Fundamental Theories of Physics

Volume 190

The international monograph series "Fundamental Theories of Physics" aims to stretch the boundaries of mainstream physics by clarifying and developing the theoretical and conceptual framework of physics and by applying it to a wide range of interdisciplinary scientific fields. Original contributions in well-established fields such as Quantum Physics, Relativity Theory, Cosmology, Quantum Field Theory, Statistical Mechanics and Nonlinear Dynamics are welcome. The series also provides a forum for non-conventional approaches to these fields. Publications should present new and promising ideas, with prospects for their further development, and carefully show how they connect to conventional views of the topic. Although the aim of this series is to go beyond established mainstream physics, a high profile and open-minded Editorial Board will evaluate all contributions carefully to ensure a high scientific standard.

More information about this series at http://www.springer.com/series/6001

Edward Anderson

The Problem of Time

Quantum Mechanics Versus General Relativity

Springer

Edward Anderson
DAMTP
Centre for Mathematical Sciences
Cambridge, UK

ISSN 0168-1222 ISSN 2365-6425 (electronic)
Fundamental Theories of Physics
ISBN 978-3-319-58846-9 ISBN 978-3-319-58848-3 (eBook)
DOI 10.1007/978-3-319-58848-3

Library of Congress Control Number: 2017954323

Printed on acid-free paper

This Springer imprint is published by Springer Nature
The registered company is Springer International Publishing AG
The registered company address is: Gewerbestrasse 11, 6330 Cham, Switzerland

Preface

This book concerns the foundations of Quantum Gravity, in particular from a conceptual point of view. It provides a self-contained introduction to this topic, resting on particular features of the accepted Paradigms of Physics: Newtonian Physics, Special Relativity (SR), Quantum Mechanics (QM), Quantum Field Theory (QFT) and General Relativity (GR). In approaching Quantum Gravity, many conceptual issues turn out to be related to notions of time. This occurs because notions of time are substantially different across these Paradigms. A first example in which this occurs is QM versus GR. Isham and Kuchař formalized the study of such discrepancies between notions of time. They did so by giving a conceptual classification of the many time-related reasons why a wide range of attempted approaches to Quantum Gravity fail to be satisfactory, in two seminal Reviews in the early 1990's [483, 586]. The current book's titular 'Problem of Time' refers to this conceptual classification. This is a multi-faceted collection of very interesting problems which turn out to be heavily interlinked. Quite a few of these problems were first glimpsed in the pioneering works of Wheeler and DeWitt in the 1960's [237, 897, 899] on the geometrodynamical formulation of GR.

The Problem of Time is, in greater generality, a consequence of the mismatch between Background Dependent and Background Independent [12, 363] Paradigms of Physics. Newtonian Physics, SR, QM, and QFT are all Background Dependent, whereas GR is Background Independent and many approaches to Quantum Gravity expect this to be Background Independent as well. So, whereas there has been quite widespread belief among theoretical physicists that the Problem of Time is a quantum matter, this is a misconception since clearly also Classical Physics can exhibit mismatches between Background Dependent and Background Independent Paradigms. Once this is taken into account, models exhibiting classical versions of the Problem of Time turn out to provide substantial conceptual insight into the harder quantum versions of the Problem of Time.

It is thus clear that further explanation of what this book (and [483, 586]) takes the Problem of Time to consist of is best done after the following.

A) Presenting the standard Paradigms of Physics and explaining how notions of time differ across these.
B) Outlining what each of Quantum Gravity and Background Independence are.

N.B. that A)—in Chaps. 1 to 8's account of notions of time and of space and of the diversity of physical laws across accepted Paradigms in Physics—serves as a *preamble*. It is not to be mistaken for *introduction of the material which the rest of the book greatly expands upon*, which is, rather, Chaps. 9, 10 and 12 on the Problem of Time and Background Independence issues which underlie this. Chapters 1 to 8 enter, rather, into assembling check lists to test foundational and Quantum Gravitational candidate times against, to see if these merit to be called timefunctions, and toward building up toward plausible Quantum Gravitational laws in Chap. 11. While these may be somewhat unexpected and indirect uses of Chaps. 1 to 8's material, this is the intended use of these Chapters in writing this book. By way of explanation, this book's main topic happens to benefit from a preliminary presentation of the types of law and notions of time and space that each of the established theories has. This is prudent given that this book's main topic is a systematic analysis of a wide range of more speculative foundational and Quantum Gravitational programs in which *only subsets* of standard theories' laws and temporal and spatial notions are kept.

From the Accepted Paradigms of Physics to Quantum Gravity

This book thus begins by considering time and clock concepts, alongside supporting notions of space, length-measuring devices, spacetime and frames. Chapter 1 gives a largely theory-free conceptual outline of these, intended for a very wide and diverse multidisciplinary audience.

Each of the Newtonian Paradigm, SR, QM, QFT and GR are then covered in turn, in Chaps. 2 to 7. This treatment includes in outline how these Paradigm Shifts affect time, clock, space, length-measuring, spacetime and frame concepts.

This theory by theory development has the further complication of not being a linear venture: these Paradigms of Physics fan out from Newtonian Mechanics as indicated in Fig. 1.a). Three distinct theoretical developments each bring in one of the three known fundamental constants of Nature:[1] Newton's gravitational constant G, the reciprocal of the speed of light c, and Planck's constant $\hbar$, as follows.

[1]Values of—and uncertainties in—these fundamental constants are as follows [661]. The speed of light in vacuo c is *defined* to be exactly 299, 792, 458 $\mathrm{m\,s^{-1}}$ due to the metre itself being defined in terms of c (see Chap. 1.13). Planck's constant $\hbar = 1.054571800(13) \times 10^{-34}$ $\mathrm{kg\,m^2\,s^{-1}}$. Newton's gravitational constant $G = 6.67408(31) \times 10^{-11}$ $\mathrm{m^3\,kg^{-1}\,s^{-1}}$. The error analysis for the Planck units is very straightforward: since G is by far the least accurately known of the fundamental units, the error in this swamps the others.

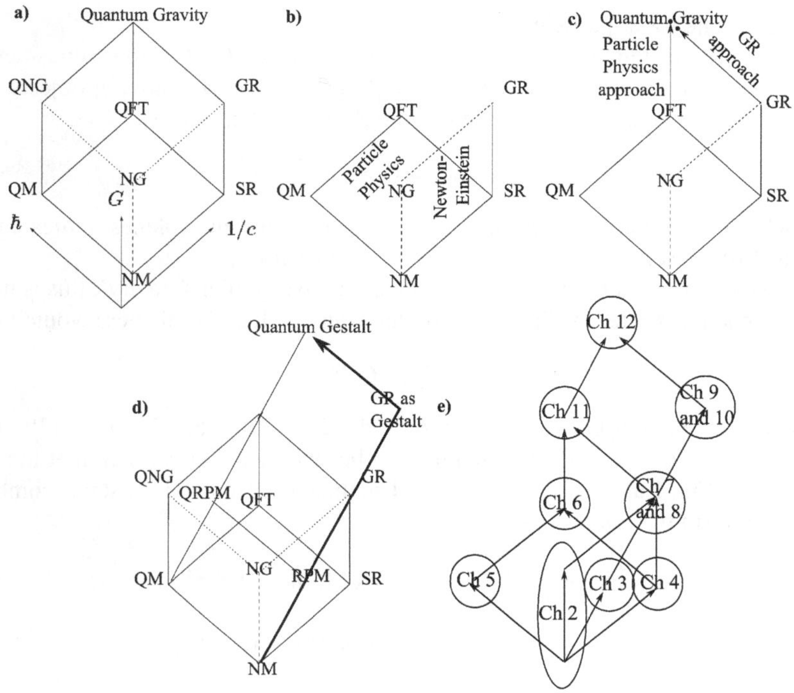

Fig. 1 **a)** Planckian cube of fundamental physical theories. Here, NM stands for Newtonian Mechanics and (Q)NG stands for (Quantum) Newtonian Gravity. **b)** indicates the 'Newton–Einstein' (alias 'classical', in the sense of 'non-quantum') plane, and the 'Particle Physics' (i.e. 'non-gravitational') plane. **c)** Gordian cube: here it is perceived that different routes along the edges to the 'final Quantum Gravity vertex' may not commute (in the algebraic sense, upon viewing its edges as maps). **d)** Cutting the Gordian cube? [Or 'thinking outside of the box' that is **a)**?] Here, (Q)RPM stands for (Quantum) Relational Particle Mechanics; see the next Subsection for an outline of what (Quantum) Gestalt means. **e)** Indicates the corresponding coverage by Chapters in Part I

1) G is significant when gravitational force is non-negligible in comparison to whichever forces dominate the physics. G was originally formulated in Newtonian Gravity, which lies within Newton's Paradigm of Physics, whereas 2) and 3) each additionally represent introducing a new Paradigm.
2) c is non-negligibly finite in SR [736]; this is relevant to objects whose velocities v are non-negligible compared to c.
3) $\hbar$ is significant in Quantum Mechanics (QM) [599], due to certain quantities coming in minimum-sized packets. For instance, angular momentum comes in $\hbar$ (or $\hbar/2$) sized packets. This is relevant in situations involving quantities comparable in size to the corresponding minimum packets.

Pairwise incorporations of these constants are furthermore as follows (Fig. 1.a).

4) Relativistic QFT [712] involves c and $\hbar$ together, corresponding to the *Compton wavelength*

$$l_{\rm C} = \hbar / m\, c \tag{1}$$

length-scale for a 'particle' of mass m.

5) GR [874]—in the sense of a Relativistic Theory of Gravitation—considers c and G together, corresponding to strongly gravitating fast-moving objects e.g. confined to around the scale given by the *Schwarzschild radius*,

$$r_{\text{Schw}} = G\,M/2\,c^2. \tag{2}$$

Each of the 'Particle Physics' and 'Newton–Einstein' planes indicated in Fig. 1.b) are self-consistent two-step Paradigm Shifts.

6) Considering $\hbar$ and G together gives 'Quantum Newtonian Gravity'; this is however much less relevant (Ex VI.0). A characteristic lengthscale here would be

$$l_{\text{g}} := \hbar^2/2\,G\,M\,m^2. \tag{3}$$

7) Finally, 'Quantum Gravity' [75, 154, 194, 237–239, 385, 471, 474, 475, 483, 485, 552, 586, 746, 845] is often held to be 'the' triple combination at the last vertex of Fig. 1.a)'s 'cube' of theories. The three fundamental constants combine here to form the *Planck units*:

$$l_{\text{Pl}} = \sqrt{\hbar\,G/c^3} \simeq 1.616228(38) \times 10^{-35}\ \text{m}, \tag{4}$$

$$t_{\text{Pl}} = \sqrt{\hbar\,G/c^5} \simeq 5.39116(13) \times 10^{-44}\ \text{s}, \tag{5}$$

$$m_{\text{Pl}} = \sqrt{\hbar\,c/G} \simeq 2.176470(51) \times 10^{-8}\ \text{kg}. \tag{6}$$

The first two of these are very small compared to 'ordinary physical quantities'. [Compare l_{Pl} with the atomic $\simeq 10^{-10}$ m and nuclear $\simeq 10^{-15}$ m lengthscales, and with the maximum precision of displacement detection in existing gravitational wave detectors corresponding to displacements of $\simeq 10^{-18}$ m. Compare also the ratio of t_{Pl} to the timescales of observational Physics with the maximally accurate clock precision of currently around 1 part in 10^{16}, as per Chap. 1.] On the other hand, m_{Pl} is very large upon considering its interpretation as a single 'fundamental particle' mass: compare e.g. the proton mass $1.672621898(21) \times 10^{-27}$ kg. By SR's $E = mc^2$, m_{Pl} corresponds to an energy scale $E_{\text{Pl}} = 1.220910(29) \times 10^{19}$ GeV. N.B. this is much larger than the 10^2 to 10^4 GeV range of the most powerful particle accelerator to date: CERN's Large Hadron Collider (LHC). Moreover, as detailed in Chap. 11, the Planck regime is expected to feature in some parts of Black Hole Physics and Early-Universe Cosmology. In particular, this book covers the Quantum Cosmology arena and simpler model arenas that exhibit features of this.

Differing Roles of Time and Space Throughout the Paradigms of Physics

Space and especially time are moreover not consistently conceived of throughout 'the Planckian cube of theories'. Due to this, consideration of which units can be built out of fundamental constants may not suffice as a conceptual framework within which to reconcile the Paradigm Shifts of Physics. Indeed, this book expounds that GR involves qualitatively distinct concepts of time, space, spacetime and frame from those used in Particle Physics.

On these grounds, this book contends that conceiving of 'Quantum Gravity' solely in terms of 'the Planckian cube' is a misleading simplification. This conceptual disparity points instead to different paths around this 'cube' not commuting (Fig. 1.c). By this disparity and the subsequent notorious difficulty with its resolution, it might be more apt to name the cube not 'Planckian' but 'Gordian': after a notorious knot of the ancient world that was presented to Alexander the Great. He is supposed to have dealt with this knot by 'thinking out of the box'. Accounts differ, however, as to whether this involved cutting it or removing it from the wooden pole it was mounted upon. Indeed, suggestions for approaching 'Quantum Gravity' differ amongst themselves as well.

A further interplay is that the Theoretical Physics literature often pays little attention to the properties entailed in calling an entity a time or a clock. This is unfortunate, because a number of such purported time quantities do not stand up against a suitable list of temporal properties. We emphasize this point in this book firstly by attributing mathematical properties to 'timefunctions', and operational character to 'clocks', in contradistinction to physical, philosophical and conceptual discussion of aspects of time.[2] Secondly, we refer to *candidate* times, timefunctions and clocks until enough properties of these have been established. There is moreover not a unique list of properties to check against, since different physical theories involve different lists of properties, as the 'Gordian cube' in Fig. 1.c foreshadows.

Background Independence in Mechanics, GR and Quantum Gravity

Following Einstein, a second perspective on the nature of—and motivation for—GR is as a freeing from absolute or background structures. From this perspective, GR is more than just a Relativistic Theory of Gravitation. Such perspectives have subsequently been dubbed Background Independence [12, 363, 483, 485, 552, 752]; contrast with how Background Dependent absolute structures pervade all six of the other non-final vertices of the cube. This book considers GR as embodying both of these perspectives at once, phrasing this in the shorthand that GR is a '*gestalt*' of a Relativistic Theory of Gravitation and of Background Independence.

[2]This book prioritizes conceptual matters concerning time over giving detailed specifics of accurate timekeeping [82, 783]. It does contain some comments on sidereal and ephemeris astronomical times, atomic clocks, and SR and GR implications for timekeeping. Some of the theoretical concepts outlined in this book, moreover, may eventually become relevant to timekeeping: space clocks, extending Earth or Solar System based reference systems to galactic and cosmological scales, and clocks in physically extreme regimes.

Some programs in Physics confine themselves to Background Dependent theories. Such approaches work for Quantum Theory within the Newtonian and Minkowskian Paradigms, while amounting to dismissing features of GR that are inconvenient in these Paradigms. This is to be contrasted with seeking new Paradigms that reconcile GR and Quantum Theory by having features of mixed quantum and GR origin! If GR's second meaning—Background Independence—is retained in passing to the quantum level, this book terms such an approach '*Quantum Gestalt*'. This book ceases to use the name 'Quantum Gravity' in this context due to it implicitly giving complete priority to the gravitational perspective on GR's identity over the Background Independent perspective. *Quantum Gestalt is a proposed family of Paradigms for Physics that encompass a number of theories and programs which implement Background Independence.* Since some kind of Background Independence is widely considered in approaches to 'Quantum Gravity', the Quantum Gestalt family includes a number of well-known examples, such as Geometrodynamics [483, 581, 899], Loop Quantum Gravity [752, 845], the Canonical Approach to Supergravity [232], and M-Theory [136, 719]. In this way, this book covers how Background Independence—and consequently the intriguing and difficult Problem of Time—are manifested in a wide range of well-known current approaches to 'Quantum Gravity'.

Both by resting upon Background Independence and by inherent interest in temporal matters, the Problem of Time is also a topic of interest more widely in Foundations of Physics and Philosophy of Physics, as well as in Theoretical Physics generally and in Background Independent Quantum Gravity in particular. None the less, this book mostly concerns Theoretical Physics rather than Philosophy of Physics.[3]

While attempting to combine QM and GR is an *example* of Background Dependence versus Independence clash, it is a *very hard* example, so it very much helps to first point out that there are other simpler examples. In particular, I) Classical Physics already exhibits such mismatches: there is also a Classical Problem of Time, which is more straightforward to resolve. II) Finite models (Minisuperspace GR and Mechanics models) already exhibit many of the Problem of Time's mismatches as well. III) Some elsewise simplified models (Midisuperspace, or, even more simply, 'slightly inhomogeneous'—perturbative—semiclassical such) exhibit all of them. I), II) and III) turn to be very insightful topics to consider prior to the Problem of Time between QM and full GR. In this vein, we outline the Classical Problem of Time in Chaps. 9 and 10 prior to its quantum counterpart in Chap. 12, and we always start by considering the above kinds of model arenas.

Returning to the Gordian cube, one can view 'freeing from absolute structures' or Background Independence as a *fourth* departure (see Chap. 3) from Newtonian Mechanics which can be adopted independently from Relativity, Gravitation and the Quantum, and their considerations of fundamental units. A simple opening here turns out to involve a Mechanics that satisfies relational criteria which arose from Leibniz's and Mach's criticisms of the Newtonian Paradigm. While no concrete

[3]For the philosophically-minded reader, most of what few such matters are mentioned in this book involve Leibniz, Mach or Broad.

such Mechanics was available in their day, Barbour and Bertotti's [105] Relational Particle Mechanics is a satisfactory such (with or without Newtonian Gravitation: Fig. 1.d). Relational Particle Mechanics is based on the following Background Independence principles.

1) *Temporal Relationalism* is that there is no meaningful time for the Universe as a whole. We shall see that this is implemented by actions which are, firstly, free of extraneous time-like quantities, and, secondly, Manifestly Reparametrization Invariant, by which there is no physically meaningful role for 'label times' either.

If time is not primary, moreover, we need to study whatever other entities that are still regarded as primary. One approach to this begins by considering configurations and configuration spaces $\mathfrak{q}$. This book is consequently also a sizeable resource on such 'spaces of shapes'[301, 539] (Appendices G–I and N).

2) *Configurational Relationalism* involves taking into account that a continuous group of transformations $\mathfrak{g}$ acting on the system's configuration space $\mathfrak{q}$ is physically irrelevant. For Mechanics, these transformations are usually translations and rotations of space, though in general Configurational Relationalism also covers physically irrelevant internal transformations, as occur in the most common types of Gauge Theory. Configurational Relationalism can be resolved, at least in principle, by *Best Matching*, which is bringing two configurations into minimum incongruence with each other by application of $\mathfrak{g}$'s group action.

Relational Particle Mechanics furthermore points to a theory of Quantum Background Independence, which, from the perspective of Quantum Gestalt, is the complement of 'Quantum Gravity' interpreted literally.

The significance of Temporal and Configurational Relationalism significantly increases upon realizing that GR itself can be recast in terms of these principles. These give two precise senses in which GR is 'Machian'. Mach's work is widely of foundational interest; for instance, some of Mach's concepts played a role in Einstein's search for GR. Moreover, the two precise senses alluded to—Mach's Time Principle and Mach's Space Principle—do not coincide with how Einstein interpreted a partly different set of Mach's ideas; also his historical route to GR ended up making at best indirect use of Machian themes. As Wheeler argued [660, 899], however, there are many routes to the same theory of GR. Some of these arrive at a dynamical formulation of GR: a theory of evolving spatial geometry: Geometrodynamics [899]. It then turns out that a more specific formulation of GR as Geometrodynamics is Machian after all ([62, 109] and Chap. 9). Finally, GR in Machian Geometrodynamics form can furthermore even be rederived from Temporal and Configurational Relationalism first principles ([62, 109] and Part II).

The proposal then is to 'cut the Gordian cube' by taking the following path (Fig. 1.d) along the 'space of fundamental theories'. a) Relational Particle Mechanics, b) the same with Newtonian Gravity, c) GR in Machian Geometrodynamics form (with SR recovered as a limiting case), d) Quantum Gestalt.

In this book, we let the physical theories themselves determine which notions of time are appropriate; see e.g. [483, 519–521, 581, 584, 586, 589, 899] for earlier such physical and conceptual approaches. With Quantum Gravity and its Quantum Gestalt subset being an unfinished subject with disputed foundations, this book considers conceptual notions to take precedence; only then is one to ask what Mathematics is required for the suitable concepts to be modelled well. This is as opposed to picking a theory for its mathematical tractability at the expense of whether it models suitable physical concepts. The Appendices provide supporting basic Pure Mathematics, Geometry and its application to configuration spaces and the Principles of Dynamics. They also provide theorems for full GR's configuration spaces and partial differential equation theorems, and various other levels of structure for Classical and Quantum Physics.

This approach is commensurate with how Physics has quite often required the development of new Mathematics that is suitable for its concepts: Calculus, Linear Algebra, Analysis.... There are moreover many tractable types of Mathematics that, however, as far as we know, Nature makes no use of. So it may be tenuous to let oneself be guided by solvability in a scientific subfield that has no experimental or observational input. This is not to be confused with how internal consistency *is* a valid guiding principle for theories. The point is that internal consistency is not a guarantee that the Universe will be as imagined, since this criterion is not by itself a guarantor of uniqueness.

The Problem of Time

We have now reached a position in which we can comment on a useful introductory breakdown of what the Problem of Time consists of. It has nine facets—closely following Isham and Kuchař [483, 586]—resulting from nine corresponding aspects of Background Independence (identified in subsequent work).

Aspect 1) Temporal Relationalism leads to the notorious *Frozen Formalism Problem*: *Facet 1)*. At the quantum level, this is the presence of an apparently frozen quantum wave equation—the Wheeler–DeWitt equation— where one would expect an equation which is dependent on (some notion of) time. This quantum Frozen Formalism Problem is very well known, but is unfortunately often confused with the entire multi-faceted Problem of Time.

Aspect 2) Configurational Relationalism leads to—in the case of GR, for which $\mathfrak{g} = \mathit{Diff}(\boldsymbol{\Sigma})$: the spatial diffeomorphisms—the *Thin Sandwich Problem*, which is *Facet 2)*. [The Thin Sandwich is a particular GR specialization of the previously mentioned notion of Best Matching.]

Each of Temporal and Configurational Relationalism moreover provides constraint equations. In the case of GR, these are the well-known Hamiltonian and momentum constraints respectively. Indeed, the above-mentioned Wheeler–DeWitt equation is the quantum Hamiltonian constraint, whereas the Thin Sandwich Problem is a particular approach to solving the momentum constraint at the classical level.

It is next natural to ask whether one has found all of the constraints: algebraic *Constraint Closure* is *Aspect 3)*. This is approached by introducing a suitable brackets structure and systematically applying the Dirac Algorithm. If the answer is in the negative, one has a *Constraint Closure Problem*: *Facet 3)*.

The objects which brackets-commute with all the constraints—or with specific subalgebraic structures thereof—are of subsequent interest. These objects—*observables* or *beables*—are useful objects due to their physical content, whereby *Aspect 4)* is *Assignment of Beables*. If obtaining a sufficient set of these to do Physics is in practice blocked—a common occurrence in Gravitational Theory—then one has a *Problem of Beables*: *Facet 4)*.

Since GR is also a theory with a meaningful and nontrivial notion of spacetime, it has more Background Independence aspects than Relational Particle Mechanics does. Indeed, the Einstein field equations of GR determine the form of GR spacetime, as opposed to SR Physics unfolding on a fixed background spacetime. From a dynamical perspective, GR's geometrodynamical evolution *forms spacetime itself*, rather than being a theory of the evolution of other fields *on* spacetime or *on* a sequence of fixed background spatial geometries. Regardless of whether spacetime is primary or emergent, there is now also need for the following.

Aspect 5) is *Spacetime Relationalism*, whereby the *diffeomorphisms* of spacetime itself, *Diff*($\mathfrak{m}$), are physically redundant transformations. Whereas this is straightforwardly implemented in the classical spacetime formulation of GR, it becomes harder to implement at the quantum level. For instance, it feeds into the *Measure Problem* of Path Integral Approaches to Quantum Gravity, so *Facet 5)* is indeed nontrivial.

Foliations of spacetime play major roles, both in dynamical and canonical formulations, and as a means of modelling the different possible fleets of observers within approaches in which spacetime is primary. Background Independence Physics is moreover to possess *Foliation Independence* : *Aspect 6)*. If this cannot be established, or fails, then a *Foliation Dependence Problem* is encountered: *Facet 6)*.

Starting with less structure than spacetime—assuming just one or both of spatial structure or discreteness—is particularly motivated by Quantum Theory [899]. Moreover, in such approaches the spacetime concept is to hold in suitable limiting regimes: *Spacetime Constructability*—*Aspect 7)*—is required. If this is false, or remains unproven, then we have a *Spacetime Construction Problem*: *Facet 7)*.

Finally, *Aspect 8)* is *Global Validity* and *Aspect 9)* is *No Unexplained Multiplicities*. These apply to all the other aspects, facets and strategies toward resolving these; contentions with these are termed, respectively, *Global Problems of Time*: *Facet 8)* and *Multiple Choice Problems of Time*: *Facet 9)*.

All in all, the Problem of Time is a multi-faceted subset of the reasons why forming 'Quantum Gravity' Paradigms is difficult and ambiguous; Further reasons are purely technical, or a mixture of both.

The classical versions of Background Independence and the Problem of Time are more straightforward, so this book presents these before their quantum counterparts. Similarly, this book makes use of the simpler Relational Particle Mechanics and Minisuperspace model arenas prior to passing to more complicated cases.

Diffeomorphisms are, moreover, crucial [483] as regards a number of Problem of Time facets, and require inhomogeneous GR models so as to feature nontrivially. Balancing this requirement, enough simplicity for calculations, and cosmological applications, this book's third choice of model arena is Slightly Inhomogeneous Cosmology: a type of perturbative Midisuperspace model. This furthermore permits investigating whether galaxies and cosmic microwave background hot-spots could have originated from quantum cosmological fluctuations [419]. Finally, this choice of model arenas amounts to concentrating on Quantum Cosmology rather than Black Hole models.

Isham and Kuchař's reviews [483, 586] on the Problem of Time are of a very high standard. The current book, however, further advances the subject in the following ways. This book's first advance is that we have enough room to trace the Problem of Time Facets back to more basic and well-known temporal concepts which reside within the Newtonian Paradigm, SR, GR, QM and QFT, by presenting the latter in prequel chapters. This book's second advance is that we present many improvements in the conceptualization of Problem of Time facets and linking them to underlying notions of Background Independence. These conceptual advances point to renaming a number of facets and aspects so as to more truly reflect their content. Consult Fig. 12.3 at the end of Part I to keep track of the evolution and end-product form of these names for the multiple parts of the Problem of Time. This book's third advance is due to those Reviews now being over 20 years old, and containing almost no mention of Supergravity, String and M-Theory, Loop Quantum Gravity, or other modern approaches to Quantum Gravity. In this way, the current book updates and expands on the range of theories considered. In particular, Canonical Supergravity turns out to be a valuable counter-example to the suggestion that passing from one GR-like theory to another leaves the Problem of Time largely unchanged. Apologies are offered as regards not covering all Quantum Gravity programs. This is beyond what can be covered in a single book. Besides, our intent is to focus on time and notions of Background Independence underlying this rather than on diversity of Quantum Gravity programs per se.[4]

Concerning This Book's Three Parts, Appendices and Epilogues

Part I is a 'first track' introduction to the above-mentioned topics that is widely accessible, including for Freshers new to a graduate school or PhD program, and for advanced undergraduates. This outlines each Fundamental Theory of Physics, alongside explaining the notion of time used in each. Conceptual outlines are subsequently provided, both of Quantum Gravity and of the nine aspects of Background Independence with the ensuing nine facets of the Problem of Time as piecemeal

[4]Readers new to Quantum Gravity may complement this book with Kiefer's [552] wide overview, in addition to introductory literature more specific to their intended research program, some of which is outlined in Chap. 11 and in Exercise Sets V and VI.

entities. Finally, Part I outlines a number of different strategies that have been suggested to deal with these facets.[5] Part I and its supporting (unstarred) Mathematical Appendices additionally contain Exercises for students to actively expand their horizons. The more challenging Exercises are marked with †, with those marked †† being considerably more challenging.

Parts II (classical) and III (quantum) concern the rather more advanced—'second track'—material necessary for a more full and up to date an account of the Problem of Time. These are supported by the starred Mathematical Appendices. Parts II and III reflect that *the Devil is in the detail.* For, as Isham and Kuchař argued, the Problem of Time facets turn out to be heavily interlinked, and none of the strategies proposed to date work when examined in sufficient detail. N.B. that this heavy interlinking takes the form that if one resolves a facet piecemeal, and then attempts to extend this resolution to resolve a second facet, then this extension has a strong tendency to spoil the resolution of the first facet. Because of this, little overall progress has arisen from treating Problem of Time facets piecemeal. This is why it is very important to list the facets together in explaining what the Problem of Time is. In particular, studying just one facet—most commonly the Frozen Formalism Problem—misses out not only the other facets but also how these interfere with each other, which is a very major part of the subject. One reason for this interference is that the facets share temporal and Background Independence roots; they have a *common origin* due to the Background Dependent and Background Independent Paradigms not fitting together. This strongly suggests that they should ultimately be approached together rather than piecemeal. Moreover, another reason for this interference is that modelling each facets' concepts brings in its own distinct type of mathematics. Indeed, this book's fourth advance is to go much further than previous authors in identifying the mathematics that modelling each facet requires.

This book's fifth advance is to *temporarily* present each facet's concepts and consequent mathematical modelling by itself, in full awareness that these approaches will subsequently need to be combined, and that the lion's share of the difficulty is in the latter. This temporary presentation of the individual facets is of pedagogical value: Part I's account, by steering clear of facet interferences, is rather probably simpler to understand than Kuchař and Isham's reviews, so it serves as an overall introduction to this subject area. This and this book's first advance combine to make Part I a useful introduction to read prior to Kuchař and Isham's reviews as well as Parts II and III.

This book's sixth advance is in demonstrating that if the mathematics needed to model each facet is taken far enough, then resolutions of different facets can be

[5] Recommended books to read beforehand or alongside this one are Lanczos [598] or Goldstein [371] for Principles of Dynamics formulations of Mechanics, Landau–Lifshitz [599] and Isham [487] for QM, Rindler [736] for SR and a brief introduction to GR, and Chaps. 1 to 4 of Peskin and Schroeder [712] for a brief introduction to QFT. Wald [874] and the entirety of Peskin and Schroeder are good books for *second* studies of GR and QFT respectively, and the first two chapters of Dirac [250] or the first four of Henneaux and Teitelboim [446] constitute a second course in specifically-constrained Principles of Dynamics.

combined after all. One caveat here is that almost all of Parts II and III concentrate on *a local* resolution of the Problem of Time, i.e. on joint treatment of all the facets bar the Global one and the Multiple Choice one. Within this restriction, we take each of few-particle Relational Particle Mechanics, Minisuperspace, Slightly Inhomogeneous Cosmology, and full GR as far as we can. Some issues considered are additionally only taken as far as the Semiclassical Quantum Cosmology regime. Within these limitations and considering the classical case first, the fourth, fifth and sixth advances can none the less be made.

Indeed, Parts II and III show *how far* the mathematics needed to model each local facet needs to be taken before it can accommodate considering multiple facets. For instance, it has long been known that replacing Euler–Lagrange actions by Jacobi actions implements Temporal Relationalism. In comparison, this book shows that to *maintain* Temporal Relationalism upon considering further facets, one needs to follow the Jacobi action up by reformulating the entirety of the Principles of Dynamics, spacetime foliation kinematics, Canonical Quantization and the Path Integral Approach to Quantum Theory. It is also very satisfying to see that some already-known key objects such as configurations, momenta, actions and constraint equations remain unchanged in the process. However, this amounts to around two orders of magnitude more work than simply replacing the Euler–Lagrange action by the Jacobi action. In this way, this book contends that the previous five decades of attempts at concurrently resolving multiple Problem of Time facets failed to get round facet interference by seldom, if ever, coupling such a level of thoroughness to mathematics which carefully fits each facet's conceptual basis.

The main 'A Local Resolution of the Problem of Time' program that this book concentrates on building up as a Combined Semiclassical, Histories and Timeless Approach. This is a Machian extension of earlier work by Halliwell, shown to hold beyond the Minisuperspace arena he explicitly investigated. The book also reviews other conceptually interesting strategies that have been suggested over the years toward attempting to resolve the Problem of Time: attempting a Klein–Gordon interpretation for Quantum GR, hidden time, matter time, and a further variety of Timeless,, Path Integral and Histories Approaches.

These results are subsequently tempered with brief accounts of Global Problems and Multiple-Choice Problems in Epilogues at the end of each of Parts II and III. Finally, further Epilogues are provided to outline Background Independence and the Problem of Time at deeper levels of mathematical structure than the usually assumed metric and differentiable structure level: the topological manifold level, topological space level, at the level of sets, and alternatives thereto. Whereas in the context of GR, Background Independence is almost always meant at the metric and differentiable manifold (diffeomorphism) levels of mathematical structure, there is no conceptual reason to stop at this level. These Epilogues represent 'third track' material, extending one's mathematical framework to include either of these would be considerably harder than the book's local resolution; these lie on, or somewhat beyond, the frontier of current research. We do however point to suitable mathematics to model the Global and Multiple-Choice Problems: stratified manifolds, sheaves, deformed cohomology, Topos Theory.... Some of these topics are supported by the double-starred Mathematical Appendices.

Since Part II and III's material is more advanced, these contain not Exercises but Research Projects. Many of the largest such—program rather than paper sized—are gathered in Part III's Conclusion and in the Epilogues. The website https://conceptsofshape.space/problem-of-time-book/ will be periodically updated to keep track of which of these projects have received significant progress. The Author can be contacted at Dr.E.Anderson.Maths.Physics@protonmail.com.

Let us end with two recommendations for readers who are interested in pursuing Quantum Gravity and are relatively new to this field. Firstly, take a wide overview.[4] Only by comparing different thoughts about conceptually similar questions—and how answers to these pan out under diverse assumptions—can one get a feel for whether the specifics of a particular Quantum Gravity program are likely to have lasting significance in humanity's understanding of Nature. Secondly, think for yourself.

Cambridge, UK Edward Anderson
2017

Acknowledgements

Foremost, I thank the five people I am close to. Without your support, I could have never written this.

For support when I was younger, I also thank in particular my father O and my friend L.

I thank Chris Isham in particular for support during the years in which I wrote this book. I also thank my PhD supervisors Malcolm MacCallum and Reza Tavakol, and subsequently Enrique Alvarez, Jeremy Butterfield, Marc Lachièze-Rey and Don Page for support with my career. For a range of discussions, book chapter proof readings, provision of useful references, invitations to speak, teaching, hosting, or support with my career, I also thank Ozgur Acik, Emily Adlam, Julian Barbour, John Barrow, Harvey Brown, Joseph Bunao, Bernard Carr, Gabriel Catren, Louis Crane, Alexis de Saint-Ours, Natalja Deng, Fay Dowker, George Ellis, Arthur Fischer, Cecilia Flori, Brendan Foster, Anne Franzen, Belen Gavela, Gary Gibbons, Mark Gotay, Alexei Grinbaum, Sean Gryb, Jonathan Halliwell, Claus Kiefer, Sophie Kneller, Tim Koslowski, Karel Kuchař, Klaas Landsman, Renate Loll, Matteo Lostaglio, Jorma Louko, Przemyslaw Malkiewicz, Flavio Mercati, Martin O'Loughlin, Niall ó Murchadha. Jonathan Oppenheim, Daniele Oriti, Brian Pitts, Alexander Polnarev, Jorge Pullin, Bryan Roberts, Carlo Rovelli, Roman Ryutin, Jan Saxl, David Schroeren, Eduardo Serna, David Sloan, Christopher Small, Rafael Sorkin, Samo Stanič, Karim Thebault, Paulo Vargas Moniz, Juan Valiente-Kroon, Hans Westman, Jimmy York Jr., and the book's technical Editors and Angela Lahee of Springer.

Much of the work this book is based upon was done with funding from Peterhouse Cambridge, the Killam Foundation, Universidad Autonoma de Madrid and the Foundational Questions Institute (fqXi.org). This work was done during my affiliation with Cambridge in the decade up to 2016. I also thank the University of Alberta and Université Paris 7 for hosting me some years, and the fqXi once again for a number of travel mini-grants. This work could have not been carried out if the Cambridge's Moore Library (Mathematics) did not have 24/7 access, for which I thank Stephen Hawking. I generally encourage 24/7 access to academic libraries elsewhere as in the best interests of foundational research actually getting done.

I also thank my friends $12A, 3B, 4C, 2D, 4E, 2H, I, 2J, 2K, 3L, 2M, 3R, 8S, T$ and W for keeping my spirits up at many points in this long journey.

Finally, I wish to pay my respects to Professors John Archibald Wheeler and David Kendall, who passed away during the years I have been working on this topic. Your works have been of great value and inspiration in my own studies, as this book makes clear.

Cambridge, UK Edward Anderson

Contents

Part I
Time in Fundamental Physics

Chapter 1
Introduction: Conceptual Outline of Time

"What then is time? If no one asks me, I know what it is. If I wish to explain it to him who asks, I do not know." Saint Augustin, 398 A.D.

There is a long history of fascination with notions of time as well as uncertainty as to their meaning. This book does not claim to give a final answer to this, but it does provide an analysis, theory by theory, as regards the observationally successful fundamental theories of Physics. The main topic of this book is how the distinct notion of time for two families of these theories—Quantum Mechanics (QM) and General Relativity (GR), or, more accurately, Background Dependent and Background Independent theories—might be pieced together in forming a theory of Quantum Gravity. We begin, however, with a largely theory-free consideration of concepts often associated with time.

1.1 Time-Related Notions

"Time present and time past
Are both perhaps present in time future,
And time future contained in time past.
If all time is eternally present
All time is unredeemable." T.S. Eliot [289].

The Universe is made up of occurrences, some of which we experience during our lives, such as ripples on a pond, cars in motion, or meeting a friend. The goal of Science is to describe and relate experiences and occurrences in the simplest possible manner [145, 632]. This is a matter of efficient codification: as regards piecing together patterns of occurrences or experiences, explaining them and assessing the likelihood of more such. Physical laws are one means of efficient codification (see Chaps. 2 to 7). Moreover, time and space concepts—and by extension time and space themselves—can also be argued to be none other than efficient codifications [632].

E. Anderson, *The Problem of Time*, Fundamental Theories of Physics 190,
DOI 10.1007/978-3-319-58848-3_1

Let us next consider various time-related notions in this light. One classification of possible occurrences is into *present*, *past* and *future*. These are *tensed* notions, in the sense that they are built into the tenses of the verbs used in our languages. The present is, the past was and the future will be. These are different and distinguishable sets of everyday occurrences. This is through experiencing a set of occurrences in the present, remembering a subset of the past that one had experienced, whereas one can neither experience nor remember the future. I.e. past, present and future differ qualitatively as regards the extent to which one perceives them.

On the other hand, *before*, *after* and *simultaneous with* are *tenseless* notions. These are useful concepts for ordering commonplace occurrences. Amongst these, 'simultaneous with' to some extent characterizes the *present now* that one experiences.

The past was before the present, which is before the future. The pairings 'past and future' and 'before and after' have an additional feature which 'present' and 'simultaneous with' do not possess. This is because 'before' and 'after' can apply in a wider range of contexts such as that the 'far past' occurred before the 'recent past', and the 'near future' will be before the 'far future'. In this way, future and past both have a notion of 'extent in time', whereby these are larger than the present, which does not possess such a notion. Thereby, each of past and future can be subdivided into portions upon which 'before' and 'after' can also act as comparatives. Adjectives such as 'far', 'recent' and 'near' are refinements of past and future that are possible due to these possessing a notion of extent in time.

Furthermore, measures of extent in time—known as *duration*—with varying precision of definition and often geared toward convenience of people's practical experience, are often used under such names as centuries, years, months, weeks, days, hours, minutes and seconds. Conventional characterization of time is by values of a single number allotted to each possible occurrence. A simple model of duration might involve taking the differences between such times for distinct occurrences.

Other simple modelling considerations are that 'a present' here corresponds to one value of a quantity, $t = t_0$, 'the future' of that present consists of all values of this number which are greater than t_0 and 'the past' of that present consists of all values which are smaller than t_0. A present is in this sense a single *instant* (instantaneous now), whereas the past and future are each comprised of many instants. The present *separates* the past from the future; it is a boundary between these that has qualities which distinguish it from either. A further idea is that which t the present corresponds to keeps on changing. So we live through a sequence of presents each of which becomes a past that might be remembered. In this manner we do eventually experience some of what was a given present's future; by this stage that present lies to our past.

1.2 Space-Related Notions Make for Useful Comparison

Space-related notions make for useful contrast with time-related ones [397, 519, 730]; moreover in some Paradigms of Physics, the two are treated jointly.

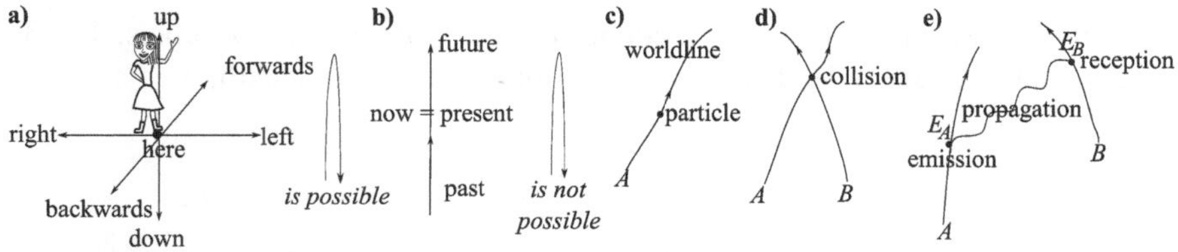

Fig. 1.1 A partial analogy between **a**) time and **b**) space, demonstrating a simple and intuitive codification of some time and space concepts. Occurrences are modelled as events: with location and dating. **c**) is the worldline of a particle or observer: its position in space at each value of time. **d**) depicts a coincidence event, such as a particle collision or two people keeping an appointment. **e**) An emission (event E_A) from one worldline A and its reception (event E_B) on another worldline B

Each 'now' is equipped with spatial properties. Intuitive space concepts here include 'up', 'down', 'forwards', 'backwards', 'to the right' and 'to the left'. These form, respectively, three 'opposite pairs' in space, in some ways (Fig. 1.1.a, b) paralleling past and future being such a pair in time. These names carry an 'on-Earth' and 'locally flat Earth' bias, in which context the vertical 'up' and 'down' pair is more distinguished from the other two 'horizontal' pairs than the latter are distinguished from each other. I.e. some of the local physics we are most accustomed to is not significantly altered if we rotate ourselves—or our coordinate system—so that e.g. what was one's notion of 'to the left' is now one's 'forward'. Gravitation near the surface of the Earth, on the other hand, acts specifically downwards. Overall, there are three independent coordinates in space, whereas there is just the one for time.

'Here' is the location at which one's own position lies, where 'up' changes status to 'down', 'forwards' to 'backwards', and 'to the right' to 'to the left'. As such, location is the counterpart of 'present' (or the present 'now'), and is a notion that likewise has Sect. 1.1's separation property. 'Above', 'below'; 'in front of', 'behind'; 'to the right of' and 'to the left of' have a similar status in each spatial direction to that of 'before' and 'after' in the temporal direction. However, one can walk to and fro in space, but one cannot move to and fro in time.

Physics moreover concerns far more than local physics on Earth. The intuitive difference about the vertical direction as compared to the two more similar horizontal directions—due to the form taken by gravity near the surface of the Earth—turns out not to be a deep fact about Nature after all. Nor does one need to use one vertical direction and two horizontal ones. One could encode the same information using combinations of these, such as a slanted diagonal that is part vertical and part horizontal, alongside two horizontal directions. Directions used may additionally vary from location to location.

Space is conventionally characterized by three numbers per constituent point. E.g. values of the familiar Cartesian coordinates $\{x, y, z\}$ (after polymath René Descartes), or spherical polar coordinates $\{r, \theta, \phi\}$. Such coordinates can be used to encode distinctions in location (at least in some patch in which they are well-defined). Moreover, change of coordinates also enters modelling of time, for all that time is characterized by *only one* coordinate.

Spatial extent quantifies of the amount of space that an entity occupies. The 3-dimensionality of space leads to distinct notions of extent in length, area and volume.

1.3 Physical Limitations on Intuitive Notions of Time and Space

Our commonplace experiences are, however, tied to living in a regime for which the bottom edge of Fig. 1.a)'s 'Planckian cube' of physical theories is a good approximation. Considering a wider region of the cube by observing, experimenting, theorizing about Fundamental Physics will cause many of these 'commonplace intuitions' to break down. We need more precision in definitions. We also need to accept that some intuitions need to change in order to fit observational facts, for sure, and also possibly as regards how one is to theorize in a more consistent manner.

1.4 Events

Following Albert Einstein [281], the sharper notion we introduce to build Physics around are events. An *event* is now taken to mean an occurrence at a specific location at a particular time. It turns out that particle and light flash concepts are useful in further development of Paradigms of Physics. Figure 1.1.c)–d)'s statement that 'this particle collision at one location occurs to the future of the emission of light at this other location' gives some inkling of the codification entailed.

1.5 Philosophical Worldviews of Time

Not all time-related concepts need be realized within a given physical or philosophical perspective, whether as primary entities or at all. This book moreover distinguishes between *philosophical worldviews* and concrete *Physical Paradigms* which may fully or partially realize given philosophical worldviews. Let us next give examples of philosophical worldviews; on the other hand, consideration of time in concrete physical theories is covered throughout the rest of this book.

For instance, *Timeless Solipsism* is a philosophical worldview in which the present now exists while the past and future do not. This position only recognizes *being*, as opposed to any further time-related notions. This worldview goes back to the fifth century B.C. perspective of Parmenides and his even more well-known student Zeno, whose 'paradoxes' were aimed at denying the occurrence of motion.[1]

[1]If a tortoise starts one unit ahead of Achilles on a line, then by the time that Achilles reaches the tortoise's starting point, the tortoise is ahead by a small amount. When Achilles reaches that new point, the tortoise is still ahead by a smaller amount, and so on. On these grounds, it may appear that speedy Achilles can none the less never catch the tortoise.... See however Appendix C.

Stark minimalistic perspective as this may be, it makes for useful contrast to many other positions, starting with the notion of *time itself flowing* as envisaged by their contemporary Heraclitus. Contrast also with philosopher Charles Broad's worldview [171, 274] in which the present and the past exist while the future unfolds in time.

It is possible as well for some temporal notions to coexist within a particular worldview but with some held to be more significant than others. E.g. in *Presentism,* the focus is on a sequence of distinct present instants which represent a distinguished notion of 'now' that 'moves forward into the future, leaving the past behind'. On the other hand, in *Eternalism* or the *Block Universe* [274, 521], all of the past, present and future are taken as given, without necessarily placing any emphasis on the present.[2] As a final example, in *McTaggart's B series* [649] Worldview, would involve the tenseless properties to entirely supplant the tensed ones.[3]

1.6 Some Properties Attributed to Time

1) Physical events $E(\underline{x}, t)$ are conventionally taken to occur *at a particular time t* and location in space coordinatized by $\underline{x}$. This can be thought of as *parametrization* of events by three spatial coordinates and one time coordinate.

   A fortiori [397], one can consider sets of events to constitute a *geometrization,* in which events are taken to be points that form some geometrical space notion of *space-time* or *spacetime.* The first of these is just a joint encoding of two geometrically separate entities: space and time, whereas the second of these is a single common co-geometrization. Chapter 2 to 7 cover examples of this including Aristotelian, Newtonian and Galilean notions of space-time, and Special Relativity (SR) 'Minkowskian' (after mathematician Hermann Minkowski) and General Relativity (GR) 'Einsteinian' notions of spacetime. Let us also use these five adjectives to refer to corresponding often-encountered Paradigms of Physics.
2) The notion of *being at a time* applies not just to events, but also to physical properties taking particular values and to questions about physical properties having particular answers.

---

[2] *Broad's Worldview* is also known as the 'growing block'. However, this book does not use this term so as to avoid giving the impression that it is a subcase of the Block Worldview; Broad's Worldview is, rather, a distinct worldview in its own right.

[3] In contrast, philosopher John McTaggart's *A series* allows for tensed notions of time, whereas his *C series* is a particular brand of Timeless Solipsism. For readers who are interested in the Philosophy of Time and are new to that field, some suggested reading is [171, 521, 730, 731, 906]. Given that the current book concerns the foundations of Quantum Gravity, let us caution that almost philosophical treatises on time pre-date or elsewise fall short of treating Quantum Gravity or even, in some cases, GR and QFT! Finally the current book makes no claim of being a *philosophical* treatment of time in Quantum Gravity.

3) *Dating* means assigning a real number—the date—to each event. This is not necessarily the aforementioned parameter notion of time, since there are other timefunctions.
4) If $t_1 = t_2$, the corresponding events E_1 and E_2 are *simultaneous with* each other, i.e. forming part of a single instant whose events are all associated with a unique value of time $t_1 = t_2$. Notions of simultaneity [521] are furthermore well-known to differ between Newtonian Physics and SR; compare Chaps. 2 and 4.
5) 'Before', 'simultaneous with' and 'after' provide an *ordering*; see Appendix A.1 for the mathematics of such in general; moreover this case is physically a *time ordering* [521, 616]. Here, in addition to the above implementation of simultaneity, if events E_1 and E_2 are at times t_1 and t_2 respectively for $t_1 < t_2$, then E_1 occurs before E_2 and E_2 occurs after E_1. This notion includes time possessing a *direction*. What we already noted about being able to move to and fro in space but not in time is relevant here. Furthermore, we appear to remember the past but not the future, whereas we can remember parts of both what lies to the right and what lies to the left.
6) *Causation* relates causes and effects; in Physics, this provides an ordering for events—*causal ordering*—with a stronger rational and physical meaning than temporal ordering's: not only preceded by but also [521, 616] influenceable by.
7) *Temporal logic* extends more basic (atemporal) logic through possessing extra "at time t_1" and "and then" constructs; see Chaps. 26 and 51 for further discussion.
8) *Duration* [83, 135, 718, 906] is a quantifier for the amount of time between two time values t_1 and t_2.
9) *Change over time* is a further notion within the perspective of time being some kind of container: a parameter of choice with respect to which change is manifest.

   Related concepts concern one state of a system "*becoming*" [731] another, undergoing *passage*, *dynamics* and *evolution*. In the last two, time plays the role of *independent dynamical variable*, with the dynamics or evolution being with respect to this variable. The notion of quantities being *conserved* under such evolution also arises at this point.

   Becoming is to be contrasted with [731] the more minimal notions of 'being' and 'being at a time'. In Timeless Solipsism—Fullt Timeless Approaches—the notion of being is all and there is no place, at least at the primary level, for the notion of becoming. While minimalism may be regarded as a virtue, it does come with the inconvenience of needing to be able to explain the *semblance of becoming* at the secondary level. Chapters 26 and 51 consider approaches to Quantum Gravity along such lines.
10) Is there time throughout the Universe? Throughout all useful models for universes? I.e. *how widespread* is the need for, and realization of, notions of time?
11) Moreover, is time *unique*?
12) Finally, the *Arrow of Time* [764, 931] concerns the apparently inevitable alignments between the directions in time of a number of areas of Physics. We postpone discussion of this until a suitable range of physical theories have been introduced so as to support these processes; cf. Ex II.12 and V.22.

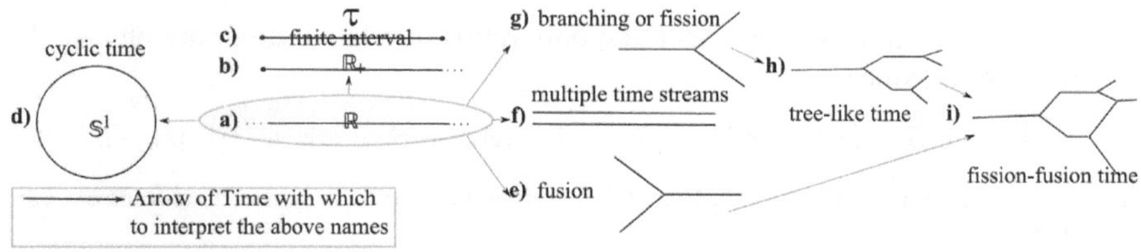

Fig. 1.2 Various topologies which have been proposed for (position-independent) notions of time. As compared to space, these are rather restricted by time being 1-d. The most obvious choice—the real line of **a**)—serves as a point of departure for a number of other alternatives, as follows. I.e. **b**) a half-infinite line for a universe with one of a beginning and no end—Big Bang to Heat Death—or an end but no beginning. **c**) A finite interval for a universe with both a beginning and an end: Big Bang to Big Crunch. **d**) One could use circular time to model a recurrent or cyclic universe: as in Hindu philosophy, the 'wheel of time', or cyclic cosmological models. Time might also branch into parallel time streams: **e**) to **g**) with subsequent compositions such as the tree **h**) and more general networks involving fusion alongside fission **i**)

1.7 Continuum Mathematics Models for Time

We next begin to consider what range of values a notion of time can take. One possibility is every value in some continuum. Conventional Mathematics builds Geometry upon assumptions of Topology. If unfamiliar with this, consult Appendices C and D.1 for an outline. The real line $\mathbb{R}$, or some interval $\mathfrak{T}$ thereof, are simple and commonly used continuum models for time. Figure 1.2 presents a fan of variants. The direction of time in these models is indicated by the arrow signs. Continuum formulations benefit from the usual continuum dynamical laws built out of the entities of Calculus: differential equations, many of which are equations for evolution in time.

1.8 Some Basic Properties of Timefunctions

Let us use the word 'timefunction' to carry mathematical connotations, in contrast to 'time' (physical and philosophical connotations) and 'clock' (operational connotations). We finally use 'timestandard' to mean a well-established timefunction that can be read off a suitably good and at least locally realizable clock.

A timefunction may implement the previous Section's item 1)'s 'time as a parameter' or have some other significance. Let us first consider how much freedom there is to be in allocating a timefunction. There are two levels of consideration here: 0) and 1).

0) Preliminarily, there is to be freedom in prescribing a timefunction as to firstly the *choice of 'calendar year zero'* and secondly of *'tick-duration'*. I.e.

$$\text{if } t \text{ is a timefunction, so is } A + B\,t \text{ for } A,\ B \text{ constants.} \tag{1.1}$$

This can be seen as a statement that only ratios of relative times are physically meaningful, since

$$\{At_1 + B - \{At_2 + B\}\}/\{At_3 + B - \{At_4 + B\}\} = \{t_1 - t_2\}/\{t_3 - t_4\}.$$

1) In some conceptualizations, the timefunction can be *reparametrizable* in excess of 0):

$$t \longrightarrow f(t). \tag{1.2}$$

2) *Being at a time* allots a specific value of the timefunction to configurations.
3) *Dating* in general involves a string of values in the manner of 2). See the Summary Fig. 12.4 for how timefunction, time and clock properties are inter-related.
4) *Field Theories* are those with space-dependent configurations as opposed to Finite Theories of e.g. point particles. Notions of time can also be *position-dependent*: $\mathrm{t}(\underline{x})$ for Field Theory in place of t.[4] Many matter theories used in the Special Relativity (SR) Minkowskian and GR Einsteinian Paradigms are Field Theories. Position-dependence gives a distinct manner from Fig. 1.2 of time acquiring a more complicated form, such as in GR's 'many-fingered' notion of times (see Chap. 8.5).
5) Timefunctions are usually taken to be a *monotonic* (rather than direction-reversing) function:

$$t \longrightarrow f(t) \quad \text{with derivative } \mathrm{d}f/\mathrm{d}t > 0 \text{ only.} \tag{1.3}$$

This makes sense as part of modelling time as having further ordering and causal properties. The notion of time as an ordering is e.g. readily implemented by the parametrized real line. One can subsequently ensure that such a model of a timefunction complies with the corresponding Paradigm of Physics' notion of causality.

Simultaneity concerns which physical events 'occur at the same time', or 'form part of a single instant'. The relation between position dependence and simultaneity is somewhat indirect; this is one way in which notions of simultaneity can become less trivial.

Monotonicity also in part underlies the Arrow of Time property, in that there is a direction involved. The further part is that the various directions are correlated. 1)'s reparametrization is now restricted to those transformations which preserve this monotonic property.

7) *Duration* can be evaluated from the timefunction.
8) The timefunction corresponds to the t which features in partial derivatives $\partial/\partial t$ in the *evolution equations* alias *equations of motion* for a physical system.
9) It may be that *the timefunction is a path function*, which depends on past history.
10) It also makes sense for a timefunction to be *operationally meaningful* (computable from observable quantities: tangible and practically accessible). While Newton did not deign to define such entities as time and space, Einstein arrived at quite distinct conceptualizations of these. Furthermore, physicist and

[4] This book's default is straight fonts for fields and slanty fonts for finite theory quantities.

philosopher of science Percy Bridgman presented an *operationalism* [168] position to crack down on unintentionally leaving gaps in how basic physical quantities are conceived of. For instance, operational considerations further enter the discussion below of clocks and rods.

11) Upon Dynamics becoming well understood, a further dynamical issue impinges upon choice of timefunction: we wish to use a notion of time in terms of which [660] *motions look simple*. More generally, notions of time, timefunctions, and candidate clocks, are to be judged not just on their own merits but also by the extent of their predictive power in the study of further dynamical systems.
12) One criterion for good timefunctions is that they be *globally valid* [483, 586]. This refers to both over time—as opposed to the half-finite and finite interval times unless there is a physical reason for this—and over space: in the case of Field Theories.
13) *Reconcileability of Multiplicity* applies in cases in which a multiplicity of elsewise-valid timefunctions occur. In general, if there is more than one plausible conceptual approach providing a timefunction, it is then interesting whether the various timefunctions are aligned. They seldom are, and this can have consequences, especially at the quantum level (see Chaps. 12, 39 and Epilogue III.A).

1.9 Non-continuum Modelling of Time

As a simple example, this could involve a finite (or countably infinite) number of discrete points. One can 'dot up' Fig. 1.2, though the dots or the 'time steps' between them might have further inherent mathematical structure. For starters, these could be the vertices and edges of a graph (see Appendix A.6 if interested). Whichever of vertices or edges can additionally be 'labelled' with structures. In this particular case, the dots would carry an order relation, so as to model time and causal orderings. Indeed, Sect. 1.6's time properties and Sect. 1.8's timefunction properties also have discrete time-dot or time-step counterparts.

Simple examples of evolution laws are now *difference equations*, or discrete time-step probabilistic models, amongst which Markov chains (see Appendix P.1) are the simplest.

In Fundamental Physics, using discrete time-steps (or similar, as per Appendix A.6) has the further issue of whether these reflect the actual form taken by Nature, rather than just our own modelling assumptions. In such a case, the 'time dots' might acquire a name that embodies this purported fundamentality, such as 'atoms of time' alias 'chronons'.[5]

A more general approach, however, involves removing continuum assumptions from one's modelling. Previous Sections mostly assumed time to be (part of) $\mathbb{R}$, which is an example of a number of continuum notions, many of which indeed orig-

[5]This name follows from that of the Ancient Greek God of time, Chronos. Hence also many other time-related words used in this book, such as chronometer, chronology, synchronization, chronogeometric, and achronal.

inated from thinking about which interesting features of $\mathbb{R}$ can be generalized. For instance, some of the properties by which the real line $\mathbb{R}$ can be modelled by Analysis (Appendix C), generalize to metric and topological spaces; $\mathbb{R}$ is additionally a manifold (Appendix D). Yet features of Nature can be modelled using *other* topological spaces, of which Mathematics provides a vast wealth. So while 'continuous versus discrete' is occasionally presented as a dichotomy, a more accurate account is that there are vastly many intermediate possibilities if one drops some part of the well-studied package of continuum assumptions exhibited by $\mathbb{R}$.

1.10 Mathematical Modelling of Space

The long-standing model for space was 3-dimensional flat Euclidean space $\mathbb{R}^3$. One way in which this can be modelled which modern Mathematics readily generalizes is in terms of a 'continuum equipped with a metric'. The Euclidean metric ('matrix') itself takes the diagonal form $\delta_{ij} = \text{diag}(1, 1, 1)$ in Cartesian coordinates. It can be arrived at as a simple reformulation of the familiar dot product; both encapsulate Euclidean Geometry's formulae for lengths and angles. The general concept of a *metric* (Appendix D.4) here is an array from which such geo*metric* information can be extracted.

The above useful generalization is to 'curved spaces'. This is not an obvious generalization, despite how curved surfaces within 3-*d* Euclidean space are immediately apparent in Nature. It is, rather, a distinct notion of Curved Geometry that exists in its own right which is relevant to the modelling of space. After all, the suggestion is that 3-*d* space itself is curved rather than involving us observing curved surfaces such as those of apples or hills within the Euclidean Geometry model of space. This generalization was envisaged by one great German mathematician, Carl Friedrich Gauss, and developed in further generality by another, Bernhard Riemann. In final form, this generalization is to a topological manifold $\boldsymbol{\Sigma}$ (Appendix D.1) equipped with a Riemannian metric $\mathbf{M}$ that is in general intrinsically curved rather than flat (Appendix D.4). This notion of geometry is furthermore meaningful in arbitrary dimension. Whereas our discussion of modelling time already mentioned manifolds, that is the 1-*d* manifold case which is uncharacteristically poor, e.g. it does not support a notion of curvature. For $d \geq 2$, however, conceiving in terms of manifolds is much richer in diversity. Gauss moreover suggested that physical space itself might be some such curved space rather than the flat Euclidean space hitherto assumed. To this end, he investigated whether the distances between three mountain tops exhibited Curved Geometry between them, but found no evidence. Such vindication had to await Einstein's work, due to the further subtleties laid out in the next Section.

1.11 Advent of Notions of Spacetime

In a separate development, Minkowski [654] pointed out that Einstein's SR can be represented as a *four*-dimensional co-geometrization of time and space: *spacetime*.

Coordinate frames in this Lorentzian sense involve specifying both time and space as coordinates. This notion of spacetime is itself flat, but its metric $\eta_{\mu\nu}$ in its simplest coordinate system is not diag$(1, 1, 1, 1)$ but diag$(-1, 1, 1, 1)$. I.e. the Minkowski metric's *signature* is nontrivial (see Appendix A.3). We shall see in Chap. 4 that this encodes distinctions in physical meaning between one of the coordinates—time t—and the other three: space x^i. I.e. the array $\eta_{\mu\nu}$ has 'chronogeometric' significance: it encodes both time intervals and lengths!

In conceiving of SR, Einstein [281] intended to create a Universal Paradigm for Physical Laws, but he then found that Gravitation resisted incorporation. He subsequently got around this by creating GR. Here spacetime not only has the signature distinction between space and time but is also in general a *curved* 4-*d* manifold. Minkowski's spacetime $\mathbb{M}^4$ is now but the special case in which there is no (or negligible) Gravitation. In this manner, Minkowskian versus Einsteinian spacetime is the indefinite-signature counterpart of Euclidean versus Riemannian Geometry. Flatness allows for SR's Lorentzian reference frames to be globally defined, but GR's more general geometry in general only permits locally defined reference frames. Both Block and Broad Worldviews can be applied to spacetime as well as to split space-time.

Within the Einsteinian Paradigm, space itself is also in general curved. Additionally, Chap. 8 demonstrates how GR can be thought of not only in terms of spacetime but also in terms of evolving spatial geometries. Moreover, in identifying curvature as a gravitational effect, it is apparent that Gauss' choice of three roughly equipotential points as regards the Earth's gravitational field was particularly unfortunate. In any case, the Earth's gravitational field is weak enough that direct detection of the curvature it induces upon space lay outside of the observational capacity of Gauss's epoch. However, curvature effects have long since been observed, alongside other vindications of Einstein's GR (see Chap. 7).

Whereas space has a richer structure than time via 3-*d* admitting far more diversity of manifold properties and of types of manifold than 1-*d* does, space is poorer than time in lacking such an ordering, causality and arrow. Spacetime possesses both of these riches. The argument about moving to and fro in space, and the Arrow of Time issues, entail that space and time are meaningfully distinct entities, regardless of their co-geometrization by spacetime. Co-geometrization by spacetime *respects* their distinction. Existence of space allows for timefunctions of the form $\mathrm{t} = \mathrm{t}(x^i)$: different times at different points, and the possibility of different observers experiencing different times. Finally, as we shall see from Chap. 5 onward, it is time—not space or length—that has a peculiar role in QM.

Dimension $d \geq 1$ is also richer than $d = 1$, as regards types of non-continuum model (see Appendix A.6 if interested). For instance, modelling with vertices and edges can now be supplemented with faces and so on. Such can be used to model spacetime as well as space. Furthermore, one of space and time being discrete does not necessarily imply that the other is. It also remains unclear whether Quantum Gravity possesses a primary notion of spacetime. Perhaps a notion along the lines of the evolving spatial geometries formulation of GR is more persistent in approaching such a regime!

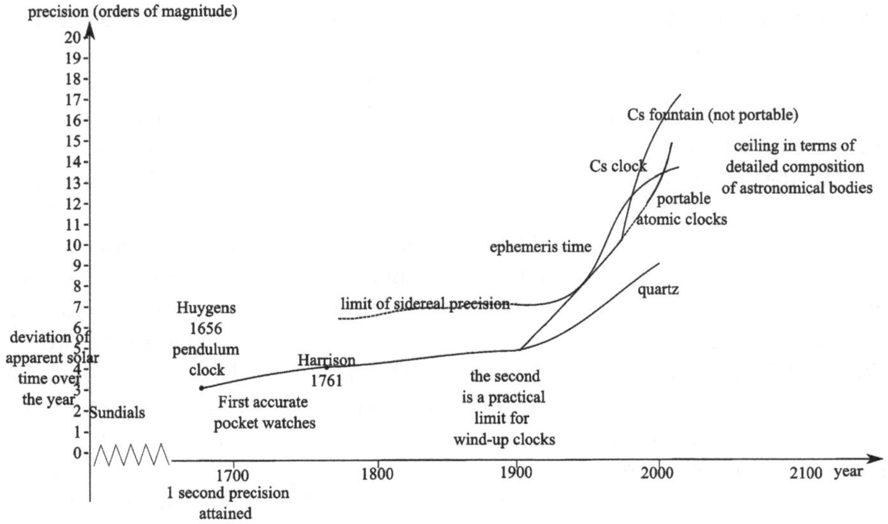

Fig. 1.3 A brief history of timekeeping

1.12 'Measuring Time': Extra Connotations in the Word 'Clock'

A *clock* is a physical entity—whether natural or artificially built *a propos*—that can be used to 'read off' 'the time'. One needs a fair amount of Physics before getting a handle on the internal workings of—and precision analyses for—some clocks. And yet the notion of what a clock is may play a foundational role in Physics.

Let us next consider some properties of clocks; these are operational and practical in character (nor is this list claimed to be complete). If a candidate object of subsystem is to be a good clock, it is likely that it will possess many, if not all, of these properties. Moreover, humankind possessed both clocks and timestandards before Newton's understanding and predictions, for these are attainable via practical considerations. Examples of these from antiquity through to the Renaissance include the hourglass, the water clock (Ex I.4.i) and the candle clock. Others are the rotation of the stars—subsequently attributed to the rotation of the Earth—and the position of the Sun or the Moon in the sky. The human pulse and the small oscillations of pendulums are further such examples. See Fig. 1.3 for an outline of the history of clocks.

1) Clocks are usually taken to *count occurrences* that are *regular* in one's notion of time, in particular *periodic*. Roughly periodic phenomena include the human pulse, the pendulum (Ex 1.4.ii), and the position of the Sun or the Moon in the sky. One type of exception to periodicity are those motions based rather on repeatable processes that need re-setting, such as the hourglass and the water clock. A uniform cross-section notched candle illustrates that regularity does not imply periodicity for a clock.
2) Clocks are to possess a suitable *reading hand*. One inspects this so as to read off the time indicated by the clock. This firstly serves to keep track of the state of the

system. Secondly, it is for convenience: accessible and swift to read. A sundial's shadow is an example of an occasionally inaccessible reading hand, since this can cease to be visible at night or in cloudy weather.

3) *Multiple clocks* can be available. This leads to questions about choosing which types of clock to rely upon.
4) If one has multiple (candidate) clocks of whatever types, one finds out which occurrences are regular or periodic by *comparison* between them. For instance, Galileo noted the superiority of pendulums over pulses as clocks. Moreover, one needs at least three candidate clocks before comparisons between them make sense.
5) Clocks vary considerably among themselves as regards how quantitatively useful they are. I.e. there are quantitative tolerance criteria that clocks need to pass before some such are allotted the task of timekeeping in a given situation of interest.

In antiquity, observation of the heavens was by far the most accurate source of timekeeping. *Sidereal time* is kept by the rotation of the Earth relative to the background of stars (of course originally interpreted the other way round). This was a successful timestandard for nearly two millennia, from Hellenic astronomer Ptolemy until the 1890s. This longstandingness is due to the Earth's rotation being reasonably stable—to 1 part in 10^8 (Ex 1.5.i). On the other hand, *apparent solar time* is defined in terms of the solar day: the interval between two successive returns of the Sun to the local meridian. Sundials approximately read off this time.

Quantitative usefulness and regularity are indeed distinct selection criteria for clocks. For instance, Chap. 3 explains how astronomical timestandards are very accurate but are none the less based on irregular motions to this level of detail. For clocks which are based on regular motions, *stability* is a meaningful quantifier. In Metrology, this refers specifically to how closely the ticks correspond to each other.

Another selection criterion for timefunctions is choice of one for which 'motion looks simple'. For example, uniform rotation was argued (e.g. by Aristotle) to be the best standard due to being the simplest to keep track of. In subsequent developments, having a firm theory of Dynamics increasing relevance of what accurately reads off this scheme's time. Finally, if one's clock reads off the time with respect to which the dynamics is simplest, then there is a sense in which it is a particularly convenient clock to use [677].

6) Clocks should actually *read the purported timefunction* rather than anything else. How this comes about (material versus spacetime property alignment) can be unclear, with standard Physics using *postulation* rather than explanation. If this substantially fails, one may have a bad clock, or a timefunction that is at best secondary in practice (if no clock can be devised that it can be directly read off from). In support of this concept, [168, 211, 317, 772, 906] consider whether the clock's read-out should correspond exactly to the dynamical time.[6] Engineering

[6]To sufficient accuracy, dynamical time diverges from Newtonian time by SR and GR corrections as per Chaps. 4 and 7.

also has a *clock bias* concept, ascertaining the practical relevance of this point. This refers to the difference between the observed ticks and the purported time-standard. In Metrology, *clock accuracy* is used to mean specifically this.

As a simple practical example, using sidereal time as the time that features in one's dynamical equations is more accurate than assuming this to be the solar time that is read off sundials.

It is also useful at this point to contrast convenient reading-hands with the separate matter of *calibration*. It may be that a clock—such as a wristwatch—only approximately marches in step with a more reliable timestandard, which however would be much more laborious to constantly monitor, such as the Solar System. One would then need to occasionally check whether the convenient reading hand's output requires updating. A reading hand can furthermore be highly stable as a distinct issue from whether it remains accurately attuned to the purported timestandard.

7) Clocks should actually *be useable within the regime of study*. This is a second kind of operational criterion.

   A classic example concerns the portability of marine chronometers. The uneven rocking of the boat offsets pendulum-based clocks. The goal is also for a type of clock that does not regularly need resetting. For instance, Magellan's expedition circumnavigating the world used multiple hourglasses with people stationed in shifts to turn them; see Ex I.4.iv) for practical difficulties with this. The reading hand of a clock should furthermore be and small and robust enough to be portable without impairment of its function. In Britain, a sizeable prize was offered for accurate determination of longitude at sea (Ex I.4.iii), which was claimed by Harrison's 1761 design of marine chronometer.

8) Simultaneity is imposed in practice by setting up a *synchronization procedure for spatially separated clocks.*

One can consider nonlocal synchronization procedures from a practical perspective [521]. In this manner, some understanding of clocks must precede—from an operational perspective—the use of position-dependent timefunctions. This is as opposed to mere mathematical consideration of SR or GR spacetime without thought as to how to populate them with actual clocks that observers read off from (see Chaps. 4 and 8 for more).

Progressive improvements as regards having stable periods occurred firstly through the introduction of quartz clocks. [These are based on *piezoelectricity*: electric currents resulting from placing certain types of crystal under stress.] Secondly, atomic clocks based on quantum oscillations of e.g. caesium (Cs) atoms were introduced, as further outlined in Sect. 5.5. See also Sect. 3.3 as regards astronomical timestandards beyond the breakdown of sidereal time.

9) *Longevity*. This is meant here in a sense other than the resistance to the regime at hand of the 'useability within the regime of study' criterion. I.e. the inclusion of self-limitations, such as the longevity of the clock's power supply or the rate at which the constituent pieces of the clock wear each other out.

10) Clock readings could *depend on their past history*.

We next begin to consider the definition of the *time unit*: the second. This was defined sidereally until the end of the 1950s. The demise of the sidereal concept due to its insufficient accuracy further carried over to a redefinition of the time unit itself. This eventually settled down into using the atomic clock timestandard's associated unit of time. This is defined as precisely 9,192,631,770 cycles of the radiation corresponding to the transition between the two hyperfine levels of the ground state of Cs-133. This definition is extrapolated to absolute zero temperature; see Sect. 7.7 for further relativistic specifications. The definition of the time unit should not be confused with the underlying conceptual entity of time itself.

Let us end by pointing to how devices used to measure time tend to have generic names. The Greek 'hora', meaning 'Goddess of the seasons', underlies the French 'horloge' and the German 'Uhr'. Moreover, the corresponding unit—the hour—was within the habitual accuracy of the Greco-Roman world (e.g. sundials). The word 'clock' itself originally meant 'bell' (from the French 'cloche'), due to church clock bells chiming hourly or quarter-hourly since the mediaeval epoch. So in each case, as the accuracy of the devices improved, the name for the generic device remained tied to the original level of accuracy. Through being free of such ties, 'chronometer' is a more meaningful name.

1.13 Measuring Length

We next consider what makes a good 'length-measuring device'. In contrast with 'time-measuring devices' being termed clocks, we do not have a generic word for 'length-measuring device'. We begin by considering features of the traditional—and still often used—measuring rods.

A first issue is the *temperature dependence of the length* of material rods. This is controlled either via fine control of the temperature or through selecting a material of low thermal expansivity (in particular Invar: iron with 36% nickel).

Secondly, rods need to be *portable*, so as to conduct measurements elsewhere than the original location of the rod. To this end, copies are made of the master standard, which are then transported elsewhere. The point of having a master standard is keeping it under carefully controlled conditions. Because of this, it itself need not be made of Invar, being chosen rather for chemical inertness and toughness to be Pt-10Ir: platinum with 10% iridium. Portable copies, on the other hand, are much more prone to accidental damage such as scratching or bending.

See also Chap. 5 for some unsatisfactorinesses with solid rods as measuring devices that are rooted in QM.

Greater accuracy can be attained by use of beams of electromagnetic radiation in conjunction with measurements made using interferometers. Astronomy involves a wealth of further means of evaluating distance (as surveyed e.g. in [888]). E.g. *parallax* based upon Trigonometry, using the width of the Earth's orbit around the Sun, or use of *standard candles*, based on the $1/r^2$ law for apparent brightness of recognizable objects with reasonably well-known absolute brightness.

In the case of length-measuring devices, the *lack* of a generic name reflects that rods or rulers suffice for most everyday purposes. This is in contrast to how clocks used for everyday purposes have undergone upgrades. Devices more accurate than rods are not widespread enough to have acquired a generic name. Rather, scientists and crafters use specialized jargon for elsewise infrequently used instruments, be these interferometers or callipers.

As regards the *unit of length*, the metre was originally (in 1790s post-revolutionary France) defined to be one ten-millionth of the distance between the North Pole and the Equator. Prior to this, each country defined its own units (some of which survive in terms such as 'yard', 'foot' and 'inch', though these are now determined from the metre via fixed conversion factors). One passed to using carefully preserved metre sticks from the pure Pt one of 1893 to its Pt–10Ir upgrade which was in use until 1960. The metre then underwent a brief period being defined as equal to 1,650,763.73 wavelengths of the orange-red emission line in the electromagnetic spectrum of the Kr-86 (krypton) atom in a vacuum, as part of the introduction of the International System of Units (SI). However, since 1983, the metre has been defined in terms of c, as afforded by SR's successes being built out of assuming that c is strictly constant. Here the metre is the path-length travelled by light in vacuum during a time interval of $1/c := 1/299,792,458$ of a second.

Let us end by pointing to Einstein's recommendation of considering the nature and non-primality of clocks and 'rods'. Sects. 1.12–1.13 might be viewed as a moderate preliminary in this direction, and this is discussed further at the end of Chap. 4.

Chapter 2
Time, Space and Laws in Newtonian Mechanics

2.1 Newton's Laws of Mechanics

Galileo conceived that all uniform motions are equally simple, as opposed to rest being simpler. This perspective enabled a major breakthrough concerning force as a notion of a body's departure from its 'natural state' due to the action of other physical entities. The particular concept of force introduced to this end by Sir Isaac Newton involves acceleration rather than velocity, in contrast to the latter having been the prevalent conceptualization since Aristotle. Galileo and Newton's conceptions were not priorly obvious due to friction being common in Nature: the rolling stone comes to rest. Moreover, rest was also associated with 'things having their place'. Feudal powers may have favoured such a concept due to its counterpart of 'people having their place', in contrast to social mobility, by which hegemonies can be challenged.

The above perspectives of Galileo and Newton can be further formalized into the first two of Newton's Laws of Mechanics, as follows.[1]

Newton's First Law. Every body continues in its state of rest, or of uniform motion in a right line unless it is compelled to change that state by forces impressed upon it.

Newton's Second Law. The change of motion is proportional to the motive force impressed; it is made in the direction of the right line in which that force is impressed.

Newton's Third Law. To every action there is always an opposing equal reaction. A further alias for this is, consequently, *Action–Reaction Principle*.

[1]This book capitalizes subject areas, specific laws of nature, specific theorems, lemmas and similar, specific principles, and the names of the Problem of Time facets, underlying Background Independence aspects and strategies for resolving these which form the main topic of this book. This is used to keep track of which phrasings refer to specific concepts that have already been introduced in the book, rather than being merely colloquial uses of the words in question.

E. Anderson, *The Problem of Time*, Fundamental Theories of Physics 190,
DOI 10.1007/978-3-319-58848-3_2

In the joint modern formulations of Calculus and vectors,[2] Newton's Second Law can furthermore be expressed as follows. Firstly, let $\underline{x}$ be the position of a Newtonian particle with velocity $\dot{\underline{x}} := \mathrm{d}\underline{x}/\mathrm{d}t$, where t is Newton's notion of time (see below). Next, Newton's notion of momentum is

$$\underline{p} := m\,\frac{\mathrm{d}\underline{x}}{\mathrm{d}t}; \tag{2.1}$$

in most applications the mass m of the particle is taken to be constant. Finally, Newton's Second Law now reads

$$\text{(Impressed force),}\quad \underline{F} := \frac{\mathrm{d}\underline{p}}{\mathrm{d}t}. \tag{2.2}$$

As one consequence of this, in the absence of external impressed forces, the momentum of a body is *conserved*: $\underline{p} = \text{constant}$.

2.2 Impact of Newtonian Mechanics

Since the inception of civilization, there has been practical demand for 'Terrestrial Mechanics'—in the form of Engineering—and for 'Celestial Mechanics': due to its timekeeping. The underlying laws for these, however, were largely not understood prior to Newton, especially as regards a unified theoretical Paradigm. Indeed, Newton's Laws of Mechanics—alongside Newton's Universal Law of Gravitation, outlined in Sect. 2.7—unified the previously separate subjects of Terrestrial and Celestial Mechanics. This Newtonian Paradigm also provided the practical means of further understanding and predicting a very wide range of phenomena.

More generally, *dynamical laws*—of which Newton's Second Law is an example—have been found to be capable of underlying substantial predictions. In the particular case of Newton's Second Law, the corresponding predictions were experimentally vindicated for over two centuries with essentially no contradictions. Substantial examples include accounting for the following.

i) Galileo's constant-acceleration model for free fall and projectiles.
ii) Uniform circular motion.
iii) Angular momentum conservation under central forces, an subcase of which recovers polymath Johannes Kepler's Second Law of planetary motion: 'equal areas swept out in equal times'. Here *angular momentum* is $\underline{L} := \underline{r} \times \underline{p}$; *total angular momentum* is $L_{\mathrm{Tot}} := L^2$.

Before briefly considering further examples of successes based on Force Laws, let us first turn to how Newton considered his Laws should be interpreted in the context of his absolute notions of time and space.

[2]Calculus was also founded by Newton, and concurrently by Leibniz. However, in the great treatise *Principia Mathematica* [676], Newton himself proved each Mechanics proposition by pictorially laid out rigorous Euclidean Geometry (including use of limiting processes).

2.3 Newtonian Absolute Space

Newton conceived of this as follows [676], presenting it in contrast with his notion of relative space. "*Absolute space, in its own nature, without relation to anything external, remains always similar and immovable. Relative space is some movable dimension or measure of the absolute spaces; which our senses determine by its position to bodies; and which is vulgarly taken for immovable space... Absolute motion is the translation of a body from one absolute place into another: and relative motion, the translation from one relative place into another.*" Newton's absolute space is continuous, infinite, imperceptible (a generalization of invisible to all senses and sensors) and cannot be acted upon. It is mathematically modelled by Euclidean $\mathbb{R}^3$ (with fixed origin and fixed axes); this also amounts to assuming well-definedness globally in space.

2.4 Newtonian Absolute Time

Newton also considered motion to occur *in* time, his principal conception of which was absolute [676]. He explained this, in contrast with his notion of relative time, as follows. "*Absolute, true and mathematical time, of itself, and from its own nature flows equably without relation to anything external, and by another name is called duration: relative, apparent and common time, is some sensible and external (whether accurate or unequable) measure of duration by the means of motion, which is commonly used instead of true time.*" Here 'equably' means 'uniformly'. Also note the use of Newton's concept of duration rather than Chap. 1's paradigm-free version. 'External' is in the sense of external to the physical entities under consideration. This includes Newtonian time being an external parameter rather than a (dependent) dynamical variable. This complies with the parametrization feature of time, but not with reparametrizability.

Newton's absolute time is likewise continuous, infinite, imperceptible and cannot be acted upon. Its infiniteness is mathematically modelled by $\mathbb{R}$ and amounts to assuming well-definedness holds globally in time itself. The last two features run against operational meaningfulness. Newtonian time is also unique enough to avoid multiplicity of times, in fact for now too strongly so, out of contravening freedom of choice of calendar year zero and of tick-duration. It is however straightforward to incorporate these features into one's practical physical calculations.

In the Newtonian Paradigm, absolute time is used to transform *kinematic* geometry into far more physically predictive *Dynamics* [96]. Within the Newtonian Paradigm, this is taken to be a universal time—one time for all the bodies and all the Laws of Physics; this precludes t being position-dependent. In fact, much of this conception of time preceded Newton, being used in the mid 1600s by Isaac Barrow, and Pierre Gassendi, and even as far back as the second century astronomer Ptolemy [521].

The Newtonian Paradigm also possesses a notion of change in time. It also possesses a notion of time as a container: a parameter of choice with respect to which

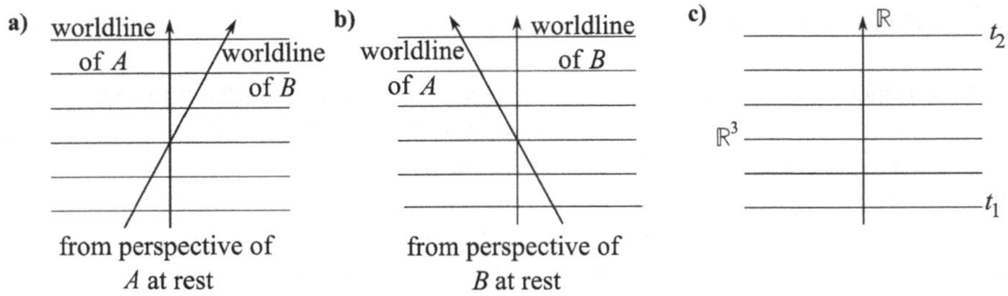

Fig. 2.1 The Aristotelian Paradigm considers **a)** and **b)** to be distinct worlds, whereas in the Galilean Paradigm they are one and the same. This is not in accord with one of **a)** and **b)** being privileged by its further identification with being at rest with respect to absolute space. **c)** contrasts the structure of Newtonian space-time. Each instant, now, or simultaneity is labelled by a value of Newtonian absolute time

change is manifest. Newton's Second Law subsequently plays the corresponding role of Dynamical Law.

Time features 1) to 11) of Sect. 1.6 are straightforwardly realized in the Newtonian Paradigm. The following four statements about these aspects of time in the Newtonian Paradigm are made to subsequently contrast with other Paradigms of Physics departing significantly from these.

1) Absolute time is taken to define a *sequence of simultaneities* representing Nature at each of its instants. Each simultaneity is here a copy of the apparent 3-*d* Euclidean Geometry of the corresponding space, containing a collection of particles (which possibly constitute extended objects).
2) *Dating procedures* are straightforward in the Newtonian Paradigm and enable the establishing of a *chronological ordering*.
3) *Causal ordering* coincides with chronological ordering here.
4) *Duration* is here indeed just the 'intuitive' *difference of datings*: $|t_2 - t_1|$.

2.5 Aristotelian, Galilean and Newtonian Paradigms Compared

Each of Aristotle and Newton put forward distinct absolute concepts for space and time (Fig. 2.1). Galileo, despite preceding Newton, made a different advance: contrast Fig. 2.1) with Newton's *unique* absolute space. In this way, the Newtonian Paradigm also involves velocity relative to absolute space, V_{abs}. On the other hand, the Galilean Paradigm is free from this, through involving instead a *privileged family* of frames moving at constant velocity v relative to one another. Note that this is a *trading* of one Absolute Paradigm for another: a unique absolute space and an absolute velocity V_{abs} for a non-unique notion of absolute space. Galileo's position did become the widely accepted one, modulo the caveat presented in Sect. 3.5.

The *Galilean transformations* are of the form

$$x \to x' = x - vt \tag{2.3}$$

for constant velocity v. One may adjoin

$$t \rightarrow t' = t \tag{2.4}$$

to this, i.e. there is just the one t, in contrast with other standard Paradigms of Physics' multiplicities. The Galilean transformations interrelate the privileged frames of reference in which Newton's First Law holds, which are termed *inertial frames*. These are at rest in absolute space or moving uniformly through it along a straight line. The Galilean transformations are the basis of Galilean Relativity. Frames related by this transformation can be envisaged as 'boats' in relative motion with constant velocity with respect to each other; these are as good as each other for the formulation of equally simple Laws of Physics. Indeed, Newton's Laws of Mechanics obey Galilean Relativity: they are invariant under Galilean transformations between inertial frames. On the other hand, in non-inertial frames, additional *fictitious forces* are perceived. Finally, contrast how Aristotle did not have any widely applicable law, without which considering simplifications in certain frames is moot.

We additionally consider additive transformations: spatial and temporal translations

$$\underline{x} \longrightarrow \underline{x}' = \underline{x} + \underline{k}, \qquad t \longrightarrow t' = t + t_0. \tag{2.5}$$

This does not incur any further complications. The second of these incorporates the desirable freedom of choice of calendar year zero. All in all, a minor modification of Newtonian time has Sect. 1.8's timefunction properties 1), 4) and 6), but not 2) or 3). As regards property 5)—operational meaningfulness—Newtonian time technically does not possess this, but via the rotation of the earth ('sidereal time') being identified in practice with Newtonian time, this property is in effect acquired.

The second Eq. (2.5) in is the corresponding freedom of choice of origin for absolute space. In fact, Newton himself identified absolute space in terms of the centre of mass of the solar system being at rest.

The issue of spatial rotations is, however, more involved; the transformation here involves a unit-determinant orthogonal matrix

$$\underline{x} \longrightarrow \underline{x}' = \underline{\underline{R}}\underline{x}. \tag{2.6}$$

The full set of translations and rotations constitute the *kinematical group* [814] of transformations between pairs of frames. Two particular cases are constant $\underline{k}$ and $\underline{\underline{R}}$ in the Newtonian kinematical group and time-dependent ones in the 'Leibnizian kinematical group' [278]. In the current setting, translations and rotations originate from modelling space as $\mathbb{R}^3$, for which these are rigid symmetries; see Sect. 2.12 for further discussion. The (infinitesimal) actions of the generators of these on a velocity vector are

$$T_V : \underline{\dot{x}} \longrightarrow \underline{\dot{x}} + \underline{V}, \qquad R_\Omega : \underline{\dot{x}} \longrightarrow \underline{\dot{x}} - \underline{\Omega} \times \underline{x}. \tag{2.7}$$

Using the above infinitesimal rotational action twice, the *real acceleration* relative to an inertial frame is related to the *apparent acceleration* in a rotating frame by

$$\underline{\ddot{x}} = \underline{\ddot{x}}_{\text{apparent}} + 2\,\underline{\Omega} \times \underline{\dot{x}} + \underline{\dot{\Omega}} \times \underline{x} + \underline{\Omega} \times \{\underline{\Omega} \times \underline{x}\}. \tag{2.8}$$

The fictitious terms here are, respectively, the *Coriolis*, *Euler*, and *centripetal* accelerations; see Ex I.4 for more.

2.6 Newton's Bucket

Rotations are less straightforward to handle and indeed led historically to complications. In particular, Newton used rotations in the argument in his Scholium [676] by which he became convinced to the reality of absolute space. *"If a vessel, hung by a long cord, is so often turned about that the cord is strongly twisted, then filled with water, and held at rest together with the water; thereupon, by the sudden action of another force, it is whirled about the contrary way, and while the cord is untwisting itself, the vessel continues for some time in this motion; the surface of the water will at first be plain, as before the vessel began to move; but after that, the vessel, by gradually communicating its motion to the water, will make it begin sensibly to revolve, and recede by little and little from the middle, and ascent to the sides of the vessel, forming itself into a concave figure (as I have experienced), and the swifter the motion becomes, the higher will the water rise, till at last, performing its revolutions in the same times with the vessel, it becomes relatively at rest in it. This ascent of the water shows its endeavor to recede from the axis of its motion; and the true and absolute circular motion of the water, which is here directly contrary to the relative, becomes known, and may be measured by this endeavor. ... There is only one real circular motion of any one revolving body, corresponding to only one power of endeavoring to recede from its axis of motion... And therefore in their system who suppose that our heavens, revolving below the sphere of the fixed stars, carry the planets along with them; the several parts of those heavens, and the planets, which are indeed relatively at rest in their heavens, do yet really move."* We return to this analysis in Chap. 3.1 with some historically-posterior arguments.

We next consider Force Laws within the Newtonian Paradigm.

2.7 Newtonian Gravity

Newton's Universal Law of Gravitation. The gravitational force between two particles with masses m_I at positions $\underline{x}_I$ is[3]

$$\underline{F}^{g}_{12} = -\frac{Gm_1m_2}{r_{12}^2}\hat{\underline{r}}_{12}, \tag{2.9}$$

where G is *Newton's universal gravitational constant*. Combining (2.9) with Newton's Second Law, gives the equation of motion for a particle in a gravitational field. In particular, this framework accounts for the following.

[3]Let us use I-indices to run over particle labels, currently 1 and 2, and $\underline{r}_{IJ} := \underline{x}_I - \underline{x}_J$, and small hats for unit vectors.

1) Kepler's other two Laws of Plantary Motion: that the planets move on ellipses with the sun at one focus and with (orbital period) $\propto$ (semi-major axis)$^{3/2}$.
2) Gravitation near the surface of the Earth. Thereby, Newton unified Terrestrial and Celestial Mechanics. Indeed, Newtonian Gravitation has considerable further success at accounting for Solar System motions, e.g. in modelling perturbations due to interactions between planets.

By the 19th century, physicists began to favour the description of forces in terms of fields pervading space. In other words, they began to consider Field Theories. From this perspective, the Newtonian gravitational potential is a scalar field: the *gravitational potential* $\phi_{12} := m_2/|\underline{r}_{12}|$ at $\underline{x}_1$ due to the particle of mass m_2 at $\underline{x}_2$. In terms of this, $\underline{F}^{g}_{12}$ may furthermore be written as $\underline{F}^{g}_{12} = -m_1\underline{\partial}\phi_{12}$. The *gravitational vector field* $\underline{g} = -\underline{\partial}\phi$ is also useful in the discussion below. Near the surface of the Earth, the magnitude of this is the familiar 'terrestrial gravity' g, whose direction is 'downwards'. Conversely, ϕ is said to be a *scalar potential* for $\underline{g}$.

Newtonian Gravity is linear, so the Superposition Principle applies as regards building up the gravitational field at each location $\underline{x}$ from each material point source. The total gravitational force due to all of a system's particles is $\underline{F}^{g}(\underline{x}) = -m\{\underline{\partial}\,\phi\}(\underline{x})$. In the Field Theoretic formulation, the combination of Newton's Second Law and (2.9) gives $\ddot{\underline{x}} = -\{\underline{\partial}\phi\}(\underline{x})$. Consider the particular case of this for two neighbouring particles at positions $\underline{x}$ and $\underline{x} + \vartriangle\underline{x}$. By subtraction and the definition of derivative, one arrives at the *tidal equation*

$$\vartriangle\ddot{\underline{x}} = -\underline{\partial}\{\underline{\partial}\cdot\vartriangle\underline{x}\} \tag{2.10}$$

for the relative acceleration of the two particles. This equation indeed accounts for the tides of the sea in terms of the position of especially the Moon and also the Sun. Moreover, this is but the most familiar of many such effects, and the relative acceleration concept is accorded further theoretical significance in Chap. 7.

A *field equation*—describing how Gravitation is sourced by masses—is also required. In differential form, this gives *Poisson's Law*

$$-\underline{\partial}\cdot\underline{g} = \Delta\phi = 4\pi G\rho, \tag{2.11}$$

where ρ is the mass density. (2.9) is then recovered as the fundamental solution [220] corresponding to the 3-*d* Laplacian operator.

A small (43 seconds of arc per century) anomalous perihelion precession of Mercury was detected in the late 19th century. At first, this was attributed to a perturbation caused by a 'planet Vulcan' (and then to a cloud of smaller bodies) in close proximity to the Sun. We shall see however in Chap. 7 that Einstein gave an entirely different explanation for this effect, and indeed no 'planet Vulcan' has ever been seen.

2.8 Electrostatics

In this book, we do not take 'Newtonian' to mean 'posited by Newton' but rather 'within Newton's Paradigm for Physics as a whole'. This covered all the Physics that was known for over two centuries after Newton's formulation and remains an excellent approximation for many practical purposes. Within this Paradigm, the next three sections consider further Force Laws that turn out to be based upon fundamental forces.

The phenomenon of static electricity, in the form of rubbing amber with cloth, has been known since the Ancient Greeks. Experimental confirmation of the corresponding force law did not however come until physicist Charles-Augustin Coulomb's work in the 18th century. Coulomb's Law for the force between two charges q_I at positions $\underline{x}_I$ is

$$\underline{F}_{12} = K \frac{q_1 q_2}{\underline{r}_{12}^2} \hat{\underline{r}}_{12}. \tag{2.12}$$

K is here Coulomb's constant; this has been subsequently interpreted as $1/4\pi\epsilon_0$ for ϵ_0 the permittivity of space (the value of this quantity is given and explained in Chap. 3.5).

The development of Field Theory was particularly significant for the study of Electricity and Magnetism and their eventual unification. Vector Calculus subsequently provided an efficient language for this. For now, Coulomb's Law can be recast as a particular case of *Gauss' Law*, in terms of a vector electric field $\underline{E}$ or a scalar potential Φ such that $\underline{E} = -\underline{\partial}\,\Phi$. The differential form of Gauss's Law is now

$$-\triangle\Phi = \underline{\partial} \cdot \underline{E} = \rho_e/\epsilon_0 \tag{2.13}$$

for ρ_e the charge density. The passage from the 3-*d* Laplacian to the inverse-square fundamental solution is just a mathematical reworking, by which Coulomb's Law is recovered analogously to how Newton's Universal Law of Gravitation is retrieved from Poisson's Law in the previous section. Working in terms of Φ superposition is again immediate, so Gauss's Law readily covers a wider range of configurations of charges.

2.9 Gravitation and Electrostatics Compared

Let us comment further here on the extent of the similarity between Coulomb's Law and Newton's Universal Law of Gravitation. Both are inverse square laws between 'charges' that feel the force in question; here these are electric charges, whereas for Gravitation they are masses. We shall see in Chaps. 3 and 7, however, that the above similarity turns out to be a coincidence of simplified regimes rather than some deep inter-relation. Moreover, electric charges come with two possible signs: positive and negative, whereas Gravitation has only one sign of 'charge': positive mass. Nor need all macroscopic bodies possess any electric charge, whereas they do all possess mass. Finally note that it is unclear at this stage that Gravitation will turn out to be very significant for further theoretical reasons beyond Newton's unification of Terrestrial and Celestial Mechanics.

Also N.B. that

$$\text{Gravitation is } 10^{40} \text{ times weaker than electrostatic attraction.} \tag{2.14}$$

This is a rough order of magnitude estimate, which holds for the range of constituent elementary particles of ordinary matter.

Moreover, mass already featured in a different manner in the conceptualization of Newtonian Mechanics. This might be interpreted as mass being a two-use concept, or as there actually being two different concepts of mass [520] that are not a priori to be assumed to be the same. I.e. inertial mass in Newton's Second Law and gravitational mass in Newton's Law of Gravitation. The latter can furthermore be split into active and passive subcases [520].

Let us next consider Newton's Second Law in a rotating rather than inertial frame. Dividing by the inertial mass m_{i}, this is [814]

$$\ddot{\underline{x}} = \underline{\mathrm{a}} + \frac{m_{\mathrm{g}}}{m_{\mathrm{i}}}\underline{\mathrm{g}} + \frac{1}{m_{\mathrm{i}}}\underline{F} - \left\{2\underline{\Omega} \times \dot{\underline{x}} + \dot{\underline{\Omega}} \times \underline{x} + \underline{\Omega} \times \{\underline{\Omega} \times \underline{x}\}\right\}, \tag{2.15}$$

since nothing can shield gravity, and where $\underline{\mathrm{a}} = \underline{\mathrm{a}}(\underline{x}, \dot{\underline{x}})$ is an acceleration field. It has additionally been noted experimentally (in e.g. 'Eötvös-type' experiments [910], named after Baron Roland von Eötvös) that $\frac{m_{\mathrm{g}}}{m_{\mathrm{i}}}$ cannot be measured, i.e. that this happens to be independent of material composition. Furthermore, elevating this from an experimental summary to a physical principle constitutes a type of *Equivalence Principle*:[4] a significant matter to which we return in Chap. 7. On the other hand, distinct electric charge-to-mass ratios are readily observed.

We end this discussion of mass with how the unit of mass—the kilogram—has been defined since 1889 as the mass of some carefully preserved lump of metal. It is presently a surface area minimizing cylinder of the Pt–10Ir alloy.

2.10 Magnetostatics

For now, take magnets to concern a further force known since antiquity to be exhibited by a few minerals such as lodestone, to be transferable unto some metals, and to be pervasive as some kind of weak background. E.g. the compass was invented in Ancient China as a tool of navigation; the background it picks up is now known to be sourced by the interior of the Earth. Magnetism was found to be sourced by electric currents; the steady-current regime case of this began to be understood as a

[4]This contradicts the Aristotelian doctrine that heavy bodies fall faster than lighter ones. On the other hand, another Ancient Greek philosopher—Epicurus—did entertain such a concept [520].

Force Law between wires due to Biot and Savart in the early 19th century. E.g. for wire elements $\underline{dl}_I$ carrying steady currents I_I at positions r_I,

$$\underline{F}_{12} = \frac{\mu_0}{4\pi} \frac{I_1 d\underline{l}_1 \cdot I_2 d\underline{l}_2}{r_{12}^2} \hat{\underline{r}}_{12}. \tag{2.16}$$

Here μ_0 is the permeability of space, which is defined to take the exact value $4\pi \times 10^{-7}$ ampère metres. Formulating the above in Field Theoretic terms gives *Ampère's Law*

$$\underline{\partial} \times \underline{B} = \mu_0 \underline{j}. \tag{2.17}$$

Equation (2.16) is now readily recovered as a particular solution of this.

Magnets have two kinds of poles—termed North and South. No pole of one kind has ever been observed in the absence of an opposite pole, e.g. bar magnets are observed to have a North pole at one end and an equal-strength South pole at the other. If a bar magnet is split into shorter bars, *each piece* has one of each kind of pole, so one cannot consider a region of space containing one kind of pole but not the other. This '*non-observation of magnetic monopoles*' is encoded as a further Law,

$$\underline{\partial} \cdot \underline{B} = 0. \tag{2.18}$$

Compare this with Gauss's Law (2.13), which in general has a non-zero source charge term on its right hand side. Magnetism has no such thing as a source pole term!

2.11 Light Flashes

For now, let us consider these just in Newtonian terms so as to enable their use as thought-experiment probe devices so as to compare [831] Newtonian Mechanics and SR in Chap. 4.

2.12 Cartesian and Curvilinear Tensors Within the Newtonian Paradigm

$\mathbb{R}^p$ can be viewed[5] as an inner product space or normed space (Appendix A.3) associated with a matrix $\mathbb{I}$ whose components are δ_{ij} (the Kronecker delta symbol). This is an efficient way to encode length of a vector $||\underline{v}||$, distance between points

[5]Most usually this is $\mathbb{R}^3$, but some issues are sufficiently well-illustrated by $\mathbb{R}$ or $\mathbb{R}^2$, and $\mathbb{R}^p$ for $p > 3$ also enter at the level of configuration space. Moreover, there is no extra complication in treating this in arbitrary dimension.

with position vectors $\underline{q}_1$ and $\underline{q}_2$: $\|\underline{q}_1 - \underline{q}_2\|$, ratios of lengths $\|\underline{v}\|/\|\underline{u}\|$, and angles between vectors, $\arccos(\frac{\underline{v}\cdot\underline{u}}{\|\underline{v}\|\|\underline{u}\|})$. One can now treat (the p-dimensional version of) Euclidean Geometry in these more modern terms.

The transformations preserving (,), $\|\ \|$ or $\mathbb{I}$ are translations $Tr(p)$, rotations $Rot(p)$ and reflections Ref. These form a group of Euclidean transformations; we focus on the case without reflections, for which we denote the group by $Eucl(p)$.[6]

Cartesian coordinate systems on $\mathbb{R}^p$ are interrelated by $\bar{x}_i = R_{ij}x_j + T_i$. For rotations R_{ij} and translations T_i, considering vectors on $\mathbb{R}^p$ is natural because inner product spaces are vector spaces; vectors are furthermore well-known to model many physical quantities. Under rotation of frames, vectors transform as

$$\bar{v}_i = R_{ij}v_j. \tag{2.19}$$

'Vector proportionality laws', such as *Ohm's Law* $\underline{\mathrm{j}} = \sigma\underline{\mathrm{E}}$ for electrical conductivity tensor σ, end up taking forms such as $\mathrm{j}_i = \sigma_{ij}\mathrm{E}_j$: laws involving 'matrix valued' quantities. This leads to one asking how such a σ_{ij} itself transforms under the group under which the vectors were already held to transform. One can keep on repeating this process for objects with increasing numbers of indices. The most well-known of these is probably $\mathrm{s}_{ij} = \mathrm{C}_{ijkl}\mathrm{e}_{ij}$ from elasticity theory, for **s** the stress, **e** the strain and **C** the elasticity, i.e. a tensorial rendition of *Hooke's Law*. Additionally, the tensor transformation law for the 2-index object is

$$\bar{\mathrm{T}}_{ij} = R_{il}\mathrm{T}_{lm}R^{\mathrm{T}}_{mj} = R_{il}R_{jm}\mathrm{T}_{lm}. \tag{2.20}$$

Tensors are of wide importance in Physics. Within this framework, the *Quotient Theorem* furthermore gives back that '(tensor) = (unknown) · (tensor)' implies that the unknown entity also transform as a tensor, justifying the above means of envisaging the need for tensors as well as just vectors.

δ_{ij} itself is a special such tensor, blessed with *isotropy*. This means that the entity is the same in all directions, in the case in which $\mathbb{R}^n$ is being interpreted as space, this property is a significant postulated attribute. This is in contrast with all other 2-tensors, which exhibit preferred directions as per the previous paragraph's argument. Having preferred directions corresponds to exhibiting *anisotropy*. A notable example of this is the electrical conductivity of graphite, which is large in a plane of directions and small in the perpendicular direction. Since observing this, it has been found out to result from a parallel-layer structure on the slightly larger than atomic scale. It is later also further insightful to identify δ_{ij} as a *metric* (indeed, it is a metric tensor and the associated distance is the basis of a metric space). This is specifically the *Euclidean metric*, and the rigid transformations $Tr(p)$, $Rot(p)$ and

[6]See Appendix A.2 for an outline of Group Theory. $Tr(p)$ and $Rot(p)$ are continuous transformations. Ref are discrete. Let us call $Eucl(p)$ as defined here the *proper* alias *special Euclidean group*, whereas including Ref would involve a *full* Euclidean group) 'Proper' and 'special' are more widely used of cases excluding discrete reflection transformations. See also Appendix B as regards other types of Flat Geometry corresponding to the preservation of other combinations of the previous paragraph's quantities.

Ref are now the corresponding *isometries*: metric-preserving transformations (see Appendix B). The word 'metric' indeed means a measurer of the basic geometrical entities: the above expressions for lengths, distances, ratios of distances and angles. The corresponding line element is

$$\mathrm{d}s^2 = \|\mathrm{d}\underline{x}\|^2 = \mathrm{d}r^2 + r^2\mathrm{d}\Omega^2, \tag{2.21}$$

for $\mathrm{d}\Omega^2$ the $\{d-1\}$-sphere line element; in particular $\mathrm{d}\Omega^2 = \mathrm{d}\theta^2 + \sin^2\theta\,\mathrm{d}\phi^2$ for a 2-sphere in 3-d.

The second form of this is in curvilinear coordinates (in this case spherical polar coordinates). Some problems (or models) match some curvilinear coordinate system well in symmetry, and some workings are solvable in particular coordinates. In general distinction between upstairs (contravariant) and downstairs (covariant) indices is required (see Appendix D.2 for more). The coordinate transformation now involves

$$\mathrm{J}^i{}_j := \frac{\partial \bar{x}^i}{\partial x^j} \tag{2.22}$$

—the *Jacobian (transformation) matrix*—in place of R_{ij}. See Appendix D.2 for a more general treatment of coordinate transformations and consequently of tensors. R_{ij} and $\mathrm{J}^i{}_j$ are but the first two cases of this encountered in this book. Each case has a concept of tensor transformation law as associated with the corresponding group of transformations. So e.g. a *Cartesian tensor* is really a $[Rot(p) = SO(p)]$*-tensor* for $SO(p)$ the d-dimensional special orthogonal group: see Appendix E). The group corresponding to $\mathrm{J}^i{}_j$ itself will be introduced in Chap. 7.

Moreover, R_{ij} maps simple cases to simple cases within the Newtonian Paradigm, but that $\mathrm{J}^i{}_j$ seldom preserves simplicity. This is useful in those few problems which admit judicious choices of simplifying coordinates. Also note the difference between 'laws are simple in these restricted frames' and 'this specific example's mathematics is simple'.

We finally point to curvilinear transformations are in general not valid over the whole of $\mathbb{R}^n$. For instance, polar coordinates are not valid at the origin since ϕ is undefined there. This is accompanied by a breakdown in the $\mathrm{J}^i{}_j$ in moving from coordinates valid in some region to coordinates invalid there. The absolute value of the determinant of $\mathrm{J}^i{}_j$—the *Jacobian* J itself—furthermore features as a factor in integrands. The familiar Vector Calculus is the most common and simple case of Tensor Calculus, and is used in both the Cartesian and curvilinear contexts.

2.13 Principles of Dynamics (PoD) formulations of Mechanics

Other than in the case of a single particle, instantaneous configuration of a system is a distinct notion from space. In the former, a system's configuration is represented by a single point in *configuration space* $\mathfrak{q}$, and the evolution of a system by a single curve therein. E.g. for N particles in d-dimensional space $\mathbb{R}^d$ the configuration space [598] is $\mathbb{R}^{dN}$ [598].

Theoreticians can moreover choose to describe the position of a particle in (absolute) Euclidean 3-space $\mathbb{R}^3$ by 3 curvilinear coordinates. It is convenience, rather than any physical reality, which underlies which particular choice is made.

Configuration spaces are a starting point for the Principles of Dynamics (below), and are also central to Presentism and Fully Timeless Approaches. Mechanical systems are usually taken to be second order, so that the initial position of the particle does not suffice to determine the motion. One requires also such as the initial velocity or the initial momentum. For N particles in $\mathbb{R}^d$, naïvely one requires the prescription of dN coordinates to describe their positions. However, the particles may not be free to move in all possible ways, e.g. some of them could be attached by means of strings, springs or rods. Such constitute *constrained* mechanical systems, which can be described in terms of less than $3n$ independent coordinates, Q^{A}.[7] Whereas one may attempt to study such particle systems directly using Newton's Laws, this may be cumbersome and requires knowledge of all the forces acting at each point in the system.

A method based on energy considerations, which is often of computational value and extends to Field Theory, was formalized by Euler and Lagrange in the 18th century: the Principles of Dynamics (Appendix J). Firstly, one considers a system's potential energy $V = V(\boldsymbol{Q})$ and kinetic energy T. A typical form for the latter is $M_{\mathsf{AB}}\dot{Q}^{\mathsf{A}}\dot{Q}^{\mathsf{B}}/2 := \|\dot{\boldsymbol{Q}}\|_{\boldsymbol{M}}^2/2$. M_{AB} is here the configuration space metric, the most common case of which is the kinetic mass metric $m_I\delta^{ij}\delta^{IJ}$ for $I = 1$ to N particles in d-dimensional space. This indicates that configuration spaces are geometrical entities: a theme developed in Appendix G. $\| \ \|_{\boldsymbol{M}}$ is a usefully concise notation here, as per Appendix A.3. One next forms the *Lagrangian* $L := T - V$: a single function, knowledge of which permits one to write down a set of equations of motion equivalent to Newton's. Consult Appendices J.1-4 as regards subsequent significant developments.

In the present case, the Principles of Dynamics' generalized momentum produces the vectorial approach's usual notion of momentum (2.1), and the Euler–Lagrange equations amount to a recovery of the vectorial approach's Newton's Second Law. Some significant Mechanics examples of Poisson brackets evaluations are the fundamental bracket

$$\left\{q^i, p_j\right\} = \delta^i{}_j, \tag{2.23}$$

and the angular momentum bracket

$$\{L_i, L_j\} = \epsilon_{ij}{}^k L_k, \qquad \left\{q^i, L_j\right\} = \epsilon^i{}_{jk} q^k, \qquad \{p_i, L_j\} = \epsilon_{ij}{}^k p_k. \tag{2.24}$$

The first Poisson bracket in (2.24) signifies that angular momentum corresponds to the $SO(3)$ group of rotations (Appendix E), and the second and third that q^i and p_i are good objects—vectors—under $SO(3)$ transformations.

---

[7]My capital sans-serif indices are general indices, many of which are reserved for particular uses in this book. This book also uses bold font for index-free presentations, so e.g. the Q^{A} are denoted more succinctly by $\boldsymbol{Q}$.

In conclusion, the Principles of Dynamics readily extends to formulation in curvilinear coordinates: these last two sections on 'useful tools' additionally combine well. These lie at the root of much efficient problem-solving within the Newtonian Paradigm. Additionally, this is not their only purpose, for they are built out of concepts that extend much further across Physics, Via a large family of Tensor Calculi and of metric geometries, and by Principles of Dynamics approaches applying to all branches of Classical Physics—involving such as fields as well as particles—such tools apply in *whichever* Paradigm of Classical Physics rather than just Newton's.

Chapter 3
Absolute Versus Relational Motion Debate

This debate has been ongoing at least since the inception of Newtonian Mechanics.[1] Newton's bucket (Sect. 2.6) thought experiment served to convinced him that absolute space was real. Moreover, observed Physics was well accounted for by Newtonian Physics until the end of the 19th century. Around then, evidence for further Physics began to accumulate and be noticed; this led to QM, SR and GR. The issue presently at stake, however, is whether Newtonian Mechanics has a conceptually and philosophically solid basis.

3.1 Two Centuries of Critique of the Newtonian Paradigm

The immovable external character of the absolute space and time—which the Newtonian Paradigm assumes—is abhorred by *relationalists*. These include the famous polymath Gottfried Wilhelm Leibniz and the noted physicist, philosopher and conceptual thinker Ernst Mach. An alternative Relational Paradigm could start along the following lines. [A mathematically precise formulation of this is postponed to Chap. 9.]

Relationalism-0) *Physics is to solely concern relations between tangible entities.*

Moreover, this is a statement universal to all of Physics rather than just concerning Mechanics. Indeed, this book uses 'tangible entities' rather than 'material objects' to include fields and 'force mediators' as well as 'matter building blocks', as befits modern Physics. Key properties of 'tangible entities' are as follows.

Relationalism-1) These *act testably and are actable upon*. [Einstein attributed this to Mach.]

[1]Note moreover that some of the literature can cause confusion on this matter due to using the word 'relative' instead of 'relational', which may subsequently be conflated with, or left open to confusion with, Relativity Theory.

E. Anderson, *The Problem of Time*, Fundamental Theories of Physics 190,
DOI 10.1007/978-3-319-58848-3_3

Things which do not act testably or cannot be acted upon are held to be *physical* non-entities. [These can still be held to be a type of thing as regards being able to *philosophize* about them or *mathematically represent* them. Absolute space is an obvious archetypal example of such a non-entity.] The intuition is that imperceptible objects should not be playing causal roles influencing the motions of actual bodies. As a first sharpening of this, in foundational physicist James L. Anderson's [13] view *"the dynamical quantities depend on the absolute elements but not vice versa"*, and an absolute object *"affects the behavior of other objects but is not affected by these objects in turn"* [60]. Background fields are intuitively fields that violate the Action–Reaction Principle.

Relationalism 2) Following Leibniz [3], any entities which are *indiscernible are held to be identical.*

I.e. Relationalism posits that physical indiscernibility trumps multiplicity of mathematical representation. Such multiplicity still exists mathematically, but the mathematics corresponding to the *true* physics in question is the equivalence class spanning that multiplicity. One would only wish to attribute physical significance to calculations of tangible entities which are independent of the choice of representative of the equivalence class. By this e.g. our Universe and a copy in which all material objects are collectively displaced by a fixed distance surely share all observable properties, so they are one and the same. The archetype of such an approach in modern Physics is Gauge Theory (see Chap. 6). This additionally carries the major insight that a mixture of tangible entities and non-entities is often far more straightforward to represent mathematically.

For now, consider separate treatments of space and instantaneous configurations on the one hand, and of time on the other. This befits the great conceptual heterogeneity between these which Chap. 1 began to present. Once this is understood, relational postulates can be stated, and a coherent subset of these are sharply mathematically implementable.

Leibniz's Space Principle is that *space is the order of coexisting things* [3].

Leibniz's Time Principle is that *time is the order of succession of things* [3]. In discussing this, Leibniz's context was whole universes and part of the point he was making is there being no meaningful notion of time with a separate existence in such a setting. I.e. the existence of the events is independent of absolute time, so that the only notions of time left are relational ones. On these grounds, one can infer the position held in this book that 'there is no time at the primary level for the Universe as a whole'. (See also [98] for somewhat similar positions, and the discussion of the Frozen Formalism Problem in Sects. 9.7, 9.10 and 12.1.) Note moreover that Leibniz acknowledged time's ordering property but not its metric property.

Leibniz's Perfect Clock is the distinct suggestion that the whole Universe is the only perfect clock. (See [104] for perfect clocks without mention of Leibniz, and around p. 41 of [906] for details.)

This is to be contrasted with Newton's position that the Universe *contains* clocks, which are regarded as substantially localized objects such as a pendulum clock. [These are placing importance respectively on calibration, and on reading-hand and stability aspects.]

Four objections to Leibniz's Perfect Clock are as follows.

1) It would be operationally impractical to use the whole Universe as a clock. This is because it would take a considerable effort to monitor the whole Universe and one only has very limited knowledge of many of its constituent parts. Including scantly known information from remote parts of the Universe would lower the accuracy of one's timestandard.
2) Suppose one were to go so far as to include the entirety of the Universe's contents in one's quest to 'perfect' one's clock. In this case, one would be treating the entirety of a closed system, at which point apparent frozennesses materializes as per Leibniz's Time Principle.
3) Adopting this principle would additionally open Pandora's box as regards how to reconcile Leibniz's meaning of 'Universe' with that of modern GR's Cosmology [702, 888]. I.e. since the Leibniz' Perfect Clock concept's intent is constructive, it would require making active use of the meaning of the word 'Universe' in its statement. [In contrast, Leibniz's Time Principle is not constructive.]
4) In any case, we shall see in Sect. 5.4 that perfect clocks are not possible in QM.

By these arguments, in this book we do not adopt this perfect clock concept. In Sects. 3.3, 5.4 and 7.7, however clocks that are considerably more extensive than a pendulum clock, pocket watch or atomic clock—such as those which are based on the Solar System—are considered, at least for calibration purposes. These can be taken to carry some vestige of the 'perfect clock' concept, but now realistically balanced with how Physics is about *precision* rather than about perfection.

Mach pointed out some flaws in Newton's bucket argument. The rotation is with respect to the 'fixed stars', pointing to the hitherto tacit inclusion of the effects of distant matter. Mach furthermore noticed that allowing for the bucket to be materially significant—*"several leagues thick"* [632]—is outside of the situation overruled by observation. This led on to various statements concerning the hypothetical origin of inertia, along the lines of 'the distribution of masses in the Universe determines inertia at each point'. Although such a *'Mach's Principle for the Origin of Inertia'* is Mach's best-known insight in the foundations of Mechanics, it plays a limited role in this book. Mach's foundational suggestions are, moreover, somewhat disjoint; the ones that this book does build upon are, rather, the following.

Mach's Space Principle is that [632] *"No one is competent to predicate things about absolute space and absolute motion. These are pure things of thought, pure mental constructs that cannot be produced in experience. All our principles of mechanics are, as we have shown in detail, experimental knowledge concerning the relative positions of motions and bodies."*

Mach's Time Principle, on the other hand, is that [632] *"It is utterly beyond our power to measure the changes of things by time. Quite the contrary, time is an abstraction at which we arrive through the changes of things."* I.e. '*time is to be abstracted from change*'. Indeed, it is change that we directly experience, and temporal notions are merely an abstraction from that, albeit a very practically useful abstraction if chosen with due care.

A further inter-connection is that Mach's Time Principle *resolves* Leibniz's Time Principle's timelessness; this is further developed in Chaps. 9, 12 and Parts II and III.

Let us end by pointing to Broad's caution [171] that time and change are "*the hardest knot in the whole of philosophy*".

3.2 Concrete Example of Relational Particle Mechanics (RPM)

Historically, however, there was a lack in viable relational theories or formulations of Mechanics. The comparatively recent *Relational Particle Mechanics (RPM)* theories, starting with that of Barbour and Bertotti (1982) [105] has made up for this deficiency. This is named after foundational physicist Julian Barbour and physicist Bruno Bertotti, and is based on the *Leibniz group* of the transformations (2.5), (2.6) with label time dependent $\underline{k}$ and $\underline{\underline{R}}$ in the role of kinematical group. See Chap. 9 for a brief outline; such models are used extensively as examples in Parts II and III.

3.3 Ephemeris Time as a Realization of Mach's Time Principle

Around the turn of the 20th century, departures from predicted positions of celestial bodies were noted, especially for the Moon. These were moreover most succinctly accounted for not by modifying lunar theory but rather by considering the rotation of the Earth to inaccurately read off the dynamical time. Physicist Willem de Sitter [231] explained this as follows. "*The 'astronomical time', given by the Earth's rotation, and used in all practical astronomical computations, differs from the 'uniform' or 'Newtonian' time, which is defined as the independent variable of the equations of celestial mechanics.*" This is a major example of clock bias and calibration.

This was then addressed by using the Earth–Moon–Sun system as providing a superior timestandard. Here, astronomer Gerald Clemence's eventual proposal in 1952 [211] involved a particular way of iteratively solving for the Earth–Moon–Sun system for an increasingly-accurate timestandard that came to be known as the *ephemeris time*. This can be viewed as an improved realization of the time of Newtonian Mechanics.

Whereas such an ephemeris time has long been in use, its Machian character has only relatively recently been remarked upon (see Chaps. 15 and 23 for details).

Ephemeris time is also interesting as a notable exception to basing clocks upon periodic motions. This is through its *incorporating* irregularities. It is also an example of a highly accurate but inconvenient primary process, as opposed to consulting a convenient 'reading hand'.

Finally, one passed from a sidereal time based *time-unit* followed suit in the late 1950s. In 1967 the time-unit was redefined so as to bring it in line with the atomic clock timestandard (Sects. 1.12 and 5.5).

3.4 Universality of Relational Thinking

Furthermore, the arguments of Leibniz and Mach are philosophically compelling enough that they should apply to not just Mechanics but to Physics as a whole. I.e these form a *universal* position over the set of laws of Physics. As subsequent Chapters shall reveal, this was a significant aspect of Einstein's thinking in developing SR and GR. See Chaps. 4 and 9 for the extent to which SR and GR succeed in addressing and resolving the absolute versus relational motion debate.

3.5 Electromagnetic Unification and the Luminiferous Aether

The last—and most historically substantial—issue to discuss arises from considering Electromognetism as unified by noted physicist James Clerk Maxwell.[2]

Let us first consider Electromagnetism in non-steady situations (i.e. beyond those in Chap. 2). The $\underline{\mathrm{E}}$ and $\underline{\mathrm{B}}$ fields are furthermore interrelated by the

$$\textit{Faraday–Lenz Law}, \quad \underline{\partial} \times \underline{\mathrm{E}} = -\dot{\underline{\mathrm{B}}}. \tag{3.1}$$

Maxwell subsequently found a *displacement current* $\mu_0\epsilon_0\,\dot{\underline{\mathrm{E}}}$, which modifies Ampère's Law to the

$$\textit{Ampère–Maxwell Law}: \quad \underline{\partial} \times \underline{\mathrm{B}} = \mu_0\,\underline{\mathrm{j}} + \mu_0\epsilon_0\,\dot{\underline{\mathrm{E}}}. \tag{3.2}$$

By this, there is a reverse coupling between the $\underline{\mathrm{E}}$ and $\underline{\mathrm{B}}$ fields. This completes electromagnetic unification. One immediate and significant consequence of this was theoretical justification of light being electromagnetic radiation in vacuo ($\rho_{\mathrm{e}} = 0$, $\mathrm{j}^i = 0$) with propagation speed

$$c = 1/\sqrt{\epsilon_0\mu_0}. \tag{3.3}$$

[Also by this relation, ϵ_0 takes an exact value,

$$8.8541878176\ldots \times 10^{-12}\,\mathrm{A}^2\,\mathrm{s}^4\,\mathrm{kg}^{-1}\,\mathrm{m}^{-3},$$

since c and μ_0 are themselves defined to be exact.] Maxwell's displacement current is clearly crucial in this regard, since it alone carries the $\epsilon_0\mu_0$ factor involved in the propagation. Indeed, that $\underline{\mathrm{B}}$ obeys the wave equation in vacuo follows from $\ddot{\underline{\mathrm{B}}} = -\underline{\partial} \times \dot{\underline{\mathrm{E}}} = -\frac{1}{\mu_0\epsilon_0}\underline{\partial} \times \{\underline{\partial} \times \underline{\mathrm{B}}\} = \frac{1}{\mu_0\epsilon_0}\triangle\underline{\mathrm{B}}$, the second equality of which involves Maxwell's displacement current. On the other hand, that $\underline{\mathrm{E}}$ obeys the wave equation in vacuo follows from $\ddot{\underline{\mathrm{E}}} = \frac{1}{\mu_0\epsilon_0}\underline{\partial} \times \dot{\underline{\mathrm{B}}} = -\frac{1}{\mu_0\epsilon_0}\underline{\partial} \times \{\underline{\partial} \times \underline{\mathrm{E}}\} = \frac{1}{\mu_0\epsilon_0}\triangle\underline{\mathrm{E}}$, where the

[2]This is the only part of Chap. 3 that is used in the standard development of Physics (Chaps. 4 to 7). The other parts feed into Chaps. 9 to 12's account of classical Background Independence and Quantum Gravity.

first equality involves Maxwell's displacement current. Both additionally contain the wave operator $\Box := -c^{-2}\partial_t{}^2 + \triangle$. Furthermore, oscillations in these fields sustain each other, so light just continues to propagate in vacuo.

One can automatically take into account the homogeneous Maxwell equations (2.18), (3.1) by formulating Electromagnetism in terms of a vector potential $\underline{A}$ such that $\underline{B} = \underline{\partial} \times \underline{A}$ and a scalar potential Φ such that $\underline{E} = -\underline{\partial}\Phi - \dot{\underline{A}}$. This leaves us with two inhomogeneous (charge or current sourced) equations (3.1), (3.2) in terms of $\underline{A}$ and Φ.

To have a full grasp of Electromagnetism, we also require a law to compare the motion of (constant mass) charged and uncharged particles in the presence of an electromagnetic field. This is provided by the *Lorentz Force Law*,

$$\ddot{\underline{x}} = \frac{e}{m}\{\underline{E} + \dot{\underline{x}} \times \underline{B}\}. \tag{3.4}$$

Note also that Maxwell's equations do not specify *with respect to which frame* c is the speed of light. Contemporary experience with other types of waves in the 19th century suggested that light should be the excitation of some medium: the '*luminiferous Aether*'.

Let us finally consider another significant consequence of Maxwell's unification of Electromagnetism, which took longer to notice and be appreciated as theoretically significant. Namely, that the set of equations (2.18), (3.1), (2.13), (3.2) have ceased to be invariant under the Galilean transformations. Instead, they are invariant under the Lorentz group, which consists of the ordinary rotations and *boosts* (see also Appendix B.2). Without loss of generality by choice of coordinate system,

$$\begin{aligned} &t \longrightarrow t' = \gamma\left\{t - vx/c^2\right\}, \qquad x \longrightarrow x' = \gamma\{x - vt\}, \\ &y \longrightarrow y' = y, \qquad z \longrightarrow z' = z \end{aligned} \tag{3.5}$$

is the boost for passing from a rest frame to one moving with constant velocity v in the x direction. Here, the *gamma factor* $\gamma := 1/\sqrt{1 - \{v/c\}^2}$. For other directions of motion, rotate the axes, apply (3.5) and then rotate back. To further establish that this change of invariance group is tied to Maxwell's displacement current, one check that the system of equations without this still possesses Galilean invariance (Ex I.11).

Assuming the existence of the Aether, its rest frame would be expected to be privileged by Maxwell's equations, by which the lack of Galilean invariance was not perceived as an immediate impasse. This led to the proposal that, out of Electromagnetism not being Galileo-invariant, experiments involving it could be used to determine motion with respect to the Aether rest frame. There was moreover speculation that this Aether rest-frame might coincide with Newton's absolute space (see e.g. p. 3 of [736]). However, the Michelson–Morley experiment (Ex I.9) gave a null result[3] for the velocity of the Earth relative to the Aether. Within the framework

[3]Today this is known to be null to 1 part in 10^{17} [447].

of Aether theory, this was in contradiction with observations of stellar aberration (Ex I.8) implying the Earth's motion through the Aether.

George Fitzgerald and Lorentz attempted to explain the above observations *constructively* in terms of the inter-particle distances for particles travelling parallel to the Aether flow being somehow contracted.

In contrast, Albert Einstein had a different, *axiomatic* strategy. This is akin [284, 285] to the more well-known case of how Thermodynamics can be based on the non-existence of perpetual motion machines. Following Einstein's approach, the outcome of the Michelson–Morley experiment can be elevated from a null result about motion and Electromagnetism to a *universal* postulate. Rather than there being Galilean invariance for Mechanics, Lorentz invariance for Electromagnetism and whatever other invariance for other branches of Physics, he gave the next Chapter's postulates.

Chapter 4
Time, Space, Spacetime and Laws in Special Relativity

4.1 Special Relativity (SR)

The *Relativity Principle* [281, 718, 736] is that all inertial frames are equivalent for the formulation of all physical laws.

This is intended to be a universal statement. There is however a source of nonuniqueness in the definition: different theoretical frameworks can have different notions of 'inertial frame'. The Relativity Principle translates to the laws of Nature sharing a universal transformation group under which they are invariant; it remains to be determined which transformation group is involved. There are two obvious physical possibilities, distinguished by whether the laws of Nature involve finite or infinite maximum propagation speed, $c_{\max}$.

If existence of absolute time is adopted as a second postulate—the *Galilean Relativity Principle*—the infinite case is selected. Universally Galileo-invariant Physics ensues, as presented in Chap. 2. On the other hand, the *Lorentzian Relativity Principle*—that light signals in vacuo are propagated rectilinearly with the same velocity at all times, in all directions, in all inertial frames—is adopted, the finite case is selected.

Moreover, the chosen speed serves universally, so it is unique (over all matter species)[1] So $c_{\max}$ can take the value c—the speed of light—without loss of generality. This gives a universally Lorentz-invariant Physics.

In the infinite case, Electromagnetism would need to be corrected, whereas, in the finite case, Newtonian Mechanics would need to be. Einstein chose the latter. N.B. that this choice involves following a Law of Nature rather than some postulated absolute structure. Also, whereas experimental evidence for Electromagnetism was already ample, that for Newtonian Mechanics was at that point confined to the low velocity ($v \ll c$) regime, in which Galilean transformations are a very good approximation to Lorentz ones. $\gamma \simeq 1 + \frac{1}{2}\{\frac{v}{c}\}^2$, as is the correction term to t, so e.g. for the motorway speed limit, $\{\frac{v}{c}\}^2$ is 1 part in 10^{14}. For the fastest planes, this is still within around 1 part in 10^{11}.

[1] See also Fig. 4.1 in this regard.

E. Anderson, *The Problem of Time*, Fundamental Theories of Physics 190,
DOI 10.1007/978-3-319-58848-3_4

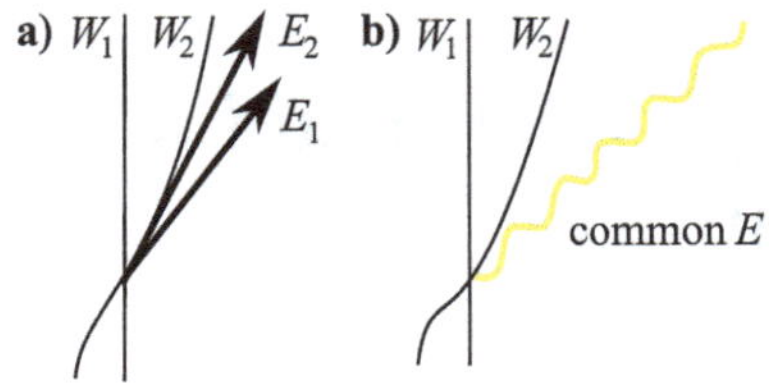

Fig. 4.1 **a)** Particle emissions as seen from an observer's frame differ between emission frames. **b)** Light emissions, however, are in each case the same. The universality of **b)** contributes to light signals being appropriate building block [349] for a universal—i.e. matter species independent —Theory of Relativity

Indeed the investigation of the high-velocity regime eventually verified Einstein's corrections to Newtonian Mechanics. This example of the great predictive power of SR is furthermore compounded by universality. In this way, for each branch of Physics, specific corrections are obtained by requiring the corresponding laws to be Lorentz-invariant. The concept of non-materially substantiated media and the proposal at the end of the previous Chapter were dismissed in this manner. Physics was subsequently rebuilt on the premise that no branch of the subject should have any room for concepts along such lines.

Galilean Relativity (Chap. 2.5) can now also be viewed as the $c_{\text{max}} = \infty$ limit of the above. *Carrollian Relativity* is the less well-known opposite limit $c_{\text{max}} = 0$: see [84, 619] and Fig. 4.4.c). This is named after Lewis Carroll, for the Red Queen's musing *"Now, here, you see, it takes all the running you can do, to keep in the same place"* [196].

These limits can be viewed as cases of group contraction (Appendix E). Finally, Bacry and Levy–Leblond's [84] generalized axiomatization leads to an extended set of *eleven* relativities (six based on relative notions of time and five on absolute time).

4.2 Invariant Interval, Indefinite Metric and Proper Time

We next introduce the metric as the invariant corresponding to the Lorentz and Poincaré transformations (named after noted physicist Hendrik Antoon Lorentz and renown mathematician Henri Poincaré). This is encoded by the Minkowski metric η with components $\eta_{\mu\nu}$, which can always be put into the form diag$(-1, 1, 1, 1)$ with respect to some basis.[2] This realizes a type of Flat Geometry. The line element takes the form

$$\mathrm{d}s^2 = -c^2\mathrm{d}t^2 + \|\mathrm{d}\underline{x}\|^2 = -c^2\mathrm{d}t^2 + \mathrm{d}r^2 + r^2\mathrm{d}\Omega^2. \tag{4.1}$$

Similarly to in Euclidean space, 'lengths' and 'angles' can be characterized in terms of a corresponding inner product. There is however now a physical distinction be-

[2]This is using the $[ct, \underline{x}]$ version of dimensionally-homogeneous coordinates.

tween time and space that is implemented mathematically via the indefinite signature of the metric. So in this context, one 'length' concept is replaced by lengths *or* times, and one 'angle' concept by angles *or* boost parameters.

Measuring the general interval, moreover, requires both rods and clocks. Thereby, spacetime co-geometrization does not extend to the operational level: devices which measure extent in space and duration in time remain distinct. However, synchronization procedures now non-trivially involve spatial measurements, by which there is some loss of independence between spatial and temporal measuring procedures.

Whereas nonzero vectors in $\mathbb{R}^3$ space always have positive norms $|x|^2$, there are three types of nonzero vector in Minkowski spacetime $\mathbb{M}^4$. Those with negative norm are called *timelike*, with zero norm, *null*, and with positive norm, *spacelike*. The existence of the three types of SR spacetime vector is central to the physical interpretation of SR. Namely (and with reference to Fig. 4.3's concept of worldline of a particle) massive particles follow timelike worldlines. Massless particles follow null worldlines. Finally, no physical form of particles follow spacelike worldlines.

Let the frame F' with coordinates t', x', y', z' be related to the frame F with coordinates t, x, y, z by uniform relative motion with velocity v such that, firstly, the x and x' axes coincide. Secondly, the primed origin runs along the unprimed x-axis. Finally, the y, y' and z, z' axes are pairwise parallel.

For a rigid rod of length $\Delta x'$ lying at rest in frame F' on the x'-axis, an observer in frame F envisages the rod to be of length

$$\Delta x = \Delta x'/\gamma \quad \text{(length contraction).} \tag{4.2}$$

Also a clock ticking out a time interval $\Delta t'$ while at rest in frame F' is observed, from another frame F, to tick more slowly: in accord with

$$\Delta t = \gamma\, \Delta t' \quad \text{(time dilation).} \tag{4.3}$$

The above two results follow from (3.5) algebraically, or graphically from Figs. 4.2.b) and c) respectively.

Proper time elapsed along an arbitrary wordline is given by

$$\Delta\tau := \int_{\tau_{\rm in}}^{\tau_{\rm fin}} \mathrm{d}\tau = \int_{t_{\rm in}}^{t_{\rm fin}} \mathrm{d}t/\gamma, \tag{4.4}$$

for t the frame in question F's time and γ built out of the velocity in F. This is just the rest-frame case of the relativistic interval.

Associated 'paradoxes' are as follows. Firstly, in the 'twin paradox', one twin stays at home and the other going on a return trip through space. Computing the proper time along each twin's worldline (Fig. 4.3.c) points to the travelling twin being younger upon arrival. But did the other twin not move in the same manner relative to the travelling twin, by which their ages should remain the same? This is the prima facie 'paradox', but it has the status of a resolved 'paradox' because under more careful consideration, it is not a symmetric situation. For the twin who

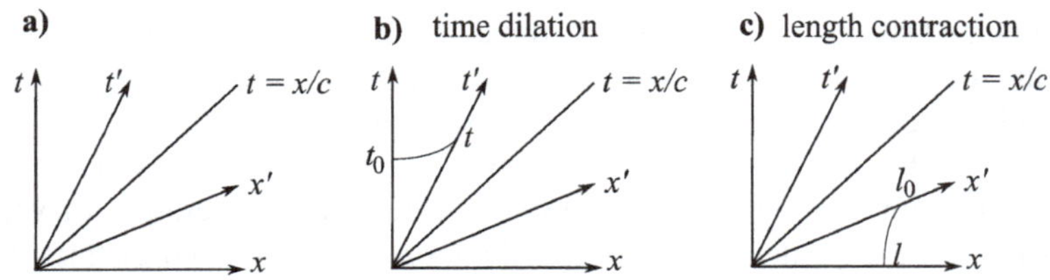

Fig. 4.2 Minkowski spacetime diagrams. **a)** Transforming between Lorentz frames. Each frame can be envisaged as populated by a fleet of observers. Subsequent thinking in terms of families rather than individual observers becomes more meaningful as notions of frame become more general. For now, the Lorentz transformation generates a boost, which corresponds to *tilting*, both of the foliation by space and also of the corresponding time direction. Moreover, it is not physically possible for tilting to extend until the spatial and temporal directions coincide at the 45 degree null line in the figure. [The 45 degree line is singled out as this physically unattainable limit of these axis rotations.] **b)** and **c)** are spacetime diagrams displaying length contraction and time dilation respectively

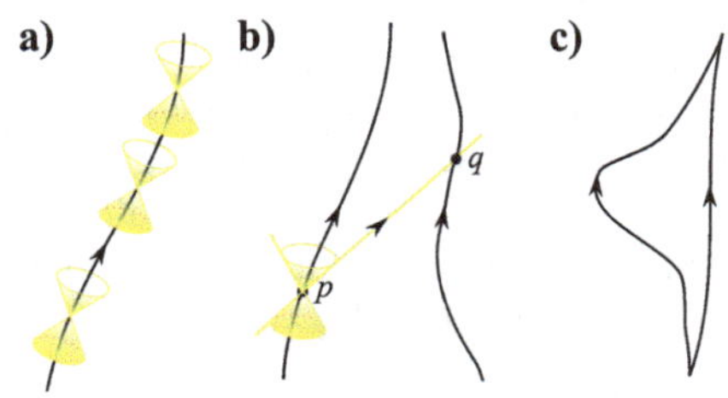

Fig. 4.3 **a)** *Worldline* concept, presented with decoration by null cones at each of its points. For extended objects, worldsheets, worldvolumes... are defined similarly. **b)** Worldlines 1 and 2 with event p on worldline 1 able to influence event q on worldline 2. **c)** The worldlines for the twin 'paradox', and the naïvely symmetric situation that does not actually model the physics experienced by the two twins

stayed at home remained in an inertial frame F, whereas the travelling twin *experienced acceleration*. Secondly, the length contraction 'paradox' can be phrased in terms of whether one can get a fast-moving long pole into a narrower garage. Non-inertialness is one key issue here again: upon the front end crashing into the back wall of the garage. The second key issue is that, signalling speed is finite, by which the back end of the pole does not yet know what has happened to the front end, so it continues for a while to move into the garage undisturbed (Ex I.7).

4.3 Minkowski Spacetime's Geometrical Structure and Its Physical Meaning

In the new *Minkowskian Paradigm*, space and time can be co-geometrized as spacetime. Time is here a coordinate on spacetime; contrast with the absolute time of Newtonian Physics's external character. Time has moreover been described as 'just another' coordinate on spacetime, indicating that time and space are less distinct in Minkowski's Paradigm than in Newton's. This has drawn some comparison (see e.g. [596]) with how the 'vertical direction' ceases to be special in passing from near-Earth to universal Gravitation. See however Sect. 4.6 as regards limitations on loss of distinction between time and space in SR.

Upon shifting from the Newtonian to the Minkowskian Paradigm, Newton's notions of absolute space and time cease to apply. For instance, one can no longer assume privileged spatial surfaces of simultaneity. The privileged surfaces are, rather, *light cones*, on which the free motion of light occurs. Moreover, these surfaces are shared by the free motion of all other massless particles, by Einstein's postulates: there is a *universal null cone structure* in Classical Physics. Additionally, massive particles are permitted only to travel from a spacetime point (event) into the interior of the future null cone of that event. Of particular significance, in free 'inertial motion' in SR, all massive particles follow timelike straight lines whereas all massless particles follow null straight lines.

Having brought in the first paragraph's co-geometrization, it makes sense to implement the Laws of Physics in terms of the 4-tensors corresponding to Minkowski spacetime $\mathbb{M}^4$ (see Sect. 4.4 for more).

From a geometrical perspective, it is notable that Minkowski spacetime $\mathbb{M}^4$ is flat; in contrast GR's notion of spacetime is in general curved. $\mathbb{M}^4$ is also absolute, in the sense of being a back-stage the Physics occurs *on*.

Causality Theory in SR [736]. All events on and within the future null cone of a spacetime point P can be influenced by P via receiving signals from P. This constitutes the *causal future* of P. This is an absolute structure in the sense that all observers agree on it. Similar statements can be made about the past null cone. For any spacetime point Q in the region not in either cone, one can always find an inertial frame in which Q is simultaneous with P. In this manner, it is appropriate to call all of this region the *causal present* of P.

Null cones indicate which events can be reached from a given event, and which can be communicated with. These are a different kind of absolute surface of significance within SR. Note also that the cones degenerate to the squashed-plane and squeezed-line limits in the cases of Galilean and Carrollian Relativities respectively [Fig. 4.4].

For the usual Lorentzian SR case, the *causal future* of a region R of spacetime $\mathfrak{m}$ is

$$\mathrm{J}^+(\mathrm{R}) := \{\mathrm{P} \in \mathfrak{m} \,|\, \exists \text{ a future directed timelike or null curve from R to P}\}.$$

The *chronological future* is

$$\mathrm{I}^+(\mathrm{R}) = \{\mathrm{P} \in \mathfrak{m} \,|\, \exists \text{ a future directed timelike curve from R to P}\},$$

$$\text{R is } \textit{achronal} \text{ if } \nexists\ \mathrm{P}, \mathrm{Q} \in \mathrm{R} \text{ such that } \mathrm{Q} \in \mathrm{I}^+(\mathrm{P}), \text{ i.e. } \mathrm{I}^+(\mathrm{R}) \cap \mathrm{R} = \emptyset.$$

The *domain of dependence* is $\mathrm{D}^+(\mathrm{R}) := \{\mathrm{P} \in \mathfrak{m} \,|$ every past inextendible causal curve through P intersects R$\}$. (The above causal notions retain their usefulness in

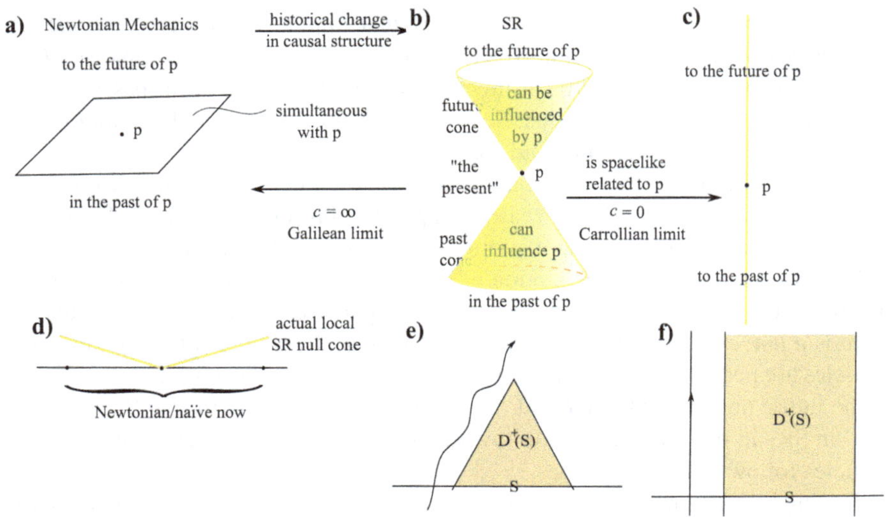

Fig. 4.4 Causality Theory. **a)** Past, present and future of an event p in Newtonian Mechanics. **b)** Past and future null cones of an event p in Minkowski spacetime $\mathbb{M}^4$. **a)** is the Galilean limit of **b)** in which the null cone is squashed into a plane. **c)** is the opposite Carrollian limit of **b)** in which the null cone is squeezed into a line. **d)** The approximate now, tied to 'almost all pairs of events observed in practice are timelike related: see p. 108 of [349]. This gives some idea as to how the Galilean view is fine for the purpose of commonplace experiences. So e.g. for the timescales of 1 ms through to 0.1 s of relevance to the 'specious present' experienced, one's SR instant can only be 600 to 60000 km across. **e)** The future domain of dependence $D^+(S)$ of a spatial region S in SR. The idea is, given a region S, to find the region that is entirely controlled by the information in S alone. The wavy arrow cannot pierce $D^+(S)$ with information from outside of S. There is no notion of domain of dependence in Newtonian theory due to $c = \infty$ meaning that any point can influence any other. However, there is a domain of dependence notion of a different shape in the $c = 0$ limit, **f)**

the more general setting of GR spacetimes [874], but can indeed already be introduced at the SR level.)

The notion of *simultaneity* (and how to set up simultaneity conventions) changes in passing from Newtonian Mechanics to SR [521]. Newtonian-type universal slices of simultaneity cease to apply; they are replaced by attributing physical significance to fixed null cones.

Isometries of Minkowski spacetime $\mathbb{M}^4$. These are those motions ('rigid motions') which leave invariant the Minkowski metric $\boldsymbol{\eta}$. They are comprised of standard rotations, boosts (Fig. 4.4.a), and space- and time-translations. The rotations and boosts themselves form the *special Lorentz group* $SO(3,1)$. Together with the space- and time-translations, they form the *Poincaré group*[3] $Poin(4)$ of transforma-

[3]Time reflections are usually physically undesirable. Space reflections are usually included (see Appendix E).

tions of the form

$$A_{\mu\nu}x^{\nu} + B_{\mu} \tag{4.5}$$

for $A_{\mu\nu}$ antisymmetric (see also Ex I.6). Since these transformations map between this Paradigm's privileged frames, they assume the role of kinematical group. The action of the generators on Minkowskian 4-vectors $X = [t, \underline{x}]$ is analogous to (2.7).

4.4 Lorentzian Tensors (Alias 4-Tensors)

Unlike multiplying by the Euclidean metric $\boldsymbol{\delta}$, multiplying by the Minkowski metric $\boldsymbol{\eta}$ changes entities by bringing in a minus sign, so this is another setting in which we need to distinguish between covariant and contravariant indices. A *contravariant Lorentzian tensor* has transformation law

$$\mathrm{T}^{\overline{\mu}\overline{\nu}\ldots\overline{\rho}} = L^{\overline{\mu}}{}_{\mu}L^{\overline{\nu}}{}_{\nu}\ldots L^{\overline{\rho}}{}_{\rho}\mathrm{T}^{\mu\nu\ldots\rho}, \tag{4.6}$$

whereas a *covariant* one has

$$\mathrm{T}_{\overline{\mu}\overline{\nu}\ldots\overline{\rho}} = L^{\mu}{}_{\overline{\mu}}L^{\nu}{}_{\overline{\nu}}\ldots L^{\rho}{}_{\overline{\rho}}\mathrm{T}_{\mu\nu\ldots\rho}. \tag{4.7}$$

In general, $\mathbf{T}$ can have a mixture of upstairs and downstairs indices, and transform in the corresponding mixed manner.

$\boldsymbol{\eta}$ itself is a special such tensor to which the Minkowskian Paradigm's chronogeometric significance is pinned. The Lorentz group is indeed the group underlying this tensor transformation law. See Chap. 7 of [736] for more about Lorentzian tensors. We finally point to SR giving a further reason for considering tensors in Physics: some hitherto 3-*d* entities can be packaged together as 4-tensors; see the next Section for examples.

4.5 Minkowskian Paradigm of Physics

First reconsider Electromagnetism in this framework. In this special case the main laws (Maxwell's equations) are already Lorentz-invariant, so they require no corrections. Moreover, these laws can be cast in an elegant spacetime notation, and the new conceptual framework greatly facilitates the study of Electrodynamics [281]. Introducing the *electromagnetic field strength tensor* $\mathrm{F}_{\mu\nu}$ (such that $\mathrm{F}_{a0} = \mathrm{E}_a$, $\mathrm{F}_{ap} = \epsilon_{apc}\mathrm{B}^c$) the Maxwell equations are

$$\partial^{\mu}\mathrm{F}_{\mu\nu} = -\mu_0 \mathrm{j}^{\mathrm{e}}_{\nu}, \tag{4.8}$$

$$\partial_{[\mu}\mathrm{F}_{\nu\rho]} = 0. \tag{4.9}$$

If one uses an *electromagnetic 4-potential* $A^\mu = [-\Phi, A^i]$ such that[4]

$$F_{\mu\nu} = 2\partial_{[\mu}A_{\nu]}, \tag{4.10}$$

then (4.9) holds trivially and one is left with

$$\Box A^\mu - \partial^\mu \partial_\nu A^\nu = -\mu_0 j^\mu, \tag{4.11}$$

where the electromagnetic current 4-vector $j^\mu := [\rho_e, j^i]$.

Next, universality required changing the forms of all the other laws of Nature.[5] For Newtonian Mechanics, Newton's Second Law and the definition of momentum are still correct, provided that proper time is employed. The relativistic laws of Nature are a great success. Indeed in many applications a major step toward proposing new laws of Physics is to consider only the Lorentz-invariant possibilities. However, Einstein found that attempting to accommodate Gravitation in this scheme presented significant difficulties (see Chap. 7).

Experimental evidence for SR comes from [736] 1) Electrodynamics, 2) nuclear power based on energy extracted as envisaged by Einstein's 'mass–energy equivalence' relation

$$E = mc^2 \tag{4.12}$$

(this is its rest frame form), 3) a large number of results on SR quantum theory that are outlined in Chap. 5. This evidence reflects that it is far more common for very small quantities of matter to attain relativistic speeds.

The Principles of Dynamics has the further virtue of readily extending to Field Theory (see Appendix K.1 for the Principles of Dynamics for fields in general). As a particular example, Electromagnetism's manifestly special-relativistic spacetime Lagrangian is

$$\mathcal{L}^{A}_{em} = -\frac{4}{\mu_0} F_{\mu\nu}F^{\mu\nu}. \tag{4.13}$$

Moreover, this can be arranged to a non-manifestly SR split spacetime Lagrangian

$$\mathcal{L} = \left\{\{\epsilon_0 \dot{\underline{A}} + \underline{\partial}\Phi\}^2 - B^2/\mu_0\right\}/2 = \left\{\epsilon_0 E^2 - B^2/\mu_0\right\}/2, \tag{4.14}$$

which is useful in dynamical and some quantum contexts (see Chap. 6.3).

4.6 More on Time and Spacetime in the Minkowskian Paradigm

1) Time as a parameter is here manifested by fixed background spacetime. Furthermore, SR replaces Newtonian Mechanics' unique timelike direction with an isotropic continuum of such [596].

[4] [ ] denotes antisymmetrization of the enclosed indices and ( ) denotes symmetrization.

[5] Since this occurred in 1905, we mean all the other classical laws of Nature known at that point.

2) Dating is more contentious in relativistic theories than in Newtonian ones, due to multiplicity of times becoming available.
3) In SR, timefunctions become locally-valued, e.g. in the sense of there being a proper time corresponding to each observer. Proper time is furthermore operationally meaningful.
4) SR has a markedly different notion of simultaneity. In contrast with Newtonian Physics (Sect. 2.4's Item 1), in SR each simultaneity does still remain a copy of the apparent Euclidean Geometry of the corresponding space, in the case of inertial frames. Accelerated frames, however, are another story entirely (see Chap. 8 for further details). SR also involves null cones—and so also the causal structure—to the forefront. For SR, chronological and causal orders are not in general coincident [596] [cf. item 2) of Chap. 2.4].
5) The SR notion of duration is not a difference of datings as per Newtonian Mechanics [item 4) of Sect. 2.4] but rather a function of the past history of the material worldline. This is clear from the twin paradox and requires update of the intuition in Sect. 1.1. Observables and histories issues carry over from the Newtonian case.
6) Global existence of timefunctions also continues to be permitted in SR. SR exhibits some multiplicity of times, albeit of a rather superficial kind: the multiplicity of inertial frames are interrelated by the Poincaré transformations.
7) Concerning the status of Minkowski spacetime $\mathbb{M}^4$, on the one hand it is often argued that in SR space and time can or must be regarded as fused into SR spacetime. Minkowski himself argued that the individual notions of time and space were "*doomed to fade away*" [654]. On the other hand, Broad [171] retorted that SR breaks only *isolation* of space and time, not their *distinction* (as separate notions, each with their own distinctive properties). E.g. signature continues to distinguish timelike and spacelike directions, whereas time retains many of its specific properties that space does not possess.

Although the above Paradigm Shifts in passing to SR are nontrivial [169, 521, 596], they are relatively minor compared to the advent of GR. In particular, time and space in SR are *also* external and absolute in the sense of SR having its own presupposed set of privileged inertial frames. Objections to absolute space of acting but not being actable upon continue to afflict SR by applying just as well to its *class of inertial frames*) [736]. Chapters 7, 9 and 10 subsequently argue that passing to GR is a more major Paradigm Shift in this sense.

As regards trading absolute structures, one has gone from separate absolute t and δ_{ij} to a unified absolute $\eta_{\mu\nu}$. We also recognize that this comes with a metric connection, and then notice that Newton and Galileo's Paradigms also happen to possess a different type of connection. So we pass from four absolute structures in Newton's own view (the fourth is V_i relative to absolute space) to three in Galileo's (V_i removed) and to a single but larger one in SR [279]. We subsequently detail how GR removes this last one.

The new privileged structures are underlied by SR's Minkowski spacetime $\mathbb{M}^4$ possessing suitable Killing vectors (see Appendix E.2 for this concept). A fortiori,

$\mathbb{M}^4$ possesses the maximal number of Killing vectors (10 in 4-*d*). These correspond to the (time and space) translation, rotation and boost generators of $Poin(4)$.

Let us end by pointing out that this Sec's argument is continued at the level of GR in Chaps. 7 to 10; see also [553] in this regard.

4.7 More on SR Clocks

Each observer has their own proper time that *depends on the past history of the clock they are carrying*. If one start with two clocks calibrated side by side, if each is taken on a trip, synchronization will not in general be maintained. So for relativistic-level accuracy—which starts at around 1 part in 10^{12} for typical macroscopic occurrences in life on Earth, such as taking a clock on a transcontinental flight – one needs to say where the clock is and how it is moving.

This accuracy moreover exceeds that which was possible when ephemeris time was introduced in 1952, but such accuracy was not attained until the late 1970s. This is to be contrasted with sub-relativistic accuracy ephemeris: in using this as deduced from the Solar System, it does not matter at which stable position therein one is allotting a timestandard to. So e.g. Clemence did not specify 'on Earth', much less 'in New York' in defining ephemeris time; the question of *where* the timestandard applies was neither posed nor practically relevant until the late 1970s.

Also 'how the clock moves' involves velocity via the time dilation formula; it would in general, and unavoidably to sufficient accuracy, also involve the effects of acceleration on the clock's internal structure. However, e.g. James L. Anderson [13] and Rindler [736] subsequently emphasized the definition of an *ideal clock* as one whose internal structure is completely unaffected by acceleration. In such a case, the clock would measure $\int \mathrm{d}\tau$, so knowledge of velocity in the relevant portion of a worldline would suffice to determine the reading on such a clock.

N.B. that synchronization includes measurement of spatial quantities as well [168, 521].

Clocks subjected to substantial accelerations have a physically understood rationale for clock bias to occur. There is also a tension between the above (SR) ideal clocks and the reality of the practical problems with portable clocks; see also Ex I.13.

Finally, we turn to a conceptual construct. *Einstein's light clocks*—based on reflecting light rays off mirrors—were useful in developing the SR, and indeed GR, notions of spacetime. The particle species independence of light pulses (see Fig. 4.1) ensures this set-up to be suitably universal. A specific construction of such a clock is due to physicist Robert Marzke and physicist and noted conceptual thinker John Archibald Wheeler [645] (Fig. 4.5) and is also discussed e.g. in [13]. See Chap. 7.7 for applications of this idea to currently planned space missions.

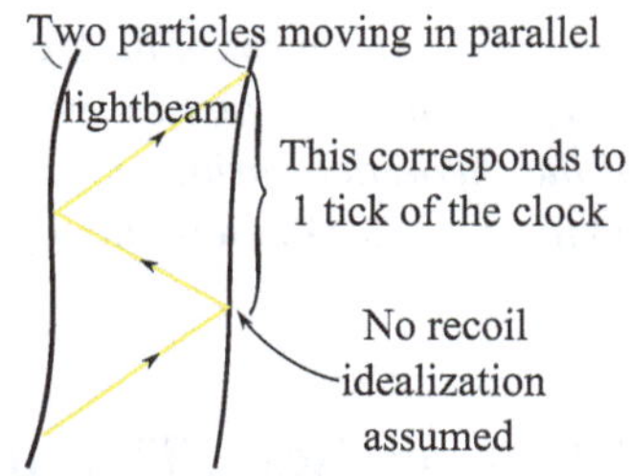

Fig. 4.5 Marzke and Wheeler's more concrete construction for a clock based on mirrors and light rays

4.8 Length Measurement in SR

Laser interferometers are preferable [349] to rods at this level; cf. also the Michelson–Morley experiment's set-up and a composition of Marzke–Wheeler clocks. Note in particular how lengthstandards passed from involving a platinum rod to a property of a light beam. None the less, the Michelson–Morley interferometer arms behave as if they were a rigid rod [13]. Space based laser rangings and future interferometers between probes such as in eLISA[6] go one step further in not having solid support for the arms along which the beams run. Moreover, realistic cases (as opposed to the Marzke–Wheeler point-particle idealization) involve finite solid contraptions at the end of each beam, emitting, reflecting or absorbing.

4.9 Einstein's Eventual Opinion on the Theoretical Status of Clocks and Rods

Einstein's eventual position on this [285] came to be *"One is struck that the theory* [SR] *... introduces two kinds of physical things, i.e. (1) measuring rods and clocks, (2) all other things ... This in a certain sense is inconsistent; strictly speaking measuring rods and clocks would have to be represented as solutions of the basic equations (objects consisting of moving atomic configurations), not, as it were; as theoretically self-sufficient entities. However, the procedure justifies itself because it was clear from the very beginning that the postulates of the theory are not strong enough to deduce from them sufficiently complete equations ... in order to base upon such a foundation of a theory of measuring clocks and rods ... But one must not legalize the mentioned sin so far as to imagine that intervals are physical entities of a special type, intrinsically different from other variables ('reducing Physics to Geometry' etc.)"* Bridgman [169] in particular laid out further support for this position. Einstein's earlier position in this regard [286] is also worth noting. *"The solid body and the clocks do not in the conceptual edifice of physics play the part of irreducible elements, but that of composite structures, which must not play any independent role in theoretical physics. But it is my conviction that in the present*

[6]This is the *European* continuation of the *Laser Interferometer Space Antenna (LISA) project*: an upcoming space mission to probe for gravitational waves; see Chap. 7 for more.)

state of development of theoretical physics these concepts must still be employed as independent concepts; for we are still far from possessing such certain knowledge of the theoretical principles of atomic structure so as to be able to construct solid bodies and clocks theoretically from elementary concepts".

4.10 Exercises I. Time in Mechanics and SR

Background Reading 1) The particularly dedicated reader might complement Chaps. 1 to 3 with philosophical accounts of time such as [730, 906]. Consider also [349]'s comparison of the Aristotelian, Galilean, Newtonian and Minkowskian Paradigms, and physicist and philosopher of physics Max Jammer's [521]'s account of simultaneity. Enthusiastic students who have not studied SR and GR in detail yet can also improve their understanding of these from [349] without investing in any more mathematics than is taught at high school. This book expands on [349]'s account—and not just in the directions already taken in [874]: toward GR—but also toward time in QM and in Quantum Gravity. It also expands on [521] by beginning to lay out other aspects of time aside from the simultaneity exposited so well there, though the current book in no way claims to provide a similar level of philosophical or historical detail. SR can be learned well by working through pp. 1–162 of [736].

Source of Projects 1) The more keen or seasoned readers of foundational or philosophical persuasion might write books comparably detailed to Jammer's on a number of aspects of time in Physics other than his treatise on simultaneity. E.g. on causality, or a spacetime structure sequel to both his book on simultaneity and his book on space [519].

Exercise 1) [Relative coordinates.] i) Take out the centre of mass by passing from point-particle coordinates $\underline{q}_I$ to relative separation vectors $\underline{r}_{IJ} := \underline{q}_J - \underline{q}_I$. ii) With i) causing one to cease to have a diagonal kinetic term, show explicitly in the 3-particle case that a such can be reinstated by taking linear combinations of the $\underline{r}_{IJ}$; these are known as *Jacobi coordinates*.

Exercise 2) Assuming Newtonian Mechanics, what shape does the surface of the water in Newton's bucket form?

Exercise 3) [Fictitious forces.] i) Show that a frame moving with velocity $V(t)$ in the Newtonian Paradigm experiences a fictitious force $-\dot{V} - V$. ii) Show that a general frame in Euclidean space experiences this, the usual rotational fictitious forces of (2.8) and a mixed term $\underline{\Omega} \times \underline{V}$. iii) Consider the hypothetical situation of invariance under the 3-*d* similarity group *Sim*(3), with infinitesimal dilational correction $\dot{x} \longrightarrow \dot{x} - \theta x$. Show that in the general *Sim*(3)-frame all of the above are experienced, alongside $\{\theta^2 - \dot{\theta}\}\underline{x} - 2\dot{\theta}\,\underline{x} + \theta\,\underline{V} + 2\theta\,\underline{\Omega} \times \underline{x}$. iv) Interpret all of the above fictitious forces.

Exercise 4) [Terrestrial timekeeping methods.] i) If a water clock is conical (with vertical axis), how should equal-time notches on it be spaced? What surface of revolution should a water clock be for the notches to be evenly spaced? [Assume water is a perfect fluid throughout.] ii) Estimate how accurately time can be kept by a pendulum which fits inside a house. iii) What was problematic about evaluating longitude at sea? Estimate the error in timekeeping, and consequently in longitude, during Magellan's voyage a) neglecting human errors and storms and b) including these.

Exercise 5) Derive the Poincaré algebra's nontrivial commutation relations (6.25)–(6.26).

Exercise 6) [SR time and length effects.] i) Estimate the time dilation for cosmic muons. ii) Derive aberration and Doppler redshift formulae for the Newtonian and Minkowskian Paradigms, including deriving the former's as a limiting case of the latter's. iii) Account for the factors of γ which occur for quantities in synchrotron physics, for instance the γ^2 in the formula for power radiated, or the γ^3 factor in the ultrarelativistic case's frequency iv) Interpret the Michelson–Morley experiment's set-up, including in terms of the *1-way* and *2-way travel times* exposited in [169]. v) How long a pole one can get into a garage of depth w by entering with it at uniform speed v? [Treat this as a 1-*d* problem; firstly assume that the shock wave travels at speed c, and next build in that these travel at speed $c_{\text{shock}} < c$].

Exercise 7) [Future improvements in measurements of G.] From the perspectives of error analysis and expected technological improvements, what arguments can be put forward for the future superiority of estimating G using interferometer methods rather than Cavendish balances?

Exercise 8) [Simpler cases of the Euler–Lagrange equations.] Work through Appendix J's 3 simplifications of the Euler–Lagrange equations, Legendre transformations in general and the specific ones involved in passage from the Lagrangian to each of the Hamiltonian and the Routhian. If L is homogeneous of degree 1, show that $H = 0$.

Exercise 9) [The Brachistochrone.] Given a fixed starting point $(x, y) = (1, 0)$ and a fixed end-point $(x, y) = (0, 1)$, calculate what shape of wire joining these is to have for a bead threaded upon it to take the least time to travel between the end-points. (Assume Newtonian Gravity in the local 'vertical' approximation.)

Exercise 10) Demonstrate that Maxwell's displacement current is crucial for the mismatch in relativities which led to Einstein's SR, by showing that the Maxwell equations in the regime without displacement current (name regime) are Galileo invariant.

Exercise 11) From first principles, find spacetime action principles for the motion of a relativistic particle, and, assuming that the particle is charged, for its motion coupled to Electromagnetism. Also show that the Lorentz Force Law (3.4) in 4-tensor notation becomes

$$\frac{d^2x^\mu}{d\tau^2} = \frac{e}{m}\mathrm{F}^\mu{}_\nu\frac{dx^\nu}{d\tau}, \tag{4.15}$$

where τ is the proper time.

Exercise 12) The *energy–momentum–stress tensor* of Electromagnetism is

$$\mathrm{T}_{\mathrm{em}}^{\mu\nu} = \frac{1}{\mu_0}\left\{\mathrm{F}^{\mu\rho}\mathrm{F}^{\nu}{}_{\rho} - \frac{1}{4}\eta^{\mu\nu}\mathrm{F}_{\gamma\delta}\mathrm{F}^{\gamma\delta}\right\}. \tag{4.16}$$

Show that this is symmetric and conserved. What statements in terms of $\underline{\mathrm{E}}$ and $\underline{\mathrm{B}}$ does this conservation encapsulate? What is the interpretation of the $\underline{\mathrm{E}} \times \underline{\mathrm{B}}$ quantity arising in this manner?

Exercise 13) [Space Chronometer.] In this case there is no constant rocking by water waves, and one would usually be able to confine travel to be away from strong astral winds. i) Suppose we lived in a Newtonian universe. Estimate the minimum time it would take for a spaceship of mass M containing ejectable gas of mass m to reach α-Centauri by application of piecewise constant accelerations of magnitude $\leq a$. What if we require the spaceship to end up there, rather than just fly past? ii) Repeat these calculations in an SR universe, computing now both the proper time aboard and how much terrestrial time would have elapsed between takeoff and the receipt of signals from the probe upon arrival α-Centauri. iii) Pass from parametrized to specific numerical estimates by use of current-era rocket engines, fuel to payload ratios and the maximum acceleration sustainable by the crew. iv) For various standard aerospace construction materials, estimate at which velocity substantial damage would be imparted by collision with a dust grain: a significant issue as pointed out e.g. in [831]. v) Estimate tolerance bounds on the accelerations and shocks sustainable by the on-board clocks hitherto used by space probes. Compare these with what crew members could sustain.

Exercise 14) Show that a Galilean frame can always be found such that $\partial_i\phi = 0$, whereas $\partial_i\partial_j\phi$ remains invariant. Interpret these results both geometrically and physically.

Chapter 5
Time and Ordinary Quantum Mechanics (QM)

If atoms were classical objects, their observed stability would be in discord with their energy losses due to electromagnetic radiation (Ex VI.5.d). Interpreting this—alongside black body radiation, the photoelectric effect and accounting for atomic spectra—pointed the way to [889] the strange new physics of Quantum Theory.

5.1 A Simple Axiomatization of QM

For now, we consider Ordinary—Nonrelativistic, Background Dependent—QM.[1]

QM Postulate I) The state of a system is now taken to be a complex-valued wavefunction ψ belonging to a *Hilbert space*, $\mathfrak{H}$ilb.[2] This is a complete (Appendix C) complex inner product space (Appendix A.3). $\langle \, | \, \rangle$ denotes this inner product; the ψ are required to be normalized with respect to this: $\langle\psi|\psi\rangle = 1$.

QM Postulate II) Any observable[3]—meaning here any physical quantity whose value can be measured at a given time—can be represented by some linear operator $\widehat{A}$ that acts on the wavefunctions and is *self-adjoint* (Appendix A.3) with respect to the inner product,

$$\langle \widehat{A}^\dagger \psi_1 | \psi_2 \rangle = \langle \psi_1 | \widehat{A} \psi_2 \rangle \tag{5.1}$$

(this acts as a guarantor of real eigenvalues).

[1]If unfamiliar with any of the non-temporal material in this Chapter, see e.g. [487, 599]; some of this Chapter's material on time in QM may however be new. See the start of Part III if interested in a more general and detailed treatment of Quantization.

[2]This is a complete inner product space (see Appendices C.2 and H.2 for a bit more mathematical context for this). This term is often used to mean the infinite version, but the current use covers the finite version as well. The lack of physical distinctions between $|\Psi\rangle$ and the phase-shifted $\exp(i\phi)|\Psi\rangle$ results in the Projective Geometry notion of a *ray* in Hilbert space is a further suitable means of modelling quantum wavefunctions.

[3]For unconstrained theories, the corresponding notion of (for now unconstrained) *classical observable* is entirely trivial: *any* function of the classical variables $A = A(\boldsymbol{Q}, \boldsymbol{P})$ is a classical observable; see Chaps. 9, 25 and 50 for the constrained case.

E. Anderson, *The Problem of Time*, Fundamental Theories of Physics 190,
DOI 10.1007/978-3-319-58848-3_5

In considering multiple classical quantities F and G, commutation $FG = GF$ trivially holds. On the other hand, *noncommutation* is a basic and essential property of the quantum world. F and G are now replaced by quantum operators $\widehat{F}$ and $\widehat{G}$ which do not in general commute; their *commutator*

$$[\widehat{F}, \widehat{G}] := \widehat{F}\widehat{G} - \widehat{G}\widehat{F} \neq 0 \tag{5.2}$$

indeed serves as a measure of their noncommutation. Additionally, the *Correspondence Principle* is that one part of Quantization may be thought of as passage from classical Poisson brackets to commutators of the corresponding quantum-mechanical operators:

$$\{F, G\} \longrightarrow \frac{\hbar}{i}[\widehat{F}, \widehat{G}]. \tag{5.3}$$

The fundamental commutator is

$$[\widehat{x}^i, \widehat{p}_j] = i\,\hbar\,\delta^i{}_j. \tag{5.4}$$

On the other hand, the angular momentum operators[4] obey

$$[\widehat{J}_i, \widehat{J}_j] = i\,\hbar\,\epsilon_{ij}{}^k \widehat{J}_k, \qquad [\widehat{J}_i, \widehat{x}^j] = i\,\hbar\,\epsilon_i{}^j{}_k \widehat{x}^k, \qquad [\widehat{J}_i, \widehat{p}_j] = i\,\hbar\,\epsilon_{ij}{}^k \widehat{p}_k. \tag{5.5}$$

I.e. the $SU(2)$ [= locally $SO(3)$: see Appendix E] structure constants and the conditions that x^i, p_i are vectors under the $SU(2)$ transformations. Note the close parallel with the algebraic form of the classical Poisson brackets (2.24). However, in the general case we would need to *construct* the objects that play an analogous role to the above operators. This is under-emphasized in most QM textbooks, which often just take the $\widehat{q}^i$, $\widehat{p}_i$ and $\widehat{J}_i$ for granted. Moreover, the algebraic structure the commutators form would not necessarily coincide with some classical precursor's Poisson one. These more general considerations are termed *Kinematical Quantization* (see [475], or Chap. 39 for an outline).

One consequence of noncommutation is that 'promoting classical quantities to quantum operators' involves choosing how to *operator-order* them. E.g. is the classical pq to be represented by $\widehat{p}\,\widehat{q}$ or $\widehat{q}\,\widehat{p}$ now that these are inequivalent, or possibly some other expression such as the symmetric operator ordering $\{\widehat{p}\,\widehat{q} + \widehat{q}\,\widehat{p}\}/2$.

Heisenberg's Uncertainty Principle (named after noted physicist Werner Heisenberg) is a fundamental limitation upon how accurately one can know a particle position and its conjugate momentum:

$$\Delta q\,\Delta p \leq \frac{\hbar}{2}. \tag{5.6}$$

One consequence of this is the replacement of the worldline concept for a particle by the more diffuse and generally unstable wavepacket concept (Fig. 5.1).

[4]See Sect. 5.2 for why we use $\widehat{J}_i$ in place of $\widehat{L}_i$ here.

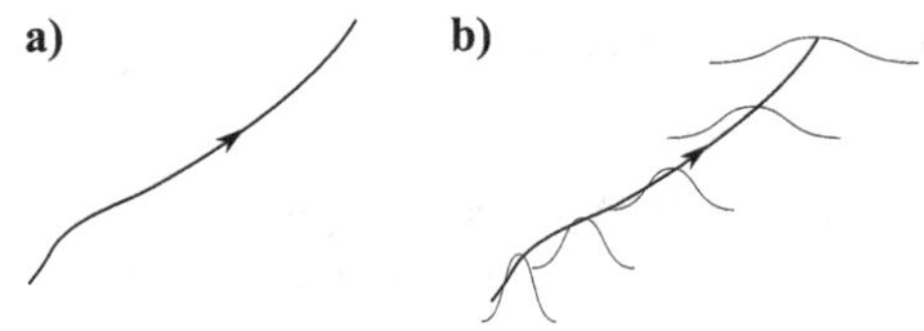

Fig. 5.1 The classical worldline **a)** becomes a spreading wavepacket **b)** [899]

The *Generalized Uncertainty Principle* is

$$\Delta A \Delta B \geq \frac{1}{2}|\langle[\widehat{A},\widehat{B}]\rangle| : \tag{5.7}$$

uncertainty is tied to noncommutativity. The Angular Momentum Uncertainty Principle $\Delta J_x \Delta J_y \geq \frac{\hbar}{2}|\langle\widehat{J_z}\rangle|$ is another well-known subcase of this.

QM Postulate III) Only inner product combinations are physically meaningful. The quantity which is physical is in general an *inner product* with operator insertion

$$\langle\psi_1|\widehat{O}|\psi_2\rangle. \tag{5.8}$$

Some special cases are the *expectation* of operator $\widehat{O}$ in state ψ_1, $\langle\psi_1|\widehat{O}|\psi_1\rangle$, and the *overlap* $\langle\psi_1|\psi_2\rangle$ between states ψ_1 and ψ_2.

It is these inner product combinations that arise from measurements, and predictions based thereupon happen to be *inherently* probabilistic. I.e. observations are consistent with QM systems not possessing certain properties, but rather the measurements one makes yield an eigenvalue that can in principle only be predicted probabilistically. In particular, if a system is in a quantum state $\psi(x) = \sum_{i=1}^{\infty} o_n\psi_n(x)$—an eigenfunction expansion form guaranteed[5] by the completeness—then the probability that a measurement produces a particular eigenvalue o_n is

$$\text{Prob}(O = o_n \,|\, \text{state is } \psi) = |o_n|^2 \quad (\textit{Born Rule}), \tag{5.9}$$

after physicist Max Born. Note that QM probabilities do not obey all of the classical probability axioms (if interested, consult Appendix P.1), in particular due to the overlaps. Negative probabilities are also at least *contemplated in setting up* Quantum Theory, even if the presence of these has been used to select against the interpretational proposals these arise for.

Moreover, the outcomes of measurements are known to be real, so if these are to be eigenvalues, it is crucial for the eigenvalues to be real, as guaranteed by the self-adjointness of physical operators.

If $\psi \to U\psi$ and $\widehat{O} \to U\widehat{O}U^\dagger$ for U a unitary transformation ($U^\dagger = U^{-1}$), then $\langle\psi_1|\widehat{O}|\psi_2\rangle \to \{\langle\psi_1|U^\dagger\}\{U\widehat{O}U^\dagger\}\{U|\psi_2\rangle\} = \langle\psi_1|\{U^\dagger U\}\widehat{O}\{U^\dagger U\}|\psi_2\rangle = \langle\psi_1|\widehat{O}|\psi_2\rangle$. Clearly an antiunitary operator—such that $U^\dagger = -U^{-1}$—also satisfies the above.

[5]The guarantee here is actually for an operator whose eigenvalues form a discrete and non-degenerate set, but one can more laboriously generalize one's way round these limitations [487].

Moreover, a Theorem of physicist Eugene Wigner's[6] guarantees there are no other possibilities.

QM Postulate IV) The wavefunction ψ obeys a *quantum wave equation*: for now, this is the *time-dependent Schrödinger equation* (after noted physicist Erwin Schrödinger)

$$i\hbar\frac{\partial\psi}{\partial t} = \widehat{H}\psi. \tag{5.10}$$

The *time-independent Schrödinger equation*

$$\widehat{H}\psi = E\psi \tag{5.11}$$

then arises when the separation ansatz $\Psi(\underline{x}, t) = \exp(iEt/\hbar)\psi(\underline{x})$ applies.

The *Evolution Postulate* is that the various probabilities involved at a given stage always sum to one; thus, this postulate is also termed *unitarity*. This evolution is in accord with the theory's time-dependent quantum wave equation.

Moreover, compliance with the quantum wave equation may alter which inner product applies. I.e. the *physical* inner product is the *final dynamical* one rather than the incipient *kinematical* one in the instance of these differing. In our case, the final inner product is Schrödinger's, with respect to which there is probability current conservation. This refers to the probability current

$$\mathrm{j}_i = \frac{\hbar}{2mi}\psi^* \overleftrightarrow{\partial_i}\psi, \tag{5.12}$$

which obeys a *conservation equation* $\partial_t\rho + \underline{\partial}\cdot\underline{j} = 0$ analogous to those of Classical Physics, corresponding to the *probability density* $\rho := |\psi|^2$. Here $\leftrightarrow$ denotes backward as well as forward action of the derivative. This in turn works via H itself being self-adjoint with respect to this, by which the inner product in question succeeds in being compatible with H.

Also note the equivalent *Heisenberg picture* in which the wavefunctions are stationary and the operators evolve according to the *Heisenberg equation of motion*

$$\frac{\mathrm{d}\widehat{O}}{\mathrm{d}t} = \frac{i}{\hbar}[\widehat{H}, \widehat{O}(t)] + \frac{\partial\widehat{O}}{\partial t}. \tag{5.13}$$

Historically this and the preceding *Schrödinger picture* with its evolving wavefunctions and stationary operators were developed separately, but were then shown to be equivalent in a manner which boils down to $\{\langle\Psi|U^\dagger\}O\{U|\Psi\rangle\} = \langle\Psi|\{U^\dagger OU\}|\Psi\rangle$.

QM Postulate V) *Collapse of the Wavefunction*. This is a second dynamical process that is held to occur in Ordinary QM despite its not being described by the evolution equation of the theory. By this, measuring $\widehat{O}$ for a system in state ψ that

[6]Any symmetry transformation in QM can be represented by an operator on the Hilbert space of states $\mathfrak{Hilb}$ that is either i) linear and unitary or ii) antilinear and antiunitary. See [885] for an accessible proof, and both that and [269, 401] for commentary.

happens to yield o_n has thrown the system into the corresponding eigenstate ψ_n. *Quantum Measurement Problem* subtleties ensue; in a nutshell, QM turns out to require an *interpretation* beyond reading off of what the theory's equations dictate (which was sufficient by itself in Classical Paradigms). The standard position here is the *Copenhagen Interpretation of QM* (after the School of renown physicist and conceptual thinker Niels Bohr). This involves a quantum subsystem being in a surrounding large; in particular the observer is treated as macroscopic and lying outside the quantum subsystem in question.

5.2 Experimental Support for QM and Examples

There is vast experimental support for QM. Some highlights are QM's explanation of the atomic spectra [599], of much of Chemistry at the molecular level [81], radioactive decay [652], and structural properties of matter [556] including explaining solid matter and why metals are shiny. Quantum Theory also explains Particle Physics [712, 885, 886] consistently with accelerator and cosmic ray data.

Example 1) Quantum harmonic oscillators are a useful model of e.g. molecular vibrations. A convenient way of studying these is in terms of creation and annihilation operators

$$\widehat{\underline{a}}^{\dagger} := \{2\, m\, \hbar\, \omega\}^{-1/2}\{m\, \omega\, \widehat{\underline{x}} - i\, \widehat{\underline{p}}\}, \qquad \widehat{\underline{a}} := \{2\, m\, \hbar\, \omega\}^{-1/2}\{m\, \omega\, \widehat{\underline{x}} + i\, \widehat{\underline{p}}\}. \quad (5.14)$$

Their actions on the nth state $|\mathrm{n}\rangle$ are $\widehat{a}^{\dagger}|\mathrm{n}\rangle = \sqrt{\mathrm{n}+1}|\mathrm{n}+1\rangle$, $\widehat{a}|\mathrm{n}\rangle = \sqrt{\mathrm{n}}|\mathrm{n}-1\rangle$—i.e. a raising and a lowering respectively—for a different n for each component of $\widehat{\underline{a}}^{\dagger}$, $\widehat{\underline{a}}$. In dimension d (often 1 or 3), one starts from a *Eucl*(d)-invariant vacuum, building up the other states by applying the creation operator.

Example 2) The rigid rotor model can be characterized in terms of $\widehat{L^2}$ and $\widehat{L}_z$ eigenvalues and eigenfunctions. $\widehat{L}_z\psi = \mathrm{m}\,\hbar\,\psi$ and $\widehat{L^2}\psi = \mathrm{l}\{\mathrm{l}+1\}\hbar^2\psi$ for l the azimuthal angular momentum quantum number and m the 'magnetic' angular momentum quantum number such that $|\mathrm{m}| \leq \mathrm{l}$. This example is closely tied to the representations of the rotation group $SO(3)$, as the first of many instances of Representation Theory (Appendix A.5) playing a significant role in Quantum Theory.

Example 3) The hydrogen atom's quantum equations (see also Ex II.2) separate into a rotor problem angularly and an extra radial equation. This produces the energy spectrum $E(\mathrm{n}) = -\hbar^2/2\,m_{\mathrm{e}}a_0^2\mathrm{n}^2$, for n now the *principal quantum number* and $a_0 := 4\,\pi\,\epsilon_0\hbar^2/m_{\mathrm{e}}e^2$ the atom's *Bohr radius*.

In some ways, it is straightforward to extend this to multi-electron atoms, in particular if one neglects electron-electron interactions to leading order. In this way a simple model of the Periodic Table can be built up.

This array of the chemical elements' row numbers have an additional factor of 2 on top of the $2\mathrm{l}+1$ factor from adding up the possible values of m for each l. Noted physicist Wolfgang Pauli posited his Exclusion Principle so as to account for this factor of 2 as a *"peculiar classically non-describable duplexity"*. This for now appended notion is called 'spin', and is a type of angular momentum. $\widehat{J}_i$ now means total angular momentum, in the sense of orbital angular momentum spin considered together, according to $\underline{\widehat{J}} = \underline{\widehat{L}} + \underline{\widehat{S}}$. The particular case of the spin of the electron can be represented by the *Pauli matrices*

$$\sigma^1 = \begin{pmatrix} 0 & 1 \\ 1 & 0 \end{pmatrix}, \quad \sigma^2 = \begin{pmatrix} 0 & -i \\ i & 0 \end{pmatrix}, \quad \sigma^3 = \begin{pmatrix} 1 & 0 \\ 0 & -1 \end{pmatrix}. \tag{5.15}$$

Heisenberg, Pauli and physicist John Slater incorporated the Pauli Exclusion Principle as wavefunction total antisymmetry, with the associated particles obeying *Fermi–Dirac statistics* (U.4).[7] Another class of particles observed obey *Bose–Einstein statistics* (U.3) associated with wavefunction total symmetry. Finally, no mixed-symmetry particles appear to be realized in Nature. This is an example of a so-called *superselection rule* [817]: specifying pure eigenstates only rather than mixed states (Appendix U.1). Also due to this, not all Hermitian operators are realized as observables.

5.3 Time in Nonrelativistic QM

Three uses of 'time' can be distinguished in Ordinary QM [185, 517].

I) *External time* is the background Newtonian time inherited from Classical Physics, occasionally also known as *laboratory time* in this context. This quantifies when the experiment is set up, the duration of the experiment, when external fields are switched on and off, and so on. Being measured by an external detached laboratory clock, it is not dynamically interconnected with the quantum entities under study in the experiment in question [185].

II) *Dynamical time* is, more relationally, determined by the quantum subsystem itself; this is also termed *internal time* in [185]. It is based on each non-stationary quantum observable $\widehat{O}$ providing its own characteristic time: that within which $\langle\widehat{O}\rangle$ changes significantly [185]. Notions of time of this kind include *time delay* in scattering experiments, *dwell time* in quantum tunnelling, and *lifetime* of unstable quantum states. These are all notions of duration in dynamical time.

III) *Observable time* carries further operational connotations, through involving the measuring apparatus as well as a quantum entity under study. An example is the notion of *time of arrival* at one's detector from a source [185].

I) to III) have more significant quantum-level distinctions than their classical counterparts do. These notions of time enter Ordinary QM in the following ways.

[7]These statistics are named after physicists Enrico Fermi, Paul Dirac, Einstein, and Satyendra Bose.

1) Whereas most physical entities are represented at the quantum level by operators, Pauli (see e.g. [701]) provided an argument placing restrictions on representing time in this manner. This led to 'time playing the role of an external parameter' in QM, though the issue is somewhat confused by the multiplicity of time concepts in QM and by Pauli's argument not being as widely encompassing as originally thought.
2) It is moreover tempting to represent *observable* times by operators, though there are a number of subtleties and impasses with this. For instance, if time is to be treated as an operator, Wigner's Theorem (in footnote 6) requires it to be an *antiunitary* operator $\widehat{\mathcal{T}}$.
3) Returning to Pauli's considerations [699], he pointed out that the commutation relation[8]

$$[\widehat{\mathcal{T}}, \widehat{H}] = i\,\hbar \tag{5.16}$$

cannot hold for physically realistic $\widehat{H}$. This is a consequence of the following result.

Stone–von Neumann Theorem *Any $\widehat{A}$, $\widehat{B}$ such that $[\widehat{A}, \widehat{B}] = i\,\hbar$ have to closely resemble $\widehat{q}$ and $\widehat{p}$.*[9] *In particular $\widehat{q}$ and $\widehat{p}$ are 'unbounded below', i.e. their spectra go down to $-\infty$. On the other hand, physically realistic Hamiltonians must be bounded below, i.e. each of these possess a ground state at finite energy.*

4) Moreover, the *Energy–Time Uncertainty Principle*

$$\Delta\mathcal{T}\Delta E \geq \tfrac{\hbar}{2} \tag{5.17}$$

has an entirely different meaning to that of the other uncertainty relations.

Firstly, it is significant that this is not stated for external time. Reasonably credible interpretations concern, rather, an internal here alias dynamical time, or an observable time. For example, $\Delta\mathcal{T}$ can be interpreted as duration in dynamical time on which each $\langle\widehat{\mathcal{O}}\rangle$ changes by the same amount as the corresponding (averaged) indeterminacy [517]. See e.g. Chap. 5.3 of [517] and [185] for further careful exposition.
5) The commutation relations that all the quantum operators are subjected to are, more precisely, *equal-time* commutation relations. This rests on the notions of 'being at a time' and of simultaneity.

---

[8]See (J.27) for the classical precursor of $\mathcal{T}$. In the case of a conservative system, this is just $t - t(0)$, which can be interpreted as t^{Newton} up to choice of calendar year zero, by which $\mathcal{T} = t^{\text{Newton}}$ and $\Delta\mathcal{T} = \Delta t^{\text{Newton}}$ in (5.17).

[9]This is a Uniqueness Theorem, named after mathematician Marshall Stone and noted mathematician and polymath John von Neumann. It is phrased for the exponentiated (alias Weyl, after mathematician Hermann Weyl) commutation relations; these have a generic unitary representation; see e.g. [407] for a commented proof. Finite Theories lie within the remit of this result, whereas Field Theories do not.

6) Sect. 5.1's *Evolution Postulate* applies again, possibly with new forms for the corresponding time-dependent wave equations.
7) Specification of an *inner product* is tied to time concepts via its use in establishing conservation of probability.
8) Sect. 5.1's collapse of wavefunction continues to hold. This is a separate manifestation of becoming, tied to measurements rather than to the evolution postulate's more habitual case of prescribing a PDE problem from which to deduce dynamical outcome. This additionally carries one of the Arrows of Time since the state of a system is markedly different 'before' and 'after' such a collapse.
9) The Copenhagen Interpretation of QM is built out of *measurements made at a particular time*. This rests on the assumption of a privileged background notion of time. In multi-measurement contexts, this is also tied to dating.

   The Copenhagen interpretation's assumption of a surrounding large is moreover a further type of Background Dependence.
10) QM Postulate II)'s notion of quantum *observable* contains an 'at a given time' clause [483], and rests on items 2) and 4) of Sect. 1.6.
11) In constructing a quantum theory's Hilbert space $\mathfrak{H}$ilb, one is to select a complete set of observables. Various time connotations are subsequently tied to this construct (see Chaps. 24 and 25). Let us note for now that these form a particular algebraic structure under equal-time commutation relations.
12) As compared to 10), a 'history' has no direct physical meaning except in so far as it refers to the outcome of the sequence of time-ordered measurements it consists of.
13) One might a fortiori seek to conceptualize in terms of histories, rather than time, at the primary level.

5.4 Clocks in QM

We next continue with Chap. 3.1's argument against perfect clocks, now at the quantum level.

1) *All quantum clocks occasionally run backwards* . Whereas background Newtonian time appears explicitly in e.g. QM's time-dependent Schrödinger equation, such a time is not precisely operationally realizable by a physical clock. Physicists William Unruh and Robert Wald [862] established this by contradiction. Suppose that there is some quantum observable $\widehat{\mathcal{T}}$ that can serve as a 'perfect' physical clock in the sense that, for some initial state, its observed values increase monotonically with the abstract time parameter t. To include the possibility that $\widehat{\mathcal{T}}$ has a continuous spectrum, decompose its eigenstates into a collection of normalizable vectors $|\tau_0\rangle, |\tau_1\rangle, |\tau_2\rangle \dots$. Here $|\tau_n\rangle$ is an eigenstate of the projector onto the interval of the spectrum of $\widehat{\mathcal{T}}$ centered on τ_n. Saying that t corresponds to a perfect clock has the following meaning.

   A) For each m, $\exists\, n > m$ and $t > 0$ such that Prob($|\tau_m\rangle$ evolves to $|\tau_n\rangle$ in Newtonian time t) $\neq 0$. This formalizes the clock having a non-zero probability of running forwards with respect to $\mathcal{T}$, and means that the physical quantity $f_{mn}(t) := \langle \tau_n | U(t) | \tau_m \rangle = 0$, for $U(t) := \exp(-i\,\widehat{H}t/\hbar)$.

B) For each m and $\forall\, t > 0$, the transition amplitude to evolve from $|\tau_m\rangle$ to $|\tau_n\rangle$ vanishes if $m > n$: the clock never runs backwards.

A) and B) are, however, incompatible with the physical requirement of positive energy (Ex II.9).

2) If $\mathcal{T}$ were to obey (5.16), an even stronger restriction would apply. This is because (5.16) implies that $U(t)|\mathcal{T}\rangle = |\mathcal{T} + t\rangle$, where $\widehat{\mathcal{T}}|\mathcal{T}\rangle = \mathcal{T}|\mathcal{T}\rangle$ as would be required for a perfect clock. However, it is well-known that self-adjoint operators which satisfy (exponentiable) representations of (5.16) necessarily have whole-$\mathbb{R}$ spectra; this result follows from the previous Section's Stone-von Neumann Theorem. So (5.16) is manifestly incompatible with requiring $\widehat{H}$ to be a positive operator.

1) and 2) amount to QM teaching us that there is a limit on global-in-time monotonicity, and bounds on accuracy criteria.

3) For a quantum clock[10] of mass M to run for a maximum interval of time T with an accuracy (here a minimal discernible time-interval) τ, the *Salecker–Wigner clock inequalities*—(named after Wigner and physicist H. Salecker) [761]

$$\text{linear spread} \quad \lambda \geq 2\sqrt{\hbar T/M}, \qquad \text{clock mass} \quad M \geq \frac{4\hbar}{c^2 \tau}\frac{T}{\tau} \tag{5.18}$$

hold. In this way, at the quantum level, a clock's tick-duration needs to be traded off against its longevity, which thus becomes a nontrivially finite concept.

5.5 Advent of Atomic Clocks

Atomic clocks greatly increased clock stability, e.g. exceeding the threshold for relativistic effects being non-negligible in the late 1970s [82]. By now, we have atomic clocks for which the stability of the primary timestandard is 2 parts in 10^{16} [621] or even better. This substantially outstrips the accuracy of contemporary astronomical timestandards.

4) 'Cleaner Clocks Principle'. The main limitations of astronomical timestandards come from limitations on the detailed knowledge of the contents of the Solar System, some of whose internal workings are speculative. For instance, details of the behaviour of the Earth's mantle enter the fluctuations which invalidate accurate use of sidereal time. In contrast, atomic clocks have simple and well-known internal constitution and physics, while being selected and further designed for being stable and well-shielded from external disturbances.

[10] This is valid for clocks that are dynamical systems that keep track of their own state, e.g. a pendulum with hands rather than just a pendulum.

Atomic clocks are themselves based upon periodic motions. By being small and localized (at least in comparison to the Solar System), atomic clocks have the following further useful properties. Firstly, they are straightforward to shield from disturbances Secondly, they are convenient as reading hands, in particular far more so than the position of the Moon. Thirdly, one need not worry about position-dependent relativistic effects within each atomic clock itself (unlike with Solar System based timestandards).

Atomic clocks, moreover, remain based on an ephemeris type conceptualization [364]. Clock bias might in principle still apply; atomic clocks still require recalibration checks. However, in the early days of atomic clocks it was determined that they read out ephemeris time to at least 1 part in 10^9 [638], which substantially eased the passage from ephemeris to atomic timestandards. None the less, this has the status of a *null experiment*, so one should keep on testing whether this premise continues to hold as precision elsewise improves....

We finally note that, a fortiori, QM underlies all time measurements. In some cases, this is by providing the atoms that emit and absorb, and in others by providing the solid state that quartz crystals, gear wheels, planets and sundials are made of.

5.6 Quantum Inputs to Measuring Lengths and Masses

Quantum Theory also underlies all length measurements, whether by providing the solid state that permits rods, reflectors and lasers or by providing the atoms which emit and absorb. Marzke and Wheeler [645] showed how quantum dependence enters a rod and an electromagnetic beam itself, and proceeded to find a way of defining length that does not involve quantum dependence. However, they did not consider that in practice Quantum Theory *also* enters their scheme as regards the structure of the 'point particles' at each end of the electromagnetic beam, so the first point stands.

How does the standard uncertainty principle limit the precision to which lengths can be measured? In practise, very accurate interferometers—such as LIGO's[11]—improve performance in this regard by employing 'squeezed states' and 'entangled beams'.

Let us next consider some quantum mechanical reasons for clocks being more fundamental than rods. Being made out of quantum matter, rods are not only complicated physical entities [349] but also are ultimately underlied by frequencies, which are clearly a temporal notion [160]. Also, by their nature and function, rods are necessarily macroscopic [761] and so interact with their environment in uncontrollable ways, while microscopic clocks (in the sense of reading hands) are possible.

Finally, following on from Chap. 2's treatment of mass, in the near future it looks likely that an exact $\hbar$ will be defined so as to be free from the 50 micrograms per century uncertainty observed in Pt–10Ir 'kilograms'.

---

[11]This is the Laser Interferometer Gravitational Wave Observatory; [839] see Chap. 7 for further discussion.

Chapter 6
Quantum Field Theory (QFT)

QFT [712, 885, 886] retains the non-commutation, evolution and measurement postulates, but involves distinct quantum wave equations and inner products. Its Compton wavelength scale l_C (after physicist Arthur Compton) arises from the balance $mc^2 = E = \hbar\omega = \hbar c/l_C$.[1]

6.1 Free Spin-0 Field

The theory for this follows from the Lagrangian $\mathcal{L} = -\partial_\mu\phi\,\partial^\mu\phi/2 - m^2c^2\phi^2/2\hbar^2$. The corresponding equation of motion is the *Klein–Gordon equation* (after physicists Oskar Klein and Walter Gordon),

$$\hbar^2\,\Box\phi = m^2c^2\phi \quad \text{for wave operator } \Box := -c^{-2}\partial_t^2 + \triangle. \tag{6.1}$$

The split Lagrangian is $\mathcal{L} = \dot{\phi}^2/2c^2 - |\underline{\partial}\phi|^2/2 - m^2c^2\phi^2/2\hbar^2$, the conjugate momentum is $\pi = \dot{\phi}/c^2$, and the Hamiltonian is $\mathcal{H} = c^2\pi^2/2 + |\underline{\partial}\phi|^2/2 + m^2c^2\phi^2/2\hbar^2$.

The equal-time commutation relations are now

$$[\widehat{\phi}(\underline{x}), \widehat{\pi}(\underline{y})] = i\,\hbar\,\delta^{(3)}(\underline{x} - \underline{y}). \tag{6.2}$$

The Klein–Gordon equation is a special-relativistic version of the time-dependent Schrödinger equation (and actually historically precedes it), which can now be recovered from the expansion

$$E\phi = mc^2\sqrt{\left\{1 + \frac{\underline{p}}{mc}\right\}^2}\,\phi = mc^2\phi - \frac{\hbar^2}{2m}\triangle\phi - \frac{\hbar^4}{8m^3c^2}\triangle\triangle\phi + \cdots . \tag{6.3}$$

[1] If unfamiliar with any of the non-temporal material in this Chapter, consult [712]; some of this Chapter's material on time in QFT (and the Wightman axioms) may however be new to the reader.

E. Anderson, *The Problem of Time*, Fundamental Theories of Physics 190,
DOI 10.1007/978-3-319-58848-3_6

The *Klein–Gordon inner product*

$$\langle \phi_1 | \phi_2 \rangle = \frac{\hbar}{2 i m c^2} \int \mathrm{d}^3 x \, \overline{\phi}_2 \overleftrightarrow{\partial_t} \phi_1 \tag{6.4}$$

ensures conservation of probability, thus amounting to a distinct form of Sect. 5.3's item 7). The candidate probability density involved, however, is negative in some places. This interpretational issue is best resolved by treating what were classical fields as themselves quantum operators. The subsequent *multi-particle interpretation of QFT* is most efficiently treated as an infinite collection of harmonic oscillators. In terms of creation and annihilation operators for these (5.14),

$$\mathcal{H} = \int \frac{\mathrm{d}^3 x}{\{2\pi\}^3} E_{\underline{p}} a^{\dagger}_{\underline{p}} a_{\underline{p}}, \tag{6.5}$$

where we have also adopted the so-called *normal (operator) ordering*: with all creation operators a_p to the left of all annihilation operators $a^{\dagger}_p$. The vacuum state is defined by $a_{\underline{p}}|0\rangle = 0$. We then build a *Fock space* (after physicist Vladimir Fock) upon this. This is the Klein–Gordon quantum state space,

$$\mathfrak{Fock} = \bigoplus_{\mathrm{n}=1}^{\infty} \bigotimes_{i=1}^{n} \mathfrak{Hilb}, \tag{6.6}$$

where $\bigotimes$ denotes tensor product, and with appropriate symmetrization subsequently incorporated to reflect that the scalar field is bosonic. The sum over n here corresponds to applying a multi-particle interpretation. Finally note for later reference that Klein–Gordon Theory straightforwardly admits a clear-cut split into positive and negative modes.

Propagators are Green's functions corresponding to the time-dependent quantum wave equation, which play a significant further role in QFT. The Klein–Gordon propagator takes the form

$$\mathcal{D}(\underline{x} - \underline{y}) \propto i \int \mathrm{d}^4 p \, \frac{\exp(-i \, \underline{p} \cdot \{\underline{x} - \underline{y}\}/\hbar)}{p^2 - m^2 c^2 + i \, \epsilon}. \tag{6.7}$$

For some purposes, one needs to specify which contour in the complex plane is to be used [712], some of which encode different implementations of causality.[2]

[2]If insufficiently familiar with Green's functions (named after 19th century mathematician George Green), consult [220]. In contrast to propagators, *correlators* are Green's functions corresponding to time-independent wave equations. Finally, examples of choices of contour used in defining propagators are retarded, advanced, Feynman and 'anti-Feynman' propagators, after noted physicist Richard Feynman. The retarded case is causal and the advanced case is anti-causal; The most usual choice for propagators is the Feynman propagator, which is the half-causal half-anticausal choice, with matter treated causally and antimatter treated anti-causally.

6.2 Free Spin-$\frac{1}{2}$ Field

The great physicist Paul Dirac discovered a linear spin-1/2 theory whose field equation is in a sense a square root of Klein–Gordon Theory's,[3]

$$\{i\,\hbar\gamma^{\mu}{}_{B}{}^{A}\partial_{\mu} - mc\,\delta_{B}{}^{A}\}\psi^{B} = 0. \tag{6.8}$$

The γ^{μ} are the *Dirac matrices*; representations of these are $i\begin{pmatrix} \mathbb{I} & 0 \\ 0 & -\mathbb{I} \end{pmatrix}$ and $i\begin{pmatrix} 0 & \sigma_i \\ -\sigma_i & 0 \end{pmatrix}$. The γ^{μ} obey the *Dirac algebra*

$$[\gamma^{\mu}, \gamma^{\nu}]_{+} = 2\,\eta^{\mu\nu}, \tag{6.9}$$

where $[\gamma^{\mu}, \gamma^{\nu}]_{+} := \gamma^{\mu}\gamma^{\nu} + \gamma^{\nu}\gamma^{\mu}$ is the *anticommutator bracket*.

There is additionally a fifth matrix $\gamma^5 := i\,\gamma^0\gamma^1\gamma^2\gamma^3 = \begin{pmatrix} 0 & \mathbb{I} \\ \mathbb{I} & 0 \end{pmatrix}$, which anticommutes with the other four; its eigenvalues ± 1 encode *chirality*, i.e. handedness.

This spin-1/2 theory can subsequently be placed on a classical Principles of Dynamics foundation. Defining the Dirac conjugate $\overline{\psi} := \psi^{\dagger}\gamma^{0}$, a Lagrangian for Dirac Theory is $\mathcal{L} = \overline{\psi}\{i\,\gamma^{\mu}\partial_{\mu} - m\}\psi$, with spacetime split form $\mathcal{L} = \overline{\psi}\{-i\,\gamma^{0}\dot{\psi} + i\,\gamma^{i}\partial_{i}\psi - m\,\psi\}$. The conjugate momentum expressions, e.g. $\pi = -i\,\psi^{\dagger}$, exhibit that 'momenta and configurations coincide' for Dirac Theory. Finally, the Dirac Hamiltonian is $\mathcal{H} = i\,\psi^{\dagger}\gamma^{i}\partial_{i}\psi + m\,\psi^{\dagger}\psi$.

Dirac Theory has the equal-time anti-commutation relations

$$[\widehat{\psi}_{A}(\underline{x}), \widehat{\psi}^{\dagger}_{B}(\underline{y})]_{+} = \delta^{(3)}(\underline{x} - \underline{y})\delta_{AB}. \tag{6.10}$$

These incorporate (but do not explain) the Pauli Exclusion Principle by implementing Fermi–Dirac statistics. The inter-relation between commutation and Bose–Einstein statistics, on the one hand, and between anticommutation and Fermi–Dirac statistics on the other, was established by physicist Markus Fierz alongside Pauli; see [885] for a distinct modern proof.

This theory also requires a multi-particle interpretation. The corresponding *Dirac inner product* is

$$\int d^3x\,\overline{\psi}_1\psi_2 = \int d^3x\,\psi_1^{\dagger}\gamma^{0}\psi_2. \tag{6.11}$$

In this case, let us denote positive energy state creation and annihilation operators by a different letter from negative energy ones; it is conventional to use $a_{\underline{p}}$ and $b_{\underline{p}}$ for these. Dirac Theory's Hamiltonian can be expanded as (once again adopting normal-ordered form)

$$\mathcal{H} = \int \frac{\mathrm{d}^3 p}{\{2\pi\}^3} \sum_{s} E_{\underline{p}}\{a^{s\dagger}_{\underline{p}} a^{s}_{\underline{p}} - b^{s\dagger}_{\underline{p}} b^{s}_{\underline{p}}\}, \tag{6.12}$$

[3]The capital Latin indices here run over 1 to 4 for 4-component spinor indices; these spinorial indices are often suppressed in this book.

for s summing over the allowed spin values. Originally, this raised issues concerning descent to arbitrarily negative energy states. However, this case's Feynman propagator,

$$\mathcal{D}(\underline{x} - \underline{y}) \propto i \int \mathrm{d}^4 p \exp(-i \underline{p} \cdot \{\underline{x} - \underline{y}\}/\hbar) \frac{\underline{\gamma} \cdot \underline{p}}{p^2 - m^2 c^2 + i\,\epsilon}, \tag{6.13}$$

has a subtle difference in interpretation [712] relative to its Klein–Gordon counterpart. This reveals that $a_{\underline{p}}$ and $b_{\underline{p}}$ correspond to a distinguishable particle–antiparticle pair.[4] There is a straightforward symmetry between antiparticle creation and particle creation, which is additionally tied to the Fermi–Dirac statistics obeyed by these particles [712, 817]. Indeed, Dirac predicted the existence of positrons—antiparticles corresponding to electrons—and those were promptly experimentally observed, in the form of deflections in an electromagnetic field corresponding to the same charge-to-mass ratio as the electron but with opposite sign. Finally, a particular feature in computing Feynman diagrams involving fermionic species is that each fermionic loop in a diagram contributes a minus sign, giving schematically an overall factor of

$$(-1)^{\mathrm{F}}. \tag{6.14}$$

This is due to [712] operator exchange in fermionic propagators such as (6.13).

6.3 Free Spin-1 Field: Electromagnetism, and Its Gauge Symmetry

We next consider a theory whose classical form is already classically well-known: Electromagnetism. As a long-ranged force, it makes sense that its mediator particle—the *photon*—is massless. Electromagnetism being linear suggests uncharged mediators, so that these do not interact electromagnetically with each other. Moreover, static forces between particles A and B require that emission and absorption of the mediator by either A or B leaves both of these in the same internal state [299]. This makes half-integer spin mediator particles impossible for such forces (here electrostatics forces).

Comparing A = B and A = $\overline{\mathrm{B}}$ (antiparticle) cases, if A and B are charged, these are same-sign and opposite-sign charges. Compute the potentials and take suitable limits. Moreover, if the mediator particle is of odd integer spin, like charges repel and opposite charges attract. This fits the bill for Electromagnetism. Take spin-1 as for now the simplest possibility (Sect. 11.7 furthermore precludes higher odd spins). Conversely, mediators of even integer spin result in universally attractive forces.

[4] By the half-causal half-anticausal choice in the Feynman propagators, some of the arrows in Feynman diagrams (Fig. 6.1.b) point backwards. This corresponds to modelling distinguishable antiparticles as if they were travelling backward in time. This convention moreover indeed works for practical purposes.

This now fits the bill for Gravitation, to which we return in Chap. 11. Electromagnetism's mediator particle is moreover massless. If it were massive, the resulting static force would go as $\exp(-mr)/r^2$, in discord with the observed inverse square law.

Following on from the Lagrangian (4.14), Electromagnetism's momentum conjugate to A_i is

$$\pi^i := \frac{\partial \mathsf{L}}{\partial \dot{\mathrm{A}}_i} = -\dot{\mathrm{A}}^i - \partial^i \Phi = \mathrm{E}^i . \tag{6.15}$$

Next note that Gauss's Law (2.13) is instantaneous: a *constraint equation*

$$\mathcal{G} := \partial_i \pi^i = 0 \tag{6.16}$$

in vacuo. It is moreover accompanied by a primary constraint (see Appendix J) $\pi_\Phi = 0$, for π_Φ the momentum conjugate to Φ. These constraints are both first-class (also in Appendix J), so they use up 2 degrees of freedom each. In this way, one passes from A_i, Φ and their conjugate momenta's redundant 4×2 phase space degrees of freedom per space point to just 2×2. This is in accord with electromagnetic waves consisting of just two modes (the transverse modes). N.B. that constraint equations become a major feature for most of the rest of this book. Electromagnetism's 'total' Hamiltonian (see Appendix J.15) is

$$\mathcal{H} = \{\pi^2 + \mathrm{B}^2\}/2 + \Phi\, \underline{\partial} \cdot \underline{\pi} \quad \text{(strictly need to include } + \lambda\pi_0). \tag{6.17}$$

Electromagnetism is moreover invariant under $U(1)$ *local gauge transformations*

$$\mathrm{A}_\mu \longrightarrow \mathrm{A}_\mu + \partial_\mu \xi \tag{6.18}$$

for any function $\xi = \xi(\vec{X})$.[5] This bears well-known relation to the above form of the constraints. Commonly useful gauge choices include each of

$$\partial_\mu \mathrm{A}^\mu = 0 \quad (\textit{Lorenz gauge}), \tag{6.19}$$

$$\partial_i \mathrm{A}^i = 0 \quad (\textit{Coulomb gauge}). \tag{6.20}$$

Electromagnetism's commutator is, in the Coulomb gauge,

$$[\mathrm{A}_i(\underline{x}), \pi_j(\underline{y})] = i\hbar\{\delta_{ij} - \triangle^{-1}\partial_i\partial_j\}\delta^{(3)}(\underline{x} - \underline{y}), \tag{6.21}$$

the combination in curly parentheses forming a transverse projector.

Maxwell's equations (4.11) play the role of wave equation. Applying a mode expansion in terms of $\underline{a_k}$ (for $\underline{k}$ each mode's momentum), the Coulomb gauge condition leads to $\underline{k} \cdot \underline{a_k} = 0$ and $\underline{k} \cdot \underline{a_k^\dagger} = 0$. The normal-ordered quantum Hamiltonian

[5] In this book, we use overhead arrows to denote 4-vector quantities.

is

$$\mathcal{H} = \int \frac{d^3k}{\{2\pi\}^3} |\underline{k}| \, a_{\underline{k}}^{\dagger} \cdot a_{\underline{k}}. \tag{6.22}$$

Building up the states with creation operators from a vacuum state works in the habitual manner for bosons.

Finally note that the photon propagator is—in the Lorenz gauge [712, 885] (after physicist Ludvig Lorenz)—

$$\mathcal{D}_{F\mu\nu}(\underline{x} - \underline{y}) \propto i \int d^4k \frac{\exp(i\underline{k} \cdot \{\underline{x} - \underline{y}\}/\hbar)}{k^2 + i\,\epsilon} \left\{ \eta_{\mu\nu} - \frac{k_\mu k_\nu}{|k|^2} \right\}. \tag{6.23}$$

More compactly, its Fourier-transformed form is

$$\frac{1}{k^2 + i\,\epsilon} \left\{ \eta_{\mu\nu} - \frac{k_\mu k_\nu}{|k|^2} \right\}. \tag{6.24}$$

6.4 Time in Quantum SR

Conventional Relativistic Field Theory has a fixed background (usually Minkowskian) spacetime structure with the field propagating with respect to this in the corresponding time [471].[6] Therein, QFT is based on selection of an inertial frame and so of a choice of time [471]. This is formalized by the first of the Wightman axioms [401, 875].[7]

Wightman-1) The theory's state are unit rays in a Hilbert space $\mathfrak{H}$ilb that carries a unitary representation of the Poincaré group *Poin*(4).

At the quantum level, SR's *Poin*(4) now manifests itself as a commutator algebra with nontrivial commutators

$$[\widehat{M}_{\mu\nu}, \widehat{P}_\rho] = i\,\hbar\{\eta_{\mu\rho}\widehat{P}_\nu - \eta_{\nu\rho}\widehat{P}_\mu\}, \tag{6.25}$$

$$[\widehat{M}_{\mu\nu}, \widehat{M}_{\rho\sigma}] = i\,\hbar\{\eta_{\mu\rho}\widehat{M}_{\nu\sigma} - \eta_{\mu\sigma}\widehat{M}_{\nu\rho} - \eta_{\nu\rho}\widehat{M}_{\mu\sigma} + \eta_{\nu\sigma}\widehat{M}_{\mu\rho}\}. \tag{6.26}$$

Since *Poin*(4) is tied to the background Minkowski metric η and its high level of symmetry, passing from QM to QFT indeed parallels passing from Newtonian Mechanics to SR as regards trading one kind of absolute time for another. Also, since *Poin*(4) is repeatedly involved in the account below, many sources of contention will arise when this background structure can no longer be assumed (see Sect. 11.3).

[6] QFT does not by itself entail SR, e.g. phonons are a *Nonrelativistic* QFT model of quanta of sound waves in materials.

[7] Moreover, the Wightman axioms (named after mathematical physicist Arthur Wightman) do not cover Quantum Gauge Theories. See e.g. [269] for a recent overview of this, and also [687] as regards an advanced consideration of their Euclidean counterpart.

Wightman-2) The 4-momentum—as defined by the action of *Poin*(4) on $\mathfrak{H}$ilb—is positive. Moreover, this can be reformulated as the *spectrum condition*: that the 4-momentum spectrum is contained within the closed future null cone.

Wightman-3) There exists a unique, Poincaré invariant state: the vacuum (thus this is also induced from background structure).

Wightman-4) The quantum fields are operator-valued distributions[8] defined on a dense (in the sense of Appendix C.6) domain $\mathfrak{D} \subset \mathfrak{H}$ilb that is both Poincaré invariant—and thus Background Dependent—and invariant under the action of the fields and their adjoints.

Wightman-5) The fields transform covariantly under the action of Poincaré transformations. This follows from Wigner's demonstration that the different types of free particle are the representations of *Poin*(4) [885]. This is outlined in Appendix E. This furthermore signifies that spin is an inherent part of SR. The linear relativistic wave equations are 'projection conditions' onto irreducible subspaces in some Hilbert space [363].

Wightman-6) At spacelike separations, quantum fields either commute or anticommute. N.B. that this depends on the fixed-background (but not necessarily highly symmetric) metric to judge what is spacelike.

The equal-time commutation relations now additionally carry the further time connotations of *microcausality* [483],

$$[\widehat{\Theta}_{\mathrm{A}}(\vec{X}), \widehat{\Theta}_{\mathrm{A}'}(\vec{Y})] = 0. \tag{6.27}$$

Here $\widehat{\Theta}_{\mathrm{A}}$ are relativistic quantum field operators at all spacelike-separated spacetime events $\vec{X}$ and $\vec{Y}$, i.e. for all pairs of points not interconnected by causal signals.

One now has distinct time-dependent wave equations and new inner products carrying time connotations via unitarity's tie to conservation of probability.

Finally, the Time–Energy Uncertainty Relation is additionally contentious in this SR setting [517]. Having to treat x, y, z differently from t on the other is 'intuitively' problematic if these are to be related by Lorentz transformations. On the other hand, the Salecker–Wigner inequalities continue to apply to SR clocks.

6.5 Interacting Field Theories, Including Quantum Electrodynamics (QED)

Free Field Theory has the following schematic form.

Free QFT I) For each field type start with the corresponding relativistic particles which are non-interacting. These correspond to classical field equations that are

---

[8]Appendix O outlines 'distributions' in the current sense of Functional Analysis. So as to not confuse this with 'probability distribution' or 'distribution of matter', this book always fully spells out these other uses.

linear in the fields and thus to quadratic Lagrangians The irreps of the *Poin*(4) symmetry group of the Lagrangian give the quantum states. Each individual particle's states form a Hilbert space.

Free QFT II) To have an arbitrary particle number scheme, one next builds the corresponding Fock space (6.6) with suitable (anti)symmetrization incorporated.

Free QFT III) Finally construct the corresponding creation and annihilation operators.

Interacting QFT then builds upon this as follows. We now need to include interaction terms in the Lagrangian; these are cubic or higher in the fields. In the Scattering Theory, the 'in' and 'out' states are free, whereas scattering in a transition region is described by a scattering matrix (*S-matrix*). If the coefficients of the interaction terms can be regarded as small, perturbation theory can be applied and useful results reasonably straightforwardly ensue. The S-matrix can be viewed in terms of the vacuum expectation values of time-ordered products of 'interpolating fields' (i.e. n-point functions) linking different-particle-number-and-species ingoing and outgoing states. At this point, one uses Wick's Theorem [712] to relate time ordering to normal operator ordering. Feynman rules arise from these considerations. As well as propagators, one is now to consider interaction vertices and S-matrix 'in' and 'out' states. Model arenas are useful in this study. As a first example, consider ϕ^4 theory, meaning that there are 4 scalar field propagators emanating from each vertex. The integrals now contain edge, external and internal vertex contributions; the *Feynman rules* [712] are an efficient prescription for computing such diagrams.

The $U(1)$ gauge symmetry of classical Electromagnetism plus a complex scalar field involves considering a local (gauged) version of the $\phi \rightarrow \exp(i\xi)\phi$ symmetry, i.e. now with $\xi(\vec{X})$ rather than just a global ξ. This requires introducing an object A_μ which transforms in opposition to ∂_μ. Letting A_μ have its own dynamics produces Electromagnetism coupled to a complex scalar field Gauge Theory. This has an $\mathrm{A}_\mu \mathrm{A}^\mu \phi^* \phi$ vertex.

One can also arrive at *QED* (*Quantum Electrodynamics*) by repeating the above procedure for a fermionic theory. This possesses a $\overline{\varphi} \gamma^a \varphi \mathrm{A}_a$ vertex (Fig. 6.1.b). This theory originated with work of Heisenberg, Pauli, Bohr and physicist Léon Rosenfeld [150, 444].

As regards model building more generally,

1) constructing Lagrangians to obey a pre-determined list of symmetries is a common procedure in Particle Physics. Including all terms with a given symmetry in the Lagrangian is an additional common premise in Particle Physics.
2) Power-counting within each Feynman diagram, the *superficial degree of divergence* of an interaction I is [885] $\Delta_I := 4 - d_I - \sum_f n_{If}\{s_f + 1\}$ for d_I the number of momentum factors and n_f the number of fields of type f with spin s_f. Theories with $\Delta_I \geq 0$ for all interactions are termed *naïvely renormalizable* (since this is for now but a 'back of the envelope' calculation). Two attitudes to non-renormalizable theories are to discard them, or to retain them in the guise of *effective theories* that are a good approximation within some particular regime.

3) The *Cluster Decomposition Principle* asserts the independence of QFT in disjoint local spacetime patches [885].

   Returning to QED, two significant features are, firstly, the involvement of a charge-to-mass ratio, which lends itself to being readjusted in detailed calculations. Secondly, the *fine structure constant* $\alpha := e^2/4\pi\hbar c \simeq 1/137$ is here available to play the role of perturbation theory's small parameter.
4) *Anomalies* [4, 139, 250] are a type of brackets algebra obstruction that specifically alters the classical symmetry group at the quantum level. In this manner, Quantum Theory refuses to accept some of what were perfectly good symmetries at the classical level. According to Dirac [250], one's Quantum Theory avoiding this problem requires 'luck'.

6.6 Yang–Mills Theory Underlying the Nuclear Forces

An early theory for the nuclear forces involved mediation by massive pions. This led to Fermi's theory (4-fermion vertices) being applied again, although this was then demonstrated to be non-renormalizable in the 1950s. [Cf. how in QED the photon splits the putative 4-fermion vertex into two QED vertices.]

There are now multiple 1-form fields to Electromagnetism's single one. Furthermore, they are coupled to each other: Yang–Mills Theory (after physicists Chen Ning Yang and Robert Mills) is nonlinear, with mediator particles now carrying charges, in contrast with Electromagnetism's photon being neutral. Yang–Mills Theory encodes these further features using larger gauge groups (Appendix E) than Electromagnetism's $U(1)$, which are furthermore noncommutative. For the modelling of the nuclear forces, the mediator particles in question are termed $W_\pm$ and Z_0 *bosons* for the weak force [$\mathfrak{g} = SU(2)$] and *gluons* for the strong force [$\mathfrak{g} = SU(3)$]; the latter carry 'red', 'green' and 'blue' colour charges. Moreover, these are just arbitrary label names [so one really has $SU(3)/\mathbb{Z}_3$]. This $SU(3)$—held to be exactly realized—should not be confused with the approximately realized *flavour* $SU(3)$ that covers up, down and strange quarks[9] The latter has been generalized to there being three generations of pairs of quarks (Fig. 6.1.a), whereas the former is the basis of the *Quantum Chromodynamics* theory of the strong force. We finally note the electroweak unification based on $SU(2) \times U(1)$ by physicists Steven Weinberg and Abdus Salam [886]. The composition without further unification of the previous two sentences' theories is known as the *Standard Model* of Particle Physics. See Fig. 6.1.a) for its remaining fundamental particle species and their relations to all the types of composite particle mentioned in this book.

α_S and α_W can at this point be defined in parallel with QED's α. However, α_S is unfortunately too large to be amenable to perturbation theory paralleling QED's.

[9] I.e. physicist Murray Gell-Mann's eightfold way explanation of the octet, singlet and decuplet patterns of the observed and predicted-and-confirmed hadrons [886].

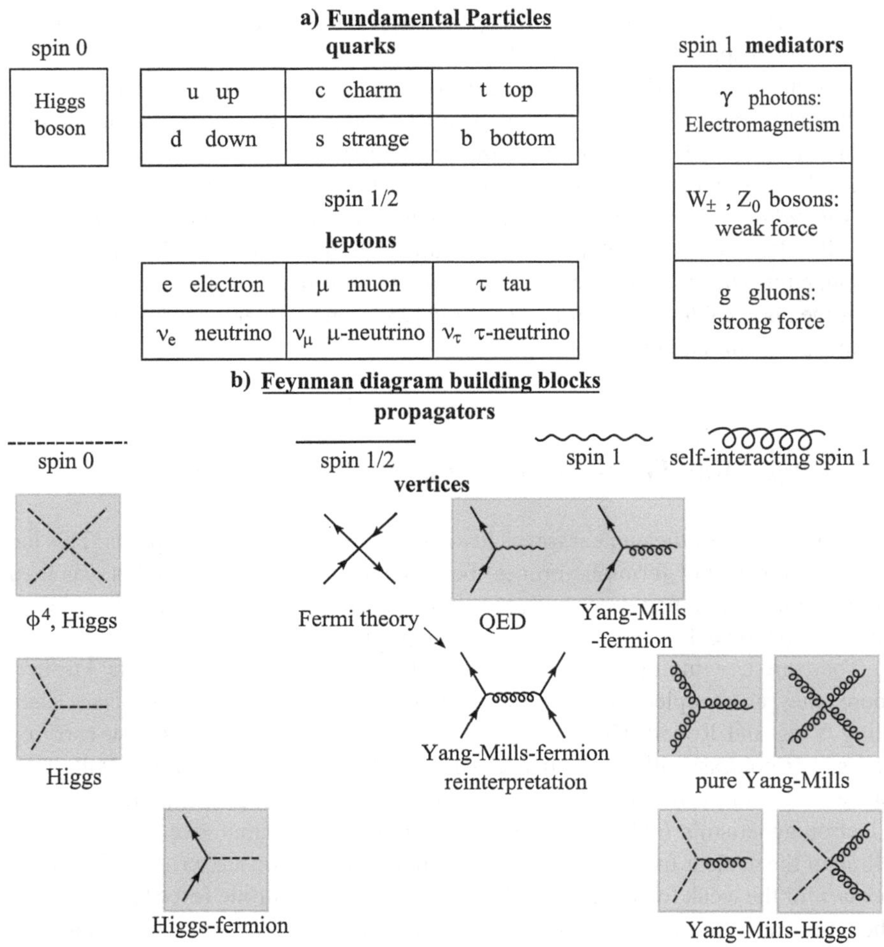

Fig. 6.1 **a)** Summary table of the building blocks of Nature: 6 quarks and 6 leptons, each representing 3 generations, alongside the mediator particles and the Higgs bosons. **b)** Notation used for the propagators and vertices in various theories

The spacetime form of the Lagrangian for (arbitrary gauge group) Yang–Mills Theory is

$$\mathcal{L} = -\frac{1}{2}\mathrm{F}^{I}_{\mu\nu}\mathrm{F}^{\mu\nu}_{I}, \tag{6.28}$$

for *Yang–Mills field strength* $\mathrm{F}^I_{\mu\nu} := \partial_\mu \mathrm{A}^I_\nu - \partial_\nu \mathrm{A}^I_\mu + i\,|[\mathrm{A}_\mu, \mathrm{A}_\nu]|^I$ and *Yang–Mills potential 1-forms* A^I_μ. The corresponding field equations are

$$0 = \mathrm{D}^\mu \mathrm{F}^I_{\mu\nu} = \mathrm{D}^\mu\left\{\partial_\mu \mathrm{A}^I_\nu - \partial_\nu \mathrm{A}^I_\mu + i\,|[\mathrm{A}_\mu, \mathrm{A}_\nu]|^I\right\}, \tag{6.29}$$

for D^μ the gauge covariant derivative (explained in Appendix F).

The 3 + 1 split of Yang–Mills Theory's Lagrangian is

$$\mathsf{L} = -\frac{1}{4}\mathrm{F}_{Iab}\mathrm{F}^{Iab} + \frac{1}{2}\{\partial_0 \mathrm{A}_{Ia} - \partial_a \mathrm{A}_{I0} + |[\mathrm{A}_a, \mathrm{A}_0]|_I\}\{\partial^0 \mathrm{A}^{Ia} - \partial^a \mathrm{A}^{I0} + |[\mathrm{A}^a, \mathrm{A}^0]|^I\}. \quad (6.30)$$

The conjugate momenta are

$$\pi^a_I := \frac{\partial \mathsf{L}}{\partial \dot{\mathrm{A}}^I_a} = \delta^{ij}\delta_{IJ}\{\dot{\mathrm{A}}^J_j - \partial_j \mathrm{A}_0 + +i|[\mathrm{A}_0, \mathrm{A}_j]|\}, \qquad \pi^\Phi_I := \frac{\partial \mathsf{L}}{\partial \dot{\Phi}^I} = 0. \quad (6.31)$$

The last equality is a primary constraint; it is partnered by the *Yang–Mills–Gauss constraint*

$$\mathcal{G}_J := D_a \pi^a_J = \partial_a \pi^a_J - \mathrm{g_c} f_{IJK} \mathrm{A}^K_a \pi^{Ia} = 0. \quad (6.32)$$

[Brackets have been evaluated here in terms of the structure constants (E.2), alongside making explicit an overall scalar factor: the *coupling constant* $\mathrm{g_c}$.] These are both first-class, and use up 2 degrees of freedom each. The Yang–Mills Hamiltonian is

$$\mathcal{H} = \pi^a_I \pi^I_a + \frac{1}{4}\mathrm{F}_{I\mu\nu}\mathrm{F}^{I\mu\nu} - \mathrm{A}^J_0 \mathcal{G}_J. \quad (6.33)$$

The Yang–Mills wave equation is just the second form of (6.29). Since it is not linear, we cannot fully treat this with mode expansions, and there are other subtleties such as Fadde'ev–Popov determinants (outlined in Chap. 52). The Standard Model further requires the Yang–Mills–Dirac Gauge Theory parallel of QED [712, 886].

Symmetry breaking is required for contact with observation. Breaking of global symmetries involves further *Goldstone boson* species (after physicist Jeffrey Goldstone) entering the physics [886]. On the other hand breaking of local i.e. gauged symmetries involves instead Higgs bosons [886], which confer mass to other particle species. [This means *inertial* mass since the 3-force Standard Model QFT within the Minkowskian Paradigm solely involves the inertial notion of mass.] Moreover, neutrinos have come to be considered to possess mass so as to explain solar observations [888].

Yang–Mills Theory's nonlinearity renders it [238, 239] somewhat more like GR than Electromagnetism is. A final feature that is common in QFTs used to approach Quantum Gravity is possession of a *mass gap*, i.e. a finite difference in energy between the vacuum state and the next lowest energy state.

6.7 Discrete Operations (Including Time-Reversal) in Quantum SR

Let T, P and C denote *time-reversal symmetry* $t \to -t$, *parity-inversion symmetry* $\underline{x} \to -\underline{x}$ and *charge conjugation symmetry* $q \to -q$ respectively (see e.g. [269,

885]). The T operator is moreover antilinear and antiunitary [885] (cf. Sect. 5.3). Note furthermore, C, P, and even CP, violations are observed experimentally; these are related to the weak force; on the other hand, even this obeys the combined CPT invariance. We finally point to C here enlarging the grouping of space-and-time that is PT, which provides yet another reason for non-closure of temporal notions by themselves.

6.8 Quantum-Level Evidence for SR

This includes the survival of muons through the atmosphere, the fine structure of atomic spectra, accounting for Nuclear Physics reactions and many observed Particle Physics processes. In particular, the predicted $W_{\pm}$ and Z_0 bosons have been confirmed, and likewise the charm, bottom and top quarks, and now the Higgs boson [838], as well as mixing angles related to the flavour changing weak decays [886].

6.9 Grand Unified Theories

These attempt to unify the electromagnetic, weak and strong forces using representations of some larger gauge group such as $SU(5)$ or $SO(10)$, within which $SU(3) \times SU(2) \times U(1)$ can be embedded: large enough and admitting complex representations. This would replace three separate coupling constants by a single one in some high-energy regime. The $SU(5)$ Grand Unified Theory itself is overruled due to non-observation of proton decay [886].

6.10 Exercises II. Time and Quantum Theory

Further Reading See e.g. [599] for an introductory account of QM, [487] for an introduction to the foundations of QM and its use of basic Linear Algebra, and [712, 885, 886] for more on QFT.

Exercise 1) Solve the n-d quantum harmonic oscillator in n-d spherical coordinates.

Exercise 2) Model the hydrogen atom's energy spectrum by considering the following equations with Coulomb potential in spherical polar coordinates. i) The time-independent Schrödinger equation. ii) The Klein–Gordon equation. iii) The Dirac equation. iv) Compare these results. v) Consider what happens to the atomic orbits in the semiclassical limit, and compare with the corresponding classical problem.

Exercise 3) [Bosonic noninteracting QFT] i) Make sure you know how this is modelled in terms of a countable collection of harmonic oscillator creation and annihilation states, and Fock space. ii) How does this work better than a single-particle interpretation of the Klein–Gordon equation?

Exercise 4) Show that Noether's Theorem links momentum to translational symmetry, angular momentum to rotational symmetry and energy to time translation symmetry. Find a Noether current and conserved quantity for a single complex Klein–Gordon scalar field and for a Dirac spin-1/2 field; both of these cases correspond to some phase symmetry.

Exercise 5) The *Feynman path integral formulation*—of conceptual and efficient computational value in QFT in particular—can be considered to be in terms of a transition probability

$$T[\boldsymbol{q}_{\rm fin}, t_{\rm fin}, \boldsymbol{q}_{\rm in}, t_{\rm in}] := \langle \boldsymbol{q}_{\rm fin}, t_{\rm fin} | \boldsymbol{q}_{\rm in}, t_{\rm in}\rangle = \langle \boldsymbol{q}_{\rm fin} | \exp\big(i\, H\{t_{\rm fin} - t_{\rm in}\}\hbar\big) | \boldsymbol{q}_{\rm in}\rangle . \quad (6.34)$$

i) By inserting a complete set of states and applying a suitable discretization and limiting procedure, and using $\boldsymbol{p}(0) := \boldsymbol{p}(\text{in})$, $\boldsymbol{q}(0) := \boldsymbol{q}(\text{in})$ and $\boldsymbol{q}(N+1) := \boldsymbol{q}(\text{fin})$, rewrite (6.34) as

$$\begin{aligned} T\big[\boldsymbol{q}(\text{fin}), t_{\rm fin}, \boldsymbol{q}(\text{in}), t_{\rm in}\big] &= \lim_{N\longrightarrow\infty} \int \prod_{\rm A=1}^{M} \frac{\mathbb{D}q_{\rm A}\mathbb{D}p^{\rm A}}{\{2\pi\,\hbar\}^M} \\ &\quad \times \exp\Bigg(\frac{i}{\hbar}\sum_{\rm B=0}^{N}\{p_{\rm A}(t_{\rm in} + {\rm B}\,\Delta t)\{q^{\rm A}\big(t_{\rm in} + \{{\rm B}+1\}\Delta t\big) \\ &\quad - q^{\rm A}(t_{\rm in} + {\rm B}\,\Delta t)\} - \Delta t\, H\big(p_{\rm A}, q^{\rm A}\big)\}\Bigg) \\ &= \int \prod_{\rm A=1}^{M} \frac{\mathbb{D}q_{\rm A}\mathbb{D}p^{\rm A}}{\{2\pi\,\hbar\}^M}\left(\frac{i}{\hbar}\int_{t_{\rm in}}^{t_{\rm fin}}\{p_{\rm A}\dot{q}^{\rm A} - H\}{\rm d}t\right). \end{aligned} \quad (6.35)$$

ii) Carry out analogous workings in the case of a scalar QFT. iii) Consider furthermore the Euclidean path integral analogue by performing a complexified spacetime coordinate transformation to imaginary time, $\tau = it$. Investigate choice of a contour in $\mathbb{C}$ ('Wick rotation', after physicist Gian-Carlo Wick) so as to move back to the Lorentzian form of the path integral. iv) Derive the form of all of this Chapter's propagators.

Background Reading 1)† Extend [349, 521, 736]'s treatments of time in Classical Physics to Quantum Theory.

Exercise 6) i) How do anti-Hermitian operators evade Pauli's point in Sect. 5.3? ii)† How else can Pauli's point be evaded? [See Sect. 41.1 for more.]

Exercise 7) [Time–Energy Uncertainty Principle.] In Ordinary QM, show that $\Delta O \Delta E \geq |\langle\, \widehat{O},\ \widehat{H}\,\rangle|/2$ for a time-independent operator O. Rearrange this to obtain a Time–Energy Uncertainty Principle for a 'characteristic evolution time'

$$t_O := \Delta O \Bigg/ \left|\frac{{\rm d}\langle\, \widehat{O}\,\rangle}{{\rm d}t}\right| ,$$

and interpret this quantity.

Exercise 8) Derive the Salecker–Wigner clock inequalities (5.18) from a suitable Uncertainty Principle.

Exercise 9) Demonstrate Unruh and Wald's contradiction (Sect. 5.4) by studying the function $f_{mn}(t)$, $m > n$, for complex t.

Background Reading 2) i) Understand Quantum Theory's transition time, life time, tunnelling time, reflection time, response time, dwell time, flight time, arrival time, pulse time, Zeno time, passage time, jump time, and coherence time as notions of time at the conceptual level [669, 670]. E.g. which of these bear which relations to each other? Which are external, internal or based on observables? Which can meaningfully enter Energy–Time Uncertainty Principles? ii)† What happens to all of these notions of time upon passing to a classical Newtonian limit? iii)† Which of these notions of time remain meaningful in QFT, and what happens to each of these upon passing to a classical Minkowskian limit?

Backgound Reading 3) Work through Chaps. 6 and 7 of [82] as regards the physics of timekeeping using atomic clocks.

Exercise 10) i) Show that all the fundamental Laws considered so far in this book are individually T, P and C invariant with the exception of the weak force. [To include Ordinary QM, consider only observable quantities, and be disposed to restrict the form of the potential.] ii) Demonstrate that even the weak force is CPT-invariant. (See also [817] for a demonstration that CPT must hold a fortiori for any local SR Field Theory.)

Background Reading 4) Consider the accounts of decoherence in the compilation [366].

Exercise 11) Demonstrate that α-tracks in bubble chambers can in fact be taken to be governed by a time-independent Schrödinger equation.

Exercise 12)†† Read Chaps. 1 to 4 of physicist Dieter Zeh's [931] on the Radiative, Thermodynamical, and Quantum (measurement: wavefunction collapse, and weak-force) Arrows of Time. Explore whether one of these is a 'Master Arrow' from which all the other Arrows follow. Does the Psychological Arrow of Time follow from any of the others?

Chapter 7
Time and Spacetime in General Relativity (GR)

GR arose historically through Einstein's hopes for universality of SR being thwarted by Gravitation alone amongst the classical laws. In the process, he recognized the importance of the Equivalence Principle: having to treat

$$\underline{a} + \frac{m_{\mathrm{g}}}{m_{\mathrm{i}}}\underline{g} \tag{7.1}$$

jointly rather than piecemeal, as per Sect. 2.9. He approached this via his *elevator thought-experiment*, in which an observer in a small enclosed laboratory is not able to discern whether they are experienced gravitational fall or rocket acceleration. This rests on the concept of *Universality of Free Fall* [287]—independence of the material composition of falling test bodies—since elsewise bodies of different compositions could be used to discern between $\underline{a}$ and $\underline{g}$. Einstein furthermore took this universality to point strongly toward there being a common underlying geometry being experienced by all the test particles. Moreover, a curved generalization of SR's notion of 4-*d* spacetime can serve this purpose. Contrast the above also with how particles with different charge-to-mass ratios move differently in an electromagnetic field: the Equivalence Principle is a statement of the non-existence of an analogous gravitational-to-inertial mass ratio.

That Gravitation in the above sense can be transformed away at any particular point is implemented by the mathematics of the spacetime affine connection.[1] By this feature, freely falling frames are but local concepts, thus deserving the name 'local frames'. The combination of Newton's Second Law and Newton's Law of Gravitation $\ddot{\underline{x}} = -\underline{\partial}\phi$ can be reformulated as a (for now affine) geodesic equation with a spacetime affine connection whose only nonzero components are

$$\Gamma^{(4)i}{}_{00} = \partial_i \phi, \tag{7.2}$$

where ϕ is the Newtonian gravitational potential.

[1] From here on, we assume the reader knows Differential, Affine and Metric Geometry; if not, take a detour to Appendix D. If elsewise unfamiliar with this Chapter's material on GR, consult [874] as preliminary reading.

E. Anderson, *The Problem of Time*, Fundamental Theories of Physics 190,
DOI 10.1007/978-3-319-58848-3_7

Indeed, connections correspond to not being able to treat an additive pair of mathematical objects in isolation from each other. This is reflected in their individual transformation laws being inhomogeneous (non-tensorial) whereas the sum of the objects does have a tensorial transformation law.[2] This transformation law can also be taken to underlie how the spacetime affine connection can be transformed away at any particular point (see Appendix D.3's normal coordinates).

Local agreement with SR is also required; a natural hypothesis here is Einstein's that SR inertial frames are global-in-spacetime idealizations of GR's local inertial frames that are attached to freely falling particles. Furthermore, in parallel with the development of SR, Einstein retained a notion of metric $\mathbf{g}$ with components $g_{\mu\nu}$[3] on spacetime to account for observers in spacetime having the ability to measure lengths and times if equipped with standard rods and clocks. I.e. the inner product characterization of length and angle carries over from SR to GR. One is consequently dealing with an in general curved *semi-Riemannian* (alias *pseudo-Riemannian*) metric. See Appendix D as regards a more general Tensor Calculus than that on $\mathbb{R}^3$ or $\mathbb{M}^4$.

The metric $\mathbf{g}$ represents Gravitation in a second sense: it replaces the single Newtonian scalar field by a geometrical decuplet of fields. Moreover, this unification of Metric Geometry and Gravitation was itself a novel physical proposal at this stage in the development of Physics. The metric connection associated with this turned out to suffice in the aforementioned role for an affine connection in the theory. As $\mathbf{g}$ reduces locally to SR's $\boldsymbol{\eta}$ everywhere locally the other laws of Physics take their SR form. One can see $\mathbf{g}$'s indefinite signature as a continuation of SR's; it is again an indefinite metric encoding the distinction between time and space by time being $--$ and space being $+++$. Indeed, notions of timelike, spacelike and null carry over to GR, as does using the first and third of these to interpret massive and massless particle based matter respectively. The straight timelike lines followed by free particles in SR's Minkowski spacetime $\mathbb{M}^4$ are bent by the gravitational field into the curves followed by relatively-accelerated freely-falling particles in the case of full GR. The straight null lines which constitute $\mathbb{M}^4$'s lightcones are similarly bent.

A natural question at this point is how one is to interpret the spacetime curvature associated with the affine connection and the metric. Another is what are suitable field equations—analogous to Maxwell's equations for Electromagnetism—and subject to the requirement of recovering the Poisson form of Newton's Law of Gravitation in a suitable limit. These field equations are to be tensorial: a realization of General Covariance.

[2]Partial derivative and connection come with transformation laws with extra compensating portions that cancel out in the covariant derivative's own tensorial transformation law: Appendix D.3. Indeed the 'compensating field' treatment of Gauge Theory in Chap. 6 can be formulated in terms of another type of connection, as per Appendix F.4.

[3]g is the spacetime metric's determinant, which is a scalar density, $\Gamma^{(4)}$ is the spacetime metric connection and ∇_μ is the spacetime covariant derivatives. While there is a role for an affine connection, it is the metric connection (indeed computable from the metric: Appendix D.4) subsumes this role in GR.

Moreover, an intermediary reformulation already reveals that Newtonian Gravitation is already a curved-space theory. Indeed [814], from (7.2)

$$\mathcal{R}^i{}_{0j0} = -\partial_i \partial_j \phi, \tag{7.3}$$

so the Newtonian tidal equation (2.10) can itself be viewed geometrically as a geodesic deviation equation (D.14), and Poisson's form (2.11) of Newton's Law of Gravitation can be further recast as

$$\mathcal{R}_{00} = -4\pi G \rho. \tag{7.4}$$

Furthermore, (7.3) is the *only* nonvanishing component. This means that this curvature tensor does not have the symmetries corresponding to Metric Geometry's Riemann tensor; the indices are clearly mismatched as well. This (non-historical) observation (and [278]) suggests that Curved Geometry is on the right track, but also that in excess of the above realization is required. SR's involvement of a spacetime metric points further toward an eventual realization of Gravitation by Metric Geometry which is free from the above intermediary geometrization's defects.

That curvature can be interpreted in terms of geodesic deviation is itself a geometrically standard fact (Appendix D.4). The above link between geodesic deviation and the Newtonian tidal equation indicates that curvature models a third aspect of Gravitation.[4] Furthermore, curvature—unlike connection—is a tensor quantity; therefore it cannot be transformed away at the point of interest. This gives a sharp mathematical basis for a substantial conceptual distinction between Gravitation in the second and third senses. I.e. local physics can be freed from Gravitation in the second sense but not in the third. [This use of 'local' requires a neighbourhood rather than a point, since curvature manifests itself though *finite-region* vector transport or geodesic deviation involving *finitely separated* geodesics.]

As compared to Gauss's investigation of the curvature of physical space (Chap. 1.10), Einstein's set-up had the good fortune that its spacetime curvature was large enough to be observed in the epoch in which he discovered GR. Moreover, Gauss and Riemann had no inkling that curvature encoded Gravitation or that space and time could be co-geometrized as 4-*d* spacetime. By these additional insights and good fortune, Einstein was able to show that Curved Geometry is relevant to modelling the Universe via some of the observations outlined in Sect. 7.5.

Returning to the issue of setting up field equations for the new Theory of Gravitation, a hypothesis that turns out to be useful and makes use of two Minkowskian Paradigm steps is as follows. I) View the source term in Poisson's Law in terms of energy.[5] II) Next extend this to sourcing by the entirety of the corresponding

---

[4]In Newtonian Gravitation formulated along the original lines with no mention made of Curved Geometry, the single word 'Gravitation' is used in all three of the above senses. These become sharply distinguished upon passing to a geometrical formulation of Newtonian Gravitation. Moreover, this distinction transcends to GR, in which setting the geometry involved is both better-behaved and more standard from a mathematical point of view.

[5]This is an SR insight: $E = mc^2$. Moreover, it has a more immediate and significant consequence: since energy gravitates and all particles have energy, everything has to couple to Gravity.

spacetime tensor: the energy–momentum–stress tensor, $\mathrm{T}_{\mu\nu}$. On these grounds, Einstein conjectured that energy–momentum–stress sources some notion of spacetime curvature and thus Gravitation (meant in the third sense). He eventually realized that this required a curvature tensor matching the properties of the $\mathrm{T}_{\mu\nu}$, i.e. with two indices, symmetric therein, and divergenceless: $\nabla_\mu \mathrm{T}^{\mu\nu} = 0$. (D.24) implies that $\mathcal{G}_{\mu\nu} := \mathcal{R}_{\mu\nu} - \mathrm{g}_{\mu\nu}\mathcal{R}/2$ is such a curvature tensor [282], and this is indeed consequently named the *Einstein tensor*.[6] Next equate this with $\mathrm{T}_{\mu\nu}$ up to proportionality as set by the Poisson equation, giving

$$\mathcal{G}_{\mu\nu} = \frac{8\pi G}{c^4}\mathrm{T}_{\mu\nu} \quad (\textit{Einstein's field equations}). \tag{7.5}$$

Various comments on curvature, the field equations and further assumptions in their derivation are now in order.

1) A cosmological constant term $\Lambda\,\mathrm{g}_{\mu\nu}$ can also be included in the left hand side, since this also fits the symmetry and conservation criteria. There are further mathematical simplicity criteria (see e.g. [859] for an account and references) that pick out this tensor (plus the cosmological constant part). The *Cartan simplicity postulates for GR*—that $\mathcal{G}^{\mathrm{trial}}_{\mu\nu}$ contains at most second-order derivatives of $\mathrm{g}_{\mu\nu}$ and is linear in these—also came to be used in axiomatizing GR. The *Lovelock simplicity postulates for GR* (after physicist David Lovelock) followed from subsequent demonstration that the linearity assumption is unnecessary in dimension $d \leq 4$ [629].
2) The number of Einstein field equations matches the number of independent components of $\mathrm{g}_{\mu\nu}$ (Appendix D.4); in the usual 4-d spacetime case, there are 10 of each. By this stroke of good fortune, the natural interpretation in which the $\mathrm{g}_{\mu\nu}$ are taken to be unknowns corresponds to a *well-determined* system: one for which the number of equations matches the number of unknowns. Since most other attempted geometrizations would face a mismatch rather than a coincidence at this point [779], this 'stroke of good fortune' is a reasonably significant further indication of GR being on the right track.
3) N.B. that the current section is not just a Paradigm Shift from Minkowski spacetime $\mathbb{M}^4$'s Flat Geometry to a generally curved notion of geometry. It is additionally a Paradigm Shift between the following two situations. i) An 'actors performing on a stage' perspective of Physics, encompassing both the Newtonian and Minkowskian Paradigms. Both Euclidean and Minkowskian geometries are rigid pre-determined background structures upon which physical events occur and physical processes unfold. ii) A 'material blobs moving around on a rubber sheet' perspective of Physics [660]. Here the distribution of the material blobs influences the shape of the rubber sheet by determining part of its curvature properties. (7.5)'s status as a 'geometry = matter' equation then means that *each of* geometry (in the form of curvature) and matter (in the form of energy–momentum–stress) *influences the other* in GR. Additionally, GR explains the

[6] Here $\mathcal{R}_{\mu\nu}$ is the Ricci curvature tensor and $\mathcal{R}$ the Ricci scalar curvature.

limited extent in practice of SR's inertial frames in terms of the sources of Gravitation, by which inertial frames cease to be structures that cannot be acted upon [736].

4) While pertinent, let us leave discussion of the well-known and yet disputed role of *Mach's Principle* [632] in the inception of GR to Chap. 9.
5) $\mathcal{G}_{\mu\nu}$ contains the same amount of information as $\mathcal{R}_{\mu\nu}$, but less than that in the Riemann curvature tensor $\mathcal{R}^{\mu}{}_{\nu\rho\sigma}$. The Weyl tensor $\mathcal{C}^{\mu}{}_{\nu\rho\sigma}$ (D.22) picks out the difference, which admits interpretation as gravitational waves. In this way, in GR, Gravitation meant in the third sense further splits into the source-controlled part governed by the Einstein field equations and a free part consisting of gravitational waves. This rests on the Weyl tensor being mathematically sharply defined as an irreducible piece of the Riemann tensor, alongside this irreducible piece covering the totality of information in the Riemann tensor which does not enter the Einstein field equations. Moreover, the conventional spacetime dimension of 4 is the minimal one supporting a nontrivial Weyl tensor and thus the general possibility of gravitational waves.

Having commented on the meaning of the Einstein field equations, we turn to some brief comments on the earlier part of this Section. Let us first expand on Sect. 2.9's coverage of Equivalence Principle concepts. The *Weak Equivalence Principle* can be taken to be just another name for Universality of Free Fall [365, 910]. On the other hand, the *Einstein Equivalence Principle* (in fact due to physicist Robert Dicke) augments the preceding with *Local Poincaré Invariance*. This consists of *Local Lorentz Invariance* and *Local Position Invariance* [910]. These are, respectively, the independence of local non-gravitational experiments from the velocity of the freely falling frame, and from where and when it is performed. Finally, the *Strong Equivalence Principle* additionally includes the effect of self-gravitation [910].

Also note for later comparison that this Chapter gives a Discover Connections and then Curvature approach, in the sense of connections arise first suggesting that the associated notions of curvature be considered as well.

Finally, once curvature is involved, the geodesic deviation equation (D.14) plays a role which is in some ways analogous to that of the Lorentz Force Law (4.15) in Electromagnetism ([660, 897] and Ex V.3).

7.1 More Systematic Formulation of GR's Mathematics

The conventional *spacetime formulation of GR* is in terms of a pair $(\mathfrak{m}, \mathbf{g})$. Here $\mathfrak{m}$ is the topological manifold (Appendix D.1) underlying each spacetime. Additionally assume that $\mathfrak{m}$ carries differentiable structure (Appendix D.2).

Spacetime diffeomorphisms are injective maps $\phi : \mathfrak{m} \rightarrow \mathfrak{m}$ which are differentiable and possess differentiable inverses are *Diff*$(\mathfrak{m})$; see Appendix D.2 for more about the mathematics of these. Because of these, 4 components' worth of information among the 10 components of $\mathbf{g}$ are unphysical.

Moreover, in the case of GR spacetime, the distinction between passive and active diffeomorphisms acquires further significance. Passive diffeomorphisms are coordinate transformations, tied to the well-known notion of Jacobian matrix (Chap. 2). On the other hand, active diffeomorphisms correspond to the moving around of points of a manifold; this is also tied to the notion of Lie derivative (Appendix D.2), which indeed provides a means of moving points around: Lie-dragging. It is the active diffeomorphisms which are the main concern in the study of GR, for the Background Independence reasons laid out in Chap. 9. Also note the step-up from Electromagnetism and Yang–Mills Theory, whose transformations occur at a fixed spacetime point (i.e. event), whereas in GR the diffeomorphism group moves points around [483].

Given that GR spacetime is also equipped with a metric, a subsequently useful notion are the *isometries*: metric-preserving injective maps. Isometries are furthermore related to both Lie derivatives and Killing forms (Appendix E.2) by

$$\pounds_{\vec{\mathrm{X}}} g_{\mu\nu} = 2\nabla_{(\mu}\mathrm{X}_{\nu)} =: (\mathcal{K}\mathrm{X})_{\mu\nu}. \tag{7.6}$$

The first equality is computational, and illustrates a common trend: that Lie derivatives can be re-expressed as covariant derivatives in the presence of sufficient structure to define the latter. Moreover, the Lie derivative notion is more minimalistic, pertaining to just Differential Geometry to the covariant derivative requiring an affine connection as well. Finally note that Killing vectors are a crucial part of the extension of the notions of symmetries and conserved quantities to more general settings than in the flat spaces of Newtonian and Minkowskian Physics; see Appendix E.2 for more.

7.2 Spacetime Action Principle for GR

The Einstein–Hilbert action for pure GR is (based on Appendix D.6's notions of density and integration)

$$\mathsf{S}^{\mathrm{GR}}_{\mathrm{EH}} = \frac{c^4}{16\pi G}\int_{\mathfrak{m}} \mathrm{d}^4x\sqrt{|\mathrm{g}|}\,\mathcal{R}(\vec{X};\mathbf{g}]. \tag{7.7}$$

One is to introduce here also a matter action S_ψ, combined additively with (7.7). This takes the $\boldsymbol{\eta} \longrightarrow \mathbf{g}$ version of its SR form, making use of minimal coupling as well (a type of local Lorentz invariance postulate). Much as one can cast all the observationally-established non-gravitational classical laws of Physics in SR form, one can cast these (now free from this non-gravitational caveat) in GR form [660].

Varying this additive combination gives Einstein's field equations for GR (7.5). The energy–momentum–stress tensor here is identified as being of the form

$$\mathrm{T}^{\mu\nu} := 2|\mathrm{g}|^{-1/2}\frac{\delta \mathsf{S}_\psi}{\delta \mathrm{g}_{\mu\nu}}. \tag{7.8}$$

Finally, to include the *cosmological constant*, pass from $\mathcal{R}$ to $\mathcal{R} - 2\Lambda$ in the action. This long was a theoretically-optional feature, but conventional ways of fitting modern cosmological observational data provide a strong argument for the practical necessity of this term.

7.3 Black Holes

Since this and the next Section consider some of GR's solutions, let us note preliminarily that the Minkowski metric η of SR (4.1) indeed also resurfaces as a solution of GR.

A first example of GR black hole metric is the *Schwarzschild solution*

$$\mathrm{d}s^2 = -\{1 - 2GM/c^2r\}c^2\,\mathrm{d}t^2 + \mathrm{d}r^2/\{1 - 2GM/c^2r\} + r^2\mathrm{d}\Omega^2 \tag{7.9}$$

(after physicist Karl Schwarzschild). This is expressed here in spatially spherical polar type coordinates. It is a vacuum solution, spherically symmetric and asymptotically flat (as a simple criterion, it is increasingly well-approximated by η in the far field). It is both *stationary*—in possession of a timelike Killing vector field—and *static*: likewise, but now additionally with the timelike Killing vector field orthogonal to the constant-time spatial hypersurfaces). These coordinates go singular at the Schwarzschild radius (2), which happens to coincide with the Michell radius (Ex V.1). One can however pass through the surface at this radius by changing coordinates. This surface furthermore has a coordinate-invariant meaning as an *event horizon*. Loosely speaking, this is a surface of no return. More specifically, in terms of Causality Theory, it is $\mathrm{H} := \dot{\mathrm{J}}^-(\mathcal{I}^+) \cup \mathfrak{m}$, where the dot denotes 'boundary of'.[7] Because the Schwarzschild solution possesses this, it contains a black hole: a region from which light cannot escape (so nothing else can escape either). Also in terms of Causality Theory, the black hole is the region $\mathrm{B} := \mathfrak{m} - \mathrm{J}^-(\mathcal{I}^+)$. On the other hand, $r = 0$ is a genuine—rather than merely coordinate—singularity, so the black hole contains a singularity. This is denoted by the jagged edge in Fig. 7.1.e).

The maximally extended Schwarzschild solution (Fig. 7.1.e) represents a black hole *and* the time reversal of a such: a *white hole*. A piece of this Schwarzschild solution—lying well outside of where the event horizon would be—is used in modelling the part of the Solar System exterior to the Sun (see the next Section). Indeed, $r_{\mathrm{Schw}} \ll r_{\mathrm{Sun}}$, so the vacuum condition ceases to apply anywhere near r_{Schw}. GR has a Newtonian limit in the sense that the correct Newtonian Physics is recovered in situations with low velocities $v \ll c$ and weak gravitational fields $\phi \ll c^2$. While G is absent from this expression, via $GM/c^2r^2 = \phi$, this amounts to $r \ll r_{\mathrm{Schw}}$. Figure 7.1.g) indicates a piece of Schwarzschild spacetime arising

[7] $\mathcal{I}^+$ here denotes *future null infinity*, $\mathcal{I}^-$ is *past null infinity* and i^0 is *spatial infinity*. These play a significant role as edges of Penrose diagrams: Fig. 7.1, where the corresponding types of geodesics can begin and end.

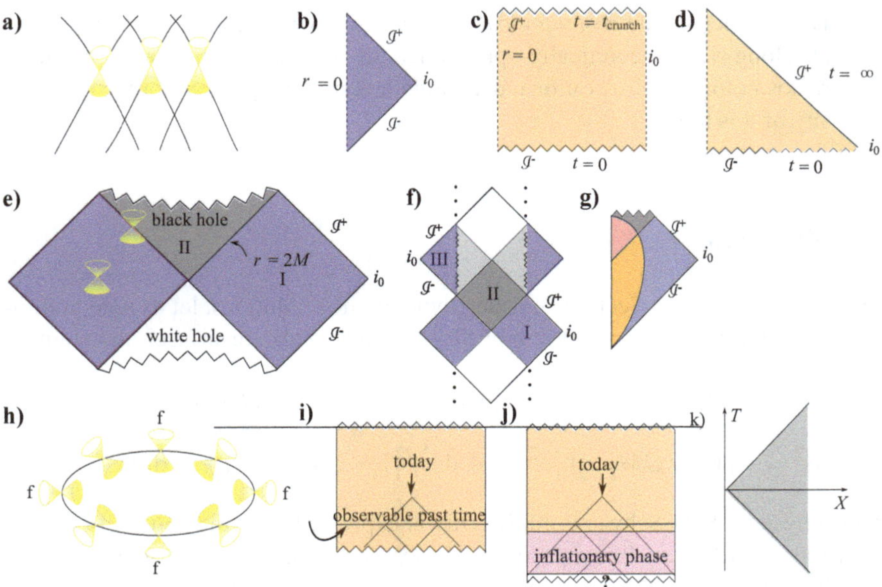

Fig. 7.1 GR spacetime. **a)** GR null cones are in general bent. Penrose diagrams (after mathematician Roger Penrose) for **b)** Minkowski spacetime $\mathbb{M}^4$, **c)** $k > 0$ FLRW, **d)** $k \le 0$ FLRW, **e)** Schwarzschild, **f)** Kerr (or Kerr–Newman) and **g)** the astrophysical truncation of Schwarzschild. Asymptotically flat regions are coloured in *blue*, black hole regions in *black*, white hole regions in *white*, cosmological dust in *brown* and stellar matter in *orange*. The jagged edges are singularities, the jagged edges with gaps are the Kerr ring singularities (traversible in some directions), and the *dashes* are mere coordinate singularities. **h)** Non-orientability in time (f indicates the future direction). **i)** The Particle Horizon Problem in Cosmology, and **j)** its resolution according to inflation by including an extended past region in *pink*. **k)** The Rindler wedge (after physicist Wolfgang Rindler): the region of Minkowski spacetime covered by Rindler coordinates T and X (this is *not* a Penrose diagram)

by stellar collapse. From balancing gravitational collapse against the maximum degeneracy pressure exertable by a Fermi gas, collapse occurs if the star's mass exceeds the *Chandrasekhar mass* (after astrophysicist Subrahmanyan Chandrasekhar) $\sim m_{\rm Pl}^3/m_{\rm p}^2 \sim 1.4 m_{\rm Sun}$, where $m_{\rm p}$ is the proton mass. Below this bound, collapse can halt in a white dwarf star or neutron star configuration.

A second example is the stationary (but not static) aximsymmetric rotating and charged black hole metric (*Kerr–Newman metric* after physicists Roy Kerr and Ted Newman)

$$\begin{aligned} ds^2 = &-\left\{\frac{dr^2}{\Delta} + d\theta^2\right\}\rho^2 + \left\{c\,dt - \frac{j}{c}\sin^2\theta\,d\phi\right\}^2\frac{\Delta}{\rho^2} \\ &- \left\{\left\{r^2 + \frac{j^2}{c^2}\right\}d\phi - j\,dt\right\}^2\frac{\sin^2\theta}{\rho^2}. \end{aligned} \tag{7.10}$$

The spatial part of the coordinates is presented here again as a type of spherical polar coordinates. Also $j := J/M$, $q^2 := GQ^2/4\pi\epsilon_0 c^4$, $\rho^2 := r^2 + j^2 c^{-2}\cos^2\theta$ and $\Delta := r^2 - 2GMr/c^2 + j^2/c^2 + q^2$. Just set Q (or q) to 0 to get the very similar uncharged rotating case (Kerr metric), whereas the charged rotating case is accompanied by an electromagnetic potential whose nonzero components are $A_t = Qr/\rho^2$, $A_\phi = -Qjc^{-2}r\sin^2\theta/\rho^2$. On the other hand, setting J (or j) to 0 gives the simpler diagonal non-rotating charged case (the Reissner–Nordström metric, after physicists Hans Reissner and Gunnar Nordström), accompanied by the electromagnetic potential whose remaining nonzero component is now $A_t = Q/r$. The physical case involves $j^2 + q^2 < 1$, for which the Kerr–Newman black hole possesses 2 distinct event horizons.

Surface gravity κ is a useful concept in Black Hole Physics. Its Newtonian Astrophysics precursor is the gravitational acceleration experienced on the surface of an astrophysical object. However, for GR black holes, this takes an infinite value; a more physically appropriate definition follows from the non-affinely parametrized geodesic equation $k^\mu \nabla_\mu k^\nu = -\kappa k^\nu$; cf. (D.10). So κ is a measure of the failure of Killing and affine agreement along the null geodesic generators of the event horizon (the k^μ are normal to the event horizon). Note furthermore that (see e.g. [874] for exposition)

$$\kappa \text{ is constant over a GR stationary black hole's horizon.} \tag{7.11}$$

Using the area of the Kerr–Newman black hole as computed geometrically, one furthermore deduces that

$$\mathrm{d}M = \frac{\kappa}{8\pi G}\mathrm{d}A + \Omega\,\mathrm{d}J + \Phi\,\mathrm{d}Q, \tag{7.12}$$

$$\mathrm{d}A \geq 0 \quad \forall \text{ black hole processes,} \tag{7.13}$$

$$\text{It is impossible to attain } \kappa = 0 \text{ by a finite number of physical processes.} \tag{7.14}$$

Equation (7.14) corresponds to the extremal black hole. Furthermore, compare (7.11)–(7.14) with the Laws of Thermodynamics in the forms (Q.1)–(Q.4) respectively, under the for now tentative identifications $\kappa \leftrightarrow cT$, $S \leftrightarrow A/8\pi c$ and $M \leftrightarrow U$, and regarding the last two terms of (7.13) as work terms. This analogy suggests that the above Laws of Black Hole Mechanics (7.11)–(7.14) are special cases of the Laws of Thermodynamics. However, the black hole is classically a perfect absorber, so κ is not conceptually a temperature. And yet Sect. 11.3 outlines how Quantum Theory removes this objective to cement this analogy. In particular,

$$0 \leq \mathrm{d}S_{\text{Total}} = \mathrm{d}S_{\text{BH}} + \mathrm{d}S_{\text{other}} \tag{7.15}$$

is a more general form for the Second Law, in accord with this law's conceptually desirable universality. For the Schwarzschild solution,

$$S = k_{\text{B}} A/4G\hbar. \tag{7.16}$$

7.4 Cosmology

The homogeneous[8] and isotropic (Sect. 2.12) metric that is most often used for this is

$$ds^2 = -c^2 dt^2 + a(t)^2\{dr^2/\{1 - kr^2\} + r^2 d\Omega^2\}: \quad (7.17)$$

the *Friedman–Lemaître–Robertson–Walker (FLRW) metric* (after physicists Alexander Friedmann, Georges Lemaître, Howard Robertson and Arthur Walker). This comes in closed (spatially $\mathbb{S}^3$: $k = +1$), open (spatially the hyperbolic space $\mathbb{H}^3$: $k = -1$) and critical (spatially flat $\mathbb{R}^3$: $k = 0$) variants. See e.g. [736] for further details of these solutions and their physical interpretation, without and with Λ. For instance, these references cover the dependence of the *scalefactor of the* (model) *universe*, $a = a(t)$'s dependence upon the equation of state $p = w\,\mathcal{E}$ of the matter content of the Universe. According to such models and their fitting to observational data,

$$\text{the age of the Universe} \simeq \{1.380 \pm 0.002\} \times 10^{10} \text{ years}, \quad (7.18)$$

while it is as yet not a foregone conclusion which sign k takes.

In isotropic model universes, *cosmic time* $t = t^{\text{cosmic}}$ is physically the time that labels the surface of homogeneity, or, dually, that is aligned with the 'Hubble flow' as followed by idealized comoving inertial observers [523, 596]. This is the proper time experienced by these observers, which roughly models our own perspective here on Earth; Chap. 20 provides further details. It is also the time in the above standard presentation of the FLRW metric.

Conformal time, on the other hand, is given by $d\eta := c\,dt/a(t)$; this puts the metric into the form

$$ds^2 = a^2(\eta)\{-d\eta^2 + dr^2/\{1 - kr^2\} + r^2 d\Omega^2\}, \quad (7.19)$$

which, for the spatially flat case, is conformal to a piece of $\mathbb{M}^4$. This is clear in the corresponding Penrose diagram Fig. 7.1.d); it is only a piece because this is *not* a valid conformal transformation along $a = 0$. The link between conformal time and causal structure becomes clearer after the following definition. *Cosmological horizons* alias *particle horizons* are the edges of where information can arrive from at a given observer's position, as per Fig. 7.1.i). Conformal time corresponds to the distance to the cosmological horizon. Moreover, different events with the same value of conformal time appear simultaneous to a comoving observer [595, 596].

See Appendix I.1 as regards further anisotropic cosmology solutions ('Minisuperspace solutions'). These are still the same pointwise over space; however, this ceases to be the case in inhomogeneous solutions; see e.g. [812] for some GR cosmologies of this form. As simpler examples, 'Midisuperspace solutions' are nontrivially spatially inhomogeneous, while retaining some nongeneric elements of sym-

[8]In this sense, 'homogeneous' means the same at each point over a mathematical space (in the present cosmological context physical 3-d space).

metry by which these are more tractable; footnote 4 points to some examples of these.

To complete this Sec's discussion of horizons, Fig. 7.1.k)'s *Rindler wedge* represents a uniformly accelerating reference frame in Minkowski spacetime. The edge of this wedge is a type of horizon at which acceleration diverges.

7.5 Evidence for GR

Gravitational redshift provides a test of the (Einstein) Equivalence Principle prior to any use of the Einstein field equations that further characterize GR. Thus such considerations do not distinguish between different metric theories of gravity. GR's general redshift formula is

$$1 + z = \sqrt{g_{tt}^{\text{obs}}/g_{tt}^{\text{source}}}. \tag{7.20}$$

The gravitational redshift between two points in the same body's gravitational field (at distances r_{source} and r_{obs} from the centre) is then approximately

$$z = \{GM/r_{\text{source}} - GM/r_{\text{obs}}\}/c^2 = r_{\text{Schw}}\{r_{\text{obs}} - r_{\text{source}}\}/2\,r_{\text{obs}}r_{\text{source}}. \tag{7.21}$$

Einstein's hypothesis in the fourth paragraph of this Chapter has since been experimentally supported [814] by Pound–Rebka type experiments (after physicists Robert Pound and Glen Rebka) [910] to better than 1 part in 10^5.

The credibility of GR itself was rapidly established by its explanation of the 43 seconds of arc per century anomalous perihelion shift of Mercury in 1915, alongside experimental verification (albeit to limited accuracy) of its prediction of the bending of light rays by the Sun in 1919. The two most common deviations from GR [910] in alternative theories are denoted by γ and β. γ quantifies variety in how much spatial curvature is produced by a unit rest mass. This enters computations for the motion of null test particles, such as calculations of the deflection of light. It also enters the more constraining Shapiro time delay by which Cassini space probe data confirms GR's value $\gamma = 1$ to 2 parts in 10^5. On the other hand, β is a measure of the nonlinear departure from the Superposition Principle applicable within Newtonian Gravity. It features in addition to γ in the massive test particle case of perihelion precession; the GR value of $\beta = 1$ has by now been confirmed to 8 parts in 10^5.

As regards cosmological developments, Hubble's Law for the recession of galaxies is $v = H_0 r$ for H_0 Hubble's constant. (This law dates from 1929; see [888] for this paragraph's original references and further details.) Moreover, astronomer Edwin Hubble only determined that the Universe was expanding (1932), rather than by how much. The first reasonable estimate of H_0 was provided in 1958, but agreement upon a value for this would not come for decades after. This is closely tied to the cosmological redshift formula

$$1 + z = a_{\text{obs}}/a_{\text{souce}} \tag{7.22}$$

for the overall recession of the galaxies. The Planck satellite mission [840] obtained H_0 to 1 part in 100, as 67.80 ± 0.77 km/s Mpc, where Mpc stands for the intergalactic length unit of megaparsecs. A second pillar of modern Cosmology—evidence of the abundances of the light elements that is tied to their genesis in the Early Universe—began with physicist Ralph Alpher's theoretical predictions in 1948, but observational secureness for it had to wait until around 1980. The third pillar—the cosmic microwave background radiation: a thermal imprint in the form of highly homogeneous and isotropic black body radiation—was first observed by physicists Arno Penzias and Robert Wilson in 1964. This observation confirmed the Big Bang scenario over its cosmological rival of that epoch: the Steady State model.

Some more modern problems with the Big Bang theory itself are why the Universe appears to be so flat, the non-observation of the monopoles that Grand Unified Theories would suggest, and Fig. 7.1.i)'s horizon problem. A candidate improvement in these regards is Inflationary Theory, which involves a period of exponential expansion. This flattens out the Universe, dilutes its monopole content and resolves the horizon problem, as per Fig. 7.1.j). Inflation was first proposed at the start of the 1980s and holds out fine against detailed modern cosmological data [122, 841] including the pattern of small inhomogeneities observed in the cosmic microwave background. Accommodation of additional supernova data [569] is a principal reason for considering universe models currently dominated by dark energy—a cosmological constant type term (or similar resultant from cosmological matter fields).

Let us now continue with Sect. 2.10's argument by pointing out that there is also a 'gravitomagnetic law' that completes the square of Inverse Force Laws and is a valid piece of a more full GR case: the weak-field regime [736]. At least in this setting, this is linked to frame dragging effects—relevant to some aspects of whether GR is Machian along the lines of the bucket argument—and has been investigated using the Gravity Probe B experiment [910] albeit here the support for GR's prediction is for now just at the 30% level.

For GPS (the global positioning system) [74, 431, 910] to attain the precision it operates at, it needs to take into account GR as well as SR effects. In particular, it is affected by both time dilation and gravitational redshift. N.B. also the importance to relativistic timekeeping of knowing the relative locations of the clocks and other equipment involved.

Einstein already predicted gravitational lensing as an extension of light deflection to optical effects with galaxies and other compact objects acting as natural lenses. See [776] for an account of this, or [431] for a briefer outline; a number of observations of gravitational lensing have by now been made [910].

Einstein also already predicted gravitational waves. For GR in weak field regimes, these have tensor modes with two polarizations: $\times$ and $+$ [874] (these are literally the distortion patters associated with each). Indirect evidence for gravitational waves comes from binary pulsar data exhibiting gravitational damping, in accord to parts in 10^3 with GR's predicted losses due to gravitational radiation [910].

In 2016, LIGO and Virgo (kilometre-scale interferometer based gravitational wave detectors based on Earth [763]) obtained the first direct evidence for gravitational waves [839]: a binary black hole merger signal.

The further proposed eLISA mission is to be an interferometer formed between three space probes forming an equilateral triangle of side 10^6 km [309].

As is clear from the difference in size between these Earth based and space based detectors, each is particularly attuned to a different part of the gravitational wave spectrum. The former are in particular for searching for gravitational waves sourced by compact astrophysical binaries[9] On the other hand, the latter are also to investigate gravitational waves of a primordial cosmological origin [226].

7.6 Notions of Time in the Spacetime Formulation of GR

1) In the spacetime formulation of GR, *time is but one of the spacetime 4-manifold's coordinates*. This clashes with Ordinary QM holding time to be a sui generis extraneous quantity. Moreover SR's signature distinction between timelike and spacelike separations remains: space and time remain distinguishable concepts.
2) The privileged frame interrelating Poincaré group of SR (or the Euclidean group of Mechanics and Ordinary QM) have been supplanted by the *spacetime diffeomorphisms* $Diff(\mathfrak{m})$ *between arbitrary coordinate systems*. As far as the Author is aware, the consequent significant increase in complexity arising from this was first pointed out by Pauli [700]. E.g. attaching significance to conserved quantities is linked to the Poincaré (or Euclidean) groups, and Chap. 11.3 presents four more examples of structures tied to $Poin(4)$. Nor does the influx of harder diffeomorphism-based mathematics end with the spacetime diffeomorphisms, as is evidenced by two further kinds of diffeomorphisms appearing in the next Chapter.
3) Furthermore, *GR's generic solutions have no Killing vectors*. In particular some time-related applications are affected by there now being *no timelike Killing vector*; this provided a privileged class of times in SR's Minkowski spacetime $\mathbb{M}^4$. In contrast, $\mathbb{M}^4$ has the *maximal* number of independent Killing vectors (10 in 4-*d*). Finally, GR is ultimately considered to be about generic solutions. In this way, much of the structure that many SR and QFT calculations are based upon is lost.

   The points made so far in the current section mean that much of the structure of Ordinary QM simply ceases to have an analogue [see Chap. 11]. As we shall see in Chaps. 9 to 12, noted quantum physicist and conceptual thinker Chris Isham [483] has attributed much of the Problem of Time to the extra subtleties brought in by the diffeomorphisms.
4) *GR's notion of simultaneity* is a straightforward extension of SR's [521].
5) *GR time retains the ordering property*.

[9] Compact astrophysical objects are white dwarfs, neutron stars or black holes. Astrophysical binaries are pairs of objects orbitally bound in close proximity to each other. Such a configuration, where each object is a neutron star or a black hole, is potentially a strong source of gravitational waves. See e.g. [296] for a pedagogical account of source counting for white dwarf stars.

6) *Causality* continues to play a major role in the spacetime formulation of GR as it did in SR, except that now *matter and gravity influence the larger-scale causal properties*. In GR, *the null cone structure is dynamical*. *Penrose diagrams* (Fig. 7.1) are useful at this point on two counts. Firstly, these are based on performing conformal transformations so as to compactify spacetimes into finite diagrams. This is based on the key underlying fact that null geodesics are conformally invariant (Ex III.11). Secondly, in Penrose diagrams, the null cones are everywhere upright, giving a very clear representation of the causal structure. Features priorly encoded in terms of null cones tipping over in other representations are clearly displayed in the Penrose diagram representation, e.g. the horizons in Figs. 7.1.e)–f).
7) We now have a further Arrow of Time to introduce: the *cosmological Arrow of Time* (Ex V.22).
8) GR spacetimes $\langle \mathfrak{m}, \mathbf{g} \rangle$ are often taken to be *time-orientable*, meaning that it is possible to divide continuously over $\mathfrak{m}$ each null cone of the metric $\mathbf{g}$ in two parts, past and future [440, 874].
9) *Closed timelike curves* exist within certain GR spacetimes. E.g. the Kerr–Newman spacetime possess such: Ex V.7). Their significance is that observers following these would experience time travel.

   One means of avoiding causality paradoxes is suppressing (regions of) solutions containing closed timelike curves. There is however also a self-consistent interpretation of closed timelike curves [666]. Solutions which are non-orientable in time are also often excised from the study of the supposedly physical GR solution space.

   See the next Chapter for notions of space within GR (as well as further aspects of time that become apparent upon performing a space–time split).
10) Energy is a substantially more complicated and unsettled concept in GR than in pre-GR Physics (Appendix K.5). This is related to time through (at least the simpler notions of) energy) being tied to time in the form of being its canonical conjugate.

7.7 GR Issues with Clocks

Chapter 4's light-and-mirror clock considerations continue to apply in GR.

Sufficiently accurate timestandards are both Specially and Generally Relativistic. We have already seen that SR confers motion-dependence to timestandards. This is a localization of the applicability of timestandards in that each 'particle' undergoing a distinct motion experiences its own timestandard. On the other hand, gravitational time dilation directly imposes location-dependence on timestandards.

So *where* in space a timestandard is to hold becomes an issue once precision exceeds SR and GR corrections. In pre-relativistic timekeeping, one need not ask *where* in, say, the Earth–Moon–Sun system the ephemeris time holds. However, once the precision exceeded around 1 part in 10^{12} in the late 1970s, relativistic timekeeping becomes relevant and differs according to where the clock is and how it is moving. Above this precision, such as 'using the Moon as a reading hand' only makes sense for specifying a timestandard in some particular localized frame. 'On the surface of the Earth' is such a qualification; due to this not being of constant radius or gravitational equipotential, 'at mean sea level' is used instead. Finally, upon introducing this standardization, the motion implicit in being 'on' the rotating Earth becomes the stipulation of 'on the rotating geoid' at mean sea level.

GR also plays a significant role in the accurate determination of positions, as is evident in the GPS system. Furthermore, since relativistic timekeeping is position-dependent, precise determination of positions is itself is a substantial input. We need to know where our clocks are in relation to one another. In this way, clock bias enters GPS considerations (see [88] and Ex V.19).

Chapter 1's definition of the second requires, for sufficiently accurate applications, additional stipulation that it is defined on the rotating geoid at mean sea level. This covers the position and motion at which it is defined.

We finally consider timestandards in more extended settings. Firstly, clocks for space travel require determination of position away from large well established approximately rigid frames such as provided by the Earth; this is already potentially an issue for LISA. One approach involves on-board clocks and frequent recalibration by signals from Earth (or similar positions with currently conventional timekeeping set up). Another possibility involves on-board clocks being calibrated by comparison with pulsar signals. These furnish an example of a case with *very* negligible gravitational interaction between a clock and the subsystem(s) it 'keeps time for'. Accurate pulsar ephemerides [310] have been computed, though for now these depend on their being observed from Earth.

Secondly, let us next consider what clocks would be suitable in the Early Universe and near black holes. We do not consider this in the sense of disturbance from accurate motion due to accelerations as in the problem of timekeeping at sea (which here has an analogue due to high spacetime curvature analogue) nor the SR conception of clocks [736]. We consider, rather the breakdown of the technology under such as high accelerations and high temperatures. Atomic clocks are vulnerable to regimes in which the primary timestandard atoms ionize. The frequency band involved is also vulnerable to [13] the Stark effect, i.e. the perturbation of a quantum system due to presence of background electric fields. This points to alternatives being required; one issue is whether black holes themselves provide any further types of notably accurate clocks, e.g. from their rotation. Finally, which clocks could—or actually did—exist in Early-Universe regimes?

7.8 Observers and Length Measurement in GR

In GR, observers are modelled as negligible energy–momentum–stress entities; contrast with how in Quantum Theory, observers are usually held to be much larger than the system in question. So the theory of 'the large' and of 'the small' is not just a direct comparison but also a comparison of each with the sizes and sensitivities assumed of its observers. Observers in GR are moreover idealized as regards their internal constitution not being posited. This is also the case for any clocks and rods involved. As such one should believe little in these idealizations once details thereof become pertinent to the physical thinking. Is the quantum way of handling these entities extendible to subsystems within the GR setting?

In GR, the actual nature of rods is ignored; idealized objects are considered instead. The concept of 'rigid bodies' is further lost at the level of GR (see e.g. p. 264 of [730]), but also one can pass to electromagnetic beam type conceptualization and technology. Marzke and Wheeler's [645] motivation for this came from Bohr and Rosenfeld's criterion for self-sufficiency of a theoretical framework [150]. They succeed in conceptualizing of the beam in quantum-free terms, but fail to remove Quantum Theory entirely from consideration due to the detailed nature of the emitting, reflecting and absorbing devices at each end. Indeed, they pointed out that length determination cannot just involve light [645], because equivalent conformally-related geometries having the same null geodesics. The 'massive particles' thus entailed are point-particle idealizations of solids, fluid bodies, or atoms, all of which are underlied by Quantum Theory.

GR has also entered the definition of the metre since 2002, due to acknowledgment that the metre is a unit of *proper length*, whose definition only applies to lengths which are sufficiently small that GR effects are negligible. One surmises that the definition would require modification on small scales if one were to be operating in a sufficiently high-curvature regime.

7.9 GR's Singularity Theorems

GR points to its own inapplicability under extreme circumstances: in the innermost part of black holes, and within a very short time interval after a cosmological Big Bang. These circumstances are furthermore likely to occur in our Universe by the Singularity Theorems of Penrose and physicist Stephen Hawking. This may be related to difficulties with combining GR and QM to form a theory of Quantum Gravity necessary for the study of these extreme regimes (see Chap. 11).

The Singularity Theorems are built using Causality Theory, conjugate points (Appendix D.3), and *trapped surfaces*: spacelike 2-surfaces both of whose null normals are converging. See [440, 784, 874] for detailed statements, proofs and examples.

Singularities can, moreover, be spacelike, timelike or null; Fig. 7.1 gives examples of the first two of these. Some singularities are inevitable within finite proper time; upon crossing the event horizon, this is true in Schwarzschild spacetime but false in Kerr spacetime (Ex V.13). With the Big Bang in mind, it may be that time runs over $\mathbb{R}_+$ rather than over $\mathbb{R}$, or over an interval $\mathfrak{T}$ if there is a Big Crunch as well.

Chapter 8
Dynamical Formulations of GR

Dynamics entails heterogeneous treatment of time and space. In particular, as Chaps. 1 and 2 indicated, Dynamics concerns configurations and momenta evolving with respect to time, and treats derivatives with respect to time differently from those with respect to space. This does not directly fit in with the SR and GR spacetime perspective, in the sense that spacetime itself neither evolves in time nor plays configuration's timeless role. Rather, GR spacetime *contains* notions of both spatial configuration and of time. Each of these can be extracted by splitting the spacetime metric up; moreover, this induces a split of GR's Einstein field equations along dynamical lines [73, 660, 899]. One may furthermore consider Dynamics to be primary, and thus ask from first principles what GR is a dynamics of, i.e. what its configurations are. In any case, GR's configurations can be taken to be spatial—i.e. positive-definite Riemannian—3-metrics, h_{ij}; in formulations in which spacetime is primary, these are furthermore spatial slices within spacetime.

In fact, as we shall see below, spatial 3-metrics are a redundant presentation; less redundantly, GR's configurations are spatial *3-geometries*: 3-metrics 'minus coordinate information'. In this way, as well as having a spacetime formulation, GR admits a dynamical formulation in terms of evolving spatial 3-metrics or 3-geometries too. Wheeler termed the latter *Geometrodynamics* [660, 897, 899].[1]

8.1 Topological Manifold Level Structure

i) In conventional dynamical formulations of GR, one first has choose a residual notion of space in the sense of a 3-surface that is a fixed topological manifold $\boldsymbol{\Sigma}$.
ii) In this book, $\boldsymbol{\Sigma}$ is usually taken to be compact without boundary for simplicity; (this book considers these to be connected as well).

[1]Wheeler is well-known for coining terminology; e.g. 'S-matrix' [895] and 'black hole' are also due to him.

E. Anderson, *The Problem of Time*, Fundamental Theories of Physics 190,
DOI 10.1007/978-3-319-58848-3_8

iii) We furthermore concentrate on specific examples with $\mathbf{\Sigma} = \mathbb{S}^3$: the 3-sphere, which is one of the simplest possibilities.

By i), a fixed $\mathbf{\Sigma}$ is to be shared by *all* the spatial configurations in a given Geometrodynamics. I.e. dynamical formulations of GR such as Geometrodynamics are built subject to the restriction of not allowing for topology change. This means that Geometrodynamics covers a more restricted range of spacetimes than the spacetime formulation of GR does: those with spacetime topology $\mathbf{\Sigma} \times \mathcal{T}$. Geometrodynamics is thus just a 'manifold topolostatics' rather than being a 'manifold topolodynamics' as well, a matter to which we return in Sect. 10.12 and Epilogue II.C. This has some superficial resemblance with Newtonian space-time, e.g. as a stringing together by labelling by time variables. However, the spatial slices involved in general differ among themselves at the metric level.

GR spacetime carries the following additional connotations.

A) Unified co-geometrization: an overall 4-metric rather than separate spatial and temporal metrics in the Newtonian case.
B) Causality structure is encoded by the indefiniteness of the 4-metric rather than Mechanics' slices being privileged surfaces of absolute simultaneity. GR's time variable is additionally highly nonunique. Different choices of this in general correspond to different foliations, each of which is valid and with the Physics involved turning out to be foliation-independent (see Chap. 10 for more).

See Chap. 9 for further motivation of ii) and iii), and Appendices C–D.1 for the meaning of i) and ii)'s technical details. Further restrictions are placed on $\mathbf{\Sigma}$ in Sect. 8.13. For now, one accepts confinement to a subset of GR's solution space so as to be able to study its dynamics within what mathematical methodology is currently known and accepted among physicists. Let us also use the notation σ in place of $\mathbf{\Sigma}$ if a 3-space is treated in isolation rather than as a slice within spacetime (in a sense made precise in Sect. 8.4). For many purposes, one can also take a finite piece of space $\mathrm{S} \subset \mathbf{\Sigma}$, rather than a whole space $\mathbf{\Sigma}$.[2]

8.2 Differential Geometry Level Structure

Let us next additionally assume that σ (or any of the preceding Sec's variants) carries differentiable structure (Appendix D.2); this is much as was considered for $\mathfrak{m}$ in Chap. 7.1. The maps preserving this level of structure are spatial diffeomorphisms, $Diff(\mathbf{\Sigma})$. Many properties of these parallel those of $Diff(\mathfrak{m})$, because at this level of structure there is not yet a metric involved whose signature distinguishes between spacetime and space. E.g. $Diff(\mathbf{\Sigma})$ are again actively interpreted (nothing in

[2]For some purposes, one does not need to concern oneself with such a piece having boundaries. These involve 'local' considerations in a sense that will be made precise at the metric level in the next Section. See e.g. [322] for the dynamics of GR including boundaries. Chapter 31 outlines arbitrary dimensional, and yet further, alternatives to this Chapter's workings.

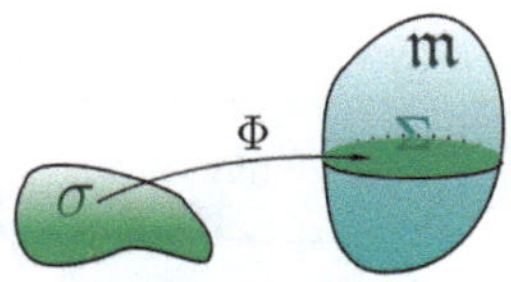

Fig. 8.1 Embedding Φ from 3-space σ to hypersurface Σ within spacetime $\mathfrak{m}$. Throughout this book, we distinguish spatial manifolds from spacetime ones by shading them *green* and *turquoise* respectively

the given active–passive argument is *signature* dependent). Also,

$$£_{\xi}\mathrm{h}_{ab} = 2\mathcal{D}_{(a}\xi_{b)} = (\mathcal{K}\,\xi)_{ab} \tag{8.1}$$

is the counterpart of (7.6) and with the same ties to Killing vectors. See Chap. 9 for further comparisons.

8.3 Metric Level Structure

A hypersurface $\langle \boldsymbol{\Sigma}, \mathbf{h}\rangle$ inherits spatiality from how it sits within the surrounding ambient spacetime $\langle \mathfrak{m}, \mathbf{g}\rangle$. To ensure that σ is indeed cast in a spatial role, moreover, this is directly equipped with a specifically Riemannian (positive-definite) 3-metric $\mathbf{h}$ with components $\mathrm{h}_{ij}(\underline{x})$.[3] It is natural to consider 3-metrics from prior consideration of Newtonian Mechanics or SR, and so as to continue to model in terms of lengths and angles.

$\mathfrak{Riem}(\boldsymbol{\Sigma})$ is GR's configuration space consisting of all the $\mathbf{h}$ on that particular fixed topological manifold $\boldsymbol{\Sigma}$; if the context in which this is used does not presuppose spacetime, the notation $\mathfrak{Riem}(\sigma)$ is used instead. The latter occurs e.g. in investigation of geometrodynamical theories in general (see e.g. Chap. 33), rather than in treatment of specifically the Geometrodynamics that is obtained by splitting GR spacetime and GR's Einstein field equations.

8.4 Single-Hypersurface Concepts

Let us next consider passing from a 3-space σ to a spatial hypersurface $\boldsymbol{\Sigma}$ embedded in a spacetime $\mathfrak{m}$.[4] More formally, a *hypersurface* $\boldsymbol{\Sigma}$ within $\mathfrak{m}$—Fig. 8.1.a)—is

---

[3]As further useful notation, h_{ij} has determinant h, inverse h^{ij} and 3-metric-compatible covariant derivative $\mathcal{D}_i$.

[4]Relativist Eric Gourgoulhon's book [382] also considers such a passage, and is carefully laid out as regards sharply distinguishing between these and other notions which involve multiple spatial hypersurfaces.

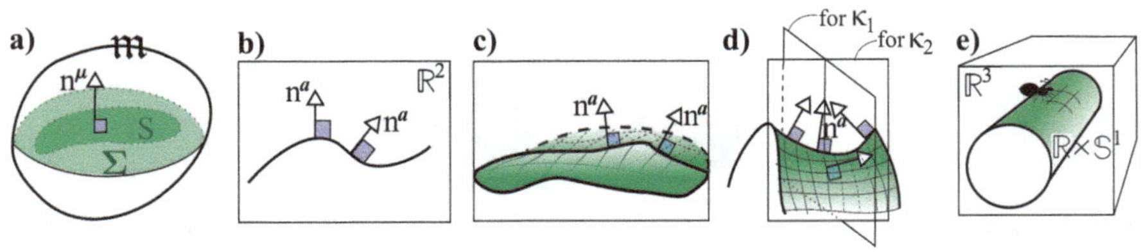

Fig. 8.2 **a)** The normal n^μ to a hypersurface, which is denoted throughout this book with a *white triangular arrow*. In this figure alone, the perpendicularity is emphasized with *blue right angles*. **b)** Extrinsic curvature of a curve in $\mathbb{R}^2$ is the rate of change of the normal along the curve. If the curve is in $\mathbb{R}^3$, then near each point p_1 the curve lies within a plane. This permits use of the preceding notion of curvature. **c)** 2-*d* surface as an example of extension of the extrinsic curvature concept to hypersurfaces with $d \geq 2$. **d)** 2-*d* surface with principal curvatures read off. **e)** An example of intrinsically flat but extrinsically curved 2-*d* surface in $\mathbb{R}^3$. The ant living on this surface only perceives 2-*d* Flat Geometry

the image of a plain spatial 3-manifold σ under a particular kind of map: an *embedding*,[5] $\mathbf{\Phi}$. This construction can also be applied locally [874]: embedding a piece S of spatial 3-surface as a piece S of hypersurface.

The notion of hypersurfaces within $\mathbb{R}^3$ is intuitively clear and well-known, e.g. a bent sheet of paper, or the surface of a globe. Hypersurfaces are more generally characterized as surfaces $\mathfrak{h}$ within a higher-*d* manifold $\mathfrak{M}$ that are of *codimension* $C := \dim(\mathfrak{M}) - \dim(\mathfrak{h}) = 1$.

Next define the *normal* n^μ to the hypersurface $\mathbf{\Sigma}$ (Fig. 8.2.a) and the *projector* $P^\mu{}_\nu := \delta^\mu{}_\nu + n^\mu n_\nu$ onto $\mathbf{\Sigma}$. The spacetime metric is furthermore said to *induce* the spatial metric on the hypersurface. This induced metric is both an intrinsic metric tensor on space, h_{ij} and a spacetime tensor $h_{\mu\nu}$. It attains such a duality by being a *hypersurface tensor*. I.e. a tensor such that for each 'independent index' $0 = \Theta_{\mu\nu\ldots\omega} n^\mu =: \Theta_{\perp\nu\ldots\omega}$ (in this context $\perp$ is pronounced 'perp', short for 'perpendicular'). Since $h_{\mu\nu}$ is symmetric, $h_{\mu\nu}n^\mu = 0$ is a sufficient condition for this to be a hypersurface tensor, and this condition is indeed met. See Chap. 31 further geometrical interpretation of the induced metric. Finally, upon Metric Geometry becoming involved, one is dealing more specifically with *isometric* embeddings.[5]

The *extrinsic curvature* of a hypersurface is its bending relative to an ambient space. For instance, a sheet of paper retains its intrinsic 2-*d* Flat Geometry when it is rolled up into a cylinder. None the less, it has nontrivial curvature relative to the ambient $\mathbb{R}^3$: Fig. 8.2.e). Extrinsic curvature can be usefully defined as the rate of change of the normal n^μ along a hypersurface,

$$\mathcal{K}_{\mu\nu} := h_\mu{}^\rho \nabla_\rho n_\nu. \tag{8.2}$$

N.B. that extrinsic (unlike intrinsic) curvature is already defined for 1-*d* $\mathbf{\Sigma}$ (curves: Fig. 8.2.b). In this case, it is a single number per point. Moreover, in $d \geq 2$, extrinsic curvature is nontrivially a tensor. E.g. for a 2-surface within a 3-*d* manifold, one

[5] If interested, consult Chap. 31 and [614] for more about embeddings.

applies the same construct to 'a basis of curves' on the surface (Fig. 8.2.c). The most convenient such are the *principal curvatures* κ_1 and κ_2 (Fig. 8.2 d). I.e. these are extrema and correspond to eigenvalues, so working with these amounts to casting the symmetric $\mathcal{K}_{ab}$ in diagonal form. The trace $\mathrm{tr}\,\mathcal{K} := \mathcal{K}$ and the determinant $\det \mathcal{K}$ are useful invariants built from $\mathcal{K}_{ab}$. These are respectively proportional to the following.

I) the *mean curvature* $:= \mathrm{tr}\,\mathcal{K}/2 = \{\kappa_1 + \kappa_2\}/2$ in 2-d; more generally $\mathrm{tr}\,\mathcal{K}/\mathrm{d}$ for a d-dimensional hypersurface.

II) The *Gauss curvature* $:= \det \mathcal{K} = \kappa_1 \kappa_2$ in 2-d.

Extrinsic curvature is symmetric and a hypersurface tensor; given the first property, $\mathcal{K}_{\mu\nu} \mathrm{n}^\nu = 0$ as follows from (8.2) suffices to establish the second.

Induced metric and extrinsic curvature are 'packaged together' as the *first* and *second fundamental forms* respectively. Between them, these contain the information about how a hypersurface is embedded in an ambient manifold.

Additionally, despite being defined in very different ways, it turns out that the intrinsic and extrinsic notions of curvature of a surface are related. For a 2-surface embedded in $\mathbb{R}^3$, the intrinsic curvature is in fact equal to (in the above convention twice) the Gauss curvature:

$$\mathcal{R} = 2\,\kappa_1 \kappa_2 \quad \text{(Gauss' Outstanding Theorem).} \tag{8.3}$$

This result furthermore substantially generalizes. For now (see Chap. 31 for yet further generalizations), allow for the embedding space itself to be curved, as well as higher-dimensional (maintaining codimension $C = 1$). The generalized result can furthermore be viewed in terms of projections of the Riemann tensor,[6]

$$\text{(Gauss equation)} \quad \mathcal{R}^{(4)}_{abcd} = \mathcal{R}_{abcd} + 2\,\mathcal{K}_{a[c}\mathcal{K}_{d]b}, \tag{8.4}$$

$$\text{(Codazzi equation)} \quad \mathcal{R}^{(4)}_{\perp abc} = 2\,\mathcal{D}_{[c}\mathcal{K}_{b]a}. \tag{8.5}$$

I.e. the left hand side is viewed here as a projection which is then computed out to form the right hand side.

The Gauss–Codazzi equations (named in part after mathematician Delfino Codazzi) admit a number of conceptually-distinct interpretations, including the following.

1) *Top-down*. Given a higher-d manifold containing a hypersurface, how do its curvature components project onto this hypersurface (as a combination of its intrinsic and extrinsic curvatures)? This involves constructing the geometry of a hypersurface within a given manifold.

[6]From here on, spacetime objects have (4) subscripts added where distinction is necessary between them and their spatial counterparts.

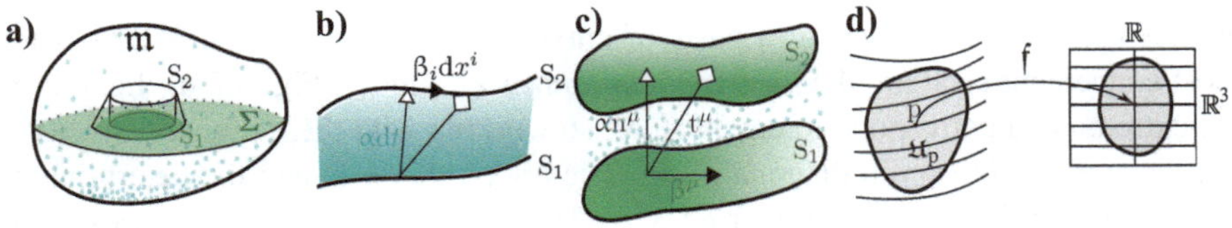

Fig. 8.3 **a)** The set-up for S_1 a local in space piece of a spatial slice. **b)** Arnowitt–Deser–Misner (ADM) 3 + 1 split of a region of spacetime, with lapse α and shift β^i, after physicists Richard Arnowitt, Stanley Deser and Charles Misner. **c)** Local presentation of t, n, β split. The *white diamond arrows* denote time flow and the flat-backed *black arrow* denotes shift along the spatial hypersurface. More generally, in this book special *black arrowheads* denote a priori spatial motions, whereas and *white arrowheads* denote motions jutting between spatial slices or through spacetime (depending on perspective). **d)** Illustrating the nature of *foliation* $\mathfrak{f}$: the rigged or decorated version of the definition of chart in Fig. D.2

2) *Bottom-up*. Given a hypersurface's intrinsic geometry and how it is bent within its ambient manifold, what can be said about the intrinsic geometry of the ambient manifold? This involves constructing the manifold locally surrounding a given hypersurface.
3) *Intrinsic to extrinsic*. Given the intrinsic geometry of both an n-d manifold and an $(n+1)$-d manifold, is there a bending by which the former can be realized within the latter as a hypersurface?

Part I makes no claims as regards these schemes' mathematical well-posedness [a concept defined in Appendix O and commented on for 1) to 3) in Chap. 31].

8.5 Two-Hypersurface and Foliation Concepts

Some notion of thin one-sided infinitesimal neighbourhood of $\mathbf{\Sigma}$ (Fig. 8.3.a) is required as regards developing a number of further concepts [382]. The notion of *foliation* [614] (Fig. 8.3.d) takes this further by considering a more extended piece of spacetime.

By considering an infinitesimal limit of two neighbouring hypersurfaces, extrinsic curvature can furthermore be cast in the form of a Lie derivative,

$$\mathcal{K}_{\mu\nu} = \pounds_{\underline{\mathrm{n}}} \mathrm{h}_{\mu\nu}/2. \tag{8.6}$$

This observation offers immediate manifest proof of its aforementioned symmetry property.

Each foliation by spacelike hypersurfaces is to be interpreted in terms of a choice of time t with an associated 'time flow' vector field t^μ. t is called a 'global timefunction' (see e.g. [874]). There are an infinity of choices for such a t. Spatial hypersurfaces here correspond to constant values of the chosen t.

For $\mathbb{M}^4$, t and t^μ already exist as fully general entities, though they are usually chosen via a global inertial coordinate system [874]; of course this ceases to exist in the case of full GR.

For dynamical formulations of GR, one usually demands the spacetime to be time-orientable so that it is always possible to consistently allocate notions of past and future.

$$\mathrm{t}^\mu \text{ is restricted by } \mathrm{t}^\mu \nabla_\mu \mathrm{t} = 1 \text{ and } \mathrm{s}^\mu \nabla_\mu \mathrm{t} = 0 \text{ for any tangential } \mathrm{s}^\mu. \tag{8.7}$$

If these hold, it is consistent to [814]

$$\text{identify } \mathrm{t}^\mu \nabla_\mu \text{ with } \partial/\partial \mathrm{t} \tag{8.8}$$

and then

$$\text{identify } \partial/\partial \mathrm{t} \text{ with } \pounds_{\underline{\mathrm{t}}}, \tag{8.9}$$

meaning an expression of the form (string of projectors) $\times\, \pounds_{\underline{\mathrm{t}}}\Theta$.

Arnowitt–Deser–Misner [73] split the spacetime metric into induced metric h_{ij}, shift β^i and lapse α pieces (Fig. 8.3.b):

$$\mathrm{g}_{\mu\nu} = \begin{pmatrix} \beta_k\beta^k - \alpha^2 & \beta_i \\ \beta_j & \mathrm{h}_{ij} \end{pmatrix}. \tag{8.10}$$

This is often presented for a foliation, though two infinitesimally close hypersurfaces suffices (or even less for some parts and weakened versions, as per Chap. 31). The corresponding split of the inverse metric is

$$\mathrm{g}_{\mu\nu} = \begin{pmatrix} -1/\alpha^2 & \beta^i/\alpha^2 \\ \beta^j/\alpha^2 & \mathrm{h}^{ij} - \beta^i\beta^j/\alpha^2 \end{pmatrix}, \tag{8.11}$$

and that of the square-root of the determinant is

$$\sqrt{|\mathrm{g}|} = \alpha\sqrt{\mathrm{h}}. \tag{8.12}$$

In the ADM formulation, t^μ is split into tangential and normal parts,

$$\mathrm{t}^\mu = \beta^\mu + \alpha\, \mathrm{n}^\mu. \tag{8.13}$$

This serves to define the *shift*, $\beta^\mu := \mathrm{h}^{\mu\nu}\mathrm{t}_\nu$: displacement in identification of the spatial coordinates between 2 adjacent slices; this is geometrically an example of *point identification map* [814]. Additionally, $\alpha := -\mathrm{n}_\gamma \mathrm{t}^\gamma$ is the *lapse*: 'time elapsed', which may be interpreted as duration of GR proper time $\mathrm{d}\tau = \alpha(t, x^i)\mathrm{dt}$.

In the ADM split, if $-\alpha^2 + \mathrm{g}_{\mu\nu}\beta^\mu\beta^\nu < 0$, the hypersurface within spacetime is spacelike and the normal direction is timelike. In particular α cannot vanish anywhere, and one is to take $\alpha > 0$ everywhere for a future-directed normal. The normal is now $\mathrm{n}^\mu = \alpha^{-1}[1, -\underline{\beta}]$. A computational form for the extrinsic curvature is

$$\mathcal{K}_{ij} = \frac{\mathrm{h}'_{ij} - \pounds_{\underline{\beta}}\mathrm{h}_{ij}}{2\alpha} = \frac{\mathrm{h}'_{ij} - 2\mathcal{D}_{(i}\beta_{j)}}{2\alpha} = \frac{\delta_{\vec{\beta}}\mathrm{h}_{ij}}{2\alpha}, \tag{8.14}$$

where $' := \partial/\partial t$ for t the coordinate time. A final useful construct at this level of structure is Canonical Quantum Gravity expert Karel Kuchař's *hypersurface derivative* [576–579],

$$\delta_{\vec{\beta}} := \frac{\partial}{\partial t} - \pounds_{\underline{\beta}}. \tag{8.15}$$

Moreover, the correction to $\partial h_{ab}/\partial t$ is (8.1) under the substitution of β^i for ξ^i.

The ADM prescription for a split of spacetime is, moreover, far from unique. The *Kaluza–Klein split* [67] (proposed by Klein alongside physicist Theodor Kaluza) parallels the *inverse* ADM split in form but uses new names and interpretations in place of the lapse and shift pieces. There is also an alternative *threading split* [440]. Here the 1-*d* temporal threads are primary rather than the 3-*d* spatial hypersurfaces; this is useful in considering observed past null cones in cosmological and astrophysical contexts. Thus it is termed a $1 + 3$ split to ADM's $3 + 1$ one. Among the many possible splits, the feature which distinguishes the ADM one is its being well-adapted to dynamical calculations. This does not just refer to its being built around the dynamical objects of GR: the spatial hypersurfaces. Additionally, it picks out four multiplier coordinates—the lapse and shift—which simplifies the dynamical equations and cleanly splits them into constrained and evolution systems. On the other hand, the threading split, is well-adapted to observational concepts such as past null cones and fluxes of gravitational waves. Finally note that the Kaluza–Klein split has a distinct main use, in $4 + 1$ dimensions, as an attempt to unify Electromagnetism and GR; see Chap. 11 for more in this regard.

8.6 Foliations in Terms of Fleets of Possible Observers

Each normalized t^μ ($t^\mu/\|t\| = \gamma[1, \underline{v}]$ for $\underline{v} = \underline{\beta}$) represents a distinct possible motion of a fleet of observers (Ex V.11.d). These are held to be combing out spacetime rather than travelling on mutually-intersecting worldlines. Elsewise, they have freedom of motion: 'rocket engines' permit each to accelerate independently of the others.

One needs to be careful at this point because of the various possible non-alignments between t^μ, n^μ and u^μ Let us start by considering the simplified situation for the *Eulerian observers* [154, 382] that correspond to each foliation, for which

$$u^\mu = n^\mu = (\text{a particular normalized } t^\mu \text{ orthogonal to the foliation}),$$

So in this case there is one thread of motion of observers per foliation, meant in a sense that is meaningfully dual to this foliation. This is in parallel to 'ray–wavefront duality' in Geometrical Optics or its configuration space analogue in the Hamilton–Jacobi formulation of Mechanics ([598] and Appendix J.14).

Two simple cases of tilted flows (Fig. 8.4.b) involve time flow aligned with each of u^μ and n^μ in turn. Now under some circumstances, the hypersurfaces to which

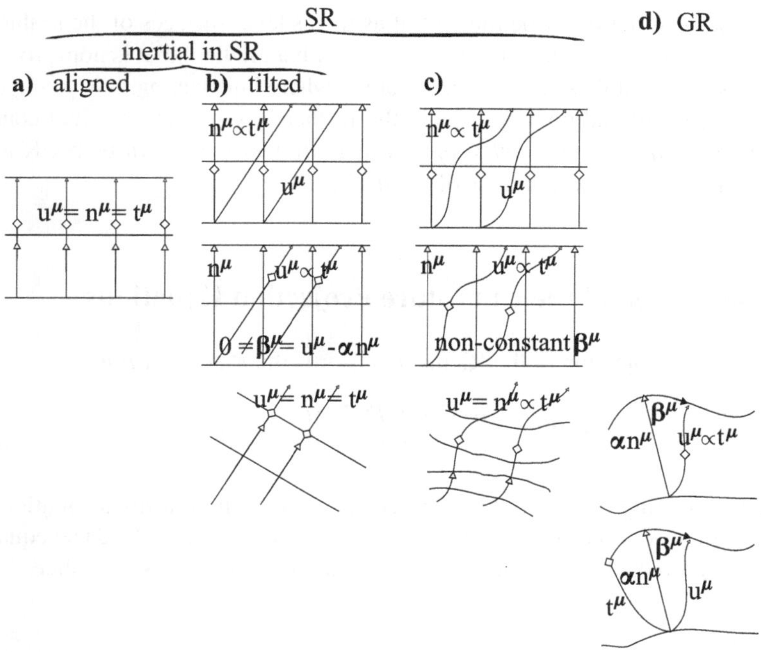

Fig. 8.4 **a)** Material flow u^μ (*small white arrows*), normal vector field n^μ and time flow t^μ all coincide for Eulerian observers (depicted here in the SR case). **b)** In this case, material flow is *tilted away* from the normals. **c)** Inhomogeneous material flow. **d)** Generic GR solution

the normal vector fields coincide have no physical significance. In the depicted SR case with homogeneous but tilted material flow, the third subfigure corresponds to dropping the initial inertial frame for a distinct material flow aligned inertial frame (rest frame). However, in the isotropic cosmology counterpart, there are hypersurfaces privileged by homogeneity, for which the preceding flexibility of changing to an equally simple frame is lost.

Consider next the more general case for which t^μ is unaligned with n^μ. Nor is it necessary for the observers to follow the flow u^μ, since 1) the predominant matter flow in the Universe is indeed not made out of observers. 2) Observers can instead be regarded as residing on independently-moving planets and rockets. 3) Such rockets can to good approximation be idealized as test particles.

In even greater generality, consider inhomogeneous material flow. Pass here from column b) to c) in Fig. 8.4, amounting to leaving the inertial frames for a more complicated general frame. On the other hand, in generic GR (Fig. 8.4.d) there are no flat hypersurfaces to begin with. The most general situation of a fleet of observers which are individually capable of undergoing arbitrary accelerations ('in rockets'), the motion of which need not be aligned with the material flow vector or the normal. So we have, overall, descended from the privileged flat foliation of flat spacetime with double alignment to a generic hypersurface in a generic spacetime with no alignments.

Foliations can moreover be thought of as the as level surfaces of the scalar field notion of time, t. t is here taken to be smooth, with a gradient that is nonzero everywhere, ensuring that these level surfaces are nowhere intersecting.

Chapter 31 will additionally explain the further 'ray–wavefront' dual concepts of *many-fingered time* and *bubble time*, which are well-known in both GR and in arbitrary spatial slice formulations of Field Theory.

8.7 Completion of the Curvature Projection Equations

The remaining projection of the Riemann tensor is the *Ricci equation*,

$$\mathcal{R}^{(4)}_{\perp a \perp b} = \frac{\delta_{\vec{\beta}}\mathcal{K}_{ab} + \mathcal{D}_b\mathcal{D}_a\alpha}{\alpha} + \mathcal{K}_a{}^c\mathcal{K}_{cb}. \tag{8.16}$$

Through containing $\partial\mathcal{K}_{ab}/\partial \mathrm{t}$, this now requires at least infinitesimal foliation concepts for its conception and manipulation. In contrast, the Gauss–Codazzi equations contain no more than $\mathcal{K}_{ab}$, which can be contemplated within a single slice.

8.8 A Further Type of Diffeomorphism: *Diff*$(\mathfrak{m}, \mathfrak{Fol})$

These correspond to foliated spacetimes. These are taken to involve *all possible* foliations $\mathfrak{Fol}$ for a given $\mathfrak{m}$. It is substantial from Chap. 9 onward to be aware that these do *not* share some of the simpler mathematical similarities common to *Diff*$(\mathfrak{m})$ and *Diff*$(\boldsymbol{\Sigma})$; see also Chap. 9.14] in this regard.

8.9 Space–Time Split of the GR Action

Under the ADM split, the Einstein–Hilbert action takes the form[7]

$$\mathsf{S}_{\mathrm{ADM}} \propto \int \mathrm{d}t \int_{\Sigma} \mathrm{d}^3x\, \mathcal{L}_{\mathrm{ADM}} = \int \mathrm{d}t \int_{\Sigma} \mathrm{d}^3x \sqrt{\mathrm{h}}\,\alpha\{\mathcal{K}_{ab}\mathcal{K}^{ab} - \mathcal{K}^2 + \mathcal{R}\}. \tag{8.17}$$

This is obtained by *decomposing* $\mathcal{R}^{(4)}$ using a combination of contractions of the Gauss and Ricci equations and discarding a total divergence since $\boldsymbol{\Sigma}$ is without boundary. Keeping a cosmological constant term just involves $-2\,\Lambda$ inside the curly parenthesis.

The result of varying with respect to this action can be recognized in terms of the three projections of the spacetime Einstein tensor. I.e. particular combinations

[7] This book uses $8\,\pi\,G = 1 = c$ units, since we are focusing on the geometrical meaning of the split involved.

of contractions of the Gauss, Codazzi and Ricci equations viewed as projection equations. [These can also be obtained by projecting the Einstein field equations themselves, though it is further useful in Canonical Approaches to decompose the underlying action instead.] One begins by considering the manifestly Lagrangian form of the action, i.e. in terms of configurations and velocities, which are here h_{ij} and $\partial \mathrm{h}_{ij}/\partial \mathrm{t}$.

8.10 The GR Action Equips $\mathfrak{Riem}(\boldsymbol{\Sigma})$ with a Metric Geometry

Let us next reformulate this action in terms of the configuration space geometry for GR. (8.17)'s kinetic term contains

$$\mathrm{M}^{abcd} := \sqrt{\mathrm{h}}\{\mathrm{h}^{ac}\mathrm{h}^{bd} - \mathrm{h}^{ab}\mathrm{h}^{cd}\} \tag{8.18}$$

contracted into $\mathcal{K}_{ab}\mathcal{K}_{cd}$ and thus into $\dot{\mathrm{h}}_{ab}\dot{\mathrm{h}}_{cd}$. M^{abcd} is a metric on the configuration space $\mathfrak{Riem}(\boldsymbol{\Sigma})$; it is termed a *supermetric* out of possessing four indices and already being built out of one preceding notion of metric, h_{ab}. Moreover, DeWitt's [237] 2-index to 1-index map $\mathrm{h}_{ab} \mapsto \mathrm{h}^A$ recasts this supermetric in the standard form for a metric: with two downstairs indices, $\mathrm{M}^{abcd} \mapsto \mathrm{M}_{AB}$. $\mathrm{T}_{\mathrm{ADM}}$ takes the form $\mathrm{M}_{AB}\delta_{\vec{\beta}}\mathrm{h}^A\delta_{\vec{\beta}}\mathrm{h}^B$. [The capital Latin indices in this context run from 1 to 6.] Pointwise, this is a $--++++$ metric, and so, overall it is an infinite-dimensional version of a semi-Riemannian metric: the *GR configuration space metric* alias *inverse DeWitt supermetric*. This 'DeWittian' indefiniteness is associated with the expansion of the Universe giving a negative contribution to the GR kinetic energy. This is entirely unrelated to the Lorentzian indefiniteness of SR and GR spacetimes themselves.

The inverse metric is

$$\mathrm{N}^{AB} = \mathrm{N}_{abcd} = \{\mathrm{h}_{ac}\mathrm{h}_{bd} - \mathrm{h}_{ab}\mathrm{h}_{cd}/2\}/\sqrt{\mathrm{h}}, \tag{8.19}$$

which is the *DeWitt supermetric* itself. DeWitt additionally studied the more detailed nature of this geometry in [237] (set as Ex V.17).

Thus, overall, the geometrical DeWitt form of the manifestly Lagrangian form of the ADM action works out to be

$$\mathsf{S}_{\mathrm{ADM}} = \int \mathrm{dt} \int_{\boldsymbol{\Sigma}} \mathrm{d}^3x\sqrt{\mathrm{h}}\,\alpha\{\mathrm{T}_{\mathrm{ADM}}/4\alpha^2 + \mathcal{R} - 2\Lambda\},$$

for

$$\mathrm{T}_{\mathrm{ADM}} = \|\delta_{\underline{\beta}}\mathbf{h}\|^2_{\mathbf{M}}. \tag{8.20}$$

8.11 GR's Momenta

Now additionally to extrinsic curvature being a characterizer of hypersurfaces, it is of further relevance due to bearing close relation to the GR momenta,

$$\mathrm{p}^{ij} := \frac{\delta \mathcal{L}_{\mathrm{ADM}}}{\delta \dot{\mathrm{h}}_{ij}} = \sqrt{\mathrm{h}}\{\mathcal{K}^{ij} - \mathcal{K}\,\mathrm{h}^{ij}\} = \mathrm{M}^{ijkl}\frac{\delta_{\vec{\beta}}\mathrm{h}_{ij}}{2\alpha}. \tag{8.21}$$

I.e. GR's momenta are a densitized version of $\mathcal{K}_{ab}$ with a particular trace term subtracted off. Finally, taking the trace,

$$\mathrm{p} = -2\sqrt{\mathrm{h}}\,\mathcal{K}. \tag{8.22}$$

8.12 GR's Constraints

The ADM–Lagrangian action encodes the

$$\textit{GR Hamiltonian constraint} \quad \mathcal{H} := \mathrm{N}_{ijkl}\mathrm{p}^{ij}\mathrm{p}^{kl} - \sqrt{\mathrm{h}}\{\mathcal{R} - 2\,\Lambda\} = 0 \tag{8.23}$$

from variation with respect to the lapse α. From variation with respect to β^i, it also encodes the

$$\textit{GR momentum constraint} \quad \mathcal{M}_i := -2\,\mathcal{D}_j\mathrm{p}^j{}_i = 0. \tag{8.24}$$

The GR momentum constraint can be straightforwardly interpreted as physicality residing not in the 3 degrees of freedom per space point choice of point-identification but rather solely in terms of the 3-metric's *other* 3, termed the *3-geometry*: the diffeomorphism-invariant information in the 3-metric. This is how GR comes to be, more closely, a dynamics of 3-geometries [237, 899] on the quotient configuration space,

$$\mathfrak{S}\mathrm{uperspace}(\boldsymbol{\Sigma}) := \mathfrak{R}\mathrm{iem}(\boldsymbol{\Sigma})/\mathit{Diff}(\boldsymbol{\Sigma}). \tag{8.25}$$

However, interpreting the GR Hamiltonian constraint is tougher. It is 'purely-quadratic in the momenta', meaning it consists of a quadratic form plus a zero-order piece but with no linear piece:

$$\mathcal{Q}\mathrm{uad} := N^{\mathsf{AB}}(\boldsymbol{Q})P_{\mathsf{A}}P_{\mathsf{B}}/2 - W(\boldsymbol{Q}) = 0. \tag{8.26}$$

We shall see in Chap. 9.10 that this property leads to the Frozen Formalism Facet of the Problem of Time.

Moreover, in terms of $\mathcal{K}_{ij}$ (and setting $\Lambda = 0$), the constraints are

$$0 = -\mathcal{H} = \mathcal{K}^2 - \mathcal{K}_{ij}\mathcal{K}^{ij} + \mathcal{R} = 2\,\mathcal{G}^{(4)}_{\perp\perp}, \tag{8.27}$$

$$0 = \mathcal{M}_i = -2\{\mathcal{D}_j\mathcal{K}^j{}_i - \mathcal{D}_i\mathcal{K}\} = 2\,\mathcal{G}^{(4)}_{i\perp}. \tag{8.28}$$

These forms of GR equations were already known to Darmois in the 1920s [227]. As indicated, the $\mathcal{K}_{ij}$ forms of these constraints serve to identify [874] these as contractions of the Gauss–Codazzi equations for the embedding of spatial 3-slice into spacetime: the *Constraint–Embedding Theorem of GR*.

GR's phase space degrees of freedom count works out as 6×2 (h_{ij} and conjugates) -3×2 (quotienting out $\mathcal{M}_i$) -1×2 (quotienting out $\mathcal{H}$) $= 2 \times 2$ degrees of freedom.[8] A more rigorous count is 10×2 (including the lapse and shift as well as h_{ij}) -3×2 (due to the shift being a Lagrange multiplier, so its momentum is zero) -1×2 (due to the lapse being a Lagrange multiplier, so its momentum is zero also) $-3 \times 2 - 1 \times 2 = 2 \times 2$. See Appendices O.5–O.6 if interested in the constraints as mathematical equations.

8.13 GR's Evolution Equations

Chapter 7 already laid down some topological restrictions such as orientability and Chap. 8.1 considered simple-product spacetimes. One may require further restrictions on the spacetime to ensure good causal behaviour. We assume $\boldsymbol{\Sigma} \times \mathfrak{T}$ preventing consideration of topology change. If S is a closed achronal set with D(S) = $\mathfrak{m}$, it is a *Cauchy surface* (named after the great mathematician Augustin Cauchy). This is where the position and velocity data for the Cauchy problem—a type of PDE problem: Appendix O—for a hyperbolic evolution (wave equation type) PDE is to be posed. A spacetime possessing a Cauchy surface is said to be *globally hyperbolic* [440, 874]. This condition allows for (local in time) determinability of GR evolution from GR initial data and excludes e.g. non-orientable spacetimes. See Fig. 8.5 for domain of dependence in the GR context. The notion of *Cauchy horizon* $\mathrm{H}^+(\boldsymbol{\Sigma}) := \overline{\mathrm{D}^+(\boldsymbol{\Sigma})} - \mathrm{I}^-(\mathrm{D}^+(\boldsymbol{\Sigma}))$ is an indicator of beyond where $\boldsymbol{\Sigma}$ fails to be a Cauchy surface. The version that is local in space builds a surface within the domain of dependence [348] of the initial $\boldsymbol{\Sigma}$. This is still a direct product at the level of topological manifolds, at least in the cases covered in this book.

The hypersurfaces $\boldsymbol{\Sigma}$ are held to be everywhere spacelike. Applying such a split entails time orientability and absence of closed timelike curves, now additionally as conditions for Cauchy surfaces to exist.

In terms of momenta, the ($\Lambda = 0$) evolution equations (ADM equation of motion) are

$$\delta_{\vec{\beta}}\mathrm{p}^{ij} = \sqrt{\mathrm{h}}\{\mathcal{R}\mathrm{h}^{ij}/2 - \mathcal{R}^{ij} + \mathcal{D}^j\mathcal{D}^i - \mathrm{h}^{ij}\triangle\}\alpha - 2\alpha\{\mathrm{p}^{ic}\mathrm{p}_c{}^j - \mathrm{p}\,\mathrm{p}^{ij}/2\}/\sqrt{\mathrm{h}}$$
$$+ \alpha\mathrm{h}^{ij}\{\mathrm{p}_{ij}\mathrm{p}^{ij} - \mathrm{p}^2/2\}/2\sqrt{\mathrm{h}}. \qquad (8.29)$$

[8]The means of carrying out this count depends on Constraint Closure and the accompanying detailed Principles of Dynamics analysis, which we postpone to Chap. 24. The degrees of freedom counts are always modulo a finite number of degrees of freedom [552] as occur e.g. in $2 + 1$ GR (Ex III.13), which still manages to have some global degree of freedom dynamical.

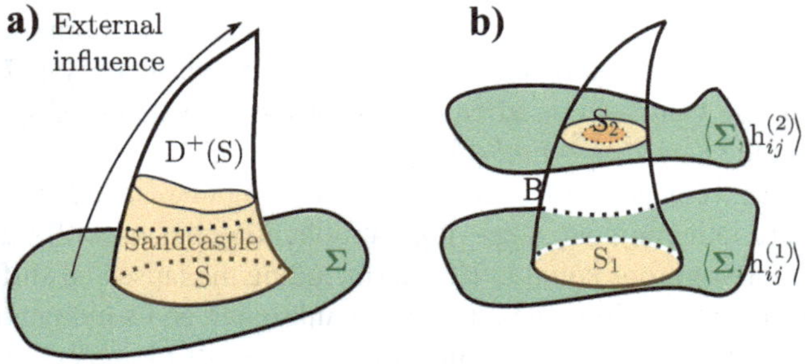

Fig. 8.5 **a)** The domain of dependence of a piece S of a spatial hypersurface $\boldsymbol{\Sigma}$, $D^+(S)$, is the portion of spacetime that is controlled solely by the physical data on S. Points on $\boldsymbol{\Sigma}$ outside S are not able to causally communicate with $D^+(S)$ (the external influence depicted). Within this is shaded an example of 'sandcastle-shaped' region for which the GR Cauchy problem results could be expected to hold. **b)** In dealing with evolutions of pieces of hypersurfaces, the pieces get smaller due to the constricting effect of the domain of dependence

In terms of the extrinsic curvature,

$$\frac{-\{\delta_{\vec{\beta}}\mathcal{K}_{ab} - h_{ab}\delta_{\vec{\beta}}\mathcal{K}\} - \mathcal{D}_b\mathcal{D}_a\alpha + h_{ab}\Delta\alpha}{\alpha} - \left\{2\mathcal{K}_a{}^c\mathcal{K}_{bc} - \mathcal{K}\mathcal{K}_{ab} + \left\{\mathcal{K}_{ij}\mathcal{K}^{ij} + \mathcal{K}^2\right\}h_{ab}/2\right\} + \mathcal{G}_{ab} = \mathcal{G}^{(4)}_{ab} = 0. \quad (8.30)$$

which form complements the constraint equations as regards forming the remaining projection of $\mathcal{G}^{(4)}_{\mu\nu}$, The three of them can also be interpreted in terms of contractions of the Gauss–Codazzi–Ricci embedding equations (thus extending the Constraint–Embedding Theorem to the *Constraint–Evolution–Embedding Theorem of GR*). It is occasionally more convenient to work instead with the following form in terms of $\mathcal{R}_{ab}$ rather than $\mathcal{G}_{ab}$:

$$\frac{\delta_{\vec{\beta}}\mathcal{K}_{ab} + \mathcal{D}_b\mathcal{D}_a\alpha}{\alpha} - \mathcal{K}\mathcal{K}_{ab} + 2\mathcal{K}_a{}^c\mathcal{K}_{bc} - \mathcal{R}_{ab} = -\mathcal{R}^{(4)}_{ab} = 0. \quad (8.31)$$

Also note the success in deriving these equations as regards removing all Riemann (and thus Weyl) tensor projections from the system of projection equations. However, some other formulations—e.g. the Threading Approach of (Sect. 36.1)—use other linear combinations which do cause some such terms to be kept. See Appendix O.7 if interested in the GR evolution equations as mathematical equations.

8.14 Other Classical Applications of Geometrodynamics

For the applications below, and others later in this book, it is generally useful to add matter to the system. For now, we add phenomenological matter; see Chap. 18 for examples of adding fundamental matter fields instead. Define $\mathcal{E} := T^{(4)}_{\perp\perp}$, $J_a :=$

$\mathrm{T}^{(4)}_{a\perp}$ and $\mathrm{S}_{ab} := \mathrm{T}^{(4)}_{ab}$. These are general matter terms which are usually prescribed as functions of matter fields that are governed by usually-separate field equations. The GR initial value problem—a type of PDE problem: Appendix O—is for the system consisting of 4 constraints [227] (via the $\mu\perp$ component of the Einstein field equations $\mathcal{G}^{(4)}_{\mu\perp} = \mathrm{T}^{(4)}_{\mu\perp}$ in suitable units):

$$\mathcal{K}^2 - \mathcal{K}_{ij}\mathcal{K}^{ij} + \mathcal{R} = 2\{\mathcal{E} + \Lambda\}, \tag{8.32}$$

$$\mathcal{D}_b\mathcal{K}^b{}_a - \mathcal{D}_a\mathcal{K} = -\mathrm{J}_a. \tag{8.33}$$

These are obtained by use of the split Einstein field equations in the doubly-contracted Gauss and the contracted Codazzi embedding equations respectively. They are constraints because they contain none of the highest time derivatives. Note that this system consists of three linear PDEs and one nonlinear algebraic equation.

The remaining 6 equations are evolution equations [227]: Eq. (8.30) with right hand side replaced by $\mathrm{S}_{ab} + \Lambda\,\mathrm{h}_{ab}$.

As PDEs, these are well supported by Analysis theorems guaranteeing their good behaviour. This work was started by the French School of Mathematical Physics. André Lichnerowicz [622] treated the GR initial value problem (constraint equations). This is viewed as a first data providing step for the GR Cauchy problem (evolution equations). The first convincing mathematical study of the latter was due to Yvonne Fourès-Bruhat [311, 312] (alias Bruhat and Choquet-Bruhat) and Jean Leray [617]. See Chap. 21 and Appendix O for further details of more up-to-date such theorems.

These PDEs can also be used to study compact astrophysical binaries using Numerical (General) Relativity [123, 202, 382, 684]. While perturbative formulations can be used to model lengthy inspirals, the 'plunge', 'merger' and 'ringdown' at the end of the process require full Numerical Relativity based on invariants of ADM's equations. A further output of such calculations is a template for the gravitational waves emitted, to be searched for within gravitational wave detector data.

8.15 Outline of Ashtekar Variables Alternative

GR admits a number of further first-order or spinorial formulations [706, 814, 874]. A particular such, which recasts GR in 'Yang–Mills like form' is the *Ashtekar Variables formulation* [75, 154] (named after physicist Abhay Ashtekar). This is in terms of a $SU(2)(\mathbf{\Sigma})$ [local $SU(2)$ on $\mathbf{\Sigma}$] 1-form $\mathrm{A}_i{}^I$. The conjugate momentum is the densitized 3-bein $\mathrm{E}^i{}_I := \sqrt{\mathrm{h}}\,\mathrm{e}^i_I$, for e^i_I the *3-bein* itself, which is related to the 3-metric by $\mathrm{h}_{ij} = -\mathrm{tr}(\mathrm{E}_i\mathrm{E}_j)$.[9] This is now a conjugate momentum, despite its relation to

[9]The capital indices here denote spinorial $SU(2)$ indices. tr denotes the trace over these. D_i is here the $SU(2)(\mathbf{\Sigma})$ covariant derivative as defined in the first equality of (8.35). $|[\ ,\ ]|$ denotes the classical Yang–Mills-type commutator. Moreover, due to the specific form of $\mathrm{A}_i{}^I$ and $\mathrm{E}^i{}_I$, h_{ij} is

the previous configurational variables h_{ab}, because a canonical transformation (Appendix J.9) has been applied. On the other hand, by involving 1) a type of mathematical unity is introduced, in the sense of all four of the fundamental forces now being associated with Gauge Theoretic connections.

A particular first-order spacetime action often used for this is (see Sect. 24.9 for generalizations)

$$\mathsf{S} \propto \int \mathrm{d}^4 x \, \mathrm{e}\, \mathrm{e}^{\mu}_{A} \mathrm{e}^{\nu}_{B} \mathrm{F}^{AB}_{\mu\nu} \tag{8.34}$$

in spacetime form. e^{μ} is here the spacetime *4-bein*, e the corresponding determinant, and $\mathrm{F}^{AB}_{\mu\nu}$ the corresponding Yang–Mills type field strength.

This formulation's constraints are

$$\mathcal{G}_A := \mathrm{D}_i \mathrm{E}^i{}_A := \partial_i \mathrm{E}^i{}_A + \big|\big[\mathrm{E}_i, \mathrm{E}^i\big]\big|_A = 0, \tag{8.35}$$

$$\mathcal{M}_i := \mathrm{tr}\big(\mathrm{E}^j \mathrm{F}_{ij}\big) = 0, \tag{8.36}$$

$$\mathcal{H} := \mathrm{tr}\big(\mathrm{E}^i \mathrm{E}^j \mathrm{F}_{ij}\big)/2\sqrt{\mathrm{E}} = 0. \tag{8.37}$$

The GR $SU(2)$ Yang–Mills–Gauss constraint (8.35) arises due to internal symmetries introduced in setting up this formulation. (8.36) and (8.37) are the polynomial forms now taken by the GR momentum and Hamiltonian constraints respectively. One can see that (8.36) is indeed associated with momentum flux since it is the condition for a vanishing (Yang–Mills–)Poynting vector. As per Geometrodynamics, this formulation's version of $\mathcal{H}$ (8.37) lacks such a clear-cut interpretation. On the other hand, it is technically simpler than Geometrodynamics' $\mathcal{H}$ since it is polynomial in this approach's canonical variables. Indeed one of the major reasons for considering Ashtekar variables formulations is their distinct and simpler form for $\mathcal{H}$.

8.16 Exercises V. Spacetime and Dynamical Formulations of GR

Exercise 1) Derive the 'Michell radius' version of $GM/2c^2$ within the Newtonian Paradigm with allowance made for c taking a finite value.

Exercise 2) i) Derive Sect. 7.5's general redshift formula for metric Theories of Gravity; deduce also that Section's various more specialized redshift formulae. ii) Estimate the redshift for a photon emitted from the Sun as observed from Earth, and one emitted down a 30-metre tower on Earth. To what extent will the time kept by a clock deviate due to gravitational redshift during an airplane flight? Finally, estimate the precision to which the ACES Earth-orbit space mission will be able to test gravitational redshift with its 1 part in 10^{16} accurate on-board atomic clock. iii) What age of the Universe corresponds to a redshift of around 10 (approximate

in fact complexified, i.e. pointwise in $GL(3, \mathbb{C})$ rather than in $GL(3, \mathbb{R})$, a point to which we return in Sect. 11.9.

maximum redshift observed in a galaxy)? What is the redshift of the surface of last scattering when the cosmic microwave background formed? [Assume a dust-filled FLRW cosmology.]

Exercise 3) i) Compare the geodesic deviation equation (D.14) and the Lorentz Force Law (4.15) at the conceptual level. ii) Derive the Newtonian tidal equation from the former. iii)† Demonstrate the absence of relativistic tidal effects in GR. Find an example of alternative theory of gravity which does admit such; compare the ratio of these relative to Newtonian tidal forces, both on Earth and for a binary pulsar.

Exercise 4) i) Use the geodesic equation in modelling Newtonian Mechanics, including consideration of non-affine parametrization. ii) Compare GR spacetime and Newtonian space-time from a conceptual point of view; in addition to the current book's intermediate geometrical formulation for the latter, consider also its Cartan-type formulation (see [776] for source material).

Exercise 5) i) Derive the FLRW solutions in the open, closed and limiting flat cases for each of dust matter and radiation matter. ii) Show that the flat dust-filled FLRW solution can be derived in purely Newtonian terms. iii) Use these solutions to estimate the age of the Universe and to posit the Particle Horizon Problem.

Exercise 6) Compare the following. a) The proper time taken for a radially infalling test particle to reach the event horizon in Schwarzschild spacetime. b) The Schwarzschild coordinate time that the particle appears to take to reach the horizon from the point of view of a distant stationary observer.

Exercise 7) i) Construct the Penrose diagram for each of the Reissner–Nordström and Kerr–Newman black holes. Comment on the geometrical form of the latter solution's singularity. ii) For a representative set of points in these solutions, consider where the wavefront of a light flash emitted from that point is after a short time interval. iii) Interpret also the surface $r = s_+ := M + \sqrt{M - a^2\cos^2\theta}$ within the Kerr solution, as well as what happens within this surface. iv) Demonstrate that the Kerr–Newman solution contains closed timelike curves; are these accessible to observers who are unwilling to traverse any event horizons?

Exercise 8) Compute the surface gravity and horizon areas for the Schwarzschild and Kerr–Newman spacetimes. Differentiate the latter to obtain the black hole form of the First Law (7.12).

Exercise 9) i) Work out the extrinsic curvature for an ellipse in $\mathbb{R}^2$, for $\mathbb{S}^2$ in $\mathbb{R}^3$ and for $\mathbb{S}^3$ in FLRW spacetime. ii) Justify the appending of

$$\pm\frac{c^2}{8\pi G}\int_{\partial\mathfrak{m}} \mathrm{d}^3x\sqrt{\mathrm{h}}\,\mathcal{K}$$

to (7.7) with $+$ for spacelike boundaries and $-$ for timelike ones.

Exercise 10) Derive the Gauss–Codazzi–Ricci equations (8.4), (8.5), (8.16) and their relations to the Einstein field equations in the ADM formulation. Show that in dimension p these have $p\{p-1\}^2\{p-2\}/12$, $p\{p-1\}\{p-2\}/3$ and $p\{p-1\}/2$ components respectively. Recover Gauss's Outstanding Theorem (8.3) as a special case.

Background Reading 1) Read [440, 874] on Causality Theory and the derivation of the Singularity Theorems. Consult also a more modern review on singularities in GR such as [210] or [784].

Background Reading 2) Read the account of Geometrodynamics in [660].

Background Reading 3) Read one of [123] or [382] on Geometrodynamics applied to the Numerical Relativity of compact binary objects.

Background Reading 4) Consider a readable introductory account of Ashtekar variables at the classical level, such as within [154].

Exercise 11) i) Obtain the ADM action from the Einstein–Hilbert action. ii) Obtain (8.27), (8.28), (8.30) from varying the ADM action. iii) Rewrite ii)'s equations in canonical form and furthermore also in terms of the DeWitt supermetric. iv) With reference to Sect. 8.6, work out $\vec{t}/\|\vec{t}\|$ as a function of $\underline{\beta}$.

Exercise 12) i) How do GR singularities differ from those elsewhere in Physics? ii)† By considering of a growing list of examples, can you come up with a concrete definition of a GR singularity? (Compare your answer with [347].)

Exercise 13) i) Show that in Lorenz gauge (6.19), Electromagnetism's Ampère–Maxwell Law is cast as the wave equation. ii) In the *harmonic gauge*

$$\bar{\gamma}^{\mu\nu}{}_{,\nu} = 0, \tag{8.38}$$

show that linearized GR's evolution equation can also be cast as a wave equation,

$$\Box\bar{\gamma}_{\mu\nu} = -16\pi\, \mathrm{T}^{(1)}_{\mu\nu}, \tag{8.39}$$

where $\bar{\gamma}_{\mu\nu}$ is the *trace reversed metric* $\gamma_{\mu\nu} - \frac{\gamma}{2}\eta_{\mu\nu}$ and $\mathrm{T}^{(1)}_{\mu\nu}$ is the first-order perturbed energy–momentum–stress tensor.

Exercise 14) Derive the forms taken by the geometrodynamical equations in i) FLRW cosmology and ii) small perturbations about the spatially spherical case thereof.

Exercise 15)† Derive the form of the spherically symmetric geometrodynamical equations. Also show that $\mathrm{d}s^2 = \{1 + GM/2c^2r\}^4 \mathrm{d}s^2_{\mathbb{R}^3}$ solves the conformally flat static version of the Lichnerowicz equation (21.6) for r a standard radial coordinate.

Exercise 16) Work through Chap. 31 on foliations.

Exercise 17)† Work through DeWitt's geometrical study of $\mathfrak{Riem}(\boldsymbol{\Sigma})$ [237]. If you are particularly interested in Applied Geometry *and* Quantum Gravity and you have a lot of time on your hands, additionally work through his study of $\mathfrak{PRiem}(\mathfrak{m})$ [241] and Kuchař's study of the space of hypersurfaces [576].

Exercise 18)† Derive the Ashtekar variables formulation's constraint equations. (Hint: you may first need to read up on the type of curved-space spinors used in this approach, e.g. in [75].)

Exercise 19) Estimate the sizes of the principal SR and GR effects in GPS timekeeping and localization (its satellites have an orbital period of 12 hours).

Exercise 20) Can pulsars serve as standard clocks for galaxy-wide timekeeping?

Exercise 21) a) Estimate the maximal tidal force which an ordinary wristwatch can withstand, alongside where within a Schwarzschild solution such tidal forces are to be found for $M_{Pl} < M < M_{\text{galactic centre}}$. b) In which ways might a classical black hole itself be used as a clock?

Exercise 22)†† Read Chap. 5 of Zeh's book [931] on the Cosmological Arrow of Time, and explore whether this is a Master Arrow. Does the analysis of which Arrows imply which others change if Quantum Gravity or Quantum Cosmology are evoked? [424].

Chapter 9
Classical-Level Background Independence and the Problem of Time. i. Time and Configuration

We now turn to the main subject of this book: Background Independence aspects and the nine ensuing facets of the Problem of Time which Isham and Kuchař identified [483, 586]. In Part I, this main subject of the current book is covered in Chaps. 9, 10 and 12. Chapter 9 and 10 demonstrate that much of Background Independence and the Problem of Time is already present at the classical level. Chapter 9 covers approaches in which one or more of space, configuration or dynamics are primary, whereas Chap. 10 covers approaches in which spacetime is primary. Chap. 11 is an introduction to Quantum Gravity, since the Problem of Time is principally motivated as a foundational issue in—or towards—Quantum Gravity. Finally, Chap. 12 gives an outline of the Problem of Time as features in sufficiently Background Independent Quantum Gravity programs. N.B. that most of the rest of this book expands on Chaps. 9 to 12 rather than directly expanding on the preliminary material in Chaps. 1 to 8.

Passage to Quantum Theory is usually from Newtonian Mechanics or SR prior to these being upgraded to GR. As per the Preface, this amounts to a Background Dependence versus Background Independence Paradigm Split, in which GR and Ordinary Quantum Theory lie on opposite sides. Historically, this situation arose by each of these two areas of Physics developing in a different direction both conceptually and technically, without enough cross-checks to keep Physics within a single overarching Paradigm. This Paradigm Split has a further practical justification which continues to apply today: that our practical experiences are of regimes that involve at most one of QM or GR. Indeed, regimes requiring both of these at once would involve the decidedly outlandish Planck units, as discussed in the Preface and Chap. 11.

Moreover, the development of GR stagnated from the 1920s through to around 1960 [910]. One knock-on effect of this was the above Paradigm Split remaining largely unaddressed. GR was subsequently revived by Wheeler's U.S. group (including ADM [73]), Zel'dovich's U.S.S.R. group, and the U.K. groups including Bondi, Sciama, Penrose and Hawking. ADM's work on the split spacetime formulation of GR toward a canonical formulation of Quantum GR did have a few significant precursors. On the one hand, the French School's work outlined in Chap. 8

E. Anderson, *The Problem of Time*, Fundamental Theories of Physics 190,
DOI 10.1007/978-3-319-58848-3_9

was significant in identifying and manipulating the GR constraints, albeit not yet in canonical form. On the other hand, in the 1950s Dirac followed up his version of classical Canonical GR with Canonical Quantization (as subsequently reviewed in [250]). Wheeler then turned attention to the conceptual underpinnings of this approach in the 1960s, envisaging some of the Problem of Time facets [897, 899]. The great Quantum Gravity pioneer Bryce DeWitt concurrently gave modern Quantum Gravity's first extensive (and last full field sweeping) treatise in the series of papers [237–239]. These cover the configuration space for the Canonical Approach and the origins of various of the strategies for addressing the Problem of Time, as well as Covariant and Path-Integral Approaches.[1] From here, Canonical and Covariant Approaches largely went their separate ways, as outlined in Chap. 11. Henceforth the number of alternative theories grew quickly to beyond what can be considered in detail in a single treatise. As further testimony to the revival of GR in the 1960s, this also included understanding the black hole concept and working out rotating black hole solutions, the birth of observational Cosmology with the detection of the cosmic microwave background, and the Hawking–Penrose Singularity Theorems. This substantially increased interest in Quantum Gravity's Planck regime as the seat for the more extreme parts of the new fields of Early-Universe Cosmology and Black Hole Physics.

GR can, moreover, be viewed as not only a Relativistic Theory of Gravitation but also as a freeing from absolute or Background Dependent structures. This is a continuation of the relational conceptualization of Mechanics outlined in Chap. 3.

Firstly, GR is often interpreted as providing a physically meaningful explanation of the privileged inertial frames of SR as being, more precisely, idealized arbitrarily large versions of GR's local inertial frames. The latter are furthermore in turn determined by the matter distribution as per Chap. 7.

A second issue concerns how Einstein , in developing GR, was influenced along these lines by Mach [288, 518], albeit not in a straightforward manner [96, 897]. Initially, he misinterpreted Mach's Origin of Inertia Principle —due to confusion between 'inertia' in the sense of 'inertial mass' and of 'inertial frames'. Moreover, he eventually abandoned his 'Machian' approach for a more indirect approach—Chap. 7's—involving spacetime frames rather than spatial frames. The resulting theory of GR can none the less be investigated as regards whether various Machian criteria apply to it. Some do, e.g. the frame dragging mentioned in Sect. 7.5. Others do not, e.g. through some GR solutions being in some sense un-Machian, such as universes with overall rotation being physically distinguishable from nonrotating ones [440].

It is useful to recollect at this point (from Sect. 3.1) that 'Machian' refers to a somewhat disjoint set of attributes that a theory might have, rather than some single coherent package that the theories being sought are to possess the entirety of. Only some parts of Mach's insights endure the passage to GR (and subsequent GR-like theories). More specifically, it is Mach's Time Principle and Mach's Space

---

[1]See Sects. 11.1–11.2 and 11.6 for these other approaches toward Quantum Gravity's own most significant precursor papers.

Principle that this book (and Barbour's work [98, 109]) draw from. Note that these are dynamical tenets, and made prior to the advent of almost all notions of spacetime largely and to Einstein's eventual correct form of the field equations of GR. This provides another sense in which Einstein's spacetime formulation of GR at most indirectly addressed Machian criteria..

To instead set up a theory of Background Independence along dynamical lines, it turns out to be rather helpful to already be familiar with the standard spacetime formulation of GR (Chap. 7) and its dynamical and a fortiori canonical reformulations (Chap. 8). The original dynamical reformulation concerns evolving spatial 3-metrics with $Diff(\mathbf{\Sigma})$ redundancy; whereas formulation in terms of $Diff(\mathbf{\Sigma})$-invariant 3-geometries is conceptually equivalent, it is the former which has the benefit of explicit computability. This Geometrodynamics is additionally a practical realization of Broad's Worldview [830], since the spacetime block grows stepwise by geometrodynamical evolution from one spatial hypersurface to the next. Sections 9.7–9.9 furthermore outline how starting from relational first principles for time and space lead to a derivation of GR in a particular geometrodynamical form [62, 98, 109], which has manifestly relational (Leibnizian and Machian) features. With more work (Sect. 10.9 and Chap. 33), GR can eventually be recovered along such lines from less structure assumed. In this way, Relationalism is not only a demonstration of the existence of a formulation in which GR is relational, but also its own route to GR (in Wheeler's sense, as per the next Section).

This is one of the ways in which the current book argues that GR succeeds in meeting Background Independence criteria as well as ones for Relativistic Theory of Gravitation. (Criteria along such lines are discussed in e.g. [40, 78, 188, 194, 250, 483, 485, 488, 552, 586, 748, 752, 795, 796, 843].) The Preface phrased this as 'GR is a gestalt theory', and pointed to this having further consequences as regards subsequent conceptualization of 'Quantum Gravity'. Indeed, from the perspective of GR being a gestalt entity, the wording 'Quantum Gravity' is itself is a misnomer since it refers solely to GR in its aspect as a Relativistic Theory of Gravitation. Whereas this does reflect what is attempted in some approaches, a number of others do consider GR as a gestalt entity. In this book, this is made clear by terming Background Independence programs not just 'Quantum Gravity' but a fortiori '*Quantum Gestalt*'; this book's 'QG' acronym then refers to the latter. Quantum Gestalt encompasses a subset of Paradigms of Physics (some of which are tentative). It also highlights the complementary possibility of studying Quantum Background Independence in the absence of any Theory of Gravitation that is compatible with Relativity[2] (see below and Chaps. 15 to 16). Moreover, in Quantum Gestalt approaches, adopting Background Independence entails the notorious

[2]As regards other names, 'Quantum GR' will not do in this role due to implying the specific Einstein field equations. Contrast with how the Quantum Gestalt position remains open-minded as to *which* Relativistic Theory of Gravitation is involved. Quantum Gestalt is also in contradistinction to 'Background Independent Quantum Gravity'. This is since the latter may carry connotations that the Background Independent and Gravitational inputs are *separate* rather than part of a coherent whole, whose classical counterpart—GR—already forms such a coherent whole.

Problem of Time as a direct consequence to be faced. This is in contradistinction to approaches beginning from a position of denying (parts of) Background Independence so as to avoid (parts of) the Problem of Time from occurring in one's scheme. This second type of approach adheres to more standard conceptualizations (usually from Quantum Theory and SR). Within these, calculations are more familiar and tractable. In contrast, Quantum Gestalt approaches involve more even-handed combinations of concepts from each of Quantum Theory and of GR viewed as both a Relativistic Theory of Gravitation and of Background Independence.

9.1 Many Routes to GR

Wheeler's works provide some useful context at this point. Firstly, Wheeler [660, 899] argued that Einstein's derivation (Chap. 7) is but the first of many routes to GR; some of the other routes are as follows.

A) and B) are the 2-way passage between the spacetime and ADM [73] split spacetime (Chap. 8) formulations of GR, of foremost relevance to this book.

C) On the other hand, in Sakharov's route, GR is conceived of as an elasticity conferred to space by Particle Physics processes. This is relevant as an example of interpreting GR as an effective theory rather than as a fundamental one; [195] reviews a number of subsequent such ideas.

D) In the Fierz and Pauli type route [300] (see also [883]), GR emerges from consideration of a spin-2 field on a fixed-background Minkowski spacetime $\mathbb{M}^4$. This is a useful perspective for Covariant Approaches to Quantum Gravity (Sect. 11.2).

See Sects. 8.15 and 11.9–11.11, and Chap. 21 for further programs which mostly postcede [660, 899] that can be argued to constitute further such routes.

9.2 Dynamics in the Great Tradition

Secondly, Wheeler alongside mathematical physicist James Isenberg furthermore argued that Physics was developed as *"dynamics in the great tradition"* [469] in the period from Galileo through to the advent of SR. Broad's point outlined in Sect. 4.6 can furthermore be expanded in this regard (a development supported also by philosopher of physics Gerald Whitrow [906]). I.e. Minkowskian and Einsteinian spacetimes are both *co-geometrizations* of space and time, rather than an end to the *actual distinction* between the two concepts. Dirac [250] also questioned spacetime's acquisition of primary status. *"One cannot, however, pick out the six important components from the complete set of 10 in any way that does not destroy the four-dimensional symmetry. So if one insists on preserving four-dimensional symmetry in the equations, one cannot adapt the Theory of Gravitation to a discussion of measurements in the way Quantum Theory requires without being forced to a more complicated description than is needed by the physical situation. This result has led me to doubt how fundamental the four-dimensional requirement in physics is."* Barbour provided further arguments for spatial or configurational primality in e.g. [101, 103].

Wheeler additionally supplied misgivings about the status of GR spacetime at the quantum level [899] (this book postpones discussion of these to Sect. 12.12). These were part of his motivation to conceive of GR as Geometrodynamics so as to take a step back from GR spacetime and return to the 'great tradition'. As an initial step, one could make the ADM split to pass to the geometrodynamical formulation of GR. However, Wheeler went further than this by asking for first principles for this without ever passing through the spacetime formulation of GR. Further on in this book, we shall encounter the Deformation Approach [454] and the Relational Approach [62, 109] which address this question. The first of these still assumes *embeddability into* spacetime, whereas the second derives that also. In this way, a geometrodynamical formulation of GR can be derived without ever passing through spacetime, i.e. never departing from the 'great tradition'. Finally N.B. that GR spacetime indeed remains as a useful reformulation; the new feature in the Relational Approach is, rather, that GR spacetime no longer plays an ontologically primary role.

9.3 Spacetime Versus 'Space or Configuration Space'

The preceding two Secs point to this dilemma of ontological primality, which can also already be seen by contrasting Chap. 7's spacetime formulation of GR and Chap. 8's geometrodynamical one.

Dynamical primality rests within the 'space or configuration space' horn of the dilemma. So do arguments for timelessness at the primary level, from Leibniz's Time Principle through to the Fully Timeless Approaches outlined in Sect. 9.12. This dilemma is moreover one of the underlying reasons for the multiplicity of Problem of Time facets (Fig. 9.1) and of strategies to deal with these (Fig. 10.2).

9.4 Configuration Spaces $\mathfrak{q}$

Now $\mathfrak{q}$ has entered consideration, it helps to supplement Sect. 2.13's outline of these with further examples, which furthermore introduce two of this book's principal model arenas.

Example 1) Scaled relational particle configurations involve just relative angles and relative separations. A theory in which just these are meaningful is *Scale and Shape Relational Particle Mechanics (RPM)* [28, 37, 100, 102, 105], alias Euclidean RPM and Barbour–Bertotti (1982) theory (reviewed in [37, 100]). On the other hand, 'pure-shape' relational particle configurations involve just relative angles and *ratios* of relative separations. *Shape RPM* [37, 45, 102] alias similarity

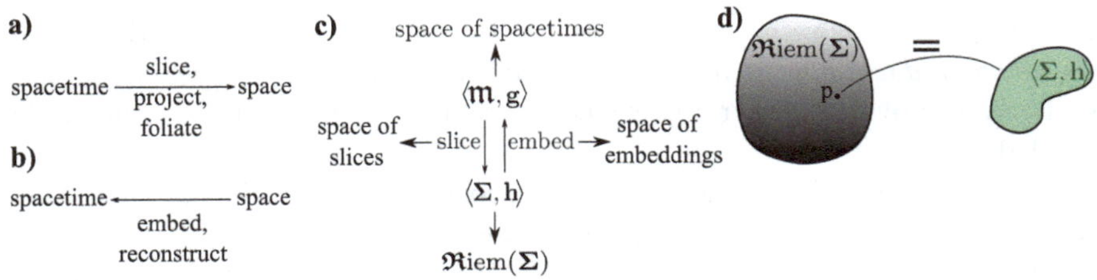

Fig. 9.1 **a)** In addition to considering each of spacetime and space, one can consider passage from spacetime to space by considering a slice and projecting spacetime entities onto it, or by foliating the spacetime with a collection of spaces. **b)** Passage in the opposite direction involves embedding rather than projecting, and is a Spacetime Construction; this is harder due to assuming less structure [41]. Cf. Wheeler's 2-way passage mentioned above; this is the basis of Facets 6) and 7). **c)** Including the spaces of each of the four preceding entities gives the eightfold that is crucial for understanding many of the facets of the Problem of Time. **d)** In particular, this book makes substantial use of 'spaces of spaces', especially configuration space $\mathfrak{q}$. In the GR case, the configuration space is $\mathfrak{Riem}(\boldsymbol{\Sigma})$, each point of which represents a 3-metric h_{ij} on the one fixed spatial topology $\boldsymbol{\Sigma}$

RPM can be viewed as a theory in which just these are meaningful. A redundant $\mathfrak{q}$ for these theories is $\mathfrak{q}(N,d) = \mathbb{R}^{dN}$ of N particles. This possesses an obvious Euclidean metric: the $\mathbb{R}^{dN}$ one rather than the spatial $\mathbb{R}^{d}$ one. See Chap. 15 and Appendix G.1 for further less redundant $\mathfrak{q}$s for RPMs; many of these turn out to have tractable and well-known geometry.

Example 2) For full GR, a redundant $\mathfrak{q}$ is $\mathfrak{Riem}(\boldsymbol{\Sigma})$ [237], as per Chap. 8.3 and further detailed in Appendix H. Figure 9.1.d) uses this example to introduce the notion of a *space of spaces*. $\mathfrak{S}$uperspace$(\boldsymbol{\Sigma})$ as per Sect. 8.12 is a less redundant $\mathfrak{q}$ for GR; if interested, see Appendix N for more about this.

Example 3) *Minisuperspace* $\mathfrak{Mini}(\boldsymbol{\Sigma})$ [657, 659] is a simpler subcase of Example 2): the space of homogeneous positive-definite 3-metrics on $\boldsymbol{\Sigma}$. These are notions of space in which every point is the same. Here full GR's $\mathrm{M}^{ijkl}(\mathbf{h}(\underline{x}))$ has collapsed to an ordinary 6×6 matrix, $M_{\mathsf{l}\mathsf{l}'}(\boldsymbol{h})$; this is an overall—rather than independently per space point—curved $(- - + + + +)$ 'minisupermetric'. Some simpler subcases nested within this are as follows.

i) Diagonal Minisuperspace involves a yet smaller 3×3 $(- - + +)$ matrix $M_{\mathsf{l}\mathsf{l}'}(\boldsymbol{h})$ [659]; Appendix I.1 further develops various subcases.
ii) Isotropic Minisuperspace: flat single-number (–) minisupermetric, for instance for $\boldsymbol{\Sigma} = \mathbb{S}^3$ with standard hyperspherical metric. This is a closed cosmological model, and simpler than i) through not modelling anisotropy.

The specific Minisuperspace models used in this book's detailed examples are spatially closed on Machian grounds. I.e. these avoid undue influence of boundary or asymptotic physics, a criterion that Einstein also argued for [286], though see also Epilogue II.C's counterpoint. The simplest choice is $\boldsymbol{\Sigma} = \mathbb{S}^3$; this is also the most conventional for closed-universe cosmologies. Unlike in Sect. 7.4, this is here to contain fundamental rather than phenomenological matter, due to having Quantization in mind [149, 433]. One needs at least 2 degrees of freedom, and Cosmology conventionally makes use of scalar fields. The simplest case brings in one minimally-coupled scalar field. The $\mathfrak{q}$ metric for this Minisuperspace is (as per Appendix I and up to a conformal factor of a^3) just 2-d Minkowski spacetime $\mathbb{M}^2$ equipped with its standard indefinite flat metric.

9.5 Configuration Spaces as Starting Point for Dynamics

Given the configurations $\boldsymbol{Q}$ indexed by A, composite objects can be built from these. In some approaches, these include velocities $\dot{\boldsymbol{Q}}$, *changes* of configuration $\boldsymbol{Q}$, or conjugate momenta $\boldsymbol{P}$. Alternatives based on 'configurational minimalism' [37] entertain the further possibility of the $\boldsymbol{Q}$ being more primary than these other objects. There are various strengths of configurational minimalism. Most stringently, one can consider just the $\boldsymbol{Q}$ in a fully timeless manner, as per the next Section; less stringently, these alongside whichever of the above notions taken to be secondary. In these approaches, further familiar useful constructs such as actions and Hamiltonians make sense as yet further composite objects; all subsequent Sections of this Chapter are of this kind. *Kinetic metrics* $\boldsymbol{M}$ with components $M_{\mathsf{AB}}(\boldsymbol{Q})$ are another type of composite object which feature in the theory's kinetic term, $T := \| \dot{\boldsymbol{Q}} \|_{\boldsymbol{M}}^2/2 := M_{\mathsf{AB}} \dot{Q}^{\mathsf{A}} \dot{Q}^{\mathsf{B}}/2$; this can furthermore be considered to equip $\mathfrak{q}$ with a metric.

9.6 Constraints Are All Versus Constraint Providers

Constraints are yet further composite objects: relations between the $\boldsymbol{Q}$ and $\boldsymbol{P}$, $\mathcal{C}_{\mathsf{C}}(\boldsymbol{Q}, \boldsymbol{P}) = 0$. These feature in particular in approaches in which one of $\mathfrak{q}$, $\mathfrak{P}$hase or Dynamics are primary, and present the following further dilemma.

A) *Constraints Unquestioned*. In this approach, constraints are merely to be prescribed ab initio regardless of what they represent. This is along the lines of Applied Mathematics' general theory of constrained systems [70, 371, 797, 805].

B) *Constraint Providers*. In this approach, one is furthermore entitled to ask why the constraints that play major roles in Fundamental Physics take their particular forms. One may then attribute further significance to how Fundamental Physics' constraints arise.

In Wheeler's words, [899], B) involves seeking 'zeroth principles' which are more primary than the constraints themselves. In particular, he asked the following question, which readily translates (Appendix J) to asking for first-principles reasons for the form of the crucial GR Hamiltonian constraint, $\mathcal{H}$.

> "*If one did not know the Einstein–Hamilton–Jacobi equation, how might one hope to derive it straight off from plausible first principles without ever going through the formulation of the Einstein field equations themselves*?" (9.1)

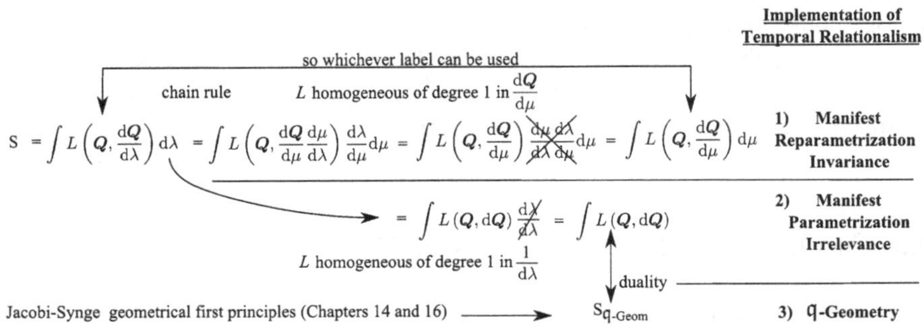

Fig. 9.2 Inter-relation of this book's three implementations of Temporal Relationalism at the level of actions

This is in the context of no longer considering just the Geometrodynamics specific to GR, but rather a multiplicity of geometrodynamical theories, and is furthermore an appeal to seek for a *selection principle* that picks out the GR case. The Deformation Approach and the Relational Approach are answers to this question.

Barbour [98, 105] further developed the idea of Constraint Providers, in the sense of underlying explanations for the form taken by Fundamental Physics' constraints. Moreover, the very well-known approach of taking a Lagrangian with particular symmetries, from which Gauge Theory—with its gauge constraints **$\mathcal{G}$auge**—ensues, can be interpreted as an example of Constraint Provider. Section 9.8 and Chaps. 14, 16, 18 consider a number of variants of this idea. In Part I, however, we first consider a different kind of Constraint Provider, as follows.

9.7 Background Independence Aspect 1: Temporal Relationalism

Temporal Relationalism Postulate. We now implement [37, 105] Leibniz's Time Principle (Chap. 3.1) in a mathematically sharp manner. The postulate itself is the following two-part selection principles for Principles of Dynamics actions.

TR-i) Include no extraneous times—such as t^{Newton}—or extraneous time-like variables—such as the ADM lapse of GR, α.
TR-ii) Include no label times either.

A first implementation of TR-ii) is for a label λ to feature in the action but be physically meaningless due to it being interchanged for any other (monotonically related) label without altering the physical content of the theory. I.e. the action in question is to be *Manifestly Reparametrization Invariant*. This requires the action to be homogeneous of degree one in its velocities $\dot{\boldsymbol{Q}} = \mathrm{d}\boldsymbol{Q}/\mathrm{d}\lambda$ (line 1 of Fig. 9.2). Further envisaging this $\mathrm{d}/\mathrm{d}\lambda$ as the Lie derivative $£_{\mathrm{d}/\mathrm{d}\lambda}$ in a particular frame—paralleling (8.9)—is useful for further reference (Sect. 10.2).

A second implementation follows from the further conceptual advance of formulating one's action and subsequent equations without use of any meaningless label at all. This gives the *Manifestly Parametrization Irrelevant* implementation in terms of *changes* $\mathrm{d}\boldsymbol{Q}$ in place of label-time velocities $\dot{\boldsymbol{Q}}$. I.e. now actions are required to be homogeneous of degree one in the changes (this is clearly equivalent by line 2 of Fig. 9.2).

Moreover, it is better still to formulate this directly, i.e. without even mentioning any meaningless label or parameter. This can be done because the Manifestly Parametrization Irrelevant implementation is, dually, a Configuration Space Geometry implementation. This final implementation provides further justification for the study of the geometry of $\mathfrak{q}$ (Chaps. 18, 21, Appendices G and H).

As a concrete example, consider Temporally-Relational but Spatially-Absolute Mechanics. An action for this is[3]

$$S_{\mathrm{J}} := \int \mathrm{d}\lambda\, L_{\mathrm{J}} := 2\int \mathrm{d}\lambda \sqrt{TW} := \sqrt{2}\int \mathrm{d}s\sqrt{W}: \tag{9.2}$$

Jacobi's action principle [598], whether the 'J' stands for the great mathematician Carl Jacobi. In the first expression, $T := \|\dot{\boldsymbol{q}}\|_{\boldsymbol{m}}^2/2$ is the kinetic energy, whose $\mathfrak{q}$ metric $\boldsymbol{m}$ is just the 'mass matrix' with components $m_I \delta_{IJ}\delta_{ij}$. Also $W := E - V(\boldsymbol{q})$ is the *potential factor*, for $V(\boldsymbol{q})$ the potential energy and E the total energy of the model universe. Moreover, Manifestly Parametrization Irrelevant formulations of this are indeed also well-known, as are dual Configuration Space Geometry formulations. The second expression in (9.2) is of this kind, now involving the *kinetic arc element* $\mathrm{d}s := \|\mathrm{d}\boldsymbol{q}\|_{\boldsymbol{m}}$. This action is indeed physically equivalent to the more familiar Euler–Lagrange action principle, though demonstration of this is postponed to Sect. 15.2.

The Manifestly Reparametrization Invariant form's conjugate momenta are $\boldsymbol{p} := \partial L_{\mathrm{J}}/\partial\dot{\boldsymbol{q}} = \sqrt{W/T}\,\dot{\boldsymbol{q}}$.

The main consequence of actions implementing TR-ii) arises via the following argument of Dirac [250]. Manifestly Reparametrization Invariant actions are homogeneous of degree 1 in the velocities. Consequently, the $k := \dim(\mathfrak{q}) = Nd$ conjugate momenta are (by the above definition) homogeneous of degree 0 in the velocities. Therefore they are functions of at most $k-1$ ratios of the velocities. So there must be at least one relation between the momenta themselves (i.e. without any use made of the equations of motion). But this is the definition of a primary constraint (cf. Appendix J.15).

Thus Temporal Relationalism indeed acts as a Constraint Provider. Moreover, the homogeneous quadratic form of the above mechanical action [98] causes the constraint it provides to also be purely quadratic:

$$\mathcal{E} := \|\boldsymbol{p}\|_{\boldsymbol{n}}^2/2 + V(\boldsymbol{q}) = E. \tag{9.3}$$

[3]Here indices I, J run over 1 to N (particle number), indices i run over 1 to d (spatial dimension) and m_I are the particles' masses.

Here $\boldsymbol{n} = \boldsymbol{m}^{-1}$, with components $\delta_{IJ}\delta_{ij}/m_I$. (9.3) is familiar from elsewhere in Physics, where it has the name and role of an *energy equation*, though as we shall see below, in the current context its interpretation is, rather, as an *equation of time*.

Finally, this approach's equations of motion are $\sqrt{W/T}\,\dot{\boldsymbol{p}} = -\partial V/\partial \boldsymbol{q}$.

We next turn to interpretational matters. Firstly, it is quite natural to ask whether there is a paradox between Leibniz' Time Principle's 'there being no time at the primary level for the universe as a whole' and our appearing to 'experience time'. Thus one is faced with having to explain the origin of the notions of time in the laws of Physics that appear to apply in the Universe.

This can be answered by pointing to discrepancies between the two situations; two preliminary such are as follows. Firstly, whereas 'time' is a useful concept for everyday experience, the nature of 'time' itself is in general less clear. Secondly, everyday experience concerns subsystems rather than the whole Universe setting of the Principle.

This book's main answer to this follows from recollecting Mach's Time Principle that 'time is to be abstracted from change'.[4] Thus timelessness for the Universe as a whole at the primary level is *resolved* by time emerging from change at the secondary level. Chaps. 15 and 23 furthermore argue that one is best served by adopting a Machian conception of time along the lines of the astronomers' ephemeris time. This is now abstracted from a 'sufficient totality of locally relevant change'.

More specifically, Temporal Relationalism provides an emergent time which can be interpreted in this Machian manner. This is the *Jacobi emergent time*, obtained by the following rearrangement of the ε constraint provided by Temporal Relationalism.

$$t^{\mathrm{em(J)}} = \int \mathrm{d}\lambda\sqrt{T/W} = \int \mathrm{d}s/\sqrt{2\,W} = \int \|\mathrm{d}\boldsymbol{q}\|_{\boldsymbol{m}}/\sqrt{2\,W}. \tag{9.4}$$

Because of this rearrangement, ε plays the role of an equation of time in the Relational Approach, rather than that of an energy equation. The third form therein—the Manifestly Parametrization Irrelevant or dual $\mathfrak{q}$-geometry form version—is furthermore manifestly an equation for obtaining time from change, so this indeed complies with Mach's Time Principle. (9.4) also ascribes to the 'choose time so that motion is simplest' tenet of Chap. 1. For, via[5]

$$* := \frac{\partial}{\partial t^{\mathrm{em(J)}}} := \sqrt{\frac{W}{T}}\frac{\partial}{\partial \lambda}, \tag{9.5}$$

it is also distinguished by its simplification of the model's momentum–velocity relations and equations of motion. In this manner, we have arrived at a recovery of Newtonian time on a Temporally-Relational footing.

[4]Chapter 12 outlines various other answers, which constitute distinct strategies for addressing the Frozen Formalism Problem.

[5]In this book, given a time variable t, its calendar year zero adjusted version $t - t(0)$ is denoted by the corresponding oversized t.

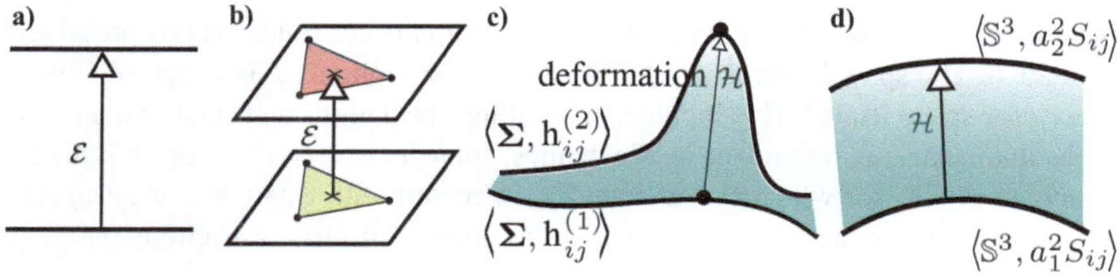

Fig. 9.3 Action of Chronos in **a)** Spatially-Absolute Mechanics, **b)** RPM, **c)** GR, for which we shall see $\mathcal{H}$ generates hypersurface deformations, and **d)** Minisuperspace whose hypersurfaces privileged by homogeneity are 3-spheres with metric proportional to the 3-sphere metric S_{ij}

In the GR counterpart of this working (Chaps. 15 and 18) we shall see that it is indeed the Hamiltonian constraint $\mathcal{H}$ *crucial to GR that arises from Temporal Relationalism as a primary constraint*, and that this amounts to a recovery of GR's version of proper time. For now, Part I offers an outline of how $\mathcal{H}$ arises in the Minisuperspace subcase of GR in Sect. 9.9.

In conclusion, we emphasize that Temporal Relationalism provides a crucial constraint whose interpretation in this context is as an equation of time, let us name the general case of the constraint provided in this manner as Chronos. $\mathcal{E}$ and $\mathcal{H}$, as arrived at within the Relational Approach, are both subcases of this. Finally Fig. 9.3 sketches how these constraints act in various models.

9.8 Aspect 2: Configurational Relationalism

Configurational Relationalism covers both of the following.

a) *Spatial Relationalism* [105] is to not ascribe any absolute properties to space.
b) *Internal Relationalism* is the post-Machian addition of also not ascribing any absolute properties to any additional internal space associated with the matter fields. This is both a useful addition and straightforward.

b) is substantially distinct though holding at a fixed spatial point whereas a) moves spatial points around. Configurational Relationalism is then approached as follows.

CR-i) One is to include no extraneous configurational structures (spatial or internal-spatial metric geometry variables of a fixed-background rather than dynamical nature).

CR-ii) Physics in general involves not only a $\mathfrak{q}$ but also a group $\mathfrak{g}$ of transformations acting upon $\mathfrak{q}$ that are taken to be physically redundant.

Since time-parametrization is really a 1-*d* metric of time, TR-i) and CR-i) reflect a single underlying relational conception of Physics: that there is to be no fixed-background Metric Geometry.

CR-ii) is a matter of practical convenience: often $\mathfrak{q}$ with redundancies is simpler to envisage and calculate with. The Internal Relationalism case of CR-ii) is a distinct formulation of Gauge Theory (as per Chap. 16) from the conventional one presented in Chap. 6. The spatial case is similar: it can also be thought of as a type of Gauge Theory for space itself.[6] This includes modelling translations and rotations relative to absolute space as redundant in Mechanics, for which $\mathfrak{q} = \mathbb{R}^{dN}$, or *Diff*($\boldsymbol{\Sigma}$) as redundant in GR, for which $\mathfrak{q} = \mathfrak{Riem}(\boldsymbol{\Sigma})$. In accord with Chap. 8.2, the *Diff*($\boldsymbol{\Sigma}$) are actively interpreted. Chaps. 14 to 19 subsequently discuss restrictions on $\mathfrak{q}$, $\mathfrak{g}$ pairings.

Best Matching implementation of Configurational Relationalism

Best Matching [105] is a substantial implementation of Configurational Relationalism at the level of Lagrangian variables (Q, $\dot{Q}$). This involves $\mathfrak{q}$, $\mathfrak{g}$ pairs so that $\mathfrak{q}$ is a space of 'configurations for which $\mathfrak{g}$ are taken to be redundant motions'. More specifically, in Best Matching $\mathfrak{g}$ acts on $\mathfrak{q}$ as a *shuffling group*. I.e. pairs of configurations are considered; one is kept fixed while the other is shuffled around—an active viewpoint—until the two are brought into minimum incongruence.

One first constructs a $\mathfrak{g}$-corrected action. For the examples considered in Part I, this involves

$$\text{replacing each occurrence of } \dot{Q} \text{ with } \dot{Q} - \stackrel{\rightarrow}{\mathfrak{g}}_g Q,$$

where $\stackrel{\rightarrow}{\mathfrak{g}}_g$ indicates group action.

Next, varying with respect to the $\mathfrak{g}$ auxiliary variables $\boldsymbol{g}$ (indexed by G) provides constraints; in view of the above, let us term these '*shuffle constraints*' and denote them by $\mathcal{S}$**huffle** (matchingly indexed by G). These arise as *secondary constraints* (defined in Appendix J.15). Being linear in the momenta, they could also be denoted by $\mathcal{L}\mathrm{in}_{\mathsf{G}}$. However, these are but a subcase of the most general linear constraints $\mathcal{L}$**in** (indexed by L rather than G), since not all possible such arise from shuffling. $\mathcal{F}$**lin** (indexed by N)—constraints which are first-class linear in the momenta—turn out to be a useful case of intermediate generality. Chap. 24 shall provide examples of these various kinds of constraints being distinct, so this emphasis of a range of names for slightly different concepts is justified.

In setting up Best Matching, the intent can be considered to be that $\mathfrak{g}$ performs the function of a gauge group. However, this intent is only known to have succeeded upon confirming Aspect 3)'s suitability of the algebraic structure between the constraints. Thus the $\mathcal{S}$**huffle** are for now *candidate* constraints associated with an *attempt* to associate $\mathfrak{g}$ with $\mathfrak{q}$. Candidate $\mathcal{S}$**huffle** constraints which succeed in the above manner belong to the more specific conceptual type $\mathcal{G}$**auge**.

Let us next address that the initial introduction of $\mathfrak{g}$ corrections appears at first sight to be a step in the wrong direction as regards freeing the physics of $\mathfrak{q}$ from $\mathfrak{g}$. This is due to its extending the already redundant space $\mathfrak{q}$ of the Q to some joint

[6] The name and concept of Gauge Theory is used here in a somewhat broader manner than that of Particle Physics, covering also e.g. Molecular Physics [624] and cosmological perturbations [110, 671].

space of Q and the $\mathfrak{g}$-auxiliary variables g. However, if **shuffle** does turn out to be of the form **gauge**, Sect. 9.14 explains that this is a type of constraint which uses up *two* degrees of freedom per $\mathfrak{g}$ degree of freedom. Each degree of freedom appended then wipes out not only itself but also one of $\mathfrak{q}$'s redundancies. Thus one indeed ends up on a $\mathfrak{q}$ that is free of these redundancies—the quotient space $\mathfrak{q}/\mathfrak{g}$—as is required to successfully implement Configurational Relationalism.

In the Best Matching procedure, one continues by taking **shuffle** in Lagrangian variables to be equations to solve for the g themselves. Next, one substitutes this extremizing solution back into the original action to obtain a reduced action on the reduced configuration space $\mathfrak{q}/\mathfrak{g}$. This action is finally elevated to be a new starting point.

Let us further clarify the nature of Configurational Relationalism and Best Matching by using Shape and Scale RPM as an example. These are taken to be fundamental rather than effective Mechanics problems, by which it makes sense for the corresponding potentials to be of the form $V(\boldsymbol{q}) = V(\underline{q}_I \cdot \underline{q}_J \text{ alone})$. This form then guarantees that auxiliary translation and rotation corrections applied to this part of the action straightforwardly cancel each other out within. The situation with the kinetic term is more complicated, because $\mathrm{d}/\mathrm{d}\lambda$ is not a tensorial operation under the *λ-dependent Euclidean group* alias *Leibniz group* which plays the role of kinematical group. This leads to the translation and rotation corrected kinetic term

$$T = \|\circ_{\underline{A},\underline{B}}\boldsymbol{q}\|_{\boldsymbol{m}}^2/2, \quad \text{for } \circ_{\underline{A},\underline{B}}\boldsymbol{q} := \dot{\boldsymbol{q}} - \underline{A} - \underline{B}\times\boldsymbol{q} \tag{9.6}$$

the '*Best Matched derivative*'. The Barbour–Bertotti action is then

$$S_{\mathrm{BB}} = 2\int \mathrm{d}\lambda\sqrt{W\,T}. \tag{9.7}$$

The momenta conjugate to the $\boldsymbol{q}$ are $\boldsymbol{p} = \sqrt{W/T}\,\{\dot{\boldsymbol{q}} - \underline{A} - \underline{B}\times\boldsymbol{q}\}$. By virtue of Manifest Reparametrization Invariance and the particular square-root form of the Lagrangian, these momenta obey a primary constraint that is purely quadratic in the momenta,

$$\mathcal{E} := \|\boldsymbol{p}\|_{\boldsymbol{n}}^2/2 + V(\boldsymbol{q}) = E. \tag{9.8}$$

Next, variation with respect to $\underline{A}$ and $\underline{B}$ give secondary constraints, respectively,

$$\underline{\mathcal{P}} := \sum_{I=1}^{N} \underline{p}_I = 0 \quad \text{(zero total momentum constraint)}, \tag{9.9}$$

$$\underline{\mathcal{L}} := \sum_{I=1}^{N} \underline{q}^I \times \underline{p}_I = 0 \quad \text{(zero total angular momentum constraint)}. \tag{9.10}$$

Note that these constraints are linear in the momenta. The first can furthermore be interpreted as the centre of mass motion for the dynamics of the whole Universe being irrelevant rather than physical. All the tangible physics is in the remaining relative vectors between particles.

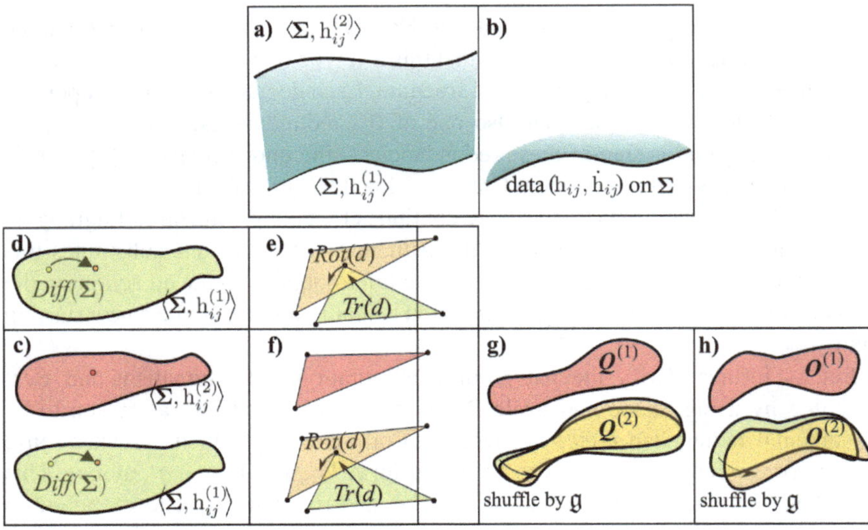

Fig. 9.4 **a)** Wheeler first contemplated a 'thick sandwich' [897]: bounding bread-slice data $h^{(1)}_{ij}$ and $h^{(2)}_{ij}$ as knowns to solve for the GR spacetime 'filling' in between. This was an attempted analogy with Feynman's path integral for quantum transition amplitudes between states at two different times [897]; however, this failed to be mathematically well-posed. **b)** Wheeler subsequently considered 'Thin Sandwich' data to solve for a local coating of spacetime [897]. This is the 'thin' limit of taking the data to be on the bounding 'slices of bread', with data h_{ij} and $\dot{h}_{ij}$. **c)** The Thin Sandwich can now be reinterpreted in terms of Best Matching $\mathfrak{Riem}(\boldsymbol{\Sigma})$ with respect to *Diff*$(\boldsymbol{\Sigma})$. This corresponds to the depicted shuffling [98, 109], in which the red space is held fixed while each point in the yellow space is moved to a new position marked in orange so as to seek out minimal incongruence between the two. See Sect. 34.1 for further details of this reinterpretation. **d)** This rests on the more basic point that *Diff*$(\boldsymbol{\Sigma})$ acts on $\boldsymbol{\Sigma}$ by moving points around. **e)** Scaled RPM's $Tr(d)$ and $Rot(d)$ actions, as another example of shuffles. **f)** then gives the analogous Best Matching for RPM triangles with respect to the rotations $Rot(d)$ and translations $Tr(d)$ [16, 98, 105]. This indicates Best Matching's applicability to a wider range of theories rather than just to Geometrodynamics or the corresponding *Diff*$(\boldsymbol{\Sigma})$. **g)** Depicts the general case of Best Matching shuffling configurations $\boldsymbol{Q}^{(1)}$ and $\boldsymbol{Q}^{(2)}$, in some shared configuration space $\mathfrak{q}$, with respect to a group $\mathfrak{g}$. **h)** Part II (Chap. 13) involves a further generalization to other levels of structure, depicted here as general objects $\boldsymbol{O}^{(1)}$, $\boldsymbol{O}^{(2)}$ in some shared space of objects $\mathfrak{O}$

The corresponding equations of motion are $\sqrt{W/T}\,\dot{\boldsymbol{p}} = -\partial V/\partial \boldsymbol{q}$.

Returning to the Best Matching procedure, the constraints (9.9), (9.10), rewritten in Lagrangian configuration–velocity variables $(\boldsymbol{q}, \dot{\boldsymbol{q}})$, are to be solved for the auxiliary variables $\underline{A}$, $\underline{B}$ themselves. This solution is then substituted back into the action, so as to produce a final *Tr*- and *Rot*-independent expression that *directly* implements Configurational Relationalism. One has the good fortune of being able to solve Best Matching explicitly for a wide range of RPMs (Chaps. 15 to 16, and Appendix G, based on [37, 539]).

Let us next consider the Geometrodynamical subcase of Best Matching. This is the so-called *Thin Sandwich*: Fig. 9.4.b) and [115, 124, 483, 586, 897]. In parallel with the above set-up for RPM, the Baierlein–Sharp–Wheeler (BSW) [89] action is[7]

$$\mathsf{S}^{\mathrm{GR}}_{\mathrm{BSW}} = \int \mathrm{d}\lambda \int_{\Sigma} \sqrt{\overline{\mathrm{T}}^{\mathrm{GR}}_{\mathrm{BSW}} \sqrt{\mathrm{h}}\{\mathcal{R} - 2\,\Lambda\}}, \quad \overline{\mathrm{T}}^{\mathrm{GR}}_{\mathrm{BSW}} := \|\delta_{\vec{\beta}}\mathbf{h}\|^2_{\mathbf{M}}. \tag{9.11}$$

Actions of this kind return the usual GR $\mathcal{H}$ as their primary constraint. See Chap. 18 as regards equivalence of the ADM and BSW actions, and more about the BSW action; this includes how this example indeed also gives rise to an emergent Machian time. In this way, Einstein's GR does happen to implement [62, 98, 109] the philosophically desirable kernel that is Mach's Time Principle.

9.9 Minisuperspace Model Arena Version

Whereas for now Part I's restriction to treating the aspects piecemeal prevents us from presenting Temporal Relationalism for full GR, we can in the meanwhile substantiate $\mathcal{H}$ arising in this manner by considering the subcase of Minisuperspace GR.

Let us first comment that modelling the Problem of Time requires a range of models which exhibit a variety of subsets of the aspect interference in GR-like theories. It is interference between Temporal and Configurational Relationalism which prevents full GR being presented in Part I. For instance, full Geometrodynamics' t^{em} depends in general on the outcome of Best Matching, and in doing so this encounters the Thin Sandwich Problem. Minisuperspace avoids this by not exhibiting Configurational Relationalism, since spatial homogeneity renders $\mathcal{M}_i$ trivial due to this constraint's dependence on spatial derivatives. On the one hand, this facilitates using Minisuperspace as a basic introductory example in Part I, but on the other hand this renders Minisuperspace too simple a model for many of Part II and III's considerations. RPM is a comparably useful model arena by, complementarily, exhibiting Configurational Relationalism, and notions of structure and thus of structure formation, which are also absent from Minisuperspace.[8] Conversely, Minisuperspace is a restriction of GR, so it inherits some features that RPMs do not possess, including kinetic metric indefiniteness and imposition of more specific restrictions on the form of the potential.

Let us next comment on this book's particular choices among Minisuperspace models, which consist of two distinct sets of modelling assumptions. 1) The matter physics is light and fast (l) as compared to the gravitational physics being heavy

[7]The overline denotes densitization, i.e. inclusion of a factor of $\sqrt{\mathrm{h}}$, which in the current case resides in $\mathbf{M}$.

[8]In RPMs, linear constraints and inhomogeneities are logically-independent features. This is in contrast with the Minisuperspace version, in which both are concurrently trivialized by homogeneity rendering the spatial derivative operator $\mathcal{D}_i$ meaningless.

and slow (h). 2) Both are h and only further degrees of freedom—anisotropy or inhomogeneity—are l.

We finally pick the isotropic minimally-coupled scalar field matter version of 1) to further illustrate Temporal Relationalism. The Misner-type [659]) action for this is

$$S = \frac{1}{2}\int \mathrm{d}\lambda\, \sqrt{\overline{T}\,\overline{W}},$$
$$\overline{T} := \exp(3\,\Omega)\left\{-\left\{\frac{\mathrm{d}\Omega}{\mathrm{d}\lambda}\right\}^2 + \left\{\frac{\mathrm{d}\phi}{\mathrm{d}\lambda}\right\}^2\right\}, \tag{9.12}$$
$$\overline{W} := \exp(3\,\Omega)\big\{\exp(-2\Omega) - V(\phi) - 2\,\Lambda\big\}.$$

Here, the *Misner variable*

$$\Omega := \ln a, \tag{9.13}$$

for a the usual cosmological scale factor. Also note that the cosmological constant term is needed to support [736] the spatially-$\mathbb{S}^3$ FLRW cosmology with scalar field matter in the case in which matter effects are presumed small. The corresponding Hamiltonian constraint is then

$$\mathcal{H} := \frac{\exp(-3\,\Omega)}{2}\big\{-\pi_\Omega^2 + \pi_\phi^2 + \exp(6\,\Omega)\big\{V(\phi) + 2\,\Lambda - \exp(-2\Omega)\big\}\big\} = 0. \tag{9.14}$$

This can then be rearranged to give the model's classical Machian emergent time,

$$t^{\mathrm{em}} = \int \sqrt{-\mathrm{d}\Omega^2 + \mathrm{d}\phi^2}\Big/\sqrt{\exp(-2\Omega) - V(\phi) - 2\,\Lambda}. \tag{9.15}$$

9.10 Temporal and Configurational Relationalism Lead to Two of the Problem of Time Facets

The most well-known *(Schrödinger-Picture) Quantum Frozen Formalism Problem* arises from elevating an equation of the form (8.26) which encompasses both GR's $\mathcal{H}$ and RPM's $\mathcal{E}$, to a quantum equation

$$\widehat{\mathcal{Q}\mathrm{uad}}|\Psi\rangle = 0. \tag{9.16}$$

Here, Ψ is the *quantum wavefunction of the* (model) *universe*. See Sect. 11.4 for the detailed form of the GR case of this equation: the so-called *Wheeler–DeWitt equation* [237, 899]. This is often viewed as the $E = 0$ case of a time-independent Schrödinger equation (5.11): a stationary alias timeless or frozen quantum wave equation which occurs in a place in which one would expect a time-dependent equation such as (5.10). On occasion, this has been interpreted at face value as a Fully Timeless Worldview arising from attempting to combine GR and Quantum Theory. See however the rest of the current Chapter, Chap. 12, and Parts II and III for further interpretations and means of bypassing such an equation arising in the first place.

Equation (5.10) is presented above in the finite-theory case for simplicity (so its given form includes just the Minisuperspace subcase of GR). The field-theoretic counterpart of (9.16) contains in place of a partial derivative $\partial/\partial Q^{\mathsf{A}}$ a functional derivative $\delta/\delta h_{ij}(\underline{x})$ (Chap. 12.1). The Wheeler–DeWitt equation arises regardless of whether from ADM's scheme that presupposes and subsequently splits spacetime, or as an equation of time *c*hronos from implementing Temporal Relationalism as per above. Moreover, from the latter perspective, the Frozen Formalism Problem already features at the classical level for the Universe as a whole; its being manifested at the quantum level is then less surprising.

On the other hand, the Lagrangian variables form of the GR momentum constraint $\mathcal{M}_i$ is the *Thin Sandwich equation* [124]; since its explicit form is rather complicated, this is postponed to Eq. (18.13). Solving this is the *Thin Sandwich Problem*, as outlined in Fig. 9.4.a)–b). This problem was furthermore identified as another of the Problem of Time facets in Isham and Kuchař 's ground-breaking reviews [483, 586].

The Thin Sandwich facet was always presented as a manifestly classical-level problem. It is indeed a problem concerning time because, firstly, it is solving for a local slab of GR spacetime immediately adjacent to $\boldsymbol{\Sigma}$. It is additionally a prerequisite for various Problem of Time strategies—including the above emergent time one and the internal time one below—due to the GR momentum constraint $\mathcal{M}_i$ interfering with resolutions of the Frozen Formalism Problem (see Part II for more). Finally, it is a major mathematical problem [115, 124]; see Chap. 18 and Appendix O.5 for an outline.

The Thin Sandwich is, moreover, sequentially generalized by the following.

a) Best Matching, which applies to a wider range of theories than just Geometrodynamics (Fig. 9.4.c–d). Because of this, sandwich- and foliation-specific concepts and nomenclature do not themselves directly generalize.
b) Configurational Relationalism, which furthermore applies to resolutions at levels other than that of the Lagrangian variables. E.g. this can also apply at the Hamiltonian level or at the level of solving the quantum equations. The most general implementation for this involves $\mathfrak{g}$ acting on one of the objects $O^{(1)}$ being compared (Fig. 9.4.f) in question, followed by an operation over all of $\mathfrak{g}$ which cancels out the dependence of $\mathfrak{g}$, such as extremization in Best Matching or group averaging (see Sect. 14.4 for details).

Most of Chap. 9 and 10's material arose from analysing which facets of the Problem*s* of Time [24, 26, 37, 483, 586] in QG already occur at the classical level [37]. This reveals that 8/9ths of the Problem of Time facets already have classical counterparts.

9.11 Other Timefunction-Based Problem of Time Strategies

Each time-dependent Schrödinger equation (5.10) can be considered to be preceded by a classical equation that is *parabolic* in the momenta [580, 582],

$$p_t = P_{\mathsf{A}} P^{\mathsf{A}} + C. \tag{9.17}$$

Here p_t is the momentum conjugate to whatever t plays the role of time in this model (Newtonian time in the most conventional case), whereas C is a function of the $\boldsymbol{Q}$ alone. On the other hand, a time-independent Schrödinger equation is preceded by an *elliptic* equation,

$$P_{\mathsf{A}}P^{\mathsf{A}} + C = 0. \tag{9.18}$$

Moreover, the GR Hamiltonian constraint $\mathcal{H}$ looks more like (9.18) than (9.17). The more general classical form

$$p_t = f(\boldsymbol{Q}, \boldsymbol{P}), \quad f \text{ homogeneous of degree } n \tag{9.19}$$

often resembles (9.17) as regards the role played by time therein; this arises in particular by solving $\mathcal{C}$hronos for p_t, with f then playing the role of Hamiltonian for the system. This form also includes Dirac-type equations.

One might next argue that the specific form of $\mathcal{H}$ (11.6) looks even more like the Klein–Gordon equation's classical precursor [581]—the hyperbolic equation

$$p_t^2 = P_{\mathsf{a}}P^{\mathsf{a}} + C, \tag{9.20}$$

where a runs over one value less than A. From this, the DeWittian indefiniteness of $\mathfrak{Riem}(\boldsymbol{\Sigma})$ might furnish a time (Chap. 20) in parallel to the Lorentzian indefiniteness of Minkowski spacetime $\mathbb{M}^n$ (Sect. 4.2). However, in the case of GR, this approach goes awry at the quantum level (Sect. 12.2).

Canonical transformations can map into and out of the general form (9.19) [580, 581]. In this case, specialize the notation to

$$p_{t^{\text{hidden}}} + H_{\text{True}}\big(Q^{\mathsf{O}}, P_{\mathsf{O}}, t^{\text{hidden}}\big) = 0, \tag{9.21}$$

for Q^{O} the theory's other configurational variables. (9.21) corresponds to [586] finding a *hidden time* candidate t^{hidden}—the conjugate to $p_{t^{\text{hidden}}}$—alongside a 'true Hamiltonian' H_{True}.

One might solve the original equation $\mathcal{Q}$uad for a particular p that is designated to be a p_t, i.e. conjugate to some candidate notion of t. However this approach has an added problem in justifying this choice. Common choices at this stage are the scale variable giving a *scale time* candidate (Chap. 20), or the matter field giving a *matter time* candidate (Chap. 22). The scale variable has the virtue of being singled out among all the other variables, since there is precisely one independent such variable. However, this is clearly not monotonic for the significant case of recollapsing universes. On the other hand, while 'the matter variable' may be unique in simple models, there are obviously multiple matter variables in more general models. It is then unclear how to pick 'the time to use' amongst these. We shall see in Sect. 12.15 that different choices of time are capable of leading to inequivalent Quantum Theories, compounding the significance of such ambiguities.

Interestingly, the *York time* candidate (after physicist Jimmy York Jr.)

$$\mathrm{t}^{\mathrm{York}} := \frac{2}{3}\mathrm{p}/\sqrt{\mathrm{h}} = -\frac{4}{3}\,\mathrm{K} \tag{9.22}$$

is monotonic in sizeable regimes. This is a type of 'dilational momentum', i.e. momentum conjugate to a scale quantity. Moreover, K is mean curvature of the extrinsic kind (cf. Sect. 8.4, and given here in the notation of a prescribed function K rather than a functional $\mathcal{K}$). So this is constant on each slice of constant York time. This approach's associated slices are therefore of *constant mean curvature (CMC)*. Chapter 21 considers further the York time candidate as an example of (9.21) in more detail.

Another proposed way forward involves introducing a matter field—e.g. the 'reference fluid' matter outlined in Chap. 22—that specifically results in a *reference matter time* candidate. In this case, (9.19) is realized as

$$p_{t^{\text{ref}}} + H_{\text{True}}\big(Q^{\text{O}}, P_{\text{O}}, t^{\text{ref}}\big) = 0, \tag{9.23}$$

where Q^{O} are now the original theory's other variables together with the other appended variables. This differs from (9.21) in that time is to be found among fields appended to one's theory rather than already hidden within. A particular case is *unimodular time*, which arises as the momentum conjugate to the cosmological constant upon elevating this to a dynamical variable.

N.B. Part I's Time, Timefunction and Clock Postulates are applied in Part II as *selection principles* to discern between the above wide range of candidates. By these and further criteria (Chap. 22), Machian emergent time 'wins out'. Moreover, not all approaches *have a time*; e.g. Sects. 9.12 and 10.6 outline further Problem of Time strategies based, rather, on timelessness or on histories in place of time.

9.12 Fully Timeless Strategies

Fully Timeless Approaches strategies[9] take Temporal Relationalism's primary timelessness at face value by addressing timeless propositions alone. This can cause at least some practical limitations, but can none the less address at least *some* questions of interest. The remaining issue is determining the extent to which the totality of Physics can be recovered from such an approach. One well-documented case [694] involves passing from the notion of 'being at a time' to timeless correlations between a subsystem configuration under study and a clock configuration. The question then arises [692] whether the notion of 'becoming' can also be supplanted, by which Physics could be taken to involve just questions about 'being' rather than about 'becoming'.

9.13 Providers, Algebraic Structure, and Beables

Figure 9.5 outlines the next two aspects of Background Independence.

---

[9]Many of these are quantum-level approaches, for which the motivation is stronger, as further detailed in Chaps. 12, 26 and 51).

$$\text{generator provider} \underset{\text{encoding}}{\overset{\text{provision}}{\rightleftarrows}} \text{generator algebraic structure} \longrightarrow \text{commutants with generators.}$$

Fig. 9.5 *Generator Providers* are a more straightforward and general concept than Constraint Providers. Generator Providers additionally cover purely timeless formulations and next chapter's Spacetime Relationalism version as well. The further structures that the rest of this chapter focuses on are, firstly, the algebraic structures formed by the generators themselves. One type of *Generator Closure Problem* arises if the generators fail to close due to unexpected brackets relations arising. Another arises if these fail to close by themselves, forcing inclusion of further generators by which a larger group is involved. Secondly, we consider entities which commute with the generators: observables or beables. The Figure's left-to-right order of these three concepts is a natural chain of decreasing primality on structural grounds. Finally, a reverse operation to Constraint Provision is the *encoding* of constraints. I.e. upon finding constraints, one aims to subsequently build auxiliary variables into the theory's action; variation with respect to this encodes the constraints required by the theory (see Chap. 33 for details)

9.14 Aspect 3: Constraint Closure

Do constraints beget more constraints? More concretely, if the constraints c_{C} vanish on a given spatial hypersurface, what can be said about $\dot{c}_{\mathsf{C}}$? If $\dot{c}_{\mathsf{C}}$ is equal to some $f(c_{\mathsf{C}})$ alone, it is said to be *weakly zero* in Dirac's sense [250], denoted by ≈ 0. There is however a lack of rigour in such 'Lagrangian' formulations of 'constraint propagation' through evaluation of $\dot{c}_{\mathsf{C}}$ from the Euler–Lagrange equations.

Let us first consider the joint space of the Hamiltonian variables $\boldsymbol{Q}$ and $\boldsymbol{P}$, as standardly equipped with the Poisson brackets algebraic structure $\{\ ,\ \}$: *phase space* $\mathfrak{P}$hase (Appendix J.11).

This formulation furthermore turns out to possess a rigorous algorithm for handling whether constraints beget further independent constraints. This is the *Dirac Algorithm* [250, 446] (laid out in Appendix J.15). This determines how classical brackets of known constraints can in general lead to further constraints, to specifier equations and to inconsistencies. Thereby, Constraint Closure is indeed a necessary check for the constraints already in hand, and one which is capable of invalidating candidate Constraint Providers. The Dirac Algorithm, moreover, serves as an archetype of how to approach this facet.

The end-product algebraic structure of constraints is, schematically,

$$\{c_{\mathsf{F}}, c_{\mathsf{F}'}\} \approx 0. \tag{9.24}$$

F here indexes *first-class* constraints, which are those that close among themselves under Poisson brackets. A constraint is *second-class* if it is not first-class. First- and second-class constraints use up 2 and 1 degrees of freedom respectively; gauge constraints are a subset of first-class constraints. See Appendix J.15 for more details, and also as regards means of removing second-class constraints.

In particular, the Dirac Algorithm can be applied to determine whether the constraints provided by Temporal and Configurational Relationalism—$\mathcal{C}$hronos and $\boldsymbol{\mathcal{S}}$**huffle** respectively—form a complete and consistent picture. A common consideration is whether the **shuffle** close among themselves in the form

$$\{\mathcal{S}\text{huffle}_{\mathsf{G}}, \mathcal{S}\text{huffle}_{\mathsf{G}'}\} = C_{\mathsf{G}\mathsf{G}'}{}^{\mathsf{G}''}\,\mathcal{S}\text{huffle}_{\mathsf{G}''}, \tag{9.25}$$

where $C_{\mathsf{G}\mathsf{G}'}{}^{\mathsf{G}''}$ are constants, so that this is a Lie algebra (Fig. 9.6.a–b). In this case, the attempt to render $\mathfrak{g}$ physically irrelevant is vindicated, insofar as this produces gauge constraints **gauge** which realize that irrelevance. [On many occasions, **gauge** = **Flin** as well, though Chap. 24 shows that this is not always the case.] If also (Fig. 9.6.c)

$$\{\textbf{shuffle}, \mathcal{C}\text{hronos}\} \quad \text{closes}, \quad \mathcal{C}\text{hronos is established as a good } \mathfrak{g} \text{ object.} \tag{9.26}$$

The final consideration is whether (Fig. 9.6.d)

$$\{\mathcal{C}\text{hronos}, \mathcal{C}\text{hronos}\} \quad \text{closes.} \tag{9.27}$$

E.g. if this were to produce a new linear constraint, it would be enforcing an enlarged $\mathfrak{g}$. If Chronos and **shuffle** do beget unexpected further constraints, then one's attempted relational formulation has a *Constraint Closure Problem*: Facet 3) of the Problem of Time.

Example 1) Electromagnetism has the Abelian algebra of constraints[10]

$$\{(\mathcal{G}|\iota), (\mathcal{G}|\mu)\} = 0. \tag{9.28}$$

Example 2) Its Yang–Mills generalization has the Lie algebra of constraints

$$\{(\mathcal{G}_I|\iota^I), (\mathcal{G}_J|\mu^J\} = f_{IJ}{}^K(\mathcal{G}_K|\iota^I\mu^J) \tag{9.29}$$

for $f_{IJ}{}^K$ the structure constants corresponding to the gauge group $\mathfrak{g}$ in question.

Example 3) As a first model involving Chronos as well, Shape and Scale RPM's constraint algebra's nonzero Poisson brackets are

$$\{\mathcal{L}_i, \mathcal{L}_j\} = \epsilon_{ij}{}^k\mathcal{L}_k, \quad \{\mathcal{P}_i, \mathcal{L}_j\} = \epsilon_{ij}{}^k\mathcal{P}_k. \tag{9.30}$$

The first of these means that the $\mathcal{L}_i$ close as a Lie algebra, which is a subalgebra of the full constraint algebra (itself a larger Lie algebra in this case). The second means that $\mathcal{P}_i$ is a 'good object'—in this case a vector—under the rotations generated by the $\mathcal{L}_i$. $\mathcal{E}$ additionally closes with these gauge constraints, in a manner that establishes it as a scalar under the corresponding transformations.

Example 4) For full GR, the algebraic structure formed by the constraints is (Fig. 9.6.e–h)

$$\{(\mathcal{M}_i\,|\,\iota^i), (\mathcal{M}_j\,|\,\chi^j)\} = (\mathcal{M}_i\,|\,[\iota, \chi]^i), \tag{9.31}$$

[10] Field Theory constraint algebraic structures are most usefully presented in terms of smearing functions, as explained in footnote 1; the undefined symbols in this Sec's presentations of Electromagnetism, Yang–Mills Theory and GR are just such smearings.

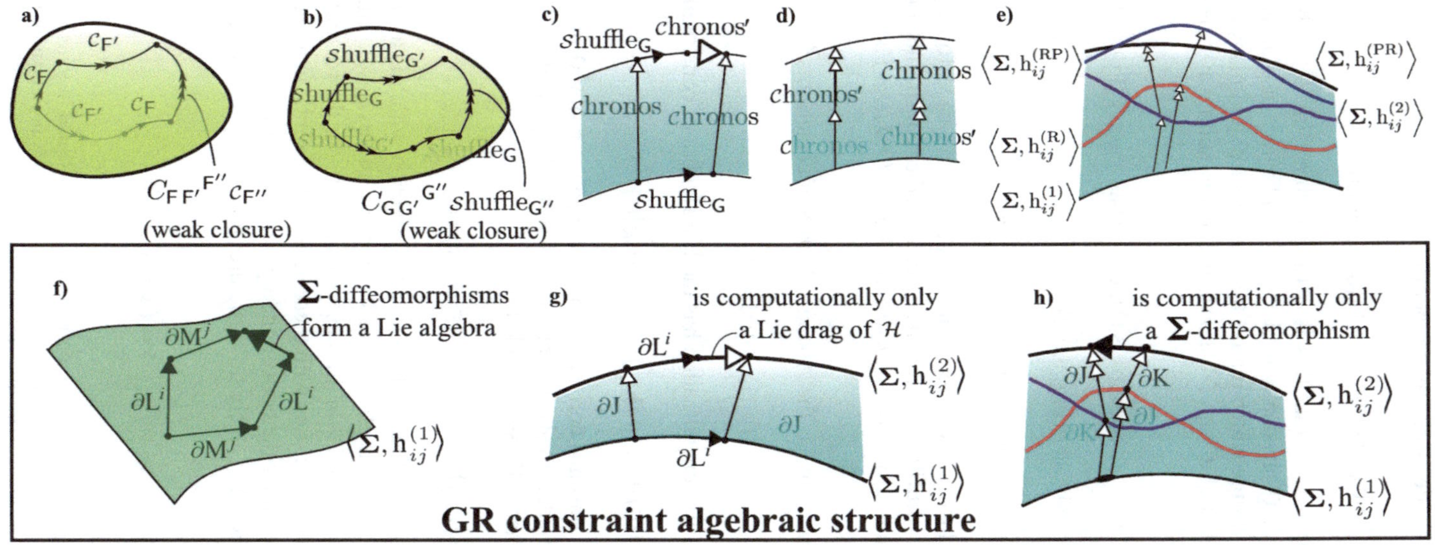

Fig. 9.6 **a)** and **b)** depict the algebraic nature of Constraint Closure for general first-class constraints and shuffle constraints respectively. [Algebra and group commutation relation diagrams are easy to pick out in this book's presentation due to being drawn upon lime-green egg-shaped spaces.] **c)** Chronos as a good **Shuffle** object: these commute up to a possible second use of Chronos. **d)** In Finite Theories, the bracket of two Chronos constraints is strongly zero. **e)** However, in Field Theories, this bracket is in general nonzero. **f)** to **h)** then depict the specific case of GR's Constraint Closure

$$\{(\mathcal{H} \,|\, \mu), (\mathcal{M}_i \,|\, \iota^i)\} = (\pounds_{\underline{\iota}}\mathcal{H} \,|\, \mu), \tag{9.32}$$

$$\{(\mathcal{H} \,|\, \mu), (\mathcal{H} \,|\, \omega)\} = (\mathcal{M}_i \mathrm{h}^{ij} \,|\, \mu \overleftrightarrow{\partial}_j \omega). \tag{9.33}$$

This closes in the sense that there are no further constraints or other conditions arising in the right hand side expressions. Therefore at the classical level for full GR, the Constraint Closure Problem is a *solved* problem.
In more detail, the first Poisson bracket means that *Diff*($\boldsymbol{\Sigma}$) on a given spatial hypersurface themselves close as an (infinite-*d*: Appendix H) Lie algebra. The second means that $\mathcal{H}$ is a good object—a scalar density—under *Diff*($\boldsymbol{\Sigma}$). Both of the above are kinematical rather than dynamical results. The third is however more complicated in both form and meaning [832]. In particular, its right hand side expression containing $\mathrm{h}^{ij}(\mathbf{h}(\underline{x}))$ has the following consequences.

i) The transformation itself depends on the object acted upon, in contrast with the familiar case of the rotations.
ii) The GR constraints form a more general algebraic entity than a Lie algebra: a *Lie algebroid*. More specifically, (9.31)–(9.33) form the *Dirac algebroid* [248, 249].[11]
iii) Also, if one tried to consider $\mathfrak{Riem}(\boldsymbol{\Sigma})$ without *Diff*($\boldsymbol{\Sigma}$) being physically irrelevant, (9.33) would in any case enforce this.
iv) Finally, by not forming a Lie algebra, the constraints—and *Diff*($\mathfrak{m}$, $\mathfrak{Fol}$)—clearly form a structure other than *Diff*($\mathfrak{m}$). Indeed, the vast difference in size between such algebras and algebroids corresponds to the variety of possible foliations.

Example 5) Minisuperspace just has the Abelian constraint algebra

$$\{\mathcal{H}, \mathcal{H}\} = 0. \tag{9.34}$$

This is much simpler than (9.31)–(9.33) because the spatial covariant derivative $\mathcal{D}_i$ now annihilates everything by homogeneity.

9.15 Aspect 4: Assignment of Beables

Given $\boldsymbol{Q}$, $\boldsymbol{P}$, one can also contemplate Taking Function Spaces Thereover. For now, in particular, $\mathfrak{U}$ consists of some class of functions $\mathit{U}(\boldsymbol{Q}, \boldsymbol{P})$ or functionals $\mathit{U}[\boldsymbol{Q}, \boldsymbol{P}]$ (indexed by U). This provides a first notion of observables. At the classical level, this

[11]See Appendix V.6 and [154] if interested in algebroids in general or the Dirac algebroid in particular. Hitherto this has usually been called 'Dirac algebra', though 'Dirac algebroid' is both more mathematically correct and not open to confusion with fermionic theory's Dirac algebra (6.9). Moreover, the Dirac algebroid is manifested even in Minkowski spacetime $\mathbb{M}^n$, upon considering arbitrary spatial hypersurfaces therein [250]; these correspond to fleets of arbitrarily accelerating observers.

is a triviality, but at the quantum level self-adjointness and Kinematical Quantization already impinge at this stage.

For constrained theories, observables are subject to 'forming zero classical brackets with the constraints'. In this way, further constrained notions of observables or beables are a logically posterior consideration to having found constraints, introduced a classical brackets algebraic structure, and Constraint Closure has been established. Observables are more useful physically than just any functions (or functionals) of $\boldsymbol{Q}$ and $\boldsymbol{P}$, due to their containing solely physical information. The Jacobi identity (E.1) applied to two constraints and one observable requires that the input notion of constraints is a closed algebraic structure: Eq. (J.42). Applied instead to one constraint and two observables, the Jacobi identity establishes that observables or beables themselves form a closed algebraic structure: Eq. (J.44). In this sense, observables form an algebraic structure that is associated with the one formed by the constraints.

Let us also use an extension from the notion of observables—which eventually carry nontrivial connotations of 'are observed'—to *beables*, which just 'are'. The latter, denoted $\mathcal{B}_{\mathsf{B}}$, are a somewhat more general notion, so as to cover a number of further 'realist' approaches at the quantum level, as explained in Chap. 50.[12]

The *unrestricted beables* $\mathcal{U}$ are the mathematically simplest notion; finding these at the classical level requires no working whatsoever. At the other extreme, *Dirac beables* alias *Dirac observables* [247] are quantities that (for now weakly classical) brackets-commute (Fig. 9.7) with all of a given theory's first-class constraints:

$$\{\mathcal{C}_{\mathsf{F}}, \mathcal{D}\} \approx 0. \tag{9.35}$$

Various interpretations proposed for these provide further colourful names for these, such as 'evolving constants of the motion' [743], or 'perennials' [404, 405, 587].

As a third alternative, Kuchař introduced [587] another type of observables, which were indeed subsequently termed *Kuchař observables* and to which this book refers to as *Kuchař beables* $\mathcal{K}$ (indexed by K). These are quantities which form zero classical brackets with all of a given theory's first-class linear constraints,

$$\{\boldsymbol{\mathcal{F}\mathrm{lin}}, \mathcal{K}\} \approx 0. \tag{9.36}$$

Whereas $\boldsymbol{\mathcal{G}\mathrm{auge}} = \boldsymbol{\mathcal{F}\mathrm{lin}}$ in the more commonly encountered cases, Chap. 24's counter-examples imply the need for a further notion of $\mathfrak{g}$-beables alias *gauge-*

[12]Moreover, this generalization does not concern a change of definition, but is rather a more inclusive context in which the entities are interpreted. The term 'beables' was originally coined by physicist John Bell [126, 127]. Both his and the Author's use of this word extend to include realist interpretations of QM. However, our uses differ as to what 'realist interpretations' we wish to include; most of those from Bell's day have ceased to be tenable positions, with the ones I discuss specifically in Part III being largely conceptually and technically unrelated to these earlier uses (Sect. 50.4). To a lesser extent, classical whole-universe modelling also motivates use of the beables concept. Finally note that these two motivations combine further in the arena of Quantum Cosmology.

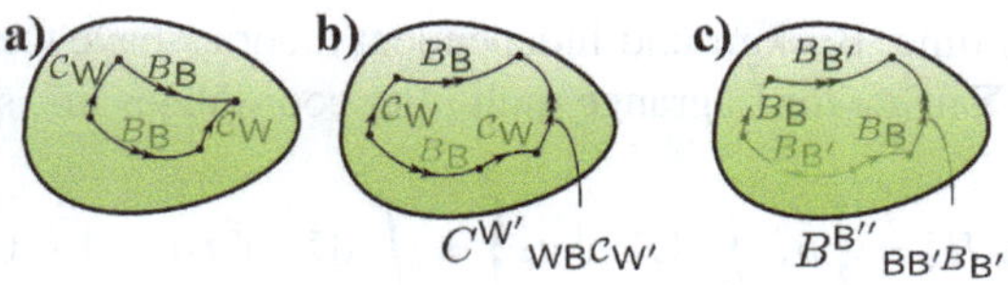

Fig. 9.7 **a)** and **b)** are respectively strong and weak beables conditions for $\mathcal{B}_\mathsf{B}$ corresponding to a constraint subalgebraic structure $\mathcal{C}_\mathsf{W}$. **c)** Beables themselves form an algebraic structure, with structure constants B. Dirac, Kuchař, gauge and Chronos beables each follow this pattern as particular subcases

invariant quantities $\mathcal{G}$ (indexed by J) obeying

$$\{\boldsymbol{\mathcal{G}}\mathbf{auge}, \mathcal{G}\} \approx 0. \tag{9.37}$$

Finally, in cases in which $\mathcal{C}$hronos closes by itself, a notion of Chronos beables $\mathcal{C}$ (indexed by H) becomes meaningful, obeying

$$\{\mathcal{C}\text{hronos}, \mathcal{C}\} \approx 0. \tag{9.38}$$

As specific examples, $\mathcal{D} = \mathcal{K} = \mathcal{G} = \mathcal{U}$ for unconstrained theories. For Minisuperspace and Spatially-Absolute Mechanics, the $\mathcal{K} = \mathcal{G} = \mathcal{U}$ are also trivially any quantities of the theory since these theories have no linear constraints at all, but the $\mathcal{D} = \mathcal{C}$ are nontrivial due to the presence of the constraint $\mathcal{C}$hronos. For Electromagnetism and Yang–Mills Theory $\mathcal{D} = \mathcal{K} = \mathcal{G} \neq \mathcal{U}$, since these just have first-class linear constraints which are gauge constraints. For RPMs more generally, we shall see that the $\mathcal{K} = \mathcal{G}$ are nontrivial, and include pure shapes and scales; these, the $\mathcal{U}$, $\mathcal{C}$ and $\mathcal{D}$ are all mutually distinct notions. For GR, $\mathcal{U} \neq \mathcal{K} \neq \mathcal{D}$, the $\mathcal{C}$ are undefined due to $\mathcal{H}$ not closing by itself, whereas the $\mathcal{K}$ include, formally, the 3-geometries themselves.

Physicist Carlo Rovelli's *partial observables* [752] do not require commutation with any constraints, so they are a particular interpretation of unconstrained observables. For a constrained system, these contain unphysical information; however one is to consider correlations between pairs of them that are physical. One often imagines each as being measured by a localized observer, so this approach usually uses the term 'observables' rather than 'beables'. Moreover, via these correlations, this scheme also eventually involves a notion of 'complete observables' that is similar to Dirac's notion of observables.

Finally, the *Problem of Beables*—more usually termed 'Problem of Observables', and which is Facet 4) of the Problem of Time—is that it is hard to construct a set of beables, in particular for Gravitational Theory. More specifically, Dirac observables or beables are harder to find than Kuchař ones (Chap. 25), and the quantum counterparts of each are even harder to find than classical ones (Chap. 50). The Dirac case are sufficiently hard to find for full GR that Kuchař [587] likened strategies relying on having already obtained a full set of these to plans involving having already caught a Unicorn. As regards this issue indeed being related to Background

Independence and time, Background Independent theories have total Hamiltonians of form $H = \int_{\Sigma} \mathrm{d}\boldsymbol{\Sigma}\, \mathrm{m}^{\mathsf{F}} \mathcal{C}_{\mathsf{F}}$ for Lagrange multiplier coordinates m^{F}, so that

$$\frac{\mathrm{d}D}{\mathrm{d}t} = \{D, H\} = \left\{D, \int_{\Sigma} \mathrm{d}\boldsymbol{\Sigma}\, \mathrm{m}^{\mathsf{F}} \mathcal{C}_{\mathsf{F}}\right\} = \int_{\Sigma} \mathrm{d}\boldsymbol{\Sigma}\, \mathrm{m}^{\mathsf{F}} \{D, \mathcal{C}_{\mathsf{F}}\} \approx 0. \tag{9.39}$$

This may appear to manifest frozenness—in the form of observables or beables being unable to change value—though Chap. 32.6 reveals straightforward interpretations along these lines to be fallacious [724].

Chapter 10
Classical-Level Background Independence and the Problem of Time. ii. Spacetime and Its Interrelation with Space

GR has more Background Independence aspects than theories of Mechanics [41]. This is because GR possesses a nontrivial notion of spacetime, which geometrizes a wider range of features than Mechanics' notion of split space-time does. The latter is far more of a composition of separate notions of space and time: multiple copies of a spatial geometry strung together by a time direction, whereas the former is a co-geometrization of space and time. GR spacetime also possesses its own versions of Generator Providing Relationalism, the corresponding Generator Closure, and observables as commutants associated with these generators.

10.1 Aspect 5: Spacetime Relationalism

Let us start afresh, now with primality ascribed to spacetime rather than (as in Chap. 9) to space, configuration or Dynamics. GR-like spacetime's own Relationalism then takes the following form.

STR-i) There are no background spacetime structures; in particular there are no indefinite-signature background spacetime metrics. Fixed background spacetime metrics are also more well-known than fixed background space metrics.

STR-ii) Consider not just a spacetime manifold $\mathfrak{m}$ but also a $\mathfrak{g}_S$ of transformations acting upon $\mathfrak{m}$ that are taken to be physically redundant.

For GR, $\mathfrak{g}_S = \mathit{Diff}(\mathfrak{m})$. $\mathfrak{m}$ can additionally be equipped with matter fields in addition to the metric. STR-i) can then be extended to include no background internal structures associated with spacetime; note the difference between these and structure on spacetime, in direct parallel to Sect. 9.8's distinction. STR-ii)'s $\mathfrak{g}_S$ can furthermore have a part acting internally on a subset of the fields. The internal part of STR-ii) is then closer to the standard spacetime presentation of Gauge Theory (Chap. 6) than the internal part of Configurational Relationalism is. On the other hand, Configurational Relationalism is more closely tied to Dirac observables or beables, since these are configuration-based notions.

E. Anderson, *The Problem of Time*, Fundamental Theories of Physics 190,
DOI 10.1007/978-3-319-58848-3_10

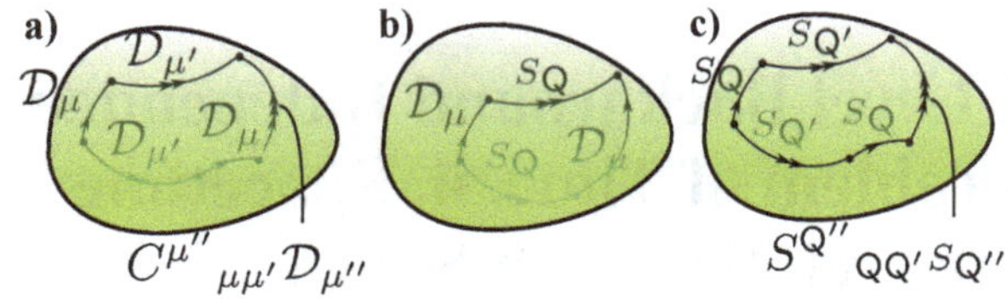

Fig. 10.1 **a)** Spacetime diffeomorphisms close as a Lie algebra. **b)** The (strong case of) spacetime observables condition. **c)** Spacetime observables themselves close as an algebraic structure

10.2 Closure of *Diff*($\mathfrak{m}$)

The *Diff*($\mathfrak{m}$) indeed straightforwardly form a Lie algebra, in parallel to how *Diff*($\boldsymbol{\Sigma}$) does:

$$|[(\mathcal{D}_\mu \,|\, X^\mu), (\mathcal{D}_\nu | Y^\nu)]| = (\mathcal{D}_\gamma \mid [X, Y]^\gamma). \tag{10.1}$$

and Fig. 10.1.a). [,] is here the differential geometric commutator of two vectors. *Diff*($\mathfrak{m}$) also shares further specific features with *Diff*($\boldsymbol{\Sigma}$), such as its right hand side being of Lie derivative form. So all three kinds of Relationalism considered up to this point are implemented by Lie derivatives.

Some differences are that whereas the generators of *Diff*($\boldsymbol{\Sigma}$) are conventionally associated with dynamical constraints, those of *Diff*($\mathfrak{m}$) are not. *Diff*($\boldsymbol{\Sigma}$)'s—but not *Diff*($\mathfrak{m}$)'s—Lie bracket is moreover conventionally taken to be a Poisson bracket.

10.3 Further Detail of This Book's Concepts and Terminology

Let us next consider the nomenclature 'absolute', 'relational' and 'background-(in)dependent'. Physicist Domenico Giulini [362], building upon James L. Anderson's precedent [12, 13], defines 'absolute' and 'Background Dependent' to be exactly the same notion; see Sect. 27.8 for details of how this is characterized. As given, this applies to what the Author terms Spacetime Nonrelationalism, though it can be extended to Spatial and Temporal Nonrelationalism as well. On the other hand, the Author identifies classical Background Independence as the multi-aspect precursor of the multi-faceted classical Problem of Time. Relationalism—viewed as the triple of Temporal, Configurational and Spacetime Relationalisms[1]—is a portion of the preceding. Because of this, the Author takes on board Giulini's conceptualization, but re-names his 'Background Dependence' as 'Nonrelationalism' (giving an 'expected synonym' absolute = nonrelational). Additionally, the Author continues to define Background (In)dependence in the more general way explained in Sects. 9 and 10.

Moreover, Giulini's definition is not straightforward, nor even claimed to be a completed item, much less one that is adhered to in other parts of the literature,

[1] See [38] for yet further notions of Relationalism.

where yet other distinctions between uses of 'absolute' and 'Background Dependent' can be found. Furthermore, let us caution that Chaps. 9 and 10's '8-aspect' classical Background Independence[2] refers to 'differentiable structure through to metric-level Background Independence', as does much other literature in making unqualified reference to Background Independence. This limitation is lifted in Epilogue II.C.

The preceding Sec's *Diff*($\mathfrak{m}$) is to be interpreted actively—point-shuffling transformations—as opposed to passively (changing coordinates). In setting up GR, Einstein originally placed stock in General Covariance, which has passive connotations. However, Kretschmann pointed out that any theory could be cast in generally covariant form. On the other hand, active diffeomorphisms continue to play a foundational role in the *physics* of GR (below). This can cause confusion because there is a *mathematical* sense in which active and passive diffeomorphisms are equivalent (see Appendix D.2). The claim, however, is that there is physical distinction between conceptualizing in terms of passive and active diffeomorphisms, with the latter being a more insightful position [483, 752]. Understanding this requires more detailed examination of the active diffeomorphisms in the GR context.

Let us start by setting $\langle \mathfrak{M}, \mathbf{m} \rangle = \langle \mathfrak{m}, \mathbf{g} \rangle$ in Appendix D.4. Then, as Isham comments, *"Invariance under such an active group of transformations robs the individual points in $\mathfrak{m}$ of any fundamental ontological significance"* [483]. To further understand what this means for GR spacetime in the presence of matter fields, Isham continues as follows. *"For example, if s is a scalar field on $\mathfrak{m}$ the value $s(X)$ at a particular point $X \in \mathfrak{m}$ has no invariant meaning"*. See also in this regard the 'hole argument' in the Philosophy of Physics literature [275, 808], though we caution that this argument has many other parts which this paragraph does not refer to.[3]

N.B. next that the Einstein field equations are invariant under the group of spacetime diffeomorphisms *Diff*($\mathfrak{m}$). As a first point, this is to be contrasted with SR, for which the Poincaré invariance group is much smaller. The main issue, however, is that for $\mathfrak{m}$ and $g_{\mu\nu}(\vec{X})$, $\widetilde{g}_{\mu\nu}(\vec{X})$ two metrics which solve Einstein's field equations, the expression that the two metrics are related by a diffeomorphism [locally $\phi : X^\mu \rightarrow \phi^\mu(\vec{X})$] is

$$\widetilde{g}_{\mu\nu}(\vec{X}) = \frac{\partial \phi^\rho}{\partial X^\mu} \frac{\partial \phi^\sigma}{\partial X^\nu} g_{\rho\sigma}(\phi(\vec{X})). \tag{10.2}$$

Active diffeomorphism invariance of the theory amounts to diffeomorphisms ϕ being guaranteed to map *solutions to solutions*. This property clearly continues to hold

[2]This is usually referred to as just metric-level Background Independence, though this does not do justice to how much of its content rests upon notions of diffeomorphism, which themselves are meaningful down to the level of differentiable structure.

[3]E.g. a hole argument was used soon after GR's inception—in Kretschmann's criticism of Einstein—so as to overrule attributing physical significance to what are now known as passive diffeomorphisms. Indeed, discussion of the hole argument started before a sharp active–passive distinction had been made, then served as one reason to make such a distinction, and only subsequently became an argument focusing upon the role of the active diffeomorphisms.

even if the theory is formulated in a coordinate-independent manner (a notion explained in Appendix D). Thus this is conceptually unrelated to spacetime coordinate transformations. So whereas any theory can be recast in a form which is invariant under passive spacetime diffeomorphisms, active spacetime diffeomorphism invariance is a property of theories themselves. This is a feature possessed by metric-level Background Independent theories such as GR, but not by dynamical Field Theories that live upon fixed backgrounds [337].

Now that diffeomorphisms and their interplay with the equations of motion have been introduced, we proceed closely following [12, 13, 362] via the following definitions.

Definition 1) An equation of motion on $\mathfrak{m}$ is *(spacetime) diffeomorphism invariant* iff (if and only if) *Diff*($\mathfrak{m}$) is a permitted invariance group for it.
Definition 2) Any field which is either non-dynamical, or whose solutions are all locally diffeomorphism equivalent, is an *absolute structure*.
Definition 3) Finally, a criterion for a theory to be *(spacetime) Background Independent* is iff its equations are *Diff*($\mathfrak{m}$)-invariant as per Definition 1), and its fields do not include absolute structures as per Definition 2). [In this book, however, we consider a more general multi-aspect notion of Background Independence.]

Let us next consider some corresponding statements about *Diff*($\boldsymbol{\Sigma}$) for Geometrodynamics. Here the role of the Einstein field equations as equations to be solved is replaced by just $\mathcal{M}_i$. Moreover, solutions are now *pairs* $(\mathbf{h}, \mathbf{K})$ or $(\mathbf{h}, \mathbf{p})$. The next parallel concerns $\phi \in$ *Diff*($\boldsymbol{\Sigma}$) mapping solutions of the form e.g. $(\mathbf{h}_1, \mathbf{K}_1)$ to solutions $(\mathbf{h}_2, \mathbf{K}_2)$. Again, statements involving this hold even if the theory is formulated in a coordinate-independent manner, so this is conceptually unrelated to spatial coordinate transformations. So whereas any theory can be recast in a form invariant under passive spatial diffeomorphisms, active spatial diffeomorphism invariance is a property of certain theories themselves. This is a feature possessed by metric-level Background Independent theories such as GR as Geometrodynamics, but not by any dynamical Field Theories upon fixed backgrounds.

We end by pointing out that the *Diff*($\mathfrak{m}$) algebra (10.1) and GR known to respect *Diff*($\mathfrak{m}$)-invariance, Spacetime Relationalism is a resolved problem at the classical level. Upon solving the Einstein field equations, the resulting Lorentzian metric on $\mathfrak{m}$ provides meaning to each of timelike-, null- and spacelike-separated, and to causality [483]. This provision of meaning holds notwithstanding of these notions themselves not being preserved by *Diff*($\mathfrak{m}$). Involving *Diff*($\boldsymbol{\Sigma}$) at the classical level is even more straightforward. As per Chap. 12, however, diffeomorphism invariance at the quantum level is not at all straightforward; indeed this is a major unresolved part of the Problem of Time.

10.4 Spacetime Observables

Diff($\mathfrak{m}$) is closely related to *spacetime observables* s_{Q} in GR. These are functions on spacetime which are manifestly *Diff*($\mathfrak{m}$)-invariant, i.e. obeying

$$|[(\mathcal{D}_\mu \,|\, \mathrm{Y}^\mu), (s_\mathsf{Q} \,|\, \mathrm{Z}^\mathsf{Q})]| = 0 \tag{10.3}$$

for smearing variables Y^μ and Z^Q. The last sentence of Sect. 10.2 furthermore implies that there is conventionally no complete spacetime analogue of the previous Chapter's notion of beables or observables. These differences stem from time being ascribed further distinction in dynamical and then canonical formulations, as compared with spacetime formulations. (10.1) is to be additionally contrasted with the Dirac algebroid (9.31)–(9.33). Clearly there are two very different algebraic structures that can be associated with GR spacetime. The first is associated with unsplit spacetime, and the second with split spacetime including keeping track of how it is split; see Sect. 27.5 for further details about spacetime observables.

10.5 Classical-Level Background Metrics

For now, let us consider the split

$$\mathrm{g}_{\mu\nu} := \eta_{\mu\nu} + k\,\gamma_{\mu\nu} \tag{10.4}$$

(or with some other background metric $\mathrm{g}^0_{\mu\nu}$ in place of $\eta_{\mu\nu}$). Here k is mathematically a perturbative ϵ and physically proportional to the fundamental constant combination $\sqrt{G}/c$. Introducing this split brings in both [477] a background topological manifold $\mathfrak{m}$ and a background metric $\eta_{\mu\nu}$ which includes a background causal structure. This approach is invariant under the infinitesimal spacetime diffeomorphisms,

$$\gamma_{\mu\nu} \longrightarrow \gamma_{\mu\nu} + 2\,\partial_{(\mu}\xi_{\nu)}. \tag{10.5}$$

One problem here is that the notion of spacelike with respect to $\mathrm{g}_{\mu\nu}$ does not in general coincide with that with respect to $\eta_{\mu\nu}$. $\mathrm{g}_{\mu\nu}$ is here split into $\eta_{\mu\nu}$ the provider of causality (which however becomes obsolete in this role) and fundamental variable $\gamma_{\mu\nu}$ (which is not however observable).

For contrast, the spatial split

$$\mathrm{h}_{ij} = \delta_{ij} + k\,\mathrm{f}_{ij} \tag{10.6}$$

introduces both a background topological manifold $\mathbf{\Sigma}$ and a background metric δ_{ij}. This approach is invariant under infinitesimal spatial diffeomorphisms

$$\mathrm{f}_{ij} \longrightarrow \mathrm{f}_{ij} + 2\,\partial_{(i}\xi_{j)}. \tag{10.7}$$

This case is less severe due to the lack of signature and causal structure.

10.6 Paths and Histories Strategies

Here one considers finite paths instead of instantaneous changes. Histories, moreover, carry further connotations than paths; for now at the classical level, these

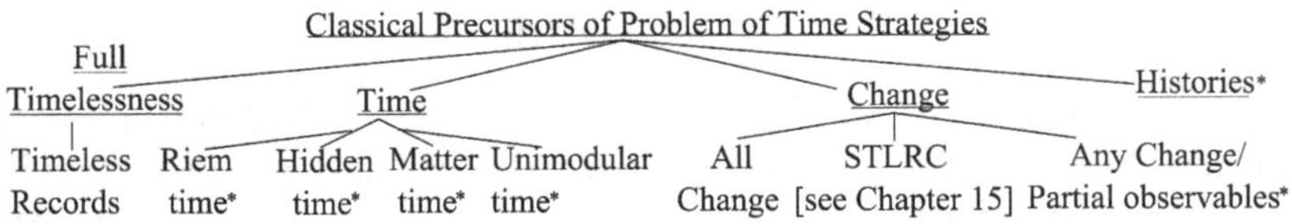

Fig. 10.2 Web of the various types of strategy and their relations. * indicates the 6 out of the 10 strategies covered by Kuchař and Isham's reviews [483, 586] that can be taken to start at the classical level. Cf. Fig. 12.1 for the full 10 of these and the quantum-level developments since

possess their own conjugate momenta and brackets. Histories Theory has a mixture of spacetime properties and canonical properties, and has more quantum- than classical-level motivation.

10.7 Web of Classical Problem of Time Strategies

Figure 10.2's branches reflect the long-standing philosophical fork between 'time is fundamental' and 'time should be eliminated from one's conceptualization of the world'. Approaches of the second sort are to reduce questions about 'being at a time' and 'becoming' to mere questions of 'being'.

A finer classification [24, 37, 483, 586] of the strategies is into Time before Quantum, Time after Quantum, Timelessness, not Time but History, and not time but change.

10.8 Aspect 6: Foliation Independence

GR spacetime admits multiple foliations. At least at first sight, this property is lost in the geometrodynamical formulation.

Foliation Dependence is a type of privileged coordinate dependence. This runs against the basic principles that GR contributes to Physics. Conversely, *Foliation Independence* is an aspect of Background Independence, and the *Foliation Dependence Problem* is the corresponding Problem of Time facet. This issue clearly involves time since each foliation by spacelike hypersurfaces is dual to a GR time-function. Moreover, *Refoliation Invariance* is encapsulated by evolving via each of Fig. 10.3.e)'s red and purple hypersurfaces giving the same physical answer as regards the final hypersurface. So whereas Foliation Independence is a matter of freedom in how to strut spatial hypersurfaces together, Refoliation Invariance instead concerns passing between such struttings.

The space–time split of GR spacetime has been shown to possess Refoliation Invariance ([573, 832], Chap. 31). By this property, GR spacetime is not just a single strutting together of spaces like Newtonian space-time is (Fig. 2.1.c). GR spacetime manages instead to be many such struttings at once in a physically mutually consistent manner (Fig. 10.3.b). Indeed, this is how GR is able to encode consistently the experiences of fleets of observers moving in whichever way they please.

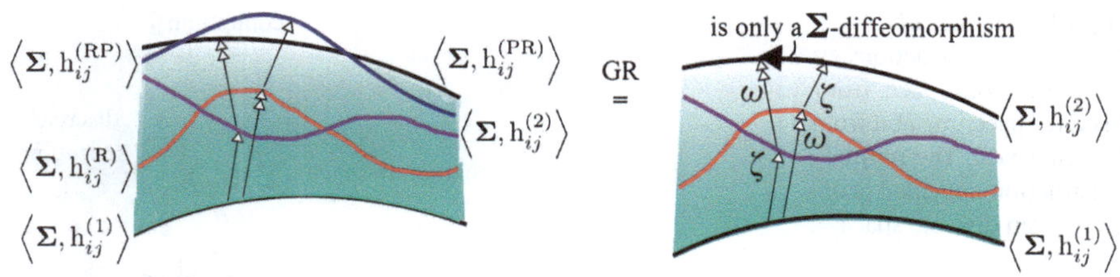

Fig. 10.3 The Foliation Dependence Problem is encapsulated by whether evolving from an initial hypersurface via the *red* (R) hypersurface produces the same final-hypersurface physical answer as evolving via the *purple* (P) hypersurface. A priori, this involves forming the *left hand side*'s 'commutator pentagon' of hypersurfaces. However for GR the two end-product hypersurfaces coincide (PR = 2 = RP) because of the form of (9.33)

Refoliation Invariance compares triples of hypersurfaces. In both cases, one starts from the same hypersurface and subsequently applies the same two operations, but in opposite orders in each case. The question is then whether the outcome of these two different orders is the same (Fig. 10.3). Moreover, one can see that this is in direct correspondence with the commutator pentagon of Fig. E.1.a). Since the individual operations involved are actions of $\mathcal{H}$, one is led to the commutator of two $\mathcal{H}$'s. Then indeed, as physicist Claudio Teitelboim pointed out [832], the form of this part (9.33) of GR's Dirac constraint algebroid guarantees Refoliation Invariance. This is achieved by the two end hypersurfaces coinciding up to a diffeomorphism of that hypersurface, as per the right hand side of (9.33). This constitutes the '*Refoliation Invariance Theorem of GR*'.

Chapter 31 further develops embeddings, slices and foliations as more advanced foundations for the ADM split (which assumes spacetime). Chapter 32 proceeds to consider a first answer to Wheeler's question (9.1) given by physicist Sergio Hojman alongside Kuchař, and Teitelboim [454]. They obtained the form of $\mathcal{H}$ by assuming as their first principles the *deformation algebroid* of the two operations in Fig. 10.3.c)–d) for a hypersurface; this takes the same form as the Dirac algebroid. This approach does however still presuppose spacetime, now in the more specific sense of embeddability into spacetime.

Chapter 32 additionally considers applications of the Foliation Formulation to observables, and as a building block for various Internal Time, Matter Time and Histories Approaches. Moreover, one most usually makes a choice to work with one of split or unsplit spacetime. A few approaches to Background Independence and Quantum Gravity, however, involve both at once [566, 768]. In this case, all of Temporal, Spatial and Spacetime Relationalism are manifested together.

10.9 Aspect 7: Spacetime Constructability

Let us next consider assuming less structure than is present in GR's notion of spacetime. In general, if classical spacetime is not assumed, one needs to recover it in a

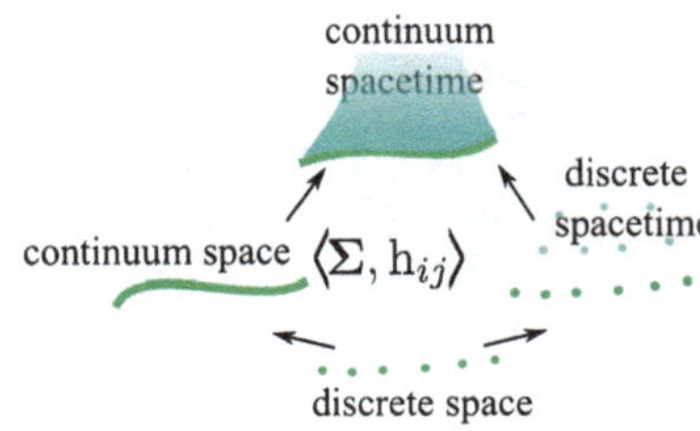

Fig. 10.4 3 types of Spacetime Construction: spacetime from space, from discrete spacetime and from discrete space. The fourth construction indicated is of space from discrete space

suitable limit. This can be a hard venture; in particular, the less structure is assumed, the harder it is. Some quantum-level motivation for this due to Wheeler [899] is outlined in Sect. 12.12. This aspect was originally known as 'Spacetime Reconstruction', though the Author takes this name to be too steeped in assuming spacetime primality. Thus we use instead the terms 'Spacetime Constructability' for the Background Independence aspect, 'Spacetime Construction Problem' for the Problem of Time facet in cases in which this is blocked, and 'Spacetime Construction' for corresponding strategies. Moreover, already at the classical level, Spacetime Constructability can be considered along two logically independent lines.

A) *From space*, as an 'embed rather than project' 'inverse problem' to the previous Section's, which is harder since now only the structure of space is being assumed.
B) *From making less assumptions about continua.*

A) and B) combine, incipiently to give a total of four construction procedures (Fig. 10.4). We further expand on this number in the sense of 'less layers of mathematical structure assumed' in Fig. 10.9.

For now, we concentrate on A). This provides a second answer to Wheeler's question (9.1). The first answer—Hojman, Kuchař and Teitelboim's Deformation Approach [454]—assumes embeddability into spacetime,. The second answer—the Relational Approach of Barbour, physicists Brendan Foster and Niall ó Murchadha and the Author [62, 109]—however, goes further by not assuming spacetime. This approach is based, rather, on 3-spaces σ in place of hypersurfaces Σ, and starts from Temporal and Configurational Relationalism first principles. It proceeds by using the Dirac Algorithm on a more general $\mathcal{H}_{\mathrm{trial}}$ obtained as the $\mathcal{C}$hronos from a more general $\mathsf{S}_{\mathrm{trial}}$ The combination of GR's particular $\mathcal{H}$ alongside local Lorentzian Relativity and embeddability into GR spacetime then arises as one of very few consistent possibilities. The few alternatives to this arising in this working differ substantially in causal structure and as regards whether they admit Refoliation Invariance. Indeed, as Chap. 33 details, these few alternatives are foundationally interesting through having, in turn, local Galilean-type Relativity, local Carrollian Relativity, and a privileged CMC foliation corresponding to York time. Note that these now arise *from the Dirac Algorithm* as the choice of factors among which one needs to vanish in order to avoid the constraint algebroid picking up an obstruction term. This is substantially different from the form of Einstein's dichotomy between universal local Galilean or Lorentzian Relativity!

Finally, emergent constructed spacetime's Relationalism, kinematics and Refoliation Invariance are cast in a Temporal Relationalism incorporating form in Chap. 34. Thereby, all facets of A Local Classical Problem of Time are addressed within the space-or-configuration primary Relational Approach.

10.10 Model Arenas, Diffeomorphisms and Slightly Inhomogeneous Cosmology

Whereas Minisuperspace exhibits '8 out of 9' of the aspects of Background Independence, we have also explained how homogeneity implies that several of the consequent Problem of Time facets are *very quickly resolved*. On the other hand, RPMs exhibit '6 out of 9' of the aspects of Background Independence [37]. This includes the Configurational Relationalism aspect that Minisuperspace does not possess, but three other facets are blocked from appearing by the absence of spacetime in this model.

Moreover, while the RPM and Minisuperspace cases are simple to calculate with, they miss the subtleties specifically associated with diffeomorphisms [483, 586]. The diffeomorphism-specific facets are the Thin Sandwich version of Configurational Relationalism, Spacetime Relationalism, further specifics about the Problem of Beables, and the Foliation Dependence and Spacetime Construction Problems.

A first arena[4] in which these appear nontrivially is Slightly Inhomogeneous Cosmology (SIC). A particular such involves inhomogeneous perturbations about the spatially-$\mathbb{S}^3$ Minisuperspace with single scalar field matter model of Sect. 9.9. At the level of Semiclassical Quantum Cosmology, this particular case becomes a *Halliwell–Hawking type model* [35, 419], named in part after physicist Jonathan Halliwell; this is the current book's choice of most complicated recurring model arena. RPMs and Minisuperspace complementarily support by one or the other having all *other* Background Independence aspects and consequent Problem of Time facets of this model. Here one considers the first few (usually two) orders of the perturbation of the metric. Each of these form a simplified $\mathfrak{q}$ in place of the full $\mathfrak{Riem}(\boldsymbol{\Sigma})$. Chapter 30 subsequently shows that SIC already exhibits the Thin Sandwich Problem—which happens to be solvable in this case—and also considers this

[4]For those interested, yet other model arenas of comparable complexity to those in this book include $2+1$ Gravity [193] and Ex III.13, the parametrized particle [586], parametrized Field Theories [584, 586], Strong Gravity—the strong-coupled limit of GR, of relevance near singularities—[472, 716, 717] and this book's index, the bosonic string as a model arena of Geometrodynamics [568, 586, 594]. Model arenas for Loop Quantum Gravity include Electromagnetism and Yang–Mills Theory [330], the Husain–Kuchař model—which has $\mathcal{G}_I$ and $\mathcal{M}_i$ without an $\mathcal{H}$—[461]

and BF theory—a type of Topological Field Theory— [330, 663] and this book's index. More complex models include various types of Midisuperspace [95]: Einstein–Rosen cylindrical gravitational waves [572], spherically-symmetric Midisuperspace models [588], and spatially $\mathbb{S}^3$ or $\mathbb{T}^3$ Gowdy cosmologies [131, 132]. The above models are in part named after physicists Viqar Husain, Nathan Rosen and Robert Gowdy; BF is named after analogues of the electromagnetic **B** and **F** fields.

model arena from a Histories Theory perspective. See Fig. 10.5 for a facetwise comparison of SIC, Minisuperspace and RPM.

10.11 Summary so Far: Seven Gates

Figure 10.6 depicts the current Chapter's Problem of Time facets as gates, expanding on a quantum-level presentation of Kuchař's[587]. For most of these gates, this and the previous Chapter also supplied a simple classical-level means of passage (at least formally, and piecemeal).

10.12 Frontiers

Facet Interference However, the Devil is in the detail. *There is a strong tendency for the Background Independence aspects—and consequent Problem of Time facets—to interfere with each other rather than standing as independent obstacles* [483, 586]. The main point of the parable of the gates is that, in dealing with time in QG, going through a further gate has a major propensity to leave one outside of gates that one had previously passed through. This is due to the Problem of Time facets bearing rich conceptual and technical relations amongst themselves due to their arising from a joint cause. I.e. the need to bridge the gap between Background Dependence and Background Independence groups of Paradigms of Physics. In [24, 37], the Author portrayed this jointness in the form of the facets being an Ice Dragon, whose different body parts which coordinate in defense when one confronts it; the 'ice' itself alludes to the Frozen Formalism Facet. By this jointness of cause, it is likely to be advantageous to indeed treat the facets as parts of a coherent package rather than disassembled into mere piecemeal problems. Moreover, addressing this matter requires many reconceptualizations, to which a Kuchař-type enchanted castle gates parable is more robust, so the current book uses the latter. These facet interference difficulties largely lie outside the scope of Part I's introductory outline; they are however the *raison d'être* for Parts II and III.

Further Aspects and Consequent Facets Some of the above aspects and corresponding facets meaningfully subdivide. For instance, the notions of closure and observables can be unpackaged as per the first four rows of Fig. 10.7. This untangles some of the various roles played by observables or beables, as well as pointing to further notions of such. For instance, there are 'Chronos beables': quantities c (indexed by H) that commute with $\mathcal{C}$hronos but not necessarily with $\mathcal{S}$**huffle**, which exist for theories in which $\mathcal{C}$hronos closes by itself as an algebraic structure.

The aspects and corresponding facets considered so far are not, however, a complete set. For instance, Chap. 9.3's 'spacetime versus space' dilemma as regards ontological primality is not exhaustive, since it tacitly favours $3+1$ formulations

Facet / Model	Temporal Relationalism	Configurational Relationalism	Problem of Beables	Constraint Closure Problem	Spacetime Relationalism	Foliation Dependence Problem	Spacetime Construction Problems	Multiple Choice Problems	Global Problems of Time
RPM	√	√	√	√	absent	absent	absent	√	√
Minisuperspace	√	trivial	Kuchař -trivial by homogeneity	trivial (single, homomogenity)	trivial (homogeneity)	trivial (homogeneity)	trivial (homogeneity)	√	√
SIC	√	√	√	√	√	√	√	√	√

Fig. 10.5 Which model arenas exhibit which Problem of Time facets. Beyond this, approaches based on mastering the diffeomorphisms are widely blocked from progress. This figure is useful at the quantum level as well

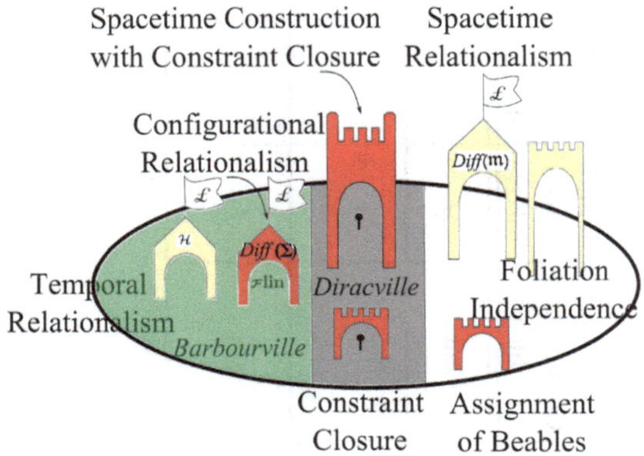

Fig. 10.6 A Local Resolution of the Problem of Time is to pass consistently through facets 1) to 7), depicted, following Kuchař, as castle gates. We elaborate on this not only as a parable but as a picture as well, by grouping and decorating the gates so as to indicate a number of significant subsets of the gates. The significant subsets of gates indicated zonally are, firstly, the *green* 'Barbourville', consisting of Temporal and Configurational Relationalism. Secondly, the *grey* cobbled 'Diracville': the domain of the Dirac Algorithm, in which the gates are additionally depicted with keyholes. Spacetime versus space is indicated with tall and short gates respectively. Relational facets are depicted as pointy-topped gates, each admitting a Lie implementation as marked by a flag. All gates have an algebraic element to either their definition [$Diff(\mathbf{\Sigma})$ and $Diff(\mathfrak{m})$ as indicated], or to their classical resolution (listed below). Four of these algebraic structures involve Poisson brackets; these are indicated by shading in *red*.
Some means of passage through classical-level gates are as follows. Emergent Machian time gets one past Temporal Relationalism. Best Matching lets one through Configurational Relationalism; Sect. 14.4 provides a more widely applicable means of passage in this case. Nontrivial termination of the Dirac Algorithm unlocks the Constraint Closure gate, and likewise in the subcase in which termination additionally provides Spacetime Construction for the taller double-gate version. This involves the Dirac algebroid, $Diff(\mathfrak{m}, \mathfrak{Fol})$, in the first case, and this being singled out from a larger family of algebroids in the second; these are the keys that fit the corresponding keyholes. Teitelboim's Refoliation Invariance depiction of the Dirac algebroid's bracket $\{\mathcal{H}, \mathcal{H}\}$ secures passage through the Foliation Dependence Problem gate. Moreover, the preceding trio are *jointly* resolved by the Dirac algebroid in the case of classical GR. The Problem of Beables is to be resolved by finding an algebraic structure of beables associated with the Dirac algebroid

over $1 + 3$ ones. A trilemma of spacetime, space–time split spacetime $(3 + 1)$ or time–space split spacetime $(1 + 3)$ primalities is more exhaustive. This extends the 8-fold Fig. 9.1 into the 14-fold Fig. 10.8. Then much as the space–time split primary formulation contributes threefold in the first three rows of Fig. 10.7, the time–space split primary formulation contributes threefold in the last three rows. An additional multiplicity arises from threadings and histories not being expected to involve the same brackets structure (Fig. 10.8). [Epilogue II.A continues this discussion by considering null splits as well.]

Aspect 8) Global Validity The underlying Background Independence aspect is that *we would prefer that all our notions of Background Independence are globally well-defined.*

		Provider	Algebraic closure	Associated algebraic structure
3 + 1	time	Temporal Relationalism → chronos, t^{em}	chronos does not close	{chronos, c} ≈ 0
	space	Configurational Relationalism → *shuffle*	*shuffle* closes	{*shuffle*, κ} ≈ 0
	space-time	Deformation-Shuffle relationalism [or just the above together]	chronos, *shuffle* do not close	{chronos and *shuffle*, D} ≈ 0
4-d	spacetime	Spacetime Relationalism	*Diff*($\mathfrak{m}$) closes	$\|[\mathcal{D}, \mathcal{S}]\|$ ≈ 0

Fig. 10.7 A further subdivision of the classification of Background Independence aspects leading to Problem of Time facets. One can also imagine a 1 + 3 split counterpart of the first three rows, which further extend this to include the further inputs envisaged in Fig. 10.8. Moreover, much as the 1 + 3 threading PDEs are distinct from the 3 + 1 ADM ones: each formulation has its own realization of conjugate momenta and brackets

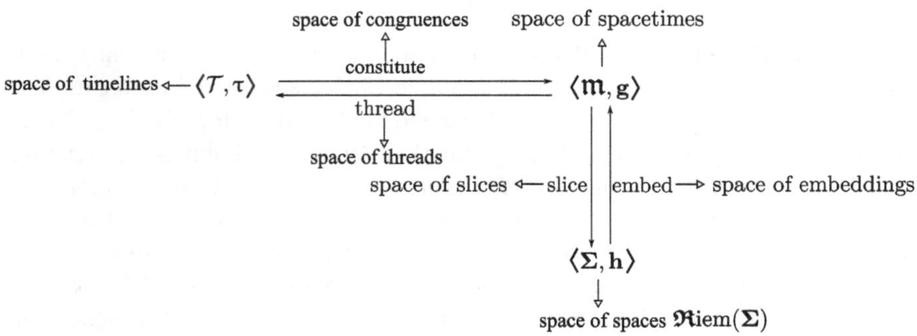

Fig. 10.8 Threading formulations augment the number of aspects and facets to 14

Wherever this fails, Facet 8) *Global Problems of Time* arises. This used to be referred to in the singular; e.g. Kuchař [586] considered the part of this directly pertaining to timefunctions; this is already visible in point 1) of Sect. 7.7. However, global issues are legion. Timefunctions may be only locally defined in space, or only locally valid in time itself; one can add 'in spacetime', 'in $\mathfrak{q}$'... to this list (Epilogue II.B). Moreover, most of the other facets *and* strategies can be beset by global issues. Another classification of Global Problems of Time is into effects involving meshing conditions of charts on manifolds versus the more involved patching of PDE solutions. The first is mathematically basic (Appendix D.2) whereas the second is not (Appendix O). Epilogue II.B considers further classifications and details of a selection of classical Global Problems. A number of open problems requiring more advanced mathematics arise from the above considerations and their even more extensive quantum-level counterparts.

Global validity approximately doubles the total number of aspects:

$$\text{(primary entities)} \times \text{(provider + algebras)} \times \text{(local + global)}$$
$$= \{3+1+3\} \times \{1+1+1\} \times 2 = 42. \tag{10.8}$$

Not all of these aspects are always present, moreover, due to some amounting to choices of perspective or only being realized for some cases of algebraic structure formed by the generators. This count does not yet include Global Problems with the strategies themselves, nor does it cover the Histories Theory counterparts of the aspects. Nor does it preclude there being dualities between some of the features of $3+1$ and $1+3$ splits.

Aspect 9) No Unexplained Multiplicities Background Independence is to have no physically meaningless ambiguities (cf. the Identity of Indiscernibles). If this fails, we have Facet 9): the *Multiple Choice Problem*. This only really becomes relevant upon making quantum-level considerations (Chap. 12). In particular, it does not refer to the multiplicity of timefunctions in a classical GR type setting, since these are well-understood as coordinates on a manifold. It refers, rather, to the effect of multiple timefunctions at the quantum level [586],

A Local Resolution of the Problem of Time This is the conceptually well-defined practical restriction to exclude Multiple Choice and Global Problems. The 'local' avoids the Global Problems of Time and the 'a' avoids the Multiple Choice Problems. In the spirit of the following remark of Isham [483], this is the arena we consider in most of Parts II and III. *"The global issues are interesting, but they are not central to the problem of time and therefore in what follows I shall assume the topology of the spacetime manifold* $\mathfrak{m}$ *to be such that global time coordinates can exist."* On the other hand, the current book also supplies a number of Global and Multiple Choice Research Projects in Epilogues II.B, III.A and III.B, alongside an outline of some of the more advanced types of mathematics that these require.

Background Independence at Further Levels of Mathematical Structure This refers to the levels of mathematical structure commonly assumed in Classical Physics; see Fig. 10.9. I.e. how far should one go in allowing these to be dynamical [43], and then subject to quantum fluctuations as Isham pioneered [480–482, 492–494, 497, 498]. First note that the Background Independence attempted—from Einstein to, 1) Ashtekar variables programs such as Loop Quantum Gravity [752, 845], and 2) to this book's main Relational Approach—is at the *metric level* (and, in fact, at the differentiable manifold level [483, 586]). Such metric and differentiable manifold level Background Independence is a very major distinguishing feature between GR on the one hand and Newtonian Mechanics, SR, QM and QFT on the other. This is as opposed to further Background Independence at the topological manifold level of mathematical level and below (or different levels, i.e. not based on layers of structure upon sets, as outlined in Sect. 12.17, Epilogue III.C and Appendix W). Moreover, humankind largely remains physically and mathematically unprepared for handling the various possible deeper levels (Epilogues II.C and III.C outline the physical status quo). The topological manifold level version of such considerations dates back to Wheeler [898]. This reflects the progression in notions of space from flat space to curved differentiable manifolds to topological manifolds. Topological spaces take this one step further.

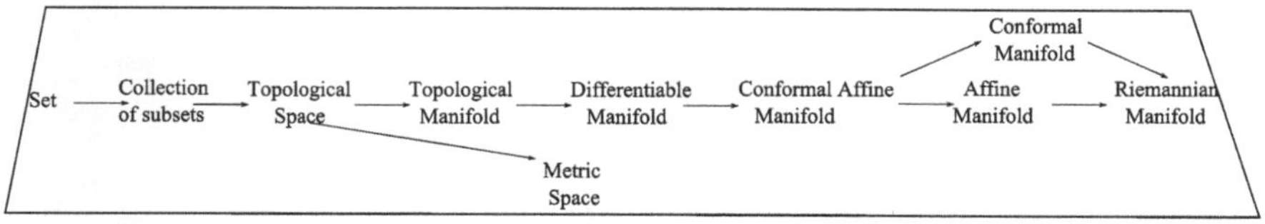

Fig. 10.9 Conventional levels of mathematical structure used in Physics to model space. As regards the upper levels, (metric information) = (conformal structure) + (localized scale). In the indefinite spacetime metric case, the conformal structure can furthermore be interpreted as causal structure

In Epilogues II.C and III.C, much of this book's 9-fold (or 21-fold) account of Background Indèpendence and the Problem of Time is argued to have a formulation which persists upon considering this descent in levels of mathematical structure. On the other hand, e.g. spacetime–space distinction varies from level to level, as does whether and how the analogues of Refoliation Invariance and Spacetime Constructability can be established. In particular, we use 'space', 'time', 'spacetime', 'slicing', 'surround', 'construct', 'thread' and 'constitute' as level-independent concepts, though some of the properties of the usual uses of these terms are level-dependent. In this manner, there is some version of the enchanted castle's gates at each level of mathematical structure.

Note also the following distinction in considering deeper levels of structure.

i) 'Single-floor' considerations: that the metric alone is dynamical (standard GR) or that the topological manifold alone is dynamical.
ii) 'Tower' considerations, in which a range of adjacent levels are dynamical, e.g. taking metrics *and* topological manifolds to be dynamical.

Overall, the single-floor case may be interpreted as removing the upper layers of structure so as to focus on the dynamical nature of the new topmost layer.

We finally point to the idea that building spacetime from 'discrete spacetime' or 'discrete space' can be generalized to far more options in terms of levels of mathematical structure. Namely, there are many intermediate mathematical formulations which assume a *subset* of the features of the continuum mathematics conventionally used in Theoretical Physics.

Chapter 11
Quantum Gravity Programs

11.1 Basic Considerations

We have now presented Background Independence and the Problem of Time for classical theories. Our goal, however, is to consider what Background Independence aspects and Problem of Time facets are exhibited by Quantum Gravity programs. The current Chapter prepares for this by providing an introduction to Quantum Gravity; Chap. 12 then meets our goal to conclude Part I.

Let us begin with some general foundational considerations about Quantum Gravity. Firstly, note that some of Nature's interactions are quantized, and consider arguments by which Quantization should apply to *all* interactions [235, 552], thus in particular applying to Gravitation as well. If Quantization did not apply to some interaction, say Gravitation, then this could be harnessed to violate Quantum Theory's Uncertainty Principle [291]. If one were to furthermore retain the classical Einstein field equations but now sourced by quantum matter,

$$\mathcal{G}_{\mu\nu} = \frac{8\pi G}{c^4} \langle \widehat{\mathcal{T}}_{\mu\nu} \rangle, \tag{11.1}$$

then there would be problems due to superpositions gravitating differently from pure states. In particular, upon performing a measurement, the passage from the former to the latter has the inconvenient feature of being instantaneous in an acausal manner [552, 874]. Moreover, such arguments do not preclude GR being too much of an effective theory to meaningfully quantize [195], much as one does not quantize e.g. the Navier–Stokes equation of Fluid Mechanics [600]. Developing this perspective faces the major difficulty that the Quantum-Gravitational analogue of the more fundamental smaller-scale theory for the constituent water molecules—itself meaningfully quantized—remain the subject of speculation. None the less, the rest of this book concentrates upon approaches which do involve Quantization.

Secondly, let us consider the qualitative similarity between gravitational and electromagnetic radiation, which Einstein was already aware of in 1916 [283]. Losses from gravitational radiation would therefore also affect atoms. Gravity is weak, however, as quantified by (2.14) redressed as an *electromagnetic* to gravitational

E. Anderson, *The Problem of Time*, Fundamental Theories of Physics 190,
DOI 10.1007/978-3-319-58848-3_11

ratio. In fact, classical collapse of the atom due to gravitational radiation would take of the order of 10^{30} years (Ex VI.5). This timescale did not affect Einstein's own argument because in that era the Universe was assumed to be infinitely old. However, Cosmology has since pointed to a finite age of the Universe (7.18) that is far below this figure, by which arguments of classical instability to gravitational radiation are rendered less pressing.

Our third point starts from Heisenberg and Pauli's proposal of QED in 1929 [444] containing the additional claim that Quantum Gravity would readily follow along similar lines. While this claim is incorrect, much can be learned from the major hole in their argument: that since gravitational charge is (gravitational) mass itself, the Equivalence Principle gives that is no gravitational analogue of an adjustable charge-to-mass ratio (cf. first paragraph of Chap. 7). This was first spotted by physicist Matvei Bronstein in 1936 [173], but he was soon executed by Stalin's regime and his work long remained forgotten. However, the same conclusion was arrived at independently in later works, such as physicist Asher Peres' study with Rosen [709] of uncertainties in the measurement of averaged Christoffel symbols stemming from the impossibility of concentrating a mass in a region smaller than its Schwarzschild radius.

Fourthly, Bronstein was probably also the first person to envisage interpreting the Planck units as characteristic scales for Quantum Gravity. In this regard, we already presented the most usual Planckian quantities in the preface as Eqs. (4) to (6); what we now provide is some further context for these.

In Black Hole Physics, the Planck mass corresponds to the (Compton wavelength) $\simeq$ (Schwarzschild radius) balance of scales. On the other hand, the cosmological balance (Compton wavelength) $\simeq$ (Hubble radius) does not involve G. Indeed, this can be rephrased as a quantum and SR energy balance $\hbar l_{\rm H} \simeq Mc^2$, for M the enclosed mass and $l_{\rm H}$ the Hubble radius of the currently observable universe. The Planck length enters Quantum Cosmology, rather, in the context of the extremely Early Universe: the era in which the 'scale of the whole Universe' was between one and a few Planck lengths.

Moreover, some quantities' fundamental units manage to depend on only a subset of the three most fundamental constants. E.g. *'Planck' force* is a purely-classical GR 'maximum tension' $F_{\rm Pl} = c^4/G = 1.210296(46) \times 10^{44}$ (kg m s^{-2} =: *Newtons*), and *'Planck' power* is likewise a purely-classical GR 'maximum power' $P_{\rm Pl} = c^5/G = 3.62851(44) \times 10^{52}$ (kg m^2 s^{-3} =: *Watts*). Indeed, 'Planck' velocity is just the pure SR c, 'Planck' angular momentum the purely quantum $\hbar$, and 'Planck' moment the quantum SR $\hbar/c$.

Some further Planck quantities of note are as follows. *Planck density* $\rho_{\rm Pl}$ comes out as, using $m_{\rm Pl}$ and $V_{\rm Pl} \sim l_{\rm Pl}^3$ in $\rho := m/V$, $c^5/\hbar G^2 = 5.15500(13) \times 10^{96}$ kg m^{-3}. Compare with 10^3 kg m^{-3} for water or 10^{17} kg m^{-3} for nuclei, with matter layers in neutron stars expected to be between slightly more dense and a factor of 10^6 less dense than nuclei. Additionally, the *Planck temperature* $T_{\rm Pl}$ is obtained by applying $E = k_{\rm B}T$—for $k_{\rm B}$ the Boltzmann constant $1.3806488(13) \times 10^{32}$ kg m^2 s^{-2} K— resulting in $\sqrt{\hbar c^5/G}/k_{\rm B} = 1.416808(33) \times 10^{32}$ K. This is well in excess of astrophysical temperatures; e.g. at the centre of the Sun, $T \simeq 10^7$ K. Finally, 'Planck' entropy is just k_B, which is a complete void in *all* of the fundamental theory dependent constants.

Moreover, if the number of theories with fundamental constants exceeded the number of independent fundamental units, then Quantum Gravity would more closely resemble Fluid Mechanics with its numerous dimensionless groups [600]. I.e. one would have a situation in which dimensional analysis does not suffice as a telling diagnostic. This could in principle not just obliterate the significance of the Planck scale but also that of the GUT 'energy desert' and of the meaningfulness of the confluence of the matter-sector coupling constants. Indeed, one might here reflect on how new Science students are taught about the potential dangers of extrapolations. For instance, Lord Kelvin's estimate of the lifetime of the Sun was out by around two orders of magnitude, due to nuclear processes not yet being known in his day.

Let us end by presenting various classifications of approaches to Quantum Gravity; some of the major approaches in the development of Quantum Gravity are then the subject of each of the current Chapter's remaining sections.

1) *Canonical* versus *Covariant* Approaches. The Covariant versus Canonical Quantization distinction can be envisaged as choosing two different paths round the 'Gordian cube' (Fig. 1.b–c). GR-centred approaches begin with classical Canonical formulations—such as Geometrodynamics or Ashtekar variables—and then set about Canonically Quantizing these (in Sects. 11.4 and 11.9 respectively). Alternatively, QFT-centred approaches (Sects. 11.2 and 11.7) consider (sufficiently weak) Gravitation as just another field on $\mathbb{M}^4$. *QFTiCS* ('in curved spacetime': Sect. 11.3) is a subsequent step, albeit limited by each curved spacetime assumed continuing to play the role of a fixed background. Path Integral Approaches (Sect. 11.6) offer additional options with QFT input. The further alternative of attempting to general-relativize Quantum Theory more thoroughly is much rarer in the literature, though mathematical physicists Klaus Fredehagen and Rudolf Haag have worked in this direction [318].
2) Whether or not to *alter one's Gravitational Theory*. Possible changes here include incorporating each of Supersymmetry and extended objects (Sects. 11.7–11.8 and 11.10–11.11).
3) Whether to incorporate *Background Independence*, and to which extent (the main subject of Chap. 12). I.e. is what one is quantizing purely a Relativistic Theory of Gravitation, or is it a Theory of Background Independence as well (*Quantum Gestalt* programs). From the latter perspective, models of just *Quantum Background Independence* then make for an interesting complement to *fixed-background Quantum Gravity* programs.
4) *Spacetime* versus *space primality* primality. Some classical roots for this ontological dichotomy have already been presented in Chap. 9, whereas Chap. 12 provides further quantum-level commentary.
5) *Top-down* approaches take a classical theory and quantize it. On the other hand, *bottom-up* approaches start from quantum-level first principles and then attempt to recover a suitably realistic classical limit (as outlined in Sect. 12.17). Almost all programs in existence either are, or were initially formulated as, top-down.

6) Finally, theories can be based on *continuum* or *discrete* mathematics. This following on from Chap. 1's sketch, and can furthermore apply in further detail as regards the level of mathematical structure that is quantized: [471, 477, 482, 492–494, 498] and Sect. 12.17), e.g. just metric-level versus differentiable manifold, topological manifold... level as well.

11.2 Covariant Approach to Quantum Gravity

This approach came historically first, as indicated in Fig. 11.5.b)'s 'family tree' for the various Quantum Gravity programs outlined in the rest of this Chapter.[1] It consists of a QFT of perturbatively small fluctuations of the metric over flat Minkowski spacetime $\mathbb{M}^4$, as studied in the 1930s by Rosenfeld [741, 742], and by Fierz and Pauli [300]. It follows on from the classical scheme outlined in Sect. 10.5.

This approach takes a particulate stance on Gravitation: a *graviton* propagating on a fixed background, which is usually $\mathbb{M}^4$. The privileged structures possessed by the Minkowskian Paradigm—as expounded in Chap. 4—then come into play. As for the electromagnetic force (Sect. 6.3), Gravity's long range leads to the corresponding mediator particle standardly being conceived of as massless. It additionally follows from Sect. 6.3 that gravitons possess even spin. Gravitons cannot moreover just be spin 0. The argument for this stems from Gravitation being sourced by the energy–momentum–stress tensor. Moreover, spin 0 couples to $\mathcal{T}_{\mu\mu}$ alone [299], $\mathcal{T}_{\mu\mu} = 0$ for Electromagnetism, and yet Gravitation is also observed to bend light. So one proceeds to consider spin 2, out of this being the next simplest case, and also due to further quite restrictive subtleties that eliminate the viability of massless particles of yet higher spin. E.g. Weinberg's papers [881, 882] detail further reasons to restrict attention to spin ≤ 2.[2] The field corresponding to spin 2 is the perturbed metric, and thus a symmetric (0, 2) tensor.

Some insights are gained from considering gravitons as a quantum-level parallel of classical linearized gravitational waves (Chap. 7.5 and Ex V.13 versus Ex VI.6). Indeed, the degrees of freedom count coincides with the 2 degrees of freedom per space point of the linearized theory.

The graviton conceptualization, alongside a few aspects of Poincaré Covariance, give back GR as a low-energy limit of a massless spin-2 QFT. Moreover, this can be arrived at by gauging the Poincaré group, resulting in the (infinitesimal form of) *Diff*($\mathfrak{m}$)-invariance.

Note furthermore that gravity gravitates, corresponding to GR-type field equations being nonlinear. Gravitons consequently form vertices with each other, for

[1] Since the approximate dates and ancestry of the various Quantum Gravity programs in this Chapter are rather nontrivial, this 'family tree' figure may quite often be a useful resource as regards outlining how these programs 'fit together' both conceptually and historically.

[2] Spin-0 and spin-2 mediators together is also a tenable possibility, as in e.g. Scalar–Tensor Theories of Gravitation.

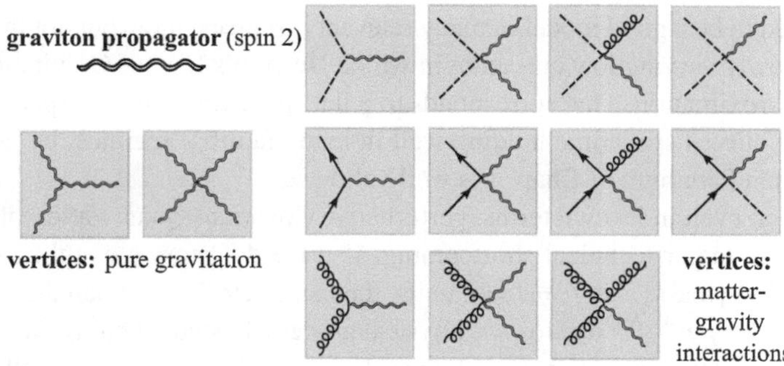

Fig. 11.1 Notation for the graviton propagator, followed by some Feynman diagram vertices involving gravitons

which Yang–Mills Theory provides some insights [238, 239]. The graviton propagator in harmonic gauge and Fourier-transformed form is

$$\frac{1}{2}\frac{\eta_{\mu\rho}\eta_{\nu\sigma} + \eta_{\mu\sigma}\eta_{\nu\rho} - \eta_{\mu\nu}\eta_{\rho\sigma}}{k^2 + i\,\epsilon}; \tag{11.2}$$

see Fig. 11.1 and Ex VI.18 for some consideration of vertices involving gravitons. One can proceed to consider Feynman diagrams involving gravitons, which began to be calculated by Feynman and DeWitt in the 1960s (see e.g. [238, 239, 297] and the review [137]), and on to effects on Particle Physics processes such as via graviton scattering. The S-matrix for graviton–graviton scattering turns out to be finite to 1 loop but infinite to 2 loops. Whereas Yang–Mills Theory is a useful intermediate more specifically due to the development of Fadde'ev–Popov (Chap. 52) and BRST (Sect. 43.1) techniques, such developments unfortunately to date fall short of being able to fully handle the case of GR.

It was furthermore established that [163] any Quantum Theory of gravitons coupled to a conserved energy–momentum–stress tensor $\mathcal{T}_{\mu\nu}$ must give the same perturbative low-energy scattering results as GR.

From a foundational point of view, Weinberg [880–882] additionally determined, firstly, that introducing interactions enforces coupling to a conserved $\mathcal{T}_{\mu\nu}$. Secondly, that this is a universal coupling, i.e. rederivation of the Equivalence Principle under the assumption of Lorentz invariance. On the other hand, the extent to which the Equivalence Principle holds at the quantum level has also remained an open question [365]. For instance, to what extent can free fall even be defined in Quantum Physics?

The graviton concept, moreover, has limited scope. With labelling particles by (inertial) mass and spin being tied to *Poin*(4), which in turn encodes the privileged $\mathbb{M}^4$ background, conceiving of Gravitation in terms of such gravitons breaks down at some point. For sure, the graviton concept is dubious [12, 471, 474] in strong field and Background Independent situations, which include in particular many of the more interesting parts of Black Hole Physics and Early-Universe Cosmology. Problems can also arise if some matters are taken too literally, such as asymptotic flatness or scattering 'in' and 'out' states being at infinity. None the less, gravitons are expected to be a good model in many relevant instances: large but not infinite regions of slowly varying low curvature in which Newton's Law of Gravitation holds to good approximation. This corresponds to a flat spacetime 'tangent space approximation' to curved spacetime holding well in many familiar regimes, by which the Minkowskian Paradigm of Chap. 1 is well validated.

Moreover, even in its own terms, perturbative Covariant Quantization of GR has the following major technical shortcoming. If a Field Theory's coupling constant has dimension $\{\text{mass}\}^D$ in $\hbar = 1 = c$ units, then an order-N Feynman diagram's integral goes like $\int p^{A-ND}\text{d}p$, for A a physical process dependent but N-independent constant. For interactions with $D < 0$, these Feynman diagram integrals all blow up beyond some $N = N_0$. Thus such interactions have problems with (at least naïve) renormalizability. In particular, this applies to GR's G, for which basic dimensional analysis gives $D = -2$ (see e.g. [193, 884]). Subsequently in the early 1970s, theoretical physicists Gerard 't Hooft and Martinus Veltman [827] carefully established that GR is non-renormalizable (see also [373] for consideration of 2-loop corrections).

Let us end by pointing to some further aspects of the conceptual interpretation of Covariant Quantization which remain unclear.

1) The background-to-perturbation split is ambiguous. Why should causality be determined by $\boldsymbol{\eta}$ rather than by the actual physical metric $\mathbf{g}$?
2) Since the physical null cone undergoes quantum fluctuations, why should microcausality with respect to $\boldsymbol{\eta}$ be involved in detail [477]? E.g. [236] the pole in the spacetime Green's function shifts upon summing Feynman diagrams which involve gravitons. In this way, the null cone actually experienced depends on the quantum state....
3) A consequence of the problem with background metrics laid out in Sect. 10.5 is that equal-time commutation relations with respect to $\mathbf{g}$ do not in general imply equal-time commutation relations with respect to $\boldsymbol{\eta}$. Moreover, this clashes both with GR's notion of coordinate time and with *Diff*$(\mathfrak{m})$-invariance, and also affects the interpretation below of the Canonical Approach.
4) QFT's involvement of time ordering of field operators clashes with the multiplicity of GR's coordinate time conception.
5) *Cosmological Constant Problem.* Effective Field Theory would have 'Λ' (i.e. the dimensionless quantity $G\,\hbar\,\Lambda/c^3$) be around 120 orders of magnitude larger. This casts doubts about whether effective Field Theory applies to the cosmological arena.
6) On the other hand, QFT's effective theory concept can be argued to devalue G's fundamentality. Quantum Theory is highly universal, whereas $1/G$ is just the coefficient of the leading term in an effective action.
7) Finally, perhaps conventional space and time only apply at scales much greater than the Planck length, from which they are separated by a phase transition [485]. In such cases, the physics of the other phase—and of the phase transition—would also need to be considered.

11.3 Quantum Field Theory in Curved Spacetime (QFTiCS)

Next consider extending study from QFT in $\mathbb{M}^4$ to QFTiCS, as started to occur in the 1970s. It is pertinent here that all uses of *Poin*(4) in the Wightman axioms 1)–5) entail assumptions of dependence upon a substantially symmetric background. Wightman-6) also relies on the background Minkowski metric η to assess what is spacelike (without relying upon any kind of symmetry). These features pose serious difficulties with extending the Wightman axioms to QFTiCS.

1) The generic spacetime possesses no symmetries, thus causing problems with extending the Wightman axioms 1) to 5) [473, 875]. In $\mathbb{M}^4$, the modes that simplify QFT are eigenfunctions of the $\partial/\partial t$ operator [325]. In particular, the construction of the Fock space (Chap. 6) is based upon the split into positive and negative modes. Standard QFT's 'natural modes' are tied to $\mathbb{M}^4$'s natural rectangular coordinates t, x, y, z, which indeed rest in turn upon *Poin*(4). Unfortunately, even in other spacetimes that have a considerable number of Killing vectors, not all the beneficial properties of these natural modes are recovered [143]. Moreover, generic GR spacetimes have no Killing vectors at all. A few instances in which conformal Killing vectors meaningfully deputize for Killing vectors are known, e.g. in relation to well-known maps between FLRW spacetimes and (pieces of) $\mathbb{M}^4$, as which result in similar-shaped Penrose diagrams as in Fig. 7.1.b)–d).[3]
2) Moreover, such spacetimes admit unitarily inequivalent Hilbert space constructions of Quantum Theory without a known means of picking out a preferred such [875]. Such an element of choice also enters the notion of 'vacuum state', yet QFTiCS in general has no unique notion of vacuum. One can furthermore expand in modes in multiple ways, which are interrelated by *Bogoliubov transformations* (see Ex VI.2 for an outline or [143, 874] for more details). E.g. in a setting with 'in' and 'out' vacua in asymptotic regimes either side of a time-dependent interaction region, the 'in' and 'out' states are related by such a transformation.
3) The spectrum condition (cf. Wightman-2) becomes compromised as follows [875]. Generic spacetimes are non-stationary and thus do not possess a time-like Killing vector, without which the total energy E is not conserved. On the other hand, the QFT definition of the energy–momentum–stress tensor $\mathcal{T}_{\mu\nu}$ requires spacetime smearing. In $\mathbb{M}^4$, since E is conserved, one can apply such a smearing of time without changing the value of E, so there is a unique and well-defined notion of E. But this possibility disappears in the generic QFTiCS case, as does any reason to expect the time-smeared E to remain positive [143]. Following Wald [875], the difficulties arising from the lack of an appropriate notion of total energy for QFTiCS can be overcome by replacing the spectrum condition by a 'microlocal spectrum condition'; see also [473]. This acts so as to restrict the singularity structure of the expectation values of the local QFT

[3]The positive norm notion itself, however, does not require Killing vectors [695].

correlators[4] (such as $\langle\,\phi(x_1)\,\ldots\,\phi(x_n)\,\rangle$) in the 'coincidence limit' ($x_1 \to x_2$ in the given example).

4) For free fields in $\mathbb{M}^4$, the notions of 'vacuum' and of 'particles' are intimately tied to the notion of positive frequency solutions. However, this rests upon the existence of a timelike Killing vector field. For QFTiCS, a notion of 'vacuum state' can be defined instead along the following lines. Call a state *quasi-free* [875] if all of its n-point correlators $\langle\,\phi(x_1)\,\ldots\,\phi(x_n)\,\rangle$—here given in the scalar field case—can be expressed in terms of the 2-point correlator by the same formula that holds for the ordinary vacuum state in $\mathbb{M}^4$. Call a state *Hadamard* (after the 19th and 20th century mathematician Jacques Hadamard) if the singularity structure of its 2-point correlator $\langle\,\phi(x_1)\phi(x_2)\,\rangle$ in the coincidence limit is the natural curved spacetime generalization of the singularity structure of $\langle 0|\phi(x_1)\phi(x_2)|0\rangle$ in $\mathbb{M}^4$. A quasi-free Hadamard state provides a 'vacuum state' notion, which leads to a corresponding notion of 'particles' in QFTiCS. Moreover, such a notion of vacuum state is highly non-unique. Indeed, for spacetimes possessing a noncompact Cauchy surface, different choices of quasi-free Hadamard states in general give rise to unitarily inequivalent Hilbert spaces [875]. I.e. in this case it is not even clear which Hilbert space of states to use. If a spacetime does not possess symmetries or other special properties, no preferred choice of quasi-free Hadamard state is in evidence. Finally note that, in Wald's view [875], seeking for a 'preferred vacuum state' in QFTiCS shares many elements with seeking for a 'preferred coordinate system' in classical GR, by which the former would also be a misguided search as well.
5) Wightman-6) alone readily generalizes to generic curved spacetimes; moreover even this becomes problematic beyond the QFTiCS regime, i.e. upon the metric becoming dynamical.
6) The labelling of particle types in $\mathbb{M}^4$ by *Poin*(4) representations is also in general a lost commodity once one passes to QFTiCS.
7) Finally, a number of aspects of requiring of Poincaré invariance in standard QFT can be replaced by requiring that the quantum fields be locally and covariantly built from GR's notion of metric. One can try to make do with *Diff*($\mathfrak{m}$), however this case has no frames in which the physics simplifies, and Representation Theory becomes much harder (see Appendix V).

In a QFTiCS setting with 'in' and 'out' regions to either side of a time-dependent interaction region, pair production near the horizon can lead to one particle falling in and the other escaping. From afar, this takes on the appearance of radiation: *Hawking radiation*. This justifies treatment of black holes in thermodynamical terms: black holes interact with their environments after all, so their having a nonzero temperature makes sense. *Black holes are, in fact, quantum-mechanically grey*. See Fig. 11.2.a) for the associated Penrose diagram, and Ex VI.3 for further details.

The *Hawking temperature* is

$$T_{\mathrm{H}} = \hbar\kappa/2\pi\, k_{\mathrm{B}} c \tag{11.3}$$

[4]See Appendices Q.9 and U.6 for a conceptual outline of these.

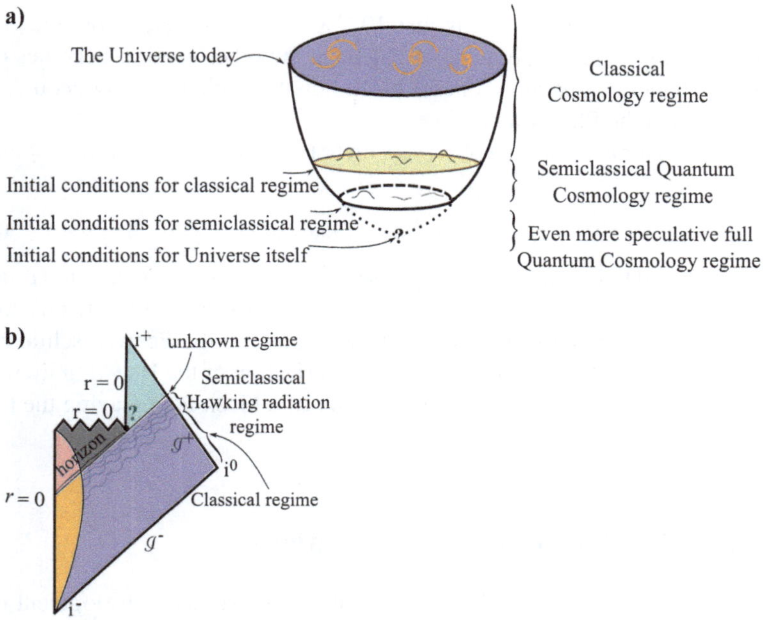

Fig. 11.2 **a)** The Early-Universe arena for Semiclassical Quantum Cosmology involves perusing a less controversial part of the Universe than the immediate vicinity of the singularity predicted by GR. Note also the expectation that there is a common Semiclassical Quantum Cosmology treatment not just for GR but for a wide range of different Gravitational Theories as well. **b)** Stellar collapse followed by black hole evaporation due to Hawking radiation. This is a modification of the Penrose diagram of Schwarzschild spacetime (Fig. 7.1.e). The extrapolated end-phase of the evaporation is, moreover, explosive; the end-products of this remain disputed. [This is marked by '?' on the diagram, and could furthermore substantially modify the diagram.] The mass of an early-universe object that would be in its end-phase of Hawking radiation today is 10^{12} kg (and no accepted physical process is known to produce such). The associated Information Paradox also remains disputed. This is between black hole radiation's supposedly perfect black body thermal spectrum carrying no information of what the black hole was formed from, versus quantum evolution being unitary. See e.g. [647] for further discussion. With there being no accepted cosmological or astrophysical processes for producing such objects, we do not particularly expect to observe Hawking radiation. Moreover, the Planck-sized universe lies well outside the scope of *Semiclassical* Quantum Cosmology. Then again, the end-phase of Hawking radiation also lies well outside the scope of a semiclassical analysis. Compare also the suggestion that the structure of any piece of spacetime may be 'foamy' if examined at a scale approaching the Planck scale. On the one hand, this is many orders of magnitude below where the averaging implied in many e.g. cosmological GR models applies. On the other hand, Planck-sized universes and end-stages of Hawking radiation are physically very special situations associated with a priori justified occurrences of very high densities and curvatures. Because of this, Quantum Gravitational effects there have further credibility than suggestions that any region of spacetime is foamy if only we could look closely enough

for κ the black hole's surface gravity. Computing κ for the Schwarzschild solution (Ex V.8) renders this explicitly in terms of Planck units,

$$T_{\rm H} = \frac{\hbar c^3}{8\pi G M k_{\rm B}} = \frac{m_{\rm Pl}}{M}\frac{T_{\rm Pl}}{8\pi}. \tag{11.4}$$

For a solar-mass black hole, this is just $10^{-8}K$, which is well below the cosmic microwave background temperature ($\simeq 3K$). On the other hand, it becomes of the order of the Planck temperature once a black hole has radiated away enough mass to itself approach the Planckian regime.

Furthermore, for accelerating observers there is an analogous *Unruh temperature* [861]

$$T_{\mathrm{U}} = \hbar a / 2\pi\, k_{\mathrm{B}} c, \tag{11.5}$$

which is predicted to be experienced by observers undergoing acceleration a in $\mathbb{M}^4$.

Finally, the Salecker–Wigner clock inequalities (5.18) extend to the GR context [113]. This can for instance be envisaged by combining the Schwarzschild radius with the inequalities. This gives an alternative derivation of the Hawking lifetime of a black hole as an upper bound on the longevity of a black hole playing the role of a clock (Ex VI.3).

11.4 Canonical Quantum Wave Equations

Chapters 8 and 9 already provided an account of the classical development of the Canonical Approach, as supplemented by Dirac's extension of the Principles of Dynamics covered in Appendix J. Developing this work and subsequently understanding it [237, 899] took until the 1960s; moreover significant steps in its Quantization entered at this point.

Let us first extend the simple scheme for Quantization presented in Chap. 5, so as to further cover constrained systems in Background Independent settings. As per Chap. 9.10, GR gives rise to a stationary wave equation—the *Wheeler–DeWitt equation* $\widehat{\mathcal{H}}\Psi = 0$—in place of a time-dependent one. These authors called this the *Einstein–Schrödinger equation*; a more detailed form for this is

$$\begin{aligned} 0 = \widehat{\mathcal{H}}\Psi := -\hbar^2 `\frac{1}{\sqrt{\mathrm{M}}}\frac{\delta}{\delta \mathrm{h}^{ij}}\left\{\sqrt{\mathrm{M}}\mathrm{N}^{ijkl}\frac{\delta\Psi}{\delta \mathrm{h}^{kl}}\right\} - \xi\, \mathcal{R}_{\mathrm{M}}(\underline{x};\mathbf{h}]'\,\Psi \\ - \sqrt{\mathrm{h}}\mathcal{R}\Psi + 2\sqrt{\mathrm{h}}\Lambda\Psi + \widehat{\mathcal{H}}^{\mathrm{matter}}\Psi. \end{aligned} \tag{11.6}$$

Here ' ' implies in general various well-definedness issues as further outlined in Chaps. 40 and 43, which are in any case absent from finite models such as (11.9), and operator-ordering issues which still partly remain for these finite models.

This is accompanied by the quantum GR momentum constraint

$$0 = \widehat{\mathcal{M}}_i\Psi = 2i\,\hbar\mathrm{h}_{ik}\mathcal{D}_j\frac{\delta}{\delta \mathrm{h}_{jk}}\Psi + \widehat{\mathcal{M}}_i^{\mathrm{matter}}\Psi, \tag{11.7}$$

and by the Klein–Gordon type inner product

$$\langle \psi_1[\mathbf{h}] | \psi_2[\mathbf{h}] \rangle = \frac{1}{2i}\prod_{x\in\Sigma}\int \mathrm{d}\mathbf{\Sigma}_{ij}\mathrm{N}^{ijkl}(\mathbf{h})\left\{\psi_1[\mathbf{h}]\,\overset{\longleftrightarrow}{\frac{\delta}{\delta \mathrm{h}_{kl}}}\,\psi_2[\mathbf{h}]\right\}. \tag{11.8}$$

This however runs into further technical problems as outlined in Sect. 12.2 and Ex VI.11.vi)

Misner's introduction of Minisuperspace was largely motivated by the wish to study more tractable versions of (11.6). His original models [657, 659] were isotropic and anisotropic, in each case without fundamental matter. For this book's main Minisuperspace model, with single minimally-coupled scalar field matter, the Wheeler–DeWitt equation reads

$$0 = \hbar^2\{\partial_\Omega^2 - \partial_\phi^2\}\Psi = \exp(6\,\Omega)\{\exp(\Omega) - 2\,\Lambda - V(\phi)\}\Psi. \tag{11.9}$$

The unreduced RPM model arena counterpart is

$$E\,\Psi = \widehat{\varepsilon}\,\Psi = -\hbar^2\triangle_M\Psi/2 + V\,\Psi; \tag{11.10}$$

compared to Minisuperspace this model permits tractable study of accompanying linear quantum constraints and of structure formation. Sect. 9.11's other classical arrangements of $\mathcal{H}$ have distinct quantum realizations—in particular, the 1970s also saw the rise of a formal scheme for Quantization based on the York time candidate [923]—though we postpone these to the next Chapter.

We end by returning to Sect. 11.2's point 3) concerning equal-times commutation relations applying once again in attempting to interpret Canonical Approaches. It is no coincidence that the formalism appearing frozen in time arises from an equation which would often be interpreted as an energy equation; this rests on energy being the quantity conjugate to time. Finally, the additional complications with the notion of energy in GR (Appendix K.5) further exasperates the already difficult position of the Time–Energy Uncertainty Principle for Quantum Gravity.

11.5 Quantum Cosmology

Consideration of this arena started with the above account of quantum Minisuperspace models in the 1960s through to the 1980s. In particular, such models with scalar field matter took off in the 1980s in relation to *Hartle–Hawking's no boundary proposal* (named in part after physicist James Hartle): that

the boundary condition for the Universe is that it has no boundary. (11.11)

Subsequently Quantum Cosmology became a possible explanation for, firstly, the origin of inflation. Secondly, for the origin of cosmic microwave background fluctuations and galaxies—the currently observed inhomogeneities—originating from quantum cosmological fluctuations [419] as amplified by inflation.

Moreover, Quantum Cosmology remains beset by conceptual difficulties [99, 101, 260, 340, 411, 413, 418, 421, 427, 428, 430, 432, 441, 442, 483, 496, 551, 552, 586, 589, 692–694, 862, 912], including the following.

1) What is the mathematical form of the quantum wavefunction of the Universe Ψ? E.g. does it really come from a frozen quantum wave equation?

2) What interpretation is to be accorded to Quantum Cosmology? E.g. the Copenhagen Interpretation of QM ceases to be possible here, since there is no longer a surrounding classical large system.
3) Quantum Cosmology has robustness issues, as regards whether ignoring certain degrees of freedom compromises the outcome of calculations [591].
4) Quantum Cosmology entails additional Arrow of Time issues. A quantum cosmological perspective might even shed further light on the origin of various other branches of Physics' arrows of time [101, 413, 433, 752]. Unfortunately this interesting question lies outside the scope of this book.

See Sect. 12.2 and Part III for further discussion of Quantum Cosmology Moreover, some of the above issues lead to previously solely philosophical contentions about Quantum Theory—in particular as applied to closed systems such as the whole Universe—to enter the realm of testable Physics.

11.6 Path Integral Approach for Gravitational Theories

This approach was initiated by Misner in 1957 [655] and begins by considering the GR version of the transition probability (6.34) type Feynman integral expression

$$\mathcal{T}[\mathbf{h}_{\text{in}}, t_{\text{in}}, \mathbf{h}_{\text{fin}}, t_{\text{fin}}] := \langle \mathbf{h}_{\text{in}}, t_{\text{in}} | \mathbf{h}_{\text{fin}}, t_{\text{fin}} \rangle = \int_{t_{\text{in}}}^{t_{\text{fin}}} \int_{\boldsymbol{\Sigma}} \mathbb{D}\mu[\mathbf{g}] \exp(i \, \mathsf{S}_{\text{EH}}[\mathbf{g}]/\hbar). \tag{11.12}$$

Here $\mathbb{D}\mu$ is a measure of integration. Also, this runs over all spacetime metric geometries on $\mathfrak{T} \times \boldsymbol{\Sigma}$ in between $\langle \boldsymbol{\Sigma}, \mathbf{h}_{\text{in}} \rangle$ and $\langle \boldsymbol{\Sigma}, \mathbf{h}_{\text{fin}} \rangle$, for time interval $\mathfrak{T} := [t_{\text{in}}, t_{\text{fin}}]$.

This approach exhibits a *Measure Problem*; see Appendix P.2 for an outline of what measures are. It is a problem because the measure involved is usually but a formal rather than explicitly-known computational object in the case of Gravitational Theories.

Some further features of the Path Integral Approach for Gravitational Theories are as follows; consult Sect. 12.8 and Chap. 52 for further details.

0) While this is not necessarily a Background Dependent pursuit, this is not a Canonical Approach.
1) It is, rather, a spacetime primary formulation. It is furthermore paths or histories that are now to be considered as primary.
2) The 'wrong sign' of the GR action causes further problems for Path Integral Approaches.
3) Moreover, these problems are ameliorated [474] by working in a Euclidean-signature sector, where

$$\langle \mathbf{h}_{\text{in}}, t_{\text{in}} | \mathbf{h}_{\text{fin}}, t_{\text{fin}} \rangle = \int_{t_{\text{in}}}^{t_{\text{fin}}} \int_{\boldsymbol{\Sigma}} \mathrm{D}\mu[\mathbf{g}] \exp(-\mathsf{S}_{\text{EH}}^{\text{Eucl}}[\mathbf{g}]). \tag{11.13}$$

Now the integration is over all Euclidean-signature metric geometries on $\mathfrak{T} \times \Sigma$ in between.

4) However, GR's action is not positive-definite, which causes further unboundedness problems.
5) Moreover, Discrete Approaches to Quantum Gravity have better-defined path integrals. Such approaches started historically with Regge Calculus in the 1960s [660] (named after physicist Tullio Regge see Ex VI.16 for more).
6) Time ordering and positive frequency impinge upon [477] the passage between Euclidean and Lorentzian sectors. At the perturbative level about flat spacetime this reduces to Wick's result from QFT (Ex II.3). However, the curved spacetime counterpart of this picks up ambiguities as regards the choice of complex contour [420].
7) Let us finally point to some approaches which involve both Path Integrals and Canonical formalism. QFT can rely on the Canonical Approach for computing what its Feynman rules are. In this way, Ordinary Quantum Theory 'in terms of path integrals' is in fact in some cases a combined Path Integral *and* Canonical Approach. Then in Quantum Gravity, such a combined scheme would pick up *both* approaches' problems.

11.7 Covariant Approaches to Alternative Theories

The 1970s also saw the beginning of a search for an extension of GR that provides a renormalizable or finite perturbation from a QFT perspective. Three types of extension are as follows.

1) Add symmetry-preserving terms.
2) Add non-unified extra degrees of freedom.
3) Alter the symmetries evoked.

Historically, early attempts involved 1), in particular Higher Derivative Theories (adding terms such as $\mathcal{R}^{(4)\,2}$ and $\mathcal{R}^{(4)}_{\mu\nu}\mathcal{R}^{(4)\,\mu\nu}$). However, physicist Kellogg Stelle [811] established the alternative that renormalizability can be attained here, *but only* wherever non-unitarity applies instead. One of these theories is the Weyl^2 Theory, with action

$$\mathsf{S}_{\mathrm{Weyl}} \propto \int \mathrm{d}^4 x \sqrt{|\mathrm{g}|}\, \mathcal{C}^{(4)}_{\mu\nu\rho\sigma}\mathcal{C}^{(4)\,\mu\nu\rho\sigma} . \tag{11.14}$$

This is conformally invariant, and also arises from gauging the conformal group, much as GR arises from gauging the Poincaré group (Ex VI.20).

The 2-*d* and 3-*d* counterparts of GR are overly simple. Riemann curvature has only 1 independent component in 2-*d*, so the Ricci scalar $\mathcal{R}^{(2)}$ carries all the geometrical information; it has only 3 in 3-*d*, so the Ricci tensor $\mathcal{R}^{(3)}_{\mu\nu}$ is all. There is consequently no Weyl tensor for $d < 4$. The physical consequences for 3-*d* GR are that there is no curvature other than where the sources are, no gravitational waves, and the degrees of freedom count per space point gives zero (though global degrees of freedom can still remain). On the other hand, in 2-d, the integral of $\mathcal{R}^{(2)}$ is a topological invariant, rendering trivial the use of such as an action. There is, however, a different way of obtaining a non-trivial Gravitational Theory in 2-d [836].

Moreover, that the integral of $\mathcal{R}^{(2)}$ is a topological invariant in 2-d and then a nontrivial action term for $d \geq 3$ is but the first of a series of objects behaving in this manner. A new member of the series becomes nontrivial upon increasing the dimension by two. The next member is $\mathcal{R}^{(d)\,2} - 4\,\mathcal{R}^{(d)}_{\mu\nu}\mathcal{R}^{(d)\,\mu\nu} + \mathcal{R}^{(d)}_{\mu\nu\rho\sigma}\mathcal{R}^{(d)\,\mu\nu\rho\sigma}$. The corresponding action integral is a topological invariant in 4-d—proportional to the well-known Euler characteristic [68] in the case of a compact oriented manifold—and becomes a nontrivial action term in 5-d and higher, for the so-called *Gauss–Bonnet* or *Lovelock Theory* [230]. While at first sight this term looks like it will produce a higher-order Gravitational Theory, careful inspection of the field equations reveals that these are actually just second-order. Indeed, the existence of this action term in $d \geq 5$ indicates the breakdown of the Lovelock simplicity postulates [629] which hold for $d \leq 4$ curvature scalars.

Theories involving connections instead of a metric can also be considered (see Appendix D), or ones in which the connection that is not necessarily the metric one. This is one way into torsion being present: as the difference of two connections, though one can also conceive of torsion existing a priori. Non-symmetric metrics can also be evoked. Non-minimal coupling can be involved as well, for instance in Brans–Dicke Theory (after physicists Carl Brans and Robert Dicke), with action

$$\mathsf{S}_{\mathrm{BD}} \propto \int \mathrm{d}^4x\sqrt{|\mathrm{g}|}\big\{\Phi\,\mathcal{R}^{(4)} - \omega\,|\partial\Phi|^2/\Phi\big\}, \tag{11.15}$$

for a parameter ω, and its Scalar–Tensor Theory generalization [910].

For later use, let us also at this point introduce *tachyons*: hypothetical matter species, for which c is not a maximum speed but rather a minimum speed. These are conceptually problematic due to ensuing causal paradoxes, so theories without these are favoured.

Unification schemes were more straightforward while Gravitation and Electromagnetism were the only known theories to unify [369]. E.g. Kaluza–Klein Theory [67] and Weyl's failed unified theory [893] were along these lines, and the second half of Einstein's life were spent looking for such theories. In particular, Kaluza–Klein theory's $U(1)$ is curled up small, came to inspire a much wider range of 'compactifications' to hide extra dimensions [385].[5] Unification became a harder proposition with the formulation of distinct and successful theories of the strong and weak forces. None the less, theoretical physicist Edward Witten considered a bigger version of Kaluza–Klein theory for this case [914]. Moreover, the GUT timescale is 10^{-35} s (GUT scales are 4 orders of magnitude out from E_{Pl} or l_{Pl}, but 8 orders of magnitude out from t_{Pl}). This modest difference in orders of magnitude suggests Gravitation could be an extra ingredient for unification.

[5] Extra spatial dimensions are relatively uncontroversial. However, considering time to have more than one dimension would, carry many technical and conceptual difficulties, starting with the difficulties with ultrahyperbolic PDEs outlined in Sect. 31.3.

Conventional GUT's, moreover, just look to unify the three non-gravitational forces. Some restrictions in combining Gravitation with the Particle Physics of the other fundamental forces of Nature are as follows.

No-Go Theorem 1 (Weinberg) *In QFT, there is an upper bound on spins, from a standard lack of currents to couple to* [885].

Note in particular that spin-2 lies within this upper bound.

No-Go Theorem 2 (Coleman–Mandula) (*after physicists Sidney Coleman and Jeffrey Mandula*). *For a QFT whose S-matrix obeys certain plausible technical conditions* [887], *if there is a mass gap then the Lie algebra of symmetries must be a direct product*

$$\text{Poin} \times \mathfrak{g}_{\text{internal}}. \tag{11.16}$$

Already in the QFT or Particle Physics context, *Supersymmetry* [887] eludes the Coleman–Mandula Theorem, because its statement does not extend to preclude Lie superalgebras. This is a transformation mapping between bosons and fermions; moreover it is a mixture of internal and spacetime symmetries. More specifically, applying the transformation twice in general leads to translation in spacetime:

$$\widehat{Q}|\text{b}\rangle = |\text{f}\rangle, \qquad \widehat{Q}|\text{f}\rangle = |\text{b}\rangle, \qquad \widehat{Q}^2|\text{b}\rangle = |\text{`b } \textit{over there'}\rangle, \tag{11.17}$$

where b are bosonic species and f are fermionic ones. In such theories, each particle has a superpartner whose spin differs by $\frac{1}{2}$, i.e. squarks and sleptons (spin 0), and gauginos and Higgsinos (spin $\frac{1}{2}$). One describes these by passing from the Poincaré group to the Poincaré supergroup, whose Representation Theory describes this enlarged set of particles.

Suggested benefits for Particle Physics in a world with Supersymmetry include the following.

1) Resolving the Hierarchy Problem (breadth of different fundamental particle masses).
2) Improving the confluence of GUT coupling 'constants' (Fig. 11.3).
3) Cancellation of anomalies is a common feature of Supersymmetry. This stems from (6.14) causing fermions to contribute Feynman diagrams of the opposite sign to the bosonic ones, with Supersymmetry furthermore fixing the proportions of these to be such that many cancellations occur.

One practical problem with Supersymmetry is that no superpartner pair has had both its species observed. This may be suggestive of Supersymmetry being a mathematical but physically spurious construct. For sure, if there is fundamental Supersymmetry, at the point at which this book was finished, there was no direct evidence of Supersymmetry being present in Particle Physics.

Moreover, one can indeed formulate Gravitational Theory in accord with Supersymmetry: Supergravity. Here the graviton has a spin-$\frac{3}{2}$ superpartner: the gravitino.

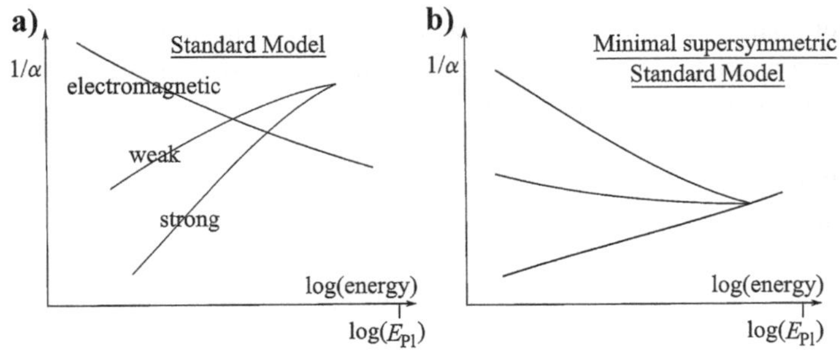

Fig. 11.3 **a)** Standard Model's inverse running coupling constants, $1/\alpha$. **b)** Minimal Supersymmetric Standard Model's inverse running coupling constants form a triple intersection

For 4-d $N = 1$ Supergravity the action is the 4-bein version of the GR Einstein–Hilbert action plus the Rarita–Schwinger action for the gravitino field ψ_μ,

$$\mathsf{S}_{\rm RS} \propto \int \mathrm{d}^4x\, \epsilon^{\mu\nu\rho\sigma}\bar{\psi}_\mu\gamma_5\gamma_\nu\mathcal{D}_\rho\psi_\sigma \tag{11.18}$$

(named after physicists William Rarita and Julian Schwinger). More extended Supergravity has multiplets running from spin 0 to spin 2, incorporating matter species as well (see Appendix V if interested). Preclusion of still larger multiplets including yet higher spins restricts the maximum size of Supergravity theories; the limiting dimension 11 materializes in this manner.

In Supergravity, further loop terms are finite [138]. Gauging *super-Poin*(d, 1) gives back Supergravity. Also note Supergravity's $N = 8$ limitation on the number of supercharges, and dimensional rigidity: at most 11-d Supergravity. String Theory is also ameliorated by passage from bosonic strings to supersymmetric strings: *superstrings*. Through Higher Derivative Theory and Supergravity, the search converged successfully to String Theory in the mid to late 1980s.

11.8 Perturbative String Theory

In this approach, point particles are replaced by strings; this turns out to smear away various of QFT's notorious problems (Fig. 11.4). A *string length* $l_{\rm S}$ subsequently enters this approach's physics; the most natural size for this is of the order of $l_{\rm Pl}$. This serves as a minimum length: $l < l_{\rm S}$ has no operational significance in perturbative String Theory. A major reason for entertaining this 'string hypothesis' is, moreover, that it underlies a well-behaved formula for graviton scattering [385].

N.B. that String Theory concerns *special-relativistic* strings. In the bosonic case, the historical *Nambu–Goto action* for a such is

$$\mathsf{S} \propto \int \mathrm{d}^2\sigma \sqrt{\{\dot{X}\cdot X'\}^2 - \dot{X}^2\cdot X'^2}. \tag{11.19}$$

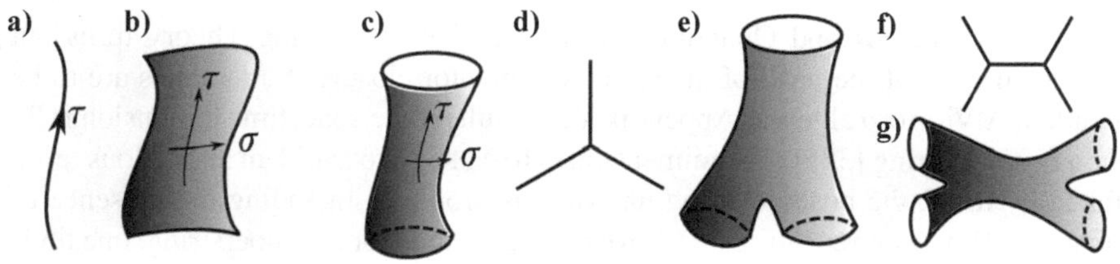

Fig. 11.4 In String Theory, SR particle worldlines **a)** become string *worldsheets*, **b)** for an *open string* and **c)** for a *closed string*. QFT vertices are smeared out, e.g. from the 3-particle vertex **d)** to the 'trousers topology' **e)**. The string theoretic analogue of Feynman diagrams are correspondingly thickened, e.g. from **f)** to **g)**. In this way, extended objects such as strings smear out Feynman diagrams, causing a number of key calculations to become better behaved (less singular)

This is however hard to work with in practice, so the equivalent *Polyakov action*,

$$\mathsf{S} \propto \int \mathrm{d}^2\sigma \, \partial_a X_\mu \partial^a X^\mu, \tag{11.20}$$

subsequently became widely used.[6]

String Theory in fact began as an unsuccessful theory of the strong force in the 1960s and 1970s. However, closed strings were found to incorporate Gravitation in the sense of necessarily including a spin-2 excitation. Consequently, by the onset of the 1980s string revolution—initiated by theoretical physicists Michael Green, John Schwarz and Edward Witten [385, 386]—String Theory was considered to be a Unified Theory including Gravitation. The above sense involves thinking in terms of the graviton—a perturbative particle concept—and associated fixed-background considerations (the above actions depend implicitly upon background metrics). The strings propagate on fixed background spacetimes, so at least this formulation of String Theory involves fixed-background SR-like notions of time, space and spacetime, rather than GR-like ones. In particular, such fixed-background spacetimes are very often confined to being static. What happens instead is that GR's Einstein field equations are emergent, due to closed strings having to contain a spin-2 excitation that one interprets to be the graviton. As a result of this, problems associated with GR might not be fundamental but refer, rather, to a D-dimensional background spacetime metric structure that the strings move in. Moreover, these perturbative and Background Dependent assumptions cease to apply in some of the subsequent developments (Sect. 11.12). One often considers D-dimensional Poincaré-invariance, though the propagation could also be on some fixed (usually also highly symmetric) curved spacetime background.

Whereas worldsheets are usually modelled as living on fixed backgrounds, a second more restrictive kind of Background Independence applies. I.e. worldsheet diffeomorphism invariance and conformal (alias Weyl) invariance hold therein.

[6]These actions are named after theoretical physicists Yoichiro Nambu, Tetsuo Goto and Alexander Polyakov.

Incorporating SR and Quantum Theory together into String Theory turns out to be restrictive at the level of quantum commutator closure if anomalies are to be avoided. A Virasoro algebra (Appendix V.3) results if the spacetime dimension is 26 for a bosonic string [385]. This dimension is fixed so as to avoid an anomalous term. Taken by itself, the bosonic string has various problems including the presence of tachyons. However, repeating the above considerations for the superstring, one finds that [385] tachyons are avoided alongside 10-d being the new anomaly-avoiding critical dimension of spacetime. Spacetime dimension 10 is more likely than 26, firstly by being closer to the physically observed dimension 4. Secondly, because Supersymmetry exhibits a pattern modulo 8 in the dimension [887], 10-d spacetime and 2-d worldsheets go well together. See Sect. 11.12 for a third reason.

String Theory does not only include a spin-2 graviton, but also has enough room, firstly, to contain the other three known forces of Nature, thus unifying the four forces. Secondly, building blocks for matter—in particular chiral fermion species—are accommodated.

One still needs to be able to shed 6 spatial dimensions in order to pass to the approximate regimes corresponding to everyday existence. The most traditional approach to this is to reuse Kaluza–Klein type compactification. More precisely, the other 6 dimensions could be [386] curled up small in the form of a particular type of complex manifold (Appendix F.1) called a *Calabi–Yau space*.[7] The geometry of this might be hoped to explain hitherto theoretically unaccounted for Standard Model parameters, such as fundamental particle masses, mixing angles, the observed Standard Model's $SU(3) \times SU(2) \times U(1)$ gauge group, or the three generations of particles observed.

Superstring Theory is highly rigid in form. As well as requiring spacetime dimension 10, Superstring Theory is a whole package in contrast to how one can add a range of extra terms to QFT actions. Moreover, there are in fact five superstring theories.

Type I, which also contains open strings and possesses $SO(32)$ symmetry.

Type IIA, with non-chiral massless fermions.

Type IIB, with chiral massless fermions.

Heterotic (meaning its right- and left-moving strings differ) String Theory with either $SO(32)$ symmetry or $E_8 \times E_8$ symmetry.[8]

All five possess the crucial closed strings whose spectra include the massless spin-2 excitations that this approach identifies with GR-type gravitons.

On the other hand, Calabi–Yau spaces, compactifications in general and other means of hiding or not perceiving extra dimensions are substantially non-unique

[7]These are named after mathematicians Eugenio Calabi and Shing-Tung Yau. See [673] for an especially accessible presentation of the various layers of structure leading to the definition of these.

[8]See Appendix E for a start on what E_8 is.

[148, 262]. So in the absence of further credible selection principles, it is very difficult to extract highly unique predictions about fundamental Particle Physics from String Theory. This is a separate issue from some of String Theory's results being applicable as mathematical methods for various much lower-energy (and thus practical) scenarios. Providing methods that can be used in multiple largely unrelated situations is not to be confused with providing a new Paradigm for Fundamental Physics as a whole. The latter would additionally involve finding a consistent physical interpretation for the mathematics, as opposed to one which shifts from application to application. Furthermore, it is not yet known what String or M-Theory are a theory *of*, as is reflected by the meaning of 'M' for now being left open. String and M-Theory remain, rather, a work in progress as regards what foundational meaning and paradigmatic interpretation they might possess.

While String Theory is free from anomalies and has finite diagrams order by order, if one considers all the orders together, there is a lack of *Borel summability*.[9] See Exercise Set VI and Sects. 11.12, 19.10 and 57.4 for further String and M-Theory concepts, results and discussions.

11.9 Ashtekar Variables and Loop Quantum Gravity

The 1986 classical-level introduction of Ashtekar variables (Sect. 8.15) was immediately followed up by Canonical Quantization. While this approach has simplified constraints as compared to Geometrodynamics, it is not equivalent to Geometrodynamics in various ways. One such is through inclusion of degenerate triads whereas Geometrodynamics did not include degenerate metrics. Another is that this approach, at least as presented so far in this book, involves *complexified* GR. To end up with real-valued GR, *reality conditions* are eventually required. Moreover, Kuchař [587] pointed out that these involve mathematics that is comparably unassailable to that which would be required for Quantum Geometrodynamics. One way around this is a real formulation involving the *Barbero–Immirzi* parameter β; this formulation has often since been used.[10]

Another of this program's early developments was to pass furthermore to $SU(2)(\boldsymbol{\Sigma})$ *holonomy variables*; these encode the geometrical effects encountered in taking a ride around each loop in $\boldsymbol{\Sigma}$. If interested, see Appendices F.4 and N for an outline of the mathematics of loops, holonomies and holonomy variables. This was followed up by considering the *loop representation* [330]. These developments amount to using a particular means of solving the $SU(2)$ Yang–Mills–Gauss constraint. Rovelli and physicist Lee Smolin [758] additionally showed that the corresponding quantum theory could be formulated in terms of Penrose's *spin networks*

[9]This construct is named after early 20th century mathematician Émile Borel; see e.g. [194, 269, 394] for further discussion of this in the context of String Theory.

[10]This is named after physicists Fernando Barbero and Giorgio Immirzi. See Sect. 24.9 for the geometrical meaning of this parameter and of the corresponding version of Ashtekar variables.

(see Sect. 43.5). One can additionally contemplate solving $\mathcal{M}_i$ by considering the *Diff*($\mathbf{\Sigma}$)-invariant counterparts of loops: *knots* (Appendix N.13).

This Loop Quantum Gravity [752, 845] has *technical advantages* over Geometrodynamics at the quantum level, as outlined in Chap. 43. On the other hand, these advantages do not by themselves alter significant Problem of Time features or strategies. A further issue is that, despite various candidates being proposed [842, 845], there is no clearly motivated, established form for Loop Quantum Gravity's quantum Hamiltonian constraint [679, 794]. Yet without this, one is dealing with mere kinematics, whereas from Sect. 2.4 onward, this book has been arguing that predictive power in Physics comes, rather, from Dynamics.

Ashtekar variables and Loop Quantum Gravity largely depend on 4-*d* spacetime features. One of these is the relation between 3-*d* space and the existence of knots. None the less, there is a 3-*d* spacetime counterpart of Loop Quantum Gravity [193]; it is the higher-*d* cases that are harder to realize.

Loop Quantum Gravity considers Quantization of area and volume. E.g. for area associated with a surface $\mathbf{\Sigma}$ (within β-real formulations)

$$\widehat{A}_{\mathbf{\Sigma}}|\psi\rangle = 8\pi\, l_{\mathrm{Pl}}^2 \beta \sum_I \sqrt{\mathrm{j}\{\mathrm{j}+1\}}|\psi\rangle, \tag{11.21}$$

where one is summing over all the lines in the spin network that thread the surface $\mathbf{\Sigma}$ (Fig. 43.1.b).

Loop Quantum Gravity also gives an expression for black hole entropy; this however depends on β; matching this with (7.16) fixes

$$\beta = \ln(2\mathrm{j}+1)/2\pi\sqrt{\mathrm{j}\{\mathrm{j}+1\}} = \ln 2/\sqrt{3}\pi; \tag{11.22}$$

the last equality is for j the minimal spin [which is 1/2 for $SU(2)$].

Finally, physicist Martin Bojowald developed Loop Quantum Cosmology [152]: Loop Quantum Gravity's counterpart of the Minisuperspace model arena. This has led to some claims about singularity avoidance, though generic such statements remain inconclusive.

11.10 Canonical Approach to Supergravity

Supergravity can also be considered from the canonical point of view; moreover, it manifests Background Independence. See [552] for an outline, or [232, 233, 314, 715, 868] for detailed accounts. This is habitually studied from a spacetime-first ontology, performing a space–time split, and then passing to a canonical formulation. To accommodate fermions, this is done as a first-order formulation rather than directly in terms of metric variables. Note also Teitelboim's alternative route [834] starting from Canonical GR and asking for a square root of the GR Hamiltonian constraint $\mathcal{H}$.

Supergravity in $3+1$ dimensions was furthermore shown [515] to be compatible with the Ashtekar variables development.

Conceptual development of Canonical Supergravity largely ceased in the 1980s. However, this has been found to differ from GR in a number of ways as regards the form of its of Background Independence [32, 36], even at the classical level. Thereby Canonical Supergravity features as a further example in Part II of this book.

11.11 Brane, Null Line, and Relational Alternatives

Passing from point particles to strings is nonunique along the following lines.

1) It admits an arbitrary-d generalization as regards which extended objects to involve, which can be followed up by the 'democratic' use of extended objects of all codimensions C. These other extended objects are called *(mem)branes* [719]. Whereas the notion of space felt by a species is the full space for a Field Theory, and just a point for a particle, it could also be a lower-d space to which a given matter species is confined, giving strings and membranes. Configurations are now the values at each point of the space of extent of an object.[11] Examples include *p-branes*, which are form fields (Appendix D.2), and *D-branes*, which are where open strings end.[12] Theoretical physicists Andrew Strominger and Cumrun Vafa [819] furthermore established the standard form of the black hole entropy (7.16) from brane microphysics, for a class of supersymmetric black holes.
2) One might instead consider null lines instead of strings. This emphasis here is on causality; this approach leads to Penrose's *Twistor Theory* (see Sect. 36.2 for a brief outline, or [707] if interested in the details).
3) Finally, one might view String Theory as following up Particle Physics' emphasis on particles by replacing these with strings. Relationalism on the other hand would ascribe reality to relations between particles. A further separate matter is modelling extended objects themselves from a relational point of view; it is not known whether this would return one to standard String and M-Theory or to an alternative.

Whereas 3)'s Relationalism was already presented in the Preface as a means of cutting the Gordian cube, Supersymmetry, strings and twistors can each be viewed as a different type of cut. Additionally, in further developing String Theory—to produce M-Theory—it is habitually 1) but not 2) or 3) which is taken on board. (Note however the twistor string theory development [918].)

[11]Branes provide further ways of hiding extra dimensions, such as 'warping', which are a further large source of phenomenological nonuniqueness.

[12]The 'D' here stands for Dirichlet boundary-value problem [220], named after 19th century mathematician Gustav Dirichlet.

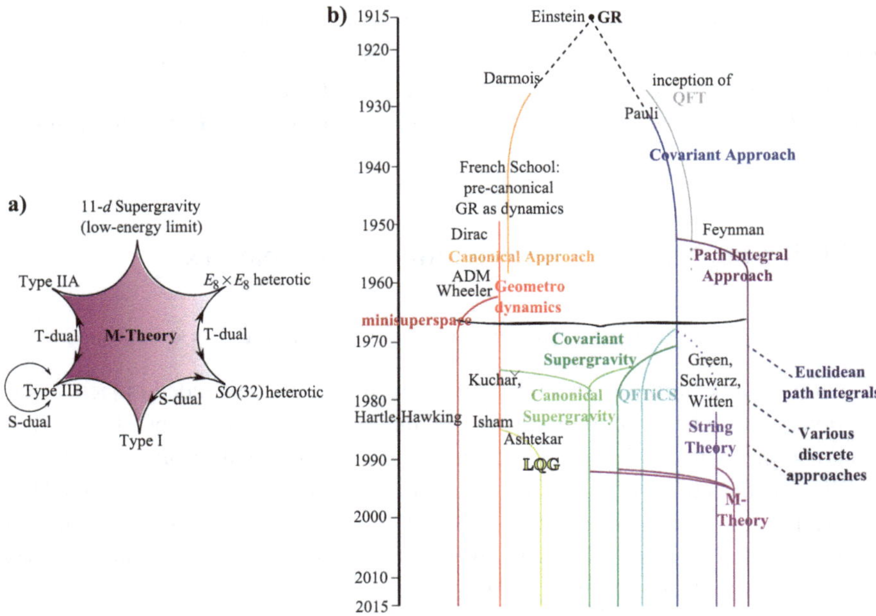

Fig. 11.5 **a)** The M-Theory web. This features not only the 5 superstring theories but also 11-*d* Supergravity as a lower-energy limit. **b)** Family tree of Quantum Gravity Programs. The underbrace marks the scope of the last all-embracing review, by DeWitt [237–239]

11.12 M-Theory

M-Theory has the following additional inputs.

A) *Dualities*, which are interrelations of the five string theories and 11-*d* Supergravity. In particular, *T-duality* relates a theory that is compactified on a circle of radius R to one on a circle of radius $1/R$ (the 'T' here stands for 'torus'). Also, *S-duality* relates the strong-coupling limit of one theory to the weak-coupling limit of another (the 'S' here stands for 'strong'). See Fig. 11.5.a) for the M-Theory web between these and 11-*d* Supergravity. So, following on from Sect. 11.8, the third argument for the stringy dimension 10 is how this can morph by T-duality into Supergravity and M-Theory's dimension 11.

One consequence of T-duality is that in M-Theory notions of space may be more complicated than elsewhere in Theoretical Physics. E.g. [782] considers both space and time from an emergent perspective in M-Theory.

M-Theory specifically possesses *M2* and *M5 branes* (spatially 2- and 5-*d* respectively). The M2 brane can be thought of as a string with an extra dimension blown up; conversely one can pass from it to a string by compactifying one dimension.

B) *Holography* concerns the possibility of a theory's degrees of freedom residing within a lower-dimensional theory on a screen (e.g. a boundary surface). This originated (see e.g. the review [917]) in 't Hooft's work on black holes. Theoretical physicist Juan Maldacena's [635] *AdS–CFT conjecture* is a subsequent major development. This concerns, on the one hand anti de Sitter space (AdS), which is maximally symmetric like $\mathbb{M}^4$ but with negative cosmological constant Λ. On the other hand, CFT is *Conformal Field Theory* [674, 719]: a type of QFT. Some forms of the conjecture concern being able to represent AdS (and asymptotically AdS) spacetimes in terms of the CFT on their boundary. This boundary can, moreover, be considered to be a type of background. If the conjecture holds, it amounts to a map between a QFT on a fixed background and a GR-like theory in the bulk interior.

We finally point to Spacetime Relationalism's axiom i) precluding perturbative String Theory from being amongst the relational theories; on the other hand, one would expect a sufficiently final form of M-Theory to comply.

11.13 Conclusion: A Family Tree Overview

We end with Fig. 11.5.b)'s approximate family tree of Quantum Gravity programs.

Chapter 12
Quantum-Level Background Independence and the Problem of Time

We culminate Part I by outlining Background Independence at the quantum level. Many of the more difficult parts of the Problem of Time [24, 26, 37, 40, 483, 552, 581, 584, 586, 589, 752, 899] occur because the 'time' of Background Independence GR and the 'time' of the ordinary Background Dependent Quantum Theory are mutually incompatible notions. This causes difficulties in trying to replace these two branches of Physics with a single framework in regimes in which neither Quantum Theory nor GR can be neglected. As explained in Chap. 11, such a replacement is required for parts of Black Hole Physics and Early-Universe Cosmology. The Problem of Time moreover is pervasive throughout sufficiently GR-like attempts at formulating Quantum Gravity, at both the quantum and classical levels. For now, we take the geometrodynamical and spacetime formulations of GR to be representative, and concentrate on these. Parts II and III subsequently comment on differences between these and Loop Quantum Gravity, Supergravity and M-Theory, as well as on Background Independence and the Problem of Time at the topological level and beyond.

12.1 Quantum Frozen Formalism Problem

The Schrödinger-picture *Frozen Formalism Problem* involves stationary alias timeless or frozen wave equations such as (5.11). These occur for GR and for model theories with Background Independence, in a setting in which in which one would expect—cf. the quantum 'evolution postulate'—time-dependent wave equations such as (5.10), or, possibly (6.1) or (6.8). This frozenness is well-known to be a consequence of the GR Hamiltonian constraint $\mathcal{H}$ being of the mathematical form $\mathcal{Q}$uad of Eq. (8.26). RPMs' $\mathcal{E}$ constraints are also of this form (9.3).

The next two paragraphs and Sects. 12.2, 12.6, 12.8, 12.9 and 12.10 introduce various strategies for this.

One of this book's main points is that the Wheeler–DeWitt equation of GR, $\widehat{\mathcal{H}}\Psi = 0$, can be traced back not only to the classical Hamiltonian constraint $\mathcal{H}$

E. Anderson, *The Problem of Time*, Fundamental Theories of Physics 190,
DOI 10.1007/978-3-319-58848-3_12

but furthermore to Temporal Relationalism. Temporal Relationalism provides constraints for the range of formulations of theories which implement this principle. These constraints are interpreted as equations of time, denoted in general by $\mathcal{C}$hronos; this interpretation provides a classical emergent Machian time resolution of the ab initio timelessness of these formulations. Both $\mathcal{H}$ and $\mathcal{E}$ can be taken to arise in this manner. Moreover, in each theory which possesses a $\mathcal{C}$hronos, this leads to an also apparently frozen quantum wave equation $\widehat{\mathcal{C}\text{hronos}}\,\Psi = 0$. Whereas $\mathcal{C}$hronos is of the form $\mathcal{Q}$uad, not all programs' $\mathcal{Q}$uad gets interpreted as an equation of time, in which case we write, rather, $\widehat{\mathcal{Q}\text{uad}}\,\Psi = 0$ at the quantum level.

Inner Product Problem alias *Hilbert Space Problem.* In Quantum Theory, the wave equation does not suffice to obtain physical answers, since are of the form $\langle\psi_1|\widehat{O}|\psi_2\rangle$, so an inner product input is also required. The Schrödinger inner product serves this purpose in Ordinary QM. Klein–Gordon Theory has its own distinct Klein–Gordon inner product; see e.g. Eqs. (6.4) and (11.8). Recollect that a Schrödinger inner product will not do in this setting because $\mathbb{M}^4$ is indefinite, which argument carries over to GR's $\mathfrak{Riem}(\boldsymbol{\Sigma})$ and minisuperspace $\mathfrak{Mini}(\boldsymbol{\Sigma})$ as well. However, a Klein–Gordon interpretation fails here too, on the further grounds outlined in Sect. 12.2 and Ex VI.11.vi). RPMs are different in this regard, since they—like Ordinary QM—have a positive-definite $\mathfrak{q}$, yielding a positive-definite inner product for which a Schrödinger interpretation is appropriate. Note moreover that the Inner Product Problem is a temporal issue—a subfacet of the Frozen Formalism Problem—due to the ties between inner products, conservation of probability and unitary evolution outlined in Sect. 5.3.

We finally point to Quantum Theory including further objects such as quantum operators and path integrals; see Chap. 52 as regards a Temporal Relationalism implementing form for these.

12.2 Timefunction-Based Strategies for Frozenness

External time is inappropriate for GR-like theories. This is firstly since it does not feature in the quantum wave equation for GR. Secondly, external time is incompatible with describing truly *closed* systems, which include in particular closed-universe quantum cosmologies. Moreover, how Quantum Theory is to be *interpreted* for whole-universe models is a recurring theme in this book. These matters leave us in need of some distinct conception of time; some possibilities for this are as follows.

A) *Emergent Time before Quantization.* If one considers there to be no time at the primary level for the whole-universe models, we have already seen subsequent classical resolution by an emergent Machian time t^{em} being abstracted from change. Unfortunately, this classical resolution fails to unfreeze the quantum equation $\widehat{\mathcal{C}\text{hronos}}\,\Psi = 0$. None the less, this can also be resolved by abstracting time from now quantum change, as follows.

B) *Emergent Time after Quantization.* A such can be considered in situations in which there are slow, heavy 'h' variables that provide an approximate time-standard with respect to which the other fast, light 'l' degrees of freedom evolve [419, 552, 586]. This occurs e.g. in the Semiclassical Approach for SIC [35, 419], h is scale (and homogeneous matter modes) and l are one or both of small anisotropies or small inhomogeneities. The Semiclassical Approach consists of the following steps.

i) Make the Born–Oppenheimer ansatz (named after Max Born and physicist Robert Oppenheimer)

$$\Psi(h,l) = \psi(h)|\chi(h,l)\rangle, \tag{12.1}$$

followed by the WKB ansatz (named after physicists Gregor Wentzel, Hendrik Kramers and Léon Brillouin)

$$\psi(h) = \exp(i\, S(h)/\hbar). \tag{12.2}$$

Each of these is accompanied by a suite of approximations, detailed in Chap. 46.

ii) We form the h-equation

$$\langle\chi|\widehat{\mathcal{Q}\mathrm{uad}}\,\Psi = 0. \tag{12.3}$$

To first approximation, this yields a Hamilton–Jacobi equation,[1]

$$\left\{\frac{\partial S}{\partial h}\right\}^2 = 2\{E - V(h)\}, \tag{12.4}$$

where $V(h)$ is the h-part of the potential. Furthermore, one way of solving this is for an approximate emergent semiclassical time $t^{\mathrm{sem}}(h)$.

iii) We next consider the l-equation

$$\{1 - |\chi\rangle\langle\chi|\}\widehat{\mathcal{Q}\mathrm{uad}}\,\Psi = 0. \tag{12.5}$$

In this initial form, this is a fluctuation equation. Moreover, it can be recast—modulo some more approximations—into an emergent-WKB-time-dependent Schrödinger equation for the l-degrees of freedom. E.g. the mechanical case of this is

$$i\,\hbar\frac{\partial|\chi\rangle}{\partial t^{\mathrm{sem}}} = \widehat{\varepsilon}_l|\chi\rangle. \tag{12.6}$$

The emergent-time-dependent left hand side arises from the cross-term $\partial_h|\chi\rangle\partial_h\psi$ (Ex VI.14). $\widehat{\varepsilon}_l$ is here the remaining piece of $\widehat{\varepsilon}$, which plays the role of Hamiltonian for the l-subsystem.

---

[1] For simplicity, we present the rest of this Section for Mechanics with one h degree of freedom; see [37] for consideration of multiple such and other generalizations.

iv) In this book's main approach, $\mathcal{Q}$uad arises as an equation of time $\mathcal{C}$hronos. t^{sem} can then be interpreted as [29, 37] a semiclassical Machian emergent time (whether for the above model arena or for GR Quantum Cosmology). To zeroth order in l, this and the classical $t_0^{\mathrm{em(J)}}$ coincide. However, this fails to be Machian in the sense of not permitting either classical or semiclassical l-change to contribute. To first order in l, however,

$$t_1^{\mathrm{sem}} = F[h, l, \mathrm{d}h, |\chi(l, h)\rangle], \tag{12.7}$$

which is now clearly distinct from the h–l expansion of (9.4),

$$t_1^{\mathrm{em(J)}} = F[h, l, \mathrm{d}h, \mathrm{d}l]. \tag{12.8}$$

This pairing of the previously known emergent semiclassical time and the classical Machian emergent time is new to the Relational Approach, as is the Machian reinterpretation of the former. Clearly also the difference between these two emergent times is in accord with the 'all changes have an opportunity to contribute' implementation of Mach's Time Principle.

As we shall see in Part II, the above derivation of a time-dependent Schrödinger equation ceases to function if the WKB scheme (ansatz and approximation). Moreover, in the quantum-cosmological context, the WKB scheme is not known to be a particularly strongly supported ansatz and approximation to make (see Chap. 46). This book props this up by combination with further Problem of Time strategies, which need to be individually developed from the classical level upwards. We return to this point in Sect. 12.9 after having surveyed the rest of the individual strategies and facets.

C) Further alternative strategies involve quantum-level continuations of Chap. 9.11's approaches. Since these involve continuing to use at the quantum level a candidate time found at the classical level, they are known as *time before quantum* approaches.

i) Riem time arises from the hyperbolic reformulation (9.20) of $\mathcal{H}$, due to the DeWittian indefiniteness. This leads to a Klein–Gordon-like quantum equation,

$$c^{-2}\partial_t^2\Psi = -\triangle_{\mathrm{True}}\Psi + C[\boldsymbol{h}]\Psi. \tag{12.9}$$

A conceptual outline of this approach is that

$$\text{perhaps the '}\triangle_{\boldsymbol{M}}\text{' in } \mathcal{H} \text{ is actually a wave operator, } \Box_{\boldsymbol{M}}. \tag{12.10}$$

One can next attempt the corresponding Klein–Gordon inner product interpretation. Unfortunately, this fails due the GR potential not being as complicit as Klein–Gordon theory's simple mass term (see Chap. 21, Ex VI.11.iv) and [581, 584] for details). This approach was traditionally billed as finding time after Quantization. It can however be set up just as well before Quantization [26].

ii) One could instead use a hidden time candidate, in terms of which a parabolic reformulation for $\mathcal{H}$ of type (9.21) arises, which is then promoted to a hidden-time-dependent Schrödinger equation

$$i\hbar\frac{\partial\Psi}{\partial t_{\text{hidden}}} = \widehat{\mathcal{H}}_{\text{True}}\Psi. \tag{12.11}$$

In particular, York time is an interesting candidate of this type, details for which are provided in Chaps. 21 and 44.

iii) If one uses the reference matter time candidate instead, another parabolic reformulation (9.23) ensues. This is then promoted to a reference-matter-time-dependent Schrödinger equation[2]

$$i\hbar\frac{\partial\Psi}{\partial t_{\text{ref}}} = \widehat{\mathcal{H}}_{\text{True}}\Psi. \tag{12.12}$$

12.3 Quantum Configurational Relationalism

We next consider Quantization for a physical theory subject to a group $\mathfrak{g}$ of physically meaningless transformations. Ab initio, there are two ways of quantizing such a theory.

1) *Reduced Quantization.* Here one first reduces out the constraints corresponding to $\mathfrak{g}$ at the classical level, and then one quantizes.
2) *Dirac Quantization.* Here one quantizes first. The constraints corresponding to $\mathfrak{g}$ are now promoted to further quantum wave equations

$$\widehat{\mathcal{G}\mathbf{auge}}\,\Psi = 0, \tag{12.13}$$

which are then solved at the quantum level.

An issue with 1) is that Quantum Theory is capable of discarding a physically-accepted $\mathfrak{g}$, by which being able to classically reduce out the classical $\mathfrak{g}$ ('Best Matching') does not necessarily imply an $\mathfrak{g}$-free quantum system (see Chaps. 42, 43, and 49 for details). On the other hand, the indirect $\mathfrak{g}$-act, $\mathfrak{g}$-all method (Sect. 9.10 and Chap. 13) continues to be applicable at the quantum level, whether as a means of formulating 2) or as an indirect means of expressing all subsequent objects required by one's theory if neither 1) or 2) can be solved.

12.4 Quantum Constraint Closure

Commutator brackets play an even more central role in Quantum Theory than Poisson brackets did at the classical level. Moreover, the quantum notion of *equal-time*

[2]The partial derivatives here are, strictly, an indication that a Minisuperspace model is for now being presented for convenience. See Chap. 47 for formal equations for this approach to full GR.

commutation relations poses significant difficulties in the context of GR. This is due to Ordinary Quantum Theory's 'equal-time' notion carrying connotations of there being a unique preassigned time, which does not fit GR's conception of time.

A first instance of equal-time commutation relations is in Kinematical Quantization (Chap. 39).

At the quantum level, constraints take the form of operator-valued equations. Moreover, passage from classical to quantum constraints is subject to operator-ordering ambiguities and well-definedness issues. One is then also to consider commutator brackets between these quantum constraints. For sure, algebraic closure of constraints is not automatically guaranteed in postulating the form these are to take at the quantum level:

$$\widehat{c}_{\mathsf{C}}\Psi = 0 \quad \not\Rightarrow \quad [\widehat{c}_{\mathsf{C}}, \widehat{c}_{\mathsf{C}'}]\Psi = 0. \tag{12.14}$$

Commutator bracket algebraic structures are furthermore not in general isomorphic to their classical Poisson brackets antecedents or approximands. Chapter 39 outlines the topological underpinnings of this discrepancy [475]. For constraint algebraic structures, this means that the quantum version is not necessarily isomorphic to the classical one. One consequence of this is that algebraic closure of classical constraints does not imply an isomorphic algebraic closure of quantum constraints, nor indeed of any other kind of quantum-level closure:

$$\{c_{\mathsf{C}}, c_{\mathsf{C}'}\} \approx 0 \quad \not\Rightarrow \quad [\widehat{c}_{\mathsf{C}}, \widehat{c}_{\mathsf{C}'}]\Psi = 0. \tag{12.15}$$

Furthermore, the latter set of constraints in general requires a distinct indexing set in place of C. It is clear from Sect. 6.5's description that anomalies are one manifestation of non-closure, and additionally a means by which a classically accepted $\mathfrak{g}$ may need to be replaced by a distinct $\mathfrak{g}'$ at the quantum level. While not all anomalies involve time or frame, a subset of them do, and these then form part of the Problem of Time. These issues are further developed in Chap. 49.

Breakdown of the closure of the constraint algebraic structure at the quantum level was termed the *Functional Evolution Problem* in [483, 586]. However, 'functional' carries field-theoretic connotations—it is the type of derivative that features in the field-theoretic form of the problem. Chapters 18 and 24 iron this out in species-neutral terms. Furthermore, to additionally include the classical case and maximally clarify the nature of this problem, we consider it better to refer to this facet as the *Constraint Closure Problem*.

Finally, this is an opportune point at which to mention that many approaches to Quantization (see [475] or Chaps. 39–42) are of at most limited value in the case of GR. This is due to the classical GR constraints forming the Dirac algebroid, whereas many an established approach to Quantization can only cope with Lie algebras.

12.5 Quantum Assignment of Beables

In Quantum Theory, observables or beables carry the further connotation of being self-adjoint operators $\widehat{B}$, by which their eigenvalues are real-valued and so can correspond to measured or experienced physical properties.

For systems which additionally possess quantum constraints, the $\widehat{B}$ are additionally to form zero quantum commutators with the quantum constraint operators. Examples of such notions are as follows.

$$\text{Quantum Dirac beables:} \quad \widehat{D} \text{ such that } \quad [\widehat{\mathcal{C}_{\mathsf{F}}}\,, \widehat{D}]\Psi = 0, \tag{12.16}$$

$$\text{quantum Kuchař beables:} \quad \widehat{K} \text{ such that } \quad [\widehat{\mathcal{F}\mathbf{lin}}, \widehat{K}]\Psi = 0, \tag{12.17}$$

$$\text{quantum } \mathfrak{g}\text{-beables:} \quad \widehat{G} \text{ such that } \quad [\widehat{\mathcal{G}\mathbf{auge}}, \widehat{G}]\Psi = 0, \tag{12.18}$$

need not coincide with the previous, and a further notion of

$$\text{quantum Chronos beables:} \quad \widehat{C} \text{ such that } \quad [\widehat{\mathcal{C}\text{hronos}}, \widehat{C}]\Psi = 0 \tag{12.19}$$

exists in cases for which $\widehat{\mathcal{C}\text{hronos}}$ closes as a subalgebraic structure. Quantum partial observables are defined as a continuation of their classical definition as well. The difference is that their capacity to 'predict numbers' now carries the inherent probabilistic connotations of Quantum Theory.

The *Problem of Quantum Beables* is that it is hard to come up with a sufficient set of these for QG theories.

Let us finally note that in the Heisenberg picture of QM, the apparent manifestation of frozenness is, rather,

$$[\widehat{H}, \widehat{B}]\Psi = 0 \tag{12.20}$$

with similar connotations to its classical antecedent (9.39).

12.6 Quantum-Level Timeless Approaches

Here the apparent timelessness of Sect. 12.1 is interpreted at face value by entertaining a Fully Timeless Worldview, and determining the extent to which Physics can be recovered therein. The familiar forms of notions such as temporal evolution, becoming and history would subsequently need to arise—as some kind of phenomenological *semblance* of dynamics or of history—from considerations of pure being. So far, this has often pointed toward adopting new interpretations of Quantum Theory. Four approaches along such lines are as follows.

A) The *Naïve Schrödinger Interpretation*. This is due to Hawking and physicist Don Page [441, 442], though its name was coined by Unruh and Wald [862]. This concerns the probabilities of being for questions about properties of the Universe, such as what is the probability that the Universe is large? Flat?

Isotropic? Homogeneous? Take note that these questions make no reference to 'when', 'how long for' or 'whether that state is attained in permanence at some point'. Answers to these questions arise by considering the probability that the Universe belongs to region R of $\mathfrak{q}$ that corresponds to a quantification of a particular such property, schematically

$$\text{Prob(R)} \propto \int_{\text{R}} |\Psi|^2 \mathbb{D}\mathbf{Q}, \tag{12.21}$$

for $\mathbb{D}\mathbf{Q}$ the volume element in $\mathfrak{q}$. This is a Timeless Approach in the sense of making no reference to time, rather than of restriction to a single instant. This approach is termed 'naïve' due to it not using any further features of the constraint equations, which limits this approach's applicability [586].[3] It is 'Schrödinger' in the sense of involving a Schrödinger inner product [439] for computing timeless relative probabilities. Finally, it is an 'Interpretation' in the sense of being an alternative to the standard Copenhagen Interpretation of QM.

B) The *Conditional Probabilities Interpretation*, due to Page alongside physicist William Wootters [694], goes further by addressing *conditioned* questions of being. An example of such a question is 'what is the probability that the Universe is flat given that it is isotropic?' See Sect. 51.2 for discussion of the particular objects that this approach is to compute. [694] moreover gives convincing arguments as regards external notions of time being incompatible with describing truly *closed* systems. Such a system's only physical states are, rather, *eigenstates* of the Hamiltonian operator, whose time evolution is essentially trivial.

C) *Records Theory* [21, 99, 101, 340, 411, 694] involves localized subconfigurations of a single instant of time. This approach requires considering—in each of space and configuration space $\mathfrak{q}$—notions of localization, probability distribution, information, correlation and pattern more generally. This book gives further novel analysis of classical and quantum Records Theory in Chaps. 26 and 51 respectively. Whether a semblance of dynamics or history can arise from this approach, however, remains an open question.

D) *Evolving constants of the motion* is a 'Heisenberg' rather than 'Schrödinger' picture of QM; see Chap. 50 for an outline and e.g. [752] for more.

C) and D) benefit from classical precursors whereas A) and B) are purely quantum approaches.

12.7 Quantum Spacetime Relationalism

Is spacetime—or any of its aspects—meaningful in QG? How does spacetime—or any originally missing aspects thereof—emerge in a suitable classical limit? Is

[3]This approach is not expected to cover all physically meaningful propositions or investigations. None the less, some elements of this approach resurface within various of the approaches below.

there a notion in QG which resembles the causality of SR, QFT and GR? If so, which aspects of classical causality are retained as fundamental, and how do the others emerge in the classical limit? Such questions lead to a Spacetime Relationalism versus Temporal-and-Configurational Relationalism debate. This in turn feeds into I) the quantum-level Feynman Path-Integral Approach versus Canonical Approach debate. II) Consideration of whether quantum-level versions of Refoliation Invariance and Spacetime Constructability aspects of Background Independence are required. Some further specific quantum level issues about spacetime are as follows.

1) Whether a hypersurface is spacelike depends on the spacetime metric **g**; however in QG this would be subject to quantum fluctuations [483]. In this way, the notion of 'spacelike' would depend on the quantum state, as would causal relations, including the microcausality condition (6.27) that is crucial for standard QFT. Most pairs of events $\vec{X}, \vec{Y} \in \mathfrak{m}$ would not be expected to be spacelike separated by at least one Lorentzian metric [318]. Moreover, if all metrics are 'virtually present' due to fluctuations, this is manifested e.g. in the path integral sum (6.27)'s right hand side being in general nonzero. In formulations in which spacetime is primary, this gives one further reason for difficulties with the notion of equal-time commutation relations in QG [483].
2) Relativity places importance upon labelling spacetime events by times and spatial frames of reference which are implemented by the deployment of physical clocks. What happens if one tries to model this using proper time at the quantum level [483]? Unfortunately, proper time intervals are built out of **g**, and thus are only meaningful after solving the equations of motion. This is rendered yet more problematic by **g**'s quantum fluctuations. On the other hand, attempting to circumvent this by casting time in the role of a quantum operator is in contravention of standard Quantum Theory for deep-seated interpretational reasons (Chaps. 5 and 6).
3) From a technical perspective, replacing *Poin*(4) by *Diff*($\mathfrak{m}$) vastly complicates the Representation Theory involved (Appendix V). Furthermore, the Representation Theory of the Dirac algebroid is *even more* difficult than that of *Diff*($\mathfrak{m}$).
4) Finally, if one's approach attempts to combine spacetime and canonical concepts, there is additional interplay as e.g. outlined Chap. 55.

12.8 Path Integral Approaches

The Problem of Time facets do not take an entirely fixed form. If one splits spacetime and works with a Canonical Approach, the Frozen Formalism Problem, Inner Product Problem and Foliation Dependence Problem: Aspect 6). occur. On the other hand, none of these occur if an unsplit spacetime formulation is used; Path Integral Approaches are along these lines. One might be tempted because of Path Integral Approaches being very successful in QFT, but these face their own set of very major problems if one attempts to apply them to Gravity. In this way, the Canonical versus

Path Integral dilemma amounts to choosing between two very different sets of hefty problems. This is rather reminiscent of Frodo offering the Ring to Galadriel, which she refuses on the grounds that her subsequently corrupted self would just become a *substantially different* kind of tyrant to Sauron [849]; due to this, the dilemma paraphrases her response.

0) 'In place of an Inner Product Problem, you will set up a Measure Problem. It shall not be limited to cases with timelike Killing vectors, but now requires *Diff*$(\mathfrak{m})$-invariance...' Having an explicit *Diff*$(\mathfrak{m})$-invariant measure is a different—but also considerable—problem, and also a reason why the Measure Problem remains time-related. More specifically, the Measure Problem is a further quantum-level part to Spacetime Relationalism. Finally, this is the most direct reason for a self-contained book on the Problem of Time to carry an outline of what measures are (Appendix P).

The long string of issues 1) to 7) of Sect. 11.6 then continues to apply, playing the analogous role of Galadriel's subsequent justification of how her corrupted self would be: "*beautiful and terrible as the Morning and the Night!... Stronger than the foundations of the earth.*" In particular, upon passage to curved spacetime, what was flat spacetime's straightforward Wick rotation from imaginary time back to real time becomes an ambiguity which is a further subfacet of the Problem of Time [193]. On these grounds, one might leave Path Integral Approaches to the QFT regime rather than entertaining their extension to Gravitational Theory, or not.

12.9 Consistent Histories Approaches

While *path* integrals are occasionally already called sum over *histories* formulations, in this book we adopt a more structured notion of history: *consistent histories*, which are paths that are furthermore decorated by projectors. See Appendix U.1 for an outline of projectors and Chap. 53 for details about consistent histories. In particular, *Histories Theory* [340, 428] gives a further family of approaches to the Problem of Time, based on the following.

Not Time but Histories Postulate. Ascribe primality to histories (rather than to time).

Moreover, Histories Theory alters some of the mathematics involved in Quantum Gravity, and is one of the ways of attempting combined Spacetime-and-Canonical Approaches. However, such programs remain largely incomplete (e.g. [566] is only a classical-level treatment).

There is also a histories before Quantization to histories after Quantization dilemma. The Isham–Linden approach [504] is of the first kind: a classical Histories Brackets Approach which, upon Quantization, becomes the Histories Projection Operator Approach. On the other hand, the older Gell-Mann–Hartle [340] version of Histories Theory is a purely quantum consideration. These two approaches differ furthermore as regards the particular form taken by the projectors, with the latter involving a discrete sequence of these, to the former's continuum; see Chap. 53 for further details.

Decoherence—a concept introduced in a more basic setting in Background Reading II.4—plays a key role in quantum-level Histories Theory, in the form of a decoherence functional; this point is further developed in Chaps. 48 and 53.

Combining the Emergent Semiclassical, Records and Histories Approaches is furthermore of particular interest [413] along the following lines. There is a Records Theory within Histories Theory [340, 411]. Histories decohering provides one possible way of obtaining a semiclassical regime in the first place. I.e. finding an underlying reason for the crucial WKB assumption without which the Semiclassical Approach does not work. What the records are will answer the also-elusive question of which degrees of freedom decohere which others in Quantum Cosmology. See [340, 411, 414] and Chap. 54 for more details about this approach.

12.10 Web of Quantum Problem of Time Strategies

This is presented in Fig. 12.1.

12.11 Quantum Foliation Independence: Aspect 6)

A Quantization of GR that retains the nice classical property of Refoliation Invariance would be conceptually sound and widely appealing. There is however no known way of guaranteeing this at the quantum level. If this property is not retained, Ψ_{in}, Kuchař argued that [586] starting with the same initial state *"on the initial hypersurface and developing it to the final hypersurface along two different routes produces inequality"*,

$$\Psi_{\text{fin-via-1}} \neq \Psi_{\text{fin-via-2}} \quad \text{(quantum Foliation Dependence criterion).} \tag{12.22}$$

Moreover, this *"violates what one would expect of a relativistic theory."*

On the other hand, association of times with foliations is expected to break down if the spacetime metric quantum-mechanically fluctuates as per the next Section's discussion.

12.12 Quantum Spacetime Constructability

Fluctuations of the dynamical entities are inevitable at the quantum level. For GR, as Wheeler pointed out [899], these are fluctuations of 3-geometry. These fluctuating geometries are, moreover, far too numerous to be embeddable within a single spacetime. Consequently, the beautiful geometrical manner in which that classical GR manages to be Refoliation Invariant breaks down at the quantum level.

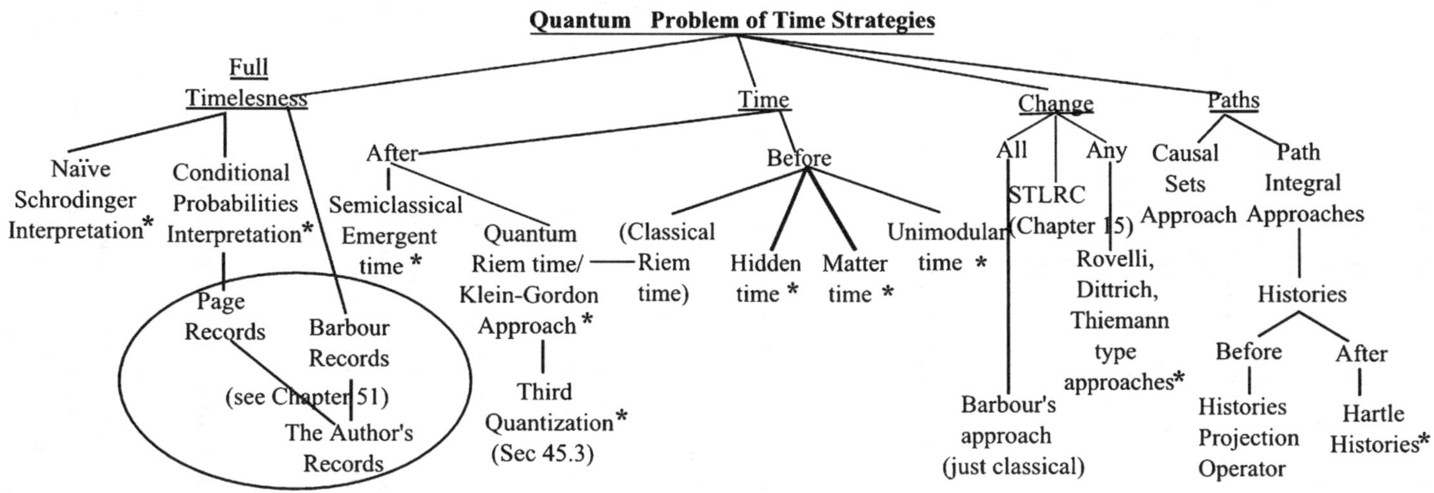

Fig. 12.1 The quantum-level extension of Fig. 10.2's web of Problem of Time strategies. * indicates the 10 strategies covered by Kuchař and Isham's ground-breaking reviews [483, 586]. Before and after are here meant relative to the Quantization procedure

Wheeler [899, 900] gave the following additional argument. Precisely-known position $\underline{q}$ and momentum $\underline{p}$ for a particle are a classical concept tied to the notion of its worldline. However, this perspective breaks down in Quantum Theory due to Heisenberg's Uncertainly Principle. In QM, worldlines are replaced by the more diffuse notion of wavepackets. Moreover, in the case of GR, the Uncertainty Principle now applies to the quantum operator counterparts of h_{ij} and p^{ij}. But by formula (8.21) this means that h_{ij} and K_{ij} are not precisely known. The idea of embeddability of a 3-space with metric h_{ij} within a spacetime is consequently quantum-mechanically compromised. Schematically,

$$\left(\begin{array}{c}\text{metric-level geometry}\\ \text{embedding data } \mathbf{h}, \mathbf{K} \text{ or } \mathbf{h}, \mathbf{p}\end{array}\right) \rightarrow \left(\begin{array}{c}\text{operators } \widehat{\mathbf{h}}, \widehat{\mathbf{p}} \text{ subject to}\\ \text{Heisenberg's Uncertainty Principle}\end{array}\right). \tag{12.23}$$

Thus Geometrodynamics (or similar formulations) would be expected to take over from spacetime formulations at the quantum level. It not then clear what becomes of notions that are strongly associated with classical GR spacetime. One such is causality. Another—if one ascribes to Wheeler's belief [897, 899] that the quantum replacement for spacetime is 'foamy'—is locality. In particular, microcausality is violated in some of these approaches [474, 477, 483].

A further issue concerns recovering continuity in suitable limits in approaches that treat space or spacetime as primarily discrete. Additional features of spacetime —e.g. its dimensionality—may require emergence in discrete (or bottom-up: see Chap. 11) approaches to Quantum Gravity [5, 151, 217, 403, 710, 804, 911]. This is not a given, since some approaches produce non-classical entities or too low a continuum dimension. Recovery of a semiclassical regime, or for the purpose of the recovery of standard Particle Physics results, has also been a long-standing difficulty with Loop Quantum Gravity [112, 320, 756, 845].

Investigation of the semiclassical and quantum commutator bracket counterpart of the classical Dirac algebroid has also begun [155]. The extension of such work to a family of such algebraic structures in parallel with Sect. 10.9 remains to be tackled.

12.13 Summary of a Local Problem of Time

Impasses arise from Background Independence versus Background Dependence clashes in attempting to put together GR and Quantum Theory are Problem of Time facets. Kuchař and Isham [483, 586] provided a classification of these very interesting foundational features. In this way, the Problem of Time is the Quantum Gravitational manifestation of the absolute versus relational motion debate.

Part I also argued that Temporal Relationalism can be approached via Machian emergent time at both classical and quantum levels, providing in particular classical and semiclassical resolutions. In response to Sect. 1.5, this approach—and many other approaches to time in QG—are concentrated principally within Broad's

Worldview (with occasional use of the Fully Timeless Worldview instead) as opposed to more widely spanning philosophers' further worldviews on time.

Classical means of handling Configurational Relationalism and Constraint Closure carry over to the quantum level (Fig. 10.6). This is subject to the caveat of not having a firm replacement for the Dirac algebroid structure of the classical GR constraints. Moreover, considering systematic approaches for finding enough observables or beables for Gravitational Theories lies outside the scope of Part I; indeed, the Problem of Beables remains unresolved for these theories. Finally, while Spacetime Relationalism, Foliation Independence and Spacetime Constructability were well-addressed at the classical level, they are harder to handle at the quantum level; even semiclassical versions of the last two remain largely unknown.

Background Independence is furthermore endemic in approaches which implement GR-like features. For instance, it is between relevant and central in Geometrodynamics, Loop Quantum Gravity, Canonical Supergravity and M-Theory (but not the perturbative part of String Theory); these represent a sizeable proportion of Quantum Gravitational research.

12.14 Aspect 8: Global Validity

We next outline the global—rather than just local—version at the quantum level. Here Chap. 11's classification by entities to be meshed together further expands into 'patching representations', 'patching functional differential equation solutions', and 'patching unitary evolutions'. Moreover, these remain little understood conceptually, let alone technically. Many of the facets and strategies have additional globality issues at the quantum level, as outlined in Epilogue III.B.

12.15 Aspect 9: No Unexplained Multiplicities

Let us first explain what we meant in Chap. 10 by the corresponding Problem of Time facet—the *Multiple Choice Problem*—only becoming relevant upon making quantum-level considerations. This is not the same as it being a quantum-level phenomenon; one is rather dealing with a four-legged beast with two legs in the classical realm and the other two in the quantum realm. More specifically, that canonical equivalence of two classical formulations of a theory need not imply unitary equivalence of the Quantization of each (Fig. 12.2). Representation of most classical canonical transformations by unitary operators cannot in fact be attained while concurrently maintaining the irreducibility of the canonical commutation relations [483]: the so-called *Groenewold–Van Hove phenomenon* [376]. In this manner, a single classical theory can lead to multiple inequivalent quantum theories. These arise since Quantization amounts to requiring some preferred subalgebraic structure of classical phase space functions to be selected [475]. See Epilogue III.A for further details and references.

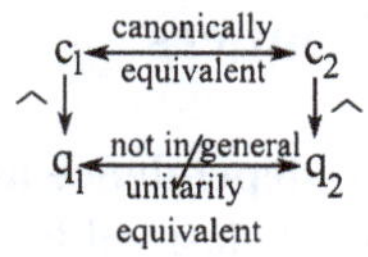

Fig. 12.2 Multiple Choice Problem: if the classical objects c_1 and c_2 are canonically related, their Quantizations q_1 and q_2 need not be unitarily related

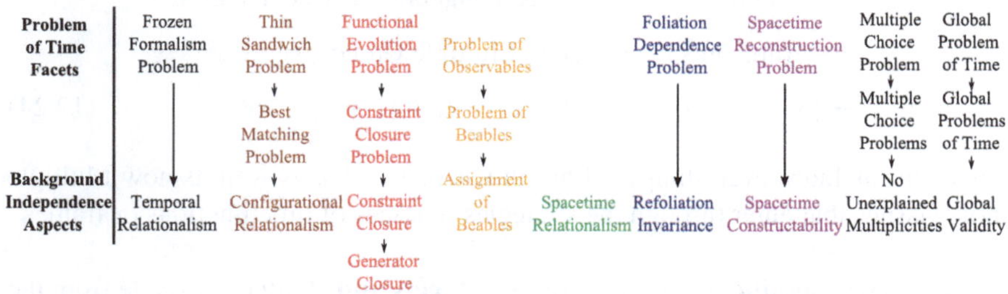

Fig. 12.3 Evolution of conceptualization and nomenclature of Problem of Time facets over the course of this book. The first row are Kuchař and Isham's [483, 586]. The last full row are the underlying Background Independence aspects arrived at in the current book. This Figure's colour scheme for the first seven is further used in Part II's presentation of each facet and of how facets interfere with each other. Moreover, 8/9ths of these aspects are already classically present: all bar the issue of physically and conceptually unaccounted-for multiplicities. Generator Closure then furthermore applies to Spacetime Relationalism, which also has its own 'generator complying' notion of observables, and, at the quantum level, a Measure Problem

Kuchař [586] characterized the Multiple Choice Problem an 'embarrassment of riches', in contrast to the Global Problem of Time for timefunctions' being an 'embarrassment of poverty'. This is since, given the Multiple Choice Problem, it is not clear which of these inequivalent Quantizations would be realized by Nature at the quantum level. For instance, this could render Loop and Geometrodynamics formulations of GR inequivalent at the quantum level; which—if any—of these would Nature choose to realize?

The above considerations of multiplicity become a temporal matter through e.g. applying to pairs of internal times, of frame variables, or, more circuitously, of classical beables subalgebraic structures selected for the purpose of Quantization. This diversity leads us to henceforth use the plural *Multiple Choice Problems* for Facet 9).

12.16 Conclusion. i. Summary Figures

Figure 12.3 summarizes the progress made in understanding of the Background Independence aspects and Problem of Time facets.

On the other hand, Fig. 12.4 juxtaposes the temporal features from earlier Chapters with some of the relations between these and Problem of Time facets. Parts II and III return to some, but by no means all, of these time features in various approaches to the Problem of Time.

12.17 ii. Quantum-Level Frontiers

1) Enlarged set of Background Independence aspects and thus of Problem of Time facets. Sect. 10.12 presented 42 classical Background Independence aspects; at the quantum level, subalgebraic structure selection additionally involves Multiple Choice aspects. The count of aspects consequently increases to

$$\begin{aligned}&(\text{primary entities}) \times ((\text{provider} + \text{algebras}) \times (\text{local} + \text{global})\\&\quad + (\text{algebras again due to Multiple Choice}))\\&= \{3 + 1 + 3\} \times \{\{1 + 1 + 1\} \times 2\} + \{1 + 1\}\} = 56. \end{aligned} \tag{12.24}$$

This count includes everything laid out in Chap. 10.12's, as well as now Multiple Choice issues that enter through the strategies' choices of timefunctions or frames.

2) Particularly significant combinations of facets and strategies. Aside from the groupings in Fig. 10.6, the following combinations of facets are of particular significance.

i) 'A Local Canonical Approach' involves facets 1) to 7) but with facet 5) confined to playing a secondary role.
ii) 'A Local Path, Spacetime or Covariant Approach' involves keeping facets 3) to 7) while dropping facets 1) and 2).
iii) 'A Local Canonical-and-Covariant Approach' involves keeping all of facets 1), 2) and 5) at once.
iv) Part II demonstrates 'factorization' into facets 1)–3), 4) and 5)–7) at the classical level, though Part III then shows that the first two factors merge at the quantum level.

N.B. that the number of facets—and of strategies for each facet—renders it highly impractical to study the Problem of Time exhaustively. Instead, in Parts II and III, I concentrate on a particularly desirable ordering of the facets, with composite strategies following suit. This ordering begins with Configurational and Temporal Relationalism, and then considers Constraint Closure as a means of Spacetime Construction. From here, one can separately deal with spacetime's own Relationalism and Refoliation Invariance on the one hand, and with Assignment of Beables on the other. The corresponding strategies are the Classical Machian Emergent Time Approach, followed by its semiclassical counterpart, which is then supported by Histories and Records Theories. The histories decohering provide a semiclassical regime, whereas the records explain which species decohere which others. The histories and the Semiclassical Approach can each provide the semblance of dynamics, with which a pure Records Approach would have difficulties. Finally, the Semiclassical Approach casts the whole scheme in a Machian framework.

3) Further theories of Gravitation. Part II shows that Supergravity has a substantially different manifestation of Background Independence from GR even at the classical level. Since Supergravity is a lower-energy limit of M-Theory, this example could furthermore be considered a prequel to investigating Background Independence and the Problem of Time in M-Theory.

4) Universal strategies. Some Problem of Time strategies are *universal* in the sense that they exist regardless of what the underlying theory is, at least for a large number of steps. For instance, universality applies to Emergent Machian Times Approaches, Timeless Approaches and Histories Approaches. A further consequence of universality is that Parts II and III are not only of interest for the specific theories used there as examples, but also of much wider interest throughout Gravitational Theory.

5) Quantum Background Independence for deeper levels of structure. Quantization of the deeper levels of mathematical structure is outlined in Epilogue III.C. At the topological manifold level, *Topological Field Theory* (*TFT*) [915, 916] is an interesting less structured variant of QFT. Along the other side of the Planckian cube, questions yet again by Wheeler [897, 898] pointed to the interesting issue of incorporating *topology change in quantum GR* [350, 351]. He envisaged this in terms of 'spacetime foam', and Path Integral Approach sums over topological manifolds and transition amplitudes between different spatial topological manifolds.

On the other hand, Isham [480–482, 497] went beyond the topological manifold level to consider Quantization at the level of topological spaces. In fact, he has also considered quantizing even more general structures that are no longer based on equipped sets. See Epilogue III.C and Appendix U for an outline, and [260, 491–494, 498] for a more full account. Within the Equipped Sets Foundational System of Mathematics, the structures involved start to simplify as one approaches the sets themselves (via collections of subsets). As regards modelling the Universe, however, this may just be a further encroachment by Background Dependent thinking. One idea here is that basing topological spaces on progressively simpler structures might be replaced by basing topological spaces on increasingly general structures, such as sheaves and topoi. See Appendix W for an outline of each; the latter also provides a replacement for the conventional set-theoretic foundations of Mathematics. This corresponds to the flat space, curved differentiable manifold, topological manifold, topological space, topos... progression in notions of space and of spacetime. It is also a wide enough arena in which to investigate the extent to which our Universe can be modelled by the conventional 'continuum' ideas [480, 481].

12.18 Exercises VI: Quantum Gravity, Background Independence and the Problem of Time

Exercise 0) Given that we have considered QFT and GR, consider the third 'double combination' vertex of the 'Planckian cube' (Fig. 1): *Quantum Newtonian Gravity*. i) Compare its characteristic lengthscale (3) with the Bohr radius and the Hubble radius, so as to produce two arguments against the physical relevance of this regime. ii) Provide a third argument based on decoherence: on what timescale would such a quantum system decohere?

Exercise 1) [A feel for the extremeness of the Planck units.] Derive each of the Planckian quantities in the Preface and in Chap. 11, alongside the order of magnitude estimates these were compared to there.

Exercise 2) [Naïve power-counting renormalization arguments.] Reconsider these (Sects. 6.5 and 11.2) in arbitrary-d.

Further Reading 1) i) Work through [143] as regards which features of $\mathbb{M}^4$ QFT are lost in various simple curved-space models. ii) Also work through [874]'s treatment of *Bogoliubov transformations* of Hawking radiation.

Exercise 3) i) Justify the Penrose diagram (Fig. 11.2.b) for a star collapsing to form a black hole which subsequently undergoes Hawking evaporation. ii) Also estimate the Hawking lifetime of a solar-mass black hole. iii) Compute the heat capacity of a black hole. iv) Show that the end-point of evaporation is explosively unstable, at least if the known Laws of Physics are taken to their logical conclusion. v) Finally give an alternative derivation of Hawking lifetime from the Salecker–Wigner clock inequalities (5.18).

Exercise 4) i) Canonically quantize Electromagnetism. Naïvely, A_μ would have 4 field-theoretic degrees of freedom. However, the Gauss constraint is present, and additionally (4.10) is invariant under the gauge transformations (6.18). Thus Electromagnetism has two degrees of freedom per space point, corresponding to light having two polarizations. Get round this feature by working in the Coulomb gauge. ii)† Consider how the Yang–Mills counterpart of a) fares.

Exercise 5) i) Show that the ansatz (10.4) leads to a wave equation and interpret the two ensuing polarizations. ii) Derive the gravitational wave formula (time-averaged luminosity) $\propto \langle \dddot{\mathrm{J}}_{ij}\dddot{\mathrm{J}}^{ij}\rangle$ for slowly-moving weakly gravitating sources. Here, $\mathrm{J} = \mathrm{I}_{ij} - \mathrm{I}\delta_{ij}/3$: a tracefree quadrupole moment, with function dependence on the $t - r$ of the asymptotic background flat spacetime. iii) Provide electromagnetic counterparts for i) and ii). iv) Give order of magnitude estimates of the timescales on which atoms modelled purely classically would succumb to collapse due to each of electromagnetic and gravitational radiation.

Exercise 6) Formulate quantum linearized GR.

Exercise 7) Work through Appendix I.1's account of the geometry of Minisuperspace.

Exercise 8) [Canonical isotropic Minisuperspace.] i) Show that the Hamiltonian constraint is now $\pi_\phi^2/2a^3 - \pi_a^2/24a6a - m^2a^3\phi^2/2 = 0$ for minimally-coupled scalar field matter; consider also the effect of adding a cosmological constant term. ii) Let the scalefactor a be an intrinsic time. Write down a classical 'true Hamiltonian' for this and promote it to a quantum wave equation. iii) Repeat with Misner time $\Omega = -\ln a$ instead. iv) Choose instead to turn the above Hamiltonian constraint into a Klein–Gordon type equation. v) Demonstrate inequivalence between this and the preceding. vi)† Show how Configurational Relationalism, Refoliation Invariance and Spacetime Construction are all trivialized in isotropic Minisuperspace, in the setting in which one adheres to the spatial hypersurfaces privileged by homogeneity.

Exercise 9) [Canonical Bianchi Minisuperspaces.] Consider

$$\mathrm{d}s^2 = \alpha^2(t)\mathrm{d}t^2 - a(t)^2 h_{ij}(t)\mathrm{d}x^i\mathrm{d}x^j$$

for $h_{ij} = \exp(-2\Omega)\beta_{ij}$ and $\beta_{ij} = \text{diag}(\beta_+ + \sqrt{3}\beta_-, \beta_+ - \sqrt{3}\beta_-, -2\beta_+)$. i) Show that the GR kinetic term forms the combination $-\pi_\Omega^2 + \pi_+^2 + \pi_-^2$. ii) Solve the corresponding quantum wave equation in the case with zero potential (this corresponds to Bianchi I spacetime). iii) For Bianchi IX, the Ricci 3-scaler potential is given by Eqs. (I.5)–(I.6); sketch this potential, and explain how Bianchi I bears relation to it in an asymptotic manner.

Exercise 10)† i) Show that preshape space—unscaled relative coordinate space—is a sphere $\mathbb{S}^{nd-1}$ in dimension d for $n = N - 1$ and N the number of particles. ii) Rederive the above results by performing Best Matching with respect to translations and dilations. iii) Calculate the Poisson bracket of ε with $\mathcal{L}$. iv) Reduce out the rotations as well, for 3 particles of equal mass in 2-d, and show that this geometry is $\mathbb{S}^2$ by finding a coordinate transformation to the standard spherical coordinates. v) Sketch where the qualitatively different types of configuration occur. In particular, where are the configurations containing tight binaries; why are these common in the Newtonian gravitational potential version of the 3-body problem? vi) What are the group orbits for the scale-invariant configuration space of 3 particles in the centre of mass frame under the action of the rotations in each of 2- and 3-d? vii) Repeat v) but now with the extra assumption of indistinguishable particles, to show that the $\mathfrak{q}$ space geometry is now closely related to $\mathbb{R}^3$. viii) Repeat v) but now without Best Matching out the dilations; what is $\mathfrak{q}$'s geometry? ix) Find the form and physical meaning of the unit Cartesian vectors in the $\mathbb{R}^3$ of scaled triangle configurations that the $\mathbb{S}^2$ of scaled triangles in 2-d can be taken to sit inside. How are these related to this model's $\mathbb{R}^4$ of unit Jacobi vectors?

Exercise 11) Show that i) in Electromagnetism the thick sandwich fails to be Wheeler–DeWitt equation (defined in Appendix O.1). ii) The given expression for York time (9.22) is the momentum conjugate to the usual 3-d Geometrodynamics' $\sqrt{\mathrm{h}}$. iii) York time is monotonic in recollapsing FLRW universes. iv) Work through [124] and [921] to [925] to gain experience of the GR initial value problem and some practise in manipulating (conformal) Killing equations. v) Explain why a Schrödinger inner product is unsuitable for an indefinite quantum wave equation. vi) Show that

$$\mathcal{E} = -\int_\Sigma \mathrm{h}_{ij}\mathrm{p}^{ij}\mathrm{d}V$$

is a conformal Killing vector for the GR configuration space metric but that this does not respect the potential term as well. [Compare its Poisson bracket with the potential term with that with its kinetic term.] What are the implications of this for a Klein–Gordon type interpretation of the equations of Quantum Geometrodynamics?

Exercise 12) i) For massless fermions, show that $\mathrm{j}^{\mu 5} := \bar{\psi}\gamma^\mu\gamma^5\psi$ is a classically-conserved current. ii) Show that quantum-level conservation is out by a term

$$-\frac{e^2}{16\pi^2}\epsilon^{\mu\nu\rho\sigma}\mathrm{F}_{\mu\nu}\mathrm{F}_{\rho\sigma}.$$

iii) Show that a δ-function derivative term arises in the corresponding brackets algebra. iv)† Interpret the above anomalous terms from a topological point of view.

Exercise 13)† Which global problems are already implicit in Chaps. 9, 10 and 12's outline descriptions of Background Independence aspects, Problem of Time facets and strategies?

Exercise 14) Work through the Semiclassical Approach's emergent time working from (12.5) to (12.6).

Exercise 15) Set up a histories formulation for i) Newtonian Mechanics. ii)† GR.

Further Reading 2) Read the introductory account to Regge Calculus in [660]. Then work through Barrett and Crane's improvement of this model [111]. Also figure out for yourself how Causal Dynamical Triangulation [6], Spin Foams [711] and the Causal Sets Approach [802, 804] differ in which structures they keep or discretize.

Exercise 16)† How is the Group and Representation Theory of $Diff(\mathbb{S}^2)$ is harder than that of $Diff(\mathbb{S}^1)$?

Further Reading 3) At an introductory level, see e.g. [552] for each of the perturbative Covariant Approach and Supergravity, [933] for String Theory, and e.g. [331] for Loop Quantum Gravity.

Further Reading 4) More advanced reading includes the overviews of Quantum Gravity in [194, 237–239, 471, 474, 478, 485, 552, 802] and Part III. More specialized books include [385, 386, 719] on String Theory, [232, 868] on Canonical Supergravity, and [154, 746, 845] on Loop Quantum Gravity.

Exercise 17)† Show that the Ashtekar variables formulation of GR fits the same set of Background Independence criteria as Geometrodynamics does. [We shall see in Part II that e.g. considering Supergravity makes a far larger difference to Background Independence and the Problem of Time.]

Exercise 18) i) Which terms in the Standard Model Lagrangian account for the Standard Model vertices (Fig. 6.1.b)? ii) Upon including Gravitation, which terms account for the further vertices in Fig. 11.1? iii) What further vertices does the Minimal Supersymmetric Standard Model exhibit? iv) Which further vertices feature in alternative theories such as Supergravity and Kaluza–Klein Theory?

Exercise 19) (Very long!) Obtain i) the equations of motion and ii)† canonical formulations corresponding to all the actions outlined in Chap. 11.

Exercise 20)† (Very long!) i) Show that gauging the Poincaré group leads to the Einstein–Hilbert action of GR. ii) Show that gauging the $\mathrm{N} = 1$ Poincaré supergroup leads to the corresponding Supergravity theory's action; also find the representations of the $\mathrm{N} = 1$ Poincaré supergroup. iii) Redo ii) for all possible values of (N, d) pairs. iv) Redo ii) now involving each of the conformal group and the $\mathrm{N} = 2$ superconformal group.

Further Reading 5) [M-Theory.] See e.g. [136] as regards a canonical formulation, and [719] for other approaches.

Further Reading 6) See [475, 919] for Geometrical Quantization, Parts II, III and [483, 586] for the Problem of Time and Background Independence, and [482] and Epilogues II.C and III.C for Background Independence and Quantization at deeper levels of mathematical structure.

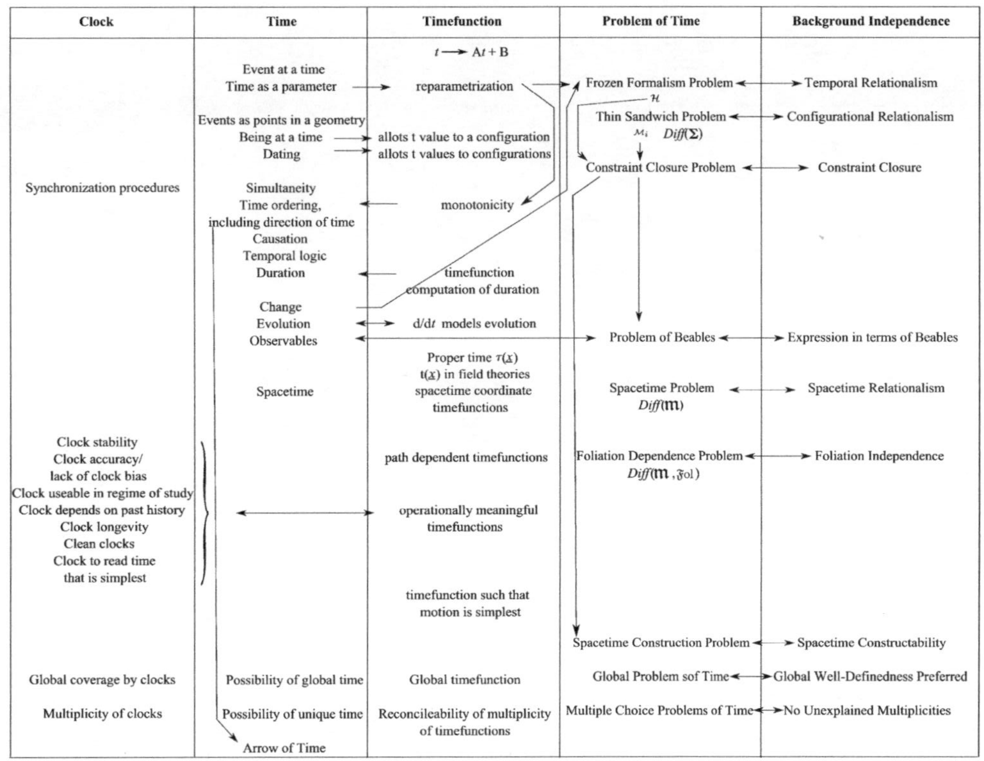

Fig. 12.4 The first eight Chapters develop the clock, time and timefunction properties in the first three columns for the standard Paradigms of Physics. On the other hand, Chaps. 9, 10 and 12 consider the Problem of Time facets and Background Independence aspects in the last two columns. These represent the main tensions between Background Dependent and Background Independent Paradigms in trying to set up theories of QG. This summary table is additionally useful through the row groupings indicating which items in each list bear relation to items in the other lists

Part II
Classical Problem of Time

This Part gives a detailed treatment of the Background Independence aspects exhibited by classical GR, alongside the ensuing Problem of Time facets. In particular, we concentrate upon interferences between these classical-level facets. We make particular use of the geometrodynamical formulation of GR, and of Relational Particle Mechanics (RPM), Minisuperspace and Slightly Inhomogeneous Cosmology (SIC) model arenas; a wider range of model arenas and alternative theories are on occasion evoked (most frequently Electromagnetism, Yang–Mills Theory, Supergravity and Ashtekar Variables approaches). In the opening chapters of Part II, we also expand on Chap. 3's absolute versus relational motion debate, in particular as regards Leibnizian and Machian features in Physics, including ephemeris-type times.

Chapter 13
Advanced Nomenclature for Facet Interference

13.1 The Various Primary Ontologies Considered

Part II begins (Chaps. 16–26) by considering worldviews in which primality is ascribed to space. Chapter 27 subsequently serves as a spacetime primality counterpoint. This primality, or features of spacetime, then enter a number of subsequent Chapters such as 31, 32, 34, which foliate spacetime, or 33, which constructs spacetime.

Spatial primality can be argued, for instance, from Mach's conceptual distinctions between space and time, Broad's point about co-geometrization not undermining conceptual distinction, and Wheeler's points about spacetime losing primary significance at the quantum level and the role of Dynamics in the development of Physics. These points suggest favouring separate treatment of space and time—(3, 1) primality—rather than spacetime primality. Epilogue II.A then chips in with subsequent arguments against null splits and for spatial surface centred $3 + 1$ splits over time centred $1 + 3$ threading splits. Configurations, configuration spaces and Dynamics are ready elements for programs with spatial primality. Indeed, Part II's account of spatial primality begins by considering configurations, configuration spaces and their Configurational Relationalism, by which this is the first Background Independence aspect to treat.

In approaches which assume space, $\boldsymbol{Q}$ and $\mathfrak{q}$, Fig. 13.1's ladder of increasing levels of structure assumed are furthermore encountered.

Rung 1) *Fully Timeless Approaches* involve a single-instant ontology in terms of the $\boldsymbol{Q}$ that form $\mathfrak{q}$ alone. We begin with this in Chap. 14, though its structural sparsity leaves one in need of making more use of what structures and concepts remain, which takes one outside of what is habitually covered in Theoretical Physics. Because of this, we postpone detailed treatment of I) to 26, first developing II) instead (Chaps. 14–25) due to its anchorage on the relatively familiar structures below.

Rung 2) '*Non Tempus sed Cambium*'—'not time, but change'—formulations assume both $\boldsymbol{Q}$ and $\mathrm{d}\boldsymbol{Q}$. Formulations in terms of $\boldsymbol{Q}$, $\dot{\boldsymbol{Q}}$ (velocities) or $\boldsymbol{Q}$, $\boldsymbol{P}$ (mo-

E. Anderson, *The Problem of Time*, Fundamental Theories of Physics 190,
DOI 10.1007/978-3-319-58848-3_13

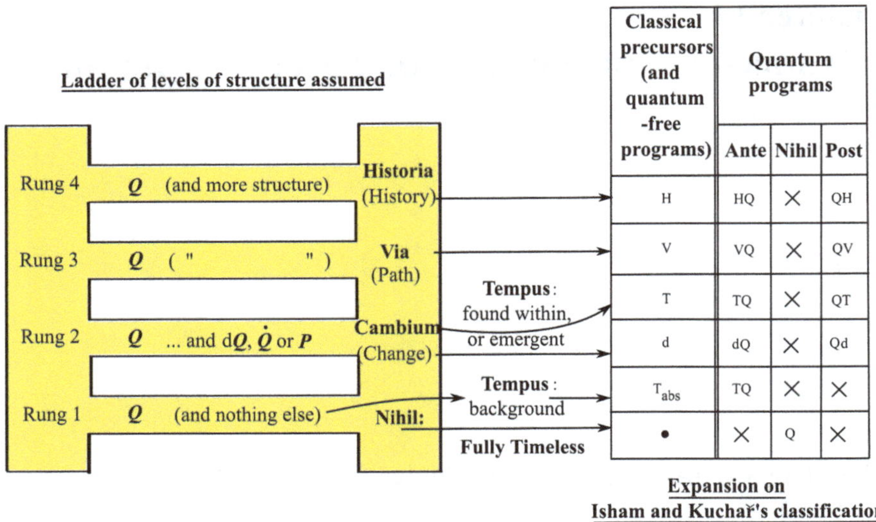

Fig. 13.1 Ladder of levels of structure assumed, including background, internal and emergent time, and subsequent extension of Isham and Kuchař's 3-fold classification—of *Tempus Ante Quantum*, *Tempus Post Quantum* and *Tempus Nihil Est*—into a 16-fold classification

menta) are also used on some occasions. The change and velocity formulations amount to extending $\mathfrak{q}$ to a tangent bundle $\mathfrak{T}(\mathfrak{q})$, whereas formulations in terms of momenta correspond to a cotangent bundle $\mathfrak{T}^*(\mathfrak{q})$. See Appendix F.4 for a general outline of Fibre Bundle Theory. These additional structures allow, for instance, for such familiar objects as actions and Hamiltonians. Formulations based on $\boldsymbol{Q}$, $\boldsymbol{P}$ and a Poisson bracket $\{\ ,\ \}$ thereupon are termed *canonical*. Finally note that, in contrast to rung I), rung II) additionally affords use of Mach's 'time is to be abstracted from change' as a source of emergent times at the secondary level.

Rung 3) '*Non Tempus sed Via*' formulations involve finite paths $\gamma : \mathfrak{I} \longrightarrow \boldsymbol{Q}(\lambda)$ rather than just infinitesimal changes on $\mathfrak{q}$.

Rung 4) Finally '*Non Tempus sed Historia*' formulations involve histories $\eta : \mathfrak{I} \longrightarrow \boldsymbol{Q}(\lambda)$, now alongside further structure than is usually ascribed to paths.

N.B. that the distinction between rungs 1) and 2) is a substantial one to emphasize. This is since both of these [and even on some occasions rungs III) and IV)] have been termed 'Timeless Approaches' alias '*Tempus Nihil Est*' [483, 586]. Adhering to 1) in place of 2) amounts to a stricter brand of timelessness. Namely, 'there is no time at the primary level so configuration is all' versus additionally allotting primality to change of configuration (or to velocity, or to momentum).

For now, let us view both rungs 3) and 4) as involving 'thick' rather than just 'thin' infinitesimal changes, in the same sense as 'thick' versus 'thin' for sandwiches in Fig. 9.4. We defer making finer distinction between rungs 3) and 4) to Chap. 28 due to parts of this resting on both canonical and spacetime primality positions.

13.2 The Cubert Classification of Quantization and Facets Ordering

In keeping track of, and presenting, Problem of Time facet orderings and interferences, a compact and clear notation for which facets are involved is much to be recommended [26]. Being clear about at which point Quantum Theory enters each approach is also crucial. CQBRT, pronounced 'Cubert' [26], generalizes a number of well-known procedural orderings, for instance not only Isham and Kuchař's *Tempus Ante Quantum*, *Tempus Post Quantum* and *Tempus Nihil Est*, but also the distinction between Dirac and Reduced Quantization. The C stands for Closure, the Q for Quantization, the B for Assignment of Beables, the R for Reduction (taking into account Configurational Relationalism), and the T for assigning a timefunction (thereby resolving Temporal Relationalism's primary timelessness).

A first obvious use is declaring (some subset of) these letters in any order; to keep this clearly distinct from surrounding text, we reserve the small upright Latin capital letters for this purpose. Then for instance TQ denotes *Tempus Ante Quantum* (more strictly, this is T ... Q, since addressing facets in between does not detract that longer ordering from being *Tempus Ante Quantum*). Likewise, Q ... T is Tempus Post Quantum, and any ordering with no T in it that is held to correspond to a complete approach is *Tempus Nihil Est* . Similarly also R ... Q is Reduced Quantization, whereas Q ... R is Dirac Quantization. N.B. that 'Reduced' is used here of the elimination of first-class linear constraints $\mathcal{F}$**lin**, and not of eliminating further non-linear constraints such as GR's $\mathcal{H}$ as well or instead. For eliminating $\mathcal{H}$ classically amounts to prescribing a timefunction, and as such is the T move. In this way, among many others, using the CQBRT notation clarifies when practitioners have been calling substantially different programs by similar names, thus avoiding mistakes of conflation as well as of misunderstood conceptual and technical content. It also permits ready identification of substantially similar programs by these having the same, or almost the same, CQBRT summary name. [Clearly programs with the same CQBRT name can have further sources of distinction, especially at the technical level.]

CQBRT is a natural successor of Isham and Kuchař's *Tempus Ante Quantum*, *Tempus Post Quantum*, *Tempus Nihil Est* classification which is in particular geared to keep track of the huge amounts of Problem of Time facet interferences in attempting to find an ordering that gets one past all the facet 'gates'. CQBRT labels *are summaries of* path orderings though the 'gates' of Kuchař's enchanted castle of Problem of Time facets. Thus they are *summaries of answers to the questions that Parts II and III are about.* Part II broadly concerns how the letters in a CQBRT path are ordered prior to hitting upon the letter Q. On the other hand, Part III concerns how each such classical string may extend to a whole CQBRT path upon following through with Quantization (and whatever further facet-addressing features are required to complete that program).

A second consideration is that even just contemplating Q and T can be envisaged to come in not 3 but 14 forms, as per the previous Sec's rungs and Fig. 13.1.

Thirdly, practical experience shows us that one often has to go some of the gates more than once. Hence there is no limit to how many B's, say, feature in a CQBRT string. Fourthly, is that partly going through a gate is allowed. For instance one might find the Kuchař beables but be ultimately interested in getting down to the Dirac beables, or one might reduce out the LQG $SU(2)(\boldsymbol{\Sigma})$ but not yet face reducing out the accompanying $\mathit{Diff}(\boldsymbol{\Sigma})$. Because of this, as well as Fig. 13.1 permitting 'T' to take four values and the void, the other letters in CQBRT are also open to taking multiple symbols and the void. For instance, we write K in place of B to signify finding the Kuchař beables. Fifthly, Part III argues the need to often 'start afresh' in dealing with quantum versions of the gates, by which most of the other letters often appear on both sides of the Q.

Finally, the reader may have noticed that CQBRT itself is only a four-facet string, to there being rather more Problem of Time facets. Let us first concede that CQBRT strings will eventually pick up Spacetime Relationalism (S), Foliation Dependence (F) and Spacetime Construction (Z) letters too, for strategies for 'A Local Resolution of the Problem of Time', and then Global and Multiple Choice letters besides for 'Full Problem of Time' resolutions. That said, RT, RTC and RTCB are all common and also widely natural as factored-out subproblems of the Problem of Time. E.g. no other facets feature until Chap. 27 in Part II or in the first thirteen Chapters of Part III). Moreover, at least at the classical level, the B and Z sequels of RTC are logically independent of each other, giving a *branching* path through the enchanted castle's gates. ZSF is finally a natural order for finishing off 'A Local Resolution of the Problem of Time', based on: 'supply the Relationalism for the Spacetime thus Constructed and finally the Foliated version of all this'. So our main classical path is of the form in Fig. 35.2.e), which is a dRCRTB subcase of CQBRT with the above obvious ZSF as a side-chain. In designing this, and its quantum counterpart, the main ambiguities to untangle reside in how to order CQBRT itself.

Chapter 14
Configuration Spaces and Their Configurational Relationalism

The opening part (Chaps. 14 to 18) on spatially primality approaches is based on a wide variety of examples of configuration spaces $\mathfrak{q}$. The current Chapter, moreover, covers the start of Chap. 13's Rung I)—configuration spaces *alone*—and does not get far enough to involve Reduction thereupon, by which no Cubert label is yet merited. We thus begin by giving, and pointing to many further, examples of configuration spaces.

14.1 Examples of Configuration Spaces

We have already seen $\mathfrak{q}(N, d) = \mathbb{R}^{dN}$ in Sect. 2.13 in the context of Newtonian Mechanics on absolute space $\mathbb{R}^d$. Section 9.4 subsequently evoked $\mathfrak{q}(N, d) = \mathbb{R}^{dN}$ again, with $\mathbb{R}^d$ now playing a fiducial role in defining redundant configurations in RPMs. Another simple example of configuration space is the space of relative coordinates *relative space*', $\mathfrak{r}(N, d) = \mathfrak{q}(N, d)/\mathrm{Tr}(d) = \mathbb{R}^{nd}$ for $n := N - 1$. In the Newtonian Paradigm, this amounts to taking out the centre of mass for convenience. In the Relational Approach, on the other hand, the centre of mass position for the whole Universe is a fortiori meaningless. In either case, setting the more usual theories of Mechanics free from overall translations is trivial. See also Appendix G.1 for various useful coordinate systems for $\mathfrak{r}(N, d)$.

See Appendices G, H, M and N for many further examples of configuration spaces with less trivial groups of physically-irrelevant transformations $\mathfrak{g}$. In the case of Mechanics, $\mathfrak{q}(N, d)$ remains a redundant configuration space, whereas the non-redundant configuration space corresponding to $\mathfrak{g}$ is the quotient space (see Appendix M) $\mathfrak{q}(N, d)/\mathfrak{g}$. These examples are clearly also tied to Configurational Relationalism, to which we next turn in greater detail.

E. Anderson, *The Problem of Time*, Fundamental Theories of Physics 190,
DOI 10.1007/978-3-319-58848-3_14

14.2 Configurational Relationalism. i. Principles Discussed

Two a priori distinct conceptualizations of Configurational Relationalism in the point particle setting are as follows.

a) $\mathfrak{g}$ acts on absolute space $\mathfrak{a}(d)$ (usually taken to be $\mathbb{R}^d$).
b) $\mathfrak{g}$ acts on configuration space $\mathfrak{q}(N, d)$, i.e. it acts, rather, on material entities of at least some physical content.

b) can be approached via a)'s consideration of the groups acting on $\mathbb{R}^d$. a) is moreover well-known: the Erlanger Program for Geometry initiated by 19th century mathematician Felix Klein, to which Appendix B's standard mathematics applies. The consequences for N-particle configurations within each geometry are outlined in Sect. 14.5, whereas Chaps. 16 and 24 introduce the corresponding RPMs.

Some useful limitations on the choice of $\mathfrak{g}$, $\mathfrak{q}$ pairs are as follows.

C) *Nontriviality*. $\mathfrak{g}$ cannot be too large; this is a degrees of freedom counting criterion. Using $k := \dim(\mathfrak{q})$ and $l := \dim(\mathfrak{g})$, a theory on $\mathfrak{q}/\mathfrak{g}$ is *inconsistent* if $l > k$, *trivial* if $l = k$ and *relationally trivial* if $l = k - 1$. Relational nontriviality is meant here in the sense of requiring at least one degree of freedom to be expressed in terms of at least one other. This is to be contrasted with degrees of freedom being meaningfully expressed in terms of some external or elsewise unphysical 'time parameter'.
B) Further *structural compatibility* is required. A simple example of this is that if one is considering d-dimensional particle configurations, then $\mathfrak{g}$ is to involve the same d (or smaller, but certainly not larger).
A) A more general structural compatibility criterion is for $\mathfrak{g}$ is to admit a group action on $\mathfrak{q}$. A group action's credibility may further be enhanced though its being 'natural'. Some further mathematical advantages are conferred from group actions being one or both of faithful or free, with the combination of free and proper conferring yet further advantages. Appendices A.2 and C.6 explain this terminology and outline the advantages; see also Appendix B.

One might additionally wish to choose $\mathfrak{g}$ for a given $\mathfrak{q}$ so as to eliminate *all* trace of extraneous background entities. The automorphism group $Aut(\mathfrak{a})$ (defined in Appendix A.2) of absolute space $\mathfrak{a}$ is an obvious possibility for $\mathfrak{g}$. Some subgroup of $Aut(\mathfrak{a})$ [560] might however also be desirable, for instance because the inclusion of some automorphisms depends on which level of mathematical structure σ is to be taken to be physically realized. In this way, $\mathfrak{g} \leq Aut(\langle \mathfrak{a}, \sigma \rangle)$ for some σ is a more general possibility. Such subgroups also comply with A) and stand a good chance of suitably satisfying criteria B) and C); see Sect. 14.5 for examples.

14.3 ii. Direct Implementation

Given a $\mathfrak{q}$, $\mathfrak{g}$ candidate pair, one seeks to represent the generators of $\mathfrak{g}$ (indexed by G) as $\mathbf{g}(\mathbf{Q}, \frac{\partial}{\partial \mathbf{Q}})$ which manifestly act on $\mathfrak{q}$, One can subsequently investigate whether some candidate objects $\boldsymbol{O}(\boldsymbol{Q})$ (indexed by v and belonging to some space of objects $\mathfrak{O}$) are $\mathfrak{g}$-invariants by explicitly checking that these are indeed preserved by the generators. For particle configurations, invariants are plentiful and intuitively clear for a wide range of $\mathfrak{g}$, as a direct consequence of the forms taken by the corresponding $\mathfrak{g}$-invariants. E.g. for $\mathfrak{g} = Eucl(d)$, separations and relative angles between particles are of this nature, and Fig. G.3 tabulates further such for other $\mathfrak{g}$.

14.4 iii. '$\mathfrak{g}$-Act $\mathfrak{g}$-All' Method: Wider Indirect Implementation

Being $\mathfrak{g}$-invariant is not the only way in which a given set of objects can be of interest. For some $\mathfrak{q}$, $\mathfrak{g}$ pairs, moreover, invariants are unknown or nonexistent. There are more general concepts of 'good $\mathfrak{g}$ objects', such as $\mathfrak{g}$-tensors (of which $\mathfrak{g}$-invariants are but one example: $\mathfrak{g}$-scalars). Further possibilities include $\mathfrak{g}$-tensor densities and 'weak $\mathfrak{g}$-tensors', meaning modulo a linear function of the generators in parallel to Dirac's notion of weak equality. Auxiliaries are here being represented in terms of $\boldsymbol{Q}$ and $\mathfrak{g}$-bundle auxiliary quantities $\boldsymbol{g}$.

The Best Matching method outlined in Chap. 9 is an example for handling non $\mathfrak{g}$-invariant quantities; it is an indirect implementation of Configurational Relationalism. It is indeed straightforward to generalize the Best Matching method to arbitrary $\mathfrak{g}$. However, since this method involves velocities or changes as well as Principles of Dynamics actions, we postpone this example to Chap. 16. Let us for now instead *further* generalize the indirect implementation of Configurational Relationalism in a formulation-independent manner. One of the many consequences of this generalization is that it applies within the current Chapter's $\mathfrak{q}$-only worldview. Another is rendering Configurational Relationalism's resolution in a more general form which can be combined with strategies for further Problem of Time facets. This is the first instance of Part II's theme of *forming General Strategies for each facet* which have enough scope that they can be *combined with each other to form A Local Resolution of the Problem of Time.*

The General Strategy for Configurational Relationalism is the *configurational relationalizing map CR*. This contains a $\mathfrak{g}$ action followed by a move that uses all of $\mathfrak{g}$, so we also refer to it more descriptively as the '*$\mathfrak{g}$-Act $\mathfrak{g}$-All Method*' [38]. This approach applies to a vast range of objects $\boldsymbol{O}$; in the current Chapter, these are composites of configurational variables alone, $\boldsymbol{O}(\boldsymbol{Q}$ alone), though further Chapters extend this as per Sect. 13.1]. Such composites cover far more than just actions S, e.g. also notions of distance (Appendices G.4 and N.8), of information and correlation (Appendix Q), and quantum operators, alongside quantum versions of the previous two notions (Chap. 42 and Appendix U). N.B. that the $\boldsymbol{O}$ need not be $\mathfrak{g}$-invariant (if they were, one would be blessed with a direct implementation of Configurational Relationalism).

In whichever case, we start by applying $\mathfrak{g}$-act; this can initially be thought of as a map $\mathfrak{O} \xrightarrow{\mathfrak{g}\times} \mathfrak{g} \times \mathfrak{O}$, $\boldsymbol{O} \mapsto \vec{\mathfrak{g}}_g\, \boldsymbol{O}$. And we end by applying $\mathfrak{g}$-all: some operation $\mathsf{S}_{g \in \mathfrak{g}}$ which makes use of all of the $\boldsymbol{g} \in \mathfrak{g}$. This has the effect of cancelling out

$\mathfrak{g}$-act's use of $\boldsymbol{g}$, so overall a $\mathfrak{g}$-invariant version of the $\boldsymbol{O}$ is produced, which we denote by

$$CR(\boldsymbol{O}) := \boldsymbol{O}_{\mathfrak{g}\text{-inv}} := \mathbb{S}_{g \in \mathfrak{g}} \circ \overrightarrow{\mathfrak{g}}_g \, \boldsymbol{O}. \tag{14.1}$$

Examples of $\mathbb{S}_{g \in \mathfrak{g}}$ include taking infs or sups, extremizing, summing, integrating and averaging; in each case 'over $\mathfrak{g}$' is meant. The group averaging subcase of this is a well-known and substantial basic technique in Group and Representation Theory. In particular, there are well-defined versions of this for finite groups (Appendix A.5) and for compact Lie groups (Appendix P.2). The sum and integral cases take the forms

$$\mathbb{S}_{g \in \mathfrak{g}} \quad \text{include} \quad \sum_{g \in \mathfrak{g}}, \quad \int_{g \in \mathfrak{g}} \mathbb{D}\boldsymbol{g}. \tag{14.2}$$

Here $\mathbb{D}\boldsymbol{g}$ denotes the group measure (in particular Appendix P.2's Haar measure for compact groups). Finally, Best Matching's own $\mathbb{S}_{g \in \mathfrak{g}}$ is extremization over $\mathfrak{g}$ (see Sect. 16.1).

'*Maps*' can furthermore be inserted between $\mathfrak{g}$-act and $\mathfrak{g}$-all to produce an even more general

$$CR(\boldsymbol{O}) := \boldsymbol{O}_{\mathfrak{g}\text{-inv}} := \mathbb{S}_{g \in \mathfrak{g}} \circ \mathit{Maps} \circ \overrightarrow{\mathfrak{g}} \, \boldsymbol{O}. \tag{14.3}$$

'*Maps*' covers a very general assortment of maps, though these are to all be $\mathfrak{g}$-invariant; if not, $\mathfrak{g}$ would act on a new type of object $\boldsymbol{O}' = \mathit{Maps} \circ \boldsymbol{O}$. '*Maps*' is useful in separating out the $\mathfrak{g}$ part for study of $\mathfrak{g}$-actions thereupon in isolation from further paraphernalia. E.g. one can view packaging of velocities into kinetic terms in this manner, alongside multiplication by $\sqrt{2\,W}$ and integration. These form a sequence of three $\mathfrak{g}$-invariant '*Maps*' which produce a Principles of Dynamics action S from the thus isolated non-$\mathfrak{g}$-invariant objects $\boldsymbol{O}$: the velocities.

As a further example, consider the Kendall-type $\mathfrak{g}$-invariant comparer between shapes

$$(\text{Kendall } \mathfrak{g}\text{-Dist}) = (\boldsymbol{Q} \cdot \overrightarrow{\mathfrak{g}_g} \, \boldsymbol{Q}')_{\mathbf{M}}. \tag{14.4}$$

Statistician David Kendall's own application of this concerns attaining 'minimal incongruence' between planar figures. Compare with Best Matching, though Kendall's notion involves just the configurations rather than any notion of change as enters Best Matching itself. See Appendix G.4 for further comparison of these and other notions of distance between shapes.

The $\mathfrak{g}$-act construct is moreover capable of storing global information. In the common case in which $\mathfrak{O}$ and $\mathfrak{g}$ are topological manifolds (the latter through being a Lie group) this takes the form of a $\mathfrak{g}$-fibre bundle.[1] In this case, $\mathfrak{g}$-all completes the computation of a particular section $\boldsymbol{O}_{\mathfrak{g}\text{-inv}}$, by a procedure that peruses all the information on each fibre so as to ensure a $\mathfrak{g}$-invariant output. This is illustrated free

[1] See Appendix F.4 for the definitions of $\mathfrak{g}$-fibre bundles and section.

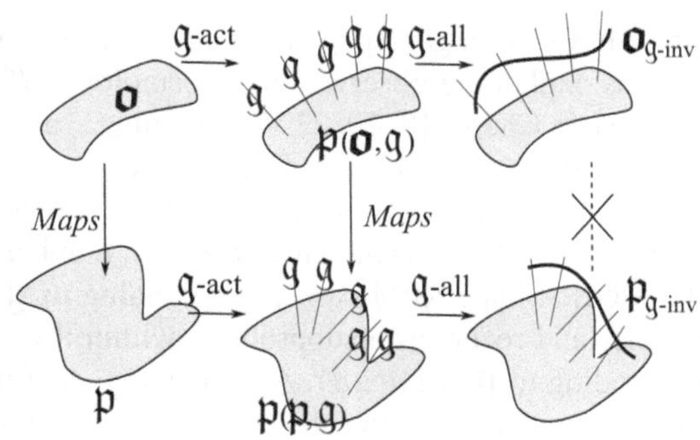

Fig. 14.1 $\mathfrak{g}$-act, $\mathfrak{g}$-all technique in the Fibre Bundles setting

of '*Maps*' on the first line of Fig. 14.1. Upon including '*Maps*' as well, because $\mathfrak{g}$-act does not act on '*Maps*' the first square commutes; the second does not, however, since $\mathfrak{g}$-all *does* in general act on '*Maps*'. Involving '*Maps*' in general clearly alters the output of summing, averaging, inf or sup taking or extremization. So the section $\boldsymbol{O}_{\mathfrak{g}\text{-inv}}$ which ignores '*Maps*' is indeed expected to differ in general from the section $\boldsymbol{O}'_{\mathfrak{g}\text{-inv}}$ which entertains some '*Maps*'.

14.5 On the Variety of Relational Configurations and RPMs

Let us next consider a variety of $\mathfrak{q}$, $\mathfrak{g}$ pairs for particle configurations for which RPM actions S can be built. N.B. that this book considers RPMs as model arenas for the geometrodynamical formulation of GR, Classical and Quantum Background Independence and ensuing Problem of Time facets [37, 45, 101, 552, 586], rather than as an attempt to model the world directly. In this way, mathematically simpler 1- and 2-d RPMs—which already exhibit many of the features of GR that are emulated by RPMs—are often preferable to the 3-d RPMs. This is because many of the extra complexities of the latter are not in line with GR's own additional complexities. The above view also leaves a number of other modelling assumptions for RPMs open, so we address them below. See Appendix G for the $\mathcal{Q}$ and $\mathfrak{q}$ corresponding to the main RPM examples considered in this book.

1) For the most habitually considered case of $\mathfrak{a} = \mathbb{R}^d$, Appendices B and E provide many $\mathfrak{g}$, according to the following sources of variety.

Is scale physically meaningless? It is not in Barbour–Bertotti's RPM [105], for which $\mathfrak{g}$ is the Euclidean group $Eucl(d)$, whereas it is in Barbour's RPM [102], for which $\mathfrak{g}$ is the similarity group $Sim(d)$.[2] One relational argument against inclusion of scale concerns this being a single heterogeneous addendum to the shapes [108]. Conceptual undesirability may also be raised as regards scale playing so dominant

[2]From here on, terming these 'Metric Scale and Shape RPM' and 'Metric Shape RPM' respectively helps get this point across while distinguishing them from yet further RPMs outlined below.

a role in Cosmology. And yet there are currently no credible alternatives to scale as regards explaining observational Cosmology [702, 736, 888]. Additionally, as we shall see in Chaps. 18, 20, 23, 46, retaining scale may enable provision of time [29]; some of these approaches in fact *rely* on the aforementioned heterogeneity. It is the scale contribution that renders the GR kinetic term indefinite, a feature not found elsewhere in Physics and which causes a number of difficulties. On the other hand, the Metric Shape RPM case corresponding to $\mathfrak{g} = Sim(d)$ is both mathematically simpler and recurs as a subproblem within the Metric Shape and Scale RPM corresponding to $\mathfrak{g} = Eucl(d)$. After making clear the nature of the choices involved, this book mostly uses the above two cases as illustrative examples.

If shears and Procrustean stretches (see Fig. B.1) are considered instead, these combine with translations and rotations to form the 'equi-d-voluminal group' $Equi(d)$, or furthermore with the dilations to form the affine group $Aff(d)$. This option ensues from preservation of the top form supported in dimension d, which is built out of the exterior product $\wedge$ instead of preserving the $\cdot$ inner product. This is the arbitrary-d generalization of the cross product formula $(\ \times\ )_3$ for area in 2-d and of the scalar triple product formula $[\ \times\ \cdot\ ]$ for volume in 3-d. Moreover, each of $Equi(d)$ and $Aff(d)$ corresponds to a further type of geometry as per Appendix E.1. Examples of the corresponding RPMs are outlined in Sects. 24.10 and in [36].

2) Are the configurations to be mirror image identified, and are the particles to be distinguishable? These translate to issues of $\mathfrak{q}$ topology, involving using not $\mathfrak{q}$ but more generally $\mathfrak{q} = \sum_{I=1}^{N} \mathfrak{a}/\mathfrak{g}'$ for $\mathfrak{g}'$ a discrete group. In particular, $\mathfrak{g}' = \mathbb{Z}_2$ for mirror image identification, $\mathbb{Z}_N$ for N indistinguishable particles and $\mathbb{Z}_2 \times \mathbb{Z}_N$ if both apply; see Appendix G.3 for examples. Finally, partial indistinguishability is also possible.
3) In cases in which RPM configuration spaces $\mathfrak{q}/\mathfrak{g}$ exhibit strata (outline Appendix M.5), various alternative approaches are pertinent, as laid out in Sect. 37.5.
4) Other models for absolute space $\mathfrak{a}$ might also be considered, such as $\mathbb{R}^d \cup \infty$, $\mathbb{S}^d$ or a more general manifold $\mathfrak{M}$. In this setting, multiple particles remain mathematically represented by multiple copies of absolute space, so the corresponding N-particle configuration space is $\times_{I=1}^{N} \mathfrak{M}$.

The case of $\mathbb{R}^d \cup \infty$—inclusion of Riemann's extra 'point at infinity' structure – additionally admits inversion in $\mathbb{S}^{d-1}$ and consequently special conformal transformations as per Appendix E.3. Now consider regarding these as physically meaningless, i.e. taking $\mathfrak{g}$ to be the conformal group $Conf(d)$ (of local angle preserving transformations) or various possible subgroups thereof [36]. The corresponding RPM is *local relative angle* alias *conformal shape mechanics* [36] and briefly outlined in Sect. 19.4.

Direct experiences in the world around us involve shape and scale. In contrast, the extra whole-universe model shears and Procrustean stretches among the equi-d-voluminal and affine transformations identify configurations under distortions of relative angles and of relative separations. Contrast also with $Conf(d)$'s additional special conformal transformations. Neither of these correspond to the totality of measurements in local everyday experience. Moreover, physical theories with scale are quite often considered that possess a distinct scale-invariant or conformally-invariant phase in an 'unbroken' higher-energy regime [395]. This is one way in which matching everyday experience is not all. Another is through Affine Geometry entering the analysis of *images* [788] of everyday experience's objects. We shall additionally see that *Conf*(d) and *Aff*(d) are useful for modelling aspects of GR and of further Theories of Gravitation (Chap. 19). This corresponds to Conformal and Affine Geometry also being two of the simplest variants at the level of Differential Geometry (Appendix D.2).

As further options, firstly $\mathbb{S}^2$ [37, 537, 539] corresponds to the geometry of 'the observed sky' in place of a 'dynamical solid geometry' $\mathbb{R}^3$. The corresponding RPM is outlined in [46]; this arises from modelling absolute space by $\mathfrak{a} = \mathbb{S}^n$. Secondly, $\mathbb{S}^3$ arises as a substitute for $\mathbb{R}^3$ space in the role of $\mathfrak{a}$ upon considering that we live within a closed GR cosmology.[3] [46] outlines the RPM corresponding to this case as well.

5) A further feature of GR which can be entertained at this point is that the notion of space $\mathfrak{M}$ itself be dynamical. This could undergo any combination of overall expansion, anisotropic change, or inhomogeneous change. A further source of variety here is whether to model particles on a dynamical $\mathfrak{M}$ [46], fields on a dynamical $\mathfrak{M}$ or just the dynamics of $\mathfrak{M}$ itself. Sections 8.3 and 8.10, as well as Appendices H and N cover various examples of this.

Many of the above sources of variety can, furthermore, be composed; see Appendices E and G. In particular, each $\mathfrak{a} = \mathfrak{M}$ has its own suite of each of metrics thereupon and of groups $\mathfrak{g}$ acting naturally on these. Section 19.8 outlines Supersymmetry as a sixth source of variety. Finally, direct considerations on some occasions succeed in finding the relational configuration space and then building a Mechanics thereupon. This is as opposed to finding a relational configuration space by having an indirectly-formulated RPM and reducing its action. These are covered in Sects. 16.7; moreover in some cases coincident theories arise from these two approaches.

Shape Statistics for the theory in question's configurations is among the structures alluded to in Sect. 13.1 as remaining available within Timeless World-views. The *Sim*(d) case of this is Kendall's well-developed 'Shape Statistics' [536, 539, 792], which motivated his own work on the corresponding notion of shape. The Author pointed out in [18, 33] that this coincides with Barbour's notion of shape [102], meaning that many further questions concerning Metric Shape RPM had already been answered by Kendall. Moreover, given the variety of notions of shape pointed out above and in Fig. G.4, and that each has a corresponding Shape Statistics, we henceforth refer to Kendall's as *Metric Shape Statistics*. See Sect. 26 and Appendices R and T for more on Shape Statistics.

[3] $\mathbb{S}^3$ is in many senses the *simplest* closed model for space. Its closest rival is the 3-torus $\mathbb{T}^3$, which, due to arising from the simplest topological identification of $\mathbb{R}^3$, is locally flat. However, this has less (global) Killing vectors, by 6 forming $SO(4)$ to 3 forming $\times_{i=1}^{3} U(1)$.

Minimal relationally nontrivial unit [36] is a particularly useful notion. This is concurrently the smallest relationally nontrivial

1) whole-universe model,
2) dynamical subsystem, and
3) Shape Statistics sampling unit.

The relational triangle (Fig. 9.4.d) is an archetype of this, as reflected both in Barbour's seminars often having involved shuffling wooden triangles and in Kendall's sampling by triangles leading to his spherical blackboard computational tool. Figure G.4 tabulates further examples, and points to 1-*d*'s limitations: 1-*d* has not only no continuous rotations and the only discrete rotation coinciding with inversion, but also no nontrivial volume forms by which it is bereft of an affine extension.

Chapter 15
Temporal Relationalism (TR)

We next allow for velocity or change at the primary level. This setting—Rung II) of Sect. 13.1's ladder—permits physical theories to take more familiar forms, e.g. in terms of an action or a Hamiltonian. As this approach obtains an emergent time at the classical level, it is of *Tempus Ante Quantum* (TQ) type. The current Chapter considers this for the 1-d version of the Metric Shape and Scale RPM model arena.

15.1 General Enough Temporal Relationalism Implementing (TRi) Strategies

Modelling closed universes from a Background Independent perspective gives Sect. 9.7's Temporal Relationalism Principle, by which there is ab initio a classical Frozen Formalism. Adopting this principle, moreover, does not make constructive use of how one defines 'universe'. This is in contrast with Chap. 3's subsequently discarded Leibniz Perfect Clock Principle. In this way Sect. 3.1's 'Pandora's box'—that our conception of the Universe has changed since Leibniz's day due to the inception of GR and the gathering of cosmological data—remains unopened.

The Temporal Relationalism Principle is useful due to admitting sharp mathematical implementations. To begin with, we work at the level of selection principles for Principles of Dynamics actions, which we build up via various natural compound object structures built upon the configurations $\boldsymbol{Q}$. In each case, Principle TR-i) (Sect. 9.7) is pre-requisite.

A first implementation of Principle TR-ii) consists of a label featuring but being physically meaningless. This is due to being able to exchange this label for any other (monotonically related) label without altering the physical content of the theory (line 1 of Fig. 9.2). An action built in this manner is *Manifestly Reparametrization Invariant*. This is the approach already followed in Part I due to its short-term advantage of affording a relatively conventional presentation of subsequent physical notions. For instance, a primary notion of velocity can then be defined as the

E. Anderson, *The Problem of Time*, Fundamental Theories of Physics 190,
DOI 10.1007/978-3-319-58848-3_15

derivative with respect to λ:

$$\text{velocity} := \frac{\text{d(configuration variable)}}{\text{d(label time}} \quad \text{i.e.} \ \frac{\mathrm{d}\boldsymbol{Q}}{\mathrm{d}\lambda}. \tag{15.1}$$

In this approach, the tangent bundle $\mathfrak{T}(\mathfrak{q})$ is realized as *configuration–velocity space*. One can next build the kinetic term $T := \| \dot{\boldsymbol{Q}} \|_{\boldsymbol{M}}{}^2/2 = M_{\mathsf{AB}} \dot{Q}^{\mathsf{A}} \dot{Q}^{\mathsf{B}}/2$. Let us assume for now that this takes the most physically standard form: homogeneous quadratic in the velocities;[1] this assumption is removed in Sect. 17.2. Finally, *action* in the sense of the Principles of Dynamics is a map $S : \mathfrak{T}(\mathfrak{q}) \rightarrow \mathbb{R}$. The current application's *Jacobi action* [598] is of the form

$$S_{\mathrm{J}}^{\mathrm{MRI}} := \int \mathrm{d}\lambda \, L_{\mathrm{J}}^{\mathrm{MRI}} = 2 \int \mathrm{d}\lambda \sqrt{T W}. \tag{15.2}$$

A second implementation for TR-ii), which also goes back historically to Jacobi [364], is *Manifest Parametrization Irrelevance*: in which no use of λ is to be made at all. One immediate consequence of this is that there is no primary notion of velocity: this has been supplanted by a 'change in configuration' differential,

$$\text{d(configuration variable)} \quad \text{i.e.} \ \mathrm{d}\boldsymbol{Q}. \tag{15.3}$$

Then similarly to $\boldsymbol{Q}$, $\dot{\boldsymbol{Q}}$ being the well-known *Lagrangian variables* for Mechanics, we term $\boldsymbol{Q}, \mathrm{d}\boldsymbol{Q}$ the *Machian variables* since these encode Mach's Time Principle allowing for change to be involved at the primary level. In either case, one's primary ontology is *a point and a tangent in configuration space*. The first case represents this as configuration–velocity space and the second as configuration–change space, but both correspond to the same tangent bundles $\mathfrak{T}(\mathfrak{q})$.

Further consequences are that kinetic energy is supplanted by *kinetic arc element*

$$\mathrm{d}s := \|\mathrm{d}\boldsymbol{Q}\|_{\boldsymbol{M}} = \sqrt{M_{\mathsf{AB}}(\boldsymbol{Q})\mathrm{d}Q^{\mathsf{A}}\mathrm{d}Q^{\mathsf{B}}}, \tag{15.4}$$

and the Lagrangian by the *physical* alias *Jacobi arc element*

$$\mathrm{d}J := \mathrm{d}s\sqrt{2\,W(\boldsymbol{Q})}. \tag{15.5}$$

[More strictly, this supplants the Lagrangian L in line 2 of Fig. 9.2.] The *formula* for action is now

$$S_{\mathrm{J}}^{\mathrm{MPI}} := \int \mathrm{d}J. \tag{15.6}$$

[1]The configuration space metric $\boldsymbol{M} = \boldsymbol{M}(\boldsymbol{Q}$ alone) used here assumed to be independent of the label time and of the velocities. Also the potential assumed is of the form $V = V(\boldsymbol{Q}$ alone); the 'potential factor' itself is of the form $W(\boldsymbol{Q}) := E - V(\boldsymbol{Q})$. These assumptions make good sense in the intended whole-universe model setting, as opposed to modelling subsystems or dissipative systems.

Lemma 1 *Equations* (15.2) *and* (15.6) *are indeed equivalent.* [This is proved in Fig. 9.2.]

A difference in formalism between the two implementations arises through (15.3) involving a *change covector* [37], which, as we shall see below, attaches 'change weights' to further Principles of Dynamics entities. These 'change weights' can furthermore be identified as a $\mathbb{Z}$-valued version of the homothetic weights subcase of the well-known conformal weights[2] In this way, the Manifestly Parametrization Irrelevant implementation modifies the Principles of Dynamics into a *homothetic Tensor Calculus* with $\mathbb{Z}$-valued exponents. This structural observation is subsequently useful in keeping track of the many further modifications (arising throughout Part II and collected in Appendix L) in passing from the habitual Principles of Dynamics to the *Temporal Relationalism implementing Principles of Dynamics (TRiPoD)*.

So for instance, kinetic energy and Lagrangian are replaced by kinetic and Jacobian arc elements respectively, which are both change covectors. (15.5) can moreover be read as these being simply interrelated by a conformal transformation. On the other hand, the *notion* of action itself remains invariant under these reformulations: it is a change *scalar*. Note also that (15.6) signifies that, in terms of the physical Jacobian arc element, dynamics takes the form of a *geodesic principle*. In this way, the problem of motion reduces to the problem of finding the geodesics of a corresponding geometry. On the other hand, in terms of the kinetic arc element,

$$S_{\mathrm{J}}^{\mathrm{MPI}} = \sqrt{2}\int \sqrt{W}\mathrm{d}s: \tag{15.7}$$

a *parageodesic principle* [659], i.e. geodesic up to a conformal factor, $\sqrt{2\,W}$. This has the advantage of being tied to $\mathfrak{q}$'s geometry, which is a unique geometry for the family of problems with different potential factors, as opposed to the preceding involving a different geometry for each of these.

The third implementation of TR-ii) is to carry out the second implementation's moves, however now regarding these as the construction of an action corresponding to a given geometry. Indeed, Jacobi historically wrote down his action principle in its aspect as a geometrical formulation of Mechanics rather than its aspect as a Timeless Worldview.

Lemma 2 *These two perspectives coincide.* [This is a mathematically straightforward, if conceptually curious, duality.]

The point of using the second implementation rather than the third is that no reference is now ever made to a parameter that is, in any case, irrelevant. More generally, it is a conceptual advance for Background Independent Physics to cease

---

[2]See Appendix D.7 for conformal and homothetic Tensor Calculi.

to use names and notions which derive from physically irrelevant or Background Dependent entities.

In more detail, Jacobi's construction associates a Mechanics to each given geometry of the homogeneous quadratic form (15.4), which covers (semi-)Riemannian geometries. This geometry conversely plays the role of configuration space $\mathfrak{q}$ for that Mechanics. Mathematical physicist John Synge subsequently generalized this [598] to a more general notion of geometry, as outlined in Sect. 17.2. One consequently postulates *geometrical Jacobi(–Synge) type actions* S_{JS} of type $S^{\mathfrak{q}\text{-Geom}}$. By a duality, these *happen* to also be Manifestly Parametrization Irrelevant and thus Manifestly Reparametrization Invariant, so the three implementations are indeed mathematically equivalent. To celebrate, we subsequently drop MRI, MPI and $\mathfrak{q}$-Geom superscripts. We often also substitute the JS subscript and its common J label subscript for the label TR for which clearly displays the conceptual type of Background Independence aspect incorporated: Temporal Relationalism.

Example 1) Jacobi's action principle for Spatially-Absolute Mechanics, for which $\mathrm{d}s = \sqrt{m_I \mathrm{d}q^I \mathrm{d}q^I}$ and $W(\boldsymbol{Q}) = E - V(\mathbf{q})$.

Example 2) Scaled 1-*d* RPM with translations trivially removed [37] is covered by another case of Jacobi's action principle. Here $\mathrm{d}s = \sqrt{\mathrm{d}\rho^i \mathrm{d}\rho^i}$ and $W(\boldsymbol{Q}) = E - V(\boldsymbol{\rho})$. This has the advantage over 1) of being a relational whole-universe model.

Example 3) Misner's action principle [659] for Minisuperspace GR is also of this form. We already presented a subcase of this in Sect. 9.9, whereas Sect. 17.1 considers a further range of such. This is furthermore a subcase of full Geometrodynamics' Baierlein–Sharp–Wheeler [89] action (9.11); the title of their paper is "*Three-dimensional geometry as carrier of information about time*", which supports the above duality. Upon subsequently passing to the relational GR action, this can be rephrased in the temporally Machian form 'geometry and change of geometry as carrier of information about time'. Moreover, GR's spatial geometries are but an example of $\mathfrak{q}$ geometry, so this can be further generalized as regards range of theories, to '*Configuration and change of configuration as carrier of information about time*'.

Example 4) We postpone nontrivially Jacobi–Synge examples to Sect. 17.2.

We keep track of TRiPoD via the end-summary Fig. L.2. The General Strategy for Temporal Relationalism consists of TRiPoD followed by TRiFol (foliations: Chap. 34), TRiCQT (Canonical Quantum Theory: Chap. 41.3) and TRiPIQT (Path Integral Quantum Theory: Chap. 52). The main virtue of this formalism is that keeping one's calculations within this formalism prevents the Frozen Formalism Problem inadvertently re-entering while one is subsequently addressing further facets. Finally note that working with TRi formulations is but the larger of two parts in handling Temporal Relationalism; approaches using this eventually need to be completed by a Machian 'time is to be abstracted form change' step.

15.2 Equivalence to the Euler–Lagrange Formulation

The familiar Euler–Lagrange formulation of Mechanics has an action of the form

$$S_{\mathrm{EL}} = \int \mathrm{d}t\, L = \int \mathrm{d}t\{T_t - V\}. \tag{15.8}$$

Here T_t is the version of the kinetic energy that is in terms of $\mathrm{d}/\mathrm{d}t$ derivatives. Also $t = t^{\mathrm{Newton}}$ in Mechanics, so this formulation immediately fails to obey TR-i). On the other hand, the previous Sec's formulations do obey both TR-i) and ii).

Lemma 3 *The previous Section's formulations are mathematically equivalent to the Euler–Lagrange formulation (subject to the latter's Lagrangian L not depending explicitly on t).*

Proof Rewrite

$$S_{\mathrm{EL}} = \int \mathrm{d}t\, L \quad \text{as } S_{\mathrm{EL}} = \int \mathrm{d}\lambda\, \dot{t}\, L(\boldsymbol{Q}, \dot{\boldsymbol{Q}}, \dot{t}): \tag{15.9}$$

λ-parametrization of the action by adjoining time to the system's configurational description (the dot here denotes $\mathrm{d}/\mathrm{d}\lambda$). The original Lagrangian's explicit t-independence means that (15.9)'s t is a cyclic coordinate. Thus Routhian reduction (Appendix J.5) applies, giving

$$L_{\mathrm{TR}}(\boldsymbol{Q}, \dot{\boldsymbol{Q}}) := L(\boldsymbol{Q}, \dot{\boldsymbol{Q}})\dot{t} - P^t\dot{t} \quad \text{for } \frac{\partial L}{\partial \dot{t}} = P^t = -E, \quad \text{constant.} \tag{15.10}$$

See Sect. 15.9 for the converse working. □

The above also serves to justify the identification of the Jacobi action's W as the combination of well-known physical entities $E - V$. The role of t in (15.8) is played by t^{Newton}, whereas (15.7) makes no primary mention of any such quantity. Indeed, the above application of Routhian reduction is termed the *parametrization procedure* in the Mechanics literature [598]. This refers to the (Nonrelational!) adjunction of the 1-d space of a time variable to the configuration space $\mathfrak{q} \longrightarrow \mathfrak{q} \times \mathfrak{T}$. This is then incorporated by rewriting one's action in terms of a label-time parameter $\lambda \in \mathfrak{T}$.

15.3 TRi Form of Conjugate Momentum

Using the standard definition of momentum conjugate to $\boldsymbol{Q}$ (J.8), the Manifestly Reparametrization Invariant form of the Jacobi action gives

$$\boldsymbol{P} = \boldsymbol{M}\sqrt{W/T} \circ \boldsymbol{Q}. \tag{15.11}$$

However, Temporal Relationalism has been argued to have no place for velocities or Lagrangians as primary entities, so a new TRi definition of momentum is required:

$$\boldsymbol{P} := \frac{\partial\, \mathrm{d}J}{\partial\, \mathrm{d}\boldsymbol{Q}}. \tag{15.12}$$

Lemma 4 *This is equivalent to the standard definition of momentum.*

Derivation This follows from the 'cancellation of the dots' Lemma which is commonplace in the Principles of Dynamics [371]. □

Since (15.12) is a ratio of changes, momentum is also revealed to be a *change scalar*. So whereas TRi places a change weight upon each use of $\mathfrak{T}(\mathfrak{q})$, it leaves $\mathfrak{T}^*(\mathfrak{q})$ invariant. Computing this gives

$$\boldsymbol{P} = \boldsymbol{M}\frac{\sqrt{2\,W}\,\mathrm{d}\boldsymbol{Q}}{\|\mathrm{d}\boldsymbol{Q}\|_M}. \tag{15.13}$$

15.4 Jacobi–Mach Equations of Motion

The usual form of the equations of motion is Euler–Lagrange's (J.3). Computing this for a Manifestly Reparametrization Invariant action gives

$$\frac{\mathrm{d}^2 Q^{\mathsf{A}}}{\mathrm{d}t^2} + \Gamma^{\mathsf{A}}{}_{\mathsf{BC}}\frac{\mathrm{d}Q^{\mathsf{B}}}{\mathrm{d}t}\frac{\mathrm{d}Q^{\mathsf{C}}}{\mathrm{d}t} = N^{\mathsf{AB}}\frac{\partial W}{\partial Q^{\mathsf{B}}}, \tag{15.14}$$

where $\Gamma^{\mathsf{A}}{}_{\mathsf{BC}}$ are the Christoffel symbols of the $\mathfrak{q}$ geometry.

However, the above make reference to times, velocities and Lagrangians. In TRi form, the equations of motion are, rather, the *Jacobi–Mach equations* that follow from Jacobi's arc element in terms of Machian variables:

$$\mathrm{d}\left\{\frac{\partial\, \mathrm{d}J}{\partial\, \mathrm{d}Q^{\mathsf{A}}}\right\} = \frac{\partial\, \mathrm{d}J}{\partial Q^{\mathsf{A}}} \quad \Rightarrow \tag{15.15}$$

$$\frac{\sqrt{2\,W}\,\mathrm{d}}{\|\mathrm{d}\boldsymbol{Q}\|_M}\left\{\frac{\sqrt{2\,W}\,\mathrm{d}Q^{\mathsf{A}}}{\|\mathrm{d}\boldsymbol{Q}\|_M}\right\} + \Gamma^{\mathsf{A}}{}_{\mathsf{BC}}\frac{\sqrt{2\,W}\,\mathrm{d}Q^{\mathsf{B}}}{\|\mathrm{d}\boldsymbol{Q}\|_M}\frac{\sqrt{2\,W}\,\mathrm{d}Q^{\mathsf{C}}}{\|\mathrm{d}\boldsymbol{Q}\|_M} = N^{\mathsf{AB}}\frac{\partial W}{\partial Q^{\mathsf{B}}}. \tag{15.16}$$

Note that (15.15) is an 'impulse formulation' of Newton's Second Law.

15.5 Differential Hamiltonian

See Appendix J.6) for an outline of the usual theory of the Hamiltonian. The Relational Approach, however, requires instead $\mathrm{d}H(\boldsymbol{Q},\boldsymbol{P}) := \mathrm{d}J(\boldsymbol{Q},\mathrm{d}\boldsymbol{Q}) - \boldsymbol{P}\,\mathrm{d}\boldsymbol{Q}$.

This is a differential change covector version of the Hamiltonian H itself, but based on the same set of (already TRi!) Hamiltonian variables $\boldsymbol{Q}$, $\boldsymbol{P}$, and so it also lives on $\mathfrak{T}^*(\mathfrak{q})$ for the homogeneous quadratic theories.

15.6 Quadratic Constraints from Temporal Relationalism

In the ADM formulation of GR, the Hamiltonian constraint $\mathcal{H}$ (8.23) arises from varying with respect to the lapse α . However, the Misner formulation has no such extraneous time-like variable, so does this case still have a $\mathcal{H}$, and, if so, where does it come from? The answer is provided by **Dirac's Lemma** (stated and demonstrated as Dirac's argument in Sect. 9.7). We now uplift this to TRi form.

Lemma 5 *Geometrical actions $S^{\mathfrak{q}\text{-Geom}}$ imply at least one primary constraint.*

Derivation $S^{\mathfrak{q}\text{-Geom}}$ is dually S^{MPI}, so it is homogeneous of degree 1 in its changes. Therefore each of its total of k momenta are homogeneous of degree 0 in the changes. Consequently, these are functions of k—1 independent ratios of changes. So there must be at least one relation between the momenta themselves without any use made of the equations of motion. But this is by definition a primary constraint. □

Moreover, the primary constraints arising from Jacobi square-root type actions also have a distinctive form. I.e. the quadratic form of this action (a single square root of a square) in turn provides a single primary constraint that is purely quadratic in the momenta. This constraint is reminiscent of the squares of the sides of a triangle in Pythagoras' Theorem, or, more usefully, of direction cosines squaring to one. In this manner, taking into account this constraint causes one to pass from considering a point and a vector in $\mathfrak{q}$ to considering just a point and a direction. Thus one has not $\mathfrak{T}(\mathfrak{q})$ but a *direction bundle* alias *unit tangent bundle*. Finally, in the current case, instead of the entities in question squaring to 1 as direction cosines do, the momenta 'square' by use of the $\boldsymbol{M}$ matrix's inner product to the 'square of the hypotenuse', $2\,W$.

Moreover, the primary constraint arising in this manner for GR is indeed $\mathcal{H}$ (Chap. 9 showed this for Minisuperspace, whereas the full GR case is in Chap. 18). *Thus in Background Independent Physics, the Hamiltonian constraint $\mathcal{H}$ that is central to the dynamics of GR $\mathcal{H}$ arises directly from the demand of Temporal Relationalism.* Its purely quadratic form is furthermore dictated by the precise (quadratic) way the action complies with Temporal Relationalism. This in turn induces $\mathcal{H}$'s purely quadratic dependence on the momenta. $\mathcal{H}$ is furthermore well-known to give the Schrödinger-picture manifestation of the most well-known Quantum Problem of Time facet: the Frozen Formalism Problem. So the Temporal Relationalism the Background Independence aspect is indeed a deeper and already classically-present replacement for the Frozen Formalism Problem.

On the other hand, for Temporally-Relational but Spatially-Absolute Mechanics following from Jacobi's action, the quadratic energy constraint

$$\mathcal{E} := \sum_{I=1}^{N} p_I^2/2m_I + V(\boldsymbol{q}) = E \tag{15.17}$$

plays an analogous role to $\mathcal{H}$. The $I = 1$ to $N \rightarrow A = 1$ to n indices, $\mathbf{q} \rightarrow \boldsymbol{\rho}$ of this does likewise for 1-*d* Metric Shape and Scale RPM ($\boldsymbol{\rho}$ here are *relative Jacobi coordinates* of Appendix G.1.) This is the first of many justifications for considering such Mechanics as useful conceptual guides to some facets of the Problem of Time for GR itself. In both cases, the quadratic form of the geometrical arc-element in the action gives rise to an equation of time,

$$\mathcal{C}\text{hronos} := \|\boldsymbol{P}\|_N{}^2/2 - W(\boldsymbol{Q}) = 0. \tag{15.18}$$

This is mathematically of the form $\mathcal{Q}$uad [cf. Eq. (8.26)] which begets a Quantum Frozen Formalism Problem. The forms of $\mathcal{E}$ and $\mathcal{H}$ are immediately recognizable as elsewise well-known equations because $\mathcal{C}$hronosis a change scalar.

Let us finally comment on the smaller case of Constraint Closure which is internal to Temporal Relationalism. Since the current Chapter's finite models only have one constraint, $\mathcal{C}$hronos, this just forms the trivial Poisson bracket with itself so Constraint Closure is attained. As we shall see in Chap. 24, however, this argument breaks down upon passing to Field Theories.

15.7 Mach's Time Principle and Its Implementations

Temporal Relationalism for the Universe as a whole is to be reconciled with our everyday experiences along the following lines.

a) Our experiences are of *subsystem* physics rather than of the Universe as a whole.
b) Mach's Time Principle—that time is to be abstracted from change—applies, whereby time emerges at the secondary level.

This sense of Machian holds for geometrical dually Manifestly Parametrization Irrelevant implementations to be in terms of *change* $\mathrm{d}\boldsymbol{Q}$ in place of label-time velocity $\mathrm{d}\boldsymbol{Q}/\mathrm{d}\lambda$. Moreover, Eqs. (15.13), (15.16) involve *ratios of changes*, as is to be expected in the Relational Approach. The next issue is which change is involved. Two diametrically opposite views on this are as follows

1) Time is to be abstracted from *any* change (a position of Rovelli's [752, 755]).
2) *The most perfect* time is to be abstracted from *all* change,

$$t^{\text{em}} = \mathcal{F}[\boldsymbol{Q},\ \text{all}\,\mathrm{d}\boldsymbol{Q}]. \tag{15.19}$$

Barbour adopted this position [98, 104], which partly rests upon taking Leibniz's Perfect Clock Principle seriously.

3) The Author's position [30, 39] represents 'middle ground' between 1) and 2), albeit with more proximity to 2). 3) is inspired by Clemence's [211] consideration of the astronomers' ephemeris time (Chap. 3.1). This was also of substantial inspiration in 2), while being free from Leibniz's Perfect Clock Principle, by which the Author was able to go further than Barbour in implementing Clemence's conceptualization. In more detail, position 3) is that

   generalized local ephemeris time (*GLET*) is to be abstracted from a sufficient totality of locally relevant change (STLRC). (15.20)

   Let us next further explore these three positions by considering the following alternatives; this can be seen as a further extension of Sect. 1.12's list of clock properties.

IA) Any subsystem will serve as a clock.

Or IB) Some subsystems are better than others as clocks, e.g. by Sect. 1.12's criteria.

IIA) Clocks are localized subsystems—'a wristwatch', as typically motivated by convenience of reading hand, and which is attuned to Newton's position that 'the Universe contains clocks'.

Or IIB) 'Ephemeris Time Principle' [211]. The changes of the actual subsystem under study are to be factored into one's timekeeping. This is generally much larger than a 'wristwatch' conception of clock: e.g. the Solar System. This is motivated by the need to at least occasionally check whether recalibration is required.

IIIA) The *less coupled* a clock and subsystem under study are, the better [335, 727]. This is usually motivated by the isolated experiment under control being a type of clean stable system upon which to base a clock.

Or IIIB) The clock and subsystem *fundamentally have* to be coupled. This is occasionally motivated from the idea that keeping time for a system requires participation. In some formulations there are difficulties as regards a clock keeping time for a subsystem if it is not coupled to it *at all*; see Sect. 47.6 for an example.

Within this last alternative, a compromise IIIC) can be reached: a small but strictly nonzero coupling can be assumed. In this case, the subsystem under study only very gently disturbs the physics of the clock. This compromise is ultimately inevitable because nothing can shield gravity, which places a bound on attaining 'clean clocks' by isolation.

In the case of time being abstracted from *any* change, IA) applies to 1). This approach also carries connotations of IIA), which is far more common among current theoretical physicists than IIB). On the other hand, the case of time being abstracted from *all* change in the Universe rests on comparison IB) and IIB) enhanced by the following.

IV) 'The more change is included the better'.

Barbour then takes this to its logical extreme, along the following lines.

V) 'Include all change'. Once again, this rests on Leibniz's Perfect Clock Principle.

Note that IV) and V) refer to recalibration, as opposed to stability or convenience of reading hand.

Much can be learnt by contrasting the extreme cases. For instance, 'any change' implements a particular sense of 'democracy'. This has been argued to be useful in generic situations (taken to be crucial in GR) in which there is no privileged timestandard. However, 'all change' implements a distinct sense of 'democracy, and STLRC does as well (here all have *the opportunity* to contribute.[3] Thus 'democracy' is not per se a clear-cut advantage of 'any change'. Moreover, IB)'s selectiveness is in close accord with humanity's history of accurate timekeeping, as per Sects. 1.12, 3.1, 5.4, 5.5 and 7.7. Both the 'all change' [104] and the STLRC [39] positions invoke this as a useful feature. This is already in evidence in Chap. 1's discussion of how sidereal time out-performed apparent solar time as regards predicting eclipses.

Philosopher Adolf Grünbaum's observation [397]—that *dynamical* facts discriminate—is of particular relevance at this point. E.g. the earth's motion slowing down and the celestial bodies speeding up are *kinematically* equivalent. However, there is a clear mechanism for the first: dynamical effects of nearby masses causing tidal friction, whereas there is no known mechanism for the latter. Also for the second multiple dynamical bodies correlatedly speed up, so multiple and correlated effects would be required, whereas for the first the earth's rotation slowing down does not require such a coincidence. This provides an argument against 'any change' leading to 'any time' approaches.[4]

The extreme position V) provides, within the context of IIB), an *incontestable* time. This is in the sense that there is no more change elsewhere that can run in concerted ways that cause of one to doubt the validity of some timestandards. However, whole-universe and perfect notions for clocks have the problems pointed out in Chaps. 3.1, 5.4 and 9. Thus only the 'any change' and STLRC positions are operationally realizable. In conclusion, the STLRC position wins out overall.

15.8 Discussion of Generalized Local Ephemeris Time (GLET)

The time abstracted from STLRC is a *generalization* [39] of the astronomers' *ephemeris time* of Chap. 3. This GLET is in accord with the clock properties of Sect. 1.12 and the choices IB), IIB), IIIC) and IV).

Taking the Earth *not* to read off a dynamical substitute for Newtonian time allowed for a substantial advance in timekeeping in the passage from sidereal to

[3]Even in generic situations, one can locally consider a ranking procedure for one's candidate times, alongside a refining procedure until a physically attainable sought-for accuracy is met. This procedure may well often give a less accurate timestandard than the familiar and highly non-generic Earth–Moon–Sun system example of ephemeris time for use on Earth. Nevertheless, it gives rise to an extremum.

[4]A similar argument can also be made as regards the greater mechanistic plausibility of expanding universes rather than of shrinking atoms.

ephemeris time. In analyzing a wider range of such situations, it is helpful to decompose postulate IIB) into two parts. The first part involves the physics of the clock itself. The second part concerns the reading of the clock not corresponding to the parameter that most simplifies the equations of motion. The latter can be more holistic—involving the system under study rather than just the convenient reading hand part of the clock itself.

There might be an 'unexpected bound' on clock precision due to the holistic effect occurring at some appreciable level. Have we seen any evidence for this to date? For sure, the *leap seconds* that some years are adjusted by are *not* of this nature. This is because these concern adjusting *civil* time to compensate for irregularities in the rotation of the Earth. Further adjustments—3 and 5 orders of magnitude smaller—between 'barycentric dynamical time' for Space Science and 'terrestrial time' for Science on Earth [783] are currently fully accounted for by standard Gravitational and Relativistic Physics. Thus no holistic realizations of IIB) are currently in evidence. This is pointed out lest the observation that the astronomers' ephemeris time notion is Machian be elsewise misunderstood to carry such holistic connotations. Nevertheless, one might keep an open mind as to whether holistic effects might show up at some level much finer than the gradual replacement of sidereal by ephemeris time in the first half of the 20th century. This suggests developing Background Independent conceptual thinking *as well as* the far more specific minutiae of yet further improving atomic clocks.

N.B. that the GLET finding procedure does not just use a change to abstract a time. This additionally checks whether using this time in the equations of motion for other changes suffices to predict these to one's desired precision. If the answer is yes, then we are done. If not, consider further locally-significant[5] changes as well or instead in one's operational definition of time. If this scheme converges without having to include the entirety of the Universe's contents, one has found a GLET that is locally more robust than just using *any* change in order to abstract a time. Furthermore, it is particularly useful to consider equations of motion in this time and propositions conditional on this time.

15.9 Emergent Jacobi Time

A particular implementation of GLET follows from how both the conjugate momenta and the equations of motion simplify by use of[6]

$$\frac{\mathrm{d}}{\mathrm{d}t^{\mathrm{em(J)}}} := * := \sqrt{\frac{W}{T}}\circ := \frac{1}{N}\circ := \frac{1}{\dot{I}}\circ = \frac{\mathrm{d}}{\mathrm{d}I}. \tag{15.21}$$

[5]This is to be judged by the criterion in Chap. 23.

[6]Here $\circ$ just denotes a more visible $\dot{}$. More generally, we could use a JS suffix or leave the notation as t^{em}.

Here the first equality is the Manifestly Reparametrization Invariant computational formula. The second relates this to finite-theory generalized lapse N (conceived of generally enough to cover both Mechanics and Minisuperspace GR). The third recasts the preceding as the *velocity of the instant* notion $\dot{I}$ [20]. The fourth further recasts this as the *differential of the instant* notion $\mathrm{d}I$. In this way, the instant I itself is identified with the emergent time $t^{\mathrm{em(J)}}$ that labels that instant. Using (15.21) amounts to finding an emergent Jacobi time $t^{\mathrm{em(J)}} = \mathcal{F}[\boldsymbol{Q}, \mathrm{d}\boldsymbol{Q}]$. This entails interpreting the quadratic constraint not as an energy constraint in the usual sense but as an equation of time. Following Mach, this is ab initio a highly dependent variable rather than the independent variable that time is usually taken to be. This is because this 'usual' situation assumes that one knows beforehand what the notion of time to use, whereas the current position involves operationally establishing that notion (see Chap. 23 for examples and discussion).

Moreover, upon finding a satisfactory such for one's error tolerance, one can pass to the more usual conceptualization in which it is taken to be an independent variable, i.e. $\boldsymbol{Q} = \boldsymbol{Q}(t^{\mathrm{em(J)}})$. Now the emergent Jacobi time [30, 37, 39, 98] formula itself belongs to the former conceptualization, being given by the change scalar

$$t^{\mathrm{em(J)}} = \int \|\mathrm{d}\boldsymbol{Q}\|_{\boldsymbol{M}} \Big/ \sqrt{2\,W(\boldsymbol{Q})}\,. \tag{15.22}$$

The Mechanics case is just the $\boldsymbol{Q} = \boldsymbol{q}$ case of the above, and 1-d RPM is the $\boldsymbol{Q} = \boldsymbol{\rho}$ case (with $\boldsymbol{M}$ going to the identity array).

The Jacobi emergent time, and modifications thereof (see Chap. 16.2) was used in this way in Barbour's work (e.g. [98, 105]).

N.B. that this emergent time is *provided by* the system; in this way, it fits Mach's Time Principle. In contrast, the usually-assumed notion of time as an independent variable is un-Leibnizian and un-Machian. However, once the above time has been abstracted from change, it *is* a convenient choice for (emergent) independent variable. At first sight, this is an 'all change' resolution, but careful examination (Chap. 23) reveals that it is in practise indeed a 'STLRC' resolution.

In the case of Mechanics, $t^{\mathrm{em(J)}}$ is a relational recovery [98] of the quantity that is more usually taken to be absolute external Newtonian t^{Newton}. Perhaps it is significant that de Sitter and Clemence both had been terming what became ephemeris time 'uniform' or 'Newtonian' time. Moreover, we further rephrase the latter as *a relational recovery of Newtonian time.*

$t^{\mathrm{em(J)}}$ does not itself unfreeze the quantum-level frozen formalism, nor does it in any other way directly give quantum equations that are distinct from the usual ones. One can view it, rather, as an object that already feature at the classical level that is subsequently to be recovered by a more bottom-up approach at the quantum level. This works because it matches up with the emergent semiclassical (WKB) time as per Chap. 46.

$t^{\mathrm{em(J)}}$ itself *starts* as the most dependent variable of all: $t^{\mathrm{em(J)}}[\boldsymbol{Q}, \mathrm{d}\boldsymbol{Q}]$. Only when it is found to sufficient accuracy is it cast as the most convenient independent variable to use, i.e. $t^{\mathrm{em(J)}}$ such that $\boldsymbol{Q} = \boldsymbol{Q}(t^{\mathrm{em(J)}}) =: t^{\mathrm{em(J)}}_{\mathrm{indep}}$. If one is to identify these with the conventional formulation of Newtonian time, this identification involves the *final* form.

There is some freedom as regards where to break the exact equality in the sequence '$t^{\mathrm{em(J)}} = I = t$ for t the Newtonian time. However, $t^{\mathrm{em(J)}}_{\mathrm{indep}} = t^{\mathrm{Newton}}$ but $t^{\mathrm{em(J)}}[\boldsymbol{Q}, \mathrm{d}\boldsymbol{Q}] \neq t^{\mathrm{Newton}}$ *until* a sufficiently good approximation is found to be used in that role once no longer regarded as a highly dependent variable. We next choose to interpret I as the *instant-labelling* role that is conceptually distinct from—but indeed ends up being 'dual to'—the Newtonian time. The role this plays is secondary to the $\boldsymbol{Q}$ and $\mathrm{d}\boldsymbol{Q}$ pertaining to the notion of space in question, so instant-labelling also has prior and posterior forms. In summary, '$t^{\mathrm{em(J)}} = I$' a priori, but '$t^{\mathrm{em(J)}} = I = t$' only holds as an a posteriori equality. The $t^{\mathrm{em(J)}} = I$ identity's—and the a posteriori $I = t$ identity's—two sides reflect the time to instant-labelling duality.

In terms of emergent Jacobi time, moreover, the momenta and equations of motion take the simpler forms[7]

$$P_{\mathsf{A}} = M_{\mathsf{AB}} * Q^{\mathsf{B}}, \tag{15.23}$$

$$D_{\mathrm{abs}}{}^2 Q^{\mathsf{A}} = ** Q^{\mathsf{A}} + \Gamma^{\mathsf{A}}{}_{\mathsf{BC}} * Q^{\mathsf{B}} * Q^{\mathsf{C}} = N^{\mathsf{AB}} \partial W / \partial Q^{\mathsf{A}}. \tag{15.24}$$

The latter is a *parageodesic equation* with respect to the kinetic metric (meaning it has a forcing term arising from the conformal W-factor). It can also be cast as a true geodesic equation with respect to the physical metric whose line element is $\mathrm{d}J$. Here, in terms of the physical $M^{\mathrm{phys}}_{\mathsf{AB}} = 2\,W\,M_{\mathsf{AB}}$, with corresponding Christoffel symbol $\Gamma^{\mathrm{phys}\,\mathsf{A}}{}_{\mathsf{BC}}$,

$$** Q^{\mathsf{A}} + \Gamma^{\mathrm{phys}\,\mathsf{A}}{}_{\mathsf{BC}} * Q^{\mathsf{B}} * Q^{\mathsf{C}} = 0, \tag{15.25}$$

for a distinct but conformally-related $*$.

N.B. that there is one kinematical geometry to serve all problems but a distinct dynamical geometry per problem. This supplies a universal reason for study of the kinematical geometry.

A final move—useful in practical calculations—involves supplanting one of the evolution equations by the emergent Lagrangian form of the quadratic constraint,

$$M_{\mathsf{AB}} * Q^{\mathsf{A}} * Q^{\mathsf{B}}/2 + W = 0. \tag{15.26}$$

[7] 'abs' here denotes the standard differential-geometric absolute derivative.

Chapter 16
Combining Temporal and Configurational Relationalisms

We now reach our first systematic combination of Problem of Time facets. Two preliminary considerations are as follows. Firstly, continuing from Sect. 9.8's discussion about postulate CR-i), at the level of standard redundant presentations, the mechanical $M_{iIjJ} = m_I \delta_{ij} \delta_{IJ}$ is unsatisfactory through involving the Euclidean metric. However, we show below that the GR counterpart succeeds in meeting this criterion. Secondly, we rephrase CR-ii) to reflect that $\mathfrak{g}$ more strictly acts not on $\mathfrak{q}$ but on some fibre bundle structure thereover, such as $\mathfrak{T}(\mathfrak{q})$ in the case of Best Matching. It additionally applies more generally to further structures based upon $\mathfrak{q}$ in the case of the wider range of examples covered by the $\mathfrak{g}$-act, $\mathfrak{g}$-all method. This Chapter, moreover, mostly concerns various generalizations and reformulations of the Best Matching implementation we already encountered in Chap. 9. We now extend this to general $\mathfrak{g}$ and then render it TRi-compatible (this is a RTQ approach, and thus in particular both a *Tempus Ante Quantum* approach and the initial portion of a Reduced Quantization scheme).

16.1 Best Matching: General $\mathfrak{g}$

Let us first summarize Best Matching for general $\mathfrak{g}$ by

$$CR(S) := S_{\mathfrak{g}\text{-free}} := \mathrm{E}_{g \in \mathfrak{g}}\big(S_{\mathrm{CR}}(\boldsymbol{Q}, \mathrm{d}\boldsymbol{Q}, \boldsymbol{g})\big). \tag{16.1}$$

The first symbol means 'the Configurationally Relational Principles of Dynamics action' (corresponding to the group action of $\mathfrak{g}$ on $\mathfrak{T}(\mathfrak{q})$ being physically irrelevant), which is alternatively known by the second symbol: the $\mathfrak{g}$-free Principles of Dynamics action. The third symbol has further computational content. Firstly, S_{CR} is a $\mathfrak{g}$-corrected form of the action S: a $\mathfrak{g}$-act move; hence its dependence on $\boldsymbol{g}$ as well as $\mathfrak{T}(\mathfrak{q})$'s Machian variables $\boldsymbol{Q}$, $\mathrm{d}\boldsymbol{Q}$. Secondly, $\mathrm{E}_{g \in \mathfrak{g}}$ denotes extremum over $\boldsymbol{g} \in \mathfrak{g}$. Lastly, $\mathfrak{g}$ and $\mathfrak{q}$ are to be a suitably compatible such pairing (or triple, including S). This is clearly a subcase of $CR(\boldsymbol{O}) = \boldsymbol{O}_{\mathfrak{g}\text{-free}}$. It furthermore encapsulates that Best Matching is a procedure for bringing pairs of configurations into

E. Anderson, *The Problem of Time*, Fundamental Theories of Physics 190,
DOI 10.1007/978-3-319-58848-3_16

'minimum incongruence' by holding one fixed and $\mathfrak{g}$-shuffling the other.[1] It has an intermediate output of note—the extremizing $g_{\text{BM}}(\boldsymbol{Q}, \mathrm{d}\boldsymbol{Q})$ themselves—as well as the final extremized $S_{\text{BM}}(\boldsymbol{Q}, \mathrm{d}\boldsymbol{Q}) = S_{\mathfrak{g}\text{-free}} = S_{\text{CR}}(\boldsymbol{Q}, \mathrm{d}\boldsymbol{Q}, g_{\text{BM}})$. Best Matching moreover involves probing with candidate generators of irrelevant motions, without yet addressing the relations between the generators, which would complete the characterization of any given group. The group relations part of Group Theory enters at a later stage, though consideration of the constraints that ensue and whether Constraint Closure applies to these (Chap. 24).

'Suitably compatible' refers here to Sect. 14.2's criteria A) to C). Within the *Non Tempus Sed Cambium* Worldview, moreover, C) ('counting'), is to furthermore take into account the presence in general of constraints that use up c degrees of freedom. The theory is then trivial if $c > k$, inconsistent if $c = k$, and relationally trivial if $c = k - 1$. Since independent constraints use up at least 1 degree of freedom each, $c \geq l + 1$ is a guaranteed least stringent bound (the 1 arising from chronos). I.e. $c = l + 1 \geq k - 1 \Rightarrow k - l = \dim(\mathfrak{q}) - \dim(\mathfrak{g}) \leq 2$ already serves to invalidate theories.

Chapter 24 shall moreover replace this rather crude bound with a tighter bound based on phase space geometry rather than just counting. This arises from considering brackets relations in addition to the group actions of generators on (some fibre bundle over) $\mathfrak{q}$; groups are, of course characterized by generators *and* relations.

We finally point to how a given $\mathfrak{q}$, $\mathfrak{g}$ pair still constitutes a substantial ambiguity as regards which action S to consider thereupon. N.B. that Relationalism does not highly uniquely control of the form that Theoretical Physics is to take.

We next consider the detailed form taken by Best Matching.

Best Matching 0) In this case of Chap. 14.4's construction, the object $\boldsymbol{O}$ is a classical Principles of Dynamics action S built upon $\mathfrak{q}$. The incipient bare action can be thought of as a map $S : \mathfrak{T}(\mathfrak{q}) \rightarrow \mathbb{R}$. For now, we interpret the tangent bundle $\mathfrak{T}(\mathfrak{q})$ as a configuration–velocity space with a product-type Jacobi action (15.7) thereupon. We next pass to the 'arbitrary $\mathfrak{g}$ frame corrected' Principles of Dynamics action by applying the basic infinitesimal group action to the incipient bare S, obtaining

$$S_{\text{CR}} = 2\int \mathrm{d}\lambda\, L = 2\int \mathrm{d}\lambda \sqrt{T\, W(\boldsymbol{Q})}, \quad T := \| \dot{\boldsymbol{Q}} - \vec{\mathfrak{g}}_{g}\, \boldsymbol{Q} \|_{M}^{2}/2. \tag{16.2}$$

Best Matching 1) We next extremize over $\mathfrak{g}$. This produces a constraint equation **shuffle** of the form **cin**, which is linear in the momenta and also a change scalar.

Best Matching 2) The Lagrangian variables $\boldsymbol{Q}$, $\dot{\boldsymbol{Q}}$ form of this constraint is to be solved for the auxiliary variables $\boldsymbol{g}$ themselves.

[1] This is modulo the Calculus of Variations not guaranteeing that extremization yields either a minimum or a unique answer. Also note that Part I's illustrative Barbour–Bertotti action (9.7) is a particular case of S_{CR}.

Best Matching 3) Substitute this solution back into the action; this is an example of Appendix J.1's multiplier elimination. This produces a final $\mathfrak{g}$-independent expression that could have been arrived at as a *direct* implementation of CR-ii).
Best Matching 4) We finally elevate this new action to serve as a new starting point.

Moreover, we shall see that in practice Best Matching 2) is often an impasse. Best Matching 2), 3) and 5) can be viewed as searching for a 'minimizer', so as to establish the minimum 'incongruence between' adjacent physical configurations (subject to footnote 1). These steps can also be viewed as a $\mathfrak{T}(\mathfrak{q})$-level reduction procedure.[2]

We also note that we can interpret $\mathfrak{T}(\mathfrak{q})$ as configuration–change space in terms of Jacobi–Mach variables $\boldsymbol{Q}$, $\mathrm{d}\boldsymbol{Q}$. Then T is supplanted by the kinetic arc element $\mathrm{d}s$ and L by the Jacobi arc element $\mathrm{d}J$, reconciling the above more detailed description with the opening summary.

Let us also now reconcile Part I's Lie derivative formulation with Part II's fibre bundle one. Begin by interpreting the Lie derivative as a point identification map [814] *between* two adjacent space slices, as opposed to a Lie dragging *within* a single such slice. Next, the Lie derivative applies due to continuous transformations being implemented; moreover all of those under consideration happen to also be differentiable, These transformations form some subgroup $\mathfrak{g} \leq \mathit{Diff}(\mathfrak{a})$ corresponding to $\mathfrak{g} \leq \mathit{Diff}(\mathfrak{a})$ at the level of Lie algebras. There is consistency between the two since one can continue to define Lie derivatives within a fibre bundle context (see e.g. [560] for more). One advantage of the Lie derivative formulation is that it provides a specific form for the infinitesimal group action. On the other hand, $\mathfrak{g}$-act does always involve this particular group action (for all that Best Matching itself does). A corresponding disadvantage is that it is a merely local treatment, whereas fibre bundles can encode further global information. See Sect. 38.1 for a further global-level advantage of the Lie derivative formulation.

Let us finally consider the smaller case of Constraint Closure that occurs internally within Configurational Relationalism: checking that the ***shuffle*** constraints arising from $\mathfrak{g}$ do indeed close among themselves. E.g. the Poisson bracket (9.30) affirms this for Metric Shape and Scale RPM, whereas (9.28), (9.29), (9.31) do so for Electromagnetism, Yang–Mills Theory and for the $\mathit{Diff}(\Sigma)$ of GR respectively. Furthermore, all of these Poisson brackets form Lie algebras.

16.2 TRi-Best Matching

The above Section, however, fails to implement Temporal Relationalism, because the $\mathfrak{g}$-correcting Lagrange multiplier coordinates breaking the Manifest Reparametrization Invariance of the action. None the less, concurrent implementation can be attained by using $\mathfrak{g}$-*correcting cyclic velocities* instead [15, 20, 37, 64,

[2]In this book, and quite commonly in the Quantum Gravity literature, reduction usually involves just linear constraints rather than quadratic constraint as well.

102]. Moreover, Manifest Parametrization Irrelevance and then its dual $\mathfrak{q}$-Geometry formulation are successive conceptual advances within the implementation of Temporal Relationalism itself. A second concurrent implementation which respects this as well uses, rather, $\mathfrak{g}$*-correcting cyclic differentials* (see also footnote 1). In this case, we form $\mathfrak{g}$-bundles in terms of $\boldsymbol{Q}$, $\mathrm{d}\boldsymbol{Q}$ and $\mathrm{d}\boldsymbol{g}$ variables: $\mathfrak{p}(\mathfrak{T}(\mathfrak{q}, \mathfrak{g}))$ with the $\mathfrak{T}$ now interpreted as 'change'. Furthermore, encoding one's $\mathfrak{g}$ auxiliary variables in either of these ways entails a nontrivial change of formalism, since it requires subsequent care with how one performs one's Calculus of Variations (see Appendix L for details). In the current Finite Theory setting, this entails so-called free end point variation (Appendix L.3). We call this method *TRi-Best Matching*.

Tri Best Matching 1) now involves an action $S_{\mathrm{TR-CR}}$ of the form (15.7) with

$$\mathrm{d}s = \|\mathrm{d}_g \boldsymbol{Q}\|_M \quad \text{for } \mathrm{d}_g \boldsymbol{Q} := \mathrm{d}\boldsymbol{Q} - \sum_{g \in \mathfrak{g}} \overrightarrow{\mathfrak{g}}_{\mathrm{d}g}\, \boldsymbol{Q}. \tag{16.3}$$

In each case, we can redo the preceding Chapter's treatment of conjugate momenta, quadratic constraint, equations of motion and the beginning of the treatment of emergent time 'by placing $\boldsymbol{g}$ suffixes' on d, I, t^{em} and $*$. Note furthermore that Temporal and Configurational Relationalism constitute *two separate Constraint Providing procedures*. These are rendered compatible with each other as per this and the next Sec. The first provides *C*hronos and the second provides ***shuffle***.

TRi Best Matching 2) now involves solving the Jacobi–Mach formulation of ***shuffle*** $= 0$ for the cyclic differential $\mathfrak{g}$ auxiliaries, $\mathrm{d}\boldsymbol{g}$:

$$CR(S) := S_{\mathfrak{g}\text{-free}} := \mathrm{E}_{g \in \mathfrak{g}}\big(S_{\mathrm{TR-CR}}(\boldsymbol{Q}, \mathrm{d}\boldsymbol{Q}, \mathrm{d}\boldsymbol{g})\big) = \{\text{extremum of } \boldsymbol{g} \in \mathfrak{g}\}$$

$$\text{of } \{S_{\mathrm{TR-CR}} \text{ built upon } \mathfrak{q}, \mathfrak{g}\} \text{ for a suitably compatible such pairing.} \tag{16.4}$$

Note that the extremization is over $\boldsymbol{g}$, rather than $\mathrm{d}\boldsymbol{g}$, since $\boldsymbol{g}$ plays the role of a cyclic differential.

On the other hand, Best Matching 2), 4) and 5) work out the same whether TRi or not [Best Matching 1) and 3)'s *mathematical procedures* are also unaffected by this change of formalism].

16.3 Emergent Jacobi–Barbour–Bertotti Time

TRi-Best Matching 3′) As a distinct application of TRi Best Matching 2)—the emergent time expression is now

$$CR(t^{\mathrm{em}}) := t^{\mathrm{em}}_{\mathfrak{g}\text{-free}} := \mathrm{E}'_{g \in \mathfrak{g}} \int \|\mathrm{d}_g \boldsymbol{Q}\|_M / \sqrt{2\,W(\boldsymbol{Q})}. \tag{16.5}$$

This is also known as Jacobi–Barbour–Bertotti emergent time, but our choice of suffices is a rather clearer reminder of this being an emergent time that complies with Configurational Relationalism. Also note how the extremization now takes an implicit form. I.e. in (16.5) a second functional—the emergent time—is subject to performing the extremization of a first functional—the relational action $S_{\text{TR–CR}}$. This means that

$$\mathrm{E}'_{g \,\in\, \mathfrak{g}} = |_{\mathrm{d}g=\mathrm{d}g_{\mathrm{BM}}(\boldsymbol{Q},\mathrm{d}\boldsymbol{Q})} \,,$$

where BM denotes the Best-Matched value. See [53] for Algebra and Fibre Bundles interpretations of this and the previous two Sections, including as cases of the $\mathfrak{g}$-Act $\mathfrak{g}$-All Method. Moreover, if one succeeds in carrying out Best Matching as e.g. per Sect. 16.7, $\mathrm{d}g$ is replaced by an extremal expression in terms of $\boldsymbol{Q}$ and $\mathrm{d}\boldsymbol{Q}$ alone. By this, both aspects of this complication are washed away and one has an expression for $t^{\mathrm{em}}_{\mathfrak{g}\text{-free}}$ paralleling that of the previous Chapter's t^{em}, albeit now in terms of the *reduced* $\mathfrak{q}$'s geometry.

16.4 TRi Configurational Relationalism in General

Let us next present the TRi $\mathfrak{g}$-Act $\mathfrak{g}$-All Method: the General Strategy for Configurational Relationalism now modified to remain within that for Temporal Relationalism.

Here

$$CR(\boldsymbol{O}) := \boldsymbol{O}_{\mathfrak{g}\text{-inv}} := \mathrm{S}_{g \,\in\, \mathfrak{g}} \circ \mathit{Maps} \circ \vec{\mathfrak{g}}_g \, \boldsymbol{O}, \tag{16.6}$$

now taking a specifically TRi-preserving form. I.e. all TRi objects $\boldsymbol{O}$ maintain TRi status upon being rendered $\mathfrak{g}$-invariant as well by this manoeuvre. A common example of this involves all three parts of the composition preserving change tensoriality, with the group action sending each change tensor to an $\mathfrak{g}$-corrected change tensor of the same rank.

We subsequently pass to giving some useful illustrative examples.

16.5 Example 1) Metric Shape and Scale RPM

In TRi form, the action for this is

$$S = \sqrt{2} \int \sqrt{E - V(\underline{\rho}^j \cdot \underline{\rho}^k \text{ alone})}\, \mathrm{d}s,$$

$$\text{for } \mathrm{d}s = \|\mathrm{d}_{\underline{B}}\boldsymbol{\rho}\|, \quad \mathrm{d}_{\underline{B}}\underline{\rho}^A := \mathrm{d}\underline{\rho}^A - \mathrm{d}\underline{B} \times \underline{\rho}^A. \tag{16.7}$$

The conjugate momenta are $\boldsymbol{\pi} = \sqrt{2\,W}\, \mathrm{d}_{\underline{B}}\boldsymbol{\rho}/\|\mathrm{d}_{\underline{B}}\boldsymbol{\rho}\|$.

These obey as a primary constraint the quadratic energy constraint

$$\varepsilon := \|\boldsymbol{\pi}\|^2/2 + V(\boldsymbol{\rho}) = E. \tag{16.8}$$

Also free end point variation with respect to $\underline{B}$ gives as a secondary constraint the linear zero total angular momentum constraint

$$\underline{\mathcal{L}} := \sum_{i=1}^{n} \underline{\rho}^A \times \underline{\pi}_A = 0. \tag{16.9}$$

The Jacobi–Mach equations are

$$\frac{\sqrt{2\,W}\,\mathrm{d}\boldsymbol{\pi}}{\|\mathrm{d}_{\underline{B}}\boldsymbol{\rho}\|} = -\frac{\partial V}{\partial \boldsymbol{\rho}}. \tag{16.10}$$

Finally, the emergent time is now

$$CR(t^{\mathrm{em}}) = \mathrm{E}'_{\underline{B} \in Rot(d)} \int \|\mathrm{d}_{\underline{B}}\boldsymbol{\rho}\| \Big/ \sqrt{2\,W(\boldsymbol{\rho})}\,. \tag{16.11}$$

The momentum-change relation and equations of motion then take the simplified forms

$$\boldsymbol{\pi} = \mathbb{I}{*}\boldsymbol{\rho}, \quad {*}\boldsymbol{\pi} = -\frac{\partial V}{\partial \boldsymbol{\rho}} \quad \text{for} \tag{16.12}$$

$$* := \frac{\mathrm{d}}{\mathrm{d}CR(t^{\mathrm{em}})} := \mathrm{E}'_{\underline{B} \in Rot(d)}\sqrt{2\,W}\frac{\mathrm{d}}{\|\mathrm{d}_{\underline{B}}\boldsymbol{\rho}\|}. \tag{16.13}$$

Note how the 1- and 2-d cases are included within the 3-d form of the auxiliaries There is no B for $d = 1$, and but a scalar B for $d = 2$. All these cases are encoded by $\underline{\rho}^A - \underline{B} \times \underline{\rho}^A$ if one allows for $\underline{B} = (0, 0, B)$ in 2-d and $\underline{B} = 0$ in 1-d (Exercise!). Correspondingly, in 1-d there is no $\underline{\mathcal{L}}$ constraint at all, while in 2-d $\underline{\mathcal{L}}$ has just one component that is nontrivially zero: $\mathcal{L} = \sum_{A=1}^{n}\{\rho^{A1}\pi_{A2} - \rho^{A2}\pi_{A1}\} = 0$.

16.6 Example 2) Metric Shape RPM

Metric Shape RPM alias similarity RPM alias Barbour 2003 RPM [102] is a mechanics in which only relative times, and *shape configurations*—consisting of relative angles and ratios of relative separations—are meaningful. (On the other hand, if the rotations are not removed, one has Kendall's notion [539] of *preshape*, cf. Appendix G.)

The action for this is

$$S = \sqrt{2}\int \sqrt{E - V(\text{ratios of } \underline{\rho}^A \cdot \underline{\rho}^B \text{ alone})}\mathrm{d}s$$

$$\text{for } \mathrm{d}s^2 = \|\mathrm{d}_{\underline{B},C}\boldsymbol{\rho}\|^2/\mathrm{I} \quad \text{and} \quad \mathrm{d}_{\underline{B},C}\underline{\rho}^A := \mathrm{d}\underline{\rho}^A - \mathrm{d}\underline{\mathrm{B}} \times \underline{\rho}^A + \mathrm{d}C\underline{\rho}^A. \tag{16.14}$$

$\mathrm{d}s^2$ is also a ratio since $I = \|\boldsymbol{\rho}\|^2$.

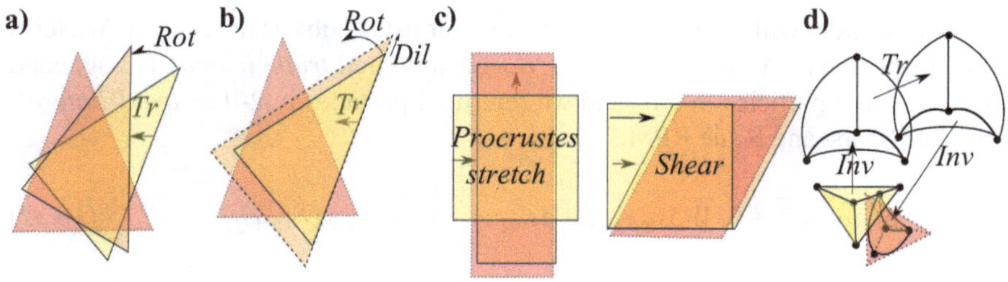

Fig. 16.1 Five notions of matching shapes (keep the *red one* fixed and perform transformations on the *yellow one* so as to bring the two into minimum incongruence). **a)** Barbour's wooden triangles, subjected to translational and rotational matching. **b)** Dilational matching, as exhibited e.g. by moving one of two parallel overhead projector slides away from the other **c)** Affine matching adds to these shears and 'Procrustean' stretches (volume-preserving in 3-d or area preserving in 2-d). These can be approximately demonstrated with two pieces of uncooked jelly. **d)** Conformal matching adds special conformal transformations instead. N.B. that the shapes depicted are in each case the minimal relationally nontrivial units for the corresponding notions of Relationalism (Fig. G.4)

The conjugate momenta are $\underline{\pi}_A = \{\sqrt{2W/I}\}\delta_{AB}\,\mathrm{d}_{\underline{B},C}\underline{\rho}^B$. The quadratic energy constraint is

$$\mathcal{E} := I\,\|\boldsymbol{\pi}\|^2/2 + V = E. \tag{16.15}$$

(16.9)'s $\mathcal{L}_i$ from variation with respect to $\underline{B}$ is now accompanied by the *zero total dilational momentum constraint*

$$\mathcal{D} := \sum_{A=1}^{n} \underline{\rho}^A \cdot \underline{\pi}_A = 0 \tag{16.16}$$

as a secondary constraint from variation with respect to C. N.B. that $\mathcal{L}_i$ and $\mathcal{D}$ are entirely independent by the 3-d case of (E.12): the former acts on shape and the latter trivializes the role of scale.

Figure 16.1 extends Best Matching from Barbour's pair of wooden triangles to a range of further cases whose minimal relationally nontrivial units are tabulated in Fig. G.4; the corresponding RPMs were introduced in Sect. 14.5 and their actions are given in Chap. 18.

Let us end by noting that one can view 'Newtonian Mechanics for island universe subsystems within RPM' as a relational route to Newtonian Mechanics.

16.7 RPM Examples of Best Matching Solved

Solving Configurational Relationalism for an RPM renders its action $\mathfrak{g}$-free and the expression for its $t^{\mathrm{em}}_{\mathfrak{g}\text{-free}}$ explicit. As outlined in Appendix G.1 and detailed in [37], this can be done in 1- or 2-d for any N and for 3 particles in 3-d, in each case with or without scale. These need particular names as whole-universe models so as to not confuse them with N-body problems in their usual subsystem context. We term these, respectively, *N-stop metroland*, *N-a-gonland* and *triangleland*. Triangleland also refers to 3 particles in 2-d, and we refer to 4 particles in 2-d as *quadrilateral-land*. For Shape and Scale RPM,

$$S_{\text{red}} = \sqrt{2}\int\sqrt{W}\,\mathrm{d}s_{\text{red}} \quad \text{for } \mathrm{d}s_{\text{red}} := \mathrm{d}s_{\text{BM}} = \|\mathrm{d}\boldsymbol{Q}\|_{\boldsymbol{M}(\boldsymbol{Q})}, \tag{16.17}$$

where the $\boldsymbol{M}(\boldsymbol{Q})$ are in-general-curved $\mathfrak{q}$ metrics.

Moreover, the above also arises by a $\mathfrak{T}(\mathfrak{q})$-level classical reduction [37], and is furthermore the basis for Reduced Quantization as per Chaps. 42–43. Upon having performed the above reduction, the RPM is in a form in which the following absolute structures of the Newtonian and Galilean Paradigms (Sect. 2.5, 4.6) are absent: not just t but also velocity $\underline{v}$ and δ. Indeed, the reduced configuration space metric is here a function of the configurations themselves rather than involving any further extraneous background entities.

Finally, note that RPMs with Newtonian Gravitation potentials have *not* freed themselves from the non-dynamical connection absolute structure of Newtonian Mechanics with Newtonian Gravitation potential [776].

16.8 Direct Implementation of Configurational Relationalism for RPMs

The idea here is to construct a $\mathfrak{g}$-invariant formulation by working directly [37] on the relationalspace.[3] This eliminates the need for any arbitrary $\mathfrak{g}$-frame variables, nor do any linear constraints arise, nor are such to be used as a basis for reduction to pass to a new action. Instead, one's action is now already directly $\mathfrak{g}$-invariant i.e. on $\widetilde{\mathfrak{q}} := \mathfrak{q}/\mathfrak{g}$ by construction. This is reflected by the more complicated form taken by the kinetic term or arc element.

This is again possible for lower-d RPMs. It involves case starting directly with the shape spaces that Kendall had determined [536, 539] in the pure-shape case, or the cones thereover in the scaled case [37]. In each case, one is associating a mechanics to the geometry which then plays the role of that mechanics' configuration space, as per the methodology of Jacobi and Synge [598] (Chaps. 14 and 17). The latter 'relationalspace formulation' indeed provides a second foundation for RPMs [18, 37], distinct from that in [102, 105]. Because the same formulation of the same mechanics is arrived at both ways—on relationalspace or by reduction—we subsequently refer to it as the *r-formulation* (denoted by an r subscript). This works out because some 1-d models and 3-particle models for $d > 1$ have particularly simple configuration spaces, as per Appendix G.1. In this way, $t^{\text{em}}_{\mathfrak{g}\text{-free}}$ is itself a t^{em} for a more directly formulated $\mathfrak{q}$ geometry.

[3]We use *relationalspace* to mean relational space for Metric Shape and Scale RPM and shape space for Metric Shape RPM, i.e. each case's non-redundant configuration space.

These identifications of RPM configuration spaces with well-known geometries are useful by allowing for very close to standard mathematical treatment at the classical and quantum levels. Moreover, this mathematics is physically interpreted in a very different manner from the standard one [37]. This can be tied to a wider range of still fairly standard mathematical methods so as to do Classical and Quantum Physics. Triangleland is the simplest model that can concurrently possess scale and nontrivial linear constraints. On the other hand, quadrilateralland [28] is more geometrically typical for an N-a-gonland than triangleland (which benefits from the $\mathbb{CP}^1 = \mathbb{S}^2$ coincidence). These models' mathematical simpleness is a triumph, because one can then carry out and check many Problem of Time calculations. [These would *not* make sense if done for atoms or molecules, say, due to the absolutist underpinnings in these latter cases.] This is just what the study of the Problem of Time needs (at least as regards the facets and strategies nontrivially manifested by RPMs).

For further use below, scaled triangleland's reduced configuration spaces possesses Cartesian coordinates that are non-obviously related to the merely-relative Jacobi coordinates. These are the 'Hopf–Dragt coordinates' [37, 266, 624] of Eqs. (G.13)–(G.15). These quantities emerge naturally from the reduction as *ubiquitous groupings* [37]. I.e. once reduction has been performed, these start to appear throughout the functional dependencies of all of the model's significant quantities. In this case, these groupings can be interpreted as a natural choices of coordinates on the reduced configuration space.

3-d-ness is moreover a well-known major complication in standard Mechanics. Nonzero total angular momentum vector $\underline{L}$ is also a substantial complication [624] for a less well-known reason (Sect. 37.3) On the other hand, there is a simple passage from the $\underline{L} = 0$ restriction of standard Mechanics to RPM. These substantial complications have hitherto blocked making mathematical progress with the status of the hypothetical extension to $\underline{L} \neq 0$ models. By this, it is not known if a more general RPM that permits $\underline{L} \neq 0$ can be constructed, nor has this been proven to not exist.

Whereas the indirect formulation involves $\mathfrak{g}$-bundles, the direct formulation's $\mathfrak{q}/\mathfrak{g}$ is more generally a stratified manifold rather than a manifold. These have but limited amenability to Fibre Bundle Methods; Sect. 37.6 outlines some more general Sheaf Methods to this end.

Dynamical study of r-formulated RPMs proceeds as follows. *Dilational momentum* p_ρ is the momentum conjugate to the scale variable ρ, whereas the momenta conjugate to the shape variables S^{a} are, schematicly, the *shape momenta* $p^{\mathrm{S}}_{\mathrm{a}}$. For Metric Shape RPM in 1-d, $\mathfrak{q}$'s isometry group provides an $SO(n)$ of conserved quantities provided that the potential respects these. These are relative dilational momentum quantities. The 2-d triangleland case has, under the same circumstances, an $SU(2)$'s worth of conserved quantities. These are a pure relative angular momentum quantity (of the base relative to that of the median: Fig. G.8), and two mixtures of relative angular momentum and relative dilational momentum [37]. [28] gives the corresponding relational interpretation of quadrilateralland's $SU(3)$ octet of conserved quantities.

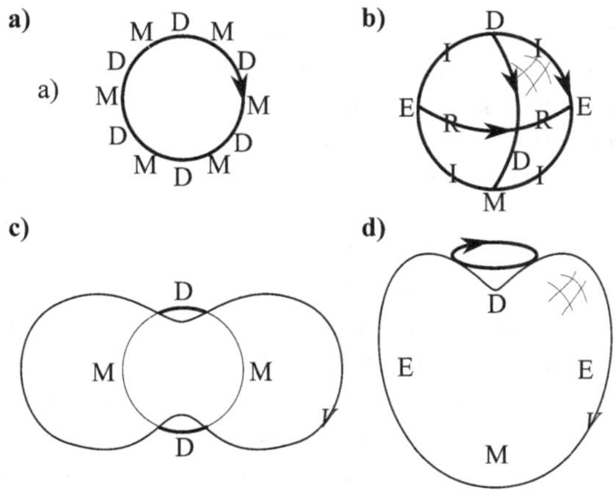

Fig. 16.2 This figure is to be interpreted using Appendix G's back-cloths and notation for configurations. **a)** and **b)** are 3-stop metroland's and triangleland's free motion geodesics. **c)** and **d)** are examples of potentials over shape space induced by harmonic oscillator potentials: the 3-stop metroland 'peanut' and the triangleland 'heart' potentials. The *thick lines* denote some of the possible classical motions within the resulting potential wells. The effective potentials in each case have an additional central skewer along the vertical axis, in the manner of a centrifugal barrier. These are relevant when the total shape momentum is nonzero. See [37] for more details and many more examples of potentials over—and dynamical trajectories along—shape spaces

Figure 16.2 subsequently presents some cases of potentials and dynamical trajectories for RPM's; these are useful as precursors and classical limits for quantum solutions used in Part III. Free problem solutions are geodesics of the shape spaces. The physical significance of $\mathbb{S}^2$'s geodesic great circles in the cases of a) 4-stop metroland and b) triangleland. The $\mathbb{S}^{N-2}$ geodesics—corresponding to N-stop metroland free motions, and the $\mathbb{CP}^{N-1}$ geodesics—corresponding to N-a-gonland free motions [28]—are also well-known. The interpretation of the configurations can then just be read off the 'back-cloth', such as in Figs. 16.2 or G.11.

16.9 Limitations of RPM Models

While RPMs have been very useful so far in the discussion, they do have some limitations. RPMs are finite rather than field-theoretic, rendering them more tractable but also not having counterparts of some more of GR's difficulties. RPMs also do not make contact with SR, with SR- or GR-like spacetime, the inherently nonlinear nature of GR that corresponds to gravity itself gravitating, or Black Hole Physics. On the other hand, RPMs are explicitly theories of Background Independence: a valuable isolation of the 'other half' of the Gestalt. They are closed systems, which is of value to Quantum Cosmology. They lie within Sect. 9.2's great tradition of Dynamics, and parallel a number of further features of Geometrodynamics [37]. Because of these features, RPMs continue to be useful as examples in a number of further Chapters.

However, due to the missing features, other theories and model arenas are clearly also required (Chaps. 17, 18 and 30). On the long run, $\mathit{Diff}(\boldsymbol{\Sigma})$ is vastly harder a $\mathfrak{g}$ than $U(1) = SO(2)$, or, indeed, any other $SO(n)$. Moreover, r-formulations are in fact seldom possible; lower-d RPMs are thus rather special in admitting such; we shall see in Chap. 18 that GR does not in general. On the other hand, as we show in Chap. 30, the r-formulation remains available to lowest nontrivial order in SIC.

Chapter 17
Temporal Relationalism: More General Geometries

17.1 Minisuperspace GR

This brief Chapter serves to introduce various further models of substantial use in this book's later discussions. We first consider $\mathrm{d}s = \|\mathrm{d}\boldsymbol{Q}\|_{\boldsymbol{M}}$ for a semi-Riemannian metric $\boldsymbol{M}$ and $W(\boldsymbol{Q}) = R - 2\Lambda$ (in the undensitized presentation). This is a Misner-type action: the Minisuperspace restriction of the BSW action. This form covers both the isotropic case with minimally-coupled scalar field [31, 149, 419] which was already introduced in Chap. 9 and the vacuum anisotropic cases [657, 659, 760] whose configuration space metric is given in Appendix I.1. These examples simpler in the sense of involving trivial $\mathfrak{g}$, and so no linear constraints. On the other hand, these examples involve *dynamical* spatial geometry. The restriction to Minisuperspace of the GR Hamiltonian constraint $\mathcal{H}$ now arises as a primary constraint, in an 'indefinite triangle' version of Sect. 15.6's Pythagorean working.[1]

In contrast, the more well-known ADM action (8.17) contains the lapse variable α. Since this lapse is an extraneous time-like variable, the ADM formulation fails to implement TR-i). It is however straightforward to show that the ADM and BSW actions are equivalent, as follows. The lapse α plays the role of a Lagrange multiplier coordinate in the ADM action. Lagrange multiplier elimination (Appendix J.1) can now be applied. In this particular case, the multiplier equation is a very simple algebraic one. I.e. $\alpha^2 = \mathrm{T}_{\mathrm{ADM-L}}/4\{R - 2\Lambda\}$, so one can readily eliminate α from the ADM action, to obtain the BSW action.[2]

The Minisuperspace cases of $\mathcal{H}$ and t^{em} are (9.14), (9.15) for the isotropic scalar field models, whereas

$$\mathcal{H} = -p_\Omega^2 + p_+^2 + p_-^2 + \exp(4\,\Omega)\big\{V(\beta_\pm) - 1\big\}, \tag{17.1}$$

[1] Most readers will already be familiar with 'indefinite triangles' from studying SR or the hyperbolic functions.

[2] This parallels the Euler–Lagrange to Jacobi equivalence. The remaining difference—that this proceeds by Routhian reduction rather than multiplier elimination—is ironed out in Sect. 18.7.

E. Anderson, *The Problem of Time*, Fundamental Theories of Physics 190,
DOI 10.1007/978-3-319-58848-3_17

$$t^{\text{em(J)}}_{\text{isotropic-MSS}} = \int \sqrt{-\mathrm{d}\Omega^2 + \mathrm{d}\beta_-^2 + \mathrm{d}\beta_+^2 \exp(\Omega)} \Big/ \sqrt{1 - V(\beta_\pm)} \tag{17.2}$$

for various vacuum diagonal anisotropic Bianchi class A models. In cases dominated by scalefactor dynamics, these amount to a relational recovery of t^{cosmic} to leading order. N.B. that—in contradistinction to the Mechanics case of Sect. 15.2—in GR λ and t *do* coincide: GR is an *already-parametrized* theory.

See e.g. [873] for an account of the dynamics of Minisuperspace, and e.g. [179] for the LQC equivalent of diagonal anisotropy.

17.2 Jacobi–Synge Relational Actions

Awareness of this second generalization started with Barbour et al. [109] pointing out two orderings in building actions: summing or integrating the squares and then taking the square root versus taking individual square roots and then summing or integrating. In Manifestly Reparametrization Invariant terms,

$$\sqrt{T_1 + T_2}\sqrt{W} \quad \text{versus} \quad \{\sqrt{T_1} + \sqrt{T_2}\}\sqrt{W}, \tag{17.3}$$

which they term a 'global square root' (as used so far in this book) and the local square root (this Section's Example 1). This is however very far from a complete characterization of the possible diversity of actions [14], as the examples below demonstrate.

Example 2) Manifestly Reparametrization Invariant actions can feature the nth root of the sum of nth power can feature instead. For instance, the configuration space version of Riemann's 'next simplest' quartic geometry (D.25) is a subcase of this:

$$S = \int \{F(Q)\mathrm{d}s^4_{\text{quartic}}\}^{1/4}. \tag{17.4}$$

In this case, primary constraint is now purely quartic in the momenta. In fact, many properties attributed to $\mathcal{H}$ for being purely quadratic transcend to this case as well, and can be traced more generally to constraints which contain no accompanying term that is linear in a further momentum variable.[3]

Example 3) Moreover, *multiple* roots and sums can also implement Temporal Relationalism, and there are other ways of making homogeneous linear functions besides, e.g. not only $f\,\mathrm{d}x$ but also e.g. $\mathrm{d}x^2/\sqrt{\mathrm{d}x^2 + \mathrm{d}y^2}$. Modulo degeneracy, at the level of Geometry this is Finsler's generalization (Appendix D.5), which was first applied to $\mathfrak{q}$ geometry by Synge [598]. These indeed still constitute a geodesic

[3]The use of 'further' here covers that $p_t = p_i p^i + p_t^2$ is entirely different in character and quantum interpretation from $p_t = p_i p^i$.

principle, just for more complicated notions of geometry. A general form for the action in this case is

$$S := \int \mathrm{d}\lambda F_{\mathrm{JS}} = \int \mathrm{d}F_{\mathrm{JS}}, \tag{17.5}$$

where JS stands for Jacobi and Synge and is homogeneous linear in the velocities or changes.

Example 4) Aside from all the previous finite relational actions in this book, Example 3) has another subcase of note:

$$S = \int \mathrm{d}\lambda \{2\sqrt{W\,T_{\mathrm{quad}}} + T_{\mathrm{lin}}\} = \int \{\sqrt{2\,W}\mathrm{d}s_{\mathrm{quad}} + \mathrm{d}s_{\mathrm{lin}}\}, \tag{17.6}$$

which is the $\mathfrak{q}$ geometry counterpart of Randers' subcase (D.26) of Finslerian Geometry. This allows for mechanics with linear 'gyroscopic' terms [598], moving charges, and is furthermore a model arena for the inclusion of spin-1/2 fermions alongside GR (see Sect. 18.11). The fermionic case additionally has disjoint quadratic and linear arc element species (i.e. a partition of the changes involved into distinct bosonic and fermionic species respectively).

Without specifying what kind of homogeneous linear function the action contains, there is no information as regards how many primary constraints there are nor about their form. For there to be an expression for *emergent Jacobi–Synge time* $t^{\mathrm{em(JS)}}$, there is to be *one* such relation, and such that $t^{\mathrm{em(JS)}}$ can be made the subject of the equation. The necessity of these caveats is demonstrated by the finite-theory local square root example having multiple quadratic constraints and by the following further example.

Example 5) Consider the free theory with action

$$S = \frac{1}{2}\int \|\dot{\boldsymbol{Q}}\|^2_{m(\boldsymbol{Q})}\mathrm{d}t.$$

The parametrized form of this does not permit Routhian reduction, since the cyclic equation is homogeneous in $\dot{t}$ and therefore cannot be used to eliminate $\dot{t}$. On the other hand, the above action clearly encodes $P_{\mathsf{A}} = M_{\mathsf{AB}}\dot{Q}^{\mathsf{B}}$ and a single quadratic constraint $\|\boldsymbol{P}\|^2_N/2 = E$. Specialize further to the case $E = 0$. Then using the momentum-velocity relation within the quadratic constraint produces an expression which is homogeneous in the hypothetical $\mathrm{d}/\mathrm{d}t^{\mathrm{em(JS)}}$, and therefore cannot be rearranged to make $t^{\mathrm{em(JS)}}$ the subject. Moreover, introducing E cures this second ailment, whereas including a potential term $V(\boldsymbol{Q})$ sorts out both.

Example 6) Next suppose that the species contained as velocities in each kinetic term square root in Example 1) are disjoint. Then the separate primary constraint arising from each square root gives its own expression for time (modulo the preceding caveat not applying to it), each in terms of the disjoint set of changes contained within the corresponding T_i. In this way, this example has multiple distinct emergent times for each such disjoint set of species, each in terms of its own species' changes alone. On some occasions, this does admit a STLRC interpretation, in which subsystem 2's disjoint changes are not relevant to a totally decoupled subsystem 1. However, it is not clear which if any of these emergent times to adopt if the kinetic metrics or the potentials couple to that species. Fortunately, this situation is not known to occur in Nature (see moreover Sect. 18.10 for specifically inhomogeneous GR, which does have a counterpart of this with a valid physical interpretation).

Example 7) The subcase of Example 4) which models fermions alongside bosons also involves a partition, albeit now into a square root of a square and a linear piece. In this case, only the quadratic piece contributes a primary constraint. However the corresponding emergent time is abstracted solely from the bosonic changes. This invalidates 'all change' approaches and also the 'all change is given the opportunity to contribute' aspect of STLRC. One way out of this is of course to doubt the physicality of *classical* fermionic theories; we shall see in Sect. 18.11 that there are also further reasons independent of Relationalism for doing so.

In summary, *the Jacobi–Synge action principle implements Leibnizian Temporal Relationalism. This does not, however, necessarily lead to a Machian resolution of this with the same conceptual features we are accustomed to from the Jacobi action itself.*

Chapter 18
Configurational Relationalism: Field Theory and GR's Thin Sandwich

18.1 Fields and Finite-Field Portmanteaux

Joint treatment of finite and field-theoretic models begins in this Chapter. This makes use of Appendix X's finite–field-theoretic portmanteau notation.[1] Q is the portmanteau configuration of finite theories' configuration $\boldsymbol{Q}$ and Field Theories' $\mathbf{Q} = \mathbf{Q}(\underline{x})$ configuration. Each of these carries the A multi-index, over one or both of particle or continuous extended object species.

We take the field theoretic kinetic metric to be *ultralocal*—i.e. having no derivative dependence—a mathematical simplicity which happens to hold over the entirety of the standardly accepted fundamental theories of Physics. The notation M_{AB} then covers both finite theories' $M_{\mathsf{AB}}(\boldsymbol{Q})$ and Field Theories' $\mathrm{M}_{\mathsf{AB}}(\mathbf{Q}(\underline{x}))$. Denote the corresponding determinant, inverse, inner product and 'norm' (indefiniteness allowed) by M, N^{AB}, $(\ ,\ )_{\mathsf{M}}$ and $\|\ \|_{\mathsf{M}}$ respectively. $\circ$ is now considered to take the *ordial derivative* form $\mathbf{d}/\mathbf{d}\lambda$: the portmanteau of ordinary and partial derivatives. Our first aim is to build relational actions, in particular in a manner rich enough to include the case of full GR. $\mathscr{L}\lfloor\dot{\mathsf{Q}}, \mathsf{Q}\rfloor$ is the portmanteau of finite theories' Lagrangian $L(t, \boldsymbol{Q}, \dot{\boldsymbol{Q}})$ and Field Theories' Lagrangian density $\mathcal{L}(\underline{x}, t, \dot{\mathbf{Q}}; \mathbf{Q}]$.[2] The latter is taken to be ultralocal in the velocities for Field Theories. One then obtains the relational action by integrating over the time portmanteau t and the notion of space portmanteau. Compliance with Manifest Reparametrization Irrelevance does make the Lagrangian portmanteau in question *look* somewhat unusual. I.e. it is not of difference-type form $\mathscr{L} = \mathsf{T} - \mathscr{V}$, but rather of *product form* $\mathscr{L} = 2\sqrt{\mathsf{T}\mathscr{W}}$. For now, we take on trust that the potential factor portmanteau $\mathscr{W} = E - \mathscr{V}$, for potential energy $\mathscr{V}$ and T is the kinetic energy.

[1] This book makes further systematic use of a number of further fonts to encode the role of the object in question (e.g. constraint or beable). If lost at any stage, consult the list of fonts in Appendix X.

[2] In fact, the first of these can be rewritten as $\mathrm{L}(t; \boldsymbol{Q}]$, which is a univariate functional due to $\mathrm{d}/\mathrm{d}t$ acting on the $\boldsymbol{Q}$ to form the velocities. However, this does not affect the types of derivatives that the theory has acting upon L, so it does not disrupt the portmanteau.

E. Anderson, *The Problem of Time*, Fundamental Theories of Physics 190,
DOI 10.1007/978-3-319-58848-3_18

To pass to a geometrical action presentation, we require rather 1) the kinetic arc element portmanteau $\mathfrak{d}\mathsf{s}\lfloor\mathbf{Q}, \mathfrak{d}\mathbf{Q}\rfloor$—of the kinetic arc element $\mathrm{d}s(\boldsymbol{Q}, \mathrm{d}\boldsymbol{Q})$ for finite theories and the kinetic arc element density $\partial \mathrm{s}(\underline{x}, \partial\mathbf{Q}; \mathbf{Q}]$ for Field Theories. 2) The physical Jacobi arc element portmanteau $\mathfrak{d}\mathscr{J}\lfloor\mathbf{Q}, \mathfrak{d}\mathbf{Q}\rfloor$—of the Jacobi arc element $\mathrm{d}J(\boldsymbol{Q}, \mathrm{d}\boldsymbol{Q})$ for finite theories and the Jacobi arc element density $\partial\mathcal{J}(\underline{x}, \partial\mathbf{Q}(\underline{x}); \mathbf{Q}(\underline{x})]$ for Field Theories.

As regards the nature of the geometries, these are now in fact infinite-dimensional generalizations of the previous Chapters' Riemannian Geometry. Moreover, the local square root does not coincide with the DeWitt-type geometry, adding extra degeneracy and functional-based issues. For convenience, let us still refer to such by the usual finite-dimensional geometries' nomenclature. I.e. we elevate names like 'Riemannian' to be finite and field-theoretic portmanteaux of the usual finite version of the notion.

In the case of Field Theories, Chap. 16.1's definitions of inconsistent, trivial and relationally trivial are recast in terms of degrees of freedom per space point. Care has to be taken now as regards nontrivial global degrees of freedom surviving.

18.2 Configurational Relationalism Including Fields

Some cases here involve augmenting $\mathfrak{a}$ to $\mathfrak{a} \times \mathfrak{i}$ for $\mathfrak{i}$ an internal space. Configurational Relationalism has hitherto in Part II rested on Mach's Space Principle. To continue to have such a supporting element, we now need to paraphrase a 'Mach-type Internal Principle' to accompany it. 'No one is competent to predicate things about gauge-dependent properties of internal space or motion thereover. These are pure things of thought, pure mental constructs that cannot be produced in experience. All our principles of Gauge Theory are, as we have shown in detail, experimental knowledge concerning gauge-independent quantities'.

Next, we consider $Aut(\mathfrak{a} \times \mathfrak{i}) = Aut(\mathfrak{a}) \times Aut(\mathfrak{i})$, or some subgroup

$$\mathfrak{g}_{\mathrm{ext}} \times \mathfrak{g}_{\mathrm{int}} \tag{18.1}$$

of this, where 'ext' standing for external transformations and 'int' for internal ones (in the same sense as in Particle Physics).

Use $\mathfrak{g}$-correcting cyclic ordial (ordinary or partial) differential portmanteau auxiliaries $\mathfrak{d}\mathbf{g}$. Encoding one's $\mathfrak{g}$ auxiliary variables in either of the above ways continues to require subsequent care with how one performs one's Calculus of Variations (Appendix L). In the portmanteau case, this entails the free end notion of space variational portmanteau.

The corresponding action is

$$\mathsf{S} = \iint_{\mathrm{NoS}} \mathfrak{d}\mathrm{NoS}\,\mathfrak{d}\mathscr{J} = \iint_{\mathrm{NoS}} \mathfrak{d}\mathrm{NoS}\sqrt{2}\mathfrak{d}\mathsf{s}\sqrt{\mathscr{W}}, \tag{18.2}$$

$$\mathfrak{d}\mathsf{s} := \|\mathfrak{d}_{\mathbf{g}}\mathbf{Q}\|_{\mathsf{M}} \quad \text{and} \quad \mathfrak{d}_{\mathbf{g}}\mathbf{Q} := \mathfrak{d}\mathbf{Q} - \sum_{\mathbf{g} \in \mathfrak{g}} \vec{\mathfrak{g}}_{\mathfrak{d}\mathbf{g}}\,\mathbf{Q}. \tag{18.3}$$

A field-theoretic update of Chap. 15's table of formulations is as follows. N.B. that

Type of variables	for which the	key portmanteau	gives the	equations of motion
Lagrangian	$\mathsf{Q}, \mathrm{d}\mathsf{Q}/\mathrm{d}t$	Lagrangian	$\mathscr{L}$	Euler–Lagrange
Machian	$\mathsf{Q}, \mathrm{d}\mathsf{Q}$	Jacobi arc element	$\mathrm{d}\mathscr{J}$	'Jacobi–Mach'
Hamiltonian	Q, P	Hamiltonian	$\mathscr{H}$	Hamilton's

the Hamiltonian formulation is unadulterated both by passing from Lagrangian to Machian variables and by bringing in portmanteau derivatives. Upon including the $\mathfrak{g}$ auxiliaries, however, there is a slight alteration to the Hamiltonian formulation. This is from the usual total Hamiltonian (Appendix J.15) to the *total* d*A-Hamiltonian* (Appendix L.6; throughout this book, A- stands for 'almost' in this same sense). Moreover, this is just reformulating the *unphysical* sector of the theory. A second table now also incorporating this expansion is as follows. The conjugate momenta

Type of variables	for which the	key portmanteau	gives the	equations of motion
Lagrangian	$\mathsf{Q}, \mathsf{m}^{\mathsf{F}}, \mathrm{d}\mathsf{Q}/\mathrm{d}t$	Lagrangian	$\mathscr{L}$	Euler–Lagrange
Machian	$\mathsf{Q}, \mathrm{d}\mathsf{Q}, \mathrm{d}\mathsf{c}^{\mathsf{F}}$	Jacobi arc element	$\mathrm{d}\mathscr{J}$	'Jacobi–Mach'
Total Hamiltonian	$\mathsf{Q}, \mathsf{m}^{\mathsf{F}}, \mathsf{P}$	Total Hamiltonian	$\mathscr{H}_{\text{Total}} = \mathscr{H} + \mathsf{m}^{\mathsf{F}}\mathcal{C}_{\mathsf{F}}$	Hamilton's
dA-total Hamiltonian	$\mathsf{Q}, \mathsf{P}, \mathrm{d}\mathsf{c}^{\mathsf{F}}$	dA-total Hamiltonian	$\mathrm{d}\mathscr{H}_{\text{Total}} = \mathrm{d}\mathscr{H} + \mathrm{d}\mathsf{c}^{\mathsf{F}}\mathcal{C}_{\mathsf{F}}$	dA-Hamilton's

are then (using the *partional derivative* portmanteau of partial and functional derivatives)

$$\mathsf{P}_{\mathsf{A}} := \partial\frac{\mathrm{d}\mathscr{J}}{\partial\,\mathrm{d}\mathsf{Q}^{\mathsf{A}}} = \mathsf{M}_{\mathsf{AB}}\sqrt{\frac{\mathscr{W}}{\mathsf{T}}}\,\mathrm{d}_{\mathsf{g}}\mathsf{Q}^{\mathsf{A}}. \tag{18.4}$$

These obey one primary constraint per relevant notion of space point, interpreted as an equation of time,

$$\mathcal{C}\text{hronos} := \mathsf{N}^{\mathsf{AB}}\mathsf{P}_{\mathsf{A}}\mathsf{P}_{\mathsf{B}}/2 - \mathscr{W}(\boldsymbol{Q}) = 0. \tag{18.5}$$

Thus it is purely quadratic in the momenta. The P also obey some secondary constraints per relevant notion of space point from variation with respect to g.

$$0 = \partial\frac{\mathrm{d}\mathscr{J}}{\partial\,\mathrm{d}\mathsf{c}^{\mathsf{G}}} := \mathcal{S}\text{huffle}_{\mathsf{G}} = \frac{\delta\{\overrightarrow{\mathfrak{g}}_{\mathrm{d}\mathsf{c}}\,\mathsf{Q}^{\mathsf{A}}\}}{\delta\,\mathrm{d}\mathsf{c}^{\mathsf{G}}}\mathsf{P}_{\mathsf{A}}; \tag{18.6}$$

these are linear in the momenta, and so are also denoted by $\mathcal{L}\mathbf{in}$.

Next, denote the joint set of these constraints by $\mathcal{C}_{\mathsf{F}}$, under the presumption that they are confirmed as first-class in Chap. 24. The indexing set designation assumes there is only one quadratic constraint, so all our examples' F ranges over G and the one quadratic value.

The corresponding Jacobi–Mach equations of motion are

$$\mathrm{d}\frac{\partial\, \mathrm{d}\mathcal{J}}{\partial\, \mathrm{d}\mathsf{Q}^{\mathsf{A}}} = \frac{\partial\, \mathrm{d}\mathcal{J}}{\partial \mathsf{Q}^{\mathsf{A}}} \quad \Rightarrow \tag{18.7}$$

$$\frac{\sqrt{2\mathcal{W}}}{\|\mathrm{d}\mathbf{Q}\|_{\mathsf{M}}}\mathrm{d}\left\{\frac{\sqrt{2\mathcal{W}}}{\|\mathrm{d}\mathbf{Q}\|_{\mathsf{M}}}\mathrm{d}\mathsf{Q}^{\mathsf{A}}\right\} + \Gamma^{\mathsf{A}}{}_{\mathsf{BC}}\frac{\sqrt{2\mathcal{W}}}{\|\mathrm{d}\mathcal{Q}\|_{\mathsf{M}}}\mathrm{d}\mathsf{Q}^{\mathsf{B}}\frac{\sqrt{2\mathcal{W}}}{\|\mathrm{d}\mathcal{Q}\|_{\mathsf{M}}}\mathrm{d}\mathsf{Q}^{\mathsf{C}} = \mathsf{N}^{\mathsf{AB}}\frac{\partial \mathsf{W}}{\partial \mathsf{Q}^{\mathsf{B}}}. \tag{18.8}$$

The previous Chapter's Best Matching procedure admits the following generalization.

Best Matching 0) Start with the 'arbitrary $\mathfrak{g}$ frame corrected' action (18.2).

Best Matching 1) Extremize over $\mathfrak{g}$. This produces a constraint equation **$\mathcal{S}$huffle** that is of the form **$\mathcal{L}$in**: linear in the momenta.

Best Matching 2) The Machian variables form of this equation, with Machian data $\mathbf{Q}$, $\mathrm{d}\mathbf{Q}$ is to be solved for the $\mathrm{d}\mathbf{g}$ themselves.

Best Matching 3) Substitute this solution back into the action: an example of d-Routhian reduction (see Appendix L). Again this produces a final $\mathfrak{g}$-independent expression that could have been directly arrived at as a *direct* implementation of CR-ii).

Best Matching 4) Finally elevate this new action to be one's new starting point.

Best Matching 3′) As a distinct application of Best Matching 2), emergent Machian times are now of the general form

$$\mathsf{t}^{\mathrm{em(Mach)}} = \mathscr{F}\lfloor \mathbf{Q}, \mathrm{d}\mathbf{Q}\rfloor, \tag{18.9}$$

a particular realization of which is t^{em} of e.g. the Jacobi–(Barbour–Bertotti) type,

$$CR(\mathsf{t}^{\mathrm{em}}) = \mathrm{E}'_{\mathbf{g}\in\mathfrak{g}}\int \|\mathrm{d}_{\mathbf{g}}\mathbf{Q}\|_{\mathsf{M}}/\sqrt{2\mathcal{W}}. \tag{18.10}$$

If one succeeds in carrying out Best Matching, moreover, both the two-functional and implicit-formulation complications are washed away. This leaves $CR(\mathsf{t}^{\mathrm{em}})$ expressed in terms of the *reduced* $\mathfrak{q}$'s geometry.

N.B. that the above expression does *not* contain a spatial integral: the field-theoretic t^{em} is *local*. Moreover, the essential line of thought of this Chapter is the *only* known approach to Configurational Relationalism that is general enough to cover the Einstein–Standard Model presentation of Physics.

The momenta in terms of the corresponding derivative $*$ are

$$\mathsf{P}_{\mathsf{A}} = \mathsf{M}_{\mathsf{AB}} * \mathsf{Q}^{\mathsf{B}}, \tag{18.11}$$

while the equations of motion now take the 'parageodesic' form

$$** \mathsf{Q}^{\mathsf{A}} + \Gamma^{\mathsf{A}}{}_{\mathsf{BC}} * \mathsf{Q}^{\mathsf{B}} * \mathsf{Q}^{\mathsf{C}} = \mathsf{N}^{\mathsf{AB}}\partial\mathcal{W}/\partial\mathsf{Q}^{\mathsf{B}}. \tag{18.12}$$

It can also be cast as a true 'geodesic' equation with respect to the physical metric whose line element is $\mathrm{d}\mathcal{J}$. Finally one of the evolution equations per relevant notion of space point can be supplanted by the emergent Lagrangian form of the quadratic 'energy-type' constraint (15.26).

18.3 Example 1) Electromagnetism Alone

Consider the space of 1-forms on $\mathbb{R}^3$. $\mathfrak{g} = Diff(\mathbb{R}^3)$ is not applied to flat-space Electromagnetism because δ_{ij} breaks this in the active sense. However, $\mathfrak{g} = Rot(3)$ can be considered. Internal Relationalism involving $U(1)$ clearly also applies.

The latter works out fine for this (including using a $\dot{\Psi}$ or $\partial\Psi$ auxiliary in place of the electric potential Φ [20]). This gives in each case the expected Gauss constraint $\mathcal{G}$. However Spatial and Temporal Relationalism is prohibitively restrictive in the case of Electromagnetism. E.g. involving $Diff(\boldsymbol{\Sigma})$ gives that the Poynting vector $\underline{\mathrm{E}} \times \underline{\mathrm{B}}$ must vanish [19]. I.e. $\underline{\mathrm{E}} = 0$, $\underline{\mathrm{B}} = 0$, or $\underline{\mathrm{E}}$ parallel to $\underline{\mathrm{B}}$ (which kills signal propagation). In any case, Electromagnetism by itself has background structures (typically the Minkowski metric $\boldsymbol{\eta}$, or the Euclidean metric $\boldsymbol{\delta}$ on flat spatial slices).

The resolution of these issues is that inclusion of GR to make the Einstein–Maxwell system frees one from these background structures and the above zero Poynting vector restriction (see Sect. 18.11). Yang–Mills Theory and the various associated Gauge Theories follow suite in these regards. More generally still, Field Theories of matter are found to not be properly supported in the absence of GR as regards attaining Background Independence.

18.4 Example 2) GR

For this particularly substantial example, $\mathfrak{q} = \mathfrak{Riem}(\boldsymbol{\Sigma})$—the space of Riemannian 3-metrics on some fixed spatial topological manifold $\boldsymbol{\Sigma}$ that is taken to be compact without boundary both for simplicity and for Machian reasons. Moreover, equipping $\boldsymbol{\Sigma}$ with $\mathbf{h}$ has a more involved form than RPMs' multiple copies of absolute space; indeed GR gives further reason to adopt Sect. 14.2's procedure b).

The group of physically irrelevant motions $\mathfrak{g}$ is usually taken to be $Diff(\boldsymbol{\Sigma})$: the diffeomorphisms on $\boldsymbol{\Sigma}$; see Sects. 21.4 and 33.7 for further alternatives.

In this case, $\mathfrak{q}/\mathfrak{g}$ is Wheeler's [237, 899] $\mathfrak{Superspace}(\boldsymbol{\Sigma}) = \mathfrak{Riem}(\boldsymbol{\Sigma})/Diff(\boldsymbol{\Sigma})$. The first action on this $\mathfrak{g}$, $\mathfrak{q}$ pair that we consider is the BSW one (9.11). This is formulated in terms of the shift β^i auxiliary which maintains contact with the earlier literature, but also fails to be Temporally Relational. We deal with this by next passing to the TRi version in terms of the *cyclic partial differential of the frame* auxiliary, $\partial\mathrm{F}^i$.

A further example of structural compatibility between $\mathfrak{q}$ and $\mathfrak{g}$ that manifests itself in Geometrodynamics is *Diff* being based upon the same underlying topological manifold $\boldsymbol{\Sigma}$ that $\mathfrak{Riem}$ is.

One could also consider less minimal $\mathfrak{q}$ than GR's $\mathfrak{Riem}(\boldsymbol{\Sigma})$, as occur in e.g. Scalar–Tensor Theories, or in Scalar–Vector–Tensor Theories [468, 516]. Finally, a diversity of actions can be constructed on a given $\mathfrak{q}$, $\mathfrak{g}$ pair. For instance, one could consider an arbitrary rather than GR-specific supermetric (as per Chap. 33), or more than just a second-order action principle. The latter could be precluded with simplicity postulates such as that the action is not to contain higher than first (or occasionally second) derivatives.

18.5 Baierlein–Sharp–Wheeler Action and the Thin Sandwich

We next consider the Baierlein–Sharp–Wheeler (BSW) action $\mathsf{S}_{\mathrm{BSW}}$ (9.11)—named after Wheeler and physicists Ralph Baierlein and David Sharp [89]. Because λ and t coincide for GR due to its status as an already-parametrized theory, the distinction between the BSW kinetic term and the ADM one is entirely conceptual rather than mathematical. The equivalence of the ADM and BSW actions for GR is then established by the multiplier elimination move done for Minisuperspace in Chap. 16 immediately carrying over to GR in general [89, 177]. Also GR's configuration space metric is indeed built out of the dynamical variables: $\mathrm{M}^{ijkl} = \sqrt{\mathrm{h}}\{\mathrm{h}^{ik}\mathrm{h}^{jl} - \mathrm{h}^{ij}\mathrm{h}^{kl}\}$, in compliance with Postulate CR-i).

The Thin Sandwich formulation [89, 897] consists of the following.

Thin Sandwich 0) Consider the BSW action.

Thin Sandwich 1) Vary this to obtain the constraint equation $\mathcal{M}_i$.

Thin Sandwich 2) Consider the 'Thin Sandwich equation', i.e. the Lagrangian-variables form of $\mathcal{M}_i$:

$$\mathcal{D}_j\left\{\sqrt{\frac{\mathcal{R} - 2\Lambda}{\{\mathrm{h}^{ac}\mathrm{h}^{bd} - \mathrm{h}^{ab}\mathrm{h}^{cd}\}\{\partial\mathrm{h}_{ab} - 2\mathcal{D}_{(a}\beta_{b)}\}\{\partial\mathrm{h}_{cd} - 2\mathcal{D}_{(c}\beta_{d)}\}}}\right.$$
$$\left.\times\left\{\mathrm{h}^{jk}\delta^l_i - \delta^j_i\mathrm{h}^{kl}\right\}\{\partial\mathrm{h}_{kl} - 2\mathcal{D}_{(k}\beta_{l)}\}\right\} = 0, \tag{18.13}$$

alongside 'thin sandwich data'

$$(\mathrm{h}_{ij}, \dot{\mathrm{h}}_{ij}), \tag{18.14}$$

as a PDE problem to be solved for the shift β^i. This equation and data jointly constitute the Thin Sandwich Problem, in the sense of 'PDE problem' explained in Appendix O.

Thin Sandwich 3.a) Construct $\pounds_{\underline{\beta}}\mathrm{h}_{ij}$: GR's $\vec{Diff}_{\underline{\beta}}\mathrm{h}_{ij}$. Then $\delta_{\vec{\beta}}\mathrm{h}_{ij} = \dot{\mathrm{h}}_{ij} - \pounds_{\underline{\beta}}\mathrm{h}_{ij}$.

Thin Sandwich 4.a) Next construct an emergent counterpart to α, $\mathrm{N} := \sqrt{\mathrm{T}_{\mathrm{BSW}}/4\{\mathcal{R} - 2\Lambda\}}$.

Thin Sandwich 5) Thin Sandwich 3.a) and 4.a) permit one to construct the extrinsic curvature $\mathcal{K}_{ij} = \mathcal{K}_{ij}(\underline{x}; \mathbf{h}, \underline{\beta}, \mathrm{N}]$ using the computational formula

$$\mathcal{K}_{ij} = \frac{\delta_{\vec{\beta}}\mathrm{h}_{ij}}{2\mathrm{N}}, \tag{18.15}$$

which is the last form in (8.14) except that BSW's emergent N has taken the place of ADM's presupposed α. This Thin Sandwich output may be considered within the Broad worldview, to a greater extent than Wheeler's interpretation [660] of the ADM split as a strutting of spacetime.

18.6 The Thin Sandwich Problem

Unfortunately the Thin Sandwich Problem, (18.13), (18.14) is hard to handle as a PDE problem. See Appendix O.5 for an outline of existence and uniqueness results for this, the most up to date of which are due to mathematical physicists Robert Bartnik and Gyula Fodor [308, 663]. Generic GR solution of this equation is, moreover, out of the question. Since the Thin Sandwich equation has a square root trapped inside the $\mathcal{D}_i$, a fairly complicated PDE ensues. Contrast how in RPM (Sect. 16.7)—and even in the SIC case (Sect. 30.4) to leading order—Best Matching gives a merely algebraic equation which is much easier to handle (at least locally).

Let us end by noting that $\mathfrak{g}$ being (along the lines of) the diffeomorphism group has the problem of blocking many an explicit construct from being more than formal.

18.7 Reparametrization-Invariant Relational Action for GR

The BSW action does succeed in being formulated free from a extraneous background time-like notion such as the GR lapse. However, this does not comply with Temporal Relationalism since the presence of the shift β^i breaks Manifest Reparametrization Invariance. None the less, Chap. 16 has laid out how to get round this deficiency.

To link between the two formalisms, $\dot{\mathrm{F}}^i$ is the *velocity of the frame*. This is numerically equal to the shift β^i. Moreover,

$$\mathsf{S}_{\text{relational}} = \int \mathrm{d}\lambda \int_\Sigma \sqrt{\overline{\mathrm{T}}_{\text{relational}}}\sqrt{\mathcal{R} - 2\,\Lambda}, \quad \overline{\mathrm{T}}_{\text{relational}} := \|\dot{\mathbf{h}} - \pounds_{\dot{\underline{\mathrm{F}}}}\mathbf{h}\|^2_{\mathbf{M}}. \tag{18.16}$$

This implements both Temporal and Configurational Relationalisms.

18.8 Geometrical Action for GR

The final action for GR as Geometrodynamics in relational form is

$$\mathsf{S}_{\text{relational}} = \iint_\Sigma \mathrm{d}^3x\, \partial\mathcal{J} = \iint_\Sigma \mathrm{d}^3x \sqrt{\mathcal{R} - 2\,\Lambda}\,\partial\mathsf{s}_{\text{relational}} \tag{18.17}$$

$$\text{for } \partial s_{\text{relational}} := \|\partial_{\underline{F}}\mathbf{h}\|_{\mathbf{M}} \quad \text{and} \quad \partial_{\underline{F}}h_{ij} := \partial h_{ij} - \pounds_{\partial \underline{F}} h_{ij}. \tag{18.18}$$

The conjugate momenta are

$$p^{ij} := \frac{\delta \, \partial \mathcal{J}}{\delta \, \partial h_{ij}} = 2\sqrt{\mathcal{R} - 2\Lambda}\, M^{ijkl} \frac{\partial_{\underline{F}} h_{kl}}{\partial s_{\text{relational}}}. \tag{18.19}$$

The GR Hamiltonian constraint $\mathcal{H}$ now follows as a primary constraint that is purely quadratic in the momenta. The GR momentum constraint $\mathcal{M}_i$ arises as a secondary constraint from variation with respect to the auxiliary *Diff*($\mathbf{\Sigma}$)-variables F^i; it is linear in the momenta. The Jacobi–Mach equations of motion are (for $\Lambda = 0$ for simplicity)

$$2\sqrt{\mathcal{R}}\frac{\partial_{\underline{F}} p^{ij}}{d s_{\text{relational}}} = \Big\{\sqrt{h}\{\mathcal{R}\, h^{ij} - \mathcal{R}^{ij} + \mathcal{D}^j \mathcal{D}^i - h^{ij}\triangle\} - \frac{2}{\sqrt{h}}\{p^{ic}p_c{}^j - p\,p^{ij}/2\}\Big\}\frac{\partial s}{2\sqrt{\mathcal{R}}}. \tag{18.20}$$

Via the Bianchi identity (D.17), these immediately propagate the above constraints without producing further conditions.

18.9 TRi Form of the Thin Sandwich

We next reiterate the Thin Sandwich procedure in the TRi formulation's manifestly temporally Machian form.

Machian Thin Sandwich 0) Consider the relational GR action (15.7) [62, 109].

Machian Thin Sandwich 1) Vary it with respect to F^i to obtain the constraint equation $\mathcal{M}_i$ [64].

Machian Thin Sandwich 2) Consider the 'Machian Thin Sandwich equation'. I.e. the Machian-variables form of $\mathcal{M}_i$

$$\mathcal{D}_j\left\{\sqrt{\frac{\mathcal{R} - 2\Lambda}{\{h^{ac}h^{bd} - h^{ab}h^{cd}\}\{\partial h_{ab} - 2\mathcal{D}_{(a}\partial F_{b)}\}\{\partial h_{cd} - 2\mathcal{D}_{(c}\partial F_{d)}\}}} \times \{h^{jk}\delta^l_i - \delta^j_i h^{kl}\}\{\partial h_{kl} - 2\mathcal{D}_{(k}\partial F_{l)}\}\right\} = 0, \tag{18.21}$$

with 'Machian thin sandwich data'

$$(h_{ij}, \partial h_{ij}), \tag{18.22}$$

for the partial differential of the frame auxiliary ∂F^i. Moreover, altering (18.13), (18.14) to (18.21), (18.22) makes no difference to the *mathematical form* of this PDE problem.

Machian Thin Sandwich 3.a) Construct $\pounds_{\partial \underline{F}}\mathrm{h}_{ij}$, and then the Best Matching corrected derivative

$$\partial_{\underline{F}}\mathrm{h}_{ij} = \partial\mathrm{h}_{ij} - \pounds_{\partial \underline{F}}\mathrm{h}_{ij}. \tag{18.23}$$

This is a distinct conceptualization of the same mathematical object as the hypersurface derivative.

Machian Thin Sandwich 4.a) Construct the *emergent differential of the instant*

$$\partial\mathrm{I} = \|\partial_{\underline{F}}\mathbf{h}\|_{\mathrm{M}} \Big/ 2\sqrt{\mathcal{R} - 2\Lambda}. \tag{18.24}$$

Machian Thin Sandwich 4′) Emergent Jacobi–Barbour–Bertotti time readily follows simply from integrating up 4.a). Moreover, 6.b) goes beyond BSW's own construction. It is GR's analogue of emergent Jacobi time as highlighted by Barbour [98]. Furthermore, it is an 'all change', or, in practice 'STLRC' implementation of Mach's Time Principle:

$$CR(\mathrm{t}^{\mathrm{em}})(\underline{x}) = \mathrm{E}'_{\underline{F} \in Diff(\Sigma)} \int \|\partial_{\underline{F}}\mathbf{h}\|_{\mathrm{M}} \Big/ \sqrt{\mathcal{R} - 2\Lambda}. \tag{18.25}$$

Another consequence of move 2) is that one can substitute the resultant extremizing F^i back into the relational GR action.

Machian Thin Sandwich 4) Take this as an ab initio new action.

Note that moves 1) to 4) constitute a reduction; with these, the Machian Thin Sandwich can be interpreted as a subcase of Best Matching.

Machian Thin Sandwich 5) is that moves 3.a) and 4.a) also permit construction of the extrinsic curvature through the computational formula

$$\mathcal{K}_{ij} = \frac{\partial_{\underline{F}}\mathrm{h}_{ij}}{2\,\partial\mathrm{I}} \tag{18.26}$$

for $\mathcal{K}_{ij} = \mathcal{K}_{ij}(\underline{x}; \mathbf{h}, \partial\underline{\mathrm{F}}, \partial\mathrm{I}]$. This subsequently enters Spacetime Construction, as further laid out in Chap. 33. See Fig. 18.1 for a summary so far, laying out the TRi modifications to the previous Thin Sandwich work following from the BSW action. Also, Chap. 31 proceeds to consider yet further completion of the thin-sandwich prescription in terms of constructing the whole of the universal hypersurface kinematics [577–579].

18.10 Comments on GR's Emergent Machian Time

Mathematical physicist Demetrios Christodoulou [208, 209] arrived at this two decades prior to Barbour, as an extension of the Thin Sandwich which he termed 'Chronos Principle'. Like Wheeler, he did not base the Thin Sandwich on relational first principles. None the less, he was certainly aware of the features that

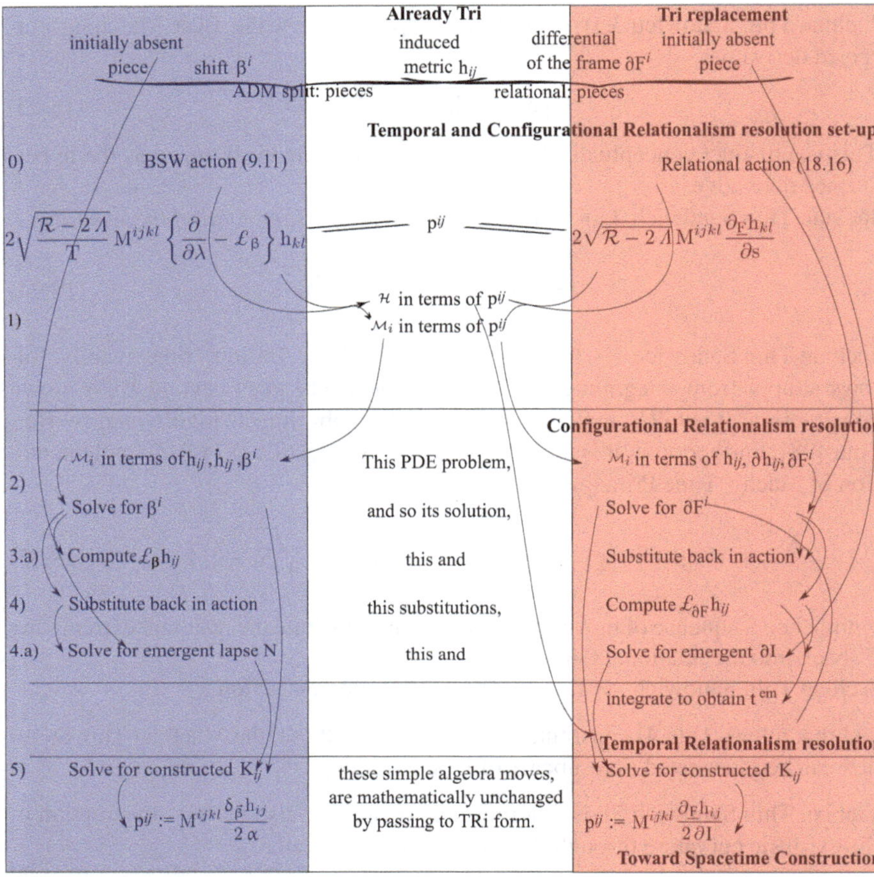

Fig. 18.1 Standard Thin Sandwich versus TRi Thin Sandwich. The current Figure can be regarded as further detail of the second floor of Fig. L.2

Barbour and the Author have argued to be a Leibniz–Mach position on relational time [39, 104]. "*They contain the statement that time is not a separate physical entity in which the changing of the physical system takes place. It is the measure of the changing of the physical system itself that is time.*" However, this work was overlooked by his contemporaries.

Let us comment further on the form of expression (18.25). It says that one has to extremize *one* functional in order to use the extremal value from that in a *second* functional. This is more complicated than the usual variational problem that involves extremizing a single functional. What happens if the time functional itself is extremized? For finite models, this is actually equivalent to the given procedure. However, for Field Theory the two become inequivalent due to factors of $\sqrt{\mathcal{W}}$ becoming entangled with the derivative operators that arise 'by parts' in the spatial integration. The necessary property that the timestandard be independent of whether

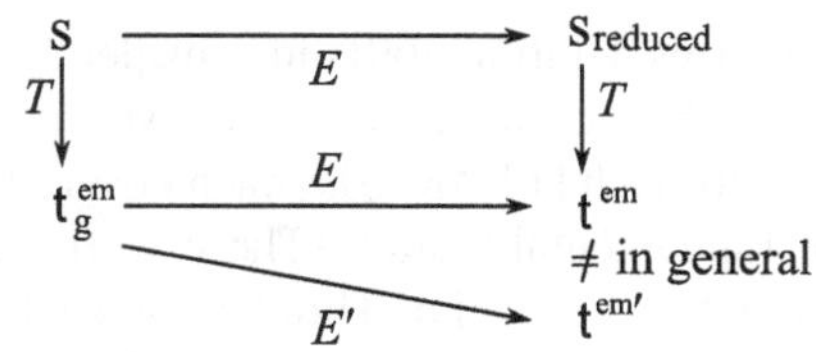

Fig. 18.2 Let us denote the map from an action to the corresponding emergent time candidate by T, We denote the map consisting of substituting in the $\mathfrak{g}$-extremum of the action by E, and the map consisting of substituting in the $\mathfrak{g}$-extremum of the timefunction itself by E' E and T naturally commute: $TE = ET$, but in general $TE' \neq ET$. This is why we use the $\mathfrak{g}$-extremum of $\mathfrak{S}$ in order to free t of $\mathfrak{g}$-dependence

the action has already been reduced then forces the extremization to take the above form (Fig. 18.2).

N.B. the local character of GR's emergent time notion. This arises from the field-theoretic use of local square roots (i.e. take the square root prior to integrating).

The final form—in the sense of Sect. 15.9—can in this case be identified as the GR proper time, now obtained as an emergent concept. In a suitable cosmological setting, this is aligned with cosmic time. We are using a distinct name and symbol I for 'instant' since this emergent entity is, most primarily, a *labeller of instants*. This ends up, very satisfactorily, being dual to the GR proper time. GR proper time is indeed a quantity which in general differs from point to point, and it is this desirable feature which arises from the field-theoretic local square root ordering (Sect. 17.2). It is in this manner that local square roots manage to be desirable in GR despite their finite theory counterparts being questionably Machian and not physically realized.

Via

$$\partial \mathrm{I} := \frac{\partial \mathrm{s}}{2\sqrt{\mathcal{R}}} = \partial\big(CR(\mathrm{t}^{\mathrm{em}})\big),$$

the $\Lambda = 0$ t^{em}–instant dual simplifies the momenta and relational equations of motion into the forms

$$\mathrm{p}^{ij} = \mathrm{M}^{ijkl}\frac{\partial_{\mathrm{F}}\mathrm{h}_{kl}}{2\,\partial \mathrm{I}}, \tag{18.27}$$

$$\partial_{\mathrm{F}}\mathrm{p}^{ij} = \left\{\sqrt{\mathrm{h}}\{\mathcal{R}\mathrm{h}^{ij} - \mathcal{R}^{ij} + \mathcal{D}^j\mathcal{D}^i - \mathrm{h}^{ij}\triangle\} - \frac{2}{\sqrt{h}}\{\mathrm{p}^{ic}\mathrm{p}_c{}^j - \mathrm{p}\,\mathrm{p}^{ij}/2\}\right\}\partial \mathrm{I}. \tag{18.28}$$

18.11 Example 3) GR with Fundamental Matter Fields

In general, the Thin Sandwich equation (18.13) is proportional to momentum flux P_i. The Thin Sandwich Problem has mostly only been considered for phenomenological matter [308, 663, 897]. Best Matching, however, concerns fundamental matter, a distinction of note since corrections to velocities do not occur in phenomenological matter terms. Giulini [360] did consider the Thin Sandwich Problem for Einstein–Maxwell Theory (including protective theorems). Christodoulou [209] and the Relational Approach [14, 58, 109] each covered this with GR coupled to a full complement of fundamental matter. [These works are at the level of the form of the equations but not at the level of Thin Sandwich Theorems.]

The specific examples below lie within the scope of Appendix H.7's configuration spaces. Π_Z are then the momenta conjugate to the matter variables ψ^Z.

Example 1) The Einstein–scalar case is useful for Cosmology, including this book's main Minisuperspace model and perturbations thereabout. Here

$$\mathsf{S}_{\text{relational}} = \iint_{\Sigma} \mathrm{d}^3x \, \partial \mathrm{s} \sqrt{\mathcal{R} - 2\Lambda - |\partial\phi|^2/2 - \mathrm{V}(\phi)} \, \partial \mathrm{s} = \left\| \partial_{\underline{\mathrm{F}}}(\mathrm{h}, \phi) \right\|_{\mathcal{M}(\mathbf{h})}, \tag{18.29}$$

for $\mathcal{M}$ as given in Appendix H.7. Moreover,

$$\mathcal{H} := \mathrm{N}_{ijkl}\mathrm{p}^{ij}\mathrm{p}^{kl} + \pi_\phi^2/2 - \overline{\mathcal{R} - 2\Lambda - |\partial\phi|^2/2 - \mathrm{V}(\phi)} = 0, \tag{18.30}$$

$$\mathcal{M}_i := -2\mathcal{D}_j\mathrm{p}^j{}_i = -\pi_\phi \phi_{,i}, \tag{18.31}$$

$$CR(\mathrm{t}^{\mathrm{em}})(\underline{x}) = \mathrm{E}'_{\underline{\mathrm{F}} \in Diff(\Sigma)} \int \left\| \partial_{\underline{\mathrm{F}}}(\mathrm{h}, \phi) \right\|_{\mathcal{M}(\mathbf{h})} \Big/ \overline{\mathcal{R} - 2\Lambda - |\partial\phi|^2/2 - \mathrm{V}(\phi)}, \tag{18.32}$$

$$\begin{aligned}
&\mathcal{D}_j \left\{ \sqrt{\frac{\mathcal{R} - 2\Lambda - |\partial\phi|^2/2 - \mathrm{V}_\phi}{\{\mathrm{h}^{ac}\mathrm{h}^{bd} - \mathrm{h}^{ab}\mathrm{h}^{cd}\}\{\partial \mathrm{h}_{ab} - 2\mathcal{D}_{(a}\partial \mathrm{F}_{b)}\}\{\partial \mathrm{h}_{cd} - 2\mathcal{D}_{(c}\partial \mathrm{F}_{d)}\} + |\partial_{\underline{\mathrm{F}}}\phi|^2/2}} \right. \\
&\qquad \left. \times \{\mathrm{h}^{jk}\delta^l_i - \delta^j_i \mathrm{h}^{kl}\}\{\partial \mathrm{h}_{kl} - 2\mathcal{D}_{(k}\partial \mathrm{F}_{l)}\} \right\} \\
&= -\sqrt{\frac{\mathcal{R} - 2\Lambda - |\partial\phi|^2/2 - \mathrm{V}_\phi}{\{\mathrm{h}^{ac}\mathrm{h}^{bd} - \mathrm{h}^{ab}\mathrm{h}^{cd}\}\{\partial \mathrm{h}_{ab} - 2\mathcal{D}_{(a}\partial \mathrm{F}_{b)}\}\{\partial \mathrm{h}_{cd} - 2\mathcal{D}_{(c}\partial \mathrm{F}_{d)}\} + |\partial_{\underline{\mathrm{F}}}\phi|^2/2}} \\
&\qquad \times \partial_{\underline{\mathrm{F}}}\phi \, \phi_{,i}.
\end{aligned} \tag{18.33}$$

Example 2) Einstein–Maxwell Theory has

$$\mathsf{S}_{\text{relational}} = \iint_{\Sigma} \mathrm{d}^3x \, \partial \mathrm{s} \sqrt{\mathcal{R} - 2\Lambda - \mathrm{B}^2/2}, \quad \partial \mathrm{s} = \|\partial_{\underline{\mathrm{F}}}\mathrm{h}, \partial_{\underline{\mathrm{F}},\Psi}\mathrm{A})\|_{\mathcal{M}(\mathbf{h})} \tag{18.34}$$

$$\mathcal{H} := \mathrm{N}_{ijkl}\mathrm{p}^{ij}\mathrm{p}^{kl} + \pi_i\pi^i/2 - \overline{\mathcal{R} - 2\Lambda - \mathrm{B}^2/2} = 0, \tag{18.35}$$

$$\mathcal{G} := \partial_i \pi^i = 0, \tag{18.36}$$

$$\mathcal{M}_i := -2\mathcal{D}_j\mathrm{p}^j{}_i = -\{\pi \times \mathrm{B}\}_i - \mathrm{A}_i\mathcal{G}, \tag{18.37}$$

$$CR(\mathrm{t}^{\mathrm{em}})(\underline{x}) = \mathrm{E}'_{\underline{\mathrm{F}},\Psi \in Diff(\Sigma)\times U(1)(\Sigma)} \int \sqrt{\|\partial_{\underline{\mathrm{F}}}\mathbf{h}\|^2_{\mathbf{M}} + |\partial_{\underline{\mathrm{F}},\Psi}\mathrm{A}|^2/2} \Big/ \sqrt{\mathcal{R} - 2\Lambda - \mathrm{B}^2/2}. \tag{18.38}$$

See [360] for consideration of its Thin Sandwich Problem, which now involves a system of 4 equations for 4 unknowns.

Research Project 1) Since the Thin Sandwich Problem plays a substantial part in the Relational Approach to the Problem of Time, investigate whether its known protective theorems extend to a wider range of theories. E.g. follow up [360, 663] by considering the Scalar–Tensor Theory, Einstein–Dirac and Einstein–Standard Model counterparts.

As some background on the Einstein–Dirac case, this has an action of the schematic form

$$\mathsf{S} = \iint_{\Sigma} \mathrm{d}^3x \big\{ \sqrt{2}\sqrt{\mathcal{W}} \partial \mathsf{s}_{\mathrm{quad}} + \partial \mathsf{s}_{\mathrm{lin}} \big\}. \tag{18.39}$$

I.e. a locally ordered Field Theoretic version of the 'Randers type' action (17.6) [14, 39]. Moreover, the species whose changes enter the quadratic and linear arc elements are disjoint: only bosonic changes enter the former, and only fermionic ones confer the latter.

Local flat space(time) frame formulations—as necessitated by the inclusion of fermionic variables—possess locally-Lorentz constraints $\mathcal{J}_{AB}$ (and conjugate, with the capital indices here being specifically 2-spinor indices, though we schematically denote this just by $\mathcal{J}$ from now on). See e.g. [232] for its explicit form. $\mathcal{J}$ can furthermore be considered to arise from local Lorentz frame Best Matching in the flat-space frames attached to each point in this formulation, alongside the usual $\mathit{Diff}(\mathbf{\Sigma})$ Best Matching that produces $\mathcal{M}_i$. In this way, (18.39) complies with Configurational Relationalism.

Being the $\mathfrak{q}$-geometry dual of a Manifestly Parametrization Irrelevant action, (18.39) furthermore complies with the incipient Leibnizian conception of Temporal Relationalism. On the other hand, spin-1/2 fermionic change $\partial \mathsf{s}^{\mathrm{trial}}_{\mathrm{lin}}$ does not subsequently enter into the Einstein–Dirac $\mathcal{H}$ or consequently into the Einstein–Dirac $\mathsf{t}^{\mathrm{em}}_{\mathfrak{g}\text{-free}}$. This has already been argued in Sect. 17.2 to be at odds with Machian resolutions of Temporal Relationalism. The above relational characterizations additionally carry over to further theories in which the spin-1/2 fermions are coupled to one or both of spin-1 gauge fields and scalar fields. Thus so far only 'time is to be abstracted from *bosonic* change' had been demonstrated [39]. One way out of this is to insist on viewing fermions as inherently quantum entities. This is a widespread view as regards placing limited weight on what preliminary classical actions can be written down for fermions. The above argument provides one more reason not to consider classical formulations of fermions within whole-universe models. We postpone addressing whether *quantum* fermionic change contributes to the *quantum* GLET to Sect. 47.5.

18.12 Example 4) Strong Gravity

For now, by Strong Gravity we mean the strong-coupled limit of GR [472]. Via the conjecture of physicists Vladimir Belinskii, Isaak Khalatnikov and Evgeny Lifshitz

[125], this is widely believed to be applicable to the primordial-cosmology universe near a singularity. Strong Gravity exists in both geometrodynamical and metrodynamical forms, with two and five degrees of freedom per space point respectively. The geometrodynamical case follows from

$$\mathsf{S} = \sqrt{2}\iint_{\Sigma} \mathrm{d}^3x\sqrt{-2\overline{\Lambda}}\,\partial \mathrm{s} \tag{18.40}$$

for ∂s the usual GR kinetic arc element (18.18), whereas the metrodynamical case has the 'bare' kinetic arc element

$$\partial \mathrm{s} := \|\partial \mathbf{h}\|_{\mathbf{M}}. \tag{18.41}$$

This accounts for the latter's three extra degrees of freedom, since the absence of F^i or β^i means that no momentum constraint $\mathcal{M}_i$ appears.

$$\mathcal{H}_{\text{strong}} := \mathrm{N}_{ijkl}\mathrm{p}^{ij}\mathrm{p}^{kl} + 2\overline{\Lambda} = 0. \tag{18.42}$$

The above notion of strong-coupled limit also readily extends to a wide range of further Gravitational Theories. E.g. whereas (17.4) cannot be a model of higher-curvature gravity due to that necessitating further pieces in order to be consistent, there is a corresponding Strong Gravity type action of this form.

Chapter 19
Relationalism in Various Further Settings

19.1 Multiple Distinct Uses of the Word 'Relational'

The word 'relational' has in fact been used in various ways by different authors. Barbour-type Relationalism [17, 37, 38, 40, 65, 98, 103, 105, 109, 398] (Chaps. 14–18) is the default in this book, by which it is often shortened to 'Relationalism'. Other meanings of 'relational' are due to e.g. Rovelli [743–745, 747, 749–752, 755] and mathematician and physicist Louis Crane [224, 225].[1] In this book, Rovelli and Crane's positions are termed '*perspectival*'. This is due to their considering sets of subsystems rather than just the whole system. This carries connotations—especially at the quantum level (Sect. 48.5)—of observed quantities depending on the particular specifics of the observer involved. Whereas Barbour-type approaches follow from more long-standing themes in the Foundations and Philosophy of Physics, the Rovelli or Crane type of approach has so far been more widely used in Quantum Gravity programs.

Some parts of the above two families of approaches are mutually compatible [37], whereas other parts require a choice to be made. The 'any change', 'all change' and STLRC fork of Chap. 15 is an example of such a choice, and also of the Author's intermediate position. Another is Rovelli's Partial Observables Approach—which is clearly based on multiple subsystems—whereas Barbour's relational position and the Author's principally concern Dirac observables or beables. Ashtekar Variables approaches use 'relational' [154, 157, 336, 845] in Rovelli's sense, though these approaches also happen to fit Barbour's and the Author's relational criteria, though this has hitherto been a not widely appreciated or used state of affairs. See [38, 39] for more detailed comparison of Barbour, Rovelli, Crane and the Author's positions on Relationalism.

---

[1]Crane's version can additionally be viewed as an early form of holography.

E. Anderson, *The Problem of Time*, Fundamental Theories of Physics 190,
DOI 10.1007/978-3-319-58848-3_19

19.2 Well-Known Theoretical Variants in Upper Layers of Mathematical Structure

Some differential-geometric level variants to GR (largely not further pursued in this book) are as follows. Not all symmetric 2-tensors are Riemannian metrics [477]. To have this property, they would need to induce a non-degenerate inner product on $\mathfrak{T}(\mathfrak{m})$. They also need to be physically ascribed (chrono)geometric significance; one can a priori also consider symmetric 2-tensors that have some entirely different significance. Some possibilities for alternative theories include one or more of the following generalizations. Ceasing to require degeneracy or symmetry, one could also have a 2-tensor or a Riemannian metric requirement (e.g. for a Finslerian generalization). One could also have more than one such entity, whether or not each is ascribed (in some perhaps partial sense) (chrono)geometric significance, as in e.g. bimetric theories.

19.3 Relationalism and Affine Geometry

On the other hand, the current book develops the affine, conformal and supersymmetric variants. This development covers, firstly, the flat space cases of these and the corresponding RPMs, secondly, the differential-geometric cases, and thirdly, the corresponding variants of the Preface's cube of physical theories.

We first turn to the case of Affine Geometry. In the flat-space case, the affine group is well-known due to being isomorphic to $GL(n, \mathbb{R})$ after taking out the centre of mass; invariants here are ratios of d-volumes in dimension d (see Appendix B.1). Using this models situations in which configurations have no *overall* meaning of either relative angle (by equivalence under global shears), or of relative ratio (by equivalence under global Procrustean stretches).

Affine Shape RPM [36] is based on $GL(n, \mathbb{R})$ Best Matching. E.g. in 2-d,

$$S = \sqrt{2}\int \mathrm{d}s\sqrt{W}, \quad \mathrm{d}s^2 = \sum_{A,B=1}^{n} \mathrm{d}_{GL}\underline{\rho}^A \times \mathrm{d}_{GL}\underline{\rho}^B \Big/ \sum_{C,D=1}^{n} \underline{\rho}^C \times \underline{\rho}^D, \tag{19.1}$$

where $\mathrm{d}_{GL}\underline{\rho}^A := \mathrm{d}\underline{\rho}^A - \mathrm{d}\underline{g}\,\underline{\underline{\underline{G}}}\,\underline{\rho}^A$ the $GL(2, \mathbb{R})$ Best Matching corrected derivative, *Best Matching corrected derivatives* for $\mathrm{d}\underline{g} := [\mathrm{d}f, \mathrm{d}e, \mathrm{d}b, \mathrm{d}c]$ auxiliaries and

$$\underline{\underline{\underline{G}}} := \left[\begin{pmatrix}1 & 0\\ 0 & -1\end{pmatrix}, \begin{pmatrix}0 & 1\\ 1 & 0\end{pmatrix}, \begin{pmatrix}0 & -1\\ 1 & 0\end{pmatrix}, \begin{pmatrix}1 & 0\\ 0 & 1\end{pmatrix}\right]^{\mathrm{T}}. \tag{19.2}$$

Also the V in this case's W is $V(-\times-/-\times-$ alone). Equation (19.1) encodes $\underline{\mathcal{P}}$ alongside Procrustean, shear, rotational and dilational constraints, which can be packaged into

$$\underline{\mathcal{G}} := \sum_{A=1}^{n} \underline{\rho}^A\, \underline{\underline{\underline{G}}}\, \pi_A. \tag{19.3}$$

Affine transformations of Minkowski spacetime $\mathbb{M}^n$ have so far been studied rather less, with 'affine QFT' receiving even less attention.

On the other hand, affine alternatives—and additions—to metric-level structure for GR are also well-known. These are based on Appendix D.3's differential-geometric considerations. E.g. one can consider metric and affine structure as independent fields to be varied. In the case of the Palatini action [75], this variation returns the metric connection and GR. There are also more generally theories, however, with an additional non-metric connection and consequently a notion of torsion. Moreover, these alternative theories have not yet been analyzed from a relational point of view.

19.4 Relationalism and Conformal Geometry

Use of Conformal Geometry is one of the simplest and most often considered variants of the cube of physical theories. In flat space, the conformal group *Conf*(d) takes the place of the Euclidean group; the corresponding invariant quantities are local angles, which are observationally very natural. The SR counterpart involves the conformal group *Conf*$(d, 1)$ in place of the Poincaré group *Poin*$(d+1)$ (Ex IV.10). The conformal transformations in this case preserve the causal structure, whereby they have been substantially studied. The Conformal Field Theory (CFT) counterpart of ordinary QFT has also been very well studied [674, 719].

Conformal Shape RPM arises by appending the further special conformal Best Matching to the $\underline{q}^I$ version of Metric Shape RPM's to give [36]

$$S = \sqrt{2}\int \mathrm{d}s\sqrt{W}, \quad \mathrm{d}s = \|\mathrm{d}_{\underline{a},\underline{b},c,\underline{k}}\mathbf{q}\|/\sqrt{I}, \tag{19.4}$$

$$\mathrm{d}_{\underline{a},\underline{b},c,\underline{k}}q^{Ia} := \mathrm{d}q^{Ia} - \mathrm{d}a^a - \left(\mathrm{d}\underline{b}\times\underline{q}^I\right)^a - \mathrm{d}c\, q^{Ia} - \left\{{q^I}^2\delta^{ab} - 2q^{Ia}q^{Ib}\right\}\mathrm{d}k_b \tag{19.5}$$

for V now of the form $V(\angle)$. Then variations with respect to $\underline{a}$, $\underline{b}$ and c yield (9.9), and the $\underline{q}^I$ counterparts of (9.10) and (16.16) respectively, whereas variation with respect to $\underline{k}$ *zero total special conformal momentum constraint*

$$\mathcal{K}_a := \sum_{I=1}^{N}\left\{{q^I}^2\delta_a{}^b - 2q^I_a q^{Ib}\right\}p_{Ib} = 0. \tag{19.6}$$

The quadratic constraint arising as a primary constraint is now

$$\mathcal{E} := I\|\pi\|^2/2 + V(\angle) = E. \tag{19.7}$$

The linear constraints now close as per the conformal algebra, and $\mathcal{K}_a$ manages to commute with $\mathcal{E}$ as well. This set-up works similarly for $d > 3$ [just requiring a different presentation for the larger *Rot*(d)].

Within GR, the spacetime conformal group continues to play an substantial role as regards causal structure. This group's status is additionally elevated to be on a par with the spacetime diffeomorphisms in e.g. in the Weyl^2 theory (11.14). On the other hand, the spatial conformal group is also significant in addressing the GR initial value problem (Sect. 21.4). Here GR is viewed as a Conformogeometrodynamical formulation associated with CMC slices, with occasional further connotations of York internal time (Chap. 21) or even of relational reformulations of GR or alternatives thereto which are partly based on conformal mathematics.

19.5 Relationalism and the Point at Infinity

A further argument for Affine Geometry is that it is a conceptually simpler geometry residing within Euclidean (and Similarity) Geometry (Appendix B.1) without appending any further structure. However, one might well argue instead that Conformal Geometry's local angles correspond naturally to directly observed quantities, whereas Affine Geometry's volume ratios do not. And yet to have a well-defined flat space theory based on local angles requires inclusion of Riemann's notion of 'point at infinity', so as to be able to formulate the inversion in the sphere and the subsequent special conformal transformation (Appendix B.1). This 'point at infinity' is open to interpretation as additional absolute structure: $\mathfrak{a}(d) = \mathbb{R}^d$ replaced by $\mathbb{R}^d \cup \infty$.

In contrast, flat space Affine Geometry does not itself require such a point. The affine or conformal fork indeed represents a choice at the level of Flat Geometry: one cannot accommodate the generators of both together by (E.35).

The above motivation of Affine Geometry, moreover, often leads to its being considered to be but a half-way house on the road to the even more simplifying Projective Geometry (Appendix B.1). Furthermore, the well-definedness of Projective Geometry itself also requires introduction of a 'point at infinity'. Projective Geometry's introduction of this structure is particularly motivated by seeking to simplify geometrical proofs. On the other hand, its invariants—the cross-ratios—also do not correspond to directly observed quantities.

Research Project 2) Does Projective Geometry admit a Relational Mechanics? [Some versions of this would not be relational *particle* mechanics due to implementing Projective Geometry ceasing to distinguish between points and lines.] As a first step, Best Matching within $\mathbb{C}^N$ was laid out in [36], alongside an outline of how cross-ratio invariance presents difficulties beyond this Chapter's other examples as regards construction of indirectly formulated actions. More generally, consider time in the projective version of the cube of theories.

19.6 The Fermionic Selection Criterion

In seeking *extensions* to standard physics, one useful criterion is the continued existence of spinors—and thus of capacity to model fermions. This is not compromised by incorporating whichever combination of conformal [706] or supersymmetric transformations, or extra affine connections. Thus this criterion is unrestrictive as regards the above-mentioned theoretical variants in the upper layers of mathematical structure. On the other hand, it can affect other extensions. E.g. at the level of the underlying topological manifold, it is well-known that some choices of $\mathbf{\Sigma}$ by themselves preclude the existence of fermions (see e.g. [673]), and that orientability of $\mathbf{\Sigma}$ is also often desired.

19.7 Relationalism in Ashtekar Variables Formulation of GR

Since Part II makes use of Fibre Bundles, it is useful to further phrase the Ashtekar variables $\mathbb{A}_i$ as a connection in the fibre bundles sense (Appendix F.4). The corresponding action (8.34) can also be more geometrically cleanly interpreted in the language of forms:

$$\mathsf{S} \propto \int \mathrm{d}^4x \, \mathrm{e}\, \mathrm{e}^{\mu}_{A} \wedge \mathrm{e}^{\nu}_{B} \wedge \mathrm{F}^{AB}_{\mu\nu} . \tag{19.8}$$

The loops themselves are moreover closely related to the connections in question. If one furthermore removes the *Diff*$(\mathbf{\Sigma})$ information from the loops, one arrives at knots. See Appendices N.12-13 for an outline of loop and knot configurations, as well as of their configuration spaces.

Knots have already been heralded as a spatially 3-*d* specific feature. Another such is complexified GR's self-duality along the lines of (F.1), as well as entering Ashtekar variables formulations, was previously already well-known due to also featuring in the Twistor Approach to (also complexified) GR (Sect. 36.2). On the other hand, the spatially 2-*d* version of Ashtekar variables gets by through its relation with Chern–Simons Theory (Appendix F.5).

Also note the following parallels between loops and preshapes. Preshapes arise by quotienting out the dilations *Dil* but not the more physically significant and mathematically harder to handle rotations *Rot*(*d*). On the other hand, Loop Quantum Gravity's loops are arrived at by quotienting out $SU(2)(\mathbf{\Sigma})$ but not the more obvious and yet mathematically harder to handle spatial diffeomorphisms *Diff*$(\mathbf{\Sigma})$. Thus, while neither are the most redundant configurations featuring in the corresponding theory, both are still partly redundant. Nor are they even 'half-way houses' in each's passage to non-redundant 'physical' kinematical variables. This is since both are prior to the *main part* of that passage, both physically and in terms of the remaining parts of each's passage being far more mathematically complex than the parts already undertaken. This has long been reflected in the former theories having been named not after preshapes but after the shapes themselves. I.e. 'Shape Geometry', 'Shape Statistics', and 'Shape Dynamics' in the sense of 'dynamics of pure shape', as per [37, 102, 536, 539] and Sect. 33.7. This suggests that it would be clearer to name the latter theory after not loops but knots. Various possibilities are Knot Quantum Gravity, Knot Quantum Gestalt and *Nododynamics*—from the Latin *nodus* for 'knot'. Let us use the last of these for the rest of this book, since it additionally makes sense at both the classical and quantum levels, just as 'Geometrodynamics' does. Indeed, Geometrodynamics itself is indeed another naming based on identifying less redundant 'physical' kinematical variables.

Relational and Background Independence criteria are held to be a substantial feature in approaches based on Ashtekar variables reformulations of GR such as Nododynamics. 'Relational' is usually meant here in Rovelli's sense, though the Author now argues that is can also be meant in the sense of Temporal and Configurational Relationalism. This is modulo a small glitch: the pure gravity case cannot be cast in Temporally Relational form at the level of the classical Lagrangian. From the point of view presupposing spacetime [14], this occurs because the lapse-uneliminated action is purely linear in the lapse. This is due to the well-known 'pure-T' character of Ashtekar's canonical action for pure GR, in contrast to the more usual $\mathrm{T} - \mathcal{V}$ form of the geometrodynamical action. This is sufficiently close to Sect. 17.2's Example 4) to preclude there being such an action. Another approach to this is that it is a subcase of $\mathcal{L} = \alpha^k \mathcal{F} + \mathcal{G}$ having Lagrange multiplier equation $\gamma \alpha^{k-1} \mathcal{F} = 0$, which, being homogeneous in α, cannot be used to eliminate α. However, it is clear that addition of matter fields, or even just a cosmological constant—a $-2\Lambda \mathrm{E} \wedge \mathrm{E} \wedge \mathrm{E}$ term in the Lagrangian—remedies this problem.

Configurational Relationalism for Ashtekar variables formulations involves $\mathfrak{g} = SU(2)(\mathbf{\Sigma}) \rtimes Diff(\mathbf{\Sigma})$. The Ashtekar Variables form of the GR momentum constraint (8.36) and Yang–Mills–Gauss constraint (8.35) follow from Appendix L.3's variation with respect to, respectively, frame auxiliaries and $SU(2)(\mathbf{\Sigma})$ auxiliaries whose meaning generalizes the previous Chapter's Electromagnetism's 'action per unit charge'. The usual approaches incorporate Configurational Relationalism by considering knot configurations. On the other hand, the relational action for the canonical formulation of GR in Ashtekar Variables lies within the Jacobi–Synge class. The Ashtekar Variables form of the GR Hamiltonian constraint ensues as a unique primary constraint (per space point), and is moreover such that a corresponding t^{em} can be isolated. So none of Sect. 17.2's caveats apply to extracting a Machian emergent time corresponding to an Ashtekar variables formulation of GR.

By now it is clear that the Barbour-type relational literature's apparent fixation upon GR in geometrodynamical form is merely at the level of choice of examples. *Temporal and Configurational Relationalism apply to Nododynamics as well.* This observation is of likely foundational importance for Nododynamics due to the additional conceptual and philosophical content of this further notion of Relationalism. On the other hand, the Husain–Kuchař model arena implements a Configurational Relationalism that is similar to Ashtekar variables GR (resulting in $\mathcal{M}_i$ and a $\mathcal{G}_I$) without possessing a Temporal Relationalism (thus this has no $\mathcal{H} = \mathcal{C}$hronos and so no emergent Machian time is based on rearranging this).

Research Project 3) Physicist Arthur Komar [563] showed that canonical transformations are capable of altering the form of the Thin Sandwich Problem. He demonstrated this in the case of a distinct canonical transformation from that used by Ashtekar. However, it does motivate investigating whether the Ashtekar canonical transformation—

$$G\,\mathrm{A}^I_a = \Gamma^I_a - i\,\mathcal{K}^I_a \tag{19.9}$$

for Γ^I_a the spin connection and $\mathcal{K}^I_a := \mathcal{K}_{ab}e^{bI}$ a close relation of the extrinsic curvature—has the further good fortune of ameliorating the Thin Sandwich Problem. Since the Ashtekar version involves six equations in six unknowns due to adjunction of $SU(2)$ gauge freedom resulting in the corresponding Yang–Mills–Gauss constraint, part of Research Project 1) is likely to be a useful precursor.

19.8 Relationalism and Supersymmetry

Supersymmetry is the third of the most immediate and often considered options to the 'standard cube' of physical theories. This corresponds to generalizing symmetry groups so as to no longer factor into spatial and internal parts. However, this split lies within the auspices of the Coleman–Mandula Theorem (11.16), i.e. *prior* to the construction of Supersymmetry so as to elude this No-Go. This version of Physics involves the Poincaré supergroup, supersymmetric QFT and replacing GR and its diffeomorphisms with Supergravity and corresponding super-diffeomorphisms.

Supersymmetric Nonrelativistic Particle Mechanics has been considered as a model arena in e.g. [321, 678, 914]. [36] subsequently supplied a simple supersymmetric RPM. The indirect formulation's incipient notion of absolute space $\mathfrak{a}$ now a *Grassmann space* $\mathbb{R}^{(d|n)}$ (after polymath Hermann Grassmann) in place of $\mathbb{R}^d$. Supersymmetry can, moreover, be regarded as a further input choice for $\mathfrak{g} \leq Aut(\langle \mathfrak{a}, \sigma \rangle)$. N.B. that physically irrelevant groups $\mathfrak{g}$ in this case can, paralleling the relativistic counterpart, cease to split into spatial and internal parts.

The supersymmetric version of Best Matching in the simple but typical case of $\mathfrak{g} = super\text{-}Tr(1)$ [36]

$$S_{\text{susy}} = \sqrt{2} \int \left\{ \|\text{d}_{a,\alpha}\mathbf{q}\| \sqrt{W} + i \sum_{I=1}^{N} \{ \bar{\theta}^I \text{d}_{a,\alpha}\theta^I - \overline{\text{d}_{a,\alpha}\theta}^I \theta^I \} \right\}, \tag{19.10}$$

for fermionic Best Matched derivatives

$$\text{d}_{a,\alpha}\theta^I := \text{d}\theta^I - \text{d}a + i\,\text{d}\alpha, \qquad \overline{\text{d}_{a,\alpha}\theta}^I := \text{d}\bar{\theta}^I - \text{d}a - i\,\text{d}\bar{\alpha} \tag{19.11}$$

and bosonic Best Matched derivatives

$$\text{d}_{a,\alpha}q^I := \text{d}q^I - \text{d}a - \bar{\theta}^I \text{d}\alpha - \text{d}\bar{\alpha}\,\theta^I. \tag{19.12}$$

The $\text{d}\alpha$ and $\text{d}\bar{\alpha}$ corrections to the fermionic species are 'Grassmann translations'. Furthermore, upon imposing Supersymmetry these also feature as corrections to the bosonic changes. Note also that $\text{d}\theta^I$ and $\text{d}\bar{\theta}^I$ are not acted at all upon by the (standard bosonic) translations; at least no intuitive natural action ties these species and transformation.

Variation with respect to a now gives a new form of 1-d zero total momentum of the Universe constraint

$$\mathcal{P}_{\text{susy}} := \sum_{I=1}^{N} \{p_I + p_{\theta I} - p_{\bar{\theta} I}\} = 0. \tag{19.13}$$

The new form just reflects that fermions also carry momentum. On the other hand, variation with respect to α and $\bar{\alpha}$ give the *zero total supersymmetric exchange momentum constraints*

$$\mathcal{S} := -\sum_{I=1}^{N} \{p_{\theta I} + i\bar{\theta}^I p_I\} = 0, \qquad \mathcal{S}^\dagger := \sum_{I=1}^{N} \{p_{\bar{\theta} I} + i\theta^I p_I\} = 0. \tag{19.14}$$

These moreover gain one piece from the fermionic sector and one piece from the bosonic sector. These constraints are accompanied by the standard quasi-bosonic $\mathcal{E}$, except that now V contains fermionic species as well:

$$\mathcal{E} := \|\mathbf{p}\|^2/2 + V(q^I, \theta^I, \bar{\theta}^I) = E. \tag{19.15}$$

Taking for now the stance of not knowing the concrete form taken by supersymmetric analogues of shape, the incipient form of V is

$$V(q^I, \theta^I, \bar{\theta}^I) = V_{\text{B}}(q^K) + \sum_{I=1}^{N} \left\{ \theta^I u_I(q^K) - \bar{\theta}^I v_I(q^K) + \sum_{J=1}^{N} \theta^I \bar{\theta}^J w_{IJ}(q_K) \right\} \tag{19.16}$$

by virtue of the automatic truncation in Grassmann polynomials afforded by the underlying anticommutativity. Demanding algebraic closure leads to the following conditions on V for this to be a function of the *super-Tr*(1) notion of shape:

$$\sum_{I=1}^{N} \bar{\theta}^I \left\{ \frac{\partial V_{\text{B}}(q^K)}{\partial q^I} + \sum_{J=1}^{N} \theta^J \frac{\partial u_J(q^K)}{\partial q^I} \right\} = 0, \tag{19.17}$$

$$\sum_{I=1}^{N} \theta^I \left\{ -\frac{\partial V_{\text{B}}(q^K)}{\partial q^I} + \sum_{J=1}^{N} \bar{\theta}^J \frac{\partial \bar{v}_J(q^K)}{\partial q^I} \right\} = 0. \tag{19.18}$$

Research Project 4) Work out the supersymmetric counterpart of Appendix G: reduced configurations, reduced configuration spaces and their isometry groups.

Whether Relationalism and Supergravity are compatible is less clear-cut. For now, let us comment that spin-3/2 fermionic change *does* enter $\mathcal{H}$, so the Machian issue with spin-1/2 does not carry over to spin-3/2; see Sect. 24.10 for more.

19.9 Supersymmetric, Conformal and Affine Combinations

One can consider each of these as a successful addition of generators to the more usual Euclidean, similarity or Poincaré groups. What happens upon attempting to make more than one of these additions concurrently?

The conformal and supersymmetric generators successfully combine to form the conformal supergroup [887]. The affine and supersymmetric generators do as well, giving the affine supergroup. On the other hand, affine and conformal generators cannot be combined for flat geometries due to the obstruction term in (E.35). Thus two distinct apex groups result from attempting to extend $Sim(d)$ further. Such obstructions preclude affine-conformal RPM, and similarly affine-conformal QFT, whereas super-affine and super-conformal RPM's [36] and QFT's are group-theoretically allowed. Indeed, Superconformal QFT has already received a certain amount of attention [368, 386].

Moreover, at the differential-geometric level, affine and conformal structures can co-exist (for all that Weyl's unified theory [893]—which failed on other grounds—is one of the best-known examples of a such). The study of Supergravity already includes consideration of affine structure due to the presence of torsion. Superconformal Supergravity—the result of gauging the superconformal group—has also been studied to some extent [529, 530]. Torsion considerations here involve a triple combination of affine, conformal and supersymmetric structure. A differential version of Projective Geometry is also well-established [688].

Research Project 5) Consider time within the super-conformal cube of theories, including setting up super-conformal RPM.
Research Project 6) Do likewise for the super-affine case.

The next most simple and most often considered option after the conformal, affine, projective and supersymmetric versions is the topological manifold one. This includes consideration of Topological Field Theory (TFT) and considerations of topology change in GR. This option and yet further descents in level of structure assumed are deferred to Epilogue II.C.

19.10 String and M-Theory Versus Relationalism

As regards perturbative String Theory, firstly recollect that Axiom i) of Spacetime Relationalism precludes this from being among the set of relational theories. Secondly, there are also some conceptual parallels between the passages from point particles to each of strings and to relational quantities alone. The latter is moreover more conservative—the relational quantities come from careful thought about the original problem rather than replacing it with a distinct problem as in String Theory. Strings, on the other hand, arose from assuming particles have material significance and then reconceiving the manner in which this material significance was realized. This is in contradistinction with the idea that only inter-particle relations have significance. This material significance provides a reason to 'string up' spacetime or space rather than auxiliary spaces from the Principles of Dynamics. Relationalism gives less new structure—leading perhaps to less mathematical richness, yet also appearing to be a safer bet as regards the 'hypotheses non fingo' tradition of Physics. This is because postulating strings builds in more assumptions whereas Relationalism removes assumptions. On the other hand, if Nature is made out of strings, relations both between strings and within a single string—*Diff*$(\mathbb{S}^1)$, *Diff*$(\mathbb{S}^1 \times \mathbb{R})$—would make sense. It would be interesting to know which aspects of Relationalism extend to this more complicated setting.

Thirdly, exporting the Configurational Relationalism or more specific Best Matching ideas to other parts of Physics would be interesting, even if not accompanied by the further demands of other kinds of Relationalism or Background Independence. E.g. one could consider what form these ideas take for target space theories of which strings on a given background.

Fourthly, note the analogy between the Nambu–Goto action (11.19) and the RPM Jacobi action. It is of the 'quadratic Jacobi' form, but the string involves *two* parameters, so this indeed remains the correct form for Manifest Reparametrization Invariance. The ensuing constraint is not to be interpreted as an equation of time, since the parametrizations of time and space on the string worldsheet are on an equal footing. In any case, the scheme is Background Dependent rather than in need of an emergent Machian time.

On the other hand, M-Theory is expected to have all three of extended objects, Supersymmetry and Background Independence.[2] Since Background Independence brings about a Problem of Time, it is exceedingly likely that M-Theory will have a Problem of Time as well. In this way, study of Background Independence and the Problem of Time is likely to also be a valuable investment from the perspective of developing and understanding M-Theory.

Moreover, many features of the Problem of Time are universal, so these will occur here as well, unless features such as Supersymmetry or extended objects specifically conspire to eliminate some facets. These features were not however designed for this, by which this is a more stringent theoretical test for these features, from the point of view that good physical theories work from multiple conceptual perspectives (cf. Wheeler's 'many routes' position in Sect. 9.1) rather than just from the one in which they were first constructed. It would thus be a pleasant surprise if, in addition to all the different usefulnesses of these structures, they also happened to cure (part of) the Problem of Time.

Geometrodynamics (which straightforwardly generalizes to arbitrary-d [845]) could be seen as a first model arena for Background Independence aspects of spatially 10-d M-Theory. A next step up might well involve using spatially 10-d Supergravity, since this is a lower-energy or 'semiclassical' limit of M-Theory.

Research Project 7) Give a relational treatment of space and time, in the presence of matter contents which span the full range of codimensions C (as opposed to being just one of point particles or fields). For now, consider the purely bosonic case.

[2]To avoid confusion, let us note here that some other uses of the term 'background (in)dependence' in the String Theory literature have a different meaning. I.e. concerning the effect of choice of vacuum on string perturbations.

Chapter 20
Other *Tempus Ante Quantum* Approaches

20.1 The Ante Postulate

This is that there is a fundamental time to be found at the classical level for the full (i.e. untruncated) classical Gravitational Theory. Various candidate implementations of this (T ... Q approaches) include the following.

A) the classical Machian emergent time already considered in Chaps. 15 to 19. The main problem with this is that it does not unfreeze the quantum GR Wheeler–DeWitt equation.
B) Riem, scale, hidden and dilational times (York time is an example of both of the previous), and matter time (including reference fluid). These are covered in Chaps. 20 and 22; many of them do unfreeze quantum GR.

This book, however, argues in favour of A) followed by its semiclassical counterpart so as to unfreeze quantum GR *after* Quantization (Chap. 46). The following turns out to often be useful in discussing B).

Timelessness-indefiniteness cancellation hypothesis. Canonical GR has an indefinite kinetic term as well as time going missing. Perhaps these two perceived deficiencies can cancel each other out.

More specifically, it is the overall scale part of the metric which causes indefiniteness in GR-like model arenas. So is overall scale, or some close relation of it (e.g. dilational momentum), tied to the isolation of, or emergence of, time? This features in a number of strategies; while this lies outside of what RPMs can model since these have positive-definite kinetic terms, it can be considered for Minisuperspace models.

20.2 Riem Time's Hyperbolic Implementation

GR's kinetic metric (inverse DeWitt supermetric) on $\mathfrak{Riem}(\boldsymbol{\Sigma})$ has indefinite signature ($- - + + + + +$ pointwise). Perhaps one could take the this indefinite direction

E. Anderson, *The Problem of Time*, Fundamental Theories of Physics 190,
DOI 10.1007/978-3-319-58848-3_20

to pick out a candidate timefunction $\mathrm{t}^{\mathrm{Riem}}$. This would be in parallel to how the indefinite direction in Minkowski spacetime $\mathbb{M}^n$ picks out a background timefunction (Chap. 4). This is a direct attempt to implement timelessness-indefiniteness cancellation. $\mathrm{t}^{\mathrm{Riem}}$ would be hoped to produce a quantum Frozen Formalism Problem resolution through this timefunction being a subcase of the more general candidate form $\mathrm{t}^{\mathrm{hyperb}}$. This is because use of this and its subsequent conjugate momentum $p_{\mathrm{t}^{\mathrm{hyperb}}}$ casts $\mathcal{H}$ into the hyperbolic form (9.20). This gives a Klein–Gordon type time-dependent wave equation (12.9) depending on the double derivative with respect to this timefunction.

The above presentation has been kept free of Configurational Relationalism by being specifically for diagonal Minisuperspace. This possesses a 3×3 $--++$ Minisuperspace metric. On the other hand, the 2×2 positive-definite anisotropy-space, $\mathfrak{Ani}$ block $\boldsymbol{M}^{+}$—a simple GR instance of shape space—has its inverse $\boldsymbol{N}_{+}$ subsequently feature in

$$H_{\mathrm{True}} = \|\boldsymbol{P}\|^2_{\boldsymbol{N}_{+}(t^{\mathrm{Riem}},\,\boldsymbol{Q})} + V\big(t^{\mathrm{Riem}},\,\boldsymbol{Q}\big). \tag{20.1}$$

This approach is more often called 'Superspace time', though this only makes sense when $\mathfrak{Riem}(\boldsymbol{\Sigma}) = \mathfrak{S}\mathrm{uperspace}(\boldsymbol{\Sigma})$ by both being equal in the case of restriction to $\mathfrak{Mini}(\boldsymbol{\Sigma})$.

The full GR case also requires the functional- rather than partial-derivative version of this working. E.g. the functional-time-dependent Schrödinger equation which is based on the functional derivative with respect to the local $\mathrm{t}(\underline{x})$, in a functional Laplacian—or Klein–Gordon operator—combination.

However, this approach fails for reasons which are already clear at the classical level. The Klein–Gordon Approach is known to work in stationary spacetimes due to the presence of a timelike Killing vector; in fact a timelike conformal Killing vector suffices as per Sect. 11.3. The GR case succeeds in paralleling the Klein–Gordon Approach in this respect, for there is a 'timelike' conformal Killing vector $\mathcal{E}$ in GR's configuration space $\mathfrak{Riem}(\boldsymbol{\Sigma})$ (see Ex VI.11.vi). For the Klein–Gordon Approach, additionally, the potential is just the constant m^2, which thus does not invalidate useability of the timelike Killing vector. On the other hand, for GR the potential term $\mathrm{V}_{\mathrm{GR}} = \sqrt{\mathrm{h}}\mathcal{R}$ does not respect the conformal Killing vector [581, 584]. I.e. one has $\{\mathcal{E}, \mathrm{T}_{\mathrm{GR}}\} = -\frac{3}{2}\mathrm{T}_{\mathrm{GR}}$ versus $\{\mathcal{E}, \mathrm{V}_{\mathrm{GR}}\} = \frac{1}{2}\mathrm{V}_{\mathrm{GR}}$, so using $\mathcal{H} = \mathrm{T}_{\mathrm{GR}} + \mathrm{V}_{\mathrm{GR}}$, $\{\mathcal{H}, \mathrm{V}_{\mathrm{GR}}\}$ does not work out right to close, answering the aforementioned Exercise. Moreover, the analogy between $\mathfrak{S}\mathrm{uperspace}(\boldsymbol{\Sigma})$ and relational space $\mathfrak{R}(N, d)$ is further strengthened by the latter also possessing a conformal Killing vector associated with scale [37].

The indefinite part of the GR supermetric is intimately related to notions of scale. This is tied to some kind of physical degrees of freedom which are closely related to *pure shape*. Thus timelessness–indefiniteness cancellation for a GR-like theory could also be expected to arise from using some kind of time that is closely related to scale, to which we next turn.

20.3 Scale Factor, Cosmic and Conformal Times

Following Sect. 9.11's outline, we now consider the multiplicity of scale variables; for GR these include the scalefactor a, $\sqrt{h}$ that goes as a^3, or the Misner variable $\Omega = \ln a$. RPMs also have a distinct meaningful set of (mass-weighted) scale variables, such as the moment of inertia I and the configuration space radius $\rho := \sqrt{I}$.[1] Make use of above GR examples as time candidates is an obvious attempt at implementing the timelessness-indefiniteness cancellation hypothesis. One major problem with using these is non-monotonicity in recollapsing universes.

It is also useful at this point to supplement Chap. 7's outline of isotropic cosmology with the Friedmann equation

$$\left\{\frac{\dot{a}}{a}\right\}^2 = \frac{8\pi G}{3}\rho + \frac{\Lambda}{3} - \frac{k}{a^2} \tag{20.2}$$

and the Raychaudhuri equation

$$\frac{\ddot{a}}{a} = -\frac{4\pi G}{3}\{\rho + 3p\} + \frac{\Lambda}{3} \tag{20.3}$$

that these obey. These follow from the Einstein Field Equations under the great symmetry reduction implied by isotropy. The dots here denote derivatives with respect to cosmic time, whereas the k is the sign of the constant curvature (± 1, or 0 in the spatially flat case). ρ and p are the density and pressure of the matter contents of the Universe, modelled phenomenologically, and we are using the customary $c = 1$ units. From the number of time derivatives featuring in each, the Friedmann equation is clearly a constraint equation (the Hamiltonian constraint), whereas the Raychaudhuri equation is an evolution equation. For dust, ρ goes as a^{-3}, and for radiation, as a^{-4}.

The cosmic time at an event E can, furthermore, be conceived of as [596]

$$\text{sup}_{\text{all proper durations}}(\text{all future-directed timelike curves ending at } E). \tag{20.4}$$

This coincides with the proper time experienced by comoving observers. To first approximation, observers on Earth are such comoving observers. Unlike scale factor variables, it is by construction monotonic along each future-directed timelike curve. Conformal time is moreover more closely linked to spacetime's causal structure. It different events for a shared value of conformal time, not cosmic time, which are seen as simultaneous by comoving observers. If cosmic time is to be taken as primary through its arising as a GLET, let us entertain the possibility of requiring a secondary timefunction for use in synchronization procedures themselves.

Both cosmic and conformal time have the disadvantage of requiring construction from multiple observational data and in a model-dependent manner. The properly

[1]This is the first of quite a few analogies between I and $\sqrt{h}$ or its global analogue, the spatial volume of the Universe, V.

normalized scale factor is more directly realized: as $\tau = t_{\text{Universe}}/\{1 + z\}$, where the numerator is the age of the Universe today.[2] However, this timefunction has the disadvantage of not coinciding with the proper times of comoving observers.

N.B. that there are limitations in the precision to which cosmic time is known. E.g. proper motions, and only approximately being able to factor velocities into proper and cosmic parts, are sources of such. Averaging assumptions—a subtle and unresolved matter in the context of GR, as outlined Sect. 30.6—are another source.

Finally note the RPM analogue of the Friedmann equation,

$$\left\{\frac{\rho^*}{\rho}\right\}^2 = \frac{2\,K}{\rho^3} + \frac{2\,R - \mathrm{s}}{\rho^4} - 2\,A + \frac{2\,E}{\rho^2}, \tag{20.5}$$

where ρ is now the RPM scale variable $\sqrt{I}$. Also, K is here an inverse square law coefficient, so that Newtonian Gravitation contributes in parallel to dust. R is a conformally invariant potential coefficient. Finally, s is the total shape momentum, which both contribute in parallel to radiation.

20.4 Parabolic and Part-Linear Implementations

We next further consider approaches which start by solving $\mathcal{Q}$uad to obtain a parabolic or part-linear form,

$$\mathsf{P}_{\mathfrak{t}^{\text{parab}}} + \mathcal{H}^{\text{True}}\lfloor \mathfrak{t}^{\text{ante}}, \mathsf{Q}^{\mathsf{O}}, \mathsf{P}_{\mathsf{O}} \rfloor = 0. \tag{20.6}$$

Here, $\mathsf{P}_{\mathfrak{t}^{\text{parab}}}$ is the momentum conjugate to a candidate classical time variable, $\mathfrak{t}^{\text{parab}}$, and the index O runs over the other coordinates. This is to play a role that closely parallels that of the external classical time of Newtonian Mechanics. Moreover, given (20.6) such a parabolic form for $\mathcal{H}$, it becomes possible to apply a conceptually-standard Quantization which yields a time-dependent Schrödinger equation (44.1). This form occurs in Parametrized Nonrelativistic Particle Mechanics model arenas [586]. See the next two Chapters for GR examples of part-linear forms.

Kuchař [573, 574, 581] developed these approaches using Parametrized Nonrelativistic and Relativistic Particle Mechanics models and Parametrized Field Theory models; see [552] for a brief introduction.

20.5 Hidden Time Approaches

A first suggested implementation of the parabolic form (9.17) of *Tempus Ante Quantum* involves the possibility of there being a *hidden* or *internal* time [483, 581, 586,

---

[2] This is, moreover, only an observed cosmological quantity modulo an unknown proper component.

922] within one's Gravitational Theory itself. The apparent frozenness would then be a formalism-dependent statement that is to be removed by applying some canonical transformation. This sends GR's spatial 3-geometry configurations to a) 1 hidden time, so $\mathrm{t}^{\mathrm{parab}} = \mathrm{t}^{\mathrm{hidden}}$. b) 2 'true gravitational degrees of freedom' (which are the form taken by the 'other variables', and are here 'physical' alias 'non-gauge'). The general canonical transformation—on GR's phase space $\langle \mathfrak{T}^*(\mathfrak{Riem}(\boldsymbol{\Sigma})), \{\,,\,\}\rangle$—is of the form

$$(\mathbf{h}, \mathbf{p}) \longrightarrow \left(\mathrm{t}^{\mathrm{hidden}}, \mathrm{p}_{\mathrm{t}^{\mathrm{hidden}}}, \mathbf{True}, \mathbf{P}^{\mathrm{True}}\right). \tag{20.7}$$

One would then seek to find a such so as to arrive at a hidden-time-dependent Schrödinger equation of form (44.1) for $\mathrm{t}^{\mathrm{hidden}} = \mathrm{t}^{\mathrm{parab}}$. One common place to seek for an hidden or internal time is among the theories' natural scalars. Let us next classify these in more detail in the case of Geometrodynamics.

1) *Intrinsic time candidates* (in the sense of intrinsic geometrical objects) though these have largely lacked in successful development [586]. One family of examples involve using some scale variable from Sect. 20.3 as a time. Note that these are based on configuration rather than on change, so they are an example of the third arrow from the bottom in Fig. 13.1.
2) *Extrinsic time candidates* (in the sense of intrinsic geometrical objects, in particular based on the extrinsic curvature). These include what is probably our best candidate so far for an internal time in GR: *York time* [483, 581, 586, 922], see Ex VI.11.iii) and the next Chapter for more detail. Elevating this to a candidate time furthermore involves exchanging the scale variable and the conjugate dilational momentum under a canonical transformation, so that the latter gains coordinate status. Dilational time candidates such as York time can be viewed as further attempts at implementing the timelessness-indefiniteness cancellation hypothesis. Unlike 1), these are clearly based on the momentum formulation of change, and so are examples of Fig. 13.1's top arrow. Finally note that internal time is not universal over all theories, though scale and dilational notions of time are universal within those theories that possess scale.

Example 2) *Einstein–Rosen time* [572, 573] is another extrinsic time candidate within the cylindrical wave Midisuperspace model arena.

20.6 Implementation by Unhidden Time

We end by pointing out that physicists David Boulware and Gary Horowitz pointed out that there is a Higher-Derivative Theory in which a natural variables set contains an already-explicit internal time candidate [162, 456].

Chapter 21
Conformal Approach and Its York Time

The York time candidate version of hidden time has been developed in particular. This approach originated with York's choice to bypass Wheeler's earlier Thin Sandwich approach on technical grounds. York's approach brings in conformal mathematics (Appendix D.7); the GR Hamiltonian and momentum constraints decouple when formulated in this manner. Wheeler moreover championed *both* of these approaches as Machian at some point [897, 901]. Finally note that various programs ([65, 103, 108, 650] and Chap. 33) use a combination of thin-sandwich and conformal mathematics to provide distinct and in some cases Machian foundations for GR.

21.1 Trace-Tracefree Irreducible Tensor Split

Begin by evoking the irreducible decomposition of symmetric second-rank tensors Σ_{ab} into their *trace part* $\Sigma := \Sigma_{ab}\mathrm{h}^{ab}$ and *tracefree part* $\Sigma^{\mathrm{T}}_{ab} := \Sigma_{ab} - \frac{\Sigma}{3}\mathrm{h}_{ab}$, since conformal weights are assigned to such irreducible pieces. In particular, if this split is applied to K_{ab}, the GR constraints (8.32) and (8.33) take the forms

$$\mathrm{K}^{\mathrm{T}}_{ij}\mathrm{K}^{\mathrm{T}ij} - \frac{2}{3}\mathrm{K}^2 - \mathcal{R} + 2\{\mathcal{E} + \Lambda\} = 0, \tag{21.1}$$

$$\mathcal{D}_b\mathrm{K}^{\mathrm{T}b}{}_a - \frac{2}{3}\mathcal{D}_a\mathrm{K} + -\mathrm{J}_a = 0. \tag{21.2}$$

21.2 Maximal and CMC Slices, and Conformal Scaling

Next, let us distinguish between Lichnerowicz's pioneering work [622] based on *maximal slices*

$$\mathrm{p} = 0, \tag{21.3}$$

E. Anderson, *The Problem of Time*, Fundamental Theories of Physics 190,
DOI 10.1007/978-3-319-58848-3_21

and York's generalization [922, 924] to *CMC slices*

$$\mathrm{K} = \text{spatial constant} \quad \Rightarrow \quad \mathrm{p}/\sqrt{\mathrm{h}} = \text{spatial constant}. \tag{21.4}$$

This generalization is significant as regards spatially-compact spacetimes, for which maximal slicing cannot be propagated (i.e. it can not be maintained throughout an extended foliation). CMC slices are moreover significant in Numerical Relativity.[1]

We subsequently consider the objects involved to be additionally conformal tensors, with conformal transformation laws

$$\mathrm{h}_{ij} \rightarrow \varphi^4 \mathrm{h}_{ij}, \qquad \mathrm{p}^{\mathrm{T}ij} \rightarrow \varphi^{-4} \mathrm{p}^{\mathrm{T}ij}, \qquad \mathrm{p}/\sqrt{\mathrm{h}};$$

$$\mathrm{J}_i, \qquad \mathcal{E}, \quad \text{and} \quad \Lambda \quad \text{are conformally invariant.} \tag{21.5}$$

See Appendix D.7 and Ex III.22) as regards the choice entailed in the first of these. The first four of these ensure that $\mathcal{M}_i$ is conformally covariant; the third also ensures conformal invariance of the CMC slice condition itself. Additionally, $\mathcal{H}$ now becomes the *Lichnerowicz equation*

$$8\,\triangle_{\mathbf{h}}\varphi = \mathcal{R}\varphi - \mathrm{p}_{ij}\mathrm{p}^{ij}/\sqrt{\mathrm{h}}\,\varphi^7 - 2\{\mathcal{E} + \Lambda\}\varphi^5 \tag{21.6}$$

for the maximal case, or its generalization the *Lichnerowicz–York equation*

$$8\,\triangle_{\mathbf{h}}\varphi = \mathcal{R}\varphi - \mathrm{p}^{\mathrm{T}}_{ij}\mathrm{p}^{\mathrm{T}ij}/\sqrt{\mathrm{h}}\,\varphi^7 + \left\{\mathrm{p}^2/6\sqrt{\mathrm{h}} - 2\{\mathcal{E} + \Lambda\}\right\}\varphi^5 \tag{21.7}$$

for the CMC case (the original versions contained no Λ term). Because $\mathcal{M}_i$ is conformally covariant, one can solve it irrespective of the subsequent solution of the conformally-transformed $\mathcal{H}$ for the physical scale φ. In this sense, *the conformal formulation decouples the GR constraints*.

Whereas (21.6), (21.7) are complicated nonlinear equations, they do benefit from quasilinearity (Appendix O.6). This allows for substantial existence and uniqueness theorems [206, 465] and Numerical Relativity applications [123], but very largely not for exact solutions. In this respect, York's formulation [922] of the GR initial value problem is in far better shape than Wheeler's earlier Thin Sandwich conceptualization. Finally, this formulation straightforwardly extends to Standard Model matter fields [64, 470].

21.3 Model Arenas

Example 1) *Minisuperspace* [483]. Here the Lichnerowicz–York equation becomes a merely algebraic polynomial equation,

$$R\,\varphi - p^{\mathrm{T}}_{ij}p^{\mathrm{T}ij}\big/\sqrt{h}\,\varphi^7 + \left\{p^2/6\sqrt{h} - 2\{\mathcal{E} + \Lambda\}\right\}\varphi^5 = 0. \tag{21.8}$$

[1] Some of the supporting PDE Analysis works alluded to in Chap. 8.14 and Appendix O involve such slices. These are used in [123, 382] for both the GR initial value problem for finding data satisfying the GR constraints and as one type of gauge fixing for the evolution equations.

This has various exactly-solvable subcases for low enough polynomial order in a suitable φ^k variable.

Example 2) *Strong Gravity*. This is also merely polynomial: (21.8) without $R\,\varphi$.

Example 3) *RPMs*. Some of these have $0 = \mathcal{D} = \sum_{A=1}^{n} \rho^A \pi_A$ as Metric Shape RPM's analogue of maximal slices. Others are based on presenting Metric Shape and Scale RPM in *scale–shape split* form (alias the metric-level cone). I.e. another type of irreducible piece tensor split in place of Sect. 21.1's trace–tracefree split, now into scale ρ and dimensionless shape variables $\underline{n}^A$. This corresponds to the quantity $D := \sum_{A=1}^{n} \rho^A \pi_A$ now not being zero by a constraint $\mathcal{D}$ but rather taking a progression of values, in parallel to how the CMC slice condition generalizes the maximal slice condition. In this formulation,

$$S = \sqrt{2} \int \sqrt{\{\mathrm{d}\rho^2 + \rho^2 \|\mathrm{d}\boldsymbol{n}\|^2\}\{E - V\}} \quad \text{and}$$
$$\mathcal{Q}\text{uad} = \{\pi_\rho^2 + \|\boldsymbol{\pi}\|^2/\rho^2\}/2 + V = E. \tag{21.9}$$

This theory's objects can additionally be allotted conformally weights. One way to do this which tightly parallels GR involves viewing (G.2) as the analogue of (H.3), whereby $\underline{\rho}^a$ scales by 1 power of RPM's analogue φ.[2] The $\underline{n}^a$ themselves are invariant. Consequently, $\mathcal{L}_i$ (and where relevant $\mathcal{D} = 0$) are automatically conformally invariant, as analogues of the conformal covariance of $\mathcal{M}_i$. This leaves $\mathcal{E}$ scaling as

$$p_\varphi^2 + \|\boldsymbol{\pi}\|^2/\varphi^2 - 2\{V(\rho_a, \varphi) - E\} = 0, \tag{21.10}$$

where a potential piece that is homogeneous of degree k in the ρ_a scales as φ^k. This is clearly always an algebraic equation as well, with simple solvable cases including $k = 2, 0, -2$ and -4 alongside various linear combinations thereof. On the other hand, the pure-shape case is analogous to the Lichnerowicz equation itself in having no p_φ^2 term, and also has V restricted to $k = 0$ which is non-scaling. This analogue—$\|\boldsymbol{\pi}\|^2/\varphi^2 = 2\{E - V\}$—is therefore always trivially algebraically solvable for φ.

21.4 Underlying Conformal Configuration Spaces

Conf$(\boldsymbol{\Sigma})$ are the conformal transformations on $\boldsymbol{\Sigma}$. *Conformal Riem*

$$\mathfrak{CRiem}(\boldsymbol{\Sigma}) := \mathfrak{Riem}(\boldsymbol{\Sigma})/\mathit{Conf}(\boldsymbol{\Sigma}): \tag{21.11}$$

the space of conformal equivalence classes of Riemannian metrics [237]; see Appendices D.7 and H.6 for further mathematical detail of this and the spaces mentioned below.

[2] This is $\rho = I^{1/2}$, to the GR φ being $\sqrt{h}^{1/6}$.

Because $\mathcal{H}$ provides 1 restriction per space point upon the 3 degrees of freedom per space point of *Superspace*, Wheeler asked what is "*2/3 of Superspace*?" In response, York evoked the following two geometrically natural possibilities [921, 922, 924, 925].

1) *Conformal Superspace* $\mathfrak{CS}(\boldsymbol{\Sigma})$ is geometrically well-defined as the space of all conformal 3-geometries (Appendix D.7) on a fixed $\boldsymbol{\Sigma}$.

$$\mathfrak{CS}(\boldsymbol{\Sigma}) := \mathfrak{S}\text{uperspace}(\boldsymbol{\Sigma})/\mathit{Conf}(\boldsymbol{\Sigma}) = \mathfrak{Riem}(\boldsymbol{\Sigma})/\mathit{Conf}(\boldsymbol{\Sigma}) \rtimes \mathit{Diff}(\boldsymbol{\Sigma}). \tag{21.12}$$

This corresponds to the maximal condition (21.3) being imposed.

2) $\{\mathfrak{CS} + \text{V}\}(\boldsymbol{\Sigma})$ [922] adjoins to this a solitary global degree of freedom—the spatial volume of the Universe. This corresponds to the CMC condition (21.4) being imposed.

Also bear in mind that the '2/3 of Superspace' picked out by each of these conditions might however not be directly related to the '2/3 of Superspace' picked out by $\mathcal{H}$ itself. Let us use $\mathfrak{T}$ruespace$(\boldsymbol{\Sigma})$ to denote the latter—the space of true dynamical degrees of freedom of GR, **True**—while acknowledging that for now this is but a formal naming rather than a space of known and understood geometry.

Conformal Riem has also been termed 'pointwise conformal superspace' [305] This name is however confusing in various ways. Firstly, it can only be understood if conformal superspace itself has already been introduced. Yet $\mathfrak{CRiem}(\boldsymbol{\Sigma})$ is a simpler space, and a strong case can be made for simpler entities to be introduced on their own terms rather than by reference to more complicated ones. Secondly, Appendix H.5 makes a *distinct* use of 'pointwise', to mean 'looking at a field at just one point', which is a very clear use. The current use, on the other hand, is along the following lines. 'Take a space that involves quotienting out $\mathit{Conf}(\boldsymbol{\Sigma})$ and $\mathit{Diff}(\boldsymbol{\Sigma})$—conformal superspace—but now do not quotient out $\mathit{Diff}(\boldsymbol{\Sigma})$ after all: pointwise'. This can however be contracted to just 'take a space that involves quotienting out $\mathit{Conf}(\boldsymbol{\Sigma})$', i.e. making no mention, rather than two implicit mentions which cancel, of the concept that is unnecessary for the definition [$\mathit{Diff}(\boldsymbol{\Sigma})$]. On these grounds, let us use instead the name 'conformal $\mathfrak{R}$iem'. We denote this by '$\mathfrak{CR}$iem', making use of '$\mathfrak{C}$' for 'conformal' paralleling the habitual use in '$\mathfrak{CS}$' for 'conformal superspace', noting that '$\mathfrak{C}$' standing for 'conformal' can *just as well* be introduced prior to any mention of superspace or the associated $\mathit{Diff}(\boldsymbol{\Sigma})$.

On the other hand, whereas passing to equivalence classes is mathematically convenient, the equivalence classes themselves can be considered to be more primary. If $\mathfrak{CRiem}(\boldsymbol{\Sigma})$ were viewed in this manner, it would make more sense for it and $\mathfrak{Riem}(\boldsymbol{\Sigma})$ to be renamed so that now $\mathfrak{Riem}(\boldsymbol{\Sigma})$'s new name derives from $\mathfrak{CRiem}(\boldsymbol{\Sigma})$'s by a 'locally-scaled' addendum. One might go as far as viewing $\mathfrak{CS}(\boldsymbol{\Sigma})$ as primary (for all that this is unlikely to be motivated by the final form of the 'true degrees of freedom' of the gravitational field). Such primality amounts to assuming not Geometrodynamics but *Conformogeometrodynamics*. In this case, a good primary name would be shape space, $\mathfrak{S}$hape$(\boldsymbol{\Sigma})$, for the conformal 3-geometry notion of shape. $\mathfrak{S}$uperspace$(\boldsymbol{\Sigma})$ would then be known as 'locally-scaled shape', $\mathfrak{CRiem}(\boldsymbol{\Sigma})$ as 'diffeomorphism-redundant shape', and $\mathfrak{Riem}(\boldsymbol{\Sigma})$ as 'locally-scaled diffeomorphism-redundant shape.

Moreover, the conventional approach to Conformogeometrodynamics involves solving the Lichnerowicz–York equation so as to pass finally from $\{\mathfrak{CS} + \mathrm{V}\}(\boldsymbol{\Sigma})$ to $\mathfrak{Truespace}(\boldsymbol{\Sigma})$ by the solution fixing a particular form of the local scalefactor φ to be the physically realized one. Whereas traditionally Conformogeometrodynamics is viewed as a convenient decoupling leading to substantial mathematical and numerical tractability, from the relational perspective, one can consider $\mathfrak{g} = Conf(\boldsymbol{\Sigma}) \rtimes Diff(\boldsymbol{\Sigma})$ or $VPConf(\boldsymbol{\Sigma}) \rtimes Diff(\boldsymbol{\Sigma})$. This approach is further motivated in Chap. 33.

Finally note the RPM to GR configuration space analogies in Fig. G.2, with preshape space and shape space analogous to $\mathfrak{CRiem}(\boldsymbol{\Sigma})$ and $\mathfrak{CS}(\boldsymbol{\Sigma})$. Limitations of this analogy are, firstly, that shape space plus scale is now a coincident analogue of $\mathfrak{Riem}(\boldsymbol{\Sigma})$ and $\{\mathfrak{CS} + \mathrm{V}\}(\boldsymbol{\Sigma})$. Secondly, conformal shape RPM's configuration space bears a tighter analogy with GR's $\mathfrak{CS}(\boldsymbol{\Sigma})$.

21.5 Canonical Twist and Definition of York and Euler Times

The *York time* candidate [483, 581, 586, 922, 923]

$$\mathrm{t}^{\mathrm{York}} := \frac{2}{3}\mathrm{h}_{ij}\mathrm{p}^{ij}/\sqrt{\mathrm{h}} = c(\lambda \text{ alone}), \qquad (21.13)$$

i.e. a spatial hypersurface constant for the CMC slice, is the canonical conjugate to the spatial volume element $\sqrt{\mathrm{h}}$. The *Euler time* candidate[3]

$$t^{\mathrm{Euler}} := D\text{: } \mathcal{D} \text{ interpreted as a time variable rather than as a constraint} \qquad (21.14)$$

is the canonical conjugate to the scale variable $\ln \rho$. These are conceptually similar in both being 'dilational momenta', i.e. canonical conjugates to scale variables. One can think of passing from considering maximal to CMC slices in GR as 'switching on' a York time variable, and similarly for the passage from pure Shape to Shape and Scale RPM as 'switching on' an Euler time variable.

In the GR case, the general true degrees of freedom–embedding variables splitting canonical transformation is now

$$(\mathbf{h}, \mathbf{p}) \longrightarrow \left(\vec{\mathcal{X}}, \vec{\Pi}, \mathbf{True}, \mathbf{P}^{\mathrm{True}}\right). \qquad (21.15)$$

This amounts to inverting the position and momentum statuses of the scale and conjugate dilational variables. Moreover, solving for the scale—as the conformal method is set up to do—isolates one momentum variable. In this way, an equation

[3]This is named in honour of Euler, due to the homogeneous and advective character of $D := \sum_I \underline{x}^I \cdot \underline{p}_I$.

of type (9.19) is formed, which generalizes the parabolic precursor to the standard-type time-dependent Schrödinger equation.

Additionally in this approach, the Hamiltonian constraint is replaced by $\mathrm{P}_{\mathrm{York}} = H^{\mathrm{True}} = \sqrt{\mathrm{h}} = \varphi^6$. The conformal scale factor φ is now interpreted as the solution of the conformally-transformed Hamiltonian constraint alias Lichnerowicz–York equation (21.7).

Viewing York time formulations as Problem of Time resolutions is however in practice hampered by, firstly, the Lichnerowicz–York equation not being explicitly solvable. Because of this, York time is in practice almost never explicitly known. Section 21.3's model arenas can, moreover, be used to investigate further features of this approach in special cases for which the time candidate is explicitly known. Secondly, the canonical transformation involved is also hard to perform in practice, and additionally suffers from the global problem outlined in Sect. 37.4.

Moreover, the York and Euler time candidates are not *exact* analogues of each other. Further consideration of this reveals that the GR and RPM families of scale variables amount to a nontrivial source of variety, since different choices lead to different formulae for conjugate dilational momenta. In turn, this affects the Internal Time Approach's procedure for solving $\mathcal{C}$hronos.

For instance, for the $f(\rho)$ family of RPM scale variables, the definition of conjugacy identifies the corresponding dilational quantities to be

$$\{f, \mathcal{D}/L_{\mathrm{D}}f\} = 1, \tag{21.16}$$

for L_{D} the linear dilational operator $\rho\,\partial_\rho$.

Example 1) It is then $\ln\rho$ which is conjugate to the Euler time candidate itself. In making the momentum conjugate to the dilational time the subject of $\mathcal{Q}$uad, the Euler time candidate produces a logarithmic 'true Hamiltonian':

$$-p_{t^{\mathrm{Euler}}} = \ln\big(F\big(p^{\mathrm{S}}_{\mathrm{a}}, S^{\mathrm{a}}, t^{\mathrm{Euler}}\big)\big). \tag{21.17}$$

For free scaled 3-stop metroland, recast $\mathcal{E}$ firstly in terms of $\ln\rho$ and $\mathcal{D}$ and then in terms of $t^{\mathrm{Euler}} = \mathcal{D}$ and its conjugate $p_{t^{\mathrm{Euler}}} = -\ln\rho$ to obtain

$$t^{\mathrm{Euler}\,2} + p_\varphi^2 = 2\,E\exp(-2\,p_{t^{\mathrm{Euler}}}). \tag{21.18}$$

Solving this equation for $p_{t^{\mathrm{Euler}}}$ gives

$$p_{t^{\mathrm{Euler}}} = -\frac{1}{2}\ln\left(\frac{t^{\mathrm{Euler}\,2} + p_\varphi^2}{2\,E}\right). \tag{21.19}$$

As we shall see in Part III, this ln is moreover quantum-mechanically inconvenient, but we can exploit the above diversity of f to replace ln with a more benevolent function.

Example 2) For instance, $f = \rho$—configuration space radius itself—has conjugate $\mathcal{D}/\rho = p_\rho$, i.e. just the radial momentum, thus furnishing the *radial dilational time candidate*, t_ρ.

Example 3) $f = 1/\rho := \upsilon$—the *reciprocal radius*—turns out to be even more useful (compare using $u = 1/r$ in Mechanics). The conjugate is now $-\rho\,\mathcal{D}$: the *reciprocal radius dilational time candidate*, t_υ. Then the free scaled 3-stop metroland $\mathcal{E}$ is

$$t_\upsilon^2 p_{t_\upsilon}^4 + p_\varphi^2 p_{t_\upsilon}^2 - 2\,E = 0. \tag{21.20}$$

Note that this is algebraic (as is any power function f), and moreover very straightforwardly solvable, giving which is solved by

$$p_{t_\upsilon} = \pm\sqrt{\frac{-p_\varphi^2 \pm \sqrt{p_\varphi^4 + 8\,E\,t_\upsilon^2}}{2\,t_\upsilon^2}}. \tag{21.21}$$

This requires $E > 0$ and thus a same-sign monotonicity sector (see two Sections down). Also the inner sign needs to be a '+' for classical consistency, and Chap. 44 furthermore requires the outer sign to be '−' as regards approximate recovery of a close-to-conventional QM.

On the other hand, for the $f(\mathrm{h})$ of GR scale variables,

$$\{f(\mathrm{h}), 2\,G(\mathrm{h})\mathrm{p}\} = 1 \tag{21.22}$$

for $G(\mathrm{h}) = 1/L_{\mathrm{D}}f(\mathrm{h})$ and linear dilational operator $L_{\mathrm{D}} := a\,\partial_a = 6\,\mathrm{h}\,\partial_{\mathrm{h}}$. It is then clear that the York and Euler time candidates are not directly analogous, but rather just one representative each from two analogous families of dilational times.

On the other hand, almost all of the other candidate times produce algebraic equations.

21.6 Dilational Time for Nontrivial $\mathfrak{g}$

Some facet interferences here are as follows.

1) The Internal Time Approach provides a conceptual form of evolutionary canonical transformation [586], which is a potentially useful resource. However, its generating function needs to be a function of the initial and final slices' metrics in the configuration representation, by which using this requires prior resolution of the Thin Sandwich Problem for $\mathcal{M}_i$ [586]. This furthermore entails solving the Lagrangian variables formulation of $\mathcal{M}_i$ for the shift β^i (Chap. 16, or the Machian variables counterpart for $\partial\mathrm{F}^i$, as per Chap. 18).
2) One may however have separate reasons to treat a different type of formulation of $\mathcal{M}_i$. Indeed, in the Conformal Approach to the GR initial value problem, it is technically preferred to solve $\mathcal{M}_i$ for the longitudinal potential ζ^i part of K_{ij} (or of the Hamiltonian variables formulation's p^{ij}). Here (e.g. in the K_{ij} version), $\mathrm{K}_{\mathrm{T}}^{\mu\nu} = \mathrm{K}_{\mathrm{TT}}^{ij} + \{\mathcal{L}\zeta\}^{ij}$, such that $\mathcal{D}_\mu \mathrm{K}_{\mathrm{TT}}^{ij} = 0$ (transverse), and where $\mathcal{L}_i$ is the conformal Killing operator: the traceless version of the Killing operator (see Appendix E.3).

1) and 2) amount to distinct formal RTQ schemes. For 2), a single scalar $\mathrm{t}^{\mathrm{York}}$ is found by solving the Lichnerowicz–York equation with the value of ζ^i from priorly solving $\mathcal{M}_i$ substituted into it. Formally, moreover,

$$\mathrm{P}_{\mathrm{t}^{\mathrm{York}}} = \varphi = \varphi\big(\widetilde{\underline{x}};\, \mathrm{t}^{\mathrm{York}}, \mathbf{True}, \mathbf{\Pi}^{\mathrm{True}}\big] = -\widetilde{\mathcal{H}_{\mathrm{True}}}. \tag{21.23}$$

In this case, the **True** arise from the conformal 3-geometry $\boldsymbol{C}$ by the standard interpretation that solving for φ breaks the conformal symmetry. This sends one from the $\mathfrak{CS}(\boldsymbol{\Sigma})$ 2/3 of $\mathfrak{S}$uperspace($\boldsymbol{\Sigma}$) to the distinct 2/3, $\mathfrak{T}$ruespace($\boldsymbol{\Sigma}$). $\widetilde{x}$ expresses local dependence in the $\mathfrak{g}$ reduced configuration space itself. Standard Quantization formally now yields the reduced York time-dependent Schrödinger equation (44.4).

3) Perhaps instead one is to use a particular case of the spacetime-vector parabolic form

$$\mathrm{P}^{\mathcal{X}}_{\mu} = -\mathcal{H}^{\mathrm{True}}_{\mu}\big(\widetilde{\underline{x}};\, \mathcal{X}^{\nu}, \mathrm{Q}^{\mathsf{O}}, \mathrm{P}_{\mathsf{O}}\big]. \tag{21.24}$$

This is a time-and-frame generalization of the single-time parabolic form (20.6). Here, $\mathsf{P}^{\mathcal{X}}_{\mu}$ are the momenta conjugate to 4 candidate embedding variables[4] $\mathcal{X}^{\mu}$, which form the 4-vector $[\mathrm{t}^{\mathrm{True}}, \mathbf{X}^i]$, where $\mathbf{X}^i$ are 3 spatial frame variables. $\mathcal{H}^{\mathrm{True}}_{\mu} = [\mathcal{H}^{\mathrm{True}}, \mathcal{M}^{\mathrm{True}}_i]$, for $\mathcal{M}^{\mathrm{True}}_i$ the true momentum flux constraint. Finally, the O index runs over the 'other variables'. Given such a parabolic form for $\mathcal{H}$, one can again quantize in a conceptually-standard manner. On this occasion, this yields a time-and-frame dependent Schrödinger equation of the form (44.5).

In the specific case of York time, this involves

$$\text{determining both } \varphi \text{ and } \zeta^i \text{ at the classical level.} \tag{21.25}$$

The four equations are, however, kept [rather than substituting the vector solution into the scalar solution as in 2)]. I.e. one has a system of the form

$$\mathrm{P}_{\mathrm{t}^{\mathrm{York}}} = \varphi = \varphi\big(\underline{x};\, \mathrm{t}^{\mathrm{York}}, \vec{X}, \mathbf{True}, \mathbf{\Pi}^{\mathrm{True}}, \underline{\zeta}\big] = -\mathcal{H}_{\mathrm{True}}, \tag{21.26}$$

$$\mathrm{P}^{X^{\mathrm{York}}}_{i} = \zeta_i = \zeta_i\big(\underline{x};\, \mathrm{t}^{\mathrm{York}}, \vec{X}, \mathbf{True}, \mathbf{\Pi}^{\mathrm{True}}\big] = -\mathcal{M}_{\mathrm{True}}. \tag{21.27}$$

This passes to the quantum equations (44.6), (44.7); it is at this level then that Configurational Relationalism is confronted in this approach. In this scheme, (21.25) is a determination of (four component) time (alias time-and-frame), rather than the role of a reduction, by which one has a TQR scheme.

Example 1) The time-and-frame part-linear form occurs for 'parametrized Field Theory' model arenas [586], which are the traditional model arenas for this approach.

[4]Chapter 31 provides further motivation for these from a spacetime primality perspective.

Example 2) For the r-formulation of 2-d RPM, one *can* reduce out the corresponding linear constraint $\mathcal{L}$. [This is due to the absolute–relative split of RPM being merely algebraic and thus simpler that the longitudinal–transverse split of GR.] The Lichnerowicz–York analogue is then (21.10). In a fair number of cases, this can be solved algebraically for $\varphi(=\rho)$ to be interpreted as $p_{t^{\mathrm{dil}}}$, which is the momentum conjugate to $t^{\mathrm{dil}}(=-p_\rho)$.

21.7 Monotonicity of Dilational Times

For both RPM and GR, it is tempting to use the singled-out scale as a time variable, however in each case one runs into monotonicity problems. These are often avoided by using as times the quantities conjugate to (a function of) the scale, such as York or Euler times, as follows.

For the Euler time candidate of RPM, $\mathcal{D} = t^{\mathrm{Euler}}$ [37] is tied to a number of substantial cases of Dynamics (see below) from the *Lagrange–Jacobi identity* alias *virial equation*

$$\dot{t}^{\mathrm{Euler}} = \ddot{I}/2 = 2\,T - \mathrm{n}\,V = 2\,E - \{\mathrm{n}+2\}V. \tag{21.28}$$

Here $\dot{}$ is viewed as $\mathrm{d}/\mathrm{d}t^{\mathrm{Newton}}$ in this context), for V homogeneous of degree n. Sums of homogeneous potentials all of which obey a common index inequality also satisfy monotonicity. Interpret n in this way from now on. This provides a fairly strong guarantee that t^{Euler} is monotonic: it is so in a number substantial sectors: $\{E \geq 0, V \geq 0, \mathrm{n} \leq -2\}$ and $\{E \geq 0, V \leq 0, \mathrm{n} \geq -2\}$ give $t^{\mathrm{Euler}*} \geq 0$, and $\{E \leq 0, V \leq 0, \mathrm{n} \leq -2\}$ and $\{E \leq 0, V \geq 0, \mathrm{n} \geq -2\}$ give, using $-t^{\mathrm{Euler}}$ as timefunction instead, $-t^{\mathrm{Euler}*} \geq 0$.

As regards whether other dilational time candidates also pass muster in this respect, we generalize (21.28) to

$$\dot{t}_F = \dot{\{}G\dot{I}\} = \ddot{I}G(I) + G'\{\dot{I}\}^2 = 2\big\{2\,E - \{\mathrm{n}+2\}V\big\}G + 4\,\mathcal{D}^2\mathrm{G}', \tag{21.29}$$

for $G(\mathrm{I}) = 1/2L_{\mathrm{D}}F$. So if G and G' are the same sign, there is monotonicity in the first two sectors above, and if they are of opposite signs, there is monotonicity in the other two.

In the case of e.g. $\Lambda = 0$ vacuum GR, maximal slicing is maintained if the lapse solves the maximal lapse fixing equation (LFE)

$$\Delta_{\mathbf{h}}\alpha = \alpha\mathcal{R} \tag{21.30}$$

albeit this is readily shown to be frozen for compact spatial topology. On the other hand, CMC slicing is maintained if the lapse solves the CMC LFE

$$2\{\alpha\mathcal{R} - \Delta_{\mathbf{h}}\alpha\} + \alpha\mathrm{p}^2/2\,\mathrm{h} = \delta_\beta\{\pi/\sqrt{\mathrm{h}}\}. \tag{21.31}$$

In the case of GR, the simplest scale variable—in the sense of having the simplest dilational conjugate π—is $2\ln a = \Omega$: the Misner variable. Moreover, upon canonically transforming according to $\mathrm{t}^{\mathrm{ss}} = p$, $p_{\mathrm{t}^{\mathrm{ss}}} = -\Omega$, it has the simplest propagation equation,

$$\delta_{\vec{\beta}}\mathrm{t}^{\mathrm{ss}} = 2\sqrt{\mathrm{h}}\{\alpha\mathcal{R}(\mathbf{x};\mathbf{h}] - \Delta\alpha\}. \tag{21.32}$$

This is an equation in the double time derivative of the scale variable a to (21.28) being that of I. (21.32) is the trace of the GR evolution equations, so the cleanest identification of the analogy is between (20.3) and (21.28). These are the dilational time candidate's propagation equations that correspond to each theory's subsequently simplest scale. Each of these constitutes a guarantee of monotonicity for certain cases. A simple such case for GR is closed Minisuperspace (for which scale variables themselves fail to be monotonic):

$$\dot{t}^{\mathrm{ss}} = 2\sqrt{h}\,\alpha\, R(\boldsymbol{h}) > 0 \tag{21.33}$$

(since $\alpha > 0$ by the definition of the lapse, $\sqrt{h} > 0$ by nondegeneracy and $R > 0$ for such closed models.

Furthermore, upon passing by canonical transformation to $t^{\mathrm{dil}} = 2\,G(h)p$, $p_{\mathrm{t}^{\mathrm{dil}}} = -f(h)$, one has the generalized dilational time candidate's propagation equation

$$\delta_{\vec{\beta}}\mathrm{t}^{\mathrm{dil}} = 2\sqrt{\mathrm{h}}\{\mathcal{R} - \{\Delta\alpha\}/\alpha\}\mathrm{G} - \mathrm{G}'\mathrm{p}^2/\sqrt{\mathrm{h}}; \tag{21.34}$$

cf. (21.28) for a further analogy. This retains monotonicity in the above closed Minisuperspace context if G and G' are of opposite signs (i.e. like in Sectors 3 and 4 of the mechanical counterpart). For $f(\mathrm{h}) = \mathrm{h}^k$, $k > 0$ guarantees this.

$f(\mathrm{h}) = \mathrm{h}^{1/2}$ is a notable subcase, for which $\mathrm{t}_{\mathrm{dil}} = \mathrm{t}^{\mathrm{York}}$, and the propagation equation is

$$\delta_{\vec{\beta}}\mathrm{t}^{\mathrm{York}} = \frac{4}{3}\left\{\mathcal{R} - \frac{\Delta\alpha}{\alpha} + \frac{\mathrm{p}^2}{4\sqrt{\mathrm{h}}}\right\}. \tag{21.35}$$

$\mathrm{t}^{\mathrm{York}}$ is, furthermore, known to have better monotonicity guarantees than in just the above closed Minisuperspace example [922]. [This falls, rather, upon the conjugate to the Misner variable.] Let us end by noting that this does *not* in general 'march in step with' t^{em}.

Chapter 22
Matter Times

22.1 Straightforward Matter Time

In Minisuperspace examples that include 1-component scalar matter, it is typical to simply isolate the corresponding momentum to play the role of the time part of the resulting time-dependent quantum wave equation. This is often taken to be 'the' alternative to scale time. However, the momenta conjugate to each of these represent two *further* possibilities. Furthermore, if canonical transformations are allowed, a whole further host of possibilities become apparent.

Using a matter scalar field as a time is often argued to be 'relational' [153]. However, this is clearly only the 'any change' sense of Machian recovery of time, so more scrutiny is due. Problems with straightforward matter time are as follows.

Problem 1) Few such models have been checked as regards their matter time candidate indeed possessing the features expected of a time.

Problem 2) In the extension to multi-component matter, why should one matter species be given the privilege of constituting the time? [I.e. is a nonuniqueness and a lack of sufficient reason.]

Problem 3) Does the Multiple Choice Problem make an appearance?

22.2 Reference-Fluid Matter Time

Here the part-linear form (20.6) is attained by use of reference fluid matter: $\mathrm{t}^{\mathrm{ante}} = \mathrm{t}^{\mathrm{matter}}$ and $\mathrm{Q}^{\mathrm{O}} = \mathrm{h}_{ij}$. This arises by extending the geometrodynamical set of variables to include matter variables coupled to these, which go on to serve as labels for spacetime events [586]. While the reference fluid matter fields are occasionally taken to be among the matter fields habitually used to model Nature, they are quite often taken to be extra fields, including instances of undetectable such. In this approach, one subsequently passes to the corresponding form of time-dependent Schrödinger equation, (44.1) or (44.5).

E. Anderson, *The Problem of Time*, Fundamental Theories of Physics 190,
DOI 10.1007/978-3-319-58848-3_22

Examples 1) and 2) are Kuchař and Torre's work on Gaussian reference fluid [586, 592] and on the reference fluid which corresponds to the harmonic gauge [593] (see also [141, 457] for null dust).

Example 3) An undetermined cosmological constant Λ as a type of reference fluid [384, 585, 862]: unimodular time.

Examples 4) to 6) A further type of Matter Time Approach by Kuchař and physicists J. David Brown, Joseph Romano and Don Marolf involves additionally forming a quadratic combination of constraints, e.g. for dust [175], more general perfect fluids [176], and massless scalar fields [590]. The point of such quadratic combinations is that they result in strongly vanishing Poisson brackets.

Example 7) Husain and Tomasz Pawlowski's more recent work [462, 463] is an example of newer such approaches using both further types or formulations of matter (now a specifically irrotational dust modification of [175]) and GR in loop form.

Problem 1) The idea of appending matter in order to have a time runs further contrary to Relationalism than hidden time due to its externalness which parallels absolute time. This applies even more so in cases in which this matter is undetectable or has no further physical function.

Problem 2) Internal and Matter Time Approaches are not aligned with Temporal Relationalism, in the sense that the time in use is a priori taken to exist in general at the classical level: *Tempus Ante Quantum* [483]). It is furthermore not in line with the 'All Change' or STLRC positions on Mach's Time Principle, since it involves using one particular change as the time for all the other changes. I.e. in this approach, the gravitational field and non-reference matter changes have no opportunity to contribute to the timestandard. Moreover, Matter Time Approaches themselves are typically additionally suspect due to issues of intangibility. Such approaches usually evoke 'reference fluids', which often have unphysical properties. [The intangibility might be taken as cover for the unphysicality, but is itself a conceptually suspect way of handling the Problem of Time along the lines suggested in this book.]

Problem 3) At least with the earlier examples [24, 483, 586], Kuchař noted that these involved having to choose between deficient notions of time and unphysical matter. I.e. matter for which either or both of violation of physical energy conditions [874] and intangibility apply. This was not however tied to any kind of No-Go Theorem and indeed Husain and Pawlowski's subsequent examples avoided this issue [462, 463]. A further advantage of their approach is that its 'true Hamiltonian' is technically simple (free from the roots which plague many other 'true Hamiltonians').

Research Project 8) Further assess the Husain–Pawlowski matter time [462, 463] from a technical point of view. Also gain a systematic understanding of which families of matter time candidates are and are not physically satisfactory.

Chapter 23
Classical Machian Emergent Time

23.1 Critique of the Previous Three Chapters' Notions of Time

These can be viewed as the most conservative family of strategies for the Frozen Formalism Problem [483, 855]. This is in the sense of seeking as soon as possible—at the classical level—among the apparent variables in one's theory for a substitute for absolute Newtonian time, or its also absolute SR replacement. This is subsequently to take over as many of absolute time's roles as possible in absolute Classical and Quantum Theory. This renders all of these roles straightforwardly conceptually resolved. Classical Machian emergent time is also obtained at the outset. This approach, however, supplants absolute time by a relational concept which itself *lives on* in an unadulterated form in more advanced Background Independent theories such as GR.

The *Tempus Ante Quantum* Approaches form an interesting counterpoint, due to being extraneous time formulations which are related by canonical transformation to formulations with no apparent such extraneous time. So renaming an apparent heterogeneous entity may not always be the 'right' answer, particularly if canonical transformations are—as standardly—allowed.

However, in seeking among the apparent variables, these *Tempus Ante Quantum* Approaches entail selecting particular variables to be those that provide 'the' time. In contrast, in the Emergent Machian Time Approach, all changes have the opportunity to contribute to the most dynamically significant notion of time.

Monotonicity and non-frozenness considerations indicate that Machian emergent time's applicability is wider than that of hidden dilational Euler time. E.g. emergent time also exists for scale-invariant models [102], characterized by $\mathcal{D} = 0$, so the Euler quantity is frozen and thus unavailable as a timefunction. Some portions of Newtonian Mechanics have monotonicity for t^{Euler} globally guaranteed. However solutions outside this portion may still possess intervals on which t^{Euler} is monotonic.

In forming time-dependent Hamiltonians, a dissipation problem arises. Time-dependent Hamiltonians are usually for subsystems that exchange energy with some

E. Anderson, *The Problem of Time*, Fundamental Theories of Physics 190,
DOI 10.1007/978-3-319-58848-3_23

other part of the Universe. On the other hand, such as (20.6) or (21.24) are whole-universe equations, causing a suspicion that dissipational mathematics is being taken outside of its physically meaningful context.

A final issue is operationality: it is highly questionable whether Tempus Ante Approaches' candidate timefunctions would actually be read off by clocks.

Classical Machian emergent time having won out, we give some comments on it and prepare its heavy–light (h–l) split for use in cosmological modelling.

23.2 Time Transformations in the Relational Approach

This involves parageodesic principle split conformal transformations (PPSCTs). This began with Misner's parageodesic formulation of Minisuperspace. The Author [22] subsequently pointed out that PPSCTs can be seen as freedom that arises in formulating relational actions. PPSCTs—laid out in Appendix L.11—underlie both the classical insights below and relational 'zeroth principles' for Misner's choice of conformal operator ordering at the quantum level (Chap. 40).

The homothetic subcase of PPSCT implements the freedom of choice of time-scale $t^{em} \longrightarrow k^2 t^{em} = t_k^{em}$, so $t_k^{em} - t_k^{em}(0) =: \mathbf{t}_k^{em} = k^2 \int \mathrm{d}s/\sqrt{2\,\mathcal{W}}$ for $t_k^{em}(0) := k^2 t^{em}(0)$.

Moreover, working through how the scaling of $\mathbf{M}$, $\mathcal{W}$ and the timefunction t conspire to cancel out at the level of the classical equations of motion reveals some interesting inter-connections. This is how one accounts for, within the Relational Approach, the non-affine transformations exhibited by Newtonian time and by the geodesic equation which enters GR's Einsteinian Paradigm. Individually, both conformal transformation and non-affine parametrization [814, 874] complicate the equations of motion (see Appendix D.2 and Ex III.11.a). Nevertheless, the equations of motion can be arranged to be preserved when both of these transformations are applied together ([22] and Ex III.11.b). In this way, in the Relational Approach, non-affine reparametrizability of time can be considered to be a consequence of a very simple property of the form of the relational action.

Affine parametrization transformations which send t_{old} to $t_{new}(t_{old})$ have the following properties.

I) Nonfreezing and monotonicity, so $\mathrm{d}t_{new}/\mathrm{d}t_{old} > 0$ which can be encoded by having it be a square of a quantity f with no zeros in the region of use.
II) This derivative, and so f, is a physically-reasonable function (to stop the change of timefunction unduly affecting the study of the motion). However, this can be recast as $\mathrm{d}/\mathrm{d}t_{new} = f^{-2}\mathrm{d}/\mathrm{d}t_{old}$, by which (with other properties matching)[1] one is free to identify this f with Ω, so any affine transformation is of a form

[1]However, conventional affine transformations are rather less smooth—C^1—than is usually assumed of conformal transformations: C^∞. In the current case, the physics involves affine transformations, which then turn out to be modelled here by conformal transformations. This motivates considering versions of the conformal transformations which are rougher than usual.

that extends to a PPSCT. If one chooses to 'complete' it to a '3-part conformal transformation' (Appendix L.11), the above calculation can be interpreted as the extra non-affine term being traded for a T term. This is by having an accompanying conformal transformation of the kinetic metric, which is then traded for $\partial^A \mathcal{W}$ by energy conservation and the compensating transformation of $\mathcal{W}$. So the freedom to affinely transform the geodesic equation on $\mathfrak{q}$ can be viewed instead as the freedom to apply a PPSCT to the system's equation of motion. *The Relational Approach's simplicity notion for equations of motion thus has the same mathematical content as prescribing an affine rather than non-affine parameter for the geodesic equation on* $\mathfrak{q}$. Thus the PPSCT-related $\vec{\mathrm{t}}$—defined in footnote 3—corresponds to 'the set of (generally) nonaffine parameters for the geodesic-like equation of motion on $\mathfrak{q}$'. (However, each is paired with a different, conformally-related $\mathbf{M}$ and $\mathcal{W}$). On, the other hand, the affine geodesic choice of emergent time function indeed remains identified with the much more restricted set of affine parameters for the geodesic equation on $\mathfrak{q}$.

Indeed, if t^{em} is a Machian emergent time, so are all the times related to it by PPSCTs. Moreover, the evolution equations that follow have a choice between two generally distinct simplifying cases.

A) use of the mechanically-natural emergent time.
B) use of geodesic rather than parageodesic form, the former corresponding to 'the dynamical curve being an affinely-parametrized geodesic on $\mathfrak{q}$'. This case has the affine geodesic choice of emergent time.

Here by 'generally distinct', we mean that the conformal factor interrelating the two is a function of the potential. If a model has constant potential, then this distinction is already included in the choice of time-scale freedom.

23.3 Examples of Mass Hierarchies and Heavy–Light (h–l) Splits

This is a classical parallel of the Born–Oppenheimer split (12.1), which originates in modelling electrons separately from the much heavier nuclei. Moreover, a form of it is commonly used in both Classical and Quantum Cosmology [552] as a means of modelling small inhomogeneities in a universe approximately modelled by a scale-factor and homogeneous matter terms.

One consideration entering 'heavy–light splits' is a mass ratio $m_l/m_h = \epsilon_{hl} \ll 1$. Moreover, this assumption is not made alone; e.g. '*sharply-peaked hierarchy*' conditions—that all the h's have similar masses $\gg$ all the similar masses of the l's—also enter at this stage:

$$\max_{i',j'} |M_{i'} - M_{j'}|/M_{i'} =: \epsilon_{\Delta \mathrm{M}} \ll 1, \qquad \max_{i'',j''} |m_{i''} - m_{j''}|/m_{i''} =: \epsilon_{\Delta \mathrm{m}} \ll 1. \tag{23.1}$$

$$\frac{m_{i''}}{M_{i'}} = \frac{\frac{m_{i''}-m}{m}m + m}{\frac{M_{i'}-M}{M}M + M} \sim \epsilon_{\text{hier}}\big\{1 + O(\epsilon_{\Delta\text{M}}, \epsilon_{\Delta\text{m}})\big\}. \tag{23.2}$$

This allows for only one h–l mass ratio to feature in subsequent approximations.

Example 1) To meet the cosmological application, this Sec's particular h–l split is aligned with the scale–shape split, whether of GR cosmology or of an RPM model arena of it.

Example 2) $m_{\text{Pl}} \gg m_{\text{inflaton}}$ is a corresponding 'gravitational mass hierarchy' which is sometimes used to motivate such approximations.

Example 3) In GR Cosmology, the scalefactor of the Universe dominates over one or both of the anisotropic or inhomogeneous modes. One could furthermore consider a two-step hierarchy which models both of these at once.

The rest of this Chapter is further motivated through being a classical precursor to the Semiclassical Approach to the Problem of Time. This sets the scene for presenting the latter as a Machian emergent time as well. It also begins to point to how some of the modelling assumptions made in the quantum cosmological version are in fact questionable. This is in the sense of being qualitatively regime-dependent and therefore requiring justification rather than just being ushered in unchallenged. This is a substantial awareness to have since many such approximations sound familiar due to having uses in other physical situations. And yet the applicability of such approximations is regime-dependent and by no means guaranteed to carry over to the quantum cosmological regime for which they have now been proposed. Let us next pass to considering examples of h–l splits.

Example 1) RPM. the action is now

$$S = \sqrt{2}\int\sqrt{E_{\text{Uni}} - V_h - V_l - I}\sqrt{\text{d}h^2 + h^2\|\text{d}\boldsymbol{l}\|^2_{\boldsymbol{M}_l}}, \tag{23.3}$$

(with $\underline{B}$'s hung on the dl's in the indirectly formulated case) for

$$V_h = V_h(\text{h alone}) = V_\rho(\rho), \qquad V_l = V_l\big(l^{\text{a}}\text{ alone}\big) = V_{\text{S}}\big(S^{\text{a}}\text{ alone}\big),$$
$$I = I\big(h, l^{\text{a}}\text{ alone}\big) = I_{\rho\text{S}}\big(\rho, S^{\text{a}}\text{ alone}\big). \tag{23.4}$$

The conjugate momenta are now (with $\Gamma = iI$ and a $\underline{B}$ hung on each $*l$ in the uneliminated case and $\Gamma = \text{a}$ in the r-case)

$$P^h = *h, \qquad P^l_\Gamma = h^2 M_{\Gamma\Lambda} *l^\Lambda. \tag{23.5}$$

The classical energy constraint is now

$$\mathcal{E} := P_h^2/2 + \|\boldsymbol{P}_l\|^2_{N_l}/2h^2 + V_h + V_l + I = E_{\text{Uni}}. \tag{23.6}$$

In the uneliminated case, this is accompanied by

$$\underline{\mathcal{L}}_l = \sum_{\text{a}=1}^{nd-1} \underline{l}^{\text{a}} \times \underline{P}^l_{\text{a}}. \tag{23.7}$$

The evolution equations are [in the same notation as Eq. (23.5)]

$$*P^h = h\|*l\|^2_{M_l} - \partial\{V_h + I\}/\partial h, \qquad *P^l_\Gamma = h^2 M_{\Lambda\Sigma,\Gamma} *l^\Lambda *l^\Sigma - \partial\{V_l + I\}/\partial l^\Gamma. \tag{23.8}$$

Here $\boldsymbol{M}_l$ is the kinetic metric on $\mathfrak{L}$: the configuration space of the light degrees of freedom. We can treat (23.6) in Lagrangian form as an equation for t^{em}_0 itself. In the current classical setting, this is coupled to the l-equations of motion; moreover we need the h-equation to judge which terms to keep. For more than one h degree of freedom, these have separate physical content. The system is in general composed of the E-equation, $k_h - 1$ h-evolution equations and k_l l-evolution equations.

The expression for the t^{em} candidate is now (with the $\underline{B}$'s and extremization thereover *absent* in the eliminated case)

$$CR(t^{\mathrm{em}}) = \mathrm{E}'_{\mathrm{d}\underline{B} \in Rot(d)} \int \sqrt{\{\mathrm{d}h^2 + h^2\|\mathrm{d}_{\underline{B}}\boldsymbol{l}\|^2_{\boldsymbol{M}_l}\}/2\{E_{\mathrm{Uni}} - V_h - V_l - I\}}. \tag{23.9}$$

[This feature carries over to GR as well, through $\mathrm{h}_{ab} =: a^2\mathrm{u}_{ab}$ leading to $\{\mathrm{d} - \pounds_{\mathrm{d}\underline{\mathrm{F}}}\}\{a^2\mathrm{u}_{ab}\} = \mathrm{a}^2\{\frac{\mathrm{da}}{a}\mathrm{u}_{ab} + \mathrm{du}_{ab} - \mathcal{D}^{\mathrm{u}}{}_{(a}\mathrm{dF}_{b)} + 0\} = \mathrm{a}^2\{\mathrm{d} - \pounds_{\mathrm{d}\underline{\mathrm{F}}}\}\mathrm{u}_{ab}$, where the 0 arises from the constancy in space of the scalefactor-as-conformal-factor killing off the extra conformal connection. Here $\mathcal{D}^{\mathrm{u}}_a$ is the covariant derivative associated with the scale-free metric u_{ab}.] By this observation, scale–shape split approximate t^{em}—and its approximate semiclassical counterpart which coincides with it—avoids the Best Matching Problem.

The h-approximation to the action (23.3) is[2] $S = \sqrt{2}\int\sqrt{\{E_h - V_h\}}\mathrm{d}h$. The conjugate momenta are then $P^h = *^h h$, the quadratic energy constraint is $\mathcal{E}_h := P^{h\,2}/2 + V_h = E_h$ and the evolution equations are $*^h P^h = -\partial V_h/\partial h$. This assumes that (using the subscript j to denote 'judging')

$$\text{(ratio of force terms)}, \quad \mathrm{F_j} := \frac{\partial I}{\partial h}\Big/\frac{\partial V_h}{\partial h} = \frac{\partial I}{\partial S}\Big/\frac{\partial V_\mathrm{S}}{\partial S},$$

$$\text{is of magnitude} \quad \epsilon_{\mathrm{sds-1j}} \ll 1, \tag{23.10}$$

$$\text{(ratio of geometrical terms)}, \quad \mathrm{G_j} := h\frac{\|\mathrm{d}\boldsymbol{l}\|^2_{\boldsymbol{M}}}{\mathrm{d}^2 h} = \rho\frac{\|\mathrm{d}\boldsymbol{S}\|^2_{\boldsymbol{M}}}{\mathrm{d}^2\rho},$$

$$\text{is of magnitude} \quad \epsilon_{\mathrm{sds-2j}} \ll 1. \tag{23.11}$$

The Author originally considered a 'scale dominates shape' approximation [37] at the level of the action, which is most clearly formulated as

$$\mathrm{F} := I/W_h = I/W_\rho, \quad \text{is of magnitude} \quad \epsilon_{\mathrm{sds-1}} \ll 1, \tag{23.12}$$

---

[2] E_h is only approximately equal to E_{Uni} since the h and l subsystems can exchange energy. $*^h := \partial/\partial t^{\mathrm{em(JBB)}}_h$. Finally, 0 and 1 subscripts denote zeroth and first-order approximations.

$$\text{G} := \frac{\|\text{d}_{\underline{B}}\boldsymbol{l}\|_M}{\text{d}\ln h} = \frac{\|\text{d}_{\underline{B}}\boldsymbol{S}\|_M}{\text{d}\ln \rho}, \quad \text{is of magnitude} \quad \epsilon_{\text{sds}-2} \ll 1. \tag{23.13}$$

Each pair—i.e. 1, 1j, and 2, 2j—are dimensionally the same but differ in further detail. However, further consideration reveals that this assumption is better justified if judged at the level of the equations of motion and thus of forces. An example of this is how the effect of Andromeda on the Solar System is not negligible at the level of the potential, but it is at the level of the tidal forces since these contain an extra two powers of 1/(distance to Andromeda).

As usual, the quadratic constraint can be taken as an equation for t^{em} via the momentum–velocity relation. The approximate emergent time candidate is

$$t^{\text{em}}_h = \int \text{d}h_0 \Big/ \sqrt{2\{E_h - V_{h_0}\}}, \tag{23.14}$$

which is of the general form

$$t^{\text{em}}_h = F[h, \text{d}h]. \tag{23.15}$$

N.B. that for this split and to this level of approximation, there is no $\mathfrak{g}$-correction to carry out. This is because the rotations act solely on the shapes and not on the scale. In other words Configurational Relationalism is trivial here.

Finally, the first approximation to the l-equations is

$$P^l_{\mathsf{a}} = h^2 M_{\mathsf{ab}} *^h l^{\mathsf{b}} * P^l_{\mathsf{a}} = h^2 M_{\mathsf{ab,c}} * l^{\mathsf{b}} * l^{\mathsf{c}} - \partial\{V_l + I\}/\partial l^{\mathsf{a}}. \tag{23.16}$$

N.B. Whenever disagreement with experiment arises, going back to the previous Machian emergent time formulation should be perceived as a possible option. Early 20th century 'anomalous lunar motions' are an archetype for this, as per de Sitter's comment given in Sect. 3.3; see also [168] in this regard.

Moreover, pure-h expressions of the general form (23.15) are unsatisfactory from a Machian perspective since they do not give l-change an opportunity to contribute. This deficiency is to be resolved by treating them as zeroth-order approximations in an expansion involving the l-physics as well. Expanding (23.9), one obtains an expression of the form (compare (23.15))

$$t^{\text{em}}_{\mathfrak{g}\text{-free 1}} = \mathcal{F}[h, l, \text{d}h, \text{d}l]. \tag{23.17}$$

More specifically,

$$t^{\text{em}}_{\mathfrak{g}\text{-free 1}} = t^{\text{em}}_0 + \frac{1}{2\sqrt{2}} \int \frac{d\rho}{W_\rho^{1/2}} \left\{ \frac{I_{\rho S}}{W_\rho} + \left\{ \frac{\text{d}S}{2\,\text{d}\ln\rho} \right\}^2 \right\}$$
$$+ O\left(\left\{ \frac{I_{\rho S}}{W_\rho} \right\}^2 + \left\{ \frac{\text{d}S}{\text{d}\ln\rho} \right\}^4 \right), \tag{23.18}$$

so one has an interaction term and an l-change term.

For comparison with the Semiclassical Approach, take note of classical adiabatic terms, which are related to the order of magnitude estimate

$$\epsilon_{\text{Ad}} := \omega_h / \omega_l = t_l / t_h, \tag{23.19}$$

for ω_h and ω_l 'characteristic frequencies' of the h and l subsystems respectively.

See [29] for a perturbative scheme for this model.

Example 2) Minisuperspace. We next consider a classical Cosmology analogue of the astronomers' ephemeris time procedure; in Part III we extend this to Semiclassical Quantum Cosmology as well. Suppose that an accurate enough time has been found for one's purposes. One can then consider an analysis in terms of Q^{l}, P_{l} as regards which features within that Universe contribute relevant change to the time-standard. This is very much expected to cover all uses of quantum perturbation theory that apply to modelling laboratory experiments. Fairly large-scale features of the Universe are expected to contribute a small amount here, in addition to the zeroth-order expansion of the Universe and homogeneous matter mode contributions. There is a limit on such ephemeris time schemes, due to their iterations being at the level of form-fitting rather than a perturbative expansion of the equations of motion themselves.

Minisuperspace models of anisotropy are one case of particular interest. For instance, diagonal Bianchi Class A models give rise to the following types of 'scale dominates anisotropic shape' correction terms: $V_{\boldsymbol{\beta}} / V_{\Omega} =: \epsilon_{\text{sds}-1} \ll 1$ (and a derivative version), and $\text{d}s_{\boldsymbol{\beta}} / \text{d}\Omega =: \epsilon_{\text{sds}-2} \ll 1$. For Bianchi IX and VII$_0$, these are both $O(\text{anisotropy})^2$, whereas for Bianchi II, VI$_0$ and VIII the latter retains a $O(\text{anisotropy})$ piece. [For Bianchi I, there is no anisotropy potential, so the second type of correction term drops out altogether.] These manifest themselves as anisotropic ephemeris time corrections to cosmic time:

$$t_1^{\text{em}} = t^{\text{cosmic}} + O\big(\text{anisotropy}^{1 \text{ or } 2}\big). \tag{23.20}$$

23.4 Problems with Classical Precursors of Assumptions Commonly Made in Semiclassical Quantum Cosmology

Classical Problem 1) Consider e.g. Newtonian Gravity or RPMs that model dust-filled GR cosmology. The corresponding regions of double collision, D, the potential has infinite abysses and peaks (Fig. 23.1's D lines are a simple example). 'Scale dominates shape' approximations are thus certainly not valid near there, so some assumptions behind the Semiclassical Approach fail in the region around these lines. So for negative powers of relative separations, the heavy approximation only makes sense in certain wedges of angle. There is also the possibility that Dynamics set up to originally run in such regions falls out from them. These considerations point to a stability analysis being required to determine whether semiclassicality is representative. I.e. there is a tension between the procedure used in Semiclassical

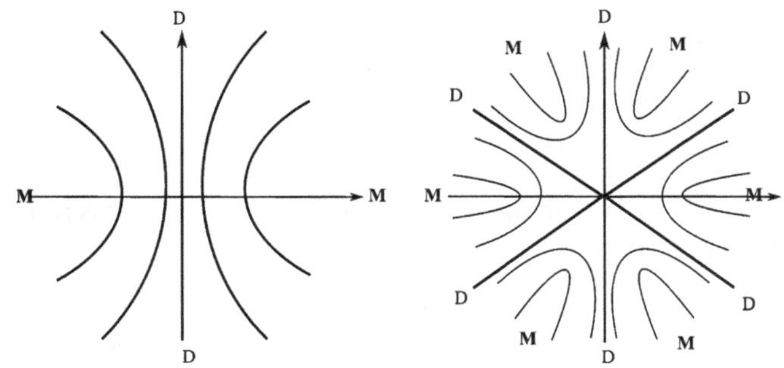

Fig. 23.1 Contours on configuration space for single and triple negative power potentials (the 1-d 3-particle case for simplicity). These have abysses along the corresponding double collision lines D and high ground in between these. (For negative-power coefficients such as for the attractive Newtonian Gravity potential.) M are merger configurations (with the third particle at the centre of mass of the other two)

Quantum Cosmology and the futility of *trying to approximate a 3-body problem by a 2-body one* [37].

Classical Problem 2) Conventional treatments so far of Semiclassical Quantum Cosmology decouples the h and l subsystems, which eases analytic solvability. As we shall see in Chap. 46, this is partly attained through neglect of the T_l term. However, the Classical Dynamics version of this (scale–shape split 1- or 2-d RPM version) involves *throwing away the central term*. I.e. the mathematical equivalent of neglecting the centrifugal barrier in the study of planetary motion. This causes unacceptable quantitative and qualitative errors (linear motion versus periodic motion along an ellipse). Furthermore, this qualitative difference indeed carries over to the RPM counterpart [37].

Research Project 9) Extend this book's local resolution of the Problem of Time to a wider range of anisotropic examples: with matter, and, especially, to cases with nondiagonal minisupermetrics.

Chapter 24
Brackets, Constraints and Closure

We now arrive at the third aspect of Background Independence: Constraint Closure (12.15). Complications and impasses with this are the corresponding third facet of the Problem of Time: the Constraint Closure Problem. The name *Functional Evolution Problem* was used by Kuchař and Isham [483, 586] in the quantum-level field-theoretic setting. Some parts of this problem, however, already occur in finite examples, for which partial rather than functional derivatives are involved. Thus the portmanteau term *Partional Evolution Problem* is more theory-independent. The further name *Constraint Closure Problem* additionally covers the classical version of the problem to some extent. At the classical level, the General Strategy for this is the Dirac Algorithm.

There are two different directions in which Constraint Closure can be taken further. On the one hand, classical-level closure can involve specifier equations as well as constraints; this reflects the full scope of possibilities in the Dirac Algorithm. Let us coin *Entity Closure* to cover this classical-level generalization of the Constraint Closure aspect; failure to attain this entails an *Entity Closure Problem* facet. On the other hand, one can pass to considering Generator Closure in situations in which group generators remain relevant but constraints do not. Closure is not a priori guaranteed because groups are determined by relations as well as generators. This provides a clearer archetype for Constraint Closure than the Dirac Algorithm, though at the classical level there are occasions in which one requires additional features of the latter. Generator Closure also covers the Timeless Approaches outlined in Chap. 14 and Chap. 27's Spacetime Relationalism. Moreover, both of these fall under the umbrella of *Equipping Objects* O *with a Brackets Structure*. We outline this in the next Section and then apply it to Constraint Closure.

Both the constraint and the spacetime generator version of this can be immediately followed up by finding a further set of objects forming zero brackets with the original set's objects: notions of observables or beables (Chaps. 25 and 27). This leads to the fourth aspect, Assignment of Beables, complications and impasses with which constitute the fourth facet: the Problem of Beables. Finally, the quantum level also involves multiple steps involving Equipping with Brackets: Kinematical Quantization (Chap. 39), quantum constraints (Chap. 49), and quantum observables or

E. Anderson, *The Problem of Time*, Fundamental Theories of Physics 190,
DOI 10.1007/978-3-319-58848-3_24

beables (Chap. 50). We consider these multiple significant uses to invite a systematic treatment of Equipping with Brackets in the Canonical Approach; Chap. 27 provides the spacetime counterpart.

24.1 General Consideration of Equipping with Brackets

Suppose that one wishes to equip a space $\mathfrak{O}$ of objects O (indexed by v) with a bracket $|[\ ,\ ]|$. A rather general possibility for the outcome of these brackets is[1]

$$|[\mathsf{O}_{\mathsf{v}}, \mathsf{O}_{\mathsf{v}'}]| = \mathsf{C}^{\mathsf{v}''}{}_{\mathsf{v}\mathsf{v}'}[\mathsf{b}_{\mathsf{u}}, c]\mathsf{O}_{\mathsf{v}''} + \Theta_{\mathsf{v}\mathsf{v}'}[\mathsf{b}_{\mathsf{u}}, c]. \tag{24.1}$$

Some significant cases are as follows.

A) $\boldsymbol{\Theta} = 0$ and $\mathbf{C} = const$ are Lie algebras (Appendix E).

B) $\boldsymbol{\Theta} = 0$ and $\mathbf{C}$ general functions are Lie algebroids (Appendix V.6, and including the case in which the $\mathbf{C}$ are operator-valued).

C) $\boldsymbol{\Theta} \neq 0$ and $\mathbf{C} = const$ admits the following further subcases.

C.1) Perhaps constants c can be fixed such that $\boldsymbol{\Theta}$ disappears. This is termed being *strongly zero*. *A*) occasionally reduces to *B*) in this manner; see Sect. 33.3 for examples.

C.2) Perhaps $\boldsymbol{\Theta}$ is an integrability that can be incorporated, by regarding $\Theta_{\mathsf{v}\mathsf{v}'}$ as $\mathsf{C}^{\mathsf{u}}{}_{\mathsf{v}\mathsf{v}'}(\mathsf{b}_{\mathsf{u}}, c)\mathsf{O}^{\mathrm{new}}_{\mathsf{u}}$ for further objects $\mathbf{O}^{\mathrm{new}}$ such that these—and any further such found recursively—can be supported by the theory in question.

C.3) Perhaps $\boldsymbol{\Theta}$ exceeds what can be supported by the theory, in which case it is a more serious *obstruction*: this kills off candidate theories, rather than just modifying them.

C.4) *Central charges* can already appear at the level of Classical Field Theory, for all that they are better-known at the quantum level in the Particle Physics literature; see Appendix V.3 for an example. $\boldsymbol{\Theta}$ here takes the form of a constant phase space functional.

C.5) $\boldsymbol{\Theta}$ can also eliminate candidates by topological means rather than by running out of degrees of freedom.

24.2 Poisson Brackets and Phase Space

As a first instance of Equipping with Brackets, consider the joint space of the $\mathbf{Q}$ and $\mathbf{P}$ alongside the classical Poisson brackets

$$\{\mathsf{F}, \mathsf{G}\} := \int_{\mathrm{NoS}} \mathrm{dNoS}\left\{\frac{\delta \mathsf{F}}{\delta \mathsf{Q}^{\mathsf{A}}}\frac{\delta \mathsf{G}}{\delta \mathsf{P}_{\mathsf{A}}} - \frac{\delta \mathsf{F}}{\delta \mathsf{P}_{\mathsf{A}}}\frac{\delta \mathsf{G}}{\delta \mathsf{Q}^{\mathsf{A}}}\right\}, \tag{24.2}$$

[1] While this remains a linear ansatz in $\mathbf{O}$, this book does not exceed this mandate. Also the $\mathbf{b}$ are 'base objects', which in this book are usually the $\mathbf{Q}$, whether or not accompanied by $\mathbf{P}$, and the c are constants.

i.e. the portmanteau of (J.24) for finite theories and (K.17) for Field Theories. This is useful because, firstly, it turns out to afford a systematic treatment of constraints, and secondly it is a preliminary step toward Quantization.

The fundamental Poisson bracket is

$$\{\mathsf{Q}^{\mathsf{A}}, \mathsf{P}_{\mathsf{A}'}\} = \delta^{\mathsf{A}}{}_{\mathsf{A}'} \tag{24.3}$$

for δ the portmanteau of the finite Kronecker δ and the product of a species-wise such with a field-theoretic Dirac $\delta^{(d)}(\underline{x} - \underline{x}')$. This bracket being established for all the $\mathbf{Q}$ and $\mathbf{P}$ means that brackets of all once-differentiable quantities $\mathscr{F}\lfloor \mathbf{Q}, \mathbf{P}\rfloor$ are established as well.[2]

Equipped in this manner, this joint space is known as *phase space*, $\mathfrak{P}$hase. The corresponding Poisson brackets preserving morphisms are the *canonical transformations*, *Can* (Appendix J.9).[3]

24.3 Lessons from the Dirac Algorithm

As a second instance of Equipping with Brackets, let the $\mathbf{O}$ be constraints c_{C}; these are already nontrivial to handle at the classical level. The $c := \mathscr{F}\lfloor \mathbf{Q}, \mathbf{P}\rfloor$: a portmanteau of $\mathrm{F}(\boldsymbol{Q}, \boldsymbol{P})$ for Finite Theories and $\mathcal{F}(\underline{x}; \mathbf{Q}, \mathbf{P}]$ for Field Theories. The combination of working in Hamiltonian variables $\mathbf{Q}$ and $\mathbf{P}$ and making use of the classical Poisson brackets turns out to allow for a *systematic* treatment of constraints: the Dirac Algorithm [250]. This is the General Strategy for addressing Constraint Closure at the classical level. Consult Appendix J.15 if you are not yet familiar with this; Appendix J.17 and Fig. 24.1 provide some additional geometrical interpretation.

---

[2]Whereas this statement may look innocuous, in Part III we shall see that its quantum counterpart is not. Moreover, classical-level considerations themselves need to justify why Poisson brackets are in use rather than e.g. i) Lagrange, ii) Peierls, iii) Schouten–Nijenhuis, iv) Nambu and associator brackets. Brief answers are as follows [32]. Poisson brackets are more convenient than i) and ii). We do not consider iii) since this corresponds to multisymplectic formulations, in which time and space are treated on an even more common footing than spacetime co-geometrization. I.e. here the usual time derivative based momenta are accompanied by spatial derivative based analogues. This is motivated by the introduction of the notion of spacetime being taken to imply necessity of joint treatment of further spatial and temporal notions. However, this motivation runs contrary to Broad's point that a formulation breaking isolation between space and time does not imply an end to the distinction between these notions. Once this is understood, the advent of spacetime clearly does not imply any necessity to replace the standard temporally-distinguished symplectic formulation with a time-and-space covariant multisymplectic one. Finally, not using iv) follows from second-order theory alongside noncommutative but associative Quantum Theory sufficing for most purposes.

[3]*Rigged phase space* $\mathfrak{R}$ig-$\mathfrak{P}$hase is a more minimalistic alternative at this stage [37]. This corresponds to specifying that the physical $\mathbf{Q}$ are distinguishable from the corresponding $\mathbf{P}$. Moreover, the nontrivial ($\mathbf{P}$ and $\mathbf{Q}$ mixing) canonical transformations do not preserve this additional structure. The corresponding morphisms are now the group *Point* of $\mathfrak{q}$-morphisms, which is much smaller than *Can*. This does not affect the type of brackets in use.

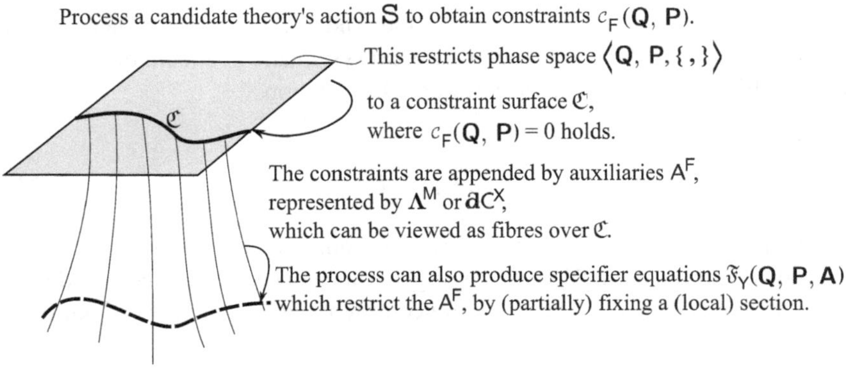

Fig. 24.1 Geometrical sketch of the outcome of Dirac-type procedures for classical Constraint Closure. We omit a loop back to readjusting some of the phase space structures—$\mathbf{Q}$, $\mathbf{P}$ and classical bracket—in the event of second-class constraints arising. While the Dirac Algorithm's Lagrange multipliers Λ^M can be incorporated by extending the $\mathbf{Q}$, the TRi-Dirac-type Algorithm's cyclic differentials $\mathfrak{a}\mathsf{c}^X$ are more heterogeneous. In any case, envisaging the auxiliary variables used in Constraint Appending as fibres keeps them separated out from the phase space which contains actual physical information. [This Figure is also eventually to be compared with the simpler assessment of Constraint Closure at the quantum level in Fig. 49.1]

We phrase this approach as starting with a *trial action* $\mathsf{S}^{\text{trial}}$ producing *trial constraints*. In cases in which Constraint Closure is completed, 'trial' names and labels are promoted to 'CC' ones, standing for 'Closure completed' as well as for the Constraint Closure aspect and facet name.

Note in particular that, while still in the process of investigating a physical theory's constraints, one does not yet know which are first-class. This is because a given constraint may close with all the constraints found so far but *not* close with some constraint still awaiting discovery. Thus one's characterization of constraints needs to be updated step by step until either of the following apply.

a) The constraint-finding procedure is complete.
b) The system under study has been demonstrated to be trivial or inconsistent. Indeed, Dirac's Algorithm is capable of rendering a candidate theory trivial by the equations it produces using up all of its degrees of freedom. Inconsistency itself arises when either even more degrees of freedom are used up, or when inconsistent equations are produced (see Counter-example 1 of Appendix J.15).

Teitelboim and physicist Marc Henneaux emphasize and largely illustrate [446] how all combinations of first and second class constraints and primary and secondary constraints are possible in whichever steps of the Dirac Algorithm.

Whereas b) is the most extreme form of Constraint Closure Problem, note that Sect. 24.4 considers milder versions involving one's candidate theory not in fact implementing an intended symmetry.

The possibility of specifier equations stems from the Dirac Algorithm involving an *appending procedure*. Let us illustrate this by the well-known case of the *total Hamiltonian*, which is formed from an incipient 'bare' Hamiltonian by adding on constraints, each multiplied by an auxiliary Lagrange multiplier coordinate. For instance, in GR, $\mathscr{H}_{\text{Total}} = \alpha\mathcal{H} + \beta^i\mathcal{M}_i$ (more strictly $+\mu p^{\alpha} + \lambda^i\pi_i^{\beta}$ for further Lagrange multipliers μ and λ^i, and π^{α}, p_i^{β} the momenta conjugate to the lapse and shift). Specifier equations can consequently appear as relations imposed upon the auxiliaries used in the appending procedure; cf. the lapse fixing equation in GR (21.30), (21.31). However, if brackets structures are imposed in ways which have no appending procedure have nothing to specify and thus no specifier equations; this is relevant to Group Theory and to imposing quantum-level brackets.

Also note the distinction between auxiliaries used for appending and smearing variables. The latter are more widely applicable since their job—'multiplication by a test function'—is to render rigorous a wider range of 'distributional' manipulations provided that these occur within an integral (Appendix P.5). In particular, this applies to classical Field Theories' Assignment of Beables (for which there is no appending procedure).

On the other hand, what we have learnt above about second-classness straightforwardly generalizes to whichever other use of bracket structures. Firstly, given a set of generators, there is no a priori guarantee that the set closes under the bracket:

$$|[\mathsf{O}_{\mathsf{V}}, \mathsf{O}_{\mathsf{V}'}]| \not\approx 0. \tag{24.4}$$

Secondly, rewrite

$$\Theta_{\mathsf{VV}'} = \mathsf{C}^{\mathsf{V}''}{}_{\mathsf{VV}'}[\mathsf{b}_{\mathsf{D}}, c]\mathsf{O}^{\text{new}}_{\mathsf{V}''} + ho_{\mathsf{VV}'}[\mathsf{b}_{\mathsf{D}}, c], \tag{24.5}$$

where the 'residue' $ho_{\mathsf{VV}'}$ consists of second-class objects alongside topological pieces (and specifier equations in the case of classical constraints; nor are these three cases necessarily distinct). Second-classness can moreover be taken to add further meaning to 3.ii)'s words: 'supported' is now meant in one of the following two senses.

α) The space of base objects $\mathfrak{BO}$ can get extended as per the effective formulation [121].

β) Alternatively, the bracket in question can get modified—in a geometrically-significant manner—to the Dirac bracket [250, 446].

In general, α) and β) are both permissible since their structures can be freed from any connotations of classical constraints, a point which enters subsequent discussion of Quantization procedures in Chap. 49.

Finally, for later convenience, we express the $\mathcal{F}$**lin** in manifestly homogeneous linear form:

$$\mathcal{F}\text{lin}_{\mathsf{N}} = \mathcal{F}[\mathbf{Q}]^{\mathsf{A}}{}_{\mathsf{N}}\mathsf{P}_{\mathsf{A}}. \tag{24.6}$$

This includes the possibility of $\mathcal{F}$ being differential operator-valued so as to accommodate Electromagnetism, Yang–Mills Theory and GR.

24.4 Some Temporal, Configurational and Closure Facet Interferences

Temporal Relationalism provides a constraint $\mathcal{C}$hronos of the form $\mathcal{Q}$uad and Configurational Relationalism implemented as Best Matching provides candidate ***shuffle*** constraints of the form $\mathcal{L}$**in**. One is then to use the Dirac Algorithm to see whether Constraint Closure is met or the Constraint Closure Problem arises. There are subsequent meaningful corresponding notions of whether each of $\mathcal{C}$hronos and the ***shuffle*** are self first-class, and whether they are mutually first-class. Moreover, the behaviour of ***shuffle*** by itself, and its interplay with $\mathcal{C}$hronos, also require scrutiny as regards whether Configurational Relationalism has succeeded. Addressing this requires a detour to $\mathfrak{g}$-specific versions of Equipping with Brackets. This composition of three facets involves dealing with multiple kinds of facet interferences, which we give individually below. We point to the useful end summary Fig. 24.2 as regards keeping track of how these various facet interferences fit together.

1) Let us first consider a joint treatment of Constraint Closure and Configurational Relationalism without any further source of constraints.
1.i) The action is now of the conceptual form $\mathsf{S}^{\text{trial}}_{\text{CR}}$.
1.ii) The ***shuffle*** constraints arise as per usual and are now put through the Dirac Algorithm. If these do not immediately close, our $\langle \mathfrak{q}, \mathfrak{g}, \mathsf{S}^{\text{trial}}_{\text{CR}} \rangle$ triple will not do. If these do close, ***shuffle*** is promoted to $\mathcal{G}$**auge** status.
1.iii) Finally Lagrange multiplier mediated Best Matching is accorded further definiteness:

$$CR(\mathsf{S}) := \mathbb{E}_{\mathsf{g} \in \mathfrak{g}}\{\mathsf{S}_{\text{TR--CR}} \text{ built upon } \mathfrak{q}, \mathfrak{g}\}, \tag{24.7}$$

where $\mathbb{E}_{\mathsf{g} \in \mathfrak{g}} = \{\text{extremum of } \mathsf{g} \in \mathfrak{g}\}$, $\mathsf{S}_{\text{TR--CR}}$ involves a suitable group action of $\mathfrak{g}$, and the whole construct gets past the Dirac Algorithm. With reference to Sect. 16.1's criteria A) to C), the last clause is an additional structural criterion D) (for 'Dirac') on top of the group action criterion A). Moreover, there is also a more stringent bound than Chap. 16.1's counting criterion C), thus supplanting it. [It manages to be a better bound through carrying out a Dirac-type algorithm determining the specific geometry of the theory's constraint surface.] Indeed, criterion D) includes C)'s worst-case scenarios: relational triviality, triviality or inconsistency as bounding cases. Furthermore, $\mathfrak{q}$—the entity taken to have some tangible physical content—has the *a posteriori right to reject* [17, 19, 109] a proposed $\mathfrak{g}$ by triviality or inconsistency. Unlike A) and C), D) depends on the choice of action S as well as of $\mathfrak{q}$ and $\mathfrak{g}$, through depending on all the constraints that this encodes. These comments apply in greater generality to schemes 3) and 5) below.

2) We next consider what it takes to have a joint treatment of Temporal Relationalism and Constraint Closure in the absence of a further source of constraints.
2.i) We now start with an action of the conceptual form $\mathsf{S}^{\text{trial}}_{\text{TR}}$.
2.ii) As usual, this provides a trial $\mathcal{C}$hronos as a primary constraint. However, one cannot just put this through the Dirac Algorithm since this is not itself TRi. So

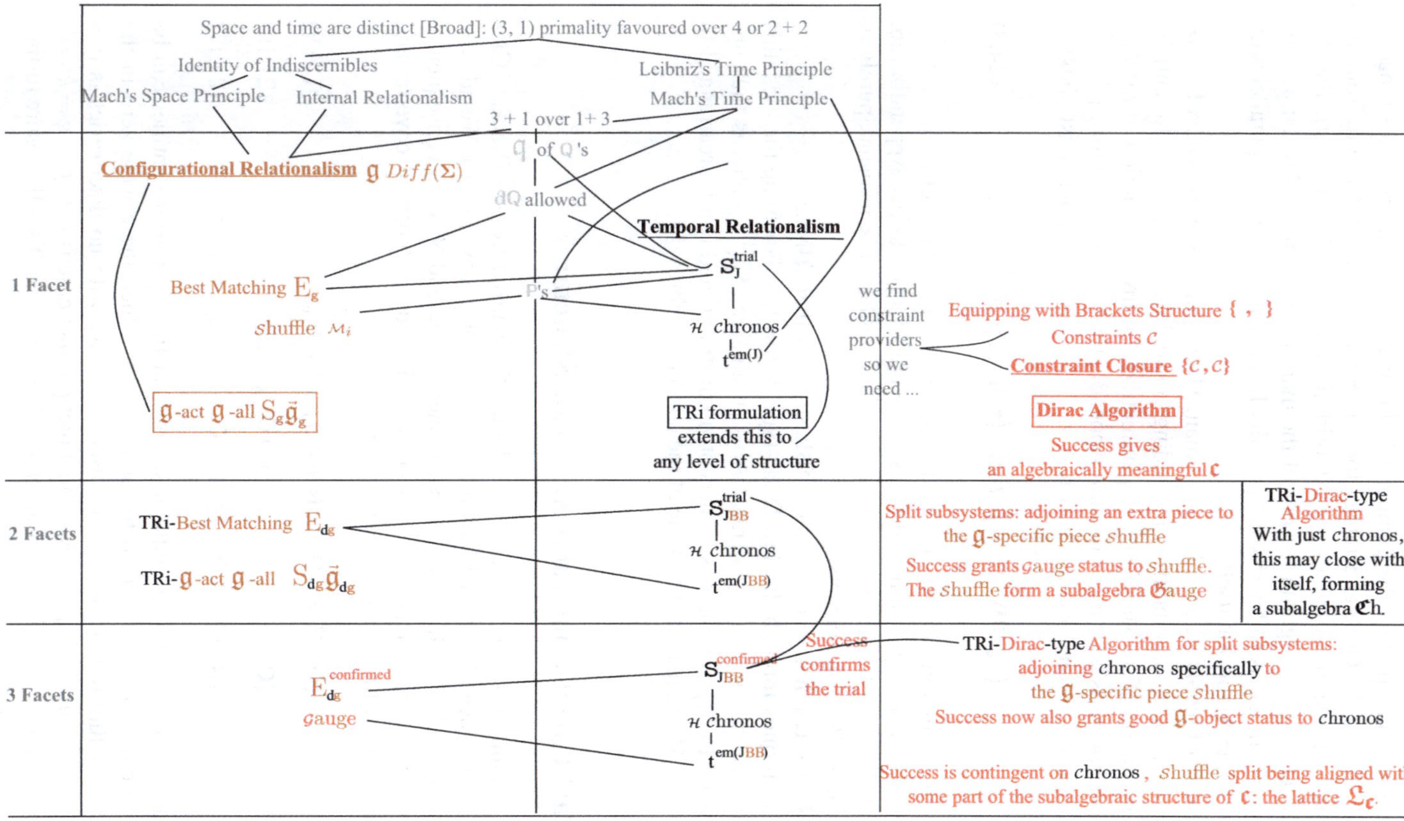

Fig. 24.2 A 'technicolour guide' to how the various facet interferences of Constraint Closure (*red*) with Temporal Relationalism (*black*) and Configurational Relationalism (*brown*) fit together

a *TRi Dirac-type Algorithm* is first required, so as to monitor Constraint Closure without spoiling the Temporal Relationalism. In this vein, we broaden our view of the General Strategy for addressing Constraint Closure from the Dirac Algorithm to Dirac-type Algorithms, thus including in particular the TRi Dirac-type Algorithm. Four of the main features of the TRi Dirac-type Algorithm are as follows; see Appendix L for a more thorough parallel with the Dirac Algorithm itself.

2.iii) Instead of involving a total Hamiltonian, TRi requires (see Appendix L) a *total* $\mathfrak{d}$*A-Hamiltonian* built by appending constraints with cyclic differentials. [The intermediate Manifestly Reparametrization Invariant notion of *total A-Hamiltonian* is built by Constraint Appending with cyclic velocities.]

2.iv) For Field Theories, we also now require *TRi-smearing* for our constraints:

$$\left(\mathcal{C}_{\mathsf{W}} \,\middle|\, \partial \mathrm{W}^{\mathsf{w}}\right) := \int d^3 z \, \mathcal{C}_{\mathsf{W}}\big(\underline{z}; \mathbf{h}, \psi, \mathbf{p}, \pi^{\Psi}\big] \, \partial \mathrm{W}^{\mathsf{w}}(\underline{z}). \tag{24.8}$$

Such expressions are then inserted inside the classical brackets.

2.v) We also now require a mixed Poisson–Peierls bracket, as per Appendix L.6. Note however that the physical part of this formulation is, as usual, purely in terms of Poisson brackets.

2.vi) TRi also requires working on $\mathfrak{d}$A-$\mathfrak{P}$hase rather than $\mathfrak{P}$hase.

2.vii) Put $\mathcal{C}$hronos through the TRi Dirac-type Algorithm. For finite theories, this trivially closes as a 1-generator Abelian algebra. $S_{\mathrm{TR}}^{\mathrm{trial}}$'s status is thus upgraded to $S_{\mathrm{TR}}^{\mathrm{CC}}$, $\mathcal{C}$hronos is an accepted constraint, and can be rearranged to form a Machian emergent time of the conceptual form $t_{\mathrm{CC}}^{\mathrm{em}}$.

24.5 Partitioned Constraint Algebraic Structures

We next attempt to *maintain* one set of objects O (indexed by u)'s Brackets Closure in the presence of a further disjoint set of them N (indexed by u). We address this by taking (24.1) with its objects partitioned into O and N. Θ is consequently split into three by the extension of (E.4)–(E.6) to include both new discoveries and obstructions, according to the schematic form

$$\{\mathsf{O}, \mathsf{O}\} = a\,\mathsf{O} + b\,\mathsf{N} + c\,\mathsf{O}_{\mathrm{new}} + d\,\mathsf{N}_{\mathrm{new}} + e, \tag{24.9}$$

$$\{\mathsf{O}, \mathsf{N}\} = f\,\mathsf{O} + g\,\mathsf{N} + h\,\mathsf{O}_{\mathrm{new}} + i\,\mathsf{N}_{\mathrm{new}} + j, \tag{24.10}$$

$$\{\mathsf{N}, \mathsf{N}\} = k\,\mathsf{O} + l\,\mathsf{N} + m\,\mathsf{O}_{\mathrm{new}} + n\,\mathsf{N}_{\mathrm{new}} + o. \tag{24.11}$$

[In all cases below, we take it without saying that further nonzero entities can be strongly removed as another path to each case for which these were zero in the first place.] Within this quite general ansatz, the case with no discoveries is $c = d = h = i = m = n = e = j = o = 0$, the direct product case is $b = f = g = k = 0$, with the orientation of semidirect product which respects the O's self-closure further allowing for $g \neq 0$.

24.6 The Remaining Temporal, Configurational and Closure Facet Interferences

3) Let us next think of the above partition as adjoining further constraints **N** to **O** $=$ ***shuffle***.

3.i) This could be considered ab initio, or resting on a Lagrangian of the form $\mathcal{L}_{CR} + \mathsf{m}^{\mathsf{u}} c_{\mathsf{u}}$ for Lagrange multipliers m^{u}.

3.ii) The ***shuffle*** are a candidate representation for $\mathfrak{g}$. This could succeed (***shuffle*** self-closure), give rise to the $C^{\mathsf{v}''}{}_{\mathsf{v}\mathsf{v}'}$ of an algebraic structure distinct from $\mathfrak{g}$, or one of C.1)–3) could occur. Moreover, now C.2) amounts to the group in question being extended away from $\mathfrak{g}$. ***shuffle***, c_{u} mutual closure then furthermore amounts to the c^{u} being good $\mathfrak{g}$ objects: tensors or tensor densities for the $\mathfrak{g}$ that ***shuffle*** represents. Finally, to attain Constraint Closure, the c_{u} need to close among themselves (possibly modulo *shuffle* terms). All candidate closures are monitored here using the Dirac Algorithm itself.

One simple way for such a candidate partition to succeed with its Brackets Closure is in the direct product form I) of Appendix E's split Lie relations. In this case, we have $\mathfrak{O} \times \mathfrak{N}$ for $\mathfrak{O}$ representing $\mathfrak{g}$ and $\mathfrak{N}$ a space of $\mathfrak{g}$-invariant quantities. A more general success takes the semidirect product form II) of Appendix E: $\mathfrak{N} \rtimes \mathfrak{O}$, for $\mathfrak{O}$ representing $\mathfrak{g}$ and $\mathfrak{O}$ some $\mathfrak{g}$-tensor. In the context of 3), we also formalize and extend the previous Sec's collection of subcases as follows.

Case *I*) b or $d \neq 0$ means the class of O-objects does not close as a subalgebraic structure.
Case *II*) $c \neq 0$ signifies that $\mathfrak{g}$ was chosen too small for $\mathfrak{O}$ to represent it.
Case *III*) $e \neq 0$ means that $\langle \mathfrak{O}, \mathfrak{g}, |[\ ,\ ]| \rangle$ can be manipulated. This requires either extending $\mathfrak{O}$ or modifying $|[\ ,\ ]|$ such that some of the hypothesized generators of $\mathfrak{g}$ are now absent. This case has an issue with what form its reduction of $\mathfrak{P}$hase takes, and also whether to modify $\mathfrak{O}$ or the space of base objects, $\mathfrak{BO}$. One can also choose to abandon ship if this occurs....
Moreover, if we use the algorithm of ensuring the **O** represent the purported $\mathfrak{g}$ prior to bringing in the **N**, all of the above are moot.
Case *IV*) If $j \neq 0$, this may indicate that the **N** are incompatible with the **O**'s $\mathfrak{g}$-invariance, to be resolved by the same methods as in III) but now treating the **O** and **N** together.
Case *V*) h or $m \neq 0$ indicate that adjoining the **N** to the **O** forces $\mathfrak{g}$ to be extended.

4) We can also consider the partition to model adjoining some constraints $\mathbf{N}_{\mathsf{v}} = c_{\mathsf{v}}$ to the *chronos* $=$ O supplied by Temporal Relationalism. This case consists of testing the three types of brackets formed from *chronos* and c_{v} using the TRi Dirac-type Algorithm.

5) Of course, our main interest for now is in concurrently taking $O =$ ***shuffle*** and $N =$ *chronos* so as to handle the first three facets at once.

5.i) We now start with an action of the conceptual form $\mathsf{S}^{\mathrm{trial}}_{\mathrm{TR-CR}}$.

5.ii) Temporal and Configurational Relationalism produce candidate $\mathcal{C}$hronos and $\mathcal{S}$**huffle** respectively. This gives three types of brackets entries for the Tri Dirac-type Algorithm to check.

For instance, Sect. 9.14's (Metric Shape and Scale) RPM Constraint Closure is of Appendix E's type II). As a more major example, the GR case has one further interrelation, which amounts to the **O** providing integrabilites for the **N**; this case is denoted by Appendix E's type III) 'Thomas' integrability algebraic form, $\mathfrak{N} \ominus \mathfrak{O}$. GR furthermore has $l = 0$, but also $k \neq 0$ taking a functional form. In conventional working theories, the outcome of this is that the $\mathfrak{g}$-constraints $\mathcal{S}$**huffle** self-close, whereas $\mathcal{C}$hronos is a good $\mathfrak{g}$ object (tensor and/or density), all of which is but a kinematical and representation-theoretic matter. The further requirement that $\mathcal{C}$hronos close with itself is moreover capable of rendering candidate theories trivial or inconsistent. However, this does not occur in RPM or GR (whereas Electromagnetism and Yang–Mills Theory trivially get by due to not even having a Chronos in the first place).

It is n straightforward to illustrate 2.iii) by examples. Firstly, for Metric Scale-and-Shape RPM, $\mathrm{dA}_{\mathrm{Total}} := \mathrm{d}I\, \mathcal{E} + \mathrm{d}B^i \mathcal{L}_i$ (strictly $+\mathrm{d}\lambda^i P_i^B$), and $\mathrm{A}_{\mathrm{Total}} := \dot{I}\, \mathcal{E} + \dot{B}^i \mathcal{L}_i$ (strictly $+\dot{\lambda}^i P_i^B$). Secondly, for GR, $\mathcal{A}_{\mathrm{Total}} = \dot{\mathrm{I}}\, \mathcal{H} + \dot{\mathrm{F}}^i \mathcal{M}_i$ (strictly $+\dot{\lambda}^i \mathrm{p}_i^\beta$), and $\partial \mathcal{A}_{\mathrm{Total}} = \partial \mathrm{I}\, \mathcal{H} + \partial \mathrm{F}^i \mathcal{M}_i$ (strictly $+\partial\lambda^i \mathrm{p}_i^\beta$). Section 24.10 populates the rest of the current's section's considerations with specific examples.

5.iii) If Constraint Closure succeeds, $\mathsf{S}^{\mathrm{trial}}_{\mathrm{TR-CR}}$ is promoted to $\mathsf{S}^{\mathrm{CC}}_{\mathrm{TR-CR}}$ status, $\mathcal{S}$**huffle** to $\mathcal{G}$**auge** with the corresponding Best Matching now having the status

$$C R(\mathsf{S}^{\mathrm{CC}}_{\mathrm{TR-CR}}) := \mathrm{E}_{\mathbf{g} \in \mathfrak{g}}(\mathsf{S}^{\mathrm{CC}}_{\mathrm{TR-CR}} \text{ built upon } \mathfrak{q}, \mathfrak{g}) \tag{24.12}$$

Here, $\mathrm{E}_{\mathbf{g} \in \mathfrak{g}}$ = {extremum of $\mathbf{g} \in \mathfrak{g}$}, $\mathsf{S}^{\mathrm{CC}}_{\mathrm{TR-CR}}$ involves a suitable group action of $\mathfrak{g}$, and the whole construct gets past the TRi-Dirac Algorithm.

Finally, the successful $\mathcal{C}$hronos is rearranged to form a Machian emergent time of the conceptual form

$$C R(\mathrm{t}^{\mathrm{em}}_{\mathrm{CC}}) := \mathrm{E}'_{g \in \mathfrak{g}} \mathrm{t}^{\mathrm{em}}_{\mathrm{trial}-\mathfrak{g}} \tag{24.13}$$

for

$$\mathrm{t}^{\mathrm{em}}_{\mathrm{trial}-\mathfrak{g}} := \int \|\mathrm{d}_g\, \boldsymbol{Q}\|_{\mathcal{M}} / \sqrt{2\, W(\boldsymbol{Q})},$$

and where E' is now likewise protected by $\mathcal{G}$**auge** closure.

24.7 Seven Strategies for Dealing with Constraint Closure Problems

If the severe form of the problem strikes, one may have to entirely abandon the candidate theory's triple $\langle \mathfrak{T}(\mathfrak{q}), \mathfrak{g}, \mathfrak{S} \rangle$. I.e. the Machian variables, a group acting

Strategy: Fix	**Notes and Examples**
0) all three: probe with nothing	No new discoveries.
1) $\mathfrak{g}$ and $\mathfrak{q}$	Adding action terms in Gauge Theory (Particle Physics). Includes strong fixings.
2) $\mathfrak{g}$ and S	Pure extension or restriction of $\mathfrak{q}$ to fit $\mathfrak{g}$: Principles of Dynamics and Gauge Theory.
3) $\mathfrak{q}$ and S	Probing by group generators: the Relational Approach that Part II opens with.
4) $\mathfrak{g}$	Gauge Theory: kill terms in S incompatible with $\mathfrak{g}$ or keep them by extending $\mathfrak{q}$ or $\mathfrak{P}$hase. Includes strong fixings.
5) $\mathfrak{q}$	Relational Approach using families of actions in Chapter 33. Includes strong fixings.
6) S	When one has a first principle or mathematically meaningful form for S, and is disposed to probe for $\mathfrak{q}$ and $\mathfrak{g}$ which realize it.
7) nothing: probe with all	A valid method - and the most general - but with no guiding principle.

Fig. 24.3 Seven strategies with some capacity for generating new theories from what is allowed by Constraint Closure. In each case, the structures which remain fixed act as a guiding principle. [Keeping a given $\mathfrak{g}$, what physics ensues? What about with a given $\mathfrak{q}$? A given S?] The unfixed complement structures correspond to types of probing for new theories. Note how this reasoning pitches the Relational Approach as a complementary method to Gauge Theory. Also, paralleling how Gauge Theory can be attempted with extra terms in the action which are then ruled out by lack of $\mathfrak{g}$ compatibility, the Relational Approach comes in a larger version in which whole families of candidate theories are treated at once (Chap. 33)

thereupon and the Jacobi–Synge geometrical action.[4] In some cases, however, modifying one or more of these may suffice to attain consistency; an interesting array of strategies for this is presented in Fig. 24.3, and supported by the comments below.

Remark 1) Fig. 24.3's strategic diversity continues to apply if $\mathfrak{P}$hase and an integrated ($\mathfrak{d}$A-)Hamiltonian—or its constituent set of constraints in whole-universe theories—are considered in place of $\mathfrak{q}$ and S. Similar considerations apply in spacetime formulations of S with $\mathfrak{g}_{\mathrm{S}}$ acting thereupon, and at the quantum level (further extending the Hamiltonian presentation).

Remark 2) Preserving a particular $\mathfrak{g}$ in Particle Physics includes insisting on a particular internal gauge group, or on the Poincaré group of SR spacetime.

Remark 3) *Strong vanishing* involves fixing hitherto free constants in $\mathfrak{S}$ so as to avoid the problem.

Remark 4) One consequence of adopting strategies permitting extension or reduction of $\mathfrak{q}$ or $\mathfrak{P}$hase is that formulations with second-class constraints are ultimately seen as half-way houses to further formulations which are free thereof. This is

[4]One might augment this to a quadruple by considering varying the type of group action of $\mathfrak{g}$ on $\mathfrak{T}(\mathfrak{q})$.

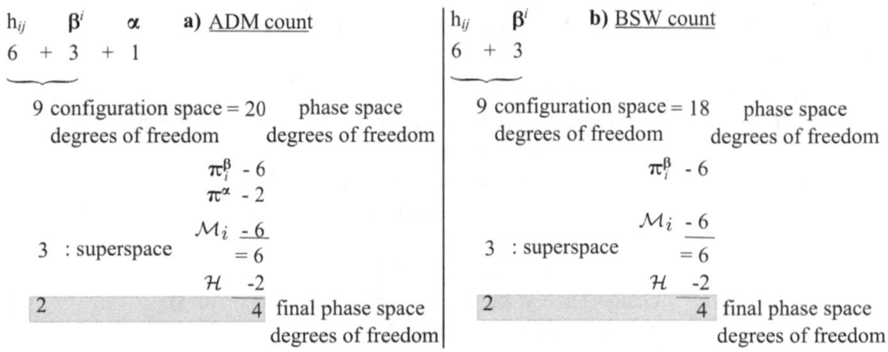

Fig. 24.4 Comparison of **a)** the ADM canonical procedure and **b)** the BSW one. The relational case is like **b)** but with F^i in place of β^i. One can consider also the ADM formulation of Geometrodynamics to involve 4 configuration space degrees of freedom due to not yet eliminating p^α; this 4 consists of $\mathfrak{G}^{(3)}$ alongside α. This corresponds to first reducing out *Diff*$(\boldsymbol{\Sigma})$ and only then engaging with α and $\mathcal{H}$

largely the context in which both the effective formulation and the Dirac bracket formulation were developed, with $\mathfrak{P}$hase getting extended in the former and reduced in the latter.

One can also consider the above strategies for the triple $\langle\mathfrak{P}\mathrm{hase}, \mathfrak{g}, \mathfrak{a}_{\mathrm{Tot}}\rangle$, where $\mathfrak{g}$ now acts on $\mathfrak{P}$hase and $\mathfrak{a}_{\mathrm{Tot}}$ is the integrated total $\mathfrak{d}$-almost Hamiltonian.

We next turn to justifying this book's presentation and development of theoretical concepts concerning types of constraint, constraint algebraic structure and Constraint Closure Problem with a series of illustrative concrete examples.

24.8 Examples of Distinctions Between Types of Constraint

We first justify the finer distinctions between types of constraint made in Sect. 9.14.

Example 1) The equivalence of canonical workings based on the ADM, BSW, relational and TRi split formulations of GR is presented in Fig. 24.4. This also serves as an example of the primary–secondary distinction of constraints being formulation-dependent. This is since $\mathcal{H}$ is secondary in the first and fourth of these formulations and primary in the second and third.

Example 2) The constraints considered so far in this book—in particular $\mathcal{L}$, $\mathcal{P}$, $\mathcal{E}$, $\mathcal{G}$, $\mathcal{G}_I$, $\mathcal{M}_i$, $\mathcal{H}$—are all first-class, so it is useful to now provide examples of second-class constraints. Firstly, in the $\boldsymbol{Q} = \mathrm{A}_i, \Phi$ formulation [715] of the massive analogue of Electromagnetism (alias *Proca Theory* after physicist Alexandru Proca),

$$\mathcal{C} := \partial_i \mathrm{E}^i + m^2 \Phi = 0. \tag{24.14}$$

This indeed uses up only one degree of freedom, so this theory has one more physical mode than Electromagnetism itself. Gravitational Theories with second-class constraints include Einstein–Dirac Theory (i.e. GR with spin-1/2 fermion matter) [232] and Supergravity [232, 314, 715, 834].

Example 3) [of relational recovery of Gauge Theory]. With $\mathfrak{g}$ being a candidate group of physically irrelevant motions, in general it remains to be ascertained whether the $\mathcal{S}$**huffle** provided by Best Matching is a gauge constraint $\mathcal{G}$**auge** which corresponds to $\mathfrak{g}$. Moreover, 'gauge' is here meant in Dirac's sense, which is relevant since 'gauge' can differ in interpretation with the choice of base objects of one's theory. E.g. **Q** alone, **Q** and d**Q**, **Q** and **P**, whole paths, or histories, are some such choices leading to distinct notions of 'gauge'. N.B. the distinction between different *notions of* Gauge Theory, as opposed to the more familiar issue of making particular *choices of* gauge within the one notion of gauge. E.g. using the Lorenz gauge for Electromagnetism (6.19) is an example of the latter, whereas two examples of the former are as follows. Dirac considered [247, 250] a notion of Gauge Theory which concerns *data at a given time*: *data-gauge*. On the other hand, physicist Peter Bergmann considered [133] a notion of Gauge Theory concerning *whole paths* (dynamical trajectories): *path-gauge*, as features further in Chaps. 27 and 32. For now, we note that it is fitting for *Configurational* Relationalism to be associated with a *data*-gauge notion.

Whether there is $\mathfrak{g}$ compatibility can at least in part be investigated prior to consideration of constraints. This is since the $\{V, P_\mathfrak{g}\}$ in $\{\mathcal{C}\text{hronos}, P_\mathfrak{g}\}$ can already be examined prior to constraints: adopting a $\mathfrak{g}$ comes with Equipping with Brackets. On the other hand, one does not assess $\mathfrak{T}(\mathfrak{q}, \dot{\mathfrak{q}})$ itself, which is tied to constraints being more simply and systematically handled in Hamiltonian-type formulations. The ensuing action can be viewed as a map from a structure that is a fibre bundles twice over: both a tangent bundles *and* $\mathfrak{g}$-fibre bundles. Specifically, it is $\mathfrak{p}(\mathfrak{T}(\mathfrak{q}), \mathfrak{g})$, rather than $\mathfrak{T}(\mathfrak{p}(\mathfrak{q}, \mathfrak{g}))$ due to the nontrivial part of the $\mathfrak{g}$ action being on the tangent bundles' fibres. Moreover, this being a $\mathfrak{g}$-fibre bundle mathematically may require excision of certain of the degenerate configurations (Appendix G), which in turn is not relationally bona fide.

Counter-example 4) Despite Dirac's conjecture, $\mathcal{F}$**lin** $\not\Rightarrow$ $\mathcal{G}$**auge** by the following technically constructed but not physically motivated counter-example given by Henneaux and Teitelboim [446]. The Lagrangian $L = \exp(y)\dot{x}^2/2$ gives a constraint $p_x = 0$ which is first-class but not associated with any gauge symmetry.

Example 5) Whereas $\mathcal{L}_i$, $\mathcal{G}$, $\mathcal{M}_i$ are uncontroversially gauge constraints, the gauge status of $\mathcal{H}$ and $\mathcal{E}$ remains disputed. Some arguments of note in this regard have been given by Kuchař, Barbour and Foster [106, 587, 589]; see also Sects. 32.5–32.6 for further arguments. This point is, moreover, directly at odds with [446], which *transform* to and from constraints of the form $\mathcal{Q}$uad. The Author pointed out [32] that this discrepancy is due to, on the one hand, [446] allowing for t-dependent canonical transformations, Can_t (see Appendix J.9). On the other hand, the relational whole-universe context has no primary-level t, by which it is not licit to adopt Can_t in this worldview. Consequently $\mathcal{C}$hronos and $\mathcal{G}$**auge** are qualitatively distinct in the relational context. The relational context furthermore makes distinction between Constraint Providers for, firstly, $\mathcal{S}$**huffle** candidates for $\mathcal{G}$**auge**, and, secondly, $\mathcal{C}$hronos.

Counter-example 6) Sect. 30.5 presents a case of $\mathcal{G}$**auge** $\not\Rightarrow$ $\mathcal{F}$**lin**.

24.9 Examples of Constraint Algebraic Structures

Example 0) Cases with reduction at any classical level explicitly attained are aided in the matter of closure by one of the following means.

a) Having less constraints to form brackets from.
b) Making use of single finite-theory classical constraints always closing with themselves by symmetric entries into an antisymmetric bracket.

In particular, for diagonal Minisuperspace, the schematic form

$$\{\mathcal{H}, \mathcal{H}\} = (\text{structure function})^i \times \mathcal{M}_i \tag{24.15}$$

of (9.33) means that no $\mathcal{M}_i$ implies closure and as a mere Abelian algebra rather than an algebroid:

$$\{\mathcal{H}, \mathcal{H}\} = 0. \tag{24.16}$$

This simplification rests on homogeneity.

Moreover, b) also renders reduced RPM trivial in this regard.

Example 1) See Sect. 9.14 for Electromagnetism and Yang–Mills Theory.

Example 2) That Sec also covered Metric Shape and Scale RPM, although in Part II the version with translations already quotiented out is preferable. Section 21.3 subsequently introduced Metric Shape RPM's extra constraint. Note furthermore that shape and scale transforming independently can be expressed as $\{\mathcal{L}_i, \mathcal{D}\} = 0$; cf. (E.13). These RPM models can be summarized by

$$\text{RPMs realize the } \varepsilon \times \mathfrak{g}\textbf{auge} \text{ subcase of } \mathcal{C}\text{hronos} \times \mathfrak{g}\textbf{auge}. \tag{24.17}$$

Example 3) Spatial diffeomorphisms form among themselves an infinite-d Lie algebra, the TRi-smeared form of which is

$$\{(\mathcal{M}_i \,|\, \partial \mathrm{L}^i), (\mathcal{M}_j \,|\, \partial \mathrm{M}^j)\} = (\mathcal{M}_i \,|\, |[\partial \mathrm{L}, \partial \mathrm{M}]|^i). \tag{24.18}$$

Example 4) In the case of full GR, the constraints' Poisson brackets form the Dirac algebroid (9.31), (9.32), (9.33) [250]. In the Theoretical Physics literature, this was long termed 'Dirac algebra', but Bojowald has more recently brought to attention that it is mathematically an algebroid, and spelled out some consequences of this in his book [154].

Moreover, we now need the TRi-smeared version of the Dirac algebroid: (24.18)

$$\{(\mathcal{H} \,|\, \partial \mathrm{K}), (\mathcal{M}_i \,|\, \partial \mathrm{L}^i)\} = (\pounds_{\partial \underline{\mathrm{L}}} \mathcal{H} \,|\, \partial \mathrm{K}), \tag{24.19}$$

$$\{(\mathcal{H} \,|\, \partial \mathrm{J}), (\mathcal{H} \,|\, \partial \mathrm{K})\} = (\mathcal{M}_i \,|\, \partial \mathrm{J} \overleftrightarrow{\partial}^i \partial \mathrm{K}) = (\mathcal{M}_i \mathrm{h}^{ij} \,|\, \partial \mathrm{J} \overleftrightarrow{\partial}_j \partial \mathrm{K}). \tag{24.20}$$

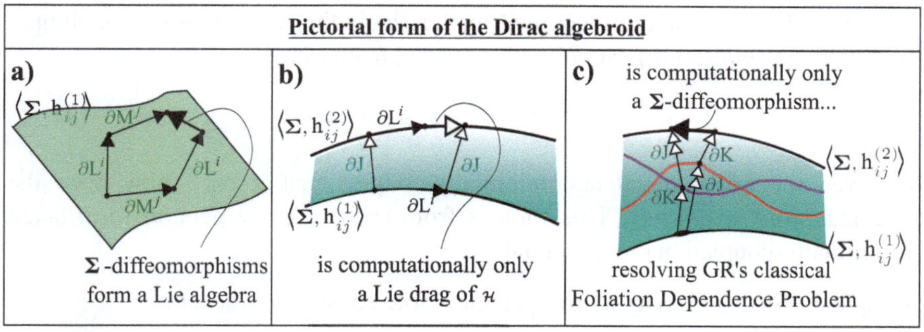

a) $\{(\mathcal{M}_i \,|\, \partial \mathrm{L}^i), (\mathcal{M}_j \,|\, \partial \mathrm{M}^j)\} = (\mathcal{M}_i \,|\, |[\partial \mathrm{L}, \partial \mathrm{M}]|^i)$

b) $\{(\mathcal{H} \,|\, \partial \mathrm{K}), (\mathcal{M}_i \,|\, \partial \mathrm{L}^i)\} = (\pounds_{\partial \underline{\mathrm{L}}}\mathcal{H} \,|\, \partial \mathrm{K})$

c) $\{(\mathcal{H} \,|\, \partial \mathrm{J}), (\mathcal{H} \,|\, \partial \mathrm{K})\} = (\mathcal{M}_i \,|\, \partial \mathrm{J} \overleftrightarrow{\partial}^i \partial \mathrm{K})$

Algebraic form of the Dirac algebroid

Fig. 24.5 Geometrical significance of the form of the Dirac algebroid formed by the constraints of GR. **a)** $\mathcal{M}_i$ generates stretches *within* $\mathbf{\Sigma}$: is just the usual Lie algebra relation in the case of $\mathfrak{g} = Diff(\mathbf{\Sigma})$. **b)** $\mathcal{H}$ as a good $Diff(\mathbf{\Sigma})$ object—a scalar density. **c)** $\mathcal{H}$ acts on spatial hypersurface $\mathbf{\Sigma}$ by deforming it into another hypersurface (*red line*) [454, 576–579, 832]. This is Teitelboim's Refoliation Invariance construct that holds at the classical level thanks to the form of the Dirac algebroid. Consider going between $\langle \mathbf{\Sigma}, \mathbf{h}^{(1)} \rangle$ and $\langle \mathbf{\Sigma}, \mathbf{h}^{(2)} \rangle$ through 2 different orderings of deformations: the first via the red hypersurface and the second via the purple one. We *only* then need the stretch associated with the commutator of ∂J and ∂K to compensate for the non-commutativity of the two deformations involved

Additionally, (24.18)–(24.20) already apply in the case of Minkowski spacetime $\mathbb{M}^n$ so as to model fleets of accelerated observers therein. This is in fact the context in which Dirac first found this algebroid [248] though he subsequently considered the GR case in [249].

Note in particular that the Poisson bracket of $\mathcal{C}$hronos $= \mathcal{H}$ with itself (24.20) gives rise to **shuffle** constraints $\mathcal{M}_i$. This indicates a greater amount of 'togetherness' between Temporal and Configurational Relationalism than the RPM model arena exhibits. Consequently, in the GR setting, Temporal Relationalism cannot be entertained without Configurational Relationalism. This is in contrast with how the two can be treated piecemeal in RPM. As an integrability, this is analogous to Thomas precession (Appendix E):

$$\text{GR manifests the } \mathcal{H} \ominus \mathcal{M}_i \text{ subcase of } \mathcal{C}\text{hronos} \ominus \mathbf{gauge}. \qquad (24.21)$$

Moreover, this is now in the form of an algebroid, as required to encode the multiplicity of foliations. In more detail, there is a parallel between composing two boosts producing a rotation: Thomas precession, and composing two time evolutions producing a spatial diffeomorphism: *Moncrief–Teitelboim on-slice Lie dragging* [662]. Finally, Fig. 24.5 elevates Fig. 9.6's pictorial explanation of the Dirac algebroid to TRi form.

Upon including minimally-coupled matter (including no curvature couplings), the *Teitelboim split* [835] for minimally-coupled matter is

$$\mathcal{H} = \mathcal{H}^{\mathrm{g}} + \mathcal{H}^{\Psi}, \qquad \mathcal{M}_i = \mathcal{M}_i^{\mathrm{g}} + \mathcal{M}_i^{\Psi}. \tag{24.22}$$

Moreover, the gravitational and minimally-coupled matter parts now obey the Dirac algebroid *separately*. This follows from (H.6) and the general form taken by minimally-coupled matter potentials.

Example 4) Strong Gravity metrodynamics' sole constraint bracket is

$$\{(\mathcal{H} \,|\, \partial \mathrm{J}), (\mathcal{H} \,|\, \partial \mathrm{K})\} = 0. \tag{24.23}$$

Its geometrodynamical counterpart is further supplemented by (24.18), (24.19).

Example 5) The Ashtekar variables constraint algebraic structure [75] is much like that of Geometrodynamics, but with an extra Gauss-type constraint included. Schematically, the new Yang–Mills–Gauss constraint $\mathcal{G}_I$ self-close as per usual, commute with $\mathcal{H}$ indicating that to be an $SU(2)(\boldsymbol{\Sigma})$ scalar and form a relation $\{\mathcal{M}_i, \mathcal{G}_I\} \sim \mathcal{G}_I$. This is commensurate with Sect. 19.7's semidirect product structure. On the other hand, the 3 geometrodynamical constraints brackets carry over with (9.33) picking up a $\mathcal{G}_I$ term without becoming free from spatial geometry-dependent structure functions. Thereby, the main interesting features and complications of the geometrodynamical case carry over. See e.g. [75] for the detailed form of this Ashtekar–Dirac algebroid of constraints.

Moreover, the above original and a priori simplest Ashtekar variables formulation being of *complex* GR, reality conditions need to be imposed in order to extract physical answers. These can be positioned at the end of one's calculations, so they are not an immediate procedural obstruction, but are none the less very hard to handle once one finally gets to them [552, 587]. One way around this is to consider instead the a priori more complicated Ashtekar variables formulation [94]. Here the integrand of the action (19.8) picks up a topological term,

$$\frac{1}{2\beta} \mathrm{e}\,\epsilon^{IJ}{}_{KL}\mathrm{e}^{\mu}{}_{I} \wedge \mathrm{e}^{\nu}{}_{J} \wedge \mathrm{F}_{\mu\nu}{}^{KL}, \tag{24.24}$$

to form the *Holst action* [455]. The relative proportionality factor thus introduced is the Barbero–Immirzi parameter, β. This also arises as a replacement for the $-i$ factor in (19.9), and subsequently features in an extra additive term in the Ashtekar $\mathcal{H}$ (8.37):

$$-\frac{\beta^2+1}{\beta^2}\frac{E^a_{[I}E^b_{J]}}{\sqrt{E}}\{GA^I_a - \Gamma^I_a\}\{GA^J_b - \Gamma^J_b\}, \tag{24.25}$$

Moreover, β can be chosen so that this is a real-variables formulation in the first place, thus not necessitating any reality conditions imposition; [94] gives the corresponding constraint algebroid. This loses some of the initial advantage in maximally simplifying the form of $\mathcal{H}$. Another issue with this is whether Nature, rather than the theoretician, fixes the value of β. This is relevant in particular due to β featuring in Nododynamics' formula for the black hole entropy [cf. Eq. (11.22)]. On the other hand, the Loop Approach's [330] results hold for *whichever* β and can thus be carried over from the original complex formulation to the real formulation as well. Loop formulations are largely disregarded in Particle Physics due to not working well for Yang–Mills Theory in that context. However, the claim is that the Loop Approach works *better* in the spatial diffeomorphism-invariant case of relevance to GR.

Given that we shall be giving some further conceptual arguments for complex formulations in Chaps. 27 and 31, it is worth bearing in mind the following. i) There are other ways in which real variables are more technically convenient to work with (Sect. 43.5). ii) The complex formulations do not only include the original formulation of Ashtekar, but also various more modern and advanced treatments of the complex case as exposited by physicist Thomas Thiemann [845].

Finally, in contrast, the Husain–Kuchař model arena has $\mathcal{G}_I$ and $\mathcal{M}_i$ alone, which indeed close as a Lie algebra.

Example 6) In the bein formulation of Canonical GR, the configuration space is now $\mathfrak{Bein}(\boldsymbol{\Sigma})$ in place of $\mathfrak{Riem}(\boldsymbol{\Sigma})$. This has local Lorentz frame rotations being physically irrelevant, leading to a local Lorentz constraint $\mathcal{J}$ (with indices suppressed).

The framed formulation of Geometrodynamics now extends to $Diff(\boldsymbol{\Sigma}) \ominus SO(3)(\boldsymbol{\Sigma})$. Correspondingly,

$$\mathcal{M}_i \ominus \mathcal{J}. \tag{24.26}$$

Additionally demanding Temporal Relationalism provides $\mathcal{H}$, in which case we find Constraint Closure in the form

$$\mathcal{H} \ominus \mathcal{M}_i \ominus \mathcal{J}. \tag{24.27}$$

Framed metrodynamics proceeds without the above extension. In this case, demanding Temporal Relationalism provides $\mathcal{H}$, from which the Dirac Algorithm produces $\mathcal{M}_i$ from $\mathcal{H}$, by which one also arrives at the same final form of Constraint Closure.

Einstein–Dirac Theory follows suite in each of the above approaches, which are furthermore useful preparation for the subsequent major example of Supergravity.

Example 7) In $1+1$ dimensions on $\mathbb{S}^1$ spatial topological manifold manages to be a *well-known* Lie algebra (albeit infinite-d), as per Appendix V. Moreover, GR in dimension $2+1$ and higher is not simple in this manner.

24.10 Examples of Constraint Closure Problems

Example 1) [of case *II*] Best Matching with respect to translations and special conformal transformations, but not with respect to rotations or dilations, fails because the first two do not form a group without the last two: Eq. (E.31).

Example 2) [of case *III*] This is a valid problem in the absence of $\mathcal{H}$, so attempting to impose $U(1)$ symmetry on Proca Theory suffices. A constraint (24.14) arises, but this is second-class so it only uses up 1 degree of freedom. One way out involves considering that Proca Theory rejects quotienting by $U(1)$ (Strategy 3). Another way out, if one insists on retaining $U(1)$ is to consider $m = 0$ to arise as a strong condition (Strategy 1). This gives a longer route to the exclusion of mass terms from $U(1)$ symmetric 1-form actions. Proca Theory can indeed be handled with Dirac brackets *or* the effective method (Exercise!). N.B. how both of these methods and the preceding strong condition all offer distinct minimalistic ways of dealing with a mismatch in an original candidate triple $\langle \mathfrak{g}, \mathfrak{q}, \mathsf{S} \rangle$.

Example 3) [of case IV] Consider Best Matching with respect to the affine transformations *Aff*(d) within a Euclidean-norm kinetic arc element [36]. This produces a constraint $\mathcal{E}$ which is incompatible with *Aff*(d), which defect can be traced from $\mathcal{E}$ possessing a Euclidean norm back to the kinetic arc element assumed. In this case, however, the answer is not to use extension or Dirac bracket, but rather to acknowledge that one needs to build an arc element free from any residual Euclidean prejudices. Here one 'abandons ship', in the sense of forfeiting a type of S for all that one can pass to a different type of S which works [36]. The 2-*d* working theory is given, rather, by (19.1). This example's required alteration so as to attain consistency is however unrelated to changing any of $\mathfrak{g}$, $\mathfrak{q}$ or $\{\, ,\, \}$.

Example 4) Suppose we try to impose a $\mathfrak{g}$ including both the $GL(d, \mathbb{R})$ transformations and the special conformal transformations acting on flat space. Then both (19.3) and (19.6) arise, but these cannot be group theoretically combined due to (E.35). This example illustrates Best Matching being sunk by a Constraint Closure Problem, due to the former being a piecemeal consideration of generators whereas the latter involves relations as well as generators.

Example 5) [of mutual second-classness across the partition of the constraint algebraic structure]. The linear constraint $\mathrm{p} = 0$ or $\mathrm{p}/\sqrt{\mathrm{h}} = \mathrm{const}$ of Conformogeometrodynamics is second-class with respect to $\mathcal{H}$ as per Chap. 33. A extension strategy for this is outlined in e.g. [650]; on the other hand, the Dirac brackets approach remains untried in this case.

Example 6) Chap. 33 is not only based on Strategy 5) but is also a rich source of further examples. In attempting to set up an elsewise GR-like metrodynamics, the Poisson bracket of two $\mathcal{H}$'s continues to read (9.33). So ab initio (24.21) continues to arise [662], and so case V) with $m \neq 0$ ensues. This is also an example of Sect. 16.1's point that a natural action of $\mathfrak{g}$ on $\mathfrak{q}$ existing does not guarantee that $\mathfrak{g}$ represents the totality of physically irrelevant transformations. The above model's enlargement amounts to being forced to pass from $\mathfrak{g} = id$ to the *Diff*$(\mathbf{\Sigma})$ that corresponds to $\mathcal{M}_i$. All in all, $\mathcal{Q}$uad can have its own say as to what form (part of) the $\mathcal{F}$**lin** have to take.

On the other hand, such a $\mathfrak{g}$-Closure Problem does not occur in attempting metrodynamical Strong Gravity; this remains consistent with just the one constraint (18.42). Finally, Chap. 33 also contains many examples of strong vanishing.

Example 7) Next consider recombining N and O objects to reveal a simpler split form, such as in realizing the $so(3, 1) \cong so(3) \times so(3)$ accidental relation (Appendix E and Ex IV.7). However, this amounts to abandoning one's originally declared partition. [Even if it is the right $\mathfrak{g}$, the partitioned method of obtaining the O ends up not being how the $\mathfrak{g}$ acts. The partition of methods of provision consequently remains a false premise.]

24.11 The Further Example of Supergravity

First of all, one here needs to generalize the Poisson brackets to accommodate mixtures of bosonic and fermionic species. Appendix J.20 outlines the subsequent Casalbuoni brackets for finite theories; the further generalization to Field Theories is straightforward.

Canonical Supergravity has additionally the expected $\mathcal{M}_i$ and $\mathcal{J}$ constraints alongside a specifically supersymmetric $\mathcal{S}_A$.[5] All we need to know for this book about $\mathcal{J}$ and $\mathcal{S}$ are that each of these constraints is linear in the momenta, and the schematic form of the subsequent constraint brackets. See e.g. [232, 868] for further details of what is known of the constraint algebraic structures for Einstein–Dirac Theory and Supergravity. In particular, in Supergravity a subset of the linear constraints—the $\mathcal{S}$—have the Supergravity counterpart of the quadratic $\mathcal{H}$ as *their* integrability [232, 834]:

$$\{\mathcal{S}, \mathcal{S}\}_{\mathrm{C}} \sim \mathcal{H} + \mathcal{M}. \tag{24.28}$$

Indeed, Teitelboim [834, 835] presented $\mathcal{S}$ as the square root of $\mathcal{H}$

$$\text{'Supersymmetry} = \sqrt{\text{Chronos}}\,\text{'}. \tag{24.29}$$

This is analogous to the Dirac operator being the square-root of the Klein–Gordon one.

By this integrability and $\mathcal{M}_i$ still being an integrability of $\mathcal{H}$, Supergravity's $\mathcal{F}$**lin** to $\mathcal{H}$ split is of the more general two-way integrability type (Appendix E). I.e.

$$\text{Supergravity's constraint algebroid is of the form} \quad \mathcal{C}\text{hronos} \ominus \mathcal{F}\textbf{lin}; \tag{24.30}$$

contrast with (24.21) for GR. An underlying reason for this difference is that as well as space–time split GR having a Thomas-type algebraic structure $\mathcal{H} \ominus \mathcal{M}_i$, Supersymmetry by itself does as well. This takes the form $S_A \ominus P_i$ in Nonrelativistic Mechanics, or $S_A \ominus P_\mu$ in Relativistic Mechanics or Field Theory: the Poincaré superalgebra. However, in considering a supersymmetric version of $\mathfrak{q} = \mathfrak{B}\text{ein}(\boldsymbol{\Sigma})$, both of these integrabilities feature at once.

In this manner, Supergravity may merit *shifting* Wheeler's question (9.1) from concerning $\mathcal{H}$ to concerning $\mathcal{S}$. I.e. why this takes the form it does, and thus what is

[5] Again, we drop the spinorial index in this book's schematic presentation. This constraint is also accompanied by a conjugate constraint.

its underlying 'zeroth principles' or Constraint Provider. It should however be cautioned that, whereas the fermions corresponding to Dirac's square root were subsequently observationally vindicated, this is not the case to date as regards superpartner particles. This can be taken as a limitation on arguing against the fundamentality of quadratic constraints like $\mathcal{H}$ on the grounds of their being supplanted in supersymmetric theories.

Lemma *If the* $\mathcal{F}$**lin** *constraints do not form a subalgebraic structure, the following consequences ensue.*

i) $\mathcal{F}$**lin** *cannot be quotiented out as a block.*
ii) *Due of Chronos's ties to Temporal Relationalism, and* $\mathcal{F}$**lin***'s to Configurational Relationalism, change of status as regards which of these constraints can be treated separately from which others also affects how Relationalism can be viewed. In particular,* $\mathcal{F}$**lin** *non-closure heralds a breakdown in Best Matching being a separate provider of linear constraints.*

By i), Supergravity does not have a counterpart of Wheeler's Superspace. By ii) and Supergravity being of the form (24.30), neither Temporal nor Configurational Relationalism can be considered alone therein. Thus approaches in which constraint provision is meaningfully subdivided along such lines are thwarted by Supergravity.

One way to deal with this is to accept that Best Matching's extremization need not be over a closed group. This is still consistent without the $\mathcal{H}$ integrability adjoined to the system of equations. This is because $\mathcal{H}$ has been substituted into the system and the system's solution is subsequently used to solve $\mathcal{H}$. In this way, the further equation implied is being considered within a Best Matching formulation.

Let us next comment on some other constraint subalgebraic structures which are realized in Supergravity.

Firstly, (24.26) recurs here, now as Supergravity's *non-supersymmetric first-class linear constraint algebra* $\mathcal{NSF}$**lin** $= \{\mathcal{J}, \mathcal{M}_i\}$. Best Matching with respect to these is a success. At least formal quotienting out by the group in question is permitted, giving 'non-supersymmetric superspace' [42]. However placing too much stock in this procedure and quotient space may run against the grain of Supersymmetry, in the sense of applying a non-supersymmetric reduction to a supersymmetric theory.

Secondly, (24.27) also recurs here, now as Supergravity's *non-supersymmetric total constraint algebroid,* $\mathcal{NSC} = \{\mathcal{J}, \mathcal{M}_i, \mathcal{H}\}$.

Further interpretations of Constraint Closure for Supergravity along the lines of 'Constraint Providers' are as follows.

Approach 1) consists of the following.

a) Extend local $SO(3, 1)$ to (24.26).
b) Include further generators associated with Supersymmetry without demanding group closure.
c) Demand Temporal Relationalism as well. Then $\mathcal{H}$ is provided and Constraint Closure is attained.

Approach 2) Consider just a) and b). Then $\mathcal{S} \ominus \mathcal{H}$ is in any case discovered, so that Constraint Closure is once again attained. Since $\mathcal{H}$ can still be rearranged to give t^{em}, it follows that in this approach t^{em} arises from *assuming Supersymmetry* in a context in which $\mathfrak{q}$ includes $\mathfrak{Riem}(\boldsymbol{\Sigma})$. This context is specified since the Nonrelativistic Mechanics case *does not* have $\mathcal{S} \ominus \mathcal{E}$ leading to t^{em}.

Approach 3) Consider just b). Then $\mathcal{S}_A$ (and conjugate) are provided, and the Dirac Algorithm gives $\mathcal{H}$ and $\mathcal{M}_i$ according to $\mathcal{S}_A \ominus \mathcal{H} \ominus \mathcal{M}_i$. In this way, assuming just the obligatory locally Lorentz frame part of Best Matching and the supersymmetry generators in a context in which $\mathfrak{q}$ includes $\mathfrak{Bein}(\boldsymbol{\Sigma})$ leads to both t^{em} and to *Diff*$(\boldsymbol{\Sigma})$ being obligatory. This has not invoked either Temporal Relationalism or Best Matching with respect to *Diff*$(\boldsymbol{\Sigma})$; these have been supplanted in this case as Constraint Providers by involvement of the supersymmetry generator. Note however that *Diff*$(\boldsymbol{\Sigma})$ being obligatory here arises in the same way in the metrodynamics assumed route to GR: $\mathcal{H} \ominus \mathcal{M}_i$. Consequently Supersymmetry is not in this case a separate source of enforcing *Diff*$(\boldsymbol{\Sigma})$ Best Matching, and Approach 3) has little more content than Approach 2).

All in all, we have established that Supergravity is far more classically distinct from Geometrodynamics than Nododynamics is. The current Relationalism or more extended Background Independence program can readily be extended to the latter but not to the former! Consequently, the Problem of Time is substantially different for Supergravity, rendering it an arena of considerable interest for future investigations into the nature of time and the foundations of QG.

Research Project 10) Does this distinct form of Background Independence taken by Supergravity impinge upon the 'space and configuration' versus spacetime primality debate?

See Chaps. 30 and 33 for yet further examples of Constraint Closure subtleties.

Research Project 11) How do Dirac-type Algorithms generalize if the Principles of Dynamics is reworked with further types of classical brackets? Work on this at least on an example by example basis; the last four examples in footnote 2 may be suitable.

24.12 Lattice of Constraint Subalgebraic Structures

Each individual constraint algebraic structure is a *Poisson algebra* in the obvious sense; see e.g. [610] for an introduction to these.

Figure 24.6 provides conceptual names for each of the constraint subalgebraic structures covered in this Chapter.

Additionally, Fig. 24.7.c) illustrates that a theory's constraint algebraic structures form a generally bounded lattice $\mathfrak{L}_\mathfrak{C}$ rather than just bounded total orders (Appendix A.1) such as in Fig. a), b) or d). See Appendix S.4 for a technical outline of lattices. Each subfigure can readily be converted [51] to a so-called Hasse diagram,

Constraints	**Constraints algebraic structures**	**Beables**	**Beables algebraic structures**
$\emptyset$	id	$\mathcal{U}_{\mathsf{U}}$ unrestricted beables	$\mathfrak{U} = c^n(\mathbb{R}^{2k})$
$\mathcal{C}$hronos	$\mathfrak{Ch}$: a 1-*d* Abelian algebra	$\mathcal{C}_{\mathsf{H}}$ Chronos beables	$\mathfrak{C}$
$\mathcal{G}\mathrm{auge}_{\mathsf{G}}$	$\mathfrak{Gauge}$ = Lie algebra $\mathfrak{g}$	$\mathcal{G}_{\mathsf{V}}$ Gauge or G-beables	$\mathfrak{G}$
$\mathcal{F}\mathrm{lin}_{\mathsf{N}}$	$\mathfrak{Flin}$	$\mathcal{K}_{\mathsf{K}}$ Kuchař beables	$\mathfrak{k}$
$\mathcal{C}_{\mathsf{F}}$	$\mathfrak{FC}$ denoting 'first class'	$\mathcal{D}_{\mathsf{D}}$ Dirac beables	$\mathfrak{d}$
arbitrary $\mathcal{C}_{\mathsf{W}} \leq \mathcal{C}_{\mathsf{F}}$	$\mathfrak{C}_{\mathsf{W}}$	$\mathcal{A}_{\mathsf{X}}$ A-beables	$\mathfrak{A}_{\mathsf{X}}$
full collection of $\mathcal{C}_{\mathsf{W}} \leq \mathcal{C}_{\mathsf{F}}$	$\mathfrak{L}_{\mathfrak{c}}$ (bounded lattice)	corresponding collection of $\mathcal{A}_{\mathsf{X}} \leq \mathcal{U}_{\mathsf{U}}$	$\mathfrak{L}_{\mathfrak{b}}$ (bounded lattice)

Fig. 24.6 For constraint subalgebraic structures, this gives the notation for the constraints and the algebras themselves. We next provide the corresponding associated notions of beables and of the beables subalgebraic structures themselves. One suggestion behind this presentation is that the variety of notions of constraint subalgebraic structures deserve comparable attention to the hitherto more widely considered notions of beables

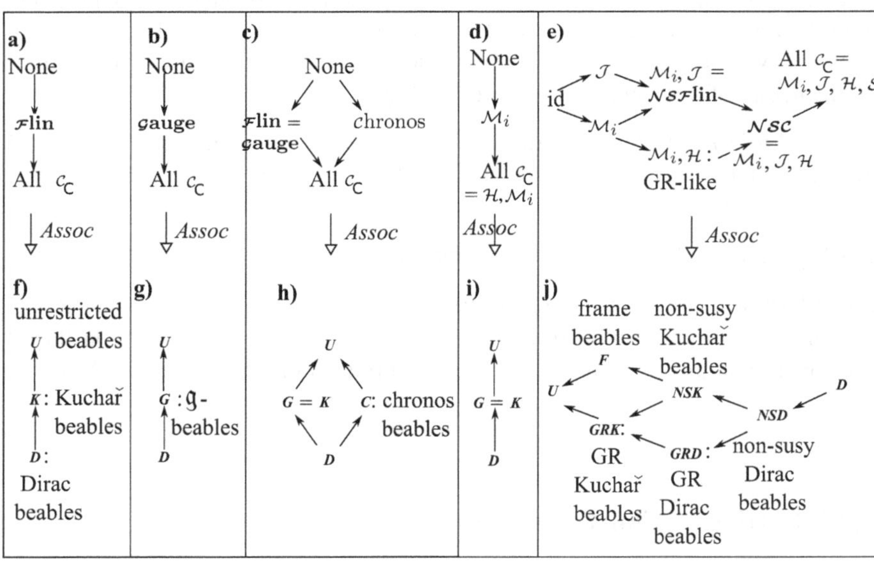

Fig. 24.7 **a)** Gives the incipient position on constraint algebraic structures. **b)** With Dirac's conjecture proving to be false, more thought in general needs to be put in as regards whether the intermediate algebraic structure is first-class linear or gauge. **c)** In RPMs $\mathcal{C}$hronos also closes separately, whereas **d)** in GR it does not. **e)** Supergravity illustrates that intermediate subalgebraic structures can be distinct from $\mathcal{F}$**lin**, $\mathcal{G}$**auge** or $\mathcal{C}$hronos. The Supergravity case has, correspondingly, intermediate beables subalgebraic structures distinct from $\mathcal{K}$, $\mathcal{G}$ or $\mathcal{C}$. **f)** to **j)** display the corresponding notions of beables. Figure E.3 can also be interpreted in terms of constraints corresponding to the generators in question, and Fig. G.3 as beables corresponding to the invariants. In this way, one can envisage $\mathfrak{L}_{\mathfrak{c}}$ and $\mathfrak{L}_{\mathfrak{b}}$ for a wider range of RPMs; see [51] for this explicitly

which are a standard presentation for lattices (and for posets more generally). In the first row of subfigures, the arrows indicate inclusion of further generators, in ways allowed by the integrability relations between these.

Supergravity provides one substantial motivation for Fig. 24.7's introduction of lattices; a second comes from the following implications for the theory of beables. The association map *Assoc*, corresponding to the 'forms zero brackets with' operation, leads to a further lattice $\mathfrak{L}_{\mathfrak{b}}$ of notions of beables for each theory, as displayed in the second row f) to j) of Fig. 24.7. *Assoc* is an order-reversing lattice morphism $\mathfrak{L}_{\mathfrak{c}} \longrightarrow \mathfrak{L}_{\mathfrak{b}}$; in fact, some coarsening of the ordering is allowed [51].

With many constraint subalgebraic structures under consideration, let us also generalize Dirac's weak vanishing notion and notation to that of $\overset{c_{\mathrm{W}}}{\approx}$, meaning vanishing weakly up to the constraints in the indicated constraint subalgebraic structure c_{W}.

Chapter 25
Taking Function Spaces Thereover: Beables and Observables

25.1 In the Absence of Facet Interference

Section 9.15 listed various notions of observables or beables—Dirac D, Kuchař K, and unrestricted U—that are common in the literature, complemented these by introducing Chronos C, and discerned between K and gauge G. The current Chapter's account, however, makes more subtle conceptual distinctions.

Let us first continue here by appreciating the U as not only the simplest case but also the only case that is free from facet interference. The fourth aspect of Background Independence is, in its purest form, to pass from considering Q and P to Taking Function Spaces Thereover, consisting of functions or functionals of the Q and P, such as $\mathbf{F}(\boldsymbol{Q}, \boldsymbol{P})$ or $\mathcal{F}(\underline{x}; \mathbf{Q}, \mathbf{P}]$ ($B := \mathscr{F}\lfloor \mathsf{Q}, \mathsf{P} \rfloor$ in portmanteau form). This is classically straightforward, since Poisson brackets can be allotted consistently to such functions. Using this criterion implies that 'thereover' means over phase space $\mathfrak{Phase}$, i.e. the space of the Q and P as equipped by the Poisson bracket $\{\ ,\ \}$. It also implies a minimum standard of differentiability for the function space $\mathfrak{U}$ of the U, and that this function space is, a fortiori, a Poisson algebra [610].

25.2 The First Great Decoupling of Problem of Time Facets

While the preceding Sec is classically straightforward, understanding this point bears three fruit. Firstly, it supplies the considerably less straightforward Kinematical Quantization with a clear-cut underlying classical counterpart. This permits envisaging Kinematical Quantization as starting one's quantum scheme by taking on the fourth aspect of the Problem of Time. Secondly, for constrained systems, further classical notions of beables and observables such as D, K, G and C

A) remain instances of Taking Function Spaces Thereover.
B) They are also subjected to a number of Constraint Closure restrictions and Temporal and Configurational Relationalism stipulations.

E. Anderson, *The Problem of Time*, Fundamental Theories of Physics 190,
DOI 10.1007/978-3-319-58848-3_25

A) justifies our claim of what the pure form of the fourth aspect is, while B) is the manifestation of a number of facet interferences between the corresponding fourth facet and the previous three.

C) This is moreover a useful classical precursor to understanding how one's initial resolution of the fourth aspect at the quantum level is then disrupted by tackling Facets 1 to 3. This leaves us in subsequent need to face the beables facet once again. The passage from the kinematical Hilbert space $\mathfrak{Kin}$-$\mathfrak{Hilb}$ to the physical Hilbert space $\mathfrak{Dyn}$-$\mathfrak{Hilb}$ requires us to find afresh what is a suitable algebraic structure of quantum operators thereover.

The notion of 'Taking Function Spaces Thereover' additionally encapsulates a type of decoupling between the Problem of Time facets. Firstly note how Constraint Closure is capable of disrupting Temporal and Configurational Relationalism as per Fig. 24.2. Resolving this triplet of facets gives a consistent phase space $\mathfrak{Phase}$. Taking Function Spaces *Thereover* entails addressing a *subsequent* mathematical problem *on* $\mathfrak{Phase}$ rather than imposing some further conditions on whether $\mathfrak{Phase}$ is adequate. Within 'space, configuration or Dynamics is primary' approaches, this imposes a strong ordering as regards proceeding through the first four gates of the enchanted castle. Classically,

1) one keeps on going through the Configurational, Temporal and Closure gates until one has consistently got past all three.
2) Only then does one turn to the fourth gate whose true name is Taking Function Spaces Thereover.

25.3 Sources of Variety Among Classical Notions of Beables

It remains to give a detailed treatment of constrained theories' beables. We begin by considering the sources of diversity among notions of beables, alongside how some of these can be traced back to facet interferences. In constrained theories, classical observables or beables [32, 133, 134, 247, 250–252, 255, 353, 446, 483, 586, 587, 724, 752, 802, 845] are further objects whose 'classical brackets' with 'the constraints' are 'equal to' zero:

$$|[c_{\mathsf{C}}, B_{\mathsf{B}}]| \text{ `='} \, 0, \tag{25.1}$$

These are motivated by being more physically useful than just any $\mathbf{Q}$'s and $\mathbf{P}$ (or functionals thereof) due to containing a higher proportion of physical information.

A consequence of (25.1) is that there are a number of different possibilities as regards which constraints, types of brackets, and notions of equality, are involved. The previous Chapter covered the first two. The bracket selected for this purpose needs to match that carried by the underlying phase space. The set of constraints selected for this purpose furthermore need to close algebraically by Appendix J.18's Lemma 1, and is among those described by Fig. 24.6. This further supports that the

Problem of Beables can only be addressed once the Constraint Closure Problem has been resolved. This is another example of ordering passage through the Problem of Time facet 'gates'. Finally, Part I already depicted in Fig. 9.7 the usual assumption of weak equality, alongside the possibility of strong equality. Algebraically, these are respectively,

$$|[c_\mathsf{C}, B_\mathsf{B}]| \approx 0 \quad \text{i.e. } |[c_\mathsf{C}, B_\mathsf{B}]| = C^{\mathsf{C}'}{}_{\mathsf{CB}}\, c_{\mathsf{C}'}, \tag{25.2}$$

$$|[c_\mathsf{C}, B_\mathsf{B}]| = 0. \tag{25.3}$$

In this way, one has weak or strong versions of each of the D, K....

See Chaps. 27 and 32 for yet further notions of observables in the spacetime setting, whether based on *Diff*$(\mathfrak{m})$ or on more subtle considerations initiated by Bergmann [133, 134], and Chap. 28 for yet more in the histories-theoretic setting.

25.4 Posing Concrete Mathematical Problems for Beables

Comparing (9.24) and (25.1) implies that the c_F are themselves in some sense beables. However, since we already knew that $c_\mathsf{F} \approx 0$, in studying beables we are really looking for further quantities outside of this trivial case. Let us call these other quantities *proper beables*; the rest of the book will always take 'beables' to mean this.

The first major mathematical observation is that (25.1) are $\eth$DEs, i.e. a portmanteau of PDEs

$$\sum_\mathsf{A} \left\{ \frac{\partial c_\mathsf{C}}{\partial Q^\mathsf{A}} \frac{\partial B_\mathsf{B}}{\partial P_\mathsf{A}} - \frac{\partial c_\mathsf{C}}{\partial P_\mathsf{A}} \frac{\partial B_\mathsf{B}}{\partial Q^\mathsf{A}} \right\} \text{`='}\, 0 \tag{25.4}$$

in the finite case, and FDEs (functional differential equations: Appendix O.8)

$$\int \mathrm{d}^n z \sum_\mathsf{A} \left\{ \frac{\delta(c_\mathsf{C}|\partial\xi^\mathsf{C})}{\delta Q^\mathsf{A}(z)} \frac{\delta(B_\mathsf{B}|\partial\chi^\mathsf{B})}{\delta P_\mathsf{A}(z)} - \frac{\delta(c_\mathsf{C}|\partial\xi^\mathsf{C})}{\delta P_\mathsf{A}(z)} \frac{\delta(B_\mathsf{B}|\partial\chi^\mathsf{B})}{\delta Q^\mathsf{A}(z)} \right\} \text{`='}\, 0 \tag{25.5}$$

in the field-theoretic case.[1]

The second major observation is that, by Appendix J.18's Lemma 2, the B themselves form a closed algebraic structure: (J.43). Thus one is not looking for individual solutions of the beables $\eth$DEs, but a fortiori for whole algebras of solutions: closed among themselves and large enough to span all of a physical theory's mathematical content. This gives the concept of 'finding basis beables'. A useful result at this point (Lemma 3 of Appendix J.18) is that if B are beables, then so are $\mathscr{F}\lfloor B \rfloor$, by which the number of individual combinations of $\mathbf{Q}$ and $\mathbf{P}$ that one needs to find may not be very large.

[1] Here $\partial\mu^\mathsf{B}$ is a smearing function. This is given in TRi form since first-class constraints are both trivially weak beables and TRi-smeared, pointing to all beables requiring TRi-smearing.

If strong beables are considered, then (25.3) and the strong case of (J.43) form a direct product $\mathfrak{c} \times \mathfrak{b}$, where $\mathfrak{b}$ denotes the algebraic structure of proper beables. On the other hand, if weak beables considered, then (25.2) and the weak case of (J.43) form a semidirect product $\mathfrak{c} \rtimes \mathfrak{b}$.

25.5 Notions of Beables: Examples and Lattice Structure

Example 1) If we just take the trivial subalgebraic structure of constraints, id, we recover the *unrestricted beables* U.

Example 0) The opposite extreme involves taking the full final set of first-class constraints c_{F}. This now corresponds to the *classical Dirac beables* (Chap. 25.8). I.e. functionals $D = \mathscr{F}\lfloor \mathbf{Q}, \mathbf{P} \rfloor$ that 'final classical brackets commute' with the c_{F}: the $|[c_{\mathsf{F}}, D]| \approx 0$ version of (9.35). In this case, Sect. 25.3's motivational adage takes its strongest form: Dirac beables, *solely* contain physical information, a property that, at the very least, is required in phrasing final answers to physical questions about a theory.

Let us now add to Sect. 9.15's list of (possibly interpretation-dependent) aliases or closely related notions *True* [743–745] alias *complete observables* [750, 751, 845]. These involve operations on a system, each of which produces a number that can be predicted if the state of the system is known.

We next turn to the general case. Take any constraints c_{W} which form some constraint subalgebraic structure $\mathfrak{c}_{\mathsf{W}}$. The corresponding notion of *A-beables* A_{X} obey

$$|[c_{\mathsf{W}}, A_{\mathsf{X}}]| \text{ `='} 0. \tag{25.6}$$

The 'A' here stands for 'algebraic substructure', since each such notion of A-beables indeed forms its own algebraic substructure $\mathfrak{a}_{\mathsf{X}}$. The totality of these algebraic substructures furthermore forms the bounded lattice of notions of beables, $\mathfrak{L}_{\mathfrak{b}}$. So the above use of 'the opposite extreme' is indeed in the sense of the unrestricted and Dirac beables corresponding, respectively, to the unit and zero of $\mathfrak{L}_{\mathfrak{b}}$. Let us also introduce the notion of *non-extremal A-beables*, meaning all those notions which are not D or U. The other examples of beables given in Sect. 9.15 are then just some of the subcases of this which are realized in some theories but not others, to which we now return.

Example 2) The classical *Kuchař beables* [106, 272, 566, 587, 589, 920] are functionals $K = K\lfloor \mathbf{Q}, \mathbf{P} \rfloor$ that 'final classical brackets commute' with all first-class *linear* constraints, i.e. the $|[\ ,\ ]|$ '=' version of (9.36). It is also possible for the $\mathcal{F}$**lin** not to close for some theories, such as for Supergravity (cf. Sect. 24.10). A further Consequence 3) of this is that K are not well-defined for Supergravity, by the Casalbuoni brackets extension of working (J.42). Thereby, much of the focus on K in earlier literature is now to be tied instead to the fully generally applicable notion of the A_{X}.

Example 3) The falseness of Dirac's conjecture—by which the $\mathcal{F}$**lin** would always coincide with the $\mathcal{G}$**auge** for some $\mathfrak{g}$—means that the $\mathcal{G}$**auge** have a separate possibility to form a closed subalgebraic structure and thus to support a corresponding notion of beables. I.e. the notion of $\mathfrak{g}$-beables G which commute with $\mathcal{G}$**auge**: the final classical brackets '=' version of (9.37).

Example 4) $\mathcal{C}$hronos closes by itself as a subalgebraic structure for some theories, e.g. for RPMs, theories of Strong Gravity, and Minisuperspace, but not for full GR itself. In cases for which this closure does occur, the corresponding notion of *Chronos beables*, denoted by C, is mathematically well-defined; this obeys the final classical brackets '=' version of (9.38).

25.6 Strategies for the Problem of Beables

The Problem of Beables (or Observables) [26, 37, 250, 483, 586, 752] is that it is hard to construct a sufficiently large set of these to describe all physical quantities. This applies in particular to Gravitational Theory. Strategies for dealing with the Problem of Beables include the following.

Strategy 1) *Use Unconstrained Beables*, U [251, 252, 752], entailing no commutation condition at all.

Strategy 0) *Insist on Constructing Dirac Beables*, D.

Insist on Constructing Dirac Beables has the conceptual and physical advantage of employing all the information in the final set of constraints of the theory, which are all first-class. It has the mathematical disadvantage that finding Dirac beables—much less a basis set for each theory in question—is a very hard venture, especially in the case of Gravitational Theories. Use Unconstrained Beables is diametrically opposite in each of the above regards. On some occasions the latter is moreover used as a stepping stone toward the former.

Using classical *partial observables* is a distinct point of view about the U. This began with Rovelli's works [191, 192, 743–745]; one might consider [235, 694] as forerunners in some respects; see also [251, 252, 750–755] and the reviews [752, 845]. Following on from Sect. 9.15's outline, partial observables involve classical or quantum operations on the system that produces a number that can be measured but possibly totally unpredictable, even if the state is perfectly known; contrast with the definition of total or Dirac observables. The physical content lies, rather, in considering pairs of these objects, with correlations between them encoding purely physical information which can be extracted. I.e. correlations of two partial observables are predictable. In particular the value of a partial observable o_1 subject to another partial observable o_2 taking a particular value is predictable; one can then consider partial observable o_2 as playing a 'clock' role. In this sense, the name 'partial' is apt: one needs to consider multiple parts before this approach starts to work. It is not however clear which partial observables in particular correspond to realistic and accurate clocks. Nor is it clear how a number of other facets of the

Problem of Time can be addressed in this approach [26, 37, 483, 586, 587]. The Partial Observables Approach's correlations, moreover, are themselves functions on the constraint surface and commute with the constraints. As such they furnish complete or Dirac observables or beables, according to one's interpretation.[2] In such a way, partial observables can also be viewed as a stepping stone toward the construction of Dirac beables, e.g. via methods developed by physicist Bianca Dittrich and Thiemann [251, 252, 845].

Moreover, Insist on Constructing Dirac Beables amounts to concurrently addressing the unsplit totality of constraints (Constraint Closure facet) and Taking Function Spaces Thereover. As a two-facet venture (Fig. 25.1), it is unsurprisingly harder than Use Unconstrained Beables, which only contemplates the latter.

Strategy 2) *Consider Kuchař Beables, κ, to Suffice*. Entertaining κ can entail treating Quad distinctly from the **Flin**. Three possible underlying reasons for making such a distinction [106, 272, 566, 586, 587, 589, 920] are Relationalism, primary-level timelessness, and the hidden nature of the associated Refoliation Invariance symmetry. A fourth, now pragmatic, reason is that κ are simpler to find than D. Moreover, if one looks more closely, some of these motivations are actually tied to the G in cases in which these and the κ are distinct. We take this on board by providing a further distinct strategy.

Strategy 3) *Consider the $\mathfrak{g}$-beables, G, to suffice*. It is more generally these—rather than the κ—which arise from the $\mathfrak{g}$-act, $\mathfrak{g}$-all construction in cases in which the candidate **Shuffle** is confirmed as a **Gauge**. Also, theories having either trivial Configurational Relationalism—or Best Matching resolved—have as a ready consequence a known full set of classical G. The G furthermore happen to coincide with the κ in many of the book's concrete examples; in these cases, the above gives a known full set of classical κ. This takes into account the triple combination of Configurational Relationalism, Constraint Closure and Taking Function Spaces Thereover aspects (Fig. 25.1). Complementarily, finding the κ is uncontroversially a timeless pursuit due to the absence of Chronos (or underlying Temporal Relationalism) from the workings in question.

Strategy 4) *Consider Chronos Beables, C*. This takes into account the triple combination of the Temporal Relationalism, Constraint Closure and Taking Function Spaces Thereover aspects. In theories with nontrivial $\mathfrak{g}$ or some further first-class constraints, this is probably best viewed as a halfway house that is available if Chronos indeed constitutes a constraint subalgebraic structure, $\mathfrak{Ch}$.

Strategy 5) *Consider split Dirac Beables*. Finally, considering Dirac beables once again, but now within the context that the underlying first-class constraints are meaningfully split according to whether they are provided by Temporal or Con-

[2]The Author does not choose to base this book on partial observables due to these carrying 'any change' connotations. Elsewhere, the Partial Observables Approach has also be considered in combination with Internal Time Approaches. Here, internal time can provide the timestandard in schemes requiring a such if one is able or willing to pay the price for the usual inconveniences of a such; see e.g. [251, 252].

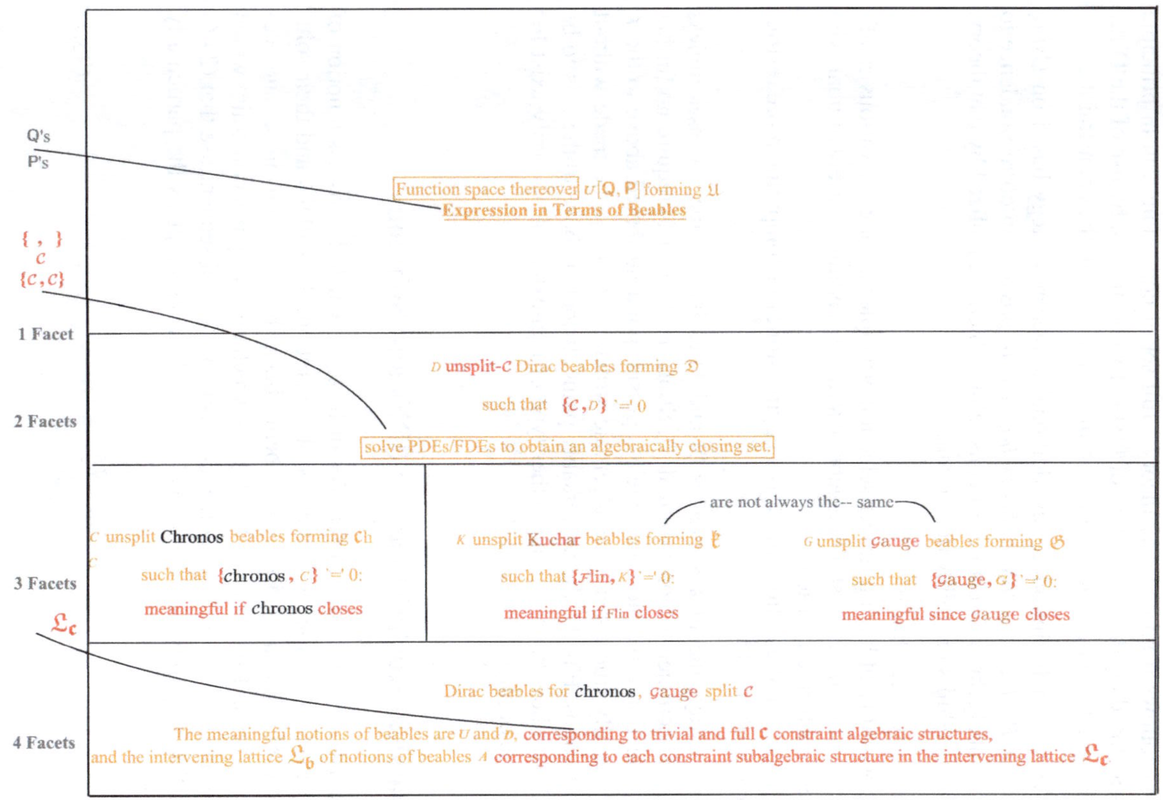

Fig. 25.1 This continuation of the 'technicolour guide' to the Problem of Time sits to the left of Fig. 24.2, with the new Taking the Function Space Thereover content highlighted in orange

figurational Relationalism. In this second-deepest level of Fig. 25.1, all four of the aspects considered so far enter.

(Non)universality arguments are pertinent at this point. Using $\mathcal{U}$ or $\mathcal{D}$ is always in principle possible. The first of these follows from no restrictions being imposed. The second follows from how any theory's full set of constraints can in principle be cast as a closed algebraic structure of first-class constraints, by use of the Dirac bracket, or the effective method, so as to remove any second-class constraints.

Strategy A) We finally introduce an additional universal strategy based on *Using Whichever A-Beables*, $\mathcal{A}_\mathsf{X}$, *that a theory happens to possess*, in correspondence to the closed subalgebraic structures of constraints which are realized by that theory. This strategy admits the following variants.

1) Making any such choice.
2) The possibility of having further selection principles among the various candidate theories whose constraint algebraic structures admit a variety of nontrivial proper subalgebraic structures.
3) Using one or more notions of $\mathcal{A}_\mathsf{X}$ as stepping stones in computing more restrictive types of beables.

Strategy A) always includes Strategies 0) and 1) as its extreme subcases, unless the strategy is confined to non-extremal $\mathcal{A}_\mathsf{X}$. Strategies 0) and 1) require no further choice or selection principle. The timeless pursuit point attributed above to the $\mathcal{K}$ is more generally true for any subalgebraic structures that can be made without involving *c*hronos. Finally, the universality point attributed above to the $\mathcal{D}$ is to be contrasted with how the $\mathcal{K}$—or any other type of non-extremal $\mathcal{A}_\mathsf{X}$—only exist for certain subsets of physical theories!

25.7 Classical Kuchař Beables: ∂DEs and Solutions

We next consider some simple examples of theories which do possess a notion of Kuchař beables in theories as examples of concrete beables ∂DEs and their solutions. These ∂DE's benefit firstly from being linear in the unknowns $\mathcal{K}$, and secondly from coming from Poisson brackets in which one entry is a constraint which depends at most linearly on the momenta. Also note for subsequent use that (24.6) and restriction to purely configurational Kuchař beables $\mathcal{K}_\mathsf{C}$ gives the particularly simple ∂DE

$$\mathcal{F}^{\mathsf{A}}{}_{\mathsf{N}}\frac{\partial \mathcal{K}}{\partial \mathsf{Q}^{\mathsf{A}}} \text{ `='} \, 0. \tag{25.7}$$

Example 1) RPMs. $\{\underline{\mathcal{P}}, \mathcal{K}\}$ '=' $0 \Rightarrow$ the PDE

$$\sum_{I=1}^{N} \frac{\partial \mathcal{K}}{\partial \underline{q}^{I}} \text{ `='} \, 0. \tag{25.8}$$

This is solved by the relative interparticle separation vectors and linear combinations thereof, mot conveniently the relative Jacobi coordinates ρ^A.

$\{\mathcal{L}, \kappa\}$ '=' $0 \Rightarrow$ the PDE

$$\sum_{I=1}^{N} \left\{ \frac{\partial \kappa}{\partial \underline{p}^I} \times \underline{p}_I + \underline{q}^I \times \frac{\partial \kappa}{\partial \underline{q}^I} \right\} \text{ `=' } 0. \tag{25.9}$$

This is solved by various dot products. In particular the κ_{c} obey

$$\sum_{I=1}^{N} \underline{q}^I \times \frac{\partial \kappa_{\mathrm{c}}}{\partial \underline{q}^I} \text{ `=' } 0, \tag{25.10}$$

which is solved by $\underline{q}^I \cdot \underline{q}^J$. Also, the pure-momentum case (κ_{p}) equation is in this case just the $\underline{q}^I \leftrightarrow \underline{p}^I$ of the previous equation, and so is solved by $\underline{p}^I \cdot \underline{p}^J$. The full equation is *not* solved by $\underline{q}^I \cdot \underline{p}^J$ but is solved by $\underline{q}^I \cdot \underline{p}^J + \underline{p}^I \cdot \underline{q}^J$: the outcome of applying the product rule to $\underline{q}^I \cdot \underline{q}^J$. Moreover, norms and angles are particular cases of functionals of the above, which are an allowed extension by Appendix J.18's Lemma 3.

For RPMs in which both (25.8) and (25.9) apply, by Appendix J.18's Composition Principle the solutions are dots of differences of position vectors, or, often more usefully, dots of relative Jacobi vectors. These are the κ [32] for Metric Shape and Scale RPM. The Hopf–Dragt coordinates (G.13)–(G.15) furthermore form a geometrically simple basis of Kuchař beables.

The classical Kuchař beables condition $\{\mathcal{D}, \kappa\}$ '=' $0 \Rightarrow$ the PDE

$$\sum_{I=1}^{N} \left\{ \frac{\partial \kappa}{\partial \underline{p}^I} \cdot \underline{p}_I + \underline{q}^I \cdot \frac{\partial \kappa}{\partial \underline{q}^I} \right\} \text{ `=' } 0. \tag{25.11}$$

This is an Euler homogeneity equation of degree zero. Therefore its solutions are ratios. The κ_{c} subcase obeys

$$\sum_{I=1}^{N} \underline{q}^I \cdot \frac{\partial \kappa_{\mathrm{c}}}{\partial \underline{q}^I} \text{ `=' } 0, \tag{25.12}$$

and the κ_{p} case is once again the $\underline{q}^I \leftrightarrow \underline{p}^I$ of the preceding. Finally, by the Composition Principle, if (25.8) and (25.11) both hold, the κ are ratios of differences, if (25.9) and (25.11) both hold, ratios of dots, and if all three apply, ratios of dots of differences. The last of these are the κ [32] for Metric Shape RPM.

For Affine Shape RPM, $\{\mathcal{S}, \kappa\}$ '=' $0 \Rightarrow$ the PDE

$$\sum_{I=1}^{N} \left\{ \frac{\partial \kappa}{\partial \underline{p}^I} \underline{\underline{S}}\, \underline{p}_I - \underline{q}^I \underline{\underline{S}} \frac{\partial \kappa}{\partial \underline{q}^I} \right\} \text{ `=' } 0. \tag{25.13}$$

This is solved in 2-d by areas between pairs of vectors, in 3-d by volumes of parallelepipeds formed by triples of vectors, and in arbitrary d by the top form supported by the dimension in question, which are formed by d-tuplets of vectors (cf. Appendix B). The κ_c case is

$$\sum_{I=1}^{N} \underline{q}^I \underline{\underline{\underline{S}}} \frac{\partial \kappa}{\partial \underline{q}^I} \text{`='} 0, \tag{25.14}$$

and the κ_p case is the $\underline{q}^I \leftrightarrow \underline{p}^I$ of this. The Composition Principle continues to apply here, so top forms of differences, ratios of top forms and ratios of top forms of differences are the forms taken by various theories' κ. The last of these corresponds to Affine Geometry and the first of these to 'equi-top-form-al' geometry (equiareal [222] in 2-d).

For Conformal Shape RPM's $\{\mathcal{K}_a, \kappa\}$ '=' $0 \Rightarrow$ the PDE

$$\sum_{I=1}^{N} \left\{ 2\{2\, p_{I[i}q_{j]I} - (\underline{q} \cdot \underline{p})\delta_{ij}\} \frac{\partial \kappa}{\partial p_{Ij}} - \{q_I^2 \delta^{ij} - 2q^{iI}q^{Ij}\} \frac{\partial \kappa}{\partial q^{Ij}} \right\} \text{`='} 0. \tag{25.15}$$

The κ_c case is

$$\sum_{I=1}^{N} \{q_I^2 \delta^{ij} - 2q^{iI}q^{Ij}\} \frac{\partial \kappa}{\partial q^{Ij}} \text{`='} 0, \tag{25.16}$$

whereas the κ_p case is now

$$\sum_{I=1}^{N} 2\{2\, p_{I[i}q_{j]I} - (\underline{q}^I \cdot \underline{p}_I)\delta_{ij}\} \frac{\partial \kappa}{\partial p_{Ij}} \text{`='} 0. \tag{25.17}$$

This provides a first example lacking symmetry under $\underline{p}_I \leftrightarrow \underline{q}^I$.

Example 2) Electromagnetism. $\{(\mathcal{G}|\partial\xi), (\kappa_\mathsf{K}|\partial\chi^\mathsf{K})\}$ '=' $0 \Rightarrow$ the FDE

$$\underline{\partial} \cdot \frac{\delta \kappa}{\delta \underline{\mathrm{A}}} \text{`='} 0. \tag{25.18}$$

This is solved by $\underline{\mathrm{E}}$ and $\underline{\mathrm{B}} = \underline{\partial} \times \underline{\mathrm{A}}$, and thus by a functional $\mathcal{F}[\underline{\mathrm{B}}, \underline{\mathrm{E}}]$ by Appendix J.18's Lemma 3. These are not however a *conjugate* pair. Since this looks to be a common occurrence in further examples, let us introduce the term 'associated momenta' to describe it.

$\mathcal{F}[\underline{\mathrm{B}}, \underline{\mathrm{E}}]$ can also be written in the integrated version in terms of fluxes:

$$\mathcal{F}\left[\iint_\mathrm{S} \underline{\mathrm{B}} \cdot \mathrm{d}\underline{\mathrm{S}}, \iint_\mathrm{S} \underline{\mathrm{E}} \cdot \mathrm{d}\underline{\mathrm{S}}\right] = \mathcal{F}[W_\gamma, \Phi_S^\mathrm{E}] \tag{25.19}$$

for electric flux $\Phi^{\mathrm{E}}_{\mathrm{S}}$ and loop variable (N.1). This is by use of Stokes' Theorem with $\gamma := \partial S$ and subsequent insertion of the exponentiation function subcase of Appendix J.18's Lemma 3; this ties the construct to the geometrical notion of holonomy. Moreover, these are well-known to form an *over*-complete set: there are so-called *Mandelstam identities* between them [330].

All of the above furthermore carries over to Yang–Mills Theory [47].

Here $\{(\mathcal{G}_I | \xi^I), (\kappa | \chi^{\mathsf{K}})\} \approx 0 \Rightarrow$ the FDE

$$\mathsf{D}_{aJ} \cdot \frac{\delta\kappa}{\delta \mathrm{A}_{aJ}} \approx 0. \tag{25.20}$$

This is solved by $\underline{\mathrm{E}}_I$ and $\underline{\mathrm{B}}_I$, so $\mathcal{F}[\underline{\mathrm{E}}_I, \underline{\mathrm{B}}_I]$ is also a solution. Once again, this can be rewritten as $\mathcal{F}[W_\gamma, \Phi^{\mathrm{E}}_{\mathrm{S}}]$, now for loop variable (N.2).

Example 3) GR as Geometrodynamics. $\{(\mathcal{M}_i \,|\, \partial \mathrm{L}^i), (\kappa \,|\, \partial\chi^{\mathsf{K}})\} \Rightarrow$

$$\left(\left\{ \pounds_{\partial \underline{\mathrm{L}}} \mathrm{h}_{ij} \frac{\delta}{\delta \mathrm{h}_{ij}} + \pounds_{\partial \underline{\mathrm{L}}} \mathrm{p}^{ij} \frac{\delta}{\delta \mathrm{p}^{ij}} \right\} \kappa_{\mathsf{K}} \,\middle|\, \partial\chi^{\mathsf{K}} \right) \text{'='} \, 0. \tag{25.21}$$

This corresponds to the unsmeared FDE

$$2\,\mathrm{h}_{jk} \mathcal{D}_i \frac{\delta\kappa}{\delta \mathrm{h}_{ij}} + \left\{ \mathcal{D}_i \mathrm{p}^{lj} - 2\,\delta^j{}_i \left\{ \mathcal{D}_e \mathrm{p}^{le} + \mathrm{p}^{le} \mathcal{D}_e \right\} \right\} \frac{\delta\kappa}{\delta \mathrm{p}^{lj}} \text{'='} \, 0. \tag{25.22}$$

In the weak case, we can furthermore discard the penultimate term. The κ_{c} subcase solve

$$2\,\mathrm{h}_{jk} \mathcal{D}_i \frac{\delta\kappa}{\delta \mathrm{h}_{ij}} \text{'='} \, 0. \tag{25.23}$$

These are, formally, 3-geometry quantities '$\mathfrak{G}^{(3)}$' by (25.23) emulating (and moreover logically preceding) the quantum GR momentum constraint (11.7). Finally note that, explicit 'basis beables' are not known in this case.

On the other hand, the FDE for the κ_{p} (formally '$\Pi^{\mathfrak{G}^{(3)}}$') is

$$\left\{ \mathcal{D}_i \mathrm{p}^{lj} - 2\delta^j{}_i \left\{ \mathcal{D}_e \mathrm{p}^{le} + \mathrm{p}^{le} \mathcal{D}_e \right\} \right\} \frac{\delta\kappa}{\delta \mathrm{p}^{lj}} \text{'='} \, 0. \tag{25.24}$$

Example 4) GR in Ashtekar variables.

$$\{(\mathcal{M}_i \,|\, \partial \mathrm{L}^i), (\kappa_{\mathsf{K}} \,|\, \chi^{\mathsf{K}})\} \text{'='} \, 0, \qquad \{(\mathcal{G}_I \,|\, \partial\xi^I), (\kappa_{\mathsf{K}} \,|\, \chi^{\mathsf{K}})\} \text{'='} \, 0. \tag{25.25}$$

The classical Kuchař beables are, formally functionals of the knots and associated momenta alone: $\mathcal{F}[\mathfrak{K}\mathrm{not}, \Pi^{\mathfrak{K}\mathrm{not}}]$. I.e. to commute with $\mathcal{M}_i$ in addition to with $\mathcal{G}_I$, one need to consider diffeomorphism-invariant classes of loops, which are knot classes (Appendix N.13). Moreover, it is loops with no intersections which satisfy the 'observables' condition, and these are very limited in non-Abelian models [330].

25.8 Examples of Classical Dirac Beables

Example 1) In Temporally-Absolute Configurationally-Relational Mechanics, Electromagnetism, Yang–Mills Theory, and the Husain–Kuchař model the absence of any non-linear constraints means that $\mathcal{D} = \mathcal{K}$, so that Sect. 25.7 suffices as a treatment of Dirac beables.

On the other hand, for each of the below, one further $\eth$DE restriction now applies in addition to each of the preceding Section's systems of equations.

Example 2) For Spatially-Absolute Mechanics, for the theories involving whichever subgroup-forming combination of $\mathcal{P}$, $\mathcal{L}$, $\mathcal{D}$ and $\mathcal{K}$, the following extra PDE applies.

$$\{\mathcal{E}, \mathcal{D}\} \text{`='} 0 \Rightarrow \sum_{I=1}^{N} \left\{ \frac{\partial V}{\partial \underline{q}^I} \frac{\partial \mathcal{D}_{\mathsf{D}}}{\partial \underline{p}_I} - \underline{p}_I \frac{\partial \mathcal{D}}{\partial \underline{q}^I} \text{`='} 0 \right\}. \tag{25.26}$$

A distinct equation is required in the other cases. E.g. the equiareal and 2-d affine cases each require distinct $\mathcal{E}$'s built from cross products rather than from Euclidean norms.

Example 3) For Minisuperspace, $\{\mathcal{H}, \mathcal{D}_{\mathsf{D}}\}$ '=' 0 applies. See e.g. [79] for direct construction of classical Dirac beables for Minisuperspace, or Chap. 29 for another method that uses histories-theoretic intermediates.

Example 4) Geometrodynamics' $\mathcal{D}$ require an extra FDE [47] $\{(\mathcal{H} \,|\, \partial \mathrm{J}), (\mathcal{D}_{\mathsf{D}} \,|\, \partial \chi^{\mathsf{D}})\}$ '=' $0 \Rightarrow$

$$\left(\left\{ \frac{\delta \mathcal{D}_{\mathsf{D}}}{\delta \underline{\mathrm{p}}} \{\underline{\mathrm{G}} - \underline{\underline{\mathrm{M}}}\, \underline{\mathrm{D}}^2\} - \frac{\delta \mathcal{D}_{\mathsf{D}}}{\delta \underline{\mathrm{h}}} \underline{\underline{\mathrm{N}}}\, \underline{\mathrm{p}} \right\} \partial \mathrm{J} \middle| \partial \chi^{\mathsf{D}} \right) \text{`='} 0. \tag{25.27}$$

This features the DeWitt vector quantities

$$\underline{\mathrm{G}} := \frac{2}{\sqrt{\mathrm{h}}} \{\mathrm{p}^{ia} \mathrm{p}_a{}^j - \frac{\mathrm{p}}{2} \mathrm{p}^{ij}\} - \frac{1}{2\sqrt{\mathrm{h}}} \left\{ \mathrm{p}^{ab} \mathrm{p}_{ab} - \frac{\mathrm{p}^2}{2} \right\} \mathrm{h}^{ij} - \frac{\sqrt{\mathrm{h}}}{2} \{\mathrm{h}_{ij} \mathcal{R} - 2\mathcal{R}^{ij}\} + \sqrt{\mathrm{h}}\, \Lambda\, \mathrm{h}^{ij} \tag{25.28}$$

which are already familiar from the ADM equations of motion (8.29), and also $\underline{\mathrm{D}}^2 := \mathcal{D}^i \mathcal{D}^j$. In unsmeared form, (25.27) is the FDE

$$\{\underline{\mathrm{G}} - \underline{\underline{\mathrm{M}}}\, \underline{\mathrm{D}}^2\} \frac{\delta \mathcal{D}_{\mathsf{D}}}{\delta \underline{\mathrm{p}}} \text{`='} 2\, \underline{\mathrm{p}}\, \underline{\underline{\mathrm{N}}} \frac{\delta \mathcal{D}_{\mathsf{D}}}{\delta \underline{\mathrm{h}}}. \tag{25.29}$$

Some examples of $\mathcal{D}$ in GR for more specialized highly symmetric cases can be found in e.g. [640, 641, 856].

Toward some further resolutions for GR beyond Minisuperspace,

i) see Chap. 30 for examples of Dirac beables $\mathcal{D}$ in SIC.
ii) Some $\mathcal{D}$ are also explicitly known [856] for some of the Gowdy Midisuperspace models.
iii) Appendix O.8 outlines two early No-Go Theorems by Kuchař and by Torre about the $\mathcal{D}$.

iv) Chap. 29 considers a further method which could extend beyond Minisuperspace.
v) Dittrich's general formal power series expansion objects for GR are of the form

$$D_{\Xi} = \sum_{n=1}^{\infty} \frac{1}{n!} \{\mathrm{V}\}^n \{\Xi, \overline{\mathcal{C}}\}_{(n)}. \tag{25.30}$$

Here Ξ denotes the dynamical fields (ψ and $\mathbf{h}$) and $\mathrm{V}^{\mu} := X^{\mu} - \mathrm{Y}^{\mu}(X^{\mu})$ is a gauge-fixing equation for Y^{μ} spacetime scalar functions. Also $\overline{\mathcal{C}}_{\mu}$ are particular linear combinations of the GR constraints [446]. Furthermore $\{\ ,\ \}_{(n)}$ is an n-fold iterated Poisson bracket, i.e. n Poisson brackets nested inside each other. Each $\mathcal{C}_{\mu}$ is contracted with that on one power of V^{μ}. This approach has already been covered contemporarily in [251, 252, 724, 845], so we detail it here no further. See [251, 252] for the conceptually relevant points of how this construct proceeds via a partial observables stepping stone, and of how some partial observables act as clock in support of the system's other dynamical variables.

Dittrich and Thiemann's approach [251, 845] gets round Torre's No-Go Theorem by [722, 724] involving series of Cauchy data derivatives that are in principle up to infinite order.

25.9 Examples of Further Notions of Beables

Explicit ∂DEs for A-beables can also be formulated, and the equations for those A-beables corresponding to subalgebraic structures of the $\mathcal{F}$**lin** have similar mathematical properties to those in Sect. 25.7.

Example 1) Bein-Geometrodynamics and Supergravity both have a simple notion of *locally Lorentz beables*, corresponding to

$$\{(\mathcal{J}_{AB} \,|\, \partial \mathrm{F}^{AB}), (\mathcal{B}_{\mathrm{B}} \,|\, \partial \mathrm{G}^{\mathrm{B}})\} \text{ `='} \, 0, \tag{25.31}$$

which is widely guaranteed even in circumstances in which other notions of beables break down. This applies e.g. to the bein formulation of GR, to Einstein–Dirac Theory and to Supergravity.

Example 2) In Supergravity, there is a notion of GR-like beables corresponding to the GR-like constraint subalgebraic structure. There also is a notion of *non-supersymmetric Kuchař beables* NSK in Supergravity, corresponding to (25.31) alongside

$$\{(\mathcal{M}_i \,|\, \partial \mathrm{L}^i), (\mathcal{B}_{\mathrm{B}} \,|\, \partial \mathrm{G}^{\mathrm{B}})\} \text{ `='} \, 0. \tag{25.32}$$

This is based on the subalgebraic structure (24.26) of the $\mathcal{NSF}$**lin** closing.

Example 3) There is also a notion of *non-supersymmetric Dirac beables* NSD corresponding to (25.31), (25.32) and

$$\{(\mathcal{H} \,|\, \partial \mathrm{J}), (\mathcal{B}_{\mathrm{B}} \,|\, \partial \mathrm{G}^{\mathrm{B}})\} \text{ `='} \, 0. \tag{25.33}$$

This is based on the subalgebraic structure (24.27) of the $\mathcal{NSC}$ closing.

Model	Constraint Provider	Algebraic Closure	Associated algebraic structure
Minisuperspace			
time	Temporal Relationalism → chronos = $\mathcal{H}$ ↓ t^{em}	✓	{chronos, $\mathcal{D}$} ≈ 0 chronos = Dirac beables
RPM			
time	Temporal Relationalism → chronos = $\mathcal{E}$ ↓ t^{em}	✓	{chronos, $\mathcal{C}$} ≈ 0 chronos beables
space	Configurational Relationalism → **shuffle** = $\mathcal{L}_i$	✓	{$\mathcal{F}$lin, $\mathcal{K}$} ≈ 0 Kuchař beables
space-time	[the above together] $\mathcal{C}_{\text{C}} = \mathcal{E}, \mathcal{L}_i$	✓	{$\mathcal{C}_{\text{F}}$, $\mathcal{D}$} ≈ 0 Dirac beables
GR			
time	Temporal Relationalism → chronos = $\mathcal{H}$ ↓ t^{em}	✗	✗
space	Configurational Relationalism → **shuffle** = $\mathcal{M}_i$	✓	{$\mathcal{F}$lin, $\mathcal{K}$} ≈ 0 Kuchař beables
space-time	Deformation-shuffle Relationalism [or just the above together] $\mathcal{C}_{\text{C}} = \mathcal{H}, \mathcal{M}_i$	✓	{$\mathcal{C}_{\text{F}}$, $\mathcal{D}$} ≈ 0 Dirac beables
spacetime	Spacetime Relationalism → $\mathcal{D}$	✓	$\|[\mathcal{D}, \mathcal{S}]\|$ ≈ 0 spacetime observables
Supergravity			
time	Temporal Relationalism → chronos = $\mathcal{H}$	✗	✗
space	Configurational Relationalism → Non-susy **shuffle** = $\mathcal{M}_i, \mathcal{J}_{AB}$	✓	{$\mathcal{NSF}$lin, $\mathcal{NSK}$} ≈ 0 non-susy Kuchař beables
susy	Supersymmetry itself → $\mathcal{S}_A$	✗	✗
space-time	[the first two together] $\mathcal{NSC} = \mathcal{M}_i, \mathcal{J}_{AB}, \mathcal{H}$	✓	{$\mathcal{NSC}$, $\mathcal{NSD}$} ≈ 0 non-susy Dirac beables
spacetime	Spacetime Relationalism → $\mathcal{D}$	✓	$\|[\mathcal{D}, \mathcal{S}]\|$ ≈ 0 spacetime non-susy observables
space, susy	[the second and third]	✗	✗
time, susy	[the first and third]	✗	✗
space, time, susy	[all three] $\mathcal{C}_{\text{C}} = \mathcal{M}_i, \mathcal{J}_{AB}, \mathcal{H}, \mathcal{S}_A$	✓	{$\mathcal{C}_{\text{F}}$, $\mathcal{D}$} ≈ 0 Dirac beables
spacetime, susy	[Spacetime Relationalism and Supersymmetry] $\mathcal{S}, \mathcal{D}$	✓	$\left\|\left[\begin{smallmatrix}\mathcal{S}\\ \mathcal{D}\end{smallmatrix}, \mathcal{S}\right]\right\|$ ≈ 0 spacetime susy beables

Fig. 25.2 Expansion on the 12 Background Independence aspects laid out in Fig. 10.7, indicating which are realized in Minisuperspace, RPM, full GR and full Supergravity. The last three reflect the progression in algebraic complexity ×, ⊖, ⊖ (explained in Appendix E)

Example 4) Finally, Chap. 30's SIC is another model for which further notions of beables are required. This also provides a concrete example of $\mathfrak{g}$-beables which do not coincide with Kuchař beables.

We are now in a position to summarize in Fig. 25.2 which of the Temporal and Configurational Relationalism, Constraint Closure and Beables aspects of Background Independence are realized in the four main theories or model arenas that have been considered in this book so far.

Research Project 12)† Work out what the concrete spaces of beables found in this Chapter are topologically, geometrically and algebraically. Also write out concrete 𝔡DE's for other types of A-beables. Solve these and elucidate the topological, geometrical and algebraic nature of the space of these.

Research Project 13) [Long-standing.] Resolve the Problem of Beables (or Observables) for Classical GR. E.g. mastery of the cleanly geometrically motivated FDE for '3-geometry momenta' sitting within (25.22) should be attainable, since this equation is concise, linear in its unknown and just second order in its derivatives.

Research Project 14) Consider further the status of beables or observables concepts in classical Supergravity, illustrated with concrete examples.

Chapter 26
Fully Timeless Approaches

This Chapter returns to rung I) of Part II's opening ladder of ontological structure: the most minimalistic approaches. At the primary level, these involve just configurations rather than any further velocities, changes, momenta, dynamical evolution, paths or histories. Minimalism is itself one motivation for these fully Timeless Approaches: some aspects of Nature can be explained without evoking further structure. Another motivation follows from these being approaches which, out of not presupposing an Arrow of Time, might be able to derive a such. Among these approaches, let us further distinguish between the following.

a) Approaches which consider the entirety of a single configuration for the Universe [99, 101].
b) The more often considered version instead involves localized subconfigurations within a single instant [21, 33, 692–694].

26.1 Propositions in the Classical Context

On major way of thinking about Timeless Approaches is to focus on *timeless propositions* or *atemporal questions*, and seek means of supplanting what are usually regarded as temporal questions. Types of atemporal questions include *questions of simple being* of the generic form 'what is the probability that an (approximate) (sub)configuration has some particular property P_1?'.

Part I already illustrated atemporal questions at the quantum level within the Naïve Schrödinger Interpretationin the cosmological setting. Let us also now point to geometrically interesting RPM model universe counterparts: finding the probability that the triangle is approximately equilateral? Isosceles? Regular?[1] Atempo-

[1]See Appendix G for the nomenclature of RPM configurations, and for what regions of configuration space they correspond to, by which valuations of propositions are realized as well-defined regions of geometrically well-understood configuration spaces. The above examples could be based, in reverse order on $|ellip| < \epsilon$, $|aniso| < \epsilon$ and $\sqrt{ellip^2 + aniso^2} < \epsilon$ for ϵ some small tolerance parameter.

E. Anderson, *The Problem of Time*, Fundamental Theories of Physics 190,
DOI 10.1007/978-3-319-58848-3_26

ral propositions also already make sense for Minisuperspace model universes. For instance, one can consider Prob(anisotropy is small) with $|\boldsymbol{\beta}| < \epsilon$ quantification. Other questions involve inhomogeneity, for which Chap. 30's SIC provides a concrete model arena. Moreover, RPMs already model some aspects of inhomogeneity. Tall triangles have a tight binary relative to the distance to the third particle: a type of 'universe contents inhomogeneity' of the subclusters involved. One can similarly consider the ratios of relative separations of multiple pairs of particles within an N-body configuration [59]. Finally, notions of uniformity are of considerable interest in Cosmology; e.g. the $area$ variable for a triangle is a such, which is maximized by equilaterality.

The above allusion to approximation reflects the practicalities of imperfect knowledge of the world. Appendix Q.6 presents both grainings and subconfigurations to this end. Note that Classical Probability Theory (Appendix P.1) suffices at the classical level; this reflects imperfect knowledge of the system. Finally, see Chap. 51 for discussion of the even more interesting quantum counterparts of these structures.

Contrast also with *questions of conditioned being*, which concern two (or more) properties within a single instant. 'What is the probability that, given that an (approximate) (sub)configuration Q_1 has property P_1, it (or some other Q_2 within the same instant) has property P_2? Recollect here Part I's Minisuperspace example of 'what is the probability that the Universe is flat given that it is isotropic'? A RPM model universe example is 'what is the probability that the triangle is approximately regular given that it is approximately isosceles'? N.B. that such questions can have significant scientific content through forming a means of predicting and testing unknowns given knowns.

The totality of atemporal questions form an 'atemporal logic'.[2]

Moreover, in the presence of a meaningful notion of time, one can consider the following additional types of *temporal questions* to be meaningful at the primary level. [These make for useful contrast, and are also the propositions that the current Chapter considers strategies for supplanting of these.]

1) *Questions of being at a particular time* involve Prob(Q_1 has property P_1 as timefunction t takes a fixed value t_1).
2) *Questions of becoming*, moreover, have the further features of Prob(state S_1 dynamically evolves to form state S_2).

Some lucid examples of questions of becoming are whether highly homogeneous universes become more inhomogeneous, or whether highly homogeneous universes become populated by supermassive black holes. The following specific example of a temporal question is, moreover, integral to the scientific enterprise itself.

[2] 'Atemporal' does not refer here to a *mathematical type* of logic, but rather to a philosophical or interpretational type. In particular, temporal logic presents more complications than atemporal logic at the philosophical and interpretational levels due to the extra temporal constructs present.

If an experiment is set up with particular initial conditions,
what final state does its initial state become? (26.1)

Note that without the qualification of 'at a time', questions of 'being' can appear to be vague. Compare e.g. 'is the Universe homogeneous?' to the same but qualified with 'today' or 'at the time of last scattering'.

Strategy A) *Rearrange* Temporal Questions in purely timeless terms, at least in principle. This usually proceeds in two steps.

1) Suppress 'being at a time'.
2) Suppress 'becoming itself'.

If these rearrangements are used in one's conception of the world and of Science e.g. (26.1), one would expect [21] the resultant formulation to strongly reflect the structure of atemporal logic.

Strategy B) *Accept* Temporal Questions. The totality of these furthermore forms a *temporal logic*.[2]

26.2 Fully Timeless Approaches

The current Chapter addresses the Rearrange Temporal Questions Strategy at the classical level. Adopting a Fully Timeless Worldview avoids the difficult issue of trying to define time as outlined in Sects. 1.8 and 1.12. However, this is replaced by three other problems, which we refer to as 'brier patches' so as to keep them grouped despite presenting the classical one here and the other two purely quantum ones in Chap. 51.

Brier patch 1) is a case of choosing to face either Saint Augustin's 'what is time' or the more recent alternative of 'how to cope in Physics without there being a time'. This is because by possessing a time, to be treated in a sui generis fashion, the former complies with the structural expectations of Ordinary QM.

As regards matters of principle, this is an issue of economy (why use a time if it is not necessary) and of primality. I.e. the Machian 'time is an abstraction from change' versus the absolutist 'time is a prerequisite for understanding change'.

Fully Timeless Postulate. One now aims to supplant 'becoming' with 'being' at the primary level (see [21, 99, 101, 340, 411, 413, 414, 418, 421, 692, 693, 731] for partial antecedents). [In this sense, these Timeless Approaches consider the instant or space as primary and change, Dynamics, history or spacetime as secondary.]

26.3 Supplanting Questions of Being at a Time

In this regard, one can pass from correlations between configurations, to clock states being amongst the configurations involved. In this manner, 'being at a time' reduces to timeless correlations between configurations.

26.4 Classical Timeless Structures. i. Good $\mathfrak{g}$ Quantities

These are a configuration space function subset of the gauge-invariant quantities ($\mathfrak{g}$-beables, equal to Kuchař beables in the more commonly used theories). They can be defined from the action of $\mathfrak{g}$ on $\mathfrak{q}$, i.e. without ever involving momenta, constraints or Poisson brackets. This restricted notion of beables can moreover be phrased without constraints (but with generators). This adds a further reason to the generator presentation of algebraic structures entertained in Chap. 9. It also conflates Configurational and Spacetime Relationalisms. In this approach, Closure is more primitive than Configurational Relationalism, since Closure involves just $\mathfrak{g}$ by itself whereas Configurational Relationalism involves $\mathfrak{g}$'s group action on $\mathfrak{q}$ as well. Let us end by noting that whereas reduced approaches proceed via further Principles of Dynamics structures, the relationalspace formulation [37] arrives elsewise at the relational $\tilde{\mathfrak{q}}$.

26.5 ii. Sub- and Super-structures for $\mathfrak{q}$

At this level, one only has configurations and the configuration spaces they form. Various sub- and super-structures of these are, moreover, also implied.

Records are subconfigurations of a single instant that are localized in space (see Appendices G.4 and N.8 for notions of distance upon which notions of localization can be based). This is partly so that records can be controlled, and partly so that one can have more than one such to compare. This also negates signal times within conventional frameworks in which such are relevant. However, this is far from necessarily the basis for a criticism; e.g. pp. 225–226 of [906] points out that Relativistic Theories make use of a similar notion of locality.

Required structures include both sub-statespaces (Appendix Q.3) corresponding to subsystems—the actually observed entities, and unions of statespaces (Appendix Q.4)—due to not knowing the exact contents of the Universe. Not knowing one's configuration beyond a certain level of precision is modelled with coarse-graining (Appendix Q.6). Let us end by noting that remembering is far from a perfect process, and different people usually have experienced at least partially different past occurrences.

26.6 iii. Information, Correlation and Patterns in $\mathfrak{q}$

Records contain information, correlations and patterns more generally (see Appendices Q and R) for technical details of these structures). More specifically, Records Theory requires *useful* information, and *meaningful* correlations and patterns. The way to view this so as to make progress is *systematic means of assessing whether a pattern is random or due to an underlying cause*.

Moreover, a caution against fully Timeless Approaches might be found in the following adage:

$$\text{correlation does not imply causation.} \tag{26.2}$$

This may continue to apply when correlation and causation pick up specific mathematical and physical meanings. [In particular, causation picks up physical theory dependent temporal content, whereas it is specifically timeless correlations that are under investigation.] This may be a reason for the more minimalist programs documented in this book to not suffice by themselves.

26.7 iv. Formalization by Stochastic Mathematics on $\mathfrak{q}$

The key to progressing with Timeless Records Approaches is to consider significant patterns in the sense of *statistically significant*, which is to be underlied by use of Stochastic Mathematics. Let the configuration space $\mathfrak{q}$ now be cast in the role of probability space upon which probability distributions are defined (Appendix P.1). Sharp notions of information, correlation and pattern assessment ensue. Statistical Mechanics is moreover one possible source of notions of information (via the 'negentropy is information' inter-relation of Appendix Q.8).

Many of Appendix R's structures applied to RPMs come from Kendall's work on the statistical theory of shape. I.e. distances between shapes based on Shape Geometry, ϵ-collinearity, and notions of information and correlation for shapes. It is the last of these that allows for diagnostics for whether a given configuration is a record. The Author's point of view is that not all configurations are records. Only those with patterns in significant excess of what is to be expected from randomness are held to be records.

Let us furthermore seek *relational versions* of the above structures. Residing within timeless instants, 'relational' boils down to Configurational Relationalism, of which there are two types of implementation as per Chap. 16: direct and indirect.

Example 1) Consider constellations of points in 1-d, modelling e.g. RPM configurations [33]. The inhomogeneous notion of clumping is already meaningful in 1-d; this is a ratios of relative separations notion. E.g. Roach [738] gives a discrete study of this, which can be interpreted as a coarse-graining of particle model configurations.

Example 2) Consider constellations of points in 2-d, modelling e.g. RPM configurations [33]. These have additional shape information in the form of relative angles. This can be handled in particular by Kendall's approach [536, 537, 539]. In particular, this probes for *collinearity in threes*, triples of points being the minimal relational unit for the similarity geometry in question. By sampling in threes, the relational triangle itself corresponds to sample size 1 for which no statistical analysis is possible. On the other hand, the quadrilateral allows for sampling up to 4 triangles. Kendall's approach furthermore involves placing a probability measure on triangleland's configuration space, which is geometrically (a portion of)

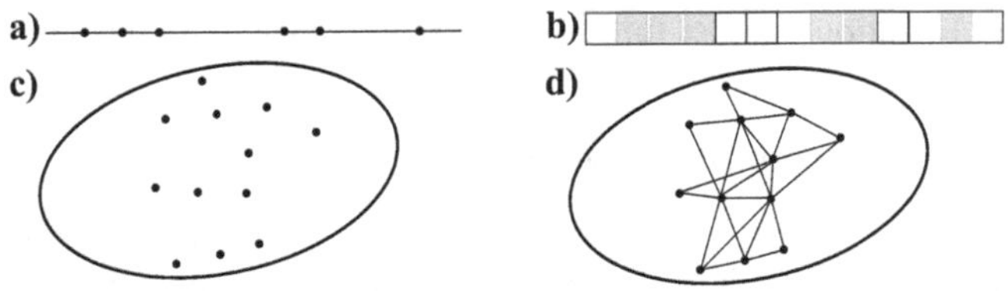

Fig. 26.1 **a)** Clumping in 1-d and **b)** a discrete model of it as per [738]. **c)** Shape data in 2-d. This could consist of e.g. standing stones or of the particle positions in an RPM. **d)** Can the number of almost-collinear triangles involved be accounted for by coincidence or is it a statistically significant pattern? This is a schematic version of the standing stones problem [539, 792], whether or not within a convex (Appendix H.2) polygon edge

the shape sphere (Fig. G.11). This is a subcase of Geometrical Probability Theory (Appendix R) and, more concretely, a Shape Statistics (see Appendix R). It tests an ϵ-collinearity (Fig. Q.2) 'tolerance parameter' quantifying the extent of deviation from collinearity that passes the test.

This is an example of using a subconfiguration space to analyze N-body configurations for N rather larger than 3. For use in Records Theory, note that this has the spherical metric on it as a notion of distance. One can then define probability distributions thereupon, and consequently such as co-variance and notions of information.

Kendall's original modelling situation for the above was the standing stones problem [536, 538, 539] (cf. Fig. 26.1.d). One of Kendall and collaborators' approaches to this built in the assumption that the standing stones lie within a compact convex polygon: 'the Cornish coastline' for the Land's End standing stones problem itself (Fig. 26.1.d). This was done so as to cope with uniform independent identically-distributed probability distributions (much as quantum physicists often perform 'normalization by boxing'). However, details of the polygon enter the test's outcome, and in an RPM setting this would be viewed as an absolute imprint. However, if one uses probability distributions that tail off, it ceases to be necessary to involve a boundary, so these method carries over to the RPM setting.

Example 3) Scaled triangles form the configuration space $\mathbb{R}^3$. $||Dra||^2$ is a Euclidean notion of distance for this, after which standard Probability and Statistics apply, including standard notions of co-variance and of information.

As regards specific Mechanics problems to be settled by Shape Statistics, clumping questions include whether a given Celestial Mechanics configuration exhibits a shape statistically significant number of tight binaries? Does it contains a shape statistically significant globular cluster? Do two images of globular clusters have sufficiently similar clustering detail to be of the same system or to be members of similarly formed populations? A relative angles question is whether a configuration exhibits a shape statistically significant number of eclipses.

Research Project 15)† Work out the smallest relationally nontrivial conformal shape space's topology and geometry, and the form taken by statistics thereupon [36].

Research Project 16)† Ditto for the affine case [36], which is likely to be interesting also from the point of view of Image Analysis [788].

Research Project 17)† Ditto for the projective case [36], with side benefits in Image Analysis [788] and in quantitative assessment of perspective drawings [815]. [Consult [698] for an outline of the forms taken by a wider range of affine and projective shape spaces.]

Example 4) The Shape Geometry of $\mathfrak{Mini}$ and $\mathfrak{a}$ni are known. As for Kendall's shape spaces, these are finite-dimensional. [For all these do not have locality in space, they are a useful precursor to the situation with perturbatively small inhomogeneities in vacuo.]

The Euclidean $\|\mathrm{d}\boldsymbol{\beta}\|^2 := \mathrm{d}\beta_-^2 + \mathrm{d}\beta_+^2$ is a useful notion of distance on $\mathfrak{a}$ni for Bianchi class A. Here one now takes the anisotropy parameters to be random variables. This involves *standard* Probability and Statistics, so standard co-variance $\mathrm{Cov}(\boldsymbol{\beta}, \boldsymbol{\beta}')$ and the n-point function in $\mathfrak{a}$ni make sense, and standard notions of information also apply.

On the other hand, Probability Theory on Minkowski spacetime $\mathbb{M}^n$ has been covered in e.g. [315] A remaining caveat preventing just uplifting some techniques to configuration spaces which carry the Minkowski metric η is that the latter come paired with specific potential functions. This is in contrast to traditional roles for Probability and Statistics on $\mathbb{R}^n$ and $\mathbb{M}^n$ involving just Metric Geometry.

Research Project 18)†† To what extent can statistical analysis of this simple notion of shape be extended to statistical analysis of the conformal-geometric notion of shape in GR? I.e. do any of the above-mentioned ideas carry over from shape space (and its cone) to $\mathfrak{CS}(\boldsymbol{\Sigma})$ and $\{\mathfrak{CS} + \mathrm{V}\}(\boldsymbol{\Sigma})$? What about the $\mathfrak{s}$uperspace$(\boldsymbol{\Sigma})$ version?

26.8 Supplanting Questions of Becoming by a Semblance of Dynamics

For the previous four sections' 'pre-records' structures to amount to a Timeless Records Theory, one has to be able to extract a semblance of dynamics or history from these timeless correlations between same-instant subconfiguration records. Dynamics or history are now to be *apparent notions* to be constructed from the instant [340, 411, 413, 418] by somehow implementing the following postulate. In the absence of change at the primary level, one can only resort to timeless correlations so as to attempt to obtain a semblance of dynamics, change, path or history. Various different approaches to this have been proposed.

Page's Records Approach In this approach [691–693], there is one present instant containing both the subsystem of primary interest and memories or other recorded evidence of 'past instants'. This is as opposed to the conventional sequence of instants. For example, one might view such a present instant configuration as researchers with data sets with memories of how they set up the corresponding experiment. Further specifics of this approach have to await quantum treatment in Chap. 51.

Limitation 1) It can readily be anticipated that Page's Records Approach is highly speculative and very hard to do calculations with.

Limitation 2) While Page's Records Approach is very general in scope, it is not all-encompassing insofar as it cannot explain the appearance of the Arrow of Time. This is because the single instant employed corresponds to a sufficiently *late* stage of the investigation, which amounts to an inbuilt direction of (the ensuing emergence of) time. Experimental set-up also gives a distinguished role to the first relevant instant, in terms of the experiment starting with the apparatus in a controlled initial state.

Barbour's 'Time Capsules' Approach Barbour terms his notion of record *'time capsules'* [99, 101]. More concretely, he considers these to be those configurations from which a semblance of dynamics can be extracted. He also considers records to be whole-universe configurations (unlike any other programs in this Chapter, which consider records to be localized subconfigurations). He furthermore conjectured a selection principle for 'time capsules'; again, this requires Quantum Theory to fully formulate, though some parts of it can already be discussed classically.

Bubble Chamber Arguments How α-particle tracks form in a bubble chamber can be modelled using just the time-independent Schrödinger equation is an underlying inspiration for various Timeless Approaches. Bell [126] subsequently suggested that perhaps Quantum Cosmology could be studied analogously. Barbour and Halliwell [413, 414] each subsequently developed their own version of such a scheme, though presenting the latter is postponed until Histories Theory has been discussed. It is useful to point out at this stage that in Quantum Cosmology, the tracks should be in configuration space $\mathfrak{q}$ rather than within a localized region of space (a 'bubble chamber'). The idea of modelling Cosmology as tracks in $\mathfrak{q}$ itself goes back to Misner [659], which further extends to viewing these tracks as 'in' and 'out' states of a scattering problem in $\mathfrak{q}$ (cf. Fig. 37.1.h).

Barbour's Conjecture 1) There are some distinctive places in the relational $\mathfrak{q}$ (possibly linked to its asymmetries or to the most uniform state).

Barbour's Conjecture 2) The quantum wavefunction of the Universe peaks around Conjecture 1)'s places—described as 'mist concentrating' [101]—by which these are probable configurations.

Barbour's Conjecture 3) These places in $\mathfrak{q}$ contain 'time capsules', from which a semblance of dynamics can be extracted.

Problem 1) When he made this conjecture [98, 101], he also suggests the wedge shape of his representation of triangleland could play a role. This feature is however merely representation-dependent [37]. So it is rather likely that his further conjectures were made without a concrete and correct mechanism in mind. Conjecture 1) can however be reinterpreted in terms of irreducible features such as a particularly uniform configuration, the maximal collision itself, or, more generally, the presence of strata. The last two of these features are laid out in Appendix G.

Problem 2) Conjecture 2) is open to assessment due to RPM quantum wavefunctions having been computed, from which quantum probability density functions can readily be extracted. While we postpone this to Part III, we comment here that using classical geometrical probability distributions and Shape Statistics to analyze the classical dynamics of RPMs is a useful classical precursor. For instance, tight binaries *are* a likely outcome from the Classical Dynamics of 3 bodies, which could then be paralleled by atom-like constructs at the quantum level. Kendall's work involving collinearity in threes furthermore features records peaking about the $\mathbb{RP}^k$ of collinear configurations of the full shape space. However, considering gross over-representation of any other angle, one concludes that classical records need not be tied to maximally physically significant zones of $\mathfrak{q}$. Another limitation on the original form of Conjecture 2) which can even more readily be envisaged at the classical level is that $\mathfrak{q}$ geometry is but one of the major players in determining confinement and concentration, the other being the potential function. In particular, potentials can confine systems to regions of $\mathfrak{q}$ which are away from the geometrically distinguished features of $\mathfrak{q}$. Of course, Conjecture 2) can be modified to encompass potential effects in tandem with $\mathfrak{q}$ geometry.

Another issue is which concrete features distinguish time capsules from just any instants. The notions of information content and pattern laid out in this Chapter are likely relevant here. Passing to a localized subconfiguration conception may help as well.

Suppose we view records as subconfigurations and subsequently approach Barbour's conjecture in this setting. The following further issue then arises. In the study of *branching processes*, one learns that 'how probable a subconfiguration is' can depend strongly on the precise extent of its contents. As an example (closely paralleling Reichenbach [731]), suppose we see two patches of sand exhibiting hoof-shaped cavities. Here there are past interactions of these two patches of sand with a third presently unseen subsystem—a horse that has subsequently become quasi-isolated from the two patches. By these, there is a clear capacity of rendering the individually improbable (low entropy and so high information) configurations of each patch of sand collectively probable (high entropy, low information) for the many sand patches–horse subsystem. This still does not explain why useful records appear to be common in Nature: a separate argument would be required to account for why branching processes are common. Whitrow [906] argued such approaches to be problematic due to their dependence on the entropy of a 'main system'. Moreover, one runs out of being able to resort to such an explanation as one's increases in subsystem size tend to occupying the whole Universe.

The Author's Records Based on Shape Statistics This approach is based, firstly, on the pre-records structures of the previous four sections. Secondly, following statistician Simon Broadbent [172], let us next consider making use of a range of different values for Shape Statistics' tolerance parameter ϵ. E.g. in the case of the standing stones problem, significant results for $\epsilon \leq 10$ minutes of arc would suggest that the stones were laid out skillfully by the epoch's standards for e.g. astronomical or religious reasons. On the other hand, $\epsilon \leq 1$ degree would suggest that they just the markers of paths or fences between plots of land. So values of tolerance criteria required to detect significant patterns can, at least qualitatively, contribute to a semblance of history or of dynamics [33]. Thereby, Shape Statistics is not only a mathematically advanced formulation of pre-records but also a candidate Records Theory in its own right.

One may view this approach as replacing the conjecture that some global configurations have special 'time capsule' properties by a concrete mathematical approach—Shape Statistics—of judging which local subconfigurations have significant pattern and semblance of dynamics properties. In contrast to Barbour's and Page's Records Approach, this approach is already viable at the classical level rather than depending on evoking specifically quantum-mechanical machinery. Finally, it is free from the bias of Conjectures 1) and 2)'s association of records with geometrically distinct regions of $\mathfrak{q}$. $\mathfrak{q}$'s geometry clearly still enters this approach though underlying the theory of Geometrical Probability and Shape Statistics thereupon, just not in a manner which necessarily privileges geometrically distinguished regions. This is because placing probability distributions on a geometrical space has a life of its own, beyond the study of that geometry.

Hartle's Information Gathering and Utilizing Systems [432] This is a model arena for a recording device or a simple robotic observer (taken to mean 'non-sentient'). As well as being interesting in its own right, this may eventually confer computational capacity to Page's Records Approach. Since this approach can also be taken to be rooted in $\mathfrak{q}$ geometry, considerations along the lines of Shape Geometry and Shape Statistics also apply here.

26.9 Cambium Records

This Chapter's modelling of records can also be conducted within less sparse worldviews. Once change is primary, abstracting time from it becomes a viable competitor to the semblance of dynamics from Fully Timeless Approaches. One can also *assume* paths or histories (Chaps. 28 and 29) and proceed to study Timeless Records Theory therein without needing to extract a semblance of history.

These less sparse worldviews come with their own versions of the sub- and super-statespace schemes. See Appendix Q.6 for grainings of $\mathfrak{T}(\mathfrak{q})$ and $\mathfrak{T}^*(\mathfrak{q})$, and for probability schemes in phase space are more familiar from Statistical Mechanics.

Let us end by arguing against what is on some occasions presented as another motivation for Timeless Approaches, namely that 'now is what we experience'. This is however a notion of psychological experience which really involves the *specious present* notion (Fig. 4.4), which in fact does have some extent in time. In this manner, approaches which make reference to consciousness are not accurately implemented by timeless instants.

Chapter 27
Spacetime Relationalism

Let us now start afresh with spacetime $\langle \mathfrak{m}, \mathbf{g} \rangle$ assumed as the primary ontology. We consider this in particular for theories within GR's Einsteinian Paradigm, and in contrast to Part II's approaches hitherto which are based on space as primary ontology.

27.1 Implementation of Spacetime Relationalism

At the classical level, Spacetime Relationalism is a resolved problem. In particular for GR, the group of physically meaningless transformations acting on spacetime is $\mathfrak{g}_{\text{S}} = \mathit{Diff}(\mathfrak{m})$. $\mathit{Diff}(\mathfrak{m})$-invariance is a central and physically sensible property of GR, and this binds together much of the Problem of Time [483]. For GR with fundamental matter sources, $\langle \mathfrak{m}, \mathbf{g} \rangle$ is augmented to $\langle \mathfrak{m}, \mathbf{g}, \psi \rangle$ for fundamental matter fields ψ and $\mathfrak{g}_{\text{S}}$ is augmented to $\mathfrak{g}_{\text{int}} \times \mathit{Diff}(\mathfrak{m})$—of the form (18.1)—if the ψ possess an internal gauge group $\mathfrak{g}_{\text{int}}$. This is moreover an extension of Spacetime Relationalism to 'Spacetime-and-Internal Relationalism'. $\mathbf{g}$ passes Sect. 10.1's STR-i) by being a solution of the Einstein field equations, which in the latter case are in turn influenced by the matter content of the Universe ψ's energy–momentum–stress tensor $\mathbf{T}$.

As regards model arenas, RPMs space-time does not have additional spacetime structure or any counterpart of $\mathit{Diff}(\mathfrak{m})$. Minisuperspace considered in terms of its spaces that are privileged by homogeneity also presents a drastic simplification. These limitations motivate adding SIC (Chap. 30) to the repertoire of model arenas that this book draws its detailed examples from.

On the other hand, SR fails to obey STR-i) due to $\eta_{\mu\nu}$ being a fixed background metric; perturbative String Theory fails likewise. It is M-Theory—or at least some limiting classical action for this—which would be expected to have further Background Independence aspects.

E. Anderson, *The Problem of Time*, Fundamental Theories of Physics 190,
DOI 10.1007/978-3-319-58848-3_27

27.2 *Diff*($\mathfrak{m}$)'s Brackets and Algebraic Structure

GR involving $\mathfrak{g}_S = \mathit{Diff}(\mathfrak{m})$ is contingent on this complying with Generator Closure under a suitable bracket. Moreover, unlike with Chap.'s 18 *Diff*($\boldsymbol{\Sigma}$) auxiliaries $\mathcal{M}_i$, these $\mathcal{D}_\mu$ are not associated with dynamical constraints, nor is the suitable bracket in question a Poisson bracket.

$$|[(\mathcal{D}_\mu \,|\, X^\mu), (\mathcal{D}_\nu | Y^\nu)]| = (\mathcal{D}_\gamma \,|\, [X, Y]^\gamma), \tag{27.1}$$

so this matter is resolved in the form of an (infinite-d) Lie algebra in parallel to that of the *Diff*($\boldsymbol{\Sigma}$). This is also to be contrast with the much larger and harder structural form taken by the Dirac algebroid.

This case requires instead the Generator Closure notion. I.e. a consistency check on whether a given set of generators close up without having to introduce further generators. There is a lattice of cases for which closure is attained paralleling Fig. 24.6. This version of closure that the spacetime setting requires is simpler than the Canonical Approach's Dirac Algorithm. This is due to the lack of appending, so that no specifier equation based complications can materialize.

27.3 The Space of Spacetimes and of GR Solutions

The space of pseudo-Riemannian metrics (10 independent components in 4-d) on a fixed topological manifold $\mathfrak{m}$ was termed $\mathfrak{P}\mathfrak{R}$iem($\mathfrak{m}$) by Isham [477]. The space of pseudo-Riemannian geometries (6 independent component entities) is

$$\mathfrak{S}\text{uperspacetime}(\mathfrak{m}) := \mathfrak{P}\mathfrak{R}\text{iem}(\mathfrak{m})/\mathit{Diff}(\mathfrak{m}). \tag{27.2}$$

This nomenclature parallels that of Wheeler's Superspace; $\mathfrak{S}$uperspacetime($\mathfrak{m}$) was furthermore considered by Stern (reported in [301]) and by Isham [477].

One also requires the *solution space* of those pseudo-Riemannian metrics that additionally solve the Einstein field equations of GR, which we denote by GR-$\mathfrak{S}$ol($\mathfrak{m}$). Let us take this to be the version without *Diff*($\mathfrak{m}$) quotiented out, while using $\mathfrak{T}$ruespacetime($\mathfrak{m}$) to denote the more interesting but harder to handle version for which *Diff*($\mathfrak{m}$) is also quotiented out. See Fig. 27.1 for comparison with the $3 + 1$ split version's main sequence of spaces.

The space of conformal spacetime metrics (9 independent component entities) is

$$\mathfrak{C}\mathfrak{p}\mathfrak{R}\text{iem}(\mathfrak{m}) := \mathfrak{P}\mathfrak{R}\text{iem}(\mathfrak{m})/\mathit{Conf}(\mathfrak{m}); \tag{27.3}$$

this attains further significance as the space of causal structures. Finally, the space of conformal spacetime geometries (5 independent component entities) is

$$\mathfrak{C}\mathfrak{S}\mathfrak{S}(\mathfrak{m}) := \mathfrak{C}\mathfrak{p}\mathfrak{R}\text{iem}(\mathfrak{m})/\mathit{Diff}(\mathfrak{m}) = \mathfrak{P}\mathfrak{R}\text{iem}(\mathfrak{m})/\mathit{Conf}(\mathfrak{m}) \rtimes \mathit{Diff}(\mathfrak{m}), \tag{27.4}$$

which we term *Conformal Superspacetime* in analogy with Conformal Superspace.

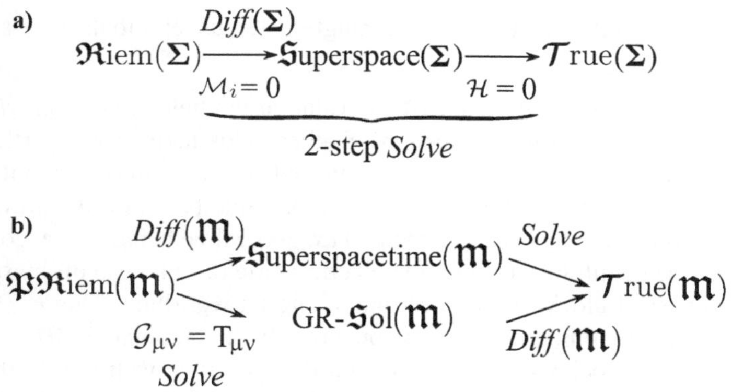

Fig. 27.1 **a)** Space–time split GR's main sequence of spaces, in contrast to **b)** this Sec's unsplit spacetime's diamond of spaces

27.4 The Path (Via) Alternative

Let us next contemplate rung III) on Part II's opening ladder of ontological structure.

Paths Postulate (alias *Non Tempus sed Via*) Perhaps it is paths in $\mathfrak{q}$ that are primary. In the finite case, denote these by $\gamma := \mathcal{Q}^{\mathsf{A}}(\lambda)$; these are taken to be labelled over the entirety of an interval, $\lambda \in \mathfrak{I}$ (so they are 'thick' rather than 'thin'). The field-theoretic counterpart is straightforward enough, with the emergent-time version now carrying a local or multi-fingered label.

Also use $\mathfrak{p}\text{ath}(\mathfrak{q})$ to denote the space of paths on $\mathfrak{q}$, e.g. $\mathfrak{p}\text{ath}(\mathfrak{Riem}(\boldsymbol{\Sigma}))$ for redundantly formulated GR.

The above are hitherto assumed to be labelled by a continuum notion of λ; one can however consider discrete time-step versions of paths as well. Note moreover that the coarse- and fine-graining notions introduced in Chap. 26 additionally apply to discrete parametrizations of paths that capture differing amounts of detail about the path.

A major inter-relation with spacetime is that 'paths in Riem' can additionally be interpreted as spacetimes. In fact this requires paths in $\mathfrak{T}(\mathfrak{Riem}(\boldsymbol{\Sigma}))$ or $\mathfrak{T}^*(\mathfrak{Riem}(\boldsymbol{\Sigma}))$, so as to know how the 3-metrics on each spatial slice fit together: i.e. including extrinsic curvature or GR momentum information. This is developed further in Chap. 31.

27.5 Spacetime Observables

Having considered $\mathfrak{m}$, we next Take Function Spaces Thereover. While there is conventionally no complete spacetime analogue of Chap. 25's beables, the notion of *Diff*$(\mathfrak{m})$-invariant quantities given by objects s_{Q} such that

$$|[(\mathcal{D}_\mu \,|\, \mathrm{Y}^\mu), (s_{\mathsf{Q}} \,|\, \mathrm{Z}^{\mathsf{Q}})]| = 0 \tag{27.5}$$

indeed remains useful and physically meaningful.[1] Let us refer to these s_{F} as *space-time observables*.

Example 1) For ϕ a scalar field on $\mathfrak{m}$ the value of the field *coincident with* some particular particle—or some collection of other fields taking on a particular set of values—are *Diff*($\mathfrak{m}$)-invariant statements and so is an observable in the above sense. This covers both matter time and internal time based on the gravitational field. Moreover [483], this is a statement *in excess* of time being a local coordinate on a spacetime manifold. Consider the practical case of time read off by spatially-localized physical clock: the proper time along its worldline. This is *Diff*($\mathfrak{m}$)-invariant if the group in question acts concurrently on the points in $\mathfrak{m}$—and consequently on the clock's worldline—and on the spacetime metric **g**. In this manner, if the initial and final points of the worldline are labelled using the matter or gravitational internal time coordinate in question, the proper time along the interconnecting geodesic would attain an intrinsic character.

Example 2) *Spacetime integrals* are a particular subcase. For instance,

$$\mathcal{F}[\mathbf{g}] := \int_{\mathfrak{m}} \mathrm{d}^4 X \sqrt{|\mathbf{g}|}\, \mathcal{R}^{\mu\nu\rho\sigma}(\vec{X};\mathbf{g})\,\mathcal{R}_{\mu\nu\rho\sigma}(\vec{X};\mathbf{g}]. \quad (27.6)$$

Such observables are furthermore intrinsically *non-local*.

Example 3) The Weyl scalars are *spacetime* scalars built out of the spacetime Weyl tensor by various products and contractions. Their specific forms were worked out by physicists Robert Debever and Jules Géheniau [338]. The Weyl scalars are furthermore foundational in the Newman–Penrose formulation of GR [706] (named in part after relativist Ted Newman). Finally, the Weyl scalars are of potential practical importance through being observable in principle in a local and convenient manner, such as by use of mathematical physicist Peter Szekeres' gravitational compass [825].

27.6 Use *Diff*($\mathfrak{m}$) or Some Larger Group?

$$\textit{Diff}(\mathfrak{m}) = \{\epsilon^{\mu}(\vec{X})\} \quad (27.7)$$

corresponds to *coordinate* alias *point transformations*[2] $\vec{X} \rightarrow \vec{\tilde{X}} = \vec{f}(\vec{X})$. Bergmann's path-based notion of gauge [133] is appropriate in this context. More-

[1] One might moreover consider a notion of weak equality instead, now meaning up to terms containing the generators. Also, one is only to select subsets of generators which close algebraically. The Jacobi identity applying to all Lie brackets, these s_{Q} are also guaranteed to close as a Lie algebra.

[2] Infinitesimal transformations $\vec{X} \rightarrow \vec{\tilde{X}}$ can be written as $\vec{X} - \vec{\tilde{X}} = \vec{\epsilon}$. Viewed as solutions in terms of Hamiltonian variables, the right hand side functions here are so-called *descriptors*: a fairly standard Gauge Theoretic notion, see e.g. [13, 134]). For GR, descriptors are of the particular form $\vec{\mathrm{v}}(\vec{X}; \Xi(\vec{X})]$. Here Ξ denotes the set of dynamical fields ψ and **h**; note that this specifically excludes the lapse α and shift β^i. Dittrich's v^{μ} in (25.30) and Chap. 27.5's Weyl scalars can be viewed as particular cases of descriptors.

over, GR is invariant under a larger group [134]: the *diffeomorphism-induced gauge group*[3]

$$Digg(\mathfrak{m}) := \{\epsilon^{\mu}(\vec{X}; \theta(\vec{X})]\}. \tag{27.8}$$

Here $\theta(\vec{X})$ denote the fields in one's theory: the metric $\mathbf{g}(\vec{X})$ alongside the matter fields $\psi(\vec{X})$].

27.7 Relationalism as Alternative Route to Physical Theories

N.B. that this book's main Relational Approach does not in general concur that constructing Lagrangians to obey a pre-determined list of symmetries is the only approach to Physics. We shall see in Chap. 38 that Wheeler wrote in favour of a wider perspective. This point of view is furthermore entailed by two of the answers to his question (9.1), as laid out in Chaps. 32 and 33. This book emphasizes, rather, looking for $\langle \mathfrak{q}, \mathfrak{g}, \mathsf{S} \rangle$ triples in space–time split approaches, or for $\langle \mathfrak{m}, \mathfrak{g}_{\mathsf{S}}, \mathsf{S} \rangle$ triples in spacetime approaches.

If such a triple fails, one is free to reconsider *any* part of the triple. In particular this means that we do not necessarily choose the *largest mathematically possible* group that a $\mathfrak{m}$ (or $\mathfrak{q}$) can take as being the physical one.[4] Nor do we necessarily choose to include the totality of terms in S that are compatible with a given $\mathfrak{g}_{\mathsf{S}}$ acting on a fixed $\mathfrak{m}$ (or a given $\mathfrak{g}$ acting on a fixed $\mathfrak{q}$). Both the possibility of including less terms are open to us, as well as that of considering more (by reconsidering one or both of the pair).[5]

All in all, *consistency* determines which theories are allowed, whether in analyzing the closure of the spacetime brackets or by the a Dirac-type Algorithm in the Canonical Approach. The second of these is demonstrated to be a comparably restrictive filter to standard procedures in Chap. 33. Some major theories—in particular standard Gauge Theory and GR minimally-coupled to fundamental matter

---

[3]See [721] for the sense in which this is 'induced'. $Digg(\mathfrak{m})$ might also be denoted $BK(\mathfrak{m})$ after Bergmann and Komar [134], though they themselves referred to it as the 'Q-group'. Physicists Josep Maria Pons, Donald Salisbury and Kurt Sundermeyer prefer to use the Bergmann–Komar name for the subsequent projected version of this group that introduced in Chap. 32.4. Incidentally, the existence of this larger invariance group does not by itself dictate that it is *the* gauge group. [This is one example of Sect. 27.7's matter, as well as the more specific reason to use the projected version.]

[4]For suppose that a model, with maximal symmetry group $\mathfrak{g}_{\max}$, fails to capture some physically meaningful features which themselves do not respect $\mathfrak{g}_{\max}$. Then there is a smaller choice group, $\mathfrak{H}$, which happens to be more physical. Moreover, one way of finding it and the missing physically meaningful features is to consider not just $\mathfrak{g}_{\max}$ for the original model, but rather it and all its subgroups. This is out of the possibility that one or more of these are *more* physically valuable than $\mathfrak{g}_{\max}$ itself.

[5]Moreover, extending $\mathfrak{q}$ or its spacetime equivalent may further enlarge the largest symmetry group $\mathfrak{H}$, break $\mathfrak{H}$ or both at once. I.e. extend the subgroup that survives a breaking in a different way from the original $\mathfrak{H}$.

fields—can indeed be arrived at by either approach. Furthermore, Chap. 33 gives examples of how such schemes can produce more than just the 'usual' theories. Thus Relationalism offers a distinct means of seeking and filtering out theories.

It is finally worth mentioning here that neither of the conventional approach's premises of Lorentz nor General Covariance are in fact guarantors of consistency, by counter-examples 2) and 3) of Appendix J.15.

27.8 Contrast Between Spacetime and Temporal and Configurational Relationalisms

The well-known Einstein–Hilbert action of the spacetime formulation of GR is built from a spacetime scalar. Thus in the spacetime formulation, no manifest corrections are required at the level of the action. This is to be compared with the $\mathbf{d}\mathfrak{g}$ that entered Chap. 18's actions, since their split-off kinetic terms do not constitute good $\mathfrak{g}$ objects without such corrections.

On the other hand, corrected-derivative entities analogous to Best Matching do still exist for spacetime. These are now interpreted in terms of an auxiliary, rather than physically-realized, higher-dimensional-manifold. E.g. Generally-Relativistic perturbation theory can be cast as an unphysical but technically-useful stack of spacetime 4-geometries that are interrelated through Lie derivative terms [814]: another example of point identification map. On this occasion, the Lie derivative is with respect to a spacetime 4-vector, rather than the Best Matching case's spatial 3-vector. This particular example can be considered as a *local region* within $\mathfrak{PR}\mathrm{iem}(\mathfrak{m})$, centred about the (usually highly symmetrical) unperturbed solution.

Thus all three Relationalisms—Temporal, Configurational and Spacetime—are *implemented by Lie derivatives* [part ii) of each, whereas part i) of each is an absence of background structures]. However, the manner in which each type of Relationalism is subsequently resolved differs amongst the three. While a 'principle of minimum incongruence' can be applied to spacetimes, there is not an underlying action object unlike for Configurational Relationalism. As regards Chap. 3's 'acts but cannot be acted upon' criterion, that the background enters dynamical field equations but the dynamics does not enter background structures is of some relevance, but suffers from issues of formulation dependence. See the discussion of the '1-orbit criterion' (group orbit) in [362] for a more advanced point of relevance at this stage.

Note moreover the absence of an underlying Machian principle as compared to each of Temporal and Spatial Relationalism. This is not just because the advent of spacetime postceded Mach's work, but also because of space and time's conceptual heterogeneity. A partial paraphrasing of Mach's Space Principle—a 'Mach-type Spacetime Principle' —is possible, along the following lines. 'No one is competent to predicate diffeomorphism dependent things about spacetime. These are pure things of thought, pure mental constructs that cannot be produced in experience. All our principles of GR spacetime are, as we have shown in detail, experimental knowledge concerning diffeomorphism-invariant quantities'. One can also extend to (spacetime) × (internal space of fields on spacetime) in parallel to Sect. 18.2. However, the above has a lack of temporal inputs despite Broad's point on time and space's co-geometrization as spacetime not overriding their conceptual distinction. Whoever takes sufficient issue along these lines is of course free to continue using separate Mach Time and Space Principles in a geometrodynamical setting.

More usually, one makes a choice to work with one of split or unsplit spacetime. However, detailed study of QFT often makes *joint* use of both Canonical and Path-Integral methods: deriving Feynman rules canonically and then manipulating complex calculations using path-integral methods. This suggests that Canonical versus Path Approaches is not a strict alternative, with combined Canonical-and-Path Approaches presenting a third alternative. Thereby, one might be cautious about ascribing primary ontology exclusively to only one of configurations, configurations and changes, paths, or the histories of the next Chapter. A case in which configurations and one of paths or histories have coprimality is also plausible.

Furthermore, a few approaches to Background Independence and Quantum Gravity do combine spacetime and split spacetime concepts. Consequently, these manifest all of Temporal, Spatial and Spacetime Relationalism at once, and require consideration of all of $\mathit{Diff}(\boldsymbol{\Sigma})$, $\mathit{Diff}(\mathfrak{m})$, $\mathit{Diff}(\mathfrak{m}, \mathfrak{F}\mathrm{ol})$, or of Canonical-and-Histories-Canonical formulations as per the next Chapter. Note additionally that the Bergmann spacetime primality, path-gauge notion and consequent notion of observable form a consistent combination.

With Canonical Supergravity—unlike that of GR—not admitting separation into Configurational and Temporal Relationalism, perhaps Supergravity is indicatory of spacetime primality. Moreover, it is usually in the spacetime setting in which Supersymmetry arises from seeking a viable violation of the direct product of internal and spacetime symmetries (11.16). Again, the unsplit *super-Diff*$(\mathfrak{m})$—e.g. from gauging the Poincaré supergroup—is more straightforward to handle than its split counterpart. We denote the additional generators here by $\boldsymbol{\mathcal{S}}$.

Finally, as regards Nododynamics, Samuel [762] pointed out that the canonical version of this is only formulated as the pull-back of a spacetime connection in the original complex case with $\beta = \pm i$. Suppose that formulation in terms of connections is taken to be central to such approaches. Additionally suppose that spacetime structure is indispensable, whether as part of Background Independence in GR-type theories or due to evoking a worldview with spacetime primality or canonical-and-spacetime coprimality. Then if these aspects are to be combined in a simple and natural way by use of connections which are geometrically spacetime objects, the original complex case would be chosen over the real variables case.

Chapter 28
Classical Histories Theory

28.1 $\mathfrak{g}$-Free Case

We finally turn to the last rung IV) of Part II)'s ladder of ontological structure. Classical histories $\boldsymbol{Q}(\lambda)$—where for now λ is a label time—are in part paths as in the previous Section. Evoking these is clearly a Historia Ante Quantum approach. Let us use ($\mathfrak{Hist}$, *Hist-Point*) to denote the space of histories and the corresponding morphisms. Isham and physicist Noah Linden [504] pursue this by considering histories to have a similar status to configurations. Further canonical structure is therefore to be ascribed to histories. In other words, Histories Theory is taken to possess the following structures.

1) Conjugate *histories momenta* $\boldsymbol{P}(\lambda)$.
2) *Histories brackets*

$$|[F(\lambda'), G(\lambda'')]|_{\mathbf{H}} := \sum_{\mathrm{A}} \left\{ \frac{\partial F(\lambda')}{\partial Q^{\mathrm{A}}(\lambda)} \frac{\partial G(\lambda'')}{\partial P_{\mathrm{A}}(\lambda)} - \frac{\partial F(\lambda)}{\partial P_{\mathrm{A}}(\lambda)} \frac{\partial G(\lambda'')}{\partial Q^{\mathrm{A}}(\lambda)} \right\}. \tag{28.1}$$

The fundamental histories bracket is, in the continuum label case,

$$\{Q^{\mathrm{A}}(\lambda), P_{\mathrm{A}'}(\lambda')\}_{\mathbf{H}} = \delta^{\mathrm{A}}{}_{\mathrm{A}'}\delta(\lambda, \lambda'). \tag{28.2}$$

Note how even a finite particle model gives a field-theoretic bracket here. The discrete-labelled version remains finite in this sense (this has $\lambda, \lambda' \longrightarrow \lambda_1, \lambda_2$ and Dirac $\delta \longrightarrow$ Kronecker δ).

3) $\langle \boldsymbol{Q}(\lambda), \boldsymbol{P}(\lambda), |[\ ,\ ]|_{\mathbf{H}} \rangle =:$ $\mathfrak{Hist}$-$\mathfrak{P}$hase: *histories phase space*. *Hist-Can* are the corresponding histories canonical transformations.

It is due to the above development that the Author [26] separates out *Non Tempus sed Historia* approaches from Isham and Kuchař's version of *Tempus Nihil Est*. This further structure is also in excess to that possessed by a theory based on clas-

E. Anderson, *The Problem of Time*, Fundamental Theories of Physics 190,
DOI 10.1007/978-3-319-58848-3_28

sical paths.[1] Isham and Linden's particular approach can now on clear grounds be dubbed a Histories Brackets Approach. It involves, moreover, continuous rather than discrete time-steps. Finally, now that we have made the above distinction between histories and paths, we can state the following postulate.

Histories Postulate (alias *Non Tempus sed Historia*): one's primary entities are to be histories (rather than configurations or paths).

Such histories primality in the sense used in this book is a perspective first brought to the GR context by the *Histories Theory* of Gell–Mann and Hartle [340, 428]. This is moreover a Historia Post Quantum approach, so discussion of it is deferred to Part III. Isham and Linden's own histories brackets work was further developed by them and their collaborators: physicists Ntina Savvidou, Charis Anastopoulos and Ioannis Kouletsis [11, 503, 504, 566, 767–771].

A case can be made that the above postulate, like spacetime primality, goes against the tradition of basing Physics on Dynamics. On this occasion, however, more similarities with Dynamics are preserved, as is clear from the paths momenta, histories brackets and paths phase space' names. One issue is whether paths or histories have as much primary operational meaningfulness as configurations do. I.e. whether they are entities that one can directly measure. Moreover, within Histories Approaches, measurements at one instant of the history constitute records, and the ontology allows for measurements at distinct instants of the history.

Marolf [639] proposed a distinct way of obtaining histories brackets. Here the Hamiltonian is viewed as an extra structure by which the Poisson bracket is extended from being a Lie bracket on phase space to a Lie bracket on the phase space of paths. This is in contrast with the more usual approach, in which one puts the equal-time formalism aside at this stage and introduces a new phase space $\mathfrak{Hist}\text{-}\mathfrak{Phase}$ in which the Poisson bracket is defined ab initio over the space of histories $\mathfrak{Hist}$.

One also has a notion of *histories constraints*, e.g.

$$c_{\mathsf{C}}^{\lambda} = \int \mathrm{d}\lambda \, c_{\mathsf{C}}(\lambda) \tag{28.3}$$

in the averaged sense. For now, an example of this in the $\mathfrak{g}$-free case is

$$\mathcal{Q}\mathrm{uad}^{\lambda} = \int \mathrm{d}\lambda \, \mathcal{Q}\mathrm{uad}(\lambda). \tag{28.4}$$

Sub-histories can be pieces of a history or the history traced in a subconfiguration space. Notions of coarse- and fine-graining of histories are furthermore clearly supported on this set of structures. Let us use $C_{\bar{\gamma}}$ to denote coarse-graining of paths where $\bar{\gamma}$ is a path consisting of a subsequence of the path γ's instants. So path formulations possess, in addition to the coarse-graining criteria in Appendix Q.3, *coarse-graining by probing at less times*.

[1]There are further quantum-level structures that distinguish a path from a history in the sense of Consistent Histories Theory (Chap. 53).

Histories have a list of desirable properties, much as time has (as laid out in Part I). For instance, histories have some form of time ordering and causality notions as well, albeit conceptualized with the notion of history replaces the notion of time.

Examples covered by the current section include Temporally-Relational but Spatially-Absolute Mechanics, 1-d scaled RPM, and the vacuum anisotropic and minimally-coupled scalar field isotropic Minisuperspaces.

Finally N.B. that histories have matching structural levels to Appendix Q's; in fact many of the latter are better known in the Theoretical Physics literature.

28.2 $\mathfrak{g}$-Nontrivial Classical Histories

Presenting this in the emergent-time case, the preceding Sec's histories constraint $\mathcal{Q}\mathrm{uad}^{\lambda}$ is now accompanied by

$$\mathcal{F}\mathbf{lin}^{\lambda} := \int \mathrm{d}\lambda \, \mathcal{F}\mathbf{lin} = 0. \tag{28.5}$$

Research Project 19) Formulate the $\mathfrak{g}$-nontrivial classical Histories Theory versions of the layers of structure of physical theory whose configurational counterparts are given in Appendix Q.

Examples covered by the current section include triangleland RPM [25] and full GR [428]. For the r-formulation of scaled triangleland, the histories can be taken to be sequences of Dragt-type coordinates $\boldsymbol{Dra}^{\lambda}$.

28.3 Classical Histories Constraint Closure and Beables

One subsequently encounters the obvious histories phase space generalizations of the notions of first- and second-class constraints, Dirac brackets, extended phase space, effective constraints and constraint algebraic structure. In cases in which second-class histories constraints are initially present, the histories version of the Dirac bracket or reformulation in terms of histories effective constraints are required. The algebraic structure obtained by the closure of the constraints is distinct from that involved in spacetime or in split spacetime. The seven strategies in Fig. 24.3 all have histories theoretic counterparts, albeit these largely remain to be studied. Finally, all constraint and observables subalgebraic structures in Fig. 24.6 pass over to histories versions $\mathfrak{L}^{\mathrm{H}}_{\mathfrak{c}}$ and $\mathfrak{L}^{\mathrm{H}}_{\mathfrak{b}}$ in conjunction with histories brackets denoted with a $\mathbf{H}$ suffix: $\mathbf{|[\ ,\ ]|_H}$ These are named by adjoining the prefix 'histories', as in 'histories Dirac observables' etc.

Chapter 29
Classical Machian Combined Approach

We next combine the classical level's Machian Scheme (Chaps. 15 and 23), Timeless Records (Chap. 26), and Isham–Linden type Histories Theory (Chap. 28). Pairwise one has the following.

1) Records within the classical Machian Emergent Time Approach (Sect. 26.9).
2) Histories within the Classical Machian Emergent Time Approach (Sect. 29.1).
3) Records within the classical Isham–Linden Histories Theory (Sect. 29.2).

A triple combination was furthermore given by Halliwell [413] for mechanical and Minisuperspace models; the Author extended this to $\mathfrak{g}$-nontrivial models and gave it a Machian interpretation [25, 37]. Most of the value of the Combined Approach is at the semiclassical quantum level (Chap. 54). For now, classical motivations include the following.

Motivation 1) This avoids purely Timeless Records Theories' need for a semblance of dynamics.
Motivation 2) A Records Theory sits within each Histories Theory.
Motivation 3) We shall see in Sect. 29.3 that Histories Theory helps with construction of beables.

29.1 Classical Machian Histories

Chapter 28 can be re-run with t^{em} in place of λ. Classically, one can use either label or emergent versions without any supporting relation linking the two. However—as we shall see in Chap. 54—semiclassically, the label version is to provide the WKB regime that produces the emergent time with respect to which the histories are to run.

There are three different accuracies of t^{em} to be considered as inputs at the classical level; Sect. 54.2 provides four more at the quantum level.

Type 1) t^{em}: the final output of the Classical Machian Emergent Time Approach.
Type 2) $\mathrm{t}^{\mathrm{em}}_0$: its un-Machian zeroth approximand.
Type 3) $\mathrm{t}^{\mathrm{em}}_1$: its Machian first approximand.

E. Anderson, *The Problem of Time*, Fundamental Theories of Physics 190,
DOI 10.1007/978-3-319-58848-3_29

Example 1) A Machian version of [11]'s Histories Theory for Minisuperspace was outlined in [31], the most salient difference being that emergent Machian time features in place of λ.

Example 2) For scaled triangleland in the r-formulation, we use that this shares the same mathematics in conformally-transformed $\mathfrak{q}$ as for 3-d mechanics in space, which is covered in [510, 765]. Physically, the classical paths are now $Dra^{\Gamma}(t^{\text{em}})$, with conjugate paths momenta $P_{\Gamma}^{Dra}(t^{\text{em}})$ (these are $\mathfrak{g}$-free emergent times). The nonzero part of the histories brackets algebra is

$$\{Dra^{\Gamma}(t_1^{\text{em}}), P_{\Lambda}^{Dra}(t_2^{\text{em}})\} = \delta^{\Gamma}{}_{\Lambda}\delta(t_1^{\text{em}} - t_2^{\text{em}}), \tag{29.1}$$

$$\{\mathsf{S}_{\Gamma}(t_1^{\text{em}}), \mathsf{S}_{\Lambda}(t_2^{\text{em}})\} = \epsilon_{\Gamma\Lambda}{}^{\Sigma}\mathsf{S}_{\Sigma}(t_1^{\text{em}})\delta(t_1^{\text{em}} - t_2^{\text{em}}), \tag{29.2}$$

$$\{Dra^{\Gamma}(t_1^{\text{em}}), \mathsf{S}_{\Lambda}(t_2^{\text{em}})\} = \epsilon^{\Gamma}{}_{\Lambda\Sigma}Dra^{\Sigma}(t_1^{\text{em}})\delta(t_1^{\text{em}} - t_2^{\text{em}}), \tag{29.3}$$

$$\{P_{\Gamma}^{Dra}(t_1^{\text{em}}), \mathsf{S}_{\Lambda}(t_2^{\text{em}})\} = \epsilon_{\Gamma\Lambda}{}^{\Sigma}P_{\Sigma}^{Dra}(t_1^{\text{em}})\delta(t_1^{\text{em}} - t_2^{\text{em}}). \tag{29.4}$$

The last two of these signify that the paths and their conjugate momenta are $SO(3)$ or $SU(2)$ vectors.

The histories energy constraint is in this case

$$\mathcal{E}^{t^{\text{sem}}} := \int \mathrm{d}t^{\text{sem}}\mathcal{E}(t^{\text{sem}}) \tag{29.5}$$

with $\mathcal{E}$ given by (9.8). Again, since there is only one histories constraint, the histories constraint algebra is trivial (the histories-Dirac Algorithm produces no unexpected secondary histories constraints).

On the other hand, in the unreduced formulation, the classical histories are $\rho^{i\mu}(t^{\text{em}})$ with conjugate paths momenta $p_{i\mu}(t^{\text{em}})$ (for now for implicitly $Rot(d)$-dependent emergent time). The nontrivial part of the histories brackets algebra is

$$\{\rho^{i\mu}(t_1^{\text{em}}), p_{j\nu}(t_2^{\text{em}})\} = \delta^i_j\delta^{\mu}_{\nu}\delta(t_1^{\text{em}} - t_2^{\text{em}}). \tag{29.6}$$

This now comes with the histories total zero angular momentum constraint

$$\mathcal{L}^{t^{\text{em}}} := \int \mathrm{d}t^{\text{em}}\mathcal{L}(t^{\text{em}}) = 0 \tag{29.7}$$

and the histories energy constraint (29.5). $\mathcal{E}^{t^{\text{em}}}$ is as per (29.5) with $\mathcal{E}$ given by the triangleland case of (9.8).

The histories constraint algebra is now Abelian. Moreover,

$$\{\mathcal{E}^{t^{\text{em(JBB)}}}, \mathcal{L}^{t^{\text{em(JBB)}}}\} = 0, \tag{29.8}$$

by which $\mathcal{L}^{t^{\text{em(JBB)}}}$ is a histories-conserved quantity.[1]

[1]We do not term this an 'observable' out of its having no operational meaning as an entity which is fully determinable within a given (even specious) instant.

While this example is useful as a formal illustration, there is an implicitness problem with it since $t^{\text{em}}_{Rot(2)\text{-free}}$ is not known unless the Best Matching Problem is solved. Moreover, if it is solved, the preceding Sec's reduced approach is rather more natural.

Example 3) In the case of full GR, this Sec's double combination furthermore requires labelling one's histories with a *many-fingered* emergent time. [One can also contemplate using t^{York} in GR—or its analogue t^{Euler} in RPM—as a label within a histories scheme.]

29.2 Records Within Classical Histories Theory

The need for a semblance of dynamics can readily dissolve if more structure is assumed, for instance at least one of Histories Theory or the Classical, or Semiclassical, Machian Emergent Time Approach are assumed. *The three of these interprotect particularly well*; however since most of this interprotection is motivated at the quantum level, we defer discussion of this to Chap. 54.

Gell-Mann–Hartle [340] and Halliwell [411, 413, 414, 421] found and studied records contained within Histories Theory (see the next Chapter for details of this Histories-and-Records Approach). At the classical level, on could consider records within the classical part of Isham–Linden's [504] reformulation of Histories Theory. These use the Isham–Linden classical histories formulation to make sense of this at the classical level.

Halliwell [413, 414] furthermore established some computationally valuable expressions for timeless probabilities. In particular, he considered an implementation of timeless propositions

$$\text{Prob}_{\text{R}} := \text{Prob(classical solution will pass through a region R of } \mathfrak{q}). \tag{29.9}$$

He approached this by considering probability distributions, firstly on the classical $\mathfrak{P}$hase, $\text{w} = \text{w}(\boldsymbol{q}, \boldsymbol{p})$ and then at the semiclassical level (Chap. 54). For now, let us note that the classical w is constant along the classical orbits,

$$0 = \frac{\partial \text{w}}{\partial t} = \{H, \text{w}\}. \tag{29.10}$$

Moreover, some applications involve a generalization of (29.9) to $\mathfrak{P}$hase [7, 416].

In terms of the characteristic function of the region R, denoted $\text{Char}_{\text{R}}(\boldsymbol{q})$,

$$\begin{aligned} \text{Prob}_{\text{R}} &= \int_{-\infty}^{+\infty} \text{d}t\, \text{Char}_{\text{R}}\big(\boldsymbol{q}^{\text{cl}}(t)\big) \\ &= \int \mathbb{D}\boldsymbol{q}\, \text{Char}_{\text{R}}(\boldsymbol{q}) \int_{-\infty}^{+\infty} \text{d}t\, \delta^{(k)}\big(\boldsymbol{q} - \boldsymbol{q}^{\text{cl}}(t)\big) \\ &=: \int \mathbb{D}\boldsymbol{q}\, \text{Char}_{\text{R}}(\boldsymbol{q})\, \mathcal{A}_{\text{R}}(\boldsymbol{q}, \boldsymbol{q}_0, \boldsymbol{p}_0): \end{aligned} \tag{29.11}$$

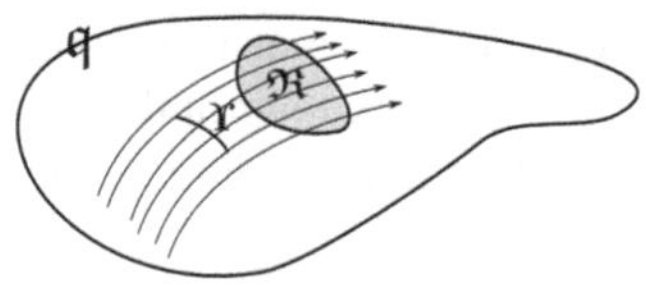

Fig. 29.1 Halliwell's approach considers propositions corresponding to fluxes through pieces of hypersurfaces Υ within configuration space $\mathfrak{q}$, and probabilities of dynamical trajectories entering regions R

the 'amount of time t' the trajectory spends in R (Fig. 29.1).[2] Furthermore, in terms of the step function θ, which serves to mathematically implement the restriction the entirety of $\mathfrak{P}$hase that is being integrated over to that part in which the corresponding classical trajectory spends time $> \epsilon$ in region R,[3]

$$\text{Prob}_{\text{R}} = \iint \mathbb{D}\boldsymbol{p}_0 \mathbb{D}\boldsymbol{q}_0 \, \text{w}(\boldsymbol{q}_0, \boldsymbol{p}_0) \, \theta\left(\int_{-\infty}^{+\infty} \text{d}t \, \text{Char}_{\text{R}}\big(\boldsymbol{q}^{\text{cl}}(t)\big) - \epsilon\right). \tag{29.12}$$

An alternative computational expression is for the flux through a piece of a $\{k-1\}$-dimensional hypersurface within $\mathfrak{q}$,

$$\begin{aligned} \text{Prob}_{\Upsilon} &= \int \text{d}t \int \mathbb{D}\boldsymbol{p}_0 \mathbb{D}\boldsymbol{q}_0 \, \text{w}(\boldsymbol{q}_0, \boldsymbol{p}_0) \int_{\Upsilon} \mathbb{D}\Upsilon(\boldsymbol{q}) \;\, \boldsymbol{\nu} \cdot \mathbf{M} \cdot \frac{\text{d}\boldsymbol{q}^{\text{cl}}(t)}{\text{d}t} \;\, \delta^{(k)}\big(\boldsymbol{q} - \boldsymbol{q}^{\text{cl}}(t)\big) \\ &= \int \text{d}t \int \mathbb{D}\boldsymbol{p}' \int_{\Upsilon} \mathbb{D}\Upsilon(\boldsymbol{q}') \;\, \boldsymbol{\nu} \cdot \boldsymbol{p}' \;\, \text{w}(\boldsymbol{q}', \boldsymbol{p}'), \end{aligned} \tag{29.13}$$

the latter equality being by passing to $\boldsymbol{q}' := \boldsymbol{q}^{\text{cl}}(t)$ and $\boldsymbol{p}' := \boldsymbol{p}^{\text{cl}}(t)$ coordinates at each t.

This is also a useful point at which to introduce the notions of *imperfect records*—the subject of Halliwell's [411]—as well as of *deteriorated, or doctored, records*. The main point is that information can be lost from a record 'after its formative event'—the word "stored" in (54.1). For instance, photos yellow with age and can also be doctored. As another example, some features of the 'cosmic microwave background' radiation that we observe have in part been formed since last scattering. The observed data has, for instance, contributions from foreground sources and as a result of the integrated Sachs–Wolfe effect [215].

[2]This is, for now, absolute time in an ordinary Mechanics model, though this Chapter can be redone in terms of label time λ for a wider range of examples; see also the next Section; finally recollect that $k = \dim(\mathfrak{q})$.

[3]ϵ itself is some small positive number that tends to 0, which is included to avoid ambiguities in the θ-function at zero argument.

Machian Nontrivial-$\mathfrak{g}$ Records Within Histories Theory Reconsider the previous Sec's example under

$$
\begin{aligned}
&q \to K \quad \text{(a configurational basis of Kuchař beables)}, \\
&p \to \pi_K \quad \text{and} \quad t \to t^{\text{em}}_{\mathfrak{g}\text{-free}}, \quad k \to r.
\end{aligned}
\tag{29.14}
$$

Some further complications are that curved measures occur at each stage and that PPSCT invariance needs to be checked as per [25].

Example 1) For the r-formulation of scaled triangleland, the curved space effects are alleviated by the conformal flatness. We then just have the

$$
q \to Dra, \qquad p \to \pi_{Dra} \quad \text{and} \quad t \to t^{\text{em}}_{Rot(2)\text{-free}}, \quad k \to 3 \tag{29.15}
$$

case of the preceding Absolute Mechanics example's results.

29.3 Beables in the Combined Approach

Halliwell's Construct for Chronos Beables in the $\mathfrak{g}$-Free Case For a simple k-d Particle Mechanics model, this is of the form [413, 421]

$$
\mathcal{A}_{\text{R}}[\boldsymbol{q}, \boldsymbol{q}_0, \boldsymbol{p}_0] = \int_{-\infty}^{+\infty} \text{d}t \, \delta^{(k)}\big(\boldsymbol{q} - \boldsymbol{q}^{\text{cl}}(t)\big). \tag{29.16}
$$

This has the property that

$$
\{\mathcal{H}, \mathcal{A}_{\text{R}}\} = 0. \tag{29.17}
$$

Thus, for $\mathfrak{g}$-trivial theories, we can write $\mathcal{A}_{\text{R}} = \mathcal{C} = \mathcal{D}$. While this construct is given above for an ordinary Mechanics example, it extends to Minisuperspace models as well [413, 414]. Finally, note that these are indeed beables—rather than histories observables—since the t has been integrated over.

The $\mathfrak{g}$-Nontrivial Version Produces Dirac Beables The Author subsequently considered the case in which $\mathcal{K}$ are a 'basis set' of configurational Kuchař beables [under the substitutions (29.14)]. (Construction of the full set of Kuchař beables would involve the extension of Halliwell's construct to such as regions of $\mathfrak{P}$hase since these have the more general dependence $\mathcal{K} = \mathcal{F}[\boldsymbol{Q}, \boldsymbol{P}]$.) For relational whole-universe models, the classical Machian emergent time $t^{\text{em}}_{\mathfrak{g}\text{-free}}$ arises to fill in the role of t and PPSCT invariance is held to apply as per Appendix L.11. Thereby, this Chapter and its descendants are Combined Machian Emergent Time, Records and Histories Approaches.

$\mathcal{A}_{\text{R}}[\boldsymbol{K}, \boldsymbol{K}_0, \boldsymbol{p}^K_0]$ now obeys $\{\mathcal{F}\mathbf{lin}, \mathcal{A}_{\text{R}}\} \approx 0$ because the $\boldsymbol{K}$ and $\boldsymbol{p}^K_0$ are Kuchař beables, to which Lemma 3 of Appendix J.18 applies. Also (29.17) still applies

(Halliwell demonstrated this to be robust to curved configuration space use [413, 414]). Therefore the $\mathcal{A}_{\mathsf{R}}[\boldsymbol{K}, \boldsymbol{K}_0, \boldsymbol{p}_0^K]$ are indeed Dirac beables, $\mathcal{D}$. To this extent, one has a formal construction of Kuchař's Unicorn (Sect. 9.15), at least for the range of theories to which Halliwell's construct can be applied. Also, as regards the various No-Go Theorems, the above avoids Kuchař's by not being of form (O.9) and Torre's by not being local in space or time.

Example 1) Since the classical Kuchař beables are known for 1- and 2-d RPMs, pure-shape or scaled, we have Dirac beables for these. To be concrete, consider the r-formulation of scaled triangleland. The extra geometrical factors vanish in this case due to configuration space flatness, and we obtain $\mathcal{A}_{\mathsf{R}}[\boldsymbol{Dra}, \boldsymbol{Dra}_0, \boldsymbol{P}_0^{Dra}]$ as follows. Additionally,

$$\mathrm{Prob}_{\mathrm{R}} = \int \mathrm{d}t^{\mathrm{em}} \int \mathbb{D}\boldsymbol{P}^{Dra} \int_{\mathrm{R}} \mathbb{D}\Upsilon(\boldsymbol{Dra})\, \mathbf{n}^{Dra} \cdot \boldsymbol{P}^{Dra}\, w(\boldsymbol{Dra}, \boldsymbol{P}^{Dra}), \qquad (29.18)$$

for w a probability distribution on the classical $\mathfrak{P}$hase, Υ a hypersurface in configuration spaces with normal n. Furthermore,

$$\mathcal{A}[\boldsymbol{Dra}, \boldsymbol{Dra}_0, \boldsymbol{P}_0^{Dra}] := \int_{-\infty}^{+\infty} \mathrm{d}t^{\mathrm{em}}\, \delta^{(3)}(\boldsymbol{Dra} - \boldsymbol{Dra}^{\mathrm{cl}}(t^{\mathrm{em}})), \qquad (29.19)$$

commutes with the classical constraints.

Alternative Indirect $\mathfrak{g}$-act, $\mathfrak{g}$-All Extension Here we make use of

$$\mathcal{A}^{\mathfrak{g}\text{-free}}[\boldsymbol{\rho}, \boldsymbol{p}_0, \boldsymbol{\rho}_0] = \int_{g \in \mathfrak{g}} \mathbb{D}g\, \vec{\mathfrak{g}}_g \left\{ \int_{-\infty}^{+\infty} \mathrm{d}t^{\mathrm{em}}_{\mathfrak{g}\text{-free}}\, \delta^{(k)}\left(\boldsymbol{\rho} - \boldsymbol{\rho}^{\mathrm{cl}}(t^{\mathrm{em}}_{\mathfrak{g}\text{-free}})\right) \right\}. \qquad (29.20)$$

Comments More generally, whenever an algebraic structure of $\mathcal{F}$**lin** can be extended by a $\mathcal{Q}$uad, Halliwell's method can be adapted to provide a construction of the extended case's A-beables from the unextended case's. From this point of view,

1) Halliwell's original working extending *id* by $\mathcal{Q}$uad to construct Chronos beables $\mathcal{C}$ from unconstrained beables $\mathcal{U}$.

For theories which are elsewise unconstrained, the $\mathcal{C}$ are Dirac beables $\mathcal{D}$. Then note that this construction carries over to

2) extending an algebraic structure of $\mathcal{F}$**lin** by a $\mathcal{Q}$uad, by which Dirac beables $\mathcal{D}$ are constructed from Kuchař beables $\mathcal{K}$.

The general formulation is more useful than 2) since not all theories' constraint algebraic structure are of the form 2); for instance Supergravity's is not. In this case, the generalized method promotes the $\mathcal{NSK}$ to $\mathcal{NSD}$, but this does not attain the goal of constructing the $\mathcal{D}$ themselves.

A final issue is that this method constructs individual beables, rather than being guaranteed to produce an entire algebraic structure's worth of these. For instance, it remains unclear whether the form of the construct closes under taking the brackets of two beables obtained by the construct.

Chapter 30
Slightly Inhomogeneous Cosmology (SIC)

Let us next develop this further and cosmologically-significant arena, for which this book's main Machian Strategy for A Local Resolution of the Problem of Time can also be worked out in detail. The particular version considered here [35] is kinematically similar to that considered by Halliwell and Hawking [419]. In particular, both are low-order inhomogeneous perturbations about $\mathbb{S}^3$ for the modelling reasons given in Sect. 9.9, include a single minimally-coupled scalar field ϕ, and are treated at most semiclassically. Moreover, the scheme for SIC presented differs from [419] in various further details on TRi and technical grounds (Fig. 30.1).

Whereas Minisuperspace is trivial as regards the Configurational Relationalism, Constraint Closure, Foliation Independence and Spacetime Constructability aspects of Background Independence, and RPM is trivial for the Foliation Dependence and Spacetime Relationalism and Construction aspects, SIC has all nine aspects nontrivial. Thus it serves as the successor of both of Minisuperspace and RPM model arenas' qualitative insights, and as a first port of call for nontrivial investigations of the Foliation Independence and Spacetime Constructability aspects. In particular, SIC still admits a solvable Thin Sandwich and consequently an explicit classical Machian emergent time and explicitly constructible beables. In fact, for SIC the Thin Sandwich procedure turns out to be only a partial reduction which needs completing by further means.

30.1 Relational Action for SIC

SIC's configurations, $\mathfrak{q}$ and $\mathfrak{g}$ are presented in Appendix I.2. The relational action is now

$$\mathsf{S}_{\text{relational}} = \sqrt{2}\int \partial \mathsf{s}_{0,1,2}\sqrt{\mathcal{W}_{0,2}}, \tag{30.1}$$

where $\partial \mathsf{s}_{0,1,2} = \sqrt{\sum_{\mathrm{n}}[\mathrm{d}_{\underline{\mathrm{F}}} f_{\mathrm{n}}, \mathrm{d}\phi, \mathrm{d}\Omega, \mathrm{d}_{\underline{\mathrm{F}}} x_{\mathrm{n}}][M_{\psi} \oplus M_{\text{grav}}][\mathrm{d}_{\underline{\mathrm{F}}} f_{\mathrm{n}}, \mathrm{d}\phi, \mathrm{d}\Omega, \mathrm{d}_{\underline{\mathrm{F}}} x_{\mathrm{n}}]^{\mathrm{T}}}$. The non-auxiliary part of this can be read off Fig. I.2. Furthermore, the auxiliary terms

E. Anderson, *The Problem of Time*, Fundamental Theories of Physics 190,
DOI 10.1007/978-3-319-58848-3_30

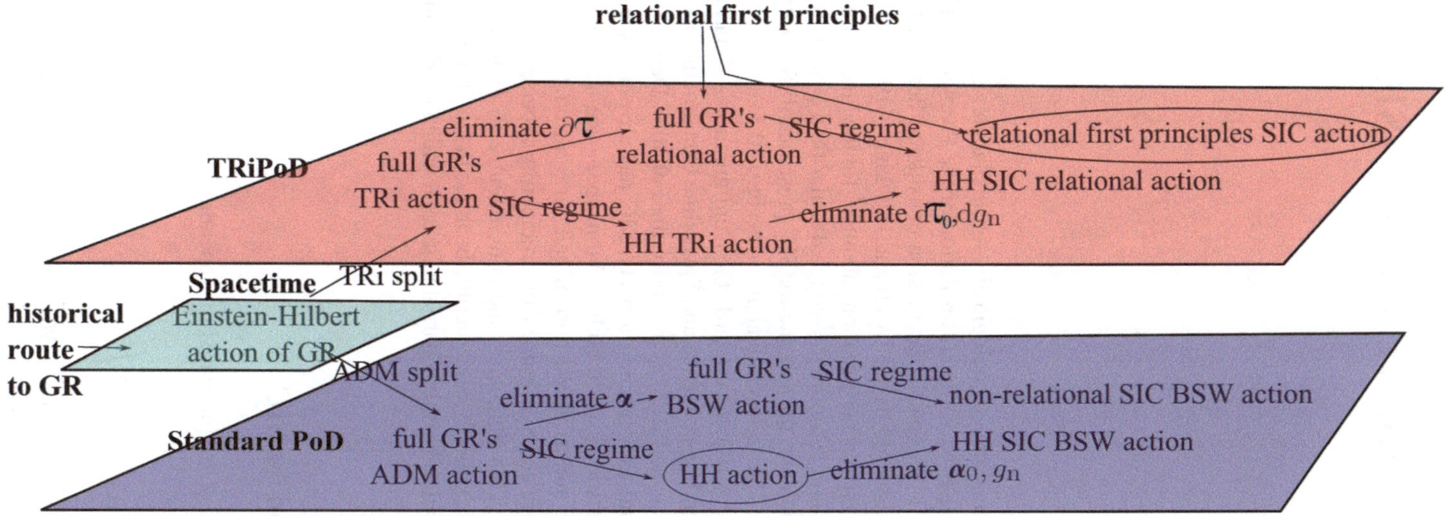

Fig. 30.1 The differences between the relational and ADM–Halliwell–Hawking (HH) formulations of perturbative SIC; PoD stands for Principles of Dynamics. The differences between the 'floors' in this diagram are in the auxiliaries used. The *arrow* to the *top encircled action* is this book's procedure, whereas the lower circled action is Halliwell and Hawking's built on the ADM split. Each of the *upstairs* and *downstairs squares* 'do not commute', so that the *two encircled actions* are *not quite* equivalent. They differ as regards how time is treated. [See Chap. 34 for what TRi split GR and $\mathcal{T}$ are.] None the less, they produce all of the same constraint equations in suitable Hamiltonian-type formulations (∂A-Hamiltonian upstairs and Hamiltonian downstairs). This is with the exception of the ADM–Halliwell–Hawking case containing a linear Hamiltonian constraint contribution from variation with respect to the perturbation of the lapse

match [419]'s Lagrangian's under the correspondence $j_n/N_0 \rightarrow \mathrm{d}j_n$, $k_n/N_0 \rightarrow \mathrm{d}k_n$; here N_0 is the zeroth approximation to the ADM lapse. In full,

$$\mathrm{d}s^{n\,2}_{0,1,2} = \frac{\exp(3\,\Omega)}{2}\left\{-\mathrm{d}a_n^2 + \frac{n^2-4}{n^2-1}\mathrm{d}b_n^2 + \{n^2-4\}\mathrm{d}c_n^2 + \mathrm{d}d_n^2 + \mathrm{d}f_n^2 + 6a_n\mathrm{d}f_n\mathrm{d}\phi + \frac{2}{3}\mathrm{d}A_n\mathrm{d}\Omega + A_n\{\mathrm{d}\Omega^2 - \mathrm{d}\phi^2\}\right\} - \exp(2\Omega)\left\{\{n^2-4\}\mathrm{d}c_n\mathrm{d}j_n + \left\{\mathrm{d}a_n + \frac{n^2-4}{n^2-1}\mathrm{d}b_n + 3f_n\mathrm{d}\phi\right\}\frac{\mathrm{d}k_n}{3}\right\} + \frac{\exp(\Omega)}{2}\left\{\{n^2-4\}\mathrm{d}j_n^2 - \frac{\mathrm{d}k_n^2}{3\{n^2-1\}}\right\}. \tag{30.2}$$

This is via the useful combination A_n defined in Eq. (I.14).

Equation (30.2) can be split up into S, V, T pieces [345, 623], and into zeroth, first and second order pieces. These pieces are readily visible above due to S being labelled by a_n, b_n and f_n factors, V by c_n factors, T by d_n factors and perturbative orders by how many powers of $\mathrm{d}y_n$, $\mathrm{d}u_n$ each term contains.

Also $\bar{W}_{0,2} = \bar{W}_0 + \sum_n \bar{W}_2^n$ for W_0 given by (9.12) with zero subscripts added and

$$\bar{W}_2^n = \frac{\exp(\Omega)}{2}\left\{\frac{1}{3}\left\{n^2 - \frac{5}{2}\right\}a_n^2 + \frac{\{n^2-7\}}{3}\frac{\{n^2-4\}}{n^2-1}b_n^2 + \frac{2}{3}\{n^2-4\}a_nb_n - 2\{n^2-4\}c_n^2 - \{n^2+1\}d_n^2\right\} + \frac{\exp(3\,\Omega)}{2}\left\{-m^2\{f_n^2 + 6a_nf_n\phi\} - \exp(-2\Omega)\{n^2-1\}f_n^2 - \{m^2\phi^2 + 2\,\Lambda\}A_n\right\}. \tag{30.3}$$

The first line here is the second-order part of the Ricci 3-scalar, $\bar{R}_2^n$, whereas the second comprises the matter potential and cosmological constant contributions.[1]

30.2 Constraints for SIC

$\mathcal{H}$ now gives (9.14) at zeroth order, and, at second order,

$$\mathcal{H}_2 = \sum_n \{{}^{\mathrm{S}}\mathcal{H}_2^n + {}^{\mathrm{V}}\mathcal{H}_2^n + {}^{\mathrm{T}}\mathcal{H}_2^n\} \quad \text{for} \tag{30.4}$$

[1]We have added a cosmological constant term omitted in [419] itself, since this is useful in subsequent cosmological modelling. This approach also immediately extends to multiple minimally-coupled scalar fields.

$$
{}^{\mathrm{S}}\mathcal{H}_2^{\mathrm{n}} = \frac{\exp(-3\,\Omega)}{2}\Bigg\{\left\{\frac{1}{2}a_{\mathrm{n}}^2 + 10\frac{\mathrm{n}^2-4}{\mathrm{n}^2-1}b_{\mathrm{n}}^2\right\}\pi_\Omega^2 + \left\{\frac{15}{2}a_{\mathrm{n}}^2 + 6\frac{\mathrm{n}^2-4}{\mathrm{n}^2-1}b_{\mathrm{n}}^2\right\}\pi_\phi^2
$$
$$
-\pi_{a_{\mathrm{n}}}^2 + \frac{\mathrm{n}^2-1}{\mathrm{n}^2-4}\pi_{b_{\mathrm{n}}}^2 + \pi_{f_{\mathrm{n}}}^2 + 2a_{\mathrm{n}}\pi_{a_{\mathrm{n}}}\pi_\Omega + 8b_{\mathrm{n}}\pi_{b_{\mathrm{n}}}\pi_\Omega - 6a_{\mathrm{n}}\pi_{f_{\mathrm{n}}}\pi_\phi\Bigg\}
$$
$$
-\frac{\exp(\Omega)}{2}\Bigg\{\frac{1}{3}\left\{\mathrm{n}^2 - \frac{5}{2}\right\}a_{\mathrm{n}}^2 + \frac{\{\mathrm{n}^2-7\}}{3}\frac{\{\mathrm{n}^2-4\}}{\mathrm{n}^2-1}b_{\mathrm{n}}^2
$$
$$
+\frac{2}{3}\{\mathrm{n}^2-4\}a_{\mathrm{n}}b_{\mathrm{n}} - \{\mathrm{n}^2-1\}f_{\mathrm{n}}^2\Bigg\}
$$
$$
+\frac{\exp(3\,\Omega)}{2}\Bigg\{m^2\{f_{\mathrm{n}}^2 + 6a_{\mathrm{n}}f_{\mathrm{n}}\phi\} + \{m^2\phi^2 + 2\,\Lambda\}\left\{\frac{3}{2}a_{\mathrm{n}}^2 - 6\frac{\mathrm{n}^2-4}{\mathrm{n}^2-1}b_{\mathrm{n}}^2\right\}\Bigg\}, \tag{30.5}
$$

$$
{}^{\mathrm{V}}\mathcal{H}_2^{\mathrm{n}} = \frac{\exp(-3\,\Omega)}{2}\left\{\{\mathrm{n}^2-4\}c_{\mathrm{n}}^2\{10\pi_\Omega^2 + 6\pi_\phi^2\} + \frac{\pi_{c_{\mathrm{n}}}^2}{\mathrm{n}^2-4} + 8c_{\mathrm{n}}\pi_{c_{\mathrm{n}}}\pi_\Omega\right\}
$$
$$
+\{\mathrm{n}^2-4\}c_{\mathrm{n}}^2\{\exp(\Omega) - 3\exp(3\Omega)\{m^2\phi^2 + 2\,\Lambda\}\}, \tag{30.6}
$$
$$
{}^{\mathrm{T}}\mathcal{H}_2^{\mathrm{n}} = \frac{\exp(-3\,\Omega)}{2}\{d_{\mathrm{n}}^2\{10\pi_\Omega^2 + 6\pi_\phi^2\} + \pi_{d_{\mathrm{n}}}^2 + 8d_{\mathrm{n}}\pi_{d_{\mathrm{n}}}\pi_\Omega\}
$$
$$
+d_{\mathrm{n}}^2\left\{\frac{\mathrm{n}^2+1}{2}\exp(\Omega) - 3\exp(3\Omega)\{m^2\phi^2 + 2\,\Lambda\}\right\}. \tag{30.7}
$$

Also $\mathcal{M}_{1i} = [{}^{\mathrm{S}}\mathcal{M}_1^{\mathrm{n}}, {}^{\mathrm{V}}\mathcal{M}_1^{\mathrm{n}}]$ is the vector corresponding to the vector of auxiliaries $[\mathrm{d}k_{\mathrm{n}}, \mathrm{d}j_{\mathrm{n}}]$. $\mathcal{M}_i$ vanishes at zeroth order, and has S and V parts to first order:

$$
{}^{\mathrm{S}}\mathcal{M}_1^{\mathrm{n}} = \frac{\exp(-3\,\Omega)}{3}\left\{-\pi_{a_{\mathrm{n}}} + \pi_{b_{\mathrm{n}}} + \left\{a_{\mathrm{n}} + 4\frac{\mathrm{n}^2-4}{\mathrm{n}^2-1}b_{\mathrm{n}}\right\}\pi_\Omega + 3f_{\mathrm{n}}\pi_\phi\right\}, \tag{30.8}
$$
$$
{}^{\mathrm{V}}\mathcal{M}_1^{\mathrm{n}} = \exp(\text{-}\Omega)\{\pi_{c_{\mathrm{n}}} + 4\{\mathrm{n}^2-4\}c_{\mathrm{n}}\pi_\Omega\}. \tag{30.9}
$$

In Hamiltonian variables, the $\mathcal{H}_2$ pieces and the $\mathcal{M}_1$ pieces moreover coincide for the relational and ADM–Halliwell–Hawking approaches. As a first step toward whether the system is well-determined, for now we have 10 degrees of freedom per mode value n and 6 constraints imposed per n. A Principles of Dynamics level treatment is required for further details, which go beyond a mere matter of counting, due to involving the particular geometrical form of $\mathfrak{P}$hase.

30.3 Constraint Closure Posed

Note the distinction between leading-order inhomogeneity brackets and subsequent mode-split brackets. In the first case, the only zeroth-order bracket is (24.16) with zero suffices on each object, no other zeroth-order brackets can be defined. Moreover, (|) reduces to the trivial when between two zeroth-order objects; it is here included only for order-by-order consistency in the presentation.

The first-order brackets are

$$\{(\mathcal{H}_0 \,|\, \mathrm{d}J_0), (\mathcal{H}_1 \,|\, \mathrm{d}K_0)\} = 0, \qquad \{(\mathcal{H}_0 \,|\, \mathrm{d}J_0), (\mathcal{H}_0 \,|\, \partial \mathrm{K}_1)\} = 0, \tag{30.10}$$

and the $J \leftrightarrow K$ of the preceding. No other first-order brackets can be defined.

The second-order brackets are

$$\{(\mathcal{H}_1 \,|\, \partial \mathrm{J}_1), (\mathcal{H}_0 \,|\, \mathrm{d}K_0)\} = (\mathcal{M}_{i\,1} \,|\, h_0^{ij} \mathrm{d}K_0 \overleftrightarrow{\partial_j} \partial J_1) \tag{30.11}$$

and the $J \leftrightarrow K$ of this. Also

$$\{(\mathcal{H}_0 \,|\, \mathrm{d}J_0), (\mathcal{M}_{i\,1} \,|\, \partial \mathrm{L}_1^i)\} = (\pounds_{\partial \mathrm{L}_1} \mathcal{H}_1 \,|\, \mathrm{d}J_0), \tag{30.12}$$

$$\{(\mathcal{M}_i \,|\, \partial \mathrm{L}^i), (\mathcal{M}_j \,|\, \partial \mathrm{M}^j)\} = 0 \tag{30.13}$$

up to fourth order. These equations mean that in this regime, the 3-diffeomorphisms have lost their group relations, whereas the deformations retain nontrivial such.

The Poisson brackets with only up to second-order right hand sides are

$$\{(\mathcal{H}_{0,1,2} \,|\, \partial \mathrm{J}_{0,1,2}), (\mathcal{H}_{0,1,2} \,|\, \partial \mathrm{K}_{0,1,2})\} = (\mathcal{M}_{i\,1} \,|\, \mathrm{d}K_0 \partial^i \partial \mathrm{J}_1 - \mathrm{d}J_0 \partial^i \partial \mathrm{K}_1) + O(y_{\mathrm{n}}^3), \tag{30.14}$$

$$\{(\mathcal{H}_{0,1,2} \,|\, \partial \mathrm{J}_{0,1,2}), (\mathcal{M}_{i\,0,1,2} \,|\, \partial \mathrm{L}^i_{0,1,2})\} = (\pounds_{\partial \mathrm{L}_1} \mathcal{H}_1 \,|\, \mathrm{d}J_0) + O(y_{\mathrm{n}}^3), \tag{30.15}$$

$$\{(\mathcal{M}_{i\,0,1,2} \,|\, \partial \mathrm{L}^i_{0,1,2}), (\mathcal{M}_{i\,0,1,2} \,|\, \partial \mathrm{M}^i_{0,1,2})\} = 0 + O(y_{\mathrm{n}}^4). \tag{30.16}$$

$\mathit{Diff}_1(\mathbf{\Sigma})$ is Abelian (up to fourth order, which is beyond where the current modelling goes). Also, $\mathit{Diff}_1(\mathbf{\Sigma})$ acts on $\mathcal{H}$ in its most nontrivial manner to second order: $\mathcal{H}$ up to first order is a Diff_1(deformed $\mathbb{S}^3$) scalar density. Additionally, $\mathit{Diff}_1(\mathbf{\Sigma})$ is $\mathit{Diff}_1(\mathbb{S}^3)$ as regards its action in the bracket of two $\mathcal{H}$'s. Finally, to second order, $\mathrm{h}^{ij} = h_0^{ij}$ are still structure functions for $\mathbb{S}^3$, so one still has an algebroid.

In the action, $\mathrm{d}h_{ij\,0} + \partial h_{ij\,1} + \partial h_{ij\,2} - \pounds_{\partial F_1}\{h_{ij\,0} + h_{ij\,1}\}$. So $\mathfrak{g} = \mathit{Diff}_1(\mathbf{\Sigma})$, which leads to $\mathcal{M}_{i\,1}$. Also (9.33) has a second-order right hand side. From this, one deduces that $\mathcal{H}_0$ cannot produce $\mathcal{M}_{i\,1}$. Finally,

$$\{(\mathcal{H} \,|\, \partial \mathrm{J}), (\mathcal{M}_i \,|\, \partial \mathrm{L}^i)\} = (\pounds_{\partial \mathrm{L}} \mathcal{H} \,|\, \partial \mathrm{J}). \tag{30.17}$$

The entry in the second slot is forced to be (first order) × (first order). Comparing with the right hand side, this has a first-order $\partial \mathrm{L}^i$ factor, so it needs the corresponding $\mathcal{H}$ to be first order as well, so that $\pounds$ acts upon it nontrivially. So on this occasion it is a bracket between $\mathcal{H}_0$ and $\mathcal{M}_1$ that produces $\mathcal{H}_1$.

SIC's modewise constraints require no smearing because they are finite block by block. By straightforward computation S, V, T cross-brackets are straightforwardly zero or weakly zero to second order. Brackets between blocks of different n are also straightforwardly zero. All this is saying is that in calculations which are at most second order, the modewise split into different values of n and the S–V–T split are preserved under the brackets operation. Consequently, at most to second order, each S, V, T piece for each value of n can be treated as a separate finite system in its own right, i.e. without reference to the other such systems.

With each block being finite, self-brackets are moreover all zero therein. This gives immediately that $\{\mathcal{H}, \mathcal{H}\}_2 = 0$ for each S, V, T piece and each n, and $\{\mathcal{M}, \mathcal{M}\}_2 = 0$ likewise.

30.4 Outcome of Dirac Algorithm and Thin Sandwich

The preceding Sec leaves just two cross-brackets to evaluate: $\{{}^{\mathrm{S}}\mathcal{H}, {}^{\mathrm{S}}\mathcal{M}\}$ and $\{{}^{\mathrm{V}}\mathcal{H}, {}^{\mathrm{V}}\mathcal{M}\}$. Straightforward evaluation of these produces nontrivial right hand side terms, in a manner which implies that ${}^{\mathrm{S}}\mathcal{M}$ and ${}^{\mathrm{S}}\mathcal{H}$ are not first-class constraints with respect to each other, and neither are ${}^{\mathrm{V}}\mathcal{M}$ and ${}^{\mathrm{V}}\mathcal{H}$.

$\{{}^{\mathrm{S}}\mathcal{H}, {}^{\mathrm{S}}\mathcal{M}\}$ and $\{{}^{\mathrm{V}}\mathcal{H}, {}^{\mathrm{V}}\mathcal{M}\}$ do not close. The corresponding phase space degrees of freedom count is depicted in Fig. 30.2.

This counting resolution in turn reveals [419] to be using rather questionable assumptions in its Quantization. Closer to the current stage of development, it also affects how one carries out the Thin Sandwich resolution and reduced configuration space treatment in the next two Secs and in Appendix I.2.

All the momentum constraint components (30.8), (30.9) are manifestly algebraic, and, being linear, manifestly solvable. To address the Thin Sandwich Problem, recast these constraints in the Jacobi–Mach formulation. By SIC's momentum-change relations, (30.8), (30.9) become the SIC Thin Sandwich equations

$$\mathrm{d}a_{\mathrm{n}} + \frac{\mathrm{n}^2 - 4}{\mathrm{n}^2 - 1}\mathrm{d}b_{\mathrm{n}} + \frac{\exp(-\Omega)}{\mathrm{n}^2 - 1}\mathrm{d}k_{\mathrm{n}} + 3f_{\mathrm{n}}\mathrm{d}\phi = 0, \qquad \mathrm{d}c_{\mathrm{n}} - \exp(-\Omega)\mathrm{d}j_{\mathrm{n}} = 0. \tag{30.18}$$

These are to be interpreted as to be solved for the first-order auxiliary variables $\mathrm{d}j_{\mathrm{n}}$ and $\mathrm{d}k_{\mathrm{n}}$. The actual solving is in this case immediate. The above two equations are decoupled and individually *well-determined*, as follows. There are even and odd $\mathrm{d}j_{\mathrm{n}}$ but also even and odd $\mathrm{d}c_{\mathrm{n}}$, while everything in the equation for the single $\mathrm{d}k_{\mathrm{n}}$ comes in a single copy. The two are, furthermore, algebraically trivial as regards making whichever of its terms the subject. In particular, the Thin Sandwich Problem choice of solutions is

$$\mathrm{d}k_{\mathrm{n}} = -\exp(\Omega)\big\{\{\mathrm{d}a_{\mathrm{n}} + 3f_{\mathrm{n}}\mathrm{d}\phi\}\{\mathrm{n}^2 - 1\} + \{\mathrm{n}^2 - 4\}\mathrm{d}b_{\mathrm{n}}\big\}, \qquad \mathrm{d}j_{\mathrm{n}} = \exp(\Omega)\,\mathrm{d}c_{\mathrm{n}}. \tag{30.19}$$

The bulk of the Thin Sandwich or Best Matching approach's work is however in the subsequent elimination. One can readily substitute (30.19) back into the kinetic metric, Appendix N.10's examination of this reveals that only a partial elimination has occurred. For, in performing this elimination of shift variables, the line element does not lose a full shift's contingent of further partner variables so as to complete

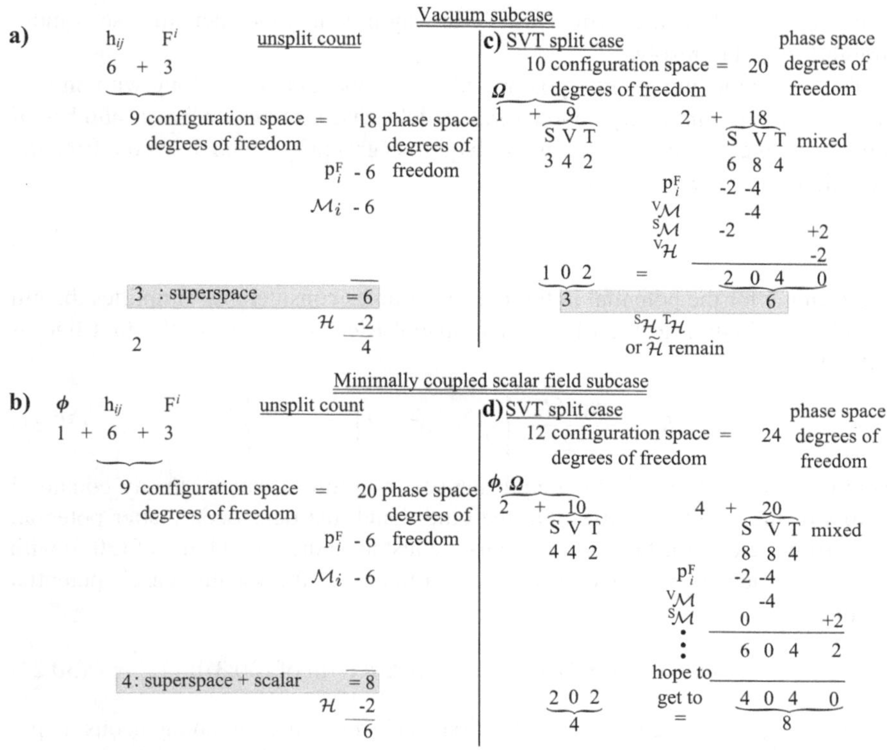

Fig. 30.2 **a)** We here juxtapose the usual relational formulation analysis for vacuum GR of Fig. 24.4 for useful comparison. **b)** is the version additionally including a minimally-coupled scalar field. **c)** and **d)** consider what happens to the vacuum and scalar field case respectively, upon specializing to SIC and applying the SVT split. Note the appearance in both **c)** and **d)** of the non SVT split variable A_n, alongside how the momentum constraints fail to quotient out a *Diff*$(\mathbb{S}^3)$'s worth of degrees of freedom by case-dependent amounts. Some of that freedom has now been transferred to the multiple Hamiltonian constraints. We concentrate upon the descent to modewise $\mathfrak{S}$uperspace$(\mathbb{S}^3)$; attaining this is marked with grey rectangles for clarity; this objective is presently attained for **c)** but not for **d)**. This is since the vacuum case **c)**'s ${}^V\mathcal{H}$ can be used to eliminate the non SVT split variable A_n. So in this case A_n attains only a temporary significance in performing successive reductions. That the two cases **c)** and **d)** behave differently in this regard is an interesting result in its own right, as regards the theory of the Thin Sandwich Problem facet

the loss of a *Diff*$(\mathbb{S}^3)$'s amount of variables. This reflects that the constraint equations involved in the reduction cannot all be first-class. So, whereas the algebraic nature of SIC's (30.19) parallels that of RPM, SIC has the further feature of not involving solely first-class constraints. This is of further interest since these constraint equations arose from an object $\mathcal{M}_i$ which, in its original unsplit form, is widely known to be first-class.

It furthermore turns out that the amount by which first-classness fails differs between the vacuum and scalar field cases. In particular, the vacuum case has the further feature that thin sandwich elimination causes the scalar mode a_n to drop out from the reduced kinetic term. This is one way in which the vacuum case is more straightforward to resolve.

The corresponding geometry is provided in Appendix N.10, along with an outline of how the minimally-coupled scalar field case's reduction has a number of further 'counter-intuitive' features. A ubiquitous grouping which drops out from the reduction is the *scalar sum variable* is

$$s_{\mathrm{n}} := a_{\mathrm{n}} + b_{\mathrm{n}}. \tag{30.20}$$

The formula for the potential in these new variables considerably simplifies the nth mode's contribution to the (densitized) spatial Ricci scalar: from the first line of (30.3) to

$$\bar{R}_{\mathrm{n}} = \frac{\exp(\Omega_{\mathrm{n}})}{2}\left\{s_{\mathrm{n}}^2 - \{\mathrm{n}^2 - 1\}d_{\mathrm{n}}^2 + \frac{A_{\mathrm{n}}}{3}\right\}. \tag{30.21}$$

Note the Ω_{n} to Ω and ϕ_{n} to ϕ equivalences in products which already contain 2 factors of the small quantities. On the other hand, the densitized matter potential and cosmological constant contributions are just as in the second line of (30.3) with n-indices attached. We make later use in particular of the vacuum case's potential combination

$$w_{\mathrm{n}} := \exp(3\Omega) \times \{(30.21) + \text{ last term of } (30.3)\}. \tag{30.22}$$

$w_2^{\mathrm{n}}/W_0 = \epsilon_{\mathrm{sds}-1} \ll 1$ gives moreover a first 'scale dominates inhomogeneous shape' small quantity, with further such arising by considering derivative versions.

Physicist Sumio Wada also considered reduction of SIC, albeit at the level of the quantum equations (i.e. within a Dirac Quantization scheme). This and Reduced Quantization as eventually attempted here are in general inequivalent. None the less, a number of constructs that worked for Wada can be derived from the Author's reduced approach to the classical $\mathfrak{q}$ geometry. Wada also introduced a scalar sum variable (our s_{n} up to proportionality). In fact, Wada performed reduction to three different extents.

Reduction 1) In [870], he only eliminates ${}^{V}\mathcal{M}_{1i}$.

Reduction 2) In [872] and the later more detailed paper with physicist Ikuo Shirai [789], he eliminates precisely all of $\mathcal{M}_{1i}$. This is the case which corresponds most closely to the Thin Sandwich.

Reduction 3) In [871], he eliminates all of $\mathcal{M}_{1i}$ alongside $\mathcal{H}_1$ (which in the Relational Approach never arises since there is no primary lapse to be independently varied).

In each case Wada corrects Ω by a distinct difference-of-squares variable. 1) uses the identification $\sum_{\mathrm{n}}{}^{\mathrm{S}}A_{\mathrm{n}}$, 2) uses $\sum_{\mathrm{n}}{}^{\mathrm{S,V}}A_{\mathrm{n}}$ and 3) uses the whole of $\sum_{\mathrm{n}} A_{\mathrm{n}}$. He qualifies the T-term within A_{n} as 'not necessary' for the quantum-level simplifications incurred. However, it is necessary as regards the combination including this turning out to be inherited from the classical $\mathfrak{q}$ geometry, as per Appendix I.2. Wada also corrects ϕ in each case by subtracting off $3\sum_{\mathrm{n}} b_{\mathrm{n}} f_{\mathrm{n}}$.

On the other hand, in the vacuum case, further reduction[2] eliminating ${}^{\mathrm{V}}\mathcal{H}$ is possible (Appendix N.10), leaving one with

$$\mathrm{d}s_{\mathrm{n}}^2 = -\mathrm{d}\zeta_{\mathrm{n}}^2 + \|\mathrm{d}v_{\mathrm{n}}\|^2. \tag{30.23}$$

Finally, the corresponding configuration space of the modewise inhomogeneities themselves is also clearly flat $\mathbb{R}^3$, with the $\underline{v}_{\mathrm{n}}$ playing the role of Cartesian coordinates:

$$\mathrm{d}s_{\mathrm{n}}^2 = \|\mathrm{d}v_{\mathrm{n}}\|^2. \tag{30.24}$$

$\mathrm{d}s_{\mathrm{n}}/\mathrm{d}\Omega = \epsilon_{\mathrm{sds}-2} \ll 1$ is the further kinetic 'scale dominates inhomogeneous shape' approximation.

The reduced action built as per above has conjugate momenta

$$p_{\mathrm{n}}^{\xi} = -\frac{\sqrt{\widetilde{w}_{\mathrm{n}}}\mathrm{d}\xi}{\mathrm{ds}}, \qquad \underline{p}_{\mathrm{n}}^{v} = \frac{\sqrt{\widetilde{w}_{\mathrm{n}}}\mathrm{d}\underline{v}_{\mathrm{n}}}{\mathrm{ds}}; \tag{30.25}$$

the tildes here denote reduction. The corresponding Hamiltonian constraint is

$$\widetilde{\mathcal{H}}_{\mathrm{n}} = \frac{1}{2}\{-p_{\mathrm{n}}^{\xi\,2} + \|\underline{p}_{\mathrm{n}}^{v}\|^2\} - \widetilde{w_{\mathrm{n}}/2}(\xi_{\mathrm{n}}, \underline{v}_{\mathrm{n}}). \tag{30.26}$$

N.B. that this is formed not only by solving the Thin Sandwich, but also requires additional steps so as to cut out all the *Diff*$(\boldsymbol{\Sigma})$ information.

The classical Machian emergent time is

$$CR(\mathrm{t}_{\mathrm{CC}}^{\mathrm{em}}) = \int\sqrt{\{\mathrm{d}\xi_{\mathrm{n}}^2 - \|\mathrm{d}\underline{v}_{\mathrm{n}}\|^2\}/\widetilde{w}_{\mathrm{n}}}. \tag{30.27}$$

Finally the reduced equations of motion are

$$\sqrt{\widetilde{w}_{\mathrm{n}}}\mathrm{d}p_{\mathrm{n}}^{\xi}/\mathrm{ds}_{\mathrm{n}} = -\frac{1}{2}\frac{\partial\widetilde{w_{\mathrm{n}}/2}}{\partial\xi_{\mathrm{n}}}, \qquad \sqrt{\widetilde{w}_{\mathrm{n}}}\mathrm{d}\underline{p}_{\mathrm{n}}^{v}/\mathrm{ds}_{\mathrm{n}} = -\frac{1}{2}\frac{\partial\widetilde{w_{\mathrm{n}}/2}}{\partial\underline{v}_{\mathrm{n}}}; \tag{30.28}$$

the reduced system still possesses an ST split.

As regards the classical Machian approach to SIC, expanding $CR(t_{\mathrm{CC}}^{\mathrm{em}})$ in the manner of Chap. 23 gives the approximation

$$\mathrm{t}^{\mathrm{em}} = t_{\mathrm{hom}}^{\mathrm{em}} - \frac{1}{2}\int\frac{\mathrm{d}\xi}{\sqrt{W_0(\xi)}}\left\{\left\|\frac{\mathrm{d}v}{\mathrm{d}\xi}\right\|^2 + \frac{W_1(\xi,\underline{v})}{W_0(\xi))}\right\} = t_{\mathrm{hom}}^{\mathrm{em}} + O(\text{inhomogeneity}^2). \tag{30.29}$$

I.e. the usual isotropic cosmic time now picks up $O(\text{inhomogeneity}^2)$ corrections, of the order of 1 part in 10^{10}.

---

[2]Clearly in this application, Sect. 2's default—of reduction signifying removal of linear constraints only—is overridden due to falling short of implementing the desired $\mathfrak{g} = \mathit{Diff}(\boldsymbol{\Sigma})$.

30.5 Beables for SIC

In the Problem of Beables for SIC, Kuchař beables $\boldsymbol{K}$ are merely functions of 'particular $\mathbb{S}^3$ mode variables and associated momenta'. This means that the small bumps in question are modes of $\mathbb{S}^3$-sphere itself, rather than of the deformed sphere. However, working overall only to second order, the sandwich is restricted to an algebraic equation, and furthermore one which is (locally) soluble. This simplification goes hand in hand with the Kuchař beables being simple to handle to this order:

$$\{(\mathcal{M}_{1i} \,|\, \partial \mathrm{L}_1^i), (\kappa_{0\mathrm{K}} \,|\, \chi_0^{\mathrm{K}})\} \approx 0, \tag{30.30}$$

so $\kappa_{0\mathrm{K}}(t \text{ alone})$ are Kuchař beables. Dirac beables are more complicated since $(\mathcal{H} \,|\, \partial \mathrm{J})$ has zeroth and first order contributions. Consequently, $(\mathcal{D}_{\mathrm{D}} \,|\, \chi^{\mathrm{D}})$ has nontrivial second and first order contributions, which are not merely functions of t alone.

Let us start by considering configurational Kuchař beables in the vacuum GR case.

$$d_{\mathrm{n}} \quad \text{and} \quad s_{\mathrm{n}} \in \mathfrak{k} \quad \text{(the space of Kuchař beables)}. \tag{30.31}$$

On the other hand, $A_{\mathrm{n}} \notin \mathfrak{k}$, which is a second difference between it and the relational triangle's ellip variable. This can be spotted since whereas ellip contains a rotationally-invariant difference of two squares, 'the other gravitational modes' in A_{n}'s own difference of sum of squares include the unphysical vector modes c_{n}. Nor does the A_{n} quantity's structure respect the status of s_{n}, rather than of the individual a_{n} and b_{n}, as an invariant quantity. [The final form of the inhomogeneous part of the vacuum potential is another quadratic difference, which does specifically compare the amount of *Diff*($\mathbb{S}^3$) invariant scalar inhomogeneity s_{n} with the amount of *Diff*($\mathbb{S}^3$) tensor inhomogeneity d_{n}. However, in this case the individual pieces are already Kuchař beables before taking the difference, so we do not add this quantity to our list of beables.] Moreover,

$$\Omega_{\mathrm{n}} = \Omega - A_{\mathrm{n}}/3 \in \mathfrak{k}, \tag{30.32}$$

as well as its arising as a blockwise simplifier of the geometry (Appendix I.2).

For the case with a minimally-coupled scalar field as well,

$$f_{\mathrm{n}} \text{ and } \phi_{\mathrm{n}} := \phi - 3\, b_{\mathrm{n}} f_{\mathrm{n}} \in \mathfrak{k}. \tag{30.33}$$

Superspace's other block reduction variable (N.13) is out be out by a factor of 1/2 in its correction to ϕ from that in the beable. Let us also note that Wada's version coincides with the latter at the level of the functional dependence of the quantum wavefunction. Additionally, much as Ω is corrected by subtracting off an ellip type variable, ϕ is corrected by subtracting off an aniso type variable; again, the resulting difference is not itself a Kuchař beable.

We next point out that the momenta associated with these that are also beables are no longer all conjugate momenta. I.e. the two $\pi_{d_{\mathrm{n}}}$,

$$\pi_{2s\mathrm{n}} := -\pi_{a_{\mathrm{n}}} + \frac{1}{4}\frac{\mathrm{n}^2 - 1}{\mathrm{n}^2 - 4}\pi_{b_{\mathrm{n}}} \tag{30.34}$$

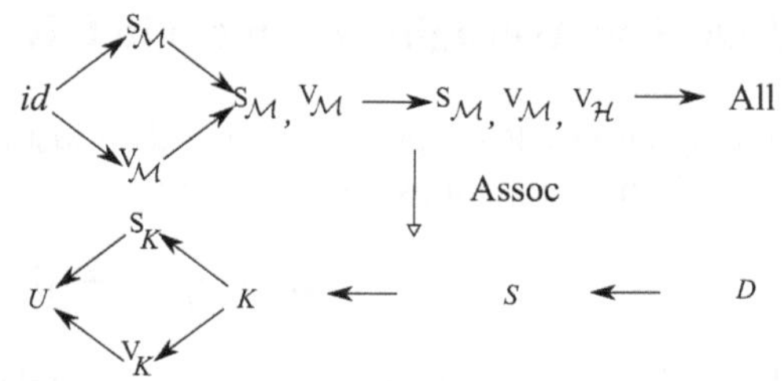

Fig. 30.3 A range of interesting notions of constraint subalgebraic structure and of A-beables for SIC

and π_Ω (which is specifically a weak Kuchař beable). Functionals of these 'basis beables' are also classical Kuchař beables. The algebra formed by the 'basis beables' has 4 pairs of brackets giving 1's in the manner of the Heisenberg algebra, alongside the one further interlinking relation,

$$\{\Omega_{\mathrm{n}}, \pi_{2s_2}\} = s_{\mathrm{n}}. \tag{30.35}$$

In the case with a minimally-coupled scalar field as well, the momenta associated with the above are $\pi_{f_{\mathrm{n}}}$ and the scalar dilational momentum

$$\pi_{2\phi} := \phi\pi_\phi + f_{\mathrm{n}}\pi_{f_{\mathrm{n}}}. \tag{30.36}$$

The further brackets are the fundamental bracket's 1 for the f_{n},

$$\{\phi_{\mathrm{n}}, \pi_{2\phi}\} = \phi_{\mathrm{n}}, \tag{30.37}$$

and the interlinking relations

$$\{f_{\mathrm{n}}, \pi_{2\phi}\} = f_{\mathrm{n}}, \qquad \{\pi_{2\phi}, \pi_{f_{\mathrm{n}}}\} = \pi_{\mathrm{n}}, \qquad \{\phi_{\mathrm{n}}, \pi_{2\mathrm{n}}\} = -\frac{3}{4}\frac{\mathrm{n}^2-1}{\mathrm{n}^2-4} f_{\mathrm{n}}. \tag{30.38}$$

Moreover, these are merely associated momenta rather than conjugate momenta: satisfying the Kuchař beables condition can require variables other than the conjugate ones.

In the vacuum case, the constraint algebraic structure is as in Fig. 30.3.a) and the corresponding notions of beables are as in Fig. 30.3.b). So in SIC, Kuchař beables $\kappa \neq G$ ($\mathfrak{g}$-beables for $\mathfrak{g} = \mathit{Diff}(\mathbb{S}^3)$; let us term the latter, more restrictive notion *Superspace beables* su. The κ do not now suffice, in the sense that eliminating $\mathcal{F}$**lin** does not in this case send one to $\mathfrak{q}/\mathfrak{g}$ (as is clear from 30.4). So eliminating $\mathcal{F}$**lin** under this symmetry restriction has ceased to coincide with gauging out the entirety of $\mathit{Diff}(\boldsymbol{\Sigma})$.

Note also that this SIC model has additional notions of A-beables, for which we suggest the names 'S-Kuchař beables' and 'V-Kuchař beables', meaning that they commute with just $^{\mathrm{S}}\mathcal{M}$ and $^{\mathrm{V}}\mathcal{M}$ respectively. Taking out just $^{\mathrm{V}}\mathcal{M}$ has a minimally-coupled scalar field case parallel; this underlies an even more partial reduction of Wada's [870].

30.6 The Averaging Problem in GR

Averaging in GR (in general, rather than just perturbative) is compromised by the Einstein field equations' nonlinearity:

$$\mathcal{G}_{\mu\nu}(\langle \mathrm{g}_{\rho\sigma}\rangle) \neq \langle \mathcal{G}_{\mu\nu}(\mathrm{g}_{\rho\sigma})\rangle. \tag{30.39}$$

The difference between these might get confused with a further source's $\mathrm{T}_{\mu\nu}$ (whether a bona fide material energy–momentum–stress tensor, or pseudo-tensor paralleling Appendix K.5). This has the status of a further unresolved problem. Note finally that the inhomogeneous corrections of (30.29) suggested by Relationalism are distinct from those due to averaging.

30.7 SIC Records

The Euclidean $\|\mathrm{d}\underline{v}\|^2$ is a useful notion of distance on $\mathfrak{Modespace}$ (defined in Appendix I.2). One can furthermore extend the scope of Sect. 26.1's propositions by 'what is Prob(inhomogeneity is small) quantified by $|\underline{v}| < \epsilon$'. On the other hand, modewise SIC is itself a local-in-time slab $\mathfrak{T} \times \mathbb{R}^3$ within Minkowski spacetime $\mathbb{M}^4$; restriction to this slab affects the detailed form of probability distributions thereupon and of subsequent statistical tests.

Research Project 20) Pass from considering Probability and Statistics on Minkowski spacetime $\mathbb{M}^4$ to considering these on a slab $\mathfrak{T} \times \mathbb{R}^3$ therein.

Probability distributions on 3-*d* flat space are straightforward, and probabilistic studies on flat indefinite spaces were already pointed to in Sect. 26.7. As regards correlations, $\mathrm{Cov}(v, v')$ and the n-point function in the $\mathfrak{Modespace}$ of the v's make sense. Finally, notions of information can be based on the $\rho \log \rho$ combination (Appendix Q.8) built from a probability density on the Euclidean space of small inhomogeneities.

30.8 SIC Histories

Unreduced Histories For the Halliwell–Hawking model, the classical histories configuration variables are $\Omega(\lambda)$, $\phi(\lambda)$, $a_{\mathrm{n}}(\lambda)$, $b_{\mathrm{n}}(\lambda)$, $c_{\mathrm{n}}(\lambda)$, $d_{\mathrm{n}}(\lambda)$ and $f_{\mathrm{n}}(\lambda)$ for λ a continuous label time. Auxiliary variables are $\mathrm{d}j^{\mathrm{o}}_{\mathrm{n}}(\lambda)$, $\mathrm{d}j^{\mathrm{e}}_{\mathrm{n}}(\lambda)$ and $\mathrm{d}k_{\mathrm{n}}(\lambda)$. Let us denote the conjugate histories momenta by $\pi_\Omega(\lambda)$, $\pi_\phi(\lambda)$, $\pi_{a_{\mathrm{n}}}(\lambda)$, $\pi_{b_{\mathrm{n}}}(\lambda)$, $\pi_{c_{\mathrm{n}}}(\lambda)$, $\pi_{d_{\mathrm{n}}}(\lambda)$ and $\pi_{f_{\mathrm{n}}}(\lambda)$. This model's histories constraints are

$$^{\mathrm{S}}\mathcal{H}_\lambda = \int \mathrm{d}t^{\lambda\mathrm{S}}\mathcal{H}(\lambda), \qquad ^{\mathrm{V}}\mathcal{H}_\lambda = \int \mathrm{d}t^{\lambda\mathrm{V}}\mathcal{H}(\lambda), \qquad ^{\mathrm{T}}\mathcal{H}_\lambda = \int \mathrm{d}t^{\lambda\mathrm{T}}\mathcal{H}(\lambda), \tag{30.40}$$

$$^{\mathrm{S}}\mathcal{M}_\lambda = \int \mathrm{d}t^{\lambda \mathrm{S}}\mathcal{M}(\lambda), \qquad ^{\mathrm{V}}\mathcal{M}_\lambda = \int \mathrm{d}t^{\lambda \mathrm{V}}\mathcal{M}(\lambda). \tag{30.41}$$

One can now take $\lambda = t^{\mathrm{em}}$ as one's choice of label time [25], to consider this case within the Classical Machian Emergent Time Approach.

Reduced Histories In the vacuum case, one could use $\xi_{\mathrm{n}}(\lambda)$, $\underline{v}_{\mathrm{n}}(\lambda)$, $d_n(\lambda)$ and $d_n^{\mathrm{e}}(\lambda)$ as histories variables, with histories conjugate momenta. One could once again use the corresponding t^{em} as choice of label time λ [25]. This formulation's histories Hamiltonian constraint is

$$\widetilde{\mathcal{H}}_\lambda := \int \mathrm{d}t^{\lambda}\widetilde{\mathcal{H}}(\lambda) \tag{30.42}$$

for $\widetilde{\mathcal{H}}$ SIC's reduced Hamiltonian constraint.

Combined Approach This is relatively straightforward modewise, through having the same $\mathfrak{q}$ geometry as for diagonal anisotropic models and particles in flat spacetime. This coincides with the original scope of Halliwell-type models [413, 414]. Finally, Halliwell's construct can be extended to promote Superspace beables $\mathcal{SU}$ to Dirac beables $\mathcal{D}$.

30.9 Summary of the Model Arenas

An efficient way of handling RPM, Minisuperspace and modewise SIC equations all at once involves the standard form

$$\mathcal{C}\mathrm{hronos} = \{\mathrm{q}\, p_h^2 + \mathrm{s}\, p_l^2\}/2 - \mathrm{w} = 0, \tag{30.43}$$

and the following table of cases.

Model	q	s	w
1- and 2-d scaled RPMs [23, 29, 37, 61]	1	$\frac{1}{\rho^2}$	$2\{E - V\}$
Isotropic Minisuperspace with minimally-coupled scalar field [31]	-1	1	$\exp(6\,\Omega)\{\exp(-2\Omega) - V(\phi) - 2\,\Lambda\}$
Bianchi IX anisotropic Minisuperspace vacuum	-1	1	$\exp(3\,\Omega) \times$ (I.5)
n-modewise vacuum SIC [34]	-1	1	$\widetilde{(30.22)}$

30.10 Frontiers of Research

Research Project 21) Work out the reduced formulation, Records, Histories, and Combined Approaches for SIC with minimally-coupled scalar matter.

Research Project 22) Extend this Chapter's work to more general SIC models. E.g. consider third-order models, models with multiple matter fields, and models with small anisotropies treated alongside the small inhomogeneities.

A limitation of modewise SIC is that its perturbative split carries background split problems, even though it is not the same as the background Minkowski split that more habitually carries such problems. Nonperturbative Midisuperspace models are better in this regard. We shall see in subsequent Chapters how this affects the Foliation Independence and Spacetime Constructability aspects of Background Independence. Reasons to push to third order include that the SVT modes now start to couple.

Research Project 23)† Consider time and the Problem of Time to a similar extent to that considered in this book, but now for full, rather than perturbative, Midisuperspaces. It may help if you begin with $\mathbb{T}^3$ or $\mathbb{S}^3$ Gowdy models.

Research Project 24)† Formulate SIC for Nododynamics. (This topic remains little developed, and not due to a lack of trying. [156] may be a useful preliminary.)

SIC in Supergravity, on the other hand, has already been considered including in mode-expanded form, in [232, 555, 868].

Chapter 31
Embeddings, Slices and Foliations

This Chapter considers Foliation Independence and its realization in the case of GR by Refoliation Invariance. We approach this by generalizing Sec's 8.4 consideration of a single hypersurface $\mathbf{\Sigma}$ within a manifold $\mathfrak{M}$ from both top-down and bottom-up perspectives. Let us first remark that considering hypersurfaces picks out codimension $C = 1$, though the current Chapter's $\mathfrak{M}$ is more general than GR's 4-d $\mathfrak{m}$. Somewhat abusing notation, let us still use the $\mathbf{\Sigma}$ and σ notation in this more general setting. A is the index corresponding to $\mathfrak{M}$ and a to $\mathbf{\Sigma}$ and σ. The current Chapter also retains the split formulation of GR's bias that $\mathfrak{M} = \mathbf{\Sigma} \times \mathfrak{I}$ for some interval $\mathfrak{I}$, and that $\mathbf{\Sigma}$ is to be compact without boundary for simplicity. The idea now is to reformulate this in a more global manner [573, 576–579, 581], in the sense of improving on how previous works [73, 250] depended on choices of coordinates which in general just hold locally. This Foliation Formulation is also of foundational interest through its giving foliations a more primary status.

31.1 Single-Slice Concepts. i. Topological and Differentiable Manifold Levels

Let us introduce the following conceptual types of map. *Slice* involves identifying a particular slice $\mathbf{\Sigma}$ in $\mathfrak{M}$. On the other hand, *Project* involves keeping only information projected onto $\mathbf{\Sigma}$. Passing to σ involves forgetting that $\mathfrak{m}$ was the source of this information, now to be regarded as set up from intrinsic first principles. *Forget* involves forgetting that $\langle \mathfrak{m}, \mathbf{\Sigma} \rangle$ contains a particular hypersurface that is picked out as a slice; see Appendix A for forgetful maps in greater generality. However, since $\mathfrak{M} = \mathbf{\Sigma} \times \mathfrak{I}$, $\mathbf{\Sigma}$ and σ are for now just related by the identity map. Going in the opposite direction, *Embed* allows for $\mathbf{\Sigma}$ to be treated as a hypersurface within some ambient $\mathfrak{M}$; this map is denoted by Φ. Since $\Phi : \sigma \rightarrow \mathfrak{M}$, the embeddings in question are homeomorphisms. The injectivity residing within this statement guarantees that the spatial hypersurface $\mathbf{\Sigma}$ does not intersect itself, unlike Fig. 31.1.c). See also Fig. 31.1.a) for how all of the above maps fit together.

E. Anderson, *The Problem of Time*, Fundamental Theories of Physics 190,
DOI 10.1007/978-3-319-58848-3_31

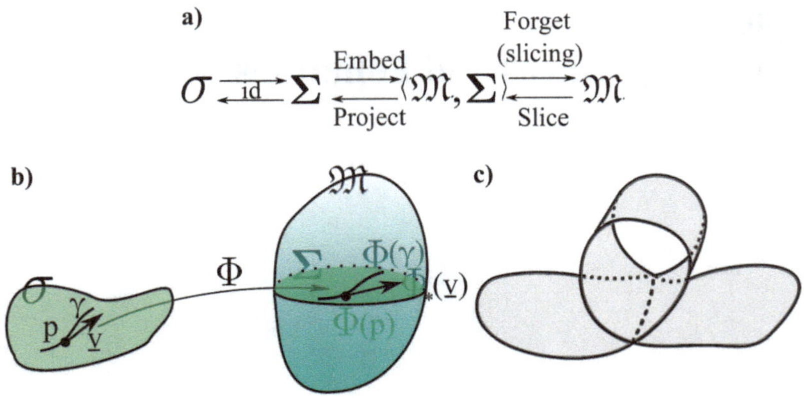

Fig. 31.1 **a)** The detailed sense in which (suitable extensions of) 'embed' and 'slice and project' form a 2-way route at the level of topological manifolds. **b)** Given an identity-and-embedding $\Phi \circ$ id (which we will take Φ to suffice to denote), and allowing for the spaces in question to possess differentiable structure, the following additional maps are established. The corresponding [382, 483] push-forward $\Phi_* \underline{v}$ of a tangent vector $\underline{v} \in \mathfrak{T}_p(\sigma)$ to a curve γ in σ is a tangent vector $\in \mathfrak{T}_{\Phi(p)}(\boldsymbol{\Sigma})$ to the image curve $\Phi(\gamma)$ in $\boldsymbol{\Sigma}$. In the opposite direction—projection-and-forgetting—Φ induces a pull-back $\Phi^* : \mathfrak{T}^*_{\Phi(p)}(\boldsymbol{\Sigma}) \rightarrow \mathfrak{T}^*_p(\sigma)$ between the space of 1-form linear maps. **c)** Embeddings are defined to preclude self-intersections [614], such as the counter-example depicted

Allowing for differentiable structure as well, Fig. 31.1.b) associates embedding and slicing with the differential-geometric notions of push-forward and pull-back (concepts laid out in greater generality in Appendix D.2).

31.2 ii. Metric Level

The situation is more complicated at the metric level. One fundamental reason for these complications is that *each* of $\mathfrak{M}$, $\boldsymbol{\Sigma}$ and σ carries its own metric, moreover with the first two bearing relation. Let us begin to consider this by splitting the Metric Geometry $\langle \mathfrak{M}, \mathbf{M} \rangle$ with respect to $\boldsymbol{\Sigma}$ into of tangential and normal parts:

$$\mathfrak{T}_{\mathrm{p}}(\mathfrak{M}) = \mathfrak{T}_{\mathrm{p}}(\boldsymbol{\Sigma}) \oplus \mathfrak{n}, \tag{31.1}$$

for $\mathfrak{n}$ the space spanned by the normal vector field n^{A}. This is normal to the corresponding *isometric embedding*: $\Phi : \boldsymbol{\Sigma} \rightarrow \mathfrak{m}$. [This is Φ_μ for some particular fixed value of μ: a signature-neutral notation for the extra dimension's coordinate with value $\mu = \mu_1$ picking out the hypersurface itself.] Using $\Phi^{\mathsf{A}} := \vec{X}^{\mathsf{A}}(\Phi(\underline{x}))$ for $\vec{X}^{\mathsf{A}}$ a coordinate system on $\mathfrak{m}$ [364, 483, 501, 502, 576], this is defined by

$$\mathrm{n}_{\mathsf{A}}\big(x^{\mathsf{c}}; \Phi\big]\, \Phi^{\mathsf{A}}{}_{,\mathsf{b}}\big(x^{\mathsf{c}}\big) = 0, \tag{31.2}$$

$$\mathrm{M}^{\mathsf{AB}}\big(\Phi\big(x^{\mathsf{c}}\big)\big)\, \mathrm{n}_{\mathsf{B}}\big(x^{\mathsf{d}}; \Phi\big]\, \mathrm{n}_{\mathsf{B}}\big(x^{\mathsf{c}}; \Phi\big] = \pm 1 \quad \forall x \in \boldsymbol{\Sigma}. \tag{31.3}$$

Here (31.2) is normality in the sense of being perpendicular to the hypersurface. Also (31.3) is a normalization condition that depends on the signature of the extra dimension: -1 is timelike and $+1$ is spacelike.

The *induced metric* can be interpreted as the pull-back $\mathbf{m} := \Phi^*\mathbf{M}$. In components, using a hypersurface-adapted coordinate system x^{c}, μ with x^{c} on $\boldsymbol{\Sigma}$ and

$$\mathrm{m}_{\mathsf{ab}}(x^{\mathsf{c}}, \mu_1) := \mathrm{m}_{\mathsf{ab}}(x^{\mathsf{c}}, \Phi_{\mu_1}] = (\Phi^*\mathbf{M})_{\mathsf{ab}}(x^{\mathsf{c}}, \mu_1) = (\Phi^*_{\mu_1}\mathbf{M})_{\mathsf{ab}}(x^{\mathsf{c}})$$
$$= \mathrm{M}_{\mathsf{AB}}(\Phi(x^{\mathsf{c}}, \mu_1))\Phi^{\mu}{}_{,\mathsf{a}}(x^{\mathsf{c}}, \mu_1)\Phi^{\mathsf{B}}{}_{,\mathsf{b}}(x^{\mathsf{c}}, \mu_1). \tag{31.4}$$

A second fundamental reason for the metric-level version's greater complexity is that one now has not only $\mathbf{h} \in \mathfrak{Riem}(\boldsymbol{\Sigma})$ to contend with, but also a notion of extrinsic curvature tensor $\mathbf{K}$, whose possible values form the space of symmetric 2-tensors, $\mathfrak{Sym}(\boldsymbol{\Sigma})$. The coordinate-free form of the definition of extrinsic curvature is the *shape operator* $\mathfrak{T}_{\mathrm{p}}(\mathfrak{M}) \to \mathfrak{T}_{\mathrm{p}}(\mathfrak{M})$ from a vector v to the variation of the normal along v (cf. Fig. 8.2). The possible extrinsic curvatures form the space. This is complicated by one needing to associate a $\mathbf{K}$ to each $\mathbf{h}$.[1]

Slice now involves identifying a particular slice $\langle \boldsymbol{\Sigma}, \mathbf{m}, \mathbf{K} \rangle$ in $\langle \mathfrak{M}, \mathbf{M} \rangle$. *Project* involves keeping only information projected onto $\langle \boldsymbol{\Sigma}, \mathbf{m}, \mathbf{K} \rangle$. Passing to σ involves forgetting that $\mathfrak{M}$ was the source of this information, including forgetting $\mathbf{K}$ since that is not intrinsic to σ. *Forget* now involves forgetting that $\langle \langle \mathfrak{M}, \mathbf{M} \rangle, \langle \boldsymbol{\Sigma}, \mathbf{m}, \mathbf{K} \rangle \rangle$ contains a particular picked-out hypersurface $\langle \boldsymbol{\Sigma}, \mathbf{m}, \mathbf{K} \rangle$ as a slice. Other reasons for this step being less trivial than at the topological level include needing firstly to derive the Gauss–Codazzi relations—(8.4)–(8.5) up to signature-dependent signs—by projections. Secondly, one needs to solve these equations in order to obtain suitable $\mathbf{m}, \mathbf{K}$ pairs. Moreover, $\langle \boldsymbol{\Sigma}, \mathbf{m} \rangle$ and $\langle \sigma, \mathbf{m} \rangle$ are *not* now in general related by the identity map. For not all the $\mathbf{m}$ that one can place on σ are necessarily isometrically embeddable into $\langle \mathfrak{M}, \mathbf{M} \rangle$. Thereby, a nontrivial *inclusion map* 'Include' is required (see Appendix A.1 for the general concept of inclusion maps. Additionally, *Embed* now involves allowing for $\langle \boldsymbol{\Sigma}, \mathbf{m} \rangle$ to be treated as a hypersurface within some ambient $\langle \mathfrak{M}, \mathbf{M} \rangle$; this map continues to be denoted by Φ.

Example 1) in the ADM GR case $\langle \mathfrak{M}, \mathbf{M} \rangle = \langle \mathfrak{m}, \mathbf{g} \rangle$: GR spacetime $\langle \boldsymbol{\Sigma}, \mathbf{m} \rangle = \langle \boldsymbol{\Sigma}, \mathbf{h} \rangle$ spatial, x^{c} = the actual spatial $\underline{x}$ and $\mu = t$. (31.4) furthermore gives an interpretation of the spatial metric as being induced by the spacetime metric, with components [576]

$$\mathrm{h}_{ab}(\underline{x}, t_1) = \mathrm{g}_{\mu\nu}(\Phi(\underline{x}, t_1))\Phi^{\mu}{}_{,a}(\underline{x}, t_1)\Phi^{\nu}{}_{,b}(\underline{x}, t_1). \tag{31.5}$$

Further Examples 2–4) Chap. 8's and the above's level of theory for $+++$ spaces interpreted as spatial hypersurfaces within $--+++$ spacetimes readily generalizes (up to sign differences only) to 2) $-++$ within $--+++$ and 3) $+++$

[1]This can be formulated in terms of a fibre bundle $\mathfrak{T}(\mathfrak{Riem}(\boldsymbol{\Sigma}))$: the tangent bundles space of extrinsic curvatures with over the base space $\mathfrak{Riem}(\boldsymbol{\Sigma})$. Another formulation involves using $\mathbf{p}$ in place of $\mathbf{K}$. In the latter case, the $\mathfrak{Sym}$ involved is a space of symmetric 2-tensor *densities* and the corresponding fibre bundles space is $\mathfrak{T}^*(\mathfrak{Riem}(\boldsymbol{\Sigma}))$.

within $++++$. They generalize again to 4) codimension $C = 1$ hypersurfaces within an n-d manifold with p timelike and $n - p$ spacelike dimensions. Note however that the physical meanings of examples 1) to 3) (with however many extra $+$'s added) are rather different. In particular, 1) is aligned with the difference between time and space and so with e.g. dynamics with respect to a time variable and the standard formulation of Causality Theory, by which it is the case of prime interest in this book. One can think of 2) as a particular slice, membrane worldsheet or 'braneworld' within a $C = 1$ bulk, and of 3) as the Euclidean counterpart of 1) [not directly physical, but of use in some programs.] Moreover, 1) to 4) diverge at the level of Analysis and consequently of supporting PDE theorems. Since the latter often involve well-posedness, this level of study is additionally relevant to the design of numerical codes. Almost all work hitherto at this level has focused on 1), for which well-posedness is attainable.

31.3 More General Examples

Further Example 5) codimension $C > 1$ surfaces can be treated with an ADM-like split with multiple special rows and columns, i.e. multiple 'lapses' and 'shifts'. Subsequently there are a matching multiplicity of extrinsic curvatures [344, 777]. $C > 2$ makes sense if the application is to the evolution of a surface layer [322] or brane that is allowed to have dynamics rather than being of a predetermined shape at all times. It is also used in Sect. 31.5's geometrical explanation of the notion of foliation.

Already at the classical level, however, there are issues with using $C > 2$ as a means of having multiple temporal dimensions. Consider for instance the ultrahyperbolic analogue of Minkowski spacetime. Here one ceases to be able to divide the structure which has replaced spacetime into future and past. Also the corresponding ultrahyperbolic analogue of the wave equation exhibits difficulties as a PDE problem (see e.g. Chap. VI.16 of [220]).

The above diversity of concepts and workings established, for the rest of this book (bar Chaps. 29 and Epilogue II.A) we restrict ourselves to Example 1) with the usual spatial dimension 3.

31.4 Spaces of Embeddings and of Slices

Here we use 'all spacetimes' to mean 'all spacetimes on a fixed topological manifold $\mathfrak{m}$, 'all hypersurfaces' to mean 'all spacetimes on a fixed topological manifold σ', and the suffixes '1' and '2' to denote particular members. Some spaces arising in these contexts are as follows.

1) $\mathfrak{PR}\mathrm{iem}(\mathfrak{m}, \mathfrak{S}\mathit{lice}(\mathbf{h}_1))$: the space of spacetime manifolds exhibiting a slice with a given intrinsic spatial geometry thereupon.

2) $\mathfrak{Slicespace}(\mathfrak{m}, \mathbf{g}_2)$: the space of all slices within a given spacetime $\mathbf{g}_2$. Kuchař considers the Differential and Metric Geometry for this in [576].
3) $\mathfrak{Emb}(\mathfrak{m}, \boldsymbol{\Sigma})$: the space of all embeddings of spatial 3-metrics on a fixed $\boldsymbol{\Sigma}$ into spacetime 4-metrics on a fixed $\mathfrak{m} = \boldsymbol{\Sigma} \times \mathfrak{T}$; e.g. Isham and Kuchař considered this in [477, 483, 501, 502].[2]

Further relevant variety arises from restricting 'spacetime' to 'spacetime solving the Einstein field equations of GR'. One would now have e.g. a version of $\mathfrak{Emb}(\mathfrak{m}, \boldsymbol{\Sigma})$ whose spacetimes are qualified as being GR ones. There are also intermediate situations, such as restricted to be 'hypersurfaces' by a property, such as being CMC; 'spacetime' and 'embedding' might also be qualified. One could also specify whether the embedding is locally or globally valid, with the latter being partnered by 'pieces' of spacetime or of space. Finally Chap. 21 bears witness to how $\mathbf{K}$ is not always left free in handling the GR constraint equations.

Let us next consider how this Sec's workings so far combine with Configurational and Spacetime Relationalisms. One can alter the output of slice or embedding schemes to 3-geometries or 4-geometries by quotienting out $\mathit{Diff}(\boldsymbol{\Sigma})$ or $\mathit{Diff}(\mathfrak{m})$ at the end of the scheme. One can alter the input by picking a representative from the equivalence class at the start of the scheme. One can envisage quotienting out $\mathit{Diff}(\boldsymbol{\Sigma})$; doing so within $\mathfrak{Slicespace}(\mathfrak{m}, \mathbf{g}_2)$ produces Kuchař's '$\mathfrak{H}$yperspace' for which he provided a topological and geometrical study of [576]. One can alternatively envisage quotienting out $\mathit{Diff}(\mathfrak{m}, \mathfrak{Slice}_1) := \mathit{Diff}(\mathfrak{m}|$ a given $\mathfrak{Slice}_1$ is preserved$)$. However, $\mathit{Diff}(\mathfrak{m}, \mathfrak{Slice}) := \mathit{Diff}(\mathfrak{m}|$ any $\mathfrak{Slice}$ is preserved$)$ is just id.

31.5 Foliation in Terms of a Decorated Chart

Let us now generalize Chap. 8's definition of *foliation* to $\mathfrak{f} = \{\mathfrak{l}_\mathsf{A}\}_{\mathsf{A}\in A}$. This denotes a decomposition of an m-dimensional manifold $\mathfrak{M}$ into a disjoint union of connected p-dimensional subsets—the *leaves* $\mathfrak{l}_\mathsf{A}$ of the foliation—such that the following holds. $\mathrm{p} \in \mathfrak{M}$ is to have a neighbourhood $\mathfrak{N}_\mathrm{p}$ in which coordinates $(x^1, \dots, x^m)$ are valid. I.e. $\mathfrak{N}_\mathrm{p} \to \mathbb{R}^m$ such that for each leaf $\mathfrak{l}_\mathsf{A}$ the components of $\mathfrak{N}_\mathrm{p} \cap \mathfrak{l}_\mathsf{A}$ are described by x^{p+1} to x^m constant: the obvious extension of Fig. 8.3.d). The codimension of the foliation is $C = m - p$.

31.6 ADM Kinematics for Foliations

For example, in ADM's 3 + 1 split of 4-d indefinite $(- - + + +)$ GR spacetime, $\mathfrak{M} = \mathfrak{m}$, so $m = 4$ and the leaves $\mathfrak{l}_\mu$ are 3-d spacelike $(+ + +)$ hypersurfaces $\boldsymbol{\Sigma}_\mathrm{t}$.

[2] They furthermore arm this with the topological structure that it inherits naturally as an open subset of $\mathfrak{C}^\infty(\boldsymbol{\Sigma}, \mathfrak{m})$. This space can be considered [426] as a manifold in the sense of Fréchet (see Appendix H.2).

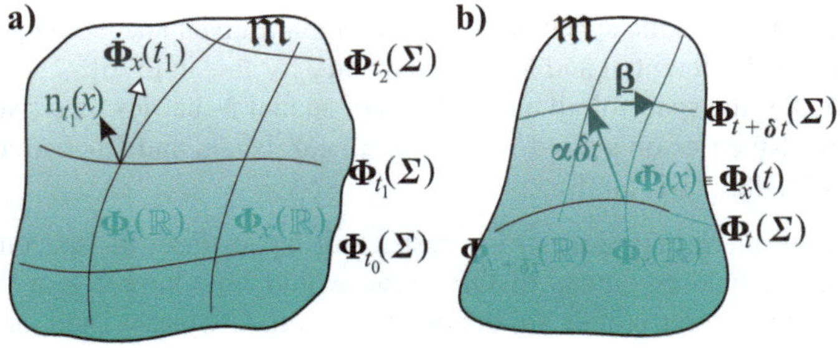

Fig. 31.2 **a)** The flow lines of the foliation of $\mathfrak{m}$ [483]. n_{t_1} is here the normal vector field on the hypersurface $\Phi_{t_1}(\boldsymbol{\Sigma}) \rightarrow \mathfrak{m}$; extending along the flow lines, one has the normal vector field n_t to the whole foliation. **b)** gives the lapse α and shift $\underline{\beta}$ in the context of a foliation [483, 501, 502]

So in this case, $p = 3$ and so $c = 1$: the temporal dimension. For GR formulated as Geometrodynamics on a fixed spatial topological manifolds $\boldsymbol{\Sigma}$, a 'global in space' foliation's leaves are each of that fixed $\boldsymbol{\Sigma}$. Henceforth we restrict attention to $c = 1$ foliations.

The foliation $\Phi_t : \boldsymbol{\Sigma} \times \mathbb{R} \rightarrow \mathfrak{m}$. Its inverse $\Phi^{-1}{}_t : \mathfrak{m} \rightarrow \boldsymbol{\Sigma} \times \mathbb{R}$ is a diffeomorphism as well, of the form [483] $\Phi_t{}^{-1}(\vec{X}) = (\sigma(\vec{X}), \tau(\vec{X})) \rightarrow \boldsymbol{\Sigma} \times \mathbb{R}$. The map $\tau : \mathfrak{m} \rightarrow \mathbb{R}$ here is a global timefunction, which provides the $t := \{\tau(\Phi_t(x)) = t \ \forall x \in \boldsymbol{\Sigma}\}$ notion of natural time parameter associated with the foliation. Isham [483] cautions, however that such a 'definition of time' is artificial (from an operational perspective: concerning its measurability and associated clock manufacture). On the other hand, $\sigma : \mathfrak{m} \rightarrow \boldsymbol{\Sigma}$ is some kind of 'space map'.

Next, for each $\underline{x} \in \boldsymbol{\Sigma}$, the map $\Phi_{\underline{x}} : \mathbb{R} \rightarrow \mathfrak{m}$ defined by $t \mapsto \Phi(\underline{x}, t) := \Phi_t(\underline{x})$ is a curve in $\mathfrak{m}$. Thus there is a corresponding a 1-parameter family of tangent vectors on $\mathfrak{m}$; let us denote these by

$$\dot{\Phi}_{\underline{x}}(t) \quad \text{whose components are } \dot{\Phi}^{\mu}_{\underline{x}}(t) = \dot{\Phi}^{\mu}(\underline{x}, t). \tag{31.6}$$

The corresponding vector field is the *deformation vector field*; Fig. 31.2 a) gives the corresponding flow lines. This exhibits a simple duality in the sense that for each $\underline{x} \in \sigma$, $\dot{\Phi}_{\underline{x}}(t)$ is a vector in $\mathfrak{T}_{\Phi(\underline{x},t)}(\mathfrak{m})$ at the point $\Phi(\underline{x}, t)$ in $\mathfrak{m}$. Isham counsels [477, 483] that this object is best regarded as an element of $\mathfrak{T}_{\Phi_t}(\mathfrak{E}mb(\sigma, \mathfrak{m}))$: the space of vectors tangent to the infinite-d manifold $\mathfrak{E}mb(\sigma, \mathfrak{m})$ of embeddings of σ in $\mathfrak{m}$ at the particular embedding Φ_t. The deformation vector field, moreover, is a reinterpretation of Sect. 8's time flow vector field t^{μ}, according to

$$t^{\mu}(\vec{X}) = \dot{\Phi}^{\mu}(\underline{x}, t)\big|_{\vec{X} = \vec{X}(\underline{x}, t)}. \tag{31.7}$$

This corresponds to viewing it as acting on a slice or leaf.

The functional derivative of $h_{ab}(\underline{x}; \Phi]$ with respect to Φ projected along n^{μ} is of value in considering dynamical evolution. Computationally, [454, 575] this takes the form

$$\mathrm{n}^\mu(\underline{x};\Phi]\frac{\delta}{\delta\Phi^\mu(\underline{x}))\mathrm{h}_{ab}(\underline{x}',\Phi]} = -2\,\mathcal{K}_{ab}(\underline{x};\Phi]\delta(\underline{x}-\underline{x}'), \tag{31.8}$$

for $\mathcal{K}$ the extrinsic curvature of the hypersurface $\Phi(\mathbf{\Sigma})$, which in this formulation is given by [cf. (31.5)]

$$\mathcal{K}_{ab}(\underline{x};\Phi] := -\nabla_\mu \mathrm{n}_\nu(\underline{x};\Phi]\Phi^\mu{}_{,a}(\underline{x})\Phi^\nu{}_{,b}(\underline{x}). \tag{31.9}$$

Also, $\nabla_\alpha n_\beta(\underline{x};\Phi]$ is here the covariant derivative obtained by parallel transporting the cotangent vector $\mathrm{n}(\underline{x};\Phi] \in \mathfrak{T}^*_{\Phi(\underline{x})}(\underline{x})\mathfrak{m}$ along the hypersurface $\Phi(\mathbf{\Sigma})$ using $\mathfrak{m}$'s metric $\mathbf{g}$.

The deformation vector can be decomposed into one piece lying along the hypersurface $\Phi_{t_1}(\mathbf{\Sigma})$ and another parallel to n_t. In the ADM formulation (31.6) can then be expanded out as (Fig. 31.2.b)

$$\dot{\Phi}^\mu(\underline{x},t_1) = \alpha(\underline{x},t_1)\mathrm{g}^{\mu\nu}\big(\Phi(\underline{x},t_1)\big)\mathrm{n}_\nu(\underline{x},t_1) + \beta^a(\underline{x},t_1)\Phi^\alpha{}_{,a}(\underline{x},t_1), \tag{31.10}$$

using $\mathrm{n}^\mu(\underline{x},t)$ as a shorthand for $\mathrm{n}^\mu(\underline{x};\Phi_t]$. From a more minimalist perspective, note that lapse α and shift $\underline{\beta}$ remain meaningful for just a pair of neighbouring slices. [This is now indexed by t_1: the value on the initial slice, and interpreted in terms of a single embedding Φ_{t_1} corresponding to this slice.] Indeed as regards $\underline{\beta}$, one can go so far as to reinterpret it in terms of changes of coordinates on a single hypersurface.

Moreover, from the spacetime perspective α and $\underline{\beta}$ depend on the spacetime metric $\mathbf{g}$ as well as on the foliation (a partly invertible relationship [483]). For a fixed foliation, α and $\underline{\beta}$ are identified with pieces of $\mathbf{g}$. E.g. these can be formulated using the pull-back $\Phi^*(\mathbf{g})$ of $\mathbf{g}$ by the foliation $\Phi : \mathbf{\Sigma}\times\mathbb{R}\rightarrow\mathfrak{m}$ in coordinates $\vec{X}^\mu$, $\mu = 0\ldots 3$, on $\mathbf{\Sigma}\times\mathbb{R}$ that is adapted to the product structure: $\vec{X}^{\mu=0}(\underline{x},t) = t$. Here [483] $\vec{X}^{\mu=1,2,3}(\underline{x},t) = \underline{x}^{\mu=1,2,3}(\underline{x})$ for x^a, $a = 1, 2, 3$ some coordinate system on $\mathbf{\Sigma}$.

The components of $\Phi^*(\mathbf{g})$ are (31.5),

$$\big(\Phi^*\mathbf{g}\big)_{0a}(\underline{x},\mathrm{t}) = \beta^b(\underline{x},\mathrm{t})\,\mathrm{h}_{ab}(\underline{x},t), \tag{31.11}$$

$$\big(\Phi^*\mathbf{g}\big)_{00}(\underline{x},t) = \beta^a(\underline{x},t)\,\beta^b(\underline{x},t)\,\mathrm{g}^{ab}(\underline{x},t) - \alpha(\underline{x},t)^2. \tag{31.12}$$

In this way, ADM's conception of the split can be replaced by one based on foliations [483, 576]. This can be interpreted as $\Phi^{\mathrm{ref}} : \mathbf{\Sigma}\times\mathbb{R}\rightarrow\mathfrak{m}$ with respect to some choice of reference foliation.

Let us end by noting that whereas $\mathcal{M}_i$ corresponds to *Diff*$(\mathbf{\Sigma})$ in an obvious manner, $\mathcal{H}$ is linked to a *hidden* invariance: Refoliation Invariance.

31.7 Spaces of Foliations

Foliations are $\langle \mathfrak{m}, \mathbf{h} \rangle \rightarrow$ (a sequence of $\mathbf{h}$'s on a given $\mathbf{\Sigma}$), i.e. curves in $\mathfrak{E}mb(\mathbf{\Sigma}, \mathfrak{m})$. Foliations map to paths in $\mathfrak{R}$iem($\mathbf{\Sigma}$) and Sym($\mathbf{\Sigma}$), or jointly in $\mathfrak{T}(\mathfrak{R}$iem($\mathbf{\Sigma}$)) or GR's phase space.

$\mathfrak{f}$oliationspace($\mathfrak{m}, \mathbf{g}_2$) is the space of all foliations within a given spacetime $\mathbf{g}_2$. Also $\mathfrak{P}\mathfrak{R}$iem(($\mathfrak{m}$), $\mathfrak{F}$ol): the space of all foliations of all spacetimes.

There are also intermediates, such as replacing each use of 'foliation' above by 'foliation with a given property'. E.g. 'foliation by CMC slices', with 'spacetime' being understood to being restricted to ones for which one or more such foliations exist. One could also specify 'globally valid foliation' or 'locally valid foliation', with the latter case being paired with 'pieces of spacetimes' rather than whole spacetimes. Also, one might replace 'spacetime' by 'spacetime solving the Einstein field equations of GR', in which case one would have e.g. GR-$\mathfrak{S}$ol($\mathfrak{m}$, $\mathfrak{F}$ol) in place of $\mathfrak{S}$uperspacetime($\mathfrak{m}$, $\mathfrak{F}$ol). This distinction reflects that one can as well foliate *any* pseudo-Riemannian spacetime, but one is interested in particular in foliations of GR spacetimes.

31.8 Refoliation Invariance

Section 10.8 already accounted well for Refoliation Invariance. Note also that the points depicted in Fig. 10.3.c) are points of *a suitable fibre bundles over* $\mathfrak{R}$iem($\mathbf{\Sigma}$) so as to represent the hypersurfaces indicated. Finally, Refoliation Invariance holding implies assuming that the *thick* sandwich [552] is satisfied which does limit the rigour and generality of the result.

31.9 Bubble Time and Its Dual: Many-Fingered Time

These are useful conceptual and technical characterizations of time in GR and in arbitrary general frame Field Theory more generally. Multiple choices of timefunction are valid herein, with each corresponding to a foliation. This validity reflects e.g. the multiplicity of GR's coordinate times and consequences of this upon performing a space–time split, giving a 'many-fingered' notion of time. In GR, moreover, time is local: in some places at a given time, a choice of finger is longer than at other places (Fig. 31.3.b). The instant of time is a slice, with a continuous sequence of non-intersecting slices forming a foliation as per Sect. 31.5.

Moreover, thinking of a particular slice as the surface of a bubble, one encounters the further notion of bubble deformation[3] under evolution due to the above local aspect of GR time.

[3]This notion of deformation indeed coincides with that of hypersurface deformation and of deformation algebroid.

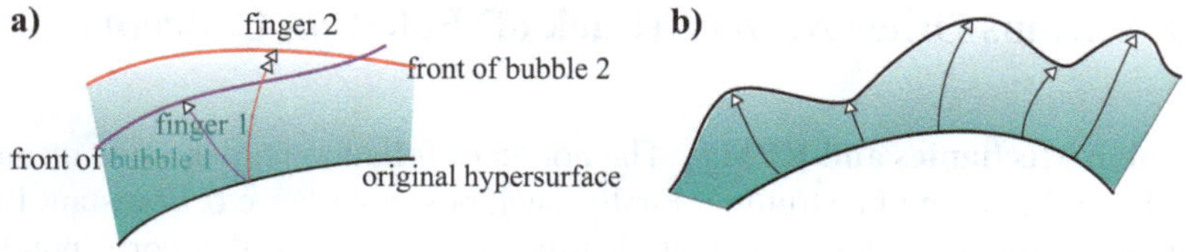

Fig. 31.3 **a)** Each finger of time corresponds to deforming the original hypersurface to a different bubble front. **b)** Note that each of these is additionally a *field theoretic* notion, meaning that the lengths of a given choice of finger in general vary from point to point. In this way, it corresponds to an outbreak of small bubble deformations

'Many-fingered time' to 'many bubble deformations time' is furthermore a 'ray to wavefront' type of duality in spacetime (Fig. 31.3.a).

As regards bubble time being more generally a *field-theoretic* feature rather than just a feature of the geometrodynamical formulation of GR, bubble time was developed in the field-theoretic context by physicists Peter Weiss [890] and Sin-Itiro Tomonaga [851]. As a functional integral method, this eventually bears close relation to quantum path integrals. Indeed, bubble time entered Canonical GR through Dirac being aware [246] of its preceding development in QFT. Bubble time's relation to geometrodynamical deformations was subsequently further developed in [454, 573, 574].

Use of bubble time in Geometrodynamics can moreover be seen as a 'covariantizing' feature—a mathematical implementation [573] of a prior insight of Wheeler's. This is attained through the formalism's many-fingered dual aspect considering all coordinate times at once. This can be envisaged as the bubble time presentation concurrently covering an infinity of ADM presentations. Each ADM presentation involves a particular local choice of coordinates, by which Refoliation Invariance ceases to be manifest; passing to a bubble time formulation restores this. Dirac's own approach avoided making a local choice of coordinates; while this does not coincide with Kuchař's bubble time approach, the two are canonically related [573]. The bubble time formulation can also be envisaged as considering all foliations at once.

Bubble time formulations are, moreover, one major way in which Parametrized Field Theory model arenas enter Problem of Time considerations [586]. Arbitrary slicing up ensues when one parametrizes a Field Theory; a salient difference in the case of GR is that (as Sect. 17.1 explained) GR comes *already-parametrized*.

31.10 Issues Involving Specific Foliations

Specifier equations arising provides one further way in which the Foliation Dependence Problem can rear its head. This is since in geometrodynamical theories, some specifier equations are foliation-fixing equations.

31.11 Various Other Arenas' (Lack of) Foliation Concepts

Newtonian Mechanics and RPMs The notion of foliation retains some meaning in RPMs, in the sense of a strutting. Firstly, such possess a lapse α or instant I like notion. Secondly, they possess a point identification map [814] that corresponds to such as the auxiliary rotations in moving along the dynamical curve in the redundant setting.

However, the GR lapse and shift (or instant and frame in the TRiPoD version) are *more* than just naïve elapsation and point-identification struts of this nature. This is in the sense that they additionally pack together with the 3-metric configurations to form a unified spacetime 4-metric, which turns out to have a number of further significant features. In addition to Chap. 27's mention of *Diff*$(\mathfrak{m})$, this now includes the following notions of space–time split GR spacetime.

1) It realizes the standard isometric embedding mathematics.
2) It possesses Refoliation Invariance, by which reslicing and consequently unrestricted recoordinatization of GR spacetime render it not just a privileged slicing evolving with respect to a single time. Indeed, it is also a theoretical framework in which each slicing evolves with respect to one of the many fingers.
3) GR spacetime's linear constraints are integrabilities of the Hamiltonian constraint (Chap. 24), so the two types of strutting are interlinked for GR but not for RPMs.
4) Subsequently GR's constraints form an algebroid versus RPMs' algebra. Moreover GR's constraints specifically form the Dirac algebroid. This firstly explains Refoliation Invariance. Secondly, it can be interpreted in terms of deformations of hypersurfaces. Indeed, the GR constraint algebraic structure is not a Lie algebra because [501, 502] the $3 + 1$ split has brought in Foliation Dependent information, and this requires a much larger algebraic structure to encode. Encoding *Diff*$(\mathfrak{m}, \mathfrak{Fol})$ for an arbitrary (rather than fixed) foliation $\mathfrak{Fol}$ can now be identified as the underlying reason for the much larger Dirac algebroid replacing unsplit spacetime's *Diff*$(\mathfrak{m})$ [since the Dirac algebroid is *Diff*$(\mathfrak{m}, \mathfrak{Fol})$].

Minisuperspace's Privileged Foliation (Diagonal) Minisuperspace spacetime has a foliation privileged by the spatially homogeneous slices. Locally and with one dimension suppressed, this gives a spherical shell version of Fig. 8.4.a). One can similarly envisage tilted cosmologies in parallel with Fig. 8.4.b). This foliation resolves Minisuperspace's foliation issues in a very straightforward manner (see however the next page's caveat).

The only surviving constraints bracket is (24.16): just an Abelian algebra. This is due to $\mathcal{D}_i$ annihilating everything; consequently, subtlety (24.21) and the algebroid structure are lost.

Minisuperspace and Minkowski Spacetimes in Arbitrary Frames Dirac [250] already pointed out that for Minkowski spacetime $\mathbb{M}^4$ in arbitrary frames— implemented by arbitrary-hypersurface foliations— the full constraint brackets (9.31)–(9.33) are required.

The above observations carry over to a certain common interpretation of Minisuperspace: the one in which the Universe's contents have to follow suit with the homogeneity. A simple example in which this is not the case are the tilted homogeneous cosmologies [812] in the sense of Fig. 8.4.b). In this case, the fluid velocity vector is not orthogonal to the group orbits. One can also consider arbitrary foliations of Minisuperspace spacetimes corresponding to the fleet of observers within now being free to accelerate as they please (cf. Fig. 8.4.c). Thus there are some situations in which spatial homogeneity is a great simplifier, though there are others for which a fully general working is needed even if a spatially-homogeneous slice foliation exists.

Finally, this carries some Equivalence Principle connotations.

Cosmic time, like any GR-type theory's timefunction, defines spatial slices by the dual level hypersurfaces that foliate the corresponding notion of spacetime. For the FLRW models, these are of constant-curvature. This gives a constant-curvature spatial hypersurface version of Fig. 8.4. (Exercise: determine which features do not carry over!)

SIC (Fig. 31.4) This has a solvable thin sandwich due to its being algebraic by the spatial derivative annihilating everything it acts on. However, this is accompanied by breakdowns of nontrivialities of some other Background Independent aspects of GR. In particular, constraints are no longer smeared, so the bracket of two $\mathcal{H}$'s is zero. With this trivialized, the Refoliation Invariance condition does not require any 3-diffeomorphisms to close up, and is independent of the 3-metric. In this particular case, this is not a very desirable simplification, since it means one remains restricted to the foliation privileged by spatial homogeneity. Indeed, one can see that this situation is the same as for *unperturbed* Minisuperspace. This also causes the constraint brackets to form a genuine algebra, thus facilitating Quantization (but at the cost of it being a restriction to a privileged foliation).

Let us next consider a model whose perturbation terms specifically depend on spatial position in a manner which specifically produces field-theoretic constraints. To zeroth order, this is as for Minisuperspace.

1 + 1 GR In the $1+1$ case, Teitelboim noted that [836] if density-normalized ($\mathrm{n}_\mu \mathrm{n}^\mu = \mathrm{h}$), undergoes the cancellation $\sqrt{\mathrm{h}}\,\mathrm{h}^{ij} = 1$, so a genuine Lie algebra ensues.

Nododynamical Counterpart of Refoliation Invariance In complex versions based on imaginary Barbero–Immirzi parameter β, the Ashtekar–Dirac algebroid sufficiently resembles the Dirac algebroid for Refoliation Invariance to follow through. On the other hand, as Giulini pointed out, in the case of real β, the connection not being a pull-back of a connection on $\mathfrak{m}$ (Sect. 27.8) has the knock-on effect of causing the Poisson bracket underlying Refoliation Invariance to not work out either [363, 364]. Thus there are two conceptual reasons—tied to two of the types of Background Independence—why to not introduce a β, or to select the original $\beta = \pm i$. In such a case, reality conditions would need to be faced by Ashtekar's original approach or one of Thiemann's more modern approaches.

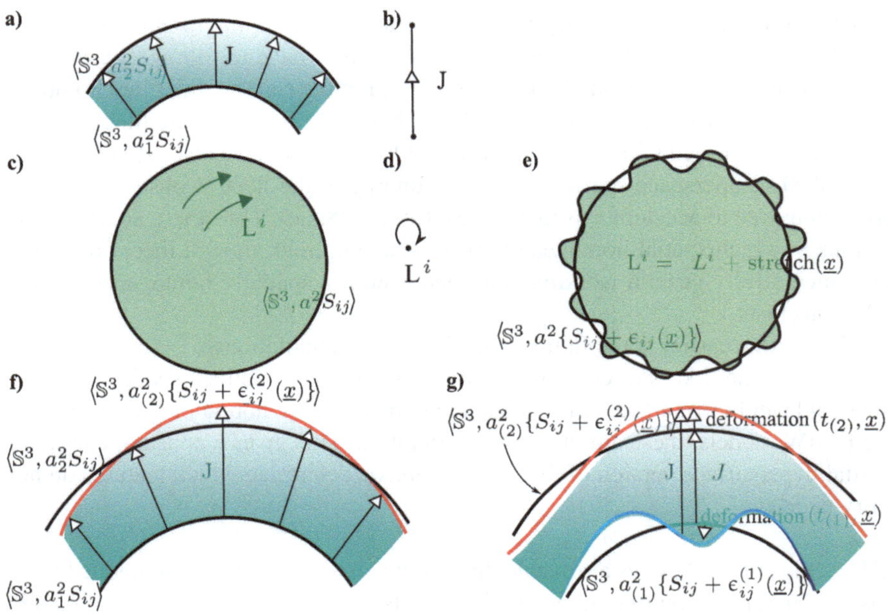

Fig. 31.4 For Minisuperspace [31] and modewise SIC, Foliation Dependence works out trivially. **a)** and **c)** are more elaborate depictions from a purely Minisuperspace perspective compared to **b)** and **d)**. For **d)**, without loss of generality, $\mathrm{d}L^i = \mathrm{id}$, due to all points being physically identical. **a)** and **c)** are moreover needed for comparison with subsequent inhomogeneous perturbations. I.e. **e)** as the slightly bumpy version of **c)** and **f)** as some indication of **e)** in the presence of small deformations. In fact the Hamiltonian constraint's action and the evolution are typically between two distinct small deformations of $\mathbb{S}^3$, as indicated in **g)**. The reader can easily imagine that the 'going via a red or purple choice of a third spatial hypersurface' extension of this progression nontrivially manifests SIC's Foliation Dependence Problem and its classical Refoliation Invariance resolution. In this way, the complicated choice of third hypersurface case is nontrivial in this example

Research Project 25) In GR, the Dirac Algebroid formed by the constraints ascertains Refoliation Invariance. Whether Supergravity possesses Refoliation Invariance is then an interesting question [868], particularly given the notable distinction (Sect. 24.10) between the Supergravity and GR constraint algebroids.

Chapter 32
Applications of Split Spacetime, Foliations and Deformations

32.1 Deformation Approach to Geometrodynamics

We next consider a first answer to one of Wheeler's principal questions: (9.1). Hojman, Kuchař and Teitelboim [454] addressed this by assuming, firstly, 3-*d* spacelike hypersurfaces described by Riemannian Geometry. More concretely, they assign primality to deformations of such hypersurfaces, subject to the further assumption—also suggested by Wheeler [899]—of their embeddability into (conventional 4-*d* semi-Riemannian) spacetime. These deformations are decomposed as per Fig. 32.1.

Evaluating Poisson brackets, these objects form the *deformation algebroid* [833]

$$\{\mathcal{S}\text{huffle}_i(\underline{x}), \mathcal{S}\text{huffle}_j(\underline{x}')\} = \mathcal{S}\text{huffle}_i(\underline{x}')\delta_{,j}(\underline{x}, \underline{x}') + \mathcal{S}\text{huffle}_j(\underline{x})\delta_{,i}(\underline{x}, \underline{x}'), \tag{32.1}$$

$$\{\mathcal{S}\text{huffle}_i(\underline{x}), \mathcal{P}\text{ure}(\underline{x}')\} = \mathcal{P}\text{ure}(\underline{x})\delta_{,i}(\underline{x}, \underline{x}'), \tag{32.2}$$

$$\{\mathcal{P}\text{ure}(\underline{x}), \mathcal{P}\text{ure}(\underline{x}')\} = \mathrm{h}^{ab}(\underline{x})\mathcal{S}\text{huffle}_j(\underline{x})\delta_{,i}(\underline{x}, \underline{x}') + \mathrm{h}^{ab}(\underline{x}')\mathcal{S}\text{huffle}_j(\underline{x}')\delta_{,i}(\underline{x}, \underline{x}'). \tag{32.3}$$

Hojman, Kuchař and Teitelboim additionally declare a *representation postulate*, by which obeying the deformation algebroid is to be entertained as a first principle for candidate Gravitational Theories. I.e. that for conventional spacetime to be produced, $\mathcal{H}^{\text{trial}}$ and $\mathcal{M}_i^{\text{trial}}$ constraints for these theories are to take a form such that they close in the same manner as the deformation algebroid of $\mathcal{P}$ure and $\mathcal{S}$huffle$_i$ (32.1)–(32.3). GR indeed satisfies this because the Dirac algebroid is of this form; the main point however is in inserting much more general ansätze and proving that the postulate narrows these down to the GR case alone.

The outcome is that the first Poisson bracket straightforwardly fixes $\mathcal{M}_i^{\text{trial}}$ to be the GR $\mathcal{M}_i$, and the second solely restricts $\mathcal{H}^{\text{trial}}$ to be a scalar density of weight 1, in each case for the reasons already familiar from Geometrodynamics (Sect. 9.14). The lion's share of the calculation involves the last Poisson bracket. Here, Hojman, Kuchař and Teitelboim evoke two subsidiary assumptions: *locality*—that the metric

E. Anderson, *The Problem of Time*, Fundamental Theories of Physics 190,
DOI 10.1007/978-3-319-58848-3_32

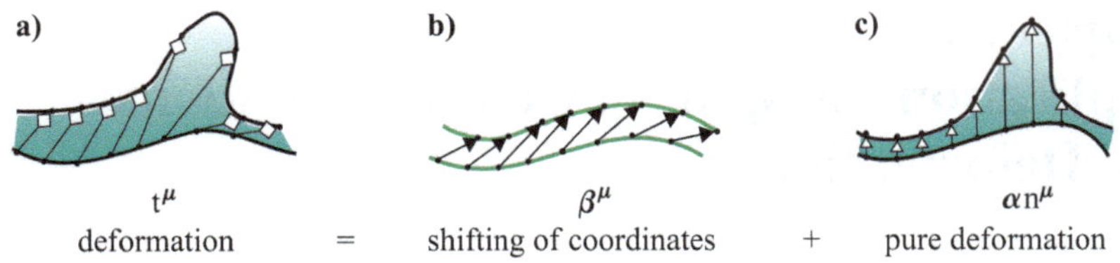

Fig. 32.1 The general deformation of a hypersurface **a)** decomposes into **b)** a shuffling of points around within that hypersurface corresponding to a constraint $\mathcal{S}\text{huffle}_i$. **c)** a pure deformation corresponding to a constraint $\mathcal{P}\text{ure}$

is to be only locally affected by a pure deformation—and 2 degrees of freedom per space point. They proceed by induction and by leaning on the 3-*d* version of Lovelock's Theorem [629] (the 4-*d* counterpart of which underlies Chap. 7's Lovelock simplicity postulates). This results in the GR form of $\mathcal{H}$ (including Λ); altering the signature also preserves the result up to at most a sign in the algebraic structure of the Poisson brackets of the deformation generators.

Teitelboim [832, 833, 835] also showed that the form of the Hamiltonian and momentum constraints $\mathcal{H}^{\text{grav-}\psi}$ and $\mathcal{M}_i^{\text{grav-}\psi}$ for GR alongside minimally-coupled fundamental matter fields ψ also fits the Deformation Approach's first principles. Here the Einstein–matter system's Hamiltonian and momentum constraints are of the form

$$\mathcal{H}^{\text{grav-}\psi} = \mathcal{H}^{\text{grav}} + \mathcal{H}^{\psi} \quad \text{and} \quad \mathcal{M}_i^{\text{grav-}\psi} = \mathcal{M}_i^{\text{grav}} + \mathcal{M}_i^{\psi}. \tag{32.4}$$

Thus the representation postulate extends additively to the constraints' *matter contributions*, so these *separately* obey the Dirac algebroid.

Teitelboim succeeded in including minimally-coupled scalars, Electromagnetism and Yang–Mills Theory in the above manner, alongside the extra postulate that $\mathcal{H}^{\text{grav-}\psi}$ is ultralocal in h_{ij}. Ultralocality holds trivially for the minimally-coupled scalar, whereas it pins down an a priori unrestricted 1-forms A_i or A_{iI} to have conjugate momenta obeying the Gauss or Yang–Mills–Gauss constraints respectively. Teitelboim furthermore argued for the gauge symmetry of the latter two being a consequence of embeddability [833, 835]. This approach leads to the further deduction that the corresponding Lie algebra is a direct sum of $U(1)$ and compact simple Lie algebras, which is in accord with the Gell-Mann–Glashow Theorem outlined in Appendix E. See also Wheeler's appraisal of Hojman, Kuchař and Teitelboim's work [900], and e.g. [17, 62, 363, 567] for subsequent commentary.

Research Project 26) Can the Deformation Approach be extended to include spin-$\frac{1}{2}$ fermions and their Gauge Theories?

Research Project 27) Is there a Conformogeometrodynamics parallel of the Deformation Approach to Geometrodynamics?

Research Project 28) Is there a Supergravity counterpart of the Deformation Approach?

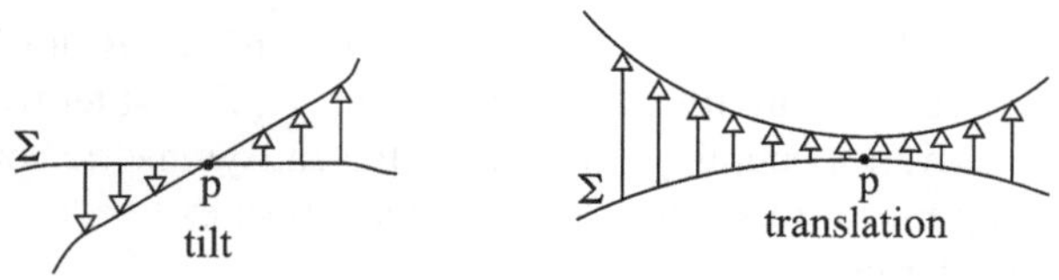

Fig. 32.2 Tilt–translation split

32.2 Universal Kinematics for Hypersurfaces in Spacetime (ADM Split Version)

Let us first observe that the split with respect to a hypersurface $\boldsymbol{\Sigma}$ of the spacetime covariant derivative ∇_μ acting on a general spacetime tensor field does not just produce the obvious spatial covariant derivative $\mathcal{D}_i$ [364, 577–579]. This is intuitively clear from spacetime derivatives involving extra components of the spacetime connection $\boldsymbol{\Gamma}^{(4)}$. The extra pieces produced by such a split can, moreover, be given the following lucid hypersurface-geometrical interpretations [576–579].

I) Hypersurface derivatives $\circ_{\vec{\beta}}$, as already encountered in Chap. 8, which implement '*shift kinematics*'.

II) *Tilts* are one part of the further translation–tilt split of pure deformations; this is a local split around each point p on the hypersurface (Fig. 32.2).

The *translation* part is such that $\alpha(\mathrm{p}) \neq 0$, $\{\partial_i \alpha\}(\mathrm{p}) = 0$.

On the other hand, the *tilt* part is such that $\alpha(\mathrm{p}) = 0$, $\{\partial_i \alpha\}(\mathrm{p}) \neq 0$. The suitability of this name [576–579] can most simply be seen in the spacetime formulation of SR, where a boosted fleet of observers on a flat spatial surface tilted at a fixed angle to the undeformed flat spatial surface. Cf. also the tilted flow in Fig. 8.4.b), in which manner tilt plays a role in Cosmology.

III) *Derivative couplings*[1] are terms linear in each of the extrinsic curvature and the tensor field itself. Absence of such terms is known as the *Geometrodynamical Equivalence Principle* [454], which is a statement concerning minimal coupling (the complement of derivative coupling).

For example, in the case of a 1-form A_μ (so this example is illustrating adjunction of matter) the 4-*d* covariant derivative's pieces decompose as follows.

$$\nabla_a \mathrm{A}_\perp = \mathcal{D}_a \mathrm{A}_\perp - \mathrm{K}_{ab}\mathrm{A}^b, \tag{32.5}$$

$$\alpha \nabla_\perp \mathrm{A}_a = -\delta_{\vec{\beta}} \mathrm{A}_a - \alpha \mathrm{K}_{ab}\mathrm{A}^b - \mathrm{A}_\perp \partial_a \alpha, \tag{32.6}$$

$$\nabla_b \mathrm{A}_a = \mathcal{D}_b \mathrm{A}_a - \mathrm{A}_\perp \mathrm{K}_{ab}, \tag{32.7}$$

$$\alpha \nabla_\perp \mathrm{A}_\perp = -\delta_{\vec{\beta}} \mathrm{A}_\perp - \mathrm{A}^a \partial_a \alpha, \tag{32.8}$$

[1]This name reflects that these feature in metric-matter kinetic cross-terms in theories with nonminimally-coupled matter fields [577–579].

Looking at the right hand sides, (32.6) and (32.8)'s first terms are hypersurface derivatives, their last terms are tilts, and (32.5) and (32.7)'s last terms and (32.6)'s second term are derivative couplings. Moreover, the antisymmetric combination of derivatives entering Electromagnetism's kinetic term ensures freedom from tilt and derivative coupling terms.

N.B. that I), II) and III) constitute a universal set of hypersurface kinematics. This is in the sense that arbitrary tensor fields can exhibit these, and no other, hypersurface kinematics features.

32.3 Thin Sandwich Completion in Terms of Hypersurface Kinematics

The Thin Sandwich treatment of Chap. 18.9 can be extended as follows by taking universal hypersurface kinematics into consideration.

Thin Sandwich 6) Thin Sandwich 4) permits one to construct an emergent version of the tilt, $\partial_b \mathrm{N}$.

Thin Sandwich 7) Thin Sandwich 3.a), 4.a) and 6) form the set of universal kinematics for split spacetime tensor fields, by which their construction further amounts to being able to construct a wide range of spacetime objects. These include the spacetime connection (Sect. 34.5), and, in the extension of the thin sandwich to include tensor field matter, the spacetime form taken by this tensor field matter.

32.4 Space–Time Split Account of Observables or Beables

We next continue Sect. 27.6's considerations of whether $\mathit{Diff}(\mathfrak{M})$ is a large enough group, by now allowing furthermore for space–time split formulations. A third group is

$$\mathit{Data}(\mathfrak{m}) := \left\{\epsilon^\mu(\vec{X};\, \Xi_{\mathrm{CD}}]\right\}, \tag{32.9}$$

where 'CD' denotes dependence on the fields only through the Cauchy data on a spatial hypersurface $\boldsymbol{\Sigma}$. The Author chooses this name for the group to emphasize its relation to the notion of data-gauge.[2] The associated *Data*-invariance involves

[2]Both notions are appropriate in canonical formalisms, hailing back to Dirac's formulation of Theoretical Physics in terms of Poisson brackets. Emphasizing this pair of notions' similarities by giving them similar names goes back to Bergmann and Komar, who termed them 'D-group' and 'D-gauge' respectively. Moreover, their 'D' stands for 'Dirac', whereas the current book's use of 'data' is additionally descriptive of the joint underlying conceptual nature of these entities. This renaming has further practical value since, firstly, coincident authorship need not imply conceptual similarity, especially in the case of an author of Dirac's creative magnitude. Secondly, this pair of notions remains relatively unknown due to resting upon under-emphasized subtleties. Consequently, many readers will not have come across these notions before, so this book presents them under clear descriptive names.

transformations which are unchanged under 4-d coordinate transformations that reduce to it on initial $\boldsymbol{\Sigma}$ on which the canonical Cauchy data are defined. For instance any spatial 3-vector, h_{ab} or p^{ab} are *Data*-invariant.

The fourth and final group we consider in this discussion is, using footnote 2's notation,

$$PDigg(\mathfrak{m}) := \{\epsilon^{\mu}(\vec{X};\, \Xi] \in Digg(\mathfrak{m}) \,|\, \epsilon^{\mu} = \mathrm{n}^{\mu}(\vec{X})\mathrm{v}^0 - \delta^{\mu}_{a}\mathrm{v}^a\}. \tag{32.10}$$

This group is also historically due to Bergmann and Komar [134], and has on some occasions been referred to as 'the Bergmann–Komar group' [724], though this term is somewhat ambiguous and we use a descriptive name instead. The name '*PDigg*' is rooted upon this group's interpretation as a projection, which was discovered in physicist Lawrence Shepley's collaboration with Pons and Salisbury [721]. In more detail, it is the projection from configuration–velocity space $\mathfrak{T}(\mathfrak{q})$ to $\mathfrak{P}$hase (via a Legendre map) of $Digg(\mathfrak{m})$. Hence *PDigg* stands for 'projected *Digg*' (or '$\mathfrak{P}$hase counterpart of *Digg*') In effect, the whole of $Digg(\mathfrak{m})$ itself cannot be completely realized in phase space, by which adopting the smaller $PDigg(\mathfrak{m})$ instead is motivated.

The corresponding active canonical transformation is

$$\mathrm{G}_{\mathrm{v}} = \vec{\mathrm{P}}_{\mu}\dot{\vec{\mathrm{v}}}^{\mu} + \{\vec{\mathcal{H}}_{\mu} + \vec{\beta}^{\rho}\mathrm{f}^{\nu}{}_{\mu\rho}\vec{\mathrm{P}}_{\nu}\}\vec{\mathrm{v}}^{\mu}. \tag{32.11}$$

Here $\vec{\mathrm{P}}_{\mu}$ are the momenta conjugate to $\vec{\gamma}^{\mu} := [\alpha, \beta^i]$: the spacetime 4-vector of auxiliaries. Also $\mathcal{H}_{\mu}$ denotes the 4-vector of constraints $[\mathcal{H}, \mathcal{M}_i]$. Finally, the $\mathrm{f}^{\nu}{}_{\mu\rho}(\mathbf{h})$ are the Dirac algebroid's structure functions wrapped up in the spacetime tensor form corresponding to $\vec{\mathcal{H}}_{\mu}$.

Bergmann and Komar [134] furthermore posited a number of relations between the four groups. This led them to conclude that Dirac and Bergmann observables turn out to coincide, but they provide no proofs for the underlying relations and this claim has since been contested [630, 920]. We next pass to considering some examples and counter-examples of relevance to the theory of Bergmann observables.

Example 1) The Weyl scalars (Sect. 27.5) can additionally be considered as a concrete proposal [134, 720] for observables in the sense of Bergmann. Indeed Bergmann and Komar converted the spatial components of the spacetime Riemann tensor and contractions with the spatial hypersurface normal n^{μ} to be purely in terms of canonical variables. This means that this formulation has h_{ab} and p^{ab} but not lapse or shift. I.e. the Weyl scalars can be written in terms of the canonical variables, as befits many of the expectations about observables. In this application, they are to be interpreted as intrinsic coordinates, and also as 'making use of a set of scalars as a gauge fixing'.

Counter-example 2) [720] argue that Torre's No-Go Theorem proves nonexistence of constant-in-time observables, i.e. constants of the motion, built as spatial integrals. This is as opposed to its referring to *Bergmann* observables such as the Weyl scalars. On the other hand, Dittrich and Thiemann's approach [251, 845] gets round Torre's No-Go by [722, 724] involving series of Cauchy data derivatives that are in principle up to infinite order. Finally, Halliwell's classical construct avoids Torre's No-Go by not being local in space or time and avoids Kuchař's by not being of form (O.9).

Let us end by examining a few features of the *Digg* and *PDigg* formulations which are retained in a model arena that is rather more tractable than full GR.

Example. Some of this section's issues already have nontrivial counterparts for Temporally Relational Mechanics. In place of *Diff*($\mathfrak{m}$), one now has

$$\text{(reparametrizations) } R = \{\epsilon(t)\}, \tag{32.12}$$

instead of *Digg*($\mathfrak{m}$),

$$Rigg = \{\epsilon(t; \underline{q}^I]\} : \tag{32.13}$$

the *reparametrization-induced gauge group*, and instead of *PDigg*($\mathfrak{m}$),

$$PRigg = \{\epsilon(t; \underline{q}^I] \in Rigg \,|\, \epsilon = t\,\xi(t; \underline{q}^I]/\alpha\} : \tag{32.14}$$

the projective version of the previous. How complete the $(P)Rigg$ are as a model of the $(P)Digg(\mathfrak{m})$ remains to be worked out in detail.

Such models already suffice to exhibit the distinction in size of the transformation group in passing from *Rigg* to *PRigg*. On the other hand, Lee and Wald's [615] further inter-relation between *Digg*($\mathfrak{m}$) nontriviality and GR's Dirac algebroid fails to work for *Rigg*. This limits the extent to which *Rigg* functions as a model arena for *Digg*($\mathfrak{m}$).

32.5 Difference Between Hamiltonians and Gauge Generators

The next two Secs are based on works of Pons, Salisbury and Sundermeyer [722, 724]. The 'evolution' generator $\delta t\{\vec{\gamma} \cdot \vec{\mathcal{H}} + \dot{\vec{\gamma}} \cdot \vec{P}\}$ does serve to replace solutions at time t by the original solutions evaluated at $t - \underset{\cdot}{t}$. However, this is merely its action on one particular member of each equivalence class of solutions. I.e. the particular member for which the lapse and shift form the chosen explicit 4-vector function $\vec{\gamma}$. Its action on all other members of these equivalence classes generates variations different from global time translations.

In more detail, points $p \in \mathfrak{Sol}_{\Xi}$—the configuration space of the dynamical fields Ξ—are specific spacetimes (plus matter fields when relevant), i.e. solutions of the equations of motion as described in a particular coordinatization.

Let us denote by D the data for the dynamical fields on some spatial hypersurface that is labelled by 'initial time' t_0. Given a specific selection of the arbitrary functions of the dynamical variables λ^{μ}, the corresponding Dirac Hamiltonian is $\mathcal{H} = \vec{\gamma} \cdot \vec{\mathcal{H}} + \vec{\lambda} \cdot \vec{P}$. This dictates—via the Poisson brackets—the time evolution in p. In particular, for an infinitesimal $\underset{\cdot}{t}$, this Hamiltonian gives what the field data D′ are on the subsequent spatial hypersurface labelled by $t_0 + \delta t$. If we carry out this procedure for all times t, of course a 'null operation' ensues: we have remained exactly at the same point $p \in \mathfrak{Sol}_{\Xi}$. This simply reflects that the dynamics as described by a given observer takes place within a given spacetime in a given coordinatization.

Finally consider the gauge generator that, after suitable choice of the descriptors, happens to coincide in its mathematical form with the Dirac Hamiltonian at time t_0. By this coincidence, its action likewise transforms the field data D into D$'$. However, these data D$'$ are now to be interpreted at time t_0, because the notion of gauge transformations in question are *equal-time* actions. What has occurred is that we have moved from p to another, albeit gauge-equivalent, spacetime p$'$. I.e. it is mathematically another point in S, but it is physically the same. Next suppose we undertake the same procedure for any time t while continuing to assume that the descriptors at time t match up with the lapse and shift at t. Thus we end up having mapped the whole spacetime p to p$'$. Notice that the field configurations in p and p$'$ differ solely as regards their time labels. In this way, a passive diffeomorphism $\mathrm{t} \rightarrow \mathrm{t} - \delta \mathrm{t}$ renders both descriptions mathematically identical. This demonstrates that the gauge generator's capacity to mimic the Hamiltonian is *conceptually unrelated* to there being real physical evolution in a given spacetime p. Therefore dynamical evolution in p is not the same notion as gauge action on p.

32.6 'Nothing Happens' Fallacy

A common type of frozen argument is that (9.39) means that nothing happens. However, the inference that 'nothing happens' is a fallacy on the following grounds [722].

On the one hand, following from the preceding Sec, we have an 'evolved configuration' D$'$ lying to the future of an 'initial configuration' D. On the other hand, D and D$'$ are related by a gauge transformation. Since 'gauge transformations do not alter the physics', we deduce that 'the physics' in D and D$'$ is the same. So the future configuration is gauge-equivalent to the initial configuration and therefore 'nothing happens'.

The fallacy comes from each of these two hands using a single common language for two sets of things that are in fact conceptually different in each case. Firstly recollect from Chap. 24.8 the distinction between gauge transformation in Dirac's sense and in Bergmann's. Furthermore there are also two corresponding notions of 'the physics'. The second hand involves mapping solutions of the equations of motion to other such solutions. Thus it requires the 'entire field configurations' of a 'whole-path', 'whole-history' or 'whole-spacetime' physics perspective, which rests on involves Bergmann's notion of gauge. In contrast, the first hand involves 'configurations at a given time t_0' (D and D$'$). I.e. a 'time-sliced' physics' dynamical perspective in which Dirac's notion of gauge applies. In this way, each of the two hands in fact uses a distinct notions of 'gauge' and also a corresponding distinct notion of 'the physics'. Since the 'nothing happens argument' does not take these differences into account, it is rendered fallacious. See [524, 720, 723, 845] for further support of this point.

Finally, this resolution of the 'nothing happens paradox' corresponds to the distinction between time-dependent beables $D_D(t)$ for t an intrinsic coordinate scalar that constitutes a gauge fixing. This is as opposed to just a constant D_D (see Chap. 3.4 of [724] for more on this point). Clearly the former are not 'constants of the motion'!

32.7 Discussion

Let us first comment that [720, 722, 723] are supportive of the notion of partial observables. [722] does limit support in the sense of insisting that partial observables be spacetime scalars (a standard tenet of Internal Time Approaches, which we touch upon two sections down). None the less, [720] argues for Weyl scalars exemplifying partial observables (which of course *do* in this case comply with being spacetime scalars). In this way, Rovelli, Dittrich and Thiemann's works on observables acquire yet wider conceptual support and motivation.

Research Project 29) Can Pons et al.'s work [722, 724] be rendered compatible with Temporal Relationalism?
Research Project 30) Do the conceptual and technical differences between GR and Supergravity alter the outcome of Pons et al.'s work?

Let us end by noting that Bergmann's approach to observables in fact involves not only spacetime primality but also the Canonical Approach at a secondary level. Moreover, the enlargement of *Diff*($\mathfrak{m}$) to *Digg*($\mathfrak{m}$) [or *PDigg*($\mathfrak{m}$)] can now also be viewed as a particular form of Spacetime Relationalism, with some inter-relations with Canonical-and-Covariant Approaches. See also the last Section below for an outline of Histories Theory counterparts.

32.8 Foliation Considerations end Unimodular Approach to Problem of Time

Let us next remark that classical knowledge of foliations suffices to end the Unimodular Approach at the level of counting degrees of freedom, as follows. In such models, the cosmological constant Λ itself plays the role of the variable with an isolated linear momentum; cf. Eq. (20.6). At the quantum level, this would get promoted to the derivative with respect to the unimodular internal time function t^{Uni}, due to the presence of which the corresponding Quantum Theory would be unfrozen. There is however a large mismatch between this single time variable and the standard Generally-Relativistic concept of time, which is 'many-fingered' with one finger per possible foliation. This is clear from the derivative with respect to t^{Uni} being a partial derivative, whereas a GR problem of Time resolution would be expected to be in terms of a functional derivative. The geometrical origin of this mismatch is that a cosmological time measures the 4-volume enclosed between two embeddings of the associated time functional. However, given one of the embeddings, the other is far from uniquely determined by the value of t^{Uni}. This is because pairs of embeddings that differ by a zero 4-volume are obviously possible due to the Lorentzian signature and cannot be distinguished in this way.

32.9 Spacetime to Foliations to Internal Time

We next pass to considering internal time candidates within a worldview in which spacetime is presupposed [483]. In such a setting, internal time candidates are required to be spacetime scalars (this is a different manifestation of a Spacetime Construction Problem from that in the next Chapter). Functions of this form succeed, moreover, in being Foliation Independent.

In contrast, in Canonical Approaches, functionals of the canonical variables are involved. Consequently, there is no a priori reason for such to be spacetime scalar fields, so one is faced with the following dilemma.

1) Perhaps one is to find functionals which are spacetime scalars, so as to establish Foliation Independence by construction, alongside recovery of the standard spacetime interpretation.
2) Perhaps instead one is to find some new classical means of arriving at the standard spacetime interpretation. The next Chapter develops this further. We first make a detour to Histories Theory, motivated by how 1970's to 1990's Internal Time Approaches have accrued various substantial histories-theoretic descendants from the 1990's onward.

32.10 Covariant-and-Canonical Histories Theory

Such would be expected to concurrently exhibit Temporal, Configurational and Spacetime Relationalism cast in terms of histories, and both configurational and histories notions of observables or beables.

Example 1) Savvidou pointed out [11, 765, 768, 769] that Isham–Linden type Histories Theory has a distinct structure for each of two conceptually distinct notions of time.

I) A kinematical notion of time that labels the paths or histories as sequences of events. (This is a 'labelling parameter of temporal logic' taken by [11] to also mean causal ordering, though see also [503].)
II) A dynamical notion of time that is generated by the Hamiltonian.

Savvidou subsequently argued that having these two distinct notions of time allows for such a Histories Theory to be canonical and covariant at once. This is clearly of interest in understanding, and reconciling various viewpoints in, QG.

Hypothetical Example 2) One could also follow solving the GR momentum constraint at the Lagrangian Best-Matching level or at the Hamiltonian level by finding a single 'time-map' Histories Theory.

Example 3) Kouletsis and Kuchař provided a means of including the set of foliations into an extension of ADM's geometrodynamical $\mathfrak{P}$hase that is Generally Covariant. This amounts to extending $\mathfrak{P}$hase to include embeddings so as to take into account the discrepancy between the *Diff*($\mathfrak{m}$) algebra and the Dirac algebroid. (This is itself an older idea of Isham and Kuchař [501, 502].) It is implemented here by constructing of a 'time map' and a 'space map'; indeed, one can set up a Histories Theory with these features. They subsequently reduce away the 'space-map' structure to pass to a new 'time-map' Kouletsis and Kuchař [568] considered the above for the bosonic string model arena, whereas Kouletsis [566] considered it for classical Geometrodynamics.

Example 4) As a further Combined Approach, consider Kouletsis' [566] tie between Histories Theory, the Internal Time Approach and the Problem of Beables. This program follows on from the preceding in involving a space map as well as a time map for how the family of geometries along each path or history embed into spacetime.

Let us end by note that, despite being based on classical paths, the Bergmann and histories notions of observables are technically and conceptually distinct extensions of which gauge groups one can attribute to a physical theory. This is clear from the shift in basic canonical entities in the latter, with the ensuing introduction of a histories brackets algebraic structure absent in Bergmann's work and with a number of subsequent parallels to Dirac's notion of observables. None the less, Savvidou [769] showed that $\mathfrak{Hist}$-$\mathfrak{P}$hase can also carry representations of *Digg*($\mathfrak{m}$).

Chapter 33
Spacetime Construction and Alternative Emergent Structures

The Relational Approach places on a common footing theories which have the same configuration space $\mathfrak{q}$ but different physically irrelevant transformation groups $\mathfrak{g}$ (cf. Sect. 27.7). This approach permits investigation in greater generality of why some symmetries happen to be widely shared in nature. This Chapter lays out a more general version of the Relational Approach, which proceeds by considering not actions S posited to have a pre-determined list of symmetries, but rather a broader range of actions corresponding to the 'zeroth principles' of the relational postulates. We now consider families of actions providing families of constraints and we then put each family together through the TRi Dirac-type Algorithm. This reveals most of the other possibilities to be inconsistent. Furthermore, filtering theories in this way or by presupposing $\mathfrak{g}$ do not always produce the same outputs. For instance, we shall see that whereas GR with local SR arises both ways, the alternatives accompanying GR in each case differ. This demonstrates the Relational Approach's capacity to find alternative theories. The Relational Approach also re-expresses the choice between universal local theories of Relativity in algebraic terms, as various ways in which the constraint algebraic structure can close.

Spacetime Construction is most clearly explained *as a procedure* after consideration of Constraint Closure and Refoliation Invariance. This is in the sense of involving a more general range of action or constraint ansätze within which the GR case—already known to comply with these other facets—sits. However, in space and configuration space primary approaches, Spacetime Construction is logically prior to considering Spacetime Relationalism and foliations. I.e. construct spacetime first, then investigate its own Relationalism, its foliations and whether it possesses Refoliation Invariance. Correspondingly, to complete the Relational Approach, we first consider Spacetime Construction, and then re-visit Spacetime Relationalism and Refoliation Invariance in this emergent spacetime context in the next Chapter.

Since such a Spacetime Construction does not assume spacetime features at the outset, it realizes the Broad worldview to a greater extent than GR as Geometrodynamics does.

The family of actions considered provides a family of $\mathcal{H}^{\text{trial}}$ constraints, which are then restricted by the TRi Dirac-type Algorithm. In this way, a second answer

E. Anderson, *The Problem of Time*, Fundamental Theories of Physics 190,
DOI 10.1007/978-3-319-58848-3_33

to Wheeler's question (9.1) arises. This program was started by Barbour, Foster and ó Murchadha [109], was continued by the Author [15, 17, 19] and eventually finished in collaboration with physicist Flavio Mercati [62]. The Relational Approach starts from the assumption of space, which is more minimalistic than starting with spacetime. This goes beyond Wheeler's further suggestion that embeddability into spacetime is necessary, proceeding instead through Constraint Closure as an exhaustive restriction from which embeddability is deduced as one option. In this way, the Relational Approach additionally incorporates a type of Spacetime Construction [62], in the sense of constructing spacetime from space. Because this approach assumes less structure, this construction is harder than splitting space. In particular, one has less structure available than in the Deformation Approach, which did adhere to Wheeler's further suggestion.

It is now additionally natural to further ask why $Diff(\mathbf{\Sigma})$ and $\mathfrak{Riem}(\mathbf{\Sigma})$ themselves are in use. There is space in this Chapter to investigate the first of these—we also contemplate $\mathfrak{g}$ = id, $Conf(\mathbf{\Sigma}) \rtimes Diff(\mathbf{\Sigma})$ and $VPConf(\mathbf{\Sigma}) \rtimes Diff(\mathbf{\Sigma})$: the group of volume-preserving conformal transformations on $\mathbf{\Sigma}$. On some occasions the additional caveat in (24.7) is realized. Consequently, one is forced to alter one's candidate $\mathfrak{g}$ due to integrability conditions appearing. $\mathfrak{Riem}(\mathbf{\Sigma})$ itself is adopted here due to the virtues of modelling space with positive-definite metrics as outlined in Part I and Appendix D.

We finally point to the second great facet decoupling uncovered in this book: noninterference between Assignment of Beables and Spacetime Construction; Figs. 25.1 and 34.6 each follow on from Fig. 24.2 without mutual interference.

33.1 Relational First Principles Ansatz for Geometrodynamical Theories

We begin with the usual choice of $\mathfrak{q} = \mathfrak{Riem}(\sigma)$ and $\mathfrak{g} = Diff(\sigma)$. However, let us now entertain a more general geometrodynamical ansatz for a family of geometrical actions built from differentiable and metric level spatial objects [37, 62, 109],

$$\mathsf{S}^{w,y,a,b} = \iint_{\sigma} \mathrm{d}^3x \sqrt{a\,\mathcal{R} + b}\, \partial \mathsf{s}_{w,y}. \tag{33.1}$$

Here, $\partial \mathsf{s}_{w,y}$ is built out of the usual $\partial_{\underline{\mathrm{F}}}$ and the more general if still ultralocal supermetric $\mathbf{M}_{w,y}$ with components $\mathrm{M}^{abcd}_{w,y} := \sqrt{\mathrm{h}}\{\mathrm{h}^{ac}\mathrm{h}^{bd} - w\,\mathrm{h}^{ab}\mathrm{h}^{cd}\}/y$. This ansatz also assumes that the kinetic metric is homogeneous quadratic in the changes. The inverse $\mathbf{N}^{x,y}$ of $\mathbf{M}_{w,y}$ has components $\mathrm{N}^{x,y}_{abcd} := y\{\mathrm{h}_{ac}\mathrm{h}_{bd} - x\,\mathrm{h}_{ab}\mathrm{h}_{cd}/2\}/\sqrt{\mathrm{h}}$ for $x := 2w/\{3w-1\}$. The parametrization by x has been chosen such that GR is the $w = 1 = x$ case. $w = 1/3$ is excluded due to rendering $\mathbf{M}$ non-invertible.

The conjugate momenta are then

$$\mathrm{p}^{ij} = \mathrm{M}^{ijkl}_{w,y} \frac{\partial_{\underline{\mathrm{F}}}\mathrm{h}_{kl}}{2\,\partial \mathrm{I}}. \tag{33.2}$$

The quadratic primary constraint is

$$\mathcal{H}^{\text{trial}} = \mathcal{H}_{x,y,a,b} := \mathrm{N}^{x,y}_{abcd}\mathrm{p}^{ab}\mathrm{p}^{cd} - \overline{a\,\mathcal{R} + b} = 0, \tag{33.3}$$

and the secondary constraint is just the usual GR momentum constraint $\mathcal{M}_i$ again. So if one takes *Diff*(σ) to be physically meaningless, the only choice is the contraction of the Codazzi embedding equation (8.28) giving the constraint $\mathcal{M}_i$.

33.2 Geometrodynamical Consistency, Local Relativity and Spacetime Construction

Consistent Geometrodynamics Theorem *If the geometrodynamical ansatz* (33.1) *is assumed, the following four outcomes are consistent.*

i) *Recovery of GR.*
ii) *A* 1*-parameter family of geometrostatics.*
iii) *A* 1*-parameter family of strong gravity theories.*
iv) *A group of formulations and theories based upon*

$$\mathcal{D}_i\{\mathrm{p}/\sqrt{\mathrm{h}}\} = 0. \tag{33.4}$$

Consistent Relativities Theorem *Upon adding minimally-coupled matter, emergent local Relativity is Lorentzian for i), Galilean for ii) and Carrollian for iii). This is in the sense of an emergent shared propagation speed that is finite for i), infinite for ii) and zero for iii).*

Classical Spacetime Construction Theorem *In case i), GR spacetime emerges by construction from assuming of just space, Temporal and Configurational Relationalism.*

Toward establishing these theorems [15, 62, 65, 109], form the Poisson brackets of the constraints and apply the TRi Dirac-type Algorithm. This gives [62] (9.31) and (9.32) with the family of constraints $\mathcal{H}_{x,y,a,b}$ in place of $\mathcal{H}$, alongside

$$\begin{aligned}
&\big\{(\mathcal{H}_{x,y,a,b}\,|\,\partial\mathrm{J}),\,(\mathcal{H}_{x,y,a,b}\,|\,\partial\mathrm{K})\big\}\\
&= -2a\,y\,\big(\mathcal{D}^j\mathrm{p}^i{}_j + \{x-1\}\mathcal{D}_i\mathrm{p}\,\big|\,\partial\mathrm{J}\overleftrightarrow{\partial}^i\partial\mathrm{K}\big)\\
&= a\,y\,\big(\mathcal{M}_i + 2\{1-x\}\mathcal{D}_i\mathrm{p}\,\big|\,\partial\mathrm{J}\overleftrightarrow{\partial}^i\partial\mathrm{K}\big)\\
&= a\,y\,\big(\mathcal{M}_i\,\big|\,\partial\mathrm{J}\overleftrightarrow{\partial}^i\partial\mathrm{K}\big) + 2\,a\,y\{1-x\}\big(\mathcal{D}_i\mathrm{p}\,\big|\,\partial\mathrm{J}\overleftrightarrow{\partial}^i\partial\mathrm{K}\big).
\end{aligned} \tag{33.5}$$

This picks up an *obstruction term to a brackets algebraic structure* [62] in four factors:

$$2 \times a \times y \times \{1-x\} \times \big(\mathcal{D}_i\mathrm{p}\,\big|\,\partial\mathrm{J}\overleftrightarrow{\partial}^i\partial\mathrm{K}\big). \tag{33.6}$$

Each factor provides a different way in which to complete the TRi Dirac-type Algorithm (Fig. 33.1). The first three are strongly vanishing options (Sect. 33.3), whereas the fourth is a weakly vanishing option which includes cases in which the TRi Dirac-type Algorithm has further steps (Sect. 33.6). Any of these options give automatic closure, so $\mathcal{H}_{x,y,a,b}$ is rendered first-class.

33.3 Strongly Vanishing Options: GR, Strong Gravity, Geometrostatics

GR with Embeddability into Spacetime The third factor in (33.6) strongly fixes [109] the supermetric coefficient to $x = 1$; correspondingly, $w = 1$. This is indeed the DeWitt value that characterizes GR (cf. Chap. 8.10). In this case, GR spacetime is furthermore constructed as follows.

Construction I) The Machian version of the Thin Sandwich construct of Chap. 18.9 applies. One can now construct an object $\mathcal{C}_{ab}$ interpreted as an emergent object of the Machian relational form

$$\frac{\mathrm{d(change)}}{\mathrm{d(other\ change}} \quad : \quad \mathcal{C}_{ab} := \frac{\partial_{\mathrm{F}}\mathrm{h}_{ab}}{\partial \mathrm{I}}. \tag{33.7}$$

Furthermore all the geometrical change is given the opportunity to contribute to the ∂I that each individual change is compared to here, so it is a STLRC entity.

Construction II) $\mathcal{H}$ subsequently takes the form of the double contraction of Gauss' embedding equation that is the GR Hamiltonian constraint. Thus it matches the contraction of Codazzi's embedding equation that is the GR momentum constraint. Consequently, a pair of embedding equations arise [832], which constitute the 4 0μ components among the 10 components of the 4-*d* Einstein field equations. The equations of motion turn out to be a linear combination of the Ricci embedding equation (34.20), the contracted Gauss embedding equation and the metric times further contractions. In this way, these equations form the TRi version of the remaining 6 Einstein field equations. So in this approach, one *recovers* equations and makes a *meaningful grouping* of them. Contrast the decomposition into projection equations of Chap. 8, or Wheeler's suggestion of presupposing embeddability into spacetime of [454, 832] and Chap. 31.

Construction III) One can then posit an ambient metric 4-geometry that the metric 3-geometry of space is locally embedded within. This could be the conventional spacetime if its signature is indefinite alias Lorentzian: $- - + + +$, corresponding to $a > 0$. In this case, one lies within the scope of the standard Mathematical Physics of Appendix O.7 applying locally. Alternatively, it could be the counterpart whose signature is positive-definite alias Euclidean: $+ + + +$, corresponding to $a < 0$. The distinction between these is not made by the Dirac Algorithm: both are consistent. See two Sections down for a physical dismissal of this Euclidean alternative.

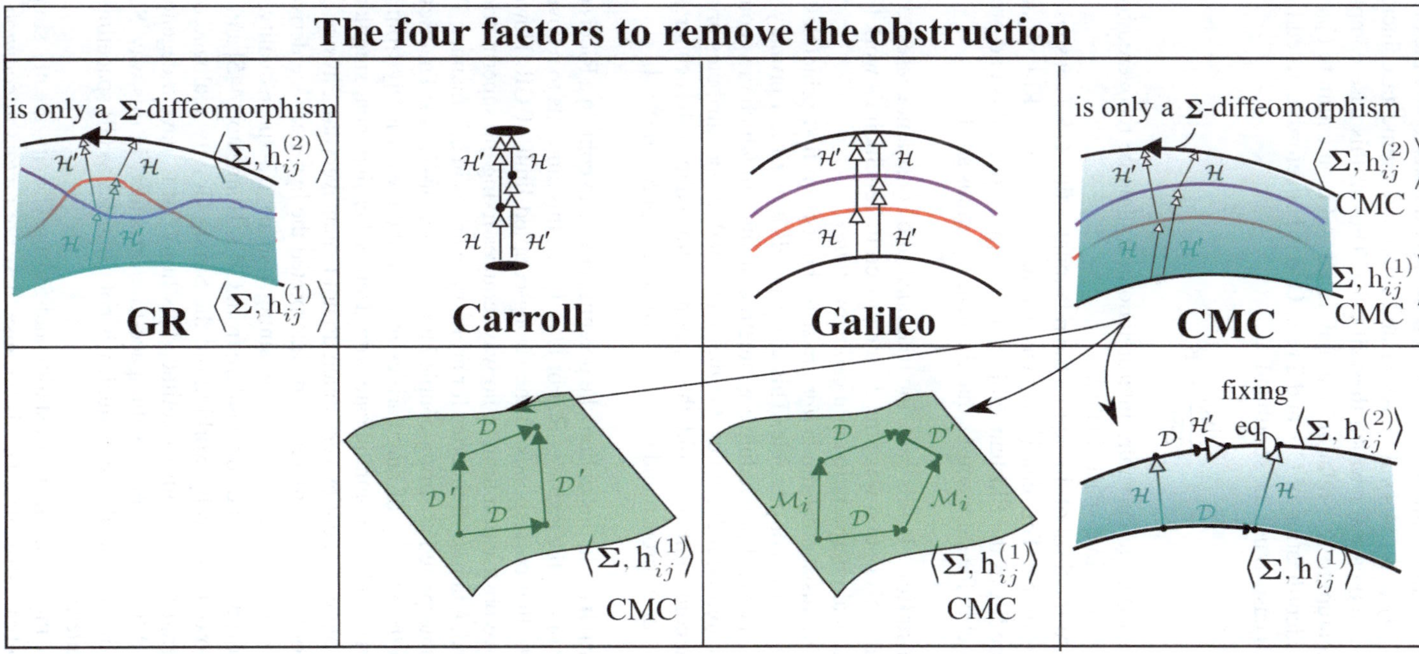

Fig. 33.1 Pictorial interpretation of the various brackets resulting from the current Chapter's geometrodynamical ansatz. The *black* and *white semicircle arrows* indicate action of $\mathcal{D}$ and the differential of the instant fixing equation respectively. See [54] for the metrodynamical analogue of this Figure. Figure 34.6 inter-relates the current Chapter's work with other facets

Construction IV) The momentum formulation is entirely unaffected by the distinction between (8.14) and (34.4). The ADM and relational momenta coincide in the $x = 1 = w$, $y = 0$ case for which they all exist. Thus in this case comparing the 'ADM-momentum to K_{ab} relation' and the 'relational momentum to $\mathcal{C}_{ab}$ relation' permit the identification of C_{ab} and $2\,\mathrm{K}_{ab}$. One is henceforth entitled in this $x = 1 = w$, $y = 0$ case to use the shorthand

$$\frac{\partial_{\mathrm{F}} \mathrm{h}_{ab}}{2\,\partial \mathrm{I}} = \mathrm{K}_{ab}. \tag{33.8}$$

The conventional extrinsic curvature interpretation can then be recovered by hypersurface tensor spacetime-space duality.

Construction V) At the level of the action, the relational action (15.7) ensues from the $x = 1$ strong fixing. This can be repackaged as, firstly, the TRi-split action (34.6), and, secondly, as the Einstein–Hilbert action. [Moreover, this end-product is a local construct in the same senses that the field equations are.]

$y = 0$: Geometrostatics In this case, the trial quadratic constraint ceases to contain a kinetic term. This option was already mentioned in Teitelboim's work [835]. It has a non-dynamical interpretation as a *geometrostatics*.

Moreover, if one insists that the *action* must be built from first principles, this geometrostatics option is not possible. This corresponds at the most primary level to the Relational Approach precluding a geometrodynamics in which the geometry indeed undergoes nontrivial dynamics. On the other hand, if one attributes primary significance to the constraint algebroid, this option is allowed in both such a restriction of the Relational Approach and in the Deformation Approach [835].

$a = 0$: Strong Gravity In this option, the potential ceases to contain a Ricci scalar since the cofactor of a in the action is $\mathcal{R}$, For $w = 1 = x$, this amounts to recovering the Strong Gravity that corresponds to the strong-coupled limit of GR. However, removing the above obstruction term in no way requires fixing the supermetric coefficient w. Instead, a family of theories for arbitrary w arises in this manner. These can moreover be interpreted as strong-coupled limits of Scalar–Tensor Theories that likewise apply in the vicinity of singularities in those theories. Clearly from each worldline only being able to communicate with itself, other than near singularities these geometrodynamical theories very much do not match everyday Physics.

Henneaux's work [445] can moreover be interpreted the hypersurface derivative or Best Matching corrected derivatives maintaining 4-space to 3-space duality, with the 4-objects involved having a distinct nature from GR's. Henneaux [445] and Teitelboim [835] followed this up by working out the Strong Gravity analogue of the geometry of hypersurfaces within spacetime. This turns out to have a degenerate-signature manifold $(0 + ++)$ for its split space-time structure. In this way, Strong Gravity serves as an example that such duality is not exclusive to GR spacetime and its Euclidean counterpart.

So it turns out that both Geometrostatics and Strong Gravity greatly simplify the constraint algebraic structure. This is because these are *factors in common* with the momentum constraint arising from the Poisson bracket of two trial-Hamiltonian constraints. In this way, $\mathcal{M}_i$ is not an integrability of the corresponding $\mathcal{H}^{\mathrm{trial}}$.

Additionally, by strongly killing off the right hand side of the Poisson bracket of two $\mathcal{H}$'s, the algebraic structure ceases to involve any structure functions. Thus it is a bona fide algebra rather than an algebroid.

All in all, (24.18), (24.19), (24.23) are obeyed [472, 835] in each of the last two cases. These correspond to entirely opposite representations of the object $\mathcal{H}$: pure potential and 'pure kinematical plus Λ-term' cases respectively.

33.4 Family Ansatz for Addition of Minimally-Coupled Matter

This extension is required for the next Section's consideration of the local Relativities corresponding to each option. This is a requirement from the perspective that these local Relativities are not a property of some container spacetime but rather of all the physical laws bar Gravitation (which is less local as per Chap. 7). Thereby, the framework requires extension to include at least two matter field laws.

To this end, we introduce fundamental bosonic matter fields ψ^{A} of unspecified tensorial rank; it turns out that a sufficient set of these can be treated all at once. These are as per Chap. 18.11 but with split-off species-wise coefficients y_ψ and a_ψ, We use

$$\partial s^{\text{grav-}\psi} = \sqrt{\partial s_{y,w}^{\text{grav}\,2} + \partial s_\psi^2} \quad \text{with } \partial s_\psi^2 := \sum_{\mathrm{z}\in\mathrm{Z}} y_\psi^{-1} \mathrm{M}_{\mathrm{zz'}} \partial\psi^{\mathrm{z}} \partial\psi^{\mathrm{z'}}$$

for configuration space metric $\mathrm{M}_{\mathrm{zz'}}$ blockwise corresponding each species ψ^{z}, taken to be ultralocal in the spatial metric $\mathbf{h}$ and with no dependence on the matter fields themselves. Also

$$\mathcal{W}^{\text{grav-}\psi} := a\,\mathcal{R} + b + \sum_\psi a_\psi \mathrm{U}_\psi .$$

This can only depend on the spatial derivatives of the spatial metric through the spatial Christoffel symbols.

For many purposes an equivalent starting point is

$$\mathcal{H}_{x,y,y_\psi,a,a_\psi,b} := \mathrm{N}^{x,y}_{abcd}\mathrm{p}^{ab}\mathrm{p}^{cd} + \sum_\psi y_\psi \mathrm{N}^{\mathrm{zz'}} \Pi_{\mathrm{z}} \Pi_{\mathrm{z'}}/\sqrt{\mathrm{h}}$$
$$\overline{- a\,\mathcal{R} + b + \sum_\psi a_{\mathrm{z}} \mathrm{U}_{\mathrm{z}}} = 0. \tag{33.9}$$

For these models, changes in all the matter degrees of freedom do have the opportunity to contribute to

$$t^{\mathrm{em}}_{\mathfrak{g}\text{-free}} = \int \partial s^{\text{grav-}\psi} / \sqrt{2\overline{W}^{\text{grav-}\psi}} . \tag{33.10}$$

The new Poisson bracket of $\mathcal{H}_{x,y,y_\psi,a,a_\psi,b}$ with $\mathcal{M}_i^{\text{grav-}\psi}$ is the obvious result of a 3-diffeomorphism Lie dragging. On the other hand,[1]

$$
\begin{aligned}
&\{(\mathcal{H}_{x,y,y_\psi,a,a_\psi,b} \,|\, \partial \mathrm{J}), (\mathcal{H}_{x,y,y_\psi,a,a_\psi,b} \,|\, \partial \mathrm{K})\} \\
&= \Bigg(ay\{\mathcal{M}_i^{\text{grav-}\psi} + 2\{1-x\}\mathcal{D}_i \mathrm{p}\} \\
&\quad + \sum_\psi \left\{ ay \left\lceil \Pi^{\mathrm{z}} \frac{\delta \pounds_{\partial \underline{\mathrm{L}}} \psi_{\mathrm{z}}}{\delta \partial \mathrm{L}^i} \right\rceil - 2\, a_\psi y_\psi M^{\mathrm{zz}'} \Pi_{\mathrm{z}} \frac{\partial \mathcal{U}_\psi}{\partial\, \partial_i \psi^{\mathrm{z}'}} \right\} \\
&\quad - 2y\left\{ \mathrm{p}_{jk} - \frac{x}{2}\mathrm{p}\, \mathrm{h}_{jk} \right\} \mathrm{h}_{il} \sum_\psi a_\psi \left\{ \frac{\partial \mathrm{U}_\psi}{\partial \Gamma^c{}_{jl}} \mathrm{h}^{ck} - \frac{1}{2} \frac{\partial \mathcal{U}_\psi}{\partial \Gamma^c{}_{jk}} \mathrm{h}^{lc} \right\} \Bigg| \partial \mathrm{J} \overset{\leftrightarrow}{\partial}{}^i \partial \mathrm{K} \Bigg).
\end{aligned}
\tag{33.11}
$$

33.5 The 3 Strong Obstruction Factors as Relativities

The GR Case of Geometrodynamics Possesses Locally Lorentzian Relativity Here $ay = a_{\mathrm{z}} y_{\mathrm{z}}$ arises, by which matter wave equations are formed between the first and second underlined terms. In this way, $c_{\mathrm{z}} = c_{\mathrm{grav}}$ is enforced: each of these matter fields ψ^{z} is forced to have the same maximum propagation speed $c_{\max}$—and consequently null cone—as Gravitation. Thereby, any pair of these matter fields ψ^{z}, $\psi^{\mathrm{z}'}$ are forced to share these entities with each other: $c_{\mathrm{z}} = c_{\mathrm{z}'}$ and a common null cone for z and z'. In this way, the Relational Approach *derives* rather than assumes the Lorentzian Relativity Principle, as a consistency condition [62, 109].

The Euclidean-signature case which also arises in this manner does not occur physically as is clear from the observed existence of finite propagation speeds.

Geometrostatics Possesses Locally Galilean Relativity Here the shared $c_{\max} = \infty$. This amounts to the local SR null cones have been squashed into planes, which is the Galilean limit of causal structure: Fig. 4.4.b). In the flat-space case, this amounts to a derivation of Galilean Relativity, in fact of an in-general curved-space geometrostatics which is a 'Galileo–Riemann' generalization [62].

Strong Gravity Geometrodynamics Possesses Locally Carrollian Relativity Here the shared $c_{\max} = 0$. Thus the null cones become squeezed into lines, so that each point can only communicate with its own worldline. This consequently possesses Carrollian Relativity (Sect. 4.1). Henneaux [445] pointed out that Strong Gravity exhibits this limit of null cones (Fig. 4.4.c). Physicist John Klauder [558] additionally studied such a zero propagation speed matter sector: 'ultralocal matter

[1]The third term's 'ceiling parenthesis' denotes the extent to which the variational derivative inside acts. The above-listed matter fields all have no Christoffel symbol terms in their potentials, so the last underlined grouping drops out.

fields'. Finally, in these last two cases (some of) the matter can be the opposite limiting case to the gravitational sector. Of course, none of the options in this paragraph are physically realistic.

Discussion As a package, the three possible strong ways of evading the obstruction term are the maximum propagation speed $c_{\text{max}} = 0$, finite and ∞ trichotomy. Local Relativity now follows from closure of the constraint algebroid rather than being postulated a priori as it was historically by Einstein. Physical observation of finite maximum propagation speed serves to select GR alone out of the above set of theories.

As an obstruction result, the three types of localized Relativity arise in a similar manner to the much-vaunted critical dimensions 26 and 10 for bosonic strings and superstrings respectively (Sect. 11.8). This is in the sense that all of these are picked out as strong impositions that produce a closed algebraic structure of constraints. Albeit this Chapter's case involves the classical Poisson brackets closure as a Dirac algebroid (9.31)–(9.33) or the simpler bona fide algebra (24.18), (24.19), (24.23). This is in contrast with String Theory's quantum commutator closure as a Virasoro algebra or its supersymmetric counterpart ([368, 385, 386] and Appendix V).

Moreover, suppose one adopts the physical choice of locally Lorentzian Relativity. This comes hand in hand with *deducing* embeddability into a GR spacetime, a formalism and worldview which has long been known to be widely insightful [440, 874].

This Chapter can furthermore be interpreted as an answer to Wheeler's question (9.1). So the GR form of $\mathcal{H}$ arises as one of but a few consistent options upon assuming just the structure of space. This answer to Wheeler's question ascribes primality to space rather than to spacetime and yet leads to spacetime emerging. Thus it provides a resolution of the classical Spacetime Construction Problem as well, in the sense of construction from space. This result can be considered to arise from local SR, GR and its spacetime structure being rigid, rather than mutable as functioning mathematical structures.

Let us finally compare Einstein's historical route to GR with the Relational Approach's. Einstein chose to consider spacetime primality instead of spatial primality. In this setting, he changed the status of frames from SR's Lorentzian inertial frames to local inertial frames that are freely falling frames. This made direct use of the spacetime connection in passing locally to freely falling frames. Considering the corresponding curvature tensors is now natural, and leads to a law relating the Einstein curvature tensor to the energy–momentum–stress tensor of the matter content. This accounts for the local inertial frames on physical grounds, and the spacetime geometry is to be solved for rather than assumed. However, this approach does not directly address Machian criteria for time and space. On the other hand, in the Relational Approach, the horn in which space is primary is chosen; time and space are conceived of separately, and Leibniz–Mach relational criteria (Chap. 3) are directly applied to each. The notion of space is broadened from that of the traditional absolute versus relational debate arena so as to include geometrodynamical theories. This approach's equations pick out a particular subcase for which an ambient spacetime manifold is implied; this is a very fruitful perspective as per Chap. 7.

So in the Relational Approach, SR arises as an idealization that holds well locally in many parts of the Universe. SR's assumed universal symmetry group and shared null cone is explained in the Relational Approach as an emergent phenomenon. Oppositely to Einstein's historical route to SR and then GR (Chaps. 4 and 7), one arrives at SR via GR.

This formulation having constructed spacetime curvature from its split space–time form, it is conversely now natural to ask whether spacetime connections also play a role in the theory: a 'Discover Curvature and then Connections' approach. One can build the latter out of elements natural to the spatially primary perspective as per (34.23). Thus in the Relational Approach one finally arrives indirectly at the identification of local inertial frames with freely falling frames. (Moreover, some—but not all—parts of the Equivalence Principle already feature in the Geometrodynamical Equivalence Principle [454].) On the other hand, in the spacetime formulation of GR, one has the Equivalence Principle modelled by connections prior to bringing in curvature: a 'Discover Connections and then Curvature' approach.

33.6 The Fourth Weakly-Vanishing Factor

Equation's (33.6) fourth factor contains a $\mathcal{D}_i\mathrm{p}$ core, the vanishing of which can be written as

$$\mathcal{D}_i\{\mathrm{p}/\sqrt{\mathrm{h}}\} = 0 \tag{33.12}$$

without loss of generality, since $\sqrt{\mathrm{h}}$ is covariantly constant. This presents a weakly-vanishing option, covering maximal slices (21.3) and CMC slices more generally. It can be formulated as

$$\mathcal{D} := \mathrm{p} - \sqrt{\mathrm{h}}\,c = 0 \tag{33.13}$$

for c spatially constant, by performing one integration. Moreover, the maximal subcase proves to be too restrictive for the combination of consistency and physical plausibility, along the lines of Chap. 21 and Sect. 33.8. These considerations eventually motivate more general realizations of solutions than the above.

33.7 Discover-and-Encode Approach to Physics

At the classical level, this amounts to trying out a $\mathfrak{g}$, finding it gives further integrability conditions that enlarge $\mathfrak{g}$ and then deciding to start afresh with this enlarged $\mathfrak{g}$.

Metrodynamics Assumed ($\mathfrak{g}$ = id) This is a more minimalist assumption [15, 682] than assuming a geometrodynamics ([683] is even more minimalistic). It leads to the following result.

Consistent Metrodynamics Theorem *Consider ansatz* (33.1) *but with* $\mathfrak{g}$ = *id, so the* $\partial \underline{F}$ *are removed. Then the following five outcomes are consistent.*

i) *Recovery of GR, through* $\mathcal{M}_i$ *appearing as an integrability condition thus forcing* $\mathfrak{g}$ = *id's enlargement to Diff*(σ).
ii) *A* 1*-parameter family of metrostatics.*
iii) *A* 1*-parameter family of Strong Gravity metrodynamics theories.*
iv) *A group of formulations and theories based upon* $\mathcal{D}_i\{\mathrm{p}/\sqrt{\mathrm{h}}\} = 0$.
v) *A 'unit-determinant geometrodynamics', corresponding to* $\mathfrak{g}$ = *id's enlargement to the group of unit-determinant diffeomorphisms, UDiff*(σ) [62].

Derivation [15, 62, 682] In this case, to start off with there is just a primary constraint $\mathcal{H}_{x,y,a,b}$. In considering the Poisson brackets this forms with itself, one no longer has an initial right to a priori 'parcel out' an $\mathcal{M}_i$. One is instead to use the first form of the right-hand side of (33.5) and define

$$\mathcal{S}_i^x := 2\{-\mathcal{D}_j \mathrm{p}^j{}_i + \{1-x\}\mathcal{D}_i \mathrm{p}\} \tag{33.14}$$

as the preliminary secondary constraint entity arising from this Poisson bracket. This is now smeared with some differential vector $\partial\iota^i$.

Equation (33.14) features unless one of $a = 0$ or $y = 0$ holds, in which case the above right hand side strongly vanishes. The Abelian constraint algebra (24.23) applies in both of these cases. Moreover, each case involves a diametrically opposite representation of the $\mathcal{H}$ object itself. I.e. the mostly kinetic $\mathcal{H}_{0,b,x,y}$ of Galileo–Riemann metrostatics: case ii) versus the zero-kinetic term $\mathcal{H}_{a,b,x,0}$ of strong metrodynamics: case iii). Note that these theories are not the same as the previous Section's, since now they have no diffeomorphism-related constraints and thus remain metrodynamical rather than geometrodynamical theories.

If $\mathcal{S}_i$ is present,

$$\{(\mathcal{S}_i \,|\, \partial\iota^i), (\mathcal{S}_j \,|\, \partial\chi^j)\} = (-2\mathcal{D}_j \mathrm{p}^j{}_i + 2\{1-x\}\{3x-2\}\mathcal{D}_i \mathrm{p} \,|\, [\partial\iota, \partial\chi]^i) \tag{33.15}$$

ensues. Comparing (33.5) and (33.15) gives that the constraint algebroid closes only if $x = 1$, $x = \frac{2}{3}$ or $\mathcal{D}_i \mathrm{p} = 0$. The last of these gives case iv) as usual.

If $x = 1$, then $\mathcal{S}_i^x$ collapses to $\mathcal{M}_i$ the generator of diffeomorphisms and therefore the main GR case i) of the working is recovered.

If $x = \frac{2}{3}$ instead, a distinct clear geometric meaning arises as follows. The corresponding

$$\mathcal{U}_a := \mathcal{S}_a^{2/3} = -2\left\{\mathcal{D}_b \mathrm{p}^{ab} - \frac{1}{3}\mathcal{D}_a \mathrm{p}\right\} \tag{33.16}$$

is the generator of *unit-determinant diffeomorphisms*: diffeomorphisms that preserve the local volume element $\sqrt{\mathrm{h}}$: case v). □

At the level of $\mathfrak{Riem}(\sigma)$, this corresponds to picking the degenerate (null signature) supermetric. This degeneracy means that case v) has no underlying relational action. This theory's exact meaning remains unknown; it is an example of a theory lying somewhere between a metrodynamics and a geometrodynamics. We introduce the names U-diffeomorphism, U-geometry and U-superspace for this theory's counterparts of the geometrodynamical entities. U-diffeomorphisms use up 2 degrees of freedom per space point, leaving U-geometry with 4.

33.8 Conformogeometrodynamics Assumed

Case 1) Maximal Slicing Obstruction We now assume from the outset that $\mathfrak{q} = \mathfrak{Riem}(\sigma)$ and $\mathfrak{g} = \mathit{Diff}(\sigma) \rtimes \mathit{Conf}(\sigma)$. Then conformal superspace $\mathfrak{CS}(\sigma) := \mathfrak{Riem}(\sigma)/\mathit{Conf}(\sigma) \rtimes \mathit{Diff}(\sigma)$. This corresponds to $\mathrm{p} = 0$.

For Conformogeometrodynamics, obtaining 'good' $\mathit{Conf}(\sigma)$-objects is somewhat more problematic than usual. If one continues to insist on second-order equations of motion, one needs a φ such that the metric and it form a simple, internally conformally-invariant pair

$$\varphi \longrightarrow \omega^{-1}\varphi, \qquad \mathrm{h}_{\mu\nu} \longrightarrow \omega^{4}\mathrm{h}_{\mu\nu}. \tag{33.17}$$

Thus $\mathrm{h}_{\mu\nu}\varphi^4$ is $\mathit{Conf}(\sigma)$-invariant, and the action is to be built out of this. Writing this is a *local* scale–shape split (since φ depends on position). This is implemented by the action[2]

$$\mathsf{S} = \iint_{\sigma} \mathrm{d}^3x\, \varphi^6 \sqrt{\varphi^{-4}\{\mathcal{R} - 8\triangle_{\mathbf{h}}\varphi/\varphi\}}\partial\mathrm{s}. \tag{33.18}$$

Here, following from Appendix D.7's consideration of conformally-covariant combinations,

$$\partial\mathrm{s} = \|\partial_{\underline{\mathrm{F}}}\mathbf{h}_{ij} + 4\,\mathbf{h}\,\partial\varphi/\varphi\|_{\mathbf{M}} \tag{33.19}$$

(for the undensitized form of $\mathbf{M}$).

This action leads to the vacuum $\Lambda = 0$ case of the Lichnerowicz equation (21.6) as a primary constraint. It also gives $\mathcal{M}_i$ as a secondary constraint from F^i-variation, and the maximal slice condition (21.3) from a part of the free end spatial hypersurface φ-variation. However the other part of this last variation is the corresponding specifier equation that entails frozenness for the compact without boundary σ's of interest (Chap. 21).

Case 2) Global Ratio Action A first way around the above frozenness is to consider a further pure-ratio action [64, 107]. This was obtained by dividing (33.18) by the homogenizing power of the volume $\mathrm{V}^{2/3}$ by the Author, Barbour, Foster and ó Murchadha. Note the similarity with forming the RPM actions (16.14) and (19.5)

[2]For simplicity, the Λ term is omitted; see [64] for its inclusion. Also, [107] considered this using 2 separate multipliers instead of a single more general auxiliary whose velocity also features in the action and has to be free end spatial hypersurface varied [64].

by dividing by $\sqrt{I}$. Moreover, this does not give GR but rather an alternative theory, and a questionable such at that, on grounds of action at a distance and no apparent means of having a viable cosmology [64].

Case 3) Generalization to CMC Slices A second way of avoiding the above obstruction was considered, now paralleling York's generalization [922] of Lichnerowicz's work [622] on the GR initial value problem: from maximal slices to CMC slices. Conformal Relationalism proceeds by considering (global) volume-preserving conformal transformations (VPCTs).

One candidate implementation for VPCTs is [65]

$$\text{case 3.a):} \quad \mathrm{h}_{ab} \longrightarrow \varphi^{\text{fin-int}\,4}\mathrm{h}_{ab} = \frac{\varphi^4}{\langle\varphi^6\rangle^{2/3}}\mathrm{h}_{ab}; \tag{33.20}$$

here the denominator is an average (N.5); this implementation was originally used [65] to provide an action from which York's initial value formulation follows. Another candidate is [16]

$$\text{case 3.b):} \quad \mathrm{h}_{ab} \longrightarrow \varphi^{\text{fin-diff}\,4}\mathrm{h}_{ab} = \{1 + \Delta\zeta\}^{2/3}\mathrm{h}_{ab} \tag{33.21}$$

Here 'fin' stands for finite, 'int' for integral and 'diff' for differential; $\varphi^{\text{fin-int}}$ was originally denoted $\hat{\varphi}$ in the literature. However, all $\mathfrak{g}$ considered so far have corresponded to infinitesimal transformations. Whereas the above two formulations each implement individual VPCTs, they do not implement the VPCTs as a group, at least not with a multiplicative action.

I.e. the composition of two transformations of the form (33.20) is not itself of that form, nor does the composition of two transformations of the form (33.21) share that form.

One possibility is to consider infinitesimal formulations instead [48]:

$$\text{case 3.c):} \quad \mathrm{h}_{ab} \longrightarrow \{1 + 4\{\chi - \langle\chi\rangle\} =: \{1 + 4\overline{\chi}\}\mathrm{h}_{ab} \tag{33.22}$$

—i.e. correction by a small contrast-type object (N.6)—and [16, 48]

$$\text{case 3.d):} \quad \mathrm{h}_{ab} \longrightarrow \varphi^{\text{inf-diff}\,4}\mathrm{h}_{ab} = \left\{1 + \frac{2}{3}\Delta\chi\right\}\mathrm{h}_{ab}. \tag{33.23}$$

Only then do $\mathit{VPConf}(\boldsymbol{\Sigma})$, and consequently $\mathit{Diff}(\boldsymbol{\Sigma}) \rtimes \mathit{VPConf}(\boldsymbol{\Sigma})$, $\mathfrak{Riem}(\boldsymbol{\Sigma})/\mathit{VPConf}(\boldsymbol{\Sigma})$, and $\mathfrak{Riem}/\mathit{Diff}(\boldsymbol{\Sigma}) \rtimes \mathit{VPConf}(\boldsymbol{\Sigma})$ realizations of $\{\mathfrak{CRiem} + \mathrm{V}\}(\boldsymbol{\Sigma})$ and $\{\mathfrak{CS} + \mathrm{V}\}(\boldsymbol{\Sigma})$ respectively make sense. The two integral implementations both encode a version of (21.4) with a particular kind of functional in place of (21.4)'s straightforward constant:

$$\mathrm{p}/\sqrt{\mathrm{h}} = \langle\mathrm{p}/\sqrt{\mathrm{h}}\rangle, \quad \text{i.e.} \quad \mathcal{D}' := \overline{\mathrm{p}/\sqrt{\mathrm{h}}} = 0. \tag{33.24}$$

On the other hand, each of the two differential implementations encodes

$$\triangle(\mathrm{p}/\sqrt{\mathrm{h}}) = 0, \text{ among the solutions of which are the CMC slices } \mathrm{p}/\sqrt{\mathrm{h}} = \text{const.} \tag{33.25}$$

Indeed, this is similar to the manner in which the Conformal Relationalism option arises as the fourth factor in (33.6). Moreover, each of the two equations above has the same structural form as the corresponding encodement of VPCT in the corresponding infinitesimal and consequently group-theoretically successful case.

Equation (33.24) and the outcome of (33.25) is to be interpreted as a constraint rather than as a slice equation in this context. Smear $\mathcal{D}$ with some partial differential scalar $\partial\sigma(z)$. This obeys the following Poisson brackets.

$$\big\{(\mathcal{D}\,|\,\partial\sigma), (\mathcal{D}\,|\,\partial\iota)\big\} = 0, \tag{33.26}$$

so $\mathcal{D}$ forms an Abelian subalgebra.

$$\big\{(\mathcal{D}\,|\,\partial\sigma), (\mathcal{M}_i\,|\,\partial \mathrm{L}^i)\big\} = (\pounds_{\partial\underline{\mathrm{L}}}\mathcal{D}\,|\,\partial\sigma) \approx 0, \tag{33.27}$$

so $\mathcal{D}$ is a scalar density as regards the 3-diffeomorphisms (just as $\mathcal{H}$ is).

$$\big\{(\mathcal{H}_{x,y,a,b}\,|\,\partial\mathrm{J}), (\mathcal{D}\,|\,\partial\sigma)\big\} = (\mathcal{L}\,\partial\mathrm{J}\,|\,\partial\sigma) - \frac{3}{2}(\mathcal{H}_{x,y,a,b}\,|\,\partial\sigma\,\partial\mathrm{J}) \tag{33.28}$$

for $\mathcal{L}\,\partial\mathrm{J} = 0$ a fixing equation for the partial differential of the instant $\partial\mathrm{J}$. I.e. another example of the TRi Dirac-type Algorithm's specifier equation. The corresponding second-order linear differential operator takes the form

$$\mathcal{L} := 2a\,\triangle - \{2a\,\mathcal{R} + 3b\} - y\left\{\frac{3}{2}x - 1\right\}c\,\mathrm{p}/\sqrt{\mathrm{h}}.$$

Furthermore, this is a TRi reformulation and generalization—by the presence of coefficients a, x, y—of the well-known CMC lapse fixing equation (21.31). By this arising on the right hand side of the last Poisson bracket, $\mathcal{D}$ knocks $\mathcal{H}$ off its perch of being hitherto first-class: these are now second-class with respect to each other. Thus the degrees of freedom count is now $6 \times 2 - 3 \times 2 - 2 \times 1 = 2 \times 2$. I.e. it is unaffected overall by there being a new constraint but its presence alters the status of an already-present constraint.

Next, consider $\mathcal{D}'$ likewise. The first two Poisson brackets are unaffected by the same geometrical interpretations applying, whereas the nonzero Poisson bracket is

$$\big\{(\mathcal{H}_{x,y,a,b}\,|\,\partial\mathrm{J}), (\mathcal{D}'\,|\,\partial\sigma)\big\} = (\mathcal{L}_1\partial\mathrm{J}\,|\,\overline{\partial\sigma}), \tag{33.29}$$

where now

$$\mathcal{L}_1 := 2a\,\triangle - \{2a\,\mathcal{R} + 3b\} - y\left\{\frac{3}{2}x - 1\right\}\langle\mathrm{p}/\sqrt{\mathrm{h}}\rangle\,\mathrm{p}.$$

Case 4) Reinterpreting $\mathfrak{CS}$ + V as $\mathfrak{CS}(\Sigma)$ A third way around the obstruction was subsequently provided by Barbour and ó Murchadha [108]. This is based on reinterpreting York's use of $\{\mathfrak{CS} + \mathrm{V}\}(\Sigma)$ as containing a global scaling redundancy, by which CMC slices are argued to be tied, rather, to $\mathfrak{CS}(\Sigma)$. This is due to identifying an overlooked homothety on the space of the geometrodynamical data,

$$(\mathrm{h}_{ij}, \mathrm{K}_{ij}) \longrightarrow (\widetilde{\mathrm{h}}_{ij}, \widetilde{\mathrm{K}}_{ij}) = \big(k^2 \mathrm{h}_{ij}, k\,\mathrm{K}_{ij}\big). \tag{33.30}$$

Since this approach does not introduce VPCTs, it avoids case 4.c)'s problems.

Comments The 0–finite–∞ propagation speed trichotomy of Carrollian, Lorentzian and Galilean Relativities arises with an unexpected fourth conformogeometrodynamical CMC slicing partner which is almost equal (from the same obstruction term, but now as a weakly-vanishing factor). This is a joint packaging that has not been seen before in Theoretical Physics; it is furthermore interesting since the last option can be interpreted as a notion of absolute simultaneity lying hidden within GR.

Some of the variants of this last option are known as 'shape dynamics' [650]. This program has so far mostly developed branches 3.a) and 4). However, this Sec's analysis suggests changing focus to branches 3.d) and 4, with some consideration of 3.c) as well. Each choice of 3.a)–d) alters the entirety of the detailed calculations of the ensuing 'shape dynamics' (i.e. the ones explicitly involving φ's). On the other hand, the 'core $\varphi = 1$ equations' of the schemes, such as the Hamiltonian constraint, momentum constraint and CMC slice condition, continue to coincide. In this way, all of the schemes in question make contact with CMC-sliced, GR paralleling formulations used in the GR initial value problem. Moreover, in infinitesimal interpretations, one needs to take care not to over-implement the infinitesimalness prior to variation. Indeed, enough use of approximations prior to variation leads instead to the *linearized* Lichnerowicz–York equation being encoded.

33.9 Simpler Cases of Spacetime Constructability

Example 1) Newtonian Mechanics and RPMs have no such notion, due to these having no notion of spacetime in the first place. The analogue of this Chapter's working for these gives that their simpler constraint brackets exhibit no corresponding obstruction terms.

Example 2) Minisuperspace modelled within the foliation by hypersurfaces privileged by homogeneity. This greatly simplifies the study of spacetime. Moreover, in this case the obstruction term vanishes, since it contains a spatial derivative operator factor $\mathcal{D}_i$ that now acts upon a homogeneous entity. Minisuperspace additionally has no $\mathcal{M}_i$. The constraint algebraic structure consequently ends up being just $\mathcal{H}$ commuting with itself (for any a, b, w, y). Thus generalized Minisuperspace is *not* restricted by constraint algebraic structure consistency. This is very similar in content to the metrodynamical case of Strong Gravity. [In the former case, R is just a spatial constant, so $aR + b$ behaves just like $0 + b'$ for a new constant b'.]

Example 3) SIC. Here the sandwich can be solved because it is algebraic. This simplification, albeit rooted in the leading-order homogeneity, unfortunately comes hand in hand with a trivialization of other Problem of Time facets, including trivialization of Refoliation Invariance by a restriction to a privileged foliation. This takes effect through finite algebraic constraints coming without smearing variables, so the self-bracket of each scalar constraint must be zero. The self-bracket of $\mathcal{H}$ is of this kind, and this argument clearly extends to the self-bracket of $\mathcal{H}_{a,b,x,y}$, on which the current Chapter focuses. Thus the current Chapter's obstruction term also collapses to triviality for this book's choice of SIC model arena. In a sense, this model has already assumed spacetime structure due to the privileged foliation involved. Furthermore, changing the coefficients in the supermetric is not restricted by the Dirac Algorithm in this regime, unlike for full GR.

We consequently need to consider *more* inhomogeneous models than this one before the obstruction term takes effect. For example, in the perturbative scheme involving specifically spatially-dependent constraints (using full rather than mode-wise expression for the constraints):

$$
\begin{aligned}
&\big\{(\mathcal{H}_{x,y,y_\Psi,a,a_\Psi,b} \,|\, \partial \mathrm{J}), (\mathcal{H}_{x,y,y_\Psi,a,a_\Psi,b} \,|\, \partial \mathrm{K})\big\}_2 \\
&\quad = \left(ay\big\{\mathcal{M}_{1i}^{\text{grav-}\Psi} + 2\{1-x\}\exp(-2\Omega)\mathcal{D}_i \mathcal{D}_1\big\} + ay \left[\pi_1^\Psi \frac{\delta \pounds_{\partial \underline{\mathrm{L}}} \Psi}{\delta \partial \mathrm{L}^i} \right] \right. \\
&\qquad \left. - 2a_\Psi y_\Psi \exp(-3\Omega)\, \pi_0^\Psi \frac{\partial \mathcal{U}_{1\,\Psi}}{\partial\, \partial_i \Psi} \,\middle|\, S^{ij}\{\partial \mathrm{K}_0 \partial_j\, \partial \mathrm{J}_1 - \partial \mathrm{J}_0 \partial_j\, \partial \mathrm{K}_1\} \right). \qquad (33.31)
\end{aligned}
$$

In this way, Spacetime Construction from space is fully functional to second order in SIC.

Additionally, the second-order contribution $\{(\mathcal{H}0_{x,y,y_\Psi,a,a_\Psi,b} \,|\, \partial \mathrm{J}_1), (\mathcal{D}_1 \,|\, \partial \mathrm{J}_1)\}$ gives the term $(\mathcal{L}\, \partial \mathrm{J}_1 \,|\, \partial \sigma_0)$ with surviving $\triangle$ contribution. Thus the $\partial \mathrm{I}$ fixing equation features as the usual kind of initial-value PDE problem for GR. SIC taken to second order is consequently a nontrivial model arena for further investigating these formulations.

Research Project 31) Is there an Ashtekar variables analogue of this Spacetime Construction from assuming just the spatial level of structure?

33.10 Caveats on Further Matter Results

[109] contains further claims that matter uniqueness and the Equivalence Principle can be derived from Temporal and Configurational Relationalism postulates alone. These claims were however subsequently shown to rely on further mathematical simplicities. This is because the matter term ansätze of [58, 109] were insufficiently diverse. Canonical formulations which presuppose spacetime [577–579] can be readily seen contain further terms, and on occasion these terms also manage to comply with the relational postulates [17, 19]. Some extra terms can be

cast into TRi form by linearities other than the square root of a quadratic kinetic term (Sect. 17.2). Some extra terms include metric–matter kinetic cross-terms.

Matter uniqueness claims concerning deriving Gauge Theory just from relational first principles fail since various other vector Field Theories can be cast into relational form [17, 19]. One such example is Proca Theory. Further examples are a subset of physicist Ted Jacobson's Einstein–Aether theories [516], which are Scalar–Vector–Tensor Theories for which the vectors are unit vectors. The latter examples of relational theories also violate the local SR null cones.

The claim that the Equivalence Principle can be derived from relational first principles was also overturned. For instance, it was found [17] that [109]'s ansätze tacitly assumed what had been known to be a 'geometrodynamical' form of Equivalence Principle since [454]. This tacit assumption entered through precluding the presence of the metric–matter kinetic cross-terms that are kinematically appropriate for nonminimally-coupled tensor matter field. The final version of [109] already conceded that Brans–Dicke Theory can also be cast in relational form; this extension contains cross-terms, and limits some GR uniqueness and Equivalence Principle derivation considerations. Furthermore, different values of the Brans–Dicke parameter corresponding to $w \neq 1$ become consistent via involvement of metric–matter cross-terms. In this way, Brans–Dicke theory and other more complicated Scalar–Tensor Theories are available not only as a strong gravity limit resolutions of Spacetime Construction but also as finite propagation speed alternatives to the GR outcome [15].

Inclusion of fermions [14, 37] requires a linear kinetic term $\mathrm{T}_{\mathrm{lin}}$ being additively appended to the product of square roots as per Sect. 18.11.

Moreover, if Temporal and Configurational Relationalism requirements are extended to a theory of Background Independence, uniqueness results are strengthened For instance, some of the other Background Independence postulates can be used to exclude all the Einstein–Aether theories . As such, there is at least some chance that a subset of the matter claims that RWR attempted to rest upon Temporal and Configurational Relationalism alone could rest on a more complete notion of Background Independence instead.

Research Project 32) Work out the a priori distinct affine and metric structure counterpart of Spacetime Construction.

Research Project 33)† Work out the full Supergravity counterpart of Spacetime Construction. This could be based on an action ansatz or a constraint ansatz.

Chapter 34
TRi Foliation (TRiFol)

We next consider an alternative version of Chap. 8's Geometrodynamics and Chaps. 31–32's Foliation Formulation. All these formulations are spacetime-assumed positions; the current Chapter's distinction is in having not ADM but TRi kinematics. This is a suitable follow-up on the previous Chapter's derivation of spacetime within the Relational Approach via the TRi Dirac-type Algorithm's Spacetime Construction.

34.1 TRi-Split Version of Geometrodynamics

At the geometrodynamical level, the differences between the two formulations are summarized in Fig. 34.1. The underlying splits are depicted in Fig. 34.2.

The *partial differential of the frame* [48] is $\partial X^\mu = h^{\mu\nu}\partial s_\nu$. This is another formulation of the point identification map. It is also a hypersurface tensor by $n_\mu \partial X^\mu = n_\mu h^{\mu\nu}\partial s^\nu = 0$, so it can be written $\partial\vec{X}$ or $\partial\underline{X}$; the Relational Approach version is $\partial\underline{F}$.

The *partial differential of the instant* [48] is $-n_\mu \partial s^\mu = \partial\mathfrak{T}$. This is another formulation of duration of GR proper time, now in the additionally simplified form of identifying τ with $\mathfrak{T}(t, x^i)$ itself. [$\partial\mathfrak{T}$ is not conceptually the same as ∂I, since the former is assumed but the latter is emergent.]

In parallel with Chap. 31 rather than Chap. 8, these can be set up by considering the TRi analogue of the time flow vector field t^μ: a change-covector ∂s^μ. The decomposition (8.13) needs to be replaced by

$$\partial s^\mu = \partial X^\mu + \partial\mathfrak{T}\, n^\mu. \tag{34.1}$$

In this way, TRi-foliations are interpreted in terms of a choice of time t and an associated time flow vector field change-covector ∂s^μ.

Next, in place of (8.7), ∂s^μ and t are restricted by

$$\partial s^\mu \nabla_\mu t = 1 \quad \text{and} \quad \partial w^\mu \nabla_\mu t = 0 \quad \text{for any tangential } \partial w^\mu. \tag{34.2}$$

E. Anderson, *The Problem of Time*, Fundamental Theories of Physics 190,
DOI 10.1007/978-3-319-58848-3_34

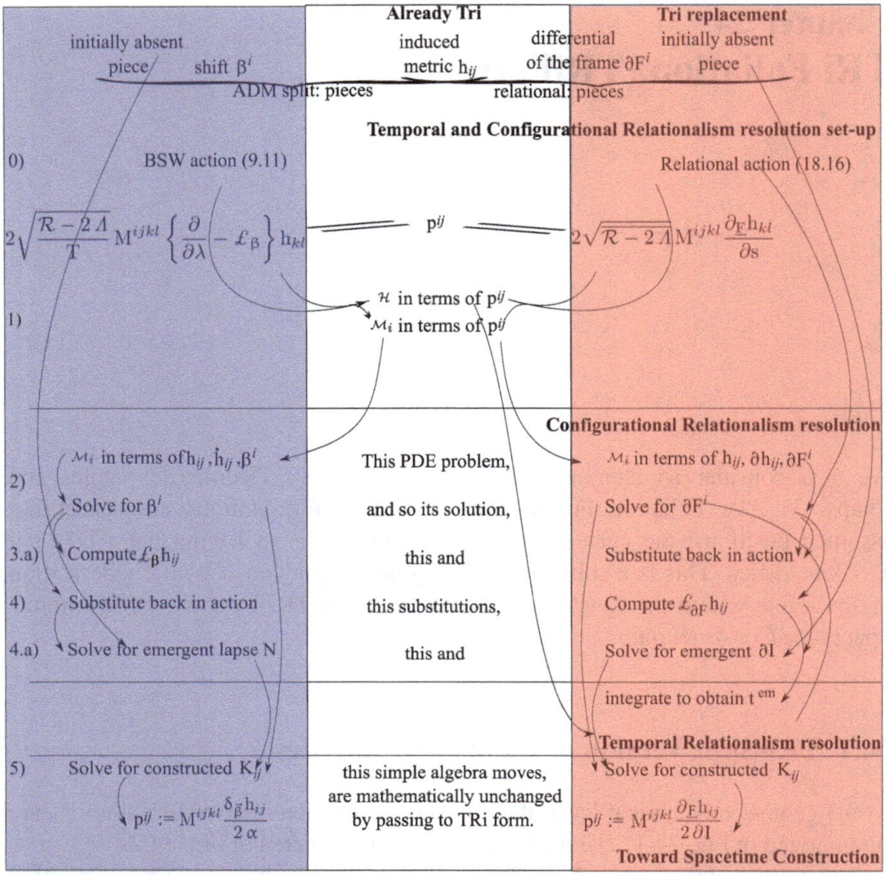

Fig. 34.1 ADM and TRi splits compared. The darker shading is for rank-± 2 objects to the lighter shading being for rank-± 1 ones

The TRi formulation furthermore admits a decomposition into change-covariant time flow and tangential parts. If (34.2) holds, it is consistent to identify $\partial s^\mu \nabla_\mu$ with both $\partial/\partial t$ and with $\pounds_{\partial \underline{S}}$. On the other hand, n^μ remains a change-scalar. Indeed, rearranging (34.1) to make n^μ the subject,

$$n^\mu = \left[\partial n^0, -\partial X^i\right]/\partial \tau \tag{34.3}$$

and ∂s^0 is numerically 1 but remains none the less a change-covector.

Next note that the TRi split of the spacetime metric $\mathbf{g}$ has pieces of three different change weights: the change-scalar induced metric, the change-1-form g_{0i} and the change-2-form g_{00}. The corresponding split of the inverse metric $\mathbf{g}^{-1}$ has three different opposing change weights: a change-scalar, a change-vector and a change-2-tensor.

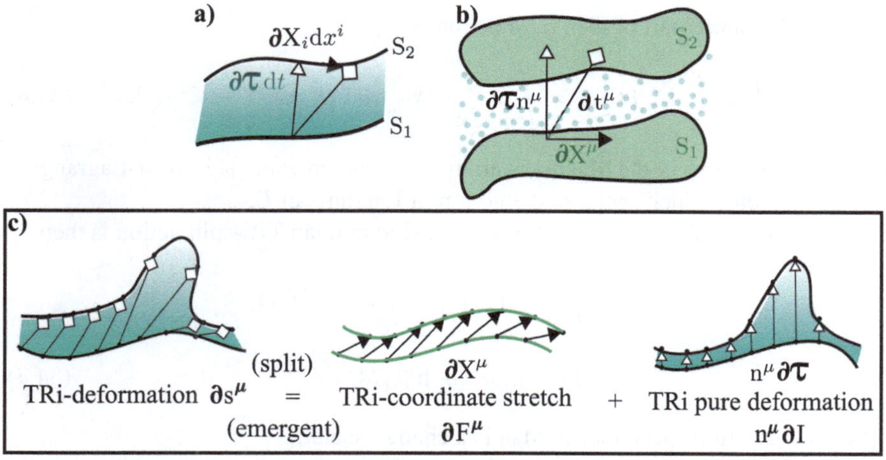

Fig. 34.2 TRi counterpart of Fig. 8.3.b)–c). **a)** $3+1$ split of a region of spacetime, with differential of the instant ∂T and differential of the frame ∂X^i. **b)** Local presentation of the split of ∂s^μ into n^μ and ∂X^μ. **c)** is the TRi version of the decomposition of deformations, in parallel to Fig. 32.1

The TRi explicit computational formula for extrinsic curvature is, in more detail,

$$\mathcal{K}_{ij} := \frac{\partial_{\underline{X}} h_{ij}}{2\,\partial\tau} = \frac{\partial h_{ij} - \pounds_{\partial \underline{X}} h_{ij}}{2\,\partial\tau}. \tag{34.4}$$

Note that this is now in terms of the Best Matching corrected derivative (18.23), whose general form is in terms of a Lie derivative correction on account of the shuffling procedure entailed. By the relation between the Lie derivative of the metric and the Killing form (E.20), and under passing from ADM split quantities to TRi split ones, this can be seen to be mathematically equivalent to the hypersurface derivative (8.14). This is underlied by Barbour's Best-Matching covariant derivative (18.23) being the 3-space dual interpretation of Kuchař's hypersurface derivative (8.15).

The Gauss–Codazzi equations are change-scalars, so their form carries over from Chap. 8. However, i) in the previous Chapter's application, due to not yet presupposing the signature of spacetime, each equation's $\mathcal{K}$ term has an extra factor of ε: the normalized version of $-a$ for ordinary Lorentzian ($\varepsilon = -1$) and Euclidean ($\varepsilon = 1$) GR. ii) These equations are in that case interpreted specifically in the 'embedding into spacetime deduced' setting as opposed to the 'projections of an assumed spacetime onto a spatial hypersurface' setting.

On the other hand, the new TRi split formulation's version of the Ricci embedding equation is

$$\mathcal{R}^{(4)}_{\perp a \perp b} = \frac{\partial_{\underline{X}} \mathcal{K}_{ab} + \mathcal{D}_b \mathcal{D}_a \partial\tau}{\partial\tau} + \mathcal{K}_a{}^c \mathcal{K}_{cb}. \tag{34.5}$$

The TRi counterpart of the ADM action is

$$\mathsf{S}_{\text{TRi}} = \int \mathrm{d}t \int_{\Sigma} \mathrm{d}^3x\, \partial\mathcal{C}_{\text{TRi}} := \int \mathrm{d}t \int_{\Sigma} \mathrm{d}^3x \sqrt{\mathrm{h}}\, \partial\mathfrak{T}\{\mathcal{K}_{ab}\mathcal{K}^{ab} - \mathcal{K}^2 + \mathcal{R}\}. \tag{34.6}$$

This is now in terms of the Machian variables h_{ij} and dh_{ij}; $\partial\mathcal{C}_{\text{TRi}}$ is the ∂-Lagrangian change-covector which replaces the notion of Lagrangian $\mathcal{L}$.

The geometrical form of the manifestly ∂-Lagrangian TRi-split action is then

$$\int \mathrm{d}t \int_{\Sigma} \mathrm{d}^3x\, \partial\mathfrak{T}\{\overline{\mathrm{T}}_{\text{TRi-JD}} + \overline{\mathcal{R} - 2\Lambda}\},$$

$$\text{for } \overline{\mathrm{T}}_{\text{TRi-JD}} = \|\partial_{\underline{\mathrm{X}}}\mathbf{h}\|^2_{\mathbf{M}}/4. \tag{34.7}$$

[The configuration space metric M_{AB} is a change-scalar.]

Whereas $\mathcal{H}$ and $\mathcal{M}_i$ are change-scalars, the TRi-split evolution equations in terms of momenta take the new form (in vacuo for simplicity)

$$\frac{\partial_{\underline{\mathrm{X}}}\mathrm{p}^{ij}}{\partial\mathfrak{T}} = \sqrt{\mathrm{h}}\left\{\frac{\mathcal{R}}{2}\mathrm{h}^{ij} - \mathcal{R}^{ij}\right\} + \frac{\sqrt{\mathrm{h}}\{\mathcal{D}^j\mathcal{D}^i - \mathrm{h}^{ij}\triangle\}\partial\mathfrak{T}}{\partial\mathfrak{T}}$$
$$- \frac{2}{\sqrt{\mathrm{h}}}\left\{\mathrm{p}^{ic}\mathrm{p}_c{}^j - \frac{\mathrm{p}}{2}\mathrm{p}^{ij}\right\} + \frac{2}{\sqrt{\mathrm{h}}}\mathrm{h}^{ij}\left\{\mathrm{p}_{ij}\mathrm{p}^{ij} - \frac{\mathrm{p}^2}{2}\right\}. \tag{34.8}$$

In this case, recasting them in terms of extrinsic curvature does not keep one within a change-scalar form due to their higher-derivative multi-slice status. Alternatively, in terms of $\mathcal{K}_{ab}$,

$$\frac{-\{\partial_{\underline{\mathrm{X}}}\mathcal{K}_{ab} - \mathrm{h}_{ab}\partial_{\underline{\mathrm{X}}}\mathcal{K}\} - \{\mathcal{D}_b\mathcal{D}_a - \mathrm{h}_{ab}\triangle\}\partial\mathfrak{T}}{\partial\mathfrak{T}}$$
$$- \left\{2\mathcal{K}_a{}^c\mathcal{K}_{bc} - \mathcal{K}\mathcal{K}_{ab} + \frac{\mathcal{K}_{ij}\mathcal{K}^{ij} + \mathcal{K}^2}{2}\mathrm{h}_{ab}\right\} + \mathcal{G}_{ab} = \mathcal{G}^{(4)}_{ab} = 0. \tag{34.9}$$

The three TRi-split Einstein field equations can all be interpreted in terms of contractions of the Gauss–Codazzi–Ricci embedding equations. In this way, one passes from the Constraint–Embedding Theorem of GR to the Constraint–Evolution–Embedding Theorem of GR. The latter has fine distinction between an ADM- and TRi-split form through the last piece—the GR equations of motion—not being already-TRi.

34.2 TRiFol Itself

Let us next reconsider Chap. 31's foliation upgrade of the ADM formulation, but now for TRi kinematics as suits working within spacetime emergent from relational first principles [48], as summarized in Fig. 34.3.

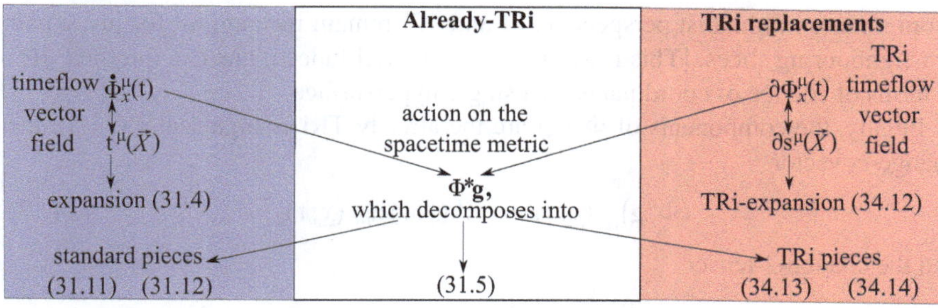

Fig. 34.3 TRiFol: foliations using TRi kinematics compared with Chaps. 31–32's foliations using ADM kinematics

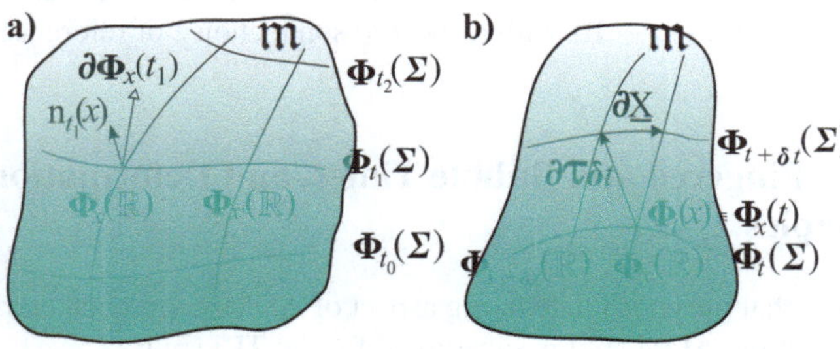

Fig. 34.4 **a)** The flow lines of the TRi foliation of $\mathfrak{m}$. **b)** The differentials of the instant and of the frame in the TRi Foliation Formulation

The basic account's time flow vector field is now the change-covector

$$\partial\Phi_{\underline{x}}(t) \text{ whose components are } \partial\Phi^{\mu}_{\underline{x}}(t) = \partial\Phi^{\mu}(\underline{x}, t). \tag{34.10}$$

The *change-covector version* [48] *of* the deformation vector field is the corresponding change-covector; Fig. 34.4.b) depicts the corresponding flow lines. N.B. in this Sec's approach, $\partial \mathrm{s}^{\mu}$ is more primary than $\partial \mathrm{X}^i$ and $\partial\mathfrak{T}$ (or $\partial \mathrm{F}^i$ and $\partial \mathrm{I}$). Thus it is the first object to be allocated nontrivial change weight and to set the 'vector or covector' sign convention and the size convention for the unit weight [44]. The manifestation of duality is now that for each $\underline{x} \in \sigma$, $\partial\Phi_{\underline{x}}(t)$ is a vector in $\mathfrak{T}_{\Phi(\underline{x},t)}(\mathfrak{m})$ at the point $\Phi(\underline{x}, t)$ in $\mathfrak{m}$. Moreover, the change-covector deformation vector field is a reinterpretation of the change-covector time flow vector field $\partial \mathrm{s}^{\mu}$ according to

$$\partial \mathrm{s}^{\mu}(\vec{X}) = \partial\Phi^{\mu}(\underline{x}, t)|_{\vec{\mathrm{X}}=\vec{\mathrm{X}}(\underline{x},t)}. \tag{34.11}$$

This corresponds again to viewing this as acting on a slice or leaf. Consequently, in the Deformation Approach the previous comment about the time flow as the first nontrivial TRi homothetic weight quantity to be encountered carries over to the likewise primary change-covector deformation vector.

The change-covector (34.10) now expands out as

$$\partial\Phi^{\mu}(\underline{x}, t_1) = \partial\mathfrak{T}(\underline{x}, t_1)\, \mathrm{g}^{\mu\nu}\big(\Phi(\underline{x}, t_1)\big)\, \mathrm{n}_{\nu}(\underline{x}, t_1) + \partial \mathrm{X}^{a}(\underline{x}, t_1)\, \Phi^{\mu}{}_{,a}(\underline{x}, t_1). \tag{34.12}$$

From a more minimalist perspective, $\partial \mathrm{I}$ and $\partial \mathrm{F}^i$ remain meaningful for just a pair of neighbouring slices. [This is again indexed by t_1.] Indeed, one can interpret $\partial \mathrm{F}^i$, in terms of change of coordinates on a single hypersurface.

Finally, the components of $\Phi^*(\mathbf{g})$ are the already-TRi change-scalar (31.5), the change-covector

$$\left(\Phi^*\mathbf{g}\right)_{0a}(\underline{x}, t) = \partial \mathrm{X}^b(\underline{x}, t)\, \mathrm{h}_{ab}(\underline{x}, t), \tag{34.13}$$

and the change-2-tensor

$$\left(\Phi^*\mathbf{g}\right)_{00}(\underline{x}, t) = \partial \mathrm{X}^a(\underline{x}, t)\, \partial \mathrm{X}^b(\underline{x}, t)\, \mathrm{g}^{ab}(\underline{x}, t) - \partial \tau(\underline{x}, t)^2. \tag{34.14}$$

In this way, the TRi-split conception can be replaced by one based on TRi-foliations. This parallels the replacement of the ADM split by one based on foliations in [483, 576]. This is $\Phi^{\mathrm{ref}} : \boldsymbol{\Sigma} \times \mathbb{R} \rightarrow \mathfrak{m}$ with respect to some choice of reference foliation.

34.3 Many-Fingered and Bubble Times, and Deformation First Principles

Let us first note that the coordinate fixing aspect of ADM's approach carries over to the case in which the ADM split is substituted for the TRi split.

The bubble time reformulation can also be used to free the TRi version [20] of ADM's approach from its own aspect as a coordinate fixing. In particular, [573] eliminates lapse α and shift $\underline{\beta}$, which corresponds to another formulation of the Thin Sandwich. The TRi counterpart of this instead eliminates the cyclic partial differential of the instant and of the frame, corresponding to another formulation of the TRi Machian Thin Sandwich.

TRi Deformation First Principles for GR were provided in Sect. 32.1 in unsmeared form. While this was implicitly derived with the plain smearings corresponding to the ADM formulation, it can clearly be rederived using TRi-smearings. A TRi version of the Deformation Approach (also in Sect. 32.1) ensues. This is not however used in the Relational Approach since those results *still* presuppose spacetime. In contrast, the Relational Approach has its own procedure as per Chap. 33 which is additionally a bona fide Spacetime Construction. Finally, Teitelboim's matter results (Sect. 32.1) also readily admit TRi counterparts.

34.4 Machian Hypersurface Kinematics

There are still three types of hypersurface kinematics, but their interpretations need to be reworked [62] from the perspective of spatial primality and this leads to two of them being renamed as well. See Fig. 34.5 for a summary.

1) *Best Matching corrected derivatives* $\partial_{\underline{\mathrm{F}}}$ have now replaced the hypersurface derivatives $\delta_{\vec{\beta}}$ they are dual to as per Sect. 34.1.

	Already-TRi	TRi replacements
1) hypersurface derivative $\delta_{\vec{\beta}}$		$\partial_{\underline{\mathrm{F}}}$ Best Matching derivative
2) tilt $\partial_i \alpha$		$\partial_i \partial \mathrm{I}$ Spatial gradient of the change of the instant
	3) derivative coupling K_{ij}	

Fig. 34.5 Universal hypersurface kinematics in standard and Machian TRi forms compared

2) *Spatial gradients of the change of the instant* $\partial_i \partial \mathrm{I}$ have now replaced the tilts $\partial_i \alpha$. The renaming involved here is in part due the name 'tilt' is itself inspired by spacetime geometry. We consequently reserve this terminology for the spacetime setting. This is one part of the spatial gradient of the change of the instant–translation split.

   The TRi version of the translation is the part such that $\partial \mathrm{I}(\mathrm{p}) \neq 0$, $\{\partial_i \partial \mathrm{I}\}(\mathrm{p}) = 0$. On the other hand, the spatial gradient of the change of the instant part such that $\partial \mathrm{I}(\mathrm{p}) = 0$, $\{\partial_i \partial \mathrm{I}\}(\mathrm{p}) \neq 0$. Subsequently the 'proper time labels instants' duality converts this to the usual notion of tilt as recognizable from the above simple SR realization. This is linked by the proper time of the observers' clocks exhibiting a constant gradient over space on the flat hypersurface tilted at a fixed angle.
3) *Derivative couplings* are as before, except that now the underlying formula for $\mathcal{K}_{ij}$ is (34.4) rather than (8.14). I.e. in Spatially Relational primary terms $\mathcal{K}_{ij}$ is now interpreted as *comparison of each geometrodynamical change with the STLRC*, $\partial \mathrm{I}$.

The above can also be used to reformulate the range of theories of matter considered in [468, 577–579]. References [14, 16, 19] also made use Kuchař's hypersurface kinematics based on the ADM split followed by the analogue of the BSW multiplier elimination of α as a guide to which theories can be formulated from relational first principles. Finally, using the X^i, τ version of the above Machian TRi hypersurface kinematics corresponding to the TRi split of the spacetime metric followed by ∂-Routhian reduction of τ returns one to the relational formulation.

34.5 Machian Thin Sandwich Completion

Consistently accommodating a spatial tensor field in a space–time TRi-split formulation in general requires associating it with further tensor fields in combinations such as

$$-\partial_{\underline{\mathrm{F}}} \mathrm{A}_a - \partial \tau \, \mathcal{K}_{ab} \mathrm{A}^b - \mathrm{C} \, \partial_a \partial \tau . \tag{34.15}$$

Subsequently at the level of constructed spacetime level, this is interpreted as $\partial \tau \, \nabla_\perp \mathrm{A}_a$ with C recast in the role of the further spacetime form's component $\mathrm{A}_\perp$.

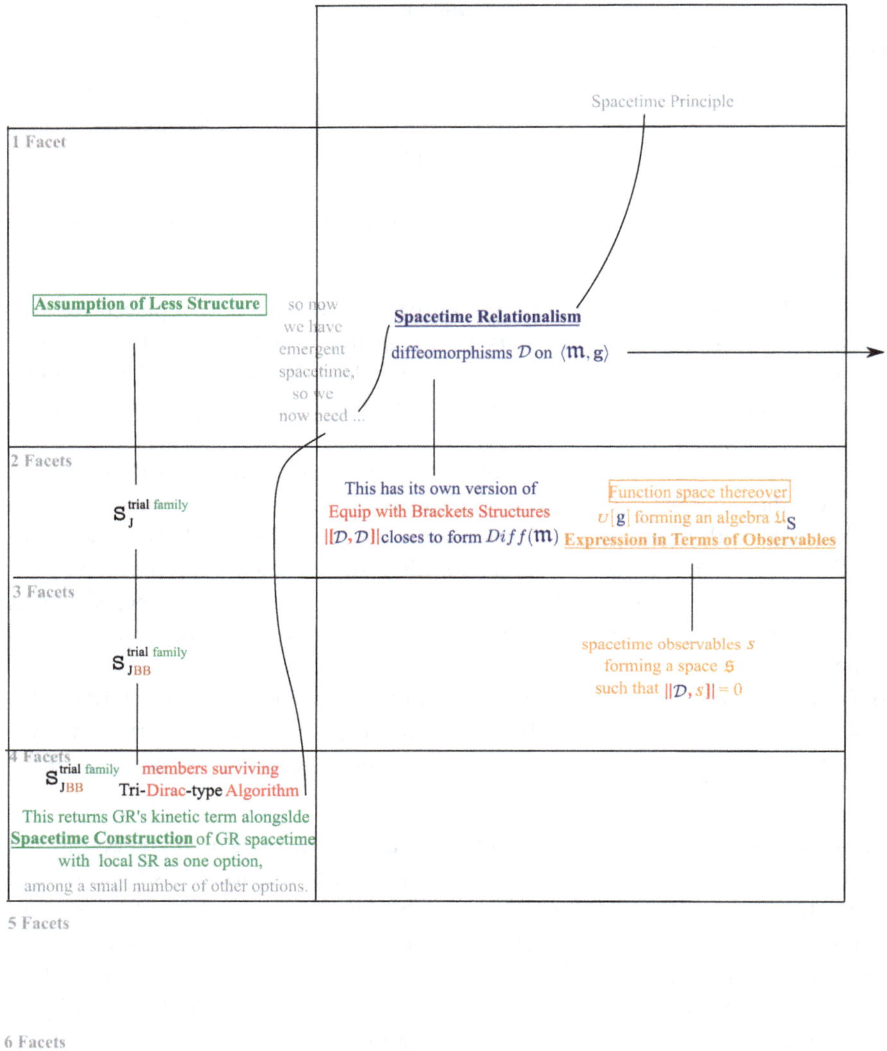

Fig. 34.6 The last part of the 'technicolour guide' to the Problem of Time: Spacetime Construction, Spacetime Relationalism and Refoliation Invariance as combined with previous Problem of Time facets. The end of Fig. 24.2 branches, so that both Fig. 25.1 and the current Fig are distinct continuations of this

Consequently, Machian Thin Sandwich 3.a) is a construction of the Best-Matched derivative, whereas Machian Thin Sandwich 5) constructs $\mathcal{K}_{ij}$. As such, postulating the following move completes the synthesis of hypersurface kinematics.

Machian Thin Sandwich 6) Machian Thin Sandwich 4.a) further permits one to construct an emergent version of the spatial gradient of the change of instant, $\partial_b\, \partial \mathrm{I}$.

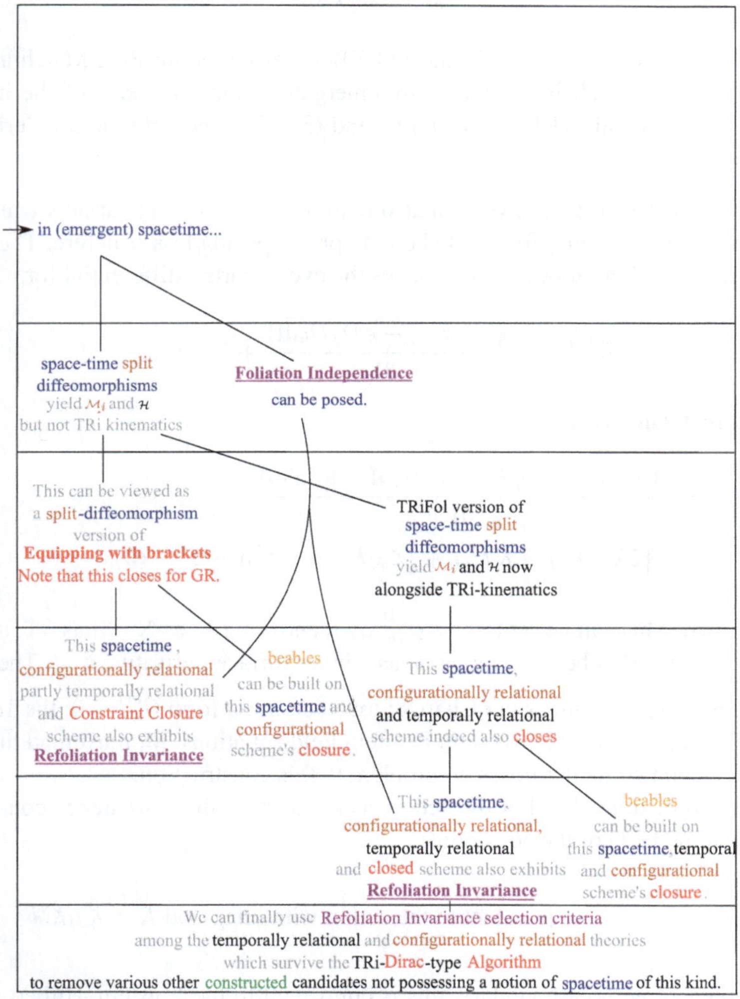

Fig. 34.6 (Continued)

In this way, one can once again construct a wide range of spacetime geometrical objects as per above.

Example 1) Re-running Sect. 32.2's example gives the following two pairs, each consisting of one change-scalar and one change-covector

$$\nabla_a \mathrm{A}_\perp = \mathcal{D}_a \mathrm{A}_\perp - \mathcal{K}_{ab} \mathrm{A}^b, \tag{34.16}$$

$$\partial \text{I} \, \nabla_{\perp} \text{A}_a = -\delta_{\vec{\partial \text{F}}} \text{A}_a - \partial \text{I} \, \mathcal{K}_{ab} \text{A}^b - \text{A}_{\perp} \partial_a \partial \text{I}, \tag{34.17}$$

$$\nabla_b \text{A}_a = \mathcal{D}_b \text{A}_a - \text{A}_{\perp} \mathcal{K}_{ab}, \tag{34.18}$$

$$\partial \text{I} \, \nabla_{\perp} \text{A}_{\perp} = -\delta_{\vec{\partial \text{F}}} \text{A}_{\perp} - \text{A}^a \partial_a \partial \text{I}, \tag{34.19}$$

In the right hand sides, (34.17) and (34.19)'s first terms are Best Matching corrected derivatives. Their last terms are emergent spatial gradients of the instant. Finally, (34.16) and (34.18)'s last terms and (34.17)'s second term are derivative couplings.

The Spacetime Construction version also requires the following variants which are somewhat more generally formulated due to presupposing less structure. The Ricci embedding equation, moreover, now takes the cyclic partial differential form

$$\mathcal{R}^{(4)}_{a \perp c \perp} = \frac{\{-\partial_{\text{F}} \mathcal{K}_{ab} - \varepsilon \mathcal{D}_a \mathcal{D}_b \partial \text{I}\}}{\partial \text{I}} + \mathcal{K}_a{}^c \mathcal{K}_{cb}. \tag{34.20}$$

In the Lorentzian GR case,

$$0 = \frac{\{\partial_{\text{F}} \mathcal{K}_{ab} - \text{h}_{ab} \partial_{\text{F}} \mathcal{K} - \mathcal{D}_b \mathcal{D}_a \partial \text{I} + \text{h}_{ab} \Delta \partial \text{I}\}}{\partial \text{I}} - \left\{2 \mathcal{K}_a{}^c \mathcal{K}_{cb} - \mathcal{K} \mathcal{K}_{ab} + \left\{\mathcal{K}_{ij} \mathcal{K}^{ij} + \mathcal{K}^2\right\} \text{h}_{ab}/2\right\} + \mathcal{G}_{ab} \tag{34.21}$$

is discovered. This can be set to be a $\mathcal{G}^{(4)}_{ab}$ by identifying F^i as X^i, ∂I as $\partial \tau$, and recasting the Best Matched derivative ∂_{F} as a hypersurface derivative $\delta_{\vec{\partial \text{X}}}$. The other two pieces of this remain as per Chap. 8; together, these form $\mathcal{G}^{(4)}_{\mu\nu} = 0$: the Temporally Relational form of the vacuum Einstein field equations. Moreover, addition of minimally-coupled matter poses no problem to this construction.

More simply at the level of the action, as fixed up by the consistency conditions imposed at the level of the constraints,

$$\iint_{\sigma} \text{d}^3 x \sqrt{\overline{\mathcal{R}}} \text{ds} \quad \text{is equivalent to the action with integrand } \mathcal{R} + \mathcal{K}_{ij} \mathcal{K}^{ij} - \mathcal{K}^2. \tag{34.22}$$

For $\boldsymbol{\Sigma}$ compact without boundary, this is equivalent to the Einstein–Hilbert action with $\text{g}_{\mu\nu}$ subjected to the TRi split.

Example 2) In support of Chap. 33's Spacetime Construction, the spacetime connection arises in 3-space terms as (e.g. in $\perp$ formulation)

$$\Gamma^{(4)c}{}_{ab} = \Gamma^c{}_{ab}, \qquad \Gamma^{(4)\perp}{}_{ab} = \mathcal{K}_{ab}, \qquad \Gamma^{(4)a}{}_{b\perp} = -\varepsilon \, \mathcal{K}_b{}^a, \tag{34.23}$$

$$\Gamma^{(4)\perp}{}_{\perp b} = 0, \qquad \Gamma^{(4)a}{}_{\perp\perp} = \varepsilon \{\partial_b \, \partial \text{I}\}/\partial \text{I}, \qquad \Gamma^{(4)\perp}{}_{\perp\perp} = 0; \tag{34.24}$$

these can be identified as given and packaged together to form $\Gamma^{(4)\mu}{}_{\nu\rho}$.

34.6 TRi Refoliation Invariance

We here we pass from the usual presentation of the Dirac algebroid (9.31)–(9.33) to the TRi-smeared version (24.18)–(24.20). Consequently, Fig. 24.5.b) replaces Fig. 10.3.b). Figure 34.6 summarizes the facets dealt with in Chaps. 33 and 34.

Chapter 35
Classical-Level Conclusion

35.1 Classical Machian Emergent Time Approach

Part II has concentrated on a *classical Machian Emergent Time Approach*. This resolves Temporal Relationalism's primary-level timelessness by Mach's 'time is to be abstracted from change'. More specifically, it is a 'GLET is to be abstracted from STLRC' realization, where STLRC stands for 'sufficient totality of locally relevant change', and GLET for 'generalized local ephemeris time' (Chaps. 15 and 23). In various settings, this is an emergent version of Newtonian time, proper time and cosmic time. We have argued at the conceptual and technical levels for such a time rather than various other candidates such as scale time, hidden time, matter time and unimodular time. We subsequently provided (Chap. 23) an approximation scheme for the classical Machian emergent time, in which the configuration variables are split into slow heavy 'h' and light fast 'l'. This covers all of 'island universe' models, and Classical and Quantum Cosmology. Moreover, this scheme is only fully Machian once one passes from the zeroth-order emergent times—whose form is $F[h, \mathrm{d}h]$—to at least first-order emergent times, of form $F[h, l, \mathrm{d}h, \mathrm{d}l]$. I.e. in the latter setting the l degrees of freedom are given the opportunity to contribute, which is itself a desirable relational premise.

35.2 Denizens of Each Problem of Time Facet

For the denizens of each facet treated piecemeal, read off the monochromatic objects in the three 'Technicolour Guide' Figs. 24.2, 25.1 and 34.6. Presenting these requires a number of Part II's mathematical developments that go beyond the scope of Part I.

1) More well-adapted and general implementations of Temporally Relational actions $\mathsf{S}_{\mathrm{rel}}$, in particular the geometrical Jacobi(–Synge) formulation (Chaps. 15 and 17).

E. Anderson, *The Problem of Time*, Fundamental Theories of Physics 190,
DOI 10.1007/978-3-319-58848-3_35

2) The $\mathfrak{g}$-Act $\mathfrak{g}$-All Method (Chap. 14) for implementing Configurational Relationalism.
3) Split algebraic structures (Appendix E and Chap. 24) are used in addressing Constraint Closure.
4) Lattices of constraint and beables subalgebraic structures $\mathfrak{L}_\mathfrak{c}$ and $\mathfrak{L}_\mathfrak{b}$ (Appendix S.4 and Chaps. 24, 25.
5) Trial families and how their resultant obstruction terms are handled (Chap. 33 in approaching Spacetime Construction and the recovery of local SR.
6) The Foliation Formulation of split spacetime in more detail, as per Chaps. 31–32.

35.3 Interferences Between Classical Problem of Time Facets. i

Part II's main theme of interference between local Problem of Time facets at the classical level can then be perused by considering the polychromatic objects in Figs. 24.2, 25.1 and 34.6.

One of the main features of this is that adhering to the TRiPoD and TRiFol formulations (Temporal Relationalism implementing Principles of Dynamics and Foliations respectively: Appendix L and Chap. 34) keeps us from re-incurring Temporal Relationalism issues in considering the other facets.

As a first instance of this, implementing Configurational Relationalism by Best Matching with cyclic differentials gives the TRi-compatible Machian form of the GR Thin Sandwich (Chaps. 16 and 18). Figure 35.2.a) depicts Configurational and Temporal Relationalism's 'loop the loop' of Chaps. 14 to 18 (Fig. 24.2). In this way, a geometrical best-matched action $\mathsf{S}_{\mathrm{rel}}$ is produced on a configuration space $\mathfrak{q}$ for which the group $\mathfrak{g}$ of transformations is held to be physically irrelevant.

Temporal and Configurational Relationalism are moreover only mathematically well-defined in cases in which *Constraint Closure* (Chap. 24 and Fig. 24.2) additionally applies. The triple $\langle \mathfrak{q}, \mathfrak{g}, \mathsf{S}_{\mathrm{rel}} \rangle$ is, rather, a candidate 'key' with which to attempt to open the 'gate' of Constraint Closure, in accord with the TRi Dirac-type Algorithm given in Chap. 24. In successful cases, the ***shuffle*** are confirmed to be gauge constraints $\boldsymbol{\mathcal{G}}$**auge**; first-class linear constraints $\boldsymbol{\mathcal{F}}$**lin** are often, but not always, equivalent to $\boldsymbol{\mathcal{G}}$**auge**. For GR, these constraints are $\mathcal{H}$ and $\mathcal{M}_i$ respectively. In this case, Fig. 35.2.a)'s path works out since $\mathcal{H}$ and $\mathcal{M}_i$ close. These furthermore close in the form of the Dirac algebroid, by which they can be treated in a split manner with $\mathcal{M}_i$ but not $\mathcal{H}$ supplying a subalgebraic structure.

Once the consistency check of Constraint Closure is passed, it makes sense to contemplate the further mathematical problem of Assignment of Beables (Fig. 25.1). We argued that Dirac beables $\mathcal{D}$ are the most interesting notion, since these take all the constraints into account. We showed furthermore that the notion of Kuchař beables $\mathcal{K}$—which commute with the $\boldsymbol{\mathcal{F}}$**lin**—only applies to a limited range of theories, due to the $\boldsymbol{\mathcal{F}}$**lin** not universally closing by themselves. This notion is are meaningful for GR. Moreover, this book's model arenas and Supergravity differ in these details, which thus pick up a theory-dependent character. A more general

	rotations	diffeomorphisms	split diffeomorphisms
group	$SO(d)$ is a finite-d compact Lie group	$Diff(\Sigma)$ is an ∞-d Lie group; not compact	$Diff(\mathfrak{m}, \mathfrak{Fol})$ is a groupoid; not compact
infinitesimal form	$\delta_{ij} - \epsilon_{ijk}\theta_k$	$\pounds_\xi$	$\pounds_\xi$, $\mathcal{H}_\perp$
finite form	$R_{ij}(\theta)$	not explicitly available	not explicitly available
corresponding constraint	$\mathcal{L}_i = \sum_I r_I \times p_I$	$\mathcal{M}_i = -2\,D_j p^j{}_i$	$\mathcal{S}\text{huffle}_i = \mathcal{M}_i$, $\mathcal{P}\text{ure} = \mathcal{H}$
group actions	independent of what is acted upon	dependent on which Σ is acted upon	dependent on which Σ and h_{ij} are acted upon
measures	Has a Haar measure available	Only a formal measure	Even less is known
Representation Theory	Has a basic such	Has been worked out in a few subcases	Requires a more general notion of rep; Representation Theory little developed

Fig. 35.1 Features of rotations versus diffeomorphisms and split diffeomorphisms

notion of A-beables A has been argued to replace the notion of Kuchař beables; 'A' here stands for 'algebraic substructure'; A-beables can be defined by association with each closed algebraic substructure of a theory's constraints.

35.4 ii. Supporting Model Arenas

Model arenas exhibit a number of simplifications. These models have helped substantially in working out the full GR case. Indeed, Part II demonstrated how the book's principal model arenas' Background Independence aspects' summary Fig. 10.5 arises.

For metric *Relational Particle Mechanics (RPMs)*, Configurational Relationalism is trivial in the scaled 1-d case and resolved for the pure-shape 1-d [59] and the scaled and pure-shape 2-d [37]. A first consequence is that Temporal Relationalism is resolved for these models at the classical level. Constraint Closure also works out straightforwardly, with **$\mathcal{S}$huffle** = **$\mathcal{G}$auge**. A second consequence is that the $\mathfrak{g}$-beables G (commuting with **$\mathcal{G}$auge**) coincide here with the Kuchař beables K. These are functions of shapes, scale and their conjugate momenta, and so are known and well-understood. Finally, there is no GR-like notion of spacetime. Consequently none of Spacetime Relationalism, Spacetime Construction or Foliation Independence arise. See Fig. 35.1 for a summary of many of the key differences arising from rotations being mathematically simpler than diffeomorphisms.

Example 2) For *Minisuperspace*, homogeneity render trivial the GR momentum constraint $\mathcal{M}_i$. A first consequence of this is that Configurational Relationalism is trivial, so Temporal Relationalism is resolved at the classical level as well. A second consequence is that the only remaining constraint is the Hamiltonian constraint $\mathcal{H}$, and finite theories with just one constraint trivially obey Constraint Closure. A third consequence is that any functions of the coordinates and the momenta will do for $\mathcal{G} = \mathcal{K}$ in Minisuperspace. Finally, using the foliation by spatial hypersurfaces which are privileged by homogeneity renders both Foliation Dependence and Spacetime Construction trivial.

Example 3) *Slightly Inhomogeneous Cosmology (SIC)* exhibits the following useful features (in perturbatively small form): spacetime, foliation, and construction features are nontrivial, as are all diffeomorphism-specific features. This is spelled out in Chaps. 30, 32 and 33. Also in SIC, the constraint algebraic structure has some surprises in store; in particular, the momentum constraint ceases to carry all the $Diff(\boldsymbol{\Sigma})$ information. Thereby, Kuchař beables $\mathcal{K}$ and Superspace beables $\mathcal{SU}$ (this model's $\mathcal{G}$)—are now distinct notions.

35.5 iii. Further Facets in the Case of GR

We subsequently retraced Sect. 35.3's steps for a trial family of theories encoded by $\mathsf{S}_{\text{trial}}$, or, alternatively, by $\mathcal{H}^{\text{trial}}$ with or without $\mathcal{M}_i$ assumed. This larger trial family approach is outlined in Fig. 24.3 and is depicted in Fig. 35.2.b) by use of a bunch of keys (representing Chap. 33's ansatz) rather than an individual key. One of these keys amounts to a recovery of the GR form of $\mathcal{H}$ (the DeWitt supermetric coefficient and enforcement of Ricci 3-scalar potential), alongside recovery of embeddability into spacetime and local recovery of SR (Chap. 33). In this way, Constraint Closure can come hand in hand with Spacetime Construction for geometrodynamical theories.

Spacetime having been recovered, it has $Diff(\mathfrak{m})$-based Spacetime Relationalism as per usual. Finally, this recovered spacetime can be foliated as per usual, except that it is more consistent to use TRiFol, which is kinematically natural within the Relational Approach. Because this paragraph's considerations are independent from those in the last paragraph of Sect. 35.3, the route through the Problem of Time facet 'gates' depicted in Fig. 35.2.e) branches in its latter portion. In this way, we have completed A Local Resolution of the Problem of Time at the classical level. In the case of GR, this is modulo one limitation in current mathematical knowledge and two formal steps as indicated below.

In a wider range of programs, for classical GR, Constraint Closure, Refoliation Invariance and Spacetime Construction all follow from either the Dirac algebroid, or from a family of algebroids which collapses to the Dirac algebroid as one of a small number of consistent possibilities as per Chap. 33.

Research Project 34)† The mathematics of the Dirac algebroid, moreover, remains to be fully developed. Improve on this while continuing to pay attention to its geometrical and physical interpretations. [See e.g. Appendix V.6 for an outline and references both for this and for the mathematics of algebroids more generally.]

Formality 1) The Thin Sandwich—the full GR case of the Best Matching implementation to Configurational Relationalism—is not an explicitly solved problem, though, at least locally, it has good existence and uniqueness properties. In particular, Configurational Relationalism needs to be resolved so as to have a practical, rather than merely formal, expression for the classical Machian emergent time.

Formality 2) The classical Problem of Beables is not in general explicitly resolved either, for sufficiently GR-like theories.

Moreover, particular cases in which Formality 1) for $Diff(\mathbf{\Sigma})$—or its Best Matching generalization for $\mathfrak{g}$ more generally—is resolved, have two further useful features.

A) The Constraint Closure Problem is now resolved by the reduced formulation possessing only one constraint (per space point in the field-theoretic case): the reduced $\widetilde{\mathcal{Q}\text{uad}}$, which in many cases straightforwardly closes with itself.
B) $\mathfrak{g}$-beables G are now automatically available at the classical level.

35.6 Further Orders of Passage Through the Problem of Time's 'Gates'

For comparison, Fig. 35.2.c) depicts the Arnowitt–Deser–Misner (ADM) approach. A spacetime ontology is assumed here. The ADM split of the metric leads to $\mathcal{H}$ and $\mathcal{M}_i$ arising together from the corresponding split of the Einstein–Hilbert action and its subsequent variation. [One could moreover treat $\mathcal{M}_i$ as secondary here since it is an integrability of $\mathcal{H}$; this orders rather than splits this path.] Constraint Closure is now attained in the usual Dirac algebroid form.

Figure 35.2.d) provides further contrast with the Deformation Approach of Chap. 32. This approach also assumes a spacetime ontology, and proceeds to split this with respect to a foliation. This produces the deformation algebroid of constraints (which has the same mathematical form as the Dirac algebroid). One next deduces circumstances under which the generators of this must take the forms of $\mathcal{M}_i$ and $\mathcal{H}$. Finally, since c) and d) are approaches which assume a spacetime ontology, these figures have Spacetime Construction crossed out as unnecessary.

Approaches with (unsplit) spacetime primary ontology (Chap. 27) are tied to spacetime diffeomorphism invariance, Generator Closure in place of Constraint Closure, and observables which commute with the generators. Some such approaches make use of larger diffeomorphism *induced* gauge groups, *Digg*. On the other hand, some *Histories Approaches* (Chap. 28) involve a histories phase space $\mathfrak{Hist}$-$\mathfrak{P}$hase that is considerably larger than the usual Canonical Approach's phase space $\mathfrak{P}$hase. These approaches indeed possess notions of history-dependent spatial diffeomorphisms, histories constraints, and histories observables which histories brackets commute with the histories constraints. Neither of these notions of observables are to be confused with those arising in conventional Canonical Approaches. Moreover, the Bergmann notion of observables has been shown to not only feature in a spacetime-primary setting but also to have subsequent ties to space–time split formulations.

The classical Timeless Approach that we principally focused on is *Timeless Records Theory*. Very detailed explicit examples are available for this by virtue of developments in *Shape Statistics* [539], which arose entirely independently from the Relational and QG literatures.

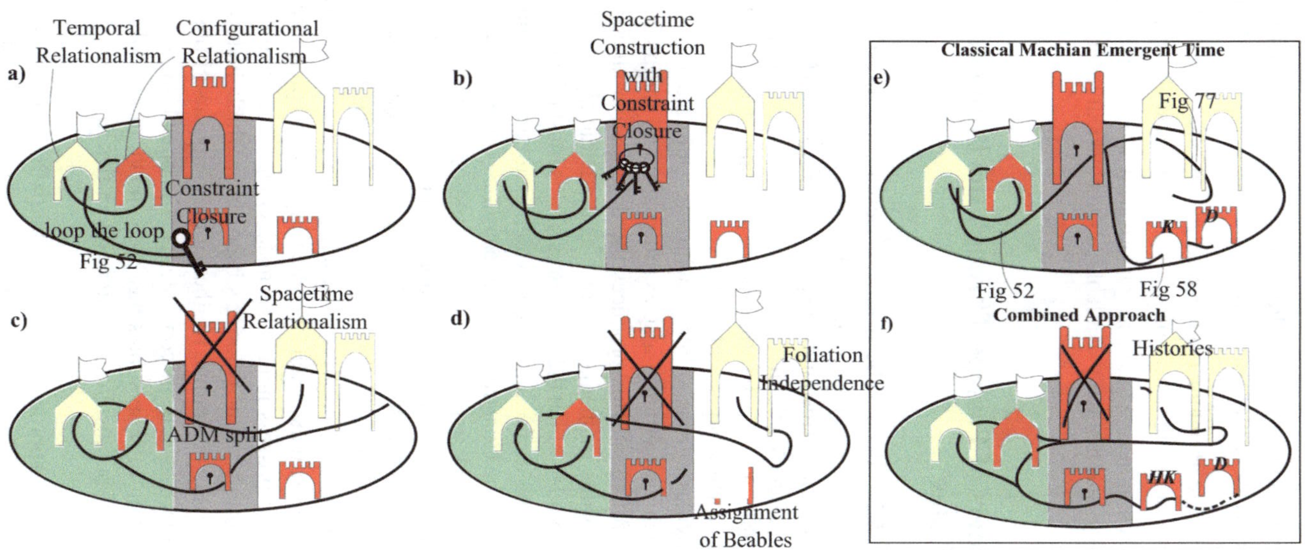

Fig. 35.2 Various orderings through the classical gates/facets of Background Independence [42]. **e)** is the book's recommended path for classical GR. See Sect. 35.6 for interpretation of **a)** to **e)**

The *Combined Approach*—consisting of emergent Machian time, histories and records—(Chap. 29 and Fig. 35.2.f) furthermore provides a construction of Dirac beables from Kuchař beables by *Halliwell's method* [413, 414]. This is depicted as the dashed path in the Figure. The rest of the path depicted, on the other hand, is a histories precursor to the Classical Machian Emergent Time approach. Because this is a histories rather than time-based approach, Temporal Relationalism is crossed out here, and Spacetime Construction is also circumvented. This approach is somewhat spurious at the classical level but we shall see in Part III that it has a descendant which is much more significant at the semiclassical level. While the classical Combined scheme has relatively few interprotections, it is a classical precursor of a semiclassical scheme which makes rather more equable use of each of histories, timeless records and emergent Machian time.

We finally pointed out that for Supergravity, the $\mathcal{F}$**lin** do not close by themselves, so a number of GR's notions, such as $\mathfrak{S}$uperspace($\boldsymbol{\Sigma}$), the possibility of treating $\mathcal{F}$**lin** and $\mathcal{Q}$uad separately, and the use of Kuchař beables, cease to be well-defined. This leaves open the possibility of further theories whose classical manifestation of Background Independence is markedly different form GR's.

35.7 Ties Between Time and Other Concepts

As well as ties between time and space, there are ties between time and configuration, time and energy, time, time, space *and* charge (as in CPT), and time and Supersymmetry. In this manner, studies of time alone are subject to open-endedness. This explains why this book covers space and configuration as well. On the other hand, energy, charge and supersymmetry's inter-connections with the Problem of Time have so far been studied at most in outline. Some Research Projects in this direction are suggested below and in Part III.

Research Project 35)† GR exhibits deep conceptual and technical Problems of Energy [823, 824] as well as Problems of Time. While a few such pairs of problems are known to be inter-related, a systematic joint study remains to be attempted.

Research Project 36)† If fundamental Supersymmetry occurs in Nature, how does this alter theorization about time? [Sections 19.8, 24.10 and 27.8 contain some preliminary considerations; this Project is intended to have further scope than Projects 10), 12), 27), 28), 30) and 33).]

Chapter 36
Epilogue II.A. Threading and Null Formulations

36.1 The 1 + 3 Threading Formulation

Spacetime primality can be envisaged as involving a 4-manifold before a (3, 1) split into space and time. Within (3, 1) primality views, moreover, one can either view space as primary: 3 + 1 approaches, or time as primary: 1 + 3 approaches. In Part II's study so far, the base objects have been spacetimes or spatial hypersurfaces: 3 + 1 approaches. We next consider timelike threads instead (see also Sect. 8.5): a 1 + 3 approach. This *threading formulation*'s analogue of foliation is *filling by a congruence of threads*: a space-filling non-intersecting collection of such threads.

Research Project 37) Some specific types of spatial slice and foliation—such as CMC—turn out to be mathematically tractable, geometrically significant and possibly of physical significance. Do any particular types of threads, or of filling congruences, have comparably fruitful developments?

Given Chap. 28 and Epilogue II.C's Canonical Approaches for a wider range of entities than just the usual configurations, is a Canonical Theory of threads conceptually sound, and if so, what form does this take? Given that the structure of *Diff*($\mathfrak{m}$) is substantially altered by a hypersurface split, what happens under a threading split?

Also assess whether there is a 'Filling Congruence Dependence Problem' analogue of the Foliation Dependence Problem. Is this classically resolved by some 'Rethreading Invariance' analogue of Refoliation Invariance? Furthermore, given Spacetime Construction from a spatial slice, how well-behaved is the analogous 'Spacetime Constitution' from threads as a PDE problem? Finally, to what extent is there threading-foliation duality not only at the level of objects but also at the level of splits of the Einstein field equations?

Research Project 38) The threading formulation gives a further setting for 'path observables' concepts. Investigate in more detail, including comparison with Bergmann and histories observables.

E. Anderson, *The Problem of Time*, Fundamental Theories of Physics 190,
DOI 10.1007/978-3-319-58848-3_36

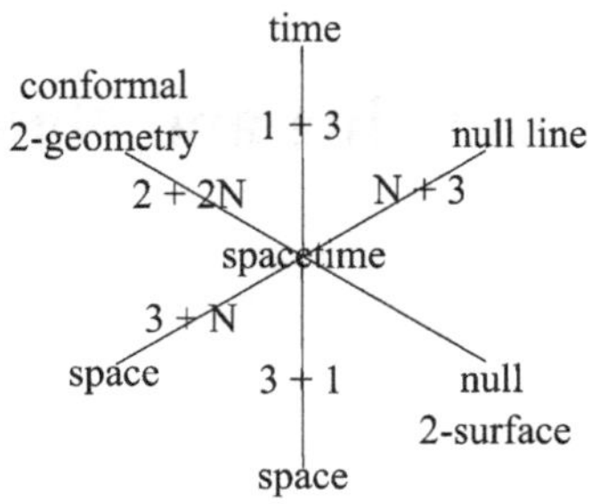

Fig. 36.1 There are 3 types of split depending on whether 0, 1 or 2 null directions are involved. While could be used in place of space or time directions, passing to null directions while leaving the time direction untouched causes additional complications and is not usually attempted. Primality can also be ascribed to each of the six entities in the figure in turn. Each of the six notions depicted as vertices has its own space of spaces, and each edge in the diagram has corresponding maps running in both directions. In these ways, the current Chapter's considerations substantially increase the number of Problem of Time facets. Whereas the 2 instances of space in the diagram are the same object, the associated PDE systems are different in the case of each split

36.2 Characteristic, 2 + 2 and Twistor Formulations

The sense in which 'characteristic' is meant here is explained in footnote 3. There are moreover two types of characteristic formulations for 4-*d* GR spacetime. The first involves one null direction. The second involves two, which form null 2-surfaces; in this case, spatial conformal 2-geometries play the role of configurations. Indexing a foliation by values of a coordinate indeed continues to make sense in the case of a null coordinate; see e.g. [874] for a brief introduction to null congruences. As for space–time splits, each of these approaches have two lobes as regards which part of the split is taken to be primary (Fig. 36.1).

Development 1) Some versions of the above are useful for causal or observer-based considerations, in contrasted with the dynamical considerations modelled by the ADM split. What a given conformal 2-geometry is causally interconnected with is a partial analogue of Geometrodynamics' consideration of which sequence of 3-geometries a given 3-geometry evolve to become.

Development 2) The above involves substantially different PDE Theory from the 3 + 1 split; see e.g. [814] for the characteristic case and [243] for the double-null case. These approaches build on the insight that characteristic directions are different, which can already be understood for simple linear wave equations (consult e.g. [220] if interested).

Research Project 39) Can spacetime be constructed from the assumption of 2-space geometries? Is the 2 + 2 approach to GR approach more heavily dependent on presupposing spacetime than the 3 + 1 approach is?

Development 3) *Twistor Theory* is an approach first developed by Penrose [707]; it has a spacetime primary ontology and is further based on the null split. The null structure is emphasized by taking null cones to be sharply defined—which results in points being quantum-mechanically fuzzy—as opposed to working with sharp points, which results in fuzzy cones. Twistor Theory furthermore involves $\mathbb{C}$ mathematics, formulation in terms of spinors, and Projective Geometry. More specifically, Ex IV.16 gives a useful preliminary indication of how Minkowski spacetime $\mathbb{M}^4$ can be reinterpreted in terms of $\mathbb{C}$ mathematics. Moreover, complex spinorial entities form their own configuration spaces—'twistor spaces', most of which are structures rooted in Projective Geometry.

Twistor Theory furthermore involves split formulation entities of note, such as 'hypersurface twistor spaces'.

Finally, note various parallels between twistor and Ashtekar variables approaches. Among these, we have already mentioned complexified GR, and that the use of self-duality is strongly tied to spacetime dimension 4. A further parallel is that both involve providing new phase space coordinates; moreover, these are nonlocal in the case of twistors, which might therefore be viewed as more radical.

36.3 Summary and Machian Evaluation

Recollect Broad's point that space and time's co-geometrizability does not preclude their being conceptually separate entities. From a Machian position, moreover, it is specifically time that is a derived entity. From this point of view, null directions are *composites* of one entity which is derived and one which is not (space). On the other hand, splitting out time is directly aligned with singling out the derived entity for distinct treatment. Using $3+1$ rather than $1+3$ furthermore builds in the position of time specifically being the derived entity as per Mach's Time Principle. In this way, space–time splits are picked out, as is spatial primality therein. In this manner, we return to this book's main perspective, now from within the much wider set of formalisms covered by Fig. 36.1.

Research Project 40) Elucidate which form temporal properties, Background Independence aspects, and Problem of Time facets take in the Twistor Approach.

Chapter 37
Epilogue II.B. Global Validity and Global Problems of Time

It would be preferable if all of the Background Independence aspects, resultant Problem of Time facets, and strategies to resolve these, were treated in a globally well-defined manner. We are however far from this goal, which involves many—and often distinct—uses of the word 'global', some of which have so far not even been explained in the literature. See Appendix O.1 for an outline of uses of 'global' in Classical Physics in general; on the other hand, the current Chapter develops these for the classical Problem of Time specifically.

'Global Problem of Time' could for instance refer to the global status of a notion of time itself. It could also refer to globality over space, such as the subset of points $t = \text{const}$ not being a complete 3-d submanifold of the spacetime [483]. Other possibilities are globality over $\mathfrak{q}$, $\mathfrak{T}(\mathfrak{q})$, $\mathfrak{P}$hase, spacetime, or some space of spacetimes. Yet further possibilities concern constructs on—or maps between—some of the preceding, for instance paths on $\mathfrak{q}$, embeddings, or foliations. 'Global Problem of Time' could specifically affect a timefunction—*Kuchař's Embarrassment of Poverty* [586]—whether merely due to coordinate restrictions on manifolds or due to more involved and not generally resolved locality of PDE solutions. However, some types of globality in the current section and Appendix O.1 can also affect frames, transformations, PDE solutions playing a role other than timefunctions, validity of beables or observables, foliations, Spacetime Constructability, and more. A further distinction is between the global breakdown of a mathematical entity itself—e.g. a function blowing up, ceasing to be defined or going complex—and the global breakdown of properties of that entity which are required for it to match its proposed physical role. E.g. non-frozenness or monotonicity could be only local for a candidate timefunction, likewise closure for a candidate algebraic structure of constraints, or annihilation by constraints for candidate beables.

All in all, *one should indeed use the plural 'global Problems of Time', for these are legion. Many of the facets, strategies, denizens and facet interferences are afflicted by one or more of these.* The above diversity of Global Problems of Time makes it clear that ignoring global issues places Problem of Time approaches into 'boxes of validity' which are rather smaller than one might naïvely expect. See also

E. Anderson, *The Problem of Time*, Fundamental Theories of Physics 190,
DOI 10.1007/978-3-319-58848-3_37

Epilogue III.B for yet further global Problems of Time at the quantum level. Some strategies for Global Problems of Time are as follows.

Strategy 1) *Globalize by Extension.* Some structures used locally may happen to remain globally valid.

Strategy 2) *Globalize by Replacement.* Some structures used locally may not remain globally valid but can be replaced by ones which are.

Strategy 3) *Globalize by Discarding.* Some structures used locally may be globally meaningless, and thus require discarding entirely in a global treatment. '*Earman's Principle*' [273] can also apply as a distinct reason to discard. This involves qualitative changes in behaviour upon mathematically taking a limit being disregarded if the effect does not occur as the limit is approached and the limit is not physically attainable.

Globalizations by Extension include 'patching together' regions; some such constructs, from Manifold Geometry or Fibre Bundle Theory are standard. However, some physical applications involve more general topological spaces than manifolds (Appendix M), or more general compositions than fibre bundles (see e.g. Sects. 37.5 and 37.6, and Appendices F.4 and W). Patching together of PDE solutions is moreover *not* in general a trivial matter (see Appendix O).

As regards the simpler model arenas' capacities to model Global Problems of Time, globality in time still has meaning in Minisuperspace [523].[1] RPMs additionally manifest some spatial globality issues due to possessing meaningful notions of localization or clumping, and some more due to possessing nontrivial configuration spaces and nontrivial timefunctions. A number of other types of globality span both Finite and Field Theories, as we demonstrate below.

37.1 Classical Emergent Machian Time

Let us first consider this book's main classical approach at the global level.

Problem 1) Representing motions by geodesics on configuration space $\mathfrak{q}$ is prima facie attractive. This works out locally (Chaps. 15 and 18) if one considers the physical geometry $\mathbb{d}\mathscr{J}$. Moreover, this is only really available modulo conformal transformations: a parageodesic principle in terms of the kinematical geometry $\mathbb{d}\mathrm{s}$. This has the advantage of involving just the one geometry rather than a distinct geometry per potential factor $\mathscr{W}$. None the less, basic Differential Geometry gives that such conformal transformations in general alter both the form of the geodesics and of the curvature of $\mathfrak{q}$. One way global inequivalence can arise is that the interrelating conformal factor is only local in the sense of just applying to a finite region on $\mathfrak{q}$. This possibility arises from conformal factors being required to be non-zero,

[1] It helps at this point to not confuse this use of 'local' with 'local degrees of freedom' in the field-theoretic sense. In particular Minisuperspace models do not possess any of the latter while they do possess some of the former.

finite and sufficiently smooth, whereas physical $\mathcal{W}(\mathsf{Q})$ are certainly capable of deviating from these conditions, at least in some regions of $\mathfrak{q}$. The Author terms this '*PoZIN*': *Problem of Zeros, Infinities and Non-Smoothnesses*; see also Fig. 37.1.a) and b).

Moreover, zeros correspond to qualitatively different situations in each of Mechanics and GR. This is due to the positive-definite to indefinite difference in the configuration space metrics.[2] In the positive-definite case, such zeros are 'halting points'. This is by $0 = W = \|\boldsymbol{P}\|_N^2/2 \Rightarrow \boldsymbol{P} = 0$ for $\boldsymbol{N}$ positive-definite.

Example 1) To illustrate the physical relevance of PoZIN, consider the well-known example of avoiding the zeros by staying within Hill's regions [636] for the Earth–Moon–Sun system (Fig. 37.1.e).

Example 2) On the other hand, in Minisuperspace's indefinite case [659] such zeros are 'spurious' rather than halting, since $0 = W = \|\boldsymbol{P}\|_N^2/2 \not\Rightarrow \boldsymbol{P} = 0$ for $\boldsymbol{N}$ indefinite. So generically the motion continues through the zero along $\mathfrak{Riem}(\boldsymbol{\Sigma})$'s null cone. In the case of Minisuperspace, this null cone consists of physically reasonable Bianchi I Kasner universes [659]. It is worth commenting on how this may at first be counter-intuitive given the differences between it and the much more familiar use of indefinite manifolds as spacetimes. In the latter case, the causal theory enforced by Relativity confines the paths followed by massive particles to lie within null cones, and those followed by massless particles to lie on null cones. However, $\mathfrak{Riem}(\boldsymbol{\Sigma})$'s own null cone does not carry connotations of causality, nor of allotting distinct physical interpretation to 'timelike, null and spacelike' intervals. Paths on $\mathfrak{Riem}(\boldsymbol{\Sigma})$ represent evolving 3-metrics, regardless of whether these paths are 'timelike, null or spacelike'. Indeed, such paths here are furthermore free to move between such types, so the preceding argument evoking a particular subcase of this—paths in minisuperspace going null—should raise no eyebrows.

Example 3) In favour of the physical relevance of PoZIN, the Bianchi IX spacetimes go through an infinity of such zeros as one approaches their cosmological singularity. These are a significant class of homogeneous anisotropic solutions due to Belinskii–Khalatnikov–Lifshitz's conjecture [125], by which they may be expected to be typical of GR solutions in this very same setting of 'approaching the cosmological singularity'. It has moreover been argued (see e.g. [183]) that such zeros place stringent local limitations on the use of Jacobi-type actions (local in space and local in configuration space). Such as [826] can in this light be interpreted as proposal for a geodesic patching (Fig. 37.6.a) approach to get around this.

Problem 2) Ensuing emergent time notions (cf. Chap. 15, 18 and 23's) are in general only intended to hold locally in the sense of a finite region, as per Fig. 37.1.a). One reason for this is that these emergent times are based on the relational action principle $\int \mathbf{ds}\sqrt{2\mathcal{W}}$ existing in the $\mathfrak{q}$ region in which they are defined. However,

[2]To be clear, zeros in the potential factor correspond to zeros of the kinetic term due to the quadratic constraint.

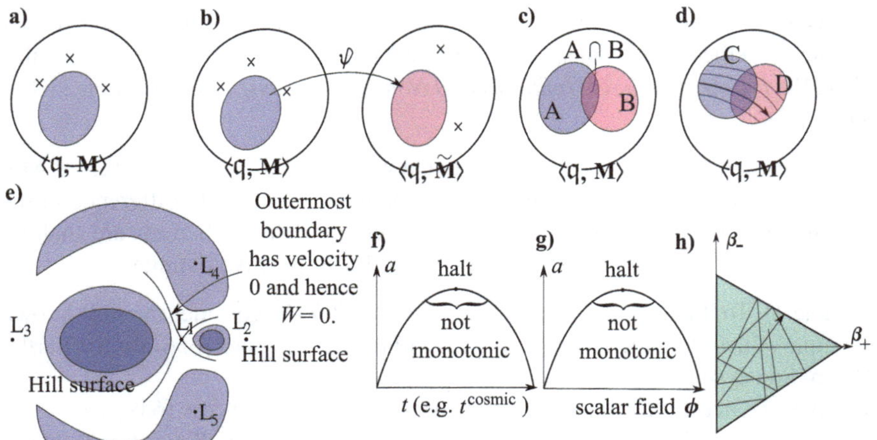

Fig. 37.1 **a)** PoZIN type restriction to a region. The crosses denote points at which zeros, infinities or non-smoothnesses of $\mathscr{W}$ occur. **b)** PoZIN type restriction applying due to the domain of definability of a function thereupon, e.g. for ψ a conformal transformation. **c)** With different approximations used in general having different domains of validity, an intersection in which all of them are valid is a smaller domain. **d)** Patching constructs extend from a region C in which one set of approximations applies to a partly overlapping region D in which some other set of approximations apply; see Sect. 59.2 for a well-known example of this. e) Equipotentials for the Earth–Moon–Sun system, illustrating Hill's regions, as an example of PoZIN being realized in a physically relevant model. The L_i here are the well-known *Lagrange points*. **f)** Closed recollapsing FLRW universe scalefactor; **g)** is a relational version of this. **h)** Bianchi IX's sequence of potential zeros; the straight line segments correspond to Kasner solution 'in' and 'out' states being scattered by the Bianchi IX potential

by immediate inspection, the formula (9.4) for the emergent Machian time candidate is itself directly disrupted by zeros in $\mathscr{W}(\mathsf{Q})$. Blow-ups in $\mathbf{ds}/\sqrt{\mathscr{W}}$—if sufficiently benevolent to permit integration thereover—correspond to blow-ups in t^{em}. Furthermore, infinities in $\mathscr{W}(\mathsf{Q})$ or zeros in $\mathbf{ds}$ correspond to frozenness in the emergent Machian time candidate.

Elsewise, if the t^{em} candidate exists for (a given portion of) a particular motion, it has the desirable property of its monotonicity being guaranteed: $\mathscr{W} > 0$, so $\mathbf{d}\mathrm{t}^{\mathrm{em}} > 0$. Note that this is the Hamilton–Jacobi case of patching together PDE solutions.

Problem 3) Chapter 23's heavy–light split is also in general only local in the sense of a finite region. In particular, consider passing from the familiar Celestial and Molecular Physics domain's use of a heavy–light approximation, with its flat mass metric, to a curved $\mathfrak{q}$ metric. In the latter case, what constitutes heavy and light modes can become a merely local condition both in space and in configuration space. The split's validating approximation can indeed break down over the course of a particular motion. One strategy here is to accept that this split is local and apply it patchwise over each of space and configuration space (Fig. 37.1.d). Moreover,

Chap. 23's other concomitant approximations need not all hold in the same region (Fig. 37.1.c).

Problem 4) Chap. 15's STLRC and GLET notions' locality is in the sense of a finite region in all of time, space and configuration space. Next consider modelling two quasi-isolated island subsystems within a universe. Here Chap. 23's approximations applied to each ensure that the details of the other's contents contribute negligibly. Each subsystem's timestandard constructed as a GLET would then be independent of the other's. Consequently, mechanical ephemeris time type constructions do not in practice by themselves appear to provide a common timestandard. Because of this, a further patching procedure is required. Moreover, the GLET concept has a physically natural means of approximate patching: examination of those changes in the Universe that can be observed by both of the patches' observers. Each has chosen a timestandard that works well for the STLRC they observe. Because of this, both GLETs work well to describe the mutually-observed change. Consequently the two patches' GLETs are reasonably attuned in their overlap.
Pulsars (Sect. 7.7 and Ex V.15) could well often furnish mutually observed subsystems. This may be the best practical answer as regards prescribing the time of meeting in terms of pulsar information verifiable from other Solar Systems or galaxies. Such a patching could also be based on each observer forming a similar notion of cosmic time, albeit this is a substantially less accurate one. Patching together GLETs may become relevant to space programs requiring multiple reference frames over space.
Let us finally note that, whereas STLRC locality bears no logical relation to PoZIN locality, it is closely related to heavy–light split locality.

37.2 Scale Times

Scale times cannot in general be used 'globally in time' in recollapsing universes, since these clearly do not have scale behave monotonically: Fig. 37.1.f).

This monotonicity problem can in some cases be countered by passing to *dilational time candidates*, though see Sects. 37.4–37.5 and 37.13 for global limitations on these as well.

37.3 $\mathfrak{g}$ Nontrivial. i. Monopoles in Configuration Space

Let us begin by considering the Dirac monopole [245] that is more familiar from QFT (Fig. 37.2.a).

2-*d* triangleland with distinguishable mirror images turns out to exhibit the Dirac monopole structure *on configuration space*, $\mathfrak{q} = \mathbb{R}^3$. On the other hand, 3-*d* (or 2-*d* with mirror images identified) triangleland—$\mathfrak{q} = \mathbb{R}^3_+$—exhibits the Iwai monopole behaviour [513] of Fig. 37.2.b). Finally, a distinct monopole also due to Iwai [512]—now in 3-*d* with $\underline{L} \neq 0$ acting as a vectorial analogue of the scalar monopole

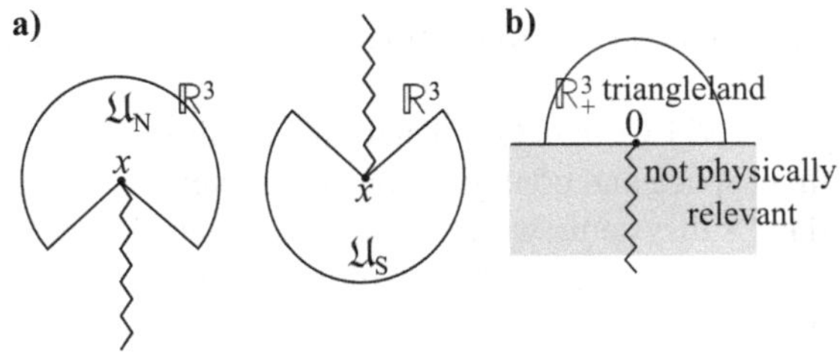

Fig. 37.2 **a)** *Dirac monopole* in $\mathbb{R}^3$. One cannot go all the way around the point x without encountering the Dirac string that emanates from it, though there is complete freedom as regards which direction this lies in. It can therefore be avoided by using a pair of local coordinate charts, as indicated. This then also applies under the independent $\mathbb{R}^3 = \mathfrak{q}$ for triangleland. **b)** The *Iwai monopole* (named after physicist Toshihiro Iwai) on the 3-*d* triangleland problem's configuration space $\mathfrak{q} = \mathbb{R}^3_+$ (half-space), on the other hand, can get by with a single coordinate chart. This is by taking the 'Iwai string' emanating from the origin 0 to lie outside of the physically relevant half-space

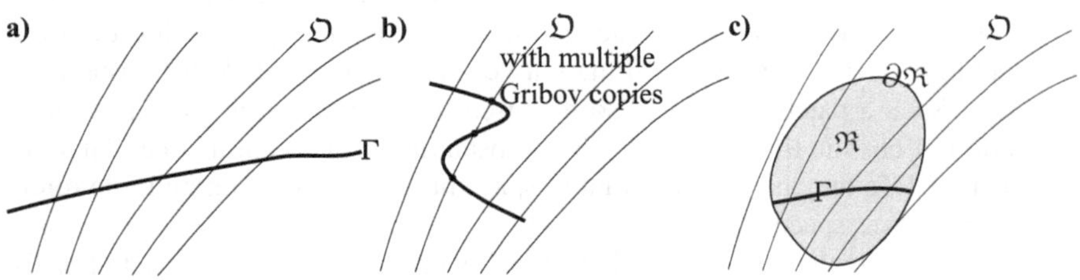

Fig. 37.3 **a)** A global section ('gauge choice') Γ cuts each gauge orbit $\mathfrak{O}$ precisely once. **b)** Extending a local section can on some occasions lead to the inevitable appearance of multiple Gribov copies due to some gauge orbits being cut more than once. **c)** Extending a local section may in any case not be possible outside of some 'Gribov region' $\mathfrak{R}$ bounded by a 'Gribov horizon' $\partial\mathfrak{R}$

charge—occurs in Sect. 16.8's modelling. All these cases are covered by standard Fibre Bundles.

37.4 ii. Gribov Phenomena

The *Gribov phenomenon* [389, 446, 790] (named after physicist Vladimir Gribov) is familiar from Gauge Theoretic QFT. This is, moreover, indeed a classical topological effect. It is an obstruction (Fig. 37.3) to there being a global section to a non-Abelian Gauge Theory's principal fibre bundles [490], which is further characterized by de Rham cohomology. In its original setting, this is unrelated to time, but that ceases to be the case in the applications below.

Problem 1) $\mathfrak{S}$uperspace($\mathbf{\Sigma}$) does not in general possess a global *Diff*($\mathbf{\Sigma}$) section [482].

Problem 2) Candidate notions of 'true dynamical degrees of freedom' require checking whether they are defined throughout both space and time. This can ad-

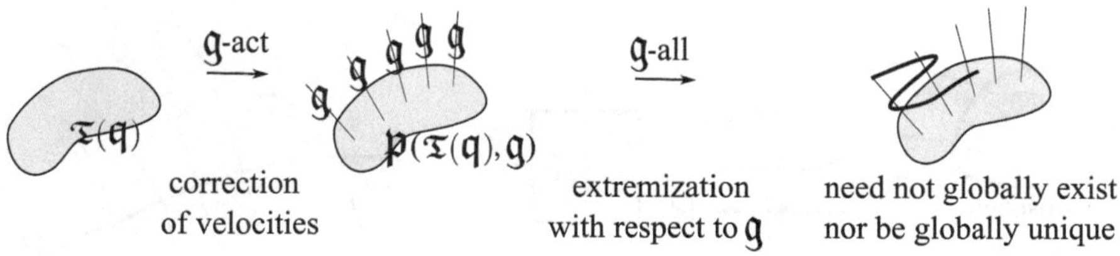

Fig. 37.4 $\mathfrak{g}$-act, $\mathfrak{g}$-all is not in general globally well-defined, for reasons covered by standard Fibre Bundle Theory

ditionally be phrased as whether the canonical transformation that separates these out exists globally in $\mathfrak{P}$hase.

Example 1) There is a Global Problem of Time in the canonical transformation separation into true and embedding (space frame and timefunction) variables in the York time version of the Internal Time Approach. This is known as the *Torre Impasse* [852]. Physicists Petr Hájíček and Jerzy Kijowski furthermore established that such a map is not unique [406].

Problem 3) Best Matching is a case of Calculus of Variations extremization, a procedure which is known to be capable of producing no, or nonunique, answers. This leads to Gribov ambiguities upon formulating its Best Matching subcase in terms of fibre bundles. Finally, in cases for which the collection of objects $\mathfrak{o}$ and the group $\mathfrak{g}$ are both topological manifolds, the Gribov ambiguity in Fig. 37.4 applies quite generally.

Counter-example 2) Problem 3) does not occur for group averaging over a compact group, since that operation is well-defined (using Appendix P.2's Haar measure) and produces a unique answer. In fact, this accounts for why fibre bundle presentations do not enter conventional accounts of group averaging.

Problem 4) Suppose $\mathfrak{g}$-act $\mathfrak{g}$-all is applied to objects O, the space of which, $\mathfrak{o}$, is not a topological manifold. In this case, a global account very likely remains relevant *but lies outside of the scope of standard Fibre Bundles*. See the next two Secs for geometrical reasons for this and for a more advanced approach to it.
Moreover, one can avoid the Torre impasse [852] (Fig. 37.6.c)—or Gribov-type issues more generally – if one uses gauge-invariant quantities rather than working with gauge-dependent quantities, as in e.g. Chap. 16's r-formulation. On the other hand, entering the r-formulation can itself be affected globally (in space or configuration space) e.g. by 1)–3) above; the r-formulation also has to contend with the previous Sec's monopoles.

Problem 5) The spatial diffeomorphisms admit an infinitesimal formulation in terms of the Lie derivative, but not a finite one (contrast e.g. with the rotations).

37.5 iii. Stratification and Its Consequences

Stratified manifolds arise very generally from reduction procedures in Physics (see for instance Chap. 16). Consequently configuration space $\mathfrak{q}$ has issues with singu-

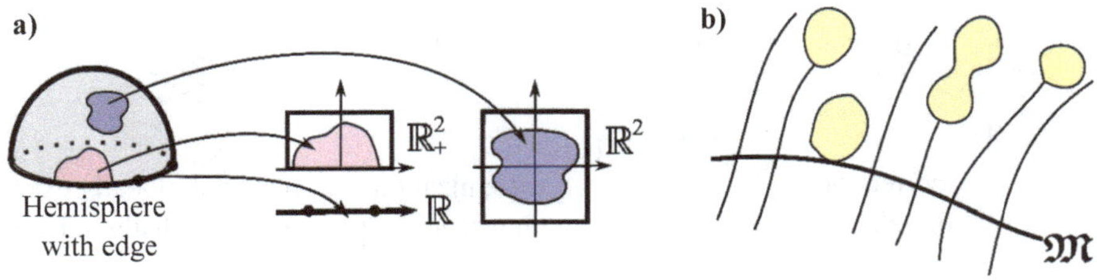

Fig. 37.5 a) This configuration space—a manifold with boundary—has three types of chart. On the other hand, double collision (D) and other collinear (C) configurations have the same isotropy group (defined in Appendix A.2), so the D's do not constitute distinct strata. b) Conceptual depiction of a general bundle or (the beginnings of) a sheaf structure

larities and edges between strata (Appendix M); one should take care not to confuse these with singularities in the potential factor. Stratified examples include the following; see Appendices G and N for more detailed accounts.

Example 1) In 3-d N-body problems' configuration spaces, collinear configurations are represented by a non-principal stratum. This is geometrically distinct because these configurations do not have an invertible inertia tensor.
Example 2) Maximal collisions in N-body problems are more severe than the preceding, both in physical space and at the level of $\mathfrak{q}$ geometry.
Example 3) Gauge group orbit spaces $\mathfrak{O}$ can also exhibit nontrivial stratification.
Example 4) Metrics which possess Killing vectors form non-principal strata within $\mathfrak{S}$uperspace($\boldsymbol{\Sigma}$).

If one sets about studying stratified manifolds using the tools of Differential Geometry, one requires multiple notions of types of chart, as illustrated by Fig. 37.5.a). Three strategies for dealing with stratified manifolds are as follows. This distinction is already modelled by the 3-body problem (Fig. G.11). Our detailed treatment of strategies is restricted to the Hausdorff second-countable case, which is habitual in Mechanics and Geometrodynamics. The remaining issue is the lack of local Euclideanness, with its associated variation in dimension from point to point (already visible in Fig. G.11).

Strategy A) *Excise Strata.* This consists of discarding all bar the principal stratum. While this simplifies the remaining mathematics to handle, it is a crude approximation and an unphysical manoeuvre. This strategy is e.g. often used in the context of removing the collinearities from the 3-d N-body problem.
Strategy B) *Unfold Strata.* Here non-principal strata are unfolded so as to end up possessing the same dimension as the principal stratum. This was considered e.g. by mathematical physicist Arthur Fischer [302]. One may however then question such an unfolding physically meaningful and mathematically unique?
The mathematical advantages of the excise and unfold strategies consist of remaining within Manifold Geometry and Fibre Bundles.
Strategy C) *Accept All Strata.* Prima facie, this is the strategy which is accord with Leibniz's Identity of Indiscernibles. Moreover, this points to harder mathematics

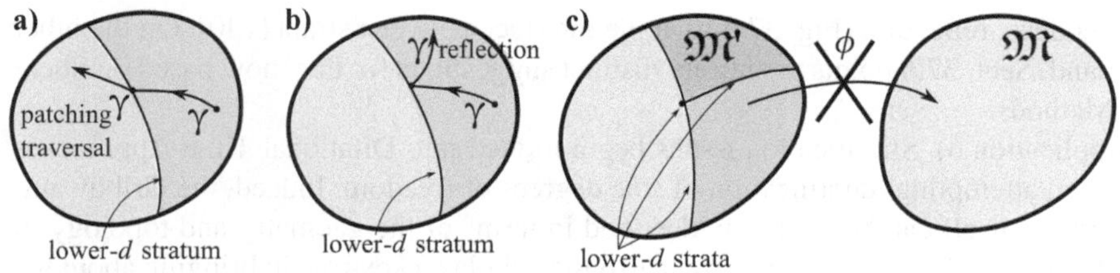

Fig. 37.6 For a geodesic γ running into another stratum, with **a)** traversing (e.g. by patching) and **b)** reflecting possibilities exhibited. **c)** Schematic picture underpinning the Torre impasse [852]: manifolds cannot be diffeomorphic to stratified manifolds

being required: Fibre Bundles do not suffice due to heterogeneity amongst what might have elsewise been fibres. To handle this, one needs at least general bundles [464, 490], and, for a wider range of applications, sheaves. The current book's relational program favours C); further quantum-level reasons for C) are outlined in Sect. 59.5.

Application 1) The above includes a relational argument against Fischer's unfolding construct (Sect. 37.5) to mathematically compensate for metrics which possess Killing vectors.

Application 2) The Bartnik–Fodor Thin Sandwich Theorem ([115] and Appendix O.5) involves two locality (in space and in configuration space) conditions. These involve avoiding

i) potential factor zeros and
ii) metrics with Killing vectors.

Note that ii) means that this theorem is an excision result, due to which it is relationally undesirable.

[In contrast, the conformal initial value problem equations are better behaved globally (in space), as per Appendix O.6.]

Application 3) Both for the Thin Sandwich and for Best Matching more generally, cases which can be solved can, moreover, usually just be solved locally. This e.g. constitutes an argument against the excision of collinear configurations. On the other hand, Best Matching of stratified manifold configurations themselves is, more generally, a stratum by stratum process. The 3-particle model in 3-d, however, does not exhibit this: one can 'spuriously rotate around the axis of collinearity', and so still perform the 'Best Matching' all in one go.

Application 4) Furthermore, the more general $\mathfrak{g}$-act, $\mathfrak{g}$-all method retains Best Matching's global contentiousness [i) above extends to the general Best Matching problem].

Application 5) Stratified configuration spaces $\mathfrak{q}$ readily admit stratum by stratum affine and metric structures [713]. Consequently, it is straightforward to have a local concept of (para)geodesics therein. In modelling Dynamics by a (para)geodesic motion, moreover, one would additionally like to know what occurs to the motion upon its striking a boundary between strata. An early idea was to consider geodesic reflection (Fig. 37.6.b) at the interface between strata [240]. On the other hand, Sect. 37.6 points to entirely distinct suggestions for this, now based on Sheaf Methods.

Application 6) Stratification issues begin to feed into Dilational Time Approaches upon attempting identification of true degrees of freedom. Indeed, the Gribov ambiguity itself can be further understood in terms of the geometry and topology of the principal stratum [759]. Strata also indeed play a key role in bringing about the Torre impasse: the embedding side of (21.15) is a manifold while the configuration space side is not. This is because in the latter case, Killing vectors exist in some places. Finally note that because the Torre impasse is a stratification effect, there is no counterpart of it for the stratification-free 1- and 2-d Metric RPMs.

We next point to the gap in the assumption made so far that unfolding is bereft of physical content.

Strategy D) *Unfold Strata Purely by Enhanced Physical Modelling.*

For example, one could take the point particles are but modelling approximations for more general bodies of finite extent. In this case, alignment of their centres of mass does not alter the isotropy group in question. However, enhanced physical modelling would not be expected to get round how quotienting in general does not preserve local Euclideanness (or Hausdorffness or second-countability). I.e. there is no guarantee that increasing modelling accuracy will be reflected by a successful unfolding of the reduced configuration space stratified manifold into a manifold.

37.6 iv. Sheaf Methods

We now address the breakdown of the scope of Fibre Bundle Methods for global results by considering more general Sheaf Methods. These are rather new within the range of theories and model arenas considered in this book, so we provide a conceptual outline below, and a brief technical outline in Appendix W. *Sheaves are tools for tracking locally defined entities by attachment to open sets within a topological space.* They generalize fibre bundles along the following lines.

1) Whereas the fibres attached to each base space point within a given fibre bundle are all the same, sheaves allow for heterogeneous attached objects (Fig. 37.5.b). Indeed, imagine a variety of shapes and sizes of 'grain' attached to a 'stem', in analogy with a 'sheaf of wheat'.
2) Sheaves admit a generalization of Fibre Bundles' notion of section, which is very useful in global considerations. In particular, obstructions to the existence of global sections can be modelled cohomologically.
3) For some purposes, the simpler notion of *presheaf* suffices. These are based on a mathematical reconceptualization in which restriction maps play a central role; see Appendix W.2 for details.
4) Sheaves themselves offer further global methodology by possessing two additional notions: a 'local to global' gluing and a 'global to local' condition; these are spelled out in Appendix W.3.

Also note the intermediate *Bundle Theory* generalization of Fibre Bundles (Appendix F.4) which admits features 1) and 2) in the absence of reconceptualizations 3) and 4). 4) is furtherly advantageous in establishing a wide range of global applications.

If sheaves (rather than general bundles) are in use, the notion of cohomology associated with 2) is *sheaf cohomology*, as per Appendix W.3. Assembly of local information into global information can proceed via a sheaf cohomology functor.

The above notions can be applied to stratified configuration spaces $\mathfrak{q}$ (Appendices G, M and H) and to the corresponding stratified phase spaces as well [714]. One application is in meshing together the heterogeneous types of charts possessed by a stratified manifold. Another is that sheaves can be used to define metric-level geodesics within stratified manifolds (Appendix M.5). This points toward handling paths that move into boundaries between strata, and thus e.g. to geodesic principles upon stratified manifolds. Thirdly, Kreck's *stratifold* [570] is the following pairing.

i) A particularly well-behaved type of stratified manifold.
ii) A sheaf construct thereover.

See Appendices M.6–M.7 and W.3 for more details, including for how these indeed meet the features of some of this book's configuration spaces. Page 28 of [713] gives a further link between stratified manifolds and sheaves.

Tangent, cotangent, symplectic and Poisson spaces, in each case corresponding to stratified configuration spaces can be studied using Sheaf Methods [714]. These are implicit in the previous Sec's Applications 2)–5), and 7) (see Appendix M.8). Application of Sheaf Methods to gauge orbit spaces $\mathfrak{O}$ has also begun [714].

Research Project 41) To what extent can Sheaf Methods advance our understanding of N-body Problem configuration and phase spaces?
Research Project 42) What about for GR's Thin Sandwich Problem?

Finally note that sheaves would not by themselves be expected to enforce isomorphisms between spaces shown to be inequivalent, such as in the Torre Impasse.

37.7 Brackets and Constraint Closure

Problem 1) The fundamental Poisson bracket's constant right hand side term can be interpreted as an obstruction 2-cocycle [475] whose presence necessitates a central extension. This interpretation adds insight to the subsequent Quantization procedure.
Problem 2) The depiction of the classical differential geometric commutator between two $\mathcal{H}$'s (Fig. 24.5.c) is but in the small. I.e. it is infinitesimally thin in time, and is usually extended (albeit finitely) in space, so that it is overall a local-in-time-and-space slab.
Problem 3) Constraint type can be local in space or in $\mathfrak{P}$hase (see e.g. Sects. 1.1.8 and 19.2 of [446]).

Problem 4) Different points within each of space and $\mathfrak{P}$hase can be associated with distinct constraint algebraic structures [446].

Let us finally note that Problems 3) and 4) lie within the modelling scope of Sheaf Methods.

37.8 Problem of Beables

Problem 1) Beables or observables are often presented as coordinate functions, which are not defined globally on curved manifolds such as $\mathfrak{q}$ or $\mathfrak{P}$hase but rather just in coordinate patches.

Problem 2) Notions of types of beables themselves are themselves in general only construed to hold locally. This is because they are defined by brackets, which—paralleling the previous Section's Problem 2)—in general only hold in a local-in-time-and-space slab. Moreover, writing out the defining brackets now explicitly gives ∂DEs for the beables, which in general only carry local guarantees for solutions. See Appendices O.3 and O.8 for a brief account of equations of this form. All in all, patching beables is at the level of ∂DE solutions rather than just of Differential Geometry.

Let us term observables and beables that are local in time and space, respectively *Fashionables* (as used by Bojowald et al.) [157, 158, 453] and *degradables* [37]. These are fitting nomenclature for local versions of these concepts: 'fashionable in Italy', 'fashionable in the 1960s', 'degradable within a year' and 'degradable outside of the fridge' all make sense. Additionally, fashion is in the eye of the beholder—observer-tied, whereas degradability is a mere matter of being rather than of observing.

Example 1) Patching observables together is very well aligned with the Partial Observables Approach, where the patching is of unrestricted fashionables.

Example 2) Examples of Dirac beables, Kuchař beables and A-beables, for instance in Chap. 24, are in fact mostly Dirac degradables, Kuchař degradables and A-degradables respectively.

Example 3) Dittrich's power series construct depends on the timefunction conjugate to the constraint being well-defined, which is in general only a local criterion. Consequently, this construct does not produce formal Dirac observables or beables, but, rather their local counterpart: formal Dirac fashionables or degradables.

Problem 3) A subset of Partial Observables Approaches are additionally interpreted via internal times. These run into [483] additional global issues typical of Internal Time Approaches.

Problem 4) Following up on Sect. 37.7, different points within each of space and phase space can be associated with distinct algebraic structures of beables (or degradables). This also lies within the modelling scope of Sheaf Methods

(Fig. 37.5.b).[3] An encouraging note here is that modelling observables by sheaves has already occurred in the literature at the quantum level; see Epilogue III.B for an outline.

37.9 Timeless Approaches

Global Problems of Time that correspond to timefunctions breaking down globally can be avoided by adopting a timeless point of view. However, some other global issues remain.

Problem 1) Elements of some Timeless Approaches can only be defined locally (e.g. in space or in configuration space).

Problem 2) Some Timeless Approaches concern localized records: these are theoretically desirable entities, but do they cover all aspects of Physics? E.g. does this perspective fail to encode topological information that is actually physically realized?

Problem 3) Moreover, records are *meant* to be localized in space. Different observers have access to different records; a consistent formulation for this however remains to be checked out.

Problem 4) There are furthermore some practical, physical and mathematical restrictions on questions of becoming. For instance, the problem need to be well-posed, including the S_1 region being extensive enough and set up to be the only significant input to the process leading to S_2. E.g. a sufficient piece S_1 of a past Cauchy surface $\boldsymbol{\Sigma}_1$ is needed to control some future piece S_2 of Cauchy surface $\boldsymbol{\Sigma}_2$. More precisely, that sufficiency is determined by $S_2 \subset D^+(S_1) \cup \boldsymbol{\Sigma}_2$, for $D^+(S)$ the future domain of dependence of set S (Fig. 8.5.b).

Problem 5) Recollect also Fig. 4.4.d)'s indication that instants are a limited modelling feature if one tries to go beyond small regions (of space or of spacetime).

Problem 6) The crucial semblance of dynamics or history may itself be a merely local construct.

Sheaves are well-motivated as regards Records Theory through being tools for tracking locally defined entities by attachment to open sets within a topological space. If all else fails, modelling Problem 3) should lie within the remit of Sheaf Methods.

37.10 Spacetime Relationalism

Issue 1) Since spacetimes are differentiable manifolds, some spacetimes cannot be covered by a single chart.

[3] Indeed, the 'Taking Function Spaces Thereover' conceptualization of observables or beables is phrased intentionally to carry Sheaf Theory connotations.

Problem 2) The point that diffeomorphisms admit an infinitesimal representation in terms of the Lie derivative—already made in the spatial case—clearly transcends to the spacetime case as well. Unfortunately, the lack of understanding of diffeomorphisms globally over manifolds carries over as well.

Problem 3) Spaces of spacetimes such as $\mathfrak{PRiem}(\mathfrak{m})$ and $\mathfrak{S}$uperspacetime($\mathfrak{m}$) have further global and topological issues; see e.g. Research Project 118).

37.11 Histories Theory

Problem 1) Notions of 'local in space' histories, of 'local in histories space $\mathfrak{Hist}$' and of 'local in histories phase space $\mathfrak{Hist}$-$\mathfrak{P}$hase' abound.

Problem 2) Histories constraints and histories observables also have global issues paralleling those in Sects. 37.7–37.8.

By building up histories as sequences of timeless records, they are also well-modelled by sheaves. Modelling of constraint and beables algebraic structures also carries over to their histories-theoretic counterparts.

37.12 Combined Approach

The Combined Approach has the following global issues in excess of the Machian Emergent Time, Histories and Records Approaches' individual global issues.

Problem 1) Its window function is assumed to fit on a single coordinate system, which places its own limitations as regards locality. This is however fine for small regions of $\mathfrak{q}$, such as 'polar caps of approximate equilaterality' on the triangleland $\mathbb{S}^2$. The method in question also continues to work approximately for examples with compact relationalspaces (which include Metric Shape RPMs as per Appendix G.1).

Problem 2) (29.12)—which ends up providing a formula for Dirac beables—requires integration along the whole history ($t = -\infty$ to $+\infty$) rather than just of segments of it. This is since the endpoints of segments contribute right hand side terms to $\{\mathcal{H}, A\}$. One doubt cast over this is that integrals over infinite time intervals are not physically realized. Additionally, since this is not a physically realized limiting case which behaves qualitatively differently, 'Earman's Principle' reinforces this doubt.

Problem 3) Finally, it is clear from Sect. 37.1 that the t^{em} in use in the Machian version of the Combined Approach does not in general run over an infinite interval. This produces further tension in Machian Combined Approaches. This globality in time is also often incompatible in practice with the global nonexistence of e.g. hidden and other emergent timefunctions.

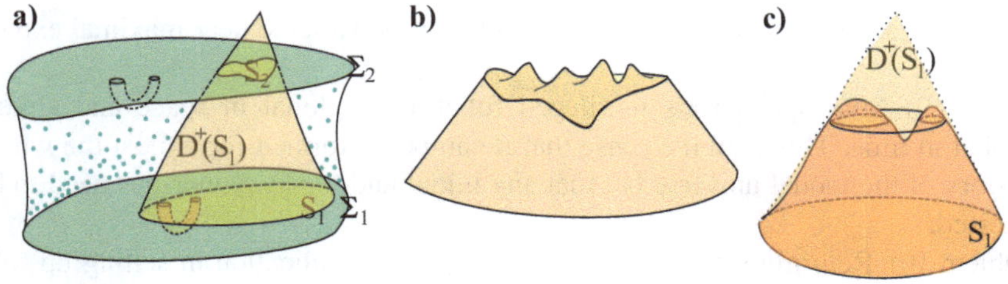

Fig. 37.7 **a)** For a space with handles, there is potentially a problem with handles leaving the domain of dependence (brown). However this book's main specific example has $\mathbf{\Sigma} = \mathbb{S}^3$, so there are no topological manifold nontrivialities, so this complication is avoided. **b)** The Cauchy evolution's spacetime volume may in fact be wiggly due to data on different parts of the initial piece of space differing in how far in time they can be evolved. **c)** So even if $D^+(S_1)$ is a wedge whose spatial topological manifold remains S_1, the capacity of the Cauchy evolution region (*orange*) to be wiggly can cause differences in the spatial topological manifold within the region that the evolution applies. For the example depicted, this differs in the manner of the 'trousers topology' for a region of spacetime

37.13 Space–Time Split, Foliations and Refoliation Invariance

Problem 1) $\mathbf{\Sigma}$ being a compact differentiable manifold means that more than one set of coordinates is required in approaches involving internal spatial functions [483].

Issue 2) The equations (8.4)–(8.5) are the starting point for Embedding Theorems with codimension $C = 1$; knowledge of such equations is not complete.

Problem 3) As Chap. 32 alluded to, Kuchař's version [577–579] of the ADM split is more global than ADM's original.

Problem 4) One is often furthermore interested in the piece of hypersurface S being reasonably regular, e.g. convex. This is a matter of efficiency: it is desirable for the piece to commandeer a reasonable-sized domain of dependence, which it would not if it had substantial concavities. One idea is to be able to locally define constructs in the interior of S, e.g. local h_{ab}, K_{ab}, α, β^a, p^{ab}. Furthermore, local PDE Theory—and construction of local solutions—are simpler and more widely applicable than global counterparts. See Appendix O for standard and global GR Cauchy Problem theorems.

Problems 5) and 6) are outlined in Fig. 37.7.a) and c).

Problem 7) Privileged foliation theories can suffer from shortcomings due to nonexistence of the corresponding foliations, whether in general or in physically significant situations. This problem can involve locality in one or both of space or in the candidate time itself.

Problem 8) Approaches making use of a particular kind of foliations can also run into global limitations. In the specific case of the York time candidate, there is no global-in-space or global-in-time guarantee of existence of a single CMC slice, of foliability of a region by such, or for the extension of this to globally cover the spacetime. [114, 203, 488, 643, 733] provide some affirmative results, whereas [114, 488, 733] contain some no-go results; see also the reviews [271, 467]. Moreover, if a global CMC foliation exists, it is unique [367]. One can also attempt to

Patch one's way out by passing to some other time variable near maximal expansion.

Problem 9) Minisuperspace's privileged foliation is global in space and almost global in time. This is in the sense that it can be performed for almost the whole history of the model universe but that any initial and final singularities need to be excised.

Problem 10) Returning to Issue 2), and its repeated application in setting up foliations, note that not all manifolds are embeddable with codimension $C = 1$. E.g. $\mathbb{RP}^2$ is not embeddable in $\mathbb{R}^3$ [614]. Occasionally this can be resolved by increasing C, or with greater generality in the higher-d space. Moreover, Global Embedding Theorems such as those of mathematician John Nash and of Whitney involve much harder mathematics and much larger lower bounds on C [387] than local ones do.

Problem 11) Teitelboim's demonstration (Fig. 10.3.b) of Refoliation Invariance is limited to lie within the aforementioned type of local-in-time-and-space slab. Furthermore, it depends on the uniqueness part of the *thick* sandwich conjecture.

Research Project 43) To what extent can Sheaf Methods advance our understanding of foliations that arise in GR-like theories?

37.14 Spacetime Constructability

Problem 1) For the principal Lorentzian $x = 1$ branch of Spacetime Construction, one is bounded to work within a localized sandcastle-shaped piece. This reflects the archetype of how solutions to Cauchy problems are only guaranteed locally in each of space and time [204, 874]. The sloping sides of this bounding region are dictated by causality (Fig. 8.5); how tall the sandcastle is, including point-by-point in space, is determined by GR's evolutionary PDEs themselves. Moreover, the above two matters are sequential: if S is a region of ill behaviour, there is limited scope for checking its chronological future $I^+(\mathrm{S})$ (cf. GR Cauchy Problem considerations in the preceding Sec). This case is further reinforced by suitable Analysis along the lines of Appendix O.7.

Problem 2) In the Strong Gravity branch of Space-time Construction, a simpler kind of Analysis suffices. This is because this case merely involves ODEs pointwise in space [716, 717]; see Appendix O.2 for protective theorems. Another consequence of this problem being pointwise in space is that the domain of dependence wedge is replaced by a solid tube. This consists of the worldlines of the points within a region of a slice S; each line within this is its own domain of dependence. Moreover, some of these lines may extend further than others, so this tube may in practice have a wiggly leading edge, much as Fig. 8.5.a)'s 'sandcastle' does.

Problem 3) Let us finally consider the Galileo–Riemann geometrostatics branch of Spacetime Construction. Since this model has no dynamics of geometry, any supporting Analysis involves equations containing only spatial derivatives. It is second-order in these, by which it is an elliptic-type problem. Note that this is unrelated to the space-time structure of the theory, which is just a stack of copies of the same 3-space: a fixed foliation the leaves of which are all equal. Moreover, given the global existence of the geometry that constitutes one copy of the leaf in question, nothing stops this simple construct extending ad infinitum in the time direction. Finally, infinite propagation speed renders the notion of domain of dependence trivial in this context (a matter already considered in Fig. 4.4.f).

Research Project 44)† Further the understanding of global Cauchy Problems in GR and in Gravitational Theory more generally. In the event of sufficiently mastering this work, applications may include i) conferring globality to Spacetime Construction, and ii) obtaining further Global Embedding Theorems.

Chapter 38
Epilogue II.C. Background Independence and Problem of Time at Deeper Levels of Structure

This is motivated [186] by Riemann's classical declaration *"Now it seems that the empirical notions on which the metrical determinations of space are founded, the notion of a solid body and of a ray of light, cease to be valid for the infinitely small. We are therefore quite at liberty to suppose that the metric relations of space in the infinitely small do not conform to the hypotheses of geometry; and we ought in fact to suppose it, if we can there by obtain a simpler explanation of phenomena."* [735] Isham [497] furthermore pointed out that Quantum Gravity is likely to require a descent of this kind. "*Nonwithstanding the current popularity of differential geometry, my strong belief is that its days are numbered, at least so far as the subject of quantum gravity is concerned. Smooth manifolds and local differential equations belong primarily to the world of classical physics and we do not believe that these are appropriate tools with which to probe the structure of spacetime (in so far as this is a meaningful concept at all) near the Planck length. At best, they are likely to be applicable in the semiclassical limit of the quantum theory of gravity (whatever that may be) and a lot more thought needs to be given to the question of which mathematical structures are really relevant for discussing the concepts of space and/or time in the "deep" quantum region.*" Thereby, Isham is not just viewing Fig. 38.1's levels of structure as the mathematics underlying physical theories (corresponding to the 'Equipped Sets' Foundational System of Mathematics) but furthermore as a sequence of structures to quantize in turn. The idea is to descend along these levels of structure; this includes making Chap. 10's distinction between 'single-floor' and 'tower' schemes. The current Chapter itself provides a *classical prequel* [43] to Isham's pioneering quantum investigations of some of the deeper levels of mathematical structure [260, 480–482, 491–494, 496–498, 508, 509] which we outline in Epilogue III.C. In this way, one passes from Wheeler's exhortation [899] to study the Superspace of Geometrodynamics to a much wider range of studies of generalized configuration spaces.

E. Anderson, *The Problem of Time*, Fundamental Theories of Physics 190,
DOI 10.1007/978-3-319-58848-3_38

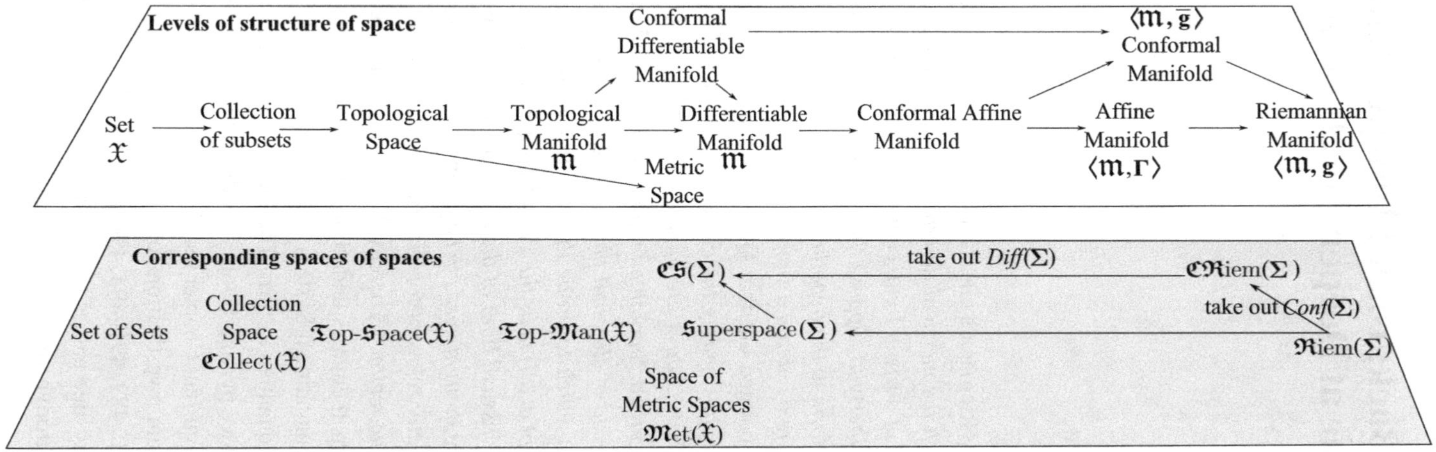

Fig. 38.1 **a)** Levels of mathematical structure commonly assumed in Classical Physics. This Chapter contemplates each of these being dynamical in turn. **b)** The spatial case version of **a)**'s corresponding spaces of spaces (omitting the spaces of affine structures)

38.1 Time, Background Independence and Problem of Time upon Descent. i. Persistent Features

First suppose that time is to be reparametrizable not only by $t' = A + Bt$ but also by any transformation that respects monotonicity. By use of continuous functions, moreover, monotonicity remains a meaningful notion down to the topological space level. Below that, monotonicity remains possible in the sense of the surviving ordering property of time.

Generalized notions of configuration Q, configuration space $\mathfrak{q}$ and change of configuration continue to make sense at all levels. The first and third of these suffice for all levels to possess a notion of action $\mathsf{S}[\mathsf{Q}, \mathrm{d}\mathsf{Q}]$. However, sufficient continuity and differentiability is required to define the usual kind of action's integral and for the standard Calculus of Variations to subsequently apply. These cease respectively at the topological space level and somewhere around[1] the differentiable manifold level. This does not however stop actions from existing or from serving as the starting point for physical calculations. For instance, defining an action sufficiently far down the levels involves a discrete sum rather than a continuous integral. One also requires a discrete counterpart of the definition of momentum and of the Calculus of Variations in order to operate beyond the level at which differentiability ceases. Subsequently, there are discrete analogues of the Euler–Lagrange equations and so on. (As some indication of the plausibility of such schemes, see e.g. the treatment by Dittrich alongside physicist Philipp Hoehn [253] for a carefully studied example of discrete Calculus of Variations including treatment of constrained systems.)

Temporal Relationalism One can consider an absence of extraneous times or time-like variables at all levels, and likewise as regards label times being meaningless. Pick an action that complies with this, and subsequently use one's notion of Calculus of Variations to procure a notion of generalized momentum. Alternatively, in Isham's categorical version [492–494, 498], generalized configurations are objects and generalized momenta the corresponding arrows alias morphisms. In such settings, a notion of generalized constraint as a relation between momenta continues to make sense, and Dirac's argument applied to Manifest Parametrization Irrelevance enforces at least one primary constraint continues to apply as well. In this way, a generalized constraint *Chronos* arises.

Since change of configuration remains available at each level of mathematical structure, a Mach's Time Principle resolution of primary-level timelessness can be based upon this, including more specifically as a STLRC implementation. Rearrangement of the generalized *Chronos* gives an expression for the emergent classical Machian time.

[1]This is left imprecise due to, for instance, stratified differentiable manifolds retaining differentiability in some neighbourhoods.

Configurational Relationalism The notion of a group of physically irrelevant transformations $\mathfrak{g}$ also makes sense at all levels of mathematical structure. In particular this could be (some subgroup of) the automorphism group of the level in question's generalized notion of space, *Aut*(NoS). Next, one passes to the corresponding quotient, e.g.

$$\widetilde{\mathfrak{q}} = \mathfrak{q}/Aut(\text{NoS}) \tag{38.1}$$

in the case involving the full $Aut(\mathfrak{q})$; compare with Eq. (8.25). Many more examples can be read off Fig. 38.2. Note that while Isham stated that the Problem of Time mostly concerns diffeomorphisms [483], this lies implicitly within the context of metric to differentiable structure level. (38.1) can indeed also be taken to be a generalization of the metric Shape Theory arising in the works of Kendall and of Barbour, and of its GR analogue.

The indirect $\mathfrak{g}$-act $\mathfrak{g}$-all implementation also extends to all levels. Let us first consider this in the single-floor context, as an extended use of the form of Sect. 14.4. This generality is based, on the one hand, the that of group actions on spaces of objects, $\mathfrak{O}$. Moreover, this generality is contingent on compatibility not yet being an issue; this awaits, rather, consideration of Constraint Closure. On the other hand, whereas not all types of $\mathfrak{g}$-all prescription cover all cases, each is very extensive in scope, and these cover at least all of the standard levels of mathematical structure used in Theoretical Physics. For instance, summing and scalar multiplication operations work in linear spaces such as vector spaces and modules, integration in measurable spaces, and inf and sup apply e.g. to normed spaces and metric spaces. Averages work out e.g. if $\mathfrak{g}$ is finite or a finite Lie group; these are 'normalizability' criteria, in the sense familiar from Probability Theory or Quantum Theory.

Secondly, in the case of towers, make use of

$$\mathsf{S}_{g_1 \in \mathfrak{g}_1} \mathsf{S}_{g_2 \text{ in each } \mathfrak{g}_2 \text{ corresponding to } g_1} \circ Maps \circ \vec{\mathfrak{g}}_{g_2} \circ Maps \ \vec{\mathfrak{g}}_{g_1} \ O. \tag{38.2}$$

See e.g. Sect. 38.3 for a more commonly encountered example of this.

As regards types of $\mathsf{S}_{g \in \mathfrak{g}}$, as one descends the levels, one needs to restrict to discrete versions: sum, discrete average, inf and sup. Principles of Dynamics actions constructs give rise to ***shuffle*** constraints in association with the Best Matching subcase of the preceding; in this way, ***shuffle*** is in general associated with *Aut*(NoS).

Next allot corresponding generalized classical brackets. Moreover, it is even more straightforward to envisage that, once at the quantum level, commutator brackets persevere.

Constraint Closure thus ultimately transcend all the way down the levels as well, based on a Dirac-type Algorithm and forming a constraint algebraic structure $\mathfrak{C}$. *Taking Function Spaces Thereover* and constrained *Assignment of Beables* forming zero brackets with constraints continue to make sense. These beables once again form an algebraic structure $\mathfrak{b}$, as per Chap. 24.

The linear–quadratic (or more generally linear–nonlinear) constraint distinction may however no longer apply, without which the intermediate notion of Kuchař

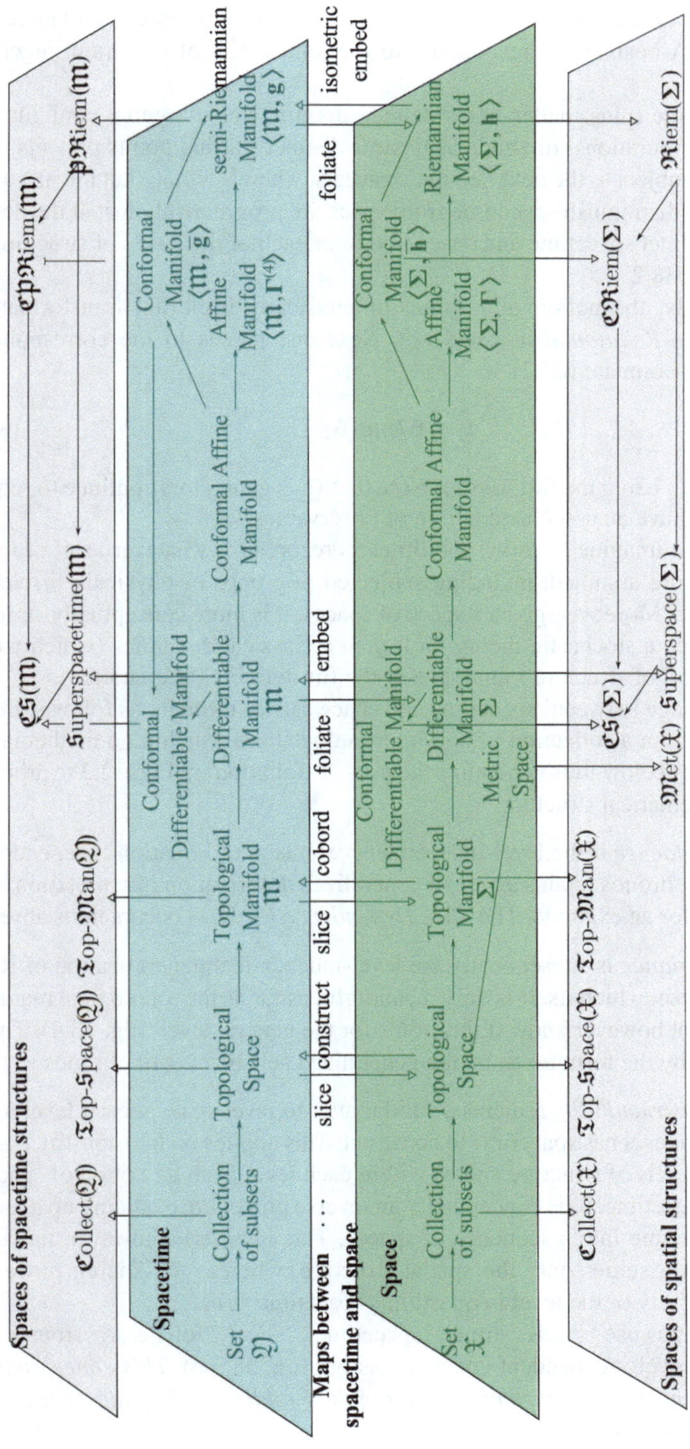

Fig. 38.2 Specific spacetime and space versions of the levels of mathematical structure, interlinked with maps and including corresponding spaces of spaces. Note in particular the advent of stratified manifolds

beables ceases to make sense. In general, what a theory possesses is a lattice $\mathfrak{L}_\mathfrak{b}$ of notions of A-beables, corresponding to the lattice $\mathfrak{L}_\mathfrak{c}$ of constraint algebraic substructures.

The spacetime (plus matter fields defined upon it) versus spatial configuration (plus field configurations) dilemma as regards distinct starting points prevails to all levels. This is subject to the next Section's caveats, whereby some (but not all) of the features which distinguish spacetime from space are progressively lost at the deeper levels. The distinct spacetime and space floors for each of the levels of structure are laid out in Fig. 38.2.

Consequently, the notion of a group of physically irrelevant transformations $\mathfrak{g}_\mathrm{S}$—*Spacetime Relationalism*—survives. Next one passes to the corresponding quotient, e.g.—compare (27.2)—

$$\widehat{\mathfrak{S}} = \mathfrak{S}/Aut(\mathfrak{S}) \tag{38.3}$$

in the case of $\mathfrak{g}_\mathrm{S}$ being the full $Aut(\mathfrak{S})$. Note that $\mathfrak{g}_\mathrm{S}$'s generators continue to require closure and to have an associated notion of observables.

One can also imagine histories and timeless records for whatever level of mathematical structure assumed, including subjected to groups of physically irrelevant transformations. Moreover, given a space of spaces, it is more conceptually straightforward to place a stochastic theory on it than a classical dynamics (which would require that level of structure's analogue of the Einstein Field Equations!).

'2-way passage between' spacetime and space carries through, as follows. Let us use 'slice' to mean a collection of non-intersecting slices which fill a mathematical space; the terms below thus generalize the uses of 'foliation' in Fig. 12.3 to arbitrary levels of mathematical structure.

Slicing Independence is the level-independent extension of Foliation Independence. In this setting, $\mathcal{C}$hronos retains a type of generalized deformation interpretation, $\mathcal{D}$ef; see Sect. 38.3 for an example. The *Slice Dependence Problem* occurs in its absence.

Reslicing Invariance is subsequently the level-independent generalization of Refoliation Invariance, which itself is only meaningful as far as the topological manifold level. We do not however know if this holds for the general level (Fig. 38.4). This is to be resolved by the form taken by the generalized self-bracket of $\mathcal{C}$hronos $= \mathcal{D}$ef.

Spacetime Constructability remains a valid matter to investigate at each level to the extent that each level has spacetime to construct. This applies both to construction by a) increasing levels of structure and b) within each level from its notion of space to its notion of spacetime. b) is conceptually an inverse procedure to slicing up a level's notion of spacetime into a sequence of spaces. This inverse is, however, harder to handle since it assumes only the spatial structure, whereas the slicing move can assume the entirety of the level in question's spacetime structure.

All in all, let us use 'space', 'time', 'spacetime', 'slice', 'foliate', 'surround' and 'construct' as level-independent concepts (as per Fig. 38.3.b). *This conceptualization indeed points to many further versions of the Problem of Time facets at each of these levels of structure.*

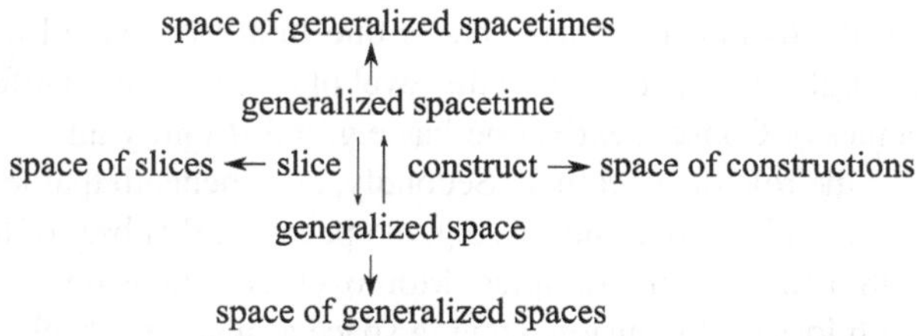

Fig. 38.3 The general level upgrade of Fig. 9.1.c)

Globally Validity and Operationally Meaningfulness Finally, these are desirable for timefunctions and all other steps involved in the above exposition, at any level of structure.

Research Project 45) How far down the levels of mathematical structure do fermions—or the more general anyons [813]—remain meaningful? This is, moreover, less of an issue in tower schemes, since one can have the required fermions upstairs *and* allow more than the usual levels to quantum-mechanically fluctuate.

38.2 ii. Losses in Earlier Stages of Descent

Loss 1) Newtonian Theory involves spatial and temporal metric Background Dependence, whereas SR involves spacetime metric Background Dependence.

Loss 2) *Rot*(d), the Euclidean group *Eucl*(d), the Lorentz group $SO(d, 1)$ and the Poincaré group *Poin*(d) are metric-level structures since they are *isometry* morphisms.

Loss 3) Geodesics are a metric—*or* affine-level structure. Consequently, the notion of (para)geodesic principle entering formulations of Temporal Relationalism is not expected to remain meaningful below these levels (or their conformal counterparts).

Loss 4) The metric level's motivation to keep actions at most quadratic may cease to apply at the lower levels (though the means of passing to quadratic formulations continue to exist withing discrete counterparts). This has the knock-on effect of chronos not necessarily retaining the GR $\mathcal{H}$'s quadraticity.

Loss 5) Spacetime signature enters a number of aspects of SR and GR, including allotting GR coordinate time. Differentiable and affine manifolds—without a metric defined thereupon—do not carry space–time distinction or relativistic notions of causality. The latter is required if one is to consider whichever of the following.

a) Time-orientability.
b) Closed timelike curves (including discarding spacetimes containing such).
c) Foliations with respect to spacelike surfaces that represent sequences of instants of time as experienced by arbitrarily moving fleets of observers. In fact, signature-dependent differences between spacetime and space are lost beneath the level of conformal metric structure.

However, some distinction remains. This is due firstly to overall dimension remaining a meaningful concept down to the level of topological manifolds (by their local Euclideanness). Consequently, one has e.g. a 3-*d* entity and a 4-*d* entity of the same type in the role of spacetime. Secondly, even beneath that level, there are still distinct larger and smaller entities, e.g. a 'space slice' subset of a 'spacetime' set. See Sect. 38.5 for further examples. Moreover, codimension $C = 1$ becomes meaningless with loss of dimension, leaving space as a strict subset of spacetime.

Loss 6) We already know from Chap. 27 that the three Relationalisms descend as far as the differential-geometric level due to being based upon the Lie derivative. In fact both the availability of a Lie derivative and this argument transcend to the level of Conformal Differential Geometry, but no further. Beyond there, a different type of implementation is required.

38.3 Topological Manifold Level

The range of model arenas in Sect. 10.10, and Chaps. 11, 14 and 19 is extensive enough for the rest of this book, but requires complementing by the rest of the current Chapter as regards probing deeper layers of mathematical structure.

In the topological manifold level variant of the cube of theories, the analogue of QFT is *Topological Field Theory* (TFT). In particular, a metric-free version of this is Chern–Simons Theory [916],

$$\mathsf{S} \propto \int_{\mathfrak{m}} \mathrm{Tr}\Big(\mathrm{A} \wedge \mathrm{dA} + \frac{2}{3}\mathrm{A} \wedge \mathrm{A} \wedge \mathrm{A}\Big), \tag{38.4}$$

where A is a 1-form field, d is an exterior derivative and $\mathfrak{m}$ is a $(2+1)$-*d* manifold. This has often been studied as a natural second variant following on from Conformal Field Theory (CFT).

Extending GR by considering change of spatial topological manifold began with Wheeler's envisaging of spacetime foam [897–899]. Indeed, considering not only the metric level but also topological manifold level Background Independence originates in these works. *This now involves questioning fixed spatial topological manifolds as fixed background structures.* In incorporating topology change into Geometrodynamics, singular metrics need to be included and one considers the Lorentzian-signature version of cobordisms (Appendix S.2) between spatial topologies. See e.g. [161] in this regard, and Appendix S.2 for simpler model arena versions of this (variable particle number RPM and 2-*d* geometry).

The distinction between single-floor topological manifold structures and the topological-and-differentiable manifold structures is superbly covered by two of mathematician John Lee's books, [613] and [614] respectively. The latter include e.g. de Rham cohomology (Appendix F.3): Algebraic Topology specific to differentiable manifolds. The considerations of hypersurfaces, embeddings and foliations in Chaps. 8 and 31 also come in purely topological manifold and tower versions.

Finally, the Hodge-* (Appendix F.2) is an example of a structure with metric and differentiable structure level inputs.

Some Background Independence issues at the topological manifold level are as follows.

1) The fixed spatial topological manifold $\boldsymbol{\Sigma}$ that pervades Geometrodynamics and Nododynamics looks to be an undesirable absolute structure.
2) Cobordisms—outlined in Appendix S.2—are a means of matching or comparing distinct topological manifolds; these involve 'ripping' operations rather than just continuous maps. One can now consider cobordism in terms of a parameter as a Manifestly Reparametrization Invariant implementation of *Temporal Relationalism*, and pass furthermore to a Manifestly Parametrization Irrelevant formulation. The action is

$$\sum_{\boldsymbol{\Sigma}} \mathrm{d}S(\boldsymbol{\Sigma}, \mathrm{d}\boldsymbol{\Sigma})$$

for a single-floor theory of topological manifolds alone, or

$$\sum_{\boldsymbol{\Sigma}} \int_{\mathfrak{m} \in \mathfrak{Riem}(\boldsymbol{\Sigma}) \text{ for each } \tau} \mathrm{d}S(\boldsymbol{\Sigma}, \mathbf{m}, \mathrm{d}\boldsymbol{\Sigma}, \mathrm{d}\mathbf{m})$$

for a tower thereupon;[2] here the $\boldsymbol{\tau}$ are variables at the topological manifolds level and $\mathbf{m}$ are at the Metric Geometry level. $\sum_{\boldsymbol{\Sigma}}$ is a usually called a 'sum over all topologies', though this really means 'topological manifolds', and it is either a merely formal sum or a sum over *a subset of* topological manifolds. Restrictions in what is summed over are indeed commonplace, with Appendix S.2 listing five such.
3) Appendix S.2 also outlines what little is known about the single-floor configuration spaces. The corresponding tower configuration spaces are of the form

$$\mathfrak{BigRiem} = \prod_{\boldsymbol{\Sigma}} \mathfrak{Riem}(\boldsymbol{\Sigma}), \tag{38.5}$$

usually subject to the fixed dimension restriction, and to further restrictions on $\boldsymbol{\Sigma}$ as well. A major technical complication with considering multiple topological manifolds dynamically is that transitions between them involve singular spaces. Thereby, one would like to take $\mathfrak{BigRiem}$ to be not only a collection of the usual $\mathfrak{Riem}(\boldsymbol{\Sigma})$ but also including less mathematically tractable spaces.
4) Returning to Temporal Relationalism, the τ (and m, if present) have opportunity to enter the changes of relevance to an emergent timestandard.
5) *Configurational Relationalism.* The Best Matching implementation is restricted to be differentiable manifold level construct due to its use of the Lie derivative. However, some of the concepts involved—such as the analogy between

[2]See [55] for limitations on tower versions of such formulations of Temporal Relationalism.

sandwiches and quantum path integrals and the idea of Best Matching pairs of configurations—continue to apply at the topological manifold level and below.

6) The $\mathfrak{g}$-act, $\mathfrak{g}$-all implementation, moreover, works at all levels of mathematical structure. Use

$$\mathbb{S}_{\tau \in \mathfrak{g}_{\mathrm{T}}} \circ \mathit{Maps} \circ \vec{\mathfrak{g}}_{\mathrm{T}}\, \mathsf{O} \tag{38.6}$$

for a single-floor theory of topological manifolds alone, or

$$\mathbb{S}_{\tau \in \mathfrak{g}_{\mathrm{T}}} \mathbb{S}_{g \in \mathfrak{g}} \circ \mathit{Maps} \circ \vec{\mathfrak{g}} \circ \mathit{Maps}\ \vec{\mathfrak{g}}_{\mathrm{T}}\, \mathsf{O} \tag{38.7}$$

for a tower. In each case, $\mathbb{S}_{\tau \in \mathfrak{g}_{\mathrm{T}}}$ could once again be a (perhaps formal) sum over some topological manifolds, or an extremization or taking an inf or sup. However, it is far from clear whether one can meaningfully assume that the Universe has three holes were the action for a three-holed universe *somewhat* larger than that for any other number of holes! None the less, Epilogue III.C gives a standard quantum-level amelioration of this issue.

Moreover, in 2) and 6)'s indirect formulations, given each particular topological manifold, it is straightforward enough to consider metric structure—or action by $\mathfrak{g}$—thereupon.

7) Let us next consider treating the single-floor case, now subject to its own non-trivial $\mathfrak{g} = \mathit{Homeo}(\boldsymbol{\Sigma})$, so that one is considering the topological manifolds up to homeomorphism,

$$\boldsymbol{\Sigma}/\mathit{Homeo}(\boldsymbol{\Sigma}). \tag{38.8}$$

This has its own '2-level formula' in the sense that it is now topological manifolds up to homeomorphism class that are subjected to 'rippings'.

8) Appendix S.2's restriction iv)—restriction to $\boldsymbol{\Sigma}$ compact without boundary—has Machian underpinnings of a traditional type going back to Einstein. On the other hand, Fig. S.1.d-e) illustrates that there is no observational basis for assuming that the Universe's spatial topology to be open or closed. Background Independence is, moreover, more widely suggestive of seeing what happens if one lifts Appendix S.2's five restrictions.

9) Spatial 3-topology $\boldsymbol{\Sigma}$ and spacetime 4-topology $\mathfrak{m}$ are related by the topological analogue of 3-metric manifolds *embedding* into 4-metric manifolds through involving 3-topologies that are cobordant to 4-topologies. The specific case of the tower involves an additional distinction between Euclidean and Lorentzian cobordisms. This level of structure's generalized deformation is a type of 'ripping' operation. The parameter corresponding to a Lorentzian cobordism might furthermore be interpreted as a time coordinate.

10) For the Morse spacetimes outlined in Appendix F.5, the Morse function f also serves as a global timefunction. By possessing this, these spacetimes preclude time non-orientability and the presence of closed timelike curves [161].

Research Project 46) Gain further understanding of time and Background Independence at the topological manifold level of structure. For instance, consider in detail which aspects of Background Independence transcend to Chern–Simons Theory.

11) *Records Theory*. The following examples can be considered in terms of obtaining topological information.

Example 1) Multiple images [597, 618]: if the Universe is small enough we would see multiple copies of the same astrophysical objects, allowing for these images to correspond to different times.

Example 2) Circles in the sky [219], allowing for the Universe to close up on a scale bigger than Hubble radius and yet still imprint evidence of closing up within the cosmological particle horizon.

Example 3) Mathematicians Partha Niyogi, Stephen Smale and Shmuel Weinberger have considered sampling unknown topology in a more general setting [681], using Čech cohomology techniques outlined in Appendices F.3 and T.3.

Moreover, in more detail, in practical terms, each of the above really concern the 'large scale shape' rather than the topology (a distinction made in Appendix S.2). See Appendix T.3 for various ways of modelling Probability and Statistics on topological manifolds.

38.4 Metric Space and Topological Space Levels

The topological manifold (Appendix D.1) is both a particularly interesting case among topological space structures (Appendix C.6) and also a very substantial restriction on the topological spaces' diversity (Fig. S.2). Arguments for adopting this package based on mathematical practicality are well-known. The question how is, however, what are the relational and Background Independence grounds for doing so.

We first consider which temporal properties and Background Independence aspects carry over to the metric and topological spaces levels of structure. The position-dependence of most notions of time in Field Theories carries over to such as spacetime lattice or space-continuum discrete-time models. This also applies to evolution laws: such as difference-equation and stochastic versions of these exist in the absence of enough structure to do ordinary Calculus. The notion of duration is moreover a metric space concept but not a topological space one.

Reduced configuration spaces $\widetilde{\mathfrak{q}}$ are generically stratified manifolds in reduced formulations, and these occur 'further down' the levels of mathematical structure than topological manifolds do. These arise since quotienting (here by $\mathfrak{g}$) does not in general preserve a number of topological properties including the three that constitute manifoldness: second-countability, Hausdorffness and local Euclideanness. *Thus Configurational Relationalism provides a second reason for considering more than the usual range of mathematical structures to Isham's first reason of 'quantizing further down' the levels of mathematical structure.*

Since the standard notion of coordinates does not descend beyond the topological manifolds level of structure, the GR feature that time is among the coordinates ceases to apply. Stratified manifolds still possess a local notion of coordinates in some regions—within a given stratum—but charts more generally differ in dimension and can have multiple strata contributing to their structure (Fig. 37.5.b). Each of the notions of orientability, curves and foliations do not descend beyond the level of topological manifolds either (though to some extent these continue to be meaningful, e.g. locally in the case of stratified manifolds). These considerations also illustrate that dimension can vary beneath the topological manifold level of structure; indeed the 'physically usual' notion of dimension is in general meaningless here.

Dispensing with basing Physics upon topological manifolds [480–482], frees one to investigate whether *topological manifold genesis* occurs. [One might extend this consideration to topological stratified-manifold genesis.] One could also consider such questions piecemeal: 'second-countability genesis', 'Hausdorffogenesis' and 'locally-Euclidean-genesis'. And which of connectedness, compactness, orientability... are emergent phenomena?

So, are each of these topological properties actually conceptually desirable for Background Independence Theoretical Physics? This is hitherto largely unexplored. E.g. Topological manifolds' second countability is a superset of 'operationally physical', since Physics in practice concerns only finite entities due to the nature of observations. Thus the second countability property may be open to weakening. Finally, whereas Hausdorffness is a great enabler of Analysis, unfortunately Appendix M outlines how some of the quotient spaces arising from Physics are not Hausdorff.

Research Project 47) The balance between spaces being 'too large' or 'too small' for doing Analysis is usually attained by considering spaces which are second-countable and Hausdorff. Does Background Independence—or physical modelling more generally—give any reason to shift this balance point? For instance, Hausdorffness is a notion of separation, and there are many other notions of separation [672, 809], including some which are only slightly distinct concepts from Hausdorffness.
Isham's choice of setting of the space of topological spaces on a fixed finite set $\mathfrak{X}$ is a first model arena for this. This space of topological spaces Top($\mathfrak{X}$) is a lattice $\mathfrak{L}_{\mathfrak{T}}$ (Appendix S.5). Moreover, since lattices can widespreadly be equipped so as to be metric spaces, different topologies τ_1 and τ_2 in Top($\mathfrak{X}$) can be compared in the format of Dist(τ_1, τ_2). In this setting, $\mathfrak{g} = Homeo$ is a particularly natural choice. In this model arena, however, finiteness renders second countability inbuilt, precludes local Euclideaness, and trivializes the manner in which Hausdorffness is distributed. I.e. on a finite set, the discrete topology alone is Hausdorff: Ex III.7.i). On the other hand , this model remains suitable as regards investigating 'separogenesis' for some of the least structured notions of separation [490, 809]: the so-called Kolmogorov and symmetric separation properties.

Moreover, the fixed nature of the above set may itself be viewed as a background structure.

As regards Timeless Records Theory, its locality criterion survives passage from metric space level considerations to topological space ones (Sect. T.1). The Čech cohomology technique mentioned in the previous Section also transcends to this level of structure, giving a 'Čech Records Theory'. Furthermore, (W.2) points to the further generalization of a Records Theory based on Sheaf Cohomology. Basing Probability and Statistics on sheaves has the advantage of descending to the topological space level as well. Kendall's theory of random sets (Appendix T.4) can also be viewed as a further example of a classical Records Theory; Isham already brought this work to attention in the QG literature in a less specific manner in [476].

Research Project 48) A somewhat less radical version—also considered by Isham at the quantum level [482]—involves forfeiting a fixed metric space. Under the caution that this need not encapsulate stratified manifolds or lead to differentiable structure and the higher levels of structure built thereupon, gain a further classical understanding of time and Background Independence at this level of structure. [One natural choice for $\mathfrak{g}$ here are the metric space isometries $Isom(\mathfrak{X})$.] What if one keeps only some of the axioms of distance?

38.5 Yet Deeper Levels of Structure

Research Project 49)† Consider *topological-space-genesis*: emergence from more general collections of subsets within some space $\mathfrak{q} = \mathfrak{Collect}(\mathfrak{X})$. Do one or both of dynamical or probabilistic considerations give collections of subsets a propensity to be, more specifically, topological spaces? [See Research Project 122) for further consideration of the latter.] Also gain further understanding of time and Background Independence at the level of sets and of collections of subsets.

Within the QG literature [628, 685], quite a lot of the work done to date with such structural sparsity is within Sorkin's *Causal Sets Approach* [801, 802]. Here, in contradistinction to much of this book, one chooses to keep the causal relation and causal ordering aspects of SR and GR spacetime, albeit now also in the absence of assuming manifoldness. Its quantum-level treatment is now in terms of path integrals. This approach makes one further hypothesis: that spacetime is fundamentally discrete [804] (so-called 'spacetime atoms'). The first and third of these assumptions are already held to apply at the classical level, and are jointly modelled by posets (Appendix A.1). These replace both the metric and the underlying topological manifold structure all in one step. In this approach, label invariance and growth order invariance take the place of coordinate invariance for causal sets. Here the passage of time is an unceasing cascade of birth events [803]; physicist Fay Dowker has furthermore argued for these to lie within Broad's Worldview [263].

The Causal Sets Approach provides a further example of space as a slice within spacetime retaining some meaningful identity in structurally sparse conditions. In this setting, slices are *maximal antichains*.[3]

As regards Spacetime Constructability, at the current level this requires an entity which becomes—or approximately resembles—a manifold in a suitable limit. See

[3]Antichains are subsets of a poset such that the elements within each of which bear no ordering relations. Maximal antichains are the largest possible ones, in some ways analogous to global slices or Cauchy surfaces in Geometrodynamics.

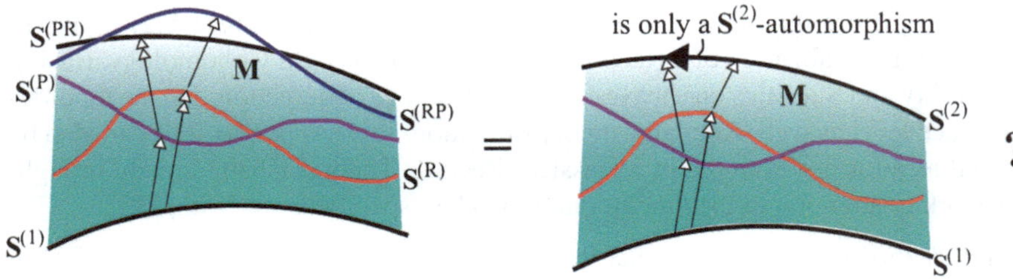

Fig. 38.4 Reslicing Invariance posed. Consider the larger notion of mathematical space **M** in the role of spacetime and the smaller notion of mathematical space **S** in the role of space. Is the *solid arrow's* transformation due to difference between passing from $\mathbf{S}^{(1)}$ to $\mathbf{S}^{(2)}$ via the *red* and *purple* intermediate **S**'s always an automorphism of $\mathbf{S}^{(2)}$?

e.g. [734] by physicists David Rideout and Petros Wallden for some advances in Spacetime Construction within the Causal Sets Approach.

Going yet another level down, one might let the cardinality of the set that the topological spaces are based upon itself classically evolving and quantum-mechanically fluctuating. Here the notion of slicing becomes a partition of a set into a bunch of 'equal-time' sets, which can be tied to at least some notions of simultaneity [521]. Spacetime Construction (including in approaches assuming just a discrete replacement for space) would be expected to be even less straightforward in this case; the less structure is assumed, the harder such a scheme is.

We finally return to Isham's quotation at the start of this Epilogue. He is in part concerned that point-set theory is used even when points are held to be physically meaningless. From this perspective he subsequently went on to ask [491] why Nature should be modelled using $\mathbb{R}$ and physical probability values should belong to the real interval [0, 1]?

Research Project 50) Provide the Principles of Dynamics at each level of mathematical structure. TFTs are a well-established arena to begin this study with. (See e.g. Sect. 38.1 and [253].)

Research Project 51)† Amongst the topics of Research Projects 45–48), let us highlight that the Reslicing Invariance generalization of Refoliation Invariance can be posed, as per Fig. 38.4. Work out in which cases this is satisfied. I.e. is Teitelboim's resolution at the level of Differential Geometry a fortunate feature, or even a selection principle on what level of structure to consider, or does it in fact reflect a level-independent state of affairs?

In conclusion, *the space of spaces construct kicks one out of the hierarchy of mathematical structures that theoretical physicists are accustomed to.* Also *in making a descent down the levels of mathematical structure, a few temporal concepts cease to be meaningfully implemented, and a few Background Independence aspects and consequent Problem of Time facets cease to exist or to be distinct. However, most persevere.* Finally, such a level by level descent also invites *transcending from a 'differential geometric' treatment of Background Independence and Problem of Time—manifolds, Lie groups, fibre bundles—to a 'categorical' treatment.*

Part III
Quantum Problem of Time

We now further explore the accounts of Quantum Gravity and Quantum Background Independence in Chaps. 11 and 12. When aspects of Quantum Background Independence are in contention—or fail to be realized by a proposed scheme—Quantum Problem of Time facets ensue. Part III concentrates in particular on quantum-level interferences between Problem of Time facets.

Chapter 39
Geometrical Quantization. i. Kinematical Quantization

Let us begin by extending Part I's outline of Canonical Quantization to more general cases, so as to include GR and models encapsulating some features thereof.

The *Deformation Quantization* approach was initiated by Weyl [894]. 'Deformation' is meant here in an algebraic sense, as in the introduction of a new parameter. This results in noncommutativity where there was none before; this accounts for the progression from classical brackets to quantum commutators. Physicist Hilbrand Groenewold [392] subsequently pointed to the significance in this scheme of the following 'star product' on $\mathfrak{P}$hase:

$$F \star G := F \exp\left(\frac{i\hbar}{2} P^{\mathsf{KK}'} \overleftarrow{\partial_{\mathsf{K}}} \overrightarrow{\partial_{\mathsf{K}'}}\right) G. \tag{39.1}$$

This came to be known as the *Moyal star product* (after applied mathematician José Enrique Moyal); see Appendix V.7 for a general outline of 'star product' operations.] $P^{\mathsf{KK}'}$ denotes the Poisson tensor outlined in Appendix J.12.

A further useful concept at this stage is *polarization*: the choice of a suitable [919] half-set among the Hamiltonian variables $\boldsymbol{Q}$, $\boldsymbol{P}$. One suitability criterion here is brackets closure, which is a type of integrability condition. I.e. if O_1 and O_2 are part of a polarization, then so is $|[O_1, O_2]|$. A well-known example of polarization is choosing the $\boldsymbol{Q}$ half of the variables: the *configuration representation* generalization of the position representation. Another involves choosing the $\boldsymbol{P}$ half: the momentum representation. The complex representation in terms of $\boldsymbol{Z} = \frac{1}{\sqrt{2}}\{\boldsymbol{P} + i\,\boldsymbol{Q}\}$ and $\boldsymbol{Z}^* = \frac{1}{\sqrt{2}}\{\boldsymbol{P} - i\,\boldsymbol{Q}\}$ lends itself to further polarizations known as 'Bargmann polarizations' after physicist Valentine Bargmann; see e.g. [75] for applications.

This book concentrates however on a subsequent development:[1] *Geometrical Quantization* [475, 919]. This is approached via choosing a polarization of $\mathfrak{q}$, in which manner it naturally partners Part II's $\mathfrak{q}$-geometry based Jacobi–Synge approach. Geometrical Quantization is general enough to embrace most of this book's

[1]Both Deformation and Geometrical Quantization have continued to evolve since their inception; see [605] for an excellent review by mathematical physicist Nicolaas Landsman.

E. Anderson, *The Problem of Time*, Fundamental Theories of Physics 190,
DOI 10.1007/978-3-319-58848-3_39

quantum treatment of its selection of model arenas and theories. This is outlined in the next two Chapters using trivial-$\mathfrak{g}$ Finite Theory concrete examples; these restrictions are lifted in Chaps. 42–43.

39.1 Unconstrained Beables Come First in Geometrical Quantization

Kinematical Quantization involves a notion of observables or beables: the simplest unconstrained such, $\widehat{\mathcal{U}}$. Quantum Theory builds on this, by which this part of Assignment of Beables is required at an earlier stage than in the classical development. Indeed, in Geometrical Quantization, finding suitable $\widehat{\mathcal{U}}$ is the first part of Quantization: Kinematical Quantization. This means that U is the first stage of Q in quantum-level Cubert facet orderings, with quantum-level considerations of constraints, reduction, and assignation of a time occurring after assigning $\widehat{\mathcal{U}}$. Some of these subsequent quantum steps are then known as Dynamical Quantization, or as *Tempus Post Quantum* assignation of a time.

The kinematical operators require a space of wavefunctions to act upon. This is mathematically a Hilbert space, $\mathfrak{H}$ilb. However, it is not the particular physically realized Hilbert space, due to dynamical (and, in the next Chapter, $\mathfrak{g}$) considerations not having yet been taken into account. Let us denote Hilbert spaces in the former role by $\mathfrak{K}$in-$\mathfrak{H}$ilb for '*kinematical Hilbert space*', and those in the latter role by $\mathfrak{D}$yn-$\mathfrak{H}$ilb for '*dynamical Hilbert space*'. The corresponding morphisms are unitary transformations. Four more specific features are as follows.

I) The examples in Sect. 39.5 explain how each Kinematical Quantization operator obeys suitable global continuity conditions, which reflect Quantum Theory's sensitivity to the topology of $\mathfrak{q}$ and $\mathfrak{P}$hase.

Before proceeding further, we recollect some classical trivialities. We can consider all $\mathcal{U} = \mathbf{F}(\boldsymbol{Q}, \boldsymbol{P})$ over a given phase space $\mathfrak{P}$hase of the $\boldsymbol{Q}$ and $\boldsymbol{P}$ are the classical level. These $\mathcal{U}$ furthermore form a Poisson brackets algebraic structure, $\mathfrak{U}$ We would expect the $\widehat{\mathcal{U}}$ to form a brackets algebraic structure as well. The next two sections consider maps between bracket structures in anticipation of this. Simple or optimistic hopes that $\mathfrak{U} = \widehat{\mathfrak{U}}$ shall be shown not to materialize; Sect. 39.4 explains why Kinematical Quantum algebraic structures are much smaller than classical $\mathcal{U}$ algebraic structures.

II) It is then a nontrivial issue that the selected set of Kinematical Quantization operators is to contain all the relational information.
III) Kinematical Quantization operators are however allowed to have some classical redundancy.[2]
IV) Kinematical Quantization operators close under the quantum commutation relations, thus indeed forming an algebraic structure $\widehat{\mathfrak{U}}$.

[2]I.e. there are more relational functions than there are independent pieces of relational information; this is not to be confused with including unphysical, gauge or nonrelational information.

39.2 Brackets Map Between Spaces of Objects

We follow up on Sect. 24.1's introduction of brackets algebraic structures, by now considering maps between two such:

$$m : \langle \mathfrak{O}_1, |[\ ,\ ]|_1 \rangle \longrightarrow \langle \mathfrak{O}_2, |[\ ,\ ]|_2 \rangle. \tag{39.2}$$

This is not per se a quantum construct, though all applications of it in this book do happen to be promotions of classical brackets to quantum commutators.

A priori and in the notation of Sect. 24.1, one might expect

$$|[\mathsf{O}_{1\mathsf{v}}, \mathsf{O}_{1\mathsf{v}'}]| = C_1{}^{\mathsf{v}''}{}_{\mathsf{v}\mathsf{v}'}\mathsf{O}_{1\mathsf{v}''} \tag{39.3}$$

to be induced upon the algebraic structure of the O_2. This occurs if m is an isomorphism. It turns out to be insightful, however, to ask *what other forms* a brackets map can take.

Outcome i) Inequivalent structure constants $C_2 \neq C_1$ may arise.
Outcome ii) The O_2 objects may pick up (24.1)'s Θ term.
Outcome iii) If prescribing m is ambiguous, different m would be expected to give different outcomes.
Outcome iv) If m does not preserve dimension (as many homomorphisms do), perhaps restriction to a proper subalgebra $\mathfrak{S}_1 < \mathfrak{O}_1$ would manage this, or even provide an isomorphism.
Outcome v) The brackets map might be a functor between two categories (Appendix W). In general, however, Quantization itself is no such well-behaved functor, as explained in Sect. 43.7.

39.3 Specifically Quantum Attributes of Brackets

1) Now $m : \langle \mathfrak{O}, |[\ ,\ ]| \rangle \longrightarrow \langle$some space of operators $\widehat{\mathfrak{O}}, [\ ,\ ]\rangle$.
2) The recipient space is more globally sensitive than the original space.
3) Quantization (classical to quantum) maps involve *operator ordering*. This is due to one's system's quantum operators in general not commuting.
4) Moreover, consistency of another kind—

$$\text{large enough sets of quantum versions of classical quantities can be inconsistent with each other} \tag{39.4}$$

—pushes one into considering a subalgebraic structure $\mathfrak{S} < \mathfrak{O}$ to be promoted, rather than the whole of $\mathfrak{O}$. This is a specifically quantum-level issue. The Groenewold–van Hove phenomenon (schematically in Fig. 12.2, with details and examples in Epilogue III.A) is a result of this kind. Finally, choice of $\mathfrak{S}$ is certainly capable of being nonunique, being the most prominent source of Multiple Choice Problems.

5) The form taken by Outcome i) is

$$|[\mathsf{S}_\mathsf{S}, \mathsf{S}_{\mathsf{S}'}]| = C_1{}^{\mathsf{S}''}{}_{\mathsf{S}\mathsf{S}'}\mathsf{S}_{\mathsf{S}''} \quad \longrightarrow \quad [\widehat{\mathsf{S}}_\mathsf{S}, \widehat{\mathsf{S}}_{\mathsf{S}'}] = C_2{}^{\mathsf{S}''}{}_{\mathsf{S}\mathsf{S}'}\widehat{\mathsf{S}}_{\mathsf{S}''}. \tag{39.5}$$

This is in accord with attribute 2), by which ranges of validity can shape the primed algebraic structure to a greater extent than the unprimed one. Also different operator orderings often lead computationally to different right hand sides.

On the other hand, Outcome ii) gives

$$|[\mathsf{S}_\mathsf{S}, \mathsf{S}_{\mathsf{S}'}]| = C_1{}^{\mathsf{S}''}{}_{\mathsf{S}\mathsf{S}'}\mathsf{S}_{\mathsf{S}''} \quad \longrightarrow \quad [\widehat{\mathsf{S}}_\mathsf{S}, \widehat{\mathsf{S}}_{\mathsf{S}'}] = C_2{}^{\mathsf{S}''}{}_{\mathsf{S}\mathsf{S}'}\widehat{\mathsf{S}}_{\mathsf{S}''} + \Theta_{\mathsf{S}\mathsf{S}'}. \tag{39.6}$$

6) Consequently,

$$\text{Classical Brackets Closure} \not\Rightarrow \text{Quantum Brackets Closure;} \tag{39.7}$$

this generalizes (12.15) from constraints to general entities.

7) If item 6) occurs, one strategy is to see whether this arose from the choices made of operator ordering and of $\mathfrak{S}$. Can some operator ordering of some choice of $\mathfrak{S}$ at least maintain closure?

Subcase A) Insist on preserving $\langle \mathfrak{S}, |[\ ,\ ]| \rangle$.
Subcase B) Identify which $\widehat{\mathsf{S}}_\mathsf{S}$ cause Θ terms to appear, and excise these; this might be used in arguing for a fixed view on how to operator-order.
Subcase C) Strong vanishing of the Θ terms may also occasionally be feasible.

Of course, since these subcases add restrictions, they are less likely to be solvable than the general case; some programs moreover hinge upon the solvability of such a subcase.

39.4 The Groenewold–Van Hove Phenomenon

The reason why we cannot promote whichever combination of—or all of—the $\upsilon = \mathbf{F}(\boldsymbol{Q}, \boldsymbol{P})$ to quantum operators is that global obstruction by the Groenewold–van Hove phenomenon applies. This gives one reason why a preferred set of $\mathbf{F}(\boldsymbol{Q}, \boldsymbol{P})$ is to be selected for promotion to quantum operators $\widehat{\upsilon} = \widehat{\mathcal{F}}[\widehat{\boldsymbol{Q}}, \widehat{\boldsymbol{P}}]$. A fortiori, the preferred set is to algebraically close under the classical Poisson bracket $\{\ ,\ \}$, and so form a subalgebraic structure. On the other hand, the latter are to algebraically close under the commutator bracket $[\ ,\ ]$. The previous two Secs have outlined why the passage between these is far from necessarily a straightforward matter. I.e. the classical Poisson brackets subalgebraic structure leads to some commutator algebra, $\mathfrak{Com}$-$\mathfrak{Al}$ of $\widehat{\mathcal{F}}[\widehat{\boldsymbol{Q}}, \widehat{\boldsymbol{P}}]$ that close under $[\ ,\ ]$, which does not have to be isomorphic to it. The latter is accompanied by commutator-preserving morphisms M in place of the classical-level canonical transformations, Can, which preserve the Poisson brackets. Because of the Groenewold–van Hove phenomenon, we add the following to Sect. 39.1's list of features.

V) Excessive polynomiality is to be avoided if at all possible due to the threat of the Groenewold–Van Hove phenomenon.

Note that cubic combinations are already afflicted by this, so one is left considering quadratic polynomials. Another consideration—balancing II) and IV)—is that one needs a large enough set of beables to express every other beable as a function(al) of these, i.e. a set of 'basis beables'. The corresponding irreps subsequently play a significant role [482].

39.5 Examples of Kinematical Quantization

These illustrate how Quantum Theory has greater sensitivity to global structure. For now, the first five of these examples are to be physically interpreted as absolutist Quantizations.

Example 1) For a particle in $\mathbb{R}$ (interpreted to be $\mathfrak{q}$), the conventional selection is just x, p. This followed by the promotion $x \longrightarrow \widehat{x}$, $p \longrightarrow \widehat{p}$, which can be represented by

$$\widehat{x} = x \quad \text{and} \quad \widehat{p} = -i\,\hbar\,\frac{\partial}{\partial x}. \tag{39.8}$$

These are the objects $\widehat{K_Z}$, which constitute the *kinematical Quantization algebra* $\mathfrak{K}$.

Example 2) For a particle in $\mathbb{R}^2 = \mathfrak{q}$, the Kinematical Quantization's selection involves not only x^i and p_i but angular momentum J as well. Upon promoting these to quantum operators, they admit the representation

$$\widehat{x^i} = x^i, \qquad \widehat{p_i} = -i\,\hbar\,\frac{\partial}{\partial x^i} \quad \text{and} \quad \widehat{J} = -i\,\hbar\left\{y\,\frac{\partial}{\partial x} - x\,\frac{\partial}{\partial y}\right\} = -i\,\hbar\,\frac{\partial}{\partial \phi}. \tag{39.9}$$

Example 3) For a particle in $\mathbb{R}^3 = \mathfrak{q}$, the Kinematical Quantization's selection now involves x^i, p_i and J_i. These are promoted to quantum operators which can be represented by

$$\widehat{x^i} = x^i, \qquad \widehat{p_i} = -i\,\hbar\,\frac{\partial}{\partial x^i} \quad \text{and} \quad \widehat{J_i} = -i\,\hbar\,\epsilon_{ijk}x^j\,\frac{\partial}{\partial x^k}. \tag{39.10}$$

In each case,

$$\mathfrak{Kin}\text{-}\mathfrak{Hilb} = \mathfrak{L}^2\left(\mathbb{R}^p, \prod_{i=1}^{p} \mathrm{d}x^i\right), \tag{39.11}$$

i.e. the square-integrable functions on $\mathbb{R}^p$ with respect to the standard measure.

Specifying $\mathfrak{K}$ also requires giving the bracket relations. The classical Poisson brackets and quantum commutators are an isomorphic pair for each d. In 1-d, this is the Correspondence Principle

$$\text{from } \{x, p\} = 1 \quad \text{to } [\widehat{x}, \widehat{p}] = i\,\hbar, \tag{39.12}$$

whereas in higher-d each of the pair has additional nontrivial brackets involving angular momentum. E.g. in 3-d (39.12) is accompanied by passage from (2.24) to (5.5). The 3-d angular momenta can also be cast in the dual form,

$$\widehat{J}_{ij} = i\,\hbar\left\{x^i\frac{\partial}{\partial x^j} - x^j\frac{\partial}{\partial x^i}\right\},$$

which presentation extends to n-d; Poisson brackets and commutators in this form are left as an Exercise. The resulting $\mathfrak{K}$ in each of the above examples is a Heisenberg algebra, *Heis*(n). Moreover, at the classical level one might well consider Poisson brackets of classical quantities other than x^i, p_i and J_i, or not think to allot co-primary status to the J_i.

The above examples are however misleadingly simple in some ways. They correspond to problems for which Kinematical Quantization is resolved by the $\widehat{x}^i$, $\widehat{p}_i$, and $\widehat{J}_i$ operators, which have come to be widely known and so are adopted without second thought. Yet the applicability of these operators reflects the underlying role of $\mathbb{R}^n$ spaces, which are all of flat, without boundary, topologically trivial, and vector spaces. Geometrical Quantization, moreover, addresses what procedure one is to adopt if a *general manifold* $\mathfrak{M}$ enters at this stage instead. In fact, Geometrical Quantization can deal with even more general possibilities than manifolds, such as stratified manifolds. Some more general examples are as follows.

Example 4) [of I) and redundancy in III).] For a particle on $\mathbb{S}^1 = \mathfrak{q}$, ϕ and p_ϕ are classically useful variables: the general coordinate q, p pair extension of the x, p pair. This p_ϕ can furthermore be interpreted as J: 2-d rotations' sole component of angular momentum. However, if we try to select ϕ and J for promotion to quantum variables, this fails due to discontinuity of ϕ at the endpoints of its range (cf. I) (this can be taken e.g. to be from 0 to 2π). An obvious next attempt is to select e.g. $\sin\phi$ instead of ϕ so as to meet this continuity condition (and smoothly so, from the sine function's periodicity). However,

$$[\sin\phi, \widehat{J}] = i\,\hbar\cos\phi, \tag{39.13}$$

so this choice does not close by itself as an algebraic structure. Thus $\cos\phi$—which was also always a valid alternative choice due to its own periodic property—has to be selected as well. This finalizes the closure, since the further nontrivial commutator

$$[\cos\phi, \widehat{J}] = -i\,\hbar\sin\phi \tag{39.14}$$

does not give any further right hand side terms. These operators are self-adjoint on

$$\mathfrak{K}\text{in-}\mathfrak{H}\text{ilb} = \mathfrak{L}^2(\mathbb{S}^1, \mathrm{d}\phi). \tag{39.15}$$

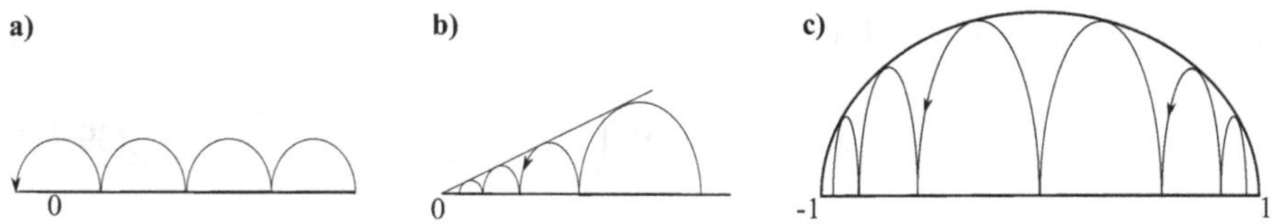

Fig. 39.1 A heuristic picture of **a)** unsuitable and **b)** suitable operator actions as regards representing momentum on the real half-line. **c)** Suitable operator actions as regards representing momentum on the interval

Moreover, clearly a change in dim($\mathfrak{K}$) is caused by passing from considering $\mathbb{R}$ to $\mathbb{S}^1$, which is topologically distinct: its fundamental group (cf. Appendix F)

$$\pi_1(\mathbb{S}^1) = \mathbb{Z} \neq id = \pi_1(\mathbb{R}). \tag{39.16}$$

Furthermore, dim($\mathfrak{K}$) has ceased to coincide with dim($\mathfrak{P}$hase($\mathbb{S}^1$)). This renders it a more striking counter-example to quantum commutators being isomorphic to underlying classical Poisson brackets.

Example 5) [of I) and of m not being an isomorphism.] For a particle in $\mathbb{R}_+ = \mathfrak{q}$ [475], suppose one were to try to represent x and p by

$$\widehat{x} = x, \qquad \widehat{p} = -i\,\hbar\,\frac{\partial}{\partial x}. \tag{39.17}$$

The latter, however, is not essentially self-adjoint since it does not respect the end-point of the $\mathbb{R}_+$ by continuing to generate a translation past it (Fig. 39.1.a). To avoid this, one uses instead

$$\widehat{p} = -i\,\hbar\,x\,\frac{\partial}{\partial x} \tag{39.18}$$

(Fig. 39.1.b). This firstly illustrates that the familiar representation of momentum—summarized by

$$\mathbf{P} = -i\,\hbar\,\frac{\boldsymbol{\partial}}{\boldsymbol{\partial}\mathbf{Q}} \tag{39.19}$$

—is in fact not universally applicable.

Secondly, the quantum commutator takes the form

$$[\widehat{x}, \widehat{p}] = i\,\hbar\,\widehat{x}: \tag{39.20}$$

the *affine commutation relation*.[3] Thus this example also illustrates a distinct algebraic structure arising upon quantizing [the corresponding classical Poisson bracket is still being (2.23)]. This change is due to Quantum Theory's greater sensitivity to topological structure; in particular, while $\mathbb{R}_+$ is contractible, it is however

[3]Exercise: show how this is indeed related to Appendix B's notion of Affine Geometry.

not a vector space [475]. These operators are self-adjoint on

$$\mathfrak{Kin}\text{-}\mathfrak{Hilb} = \mathfrak{L}^2\left(\mathbb{R}_+, \frac{\mathrm{d}x}{x}\right). \tag{39.21}$$

Note in particular that the $\widehat{p}$ constructed here is self-adjoint with respect to this and not with respect to 'the usual' dx.

Example 6) For a particle on an interval $\mathfrak{I} = \mathfrak{q}$ there are two endpoints to respect. Without loss of generality, place these at ± 1. Moreover,

$$x, \qquad \sqrt{1 - x^2} \quad \text{and} \quad - i\,\hbar\sqrt{1 - x^2}\frac{\partial}{\partial x} \tag{39.22}$$

are self-adjoint on

$$\mathfrak{Kin}\text{-}\mathfrak{Hilb} = \mathfrak{L}^2\left(\mathbb{R}_+, \frac{\mathrm{d}x}{\sqrt{1 - x^2}}\right). \tag{39.23}$$

These can furthermore be recast as

$$\sin\theta, \qquad \cos\theta \quad \text{and} \quad - i\,\hbar\frac{\partial}{\partial\theta}, \tag{39.24}$$

though the interpretation these receive here is different from that in Example 4), due to -1 and 1 now not being identified.

Let us next consider a method which works for quite a wide range of examples including some of this book's more widely used ones. It is due to noted mathematician George Mackey, and has been much used in Isham's works [475, 482, 491, 497]. It applies when $\mathfrak{q}$ takes the form of a homogeneous space $\mathfrak{g}_1/\mathfrak{g}_2$ ($\mathfrak{g}_2$ is a subgroup of $\mathfrak{g}_1$).

In this case, the corresponding kinematical quantum algebraic structure $\mathfrak{K}$ can be decomposed as semidirect products (Appendix A)

$$\mathfrak{v}^*(\mathfrak{q}) \rtimes \mathfrak{g}_{\text{can}}(\mathfrak{q}). \tag{39.25}$$

Here, $\mathfrak{g}_{\text{can}}(\mathfrak{q})$ is the *canonical group* and $\mathfrak{v}^*$ is the dual of a linear space $\mathfrak{v}$. This is natural due to carrying a linear representation of $\mathfrak{q}$ such that there is a $\mathfrak{q}$ group orbit in $\mathfrak{v}$ which is diffeomorphic to $\mathfrak{q}/\mathfrak{g}$ [475]. Furthermore,

$$\mathfrak{K} = \langle \mathfrak{v}^* \rtimes \mathfrak{g}_{\text{can}}(\mathfrak{q}), [\ ,\ ] \rangle. \tag{39.26}$$

Also $\mathfrak{v}^* = \mathfrak{v}$ for finite examples, while $\mathfrak{g}_{\text{can}}(\mathfrak{q}) = Isom(\mathfrak{q})$ for many of the examples in this book. Semidirect product groups have the further good fortune that the powerful techniques of Mackey Theory (Appendix V.1) are available to set up the corresponding Representation Theory.

Let us now return to Examples 1) to 3), as regards a sense in which these are *less simple* than Examples 4) and 5). This follows from the central extension already observed at the classical level (Sect. 37.7) also affecting the form of $\mathfrak{K}$. By this e.g.

the $\mathfrak{q} = \mathbb{R}$ case's $\mathfrak{K}$ picks up an extra $\mathbb{R}$—the 1—in addition to the $\mathbb{R}$ of q in the $\mathfrak{v}$ and the $\mathbb{R}$ of p which forms $\mathfrak{g}_{\text{can}}(\mathbb{R})$. In contrast, Example 5)'s

$$\mathfrak{K} = \mathbb{R} \rtimes \mathbb{R}_+ = Aff(1)$$

involves no central extension term. Furthermore, this case does not involve $\text{Isom}(\mathbb{R}_+)$ either. This is because the 1-d isometry group is formed here by the Killing vector $\frac{\partial}{\partial x}$. On the one hand, $\frac{\partial}{\partial x}$ arises from locally solving the Killing equation, but on the other hand this expression does not comply with the required essential self-adjointness to be part of $\mathfrak{K}$. The first attempt at Example 4) above can be interpreted as [475] $\mathfrak{K} = \mathbb{R} \rtimes SO(2)$ being obstructed by the nontrivial cohomology group (Appendix F)

$$H^1(\mathbb{S}^1, \mathbb{R}) = \mathbb{R}. \tag{39.27}$$

This is resolved by bringing in multiple coordinate charts; two will do, corresponding to use of

$$\mathfrak{K} = \mathbb{R}^2 \rtimes SO(2) = Eucl(2).$$

This $\mathfrak{K}$ is moreover shared by Example 5) as well.

Example 7) On $\mathbb{S}^k = SO(k+1)/SO(k)$, $\mathfrak{g}_{\text{can}} = Isom(\mathbb{S}^k) = SO(k+1)$, acting upon $\mathfrak{v} = \mathbb{R}^{k+1}$. The $\mathbb{R}^{k+1}$ objects here obey $\sum_{i=1}^{k+1} u_i^2 = 1$, which can be represented by the (hyper)spherical polar coordinate unit vectors. These are self-adjoint on

$$\mathfrak{Kin}\text{-}\mathfrak{Hilb} = \mathfrak{L}^2\left(\mathbb{S}^k, \sin^{nd-1-\mathsf{A}}\theta_{\mathsf{A}} \prod_{i=1}^{A-1} \sin^2\theta_i \, \mathrm{d}\theta_i \mathrm{d}\theta_{\mathsf{A}}\right) \tag{39.28}$$

and form

$$\mathfrak{K} = \mathbb{R}^{k+1} \rtimes SO(k+1) = Eucl(k+1). \tag{39.29}$$

Example 8) For N-particle 1-d Metric Shape and Scale RPM, $\mathfrak{q} = \mathbb{R}^n$ for $n := N-1$. Thus this shares

$$\mathfrak{Kin}\text{-}\mathfrak{Hilb} = \mathfrak{L}^2\left(\mathbb{R}^n, \prod_{i=1}^{n} \mathrm{d}x^i\right) \tag{39.30}$$

and

$$\mathfrak{K} = Heis(n) \tag{39.31}$$

mathematics with the absolutist version of $\mathbb{R}^n$. What differs now is the physical interpretation of $\mathfrak{K}$'s objects. These are now relative separations ρ^i, their conjugate momenta

$$\widehat{\pi}_i = -i\,\hbar \frac{\partial}{\partial \rho^i} \tag{39.32}$$

and relative dilational momenta

$$\widehat{D}_{ij} = -i\hbar\left\{\rho_i \frac{\partial}{\partial \rho^j} - \rho_j \frac{\partial}{\partial \rho^i}\right\}. \tag{39.33}$$

This is supported by

$$\pi_1\big(\mathfrak{s}(N,1)\big) = \pi_1\big(\mathbb{S}^{N-2}\big) \quad \text{being trivial} \tag{39.34}$$

for all the relationally nontrivial cases ($N \geq 4$).

Example 9) For closed single scalar field Minisuperspace, working with Misner's scale variable Ω exhibits *Poin*(2) as isometry group. The mathematical outcome of Kinematical Quantization is thus fairly standard, albeit again with non-standard physical interpretation due to occurring in a configuration space—Minisuperspace—rather than in physical spacetime. This involves the isometry group consisting of a single 'boost' K and a 'translational' 2-vector $P = (p_\Omega, p_\phi)$ associated with one Misner scale variable coordinate and one scalar field coordinate. Moreover, this acts upon another Minisuperspace 2-vector $X = (\Omega, \phi)$. These are all self-adjoint on

$$\mathfrak{K}\text{in-}\mathfrak{H}\text{ilb} = \mathfrak{L}^2\big(\mathbb{M}^2, \mathrm{d}\Omega\,\mathrm{d}\phi\big), \tag{39.35}$$

and form

$$\mathfrak{K} = \mathbb{M}^2 \rtimes Poin(2): \tag{39.36}$$

the indefinite $2-d$ counterpart of the familiar Heisenberg group. Furthermore, this lies within the $\mathfrak{v}^*(\mathfrak{q}) \rtimes \mathfrak{g}_{\text{can}}(\mathfrak{q})$ form. [The objects acted upon here obey $c^2 - s^2 = 1$.] However, if one works with the scale variable a, a distinct algebra with some $\mathbb{R}_+$ type affine features arises [50]. [a versus Ω amounts to a change in the topology of the domain involved. This inequivalence is thus rooted in Quantum Theory's global sensitivity.] In this way, choice of scale variable leads to inequivalent Quantizations: a Multiple Choice Problem. This additionally becomes a Multiple Choice Problem of Time if the scale variable chosen is to play a temporal role.

Example 10) For vacuum diagonal Bianchi IX anisotropic Minisuperspace, working in $\Omega, \beta_\pm$ variables exhibits *Poin*(3) as isometry group. This now consists of a rotation J in $\mathfrak{A}$ni, 2 boosts K mixing anisotropy and Misner scale variable, and a 'translational' Minisuperspace 3-vector $P = (p_\Omega, p_{\beta_+}, p_{\beta_-})$. Moreover, this group acts upon another Minisuperspace 3-vector $X = (\Omega, \beta_+, \beta_-)$. These objects are all self-adjoint on

$$\mathfrak{K}\text{in-}\mathfrak{H}\text{ilb} = \mathfrak{L}^2\big(\mathbb{M}^3, \mathrm{d}\Omega\,\mathrm{d}\beta_+\mathrm{d}\beta_-\big), \tag{39.37}$$

and form

$$\mathfrak{K} = \mathbb{M}^3 \rtimes Poin(3). \tag{39.38}$$

Again, working with a instead of Ω gives a distinct Kinematical Quantization algebra.

39.6 ii. Further Global Nontriviality

Further global effects can stem from nontriviality of $\mathfrak{q}$'s Chern classes (outlined in Appendix F.5). E.g. the first Chern class classifies the twisted representations [475], a notion bearing close relation to fermions.

Example 1) For the action of $SO(2)$ on $\mathbb{Z}$, the cohomology groups are

$$H^p(\mathbb{S}^1, \mathbb{Z}) = 0, \tag{39.39}$$

by which these effects are trivialized.

Example 2) On the other hand,

$$\text{for } k > 1, \quad H^p(\mathbb{S}^k) = 0 \quad \text{for } p < k \quad \text{but not for } p = k, \tag{39.40}$$

so a few of the more subtle effects persevere. These $k > 1$ results are furthermore directly relevant to Metric Shape RPMs since $\mathfrak{S}(N, 1) = \mathbb{S}^{N-2}$ for $N \geq 4$.

39.7 Conceptual Outline of the Kochen–Specker Theorem

This [562] is a further nontriviality which can be considered to be part of subalgebraic structure selection. Discussing this requires first considering *valuation functions*. For a state S and a physical quantity A, a classical valuation function is just $V_S(A) = f_A(S)$, which allots the value of f_A in the state S. f_A is itself a function from the state space to $\mathbb{R}$ [487]. On the other hand, a quantum valuation function $V_\psi(A)$ is a more involved construct (for ψ the quantum wavefunction);

$$\text{function}\big(\text{valuation}(A)\big) = \text{valuation}\big(\text{function}(A)\big) \tag{39.41}$$

would appear to be a natural condition to impose on this, for any function: $\mathbb{R} \longrightarrow \mathbb{R}$. However, (39.41) implies the following conditions [487].[4]

i) $V_\psi(O_1 + O_2) = V_\psi(O_1) + V_\psi(O_2)$ (additivity).
ii) $V_\psi(O_1 O_2) = V_\psi(O_1) V_\psi(O_2)$ (multiplicativity).
iii) $V_\psi(\mathbb{I}) = 1$, for $\mathbb{I}$ the physical quantity corresponding to the unit operator $\widehat{\mathbb{I}}$, holds so long as there is at least one O such that $V_\psi(O) \neq 0$ (unit property).
iv) $V_\psi(0) = 0$ (zero property).

[4]Furthermore, by using projectors $\widehat{P}_i$ to represent quantities playing the role of the A_i, and the Proposition–Projector Association of Sect. 51.1, this becomes a matter of truth valuations. $V_\psi(\widehat{P}_i) = 0$ or 1 (projector truth valuation). Moreover, $\sum_{i=1}^n V_\psi(\widehat{P}_i) = 1$ for a set of projectors resolving the identity $\big(\sum_{i=1}^n \widehat{P}_i = 1$ for $\widehat{P}_i \widehat{P}_j = 0$ if $i \neq j\big)$ is guaranteed by i) and iii), though the preceding sentence forces this to be realized by a single 1 and $n-1$ zeros. Proof of this footnote's statements is left as an Exercise.

The *Kochen–Specker Theorem*, however, is that

$$\text{there is an obstruction to function}\big(\text{valuation}(O)\big)$$
$$= \text{valuation}\big(\text{function}(O)\big) \text{ for Hilbert spaces of dimension} > 2. \qquad (39.42)$$

See e.g. [708] for an updated simpler proof, complete with comments and references to previous proofs.

Chapter 40
Geometrical Quantization. ii. Dynamical Quantization

The next step is Dynamical Quantization. For GR (and a wider range of Background Independent theories), Dynamical Quantization moreover reveals frozenness. We argued in Part I that this is already classically present and resolved at that level by time being abstracted from change. The quantum level, however, requires a new resolution, as can be envisaged from quantum change in general differing from classical change. That these two resolutions differ is particularly clear from the position that 'all change is given the opportunity to contribute'. This approach, however, is to await Chap. 46, so that we first map out the effects of generalization to nontrivial $\mathfrak{g}$, Field Theory, GR, and *Tempus Ante Quantum* options.

Dynamical Quantization 1) Represent $\mathcal{Q}$uad by some $\widehat{\mathcal{Q}\mathrm{uad}}$ functional of Kinematical Quantization operators. The outcome of this procedure is a quantum wave equation. One nontriviality with this is that it entails a *dynamical operator ordering* ambiguity. Another is whether $\mathcal{Q}$uad is well-defined, on account of involving compositions of operators which do not in general come with Functional Analysis reasons to be well-behaved.

Dynamical Quantization 2) Solve $\widehat{\mathcal{Q}\mathrm{uad}}\,\Psi = 0$ so as to pass from $\mathfrak{Kin}$-$\mathfrak{Hilb}$ to $\mathfrak{Dyn}$-$\mathfrak{Hilb}$ itself: the Hilbert space therein which is annihilated by $\widehat{\mathcal{Q}\mathrm{uad}}$. One finally interprets the spectra of a sufficient set of self-adjoint operators acting on this.

40.1 Operator Ordering

We next need to pay attention to how $\mathcal{Q}$uad's classical product combination,

$$\mathsf{N}^{\mathsf{AB}}(\mathbf{Q})\mathsf{P}_{\mathsf{A}}\mathsf{P}_{\mathsf{B}}, \tag{40.1}$$

gives rise to an operator ordering ambiguity upon quantizing.[1] This has consequences for the physical predictions of one's theory. Moreover, there is no estab-

[1] We include field versions at no extra cost, though we postpone treating the full GR quantum wave equations until Chap. 43, i.e. until after $\mathfrak{g}$ has been introduced.

E. Anderson, *The Problem of Time*, Fundamental Theories of Physics 190,
DOI 10.1007/978-3-319-58848-3_40

lished way to prescribe the operator ordering in the case of (model arenas of) Quantum Gravity. One approach which picks out certain operator orderings is based on an extension of the General Covariance Principle to configuration space $\mathfrak{q}$: *DeWitt's General Covariance Principle* [234].

$$\mathsf{N}^{\mathsf{AB}}(\mathbf{Q})\mathsf{P}_{\mathsf{A}}\mathsf{P}_{\mathsf{B}} \longrightarrow \frac{1}{\sqrt{\mathsf{M}}}\widehat{\mathsf{P}}_{\mathsf{A}}\{\sqrt{\mathsf{M}}\mathsf{N}^{\mathsf{AB}}(\mathbf{Q})\widehat{\mathsf{P}}_{\mathsf{B}}\} \tag{40.2}$$

is one notable implementation of this. This is working within the configuration representation. Moreover, if $\widehat{\mathsf{P}}_{\mathsf{A}}$ is additionally representable by (39.19), this construct is proportional to the Laplacian: $-\hbar^2\triangle_{\mathsf{M}}$. Thereby, it is termed the *Laplacian operator ordering*. Nor is this a unique implementation of DeWitt's General Covariance Principle, for one can include a Ricci scalar curvature term so as to have the ξ-*operator ordering*

$$\triangle^{\xi}_{\mathsf{M}} := \triangle_{\mathsf{M}} - \xi\,\mathscr{R}_{\mathsf{M}} \tag{40.3}$$

[234, 409, 441], for any real number ξ, is a more general such.[2] N.B. this gives inequivalent physics for each value of ξ, though the following limitation applies in Semiclassical Approaches.

Lemma 1 *The physics for all ξ-orderings coincides to one loop, i.e. to $O(\hbar)$.*
This was established by physicist Andrei Barvinksy in [117, 118, 120], *so we term it 'Barvinsky's first equivalence'.*

Misner [659] furthermore pointed to a unique choice among the ξ-orderings: *conformal operator ordering*

$$\triangle^{\mathrm{c}}_{\mathsf{M}} := \triangle_{\mathsf{M}} - \xi^{\mathrm{c}}\mathscr{R}_{\mathsf{M}} := \triangle_{\mathsf{M}} - \frac{k-2}{4\{k-1\}}\mathscr{R}_{\mathsf{M}}. \tag{40.4}$$

This is the unique conformally-invariant choice for each configuration space dimension $k > 1$ (which also corresponds to the relationally-meaningful values) provided that the Ψ it acts upon itself transforms as [874]

$$\Psi \longrightarrow \overline{\Psi} = \Omega^{\{2-k\}/2}\Psi. \tag{40.5}$$

Let us next pinpoint the nature of the conformal invariance referred to in conformal operator ordering. This does not apply to space itself; *Misner's identification* [659] is that it applies, rather, to the Hamiltonian constraint under scaling transformations.

$$\mathcal{H} = 0 \longrightarrow \overline{\mathcal{H}} = 0. \tag{40.6}$$

This can be generalized to the invariance

$$\mathcal{C}\text{hronos} = 0 \quad \longrightarrow \quad \overline{\mathcal{C}\text{hronos}} = 0. \tag{40.7}$$

[2] An underlying simplicity here is the exclusion of no more complicated curvature scalars, i.e. no higher-order derivatives or higher-degree polynomials in the derivatives.

The nature of the conformal invariance was moreover taken once step further back [22], to the level of actions. This reveals that Misner's identification is underlied by the PPSCT invariance (L.21) of the relational product-type parageodesic-type action. In this way, it is clear that Misner's conformal invariance applies to the kinetic arc element $\mathrm{d}s$ alongside a compensatory conformal invariance in the potential factor W. This reflects that the combination which actually features in the action $\mathrm{d}\widetilde{s}$ is not physically meaningfully splittable into kinetic and potential factors, as per Appendix L.11. PPSCTs are further appropriate in the whole-universe setting for Quantum Cosmology. In this way, demanding conformal operator ordering can be seen as choosing to retain this simple and natural invariance in passing to the quantum level.

We end by pointing to some useful simpler cases of ξ-operator ordering.

Simplification 1) For models with 2-d configuration spaces the conformal value of $\xi^{\mathrm{c}} = \{k - 2\}/4\{k - 1\}$ collapses to zero. Now conformal and Laplacian operator orderings coincide.

Simplification 2) For models with zero Ricci scalar, all of the ξ-orderings reduce to the Laplacian one.

Simplification 3) Suppose a space has constant Ricci scalar. Then the effect of a $\xi \mathscr{R}_{\mathsf{M}}$ term, conformal or otherwise, can simply be absorbed into redefining the mechanical energy or the GR cosmological constant.

40.2 Quantum Wave Equations

We next arrive at the crucial time-independent Schrödinger equation; if the above family of ξ-operator orderings is used, this takes the form

$$\widehat{chronos}\,\Psi = 0 \quad \Rightarrow \quad \triangle^{\xi}_{M}\Psi = 2\{V - E\}\Psi/\hbar^2. \tag{40.8}$$

Example 1) For scaled N-stop metroland,

$$\triangle_{\mathbb{R}^n}\Psi = \triangle^{\mathrm{c}}_{\mathbb{R}^n}\Psi = 2\{V - E\}\Psi/\hbar^2. \tag{40.9}$$

Example 2) For the generalized Klein–Gordon-type equation,

$$\Box_{M(Q,t)}\Psi = 2m(Q, t)\Psi/\hbar^2. \tag{40.10}$$

For now, this covers Minisuperspace examples, both the current Chapter's and as found in e.g. [149, 659, 760].

N.B. the first equality is due to Simplification 2); for $N = 3$ and for the isotropic Minisuperspace example, $\dim(\mathfrak{q}) = 2$, so Simplification 1) could also be evoked.

40.3 Addendum: $\mathfrak{q}$-Primality at the Quantum Level

Quantum Theory unfolding on configuration space $\mathfrak{q}$ has often been argued for, though often in the context of also arguing for Quantum Theory unfolding equally well on whichever polarization within $\mathfrak{P}$hase, of which $\mathfrak{q}$ is but one.

A further argument for $\mathfrak{q}$ primality is how Geometrical Quantization procedures [475, 633] such as the above are centred around the structure of $\mathfrak{q}$. Therein, the canonical group $\mathfrak{g}_{\text{can}}$ arises from $\mathfrak{q}$, and $\mathfrak{v}^*$ is a space this acts upon. Additionally, the point transformations *Point* remain relevant to $\mathfrak{K}$in-$\mathfrak{H}$ilb, and for the operator orderings of the constraints as well.

Of course, some other approaches involve the canonical transformations, *Can*. That classical equivalence under canonical transformations is in general broken in the passage to Quantum Theory due to the Groenewold–Van Hove phenomenon provides a further reason to question the licitness of all canonical transformations. Moreover, using just *Point* does *not* resolve out all aspects of unitary inequivalence. A deeper question is which weakening of the *Can*'s or *Point*'s at the classical level form up into classes that are preserved as unitary equivalence at the quantum level. On the other hand, the solution of the currently posed problem is likely to involve a rather larger proportion of the canonical transformations. Holding canonical transformations in doubt affects, firstly, Internal Time and Histories Theory approaches to the Problem of Time, and, secondly, Ashtekar variables approaches.

This points to use of *positive operator-valued measures* on $\mathfrak{P}$hase. These generalize both *projection-valued measures* on configuration space (in terms of which Mackey's results can be reformulated) and self-adjoint operators on Hilbert space (while remaining physically meaningful in at least some contexts). See [605] for a sharp outline of the above two terms in italics and for the reformulation in question, which is based on a combination of Appendix P.3's Borel subsets and Appendix V.7's $\mathfrak{C}^*$ algebras.

Since the above comes out frozen, there is a need to finish Temporal Relationalism off at the quantum level, which we gradually address below.

Chapter 41
Further Detail of Time and Temporal Relationalism in Quantum Theory

41.1 Time in Quantum Theory Revisited

Many of Part I's basic considerations of time in QM remain to be revisited.[1] We return to some of these in the current Chapter; further such are covered in Chaps. 42 and 48.

Two lines of discourse concerning the limited validity of Pauli's argument [item 3) of Sect. 5.3] are as follows.

Firstly, there are well-documented technical limitations with this [327]. These considerations are moreover consistent with entertaining more general notions of quantum observables. In particular, positive operator-valued measures can also be applied here, now in postulating dynamical time operators corresponding to a number of the notions of time in QM [178, 327, 328]. These match how some experiments record types of quantum time.

Secondly, Sect. 39.4's consideration of Kinematical Quantization points to a further kind of limitation. This is due to Kinematical Quantization requiring a greater diversity of commutation relations rather than the only—if most commonly considered—one which enters Pauli's argument. This follows from sensitivity to the underlying topology, whereas the topology of time itself is an unsettled matter (Fig. 1.2). Because of this, the time–Hamiltonian commutation relation does not necessarily need to take the naïve form (5.16) that is involved in Pauli's argument. In particular, for quantum GR, the 'standard' commutation relations are supplanted by affine-type ones. In cosmological models, if one or both of a Big Bang or Big Crunch hold, time forms but a half-line or an interval.

---

[1]Contrast with how Part I's considerations of time in QFT through the Wightman axioms—which capture many of these points on time—were already argued in Chap. 11 to very largely breaks down even at the level of QFTiCS. The exceptions to the 'many' are the evolution and inner product postulates running up against the Frozen Formalism Problem and the Inner Product Problem respectively in QG. These were revisited in Sect. 12.1, though the latter is further considered at the end of the current Chapter as well.

E. Anderson, *The Problem of Time*, Fundamental Theories of Physics 190,
DOI 10.1007/978-3-319-58848-3_41

Such counter-arguments moreover affect items 1) and 2) of Sect. 5.4 as well, concerning quantum clocks at the level of QG. Bearing in mind the above discrepancies, the following complementary questions are of interest.

Research Project 52) Reassess the Salecker–Wigner clock inequalities within the full QG setting.

Many of the conceptual types of time in Quantum Theory (Chap. 5 and [669, 670]) are moreover realized in QG, including further and nontrivially distinct variants of these notions arising.

Research Project 53) Match up time operator proposals in Ordinary QM with the roles ascribed to time in various Problem of Time approaches. Which time concepts of significance in Quantum Theory (Background Reading 2 in Exercises II) remain meaningful in Quantum Cosmology and in general QG?

Research Project 54) Moreover, Research Project 53)'s considerations in turn affect how Energy–Time Uncertainty Principles are to be interpreted in the QG context. GR furthermore begets many subtleties in conceptual meaning and technical realization of notions of energy. Overall, 'QG Energy–Time Uncertainty Principles' remain *very far* from being understood. Treat this more comprehensively.

41.2 The Quantum Frozen Formalism Problem

This arises at the level of the quantum quadratic constraint—the Wheeler–DeWitt equation in the case of GR—as per Chaps. 9 and 12. Chapters 9 and 15 to 18 argued moreover that this has a classical precursor which is more readily handled. There are *Tempus Ante Quantum* , *Tempus Post Quantum* and *Tempus Nihil Est* approaches to this, which are expanded upon from Chap. 12's outline in, respectively, Chap. 44, Chaps. 45–47 and Chap. 51. On the other hand, Chaps. 52 and 53–54 bypass this matter by use of Quantum Path Integrals and Histories Theory respectively. In this way, all the main classical options of Part II (cf. Fig. 13.1) are further pursued in Part III at the quantum level; a few further purely quantum variants arise as well.

Part II's main approach at the classical level involves, firstly, adopting primary-level timelessness. Secondly, abstracting time from change as Mach suggested, by which we termed this output a classical Machian emergent time. In this way, there is reconciliation with notions of time which do appear to play major roles in the local physics familiar from experience. Thirdly, we extended TRi actions to a full TRiPoD (Appendix L) and TRiFol (Chap. 34, by which Temporal Relationalism remains implemented upon incorporating each of the other local classical aspects of Background Independence.

41.3 Temporal Relationalism Implementing Canonical Quantum Theory (TRiCQT)

Part III then follows suit; a major quantum-level continuation of TRiPoD is TRiCQT (Canonical Quantum Theory). This is entirely already-TRi; it is the Semiclassical Approach and Path Integral Quantum Theory formulations that require supplanting some entities by TRi versions. It consists of the following.

1) The kinematical quantum algebraic structure's operators $\widehat{\mathbf{K}} \in \mathfrak{K}$—promoted from classical objects selected from among the $\mathcal{U}$. [So at the quantum level, one in general starts with a subalgebraic structure of the 6)'s which are functions of the 1)'s, rather than with the 1)'s themselves.]
2) Quantum commutators $|[\ ,\ ]|$: descendents of the Poisson brackets or their modification to the Dirac brackets.
3) The quantum Hamiltonian operator $\widehat{\mathscr{H}}(\widehat{\mathbf{K}})$ and the time-independent wave equation $\widehat{\mathscr{H}}(\widehat{\mathbf{K}})\Psi = E\,\Psi$ (possibly $= 0$) formed from it. In theories with zero bare Hamiltonian, this role can be taken to be adopted by a quadratic quantum constraint $\widehat{\mathcal{Q}\mathrm{uad}}$ such as the quantum Hamiltonian constraint $\widehat{\mathcal{H}}$ of GR. This is under the proviso that Quantum Theory for the whole Universe produces a timeless wave equation of the Wheeler–DeWitt type. The wave equations here making no reference to time, no time enters the wavefunctions of the Universe which solve them. Nor would it be natural for a Hilbert space inner product associated with these to refer to a time.
4) Any supplementary wave equations $\widehat{\mathcal{F}\mathrm{lin}_{\mathrm{N}}}(\widehat{\mathbf{K}})\Psi = 0$ resulting from quantum constraints $\widehat{\mathcal{F}\mathrm{lin}_{\mathrm{N}}}$, as considered in Chaps. 42–43.

   With the quantum constraints and the commutator both being change-scalar objects, the standard forms of all of
5) the quantum constraint algebraic structure,
6) the quantum beables defining equations and
7) the quantum beables algebraic structures are a priori change-scalars.

See Chaps. 49 and 50 for more about 5) to 7).

Part III's UQ scheme's need for a quantum-level T step is subsequently contemplated in this and the next thirteen Chapters, including interplay with nontrivial $\mathfrak{g}$, closure and beables as well as Timeless and Histories alternatives to T. Classical Machian emergent time itself does not survive passage from the classical to the quantum. However, as we shall see in detail in Chaps. 46 and 47, a semiclassical Machian emergent time springs up to take its place. Indeed, the main approach proposed in this book is a T . . . Q scheme with a semiclassical Machian emergent time.

41.4 Do Absolute and Relational Mechanics Give Distinct QM?

RPMs Versus Standard QM Models This follows on from the historical interest in freeing Mechanics from absolute structure. Asking whether a different sort of

QM (or generalization thereof) would have arisen if Mechanics had been cast in relational form prior to the advent of QM, is of subsequent interest [37, 746, 793]. Note that this question is not rendered obsolete by passing to special-relativistic QM or QFT on flat spacetime, by arguments such as Chaps. 6 and 11's. In this way, Minkowski spacetime $\mathbb{M}^n$'s privileged timelike Killing vector that effectively re-assumes Newtonian time's absolute role.

So, have traditional absolutist Paradigms of Physics—whether Newtonian or Minkowskian—been misleading us as regards the form taken by Quantum Theory? Or does a highly coincident theoretical framework arise from the Relational Approach? At least for the RPMs considered in detail in this and the next Chapter—Metric Shape RPMs with and without scale—it is the latter situation which largely applies.

For now, we just consider the effect of incorporating Temporal Relationalism. One advantage of similarities between Relational and Absolute QM is the ready availability of methods of calculation. However, differences in interpretation of the mathematical objects common to both cases is also pertinent. So, for instance, harmonic oscillator and nonrelativistic hydrogen mathematics recur in RPMs [37], but now the variables involved have relational whole-universe significance. In this manner, RPMs are *more useful than* harmonic oscillators or hydrogen as model arenas of whole-universe Quantum Cosmology. RPMs' technical similarity to the former is a *large bonus* since by it well-known types of calculation can be carried over to a new setting which has whole-universe quantum cosmological significance. The usefulnesses and limitations of Atomic and Molecular Physics methods and concepts are exported in this way to Quantum Cosmology in [37]. Let us consider scaled N-stop metroland [37] as a concrete $\mathfrak{g}$-free example. The free problem for this gives Bessel functions (Fig. 41.1.b). Adding a harmonic oscillator to this gives Laguerre functions (these arise in Ex II.1 and Fig. 41.1.b), which are indeed also familiar from the radial equation for hydrogen and the isotropic harmonic oscillator.

RPM models can furthermore have potentials selected in accord with the Cosmology–Mechanics analogy (Sect. 20.3). This focuses attention on cases with further resemblances with standard cosmological models.

One source of differences between Absolute and Relational QM is in closed universe effects diminishing the extent of eigenspectra by collapse in the number of admissible combinations of quantum numbers [237]. One example of this is *energy interlocking*: suppose that the total energy of a model universe with two subsystems is fixed to be E_{Uni}, and the first subsystem takes one of its energy eigenvalues E_1. The second subsystem must then have energy $E_{\mathrm{Uni}} - E_1$, which will often not be an eigenvalue, in which case E_1 ceases to be part of the first subsystem's eigenspectrum. This is a type of closed-universe feature which was already known to DeWitt [237]. These collapses indeed violate the Cluster Decomposition Principle (Sect. 6.5). This occurs furthermore for a lucid reason: whole closed universes lie outside of the conceptual remit of the flat spacetime QFT framework realm of this principle. Finally note that such collapses need not apply in settings in which E_{Uni}

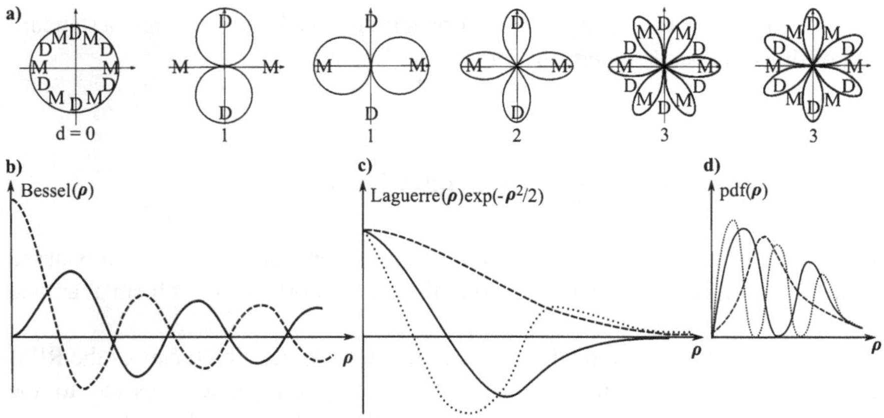

Fig. 41.1 **a)** Pure-shape—or separated-out shape part of—some of the quantum wavefunctions for 3-stop metroland in some particularly shape theoretically meaningful cases. These are clearly the simpler 2-*d* analogue of the spherical harmonics orbitals. The solutions here are flowers of 2d petals, for d the dilational quantum number. **b)** Bessel functions and **c)** Laguerre functions, as occur in the 'radial' scale ρ part of this problem in the free and harmonic oscillator cases respectively. For the latter,the corresponding confined probability density functions are sketched in **d)**

takes a range of values, which it could in some multiverse interpretations. [Yet further differences are best discussed after increasing the repertoire of examples to $\mathfrak{g}$-nontrivial theories, and so are deferred to the next Chapter.]

Interpreting Quantum Theory in Closed Universes The usual Copenhagen interpretation of QM cannot apply to the whole-universe models. In Ordinary QM, one presupposes that the quantum subsystem under study has an ambient classical world, crucial parts of which are the observers and the measuring apparatus. In familiar situations, Newtonian Mechanics turns out to give an excellent approximation for this classical world. However, there are notable conceptual flaws with extending this 'Copenhagen' approach when applied to the whole Universe. This is because observers and measuring apparatus are themselves made out of quantum-mechanical matter, and are always coupled at some level to the quantum subsystem. Treating them as such requires further observers or measuring apparatus so the situation repeats itself. But this clearly breaks down upon consideration of the whole Universe.

The *Many-Worlds Interpretation* is an alternative, though we do not discuss it further in this book (see e.g. [877] for some pros and [545] for some cons). A number of further replacements are tied to various Problem of Time strategies such as the Conditional Probabilities Interpretation, Histories Theory and Records Theory, as covered in subsequent Chapters.

Solutions of Simple Minisuperspace Model Wave Equations These also have tractable mathematics with some common ground with Ordinary QM [149, 433, 659, 760]. Equation (40.10) is elementary to solve for $V(\phi) = 0$, and $\exp(6\,\Omega)V(\phi)$

can subsequently be treated as an interaction term to which one can apply a standard form of time-independent perturbation theory.

41.5 Inner Product and Adjointness Issues

This Section is but a technical addendum of subsequent use in the book; it is based on the Inner Product Problem being part of the Frozen Formalism Problem as argued in Chap. 12.

Firstly, the definiteness-indefiniteness difference (Appendix O) causes the RPM time-independent Schrödinger equation to be elliptic-type, as opposed to the hyperbolic-type Wheeler–DeWitt equation of GR. In this way, the techniques required for RPMs concern solution of one or both of constrained and curved-space elliptic-type equations.

Secondly, we know from Part I that a Schrödinger inner product suffices for RPM. One might then hope that a Klein–Gordon inner product would suffice for Minisuperspace. However, Klein–Gordon type equations with non-constant mass terms in general give rise to further significant complications; see Sect. 12.2 and Ex VI.11.vi)

Thirdly, Minisuperspace models are on some occasions also presented in terms of a first-order square-rooted equation in place of a second-order Klein–Gordon equation. Moreover, the latter is inequivalent in the general time-dependent case. This is due to the first square-root operator $\widehat{\mathcal{S}}$ in general acting nontrivially upon the second in $\widehat{\mathcal{S}}\widehat{\mathcal{S}}\Psi = 0$ [149]. Consequently

$$\widehat{\mathcal{S}}\widehat{\mathcal{S}}\Psi = 0 \text{ is not equivalent to } \widehat{\mathcal{S}}\Psi = 0\,. \tag{41.1}$$

Fourthly, consider sending $\widehat{chronos}\,\Psi = E\,\Psi$ to the conformally related $\overline{chronos}\,\overline{\Psi} = \overline{E}\,\overline{\Psi} = \{E/\Omega^2\}\overline{\Psi}$. In this case, the eigenvalue problem has a weight function Ω^{-2} which features in the inner product:

$$\int_{\mathfrak{q}} \overline{\Psi_1}^*\overline{\Psi_2}\Omega^{-2}\sqrt{\mathsf{M}}\,\mathrm{d}^k x. \tag{41.2}$$

In the RPM case, $\langle\mathfrak{q}, \mathsf{M}\rangle$ denotes the relationalspace portmanteau at the level of Riemannian Geometry. This inner product additionally succeeds in attaining PPSCT invariance. It is equal to

$$\int_{\mathfrak{q}} \Psi_1^*\Omega^{\{2-k\}/2}\Psi_2\Omega^{\{2-k\}/2}\Omega^{-2}\sqrt{\mathsf{M}}\,\Omega^k\mathrm{d}^k x = \int_{\mathfrak{q}} \Psi_1^*\Psi_2\sqrt{\mathsf{M}}\,\mathrm{d}^k x \tag{41.3}$$

in the PPSCT representation which is mechanically natural in the sense that E comes with the trivial weight function, 1. (Cf. [659] for the Minisuperspace case.)

Let us finally introduce the *Rieffel induced inner product* [434] (after mathematical physicist Marc Rieffel)

$$\int_{\mathfrak{q}} \mathrm{d}x \, \Psi_1^* \Psi_2 = \delta(0)(\Psi_1^* \circ_I \Psi_2), \tag{41.4}$$

for $\circ_I$ an induced product [603]. (41.4) can additionally be thought of as a renormalized inner product. This is an alternative to the Klein–Gordon-type inner product as regards Minisuperspace calculations, and a useful means of handling quantum Hamiltonians which have continuous spectra.

Chapter 42
Geometrical Quantization with Nontrivial $\mathfrak{g}$. i. Finite Theories

42.1 Configurational Relationalism at the Quantum Level

We next bring in a third Problem of Time facet, choosing Configurational Relationalism since at the classical level this complements Temporal Relationalism as the other Constraint Provider. In particular, Configurational Relationalism at the classical level produces $\mathcal{S}$**huffle** constraints, to Temporal Relationalism's $\mathcal{C}$hronos constraint. The former are subsequently checked to be of the form $\mathcal{F}$**lin** at the classical level, by means of a Dirac-type Algorithm.

On the one hand, Dirac Quantization [250] involves tackling these constraints at the quantum level: Q ... R. On the other hand, Reduced Quantization, involves solving the $\mathcal{F}$**lin** classically prior to initiating Quantization R ... Q. We follow [483, 586] here in terming procedures in which $\mathcal{C}$hronos is also solved at the classical level '*Tempus Ante Quantum* ' rather than a further type of reduction; T ... Q is a briefer summary name. Finally, the direct approach to Configurational Relationalism gives a reduced $\widetilde{\mathcal{C}\text{hronos}}$ without passing through an elimination of any $\mathcal{F}$**lin**.

In the direct approach—or in simple modelling for which $\mathfrak{g} = id$ so the $\mathcal{F}$**lin** are absent in the first place such as in the previous Chapter—the notions of Dirac and Reduced Quantization coincide. On the other hand, when the $\mathcal{F}$**lin** are present, the outcomes of Dirac and Reduced Quantization are quite generally distinct [76, 117, 118, 740], since

$$\text{reduce and quantize operations are not expected to commute.} \tag{42.1}$$

At least on some occasions, however, operator orderings can be chosen by which these do match up.

In the reduced approach, one hopes that classically resolved Configurational Relationalism remains resolved at the quantum level, or, more generally, that a classically consistent $\langle \mathfrak{q}, \mathfrak{g}, \mathsf{S} \rangle$ triple produces a consistent quantum theory for the same $\mathfrak{g}$. This can readily be arranged for the RPMs considered, but can arise more generally as some of the of the Quantum Constraint Closure Problems postponed to Chap. 49.

E. Anderson, *The Problem of Time*, Fundamental Theories of Physics 190,
DOI 10.1007/978-3-319-58848-3_42

If one is considering an r-formulation, one can use the standard notions for each quantum-level structure in turn. If not, one can at least formally consider a $\mathfrak{g}$-act $\mathfrak{g}$-all approach at the quantum level. E.g. given a quantum operator $\widehat{\mathcal{O}}$ that is not $\mathfrak{g}$-independent,

$$\widehat{\mathcal{O}}_{\mathfrak{g}\text{-free}} := \int_{g \in \mathfrak{g}} \mathbb{D}g \exp\Big(i \sum_{g \in \mathfrak{g}} \overrightarrow{\mathfrak{g}}_g\Big) \widehat{\mathcal{O}} \exp\Big(-i \sum_{g \in \mathfrak{g}} \overrightarrow{\mathfrak{g}}_g\Big) \tag{42.2}$$

for $\overrightarrow{\mathfrak{g}}_g$ a Configurationally Relational counterpart. This particular $\mathfrak{g}$-act, $\mathfrak{g}$-all procedure is now based on the exponentiated adjoint group action followed by integration over the group (see also [641]). One can similarly pass from kets $|\Psi\rangle$ that are not $\mathfrak{g}$-independent to

which are not Configurationally Relational to ones which are:

$$|\Psi_{\mathfrak{g}\text{-free}}\rangle := \int_{g \in \mathfrak{g}} \mathbb{D}g \exp\Big(i \sum_{\mathfrak{g} \in \mathfrak{g}} \overrightarrow{\mathfrak{g}}_g\Big) |\Psi\rangle. \tag{42.3}$$

[A further possibility is that the averaging might also be attempted at a subsequent level of structure such as the actually physically meaningful expectations.] Other applications of operators and wavefunctions such as finding operator eigenvalues or constructing a wavefunction basis have enough partial meaning to on occasion be considered for individual $\mathfrak{g}$-act $\mathfrak{g}$-all moves.

Such *Group-averaging Quantizations* are—at least conceptually—variants on Dirac Quantization, and are valid quantum implementations of Configurational Relationalism.

In passing to the quantum level, the Relational Approach has, furthermore, shed most of its distinctive TRi features, as per Sect. 41.3.

42.2 Dirac Quantization of Finite Models

For $\mathfrak{g}$-nontrivial cases, one quite often solves the $\widehat{\mathcal{F}\mathbf{lin}}\,\Psi = 0$ prior to handling the $\widehat{\mathcal{Q}\mathrm{uad}}\,\Psi = 0$. This amounts to passing from $\mathfrak{Kin}$-$\mathfrak{Hilb}$ to $\mathfrak{Dyn}$-$\mathfrak{Hilb}$ via a $\mathfrak{g}$-$\mathfrak{Hilb}$ of quantum wavefunctions that solve just $\widehat{\mathcal{F}\mathbf{lin}}\,\Psi = 0$. I.e. the space of those $\mathfrak{Kin}$-$\mathfrak{Hilb}$ states which are annihilated by $\widehat{\mathcal{F}\mathbf{lin}}$ but not necessarily by $\widehat{\mathcal{Q}\mathrm{uad}}$. Another alternative is to solve $\widehat{\mathcal{F}\mathbf{lin}}\,\Psi = 0$ and $\widehat{\mathcal{Q}\mathrm{uad}}\,\Psi = 0$ together as a package. One way in which this is relevant is via not all theories' constraints being tractable as split up into independent $\mathcal{F}\mathbf{lin}$ and $\mathcal{Q}\mathrm{uad}$ packages, due to the presence of integrability conditions (Chap. 24).

Promoting $\mathcal{F}\mathbf{lin}$ to $\widehat{\mathcal{F}\mathbf{lin}}$ can on some occasions entail operator-ordering ambiguities. The following result is useful on some occasions.

Lemma *Suppose that one is in the configuration representation in cases in which the momenta can additionally be represented by* (39.19). *Then if the* $\mathcal{F}\mathbf{lin}$ *are* *operator-ordered with* P *to the right,* $\Psi = \Psi[\kappa_{\mathsf{c}}$ *alone*], *where the* κ_{c} *are classical configurational Kuchař beables.*

Proof With this momentum representation and operator ordering, the classical expression (24.6) for the $\mathcal{F}$**lin** becomes

$$\widehat{\mathcal{F}\mathbf{lin}} = \frac{\hbar}{i} F^{\mathsf{A}}{}_{\mathsf{N}} \frac{\partial}{\partial \mathsf{Q}^{\mathsf{A}}} \Psi = 0. \tag{42.4}$$

This furthermore imposes an equation on Ψ which coincides (up to proportionality) with the classical configurational Kuchař beables equation (25.12). □

For RPMs, the unreduced $\mathfrak{q} = \mathbb{R}^{nd}$, for which

$$\mathfrak{K} = \mathfrak{v} \rtimes \mathfrak{g}_{\mathrm{can}} = \mathbb{R}^{nd} \rtimes Eucl(nd) = Heis(nd). \tag{42.5}$$

The selected objects are now $\widehat{\underline{\rho}}^A$, $\widehat{\underline{p}}_A$ and the shape conserved quantities $\mathrm{s}^{\Gamma\Delta}$, which together obey the obvious standard commutation relations.

Example 1) For Metric Shape and Scale RPM,

$$\widehat{\mathcal{L}}\Psi = \frac{\hbar}{i}\sum_{A=1}^{n} \underline{\rho}^A \times \frac{\partial \Psi}{\partial \underline{\rho}^A} = 0 \tag{42.6}$$

holds [746, 793]. This means that $\Psi = \Psi(- \cdot -)$ (in Fig. E.4's notation). Then finally

$$\widehat{\mathcal{E}}\Psi = -\frac{\hbar^2}{2}\Delta_{\mathbb{R}^{nd}}\Psi + V(- \cdot -)\Psi = E\,\Psi. \tag{42.7}$$

Example 2) Metric Shape RPM has the linear constraints (42.6) and [37]

$$\widehat{\mathcal{D}}\Psi = \frac{\hbar}{i}\sum_{A=1}^{n} \underline{\rho}^A \cdot \frac{\partial \Psi}{\partial \underline{\rho}^A} = 0. \tag{42.8}$$

Together, these enforce $\Psi = \Psi(- \cdot -/- \cdot -)$. Furthermore,

$$\widehat{\mathcal{E}}\Psi = -\frac{\hbar^2}{2}\Delta_{\mathbb{R}^{nd}}\Psi + V(- \cdot -/- \cdot -)\Psi = E\,\Psi. \tag{42.9}$$

Example 3) 3-d Conformal Shape RPM has [36]

$$\widehat{\mathcal{K}}_i\Psi = \frac{\hbar}{i}\sum_{I=1}^{N}\{|\underline{q}^I|^2\delta_i{}^j - 2q_i{}^I q^{jI}\}\frac{\partial\Psi}{\partial q^{jI}} = 0, \tag{42.10}$$

which in conjunction with

$$\widehat{\underline{\mathcal{P}}}\Psi = \frac{\hbar}{i}\sum_{I=1}^{N}\frac{\partial\Psi}{\partial \underline{q}^I} \tag{42.11}$$

and the $\boldsymbol{q}$-versions of (42.6) and (42.8) signify that $\Psi = \Psi(\angle)$. Finally,

$$\widehat{\mathcal{E}}\,\Psi = -\frac{\hbar^2}{2} I\, \triangle_{\mathbb{R}^{Nd}} \Psi + V(\angle)\Psi = E\,\Psi. \tag{42.12}$$

Example 4) 2-d Affine Shape RPM has [36] linear constraints

$$\widehat{\underline{\mathcal{G}}}\,\Psi = \frac{\hbar}{i} \sum_{A=1}^{n} \underline{\rho}^A \underline{\underline{G}} \frac{\partial \Psi}{\partial \underline{\rho}^A} = 0. \tag{42.13}$$

Thus $\Psi = \Psi(- \times - / - \times -)$. There is moreover a further constraint $\mathcal{E}$, which, in the case of 4-particle Affine Shape RPM, takes the form

$$\widehat{\mathcal{E}}\,\Psi = -\frac{\hbar^2}{2} \sum_{\text{cycles } C,D=1}^{3} \left(\underline{\rho}^C \times \underline{\rho}^D\right)_3 \sum_{\text{cycles } A,B=1}^{3} \left(\frac{\partial}{\partial \underline{\rho}^A} \times \frac{\partial}{\partial \underline{\rho}^B}\right)_3 \Psi$$
$$+ V(- \times - / - \times -)\Psi = E\,\Psi. \tag{42.14}$$

Example 5) As a simple case of quantum-level $\mathfrak{g}$-averaging, for the Dirac Quantization approach to triangleland, $SO(2) = U(1)$, rotation-averaged operators are given by

$$\mathsf{S}_{g \in \mathfrak{g}} = \int_{g \in \mathfrak{g}} \mathbb{D}\boldsymbol{g} \times = \int_{\zeta \in \mathbb{S}^1} \mathbb{D}\zeta \times = \int_{\zeta=0}^{2\pi} \mathrm{d}\zeta \times. \tag{42.15}$$

Here, $\vec{\mathfrak{g}}_g$ is the infinitesimal 2-d rotation by the matrix $\underline{\underline{R}}_\zeta$ acting on the vectors of the model, where ζ here denotes angle of absolute rotation.

42.3 Reduced Quantization of Finite Models

RPM Examples: Kinematical Quantization

Example 1) For Metric Shape RPMs in 1-d, $\mathfrak{q}^{\mathrm{r}} = \mathbb{S}^{n-1}$, so the mathematics of Example 7) of Sect. 39.5 carries over as regards Kinematical Quantization.
Example 2) In 2-d,

$$\mathfrak{q}^{\mathrm{r}} = \mathbb{CP}^{n-1} \cong \frac{SU(n)}{SU(n-1) \times U(1)},$$

so this also lies within the remit of Sec's 39.5 general method. In this case, evoking the homogeneous polynomials of degree 2 in $\mathbb{C}^n$ [61],

$$\mathfrak{K} = \mathrm{HomPoly}\big(\mathbb{C}^n, 2\big) \rtimes SU(n). \tag{42.16}$$

Note moreover that these succeed in complying with the opposing restrictions imposed by Sect. 39.1's criterion II) and Sect. 39.4's criterion V).

Example 3) The special triangleland subcase of this yields

$$\mathfrak{K} = \mathbb{R}^3 \rtimes SO(3) = Eucl(3) \tag{42.17}$$

by some well-known maps (Exercise!). The Kinematical Quantization operators here are 1) the Dragt coordinates Dra^Γ, and 2) the $\widehat{S}^\Gamma$, which are relative angular momenta and mixed relative dilational and relative angular momenta [28].

Example 4) See [61] for a detailed account of the more typical case of the quadrilateral, for which $\mathfrak{q}^\mathrm{r} = \mathbb{CP}^2$. The above results are supported by how the fundamental groups

$$\pi_1\big(\mathfrak{S}(N,2)\big) = \pi_1\big(\mathbb{CP}^k\big) \quad \text{being all trivial.} \tag{42.18}$$

Example 5) For Metric Shape and Scale RPMs, the 1-d case requires no reduction, and so was already covered in Sect. 39.5.

Example 6) The 2-d case has $\mathfrak{q}^r$ be the cones $\mathrm{C}(\mathbb{CP}^{n-1})$, which are rather less thoroughly explored spaces.

Example 7) The special triangleland case has

$$\mathfrak{K} = \mathbb{R}^3 \rtimes Eucl(3) = Heis(3). \tag{42.19}$$

The Kinematical Quantization operators here are the Dra^Γ, their conjugates the P^{Dra}_Γ, and the shape conserved $SO(3)$ quantities $\widehat{S}^\Gamma$. This example also exhibits low-order polynomiality and trivial second Chern classes.

Laplacians I denote the general shape space Laplacian by $\triangle_{\mathfrak{S}(N,d)}$.

Example 1) For Metric Shape RPM in 1-d,

$$\triangle_{\mathbb{S}^{n-1}} = \frac{1}{\sin^{nd-1-\mathsf{A}}\theta_\mathsf{A} \prod_{i=1}^{\mathsf{A}-1} \sin^2\theta_i} \frac{\partial}{\partial \theta_\mathsf{A}} \left\{ \sin^{nd-1-\mathsf{A}}\theta_\mathsf{A} \frac{\partial}{\partial \theta_\mathsf{A}} \right\}. \tag{42.20}$$

Example 2) In 2-d,

$$\begin{aligned}\triangle_{\mathbb{CP}^{n-1}} = \frac{\{1+\|R\|^2\}^{2n-2}}{\prod_{p=1}^{n-1} R_p} &\left\{ \frac{\partial}{\partial R_p} \left\{ \frac{\prod_{p=1}^{n-1} R_p}{\{1+\|R\|^2\}^{2n-3}} \{\delta^{pq} + R^p R^q\} \frac{\partial}{\partial R_q} \right\}\right. \\ &\left. + \frac{\partial}{\partial \Theta_{\tilde{\mathrm{p}}}} \left\{ \frac{\prod_{p=1}^{n-1} R_p}{\{1+\|R\|^2\}^{2n-3}} \left\{ \frac{\delta^{\tilde{\mathrm{p}}\tilde{\mathrm{p}}}}{R_{\tilde{p}}^2} + \mathbb{I}^{\tilde{\mathrm{p}}\tilde{\mathrm{p}}} \right\} \frac{\partial}{\partial \Theta_{\tilde{\mathrm{p}}}} \right\} \right\}.\end{aligned} \tag{42.21}$$

This expression makes use of the multipolar form for the inhomogeneous coordinates $Z_q =: R_q \exp(i\Theta_q)$.

Example 3) The special triangleland case of the above can furthermore be cast as a the standard spherical Laplacian time-independent Schrödinger equation, albeit in the relational Θ, Φ coordinates (Appendix G.3).

Examples 4) to 6) Metric Shape and Scale RPMs, using the scale–shape split form and (G.5),

$$\triangle_{\mathfrak{R}(N,d)} = \triangle_{\mathrm{C}(\mathfrak{S}(N,d))} = \rho^{k(N,d)}\partial_\rho\{\rho^{-k(N,d)}\partial_\rho\} + \rho^{-2}\triangle_{\mathfrak{S}(N,d)}. \quad (42.22)$$

Building the scaled RPM Laplacians corresponding to all the above pure-shape ones is straightforward.

Conformal Laplacians By the configuration space dimension being 2 in each case, $\triangle^{\mathrm{c}}_{\boldsymbol{M}} = \triangle_{\boldsymbol{M}}$ for pure-shape 4-stop metroland and triangleland, and for scaled 3-stop metroland. Simplification 3) applies to pure-shape N-stop metrolands and N-a-gonlands. As regards specific values of the constants (Exercise!) for pure-shape N-stop metroland,

$$\triangle^{\mathrm{c}}_{\mathbb{S}^{n-1}} = \triangle_{\mathbb{S}^{n-1}} - \{n-1\}\{n-3\}/4. \quad (42.23)$$

Pure-shape N-a-gonland has instead

$$\triangle^{\mathrm{c}}_{\mathbb{CP}^{n-1}} = \triangle_{\mathbb{CP}^{n-1}} - 2n\{n-1\}\{n-2\}/\{2n-3\}. \quad (42.24)$$

Finally, for scaled N-a-gonland,

$$\triangle^{\mathrm{c}}_{\mathrm{C}(\mathbb{CP}^{n-1})} = \triangle_{\mathrm{C}(\mathbb{CP}^{n-1})} - 3n\{2n-3\}/4\{n-1\}\rho^2. \quad (42.25)$$

We subsequently use the following notation for the numerical coefficient of this subtracted-off term:

$$c(N,d) := 0 \quad \text{for } d = 1 \quad \text{and} \quad 3n\{2n-3\}/4\{n-1\} \quad \text{for } d = 2. \quad (42.26)$$

Time-Independent Schrödinger Equations and Simple Solutions For Metric Shape RPM,

$$\triangle^{\mathrm{c}}_{\mathfrak{S}(N,d)}\Psi = 2\{V - E\}\Psi/\hbar^2, \quad (42.27)$$

whereas for Metric Shape and Scale RPM,

$$\triangle^{\mathrm{c}}_{\mathfrak{R}(N,d)}\Psi = 2\{V - E\}\Psi/\hbar^2. \quad (42.28)$$

The classically-reducible scaled-RPM series' time-independent Schrödinger equations are therefore encapsulated by the $d = 1, 2$ [61] cases of

$$-\hbar^2\{\partial_\rho^2 + k(N,d)\rho^{-1}\partial_\rho + \rho^{-2}\{\triangle_{\mathrm{S}(N,d)} - c(N,d)\}\Psi = 2\{E_{\mathrm{Uni}} - V(\rho, S^{\mathrm{a}})\}\Psi. \quad (42.29)$$

Equation (42.29) furthermore separates into scale and shape parts for a number of suitable V. The scale part of this has been solved for the general free and isotropic harmonic oscillator potential cases [37, 61], giving Bessel and Laguerre functions (Fig. 41.1.b–d). On the other hand, some case of the shape part of this have been solved in [37, 59, 61]. This gives spherical harmonics mathematics for

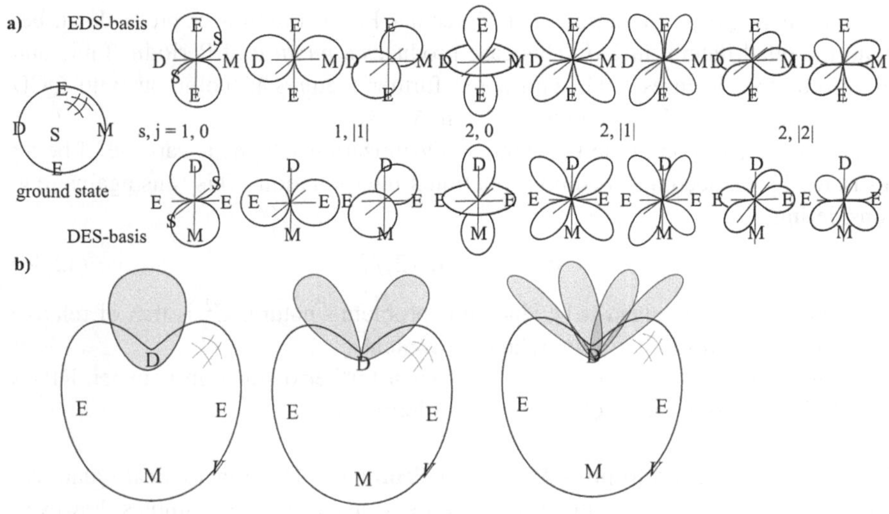

Fig. 42.1 **a)** Wavefunctions for triangleland for the bases with ordered axes EDS and DES (defined in Appendix G). The solutions here are spherical harmonics, corresponding to total shape momentum s with MD direction ('z') component j. **b)** The corresponding harmonic oscillators have quantum wavefunctions confined in Fig. 16.2's well. See [37] for sketches of the types of function involved in the 'radial' scale factor of the wavefunction

4-stop metroland and triangleland (Fig. 42.1), hyperspherical harmonics for N-stop metroland more generally, and products of Jacobi polynomial and Wigner D-function terms [1] for quadrilateralland [61].

RPMs Versus Ordinary QM Revisited Following on from Sects. 41.4 and 42.3, mathematics familiar from Absolute QM clearly persists in $\mathfrak{g}$-nontrivial RPMs as well [37]. One underlying reason for close analogies between simple Absolute and Relational QM is that both involve $SO(n)$ mathematics. For while RPMs have zero total (absolute) angular momentum, they still possess relative angular momentum in $d \geq 2$. Even in $1 - d$ they still possess relative dilational momentum which has the same mathematics as angular momentum. Many aspects of the quantum theory of $SO(n)$ are moreover independent of the physical interpretation. E.g. addition of angular momentum applies just as well to relative angular momentum as it does to absolute angular momentum, and indeed also to relative dilational momentum and to mixtures of relative angular and dilational momenta.

Energy interlocking is now accompanied by *angular momentum counter-balancing* [37]. I.e. so as to comply with $\underline{\mathcal{L}} = 0$, the relative angular momenta of the constituent parts of a system must have angular momentum quantum numbers which cancel out. Furthermore, effects solely arising from the inclusion of the entirety of an infinite system are to be doubted by 'Earman's Principle', whereas inclusion of the entirety of a finite system is a non-issue. As such, there is no problem with finite model whole-universe effects along these lines, but there may be problems with them in Field Theory models, including SIC.

Tight analogies with Atomic and Molecular Physics moreover break down beyond the quadrilateral in 2-*d*. This case produces a *mixture* of Periodic Table and Eightfold Way features [61], with some further features in common with QCD 'colour' since $Isom(\mathbb{CP}^2)$ is, in more detail, $SU(3)/\mathbb{Z}_3$.

The $\mathfrak{v}^*$ acted upon in the Kinematical Quantization scheme is capable of being restricted by its needing to admit a relational interpretation. This runs against e.g. considering

$$\mathfrak{K} = \mathbb{C}^2 \rtimes SU(2)/\mathbb{Z}_2 \tag{42.30}$$

for pure-shape triangleland [61], since this problem's natural $\mathbb{C}^2$ is that of relative coordinates which still carry nonrelational content.

Finally Problem 1) of Sect. 42.4 provides a further distinction between RPMs and Ordinary QM at the level of operator ordering.

The Reduced SIC Example's Modewise Problem The Kinematical Quantization for this involves $\mathbb{M}^4$. Unlike Sect. 39.4's Minisuperspace examples, however, one is now restricted to a local-in-'time' slab $\mathbb{R}^3 \times \mathfrak{T}$ (for $\mathfrak{T}$ a time interval), and this feature is known to impinge on the outcome of Kinematical Quantization as per Example 6) of Sect. 39.5. Kinematical Quantization operators are now[1]

$$\begin{aligned} &v_{i\mathrm{n}}, \quad \zeta_{\mathrm{n}}, \quad \sqrt{1-\zeta_{\mathrm{n}}^2}, \quad \frac{\partial}{\partial v_{i_{\mathrm{n}}}}, \quad \sqrt{1-\zeta_{\mathrm{n}}^2}\frac{\partial}{\partial \zeta_{\mathrm{n}}}, \\ &v_{j\mathrm{n}}\frac{\partial}{\partial v_{i_{\mathrm{n}}}} - v_{i\mathrm{n}}\frac{\partial}{\partial v_{i_{\mathrm{n}}}}, \quad \sqrt{1-\zeta_{\mathrm{n}}^2}\frac{\partial}{\partial v_{i_{\mathrm{n}}}}, \quad \zeta_{\mathrm{n}}\frac{\partial}{\partial v_{i_{\mathrm{n}}}}. \end{aligned} \tag{42.31}$$

These are self-adjoint on

$$\mathfrak{Kin}\text{-}\mathfrak{Hilb} = \mathfrak{L}^2\left(\mathbb{R}^3 \times \mathfrak{T},\ \mathrm{d}s_{\mathrm{n}}\mathrm{d}d_{\mathrm{n}}^o\mathrm{d}d_{\mathrm{n}}^e\frac{\mathrm{d}\zeta_{\mathrm{n}}}{\sqrt{1-\zeta_{\mathrm{n}}^2}}\right). \tag{42.32}$$

The corresponding quantum wave equation additionally lies within the remit of (40.10), albeit now for a somewhat more complicated form of the Laplacian due to its being built out of a less straightforward explicit form for $\widehat{p}_{\zeta_{\mathrm{n}}}$:

$$-\hbar^2\left\{-\sqrt{1-\zeta_{\mathrm{n}}^2}\frac{\partial}{\partial \zeta_{\mathrm{n}}}\sqrt{1-\zeta_{\mathrm{n}}^2}\frac{\partial}{\partial \zeta_{\mathrm{n}}} + \triangle_{v_{\mathrm{n}}}\right\}\Psi + V(\zeta_{\mathrm{n}}, v_{i\mathrm{n}})\Psi = 0. \tag{42.33}$$

This amounts to a further correction on [34]'s sketch of a wave equation. Both this and Halliwell and Hawking's earlier quantum wave equation—based on supposing that the SVT split pieces of the classical constraints remain first-class constraints, which they do not by Chap. 30—are of this schematic form as well. The latter is

[1]This ζ_{n} is linearly scaled relative to the originally introduced ζ_{n}, so that the two ends of the slab are set to ± 1. This just amounts to use of the freedom of tick-duration and of calendar year zero, and is helpful as regards casting subsequently encountered mathematical entities into standard form.

clearly additionally supplemented by a number of linear quantum constraints as befits a Dirac Quantization. In this manner, study of (40.10) covers SIC in addition to Sect. 40.2's variable mass Klein–Gordon and Minisuperspace applications, making it a particularly worthwhile family of models to study (Chaps. 43–46).

N.B. that (42.33) continues SIC schemes' feature of splitting into scalar S (i.e. now scalar sum) and tensor T parts; each of these is coupled to the scale variable but not directly to the other).

42.4 Three More Operator Ordering Problems

1) Discrepancy Between Quantum Cosmology and Molecular Physics This is between the 'relational portion' of Newtonian Mechanics and that of r-approaches. The former possesses a chain of transformations as conventionally used in Molecular Physics [624]. I.e. from particle positions to relative Jacobi coordinates to spherical coordinates to Dragt-type coordinates, in each case plus absolute angles [513]. This scheme consequently has an absolute block in its configuration space metric. This participates in the formation of the overall volume element $\sqrt{M_{\text{abs-rel}}}$, which subsequently enters the *relational* block's part of the Laplacian. For, since $\sqrt{M_{\text{abs-rel}}}$ sits inside the $\partial/\partial Q_{\text{rel}}$ derivative, by this means, if absolute space is assumed, it leaves an imprint on the 'relational portion'. This mechanism is however absent if one considers a relationally-motivated Lagrangian as in r-approaches.

In any case, it is reasonable from the relational perspective that modelling a molecule in a universe differs from modelling particles as a whole-universe model arena. This is because the first of these possesses an inertial frame concept due to the rest of the model universe. Suppose that one takes the RPM–Geometrodynamics analogy as primary rather than trying to describe reality in terms of a few (or even very many) non-relativistic particles. This gives serious reason *not* to use Molecular Physics' quantum equations in the whole-universe model context. This mathematical analogy ends with the Classical Dynamics and the quantum *kinematics*.

In more detail [37], this difference is a consequence of the nontriviality of the rotations. In the case of triangleland,

$$\partial_{\mathrm{I}}^2 + g\,\mathrm{I}^{-1}\partial_{\mathrm{I}}. \tag{42.34}$$

arises in each context, but with $g = 2$ in a purely relational formulation, $g = 3$ within 2-*d* absolute space, and $g = 5$ within 3-*d* absolute space. These correspond to $SO(2)$ imprinting an extra 1, and $SO(3)$ an extra 3, on top of the natural spherical polar coordinates' 2 that occurs within each of these cases. The quantum cosmologically inspired conformal operator ordering's curvature term is also a further source of discrepancies between such formulations.

2) Dirac–Reduced Inequivalence Laplacians do not in general map to Laplacians under reduction, due to extra hypersurface geometry terms appearing [37]. Conformal operators do not in general map to conformal operators both for this reason and because the reduction diminishes the dimension, so it moves between cases with different conformal correction coefficients. By these geometrical effects, reducing and selecting a Laplacian or conformal operator ordering are not in general commuting operations.

Because of this facet interference, specifying Laplacian or conformal operator ordering is by itself unsatisfactory. Such a statement needs to be supplemented by whether it is to be carried out before or after reduction. The suggested way out follows from DeWitt's General Covariance of $\mathfrak{q}$ argument making best sense in the case involving solely true degrees of freedom, rather than these mixed with gauge ones. Such an operator ordering prescription should be applied to the most reduced configuration space. [Unfortunately, this leaves one stuck in the general case, since this has no explicit form for the most reduced configuration space.]

There is however a modicum of approximate protection from this noncommutation of procedures, this is due to the following result.

Lemma (Barvinsky's [117, 118] second approximate equivalence). *To* 1 *loop—i.e. to* $O(\hbar)$*—the Laplacian operator ordering coincides for reduced and Dirac schemes.*

Furthermore, to this order, all ξ-operator orderings coincide by Barvinsky's first equivalence. In particular Barvinsky's second equivalence additionally applies to conformal operator orderings.

3) Breakdown of Conformal Operator Ordering in Descent of Level of Structure E.g. in Affine Shape RPM, the form of (42.14) does not produce a Laplacian (or ξ- or conformal modification thereof) upon Quantization. So there is an additional level of mathematical structure limitation on adopting Laplacian and conformal operator orderings.

Chapter 43
Geometrical Quantization with Nontrivial $\mathfrak{g}$. ii. Field Theories and GR

43.1 Further QFT Subtleties

This expands on Chap. 6 as regards further Background (In)dependence issues. QFT techniques usually do not carry over to Background Independent settings, yet there is still value in learning why.

Subtlety 1) The S-matrix is a Background Dependent notion. This is a significant point to make due to the S-matrix playing a very major role in QFT and perturbative String Theory.

Subtlety 2) *Regularization* is also a Background Dependent notion. The mathematical meaningfulness of QG FDEs is subject to (O.11) to be resolved by regularization (O.12). Different methods for this exist; end results are moreover required to be method-independent.

Example 1) *Pauli–Villars regularization* involves

$$\mathcal{T}^{\text{phys}} = \lim_{M \to \infty} \big(\mathcal{T}(m) - \mathcal{T}(M)\big), \tag{43.1}$$

where M is a 'regulator mass and $\mathcal{T}$ denoting the transition amplitude.

Example 2) In *dimensional regularization*, if one's intent is to calculate in dimension 4, one works with dimension $n > 4$ and then takes $\lim_{n \to 4}$.

Example 3) In *point-splitting regularization*, products of operators at a coincident point are resolved according to

$$\widehat{\mathcal{S}}(x)\widehat{\mathcal{O}}(x) \to \widehat{\mathcal{S}}(x + \epsilon/2)\widehat{\mathcal{O}}(x - \epsilon/2) \tag{43.2}$$

with $\lim_{\epsilon \to 0}$ taken at the end of the calculation (cf. 'coincidence limit' in Sect. 11.3).

Example 4) The more modern ζ-function and heat kernel regularizations [438] are more widely applicable, e.g. for de Sitter QFTiCS, black hole event horizons, as well as in CFT and consequently in perturbative String Theory.

E. Anderson, *The Problem of Time*, Fundamental Theories of Physics 190,
DOI 10.1007/978-3-319-58848-3_43

Subtlety 3) The so-called *BRST Quantization*[1] method is successful for the flat spacetime Yang–Mills Gauge Theories used to model Particle Physics; see [446] for an excellent introductory account. However, neither the BRST procedure nor subsequent generalizations such as the Batalin–Vilkovisky approach [446, 886] succeed in the case of GR; non-renormalizability and diffeomorphism invariance underlie some of the problems here.

Subtlety 4) In Gauge Theoretic QFTs, there is a further θ-sector [473] global subtlety. This is based on a θ-vacuum featuring a real-valued parameter θ, which features as a phase angle. This is classified by the homotopy group $\pi_0(\mathfrak{g})$ for $\mathfrak{g}$ the Yang–Mills gauge group.

Subtlety 5) Some more mathematically conscientious approaches set out to model quantum operators using $\mathfrak{C}^*$-algebras or similar; see Appendix V.7 for an outline of the technicalities.

43.2 Unconstrained Examples

Example 1) QFT correlators $\langle O_1 \ldots O_n \rangle$ are observables in the context of QFT, but do not extend to *Diff*$(\mathbf{\Sigma})$-invariant expressions. [Perhaps one could work instead with boundary observables.]

Example 2) $\mathfrak{C}^*$- and $\mathfrak{W}^*$-algebras are used furthermore in modelling local observables in Ordinary QM, QFT and QFTiCS [401, 473, 603]. This incorporates at least the SR notion of causality; it is furthermore an approach which in part makes pioneering use of presheaf mathematics [401].

Example 3) More general operators can also be contemplated here, by once again evoking positive operator-valued measures.

See also Chap. 52 for further QFT subtleties, now within the context of path integral formulations.

43.3 Dirac Quantization of Geometrodynamics. i. Kinematical Quantization

Let us begin with a fully general pointwise treatment, as befits Minisuperspace or Strong Gravity. For full Minisuperspace, $\mathfrak{q} = \mathfrak{S}\mathrm{ym}^+(3, \mathbb{R})$, which takes the mathematical form (H.2). In this case, a simple candidate for Kinematical Quantization is

$$\mathfrak{K} = \underline{\mathfrak{S}\mathrm{ym}}(3, \mathbb{R}) \rtimes \mathfrak{S}\mathrm{ym}^+(3, \mathbb{R}), \tag{43.3}$$

[1]These techniques are named after physicists Carlo Becchi, Alain Rouet, Raymond Stora and Igor Tyutin, and Igor Batalin and Grigori Vilkovisky, respectively.

formed by adjoining the conjugate momenta p^{ij}: densitized 3×3 symmetric matrices which form the space $\underline{\mathfrak{S}\text{ym}}(3, \mathbb{R})$. Moreover, naïve commutation relations

$$\left[\widehat{h}_{ij}, \widehat{p}^{kl}\right] = 2\,i\,\hbar\,\delta_{(i}{}^{k}\delta_{j)}{}^{l}. \tag{43.4}$$

fail to incorporate the positive definiteness condition, $\det h > 0$. Isham and physicist A.C. Kakas' affine commutation relations are required instead [499, 500]:

$$\left[\widehat{h}_{ij}, \widehat{p}_{k}{}^{l}\right] = 2\,i\,\hbar\,\widehat{h}_{k(j}\delta_{i)}{}^{k}. \tag{43.5}$$

The resemblance with the previous Chapter's Quantization of $\mathbb{R}_+$ is more than just superficial since $\mathbb{R}_+$ can be viewed as $GL^+(1, \mathbb{R})$. (43.3)'s candidate $\mathfrak{K}$ is furthermore likely to be too small due to $GL^+(3, \mathbb{R})$ possessing isometries which are not yet incorporated into this selection; due to this, we consider this candidate with a pinch of salt. Note moreover that this candidate and the previous Chapter's examples of diagonal Minisuperspace $\mathfrak{K}$ do not form a consistent whole. The simplest way of viewing the two together involves recasting the general case's diagonal as a new vector of quantities. Additionally, the current chapter uses a rather than Ω for scale variable; on the other hand, the previous Chapter does find and incorporate the simpler examples' extra isometries. Finally, the above generalizes working with $a \geq 0$ rather than the $\Omega \in \mathbb{R}$ version of the Quantization of diagonal Minisuperspace in Sect. 39.5.

Isham [475] went on to propose

$$\mathfrak{K} = \mathfrak{C}^{\infty}\big(\boldsymbol{\Sigma}, \underline{\mathfrak{S}\text{ym}}(3, \mathbb{R})\big) \rtimes \mathfrak{C}^{\infty}\big(\boldsymbol{\Sigma}, \mathfrak{S}\text{ym}^{+}(3, \mathbb{R})\big) \tag{43.6}$$

as a model for the full GR case. The second factor corresponds to the space of the $\mathbf{h}$, $\mathfrak{Riem}(\boldsymbol{\Sigma})$: the collection over $\boldsymbol{\Sigma}$ of the $\mathfrak{S}\text{ym}^{+}(3, \mathbb{R})$. The first factor corresponds to the space of the $\mathbf{p}$: the collection over $\boldsymbol{\Sigma}$ of the $\underline{\mathfrak{S}\text{ym}}(3, \mathbb{R})$.

Problem 0) Equal-time commutation relations depend on there being a background manifold $\boldsymbol{\Sigma}$.

Once again, naïvely Geometrodynamics' h_{ij} and p^{ij} might be expected to resemble (39.12), in being promoted to $\widehat{\mathrm{h}}_{ij}$ and $\widehat{\mathrm{p}}^{ij}$ which obey

$$\left[\widehat{\mathrm{h}}_{ij}(\underline{x}), \widehat{\mathrm{p}}^{kl}(\underline{x}')\right] = 2\,i\,\hbar\,\delta_{(i}{}^{k}\delta_{j)}{}^{l}\delta^{(3)}(\underline{x}, \underline{x}'). \tag{43.7}$$

Problem 1) There is however a classical-level inequality on the determinant

$$\det \mathrm{h} > 0 \quad \text{(nondegeneracy condition)}. \tag{43.8}$$

The Affine Geometrodynamics [476, 479, 499, 500, 559] commutation relations which take (43.8) into account are, rather,

$$\left[\widehat{\mathrm{h}}_{ij}(\underline{x}), \widehat{\mathrm{p}}_{k}{}^{l}(\underline{x}')\right] = 2\,i\,\hbar\,\widehat{\mathrm{h}}_{k(j}\delta_{i)}{}^{k}(\underline{x})\delta^{(3)}(\underline{x}, \underline{x}'). \tag{43.9}$$

N.B. that this point postcedes some of the literature and has very largely been overlooked or ignored since.

RPMs moreover provide modes for Affine Quantization of intermediate complexity between $\mathbb{R}_+ = GL^+(1, \mathbb{R})$ corresponding to length > 0 and (full Minisuperspace) $= GL^+(3, \mathbb{R})$ corresponding to volume > 0. I.e. two of the variants of triangleland involve area > 0, whereas the same with the zero edge of collinearities included corresponding to area ≥ 0.[2] In all of these models, an underlying issue is that since *inequality constraints* are a subcase of nonholonomic constraints, they are not covered by the Dirac Algorithm. Because of this, a separate procedure—Affine Quantization's well-defined kinematical quantum operator selection—is required *in addition to* the Dirac Algorithm.

Problem 2) GR's purely configurational commutator

$$\left[\widehat{\mathrm{h}}_{ij}(\underline{x}), \widehat{\mathrm{h}}_{kl}(\underline{x}')\right] = 0 \tag{43.10}$$

is entangled in a further conceptual issue. In QFT, the commutator of two configuration variables being zero means that these can be measured simultaneously at two points $\underline{x}, \underline{x}'$ in the notion of space they are defined upon. But in the case of GR's metric field, there is not an a priori notion of simultaneity. This moreover interferes with extending the $\mathfrak{C}^*$-algebra approach to QFT to the case of QG [483].

Problem 3) Given that doubts were already raised as regards the complete characterization of the pointwise version (43.3) of (43.6), we take (43.6) itself with *two* pinches of salt.

Problem 4) The model (43.6) points to GR requiring a large step-up in difficulty of the ensuing Representation Theory [475]. Geometrodynamical approaches indeed remain gridlocked around this point.

43.4 ii. Dynamical Quantization

GR as Geometrodynamics additionally has quantum constraints to [501, 502]. Equation (11.7) is a commonly used form for GR's quantum linear momentum constraint, which involves the operator ordering with P to the right [662].

Additionally, GR's Hamiltonian constraint becomes the Wheeler–DeWitt equation (11.6).

Problem 5) The Wheeler–DeWitt equation is additionally a second-order FDE. While, prima facie, this is a type of equation which is well-known in QFT, therein regularization methods for such equations are unfortunately based on fixed backgrounds. For instance, the regulator mass is a background structure, whereas the point-splitting vector also relies on the background structure. Thus none of these methods are applicable in the Wheeler–DeWitt equation's own Background Independent context, for which it remains unclear how to proceed with regularization. Since regularization is specifically a FDE issue, moreover, both Minisuperspace and RPMs are free of it through merely being PDE problems.

[2]This is mass-weighted area, so this is as yet another analogy between I and $\sqrt{\mathrm{h}}$.

Example 7) The special triangleland case has

$$\mathfrak{K} = Eucl(3) \times Aff(1). \tag{43.11}$$

The Kinematical Quantization operators here are the dra^{Γ}, the shape conserved $SO(3)$ quantities $\widehat{S}^{\Gamma}$, the hyperradius ρ and its conjugate $\widehat{p}_{\rho} = \rho\partial/\partial\rho$. This example also exhibits low-order polynomiality.

While we have argued for a conformal operator ordering resolution of this, there are various further obstacles to using this resolution for full GR. A preliminary minor snag is that k is now infinite so the conformally-transformed quantum wavefunction becomes ill-defined. None the less, the k's in the working (41.3) continue to formally cancel, and it is the outcome of this working, rather than Ψ itself, that has physical meaning. This gives the *conformal-ordered Wheeler–DeWitt equation* [22]

$$\hbar^2 \text{`}\left\{\frac{1}{\sqrt{\mathrm{M}}}\frac{\delta}{\delta \mathrm{h}_{ij}}\left\{\sqrt{\mathrm{M}}\mathrm{N}^{ijkl}\frac{\delta}{\delta \mathrm{h}_{kl}}\right\} - \frac{1}{4}\mathcal{R}_{\mathbf{M}}(\underline{x};\mathbf{h}]\right\}\text{'}\Psi + \sqrt{\mathrm{h}}\{\mathcal{R} - 2\Lambda\}\Psi = 0. \tag{43.12}$$

Three greater snags with this are given as the next three Problems.

Problem 7) While (43.12) manifests conformal operator ordering, this equation is but at the level of a pre-regularization nicety.

Problem 8) Sect. 42.4's Problem 2)—that conformal operator ordering should in any case apply to the reduced version of $\mathcal{Q}$uad—continues to apply here. Furthermore, a reduced $\mathcal{H}$ is unfortunately unavailable for Geometrodynamics.

Problem 9) The usual form in which $\widehat{\mathcal{H}}$ and $\widehat{\mathcal{M}}_i$ are presented, however, amounts to not taking on board the affine representation of p^{ij}. This affects the form of the Wheeler–DeWitt equation. Furthermore, it is not even clear whether affinely representing momenta and building Laplacians are mathematically compatible. Inserting the affine representation for the momentum operator into the Isham–Kakas [499, 500] formulation gives

$$\mathcal{H}\Psi = -\text{``}\frac{\hbar^2}{4\mathrm{e}}\{\mathrm{e}_A{}^P\mathrm{e}_B{}^Q\mathrm{e}_L{}^D\mathrm{e}_M{}^F\delta_{LM}\delta_{PQ} + \mathrm{e}_A{}^P\mathrm{e}_P{}^F\mathrm{e}_B{}^Q\mathrm{e}_Q{}^D - \mathrm{e}_A{}^P\mathrm{e}_P{}^D\mathrm{e}_B{}^Q\mathrm{e}_Q{}^F\}$$
$$\times\, \mathrm{e}_R{}^A\frac{\delta}{\delta \mathrm{e}_R{}^D}\mathrm{e}_S{}^B\frac{\delta}{\delta \mathrm{e}_S{}^F}\text{''}\Psi + \mathrm{e}\{2\Lambda - \mathcal{R}(\underline{\mathrm{x}};\mathbf{e}]\}\Psi = 0. \tag{43.13}$$

This is in triad notation and with no attempt made for now as regards regularization or picking a particularly significant operator ordering. Affine Quantization additionally interferes with the benevolent properties imparted upon first-class linear constraints by operator ordering with **P** to the right.

Problem 10) In approaches to Quantum GR which set out to use $\mathfrak{g}$-averaging, in general well-definedness issues appear as regards the measure $\mathbb{D}\mathrm{g}$ over the group $\mathfrak{g}$. In particular, it is not clear how $Diff(\boldsymbol{\Sigma})$ would be treated explicitly in this manner.

Closure of quantum equations is for now being assumed in discussing the above schemes; this is checked in Chap. 49.

See also [475] for details of the counterparts of Sect. 39.6's global issues in the case of GR as Geometrodynamics. N.B. that for a theory like GR, the involvement of $\pi_0(\mathfrak{q})$ is unlikely to bear much relation to one's final quantum theory. Another is that the configuration spaces in question being homogeneous spaces carries the further implication of there being a unique group orbit, i.e. the group action is transitive. This has the good fortune of guaranteeing the straightforward applicability of Group Quantization techniques.

In the full GR setting, reduction down to $\mathfrak{S}$uperspace($\boldsymbol{\Sigma}$) is usually taken to be the same procedure as removing the linear constraints. Some works consider Quantization after reduction down to superspace [472]. On the other hand, Chaps. 24 and 30 illustrate how these two procedures are capable of being distinct.

For e.g. the spatially $\mathbb{S}^3$ case of Geometrodynamics considered in the current book, Isham [475] has shown that GR's counterpart of the θ-sector is trivial, via

$$\pi_1\big(\mathfrak{S}\text{uperspace}(\mathbb{S}^3)\big) \cong \pi_0\big(\mathit{Diff}_{\mathrm{F}}(\mathbb{S}^3)\big). \tag{43.14}$$

43.5 Dirac-Type Quantization of Nododynamics Alias LQG

This follows on from Sect. 11.9's briefer outline. It uses methods along the lines of Geometrical Quantization [254, 844]; more specifically Group-averaging Quantization, in particular *Refined Algebraic Quantization*. In this sense, Nododynamics implements Configurational Relationalism at the quantum level.

Quantum Nododynamics [75, 80, 154, 752, 845] usually evokes either 1) Dirac Quantization. Or 2) a method lying between this and Reduced Quantization. In 2), on the one hand, the $SU(2)$ Yang–Mills–Gauss constraint (8.35) is handled at the classical level by introducing holonomy variables (closely related to Wilson loop variables: Appendix N.2). On the other hand, the momentum constraint (8.36) is promoted to a quantum equation. Such variables clearly also arise in a natural way within Yang–Mills Theory itself, though their use in QFT has been very largely discontinued due to loop states being too numerous and too singular. Loop Quantum Gravity (LQG) adherents suggest however that Wilson loops be reintroduced in cases for which a conventional Gauge Theory's scope is augmented by the inclusion of the $\mathit{Diff}(\boldsymbol{\Sigma})$.[3] In this situation, the physical irrelevance of $\mathit{Diff}(\boldsymbol{\Sigma})$ greatly cuts down on the number of loop states [752].

[3] See e.g. physicists Rodolfo Gambini and Jorge Pullin's book [330] for a conceptually if not technically detailed account of the loop transformation underlying the loop representation. This in turn rests on loop space mathematics (outlined in Appendix N.12). See also e.g. Ashtekar and physicist Jerzy Lewandowski's account [77] for a more rigorous treatment. It has also been determined that the loop representation works just as well for whichever value of the Barbero–Immirzi parameter β. Thus, in particular, it works for both complex-variables and real-variables forms of Nododynamics.

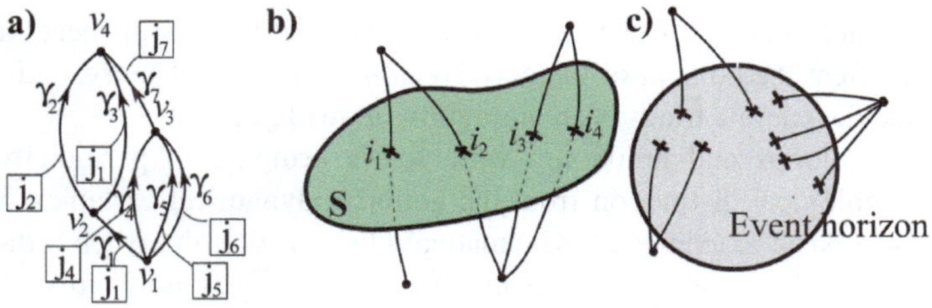

Fig. 43.1 Outline LQG figures of **a)** a spin network, with vertices v, and edges e labelled by spins j. **b)** The spin network's punctures at the intersection points i with a surface S under consideration build up the associated area spectrum. **c)** Finally the preceding is considered for an event horizon—which is a flat surface other than where quantized deficit angle is induced by punctures—in building up the LQG black hole entropy result

We are now left with i) a quantum momentum constraint

$$F^I_{ij}\frac{\delta\Psi}{\delta A^I_j} = 0, \tag{43.15}$$

which has a useful and geometrically clear-cut meaning via Sect. 42.2's Lemma. ii) A quantum Hamiltonian constraint, the precise form of which remains disputed [679, 793, 842, 845]. This is due to functional well-definedness, Barbero–Immirzi β and operator-ordering ambiguities, regularization, of the physical significance of the interaction terms therein and their mathematical realizability. Two strategies for dealing with this momentum constraint are as follows.

Strategy A) *Use Knots.* i) can be taken into account, at least formally, by use of knot states [330, 757]. See Appendix N.13 for a mathematical outline of knots. One furthermore needs to pass from formal to practically useable knot states[4]

Strategy B) *Use Nets.* This makes use of a version of *spin networks*. These are, more generally, a combinatorial approach which was originally pioneered by Penrose on its own merits. See Fig. 43.1 for general [533] and Nododynamics-specific [758] outlines. In Nododynamics, spin networks form an orthonormal basis for the quantum states; the additional 'overcompleteness' connotation in loops states being a priori too numerous is sorted out for *Diff*$[\mathbf{\Sigma})$-invariant theories.

Let us next update Part I's list of issues with Nododynamics alias LQG.

1) The classical inclusion of degenerate triads at the very least raises questions at the quantum level. Some of these arise from having seen the difference between Plain and Affine Quantum Geometrodynamics due to an imposition of an inequality in the latter case. Others arise due to physical distinctions between

[4]For instance, one might use a framing by which loops become ribbons—a further type of regularization that is possibly more amenable to theories with metric Background Independence. This is illustrative of a main approach in freeing oneself from the consequences of loops themselves being too singular.

models which include or excise lower-d strata. Distinctions of these kinds are likely to affect the content of the corresponding Quantum Theory, and whether to include degenerate triads is an ambiguity of this kind.
2) Introducing loops and holonomy variables is bringing in periodicity which causes topological distinction from the geometrodynamical variables. We then know from basic examples of Kinematical Quantization that such a difference can produce substantial differences in one's resulting quantum theory.
3) By the Multiple Choice Problem, canonically equivalent classical theories need not lead to unitarily equivalent quantum theories. So even if Nododynamics can be regarded as classically canonically equivalent by (19.9) to Geometrodynamics in all senses—which 1) and 2) may preclude—Quantum Nododynamics may still differ from Quantum Geometrodynamics. Consequently, telling which of these, if any at all, is realized by Nature would likely be impossible in the absence of experiments.
4) Recollect Kuchař's [587] well-known point that the imaginary-β version ultimately has to face reality conditions which are around as mathematically forbidding as the procedural obstructions to quantizing Geometrodynamics are. We have seen that this observation in part led to real-variables formulations of Nododynamics taking root in the 1990's. Moreover, Samuel's point about the complexified version alone having spacetime connections (Sect. 27.8) feeds into Giulini's point that this version alone also meets the classical Refoliation Invariance criterion (Sect. 31.11). As an aspect of Background Independence, this applying can furthermore be used as a filter on Background Independent Theories. In this way, just complex-variables Nododynamics gets by, and this is precisely the subset of Nododynamics to which Kuchař's quantum-level point applies.
5) A second well-known issue with Nododynamics are whether it has suitable classical and semiclassical limits. Does it support a limiting regime sufficiently like $\mathbb{M}^4$ as regards performing Standard Model calculations, and do this and a suitable Semiclassical Quantum Cosmology arena emerge from a full Quantum Nododynamics? We return to the latter in Sect. 56.1.

Finally, in the much more specific, if somewhat more heuristic LQC regime, the quantum Hamiltonian constraint now takes the form of a higher-order difference equation; see [152] for examples.

43.6 Dirac Quantization of Super-RPM and Supergravity

See [232, 868] for an account of the general case. A few simple models are as follows.

For quantum *superTr*(1),

$$\widehat{\mathcal{S}} = -\frac{\hbar}{i}\sum_{I=1}^{N}\left\{\frac{\partial}{\partial\theta^I} + i\,\bar{\theta}^I\frac{\partial}{\partial q^I}\right\}\Psi = 0, \tag{43.16}$$

$$\widehat{\mathcal{S}}^{\dagger} = \frac{\hbar}{i}\sum_{I=1}^{N}\left\{\frac{\partial}{\partial\bar{\theta}^{I}} + i\,\theta^{I}\frac{\partial}{\partial q^{I}}\right\}\Psi = 0, \tag{43.17}$$

$$\widehat{\mathcal{P}}_{\text{susy}}\Psi = \frac{\hbar}{i}\sum_{I=1}^{N}\left\{\frac{\partial}{\partial q^{I}} + \frac{\partial}{\partial\theta^{I}} - \frac{\partial}{\partial\bar{\theta}^{I}}\right\}\Psi = 0, \tag{43.18}$$

$$\widehat{\mathcal{E}}\Psi - \frac{\hbar^2}{2}D^2_{\mathbb{R}^N}\Psi + V\big(q^I, \theta^I, \bar{\theta}^I\big)\Psi = E\,\Psi. \tag{43.19}$$

with V within the form allowed by (19.18).

Supersymmetric Minisuperspace models have been substantially reviewed in [232, 868].

Supergravity is, moreover, a counter-example to Reduced Quantization—taken to mean reducing out $\mathcal{F}$**lin** but not $\mathcal{Q}$uad—not making sense for all theories, due to $\mathcal{F}$**lin** not forming a subalgebraic structure in this case.

Research Project 55)† Consider Background Independence and the Problem of Time for Canonical Quantum Supergravity in comparable detail to the current book's account of GR as Geometrodynamics.

Research Project 56) What happens in detail if a Dirac-type time-dependent wave equation is taken to model the whole Universe? This could e.g. correspond to the wavefunction of the Universe actually being a fermionic entity rather than a scalar. One might in particular pursue using a Dirac-type inner product approach to the Problem of Time in the case of Supergravity, due to first-order equations—the supersymmetry constraints—having been argued to play a more primary role in this case....

Research Project 57) Consider basing Quantum Cosmology on a *density matrix of the Universe*, i.e. with a mixed-state entity taking the place of the 'wavefunction of the Universe'.

43.7 Is Quantization Is a Functorial Prescription?

Categories—pairings of a type of mathematical objects with corresponding morphisms as outlined in Appendix W.1—can be useful in Mathematics, and some of this usefulness transcends to Mathematical Physics. Functors are maps between categories. It would be very helpful if Quantization could be viewed as a functor. This would be a map from something like (Poisson brackets algebras on phase space $\mathfrak{P}$hase as paired with canonical transformations *Can*) to some kind of (operator algebra on Hilbert space with corresponding unitary transformations *Uni*). Unfortunately, Quantization in general defies such a description [604].

It is not clear whether one can always quantize a given classical system. The lack of physical fundamentality of the classical world may, moreover, render the notion of Quantization—passage from classical to quantum—ultimately meaningless. I.e. a 'bottom-up' approach—such as formulating quantum theory from first principles followed by considering a suitable classical limit—may well be more desirable in the long run than the 'top-down' approach that is Quantization. This is reflected by QG programs often taking the following form.

A) Start by considering a classical system and a Quantization procedure.
B) One subsequently seeks for quantum first principles which lead to a Quantum Theory of the kind arrived at by A).
C) Finally consider whether a (semi)classical description of the world can be recovered in a suitable limit.

A functorial prescription *is available* in some cases, such as model arenas and particular instances of CFTs and TFTs. More generally, however, category-theoretic thinking does not point to a particular prescription for Quantization, much less one that is established to match Nature. It helps in seeing this to tease apart the various procedures that Quantization involves, e.g. the split into Kinematical and Dynamical Quantization, and the further split due to incorporating $\mathfrak{g}$ nontriviality. These more basic pieces already manifest individual difficulties—or at least ambiguities—as regards there being some kind of 'universal Quantization functor prescription'.

1) A preferred subalgebra choice is required (and this is nontrivial e.g. due to the Groenewold–Van Hove phenomenon [374]). However, this renders any functorial prescription ambiguous.
2) Global sensitivities have to be met at the quantum level, despite playing a rather lesser role at the classical level. Moreover, one could pre-emptively pass to globally careful classical modelling, so that non-functoriality would not be induced on such grounds.
3) For Geometrodynamics, promoting the constraints to the quantum level also brings in well-definedness issues.
4) In addition to these individual difficulties, the order in which some of these procedures are performed affects the outcome. E.g. following reduction by conformal operator ordering does not match up with vice versa, by Sect. 42.4 There might however be reasons to order some particular way, which one might attempt to turn into a functorial prescription.

Let us end by noting that Landsman's arguments [604] on whether Quantization is a functor extend to whether more subtle choices of categories could help toward attaining this, though this lies beyond the scope of the current book.

Chapter 44
Tempus Ante Quantum

We have found a frozen quantum wave equation and the classical approach's Machian emergent time does not unfreeze this. The classical resolution is of the form TRi ... T whereas the quantum resolution is of the form TRi ... Q ... T. In both cases, we set up TRi over a sufficient width of formalism to accommodate the other local facets, and we then finish off Temporal Relationalism by much more summarily abstracting time from change. This works out differently before and after Quantization, since these involve distinct classical and quantum changes respectively. We consider various TRi ... Q ... T schemes in more detail over the next four chapters.

It has often been suggested that [101, 412, 483, 586] successors to the Wheeler–DeWitt equation arising by supplanting GR by an alternative theory would likely persist in having a Frozen Formalism Problem. This is largely the case, for all that Sect. 20.6 points to a rare counter-example of unhidden-time-dependent Schrödinger equation.

44.1 Finite $\mathfrak{g}$-Free Models

Kinematical Quantization The argument about the time–Hamiltonian version of time–energy commutation relations in Sect. 39.5 involves interpreting time as an adjoined configurational variable. In the Internal Time Approach, however, time arises from within the system itself, accompanied by a 'true Hamiltonian'. This may alter the status and form of the corresponding commutation relations.

Dynamical Quantization After casting a formerly purely-quadratic constraint Quad in the general classical parabolic form (20.6), one obtains a time-dependent Schrödinger equation of the form

$$i\hbar\frac{\partial\Psi}{\partial t_{\text{ante}}} = \widehat{H}_{\text{True}}\big(t_{\text{ante}}, Q^{\mathsf{O}}, \widehat{P}_{\mathsf{O}}\big)\Psi. \tag{44.1}$$

The O index here runs over the other variables. One can follow up this unfreezing with the following considerations.

E. Anderson, *The Problem of Time*, Fundamental Theories of Physics 190,
DOI 10.1007/978-3-319-58848-3_44

1) Evoke the obvious Schrödinger inner product (at least formally) associated with (44.1).
2) A relatively standard interpretation of QM ensues [586], leaning upon the classical hidden time candidate in parallel to Ordinary QM being propped up by Newtonian time (Chap. 5).

A particular formal example of note is the York-time-dependent Schrödinger equation for Minisuperspace,

$$i\hbar \frac{\partial \Psi}{\partial t_{\text{York}}} = \widehat{H_{\text{True}}}\big(t_{\text{York}}, \boldsymbol{True}, \widehat{\boldsymbol{P}}^{\text{True}}\big)\Psi. \tag{44.2}$$

A first caveat with taking a classical time for dynamical QM use, moreover, is apparent upon examining physicist Paul Busch's [185] trichotomy of notions of time in Ordinary QM (Sect. 5.3). In particular, Quantum Theory's notion of dynamical time is itself quantum-mechanical, and, as such, is subject to quantum fluctuations [517].

44.2 Nontrivial $\mathfrak{g}$ Models

An argument for the reduced facet ordering R somewhere prior to Q being the physical ordering is as follows. One would not expect that appending unphysical fields to the reduced description should change any of the physics of the true dynamical degrees of freedom. So in cases for which these two approaches do differ, one should adhere to the reduced version.

Let us next consider various specific strategies for nontrivial-$\mathfrak{g}$ models. In RTQ schemes, the general classical parabolic form (20.6) yields the time-dependent Schrödinger equation for the reduced configuration space evolution,

$$i\hbar \frac{\partial \Psi}{\partial \mathrm{t}^{\text{ante}}} = \widehat{\mathscr{H}_{\text{True}}}\big\lfloor \mathrm{t}^{\text{ante}}, \widetilde{\mathbf{Q}}^{\mathrm{O}}, \widehat{\widetilde{\mathbf{P}}}_{\mathrm{O}}\big\rfloor\Psi. \tag{44.3}$$

A particular example of such is provided by the reduced York time dependent Schrödinger equation

$$i\hbar \frac{\delta \Psi}{\delta \mathrm{t}^{\text{York}}} = \widehat{\mathcal{H}_{\text{True}}}\big(\widetilde{x};\, \mathrm{t}^{\text{York}}, \mathbf{True}, \widehat{\widetilde{\mathbf{P}}}^{\text{True}}\big]\Psi. \tag{44.4}$$

This is contingent on having classically solved the accompanying $\mathcal{M}_i$ as part of ensuring that **True** are the true gravitational degrees of freedom.

On the other hand, in classical internal or matter time-and-frame finding TQR schemes, the generic spacetime-vector time-and-frame-dependent Schrödinger equation is

$$i\hbar \frac{\delta \Psi}{\delta \mathcal{X}^{\mu}} = \widehat{\mathscr{H}_{\text{True}\,\mu}}\big\lfloor \mathcal{X}^{\nu}, \mathbf{Q}^{\mathrm{O}}, \widehat{\mathbf{P}}_{\mathrm{O}}\big\rfloor\Psi. \tag{44.5}$$

The York time candidate case of this is

$$i\,\hbar\,\frac{\delta\Psi}{\delta \mathrm{t}^{\mathrm{York}}} = \widehat{\mathcal{H}_{\mathrm{True}}}\big(\underline{x};\, \mathrm{t}^{\mathrm{York}}, X^i, \mathbf{True}, \widehat{\mathbf{P}}^{\mathrm{True}}, \zeta^i\big], \tag{44.6}$$

$$i\,\hbar\,\frac{\delta\Psi}{\delta \mathrm{X}^i_{\mathrm{York}}} = \widehat{\Pi^{\mathrm{True}}}_i\big(\underline{x};\, \mathrm{t}^{\mathrm{York}}, X^i, \mathbf{True}, \widehat{\mathbf{P}}^{\mathrm{True}}\big]. \tag{44.7}$$

Let us end by noting that Reference Matter Approaches usually also end up with a quantum equation of the general form (44.5).

44.3 Problems with These Approaches

For full GR, moreover, the functional dependence of $\mathscr{H}_{\mathrm{True}}$ on the other variables is either unknown, or just known implicitly through the solution of such a partial differential equation (Chap. 21). Thereby, the quantum 'true Hamiltonian' $\widehat{\mathscr{H}_{\mathrm{True}}}$ cannot be explicitly defined as an operator. Even if it could be, it would be fraught with severe operator-ordering ambiguities and well-definedness issues. The explicit elimination of $\mathcal{M}_i$ is just formal as well (if not so technically hard, as per Sect. 21.6).

Furthermore, simpler models such as RPMs [37] and Minisuperspaces[1]—which avoid the implicit dependence impasse by having a solvable algebraic equation in place of the Lichnerowicz–York quasilinear elliptic PDE—run into further impasses. For instance, their quantum equations *still* have major operator ordering ambiguities, well-definedness issues, and, additionally, look very little like standard formulations of QM in the case of models for which these are also available. Because of this, the '*Tempus Ante Quantum* ' insertion of a T assignment procedure prior to a Quantization Q results in mathematics is highly unlike the rest of Part III's. Even in cases where approximations yield familiar equations, these equations do not resemble the standard quantum equations for the system in question.

Let us comment further on the well-definedness issues which transcend the model arenas. Most of these candidate true Hamiltonians involve multiple layers of roots; recollect e.g. (21.21). Moreover, many of the radicands in question contain mixed-sign terms. Consequently nice guarantees of being able to promote such to an operator do not apply. (Occasionally, protection is offered by a Spectral Theorem [729] for the well-definedness of positive combinations under square root signs.)

On the other hand (21.17)'s candidate true Hamiltonian corresponding to the Euler time candidate for RPM involves the logarithm of a combination of quantum operators. This gives a 'logarithmic impasse' as regards the following.

[1] Section 4.2.3 of Isham's [483] can also be taken as the starting point for a Minisuperspace model analysis similar to the current Section's.

1) Ensuring a rigorous Functional Analysis underpinning for the corresponding candidate true Hamiltonian quantum operator.
2) A hefty and unusual operator-ordering ambiguity is incurred.

Moreover, logarithmic candidate true Hamiltonians are fortunately rare, and can be avoided by making other choices of candidate dilational time.

Both simple and subtle Multiple Choice Problems can already arise in such model arenas; see Epilogue III.A for more.

Let us end with more details about the above-mentioned approximation schemes. One method for these is to approximate in a series at the classical level, and only subsequently promote the outcome of that to quantum operators. These are rather better-defined and less ambiguous equations, with some parallels to approximate treatments of relativistic wave equations [143, 552]. Scaled 3-stop metroland examples suffice to illustrate these points.

Example 1) Using the conjugate to reciprocal radius υ as a candidate time, (21.21) is promoted to the υ-time-dependent Schrödinger equation

$$i\hbar\frac{\partial\Psi}{\partial t_\upsilon} = -\sqrt{\frac{\hbar^2\partial^2/\partial\varphi^2 + \sqrt{\hbar^4\partial^4/\partial\varphi^4 + 8\,Et_\upsilon^2}}{2t_\upsilon^2}}\,\Psi. \tag{44.8}$$

Example 2) On the other hand, working with the Euler time candidate, (21.19) is promoted to the Euler-time-dependent Schrödinger equation

$$i\hbar\frac{\partial\Psi}{\partial t^{\mathrm{Euler}}} = \frac{1}{2}\ln\left(\frac{t^{\mathrm{Euler}\,2} - \hbar^2\partial^2/\partial\varphi^2}{2\,E}\right)\Psi. \tag{44.9}$$

Examples 1) and 2) already demonstrate unusual quantum wave equations arising, as well as the presence of nested root and logarithms which lead to well-definedness issues. Also giving this pair of examples for different choices of candidate dilational time for the same underlying classical model illustrates nontrivial multiplicity. Both examples, moreover, have unambiguous operator ordering. This is because the radicands contain mutually commuting variables alone (p_φ and t_υ), whereas the same can be said for the argument of the logarithm (p_φ and t^{Euler}). Furthermore, neither of these equations looks anything like the ulteriorly exactly solvable scaled 3-stop metroland free problem. We next reconsider these models under application of classical expansion in powers of the momenta prior to attempting Quantization; each example is now up to $O(\hbar^4) \times \partial^4\Psi/\partial\varphi^4$ corrections.

Example 1) gives

$$i\hbar\frac{\partial\Psi}{\partial t_\upsilon} = -\frac{\hbar^2}{4\{2\,E\}^{1/4}t_\upsilon^{3/2}}\frac{\partial^2\Psi}{\partial\varphi^2} - \frac{\{2\,E\}^{1/4}}{t_\upsilon^{1/2}}\Psi \tag{44.10}$$

The rectifying transformation[2] $t^{\mathrm{Rec}} = -1/\{2\,E\}^{1/4} t_\upsilon^{1/2}$ sends this to an ordinary time-dependent Schrödinger equation with a particular (rectified) time-dependent potential,

$$i\,\hbar\frac{\partial\Psi}{\partial t^{\mathrm{Rec}}} = -\frac{\hbar^2}{2}\frac{\partial^2\Psi}{\partial\varphi^2} - \frac{2}{t^{\mathrm{Rec}\,2}}\Psi. \tag{44.11}$$

This admits a straightforward solution under separation of variables, which, recasting in terms of t_υ and making use of Sect. 41.4)'s dilational quantum number, is of the form

$$\Psi \propto \exp\bigl(i\bigl\{\pm\,\mathrm{d}\varphi + \bigl\{2\{2\,E\}^{1/4}\sqrt{t_\upsilon}/\hbar + \hbar\,\mathrm{d}^2/2\{2\,E\}^{1/4}\sqrt{t_\upsilon}\bigr\}\bigr\}\bigr). \tag{44.12}$$

Example 2), on the other hand, gives

$$i\,\hbar\frac{\partial\Psi}{\partial t^{\mathrm{Euler}}} = -\frac{\hbar^2}{2\,t^{\mathrm{Euler}\,2}}\frac{\partial\Psi^2}{\partial\varphi^2} + \ln\left(\frac{t^{\mathrm{Euler}}}{\sqrt{2\,E}}\right)\Psi. \tag{44.13}$$

Picking $t^{\mathrm{Rec}} = -1/t^{\mathrm{Euler}}$ gives a an ordinary time-dependent Schrödinger equation, now with potential $-\{t^{\mathrm{Rec}}\}^{-2}\ln(\sqrt{2}\,E t^{\mathrm{Rec}})$, which can be solved to give

$$\Psi \propto \exp\left(i\left\{\pm\mathrm{d}\varphi + \frac{\hbar\,\mathrm{d}^2}{2\,t^{\mathrm{Euler}}} + \frac{t^{\mathrm{Euler}}}{\hbar}\left\{1 - \ln\left(\frac{t^{\mathrm{Euler}}}{\sqrt{2\,E}}\right)\right\}\right\}\right). \tag{44.14}$$

Moreover, neither of the above solutions look like free waves.

[2]This is a redefinition under which the time-dependent Schrödinger equation takes on a simplified form by the timefunction absorbing prefactors of the configuration variables' derivatives.

Chapter 45
Tempus Post Quantum. i. Paralleling QFT

Post Postulate In strategies in which time does not always feature at the fundamental level, it is none the less capable of emerging in the quantum regime: Q . . . T schemes.

By such an emergence, the Hilbert space structure of one's final Quantum Theory can be largely unrelated to that of one's incipient Wheeler–DeWitt equation-type Quantum Theory. Emergent strategies for the latter include the following.

45.1 Attempting a Schrödinger Inner Product

For GR, approaches based on a Schrödinger-type inner product fail due to the indefiniteness of the supermetric underlying the Wheeler–DeWitt equation (Exercise VI.11.vi). Model arenas which tractably exhibit these features include Minisuperspace and SIC. On the other hand, there is (mostly) no Inner Product Problem for RPMs, since these come with positive-definite kinetic metrics. So in this case, the corresponding natural Schrödinger inner product suffices.

45.2 Attempting a Klein-Gordon Inner Product Based on Riem Time

Next consider the Wheeler–DeWitt equation not as a time-independent Schrödinger equation, but rather as an analogue of the Klein–Gordon time-dependent wave equation (12.9) with a corresponding Klein–Gordon type inner product. Such approaches carry QFT—and thus spacetime primality—undertones.

Problem 1) Superspace null cones are not respected by superspace trajectories, which limits the analogy.

E. Anderson, *The Problem of Time*, Fundamental Theories of Physics 190,
DOI 10.1007/978-3-319-58848-3_45

Problem 2) Attempting a Klein-Gordon inner product based on Riem time fails regardless of whether the scheme is Q ... T or T ... Q. This is because the scheme's candidate inner product (6.4)[1] comes with an Inner Product Problem.

Problem 3) The positive–negative modes split of states in the usual Klein–Gordon Theory arises from the presence of a privileged time. Since there is no such privileged time in the case of full GR, however, Quantization schemes for this do not exhibit this familiar and useful feature.

Moreover, from the parallel with Klein–Gordon failing as a 'first Quantization' leading to its reinterpretation as a second-quantized QFT, one might next try the following.

45.3 'Third Quantization' Generalized and Renamed

A further suggestion is that the quantum wavefunctions $\Psi[\mathbf{h}]$ which solve the Wheeler–DeWitt equation might be elevated to operators, by which we now have an equation

$$\widehat{\mathcal{H}}\,\widehat{\Psi}\,\psi = 0. \tag{45.1}$$

This has usually been termed *Third Quantization*, in analogy with 'Second Quantization' (an older name for QFT).

The 'Second Quantization' name is however questionable [885]. This is because it is a map between ($\mathfrak{H}$ilb, *Uni*) type structures, rather than being a map from ($\mathfrak{P}$hase, *Can*) to ($\mathfrak{H}$ilb, *Uni*) one like ('first') Quantization is. This argument extends to 'Third Quantization' being questionable nomenclature.

It is useful to consider RPMs as well at this point; for these, the analogue of 'Third Quantization' is, rather, a type of 'Second Quantization',

$$\widehat{\mathcal{E}}\,\widehat{\Psi}\psi = 0. \tag{45.2}$$

This occurs because RPM is finite rather than field-theoretic. While this bears some *technical* resemblance to QFT,[2] *interpretationally* the quantum wavefunction of the finite model Universe has once again been elevated to a quantum operator. This further renders it clear that 'Third Quantization' is a Field Theory specific term.

---

[1]Isham [483] further asserted that $\mathrm{d}\boldsymbol{\Sigma}_{ab}$ needs to be spacelike with respect to the GR kinetic metric M^{abcd}, and that making this inner product rigorous is difficult. It is certainly only intended as a formal expression which has yet to take $\mathcal{M}_i$ into account. For instance, this could be formally attained by projecting the inner product down to $\mathfrak{S}$uperspace($\boldsymbol{\Sigma}$). This matter is absent in the Minisuperspace examples which are used widely in such an approach. The conceptual core, however, is clear: the expression is *"invariant under deformations of the 'spatial' hypersurface in* $\mathfrak{R}$*iem*($\boldsymbol{\Sigma}$)*"* [483]. This is (paraphrasing) the Quantum GR analogue of the normal Klein–Gordon inner product's time-independence property.

[2]However, relativistic QFT's motivation in terms of finding a satisfactory inner product is absent here, since the first Quantization's Schrödinger inner product works just fine for RPMs.

All in all, we prefer to term the approach that this Section concerns 'Operator-Valued Wavefunction of the Universe', since this covers both sides of the finite field-theoretic portmanteau as well as getting around the above nomenclatural issue.

Moreover, whereas this Sec's approach is of interest as regards various technical issues, it was not held to provide a satisfactory approach to the Problem of Time up to the early 1990s [483, 586], and, at the point of writing this book, the Author is not aware of any subsequent advances in this regard.

45.4 Problem of Time Strategies in Affine Geometrodynamics

In Affine Geometrodynamics [499, 500, 559], one has a distinct form for the unreduced Hilbert space and for the detailed structure of the Wheeler–DeWitt equation.

Tempus Ante Quantum approaches have no plain–affine distinction at the classical level at which the timefunctions are found. However, different commutation relations and operator orderings can subsequently arise at the quantum level.

The affine approach still has an analogue of Riem time, since the signature of the wave equation is unaltered by passing to the affine approach. Because the subsequent inner product issue involving the potential not respecting the conformal Killing vector can be traced back to the classical level, that a Klein–Gordon inner product cannot be used carries through to the affine case as well. The Author is not aware of 'Third Quantization'—alias the Operator-valued Wavefunction of the Universe Approach—having been tried in the affine case. However, a number of the reasons for this not appearing to be very promising as a Problem of Time resolution carry over to the affine case.

Chapter 46
Tempus Post Quantum. ii. Semiclassical Machian Emergent Time

We next turn to the semiclassical version of the Emergent Machian Time program that we argued substantially in favour of in Chaps. 14 to 19 and 23. As we shall see below, this shares some but not all features with the more well known case of Semiclassical Ordinary QM [165, 603, 605, 652].

We consider in particular the approximations, equations and regimes [23, 29, 37, 93, 119, 174, 237, 419, 483, 550–552, 554, 586, 607, 696] for the more general heavy–light split problem

$$-\hbar^2\{q(\mathrm{h})\partial_{\mathrm{h}}^2\Psi + r(\mathrm{h})\partial_{\mathrm{h}}\Psi + s(\mathrm{h})\triangle_{\mathrm{l}}\Psi\} = w(\mathrm{h}, \mathbf{l})\Psi. \tag{46.1}$$

This embraces the following cases.[1]

Model	q	r	s	w
1- and 2-d scaled RPMs [23, 29, 37, 61]	1	$\frac{k(N,d)}{\rho}$	$\frac{1}{\rho^2}$	$2\{E - V\} - \frac{\hbar^2 c(N,d)}{\rho^{-2}}$
Isotropic Minisuperspace with minimally-coupled scalar field [31]	-1	0	1	$\exp(6\Omega)\{\exp(-2\Omega) - \mathrm{V}(\phi) - 2\Lambda\}$
Bianchi IX anisotropic Minisuperspace vacuum	-1	0	1	$\exp(3\Omega) \times$ (I.5)
n-modewise vacuum SIC [34]	$\zeta_{\mathrm{n}}^2 - 1$	ζ_{n}	1	$\widetilde{(30.22)}$

$k(N, d)$ is here the shape space dimension, and $c(N, d)$ is the conformal operator ordering coefficient given by (42.26). The table also rests upon 1) the Laplacian part of this operator ordering giving $q(h) = \mathrm{N}^{hh}$ and $r(h) = \{\sqrt{\mathrm{M}}\mathrm{N}^{hh}\}_{,h}/\sqrt{\mathrm{M}}$ for this range of examples, for which the blockwise split $\mathbf{M} = \mathrm{M}_{hh}(h) \oplus \mathbf{M}'_l$ applies. 2) $\mathbf{M}'_l$ takes either the 'Cartesian split' form $\mathbf{M}_l(l)$ or the 'polar split' form $h^2\mathbf{M}_l(l)$ for this

[1] Some other earlier models—such as [93, 170, 228, 689, 690], which consider particles in absolute space as semiclassical model arenas—also fall within this ansatz.

E. Anderson, *The Problem of Time*, Fundamental Theories of Physics 190,
DOI 10.1007/978-3-319-58848-3_46

range of examples. We have also split the 'potential'[2] v into v_h, v_l and j_{hl} pieces: heavy, light and interacting.

46.1 Born–Oppenheimer Scheme

For Semiclassical Quantum Theory, this consists of ansatz (12.1) alongside the following approximations.

Firstly, let $\widehat{C}$ denote the complement of the heavy kinetic term, $\widehat{H} - \widehat{T}_h$. The $|\,\chi\,\rangle$-wavefunction expectation value (integrated over the l degrees of freedom, i.e. 'l-averaged') is

$$c(h) := \langle\chi|\widehat{C}|\chi\rangle = \int_{\mathfrak{L}} \mathbb{D}\mathfrak{L}\, \chi^*\big(h, l^{\mathrm{a}}\big)\, \widehat{C}\big(h, l^{\mathrm{a}}, p^l_{\mathrm{a}}\big)\, \chi\big(h, l^{\mathrm{a}}\big). \tag{46.2}$$

Here, the associated integration is over the configuration space $\mathfrak{L}$ of the l degrees of freedom. In particular, for RPMs it is over shape space $\mathfrak{S}(N, d)$, whereas for vacuum anisotropic minisuperpace it is over anisotropyspace, $\mathfrak{Ani}$, and for modewise SIC, it is over $\mathfrak{Modespace}$. In each case, $\mathbb{D}\mathfrak{L}$ is the measure thereover: $\mathbb{D}\mathbf{S}$, $\mathbb{D}\boldsymbol{\beta}$ and $\mathbb{D}\boldsymbol{v}_{\mathrm{n}}$ respectively. Since

$$\widehat{C}\big(h, l^{\mathrm{a}}, p^l_{\mathrm{a}}\big)\big|\chi\big(h, l^{\mathrm{a}}\big)\big\rangle = c(h)\big|\chi\big(h, l^{\mathrm{a}}\big)\big\rangle, \tag{46.3}$$

(46.2) may also be regarded as an 'h-parameter-dependent eigenvalue'.

For some purposes, $|\,\chi\rangle$ requires explicit suffixing by its quantum numbers, which we denote by a single straight Latin letter multi-index k. Thus the above c is, strictly, c_{kk} and there is an obvious off-diagonal equivalent

$$c_{\mathrm{kk'}} := \langle\chi_{\mathrm{k}}|\widehat{C}|\chi_{\mathrm{k'}}\rangle. \tag{46.4}$$

The *Born–Oppenheimer approximation* alias 'diagonal dominance condition' is that

$$\text{for } \mathrm{k} \neq \mathrm{k}', \quad |c_{\mathrm{kk'}}/c_{\mathrm{kk}}| =: \epsilon_{\mathrm{BO}} \ll 1. \tag{46.5}$$

This is the first of various limitations on dimensional analysis in approximation-making in Semiclassical Quantum Cosmology. Assuming that (46.5) holds, we consider $\langle\chi| \times$ {the time-independent Schrödinger equation} with the Born–Oppenheimer ansatz substituted in.

Secondly, the h-derivatives acting upon the Born–Oppenheimer wavefunction product ansatz produces multiple terms by the product rule, schematically

$$|\chi\rangle\partial_h^2\psi, \qquad \partial_h\psi\,\partial_h|\chi\rangle, \qquad \psi\,\partial_h^2|\chi\rangle. \tag{46.6}$$

[2]These are not potentials per se, due to, firstly, overall factors of 2 and that some of the original equations having an $\mathrm{N}^{hh}(h)$ function prefactor being divided out. Secondly, due to terms going like $\hbar^2$ originating from the conformal operator ordering being packaged into v, by which (46.1) does not necessarily display all $\hbar$ dependence.

The first term is always kept. In Born–Oppenheimer's own application of this ansatz—to Atomic and Molecular Physics—the other two terms are discarded due to being far smaller than the first. However, as detailed in Sect. 46.6, the Emergent Semiclassical Time Approach to the Problem of Time and Quantum Cosmology requires the second term to be kept, due to the qualitative effect of doing so overriding its smallness.

Equation (46.1) also in general contains a linear derivative term; both curvilinear coordinates and curved spaces are conducive toward this (Exercise!) In this case, the product rule produces two terms, schematically

$$|\chi\rangle\partial_h\psi, \qquad \psi\,\partial_h|\chi\rangle. \tag{46.7}$$

The second of these is often also discarded as small.

46.2 Discussion of Adiabatic Approximations

We next consider quantum-level adiabatic approximations [652]. We distinguish between the following two 'pure' types of these. Some quantities are small through $|\,\chi\rangle$ being far less sensitive to changes in h-subsystem physics than to changes in l-subsystem physics. On the other hand, some quantities are small through $|\chi\rangle$ being far less sensitive to changes in l-subsystem physics than ψ is sensitive to changes in h-subsystem physics. Let us name these two types of adiabaticity as, respectively, 'internal to the l-subsystem'—labelled by a(l)—and 'mutual between the h and l subsystems'—labelled by a(m). In dimensional analysis terms, classical adiabaticity's $\omega_h/\omega_l \sim l/h \sim \partial_h/\partial_l$ accounts for both a(l) and a(m). Moreover, that these three types of adiabaticity condition need not imply each other is the second example of limitation on dimensional analysis. Some wavefunctions can attain this by being very steep or wiggly even for slow processes; consider for instance the 1000th Hermite function. High wiggliness is furthermore related to high occupation number. This is due to quantum states increasing in number of nodes as one increases the corresponding quantum numbers. In turn, high occupation number is itself a characterization of semiclassicality.

A third type of quantity involves $|\,\chi\rangle$ being far less sensitive to changes in h-subsystem physics than ψ is. Of course, since three quantities only support two independent ratios, this case's smallness is the ratio of the previous two smallnesses, which can itself be small if the a(l) smallness is much smaller than the a(m) one. So whereas the above discards can be made by comparison with $|\,\chi\rangle$'s l-change, any justification of making these discards while not discarding the first of each set of terms is of the third type. Let us end by pointing out that a(l)/a(m) need not be of order 1; indeed, the Born–Oppenheimer approach to Ordinary QM conventionally regards this ratio itself as small.

46.3 WKB Scheme

This next step consists of the ansatz (12.2) for the h-wavefunction alongside the approximations below; see Sect. 46.8 for quantum cosmological caveats with this. For ease of physical interpretation, let us rewrite the principal function S by isolating a heavy mass M, $S(h) = M\,(h)$. [For 1 h degree of freedom, this is trivial; for more than 1, it still makes sense if the sharply-peaked mass hierarchy condition (23.1) holds.] The associated *WKB approximation* [165, 652] is the negligibility of second derivatives,

$$\left|\frac{\hbar}{M}\frac{\partial_h^2 \mathrm{F}}{|\partial_h \mathrm{F}|^2}\right| =: \epsilon_{\mathrm{WKB}} \ll 1. \tag{46.8}$$

The associated dimensional analysis expression is $\hbar/M\,\mathrm{F} =: \epsilon_{\mathrm{WKB}'} \ll 1$. This is to be interpreted as (quantum of action) $\ll$ (classical action) by reinterpreting S as classical action (see e.g. [598]), which has clear semiclassical connotations.

One incentive for using 1 h degree of freedom is that this trivially gets round having to explicitly solve nonseparable Hamilton–Jacobi equations. This practical problem generally plagues the case of > 1 h degrees of freedom [371].

46.4 Scale–Shape Split h- and l-Equations

In Semiclassical Quantum Cosmology, the *h-equation* is $\langle\chi| \times$ {time-independent Schrödinger equation (42.29)}, with the Born–Oppenheimer and WKB ansätze substituted in:

$$\mathrm{q}\big\{\{\partial_h S\}^2 - i\,\hbar\,\partial_h^2 S - 2\,i\,\hbar\,\partial_h S\langle\partial_h\rangle - \hbar^2\big\{\langle\partial_h^2\rangle\big\}\big\} - \mathrm{r}\big\{i\,\hbar\,\partial_h + \hbar^2\langle\partial_h\rangle\big\} - \hbar^2\,\mathrm{s}\langle\triangle_l\rangle$$
$$= \mathrm{w}_h + \langle \mathrm{w}_l\rangle + \langle \mathrm{j}_{hl}\rangle. \tag{46.9}$$

Also, the *l-equation* is $\{1-\mathrm{P}_\chi\}\times$ {time-independent Schrödinger equation (42.29)}. For now, this takes the form of a fluctuation equation

$$\{1-\mathrm{P}_\chi\}\big\{\mathrm{q}\big\{2\,i\,\hbar\,\partial_h S\,\partial_h + \hbar^2\partial_h^2\big\} + \mathrm{r}\,\hbar^2\,\partial_h + \mathrm{s}\,\triangle_l + \mathrm{w}_l + \mathrm{j}_{hl}\big\}|\chi\rangle = 0; \tag{46.10}$$

P_χ is here the projector $|\chi\rangle\langle\chi|$.

46.5 Semiclassical WKB Emergent Time

In the Semiclassical Approach to the Problem of Time and Quantum Cosmology, it is standard to assume that $\partial_h^2 S$ is negligible by the WKB approximation so as to remove the second term from the h-equation. Furthermore, by identifying S as Hamilton's function and by using the momentum–velocity relation

$$\partial_h S = p_h = \sqrt{\mathrm{h}}\,M_{hh}{*}h, \quad \text{where } {*} := \mathrm{d}/\mathrm{d}t^{\mathrm{sem}}. \tag{46.11}$$

Classical–Semiclassical Machian Emergent Time Alignment Lemma To zeroth approximation (denoted by 0 indices) $t_0^{\mathrm{sem}} = t_0^{\mathrm{em}}$, so the notation can be simplified to a new notion of t_0^{em} meaning 'classical or semiclassical zeroth order'.

This observation points to emergent WKB time furthermore being interpretable as an emergent Machian time; Chap. 23's properties and critiques extend to approximate emergent WKB time as well. This can moreover be combined with Chap. 23 to give that the approximate emergent WKB time is aligned with Newtonian, proper and cosmic times in various contexts.

Proceed by rearranging the chain rule expression $* = *h\,\partial_h + *l\,\partial_l$ to isolate

$$\partial_h = \{* - *l\,\partial_l\}/*h := \Diamond *h := \heartsuit, \tag{46.12}$$

where the $\heartsuit$ and $\Diamond$ operators are to act upon $F(t^{\mathrm{sem}}, \boldsymbol{l})$. The full—bar $\partial_h^2 S$ term neglected by the WKB approximation—Machian h-equation is [29]

$$\mathrm{q}^{-1}\{\bar{*}h - i\,\hbar\{\mathrm{r} + 2\mathrm{q}\langle\heartsuit\rangle\}\}\bar{*}h - \hbar^2\{\mathrm{q}\langle\heartsuit^2\rangle + \mathrm{r}\langle\heartsuit\rangle + \mathrm{s}\langle\triangle_l\rangle\} = \mathrm{w}_h + \langle\mathrm{w}_l\rangle + \langle\mathrm{j}_{hl}\rangle. \tag{46.13}$$

If one now neglects the second, third, fourth, fifth, sixth and eighth terms and the $*l\,\partial_l$ contributions—for which Sect. 47.1 provides various justifications—then this h-equation collapses to the standard semiclassical approach's Hamilton–Jacobi equation,

$$\{\partial_h S\}^2 = \mathrm{w}/\mathrm{q}, \quad \text{or} \quad \bar{*}h^2 = \mathrm{C}_0. \tag{46.14}$$

The second form, with $\mathrm{C}_0 := \mathrm{qw}$, arises from use of (46.11), which is particularly justified in the current context due to S being a standard Hamilton–Jacobi function. The more general case entails assuming that the quantum-corrected Hamilton–Jacobi equation has a solution which still has properties akin to those of a standard Hamilton–Jacobi function. A reformulation of the latter in the RPM cases is of use in further discussions in this book is the 'analogue Friedmann equation'—similar to (20.5)—

$$\left\{\frac{*h}{h}\right\}^2 = \frac{2\,E}{h^2} - \frac{2\,V_h}{h^2}. \tag{46.15}$$

(46.14) is furthermore solved by the emergent time expression

$$t_0^{\mathrm{em}} = \int \mathrm{d}h\sqrt{\mathrm{h}/\mathrm{C}_0}. \tag{46.16}$$

Since the densitization cancels out therein, this indeed returns (9.15), (17.2) and (30.27) for the two Minisuperspace examples and for SIC respectively. Moreover, this first approximation is rather un-Machian (in the sense of STLRC) due to deriving its change just from the scale variable.

This is to be contrasted with the account at the end of Chap. 44. Finally, the first approximation is such that the classical and semiclassical Machian emergent times coincide to zeroth order in this cosmological setting.

46.6 l-Time-Dependent Schrödinger Equation

The core of Semiclassical Quantum Cosmology's passage from a fluctuation equation to a semiclassical emergent time-dependent quantum wave equation is the Machian emergent time rearrangement

$$N^{hh} i\,\hbar \frac{\partial S}{\partial h}\frac{\partial|\chi\rangle}{\partial h} = i\,\hbar\, N^{hh} p_h \frac{\partial|\chi\rangle}{\partial h} = i\,\hbar\, N^{hh} M_{hh} {*h} \frac{\partial|\chi\rangle}{\partial h}$$
$$= i\hbar\sqrt{h}\frac{\partial h}{\partial t^{\mathrm{sem}}}\frac{\partial|\chi\rangle}{\partial h} \simeq i\,\hbar\sqrt{h}\frac{\partial|\chi\rangle}{\partial t^{\mathrm{sem}}}. \tag{46.17}$$

This uses (46.11) and the chain rule in reverse.

Moreover, the full semiclassical emergent time-dependent wave equation is

$$i\,\hbar\{1-\mathrm{P}_\chi\}\sqrt{\mathrm{h}}\,\diamondsuit|\chi\rangle = \{1-\mathrm{P}_\chi\}\left\{-\frac{\hbar^2}{2}\left\{\mathrm{s}\,\triangle_l + \mathrm{r}\heartsuit + \mathrm{q}\heartsuit^2\right\} - \{\mathrm{w}_l + \mathrm{j}_{hl}\}/2\right\}|\chi\rangle. \tag{46.18}$$

This case is making use of one of Eqs. (46.14)–(46.16) to express h as an explicit function of t^{sem}.

Equation (46.18) is furthermore usually approximated by a core semiclassical emergent time-dependent Schrödinger equation, such as

$$i\,\hbar\frac{\partial|\,\chi\rangle}{\partial t^{\mathrm{sem}}} = H_l\,|\,\chi\rangle = -\frac{\hbar^2}{2}\frac{\triangle_{\mathrm{S}(N,d)}|\,\chi\rangle}{h^2(t^{\mathrm{em}})} + V_l|\,\chi\rangle \tag{46.19}$$

for RPMs. (46.19) is, modulo the h–l coupling term, 'ordinary relational l-physics'. Thus the approximate core situation has 'the scene set' by the h-subsystem for the l-subsystem to possess dynamics. I.e. the fluctuation l-equation (46.10) can be rearranged to obtain a time-dependent Schrödinger equation with respect to an emergent time as 'provided by the h-subsystem'. This corresponds to considering (46.9) and (46.10) as a pair of equations to solve for the unknowns t^{em} and $|\chi\rangle$.

The standard and Machian formulations of semiclassical Geometrodynamics, moreover, continue to realize the Broad worldview.

46.7 Rectified Time and Its Relation to Shape Space

The general time-dependent Schrödinger equation core simplifies under passing to a rectified emergent time according to

$$\bigstar := \partial/\partial t^{\mathrm{rem}} := \sqrt{\mathrm{h}}\mathrm{s}^{-1}\partial/\partial t^{\mathrm{sem}} = \sqrt{\mathrm{h}}\mathrm{s}^{-1}*, \tag{46.20}$$

since this converts the relative coefficient of '$i\,\hbar\,\partial_t$' and $-\frac{\hbar^2}{2}\triangle_l$ to 1. This is the relative coefficient that one is accustomed to from basic QM. Additionally, in the current context, this is suggestive of t^{rem} being a simplifying and more geometrically natural timefunction to work with in quantum shape physics, given that $\triangle_l = \triangle_{\text{shape space}}$ for all of this Chapter's examples. In the RPM case, t^{sem} in contrast amounts to working with the restriction to the shape part of the relationalspace cone over shape space. Furthermore, t^{rem} is as Machian and as t^{sem} is, and in a matching GLET manner through all changes also having an opportunity to contribute to it. In fact, the two are related by a conformal transformation, which is a relationally-motivated PPSCT freedom. Moreover, if the 'calendar year zero adjusted' t^{sem} is monotonic, then t^{rem} is as well; this is elementarily true for all of this Chapter's examples. Finally, all of this Chapter's examples have nontrivial rectification, for RPMs have trivial $\sqrt{\mathrm{h}}$ but nontrivial polar s, whereas the other three examples have nontrivial cosmological $\sqrt{\mathrm{h}}$ but trivial Cartesian s.

To proceed, let us expand equation (46.12) to the useful 4-suit summary

$$\heartsuit := \frac{\diamondsuit}{\ast h} := \frac{\ast - \ast l\,\partial_l}{\ast h} = \partial_h = \frac{\star - \star l\,\partial_l}{\star h} =: \frac{\clubsuit}{\ast h} =: \spadesuit, \tag{46.21}$$

noting that the new $\spadesuit$ and $\clubsuit$ operators are to be used to act on $F(t^{\mathrm{rem}}, l)$. Introduce also $\widetilde{\mathrm{w}}_h + \widetilde{\mathrm{w}}_l + \widetilde{\mathrm{j}}_{hl} := \mathrm{w}_h/2\mathrm{s} + \mathrm{w}_l/2\mathrm{s} + \mathrm{j}_{hl}/2\mathrm{s}$. Clearly the split into h, l and hl parts is *not* in general preserved, though the corresponding ratios are. The rectified l-equation is now

$$i\,\hbar\{1 - \mathrm{P}_\chi\}\sqrt{\mathrm{h}}\clubsuit|\chi\rangle = \{1 - \mathrm{P}_\chi\}\left\{-\frac{\hbar^2}{2}\{\Delta_l + \mathrm{rs}^{-1}\spadesuit + \mathrm{qs}^{-1}\heartsuit^2\} - \widetilde{\mathrm{w}}_l - \widetilde{\mathrm{j}}_{hl}\right\}|\chi\rangle. \tag{46.22}$$

It now also makes sense to introduce the rectified h-equation

$$\begin{aligned} &\mathrm{q}^{-1}\{\mathrm{s}\star h - i\,\hbar\{\mathrm{r} + 2\mathrm{q}\langle\heartsuit\rangle\}\}\star h \\ &\quad = \hbar^2\{\langle\Delta_l\rangle + \mathrm{rs}^{-1}\langle\spadesuit\rangle + \mathrm{qs}^{-1}\langle\spadesuit^2\rangle\} + 2\widetilde{\mathrm{w}}_h + \langle\widetilde{\mathrm{w}}_l\rangle + \langle\widetilde{\mathrm{j}}_{hl}\rangle, \end{aligned} \tag{46.23}$$

so that coupled treatments are in terms of a common set of variables.

In summary, in the Machian Emergent Time Approach, the Wheeler–DeWitt equation's Quantum Frozen Formalism Problem still occurs rather than being unfrozen by the classical t^{em}. However, t^{sem} or t^{rem} can subsequently be abstracted from suitably semiclassical quantum change. This amounts to starting afresh as regards obtaining an emergent time. Moreover, this distinction is itself well founded on Machian grounds, due to requiring that *quantum change* be given the opportunity to contribute to the time being abstracted from change.

46.8 The WKB Assumption Is Crucial but Unjustified

Let us first present the form taken by more general quantum wavefunctions, so one can see how the WKB ansatz is a specialized rather than general form for a wavefunction. The S-function arises from solving an h-equation that is (at least approximately) a Hamilton–Jacobi equation. However, Hamilton–Jacobi equations have 2 solutions $S^{\pm}$.[3] Thus one would not in general expect $\exp(i\,S/\hbar)$, but rather a superposition [97, 366, 483, 586, 930, 931]

$$\psi(h) = A_{+}\exp\big(i\,S_{+}(h)/\hbar\big) + A_{-}\exp\big(i\,S_{-}(h)/\hbar\big). \tag{46.24}$$

Moreover, while the WKB regime is familiar from Ordinary QM, the circumstances of its applicability there unfortunately do not carry over to the quantum cosmological context. Namely, the assumption of the WKB ansatz rests on the preexistence of a surrounding classical large system [599]. This is part of the Copenhagen Interpretation of QM, and is no longer tenable when the whole-universe models are under consideration.

Another common argument is that the WKB ansatz encountered in Ordinary QM often corresponds to the laboratory set-up being a 'pure incoming wave'. However, laboratory preparation of a convenient quantum state has no analogue in Quantum Cosmology. Furthermore, suppose one adopts an analogue of the pure incoming wave *a priori* in an entirely theoretical Quantum Cosmology calculation. Since its wavefronts pick out a direction by orthogonality which serves as timefunction, this amounts to 'supposing time' rather than a 'bona fide emergence of time' as a Machian Problem of Time resolution would require.[4]

Moreover, this is a major issue for Semiclassical Quantum Cosmology, including as regards how time is to be interpreted therein. This is because the calculation (46.17)—by which an emergent semiclassical time-dependent wave equation (46.18) is extracted from a cross-term in the fluctuation equation (46.10)—ceases to work [97, 99, 101, 483, 586, 929–931] in the absence of the WKB ansatz. For instance, this extraction of an emergent time has no working counterpart (Exercise!) for the generalized wavefunction solution (46.24).

Finally, as we shall lay out in Chaps. 51 to 54, attempts to prop the WKB ansatz up often involve elements from additional Problem of Time strategies.

[3] For the case in which the velocities feature solely homogeneous quadratically in the kinetic term, these are $\pm$ the same expression. However, more generally, the 2 solutions are $\pm$ in the sense of being a complex conjugate pair.

[4] A further argument involves constructive interference underlying classicality [897, 899]. However this amounts to imposing, rather than *deducing* (semi)classicality. This also applies to using (semi)classicality as a 'final condition' restriction on quantum-cosmological solutions. In this manner, LQC can also be argued to not address this point either, and with further issue due to the proportion of solutions rejected on such grounds being larger than is elsewise usual in Quantum Cosmology (see Sect. 43.5). Such a restriction is also akin to how Griffiths and Omnès remove by hand the superposition states which they term "grotesque universes" [390] due to their behaviour being very unlike that we experience today.

Chapter 47
Tempus Post Quantum. iii. Semiclassical Quantum Cosmological Modelling

Section 23.4's classical considerations already revealed difficulties with some details with approximations made in Quantum Cosmology's status quo. These were envisaged by analogy with hitherto much more thoroughly studied areas of Classical Dynamics. We now extend this discussion to quantum (or at least semiclassical) approximations hitherto made in Quantum Cosmology.

47.1 Back-Reaction, Higher Derivative, and Expectation Terms

Many Approximations Problem Whereas a classical such already featured in Chap. 23, the number of approximations increases further at the quantum level. In concomitant uses of approximations, it is harder to meaningfully isolate testing whether any specific one applies; in particular, this affects testing Quantum Cosmology's crucial WKB assumption.

The rest of this Chapter provides an outline of the terms involved and of some of the simpler regimes that arise from keeping small numbers of them. Some of the first papers in this area [119, 554, 555] involved expansions in one parameter. Semiclassical Quantum Cosmology is, however, a multiple parameter problem [689]; the current Chapter lists independent small quantities involved. See e.g. [23, 29, 37] for a conceptual outline of Semiclassical Quantum Cosmology with multiple independently-small quantities.

Back-Reaction Terms The full Semiclassical Quantum Cosmology equations have the particularly interesting feature that the l-subsystem can back-react on the h-subsystem, rather than just receiving a timestandard from it. This was pioneered in [174] and briefly reviewed within [552]. See Sect. 47.3 for an outline of the qualitative difference that including back-reactions has on Semiclassical Quantum Cosmology's system of equations.

E. Anderson, *The Problem of Time*, Fundamental Theories of Physics 190,
DOI 10.1007/978-3-319-58848-3_47

Motivation 1) Back-reaction terms give the l-subsystem the opportunity to contribute to the final more accurate estimate of the emergent timefunction. This is a further part of implementing Mach's Time Principle in a STLRC manner.

Motivation 2) Including these terms mean that the h-equation ceases to take the form of a decoupled conservative system's Hamilton–Jacobi equation. Now instead, h-system energy can be interchanged with l-system energy, so the l-system can take a range of energies rather than being frozen at one particular energy. In this way, the l-system can transition between energy levels. Note that this is not only 'book-keeping, but also an essential feature for such an l-subsystem to possess if it is to describe the standard observed Quantum Theory.

Motivation 3) Back-reaction is moreover conceptually central to GR. The Einstein field equations (7.5) permit matter to back-react on geometry. Furthermore, nothing can be shielded from gravity. Finally, back-reaction is also commensurate with GR's gestalt aspect as supplanter of absolute structure.

Higher Derivative Terms Particular caution is required in neglecting higher derivatives in modelling PDEs. For instance, higher derivative terms are well-known to be qualitatively dangerous in Fluid Dynamics [600]. The Navier–Stokes equation is qualitatively very different from the Euler equation both in boundary layers and in some adjacent regions which are influenced by these. That e.g. $x^{-1}\partial_x$ and ∂_{xx} are none the less dimensionally identical alludes, moreover, to a third limitation on dimensional analysis.

Quite clearly (46.9)–(46.18) contain higher time derivatives than the $i\hbar\partial_t^{\mathrm{em}}$ term that is conventionally kept in Semiclassical Quantum Cosmology. This could well cause difficulties with the behaviour of the PDE in some regions of space and of configuration space $\mathfrak{q}$. In the quantum setting, furthermore, inclusion of such terms marks passage from a time-independent Schrödinger-type equation to a Klein–Gordon-type one. Even in the simplest models, this is known to require a different type of inner product and of interpretation of the associated Quantum Theory. It is worth reiterating here that 'Klein–Gordon-type' equations are prone to substantial extra impasses arising from their differences from actual (constant-mass) Klein–Gordon Theory [581]. All in all, keeping higher derivative terms sends the Semiclassical Quantum Cosmology time-dependent Schrödinger equation to a more general time-dependent quantum wave equation. Finally, such approximations are also made in some Internal Time Approaches; see e.g. (44.10).

Expectation or Average-Type Terms Averaged and unaveraged terms are clearly dimensionally identical, by which neglecting averaged terms is a fourth limitation on dimensional analysis. This assumption is moreover usually made in the Quantum Cosmology literature. The reason given for making it, at the physical level, is *destructive interference*; in turn, this is supported by the mathematics of the *Riemann–Lebesgue Theorem* [184].

Motivation 1) However, the analogous assumption fails to hold in Atomic and Molecular Physics. Here, neglecting such terms leads to a considerable error in predicting observed spectra. This was a major issue in the early 1930s, which was resolved by keeping such terms and proceeding via the *Hartree–Fock method* [81, 324] outlined in Sect. 47.6. The effect of dropping these averaged terms on the mathematical form of the system of equations is drastic, turning coupled integro-differential equations into much simpler differential equations. Thus, while making this assumption greatly increases the chances of dealing with familiar mathematics and analytic solutions, Atomic and Molecular Physics has taught us to be cautious about discarding such terms, alongside providing means of handling such terms if they are kept.

For now, let us provide simple examples of expectation terms which cannot be dismissed as smaller. In 3-stop metroland's analogue of the central problem, $\langle\partial_\varphi^2\rangle|\chi\rangle$ and $\partial_\varphi^2|\chi\rangle$ are of the same size since the wavefunctions in question are eigenfunctions of this operator. This type of example clearly generalizes more widely due to being based on the eigenfunctions of the Laplacian operator, which is common in both laboratory and quantum cosmological models.

Classification of the h- and l-Equations' Terms Let us finally display the qualitative types of the 14 often-neglected terms which arise in the book's Semiclassical Quantum Cosmology model system of equations (Fig. 47.1).

47.2 Solving the h-Equation for Emergent Machian Time

We now parallel Chap. 23's procedure of correcting the h-system's approximate emergent time by the leading-order corrections of an expansion. let us first consider the simple case of expanding in powers of $\hbar$:

$$t^{\mathrm{rem}} \propto \int 2\,\mathrm{d}h/\{\mathcal{P} \pm \sqrt{\mathcal{P}^2 - 4\mathcal{Q}}\}, \tag{47.1}$$

$$\mathcal{P} := i\,\hbar\,\mathrm{s}^{-1}\{\mathrm{r} + 2\mathrm{q}\langle\spadesuit\rangle\},$$

$$\mathcal{Q} := \mathrm{qs}^{-1}\{2\{\widetilde{\mathrm{w}}_h + \langle\widetilde{\mathrm{w}}_l\rangle + \langle\widetilde{\mathrm{j}}\rangle\} + \hbar^2\{\mathrm{rs}^{-1}\langle\spadesuit\rangle + \mathrm{qs}^{-1}\langle\spadesuit^2\rangle + \langle\triangle_l\rangle\}\}. \tag{47.2}$$

Equation (47.1) suffices to establish that now indeed the quantum l-subsystem change has an opportunity to contribute:

$$t^{\mathrm{rem}} = \mathcal{F}\big[h, l, \mathrm{d}h, |\chi(h,l)\rangle\big]. \tag{47.3}$$

Perturbative investigations benefit from binomially expanding t^{rem} to obtain the leading-order effects of the various terms. Using $\mathcal{Q} = \mathrm{Q}_0 + \mathcal{Q}_1$, this produces

$$t^{\mathrm{rem}} = t_0^{\mathrm{rem}} - \frac{1}{2}\left\{\int \frac{\mathcal{P}}{\mathrm{Q}_0}\mathrm{d}h + \int \frac{\mathcal{Q}_1 + \mathcal{P}^2/4}{\mathrm{Q}_0^{3/2}}\mathrm{d}h\right\} + O\left(\frac{\mathcal{Q}_1^2}{\mathrm{Q}_0^2} + \frac{\mathrm{Q}_1\mathcal{P}^2}{\mathrm{Q}_0^2} + \frac{\mathcal{P}^4}{\mathrm{Q}_0^2}\right). \tag{47.4}$$

term	$O(\hbar^x)$ $x =$	back-reaction	time derivative	higher derivative	average	diabatic	j perturbatively small	WKB approximation
$h1)\ \hbar\partial_h \mathrm{S}\langle\partial_h\rangle$	1	✓	✓		✓	l		
$h2)\ \hbar^2\langle\partial_h^2\rangle$	2	✓	✓	✓	✓	l		
$h3)\ \hbar^2 h^{-1}\langle\partial_h\rangle$	2	✓	✓		✓	l		
$h4)\ \langle \mathrm{j}\rangle$	0	✓			✓		✓	
$h5)\ \hbar^2 h^{-2}k(\xi)$	2							
$h6)\ \hbar\partial_h^2 \mathrm{S}$	1		✓	✓				✓
$l1)\ \hbar\partial_h \mathrm{S}$	1		✓			m		
$l2)\ \hbar^2\langle\partial_h^2\rangle$	2		✓	✓	✓	m		
$l3)\ \hbar^2 h^{-1}\langle\partial_h\rangle$	2		✓		✓			
$l4)\ \langle \mathrm{j}\rangle$	0						✓	
$l7)\ \mathrm{j}\lvert\chi\rangle$	0						✓	
$l8)\ \hbar^2\langle\triangle_l\rangle$	2				✓	m		
$l9)\ \hbar^2\partial_h^2\lvert\chi\rangle$	2		✓	✓		m		
$l10)\ \hbar^2 h^{-1}\partial_h\lvert\chi\rangle$	2		✓			m		

Fig. 47.1 Illustration of how the preceding qualitative types of terms are by no means mutually exclusive: some terms belong to multiple qualitative types. The h and l notation refers to whether the term in question features in the h-equation or in the l-equation

Regime 1) Even if one assumes that the expectation terms are negligibly small, there is a novel operator-ordering term $i\hbar \mathrm{s}^{-1}\mathrm{r}$ within $\mathcal{P}$ [37]. Moreover, incorporating this does *not* require coupling the Machian emergent time procedure to the quantum l-equation. I.e. it is a quantum correction to the h-physics itself rather than a Machian l-subsystem change contribution.

Regime 2) Keep $\mathcal{P}$'s other term $2i\hbar \mathrm{s}^{-1}\mathrm{q}\langle\spadesuit\rangle$, which is both an expectation and a back-reaction.
Furthermore, suppose that one or both of $\mathcal{P}$'s terms are kept; one reason for keeping both is their jointly being $O(\hbar)$. What were hitherto pairs of solutions differing only by sign at the classical level become furtherly distinct pairs. Moreover, any of these terms gives t^{rem} a *complex* correction (see Sect. 59.2 for further discussion). Finally, comparison with the classical counterpart (23.18) reveals that the $\mathcal{P}$ terms are in place of a distinct classical l-change term.

Regime 3) It also makes good sense [37] to keep $\langle\widetilde{\mathrm{j}}\rangle$, which is the expectation of an interaction potential mediated back-reaction. This has a classical counterpart, involving keeping the interaction term rather than its expectation.

47.3 Some l-Time-Dependent Schrödinger Equation Regimes

Basic and Back-Reacting Regimes This follows on both Sects. 46.6 and 47.2.

Regime A) *Decoupled time-dependent Schrödinger equation*. This makes sense in conjunction with the classical or Regime 1) treatments of the h-equation, which act solely as time provider. Such a decoupled time-dependent Schrödinger equation is furthermore analytically tractable for a range of RPMs [37], and for the SIC case [52], for which it is additionally a *free* such.

Regime B) *Interaction potential kept*. For instance, this can often be treated as a time-dependent perturbation about the previous [37], which is a well-established type of problem in basic QM [651]. This amounts to letting the h-system act upon the l-system. To some extent, this can still be paired with the classical or Regime 1) treatments of the h-equation. However, this only makes full sense—as regards book-keeping and Relationalism both—if the l-system back-reacts on the h system, noting in particular Regime 3)'s matching means of interaction. Some RPM models of this, for instance, are mathematically tractable by use of Green's functions [23, 37]. Other back-reaction mechanisms are not precluded, but remain to be investigated. One scheme involves obtaining t_0^{em} from the classical or Regime 1) Hamilton–Jacobi equation, place it in the l-equation of Regime A or B to solve for $|\,\chi_0\rangle$, and next use this in a less approximate form of the h-equation to produce a corrected t_1^{sem} and keeping on iterating in this manner.

Diabatic Regimes Physicists Serge Massar and Renaud Parentani considered including diabatic terms [646] in Minisuperspace modelling. They found expanding universe–contracting universe matter state couplings. They also observed a quantum-cosmological version of the *Klein paradox*, i.e. backward-travelling waves being generated from an initially forward-travelling wave.

Higher Derivative Term Regime Kiefer and physicist Tejinder Singh's expansion [554] treats higher derivative terms along the lines of the next-order correction (6.3) to the time-dependent Schrödinger equation from the Klein–Gordon equation. Look up 'Foldy–Wouthuysen transformation' in e.g. [144] for the underlying technique used.

Unitarity is moreover not exact in the Semiclassical Approach, due to the approximations made. Furthermore, the type of inner product that is appropriate can differ with the order of approximation. Finally note that the use of higher-order WKB approximations remains to be considered in Semiclassical Quantum Cosmology (see also Research Project 58).

47.4 Dirac-Quantized Semiclassical Schemes

We have so far incorporated Configurational Relationalism by working within Reduced Quantization. We now consider the Dirac Quantization alternative to this (see also Fig. 47.2 for a figure summarizing this and outlining how it fits in with various other Problem of Time facets). Both of these feature in the end-summary Fig. 47.2. One motivation for this is that how to carry out Reduced Quantization for full GR is not known, whereas the full $\mathcal{M}_i$ and $\mathcal{H}$ are first-class, by which Dirac Quantization remains open. Another is that this is a very natural facet ordering to consider in comparing various approaches' facet interferences.

This consideration leads to the **$\mathcal{g}$auge** constraints contributing their own h- and l-split equations [37, 552] to the Emergent Semiclassical Approach's system of equations. These further equations do not immediately enter the specification of the emergent time,[1] which arises from $\mathcal{Q}$uad rather. However, there is facet interference through the auxiliary variable $\mathbf{d}\mathrm{g}^{\mathrm{G}}$ entering the linear constraints in Lagrangian form. Additionally, via the properly auxiliary-corrected version of the momentum–velocity relation, the **$\mathcal{g}$auge** also enter the l-time-dependent Schrödinger equation:

$$i\hbar\left\{\frac{\partial}{\partial \mathrm{t}^{\mathrm{em}}} - \widehat{\mathcal{g}\mathrm{auge}}_{\mathrm{G}}\frac{\partial \mathrm{g}^{\mathrm{G}}}{\partial \mathrm{t}^{\mathrm{em}}}\right\}|\chi\rangle = \widehat{\mathscr{H}_{\mathrm{l}}}|\chi\rangle. \tag{47.5}$$

This is a semiclassical analogue of the classical Best Matching corrections. A particular case of this is [607] the GR version of the Tomonaga–Schwinger equation,

$$i\hbar\left\{\frac{\delta}{\delta \mathrm{t}^{\mathrm{em}}} - \mathcal{M}_i\frac{\delta \mathrm{F}^i}{\delta \mathrm{t}^{\mathrm{em}}}\right\}|\chi\rangle = \widehat{\mathcal{H}}_{\mathrm{l}}^{\mathrm{GR}}|\chi\rangle. \tag{47.6}$$

[1] This is rectified where suitable.

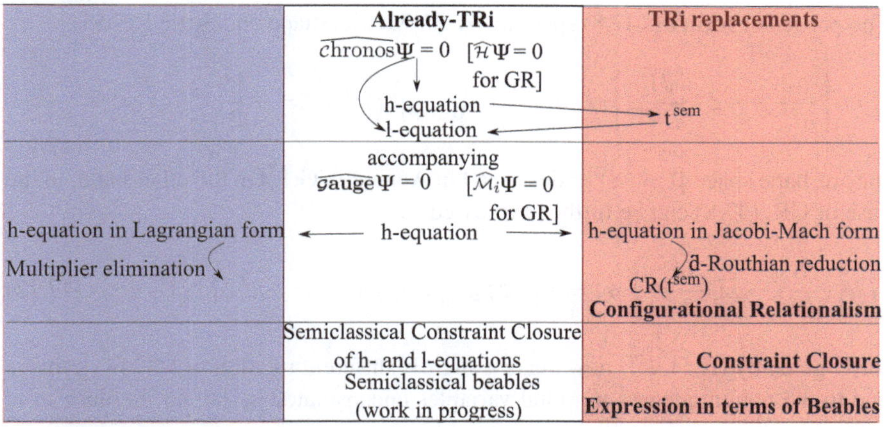

Fig. 47.2 End-summary of the TRi Semiclassical Approach

If one has succeeded in freeing oneself from the above complication, a reduced version is obtained. The corresponding time-dependent Schrödinger equation is, formally,

$$i\hbar \frac{\partial|\chi\rangle}{\partial t^{em}} = \widehat{\widetilde{\mathscr{H}}}_{l}|\chi\rangle. \tag{47.7}$$

A further form of l-time-dependent Schrödinger equation is

$$i\hbar\left\{\frac{\partial}{\partial t^{rem}} - \widehat{\mathfrak{gauge}}_{G} \frac{\partial g^{G}}{\partial t^{rem}}\right\}|\chi\rangle \propto -\frac{\hbar^2}{2}\triangle^{c}_{Preshape}|\chi\rangle + \cdots . \tag{47.8}$$

This is accompanied by 1)

$$CR(t^{rem}) \propto \mathrm{E}'_{\mathsf{g}\in\mathfrak{g}}\left(\int \|\partial_{\mathsf{g}}\mathbf{Q}\|/\{-\mathcal{P}\pm\sqrt{\mathcal{P}^2-4\mathcal{Q}}\}h^2\right), \tag{47.9}$$

where the semiclassical meaning of the extremization symbol is postponed to Sect. 49.6. This extremization is moreover unnecessary in scale–shape split RPMs and Minisuperspace but now involving an object S_{semi} whose detailed form remains to be specified in the next update of [37]. This unknown object reduces to the relational action S_{rel} in the classical limit but presumably contains quantum corrections. $\mathcal{P}$ and $\mathcal{Q}$ are generalizations of the previous specific example of forms for these.

2) *h*- and *l*-$\mathfrak{gauge}$ equations,

$$\langle\widehat{\mathfrak{gauge}}\rangle = 0, \qquad \{1-\mathrm{P}\chi\}\widehat{\mathfrak{gauge}}|\chi\rangle = 0. \tag{47.10}$$

The particular form of (47.8) specifically for Metric Shape and Scale RPM is

$$i\hbar\left\{\frac{\partial}{\partial t^{\text{rem}}} - \widehat{\underline{\mathcal{L}}}\cdot\frac{\partial \underline{B}}{\partial t^{\text{rem}}}\right\}|\chi\rangle = -\frac{\hbar^2}{2}\triangle^{\text{c}}_{\mathfrak{p}(N,d)}|\chi\rangle + \cdots = -\frac{\hbar^2}{2}\triangle^{\text{c}}_{\mathbb{S}^{nd-1}}|\chi\rangle + \cdots, \tag{47.11}$$

for preshape space $\mathfrak{p}(N,d)$ as described in Appendix G.1. On the other hand, in the case of GR, (47.8) can be further expressed as

$$i\hbar\left\{\frac{\delta}{\delta t^{\text{rem}}} - \widehat{\mathcal{M}_i}\,\frac{\delta \underline{\text{F}}^i}{\delta t^{\text{rem}}}\right\}|\chi\rangle = -\hbar^2\triangle^{\text{c}}_{\text{CRiem}(\boldsymbol{\Sigma})}|\chi\rangle + \cdots = -\hbar^2\triangle^{\text{c}}_{\text{U}}|\chi\rangle + \cdots, \tag{47.12}$$

for $\mathbf{U}$ given by (H.5). t^{rec} now bears a slightly different relation to t^{sem} [31]; this is due to the nonuniqueness in radial variables encapsulated by taking, in place of r, some $f(r)$ that is monotonic over a suitable range.

Dirac-type wave equations were considered in [419, 549]. This is modulo Chap. 30's observation that the SVT-split constraints are not first-class, though neither this nor using a reduced approach affect the harmonic oscillator type output to this (SVT uncoupled modes) level of accuracy. Wada [789, 872] also provided various partial reductions at the level of solving quantum-level constraint equations.

47.5 Extension Including Fermions

Proceed as per the classical counterpart. Now the zeroth-order l-time-dependent Schrödinger equation contains only fermionic potential V_{f}, so Ψ can be separated into bosonic $\Psi(\text{b}, t_0)$ and fermionic $\Psi(\text{f})$ factors. However, the separated-out fermion part reads (for C the constant of separation)

$$\{\text{V}_{\text{f}} - C\}\Psi_{\text{f}} = 0, \tag{47.13}$$

which at most has algebraic polynomial roots for solutions. Thus fermion-l-nontriviality does not feed into the first-order system. So there is a breakdown of 'giving an opportunity to all species'. However, the interest in Nature of linear theories of fermionic species is *field-theoretic*. In this case, (47.13) is a PDE due to the potential containing spatial derivatives. Therefore Ψ does pick up nontrivial l-fermion dependence at zeroth order, so one continues to reside within the GLET is to be abstracted from STLRC interpretation that this book expounds.

47.6 Variational Methods for Quantum Cosmology

Variational Principles Suitable such are part of the standard textbook knowledge of QM (see e.g. [652]). Variational principles for time-independent QM are the most commonly encountered. Firstly, the *Ritz Principle* (after physicist Walther Ritz) is

$$\mathcal{E}[\psi, \psi^*] = \langle\psi|H|\psi\rangle/\langle\psi|\psi\rangle. \tag{47.14}$$

Varying this with respect to ψ^* returns the time-independent Schrödinger equation. Secondly,

$$\mathcal{J}[\psi, \psi^*] = \langle\psi|H|\psi\rangle - E\,||\psi||^2, \tag{47.15}$$

for E a Lagrange multiplier encoding the normalization condition $||\psi||^2 = 1$. We now require a combination of four extensions in order to model various Semiclassical Quantum Cosmology regimes.

Extension 1) is to one or both of curved and indefinite configuration spaces. This is unproblematic; see e.g. [27] for examples.

Extension 2) is to models with **gauge** constraints. This is again unproblematic.

Extension 3) is to time-dependent quantum variational principles. Atomic and Molecular Physics is also a source for some aspects of this. The time-dependent successors of the Ritz and multiplier principles are moreover distinct [146], and it is the latter that is more suitable to our purposes. I.e.

$$\mathcal{I}[\psi, \psi^*] = \langle\psi|\{i\,\hbar\,\partial_t - H - \Lambda\}|\psi\rangle, \tag{47.16}$$

where Λ now corresponds both to normalization and to freedom to change phase factor.

Extension 4) is to Hartree–Fock type variational principles (named after Fock and physicist Douglas Hartree). These are also familiar from Atomic and Molecular Physics, where they are used to incorporate electron–electron interaction terms, which are a type of expectation term. This is usually approached by an extended version of (47.14) including these expectation terms. Furthermore, iterative ('self-consistent') methods—for handling the nonlinear dependence on the wavefunctions entailed by the presence of averaged terms—have been developed. In a nutshell, one iterates by substituting approximate solutions to equations into more accurate versions of those equations; this is a numerical rather than analytic method. Averaged alias expectation terms in Semiclassical Quantum Cosmology can be included in a directly analogous manner.

These extensions can furthermore be combined with each other. Time-dependent Hartree–Fock theory in the context of Atomic and Molecular Physics has been considered in e.g. [146, 244, 546]. A variant of (47.16) which encodes the semiclassical l-equation by itself is

$$\mathcal{I}[\chi, \chi^*] = \langle\chi|\left\{i\,\hbar\clubsuit - \widehat{N}_l - \frac{1}{2}\{i\,\hbar\langle\clubsuit\rangle - \langle\widehat{N}_l\rangle\} - \{\Lambda - \langle\Lambda\rangle\}\right\}|\chi\rangle, \tag{47.17}$$

using for now the t-derivative-free $\widehat{N}_l := -\hbar^2\triangle_l/2 + V_l^{\mathrm{rec}} + J^{\mathrm{rec}}$ as an approximand for $\widehat{H}_l$. Here, variation with respect to χ^* encodes the l-equation.

Variational principles similar to (47.16) are moreover known for other time-dependent quantum wave equations. So including $-\hbar^2\{\spadesuit^2 + \spadesuit\}/2$ corrections into the variational scheme is not a problem either. Just add $-\hbar^2\{\{\spadesuit^2 + \spadesuit\}/2 - \langle\spadesuit^2 + \spadesuit\rangle/4\}$ into the innermost factor of (47.17). As regards modelling Quantum Cosmology, one benefits from the Hartree–Fock method having been set up for Field Theories (see e.g. [546]). [This is mentioned in anticipation of applying such methods to inhomogeneous GR models.]

Note furthermore that variational approaches often begin by entering *zeroth order trial wavefunctions*. For Atomic and Molecular Physics, these could for instance be an antisymmetrized product of a simple hydrogenic model's orbitals. A notable feature in the SIC case is that one can split the quantum wavefunction up, firstly as an n-modewise product, and secondly via the wavefunction for each mode itself being a product of S, V and T parts:

$$\Psi = \Psi_1 \cdots \Psi_{\mathrm{n}-1}\Psi_{\mathrm{n}}\Psi_{\mathrm{n}+1}, \quad \Psi_{\mathrm{n}} = \Psi_{\mathrm{n}}^{\mathrm{S}}\,\Psi_{\mathrm{n}}^{\mathrm{V}}\,\Psi_{\mathrm{n}}^{\mathrm{T}}. \tag{47.18}$$

This 'plain product' can furthermore be interpreted as a zeroth approximation *variational trial wavefunction*. Moreover, there is no problem in applying a plain product within Hartree–Fock type methods. Indeed, the very first works on these methods (by Hartree and Slater) used the simple product, prior to Slater introducing the antisymmetrized product. The plain product ansatz is probably more suitable for Quantum Cosmology within a Hartree–Fock type scheme. So the idea is to treat inhomogeneous modes independently to zeroth approximation prior to considering interactions between them, in analogy to making use of hydrogenic orbitals prior to considering electron-electron interactions within multi-electron atoms and molecules. Furthermore, both Minisuperspace and RPMs fall short of having the requisite features to be useful model arenas for this approach.

We next meet Sect. 15.7's promise, by pointing out that if the quantum wavefunction factorizes exactly, then the Machian emergent time working (46.17) ceases to function. This requires a subsequent non-factorized level of approximation.

Straightforward classical-type variational principles modified to include expectation corrections can also straightforwardly be construed [27], e.g.

$$\mathsf{S} = \int \mathrm{d}t^{\mathrm{rem}}\big\{T - V - \langle\widehat{O}\rangle\big\}. \tag{47.19}$$

Finally, variational principles encoding the coupled h-l system of Semiclassical Quantum Cosmology are more involved; for these [27] only provides more tentative suggestions. On the one hand, in the regime in which t_0^{rem} is satisfactory, a Hartree–Fock type procedure can be applied on the l-equation with average terms kept. On the other hand, variationally encoding a coupled system which includes a Machian emergent time procedure presents further complications. One is now dealing with a hitherto mathematically unfamiliar type of mixed classical–quantum system of equations of the form

$$\begin{cases} \text{(expectation-corrected Hamilton–Jacobi equation for emergent time)} \\ \text{(emergent-time-dependent Hartree–Fock scheme).} \end{cases} \tag{47.20}$$

47.7 Perturbative Schemes

Whether one is to include the Machian emergent time determination step in the evaluation loop affects purely perturbative approaches as well as part-variational formulations. On the one hand, if this is not included, perturbation schemes are technically standard. On the other hand, if this is included, these schemes become more complicated and less standard [37] due to being quantum perturbation schemes that are coupled to classical perturbation schemes.

See [29, 37] for this perturbation scheme's equations. Including the $\hbar$ correction from operator-ordering can be combined with all the other considerations in Chap. 23, at least at the conceptual level.

47.8 Problems

Problem 1) with the Semiclassical Approach is that having invoked a Wheeler–DeWitt equation results in inheriting some of its problems [483, 586]. The RPM case of this is less severe: there is no Inner Product Problem, no functional derivatives or need for regularization.

Problem 2) How to relate the probability interpretation of the approximation with that for the underlying Wheeler–DeWitt equation itself remains unclear [483, 586]. The imprecision due to omitted terms means deviation from exact unitarity. This gives problems with the probability interpretation that is to be accorded to Quantum Theory [483, 586]. How is one to make sense of the sequence of these corresponding to increasingly accurate modelling? This is up to and including their relation with the probability interpretation for the unapproximated Wheeler–DeWitt equation itself. E.g. some of these involve other than Schrödinger equations and thus require new inner products and new probability interpretations based thereupon.] Thereby, the Hilbert space structure of the final theory may be related only very indirectly (if at all) to that of the Quantum Theory with which the construction starts.

Research Project 58) Moreover, it is in the above setting that higher-order WKB techniques are expected to become relevant [554] to Quantum Cosmology. Does the Semiclassical Approach to emergent time survive such upgrades?

Let us next compare the Semiclassical Approach with the Internal Time Approach. In the Semiclassical Approach, scale is not taken to be a time, but is rather considered to be part of some set of heavy, slow, set of variables. These go into providing an approximate time rather than into the fast, light set of variables that deal with actually-observed subsystem physics. This furthermore possesses a positive-definite kinetic term, which is far more familiar, and conceptually more satisfactory. The scale physics is moreover not heavy and slow in all regions of all models, though this is fortunately the case in Early-Universe Cosmology.

An additional difference is that these Tempus Ante schemes prevail at the quantum level, whereas emergent Machian time needs to start afresh at the quantum level.

In this way, classical Machian emergent time is identified as the precursor of semiclassical emergent time, which can indeed be interpreted in parallel as a Machian construct. Finally, the reason for having to start afresh is clear from a Machian perspective: quantum changes are different from classical ones and so differ in how they contribute to the emergent time.

Research Project 59) Provide a Mathematical Physics treatment of the equations of Semiclassical Quantum Cosmology. I.e. build up an analogue of Mathematical Relativity's treatment of classical GR for Semiclassical Quantum Cosmology's own system of equations.

This has the following suite of open problems.

1) Can the quantum-corrected h-equation be taken to comprise a semiclassical quantum equivalent of 'the Hamilton–Jacobi equation encodes all information about the h-system'? I.e. precisely how can semiclassical-corrected counterparts of equations of motion be obtained from a semiclassically corrected Hamilton–Jacobi equation? We anticipate that this is surmountable, though how uniquely is unclear and may well depend on the detailed sense in which 'semiclassical' is to be mathematically implemented [603, 605].
2) What happens at the quantum level to the role of the forces determining which approximations to make in the 'GLET is to be abstracted from STLRC' procedure advocated in Chap. 23? [Force equations are far less common at the quantum level; which formalism is one to use for these?]
3) Can Eq. (47.20) be anchored on a variational principle?

Note furthermore the expectation that not only GR, but also a wide range of different Quantum Gravity Theories as well, would have a *common* Semiclassical Quantum Cosmology treatment. However, Supergravity lies outside of these [555], thus meriting further treatment as a source of both conceptual and technical variety.

While emergent semiclassical time t^{sem} is accompanied by an 'l-Hamiltonian', it is not a priori clear whether these two quantities are to be expected to be conjugate. Firstly, these are semiclassical constructs that may not meaningfully correspond to a classical phase space extension, such as the one by which time and energy are canonically conjugate in ordinary Mechanics. Secondly, the 'l-Hamiltonian' in general depending upon t^{sem}; for time-dependent Hamiltonians even the classical conjugacy argument breaks down. Thirdly, commutation relations and uncertainties between time and energy are a delicate matter in QM, as per Sects. 5.3 and 41.1. That emergent time is not a background parameter helps in this regard. This time is, rather, internal to the QM, in the sense of the quantum time classification of Sect. 5.3.

We finally point to Isham's argument [482] that spacetime and differential geometric modelling at most applies at the semiclassical level.

Chapter 48
Semiclassicality and Quantum Cosmology: Interpretative Issues

This Chapter serves to introduce a number of further useful notions and tools for handling and understanding semiclassical regimes.

48.1 Coherent States

Schrödinger's wavepacket concept (Sect. 5.1) matured into the notion of *coherent states*. The theory of coherent states has been well developed for harmonic oscillators. Coherent states are significant through having a number of similarities with classical point-particle states and further semiclassical attributes. For instance, they realize minimal entropy production and minimal uncertainty [932]. Also, $\psi_c(t) \approx \psi_{c(t)}$ for c a classical parameter evolving according to $c = c(t)$ [605], by which knowledge of the classical motion goes far toward understanding the corresponding coherent state. See e.g. [932] for a brief and lucid account which also contains the original references.

48.2 Wigner Functionals

In outline, these are a type of quantum-mechanical probability density function on $\mathfrak{P}$hase, analogous to the classical $\mathrm{w}(\boldsymbol{q}, \boldsymbol{p})$ (Chap. 29). E.g. in the flat-configuration-space K-dimensional Cartesian case, the *Wigner functional* takes the form [907]

$$\mathcal{W}\mathrm{ig}[\boldsymbol{q}, \boldsymbol{p}] \propto \iint \mathrm{d}^K y \langle \psi(\boldsymbol{q} + \boldsymbol{y}) | \exp(2i\, \boldsymbol{y} \cdot \boldsymbol{q}) | \psi(\boldsymbol{q} - \boldsymbol{y}) \rangle. \tag{48.1}$$

A fortiori, this is a *semiclassical* analogue of $\mathrm{w}(\boldsymbol{q}, \boldsymbol{p})$, and in fact only a *quasiprobability* distribution in the sense that it can take negative values. See [91, 140, 165, 197, 450, 829] for reviews, including for further detail of its physical interpretation. The Wigner functional is such that its integral over $\boldsymbol{p}$ gives $|\psi(\boldsymbol{q})|^2$ and its integral over $\boldsymbol{q}$ gives $|\psi(\boldsymbol{p})|^2$. A further distinguishing feature is that the equation of motion for it very closely parallels the classical one [140].

E. Anderson, *The Problem of Time*, Fundamental Theories of Physics 190,
DOI 10.1007/978-3-319-58848-3_48

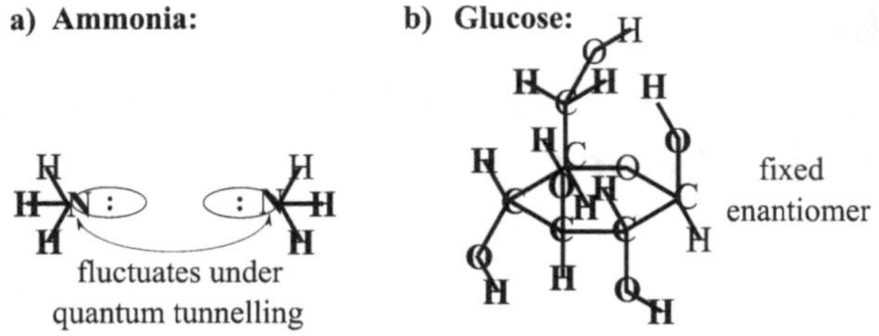

Fig. 48.1 A quantum behaviour size interface lies somewhere between the sizes of the depicted molecules

48.3 Decoherence

We next consider why we do not observe superpositions of macroscopic objects, such as of dead and live cats.[1] One of the approaches toward understanding this, initially due to physicist Dieter Zeh [928], is *decoherence*: that interaction with their environment swiftly measures such a system, reducing it to an entirely live or dead state.

On the other hand, quantum phenomena indicate that very small objects are not swiftly reduced in such a manner. For example, the ammonia molecule exhibits the properties of the superposition of states indicated in Fig. 48.1.a). This begs the question of for what molecular size do such superpositions quickly become reduced. The outcome of this is that glucose molecules (Fig. 48.1.b) are observed to stay in one chiral state, in contradistinction with ammonia molecules. Therefore, the boundary for such behaviour lies somewhere between the size of ammonia molecules and that of glucose molecules (which are still much smaller than cats!)

Decoherence in standard Quantum Theory can be taken to involve mixed state density matrices being traced over by removal of environmental modes. This suppresses interference terms between states and physically realizes a diagonalized form (at least with respect to a suitably chosen basis). See e.g. [366] and references therein for further details. [932] renders clear that coherent states and decoherence are not only nominally similar but indeed conceptually and physically related as well.

The quantum cosmological version of decoherence is moreover somewhat different [366, 551, 552] from that of Ordinary Quantum Theory. For instance, in Quantum Cosmology there may be no pre-existing notion of time in which to decohere.... Unlike most of the topics in [669, 670], decoherence has been substantially carried over to the Quantum-Cosmological setting.

We finally point to Zeh's further suggestion [929] that the matter and gravity inhomogeneities decohere the Minisuperspace degrees of freedom; see e.g. [549, 552] for more on this topic.

[1] More formally, one is dealing here with the Quantum Measurement Problem (outlined in Sect. 5.1 and [487]), which remains an unsettled, and major, area of research (see e.g. [773] for a detailed review).

48.4 Environments

We next consider what is meant above by 'the environment', with particular consideration of the strategic ambiguity that this causes in the quantum cosmological setting.

Strategy 1) *Whole Universe Models have No Environment.* E.g. consider an N-body RPM to represent the whole universe. This most idealized interpretation is moreover much less robust than absolute approaches as regards being able to assume the existence of additional particles whose contributions are traced over.

Strategy 2) *Scale Models with Shape as Environment.* Whereas this approach suffers from an overly simple original system, this feature is alleviated once one considers the shape perturbations thereabout. Such small shape environments include not only the metric notion of shape for RPMs, but also one or both of small anisotropies and small inhomogeneities in the case of GR models of perturbations about isotropic Minisuperspace. Contrast the lack of robustness in the given example of Strategy 1) with how Minisuperspace models can readily be extended to admit such perturbations. In this manner, models neglecting these degrees of freedom at the level of the dynamical equations can still cast such in the role of environment for decoherence and accompanying approximate information storage. N.B. that even one particle is capable of serving as a nontrivial environment [411].

Strategy 3) *Scale Models with Shape as both Perturbation and Environment.*| For instance, study an RPM in which a small set of the particles—say a triangle of particles—dominate over the others, which contribute to a small but non-negligible extent. One could alternatively study Cosmology using a SIC model, with scale (and any other homogeneous isotropic mode) dominating over small anisotropy and inhomogeneity modes. This is expected to most closely fit the situation in actually realized Cosmology: that decoherence occurred through coupling with small inhomogeneities' multipole terms [366, 549, 551, 552, 929].

48.5 Is Physics only About Subsystems?

We now continue Sect. 19.1's discussion of notions of Relationalism at the quantum and quantum cosmological levels. At the quantum level, Crane [224] works along the following lines (the split into pieces is the Author's own, for convenience of discussion and of proposing variants).

Crane 1) Preliminarily, Quantum Theory makes sense for subsystems, each of which has its own Hilbert space, within which the standard interpretation of Quantum Theory applies.

Crane 2) Quantum Theory solely concerns such subsystems.

Crane 3) What would elsewise be Quantum Theory for the whole universe does not possess a Hilbert space.

Crane 4) Nor does this possess the standard interpretation of Quantum Theory.

Crane 5) None the less, Quantum Theory does make sense for whole-universe models in some kind of semiclassical limit, which does possess a Hilbert space.

Let us use 'Perspectivalism 1)' as a postulate name which jointly refers to Crane 1) to 5).

Crane makes use of sub-statespaces Sub $\mathfrak{S}$ (though his notion of these does not carry the same local physical connotations as mine in Appendix Q.3). For QG, Crane further considers subsystem–environment splits of the universe in which the observer resides on the surface of this split.[2] Each of these splits has its own Hilbert space, $\mathfrak{S}$ub-$\mathfrak{H}$ilb.

One can readily imagine weakening Crane 2) to 'Quantum Theory almost always concerns subsystems'. Also, a standard alternative to Crane 3) is that Quantum Cosmology does not lack a Hilbert space (shrunken as it may be due to closed-universe effects as per Sects. 41.4 and 42.3). The scalefactor of the universe is an example of a whole-universe—rather than localized subsystem—variable that can plausibly enter one's physical propositions, so it lies outside of Crane 2). However, this can be taken to belong to the semiclassical regime, by which Crane 5) applies.

Finally, Crane 5) faces the problem that—in constituting a qualitative change arising from taking a semiclassical limit—it falls afoul of 'Earman's Principle'. Thus a more standard alternative to Perspectivalism 1) is the weakened form Perspectivalism 1W), comprising of Crane 1), the weakened form of 2), and Crane 4). This is a nontrivial addendum to relational programs because it focuses on the Quantum Theory of multiple subsystems—the entire set of Sub $\mathfrak{S}$—rather than that for whole-universe models. This will tend to lead to multiple inequivalent quantum theories, where the inequivalences can be explained by the differences in observers; we return to this idea in Epilogue III.A's strategizing about the Multiple Choice Problem.

Perspectivalism 2) Crane also allows for the Quantum Theory of *observers observing other observers observing subsystems*; Rovelli considers a similar notion in [747]). This provides a further level of structure between the many Hilbert spaces associated with all the observers in Perspectivalism 1) (or 1W).

Finally, let Perspectivalism 3) and 4) denote use of partial observables and of 'any change' leading to 'any time' respectively, each of which carry over straightforwardly to the quantum level.

[2]In fact, Crane [225] considers defining observers as boundaries of localized regions. Though clearly not all such boundaries will have observes realized upon them. Also, sizeable boundaries would need to be populated by many observers, forming a 'shell of observers' or a 'shell array of detectors'. The Information Gathering and Utilizing System concept may help as regards practically realizing such shells.

Chapter 49
Quantum Constraint Closure

We next turn to a fourth aspect—Constraint Closure—that the quantum constraints arising from Temporal and Configurational Relationalism require as a consistency check. [At the quantum level, defining the quantum constraints that these three other aspects revolve around requires preliminary Assignment of a Function Space in the form of Kinematical Quantization.]

More specifically, all first-class classical constraints c_{F} are to be promoted to operator-valued quantum constraints.[1] We provided some candidates for the quantum constraints in Chaps. 40–43. The next requirement is that these close as an algebraic structure $\widehat{\mathfrak{C}}$ under the commutator brackets. Having constraint providing principles and arguments for particular types of operator ordering in no way guarantee overcoming this hurdle as well.

Section 43.3's Problem 0) concerning dependence of the commutation relations on the background manifold $\boldsymbol{\Sigma}$ applies once again here.

N.B. also that sets of independent classical constraints are much smaller than Kinematical Quantization's input set of functionals of the Hamiltonian variables. Because of this, selection of a constraint algebraic structure comes with less variety than Kinematical Quantization's choice of subalgebraic structure.

Nor need the algebraic structure $\widehat{\mathfrak{C}}$ formed by the quantum constraints be isomorphic to that of the classical constraints $\mathfrak{C}$, as per the brackets map ambiguity (12.15). The underlying fundamental brackets being different on each side of the brackets map provides a first reason for the brackets map ambiguity. A second reason is that different operator orderings of constraints are expected to obey different commutation relations. Indeed, succeeding in obtaining quantum constraint closure can itself motivate the form of operator-ordering adopted. Constraint Closure could well also be a stringent filter on operator ordering, by which some hitherto conventional choices of operator ordering could come to be overruled.

There is moreover a quantum-level issue of why the c_{F} are chosen for promotion to operator-valued expressions rather than an equal number of functionally-

[1] This is in the context of having used extended $\mathfrak{P}$hase, Dirac bracket or reduction to free ourselves of second-class constraints, so the classical constraints in question are all first-class.

E. Anderson, *The Problem of Time*, Fundamental Theories of Physics 190,
DOI 10.1007/978-3-319-58848-3_49

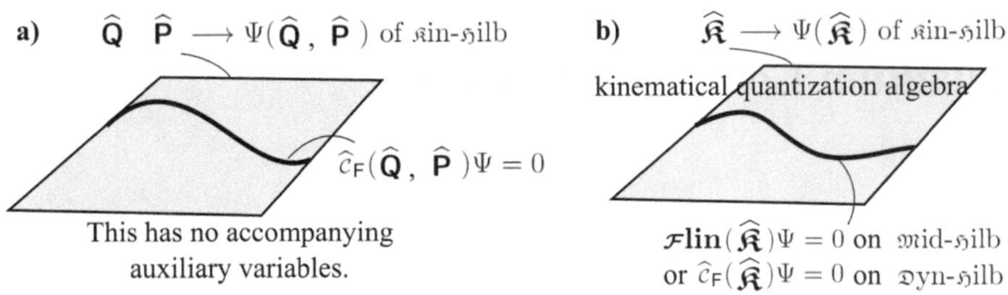

Fig. 49.1 Quantum Constraint Closure Algorithms are more straightforward than Classical Dirac–type Algorithms of Fig. 24.1 which it is tied to at the classical level. This loss of complexity is due to there no longer being any appending of constraints by auxiliary variables. The quantum-level Constraint Closure scheme is depicted in a) and b) for trivial (or specialized) and general Kinematical Quantization respectively

independent $f_{\mathsf{F}}(c_{\mathsf{F}'})$. Such a change in selection is furthermore capable of substantially change the form of the ensuing algebraic structure.

There are some ways in which Quantum Constraint Closure Algorithms are simpler than the classical Dirac Algorithm (compare Figs. 49.1 and 24.1). The quantum case is, moreover, already-TRi, in contrast to the classical case which requires development of a TRi version.

49.1 Split Quantum Constraint Structures and Nontrivial $\mathfrak{g}$

Let us next recollect Sect. 24.5's notion and notation for the partition of objects into $\mathfrak{o}$ and $\mathfrak{n}$, in particular with $\mathcal{F}$**lin** or $\mathcal{G}$**auge** in the role of the objects O of $\mathfrak{o}$. Various possibilities at the level of the brackets map m involve the classical-level structure constants C mapping to each of the following.

Outcome i) To the same C.
Outcome ii) To distinct C' and yet with $\mathfrak{g}$-compatibility preserved.
Outcome iii) Likewise but with $\mathfrak{g}$-compatibility violated.
Outcome iv) To $C'\mathsf{O} + \Theta$ and yet with $\mathfrak{g}$-compatibility preserved.
Outcome v) To $C'\mathsf{O} + \Theta$ with $\mathfrak{g}$-compatibility violated.

Outcomes iii) to v) clearly feed into the brackets map ambiguity. Furthermore, Outcomes iii) and v) entail a $\mathfrak{g}'$ distinct to the classically accepted $\mathfrak{g}$ being required at the quantum level. In such a case, clearly, Configurational Relationalism being a priori resolved at the classical level clearly does not automatically carry over to Configurational Relationalism remaining resolved at the quantum level. This is a second reason, in addition to practical inability to reduce at the classical level, to pursue Dirac Quantization approaches.

Many of the successes with classical Constraint Closure successes falter under almost every possible choice of operator ordering. This refers to the closures of $\mathfrak{g}$, of $\mathcal{F}$**lin**, of all the first-class constraints as per GR and Dirac formulations of RPMs, or, in Supergravity, of $\mathcal{NSF}$**lin** and $\mathcal{NSC}$. Two cases which are exempt from this are as follows.

Lemma 1 *If there is only one constraint and it is finite (as opposed to field-theoretic), then it commutes with itself regardless of how it is operator-ordered.*

Lemma 2 *The classical and quantum constraint algebraic structures* $\mathfrak{k}$ *and* $\widehat{\mathfrak{k}}$ *of the* $\mathcal{F}$**lin** *are isomorphic if these constraints are operator-ordered with the* P *to the right.*

Proof In this operator ordering, the quantum constraints coincide up to proportion with the classical generators. □

Caveat. This is moreover contingent on momentum being representable as (39.19), by which global sensitivity in Kinematical Quantization limit the scope of applicability of Lemma 2.

Corollary *This additionally holds for any constraint subalgebraic structure* $\mathfrak{a}_{\mathsf{X}}$ *of the* $\mathcal{F}$**lin**.

Let us end with a further consequence of $\widehat{\mathfrak{C}}$ not coinciding with $\mathfrak{C}$ is that each will in general have a distinct lattice of subalgebraic structures, $\mathfrak{L}_{\widehat{\mathfrak{C}}}$ and $\mathfrak{L}_{\mathfrak{C}}$ respectively.

49.2 (Counter)Examples of Quantum Constraint Closure

Example 1) r-formulated RPMs attain quantum Constraint Closure by Lemma 1.[2]

Example 2) Dirac-formulated RPMs have a constraint subalgebra of $\widehat{\mathfrak{gauge}}$ which is isomorphic to its classical counterpart by Lemma 2. Further tinkering with operator ordering permits closure upon inclusion of $\widehat{\mathcal{C}\text{hronos}}$ [37].

Example 3) Minisuperspace models also attain Constraint Closure by Lemma 1.

Example 4) Metrodynamical Strong Gravity can be considered pointwise, by which Lemma 1 resolves this case as well.

Example 5) Reduced SIC has at most one finite constraint per independent S, V, T sector.

Examples 6) and 7) Electromagnetism and, by Lemma 2, pure Yang–Mills Theory maintain their Lie algebra status. This straightforwardness stems from these theories possessing conventional linear gauge constraints alone.

[2]See Chap. 24 for a sequence of previous names of varying generality for the Constraint Closure Problem.

Example 8) For GR as Geometrodynamics, Moncrief and Teitelboim [662] pointed out how Lemma 2's operator-ordering for $\mathcal{M}_i$ favours brackets closure as regards brackets containing this constraint.

Let us next note the additional technical field-theoretic matter that the commutators of the c_{F} containing products of at least two functional derivatives, by which they require regularization. So far in these examples, the need for regularization has at most featured in products of two linear functional differential operators.

Example 9) For Geometrodynamics, however, the commutator of two quadratic functional differential operators

$$\big[\widehat{\mathcal{H}}(x), \widehat{\mathcal{H}}(y)\big] \quad \text{contains a product of four functional derivatives,} \tag{49.1}$$

which increases the formidability of the corresponding regularization task. This is one reason why GR quantum constraint algebraic structure remains mysterious. This situation is not ameliorated in geometrodynamical Strong Gravity, though at least the commutator computations there are somewhat simplified by the absence of Ricci 3-scalar terms.

Moreover, in Geometrodynamics interpreted with a dust matter time candidate, the issue of using functions of constraints rather than the constraints themselves is realized nontrivially. This occurs e.g. in Kuchař and Brown's [175] selection of quadratic combinations of the usual constraints. This has the particular merit of closing as a Lie *algebra* of constraints, both classically and quantum-mechanically.

Example 10) Operator ordering $\mathcal{F}$**lin** (or some subspace therein) with P to the right due to its favouring at least partial quantum commutator closure transcends to Nododynamics as well. Difficulties with regularizing—and elsewise defining—$\mathcal{H}$ of course persist here too; [155] is a useful review of contemporary progress.

Example 11) Lemma 2's caveat applies in particular in Affine Geometrodynamics.

Research Project 60)† Continue assessment of how $\widehat{\mathcal{M}}_i$ is to be operator-ordered in Affine Geometrodynamics, and whether $\widehat{\mathcal{H}}$ can continue to be ascribed a conformal operator ordering in this setting. Proceed by computing out the resultant candidate Affine Geometrodynamics quantum constraint algebraic structure.

Example 12) Other operator orderings, such as the symmetric operator ordering $\frac{1}{2}\{\widehat{x}\widehat{p} + \widehat{p}\widehat{x}\}$, change the algebraic structure. This is already the case in Particle Mechanics, whether Absolute or Relational [37].

49.3 Anomalies

Anomalies are one manifestation of quantum commutator non-closure, already mentioned in Sect. 6.5's outline, and manifested as a subcase of the previous Section's Θ term. We have already seen examples of these (and avoidance thereof) in Gauge Theory (Ex VI.12), Supersymmetry (Sect. 11.7) and String Theory (Sect. 11.8); see also e.g. [712] for further examples involving specifically non-Abelian Gauge Theory.

Anomalies were also interrelated with Configurational Relationalism in Sect. 12.4, due to their being a means by which a classically accepted $\mathfrak{g}$ may need replacing by a distinct $\mathfrak{g}'$ at the quantum level. In particular, at the level of quantum constraint commutators, Field Theory permits *Schwinger terms*

$$\Theta_{\mathsf{AB}} = \sum_{k \geq 1} \delta^{(k)}(\underline{x} - \underline{x}')\Phi_{\mathsf{AB}}, \tag{49.2}$$

where (k) here indicates the kth order derivative, of the form $\text{Array}^{i_1 \cdots i_k}(\underline{x}, t)\partial_{i_1} \cdots \partial_{i_k}$. Anomalies are, moreover, topological in origin, as outlined in Sect. 59.6. This furthermore implies that anomalies hold independently of choice of regularization, which is a metric-level issue.

Moreover, the above array contains one projector per derivative in the case based upon a general hypersurface [269]. A classical gauge group can also be rejected at the quantum level due to the quantum quadratic constraint ceasing to be compatible with it. Thus even the operator ordering with **P** to the right does not guarantee that the classical algebra of the $\mathcal{F}$**lin** remains relevant at the quantum level.

Example 1) The previous Section's Example 12) can materialize via anomalies arising. As a more specific sub-example, the so-called *conformal anomaly* can occur in finite models, for instance in a 1-particle model with conformally invariant potential term. This model does not however have much Background Independence or whole-universe meaning.

Example 2) GR in the spacetime setting exhibits [139] 'Lorentz' and 'Einstein' anomalies due to $\langle \mathrm{T}_{\mu\nu} \rangle$ being symmetric and conserved respectively. There is also a 'Weyl' alias trace anomaly to $\langle \mathrm{T}_{\mu}{}^{\mu} \rangle$ carrying over to the quantum level in cases for which $\mathrm{T}_{\mu}{}^{\mu} = 0$ holds classically (such as for Electromagnetism).

See also [4] for a wide range of cases of gravitational anomalies from a spin-by-spin Covariant Quantization perspective.

On the other hand, Kuchař and Torre [583, 857] have considered anomalies within a canonical perspective, including in particular examples of Foliation Dependent anomalies and formulations which avoid these arising, at least in some model arenas. Finally, see e.g. [332, 620] for accounts of anomalies in Nododynamics.

All in all, whereas some anomalies are related to one or both of Gravitational Theory and Background Independence, others are not. Thus only some [583, 857] anomalies become entwined in the Constraint Closure Problem aspect of Background Independence, though bearing relation to time, space or frames.

49.4 Strategies for Dealing with Quantum Constraint Closure Problem

Suppose that one has a candidate triple $\langle \widehat{\mathfrak{K}}, \mathfrak{g}, \widehat{\mathcal{C}_{\mathsf{F}}} \rangle$ which exhibits inconsistency or the appearance of unexpected extra equations more generally. One can then probe with each element of the triple in turn, much as in Fig. 24.3.

Strategy 1) Tweaking $\widehat{\mathfrak{K}}$ may either amount to selecting a different subalgebraic structure directly, or to extending or restricting $\mathfrak{q}$ or $\mathfrak{P}$hase, which may *induce* a need to select differently.
In particular, some supersymmetrization of Field Theories and of Gravitation have many successes along the latter lines, due to the fermionic terms' opposite signs (6.14) permitting cancellation.

Strategy 2) On the other hand, tweaking $\widehat{\mathcal{C}}_{\mathsf{F}}$ may either involve restricting an underlying family of classical theories, or altering how one operator-orders and regularizes the constraints. I.e. Quantization is itself a source of families of theories, even when just one theory was being considered at the classical level.
The first case includes the possibility of a strong restriction. For instance, the spacetime generator version of strong restriction is how String Theory acquires its particular dimensionalities: 26 for the bosonic string, or 10 for the superstring.

Strategy 3) Permit $\mathfrak{g}$ to be altered. One case of this is *Accepting an Anomaly*: the quantum-level loss of what had been a symmetry at the classical level. If this occurs, then success at accommodating the classical $\mathfrak{g}$'s Configurational Relationalism becomes immaterial, This leaving one in need of treating Configurational Relationalism afresh for the quantum-mechanically realized $\mathfrak{g}'$.

Finally, as per Fig. 24.3, one can proceed by altering two or even all three of the inputs.

49.5 Quantum Implications of Constraints Closing as Algebroids

Many approaches to Quantization ([475], Chaps. 39–43) is of at most limited value in GR, due to the classical GR constraints forming the Dirac algebroid instead of a Lie algebra. Moreover, the issue remains of whether the quantum GR constraints obey the same algebraic structure as the classical ones. To date, this is not even settled in the semiclassical regime, even for simplified model arenas (nor is this easy to investigate due to involving Midisuperspace features). Because of this, we do not know what the constraint algebraic structure is, but, pace matter time approaches, it is rather probably still some algebroid. All in all, algebroids and their Representation Theory are likely to be relevant at the quantum level; Appendix V.6 provides a general outline of these matters.

In some approaches, the gravitational constraint algebroid is enlarged. This occurs e.g. in LQG, Histories Theory [566] and Supergravity [232].

The possibility of Quantum Foliation Independence and Quantum Refoliation Invariance restricting the manner in which the quantum constraint algebraic structure can close is further entertained in Chap. 55.

49.6 The Semiclassical Case

Constraint Closure is less of an issue here due to Barvinsky's operator ordering coincidence Lemmas alongside the commutators themselves only needing to close up to first order in $\hbar$.

One now has

$$CS(\mathsf{S}^{\mathrm{CC}}_{\text{semi-TR-CR}}) = \mathrm{E}_{g \in \mathfrak{g}}(\mathsf{S}^{\mathrm{trial}}_{\text{semi-TR-}\mathfrak{g}}) \quad \text{built upon } \mathfrak{q}, \mathfrak{g} \tag{49.3}$$

with suitable group action of $\mathfrak{g}$ on the action S, and where the whole procedure gets past the Semiclassical Constraint Closure Algorithm. [$\mathfrak{g}$ corrections still feature at the semiclassical level due to the Hamiltonian to Lagrangian or Machian variables substitution in the emergent time provision bringing such in.]

49.7 Is There a Quantum Dirac-Type Algorithm?

Research Project 61) Let us end by asking whether constraints arising from quantum non-closure themselves use up only one degree of freedom per equation, in parallel to the phenomenon of second-class constraints at the classical level? If so, do extension and passage to Dirac brackets have quantum-level analogues? The latter would moreover involve inverting an operator-valued matrix, which may entail some formal limitations....

Chapter 50
Quantum Beables or Observables

50.1 Types of Constrained Quantum Beables

Upon passing from $\mathfrak{Kin}$-$\mathfrak{Hilb}$ to $\mathfrak{Dyn}$-$\mathfrak{Hilb}$, Chap. 39's Taking Function Spaces Thereover ceases to address the matter of Constrained Theories' Assignment of Beables. The incipient function space $\widehat{\mathfrak{U}}$ of the $\widehat{U}$, while needed for the Quantization and facet-addressing steps hitherto, does not play the role of the final theory's Associated Function Space. One requires, rather, a constrained notion of quantum beables $\widehat{B}$, and the function space formed by these, $\widehat{\mathfrak{U}}$. Par excellence, these are the quantum Dirac beables, $\widehat{D}$, which form the function space $\widehat{\mathfrak{D}}$. In this way, one needs once again to go through the gate of Taking a Function Space Thereover.

To this end, let us next consider Constrained Theories' wider range of examples of notions of beables or observables at the quantum level. For each notion of constraint subalgebraic structure $\widehat{C}_{\mathsf{W}}$ and each notion of classical A-beables $\widehat{A}_{\mathsf{X}}$ such that $|[C_{\mathsf{W}}, A_{\mathsf{X}}]|$ '$=$' 0, this classical bracket uplifts to a quantum commutator

$$[\widehat{C}_{\mathsf{W}}, \widehat{A}_{\mathsf{X}}]\Psi \text{ `$=$' } 0. \tag{50.1}$$

This is moreover an equation for $\widehat{A}_{\mathsf{X}}$ and not a further restriction on Ψ.

These notions of beables once again form a bounded lattice $\mathfrak{L}_{\widehat{\mathfrak{b}}}$ under the inclusion operation. Some common specific types of quantum beables are Kinematical Quantization's unrestricted beables $\widehat{U}$, as well as the Chronos beables $\widehat{C}$, Kuchař beables $\widehat{K}$, $\mathfrak{g}$-beables $\widehat{G}$, and Dirac beables $\widehat{D}$. That Kinematical Quantization involves selecting a subalgebraic structure $\langle \mathfrak{K}, |[\ ,\ ]| \rangle \subset \mathfrak{U}$ (the algebraic structure of the U) for promotion to a quantum kinematical algebraic structure $\langle \widehat{\mathfrak{K}}, [\ ,\ ] \rangle$ already illustrates the nontriviality of such an uplift from classical to quantum beables.

Example 1) In the absence of any linear constraints, Kuchař beables K remain a trivial matter at the quantum level. For instance, scaled 3-stop metroland is a simple model for which, classically, $K = U$ still applies. This already satisfies conditions I) to IV) of Sect. 39.1 in evidence at the level of Kinematical Quantization, via using $\sin\varphi$ and $\cos\varphi$ instead of φ. φ itself was already established to be only locally defined even at the classical level through not being defined for $\theta = 0$ or π.

E. Anderson, *The Problem of Time*, Fundamental Theories of Physics 190,
DOI 10.1007/978-3-319-58848-3_50

φ moreover fails the global continuity condition I), unlike $\sin\varphi$ and $\cos\varphi$, which are suitably periodic.

Lemma *Suppose that we operator-order the* $\widehat{\mathcal{F}\mathbf{lin}}$ *with* P *to the right. Then the classical configurational Kuchař beables* $\mathfrak{d}DE$ (25.7) *reappears as the quantum configurational beables operator.*

Proof Insert (42.4) in (12.17) to obtain $f^{\mathsf{A}}{}_{\mathsf{N}}\frac{\partial K}{\partial \mathsf{Q}^{\mathsf{A}}}\Psi = 0$. Next compare with (25.7) using the configuration representation $\widehat{\kappa} = \kappa$, which is valid since $\kappa = \kappa(\mathbf{Q}$ alone). □

N.B. that both the previous Chapter's caveat, and its generalization to other types of A-beables which correspond to algebraic structures within the $\widehat{\mathcal{F}\mathbf{lin}}$, continue to apply.

Example 2) For N-stop metroland, which has no relative angles in space to encode, v) is straightforward. The objects selected in Kinematical Quantization—$\widehat{n}^{\mathsf{A}}$, $\widehat{\mathsf{D}}$ for the pure-shape case and $\widehat{\rho}^{\mathsf{A}}$, $\widehat{p}^{\mathsf{A}}$ and $\widehat{\mathsf{D}}$ for the scaled case—are clearly sub-cubic. Each case's full set of objects can be taken as a particular basis of quantum Kuchař beables for each problem's relational quantities.

Example 3) For triangleland as formulated in terms of the $\underline{n}^i$, all selected objects are sub-cubic again. I.e. $\widehat{\mathrm{dra}}$, s^Γ in the pure-shape case, and $\widehat{\mathrm{Dra}}$, $\widehat{\mathrm{P}}^{\mathrm{Dra}}$ and $\widehat{\mathrm{S}}^\Gamma$ for the scaled case.

In each of the above Examples, the first objects listed are quantum configuration beables, which are readily accessed using the Lemma.

Example 4) By the Lemma, Electromagnetism and Yang–Mills Theory's classical configurational $G = \kappa = D$ beables bracket carries over to the quantum level.

Example 5) The simplest examples of constructing D are in cases for which $D = C$. The traditional setting in which such have been computed is for Minisuperspace models; see e.g. [79].

Examples 6–8) In the particular case of GR as Geometrodynamics, the requirements are, formally, that

$$[\widehat{\mathcal{M}}_i, \widehat{\kappa}]\Psi = 0, \tag{50.2}$$

$$[\widehat{\mathcal{M}}_i, \widehat{D}]\Psi = 0, \qquad [\widehat{\mathcal{H}}, \widehat{D}]\Psi = 0. \tag{50.3}$$

In particular, the latter constitutes a hard and very largely unsolved problem. For Plain Geometrodynamics, the Lemma gives that the configurational $\widehat{\kappa}$ lie among the classical κ, but for Affine Geometrodynamics, the caveat applies. The Lemma can also be used in (at least simple) approaches to quantizing Nododynamics, for which configurational beables are loops, or, less redundantly, knots. Also note here that Geometrodynamics and Nododynamics still lack satisfactory rigorously formulated expressions for $\widehat{\mathcal{H}}$, which has the knock-on effect of not even being able to rigorously pose the equations to be satisfied by quantum Dirac beables in these cases. Finally, note that the $\mathfrak{C}^*$-approach is limited within the context of GR [483].

50.2 Indirect Constructions for Quantum Dirac Beables $_D$

Example 1) In $\mathfrak{g}$-free theories, any operator $\widehat{O}$ can be subjected to the construction

$$\widehat{O}_D := \int \mathrm{d}t \exp(i H_0 t)\, \widehat{O} \exp(-i H_0 t) \tag{50.4}$$

for a suitable notion of time t (e.g. label time λ in Minisuperspace or Newtonian time in Mechanics), which again needs to run over all values of time rather than just some interval. Formal field-theoretic generalizations of this construct are straightforward as well. DeWitt gave an early treatment of such a construct in [235], based on (50.4) and specialized to the semiclassical case. Marolf [641] subsequently treated such objects in the case of perturbative Quantum Theory.

Example 2) Another means of attaining beables in the case of GR, developed e.g. by physicist Steve Giddings, Marolf and Hartle [353], involves integrating $\widehat{O}$ over all of spacetime. We do not follow this approach in this book because integrating over all spacetime is particularly problematic from an operational point of view.

Example 3) Suppose that a basis set of $\widehat{K}$ are known. Example 1) can now be re-run to construct Dirac beables from these.

Example 4) For instance, Halliwell's semiclassical Dirac beables construction is along the lines of Example 1) and can furthermore be extended along the lines of Example 3).

Examples 1) to 4) and the Lemma, moreover, are only addressing how to end up with examples of D and K respectively. This is as opposed to these procedures ensuring that the quantities produced have compatible commutation relations, form a closed algebraic structure, and consist of a basis (contain all the information and in a minimal non-redundant manner).

Example 5) Barvinsky's approximate equivalence Lemma 1 of Sect. 40.1 ameliorates Constraint Closure at the semiclassical level to leading order in $\hbar$.

Example 6) The classical-level $K \neq G$ of modewise SIC carries over to a $\widehat{K} \neq \widehat{G}$ distinction as well.

Example 7) Supergravity has e.g. notions of quantum non-supersymmetric Kuchař beables $\widehat{NSK}$ and quantum non-supersymmetric Dirac beables $\widehat{NSD}$, which are further examples of $\widehat{A}_X$ beyond the list of examples of notions of beables given at the beginning of this Section.

50.3 Quantum-Level Problem of Beables

A first issue with this is how to select a subalgebra of one's classical beables to promote to quantum beable operators. A second issue concerns how one is to operator-order these so as to ensure each of the following.

i) The beables close among themselves.
ii) The beables succeed in forming zero brackets with the final selection of operator-ordered and regularized quantum constraints.

The quantum Problem of Beables is that notions of beables are hard to construct for constrained quantum systems, with the most interesting quantum Dirac beables $\widehat{\mathcal{D}}$ being the hardest to construct of all.

Strategy 1) *Promote*. In this case, one takes classical beables beforehand, one can attempt to promote them to quantum ones. This might occur at the level of Kinematical Quantization [475] or be viewed as a process in addition to this. In either case, the need arises to select a classical subalgebra of objects to promote to quantum operators.

Three particular reasons why a classical resolution of the Problem of Beables may not pass over to the quantum level are as follows. Firstly, we know (from Chap. 39) that the quantum commutator algebraic structures are not necessarily the same as the classical Poisson algebraic structures for a given system. Secondly, that we need to select a subalgebra of classical objects to promote to quantum operators. Thirdly, whether a given object's brackets with the constraints contribute to equate to zero depends on the following.

a) The selection of classical subalgebraic structure to promote to form a quantum kinematical operator subalgebraic structure, which is afflicted by the Multiple Choice Problem.
b) The operator ordering involved in formulating the quantum constraints.
c) The operator ordering involved in formulating each candidate quantum beable.

Strategy 2) *Start Afresh*. One might also start afresh in the quest to find beables at the quantum level. This makes particular sense upon realizing that in general the classical and quantum brackets correspond to different algebraic structures $\mathfrak{b}$ and $\widehat{\mathfrak{b}}$; this is due to quantum level workings being more globally sensitive [475]. In general, the entities that commuted with the classical constraints with respect to one brackets structure should not be expected to result in quantum operators that commute with the quantum constraints with respect to an distinct brackets algebraic structure! Schematically, for whichever appropriate pairing of classical and quantum types of bracket,

$$|[\mathcal{C}_{\mathsf{W}}, \mathcal{A}_{\mathsf{X}}]| \text{ `='} \ 0 \quad \not\Rightarrow \quad [\widehat{\mathcal{C}}_{\mathsf{W}}, \widehat{\mathcal{A}}_{\mathsf{X}}]\Psi = 0. \tag{50.5}$$

Whether this occurs is, moreover, dependent *twice over* on operator-ordering ambiguities (in the beable operators and in the constraint operators).

In this way, classical beables can fail to be quantum beables. This parallels perfectly good classical symmetries failing to be quantum symmetries due to anomalies arising. This is in the sense that both are bracket obstructions upon passing from classical to quantum brackets. These statements are moreover dependent *twice over* on operator-ordering ambiguities (in the beables operator and in the constraint operators). This often adds to the futility of B ... Q schemes. This is since classical Problem of Beables resolutions may not straightforwardly carry over to the quantum Problem of Beables. $\mathfrak{g}$-beables are often exempt from invalidation due to the nice properties of Lie groups under Quantization schemes. However, in cases beyond this remit (Dirac beables, the classical Dirac algebroid...) one may need to solve the quantum Problem of Beables afresh. See the Halliwell-type combined scheme of Chaps. 29 and 54 for an example of separate classical and quantum implementations for Dirac-type beables. On the other hand, triangleland's Kuchař beables carry over straightforwardly to the quantum level [25].

The above complication might be avoided by making strong restrictions at the level of getting the algebraic structure of the beables to close.

The Quantum Constraint Closure Problem may cause there to be more $\widehat{\mathcal{C}}$'s than there were $\mathcal{C}$'s. The definition of quantum beables is more stringent due to requiring commutation with the additional $\widehat{\mathcal{C}}$'s.

If $\mathfrak{g}$ remains a bona fide symmetry at the quantum level, then there is no need for more Configurational Relationalism, and classical Kuchař beables remain. However each beables construction and closure success obtained at the classical level of closure may falter under operator ordering. In this way, some theories which had such a closure supporting a notion of beables at the classical level may lose that closure and with it a corresponding notion of beables at the quantum level.

Aside from the above bracket inequivalences and operator orderings, trying to promote known classical beables to quantum ones also falls afoul of the Multiple Choice Problem (see Epilogue III.A).

Research Project 62) Consider the quantum-level implications of Dittrich's approach to observables; see Sect. 25.8 for an outline and [251] for further details.

Research Project 63) [Long-standing] Resolve the Problem of Observables—or Beables—in QG.

50.4 Beables Motivated from Realist Interpretations

It is in particular at the quantum level that Bell's term 'beable' is distinct from the earlier term 'observable'. 'Observable' may be taken to imply that observers exist, and so also the measurement process and corresponding notorious Quantum Measurement Problem. Moreover, the situation worsens once it is being used for Cosmology rather than laboratory physics. In this setting, the Copenhagen Interpretation of QM can no longer apply. Beables, on the other hand, are quantities that just are. These steer clear of connotations carried by 'measurement' in QM, as well as being a better concept for whole-universe Cosmology. They entail more of a realist than an instrumentalist interpretation of QM. Some variants of such interpretations that remain alive and well—inclusion of which is a good part of why I use the term 'beables' in this book—are as follows; these all feature in subsequent chapters.

1) Formulations of decoherence (Sect. 48.3),
2) Histories Theory [340, 429, 504] (Chaps. 53 and 54), and
3) The contextual-realist approach of Doering and Isham [260] (Epilogue III.C).

Chapter 51
Fully Timeless Approaches at the Quantum Level

At the kinematical level, Fully Timeless Approaches exhibit the following triviality. Since these approaches have just configurations and no momenta at the primary level, working in the configuration representation makes it immediately clear that these approaches possess no nontrivial commutation relations at the primary level either. Nontrivial commutation relations can only enter such approaches if they succeed in providing emergent momenta at some more secondary level. Configurational notions of beables moreover remain defined at the primary level.

51.1 Quantum-Level Propositions

Let us next continue here with Sect. 26.1's chain of thought that Physics consists of questions which can be rephrased in a logical framework. While this was largely trivial at the classical level, we shall see below that the quantum version of this becomes nontrivial. This chain of though being due to mathematician George Mackey [650], we refer to it as '*Mackey's Principle*'. It was subsequently suggested by Isham and Linden [515] in the Histories Theory context. The Author then pointed to its wider applicability, in particular to Timeless Approaches as well [38]. These approaches involve timeless questions—questions of being—which are based on ($\mathfrak{q}$, *Point*)'s $\psi = \psi(\mathfrak{q})$ level of structure; in this way, such approaches lie within the $\mathfrak{q}$-primality worldview. The corresponding probabilities are timeless probabilities, requiring just the above level of structure (including a characterization of regions of $\mathfrak{q}$). This lies is within Chaps. 9, 10 and 26's classification of questions into purely timeless being, being at a time, and becoming. Let us continue by considering how purely timeless questions about being might be formulated and handled at the quantum level.

Timeless Peaking Implementation Solutions to the time-independent Schrödinger equation are interpreted here in terms of where the probability distribution function peaks. This involves integrating over some region of the classical $\mathfrak{q}$ (or

E. Anderson, *The Problem of Time*, Fundamental Theories of Physics 190,
DOI 10.1007/978-3-319-58848-3_51

in some applications, a region of the classical $\mathfrak{P}$hase). Some cases of this are as follows.

1) 'Modes and nodes', which have already been used in this book to explain whole-universe time-independent Schrödinger equations (Chaps. 39–43).
2) The Naïve Schrödinger Interpretation (Sect. 12.6 and below).
3) Barbour's Conjecture 2) of Sect. 26.8.
4) Combined Approaches (postponed until Chap. 54).

Some Problems with Timeless Approaches Let us next extend Sect. 26.2's Timeless Approach 'brier patches' to the quantum level.

Brier Patch 2) *Nonstandard Interpretation of Quantum Theory.* Fully Timeless Approaches and Histories both soon become entwined in general questions about the interpretation of Quantum Theory. In particular, this applies to whole-universe replacements for standard QM's Copenhagen Interpretation. Note here that some criticisms of Timeless Approaches [601, 604] are underlied by wishing to preserve aspects of the Copenhagen Interpretation , which may not be appropriate for Quantum Cosmology or QG.

Brier Patch 3) *Wheeler–DeWitt equation Dilemma.* Kuchař pointed out that Timeless Approaches face the following dilemma.

Either—horn 1—invoke the Wheeler–DeWitt equation, thereby inheriting some of its problems.

Or—horn 2—or do not invoke it, risking the alternative problem of one's approach being incompatible with the Wheeler–DeWitt equation. The action of the Wheeler–DeWitt operator would in this case kick purported solutions out of the physical solution space [601].

Naïve Schrödinger Interpretation The Naïve Schrödinger Interpretation is an Interpretation of Quantum Theory for whole-universe models [448, 450, 451, 897]. It concerns probabilities of 'being', as already outlined in Sect. 12.6. Moreover, one makes no attempt here to supplant all questions by questions of being.

Answers to questions of being follow here from considering Prob(the Universe belongs to region R of $\mathfrak{q}$), which corresponds to a quantification of a particular such property, as per (12.21). More precisely, one needs to address relative probabilities of this kind, out of not being able to normalize individual expressions of the above kind.

Example 1) r-formulation of RPMs. In this case, simple (and "geometrically nice") questions of being are listed in Chap. 26.1; e.g. [37] furthermore computes answers to these. E.g. for Metric Shape and Scale RPM, (12.21) becomes

$$\text{Prob}(\boldsymbol{\rho} \text{ has property P}) = \int_{A_{\text{P}} \subseteq \mathfrak{R}(\text{N, d})} |\Psi[\mathbf{S}, \rho]|^2 \mathbb{D}[\mathbf{S}, \rho], \tag{51.1}$$

where A_{P} denotes the P-affirmative subset.

Example 2) For isotropic Minisuperspace with scalar field,

$$\text{Prob}(\alpha, \phi \text{ has property P}) = \int_{A_P \subseteq \mathfrak{M}\text{INI}} |\Psi[\alpha, \phi]|^2 \text{d}\alpha\, \text{d}\phi; \tag{51.2}$$

consult Hawking and Page's work [450, 451] for various concrete such.
On the other hand, for diagonal Bianchi IX vacuum,

$$\text{Prob}(\alpha, \beta_\pm \text{ has property P}) = \int_{A_P \subseteq \mathfrak{M}\text{ini}} |\Psi[\alpha, \beta_\pm]|^2 \text{d}\alpha\, \text{d}\beta_+ \text{d}\beta_-. \tag{51.3}$$

Concrete such P correspond e.g. to quantifications of anisotropy as per Sect. 26.1.
Example 3) For SIC treated modewise,

$$\text{Prob}(\zeta_\text{n}, \upsilon_\text{n} \text{ has property P}) = \int_{A_P \subseteq \mathbf{Z} \times \mathfrak{M}\text{odespace}} |\Psi[\zeta_\text{n}, \upsilon_\text{n}]|^2 \text{d}\zeta_\text{n} \text{d}\upsilon_\text{n}, \tag{51.4}$$

for $\mathbf{Z}$ the allowed interval range of ζ_n.
Example 4) For GR as Geometrodynamics,

$$\text{Prob}(\mathbf{h} \text{ has property P}) = \int_{A_P \subseteq \mathfrak{R}\text{iem}(\mathbf{\Sigma})} |\Psi[\mathbf{h}]|^2 \mathbb{D}\mathbf{h}. \tag{51.5}$$

Concrete such P corresponding to Examples 3) and 4) include quantifications of inhomogeneity, as per e.g. Sect. 26.1 or Appendix N.8.

Problem 1) The Naïve Schrödinger Interpretation is indeed of limited use since it does not accommodate questions of being at a particular time, or of becoming [601]. While subsequent Sections provide routes around this, these routes extend beyond the Naïve Schrödinger Interpretation itself.

Problem 2) This approach is termed 'naïve' due to its not using any further features of the constraint equations. This is less severe when r-formulations are available; however it also fails to deal with Quad, which is so symptomatic of the Frozen Formalism Problem. By this, doubts should be cast on a strategy that purports to handle Frozen Formalism Problem while leaving Quad not bypassed but simply *unaddressed*.

Problem 3) In particular, this leaves the Naïve Schrödinger Interpretation menaced by horn 2 of the Wheeler–DeWitt equationWheeler–DeWitt equation Dilemma, in particular via this approach's consequently 'naïve' inner product postulation.

Problem 4) Generally non-normalizable probabilities are also involved, though, however, these support finite *ratios* of probabilities (the relative probabilities previously alluded to).

Problem 5) In the case of Geometrodynamics, time enters the scheme as an internal coordinate function of h_{ij}. Therefore it is represented by an operator. However, as pointed out in e.g. [495], there are problems with representing time as an operator.

Classical q Regions Implementations of Quantum Propositions The above propositions are associated with classical regions; consequently, these combine along the following lines. Set theory's complement ^{c}, union $\cup$, intersection $\cap$ and inclusion $\subseteq$ are a realization of conventional propositional logic's $\neg$, $\vee$, $\wedge$ and $\preceq$. Continuous regions of a manifold are a type of example of such sets, by which these implement conventional propositional logic.

Problem 6) The above logical structure for the propositions is a questionable one to use in a quantum-mechanical context due to its *classical* form.

Problem 7) (51.5) should be taken with a pinch of salt due to its involvement of the measure on $\mathfrak{Riem}(\boldsymbol{\Sigma})$. This would be furtherly problematic in any formally reduced scheme, in which the measure on $\mathfrak{Superspace}(\boldsymbol{\Sigma})$ would feature.

Moreover, Chap. 54's Combined Approach implements quantum propositions in this manner as well.

Proposition–Projector Association The aim is to implement propositions at the quantum level by projectors. This includes going beyond [515] the usual context and interpretation ascribed to these in Ordinary Quantum Theory. This implementation is preferable to representation by classical regions of integration. One reason for this is that quantum propositions do not in general combine Booleanly, whereas classical regions do.

51.2 Conditional Probabilities

Conditional Probabilities in Ordinary Quantum Theory Here for state ρ and proposition P implemented by projector $\widehat{\mathrm{P}}$, $\mathrm{Prob}(P;\,\rho) = \mathrm{tr}(\widehat{\rho}\widehat{\mathrm{P}})$. In this context, Gleason's Theorem (Appendix U.3) provides strong uniqueness criteria (see e.g. [499]) for this choice of object from its satisfying the basic probability axioms.

Note moreover that conditional probabilities in Ordinary Quantum Theory are given by [499]

$$\mathrm{Prob}(B \in b \text{ at } \mathrm{t} = \mathrm{t}_2 \,|\, A \in a \text{ at } \mathrm{t} = \mathrm{t}_1;\, \rho) = \frac{\mathrm{Tr}(\mathrm{P}^B_b(\mathrm{t}_2)\,\mathrm{P}^A_a(\mathrm{t}_1)\,\rho\,\mathrm{P}^A_a(\mathrm{t}_1))}{\mathrm{Tr}(\mathrm{P}^A_a(\mathrm{t}_1)\,\rho)}. \tag{51.6}$$

We here denote the projector—for a beable A and a a subset of the values that this can take—by P^A_a. This is given in a 2-time context, i.e. to be interpreted as *subsequent* measurements. It also follows that

$$\mathrm{Prob}(B \in b \text{ at } \mathrm{t}_2 \text{ and } A \in a \text{ at } \mathrm{t}_1) = \mathrm{Tr}\big(\mathrm{P}^B_b(\mathrm{t}_2)\,\mathrm{P}^A_a(\mathrm{t}_1)\,\rho\,\mathrm{P}^A_a(\mathrm{t}_1)\big), \tag{51.7}$$

and this extends in the obvious way to p propositions at times t_1 to t_p.

Supplant 'at a Time' This can be replaced by a timeless correlation between the subconfiguration of primary interest and the value of a particular 'clock' subconfiguration. As compared to Chap. 26, most tools for implementing this only become available at the quantum level. In the literature, this approach has been considered for each of the 'any', 'all' and 'sufficient local' versions.

Conditional Probabilities Interpretation in Timeless Quantum Theory The Conditional Probabilities Interpretation was proposed by Page and Wootters [720]; [343, 361, 601, 604, 753] contain subsequent useful comments, criticisms and variants. It extends the Naïve Schrödinger Interpretation as regards the range of questions which can be answered, by which it is an improvement as regards Problem 1) above. In particular, it addresses questions about conditioned being: conditional probabilities for the results of a pair of beables A and B concern correlations between these at a single instant in time. E.g. 'what is the probability of a triangle model universe being almost collinear given that it is almost isosceles?' Or 'what is the probability of the Universe being almost-flat *given* that it is almost-isotropic?', which can be interpreted within each of anisotropic Minisuperspace, SIC, Midisuperspace, and full GR. Moreover, both technically and as interpretations of QM, the Conditional Probabilities Interpretation and the Naïve Schrödinger Interpretation are highly distinct, e.g. for the following reasons.

1) The Conditional Probabilities Interpretation implements propositions at the quantum level by use of projectors.
2) The Conditional Probabilities Interpretation addresses questions concerning conditioned being by *postulating* the relevance of conditional probabilities, for finding B in the subset b, given that A lies in the subset a for a (sub)system in state ρ

$$\text{Prob}(B \in b \,|\, A \in a;\, \rho) = \frac{\text{Tr}(\text{P}^B_b\, \text{P}^A_a\, \rho\, \text{P}^A_a)}{\text{Tr}(\text{P}^A_a\, \rho)}. \tag{51.8}$$

   N.B. that these occur within the one instant rather than ordered in time (one measurement and *then* another measurement). Thereby this postulation lies outside of Quantum Theory's conventional formalism, for all that (51.8) superficially resembles (51.6). So Conditional Probabilities Interpretation is indeed also meant in the sense of an interpretation of QM.

The Conditional Probabilities Interpretation replaces questions of 'being at a time' by simple questions of conditioned being, along the following lines. Make use of one subsystem A as a timefunction, so that the above question about A and B can be rephrased to involve which value B takes when the timefunction-giving A indicates a particular time [495, 720]. The Conditional Probabilities Interpretation is moreover tied to considering a localized notion of clocks, and one which is configuration-based rather than change-based, as in 'the clock reads three o'clock'.

The traditional development of the Conditional Probabilities Interpretation did not extend to modelling the propositions within a logical scheme, due to pre-dating such considerations. This further layer of structure can however be added to the Conditional Probabilities Interpretation so as to comply with 'Mackey's Principle'.

Problem 1) This means of replacement of ‘being at a time’ questions has the practical limitation that A may not happen to have features rendering it suitable as a sufficiently good timefunction.

Problem 2) Supplanting ‘being at a time’ by ‘being’ is as far as the original Conditional Probabilities Interpretation goes. It is not a full resolution of Problem with Naïve Schrödinger Interpretation 1) due to not addressing questions of becoming. Page did however proceed to address this; see the next Section.

Problem 3) Kuchař pointed out that horn 1 of the Wheeler–DeWitt equation Dilemma also applies to the Conditional Probabilities Interpretation [601].

Some of Kuchař’s critiques [601, 604] of the Naïve Schrödinger Interpretation and the Conditional Probabilities Interpretation can however be interpreted as arising from not accepting a separate ‘being’ position, as opposed to constituting conceptual or technical problems once one has adopted such a position. E.g. Kuchař [601, 604] pointed to the Conditional Probabilities Interpretation leading to incorrect forms for propagators. Page’s response was that the Conditional Probabilities Interpretation is a timeless conceptualization of the world, so it does not need 2-time entities such as propagators.

Gambini–Porto–Pullin Approach Gambini and Pullin’s work with physicist Rafael Porto [341–343, 753] is built upon conditional probabilities of the following form:

$$\text{Prob}\begin{pmatrix}\text{beable } B \text{ lies in interval } \Delta B \\ \text{provided that the timefunction } \tau \text{ lies in interval } \Delta\tau\end{pmatrix}$$
$$= \lim_{T\longrightarrow 0}\left(\frac{\int_0^T \mathrm{d}t\,\mathrm{Tr}(\mathrm{P}_{\Delta_B}(t)\mathrm{P}_{\Delta t}(t)\rho_0\mathrm{P}_{\Delta t}(t))}{\int_0^T \mathrm{d}t\,\mathrm{Tr}(\rho_0\mathrm{P}_{\Delta t}(t))}\right). \tag{51.9}$$

The P(t) here are Heisenberg time evolutions of projectors P, i.e. $\mathrm{P}(t) = \exp(i\mathrm{H}t) \times \mathrm{P}\exp(-i\mathrm{H}t)$. This approach is based on the ‘any change’ notion of relational clocks (though this can be modified [30, 39] to the STLRC notion). These clocks are moreover taken to be non-ideal at the quantum level, giving rise to the following.

1) A decoherence mechanism. This is based on the argument that imprecise knowledge arises due to having to imprecise (non-ideal) clocks and rods (though that needs to be checked case by case rather than assumed [10]).
2) A modified version of the Heisenberg equations of motion. This is of the Lindblad type [862], which for finite theories and at the semiclassical level—with coherent-state connotations—takes the form

$$i\hbar\frac{\partial\rho}{\partial t} = [H, \rho] + \mathrm{r}[\rho]. \tag{51.10}$$

The form of the r-term here is $\sigma(t)[H, [H, \rho]]$, where $\sigma(t)$ is dominated by the rate of change of width of the probability distribution. By the presence of this ‘emergent becoming’ equation (51.10), this approach looks to be more promising in practice than Page's extended form of Conditional Probabilities Interpretation. This approach also has the advantage over the original Conditional Probabilities Interpretation of producing consistent propagators. By the presence of the $\mathfrak{r}$-term, the evolution is not unitary, which is all right insofar as it represents a system about which we have imprecise knowledge (cf. Appendix Q.3).

Quantum Dirac beables for this scheme are furthermore considered in the paper co-authored with physicist Sebastian Toterolo [344].

A full enough version of GR to model inhomogeneities wold require the $\delta\rho/\delta t$ version of this equation. On the other hand, both the specific Lindblad case and all explicit investigations to date involve the finite $\partial\rho/\partial t$ subcase.

Considering sets of propositions and using t^{sem} for t above, this scheme can be cast to comply with both Mach's Time Principle and Mackey's Principle. The first of these can be implemented by conditioning on t^{em}, or, more accurately, on t^{sem}. The Author views this option as possible for Gambini–Porto–Pullin schemes, unlike in the standard Conditional Probabilities Interpretation. This is due to these being less fully timeless, by being a limiting case of a temporal construct, in which there is a place in the theoretical scheme for the aforementioned propagators.

51.3 Timeless Records Theories

Quantum-level Timeless Records Approaches [21, 101, 718, 719] are a very natural successor to not only classical-level Records Approaches (Chap. 26) but to classical-level $\mathfrak{q}$ primary Relationalism as well.

Quantum Pre-records Theory This make use of quantum versions of the types of structure outlined at the classical level in Sects. 26.4 to 26.7, as follows. In the configuration representation, classical notions of localization in space and in $\mathfrak{q}$ continue to suffice, (these may be hard to obtain and use for QG in general). At least for simple model arenas, the mathematical structure of the quantum state space is also well-established [130]. See Appendices U.5–6 for outlines of quantum notions of information and of correlation; however, these also fall short if QG in general is the objective. Finally, whereas a means of assessing significant patterns in timeless records was demonstrated at the classical level in Sect. 26.7 by use of Kendall's Shape Statistics, the quantum counterpart of this remains unexplored, even for simple model arenas, never mind for full QG.

Research Project 64)† Set up the quantum-level analogue of Shape Statistics for use in simple RPM models of Quantum Records.

Next, as regards completing Pre-Records Theory to Records Theory by consideration of semblance of dynamics, four purely atemporal alternatives are as follows.

Page's Records Approach Page followed up [717–719] on the Conditional Probabilities Interpretation's success in supplanting 'being at a time' by timeless correlations between the configurations of the studied subsystem and of the clock used, by considering whether the Conditional Probabilities Interpretation computational object could be used in supplanting 'becoming' as well. In this way, he addressed correlations within a single present instant configuration which contains memories of what might otherwise be regarded as a sequence of 'previous configurations'.

By being an extension of the Conditional Probabilities Interpretation, this continues to be rooted on Proposition–Projector Association, and could be built up to be additionally in accord with 'Mackey's Principle'. This set of quantum-level details further advances Sect. 26.8's account.

Bubble Chamber Arguments Bubble chamber α-particle tracks can be explained in terms of a time-independent Schrödinger equation and can therefore be treated as 'timeless': [693] and Ex II.11. Various approaches to Quantum Cosmology—e.g. by Barbour [98, 101], Halliwell and students [420–422, 426, 429, 430], or by physicists Mario Castagnino and Roberto Laura [200, 201]—are based on extending this insight.

Barbour's 'Time Capsules' Conjectures Let us now continue Sect. 26.8's classical-level discussion with some quantum-level commentary.

As regards Barbour's Conjecture 2), his 'concentration of mist' is clearly a particular example of Timeless Peaking Interpretation. No basis for this conjecture is found however among the simple solved concrete models used in this book or in [37, 60]. RPM $\mathfrak{q}$ geometry does not drastically affect the probability distribution of Ψ to substantially peak about configurations of the required sort.

More concretely, as regards whether the quantum wavefunction peaks about the maximal collision, scaled triangleland has this point effectively excised by the potential being singular there. The wavefunctions are zero there, though a number of wavefunctions remain *centred about* thereabout. As regards whether there is heavy peaking about notions of uniform state (a concept argued to be important in e.g. [731]), pure-shape triangleland does not seem to exhibit this [37].

Addressing Barbour's Conjectures in full moreover requires the configuration space to not be conformally flat. This is because in conformally-flat spaces, curved $\mathfrak{q}$ geometry can (at least locally) recast by a PPSCT into the form of a potential factor. On the other hand, in non-conformally-flat spaces, the $\mathfrak{q}$ geometry possesses an irreducible part which can not be re-encoded as a potential effect. In this sense, triangleland does not suffice but [60]'s quadrilateralland does.

Since RPMs yield elliptic time-independent Schrödinger equations, it is relevant that elliptic equations on manifolds are capable of producing patterns which reflect the underlying shape [37, 612, 634]. On these grounds, geometrically induced patterns are more widely plausible, though this comes with no guarantees of *pronounced* peaking, or as regards any of the suggestions of *which regions* peaking occurs in.

Finally, no evidence has been forthcoming as regards Barbour's Conjecture 3).

Halliwell's Imperfect Records [419] Here one considers information in a wave pulse signal that is picked up and stored in a detector in terms of approximands or modes. A detector can be attuned to pick up the harmonics that are principal contributors to the form of the signal. In this way one, a good approximation to a signal can be obtained by storing relatively little information. Compare for instance the square wave with the almost-square wave that is comprised of the first 10 harmonics of the square wave. Even a detector that is only capable of storing one bit of information is capable of forming an imperfect record.

The Author's Records The idea here is to complete Pre-Records Theory to Records Theory by use of further attributes of Shape Statistics. Whereas this is currently classically viable, Research Project 64) would need to be tackled prior to being able to extend this approach to the quantum level.

Comparative Discussion Timeless approaches are often tied to nonstandard interpretations of QM; this is the case e.g. for the Naïve Schrödinger Interpretation, programs inspired by the bubble chamber model arena, Conditional Probabilities Interpretation, and Page's approach.

N.B. that bubble chambers are carefully selected environments for revealing tracks. However, Halliwell has emphasized that records can be imperfect. On these grounds, one may then expect records in general to be poorer than bubble chamber tracks. Perhaps they are far poorer, along the lines of physicist Erich Joos' model with Zeh of a dust particle decohering due to the cosmic microwave background photons [540]. This is a far more generic situation to encounter in Nature than a bubble chamber! In such a situation, records would be exceedingly diffuse, due to the information being dispersed by the cosmic microwave background photons and ending up spread over cosmological space. This affects Quantum Cosmology, as regards quantification of records due to these being problematic to access. The relevant information is now stored 'all over the place' and in a diffuse manner. In turn, this means that the information available may be of too poor a quality to construct history from it in any detail.

This disparity in likelihoods would leave a selection principle for 'time capsules' —envisaged as resembling bubble chamber tracks—needing to do a lot of work to compensate.

The description given above of Halliwell's notion of detectors storing imperfect records clearly middles between the Joos–Zeh and bubble chamber conceptualizations. This represents an 'average' detector, rather than one which is carefully attuned and capable of storing large amounts of relevant information. Cf. the former not being so dominant as to preclude some natural Physics running along the lines of the bubble chamber.

Barbour's approach has the further issue of not being based on Proposition–Projector Association. Additionally, quantitative detail of how to assess quantum level 'mist concentration' from a statistical perspective remains unclear. The classical Shape Statistics approach may well be a valuable guide toward a quantum-level understanding of this.

Whereas collinearities in threes and the bubble chamber tracks may look similar, Shape Statistics can be used to test any other aspect of shape, so approaches of this kind are expected to be generically applicable.

Finally, as regards Affine Geometrodynamics, the Naïve Schrödinger Interpretation should not care, in so far as this does not make use of commutation relations or of the Wheeler–DeWitt equation. On the other hand, the Conditional Probabilities Interpretation does work in a Wheeler–DeWitt framework [601], this will change in detail upon passing to Affine Geometrodynamics. Classical Records Theory is unaffected by this difference, though subsequent changes would be expected at the quantum level.

51.4 Records Approaches with More than Just Timeless Structure

Example 1) Physicist Mario Castagnino's scheme [200] builds in a time asymmetry in the choice of admitted solutions. As such, it does not contest the Arrow of Time issue. Note that this is a price to pay in passing from a purely Timeless Approach to each of the approaches below as well.

Example 2) Records Theory can be considered within the Semiclassical Machian Emergent Time Approach, whether of the 'all' or of the 'STLRC' variety. The basic idea here is to follow up a semiclassical scheme by considering timeless correlations therein. This approach started with Halliwell's follow-up [416] of Halliwell and Hawking's Semiclassical Quantum Cosmology [427]. It has the benefit of not requiring a purely timeless semblance of dynamics, since the overarching semiclassical scheme provides an emergent time. This idea can furthermore be carried over to the specifically Machian formulation of Semiclassical Quantum Cosmology of Chap. 46.

Example 3) The Gambini–Porto–Pullin scheme is also amenable to being 'embedded' in this way within a Semiclassical Approach, at the cost of expanding somewhat on Gambini–Porto–Pullin's own meaning of 'semiclassical'.

Example 4) Section 53.7's consideration of Records Theory within Histories Theory.

Example 5) Chapter 54's treatment of with the triple combination of Records Theory within Histories and Semiclassical Machian Emergent Time Approaches.

Chapter 52
Spacetime Primary Approaches: Path Integrals

52.1 Unconstrained Models

In the primary spacetime ontology, the quantum path integral takes the form

$$\int \mathbb{D}\mathbf{Q} \exp\big(i\, \mathsf{S}[\mathbf{Q}]/\hbar\big). \tag{52.1}$$

One can stay within this formulation or apply a $3+1$ split and canonical reformulation, as per Ex II.5. This Exercise's general transition amplitude $\mathcal{T}$ furthermore features as the integral kernel in the relation for passing between initial and final quantum wavefunctions,

$$\psi[q_{\text{fin}}, t_{\text{fin}}] = \int \mathrm{d}q_{\text{in}} \mathcal{T}[q_{\text{fin}}, t_{\text{fin}}, q_{\text{in}}, t_{\text{in}}]\psi[q_{\text{in}}, t_{\text{in}}]. \tag{52.2}$$

52.2 Path Integrals in Gauge Theory

Directly Formulated Case If we have the good fortune of being able to formulate our action and measure in gauge-invariant terms, one can keep on working with a close analogue of (52.1). I.e.

$$\int \mathbb{D}G_{\text{c}} \exp\big(i\, \mathsf{S}[G_{\text{c}}]/\hbar\big), \tag{52.3}$$

where the G_{c} form a basis of configurational $\mathfrak{g}$-invariant quantities.

Indirectly Formulated Case This is often necessary due to lack of availability of the former. One now inserts an expansion of unity in the following form [712]:

$$\iint \mathbb{D}\mathbf{Q}\, \mathbb{D}\mathbf{A} \det\left(\frac{\delta \mathcal{F}}{\delta \mathbf{A}}\right) \delta(\mathcal{F}_{\mathsf{G}}) \exp\big(i\, \mathsf{S}[\mathbf{Q}, \mathbf{A}]/\hbar\big). \tag{52.4}$$

E. Anderson, *The Problem of Time*, Fundamental Theories of Physics 190,
DOI 10.1007/978-3-319-58848-3_52

Therein, the *Fadde'ev–Popov factor* [295] (after physicists Ludvig Fadde'ev and Victor Popov) determinant is a subcase of Jacobian, whereas the δ-function imposes a basis set of gauge-fixing conditions $\mathcal{F}_{\mathsf{G}}$. The **A** are auxiliary fields, for now to be regarded as gauge auxiliary fields. One has the option of performing a $3+1$ split with action in the third form of (L.18), so as to pass to a canonical version.

A further separate technical matter is the inclusion of opposite-chirality 'ghost species'. These enter a modified form of the Lagrangian as well as providing extra measure factors followed by integrals thereover. Their introduction is a computational and unitarity-maintaining device; answers to physical questions are themselves free from these ghosts.

Example 1) In non-gauge models and directly gauge-invariant formulations, the Fadde'ev–Popov factor is multiplicatively trivial.

Example 2) On the other hand, Electromagnetism and Yang–Mills Theory are non-trivial in this way; indeed, the Fadde'ev–Popov approach opens up a rigourous Quantization of the latter. See [712] for a simple treatment of these examples and [446] for a more general treatment.

The field theoretic case's formula for the determinant corresponds to the array

$$\mathrm{W}_{\mathsf{GG}'}\delta(\mathrm{t}^{\mathrm{in}}-\mathrm{t}^{\mathrm{fin}}) = \frac{\delta}{\delta u(\mathrm{t}_{\mathrm{in}})}\frac{\delta}{\delta v(\mathrm{t}_{\mathrm{fin}})}\int\{\mathcal{g}\mathrm{auge}_{\mathsf{G}}[v], \mathcal{F}_{\mathsf{G}'}[u]\}\,\mathrm{dt}. \tag{52.5}$$

Example 3) For Finite Theories, this simplifies to

$$W_{\mathsf{GG}'} = \int\{\mathcal{g}\mathrm{auge}_{\mathsf{G}}, \mathcal{F}_{\mathsf{G}'}\}\mathrm{d}t. \tag{52.6}$$

In each case, one can contemplate extending from **gauge** to c_{F}, with a matching enlargement of $\mathcal{F}_{\mathsf{G}'}$ to $\mathcal{F}_{\mathsf{F}'}$.

52.3 Strategies for GR Path Integrals

Strategy 0) (11.12) is an incipient form for the spacetime formulation of GR's $\mathcal{T}$. *Diff*($\mathfrak{m}$)-invariance is subsequently formally incorporated in the indirect Fadde'ev–Popov manner.[1] This approach can be taken to manifest General Covariance.

To obtain a canonical version instead, begin by applying the ADM split. S now takes the analogous form to (J.39):

$$\mathsf{S}_{\mathrm{GR}} = \int \mathrm{d}\lambda \int_{\Sigma} \mathrm{d}\boldsymbol{\Sigma}\{\dot{\mathrm{h}}_{ij}\mathrm{p}^{ij} - \alpha\mathcal{H} - \beta_i\mathcal{M}^i\}. \tag{52.7}$$

[1] Indeed, the Fadde'ev–Popov method is on some occasions co-attributed to DeWitt [886], who applied it to the case of GR as well [238, 239].

Problem 1) Moreover, in considering the measure and Fadde'ev–Popov factors, the following trichotomy is induced depending on how $\mathcal{H}$ is to be treated.

Strategy 1) Consider just building in the *Diff*($\mathbf{\Sigma}$)-invariance corresponding to $\mathcal{M}_i$. This is uncontroversial, because $\mathcal{M}_i$ is a gauge constraint. Here, $\mathsf{A} = \underline{\beta}$ as features in the action, and a triplet $\mathcal{F}_i$ of *Diff*($\mathbf{\Sigma}$)-fixing conditions is applied. This part of the working at least conceptually parallels Yang–Mills Theory.

Strategy 2) Next, consider $\mathcal{H}$ as a gauge constraint as well. This case has a fourth gauge auxiliary variable α and a fourth gauge-fixing condition, so that now one has a quartet $\mathcal{F}_\mu$.

Strategy 3) Finally, one might not regard $\mathcal{H}$ as a gauge constraint, for the reasons laid out in Chaps. 27 and 32. In this case, $\mathcal{H}$ still has an auxiliary variable α that is paired to by multiplier-appending in the action. This can then enter the path integral measure and subsequently be integrated over. One can furthermore envisage a corresponding generalized fixing condition for first-class constrained systems, and a generalized Fadde'ev–Popov method which works for first-class constraints regardless of whether these are gauge constraints.

Notice that Strategy 3)'s schematic form looks very similar to 2), though making such a scheme concrete would take more technical and interpretational work. Also, while the above trichotomy is often ignored, it is clear from the problems below that full GR in any case remains but a very formally treated example.

In passing from a Canonical Approach to a spacetime formulated Path Integral Approach, a '*Non Tempus sed Via*' resolution of the Frozen Formalism Problem is posited. (This begs further ontological questions in addition to the above questions about the habitual technique used to handle path integrals. Moreover, this does not address any of the other facets of the Problem of Time.) One does elude the Inner Product Problem with one's time-dependent quantum wave equation, but this is at the cost of incurring a new Measure Problem. Unfortunately, much of the value attributed in Quantum Gravity to Path Integral Approaches over Canonical Approaches follows from the misunderstanding that the notorious Problem of Time *is* the Frozen Formalism Problem. Once it is clear that this is only a small subset of the problem, what Path Integral Approaches attain is not 'resolving the Frozen Formalism Problem which is the Problem of Time', but rather 'trading the Frozen Formalism Problem for further parts of the Problem of Time'.

Finally, using Strategy 2) or 3) need not automatically eschew spacetime General Covariance, in that some practitioners attempt concurrently Canonical-and-Covariant Approaches.

Problem 2) *Measure Problem.* There are, furthermore, diffeomorphism-based complications with specifying the measure part of the gravitational path integral [477]. Precisely what form these complications entail depends on which of Strategies 0) to 3) are adopted. Strategies 0) and 1) require the more moderate tasks of defining a *Diff*($\mathfrak{m}$)-invariant measure on $\mathfrak{S}$uperspacetime($\mathfrak{m}$) [477], and a *Diff*($\mathbf{\Sigma}$)-invariant measure on $\mathfrak{S}$uperspace($\mathbf{\Sigma}$), respectively. On the other hand, Strategies 2) and 3) amount to defining a *Diff*($\mathfrak{m}$, $\mathbf{\Sigma}$)-invariant measure on $\mathfrak{T}$ruespace($\mathbf{\Sigma}$). This involves yet further mathematical complications out of being an algebroid rather than an algebra. Further distinctions arise, depending on whether this space arises in a purely Gauge Theoretic context or a more general situation involving first-class but not necessarily gauge constraints. In some of these approaches, the GR path integral is viewed as more like an energy Green's function than a propagator due to the absence of an internal time [552]. For sure, all four of these strategies amount to the Inner Product Problem being traded for another Background Independence issue,[2] with Strategies 0), 2) and 3) being further specifically tied to time.

Problem 3) The indefiniteness of the GR action causes unboundedness from below upon being inserted in GR's path integral's exponential term [352]. This is clear upon conformally transforming the metric (Exercise!) There are moreover indications that the measure produces an opposite-sign term which can cancel this off [435]. In GR, passing to the Euclidean version no longer fully controls the behaviour of the action in the exponent, due to the complex rotation nullifying the Lorentzian indefiniteness of spacetime but not the DeWittian indefiniteness of the configuration space $\mathfrak{q}$.

Some model arena considerations are useful at this point.

Example 1) Minisuperspace. Diffeomorphism triviality means here that these models are not hampered by any Measure Problems. These models moreover already exhibit GR's indefiniteness and have a nontrivial role for $\mathcal{H}$. Here, Strategy 1) amounts to doing nothing, whereas Strategy 0), and some distinction between Strategies 2) and 3), are already realized. [This is subject to the limitations of privileged slices of homogeneity, and of one finite $\mathcal{H}$ trivially closing as an Abelian algebra.]

Example 2) RPMs. Strategy 1) remains nontrivial here as well. On the other hand, distinctions between Strategies 2) and 3) are clearly not tied to spacetime considerations here, nor does including $\mathcal{E}$ among the constraints substantially complicate the constraint algebra, nor is the action indefinite. This example is moreover more naturally formulated along the lines of the next Section.

Problem 4) In Quantum Theory, Complex Methods are commonly used, such as slightly deformed contour integrals in expressions for propagators; see e.g. [712] for the QFT case. This method involves making a $t \longrightarrow i\tau$ substitution on account of the Euclidean path integral being better behaved. Complications moreover arise in attempting to extend Complex Methods to the curved spacetime setting that is required for GR. Even for Minisuperspace, the contours between the real-Euclidean and real-Lorentzian 4-manifolds are ambiguous and with a substantial multiplicity [420]. Additionally, Wick rotations are non-generic (see e.g. [552]). There is little guarantee of a complex 4-d curved spacetime containing both suitable Lorentzian and Euclidean real 4-manifolds as submanifolds. In this way, a nontrivial issue arises in place of a flat-spacetime triviality (see again [552]). Whereas Euclidean approaches avoid issues with microcausality, these are however replaced by relying on analytic continuation working out [474], and in the current context this has

[2]This Measure Problem is, additionally, a technical trade-off [193], in the sense of having it instead of the Canonical Approach's Operator Ordering Problem.

become hard to establish and is on occasion demonstrably false. Let us end by noting that complex correction terms occur widely in semiclassical approaches; see e.g. [37, 157, 158, 453].
Some of the above problems are also ameliorated upon passing to Discrete Approaches [111, 253, 628, 685, 911]. However, wishing Nature to be calculable does not make Her discrete. And, in any case so far, discrete[3] approaches have a poor track record as regards the usually last to consider—and especially unsurmountable—Problem of Time facet that is the Spacetime Construction Problem.

Problem 5) Path Integral Formulations additionally exhibit a Constraint Closure Problem, since the measure term's Fadde'ev–Popov factors would be affected if Constraint Closure were not to hold.

Problem 6) Moreover, for GR the Batalin–Vilkovilsky approach to path integrals is more suitable than the Fadde'ev–Popov one [820], and yet still falls short of being satisfactory in further ways [319].

52.4 Temporal Relationalism Implementing Path Integral Quantum Theory (TRiPIQT)

This (Fig. 52.1) is a split-action almost-canonical formulation, whose action is of the third form in (L.18). The expansion of unity inserted into the measure part of the path integral now involves auxiliary variables of the form $\mathbf{d}\mathsf{A}$. The $\mathcal{F}_\mathsf{G}$ are also to be reformulated as $\mathbf{d}\mathcal{F}_\mathsf{G}$, so the TRi-Fadde'ev–Popov form of the action is

$$\iint \mathbb{D}\mathbf{Q}\,\mathbb{D}\,\mathbf{dA}\,\det\left(\frac{\delta\,\mathbf{d}\mathcal{F}}{\delta\,\mathbf{dA}}\right)\delta(\mathbf{d}\mathcal{F}_\mathsf{G})\exp\big(i\,\mathsf{S}[\mathbf{Q},\mathbf{dA}]/\hbar\big). \tag{52.8}$$

Now the array corresponding to the determinant is $\mathbf{d}\mathsf{W}_{\mathsf{GG}'}$, which depends linearly on $\mathbf{d}\mathcal{F}_\mathsf{G}$. The above covers the following examples.

Example 1) Straightforward rearrangements of Electromagnetism and Yang–Mills Theory.

Example 2) The fully TRi formulation of RPM path integrals.

Example 3) For Minisuperspace models, the TRi path integral formulation boils down to re-representing α as $\mathrm{d}I$.

Example 4) For full GR, the TRi formulation requires firstly for the action to be cast in the form

$$\mathsf{S}_{\mathrm{GR}} = \iint_\Sigma \big\{\partial \mathrm{h}_{ij}\mathrm{p}^{ij} - \partial \mathrm{F}^i\,\mathcal{M}_i - \partial \mathrm{I}\,\mathcal{H}\big\}, \tag{52.9}$$

secondly the development in this Sec's opening paragraph, and thirdly the adoption of the previous Section's Strategy 3).

[3]Or almost-discrete, e.g. involving an ancillary continuum sample space in the Causal Sets Approach or eventually taking a continuum limit in the Causal Dynamical Triangulation Approach.

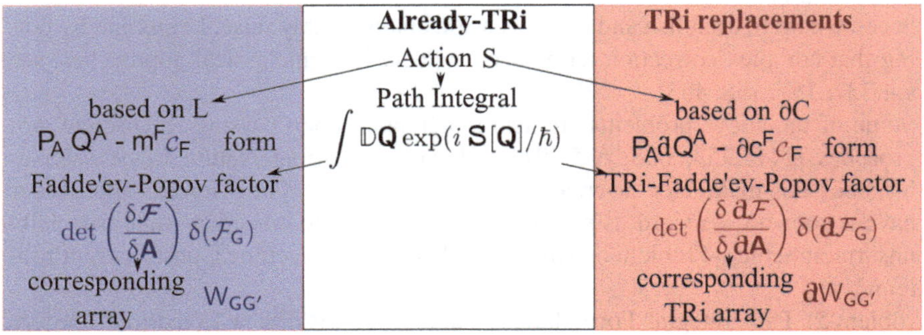

Fig. 52.1 TRiPIQT

52.5 Canonical-and-Path-Integral Approaches

E.g. QFT can be taken to involve a such in the usual Minkowski spacetime $\mathbb{M}^n$ context for which Canonical and Path-integral Approaches *support* each other. As well as such approaches inheriting *both* the canonical and the path-integral difficulties, it is not clear whether the large amount of path integral to canonical compatibility exhibited in e.g. flat-spacetime QFT continues to hold in QG. And, if not, whether *either* approach can be self-sufficient, or gain sufficiency from combination with some *third* program such as algebraic QFT [425]. If one tries to combine diverse such approaches for QG, then one would expect most of the problems of *each* to be present, including a wide range of time-related problems.

Additionally, one can proceed through Covariant-and-Canonical Approaches; such can use e.g. Peierls brackets, or take the form of Histories Approaches. Finally, a distinct covariant spacetime take on path integrals is afforded by the so-called *antifield formalism*; see e.g. [446] for an exposition.

Chapter 53
Histories Theory at the Quantum Level

At the quantum level, histories are not just paths but rather have the further distinction of being *decorated with* projectors. Thus we require more than just the previous Chapter's Feynman path integral mathematics. Moreover, we need to cover two cases.

1) The promotion of Chap. 28's 'Historia ante Quantum' scheme to the quantum level H ... Q.
2) Starting afresh with 'Historia Post Quantum' Q ... H.

53.1 Gell-Mann–Hartle-Type Histories Theory

This is a Q ... H scheme [340, 429] (see also [483, 780]), and consists of the following steps.

1) Individual histories are built out of strings of projectors $\mathrm{P}_{a_i}^{A_i}(\mathrm{t}_i)$, i = 1 to N at discrete time-steps t_i,

$$c_\eta := \mathrm{P}_{a_N}^{A_N}(\mathrm{t}_N)...\mathrm{P}_{a_2}^{A_2}(\mathrm{t}_2)\mathrm{P}_{a_1}^{A_1}(\mathrm{t}_1). \tag{53.1}$$

   N.B. that these projectors do not imply measurement; Histories Theory is, moreover, intended to have [340, 428, 483] other than the standard interpretation of QM. Indeed, assigning probabilities to histories does not work in Quantum Theory. For suppose $a(\mathrm{t})$ has amplitude $A[a] = \exp(i\, \mathsf{S}[a])$ and $b(\mathrm{t})$ has amplitude $B[b] = \exp(i\, \mathsf{S}[b])$. Then these are nonadditive since in general $|A[a] + B[b]|^2 \neq |A[a]|^2 + |B[b]|^2$.
2) Also consider notions of fine- and coarse-graining, which correspond to different levels of imperfection of knowledge. By this, families of histories are partitioned into subfamilies. Let $C_{\bar{\eta}}$ denote coarse-graining, where $\bar{\eta}$ is a subsequence of the history η's times and each projector in the new string may concern a less precise proposition. Note that this is a further type of coarse-graining—by probing at less times—in addition to the coarse-graining criteria used Chap. 51 and Appendix U.2.

E. Anderson, *The Problem of Time*, Fundamental Theories of Physics 190,
DOI 10.1007/978-3-319-58848-3_53

3) Finally consider the *decoherence functional* between a pair of histories η, η', given by

$$\mathcal{D}\mathrm{ec}[c_{\eta'}, c_\eta] := \mathrm{tr}(c_{\eta'} \rho c_\eta). \tag{53.2}$$

This is useful as a 'measure' of interference between $c_{\eta'}$ and c_η. It is zero for perfectly consistent theories, and has the following properties.

$$\mathcal{D}\mathrm{ec}[c_{\eta'}, c_\eta] = \mathcal{D}\mathrm{ec}[c_\eta, c_{\eta'}] \quad \text{(Hermeticity)}, \tag{53.3}$$

$$\mathcal{D}\mathrm{ec}[c_{\eta'}, c_\eta] \geq 0 \quad \text{(Positivity)}, \tag{53.4}$$

$$\sum_{c_{\eta'}, c_\eta} \mathcal{D}[c_{\eta'}, c_\eta] = 1 \quad \text{(Normalization)}, \tag{53.5}$$

$$\mathcal{D}\mathrm{ec}[c_{\eta'}, c_\eta] = \sum_{\eta' \in \bar{\eta}', \eta \in \bar{\eta}} \mathcal{D}[c_{\eta'}, c_\eta] \quad \text{(Superposition Property)}. \tag{53.6}$$

This scheme has the new probability postulate that

$$\mathcal{D}\mathrm{ec}[c_{\eta'}, c_\eta] = \delta_{\eta', \eta} \mathrm{Prob}(a_N \mathrm{t}_N, a_{N-1} \mathrm{t}_{N-1}, \ldots a_1 \mathrm{t}_1; \rho_0). \tag{53.7}$$

Let us end by noting that approximate consistency is held to suffice in this approach.

Research Project 65) To what extent do paths by themselves already support a notion of decoherence? E.g. the Path Integral Approach already possesses the notions of coarse-graining and finest possible graining that are more usually used in, and attributed to, Histories Theory. Also the basic notion of decoherence functional looks to already be defined at the level of paths alone, including its formulation in terms of class functionals as per the next Chapter.

53.2 Histories Projection Operator (HPO) Approach

This is Isham and Linden's [503, 504, 770, 771] quantum-level continuation of Chap. 28's classical treatment of the Histories Brackets Approach; see also e.g. [11, 766–769]. By this, it is a Q ... H scheme. Chapter 28's classical histories Poisson brackets are promoted to quantum histories kinematical commutators. This scheme's use of continuous time results in a 1-*d* QFT in the time direction. Moreover, passing from discrete to continuous time necessitates a continuum limit [816] of the tensor product enlargement from $\mathfrak{H}$ilb to $\bigotimes_{i=1}^{n} \mathfrak{H}$ilb. Indeed, H ... Q approaches can be seen as providing a second opportunity to a number of T ... Q approaches and to use of Geometrical Quantization. For instance, new Kinematical Quantization commutator and constraint-bracket algebroids ensue.

Example 1) History of nonrelativistic particles in absolute space. With the notion of a 1-*d* QFT in the time direction being relatively unfamiliar, I provide an outline this for this Example. By this, the nontrivial commutation relation is not like Ordinary QM's $[x_i, p_j] = i\,\hbar\,\delta_{ij}$, but rather takes the field-theoretic form

$$[x_{t_1}, p_{t_2}]_{\mathbf{H}} = i\,\hbar\,\delta(t_1 - t_2). \tag{53.8}$$

I.e. a continuous Dirac delta is involved in place of a discrete Kronecker delta. This is furthermore better handled in a smeared formulation; the

$$\mathfrak{L}^2(\mathbb{R}, \mathrm{d}x) \tag{53.9}$$

function space of real square-integrable functions on $\mathbb{R}$ enters at this stage [504] in the role of test functions. The nontrivial commutation relation is now

$$[x_f, p_g] = i\,\hbar \int_{-\infty}^{+\infty} f(x) g(x) \mathrm{d}x, \tag{53.10}$$

for $f, g \in \mathfrak{L}^2(\mathbb{R}, \mathrm{d}x)$. A quantum theory based on a Fock space—as is habitual in QFT—ensues; [504] provides further details.

The HPO approach additionally has a quantum histories quadratic constraint,

$$\widehat{\mathcal{Q}\mathrm{uad}}^{\lambda}\,\Psi = 0. \tag{53.11}$$

Like the Gell-Mann–Hartle approach, the HPO approach also possesses histories-theoretic notions of coarse- and fine-graining. The role of the previous Section's c_η is now undertaken by the following tensor product of the projectors:

$$c_\eta := \mathrm{P}_{a_N}^{A_N}(\mathrm{t}_N) \otimes \ldots \otimes \mathrm{P}_{a_2}^{A_2}(\mathrm{t}_2) \otimes \mathrm{P}_{a_1}^{A_1}(\mathrm{t}_1). \tag{53.12}$$

These immediately inherit the projector axioms from the individual projectors, whence the name of the program. Finally, the HPO approach also provides a form of decoherence functional; see e.g. [11] for its specific form.

53.3 Computation of Decoherence Functionals

For a given coarse-graining $\mathcal{C}_\eta$ consisting of classes $\{c_\eta\}$,

$$\langle \mathsf{Q}^{\mathrm{fin}} | \mathcal{C}_\eta | \mathsf{Q}^{\mathrm{in}} \rangle = \sum_{\eta \in \mathsf{Q}^{\mathrm{in}} c_\eta \mathsf{Q}^{\mathrm{fin}}} \exp\big(i\, \mathsf{S}[\eta]/\hbar\big). \tag{53.13}$$

This formulation for the class function can be interpreted in terms of paths η without association of projectors. Next,

$$\mathcal{D}\mathrm{ec}[\eta, \eta'] = \mathcal{N} \sum_{I,J} \mathrm{Prob}_I^{\mathrm{fin}} \langle \psi_I^{\mathrm{fin}} | \mathcal{C}_{\eta'} | \psi_J^{\mathrm{in}} \rangle \langle \psi_I^{\mathrm{fin}} | \mathcal{C}_\eta | \psi_J^{\mathrm{in}} \rangle \mathrm{Prob}_J^{\mathrm{in}}, \tag{53.14}$$

where the probabilities are initial and final inputs alongside the initial and final states.

$$\mathcal{N} := 1 \Big/ \sum_{I,J} \text{Prob}_I^{\text{fin}} |\langle \psi_I^{\text{fin}} | \mathcal{C}_S | \psi_J^{\text{in}} \rangle|^2 \text{Prob}_J^{\text{in}}, \tag{53.15}$$

where $\mathcal{C}_S$ is the sum of all the paths in (53.13). This is built in terms of a kernel path integral,

$$\langle \psi_I^{\text{fin}} | \mathcal{C}_\eta | \psi_J^{\text{in}} \rangle := \psi_I^{\text{fin}}(\mathsf{Q}^{\text{fin}}) \circ \langle \mathsf{Q}^{\text{fin}} | \mathcal{C}_\eta | \mathsf{Q}^{\text{in}} \rangle \circ \psi_J^{\text{in}}(\mathsf{Q}^{\text{in}}). \tag{53.16}$$

Here $\circ$ is some Hermitian but not necessarily positive inner product, whereas $\psi^{\text{in}} = \psi^{\text{in}}(\mathsf{Q}^{\text{in}})$ and $\psi^{\text{fin}} = \psi^{\text{fin}}(\mathsf{Q}^{\text{fin}})$ are initial- and final-condition quantum wavefunctions.

53.4 Further Theories, Structures and Problems

Research Project 66) Consider RPMs as a theoretical probe of Savvidou's 2-times and Canonical-and-Covariant Approach: how many features of this approach exists in a Background Independent but non-spacetime setting?

Research Project 67) Complete the Kouletsis–Kuchař model [568] at the quantum level. Does the histories group approach really get past the impasses encountered by the canonical group approach?

Problem 1) It is not clear whether the Gell-Mann–Hartle Approach is in correspondence with an algebra of propositions. In this approach, histories are the product of Heisenberg picture projectors. Moreover, such products are usually not themselves projectors. So this approach does not implement propositions as projectors. It does however possess a 'disjoint sum of histories' OR and a NOT operation. A further possibility is to implement propositions along Hartle's lines, from consideration of how each history intersects with a given region R in configuration space. For this purpose, Hartle uses (proper time spent in R), which is independent of canonical slicing. However, this is not to be expected to cover all physically-relevant propositions and the interrelations between them at the quantum level. This parallels Sect. 51.1's questioning the Naïve Schrödinger Interpretation 's use of classical regions to pose its quantum-mechanical questions. Additionally, this approach gives meaning to a variable which is not dynamical or subsequently quantum-mechanical, a procedure upon which Kuchař has cast some doubt. It amounts to altering a feature of standard Quantum Theory without as yet providing enough justification and interpretation for this.

The HPO Approach, however, bypasses Problem 1) by finding a different way in which to anchor the Proposition–Projector Association to Histories Theory. Indeed, the HPO Approach can be further motivated as a demonstration that a somewhat different correspondence between histories and questions can be made. The above correspondence can readily be completed with notions of negation and disjoint sum to form an orthoalgebra of propositions, $\mathfrak{up}$ (Appendix S.4). Questions about histories are moreover another case of simplified logical structure as compared to temporal logic [503]. In the HPO Approach, the decoherence functional is regarded as a functional $\mathfrak{up} \times \mathfrak{up} \longrightarrow \mathbb{C}$. Is this additional structure a significant further mathematical construct? Some partial answers to this are as follows.

A) The structure of this map parallels that of maps being representable by matrices, by which it makes sense to talk in terms of decoherence involving negligibility of off-diagonal elements.
B) Decoherence functionals are the analogues of quantum states in the parallel given by Isham–Linden's between Histories Theory and Ordinary QM.

Problems 2 to 5) Most of the Problems with Path Integral Approaches continue to apply. The extra structures that Histories Theory brings in are not known to ameliorate the problems with the underlying path integrals themselves.
Problem 6) Which degrees of freedom decohere which is unclear for GR [413].
Possible Problem 7) Is the generalization of Quantum Theory involved in Histories Theory self-consistent and meaningful? Does it reduce to Ordinary Quantum Theory in cases which are testable by experiment? In the case of a relativistic particle, Kuchař has argued [589] that it does not. Furthermore, Dowker and quantum theorist Adrian Kent [264, 265, 540, 542–544] commented that future and past can easily fall apart in this scheme. This may compromise the capacity to do Science in such a universe.

53.5 TRi Quantum Histories Theory

The input action for this can be built from t^{sem} by the last form in Eq. (L.18). This is linear in $\mathrm{d}/\mathrm{d}t^{\mathrm{sem}}$ and in $\mathrm{d}t^{\mathrm{sem}}$, by which it is Manifestly Parametrization Irrelevant, so one is in fact free to use any other label time λ if one so wishes. Because of this, there is no physical content to using the emergent time in the construction of the decoherence functional.

For now, let us consider the $\mathfrak{g}$-free case. Start from ($\mathfrak{Hist}$-$\mathfrak{P}$hase, *Hist-Can*) and apply Kinematical Quantization. The kinematical commutator algebra is obtained by selecting a subalgebra of the classical histories quantities. The constituent commutators suitably reflect global considerations, and the quantum histories constraints are some operator ordering of their classical counterparts. A kinematical Hilbert space and quantum constraint solving maps are next evoked, so to obtain (at least formally) ($\mathfrak{Hist}$-$\mathfrak{H}$ilb, *Hist-Uni*). This is meant in the enlarged sense of Isham and Linden, and is held to be useful for the following reasons.

1) Given the subsequent histories group, its unitary representations specifically permit access to the orthoalgebra $\mathfrak{up}$ of projectors as propositions about the theory's histories.
2) The Histories Approach involves a fresh set of algebraic entities. In this way, it provides a second opportunity for the corresponding Quantization scheme to work out in practice. [The price to pay, moreover, may be significant: basing a

Canonical Quantization scheme on histories rather than on configurations is less well-established, even at the conceptual level.]

The kernel path integral in (53.16) takes e.g. the Manifestly Reparametrization Invariant form $\langle \mathsf{Q}^{\text{fin}} | C_\eta | \mathsf{Q}^{\text{in}} \rangle$. This differs from the one in Hartle's review [428] due to use of the TRi differential of the instant $\mathbf{d}\mathsf{I}$ in place of the lapse α.

In terms of the answers to physical questions, moreover, there would appear to be no difference between the HPO and Gell-Mann–Hartle schemes. HPO is a technical refinement useful for establishing theorems, and also a conceptual improvement due to its manifestly fitting the Proposition–Projector Association.

Example 2) Scaled N-stop metroland's histories formulation exactly parallels the mathematics of Example 1) while casting this mathematics in a context relevant to Quantum Cosmology. One interpretational difference here is in the use of emergent time t^{em} or label time λ in place of absolute time t.

As regards the Hartle-type formulation of Histories Theory for RPMs, this approach has workings with physical outcomes paralleling those of Chap. 52's formalism, albeit with the following differences.

Difference 1) It is cast in terms of the habitual multipliers rather than in terms of TRi cyclic differentials.

Difference 2) This approach allows for time to be treated discretely. [In this case, the subsequently-attached projectors at each time form finite strings, composed by plain multiplication.]

Difference 3) Projectors having been evoked, the paths γ decorated by such are to be re-denoted as histories η.

Difference 4) Histories are here plain products of projectors and so are not themselves projectors, by which there is a lack of Proposition–Projector Association for these whole histories.

On the other hand, the HPO formulation for RPMs has histories brackets paralleling those in Sect. 53.2. Next consider tying a continuous limit of tensor product strings of projector operators to each of the paths in question. This overall object is itself a projector and therefore implements a histories proposition. These objects are therefore to be regarded as quantum histories.

53.6 Further Examples of Histories Formulations

Example 3) Minisuperspace. Anastopoulos and Savvidou [11] gave an HPO approach to isotropic Minisuperspace models with scalar field matter. The Author made a slight modification of this within the Relational Approach [31].

$\mathfrak{g}$-Nontrivial Histories Theory In indirectly formulated versions, there are also quantum first-class linear constraints

$$\widehat{\mathcal{F}\mathbf{lin}}^{\lambda}\Psi = 0. \tag{53.17}$$

In this case (or in Field Theories) quantum histories Constraint Closure can become nontrivial. In Historia ante Quantum Approaches, each classical histories constraint subalgebraic structure is promoted to a quantum operators counterpart $\widehat{\mathcal{C}}^{\lambda}_{\mathsf{W}}$. The two are, once again, not necessarily isomorphic.

Histories observables are also to be found, alongside addressing whether these are operationally meaningful. This is rather probably handled by starting afresh, rather than by promoting classical histories observables. Each notion of histories quantum A-observables $\widehat{A}^{\lambda}_{\mathsf{X}}$ obeys

$$\big[\widehat{\mathcal{C}}^{\lambda}_{\mathsf{W}}, \widehat{A}^{\lambda}_{\mathsf{X}}\big]_{\mathbf{H}}\Psi = 0. \tag{53.18}$$

Finally, in relational treatments based on Dirac-type Quantization, the TRi path integrals further differ from Hartle's presentation through containing further cyclic differentials.

Example 4) The HRQ ordered approach to RPM has a single histories constraint: the histories energy constraint, which is

$$\widehat{\mathcal{E}}_{t^{\mathrm{em}}} := \int \mathrm{d}t^{\mathrm{sem}}\widehat{\mathcal{E}}\big(t^{\mathrm{sem}}\big) \tag{53.19}$$

in the Machian approach. Compared to the previous specific examples, this in general has additional curved configuration space features.

The more specific case of scaled triangleland in the HRQ formulation gives analogous mathematics to that of the particle in 3-d, by which $\mathfrak{g}_{\mathrm{can}} = Eucl(3)^{\lambda}$. The canonical group now consists of the histories of mixed relative dilational momenta and relative angular momenta $SO(3)^{\lambda}$ objects $\mathrm{s}(\lambda)$ and the $(\mathbb{R}^3)^{\lambda}$ objects $\boldsymbol{p}^{Dra}(\lambda)$. The corresponding linear space $\mathfrak{v}^*$ is an $\mathbb{R}^3$ at each value of time, of objects $\boldsymbol{Dra}(\lambda)$ that are the histories of *ellip*, *area* and *aniso*. Again, these continue to combine as a semidirect product with no problems in considering it to be the standard $\mathfrak{g}_{\mathrm{can}} \rtimes \mathfrak{v}$ at each value of time, which again forms an associated 1-d QFT. The kinematical quantum histories algebra for this model turns out to be straightforward [37].

Equation (53.19) is here the sole constraint, and the quantum histories constraint algebraic structure is straightforward, so there is no Constraint Closure Problem here.

The quantum path integral is

$$\big\langle \boldsymbol{Dra}^{\mathrm{fin}}\big|C_{\gamma}\big|\boldsymbol{Dra}^{\mathrm{in}}\big\rangle = \int_{\eta} \mathbb{D}\boldsymbol{P}^{Dra}\,\mathbb{D}\boldsymbol{Dra}\,\mathbb{D}\mathrm{d}I\,\exp\big(iS\big[\boldsymbol{Dra}, \boldsymbol{P}^{Dra}, \mathrm{d}I\big]/\hbar\big). \tag{53.20}$$

Additionally, we can use that $I = t^{\mathrm{em}}$ in the Machian version of this scheme. The decoherence functional (53.14), (53.23) now takes the following form. $\psi^{\mathrm{in}} = \psi^{\mathrm{in}}(\boldsymbol{Dra}_{\mathrm{in}})$ and $\psi^{\mathrm{fin}} = \psi^{\mathrm{fin}}(\boldsymbol{Dra}^{\Gamma}_{\mathrm{fin}})$. The $\mathbb{D}\boldsymbol{Dra}$ that plays the role of $\mathbb{D}\mathrm{Q}$ in these equations is trivial as the corresponding configuration space is flat and the $\boldsymbol{Dra}$ play the role of Cartesian coordinates for it (Appendix G.3). The associated $\mathbb{D}\boldsymbol{P}^{Dra}$ that plays the role of $\mathbb{D}\mathsf{P}$ is likewise geometrically trivial. Use an action of the third form of (L.18), specialized to scaled triangleland in Dragt coordinates:

$$S^{\mathrm{r}}_{\Delta\text{-ERPM}} = \int \big\{ \mathrm{d}Dra^{\Gamma}\,\Pi^{Dra}_{\Gamma} - \mathrm{d}I\,\varepsilon \big\}. \tag{53.21}$$

The Fadde'ev–Popov determinant and gauge fixing factors are trivial for the r-formulation of this model.

$$\langle \boldsymbol{Dra}_{\mathrm{fin}} | C_{\eta} | \boldsymbol{Dra}_{\mathrm{in}} \rangle = \int_{\eta} \mathbb{D}\boldsymbol{Dra}\,\mathbb{D}\boldsymbol{P}^{Dra}\,\mathrm{d}I\,\exp\big(i\,\mathsf{S}\big[\boldsymbol{Dra}, \boldsymbol{P}^{Dra}, \mathrm{d}I\big]/\hbar\big); \tag{53.22}$$

$\mathcal{D}\mathrm{ec}[\eta, \eta']$ is next built from this via (53.14)–(53.16) under $\mathbf{Q} \longrightarrow \boldsymbol{Dra}$. In particular, the case of 1 particle in 3-d has $\mathfrak{q} = \mathbb{R}^3$, which corresponds to the space of the $\boldsymbol{Dra}$.

Example 5) On the other hand, the HQR ordered approach to RPMs provides a more specific example of the preceding indirect approach. This formulation more closely parallels what is formally available for full GR than Example 4) does. The configurations are now triangles redundantly described by the four components of the pair of relative Jacobi vectors. The classical histories are then built up from sequences of these and their momenta. The $\mathcal{L}^{\lambda} = 0$ constraint remains to be imposed at the quantum level.

The quantum path integral in this case is

$$\big\langle \boldsymbol{\rho}^{\mathrm{fin}} \big| C_{\gamma} \big| \boldsymbol{\rho}^{\mathrm{in}} \big\rangle$$
$$= \int_{\eta} \mathbb{D}\boldsymbol{p}\,\mathbb{D}\boldsymbol{\rho}\,\mathrm{d}I\,\mathbb{D}\mathrm{d}\underline{\mathrm{B}}\,\mathsf{D}_{\mathcal{F}}[\boldsymbol{\rho}, \mathrm{d}I, \mathrm{d}\underline{\mathrm{B}}]\,\delta\big(\mathcal{F}[\boldsymbol{\rho}, \mathrm{d}I, \mathrm{d}\underline{\mathrm{B}}]\big)\exp\big(i\,S[\boldsymbol{\rho}, \boldsymbol{p}, \mathrm{d}I, \mathrm{d}\underline{\mathrm{B}}]\big). \tag{53.23}$$

The decoherence functional (53.14) can now be computed as follows. $\psi^{\mathrm{in}} = \psi^{\mathrm{in}}(\boldsymbol{\rho}_{\mathrm{in}})$ and $\psi^{\mathrm{fin}} = \psi^{\mathrm{fin}}(\boldsymbol{\rho}_{\mathrm{fin}})$. The $\mathbb{D}\boldsymbol{\rho}$ that plays the role of $\mathbb{D}\mathsf{Q}$ in equations (53.14) is trivial as the corresponding configuration space is flat and the $\boldsymbol{\rho}$ play the role of Cartesian coordinates for it. The associated $\mathbb{D}\boldsymbol{p}$ that plays the role of $\mathbb{D}\mathsf{P}$ is likewise trivial. Use action (L.18), in particular in its second form as specialized to the triangleland case.

The TRi-Fadde'ev–Popov and gauge-fixing factors are now in general nontrivial. Explicit evaluation of these requires making a gauge-fixing choice for scaled triangleland. In triangleland, a double collision D and a merger M can be used as North and South poles respectively (Appendix G.3); this entails choosing between two overlapping charts centred about M and D. Consider first the 'triangle base = x' gauge's $\mathcal{F}_{\mathrm{M}} := \theta_1 = 0$. $\boldsymbol{\{}\mathcal{F}_{\mathrm{M}}, \mathcal{L}\boldsymbol{\}} = 1$, so the Fadde'ev-Popov determinant does not contribute any nontriviality to the integration in this gauge. Moreover, the above argument is clearly not globally valid, since the M-chart is not. We can counter this by considering also the median = x gauge's $\mathcal{F}_{\mathrm{D}} := \theta_2 = 0$. This gives the same histories bracket as above, and split the region of integration into two charts within each of which one of these two gauges is entirely valid. As regards the delta function, $\mathcal{F}_1$ can be written in Cartesian coordinates as $\arctan(\rho_y^1/\rho_x^1) = 0$. I.e. $\rho_y^1 = 0$ provided that $\rho_x^1 \neq 0$, which is guaranteed by the choice of region of integration in which this gauge-fixing is applied. So, allowing for M-indices to be rewritten as 1's and D-indices as 2's, and for the index F to run over 1 and 2,

$$\langle \rho_1 | \mathcal{C}_\gamma | \rho_2 \rangle = \int_{\substack{\gamma \text{ represented} \\ \text{in F-chart}}} \mathbb{D}\rho\, \mathbb{D}p\, \mathrm{d}I\, \mathbb{D}\mathrm{d}\underline{B}\, \delta\big(\theta_{\mathrm{F}}(\rho^{\mathrm{F}})\big) \exp\big(iS[\rho, p, \mathrm{d}I, \mathrm{d}\underline{B}]/\hbar\big). \tag{53.24}$$

Finally, $\mathcal{D}\mathrm{ec}[\gamma, \gamma']$ is built out of this through (53.14)–(53.16) using $\mathsf{Q}^{\mathsf{A}} \longrightarrow \underline{\rho}^i = \rho$ and $\mathrm{c}^{\mathsf{G}} \longrightarrow \underline{B}$.

This book's simple RPM models are moreover free of the following Histories Theory versions of the Measure Problem, Functional Evolution Problem, Thin Sandwich Problem, Diffeomorphism-specific Problems and Foliation Dependence Problem. The reduced approach to the scaled triangle is, moreover, also free of global issues (unlike the Dirac approach to the scaled triangle). So at least for this model, one looks to have a rather good resolution of the Problem of Time, upon which the next Chapter builds an even better resolution.

Example 6) SIC. This is a further modewise reduced calculation which can be investigated using the HPO Approach. This does not look to be substantially harder than the above Minisuperspace and reduced triangleland calculations. In this way, Histories Theory for SIC has entered the realm of calculable physics.

Research Project 68) Investigate quantum histories for modewise SIC. In particular, is this case's 1-*d* QFT entirely straightforward, or is it affected by a central term?

Example 7) [8] also considered the HPO Approach for simple QFT.

Example 8) Hartle already considered Electromagnetism as a model arena of constrained physics [429]; an HPO counterpart has now been provided as well [182].

Example 9) The general GR case was set as Ex VI.15.ii) along the lines of the Gell-Mann–Hartle Approach to Histories Theory. An answer, in outline, is that for full GR, the configurations are now 3-geometries redundantly described by the six components of the 3-metric $\mathbf{h}$. The classical histories are next built up from sequences of these. $\mathcal{M}_i$ and $\mathcal{H}$ remains to be imposed as quantum-level histories constraints.

The decoherence functional for GR (53.14) is further evaluated as follows. $\psi^{\mathrm{in}} = \psi^{\mathrm{in}}(\mathbf{h}^{\mathrm{in}})$ and $\psi^{\mathrm{fin}} = \psi^{\mathrm{fin}}(\mathbf{h}^{\mathrm{fin}})$. Use the action (52.9); the computation of the measure needs to be left formal. Including a minimally-coupled scalar field,

$$\big\langle \mathbf{h}^{\mathrm{fin}}, \phi^{\mathrm{fin}} \big| \mathcal{C}_\gamma \big| \mathbf{h}^{\mathrm{in}}, \phi^{\mathrm{in}} \big\rangle = \int_\gamma \mathbb{D}\mathbf{h}\, \mathbb{D}\phi\, \mathbb{D}\mathbf{p}\, \mathbb{D}\pi_\phi\, \mathbb{D}\partial \mathrm{I}\, \mathbb{D}\partial\underline{\mathrm{F}}\, \mathsf{D}_{\mathcal{F}}[\mathbf{h}, \mathbf{p}, \phi, \pi_\phi, \partial\mathrm{I}, \mathrm{d}\underline{\mathrm{F}}]$$
$$\times\, \delta\big(\mathcal{F}^\gamma\big[\mathbf{h}, \phi, \partial\underline{\mathrm{F}}^\gamma\big]\big) \exp\big(i\, \mathsf{S}[\pi, \pi_\phi, \mathbf{h}, \phi, \partial\underline{\mathrm{F}}, \partial\mathrm{I}]\big). \tag{53.25}$$

Finally, $\mathcal{D}\mathrm{ec}[\eta, \eta']$ is built out of this through (53.14)–(53.16) using $\mathsf{Q} \longrightarrow \mathbf{h}$.

See also [548] for a Histories Theory approach to Affine Geometrodynamics including a Minisuperspace example.

Research Project 69) Consider concrete quantum-level examples of Histories Theory approaches with nontrivial diffeomorphisms, from SIC to Midisuperspace.
Research Project 70) Further develop Kouletsis and Savvidou's approaches to Histories Theory [567, 769] for full GR, now at the quantum level.

53.7 Records Within Quantum Histories Theory

N.B. that a Timeless Records Theory sits within each histories theory [340, 342, 411, 413–415, 421, 422]. This is moreover independent of the Gell-Mann–Hartle versus HPO distinction. This is because records involve single-time histories, i.e. a single projector, and the ordinary and tensor products of a single projector obviously coincide and indeed trivially constitute a projector. Thus the Proposition–Projector Association can be applied and a propositional logic structure can be based on this.

Additionally, paralleling the treatment of records in Chap. 51 and Appendices Q and U, one can conceive of the following notions.

1) Subhistories, Sub-$\mathfrak{H}$ist.
2) Distances between histories $Dist(\eta, \eta')$ and a corresponding sense of localization.
3) Notions of information for quantum histories as considered by Hartle [429], Isham and Linden [505], and Kent [541].
4) Notions of correlation between histories. For instance, the decoherence functional is already a key structural element for this.
5) Histories propositions or logic remain an atemporal construct. Once again, this is an area that mostly only becomes interesting at the quantum level [489].

Since in Records Theory, the histories brackets reduce to the usual Poisson brackets, records approaches return one to the *usual* canonical group. The opening of the next Chapter's Interprotections 2) to 4) concern further interesting properties of Records Theory within Histories Theory.

The conditional probability of records $\underline{\kappa} = (\kappa_1, \kappa_2, \dots, \kappa_n)$ given the past alternatives $\underline{\eta}$ is

$$\text{Prob}(\underline{\kappa}|\underline{\kappa}) = \text{Prob}(\underline{\kappa}, \underline{\kappa})/\text{Prob}(\underline{\eta}) = \text{Tr}\big(R_{\underline{\kappa}}\,\rho_{\text{eff}}(\underline{\eta})\big), \tag{53.26}$$

$$\text{for } \rho_{\text{eff}}(\underline{\eta}) = \mathcal{C}_{\underline{\eta}}\,\rho\,\mathcal{C}_{\underline{\eta}}^{\dagger}/\text{Tr}\big(\mathcal{C}_{\underline{\eta}}\,\rho\,\mathcal{C}_{\underline{\eta}}^{\dagger}\big). \tag{53.27}$$

Here, $R_{\underline{\kappa}}$ denotes a *records projector*, which can be envisaged as the obvious subcase of the general histories projector. Note also the relation

$$\text{Prob}(\underline{\eta}) = \text{Tr}\big(\mathcal{C}_{\underline{\eta}}\,\rho\,\mathcal{C}_{\underline{\eta}}^{\dagger}\big) = \text{Tr}\big(R_{\underline{\eta}}\,\rho\,\big(t_n^{\text{sem}}\big)\big). \tag{53.28}$$

Perfect correlation between records and past alternatives is only guaranteed if Prob($\underline{\kappa}$ | $\underline{\eta}$) = 1, which only occurs if $\rho_{\text{eff}}(\underline{\eta})$ is pure. For the more general case of $\rho_{\text{eff}}(\underline{\eta})$ mixed, Prob($\underline{\kappa}$ | $\underline{\eta}$) < 1 and this correlation is imperfect. *So in general one can expect the presence of imperfect records.* In particular, imperfect records are still possible in settings with very small environments (even just 1 degree of freedom's worth) [411]. This permits Minisuperspace models, small RPMs and small detector models to remain meaningful.

Example 1) Halliwell's investigation of imperfect records [411] concerns a heavy particle moving through a medium consisting of a few light particles. The heavy particle disturbs these into motion. Subsequent instants consist of the particles' positions and momenta. It is these instants which are the records, and the dynamics or history of the large particle can subsequently be constructed from them (to some approximation). This example provided the first confirmation that a very small environment of light particles indeed suffices in order to have a nontrivial notion of imperfect record. This points to the possibility of using finite model arenas with further desirable theoretical properties, such as this book's main model arenas: RPM, Minisuperspace and modewise SIC. Finally, as regards the information content aspect of records, in this work Halliwell also conjectured that the number of bits required to describe a set of decoherent histories is approximately equal to number of bits thrown away to the environment.

Chapter 54
Combined Histories-Records-Semiclassical Approach

At the semiclassical quantum level, the Combined Approach becomes a rather more interprotected procedure than its classical precursor of Chap. 29. It is a combination of the Semiclassical Machian Emergent Time Approach (Chaps. 12 and 46), quantum Records Theory (Chap. 51), and the quantum Isham–Linden type HPO Histories Theory (Chap. 53). Pairwise, one has the following.

1) Records within the Semiclassical Machian Emergent Time Approach (Sect. 51.4).
2) Histories within the Semiclassical Machian Emergent Time Approach (Sect. 53.5).
3) Records within the HPO version of Histories Theory (Sect. 53.7).

Halliwell's study [408] of timeless correlations *within* the Semiclassical Approach arena of Halliwell–Hawking [419] is a precursor of the triple combination. Halliwell introduced the triple combination in [413] for mechanical and Minisuperspace models; the Author subsequently provided a Machian and nontrivial-$\mathfrak{g}$ in [25]. This is a particularly interesting prospect because in this combination, the individual strategies can remove weaknesses from each other along the following lines.

Firstly recollect that we already encountered three examples of this at the classical level (the three motivations at the start of Chap. 29). There is, moreover, a substantial increase in interprotections upon passing to the semiclassical level.

Interprotection 1) The Semiclassical Approach's assumption of a WKB regime requires justification. This is approached using decoherence in the form of histories decohering [552, 931].[1]

Interprotection 2) Within each Histories Theory, there is a Timeless Records Theory, as per Sect. 53.7. At the quantum level, this was pointed out by Gell-Mann and Hartle [340] and extended to imperfect cases by Halliwell [411, 413–415]. In this way, Histories Theory supports Records Theory by providing guidance as to the form a working Records Theory would take. This also allows for these two to be cast as a joint mathematical package. It also provides a common atemporal logic

[1]On the other hand, Gambini, Porto and Pullin's work in Sect. 51.2 provides a distinct non histories theoretic source of decoherence.

E. Anderson, *The Problem of Time*, Fundamental Theories of Physics 190,
DOI 10.1007/978-3-319-58848-3_54

grounding of the two [37, 260, 504], which can become nontrivial at the quantum level, i.e. 'Mackey's Principle'.

Interprotection 3) As Gell-Mann and Hartle said [340]

$$\text{“}\textit{records are somewhere in the Universe where information is stored when histories decohere}\text{”}. \tag{54.1}$$

Interprotection 4) The elusive question of 'what decoheres what' can be approached based on what the records involved are [340, 413]. In this way, Records Theory in turn supports Histories Theory. For instance, Sect. 48.3 pointed to the suggestion that small inhomogeneities decohere the Minisuperspace degrees of freedom.

Interprotection 5) By providing an underlying Dynamics or history, one or both of the Semiclassical Machian Emergent Time Approach and Histories Theory overcome present-day purely timeless Records Theory's principal weakness of needing to find a practicable construction of a semblance of dynamics or history. This goes a long way towards Records Theory being a viable part of one's worldview.

Interprotection 6) The Semiclassical Approach provides a Machian scheme for quantum histories and quantum records to reside within.

Interprotection 7) Working in a semiclassical regime helps with the computation of timeless probabilities of histories entering given configuration space regions. In particular, given an eigenstate of the quantum Hamiltonian, this approach determines what the probability is of finding the system in a region of configuration space, without reference to time [413].

Interprotection 8) Histories Theory provides a means of construction of semiclassical Dirac beables. This is also a timeless construct, and one which makes use of semiclassicality.

At the classical level, Interprotections 1), 3) and 4) are absent since they concern the purely quantum notion of decoherence. On the other hand, Interprotection 7) drops out since it concerns a purely quantum probability computation.

Let us finally propose how to order the constituent parts of the Combined Approach. *Meaningless label* histories come first, this gives the Semiclassical Approach, and the Machian emergent-time version of Histories Theory ensues. Finally, localized Timeless Approaches sit inside the last two of these. On the other hand, the Semiclassical Approach sits within the Timeless Approach, since its equations arise from the Wheeler–DeWitt equation. However the Timeless Approach can itself be taken to sit within the meaningless label time Histories Theory. So down both strands of the argument, histories are the most primary entities in the Combined Approach.

54.1 $\mathfrak{g}$-Free Models Without Machian Emergence

Let us first explain some of the features of Halliwell's approach. This considers a Gell-Mann–Hartle approach to histories, which picks up some Problem of Time issues from Chap. 53, though other such are absent from the model arenas considered. Halliwell implements propositions using regions of configuration space $\mathfrak{q}$. In fact, Hartle and Halliwell both separately considered such schemes, each using different versions of class functionals $\widehat{\mathcal{C}}_{\mathrm{R}}$ relating to $\mathrm{Prob}_{\mathrm{R}} :=$ Prob(trajectory enters a region R of $\mathfrak{q}$). The extent to which the Naïve Schrödinger Interpretation 's issue due to using regions carries over to Halliwell's approach is covered in Fig. 59.2.

Halliwell investigated Interprotection 7) in 2003 [413] for a free particle, by addressing the following question for an energy eigenstate. What is the probability of finding the system in a series of regions of configuration space $\mathfrak{q}$ without reference to time?

Halliwell subsequently investigated [414, 415] decohering histories as a possible means of constructing the probability distribution for the Wheeler–DeWitt equation. [This also uses Semiclassical Approach techniques and could be useful for avoiding problems involving how to interpret the Semiclassical Approach's Wheeler–DeWitt equation.] Some intermediate and supporting steps in this program were co-authored with his students Peter Dodd, Joerg Thorwart and Petros Wallden [418, 421, 422].

Semiclassicality is helpful with explicit construction of the class functional. First apply the WKB ansatz (12.2). Comparing with Fig. 29.1, the semiclassical flow lines now correspond to $\nabla_{\mathrm{C}} S$; to zeroth order, these coincide, but there are subsequently semiclassical corrections. Halliwell [408] furthermore established that

$$\mathcal{W}\mathrm{ig}[\boldsymbol{q}, \boldsymbol{p}] \simeq |\chi(\boldsymbol{q})|^2 \delta^{(K)}(\boldsymbol{p} - \boldsymbol{\partial} S), \tag{54.2}$$

where $\boldsymbol{p}$ would be equal to $\boldsymbol{\partial} S$ at the purely classical level, as per Hamilton–Jacobi Theory. Next [408, 410, 441],

$$\mathrm{Prob}^{\mathrm{semicl}}_{\Upsilon} \simeq \int \mathrm{d}t \int_{\Upsilon} \mathbb{D}\Upsilon(\boldsymbol{q}) \; \boldsymbol{v} \cdot \boldsymbol{\partial} S \, |\chi(\boldsymbol{q})|^2. \tag{54.3}$$

N.B. that this is 'starting afresh' as opposed to promotion of the classical precursor (29.13).

Class Functionals Halliwell's treatment continues within the standard framework of decoherent histories. The key step here is the construction of *class functionals*, which we denote by $\widehat{\mathcal{C}}_{\mathrm{R}}$, so as to model $\mathrm{Prob}_{\mathrm{R}}$. $\widehat{\mathcal{C}}_{\mathrm{R}}$ refers to a set of histories η which cross over into region R. Halliwell's approach to this [413–415] uses integrals over all time to resolve the further issue of compatibility with H, by which we choose Halliwell's version over Hartle's.

In seeking a mathematical implementation of class functionals, it makes sense to preliminarily point out that one cannot use the most obvious

$$\widehat{\mathcal{C}}_{\mathrm{R}}(\boldsymbol{q}_{\mathrm{f}}, \boldsymbol{q}_0) = \int_{-\infty}^{+\infty} \mathrm{d}t \exp(-iEt/\hbar) \times \int \mathbb{D}\boldsymbol{q}(t) \exp\big(iS[\boldsymbol{q}(t)]/\hbar\big)\, \theta\Big(\int_0^{\lambda} \mathrm{d}t \, \mathrm{Char}_{\mathrm{R}}\big(\boldsymbol{q}(t)\big) - \epsilon\Big) \tag{54.4}$$

—where the E-factor' here comes from [421] assuming Sect. 41.5's Rieffel inner product [413, 434]—due to its non-commutation with H.

However, the 'sharpened' object[2]

$$\widehat{\mathcal{C}}_{\mathrm{R}}^{\sqcap\!\sqcup} = \theta\left(\int_{-\infty}^{\infty} \mathrm{d}t\,\mathrm{Char}_{\mathrm{R}}\big(\boldsymbol{q}(t)\big) - \epsilon\right)\mathrm{P}(\boldsymbol{q}_{\mathrm{f}}, \boldsymbol{q}_0)\exp\big(i\,S(\boldsymbol{q}_{\mathrm{f}}, \boldsymbol{q}_0)\big), \tag{54.5}$$

is satisfactory, both conceptually and as regards commutation with H. This is given above for a specific example from ordinary Mechanics; Halliwell furthermore considered examples of it for Minisuperspace [413].

Moreover, semiclassicality helps at this particular point by providing an explicit construction of the class functional. N.B. also that this class functional is not the end of the story since it is technically unsatisfactory, as resolved in [414, 415].

Construction of Quantum Chronos Beables The class functionals obey

$$[\widehat{\mathcal{C}\mathrm{hronos}}, \widehat{\mathcal{C}}_{\mathrm{R}}^{\sqcap\!\sqcup}] = 0 \tag{54.6}$$

by construction. The $\widehat{\mathcal{C}}_{\mathrm{R}}^{\sqcap\!\sqcup}$ are therefore examples of quantum Chronos beables $\widehat{\mathcal{C}}$, which are additionally quantum Dirac beables $\widehat{\mathcal{D}}$ for such models as here with no further constraints.

While the semiclassical Halliwell construct is formulated in terms of histories theoretic structures, note that its output indeed consist of beables rather than histories observables. This is because of the integration over all t. N.B. that this is an example of 'starting afresh' with a new structure rather than of trying to promote Halliwell's distinct classical construct to the quantum level.

Finally, through involving use of the semiclassical quantum $\widehat{\mathcal{C}}_{\mathrm{R}}^{\sqcap\!\sqcup}$, the semiclassical Dirac beables construct also 'starts afresh' rather than being a promotion of the classical combined scheme's own Dirac beables construct.

Decoherence Functionals The class functional can now be used to re-express the decoherence functional between pairs of histories η, η'

$$\begin{aligned}\mathcal{D}\mathrm{ec}[\eta, \eta'] &= \int_{\eta}\mathbb{D}\boldsymbol{q}\int_{\eta'}\mathbb{D}\boldsymbol{q}'\exp\big(i\{\mathsf{S}[\boldsymbol{q}(t)] - \mathsf{S}[\boldsymbol{q}'(t)]\}/\hbar\big)\rho(\boldsymbol{q}_0, \boldsymbol{q}_0'), \\ &= \iiint \mathbb{D}\boldsymbol{q}_{\mathrm{f}}\mathbb{D}\boldsymbol{q}_0\mathbb{D}\boldsymbol{q}_{\mathrm{f}}'\widehat{\mathcal{C}}_{\eta}^{\sqcap\!\sqcup}[\boldsymbol{q}_{\mathrm{f}}, \boldsymbol{q}_0]\widehat{\mathcal{C}}_{\eta'}^{\sqcap\!\sqcup}[\boldsymbol{q}_{\mathrm{f}}', \boldsymbol{q}_0']\Psi(\boldsymbol{q}_0)\Psi^*(\boldsymbol{q}_0').\end{aligned} \tag{54.7}$$

This can be further reformulated using the notion of *influence functional*, $\mathcal{I}$ [298], which encodes an environmental property: the influence on a given subsystem; its

[2]Here, $\boldsymbol{q}^{\mathrm{cl}}(t)$ is the classical trajectory, $\boldsymbol{q}_0$, $\boldsymbol{p}_0$ is initial data, and θ is the step function. The corresponding cofactor is the standard semiclassical approximation to the unrestricted path integral. f_{f} is the characteristic function of region R, ϵ is a small positive number, and $S(\boldsymbol{q}_{\mathrm{f}}, \boldsymbol{q}_0)$ is the classical action between $\boldsymbol{q}_{\mathrm{f}}$ and $\boldsymbol{q}_0$. See [421] for the detailed form of the prefactor function P.

mathematical form is

$$\begin{aligned}\mathcal{I}[\boldsymbol{q}(t),\boldsymbol{q}'(t)] &= \sum_{f}\iint \mathbb{D}\boldsymbol{Q}(t)\mathbb{D}\boldsymbol{Q}(t') \\ &\quad\times \exp\big(i\big\{S_0\big[\boldsymbol{q}(t)\big]-S_0\big[\boldsymbol{Q}'(t)\big]+S_i\big[\boldsymbol{q}(t),\boldsymbol{Q}(t)\big]-S_i\big[\boldsymbol{q}'(t),\boldsymbol{Q}'(t)\big]\big\}/\hbar\big).\end{aligned} \tag{54.8}$$

Next, (54.7) gives

$$\mathcal{D}\mathrm{ec}[\eta,\eta'] = \iiint \mathbb{D}\boldsymbol{q}_{\mathrm{f}}\mathbb{D}\boldsymbol{q}_0\mathbb{D}\boldsymbol{q}'_0\widehat{\mathcal{C}}^{\sqcap\!\sqcup}_{\eta}[\boldsymbol{q}_{\mathrm{f}},\boldsymbol{q}_0]\widehat{\mathcal{C}}^{\sqcap\!\sqcup}_{\eta'}[\boldsymbol{q}_{\mathrm{f}},\boldsymbol{q}'_0]\mathcal{I}[\boldsymbol{q}_{\mathrm{f}},\boldsymbol{q}_0,\boldsymbol{q}'_0]\Psi(\boldsymbol{q}_0)\Psi^*(\boldsymbol{q}'_0). \tag{54.9}$$

If [421]'s conditions furthermore hold—modelling environment-system interactions—then $\mathcal{I}$ takes the form

$$\mathcal{I}[\boldsymbol{q}_{\mathrm{f}},\boldsymbol{q}_0,\boldsymbol{q}'_0] = \exp(i\boldsymbol{q}\cdot\boldsymbol{\Gamma}+\boldsymbol{q}\cdot\boldsymbol{\sigma}\cdot\boldsymbol{q}). \tag{54.10}$$

Here, $\boldsymbol{\sigma}$ is a non-negative matrix, $\boldsymbol{q}^- := \boldsymbol{q}-\boldsymbol{q}'$, whereas Γ_{Γ}, $\sigma_{\Gamma\Lambda}$ are real coefficients depending on $\boldsymbol{q}+\boldsymbol{q}'$ alone. Furthermore in this case,

$$\mathrm{Prob}_{\mathrm{R}} = \iint \mathbb{D}\boldsymbol{p}_0\mathbb{D}\boldsymbol{q}\,\theta\bigg(\int_{-\infty}^{+\infty}\mathrm{d}t\,\mathrm{Char}_{\mathrm{R}}\big(\boldsymbol{q}^{+\mathrm{cl}}\big)-\epsilon\bigg)\widetilde{\mathcal{W}\mathrm{ig}}\big[\boldsymbol{q}_0^+,\boldsymbol{p}_0\big], \tag{54.11}$$

where $\boldsymbol{q}^+ := \{\boldsymbol{q}_0+\boldsymbol{q}'_0\}/2$ and $\boldsymbol{q}^{+\mathrm{cl}}(t)$ the classical path with initial data $\boldsymbol{q}_0^+,\boldsymbol{p}_0$. Finally,

$$\widetilde{\mathcal{W}\mathrm{ig}}\big[\boldsymbol{q}_0^+,\boldsymbol{p}_0\big] = \int\mathbb{D}\boldsymbol{p}\exp\bigg(-\frac{1}{2}\{\boldsymbol{p}_0-\boldsymbol{p}-\Gamma\}\cdot\boldsymbol{\sigma}\cdot\{\boldsymbol{p}_0-\boldsymbol{p}-\boldsymbol{\Gamma}\}\bigg)\mathcal{W}\mathrm{ig}\big[\boldsymbol{q}_0^+,\boldsymbol{p}_0\big]. \tag{54.12}$$

is the Gaussian-smeared version of the Wigner functional, which itself is

$$\mathcal{W}\mathrm{ig}[\boldsymbol{q},\boldsymbol{p}] = \frac{1}{\{2\pi\}^k}\int\mathbb{D}\boldsymbol{q}\exp(-i\,\boldsymbol{p}\cdot\boldsymbol{q})\rho\big(\boldsymbol{q}^++\boldsymbol{q}^-/2,\boldsymbol{q}^+-\boldsymbol{q}^-/2\big). \tag{54.13}$$

Example 1) Halliwell considered this for absolutist particle models [413], for which the configuration spaces are $\mathbb{R}^n$.

Example 2) Simple Minisuperspace models for which the Minisuperspace is flat. Since these are indefinite, they is qualitatively distinguished from the previous.

54.2 Machian Time Version Sitting Within Semiclassical Approach

This involves the $t \longrightarrow t^{\mathrm{sem}} \simeq t_1^{\mathrm{sem}}$ version of the preceding three Sections. We also return here to Sect. 29.1's enumeration of types of emergent Machian time.

Type 4) $\mathrm{t}^{\mathrm{sem}}$, which is taken to be formally arbitrarily satisfactory within the premises of the semiclassical approach.
Type 5) $\mathrm{t}_1^{\mathrm{sem}}$: the distinct semiclassical quantum Machian first approximand; the two zeroth approximands coincide, so let us relabel Type 2) as $\mathrm{t}_0^{\mathrm{em}}$.
Type 6) $\mathrm{t}^{\mathrm{sem}\prime}$, which is taken to be some formal improvement on t^{em} that is not dependent on any use of semiclassical approximations.

54.3 Combined Approach for $\mathfrak{g}$-Nontrivial Theories

Many examples require generalization of the Wigner functional to curved space. This enters not only the volume elements, but also causes the sums inside the bra and ket to cease to be trivially defined. For Riemannian configuration spaces, this was resolved in [189, 402, 532, 625, 913]. See instead [860] for an approach which assumes just affine structure. Mathematicians Liu Zhang-Ju and Qian Min also extended their work [625] to principal fibre bundles over Riemannian manifolds, which meets the requirements for extending Halliwell's 2003 approach's treatment in terms of Wigner functionals. We thereby take 'Wigner functionals in curved space' in Liu and Qian's sense.

Reference [408]'s straightforward approximations in deriving (54.2) locally carry over [25], so for $\boldsymbol{K}$ a basis of Kuchař configurational beables,

$$\mathcal{W}\mathrm{ig}\left[\boldsymbol{K}, \boldsymbol{P}^{K}\right] \simeq \left|\chi(\boldsymbol{K})\right|^2 \delta^{(r)}\left(\boldsymbol{P}^{K} - \partial^{K} S\right) \tag{54.14}$$

($\boldsymbol{P}^K$ being $\partial^K S$ for classical trajectories). One next applies the Halliwell-type heuristic of replacing w by $\mathcal{W}$ig in expressions for timeless probabilities, giving

$$\mathrm{Prob}_{\Upsilon}^{\mathrm{semicl}} \simeq \int \mathrm{d}t^{\mathrm{sem}} \int_{\Upsilon} \mathrm{d}\Upsilon(\boldsymbol{K}) \boldsymbol{v}^{K} \cdot \frac{\partial S}{\partial \boldsymbol{K}} \left|\chi(\boldsymbol{K})\right|^2. \tag{54.15}$$

Decoherence Functional The decoherence functional is the Q to κ version of the first equation in (54.7) Class functionals are next inserted into the expression for the decoherence functional, giving (modulo PPSCT factors) [25]

$$\begin{aligned} \mathcal{D}\mathrm{ec}\left[\eta, \eta'\right] \\ &= \iiint \mathbb{D}\boldsymbol{K}_{\mathrm{f}} \mathbb{D}\boldsymbol{K}_0 \mathbb{D}\boldsymbol{K}'_{\mathrm{f}} \widehat{\mathcal{C}}_{\eta}^{\sqcap\!\sqcup} [\boldsymbol{K}_{\mathrm{f}}, \boldsymbol{K}_0] \widehat{\mathcal{C}}_{\eta'}^{\sqcap\!\sqcup} \left[\boldsymbol{K}'_{\mathrm{f}}, \boldsymbol{K}'_0\right] \Psi(\boldsymbol{K}_0) \Psi\left(\boldsymbol{K}'_0\right). \quad (54.16) \\ &= \iiint \mathbb{D}\boldsymbol{K}_{\mathrm{f}} \mathbb{D}\boldsymbol{K}_0 \mathbb{D}\boldsymbol{K}'_0 \widehat{\mathcal{C}}_{\eta}^{\sqcap\!\sqcup} [\boldsymbol{K}_{\mathrm{f}}, \boldsymbol{K}_0] \widehat{\mathcal{C}}_{\eta}^{\sqcap\!\sqcup} \left[\boldsymbol{K}_{\mathrm{f}}, \boldsymbol{K}'_0\right] \mathcal{I}\left[\boldsymbol{K}_{\mathrm{f}}, \boldsymbol{K}_0, \boldsymbol{K}'_0\right] \Psi(\boldsymbol{K}_0) \Psi^*\left(\boldsymbol{K}'_0\right). \quad (54.17) \end{aligned}$$

The second step is in the context that the Universe contains a classically-negligible but quantum-non-negligible environment as per Appendix 48.4, so it indeed makes sense to use the influence functional $\mathcal{I}$.

Alternative Indirect $\mathfrak{g}$-Act, $\mathfrak{g}$-All Implementation In this case,

$$\widehat{\mathcal{C}}_{\mathrm{R}}^{\sqcap\sqcup\,\text{g-free}}[\rho_{\mathrm{f}},\rho_0]=\int_{\mathrm{g}\in\mathfrak{g}}\mathbb{D}\mathrm{g}\;\vec{\mathfrak{g}}_{\mathrm{g}}\left(\theta\left(\int_{-\infty}^{+\infty}\mathrm{d}t^{\mathrm{sem}}\,\mathrm{Char}_{\mathrm{R}}\left(\rho_0^{\mathrm{f}}(t^{\mathrm{em}})\right)-\epsilon\right)\right.$$
$$\left.\times\,\mathrm{P}(\rho_{\mathrm{f}},\rho_0)\exp\left(iS(\rho_f,\rho_0)/\hbar\right)\right). \tag{54.18}$$

It is indeed physically desirable for these to be individually $\mathfrak{g}$-invariant. Next form the decoherence functional out of $\widehat{\mathcal{C}}_{\mathrm{R}}^{\sqcap\sqcup\,\mathrm{g-free}}$,

$$\mathcal{D}\mathrm{ec}[\eta,\eta']=\int_{\mathrm{g}\in\mathfrak{g}}\mathbb{D}\mathrm{g}\;\vec{\mathfrak{g}}_{\mathrm{g}}\left\{\iiint\mathbb{D}\boldsymbol{q}_{\mathrm{f}}\mathbb{D}\boldsymbol{q}_0\mathbb{D}\boldsymbol{q}_0'\,\widehat{\mathcal{C}}_{\eta}^{\sqcap\sqcup\,\text{g-free}}[\boldsymbol{q}_{\mathrm{f}},\boldsymbol{q}_0]\right.$$
$$\left.\times\,\widehat{\mathcal{C}}_{\eta'}^{\sqcap\sqcup\,\text{g-free}}[\boldsymbol{q}_{\mathrm{f}},\boldsymbol{q}_0']\underbrace{\Psi(\boldsymbol{q}_0)\Psi^*(\boldsymbol{q}_0')}\right\}\mathcal{I}[\boldsymbol{q}_{\mathrm{f}},\boldsymbol{q}_0,\boldsymbol{q}_0']\Psi(\boldsymbol{q}_0)\Psi^*(\boldsymbol{q}_0'). \tag{54.19}$$

A problem with this alternative approach is that it becomes blocked early on as regards being more than formal when $\mathfrak{g}=\mathit{Diff}(\boldsymbol{\Sigma})$.

Example 1) Scaled triangleland avoids Curved Geometry subtleties[3] [25] since this model possesses a flat presentation of configuration space. On the one hand, its r-formulation involves the q^i to Dra^{Γ} of the flat-space case, as in [25, 37]. On the other hand, for the indirect formulation, $\mathfrak{g}=SO(2)=U(1)$, so in Example 5) of Sect. 42.2's notation,

$$\int_{g\in\mathfrak{g}}\mathbb{D}g\,\times=\int_{\zeta\in\mathbb{S}^1}\mathbb{D}\zeta\,\times=\int_{\zeta=0}^{2\pi}\mathrm{d}\zeta\,\times. \tag{54.20}$$

54.4 Construction of Quantum Dirac Beables from Quantum Kuchař Beables

For theories which possess of a notion of Kuchař beables, we have class functionals

$$\widehat{\mathcal{C}}_{\mathrm{R}}^{\sqcap\sqcup}=\theta\left(\int_{-\infty}^{+\infty}\mathrm{d}t_1^{\mathrm{sem}}\mathrm{Char}_{\mathrm{R}}\left(\boldsymbol{K}\left(t^{\mathrm{sem}}\right)\right)-\epsilon\right)\mathrm{P}(\boldsymbol{K}_{\mathrm{f}},\boldsymbol{K}_0)\exp\left(i\,S(\boldsymbol{K}_{\mathrm{f}},\boldsymbol{K}_0)/\hbar\right), \tag{54.21}$$

obeying

$$[\widehat{chronos},\widehat{\mathcal{C}}_{\mathrm{R}}^{\sqcap\sqcup}]=0 \tag{54.22}$$

[3] To have this nontrivial, consider e.g. the relational quadrilateral and the non-diagonal Bianchi IX Minisuperspace.

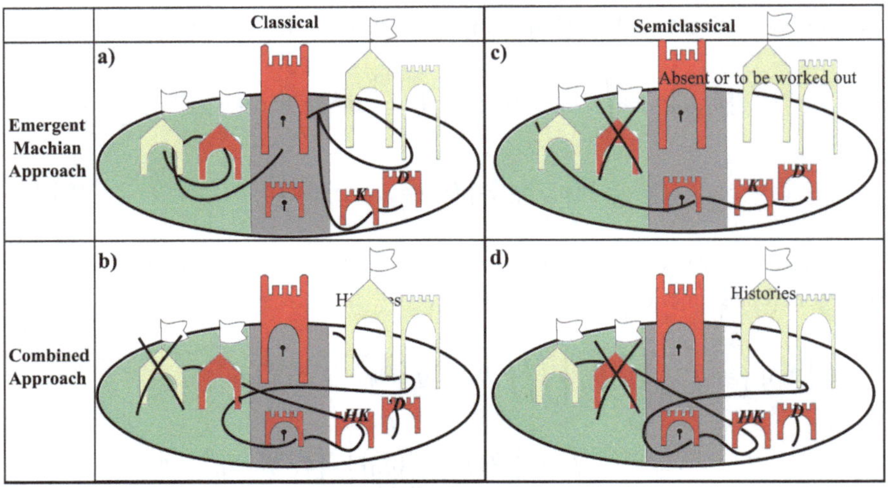

Fig. 54.1 This extends Fig. 35.2 to its quantum counterpart. While **d)**'s main loop is the same shape as **b)**'s, **d)** provides the decoherence upon which **b)**'s starting point depends. In each of **b)** and **d)**, the disjoint path provides a construction for Dirac beables so as to complete **a)** and come closer to completing **c)**

by construction, and

$$[\widehat{\mathcal{F}\mathbf{lin}}, \widehat{\mathcal{C}}_{\mathrm{R}}^{\sqcap\sqcup}] = 0 \tag{54.23}$$

due to being functionals of the Kuchař beables. These $\widehat{\mathcal{C}}_{\mathrm{R}}^{\sqcap\sqcup}$ are therefore quantum Dirac beables, $\widehat{\mathcal{D}}$.

However, the above form does suffice as a conceptual-and-technical start for RPM version of Halliwell-type approaches, and amounts to an extension of these to cases including linear constraints also.

Another caveat is that the semiclassical procedure for finding Dirac beables does not fully apply in the case of Supergravity, for the same reasons that the classical procedure does not (Sect. 29.3).

The above working clearly generalizes to restricting other A-beables subalgebraic structures by commutation with $\mathcal{C}$hronos, in cases in which the corresponding extended constraint algebraic structure itself closes. Supergravity moreover possesses various cases of this more limited construct.

54.5 Frontiers of Research

See Fig. 54.1 for comparison of this Chapter's Problem of Time strategy with the book's main previous such.

The method for constructing Dirac beables provides quantities which commute with $\mathcal{Q}$uad and the $\mathcal{F}$**lin**. Yet no ready means of checking that the quantities provided in this way can be assembled into an algebraic structure. Thus it does not look at all likely that the quantum Dirac beables constructed in this manner will form a subalgebraic structure of the algebraic structure of quantum Kuchař beables.

$\widehat{\mathcal{C}}_{\mathrm{R}}^{\sqcap\!\sqcup}$ comes with the following further issues.

2) It involves a slight spreading.
3) The $\mathrm{Prob}_{\mathrm{R}}$ notion is open to difficulties in general for chaotic reasons.
4) The θ function has harsh edges, which cause a *Quantum Zeno Problem*,[4] which is resolved by Halliwell's subsequent construction [414] to which we next turn.

Smooth Window Function Extension The above Secs' class functional, while conceptually illustrative, has technical problems, in particular it suffers from the Quantum Zeno Problem. Smoothing out Halliwell 2003's [413]'s sharp-edged window function—also done by Halliwell, now in 2009—[414] removes this, and is equivalent to taking the region in question to contain a finite potential. Here the class functional being the corresponding S-matrix and the smoothed-out case representing a softening in the usual sense of Scattering Theory (albeit now in configuration space rather than in space). This upgrade manages to remain compatible with H.

In further detail, [422], this approach makes use of

$$\mathrm{P}(N\epsilon)\ldots\mathrm{P}(2\epsilon)\mathrm{P}(\epsilon)=\exp(-iHt). \tag{54.24}$$

One furthermore now conceptualizes in terms of probabilities that the region is *never* entered. Halliwell [414] additionally applied a 'softening' result [277]; in this way, one passes to

$$\mathrm{P}(N\epsilon)\ldots\mathrm{P}(2\epsilon)\mathrm{P}(\epsilon)=\exp\bigl(-i\{H-iV_0P\}t\bigr) \tag{54.25}$$

for $\epsilon\,V_0\simeq 1$. The class functional for *not* passing through region R is

$$\widehat{\mathcal{C}}_{\not{\mathrm{R}}}^{\frown}=\lim_{t_1\longrightarrow-\infty,t_2\longrightarrow\infty}\exp(iHt_2)\exp\bigl(-i\{H-iV\}\{t_2-t_1\}\bigr)\exp(-iHt_1). \tag{54.26}$$

Such expressions do not suffer from the Quantum Zeno Problem, and they are still quantum Chronos beables $\widehat{c}$. Moreover, in the smooth window function approach [414], regions of $\mathfrak{q}$ that are large enough need no environment.[5] Sect. 48.4's 'Scale Models with Shape as both Perturbation and Environment' strategy offers a second resolution to [421]'s issue of 'losing the environment', i.e. arguing that it was hitherto negligible in the study but is nevertheless available at this stage of the study as an environment.

[4]This actually-observed quantum effect's name follows from its resemblance to *Zeno's Arrow 'Paradox'*. Here an arrow in flight is argued not to be in motion at any instant, by which it is not in motion at all. This amounts to breaking time down point-by-point rather than in intervals as in Zeno's Achilles and Tortoise 'Paradox'; it is also clearly fallacious once one has a sufficiently sturdy theory of limits. The similarity with the quantum effect is, rather, at the level of sufficiently frequent quantum observation impossibilitating the occurrence of quantum change.

[5]Here, 'large' means compared to wavelength, which is solution-dependent.

One would next apply $\lambda \longrightarrow t^{\mathrm{sem}}$'s in the Machian version of the updated Combined Approach.

Extending the smooth window function approach to the RPM arena mostly concerns defining class functionals somewhat differently, so as to get these to be better-behaved as regards the Quantum Zeno Effect. In the case of nontrivial-$\mathfrak{g}$ theories which are reducible, the $\widehat{C_{\mathfrak{R}}}$ are quantum Dirac beables $\widehat{\mathcal{D}}$.

Let us end by pointing out that the arrival time problem in Quantum Theory has similar conceptual content and mathematics [417, 423].

Research Project 71) Halliwell [413, 414] and Marolf's [641] approaches to Histories Theory remain to be compared in detail with each other. Does one of these confer greater advantages, do the two approaches produce compatible results?

Research Project 72) The smooth window function approach [414, 415] remains to be applied to examples with additional linear gauge constraints.

Research Project 73) Consider the quantum-level Combined Approach and its construction of Dirac beables for SIC.

Research Project 74) Consider the Combined Approach for Nododynamics.

Chapter 55
Quantum Foliation Independence Strategies

55.1 Constraint Algebraic Structure's Ties to Other Facets

One of the more conceptually interesting possibilities for quantum constraint nonclosure is its arising in the form of Foliation Dependent terms. For instance anomalies could result from scale being included among the physically-irrelevant variables, though the concept of 'spatial frame' now involves a conformal factor as well as a shift. This is at least known to occur in simpler models which do not receive a Background Independent interpretation: *conformal anomalies* for e.g. a nonrelativistic particle in a scale-invariant potential. A complication with this is the interplay between different operator orderings and regularizations on the one hand and Foliation Dependent effects on the other. E.g. whether suitable choices of these other matters can be made such as to avoid Foliation Dependence in the ensuing constraint algebraic structure.

Research Project 75) Are deformation first principles—successfully considered at the classical level in Sect. 32.1—also useful at the quantum level? Does the Deformation Approach's classical derivation of GR have a quantum—or at least semiclassical—counterpart?

55.2 Semiclassical Refoliation Invariance?

Given Teitelboim's classical demonstration of Refoliation Invariance as the geometrical counterpart of the key part of the Dirac algebroid, does the semiclassical constraint algebraic structure of GR imply a similar result?

Strategy 1) *Accept.* Demand classical GR's Foliation Independence and Refoliation Invariance continue to hold at the quantum level. We next need to face that the commutator algebroid at the quantum level is almost certainly distinct from classical GR's Dirac algebroid. One's quantum constraint algebraic structure may contain Foliation Dependent anomalies as well [583, 857].

E. Anderson, *The Problem of Time*, Fundamental Theories of Physics 190,
DOI 10.1007/978-3-319-58848-3_55

Strategy 2) *Discard.* Use Background Dependent or privileged slicing alternative theories. These involve times which this approach imposes sufficient significance upon that they cannot be traded for other times. These are more like the Ordinary Quantum Theory notion of time than the conventional view of time in GR. In such an approach, Foliation Dependence is built in from the start, and so the matter of Refoliation Invariance is avoided. At the classical level, this amounts to throwing away an established Background Independence aspect of GR. On the other hand, at the quantum level, Refoliation Invariance remains to be demonstrated and might conceivably not hold.

The Discard strategy should not be confused with instances of some special highly-symmetric solutions can possess geometrically-privileged foliations. Generic solutions are however central to GR, and even perturbations about highly-symmetric solutions cease to have geometrically-privileged foliations to the perturbative order of precision [589]. Additionally, even highly-symmetric solutions that admit a privileged foliation in GR are *refoliable*. In this way, the problem below with losing Refoliation Invariance is curtailed.

Kuchař argued that [586] *"The foliation fixing prevents one from asking what would happen if one attempted to measure the gravitational degrees of freedom on an arbitrary hypersurface. Such a solution amounts to conceding that one can quantize gravity only by giving up GR. I.e. to say that a quantum theory makes sense only when one fixes the foliation is essentially the same thing as saying that Quantum Gravity makes sense only in one coordinate system."*

Fixed foliations are a type of Background Dependence, which, from the relational perspective, is undesirable from the outset. For all that investigating the effects of making *just* this concession is indeed also of theoretical interest.

Research Project 76)† What is the semiclassical analogue of the Dirac algebroid? Does it guarantee Refoliation Invariance? Do its other commutator brackets close in the same pattern as GR's?

Research Project 77)†† Find a full and rigorous QG analogue of the Dirac algebroid. How unique is it? How universal are its features? Does it guarantee Refoliation Invariance?

55.3 Suitable Model Arenas for Quantum Foliation Issues

RPM and Minisuperspace models, while either trivially or fortunately possessing quantum Constraint Closure, are unsuitable for investigating foliation issues at the quantum level. Thus more complicated examples are required.

Example 1) SIC has already been presented as a more viable model for investigation of Foliation Independence. One particular quantum issue here is whether the algebraic structure of $\widehat{\mathcal{H}}$ and $\widehat{\mathcal{M}}_i$ can be formulated in a Foliation Independent manner.

Example 2) In [583], Kuchař approaches parametrized Field Theory by finding an operator ordering in which its Hamiltonian maps to that of a massless scalar field on $\mathbb{M}^2$ viewed in 'cylindrical' coordinates.

Example 3) Kuchař and Torre [594] argued that the bosonic string can be viewed as a model arena of Geometrodynamics. This application, moreover, involves a Quantization procedure which is distinct from that usually performed on a bosonic string. In this approach, the string exhibits anomalies; furthermore, these are Foliation Dependent. They considered worldsheet diffeomorphism-covariant Quantizations, among which an anomaly-free constraint algebraic structure can be found by obtaining suitable internal time 'embedding variables'. For these models, this resolves not only the Constraint Closure Problem but their Foliation Dependence Problem as well.

Example 4) Torre [857] exhibited quantum-level Refoliation Invariance in a functional time evolution approach to $1 + 1$ GR.

Example 5) Bojowald [155] has argued for quantum corrections to the Dirac algebroid maintaining a deformed notion of General Covariance.

Research Project 78) Consider the quantum-level consequences of Pons et al.'s classical level work [722, 724] (Sect. 32.4–32.6). In particular, what are the quantum-level consequences of the Hamiltonian to gauge condition distinction and of the 'nothing happens' fallacy?

55.4 Foliation Problems at the Quantum Level

Example 1) Let us now return to the form taken by a split-spacetime diffeomorphism transformation—unlike that for a rotation transformation—depending on what object it acts upon ([832] and Fig. 35.1). At the quantum level, this has the further repercussion of greatly complicating the corresponding Representation Theory.

Example 2) *Foliation Dependence Problem with Internal Times.* The bubble time version describes a hypersurface in spacetime only after the classical equations have been solved. This unclear status of foliations limits the extent to which such internal time approaches are understood [552]. Quantum use of bubble times features moreover in the Semiclassical Approach.

Example 3) *Foliation Dependence Problem with Path Integral Approach.* The path integral's definition furnishes a direct counterpart of Foliation Independence (the construction of intermediate surfaces in Fig. 10.3.b). The Path Integral Approach is an interesting place to look for whether this probably physically desirable result of classical GR extends to the quantum level.

Example 4) *Foliation Dependence Problem with Histories Theory.* This is that GR time's many-fingeredness brings in Foliation Dependence issues to Histories Theory.

Finally, some constructive examples toward the Accept strategy are as follows.

Example 5) In Savvidou's work [768–771] to the HPO Approach, she pointed out that the space of histories $\mathfrak{Hist}$ has implicit dependence on the foliation vector. This is the unit normal vector to the spatial hypersurface n^μ mentioned in Chap. 8 as being orthogonal to a given hypersurface, now re-interpreted as the foliation 4-vector [with a label running along the foliation, $\mathrm{n}^\mu(\lambda)$]. With this HPO approach admitting 2 types of time transformation, the histories algebroid turns out to be Foliation Dependent. However, the probabilities—which are the actual physical quantities—are Foliation *In*dependent. In this way, this approach avoids ending with a Foliation Dependence Problem.

Example 6) The preceding example leads to a further possibility, namely that the classical foliation vector itself be among the structures being quantized. This was considered by Isham and Savvidou [506, 507, 767]. Here,

$$\widehat{\mathrm{n}}_\mu \Psi = \mathrm{n}_\mu \Psi, \qquad \widehat{\Pi}^{\mu\nu}\Psi = i\left\{\mathrm{n}_\mu \frac{\partial}{\partial \mathrm{n}_\nu} - \mathrm{n}_\nu \partial\partial \mathrm{n}_\mu\right\}\Psi, \tag{55.1}$$

where the antisymmetric $\Pi^{\mu\nu}$ is the momentum conjugate to n^μ and satisfies the Lorentz algebra. They subsequently apply a Group-theoretic Quantization to the configuration space of all foliation vectors for the Minkowski spacetime model arena of the HPO Approach.

Example 7) Whereas *Diff*$(\mathfrak{m})$ is kinematical and shows up in Path Integral Approaches [477], in contrast, *Digg*$(\mathfrak{m})$is dynamical and shows up in Canonical Approaches. However, quantum-level consequences of this remain to be worked out.

Example 8) Canonical-and-Covariant Histories Approaches may offer further perspectives on quantum-level foliations, though this remains but tentatively outlined [769].

Chapter 56
Quantum Spacetime Construction Strategies

Construction of spacetime from space has been much less developed at the quantum level than at the classical level. Whereas attempting to recover spacetime in a suitable limit from a quantum theory with some non-continuum inputs has been more widely studied, we postpone discussion of this to Epilogue III.C. For now, let us note that such programs are indeed often designed so the recovery of a continuum with spacetime properties is the last facet to encounter. [The Cubert Z carries connotations of coming last in these strategic contexts.]

56.1 Semiclassical Spacetime Construction

Research Project 79)† Example 1) What is the quantum—or at least semiclassical—counterpart of the classical Spacetime Construction workings of Chap. 33? Paralleling Wheeler's classical considerations, this can be considered to be an investigation of why the *semiclassical* GR Hamiltonian constraint takes its given form. In this case, can the versions assuming each of metrodynamics, geometrodynamics and conformogeometrodynamics be worked out? (See [716, 717] for a start on the Strong Gravity subcase at the quantum level.) Having seen rigidities which give back GR play a significant role at the classical level, do these have a direct quantum counterpart, and are there any further rigidities which greatly cut down on choices of operator ordering and regularization?

If there is a semiclassical hypersurface kinematics (Project 78), does this furthermore admit a Machian interpretation, and is there a semiclassical counterpart of TRiFol? Try starting with the geometrodynamics-assumed (*Diff*($\boldsymbol{\Sigma}$)-invariant) but arbitrary supermetric ansatz

$$\widehat{\mathcal{H}}_{a,b,x,y} = -\frac{\hbar^2}{\sqrt{\mathrm{M}_{x,y}}}\frac{\delta}{\delta \mathrm{h}_{ij}}\left\{\sqrt{\mathrm{M}_{x,y}}\mathrm{N}^{ijkl}_{x,y}\frac{\delta}{\delta \mathrm{h}_{kl}}\right\}\Psi - \hbar^2\xi\mathcal{R}(\underline{x};\mathrm{M}_{x,y}]\Psi$$
$$+\sqrt{\mathrm{h}}\left\{a + b\mathcal{R}(\underline{x};\mathbf{h}]\right\}\Psi, \tag{56.1}$$

E. Anderson, *The Problem of Time*, Fundamental Theories of Physics 190,
DOI 10.1007/978-3-319-58848-3_56

much as Chap. 33 starts with the classical ansatz (33.3), alongside the 'momenta to the right' ordering of $\widehat{\mathcal{M}}_i$. Compute the three commutator brackets formed by these, including under Semiclassical Approach assumptions.

Example 2) Consider Spacetime Construction in the simpler, if more restrictive, SIC setting.

Example 3) In Nododynamics alias Loop Quantum Gravity, semiclassical *weave states* have been considered by e.g. Ashtekar, Rovelli, Smolin, and physicists Matthias Arnsdorf and Luca Bombelli; see [845] for a brief review and critique of these works.

Subsequent semiclassical construction work has mostly involved coherent states (Sect. 48.1) instead, as constructed in this particular case by the '*complexifier method*' [845]. See [320] for a further distinct approach to these. In Lorentzian spin foam models, semiclassical limits remain a largely open problem. E.g. [112] is a treatment of Lorentzian spin foams that also covers how Regge Calculus's action emerges as a semiclassical limit in the Euclidean case. On the other hand, almost all the semiclassical treatment in the more recent review [711] remains Euclidean. The further LQC truncation [153] does possess solutions that look classical at later times. [These are amidst larger numbers of solutions that do not, which are, for now, discarded for this reason. This is somewhat unsatisfactory due to its replacing predictivity by what amounts to a *future* boundary condition.] It also possesses further features of a semiclassical limit [151]. (This means a WKB regime with powers of both $\hbar$ and the Barbero–Immirzi parameter β being neglected. It is still subject to open questions about correct expectation values of operators in semiclassical states.)

Let us end by pointing out that Isham has argued [482] for spacetime being a meaningful entity *at most* at the semiclassical level.

56.2 Spacetime Construction in Histories Theory

For instance, recovery of microcausality at the quantum level is possible in Savvidou's [768, 769] or Kouletsis' [566] Canonical-and-Covariant Approaches. However, once again, this work very largely remains to be extended to the quantum level.

Chapter 57
Quantum-Level Conclusion

In Part III, we further considered time in 'Quantum Gestalt': Quantum Gravity treated on an equal footing with Background Independence. We began by providing further detail of what Quantization means, firstly within Canonical Quantization Programs (Chaps. 39 to 43) and secondly within Path Integral Approaches (Chap. 52). The structures used in formulating Quantum Theory have, moreover, been argued in Chaps. 39 to 43 to transcend from Ordinary QM and QFT based on Newtonian and Minkowskian absolutism to the Relational Approach as well.

Quantum Theory's global sensitivity points to adopting Affine Geometrodynamics at the quantum level, rather than the hitherto more often considered Plain Geometrodynamics. This is a consequence of the positive-definiteness of spatial 3-metrics. That said, one seldom gets far enough in non-formal detail with Problem of Time approaches to Geometrodynamics for this Affine versus Plain distinction to have started to make a difference.

57.1 The First Four Facets

Let us next sum up as regards Part III's main theme of quantum level interferences between Problem of Time facets.

Kinematical Quantization—the nontrivial quantum analogue of the trivial assignment of unconstrained beables υ at the classical level—needs to be considered first in Canonical Approaches. At the level of Problem of Time facets, this is amounts to a *preliminary* consideration of Assignment of Beables. This involves a Kinematical Quantization algebra $\mathfrak{K}$ with commutator bracket product $[\ ,\ ]$ acting on a Hilbert space $\mathfrak{K}$in-$\mathfrak{H}$ilb. Having done this, Configurational and Temporal Relationalisms can be considered; these giving constraints $\widehat{\mathcal{C}hronos}$ and $\widehat{\mathcal{F}\mathbf{lin}}$ respectively, Constraint Closure then needs to be addressed before Assignment of Beables is further amended so as to respect the constraints.

Dynamical Quantization subsequently involves operator ordering and the Wheeler–DeWitt equation exhibiting the Frozen Formalism Problem and the Inner Product Problem.

E. Anderson, *The Problem of Time*, Fundamental Theories of Physics 190,
DOI 10.1007/978-3-319-58848-3_57

As regards residing within Temporal Relationalism implementing (TRi) formalism, the general classical-level Temporal Relationalism implementing scheme (TRi-PoD: Principles of Dynamics) can readily be extended to form TRiCQT ('Canonical Quantum Theory'). This is in great part by virtue of the classical Principle of Dynamics entities that become significant in Canonical Quantum Theory—Hamiltonian variables, Poisson brackets, Hamiltonians, constraints, beables—being already change-scalar quantities, alongside there not being a quantum analogue of Dirac-appending of constraints.

Semiclassical Machian emergent time. The main classical strategy considered in Part II is the Classical Machian Emergent Time Approach. In Part II, we argued in favour of this over other candidate times such as hidden, matter and unimodular, as well as over spacetime-first and entirely-timeless and spacetime-first approaches. Part III has now provided further quantum-level arguments against the other approaches that survived Part II's classical-level considerations. One salient case of this is that hidden time candidates provide (Chap. 44) very hard quantum equations, approximands to which do not resemble the underlying problem's quantum physics (which is known in full for a few simple models). Another is that quantum arguments for purely timeless worldviews (Chap. 51) are lacking somewhat in generality and greatly as regards current practical viability.

The classical Machian emergent time t^{em} moreover fails to immediately carry over as a Frozen Formalism Problem resolution at the quantum level. It has moreover an emergent semiclassical time sequel (Chaps. 46-47). This is a reinterpretation of—and further correction of—the already well-known notion of emergent WKB time $\mathsf{t}^{\mathrm{sem}}$ [552]. We recast this in Machian form, providing a perturbative Semiclassical Machian Scheme for the RPM analogue of Halliwell–Hawking-style Semiclassical Quantum Cosmology. In this scheme, the semiclassical $\mathsf{t}^{\mathrm{sem}}$ coincides with the classical t^{em} to zeroth order but not so to higher (perturbative) order. This is clear from *quantum* change being given the opportunity to enter the former Machian time. One passes from an emergent Machian time of the form $\mathfrak{F}[\mathsf{h}, \mathsf{l}, \mathbf{d}\mathsf{h}, \mathbf{d}\mathsf{l}]$ to one of the form $\mathfrak{F}[\mathsf{h}, \mathsf{l}, \mathbf{d}\mathsf{h}, |\chi(\mathsf{h}, \mathsf{l})\rangle]$. This takes into account that the light subsystem has passed from a classical to a quantum description, by which *quantum* change is now being given an opportunity to contribute.

Some remaining issues with the Semiclassical Approach are as follows.

1) A priori motivation for use of the WKB approximation is limited. If this is dropped, moreover, the emergent time mechanism ceases to function. We also show in Epilogue III.B that this ansatz is merely local over configuration space and space.
2) The Semiclassical Quantum Cosmology literature has hitherto suffered more widely from insufficiently justified approximations. Some of these approximations are moreover inconsistent at least in fairly analogous quantum and even classical systems. For instance, Part II pointed to one such approximation implying that the 2-body problem sits stably within the 3-body problem. Another approximation amounts to the neglect of central terms. Part III itself pointed to higher derivative terms being dropped—dangerous in Fluid Mechanics—and averaged terms being dropped: dangerous in Atomic and Molecular Physics. As

such, humankind has probably only just seen the tip of the iceberg as regards making Semiclassical Quantum Cosmology calculations. The current book has outlined the physical significance of each often-omitted term and commented on a few of the simpler regimes obtained by keeping but a few of these terms. We provided more full equations keeping correction terms, and gave a start on variational methods to underlie numerical treatment of such subsystems. This is of interest as regards details of the possible quantum cosmological origin of structure: galaxies and cosmic microwave background hot-spots.

Configurational Relationalism remains resolved in a range of models: having reduced at the classical level, Quantization does not unreduce the system.

In cases in which reduction cannot be carried out in practice, emergent WKB time is, like its classical counterpart, only known implicitly due to a pending extremization over $\mathfrak{g}$ in its definition. This is one significant Problem of Time facet interference. It is moreover also a subcase of the indirectly formulated quantum-level $\mathfrak{g}$-act $\mathfrak{g}$-all method for addressing the Configurational Relationalism facet.

Two further double—and one triple—facet interferences concern how quantum constraint commutator brackets may behave differently from their classical counterparts. This may cause a difference in which group is physically irrelevant in passing from a classical theory to its quantum counterpart. This description can be linked to a subset of anomalies, by which the physically irrelevant group gets smaller: $\mathfrak{g} \longrightarrow \mathfrak{g}_{\text{QM}}$. The subset in question are those anomalies tied to time, space, frame or $\mathfrak{q}$. It may also occur that classically compatible $\mathcal{C}$hronos and $\mathcal{F}$**lin** are promoted to quantum-mechanically incompatible $\widehat{\mathcal{C}\text{hronos}}$ and $\widehat{\mathcal{F}\textbf{lin}}$. Finally, $\widehat{\mathcal{C}\text{hronos}}$ may imply new integrabilities, which render the original $\mathfrak{g}$ unsuitable.

Whereas the above facet interferences already have classical counterparts as per Fig. 35.2.a), the following further up to five-way interferences between facets very largely do not.

Firstly, the classical and quantum constraint algebraic structures $\mathfrak{C}$ and $\widehat{\mathfrak{C}}$ do not in general coincide. The subalgebraic structures admitted by each consequently differ in general as well.

Secondly, promoting classical constraints and classical beables to their quantum versions require choice of operator ordering. Due to this, the *particular notion* of beables (Chaps. 25 and 50) that one may wish to use for one's problem may differ between the classical and quantum levels. This last problem clearly does not occur if one uses the zero and unit extremes of the lattices in question: the algebraic structure of all the first-class constraints $\mathcal{C}_{\text{F}}$ corresponding to the Dirac beables $\mathcal{D}$, and the trivial algebra id of no constraints corresponding to the unconstrained beables $\mathcal{U}$. It is thus unsurprising that managing to promote classical beables to quantum ones $\mathcal{B} \longrightarrow \widehat{\mathcal{B}}$ is the exception rather than the norm. This is one reason why one often needs to start afresh at the quantum level. Another is that only a subalgebraic structure of the quantum level beables $\mathfrak{b}$ can be consistently promoted to quantum beables $\widehat{\mathfrak{b}}$, due to the Multiple Choice Problem.

Thirdly, suppose Kuchař beables exist at both the classical and quantum levels, for which Configurational Relationalism stays resolved with the same $\mathfrak{g}$, and for which momentum P can be represented as

$$\widehat{\mathsf{P}} = -i\,\hbar\,\frac{\partial}{\partial \mathsf{Q}}.$$

Then operator ordering with P to the right preserves the classical first-class linear constraint algebraic structure. Furthermore, classical Kuchař beables (or any other type of A-beables corresponding to constraint subalgebraic algebraic structures of the first-class linear constraints $\mathcal{F}$**lin**) can be promoted to quantum ones in a suitable operator ordering.

57.2 Spacetime, Timeless, Histories and Combined Approaches

At the quantum level, Spacetime Primality Approaches become Path Integral Approaches. Whereas Gauge Theory is well understood in such terms, a multitude of problems surface in the case of GR. Some of these concern the unusual form of the GR action, and others the difficult nature of the diffeomorphism group (e.g. diffeomorphism-invariant measures). The QFT use of imaginary time also undergoes severe shortcomings in the presence of curved, and generic, spacetime geometries. Finally, Temporal Relationalism implementing Path Integral Quantum Theory (TRiPIQT: Fig. 52.1) renders quantum path integral approaches compatible with Temporal Relationalism.

Quantum Histories Approaches involve not just paths but strings of projectors attached to these. Some such schemes, due to Isham and Linden, follow on from quantizing classical histories theory, whereas other schemes such as Gell-Mann and Hartle's start de novo at the quantum level.

As regards Timeless Approaches, whereas at the quantum level a wider variety of mechanisms for obtaining a semblance of dynamics have been proposed, few such are conducive to carrying out calculations. Additionally, a quantum-level successor of the most promising classical scheme—based on Shape Statistics—for now remains unformulated.

On the other hand, there are a number of approaches available in which timeless records sit within formulations in which further structure is assumed. These offer less radical but more solid ways out than being able to derive a semblance of dynamics. In particular, one can consider Records Theory within one or both of the Semiclassical Machian Emergent Time Approach or Histories Theory. Additionally, many of the strategies suggested for attempting to justify the WKB ansatz requires investing in Histories or Timeless Approaches [329, 552, 931]. In the Author's opinion, this is best approached by considering a combination of Timeless Records, Histories Theory and Machian Emergent Time Approaches. This extends a previous formulation by Halliwell [413, 414] to an explicitly Machian interpretation. Halliwell also provided a corresponding semiclassical means of constructing observables for $\mathfrak{g}$-free models. This can also be extended to the $\mathfrak{g}$-nontrivial case as a means of constructing semiclassical Dirac beables given semiclassical Kuchař beables.

57.3 State of Completion in Model Arenas

Relational Particle Mechanics (RPM) Configurational Relationalism is resolved at the classical level for a range of simple such, and can be taken to remain resolved in passage to the quantum level. Temporal Relationalism is subsequently resolved in a directly computable manner in Quantum Cosmology model arena regimes by semiclassical Machian emergent time. To leading order, this is a relational recovery of Newtonian time. Constraint Closure remains a non-issue at the quantum level for RPMs. Quantum Kuchař beables remain well-defined and are immediate to construct, and subsequently Halliwell's method for promotion to quantum Dirac beables applies. Spacetime and hence Foliation Dependence and Spacetime Construction issues are absent from RPMs. We are therefore done as regards providing A Local Resolution of the Problem of Time for these RPMs. Reference [37] covers some of the RPM models and the Problem of Time therein in further detail.

Minisuperspace This has trivial notions of Configurational Relationalism, Constraint Closure and Kuchař beables, as well as of Spacetime Relationalism, Foliation Dependence and Spacetime Construction within its foliation privileged by spatial homogeneity. Such models have a directly computable semiclassical Machian emergent time, which to leading order is a relational recovery of cosmic time. Finally, Halliwell's method was designed in the first place for these very models, by which A Local Resolution of the Problem of Time has also been attained for these.

Slightly Inhomogeneous Cosmology (SIC) [34, 50] To leading nontrivial perturbative order about an isotropic $\mathbb{S}^3$ Minisuperspace with scalar field matter, this has Configurational Relationalism resolved at the classical level, and can be taken to remain resolved in passage to the quantum level. This model also has a directly computable semiclassical Machian emergent time; once again this is a relational recovery of cosmic time to leading order. There are no known problems with quantum-level Constraint Closure or Assignment of Beables here, though these remain to be computed explicitly. Halliwell's method now permits finding quantum Dirac beables as functionals of quantum $\mathfrak{g}$-beables which no longer coincide with quantum Kuchař beables in this model arena. For this model, Foliation Dependence and Spacetime Construction remain to be worked out, even at the semiclassical level.

57.4 Research Frontiers

GR: Semiclassical and Quantum Counterparts of the Dirac Algebroid The current state of knowledge still leaves us with most Foliation Independence and Spacetime Construction matters very undeveloped, even at the semiclassical level. The extent of Quantum Refoliation Invariance, Spacetime Construction and TRiFol remains unclear. Foliation Dependent quantum constraint algebraic structures constitute yet another example of interference between Problem of time facets. Research Projects 76), 77) and 79) have been suggested to address these matters.

Loop Quantum Gravity: Quantum-Level Position This is better-defined at the quantum level than Geometrodynamics is. However, there are a number of issues with the various branches of this program, as outlined in Sects. 43.5 and 56.1.

Research Project 80)† Carry out a quantum-level survey of the Problem of Time throughout LQG alias Quantum Nododynamics.

Open Universe GR Models

Research Project 81) To what extent do Background Independence and Problem of Time ideas apply in a) asymptotically flat and b)† asymptotically AdS spacetimes? [These are natural alternatives to compact without boundary spaces, and are widely used models. Asymptotic flatness is assumed in most treatments to date of isolated Generally-Relativistic astrophysical objects, whereas AdS is very central in Holographic Approaches. Also bear in mind that whereas *metric* level Relationalism provides arguments for favouring closed (compact without boundary) spaces, *topological manifold* level Relationalism more generally encourages treatment both of open models and of open and closed models considered together. This is additionally a useful precursor to considering topological manifold level Background Independence.]

Research Project 82)†† Since the Universe we actually live in looks to be, more accurately, FLRW on larger scales, develop a suitable notion of asymptotically-FLRW. [This would be a more complicated venture than setting up asymptotically flat or AdS formulations.] Consider the extent to which Background Independence and the Problem of Time apply here.

CPT in QG?

Research Project 83)† To what extent does CPT invariance carry over to QFTiCS? To more full QG programs than this? If this remains relevant in Quantum GR, does this interrelation of charge and time impinge upon Problem of Time facets and strategies?

Arrow of Time in QG and Quantum Cosmology?

Research Project 84)† Does the quantum cosmological setting—or the full QG one—affect the Arrow of Time arguments, or provide new reasons to suggest a Master Arrow?

Supergravity This book has revealed that Background Independence and the Problem of Time for Supergravity are substantially classically distinct from Geometrodynamics. This is due to a change in the integrability structure of the constraints. This causes further changes in the interpretation of Temporal and Configurational Relationalism and in the types of notions of beables which are appropriate.

Further implications of this for the nature of time and the foundations of QG remain to be worked out. To date, the Semiclassical Approach to Supergravity has been most developed in [555]. By the anomaly-cancelling properties of the supersymmetry, there is some chance of obtaining a better-behaved quantum (or at least semiclassical) constraint algebroid, as regards Constraint Closure, Refoliation Invariance and Spacetime Construction.

Research Project 85)† Investigate Background Independence and the Problem of Time for Quantum Supergravity.

M-Theory Canonical Approaches to M-Theory include e.g. theoretical physicists David Berman and Malcolm Perry's [136], or Jonathan Bagger, Neil Lambert and Andreas Gustavsson's [85–87, 400]. Whereas a few approaches to time in M-Theory have been considered, e.g. in [782] by theoretical physicist Nathan Seiberg from an emergent spacetime perspective, such works are as yet few and far between. Supergravity is itself a kind of semiclassical limit of M-Theory. In this way, at least some semiclassical approaches to M-Theory follow from passing to Supergravity, which further motivated making use of Supergravity as an example in this book.

Moreover, some of the conceptual changes in passing from GR to Supergravity significantly change again in passing to Superconformal Supergravity. For now, very little is known about whether any of the Canonical Formulations of M-Theory are yet again distinct, including among themselves. E.g. [85–87, 400] involves a different type of brackets algebraic structure, and so of constraint algebraic structure and of notions of observables or beables [32].

Research Project 86)†† Perhaps the relational notions along the lines of those presented in this book are furthermore necessary for a notion of M-Theory which possesses GR, or Supergravity, type Background Independence. Might the elusive meaning of 'M' turn out to be resolved by 'M for Mach'?

Research Project 87)†† To what extent are Holographic Approaches Background Independent?

The established physical theories combine to give the appearance that there is no such thing as time at the primary level for whole-universe models. However, following Mach, time can be abstracted from change in a wide range of circumstances, including for all practical purposes in everyday life. That time is an abstraction from change is also the answer we give to Saint Augustin's opening question.

Why time consistently exhibits a direction, an also widely-noted and mysterious issue, remains an issue that is much more widely *not* convincingly explained to date, although this ('the Arrow of Time') was never the subject of this book.

SR, QM and GR have moreover all entered accurate measurement of length and time, and the corresponding definitions of the units for these quantities as well. This gives practical interest to further understanding Fundamental Physics: it eventually enters both the accuracy and the conceptualization of all Physics.

Returning to Einstein's sins that the Introduction ends with, Quantum Theory still implicitly enters GR's timestandards. But there is no such thing as a purely classical world; ultimately a bottom-up theory of Quantum Gravity would be expected to be consistent in this regard. For now, GR's idealizations are still doing fine for Science. . . .

Let us end with warnings and reassurances. "*I wasted time, and now doth time waste me.*" William Shakespeare [787]. Thus in theorizing, ignore time at your peril! "*The strongest of all warriors are these two—Time and Patience.*" Leo Tolstoy [850]. So when faced with the knot of Time, employ Patience, and "*all we have to decide is what to do with the time that is given us*" J.R.R. Tolkien [849]. Along such lines, if all we have is an emergent time in some regimes, then let us see how much we can do with it.

Chapter 58
Epilogue III.A. The Multiple Choice Problem

This Epilogue is an expansion of Sect. 12.15's outline of the Groenewold–van Hove phenomenon that is the main part of the Multiple Choice Problem. See [392, 866] for the original papers. More recent technical papers include [375, 376, 378–381]; these are on particular examples given by mathematical physicist Mark Gotay, some in collaboration with mathematical physicists Janusz Grabowski, Hendrik Grundling, C.A. Hurst and Gijs Tuynman. [376] itself reviews all preceding papers; see also [355, 361] for subsequent commentary.

Moreover, generic GR gives further reasons for such choices to be made. Here there is no geometrically natural choice for the internal spacetime coordinates. Classically these all have equal status [479, 483], and yet the limitations on consistent quantum algebraic structures push one toward making such a choice. Part of the Multiple Choice Problems in QG arise from this clashing combination of quantum and GR features. More widely, quantizing nonlinear theories is likely to produce this phenomenon [483]. Let us end by noting that this occurs not only for field theories but for finite models as well (a common misconception prior to [483]).

58.1 Multiple Choice Problems

Multiple Choice Problems are exhibited by not only the above choices of timefunction and of spatial frame, but also by the choice of classical beables subalgebraic structure.

The Multiple Choice Problem, moreover, perseveres even in the Semiclassical Approach when studied in sufficient detail [586]. For instance, there could be multiple regions with distinct WKB approximations holding in each, leading to distinct emergent times in each such region. There could also be a variety of different approximate emergent times within each such region. The Groenewold–Van Hove phenomenon is on occasion avoided here, due to some subalgebraic structure choices and unitary inequivalences being of negligible order in semiclassical expansions. E.g. these may be $O(\hbar)$ smaller than the smallest terms kept.

E. Anderson, *The Problem of Time*, Fundamental Theories of Physics 190,
DOI 10.1007/978-3-319-58848-3_58

Section 20.3's scale time ambiguity corresponds to selecting different entities to be among one's subalgebra of quantum beables, which is a source of the Multiple Choice Problems upon passing to the quantum level. In Internal Time Approaches more generally, the Multiple Choice Problem applies e.g. to making two different choices of internal coordinates [483].

In Path Integral Approaches, there is no longer a time whose selection causes this problem. But there often does remain a choice of frame, and always a choice of beables, and each of these is rather likely to be afflicted with a Multiple Choice Problem. These can additionally be passed on to the measures involved in gravitational path integrals [477, 586].

The above problems with path integrals carry over to Histories Theory, which also suffers from the following further problem. Histories Theory is usually approached in gauge-dependent form. This is with each gauge being equipped with an internal time. In this way, many internal time problems return, one of which is the Multiple Choice Problem in the timefunction. In this book, we get round this problem with internal time by making use of emergent Machian time instead. However, this still carries no guarantee of freedom from Multiple Choice Problems, e.g. due to the above points about the Semiclassical Approach continuing to hold in the version receiving a Machian interpretation.

Let us end by pointing to Foliation Dependence as one of the ways in which the Multiple Choice Problem can be manifested.

58.2 Strategies for the Multiple Choice Problems

1) *Total Excision.* Work only with models that do not exhibit the Multiple Choice Problem. This is however draconian, given that there is no deep-seated physical reason to reject these models, which are also for now in the majority amongst those studied (Sect. 58.3).
2) *Unconditional Acceptance* of all cases of the phenomenon as part of quantum reality. In this case, one would need experimental tests to determine which choice Nature makes.

1) and 2) are opposite extremes; the next three strategies lie somewhere in between.

3) *Patching.* For instance, one could accept that timefunctions are both multiple and local, and proceed to consider how to interpolate between such in passing between patches. This is carried out in e.g. Bojowald et al.'s works [157, 158, 452, 453]. Multiplicity of times is also inherent in Rovelli's 'any change' and 'partial observables' approaches to Relationalism, and also in the STLRC approach (where one tests one's way among the many to find which is locally-best); see also Sect. 58.2.

   This timefunction patching strategy, however, has so far not addressed the entirety of the Multiple Choice Problems. This is firstly because while each patch having its own time manifests some aspects of the Multiple-Choice Timefunction

Problem, it is not known to overcome, more specifically, the Groenewold–van Hove phenomenon itself. Secondly, this strategy has not hitherto been used to address the multi-facetedness of the Multiple Choice Problems. On the other hand, through this strategy's use of a moments expansion, it does concurrently offer a new approach to the Inner Product Problem. It additionally works toward resolving the Global Problem of Time (see the next Epilogue).

4) *Partial Excision*: remove the problem by considering a hypothetical subspace of the canonical transformations such that the square of maps of Fig. 12.2 commutes. It is not however clear [37] which nontrivial subspace would have this property!
5) *Perspectival Acceptance* of those cases of the phenomenon for which there is physical justification for the choices through their correspondence to different perspectives. This would for instance allow for different observers observing different subsystems that have different notions of time; this fits in well with the partial observables and patching approaches.

Research Project 88)† To what extent does the Multiple Choice Problem coincide with to differences in perspective between observers?

58.3 Specific Model Arenas

A common misconception [483] has been that the Multiple Choice Problem would only be of concern in the study of infinite models. It can however occur for finite models as well. Indeed concrete model studies of it have concentrated on finite models. Namely, the following phase spaces: $\mathbb{R}^{2n}$ [375], $\mathbb{S}^2$ [381], $\mathfrak{T}^*\mathbb{S}^1$ [378] which do exhibit the Multiple Choice Problem, and $\mathbb{T}^2$ [379] which does not. $\mathfrak{T}^*\mathbb{R}_+$ admits a polynomial Quantization [377]. [376] reviews all of these. See [399] for an excellent part-worked introductory example.

Some of the above cases can furthermore be interpreted in whole-universe model arena terms; for r-formulations of RPMs the current frontier of knowledge is as follows.

Metric Shape and Scale RPM

	$N=3$	$N\geq 4$
$d=1$	$\mathfrak{Phase}=\mathbb{R}^4$	$\mathfrak{Phase}=\mathbb{R}^{2n}$
	Both exhibit MCP	
$d=2$	$\mathfrak{Phase}=\mathbb{R}^6$ Exhibits MCP	$\mathfrak{Phase}=\mathfrak{T}^*(\mathrm{C}(\mathbb{CP}^{n-1}))$ MCP status unknown

Metric Shape RPM

	$N=3$	$N\geq 4$
$d=1$	$\mathfrak{Phase}=\mathfrak{T}^*\mathbb{S}^1$ Exhibits MCP	$\mathfrak{Phase}=\mathfrak{T}^*\mathbb{S}^{n-1}$ MCP status unknown but relationally trivial
$d=2$	$\mathfrak{Phase}=\mathfrak{T}^*\mathbb{S}^2$ MCP status unknown	$\mathfrak{Phase}=\mathfrak{T}^*(\mathbb{CP}^{n-1})$ MCP status unknown

This settles that Multiple Choice Problem does occur for some RPMs, which, conversely, pose a number of interesting extensions to the Multiple Choice Problem analysis so far.

Research Project 89)† How ubiquitous is the Multiple Choice Problem for Minisuperspace and modewise SIC? What is known about the Multiple Choice Problem for systems that retain linear constraints at the quantum level? What about in simple Field Theories?
E.g. do the various distinct isotropic Minisuperspace 'true Hamiltonians' in Isham's account (Sect. 4.2.3 of [483]) which correspond to different choices of time candidate have an underlying manifestation of the Groenewold–van Hove phenomenon at the level of Kinematical Quantization? If this example were to fail to exhibit this phenomenon, find other Quantizations of Minisuperspace models for which it does occur.

Chapter 59
Epilogue III.B. Quantum Global Problems of Time

59.1 Extending Classification of Global Problems of Time

The classical level considerations of meshing together manifold charts and patching together local solutions of PDEs are augmented at the quantum level to include the following patchings.

1) Various classes of operators on Hilbert spaces. Self-adjointness confers a more global character upon quantum operators than classical functions are bestowed with. One now has kinematical operators, constraint operators, dynamical operators and beables operators to contend with. Maps between classical and quantum spaces further complicate the Analysis involved due to the heterogeneity between domains and codomains.
2) How to patch together representations.
3) Since some quantum wave equations are FDEs rather than PDEs, patching their solutions together may be more intricate.
4) Inner products are required as well as wavefunctions so as to handle physical quantities. One now has to patch together quantum mechanical unitary evolutions which involve both of these features. In general, this list of quantum-level patchings is short on conceptual, let alone technical, understanding.

Some particular examples do afford a more concrete understanding. For instance, in some particular cases of CFT and TFT [916], there is a sense in which Hilbert spaces can be associated with boundaries. This can furthermore be interpreted in terms of assigning representations, and in terms of evolving through intermediary boundaries. On the other hand, in spin network approaches, vertices are labelled by representations and edges by intertwiners (defined in Appendix W.1). This gives a concrete implementation of patching representations together.

A more well-known global issue concerns viewing the quantum wavefunction as defined on a section. So in the event of a lack of global section, a fibre bundle presentation with multiple charts is required. In this setting, the corresponding quantum states belong to a cohomological—rather than just plain—Hilbert space.

E. Anderson, *The Problem of Time*, Fundamental Theories of Physics 190,
DOI 10.1007/978-3-319-58848-3_59

59.2 Semiclassical Approach

Problem 1) What happens when a wavepacket approaches a zero of the potential factor? This is the wavepacket version of the geodesic problem 1) of Sect. 37.1.

Problem 2) The forcedly local nature—in the sense of finite regions in space and in configuration space— of Sect. 37.1's Problems 2) to 4)'s heavy–light split designations carries over to the quantum level. The relevant set of approximations is now that given in Chap. 46.

Problem 3) Abstracting a GLET from a STLRC remains local in the context of Semiclassical Quantum Cosmology.

Problem 4) In Chap. 46, we already alluded to how emergent semiclassical time picks up a significant imaginary part if considered over too large a piece of $\mathfrak{q}$.

Problem 5) In particular, WKB regimes are not expected to hold globally in time, space or configuaration space. Classical–semiclassical Machian emergent time alignment extends the relevance of the issues raised at the classical level in Chap. 23 and in Epilogue II.B. Semiclassical emergent time is moreover globally limited (in space and in configuration space) by S having zeros. There is often oscillatory behaviour to one side of these and decaying behaviour to the other. In such cases, the WKB procedure is both invalid at the zero and a very poor approximation nearby.

A distinct approximation is therefore required around each zero; cf. the theory of connection formulae [667] for moving between WKB regimes in the case of ODEs (Fig. 59.1.d). Because of this, a time arising from a WKB procedure cannot be claimed to be generically applicable over $\mathfrak{q}$. One should rather expect a number of patches in $\mathfrak{q}$ in which a different regime applies; therein emergent semiclassical time is not a valid answer to the Problem of Time. This raises the interesting question of whether these patches remain timeless, or whether the connection formulae provide their own emergence mechanism. Additionally, if the zeros are sufficiently near to each other, there is no room for a WKB regime in the region between them. Applicability of the WKB procedure is thus expected to be local along such lines within any $\mathfrak{q}$ that contains many potential factors zeros. Moreover, the theory of (some generalization of) connection formulae for use in Quantum Cosmology largely remains to be developed.

Research Project 90) In the quantum cosmological setting, what form does the wavefunction take in the connecting regions? Is there any resolution of the Problem of Time in these regimes? If this holds, can this be patched with the emergent WKB time resolution in the other regions? Does some version of the connection formula procedure for patching together regions remain suitable at the level of prolonging quantum evolution?

59.3 Hidden Time Approaches

Problem 1) The quantum-level $\widehat{\mathscr{H}}^{\text{True}}$ is also questionable as regards its global well-definedness. This may well just be defined in a localized patch within $\mathrm{t}^{\text{hidden}}$ itself; this already occurred in Sects. 37.4–37.5's classical treatment.

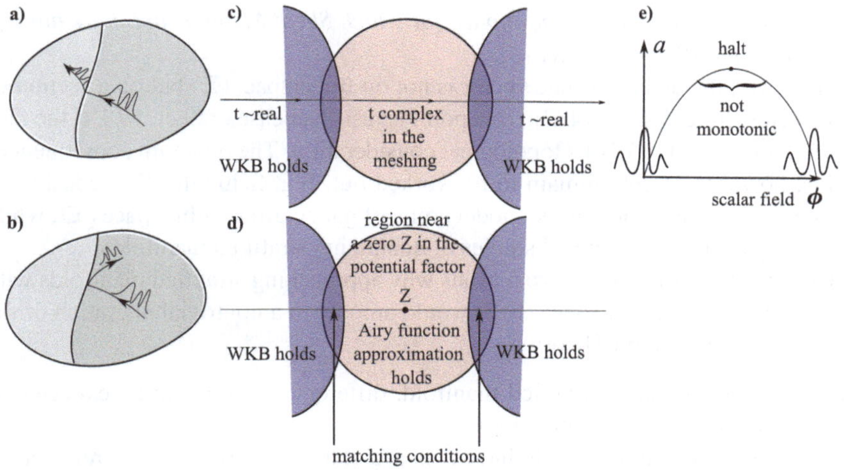

Fig. 59.1 **a)** Wavepacket moving across a stratum barrier. **b)** Wavepacket reflected on a stratum barrier. **c)** Time in a Patching Approach arising in the Semiclassical Quantum Cosmology arena. **d)** An analogous ODE problem (e.g. 1-*d* QM), for which WKB breakdown near a zero Z is patched over using the *connection formula method*. I.e. firstly, an approximation around Z is made in terms of Airy functions [1]. Secondly, matching conditions are applied to relate the solutions in each region. **f)** Global behaviour of wavepackets in isotropic Minisuperspace. The initial condition specification indicated is for scale factor time [552]. This is also to be contrasted this with the single point data prescription of the classical trajectory in Fig. 37.1.g)

Problem 2) The corresponding quantum wave equation can also have negative-probability interpretation issues in some finite regions of the state space.

59.4 Basic Monopole and Gribov Effects at the Quantum Level

The Gribov ambiguity [446] now interferes with gauge-fixing as a route to Quantization. Much of the mathematics of monopole defects [886] is intended for use at the quantum level. See e.g. [674, 675, 886] for a further range of global effects well-known in standard QFT, and [473, 475] for the less well-known and more involved extension of some of these to QFTiCS and to full GR.

On the other hand, the reduced phase approach is not affected by the Gribov obstruction.

59.5 Quantum Issues Following from Stratification

From the classical treatment, we have learned that *configuration spaces* $\mathfrak{q}$ *are not in general manifolds, nor are Fibre Bundle Methods always applicable to them. They are more generally stratified manifolds, for which Sheaf Methods are more natural and more generally applicable.*

Moreover, the quantum state vector is not on $\mathfrak{S}$uperspace(Σ) but on a section of a flat vector (fibre) bundle [482]. This point is just a subcase of Sect. 59.1's, the flatness resulting from (43.14)'s topological consideration. The quantum consequences of this, however, largely remain to be worked out. E.g. Schmidt [775] considered Geometrical Quantization in the model arena of gauge group orbit spaces $\mathfrak{O}$, while advocating further finite model studies of quantizing stratified manifolds.

The first five issues below are reasons why approaching stratified manifolds with an Unfold Strata strategy (Sect. 37.5) would amount to a nontrivial alteration of the corresponding Quantum Theory.

Issue 1) In quantizing a stratified manifold, different strata would be expected to contribute their own representations.

Issue 2) Since operator self-adjointness is a globally sensitive matter, removal of strata would likely affect the form of quantum theory in question.

Issue 3) Some types of operator ordering for the quantum wave equation, such as the Laplacian or conformal Laplacian operator orderings, have geometrical content which can only be defined on a stratum-by-stratum basis.

Issue 4) Wavefunctions on an unfolded configuration space would be expected to have a distinct probability density over the part which does not require unfolding.

Dynamical Issue 5) There is a quantum wavepacket counterpart of the classical phenomenon of geodesics hitting strata (Fig. 59.1.a–b).

Issue 6) At least in some cases, non-generic gauge orbit strata can be taken to possess an analogue of magnetic charge, whereby they can be viewed as a type of magnetic monopole. In this way, the inclusion of non-principal strata can affect a quantum theory's dynamical behaviour.

Issue 7) Stratified manifolds require General Bundle Methods, or a fortiori Sheaf Methods, in place of just Fibre Bundle Methods. Wavefunctions are now to be defined on *sheaf* sections; in the event of a lack of a global such, a multiple sheaf section presentation is required.

For now, we know that the various trianglelands in Fig. G.11 differ both at the level of Kinematical Quantization and of the forms taken by their quantum wavefunctions. In this way, the differences between these models have quantum-level consequences.

59.6 Constraint Closure

Problem 1) Paralleling the classical situation with Poisson brackets, the quantum commutator remains a local-in-time-and-space slab construct, rather than necessarily holding globally.

Problem 2) Anomalies carry topological connotations. Some anomalies are purely topological, due to which they are entirely missed by perturbative methods. In this manner, the Quantum Constraint Closure Problem is at least part global. Such anomalies can be tied to characteristic classes, cocycle conditions and Index Theorems; see e.g. [673] for an outline and [139] for a more detailed exposition including of such in Gravitational Theory.

Research Project 91) [alias Problem 3)] E.g. [443] demonstrated that further cases of anomalies can arise from nongeneric strata. Provide a topological and geometrical characterization of this phenomenon alongside a physical interpretation.

Problem 4) The algebraic structure formed by the quantum constraints can vary from point to point in space or in $\mathfrak{P}$hase. This remains moreover within the remit of Sheaf Methods.

59.7 Observables and Beables

Quantum observables or beables are in general not globally defined over all of time, space, $\mathfrak{P}$hase (or $\mathfrak{q}$ in the purely configurational case). The local operationally favoured versions of these entities are 'fashionables' and 'degradables', respectively (as per Sect. 37.8). Some global issues with this are as follows.

Problem 1) At the quantum level, it is somewhat strange for globally-defined kinematical commutator brackets algebraic structures and constraints to be associated solely with locally-commuting entities.

Problem 2) In the absence of global quantum observables or beables, it is furthermore unclear how one is to patch between different sets.

Issue 3) Consideration of localized notions of observables or beables carries over to the quantum and semiclassical levels. For instance, in Bojowald et al.'s [157, 158, 453] fashionables approach at the semiclassical level (Fig. 59.1.c), a moments expansion is used to bypass the Inner Product Problem. Whereas fashionables are real-valued, solving the constraints gives that this approach's notion of time go complex around the semiclassical regime's turning points. [Their sense of 'semiclassical' is that they neglect O($\hbar^2$) and higher moment polynomials.] In this approach, moreover, the imaginary part of a time variable becoming significant is a diagnostic for that time variable ceasing to be accurate. This is due to the onset of non-unitarity in the evolution brought about by the local nature of the Semiclassical Approach.

Time having to be complex in this program—and in other more orthodox treatments of connecting regions—may furthermore interfere with such a time candidate having enough of the list of time properties to pass muster as a time. Let us finally recollect (Chap. 52) that the use of Complex Methods for detailed path-integral or canonical type calculations has a poor track record in Gravitational Theory.

Problem 4) Sect. 37.12's argument on operational grounds against types of quantum beables which are integrals over all t, space, $\mathfrak{q}$, or $\mathfrak{P}$hase, continues to hold at the quantum level.

Problem 5) That quantum observables, beables, fashionables or degradables constitute an algebraic structure is a further hurdle for the Patching Approach, which has hitherto largely not been taken into account.

Let us finally consider modelling algebraic structures of observables or beables by sheaves. Sect. 37.8's classical-level transcends to the quantum level; this can be seen as a more mathematically advanced approach to patching which can take Problem 5)'s algebraic structure point into account. We note that Haag [401] has pioneered the modelling of quantum observables by sheaves. This is also among the many applications covered in mathematical physicist Urs Schreiber's immense review [778] of advanced topological and categorical applications to Theoretical Physics.

59.8 Timeless Approaches

Problems 1) and 2) are Sect. 37.9's 2) and 5), which are in practice mostly to be considered at the quantum level. This is due to the importance of quantum conceptualization of records and of the role of (some variant on) Quantum Theory in whatever actually-proposed mechanisms for obtaining a semblance of dynamics or of history.

Problem 3) The Wheeler–DeWitt equation has at most been studied in neighbourhoods of highly symmetric metrics such as Halliwell–Hawking's around $\mathbb{S}^3$. Giulini [358] comments that it would be interesting to know how 'far' from such a point one has to go in order to encounter singular regions and signature change. In this way, locality in configuration space $\mathfrak{q}$ enters consideration. He additionally comments on how Wheeler–DeWitt equations have apparently not been studied in the neighbourhoods of metrics with spatial Ricci scalar $\mathcal{R} < 0$ (as opposed to $\mathbb{S}^3$'s $\mathcal{R} > 0$).

Problem 4) The Global Problems concerning the heavy–light split sharpen the part of Barbour's conjecture concerning the importance of $\mathfrak{q}$'s geometry to the quantum probability distribution.

Problem 5) Sect. 51.3's argument—about conformal non-flatness being required to separate out $\mathfrak{q}$'s geometry effects from potential effects in investigating Barbour's conjecture—is only locally valid. This is because viewing the potential factor $\mathscr{W}$ as a conformal factor is in general only locally valid due to the Problem of Zeroes, Infinities and Non-Smoothnesses (PoZIN).

Issue 6) Our classical-level argument for sheaves providing a more advanced set-up for Records Theory furthermore transcends to quantum Records Theory.

59.9 Paths and Histories Approaches

Problem 1) Since nongeneric gauge orbit strata affect classical motion, they can contribute nontrivially to quantum path integrals [759]. This can lead to some quantum states being localized; [290] gives a finite mechanical example.

Problem 2) The Fadde'ev–Popov construction of path integrals is less straightforward in cases with multiple strata [759].

Problem 3) The Gribov ambiguity in the path-integral formulation of Yang–Mills Theory continues to feature in Gravitational Theory.

Problem 4) At the classical level, we already alluded to how the consequences of Histories Theory are usually considered in gauge-dependent form. Moreover, each gauge choice is tied to an internal time candidate; consequently many internal time problems recur, including quantum-level ones. Furthermore, by Histories Theory being based on paths, Problem 3) recurs here as well; e.g. the forms of Eqs. (53.24) and (53.25) are affected.

Problem 5) This book's main program, moreover, breaks contact with such internal times by using emergent Machian time instead, including within the context of Histories Theory. This program, however, has its own set of Global Problems to face as per Sect. 59.2.

Problem 6) As at the classical level, histories constraints and histories observables also have global issues, now paralleling those in Sects. 59.6–59.7.

Problem 7) Histories have the possibility of intersecting non-uniquely with spatial hypersurfaces [586].

Issue 8) By being built up as sequences of quantum timeless records, quantum histories are also well-modelled by sheaves.

Issue 9) So are quantum histories constraint and observables algebraic structures, $\mathfrak{L}^{\mathrm{H}}_{\widehat{\mathfrak{c}}}$ and $\mathfrak{L}^{\mathrm{H}}_{\widehat{\mathfrak{b}}}$.

59.10 Combined Approach

In addition to each constituent approach's problems, the Combined Approach has the following additional Global Problems. Section 37.12's Problems 1) to 3) each have obvious quantum-level analogue. In particular, Problem 3)'s semiclassical counterpart is that t^{sem} is only locally defined for even more reasons than t^{em} is. Problem 2) is the conflict between this and the requirement of integration from $-\infty$ to $+\infty$ so as to have quantum Chronos or Dirac beables. In contrast, Anastopoulos' distinct histories-based construct [9] does not require integration from $-\infty$ to $+\infty$; the end-products of this are, however, histories observables rather than beables.

Problem 4) Semiclassical emergent Machian time now only provides a local semblance (PoZIN, multiple approximations, GLET); presumably also the histories only decohere locally since the WKB regime produced is in general only local.

Problem 5) Gell-Mann and Hartle's 'somewhere' in statement (54.1) is meant locally.

Issue 6) The timeless probabilities of Halliwell concern localized regions in $\mathfrak{q}$. Consideration of multiple—or more extended—such regions is sketched in Fig. 59.2.

Issue 7) The class functional (54.5) is to be interpreted locally. Through selecting a region, using such a construct to solve the Problem of (especially chronos or Dirac)

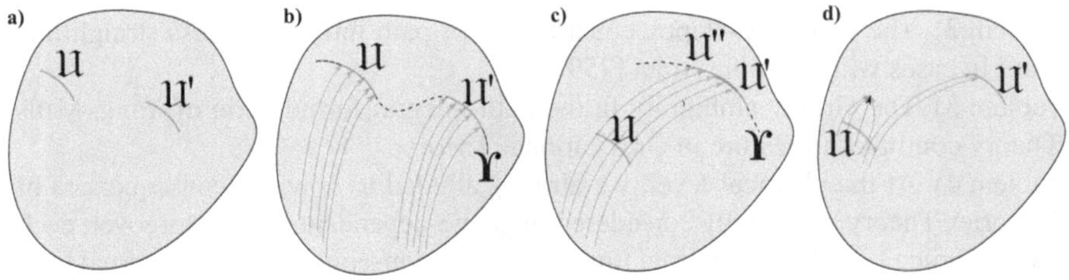

Fig. 59.2 Composing Halliwell's regions construct. **a)** Consider now two regions $\mathfrak{U}$ and $\mathfrak{U}'$ which are pieces of general hypersurfaces Υ and Υ' within the configuration space $\mathfrak{q}$. **b)** Occasionally there will exist a single hypersurface Υ orthogonal to the classical flow that includes both $\mathfrak{U}$ and $\mathfrak{U}'$. Furthermore, by (29.13), $\text{Prob}_{\mathfrak{U} \text{ OR } \mathfrak{U}'} = \text{Prob}_{\mathfrak{U} \cup \mathfrak{U}'}$ just as for the Naïve Schrödinger Interpretation. **c)** However, if $\mathfrak{U}$ and $\mathfrak{U}'$ do not lie on the same hypersurface, this is not directly analogous to the Naïve Schrödinger Interpretation. Now one can consider the flow evolution $\mathfrak{U}''$ of region $\mathfrak{U}$, say, so that it lies in a hypersurface Υ that extends $\mathfrak{U}'$. In this case, one can now compose $\mathfrak{U}'$ and $\mathfrak{U}''$ just as one did for the Naïve Schrödinger Interpretation. d) Moreover, there need not always exist a section that extends $\mathfrak{U}'$ while also containing a flow evolution image $\mathfrak{U}''$ of $\mathfrak{U}$. This case is an example of the composition of Halliwell's implementation not always reducing to a parallel of the Naïve Schrödinger Interpretation's composition. Local sections and meshing conditions between them suffice for composition, but this does not always hold either. These non-existences reflect that some flows can be pathological, e.g. exhibiting breakdowns in well-definedness or smoothness

Beables or Observables at the quantum level is indeed compatible with the basic ethos of degradables or fashionables. So in general Halliwell's method just constructs quantum chronos or Dirac degradables. Nor have such yet been built with global or even widely non-local geometrical considerations in mind. Two reasons why in general these are just degradables are as follows.

i) *h–l* splits are not in general global (in space or in configuration space).
ii) Problem 4) applies to these degradables as well. In particular, the precluded locality in time would be a desirable property for beables that can be used in practice.

Problem 8) Halliwell's scheme does not a priori use a Naïve Schrödinger inner product (see also Fig. 59.2). None the less, its implementation of propositions is also by regions of classical $\mathfrak{q}$ (or maybe a generalization to regions of $\mathfrak{P}$hase). So it still suffers from probabilities corresponding to regions composing too simply (Booleanly) for one to be able to represent all quantum propositions in such a form. E.g. quantum propositions in general compose nondistributively [484, 486, 503], whereas the composition of configuration space regions is distributive; see also Appendix S.4. See Epilogue III.C for a possible way out.

59.11 Refoliation Invariance and Spacetime Construction

Research Project 92) To date, Quantum Spacetime Construction just has local protective theorems. This locality is in space, time and configuration spaces, given that all of these already applied at the classical level. However, this global quantum version should await obtaining local quantum results.

Research Project 93) Quantum Refoliation Invariance has hitherto only been studied locally. Pass to a global version.

Chapter 60
Epilogue III.C. Deeper Levels' Quantum Background Independence and Problem of Time

We finally return to Riemann's question that Epilogue II.C opened with, now addressing it at the quantum level. Whereas commutation relations can be imagined to all levels, are *equal-time* commutation relations impaired by loss of temporal properties in the descent of levels of mathematical structure? Which aspects of Background Independence survive this descent, and which break down at each level?

Another interesting question is how far down this descent the notion of fermions remains supported. At the level at which this breaks down, the notion of Supersymmetry is compromised as well. Note here that supersymmetric TFTs are well-known [915]; on the other hand, at least some elements of topological manifold level structure may be indispensable for Supersymmetry.

Moreover, Mackey's $\mathfrak{g}/\mathfrak{H}$ example (Sect. 39.5) transcends to the generalized $\mathfrak{q}$ of other levels of mathematical structure [480, 481, 491].

Spacetime Construction attempts from assuming less mathematical structure are fairly common, usually in the guise of 'Discrete Approaches'. For instance, one investigates here how such as curvature and causality arise—at least in suitable limits—within less structured approaches. What of a single, integer-valued and sufficiently large dimension? We prefer however to consider these matters in terms of dropping some of the topological manifold or topological space assumptions that make up the conventional package of 'the continuum'. This is much more general that a continuum versus discrete dichotomy, because there is a rich range of models which are intermediate between the naïve notions of 'continuum' and 'discrete'.

While 'Discrete Approaches' have a taming effect on the path integral, Spacetime Construction Problems pose significant difficulties for these approaches. For instance, difficulties are to be expected to ensue from the Causal Sets Approach's insistence on very sparse structural assumptions. At the point this book was written, recent advances were e.g. Rideout and physicists Seth Major and Sumati Surya's account [634] for a recovery of a spacetime-like notion of topology, or Rideout and Wallden's work [734] for a (classical-level) recovery of a metric.

Isham [497] has also pointed out that functional integrals involve distributions—Quantum Theory 'roughens up' spaces so why not sum over differentiable manifolds alongside manifolds with singularities in place of just over differentiable manifolds?

E. Anderson, *The Problem of Time*, Fundamental Theories of Physics 190,
DOI 10.1007/978-3-319-58848-3_60

Twistor Theory is an area of Theoretical Physics in which Sheaf Methods have featured since the 1970s. E.g. [707] covers this point at the level of Twistor Theory's quantum wavefunctions. This is another way in which the Twistor Approach was well ahead of its day as a pioneering program, and this extends to use of sheaf cohomology as well. The Twistor Approach can furthermore be viewed as a null structure Quantization counterpart to Isham and Savvidou's quantizing of the foliation vector in Sect. 55.4.

60.1 Topological Manifold Level Considerations

We consider here Wheeler's spacetime foam conceptualization [897], in which quantum fluctuations could be expected to cause the metric to change signature and alter the topology of space. In this manner, Quantum Theory might tolerate—or even motivate—inclusion of metrics which are more general (e.g. degenerate or singular) amongst a theory's configurations. On these grounds, the elsewise-motivated classical-level inclusion of degenerate beins in Ashtekar variables formulations might come to be acceptable, and similarly the quantum-level consideration of Plain rather than Affine Geometrodynamics.

This approach originated from Wheeler's idea of applying Feynman path integrals to Quantum Gravity. So perhaps now summing over histories entails sums over topologies. One would next contemplate computing transition amplitudes for 'topology change' (meaning change in spatial topological manifold). More specifically, which topologies is one to include, and what kind of metrics are there to be thereupon (Appendix S.2)?

Problem 1) The non-classifiability of 3-d topologies renders 'GR with summing over topologies' highly formal.

Problem 2) Geroch's Theorem (Appendix S.2) entails choices which are both technically and conceptually difficult.

Model Arena 1) Topological Field Theory (TFT) [674, 916], in particular the Chern–Simons Theory following from (38.4). This is a metric-free model. Moreover, being spatially 2-d, it is free from the excessive complications caused by 3-d topologies.

Problem 3) Spatially 2-d models are sufficiently tractable that one can get far enough to espy further technical problems. Quantum wavefunctions tend to concentrate about the classical action's extremal contribution, albeit now only due to this contributing more rather than being the sole contributor as it was at the classical level. However, the topological manifold level counterpart of this involves extremization of discrete, rather than continuous, parameters in the action. At the very least, this requires an extended version of the Calculus of Variations to justify, and how to proceed in general remains unclear. Moreover, whereas the largest value of the action contributes the most probability, other cases of sizeable action contribute as well. On the one hand, this covers the well-known case of multiple saddle points in standard approaches to Semiclassical Quantum Cosmology. On the other hand, it also covers dominance by less probable but more numerous configurations, of which [193] provides a spatially 2-*d* GR example. Both of these are quantum effects which smoothen out classical extrema, and both cast some doubts on semiclassical approximations used in Quantum Cosmology.

Problem 4) If one's model allows for pinching off (Appendix S.2), a quantum-level microscopic problem ensues due to virtual pairs pinching all over the place, which might compromise some aspects of local SR physics.

Model Arena 2) There is a partial analogy between topology change in GR and change in particle number in RPMs. One can view this as modelling operations that alter the 'list of contents of the Universe' aspect of topology change in GR. Finite RPMs would involve

$$\bigcup_{\mathrm{N}\in\mathbb{N}_0} \mathfrak{q}(N,d) \tag{60.1}$$

so as to allow for particle coalescence and splitting, or creation and annihilation. Some instances of stratified $\mathfrak{q}$ for RPM can at least on some occasions be regarded as already being of this form. This analogy additionally suggests the concept of interaction terms between topologies which parallels particle non-conserving interaction terms. Moreover, this has the interpretation of 'Second Quantization' in the sense of model universe creation rather than of mere particle creation within a universe; this is far from a clear upgrade both physically and philosophically. This is the opposite of 'Second Quantization' as a model arena of 'Third Quantization' as a Problem of Time scheme (Sect. 45.3). By this analogy, the interacting theory involves topology-changing 'ripping' operations, i.e. cobordisms: Appendix S.2. The variable particle number N-a-gonlands provide a more specific variable particle number model arena for Quantum Cosmology. Here,

$$\sum_{\mathrm{N}\geq 3\ \mathrm{N\text{-}a\text{-}gonlands}} = \sum_{k\in\mathbb{N}\,\mathbb{CP}^k} \tag{60.2}$$

gives a sum of manifolds over a simple family of different dimension. This is also a far simpler proposal than summing over the manifolds for a fixed dimension $p > 2$ due to the diversity issues in Appendix G.

Let us end with an example of some progress with kinematical-level detail for topology change in GR itself. This takes the form of a selection rule for the Path Integral Approach to Quantum Gravity by Gibbons and Hawking [350, 351]: that handle creation and annihilation must involve handles in pairs.

Research Project 94) Further develop this Sec's model arenas as regards the form taken by time and Background Independence therein.

60.2 Metric and Topological Space Level

Most of the currently pursued approaches to Quantum Gravity assume continuum notions at some level or other [186]. However, beyond a certain point, use of continuum notions (on which manifolds are based): spacetime, space, Principles of Dynamics spaces, Lie groups, probabilities... becomes a presupposition of background structure. So are the choice of function spaces (most notably standard Hilbert spaces at the quantum level), and the standardly-adopted assumptions that one's mathematics can be rooted in Set Theory and that standard binary logic is to be used.

'Discrete Approaches' can be viewed as a challenge to modelling using manifolds. Moreover, metric and topological spaces include both discrete and continuum features. Isham considered quantizing at each of these levels in [479, 482, 508, 509].

In fact, the suggestion [508, 509] of quantizing distance itself goes back to Wheeler [897]. Quantizing distance requires implementing an inequality constraint, now with

$$\text{Dist}(p, q) \geq 0 \tag{60.3}$$

somewhat paralleling the $\det \mathrm{h} \geq 0$ condition encountered in Affine Geometrodynamics. The space of norms now plays the role of generalized configuration space.

This case also involves [479] a semidirect product of groups, by which once again Mackey Theory can be used to extract representations. There is also a square root operation to contend with, with the entity inside being concentrated on a curve, whereby Isham is concerned that this is singular enough for this square root to cause difficulties. The above two examples moreover turn out to have enough parallels for QFT to permit a Fock space based approach [508, 509]. In particular, analogues of creation and annihilation operators can be defined for these.

Finally in [480, 481] Isham considers various attitudes to time at the topological manifold level, including timelessness, discrete time-steps, and path integrals which incorporate transition amplitudes for change of topological space. On the other hand, Isham [497] viewed time as a continuous label, and along the lines of an internal time, at least in the semiclassical limit.

Research Project 95) Develop further this Sec's considerations of quantizing at the topological space level. For instance, the Machian Emergent Time and Records Approaches exposited in this book remain unaddressed at this level.
For now, this is procedurally obstructed due to not much being known about how to quantize stratified manifolds, which are the reduced configuration and phase spaces in question. In particular, what happens to stratified manifold 'fitting together' conditions upon quantizing? The expectation is that different strata contribute representations to the Quantum Theory in hand, but how are these to be fitted together to making a coherent, computationally viable Quantum Theory? Also, how are quantum wave equations, and wavefunctions solving these, to be patched together at the quantum level? Do particular stratified manifold pairings with sheaves (or differentiable structures) persist through to the quantum level, and do such help with the previous two questions?

60.3 Yet Deeper Levels of Structure

Quantizing causal sets has been considered in e.g. [491, 637, 800, 801, 821]. This resembles a Histories Theory in being based on paths and on presuming some aspects of spacetime (i.e. causal structure). However, it does not involve strings of projectors, so it does not strictly meet this book's specifications for a Histories Theory. Thereby, it involves a *third* approach to the implementation of propositions, at the level of regions in the space of paths, $\mathfrak{p}\mathrm{ath}(\mathfrak{q})$.

Moreover, Spacetime Construction difficulties do however follow from the Causal Sets Approach's insistence on very sparse structure. This is already problematic for this approach at the classical level (Sect. 38.5), and remains very largely unexplored at the quantum level.

A further question at the level of sets themselves is whether the cardinality of the underlying set is itself subject to quantum fluctuations. Isham [492–494, 498] entertained this possibility by quantizing at the level of sets themselves.

In the single-floor case, if only collections of subsets of a set contain physically meaningful information, what is the Quantum Theory on Collect($\mathfrak{X}$)? [I.e. on the space of collections of subsets of $\mathfrak{X}$.] On the other hand, in the tower case, what is the effect on the upper layers of structure if 'the underlying set' is allowed to quantum-mechanically fluctuate?

Research Project 96) Develop further this Sec's considerations of quantizing at the levels of collections of subsets and of sets.

Research Project 97) How well do discrete (and continuous limit of discrete) models fit within this Epilogue's considerations? E.g. what do these already cover versus what gaps this Epilogue reveals in the modelling assumptions of such theories studied so far?

Research Project 98) Investigate fully quantum-level Spacetime Constructions.

60.4 Situations with Negligible Deeper Levels of Structure

Semiclassical Quantum Cosmology does not involve fluctuations of topological manifolds or beyond; this protects us from the tower of mathematical structure for some practical purposes. Of course, the most interesting questions in QG concern more full regimes. E.g. Isham [482] has furthermore argued for modelling of space(time) base on Differential Geometry to be unlikely to apply beyond the semiclassical level.

60.5 Records and Histories

Research Project 99) Given Part II's position for timeless records based on advances involving Stochastic Mathematics, what is a quantum-level counterpart? What can be said about theories of quantum records along such lines?

Research Project 100) Formulate classical Histories Theory at each level.

Research Project 101) Formulate quantum Histories Theory at each level.

Research Project 102) Formulate the Combined Approach (Machian emergent time, histories and records) at each level.

Research Project 103) Reflect on whether there are conceptual or technical reasons to favour the standard meaning of configuration over an extended meaning that also covers such as paths in time or histories.

60.6 Quantum Theory, Categories and Topoi

Category Theory and Topos Theory offer alternatives[1] (Fig. 60.1) to the 'Equipped Sets' Foundational System for the levels of mathematical structure that is conventionally used in Theoretical Physics. *The idea here is that perhaps the conventional 'Equipped Sets' Foundational System of Mathematics is itself a fixed background structure.* Isham [492–494, 498] quantized *small categories.* (Appendix W.1)[2] Here the objects are generalized configurations and the morphisms alias arrows as generalized momenta. The subcase for which Mackey's trick applies can once again be exploited, along with creation–annihilation operator analogues.

Research Project 104) The question of spaces of spaces for Theoretical Physics can be reframed in terms of spaces of categories, or indeed of categories of categories. While the set of sets impasse carries over to the category of categories, it does not carry over to the small categories since the category of small categories is not itself small. Finally, consider all of dynamical evolution, Probability Theory and Quantization on a general category; Isham looked into the last of these in the moderately general case of small categories [492–494, 498]. The suggested Research Project concerns the first two of these, as well as generalizing the third.

As per Appendix W.3, sheaves are the basis for more general patching constructs. Presheaves (Appendix W.2) are a less structured alternative. Presheaves do not in general possess the gluing property, by which sheaves are more amenable to patching constructs.

Topoi (the plural of *topos*) can be envisaged as categories with three extra structures providing some properties similar to those of sets, as per Appendix W.4. Using topoi instead of sets leads to use of multi-valued and contextual logic. 'Multi-valued' here means extra answers in addition to YES and NO, whereas 'contextual' means that such 'valuations' can differ from place to place. These features occur because topoi are Geometrical Logics; valuations here are somewhat analogous to Differential Geometry's locally-valid charts, with binary Logic's globally-valid valuations playing the analogous role to flat space.

Topos Theory is, alongside Sheaf Methods, a potentially useful tool as regards provision of superior patching methodology. Topoi are capable of supporting patching constructs to an even greater extent than sheaves. Finally, as we shall see below,

[1]See Appendix W for a more technical outline of this Sec's notions.

[2]This is a substantial limitation on applications, since most of the categories of interest as regards the mathematics underlying Theoretical Physics are larger: at least locally small. It may also be tied to yet another assumption of background structure.

presheaves and sheaves coincide in some cases, and can occur in the same package as—rather than instead of—topos structures.

Topos approaches are likely to be useful in Quantum Theory or some replacement or generalization thereof, whether or not this ameliorates the clash with Gravitation. Isham considered using topoi to upgrade Quantization in [187, 260, 495, 496] (some co-authored with mathematical physicist Andreas Doering or philosopher of Physics Jeremy Butterfield; see [448] for Landsman alongside mathematical physicists Chris Heunen and Bas Spitters' alternative approach). N.B. this use of Topos is more subtle than 'Quantization on'. Rather, it concerns conceptual issues in (the interpretation of) Quantum Theory being usefully recastable in terms of Topos Theory.

Application 1) Isham and Doering [260, 261] succeeded in reformulating the Kochen–Specker Theorem (39.42) in terms of a presheaf on the category of self-adjoint operators. This is based on *contexts*: von Neumann subalgebras of operators which commute among themselves, so this approach is termed 'contextual realism'. This presheaf formulation furthermore points to a generalization of the definition of valuation that is more suitable to Quantum Theory. In this setting, the Kochen–Specker Theorem can be reformulated as a statement that this presheaf lacks global sections. Representing propositions as subobjects of state space in this setting requires *daseination*, which is conceptually a type of coarse-graining.

Research Project 105) and Application 2) Expand on the relation alluded to in [489] between Histories Theory [504] and the Topos approach.

Research Project 106) and Application 3) Expand on [37]'s sketch of how [260] can be interpreted as a Timeless Records Theory.

Research Project 107) Consider the Kochen–Specker Theorem for QG theories themselves.

Application 4) Schreiber [778] has pioneered the application of sheaf and topos methods to reformulating standard Classical Physics, as well as to a large collection of applications in Supergravity and in String and M-Theory.

60.7 Multiple Choice Problem Revisited

Research Project 108) The Groenewold–van Hove phenomenon itself bears the hallmarks of a global obstruction. This has not however to date received a precise mathematical characterization, which rather probably requires going beyond the range of global considerations habitually made in Theoretical Physics. Cf. the Kochen–Specker Theorem receiving interpretation in terms of presheaves and Topos Theory as an example of 'going beyond'. Moreover, these two cases are conceptually distinct enough to likely require distinct technical resolutions. I.e. the Kochen–Specker case involves patching together local contexts, whereas the Groenewold–van Hove case *truncates* to at most quadratic polynomials rather than considering multiple local patches of types of polynomial.

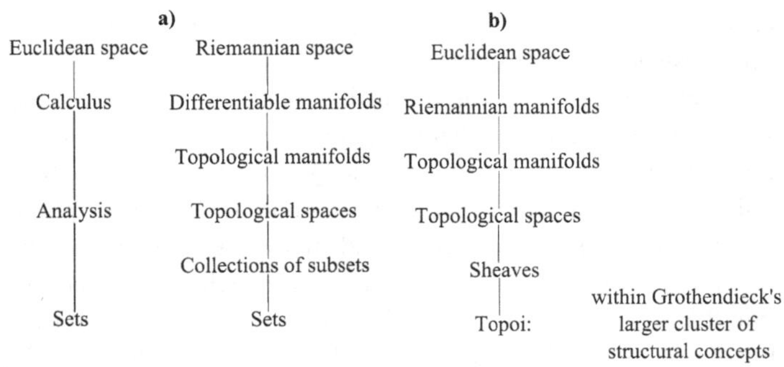

Fig. 60.1 **a)** Levels of mathematical structure commonly assumed in Classical Physics, as based on equipped sets. **b)** Progression in conceptualization of notions of space

This points to the Groenewold–van Hove phenomenon involving a codomain restriction (of the subalgebraic structure selection map) and a more severe type of global obstruction which is to be characterized rather than patched over. It looks to be an obstruction arising in mapping between two distinct cohomologies. I.e. a classical cohomology—such as de Rham cohomology, or more concretely, Poisson cohomology [458]—and a cohomology for the modelling of quantum-level operators—for which e.g. Hochschild or cyclic cohomology [627] have at least some suitable features. For the problem described here, it would be interesting to precisely characterize how simple unobstructed realizations of the first can be 'deformed' into more complicated realizations of the second. Mathematical physicist Maxim Kontsevich's work [565] on Deformation Quantization may be a useful starting point.

60.8 Background Independence, Categories and Topoi

Research Project 109) Study the classical counterpart of Isham's Quantization of small categories [492–494, 498]. What is the corresponding classical dynamics? How does one formulate Probability and Statistics on such categories?

Research Project 110) Consider time and Background Independence in the categorical setting, including e.g. an appraisal along these lines for [492–494, 498].

Research Project 111)† Consider time and Background Independence in the setting of topoi, including e.g. an assessment along these lines of Isham and Doering's work [256–259].

Research Project 112)†† Are Relationalism—or Background Independence more generally—criteria as to whether to use the standard 'equipped sets' Foundational System for Mathematics to model the deeper levels of Theoretical Physics, or to use categorical or topos-theoretic Mathematics instead? For instance, might it be that standard notions—such as giving points primary ontological status, or the conventional notion of open sets—that are artificial?

Research Project 113) One longstanding suggestion in Quantum Gravity has been termed 'field marshal covariance' [818], as in 'outranking' General Covariance due to its having an even wider scope. Heunen, Landsman and Spitters have formulated a notion of '*general tovariance*' [449]—a topos counterpart of General Covariance—which might constitute a realization of 'field marshal covariance'. As other alternatives, Noncommutative Geometry's version of General Covariance [216] is also more general than GR's, as is Supergravity's. However Topos Theory greatly further outstrips these theories as regards both mathematical generality and non-assumption of 'standard' mathematical structures in Physics that are in fact only standard due to being rooted in Background Dependent assumptions. Assess 'general tovariance' from a Background Independence perspective.

Research Project 114)† Let us further widen the scope of the preceding to '*general grovariance*'. This generalization is based on Grothendieckian mathematics having an even broader scope of structural concepts [66] than even its Topos Theory portion does, by which maybe that another part of this mathematics is required for the foundations of QG....

Appendices: Mathematical Methods for Basic and Foundational Quantum Gravity

Unstarred Appendices support Part I's basic account. Starred Appendices support Parts II and III on interferences between Problem of Time facets. Double starred ones support the Epilogues on global aspects and deeper levels of mathematical structure being contemplated as Background Independent. If an Appendix is starred, the default is that all of its sections are starred likewise; a few are marked with double stars.

Appendix A
Basic Algebra and Discrete Mathematics

A.1 Sets and Relations

For the purposes of this book, take a *set* $\mathfrak{X}$ to just be a collection of distinguishable objects termed *elements*. Write $x \in \mathfrak{X}$ if x is an element of $\mathfrak{X}$ and $\mathfrak{Y} \subset \mathfrak{X}$ for $\mathfrak{Y}$ a subset of $\mathfrak{X}$, $\cap$ for intersection, $\cup$ for union and $\mathfrak{Y}^{\mathrm{c}} = \mathfrak{X} \backslash \mathfrak{Y}$ for the complement of $\mathfrak{Y}$ in $\mathfrak{X}$. Subsets $\mathfrak{Y}_1$ and $\mathfrak{Y}_2$ are *mutually exclusive* alias *disjoint* if $\mathfrak{Y}_1 \cap \mathfrak{Y}_2 = \emptyset$: the empty set. In this case, write $\mathfrak{Y}_1 \cup \mathfrak{Y}_2$ as $\mathfrak{Y}_1 \amalg \mathfrak{Y}_2$: *disjoint union*. A *partition* of a set $\mathfrak{X}$ is a splitting of its elements into subsets $\mathfrak{p}_{\mathsf{P}}$ that are mutually exclusive and *collectively exhaustive*: $\coprod_{\mathsf{P}} \mathfrak{p}_{\mathsf{P}} = \mathfrak{X}$. Finally, the *direct* alias *Cartesian* product of sets $\mathfrak{X}$ and $\mathfrak{Z}$, denoted $\mathfrak{X} \times \mathfrak{Z}$, is the set of all ordered pairs (x, z) for $x \in \mathfrak{X}$, $z \in \mathfrak{Z}$.

For sets $\mathfrak{X}$ and $\mathfrak{Z}$, a *function* alias *map* $\varphi : \mathfrak{X} \to \mathfrak{Z}$ is an assignation to each $x \in \mathfrak{X}$ of a unique image $\varphi(x) = z \in \mathfrak{Z}$. Such a φ is *injective* alias *1 to 1* if $\varphi(x_1) = \varphi(x_2) \Rightarrow x_1 = x_2$, *surjective* alias *onto* if given $z \in \mathfrak{Z}$ there is an $x \in \mathfrak{X}$ such that $\varphi(x) = z$, and *bijective* if it is both injective and surjective. For $\mathfrak{Y} \subset \mathfrak{X}$, the corresponding *inclusion map* is the injection $j : \mathfrak{Y} \to \mathfrak{X}$ with $j(y) = y \ \forall \, y \in \mathfrak{Y}$. $\mathfrak{X}$ is *countable* if it is finite or admits a bijection $\varphi : \mathfrak{X} \to \mathbb{N}$: the natural numbers. On the other hand the set of real numbers $\mathbb{R}$ is uncountable. $|\mathfrak{X}|$ is the number of elements in the set $\mathfrak{X}$ in the finite case, or the extension of this to the notion of *cardinality* of the set more generally.

A *binary relation* R on a set $\mathfrak{X}$ is a property that each pair of elements of $\mathfrak{X}$ may or may not possess. We use $a \, R \, b$ to denote 'a and $b \in \mathfrak{X}$ are related by R'. Simple examples include $=$, $<$, $\leq$, $\subset$ and $\subseteq$. Some basic properties that an R on $\mathfrak{X}$ might possess are as follows ($\forall \, a, b, c \in \mathfrak{X}$). *Reflexivity*: $a \, R \, a$. *Symmetry*: $a \, R \, b \Rightarrow b \, R \, a$. *Antisymmetry*: $a \, R \, b$ and $b \, R \, a \Rightarrow a = b$. *Transitivity*: $a \, R \, b$ and $b \, R \, c \Rightarrow a \, R \, c$. *Totality*: that one or both of $a \, R \, b$ or $b \, R \, a$ holds, i.e. all pairs are related. Commonly useful combinations of these include the following.

1) *Equivalence relation*, $\sim$, if R is reflexive, symmetric and transitive.
2) *Partial ordering*, $\preceq$, if R is reflexive, antisymmetric and transitive, e.g. $\leq$ or $\subset$:

E. Anderson, *The Problem of Time*, Fundamental Theories of Physics 190,
DOI 10.1007/978-3-319-58848-3

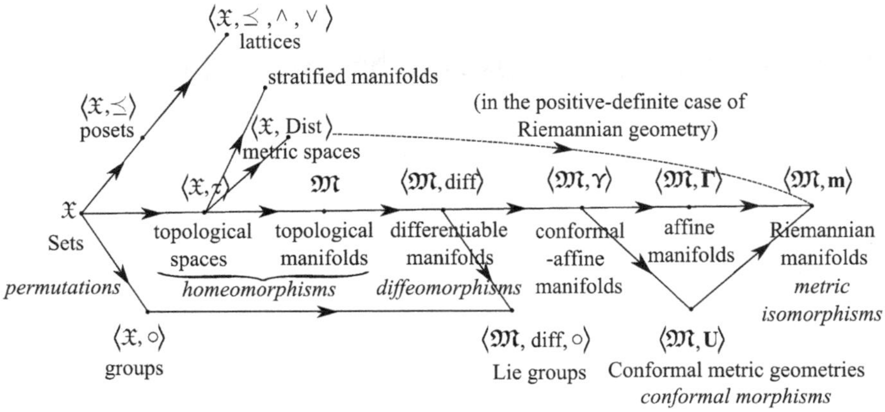

Fig. A.1 Equipping sets with a variety of structures. The main structures considered in this book are presented in *straight font*, and we use *italic font* to denote some of the associated morphisms. Reversing the *arrows* gives examples of forgetting structures. τ is a topology, Dist is a metric space metric, diff are differentiable structures, γ and $\mathbf{\Gamma}$ are connections, and $\mathbf{U}$ and $\mathbf{m}$ are Riemannian metrics; γ and $\mathbf{U}$ are conformally invariant versions

ordering by 'is a subset of'.

3) *Total ordering*, alias a *chain*, if R is both a partial order and total.

A fairly standard Foundational System for Mathematics involves equipping sets $\mathfrak{X}$ with further layers of structure ς; we denote this by $\langle \mathfrak{X}, \varsigma \rangle$. For example, a set equipped with a partial order is a *poset* $\langle \mathfrak{X}, \preceq \rangle$. One can furthermore equip an already established equipped space $\mathfrak{S}$ (rather than just a set) with extra layers of structure $\langle \mathfrak{S}, \varsigma \rangle$. Such equipping is what is meant in Chap. 10 and Epilogue II.C by 'levels of mathematical structure', within the traditional (if in some ways restrictive) context that the base level consists of Set Theory. See Fig. A.1 for further examples of equipping.

A *(homo)morphism* is a map $\mu : \mathfrak{S}_1 \rightarrow \mathfrak{S}_2$ that is structure-preserving. In particular, if a such is bijective (equivalently invertible) it is an *isomorphism*, if $\mathfrak{S}_1 = \mathfrak{S}_2$ it is an *endomorphism*, and if both of these apply, it is an *automorphism*.

A *forgetful map* is one that 'forgets' (or 'strips off') some of the layers of structure: $\phi : \langle \mathfrak{S}, \varsigma \rangle \rightarrow \mathfrak{S}$. There is corresponding loss of structure preservation in the maps associated with the latter equipped space. The obvious reversal of each of the equipping maps in Fig. A.1 readily provides an example of forgetful map.

Let us end by noting that whereas many further mathematical entities are often thought of as arising by imposing further layers of structure on a set as per above, this is not the only way of doing mathematics. E.g. Category Theory [611, 612, 631] and Topos Theory [126, 260] offer extensions and alternatives in this regard. While these are very briefly outlined in Appendix W, all detailed applications in this book can be taken to be based upon sets.

A.2 Groups

A *group* $\langle \mathfrak{g}, \circ \rangle$ is a set $\mathfrak{g}$ equipped with an operation $\circ$, such that $\forall\, g_1, g_2, g_3 \in \mathfrak{g}$,

i) $g_1 \circ g_2 \in \mathfrak{g}$ (closure),
ii) $\{g_1 \circ g_2\} \circ g_3 = g_1 \circ \{g_2 \circ g_3\}$ (associativity),
iii) $\exists\, e \in \mathfrak{g}$ such that $e \circ g_1 = g_1 = g_1 \circ e$ (identity), and
iv) $\exists\, g_1^{-1} \in \mathfrak{g}$ for each g_1 such that $g_1 \circ g_1^{-1} = e = g_1^{-1} \circ g_1$ (inverse).

Groups (see [69, 213] for further reading) are one of the most widely useful mathematical structures in Physics [526]. They encode the mathematics of transformations and symmetries. This book often uses $\mathfrak{g}$ as a shorthand for $\langle \mathfrak{g}, \circ \rangle$. $|\mathfrak{g}|$ denotes the *order* of $\mathfrak{g}$, i.e. the number of elements of $\mathfrak{g}$. If additionally $g_1 \circ g_2 = g_2 \circ g_1$ (commutativity), $\mathfrak{g}$ is said to be *Abelian* (after mathematician Niels Abel). $\mathfrak{H}$ is a *subgroup* of $\mathfrak{g}$, denoted $\mathfrak{H} \leq \mathfrak{g}$, if $\mathfrak{H} \subseteq \mathfrak{g}$, and $\mathfrak{H}$ is closed with respect to the same group operation $\circ$ that $\mathfrak{g}$ possesses while containing the identity and all its elements' inverses. If $\langle \mathfrak{g}, \circ \rangle$ and $\langle \mathfrak{K}, * \rangle$ are groups, the *(direct) product group* $\mathfrak{g} \times \mathfrak{K}$ is defined by having (g, k) as its elements: the Cartesian product of the sets of $\mathfrak{g}$ and $\mathfrak{K}$, with group operation

$$(g_1, k_1) \square (g_2, k_2) = (g_1 \circ g_2, k_1 * k_2).$$

Two groups $\langle \mathfrak{g}, \circ \rangle$ and $\langle \mathfrak{K}, * \rangle$ are *isomorphic*, denoted by $\cong$, if there is a bijection φ that preserves the group structure: $\varphi(g \circ k) = \varphi(g) * \varphi(k)\ \forall\, g \in \mathfrak{g}, k \in \mathfrak{K}$.[1]

Examples of groups include 1) $\langle \mathbb{R}, + \rangle$, 2) $\langle \mathbb{R}/\{0\}, \times \rangle$. 3) The cyclic groups $\mathbb{Z}_n$, which are Abelian and with the sole relation $g^n = e$, and 4) the permutations of a finite set $\mathfrak{X}$ form a group, $Perm(\mathfrak{X})$. 5) The automorphisms for each $\mathfrak{S}$ form a group, $Aut(\mathfrak{S})$; $Perm(\mathfrak{X})$ is the simplest case, corresponding to $\mathfrak{S} = \mathfrak{X}$ finite. 6) Lie groups (named after 19th century mathematician Sophus Lie) such as Appendix E's standard examples and Appendix V's larger ones. 7) Given a group $\mathfrak{g}$, its *centre* $Z(\mathfrak{g})$ is another group: the subgroup of $\mathfrak{g}$ formed by the elements of $\mathfrak{g}$ which commute with all the other elements: $\{c \in \mathfrak{g} \,|\, cg = gc\ \forall\, g \in \mathfrak{g}\}$. Exercise Set IV is dedicated to groups.

Groups are of yet further interest through acting on other mathematical or physical objects (sets of points, figures, physical matter...) A *group action* on a set $\mathfrak{X}$ is a map $\alpha : \mathfrak{g} \times \mathfrak{X} \to \mathfrak{X}$ (often denoted by $\vec{\mathfrak{g}}_g$ in this book) such that

i) $\{g_1 \circ g_2\}x = g_1 \circ \{g_2 x\}$ (compatibility) and
ii) $ex = x\ \forall\, x \in \mathfrak{X}$ (identity).

Subcases include *left action* gx, *right action* xg, and *conjugate action* gxg^{-1}.

By a *natural* group action, we just mean one which does not require a choice as to how to relate $\mathfrak{X}$ and $\mathfrak{g}$ due to there being one 'obvious way' in which it acts. E.g. Perm($\mathfrak{X}$) acts naturally on $\mathfrak{X}$, or an $n \times n$ matrix group acts naturally on the

[1]This is the well-known *group isomorphism* subcase of Appendix A.1's more general notion of isomorphism, and is useful in classifying groups.

corresponding n-vectors. An action is *faithful* if $g_1 \neq g_2 \Rightarrow g_1 x \neq g_2 x$ for *some* $x \in \mathfrak{X}$, whereas it is *free* if this is so for *all* $x \in \mathfrak{X}$. So free $\Rightarrow$ faithful, but not vice versa. Finally, an action is *transitive* if for every $x, y \in \mathfrak{X}$ there is a g such that $gx = y$.

For $\mathfrak{g}$ a group acting on a set $\mathfrak{X}$, and $x \in \mathfrak{X}$, the *group orbit* $\text{Orb}(x) := \{gx \mid g \in \mathfrak{g}\}$: the set of images of x, and the *stabilizer* alias *isotropy group* alias *little group* $\text{Stab}(x) := \{g \in \mathfrak{g} \mid gx = x\}$: the set of $g \in \mathfrak{g}$ that fix x. := denotes 'is defined by'.

A subgroup $\mathfrak{N}$ that is invariant under conjugate action—$g\,\mathfrak{N}\,g^{-1} = \mathfrak{N}$ for every $g \in \mathfrak{g}$—is called a *normal* subgroup of $\mathfrak{g}$, denoted by $\mathfrak{N} \lhd \mathfrak{g}$. The quotient of one group by another, $\mathfrak{g}/\mathfrak{H}$, only makes sense if $\mathfrak{H} \lhd \mathfrak{g}$. In this case, $\mathfrak{g}/\mathfrak{H} := \{g\mathfrak{H} \mid g \in \mathfrak{g}\}$ is itself a group: the *quotient group* . Finally, the *semidirect product group* $\mathfrak{g} = \mathfrak{N} \rtimes \mathfrak{H}$ is given by

$$(n_1, h_1) \,\square\, (n_2, h_2) = (n_1 * \varphi_{h_1}(n_2), h_1 \circ h_2)$$

for $\langle \mathfrak{N}, * \rangle \lhd \mathfrak{g}$, $\langle \mathfrak{H}, \circ \rangle$ a subgroup of $\mathfrak{g}$ and $\varphi : \mathfrak{H} \to Aut(\mathfrak{N})$ a group homomorphism.

A transformation is *passive* if it changes the problem in hand's coordinate description. On the other hand, it is *active* if it is held to actually move around the problem in hand's entities. This book's main use of this distinction is for diffeomorphisms (in Chaps. 7 to 10).

A.3 Linear Algebra. i. Fields and Vector Spaces

For F a set and two operations $\circ, \square$, $\langle \mathbb{F}, \circ, \square \rangle$, is a *field* if

i) $\langle \mathbb{F}, \square \rangle$ is an Abelian group.
ii) $\langle \mathbb{F}/\{e\}, \circ \rangle$ is an Abelian group for e the identity of the preceding.
iii) $\circ$ is *distributive* over $\square$: $\forall\, x, y, z \in \mathbb{F}, x \circ \{y \,\square\, z\} = \{x \circ y\} \,\square\, \{x \circ z\}$.

Ex III.0 asks the reader to investigate which of $\mathbb{N}$, $\mathbb{Z}$, $\mathbb{Q}$, $\mathbb{R}$, $\mathbb{C}$ are fields.

A *vector space* $\mathfrak{v}$ over the field $\mathbb{F}$ is an 'additive group' $\mathfrak{g} = \langle \mathbb{F}, + \rangle$ together with a product $\mathbb{F} \times \mathfrak{v} \to \mathfrak{v}$ sending the pair (p, v) to the scalar-multiplied pv, such that $\forall\, p, q \in \mathbb{F}$ and $\forall\, g, h \in \mathfrak{g}$,

i) $\{p + q\} \times g = pg + qg$ and $p \times \{g + h\} = pg + ph$ (distributivities),
ii) $p\{qg\} = \{pq\}g$, and
iii) $1\,g = g$.

See [213] for more about fields and vector spaces.

Examples of vector spaces used in Physics include 3-vectors in $\mathbb{R}^3$ and 4-vectors in Minkowski spacetime $\mathbb{M}^4$ (both for $\mathbb{F} = \mathbb{R}$), while the $\mathbb{F} = \mathbb{C}$ case is useful in Quantum Theory. Dirac's kets used in Quantum Theory are in fact often, but not always, *infinite-dimensional* vectors. These form an infinite-dimensional vector space of the Hilbert space type (see Appendix C.2 for an outline). The complementary function part of the solution of a linear ODE (which solves the corresponding homogeneous equation), and Fourier series are further widely useful examples belonging to finite and infinite vector spaces respectively.

$\mathfrak{w} \subseteq \mathfrak{v}$ is a *subspace* of $\mathfrak{v}$ if it is itself a vector space under the same two operations. For two subspaces $\mathfrak{u}, \mathfrak{w}$ of $\mathfrak{v}$, their *sum*, denoted $\mathfrak{u} + \mathfrak{w}$, is the set $\{pu + qw \in \mathfrak{v} \,|\, u \in \mathfrak{u}, w \in \mathfrak{w}, p, q \in \mathbb{F}\}$. $\mathfrak{v}$ is the *direct sum* of $\mathfrak{u}$ and $\mathfrak{w}$, denoted $\mathfrak{v} = \mathfrak{u} \oplus \mathfrak{w}$ if $\mathfrak{v} = \mathfrak{u} + \mathfrak{w}$ and $\mathfrak{u} \cap \mathfrak{w} = 0$.

For $\mathfrak{w} \subseteq \mathfrak{v}$, the *span* $\langle \mathfrak{w} \rangle$ is the intersection of all subspaces containing $\mathfrak{w}$. The vectors $v_i \in \mathfrak{v}$ ($i = 1$ to n) are *linearly independent* iff (if and only if) $\sum_{i=1}^{n} a_i v_i = 0$ only for $a_i \in \mathbb{F}$ all zero. A *basis* for a vector space $\mathfrak{v}$ is any linearly independent subset that spans $\mathfrak{v}$. The number of vectors in a basis set is the *dimension* $\dim(\mathfrak{v})$ of the vector space $\mathfrak{v}$. If $\{v_i, i = 1 \text{ to } p\}$ is a basis, each vector $v \in \mathfrak{v}$ can be expressed as $v = \sum_{i=1}^{p} a_i v_i$ for unique scalars $a_i \in \mathbb{F}$; these are termed the *components* of v with respect to this basis.

Linear maps can be set up between vector spaces: $L : \mathfrak{v} \to \mathfrak{w}$ such that $L(ax + by) = aL(x) + bL(y)$. Use $\text{Im}(L) := L(\mathfrak{v}) \leq \mathfrak{w}$ to denote the *image* of $\mathfrak{v}$ under L and $\text{Ker}(L) := \{v \in \mathfrak{v} \,|\, L(v) = 0\} \leq \mathfrak{v}$ for the *kernel* of L. If the vectors in question are written with respect to specific bases, the map in question is cast in matrix form. For $\mathbb{F} = \mathbb{C}$, one can also define *antilinear* maps: $A(az + bw) = \bar{a}A(z) + \bar{b}A(w)$ where the bars denote complex conjugate.

For $\mathfrak{u}, \mathfrak{v}, \mathfrak{w}$ vector spaces over the same $\mathbb{F}$, $\mu : \mathfrak{u} \times \mathfrak{v} \to \mathfrak{w}$ is a *bilinear map* if it is linear in each argument. μ is *degenerate* if there is either a nonzero $u \in \mathfrak{u}$ such that $\mu(u, v) = 0 \; \forall \, v \in \mathfrak{v}$, or a nonzero $v \in \mathfrak{v}$ such that $\mu(u, v) = 0 \; \forall \, u \in \mathfrak{u}$.

Two vector spaces $\mathfrak{v}, \mathfrak{w}$ are *isomorphic* if there is a bijection φ between them such that $\varphi(v_1 + v_2) = \varphi(v_1) + \varphi(v_2)$ and $\varphi(pv_1) = p\varphi(v_1) \; \forall \, v_1, v_2 \in \mathfrak{v}$ and $p \in \mathbb{F}$. They are *homomorphic* if these relations hold but bijectivity is dropped.

Some bases are more convenient than others: *bases can be chosen to maximally simplify problems*. This includes casting matrices in diagonal form, or as close to it as possible, and with as many as possible of the remaining elements being 1's (on some occasions -1's are also required). In the case of $\mathbb{F} = \mathbb{R}$, bilinear forms $\mu(\underline{x}, \underline{y})$ which are *symmetric*: $\mu(\underline{x}, \underline{y}) = \mu(\underline{y}, \underline{x})$, the corresponding matrices are always diagonalizable. One can select a basis such that the corresponding matrix takes the form $\text{diag}(\mathbb{I}_s, -\mathbb{I}_{r-s}, 0)$. Here $\mathbb{I}_p$ denotes the $p \times p$ identity matrix, with s and $r - s$ indicating the dimensionality of each block. $s + \{r - s\} = r$ is the *rank*: number of nonzero diagonal entries. Mathematicians define *signature* by $s - \{r + s\}$.[2] On the other hand, for $\mathbb{F} = \mathbb{C}$, the natural counterpart to symmetric bilinear forms are *sesquilinear Hermitian forms* (after mathematician Charles Hermite). These are linear in the first argument and antilinear in the second, and with $\mu(\underline{x}, \underline{y}) = \mu(\underline{y}^\dagger, \underline{x})$, where the Hermitian $\dagger$-symbol denotes complex conjugate transpose.

[2]Physicists often use 'signature' in a more loose but none the less related manner, mostly out of only needing to distinguish between one minus in spacetimes and no minuses in spaces. So physicists use '$- - + + +$' or 'indefinite', whereas the mathematicians' concept of signature involves the number of plusses with the number of minuses subtracted off, giving spacetime signature 2.

Multilinearity follows similarly to bilinearity; for $\mathbb{F} = \mathbb{R}$ and constituent linear spaces $\mathbb{R}^n$, this gives the mathematics of Cartesian tensors (Sect. 2.12).

The invertible linear maps $L : \mathfrak{v} \rightarrow \mathfrak{v}$ (or the matrices corresponding to these) form the *general linear group* $GL(\mathfrak{v})$. This book makes use of the $GL(n, \mathbb{C})$ and especially $GL(n, \mathbb{R})$ cases. The unit-determinant version is denoted by $SL(\mathfrak{v})$, standing for *special linear*; in this book, 'special' is furthermore taken to mean restriction to unit determinants in a wider range of contexts.

Suppose $v_1, v_2 \in \mathfrak{v}$ and μ is either symmetric bilinear (for $\mathbb{F} = \mathbb{R}$) or Hermitian sesquilinear (for $\mathbb{F} = \mathbb{C}$). Then $\mu(v_1, v_2) : \mathfrak{v} \times \mathfrak{v} \rightarrow \mathbb{F}$ can also be viewed as an *inner product* $(v_1, v_2)_\mu$ (generalizing $\mathbb{R}^p$'s dot product). A (for now finite) *inner product space* is a vector space equipped with an inner product.

We use $\|\underline{v}\|_\mu$ to denote $(\underline{v} \cdot \underline{v})_\mu^{1/2}$. This is a *norm* on $\mathfrak{v}$ if it obeys

i) $\|\underline{v}\|_\mu \geq 0$ with equality if $v = 0$ (positive definiteness),
ii) $\|\underline{u} + \underline{v}\|_\mu \leq \|\underline{u}\|_\mu + \|\underline{v}\|_\mu$ (triangle inequality alias subadditivity),
iii) $\|p\,\underline{v}\|_\mu = |p|\,\|\underline{v}\|$, $p \in \mathbb{F}$ (absolute homogeneity).

In the finite case, a *normed space* is just a vector space $\mathfrak{v}$ equipped with a norm.

A well-known example of this is the Euclidean norm and the Euclidean inner product ($\mu = id$, whose components are δ_{AB} in a suitable basis). Another is the complex norm and inner product, based on $(z \cdot w) = z\bar{w}$, where the bar denotes complex conjugation.

A *projector* is a map $P\underline{x} = (\underline{x} \cdot \underline{n})\underline{n}$, giving the component of $\underline{x}$ in the direction $\underline{n}$.

Vectors v and u are *orthogonal* with respect to a given inner product if $(v, u) = 0$. For δ_{ij} the Kronecker delta symbol, a set of vectors v_i for which $(v_i, v_j) = \delta_{ij}$ are *orthonormal*; orthonormal bases can readily be constructed by keeping on picking linearly independent vectors and removing projections onto vectors which have already been picked.

The *dual vector space* $\mathfrak{v}^*$ of $\mathfrak{v}$ consists of the linear maps $\varphi : \mathfrak{v} \rightarrow \mathbb{F}$. Finite 'transposed vectors' are a simple example. As another simple example, in Dirac's quantum notation, the bras $\langle\,\psi\,|$ are the corresponding duals to the kets $|\,\psi\,\rangle$. (More complicated examples exhibit more substantial distinction between vector spaces and their duals.) Let $\{\underline{e}_i\}$ be a basis of $\mathfrak{v}$; then the *dual basis* $\{\underline{f}_i\}$ is a basis of $\mathfrak{v}^*$ such that $(\underline{e}_i \cdot \underline{f}_j) = \delta_{ij}$. In the case of $\mathbb{F} = \mathbb{R}$, Cartesian tensor multilinearity (Sect. 2.12) follows.

The $2p \times 2p$ *symplectic matrix* is

$$\mathbb{J}_p := \begin{pmatrix} 0 & \mathbb{I}_p \\ -\mathbb{I}_p & 0 \end{pmatrix}. \tag{A.1}$$

This corresponds to casting the *symplectic quadratic form* [70] according to a convenient choice of basis.

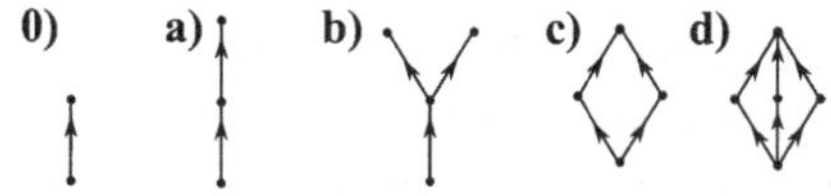

Fig. A.2 Directed graphs. Each can additionally be interpreted as a poset

A.4 ii. Rings and Modules*

A *monoid* is a set $\mathfrak{X}$ obeying the group axioms except that not all of its elements are necessarily invertible. A *ring* is a set $\mathfrak{R}$ with two binary operations such that axioms i) and iii) for fields hold, but now in place of ii), we require that $\mathfrak{R}$ is a monoid under multiplication. An *ideal* $\mathfrak{J}$ in a ring $\mathfrak{R}$ is a subgroup of $\langle \mathfrak{R}, + \rangle$ such that $\mathfrak{R}\mathfrak{J} \subseteq \mathfrak{J}$, $\mathfrak{J}\mathfrak{R} \subseteq \mathfrak{J}$ (i.e. under both right and left action). Ideals play a major role, analogous to that of normal subgroups for Group Theory. Finally, *modules* are the analogous notion to vector spaces upon replacing the input field by a ring. All that is needed to follow this book is that addition and direct sum continue to be defined here. See especially [213] for more about rings and modules.

A.5 Representation Theory

Representations are homomorphisms $\rho : \mathfrak{g} \rightarrow GL(\mathfrak{v})$ by which one can pass from handling groups to the more convenient problem of just handling matrices. The mathematics of this—*Representation Theory*—(see in particular [786]) is based on both Group Theory and Linear Algebra, and has been considerably developed. This is widely useful for Physics [526] (symmetries and Quantum Theory in particular)

A representation is *irreducible* (an *irrep*) if it cannot be broken up into a direct sum of representations.

In basic Group and Representation Theory, it is quite often useful to construct a $\mathfrak{g}$-invariant version $O_{\mathfrak{g}\text{-free}}$ of a non-inherently $\mathfrak{g}$-invariant object O for which addition and scalar multiplication are meaningful. This is by applying a $\mathfrak{g}$-action to O followed by averaging over all of $\mathfrak{g}$. E.g. for a finite group, this takes the form

$$O_{\mathfrak{g}\text{-free}} = \frac{1}{|\mathfrak{g}|} \sum_{g \in \mathfrak{g}} \vec{\mathfrak{g}}_g \, O. \tag{A.2}$$

Ex IV.1 is a worked example for gaining familiarity with this Sec's concepts.

A.6 Graphs and Generalizations

A *graph* (see especially [159]) is a set of vertices and the corresponding set of edges between vertices. Mathematical entities such as groups, knots, and Feynman diagrams can on some occasions be usefully represented by graphs. This illustrates a more general sense than the preceding Sec's of the concept of 'representation';

see Appendix W for further discussion. Graphs can additionally be generalized to versions including higher-dimensional entities starting with faces and solid blocks; see Appendices F and G for examples. Generalizations along the lines of *labelled versions* of graphs are also possible, such as directed graphs—with ordering arrows on their edges: Fig. A.2—and coloured graphs [159] to labellings by representations in spin networks: Sect. 43.5.

Appendix B
Flat Geometry

"*He who attempts natural philosophy without geometry is lost*". Galileo Galilei

B.1 Real Geometry

We approach this here from a simple Kleinian position,[1] by considering $\mathfrak{g} \leq Aut(\langle \mathbb{R}^d, \varsigma \rangle)$ for various layers of mathematical structures ς. ς could be $\cdot$ (scalar products: the Euclidean metric δ_{ij}), but also / (ratios), $-$ (differences, as features e.g. in the Euclidean notion of distance), $\angle$ (angles), and d-dimensional volumes (such as areas built out of cross products $\times$ in 2-d or volumes built out of scalar triple products $[\ \times \cdot\ ]$ in 3-d). Let us denote the last of these in the general case by $\bigwedge$, for the top form supported in dimension d. Additionally, a number of combinations of these structures are possible.

To be clear about the above shorthands' definitions, let $\underline{u}$, $\underline{v}$, $\underline{w}$, $\underline{y} \in \mathbb{R}^d$. The scalar product is a 2-slot operation $\underline{u} \cdot \underline{v}$. The Euclidean norm alias magnitude is a special case of the square root of this: $||\underline{v}|| := \sqrt{\underline{v} \cdot \underline{v}}$. Also

$$(\textit{Euclidean distance} \text{ between } \underline{u} \text{ and } \underline{w}) := ||\underline{u} - \underline{w}||, \tag{B.1}$$

i.e. the Euclidean norm of the difference between the two vectors $\underline{u} - \underline{w}$. Ratios are then a 2-slot operation acting on scalars, e.g. a ratio of two components of a vector.

$$(\text{ratio of magnitudes of } \underline{u} \text{ and } \underline{w}) := \frac{||\underline{u}||}{||\underline{w}||}, \tag{B.2}$$

$$(\text{ratio of distances}) := \frac{||\underline{u} - \underline{v}||}{||\underline{w} - \underline{y}||}, \tag{B.3}$$

[1] See [222] and [815] for further positions. Also note that this Appendix's variety of flat geometries is a major precursor of understanding the variety of differential geometries with extra structures.

E. Anderson, *The Problem of Time*, Fundamental Theories of Physics 190,
DOI 10.1007/978-3-319-58848-3

$$\text{(ratio of scalar products)} := \frac{(\underline{u} \cdot \underline{v})}{(\underline{w} \cdot \underline{y})}. \tag{B.4}$$

The *angle* between $\underline{u}$ and $\underline{w}$ is the arccos of the particular combination

$$\text{(scalar product of unit vectors } \hat{\underline{u}} \text{ and } \hat{\underline{v}} = (\hat{\underline{u}} \cdot \hat{\underline{v}})) = \frac{(\underline{u} \cdot \underline{v})}{\|\underline{u}\| \, \|\underline{v}\|}, \tag{B.5}$$

which is a product of square roots of 2 subcases of (B.4). Finally, p-volume between vectors is

$$\text{(areas of parallelograms formed by vectors } \underline{u}, \underline{v}) := (\underline{u} \times \underline{v})_3 \quad \text{in 2-}d, \quad \text{and} \tag{B.6}$$

$$\text{(volumes of parallelepipeds formed by vectors } \underline{u}, \underline{v}, \underline{w}) := \underline{u} \times \underline{v} \cdot \underline{w} \quad \text{in 3-}d. \tag{B.7}$$

Possible $\mathfrak{g}$ include the following; see Figs. B.1 and B.2 for the meanings of the types of transformations mentioned. $\mathfrak{g} = id$: a trivial limiting case corresponding to no transformations being available. $\mathfrak{g} = Aut(\langle \mathbb{R}^p, -\rangle) = Tr(p)$, the *translations* $\underline{x} \rightarrow \underline{x} + \underline{a}$, which form a p-dimensional Abelian group $\langle \mathbb{R}^p, +\rangle$. $\mathfrak{g} = Aut(\langle \mathbb{R}^p, /\rangle) = Dil$: *dilations* alias *homotheties* $\underline{x} \rightarrow k\underline{x}$, which form a 1-$d$ Abelian group $\langle \mathbb{R}^+, \cdot\rangle$. $\mathfrak{g} = Aut(\langle \mathbb{R}^p, -/-\rangle) = Tr(p) \rtimes Dil$. $\mathfrak{g} = Aut(\langle \mathbb{R}^p, \cdot\rangle) = Rot(p)$: *rotations* $\underline{x} \rightarrow \underline{\underline{B}x}$ forming the *special orthogonal group* $SO(p) := \{B \in GL(p, \mathbb{R}) \,|\, B^{\mathrm{T}}B = \mathbb{I}, \det B = 1\}$, which is of dimension $p\{p-1\}/2$ $\mathfrak{g} = Aut(\langle \mathbb{R}^p, - \cdot -\rangle) := Isom(\mathbb{R}^p) = Tr(p) \rtimes Rot(p) =: Eucl(p)$: the $p\{p+1\}/2$-dimensional *Euclidean group* of isometries, corresponding to *Euclidean Geometry* itself.[2] $\mathfrak{g} = Aut(\langle \mathbb{R}^p, \cdot/\cdot\rangle) = Rot(p) \times Dil$. $\mathfrak{g} = Aut(\langle \mathbb{R}^p, - \cdot - / - \cdot -\rangle) = Tr(p) \rtimes \{Rot(p) \times Dil\} =: Sim(p)$: the $p\{p+1\}/2 + 1$ dimensional *similarity group* corresponding to *Similarity Geometry*.

Next, using $\wedge$ for the general p-dimensional case for reasons explained in Sect. D.2 $\mathfrak{g} = Aut(\langle \mathbb{R}^p, \wedge\rangle) = SL(p, \mathbb{R})$. This is the $p^2 - 1$ dimensional special linear group, consisting of the $p\{p-1\}/2$ rotations, $p\{p-1\}/2$ shears and $p-1$ 'Procrustean stretches'. $\mathfrak{g} = Aut(\langle \mathbb{R}^p, \wedge-\rangle) = Tr(p) \rtimes SL(p, \mathbb{R})$: the $p\{p+1\} - 1$ dimensional 'equi-p-voluminal group' corresponding to 'equi-p-voluminal geometry'. (For $p = 2$, $\wedge = \times$ and this is the quite well-known [222] *equiareal geometry*.) $\mathfrak{g} = Aut(\langle \mathbb{R}^p, \wedge/\wedge\rangle) = GL^+(p, \mathbb{R})$: the p^2-dimensional general linear group, consisting of rotations, shears and Procrustean stretches now alongside dilations. $\mathfrak{g} = Aut(\langle \mathbb{R}^p, (\wedge-)/(\wedge-)\rangle) = Tr(p) \rtimes SL(p, \mathbb{R}) =: Aff(d)$ the $p\{p+1\}$-dimensional *affine group* of linear transformations, corresponding to *Affine Geometry*.

So far, the above transformations can all be summarized within the form of the equation at the top of Fig. B.1. The most general case included in this is Affine Geometry, within which all the other $\mathfrak{g}$ above are realized as subgroups.

---

[2]In a slight abuse of notation, elsewhere in this book, we use *Eucl* and *Sim* to denote the cases excluding the rotations; in the *SL* and *GL* cases, we denote the cases excluding the reflections by SL^+ and GL^+.

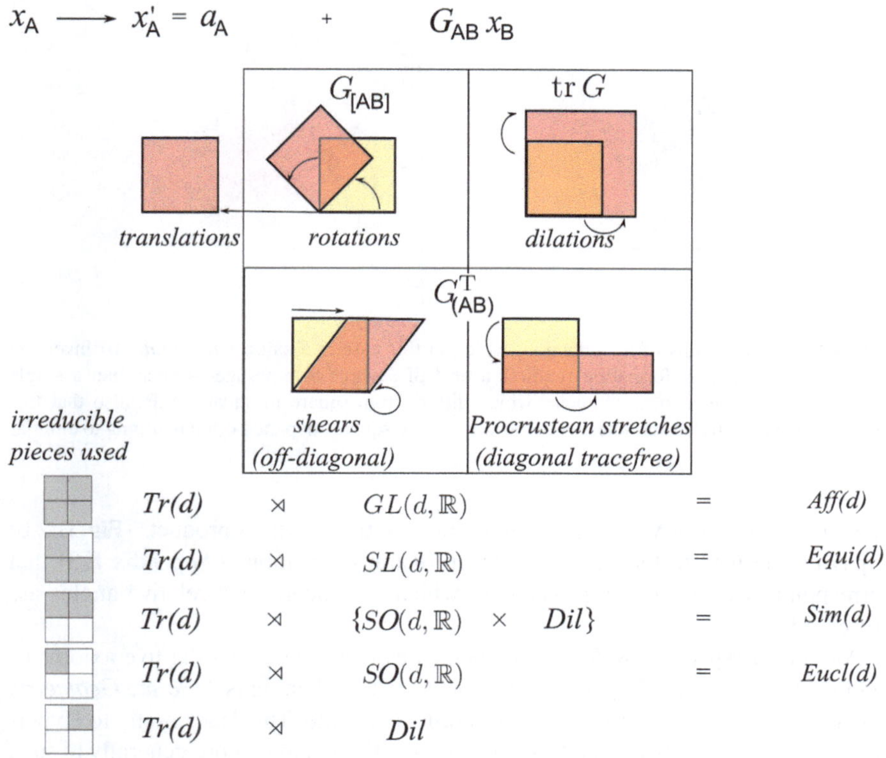

Fig. B.1 Elementary transformations, in each case from a yellow square to a red image. 2-d illustration of translation, rotation, dilation, shear, and Procrustean stretch, i.e. d-volume top form preserving stretches, in particular area-preserving in 2-d and volume-preserving in 3-d. Underneath, we also indicate the relation of the last four of these to the irreducible pieces of the general linear matrix $\underline{\underline{G}}$, and which geometrically illustrious groups these transformations form part of. The T superscript denotes 'tracefree part'. Procrustean stretches, moreover, do not respect ratios and shears do not respect angles

Reflections can also be involved in each case, about an invariant mirror hyperplane (e.g. a line in 2-d or a plane in 3-d). These are a third elementary type of isometry; unlike translations and rotations, they are a discrete operation. The case of a mirror through the origin is characterized by the unit normal $\underline{n}$; here the explicit form for the corresponding reflection is the linear transformation

$$Ref : \underline{v} \rightarrow \underline{v} - 2\,(\underline{v} \cdot \underline{n})\,\underline{n}. \tag{B.8}$$

A further direction in p-dimensional geometry arises by introducing *inversions* in $\mathbb{S}^{p-1}$

$$Inv : \underline{v} \rightarrow \frac{\underline{v}}{\|\underline{v}\|^2}. \tag{B.9}$$

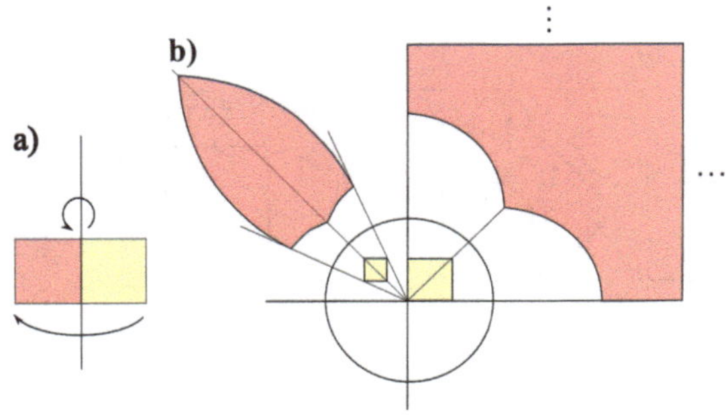

Fig. B.2 2-*d* depictions of **a)** reflection, which in this case is about a *mirror line*. **b)** Inversion in the *circle*. This transformation requires a grid of squares to envisage—rather than a single square—since it has a local character which differs from square to square. N.B. also that this can map between circles and lines, with the sides of the squares depicted often mapping to circular arcs

Inversions *also* preserve angles—but *not* other ratios of scalar products (Fig. B.2.b) —paving the way to the yet larger group of transformations (Appendix E.3) that correspond to *Conformal Geometry*, in which specifically just relative angles are preserved.

Another perspective [815] on Geometry involves weakening the five axioms of Euclidean Geometry [222]. The best-known such weakening is *Absolute Geometry*, which involves *dropping just* Euclid's parallel postulate. This leads firstly to Hyperbolic Geometry arising as an alternative to Euclid's, and then more generally to such as Riemannian Differential Geometry (Appendix D.4). In contrast, Affine Geometry *retains* Euclid's parallel postulate, and indeed places central importance upon developing its consequences ('parallelism'). This approach drops instead Euclid's right-angle and circle postulates. These two initially contrasting themes continue to play major parts in the eventual generalization to Affine Differential Geometry (Appendix D.3).

Two more primary types of geometry are, firstly, *Ordering Geometry* [222], which involves just an 'intermediary point' variant of Euclid's line postulates. By involving *neither* the parallel postulate or the circle and right-angle pair of postulates, this can be seen as serving as a *common* foundation for both Absolute and Affine Geometry [222]. Secondly, in *Projective Geometry*[3] one ceases to be able to distinguish between lines and circles, in addition to angles being meaningless and no parallel postulate holding. This now corresponds to the *projective linear group* $PGL(p,\mathbb{R}) = GL(p,\mathbb{R})/Z(GL(p,\mathbb{R}))$ [815].

For practical use within Euclidean theories of space, note in particular that 'spatial' measurements in our experience are of the forms (B.3) and (B.5), i.e. measuring tangible objects *against a ruler* and measuring angles *between* tangible entities. On the other hand, more advanced, if indirect, physical applications make use of (extensions of) the other notions of geometry above.

[3]See [222, 815] for an introduction to Flat Projective Geometry.

B.2 Minkowski Spacetime Geometries

This case has an indefinite flat metric η_{AB} on $\mathbb{M}^{p+1}$ in place of Euclidean Geometry's positive-definite metric δ_{AB} on $\mathbb{R}^p$. While the most obvious application of this is to SR spacetime in support of Part I [736], it is also used under various other circumstances such as Minisuperspaces and perturbations thereabout (Appendix I) used in Part II.

One can furthermore quite readily envisage counterparts of each of the above types of geometry in this next setting, though we restrict mention to the ones that this book makes use of. Preserving the indefinite interval takes the place of preserving the Euclidean norm. This is attained by, firstly, translations $Tr(p, 1)$, which now include both the already-familiar *spatial* translations $Tr(p)$ and the *time translations* $t \rightarrow t + k$. Secondly, rotations alongside *boosts*—the tx component analogue of rotations with the indefiniteness resulting in these taking the form of a 'hyperbolic' rather than 'standard' rotation—form the *proper orthochronous Lorentz group* $SO^+(p, 1)$. The full Lorentz group $O(p, 1) := \{M \in GL(p + 1, \mathbb{R}) \,|\, M\eta M^{\mathrm{T}} = \mathbb{I}\}$ involves also reflections not only in space but also in time; removing each of these gives rise to the qualifiers 'proper' and 'orthochronous' respectively. Finally, the second and third items form together the (proper orthochronous) *Poincaré group* $SPoin^+(p + 1) = Tr(p, 1) \rtimes SO^+(p, 1)$, whereas the (full) Poincaré group ensues from considering all three together: $Poin(p + 1) = Tr(p, 1) \rtimes O(p, 1)$.

A first instance of this arises in the $p = 3$ case as the *kinematical* transformation group for Electromagnetism, and is subsequently taken to be the kinematical transformation group for all of Classical Physics bar Gravitation. In this way, it lies beyond what *Eucl*(d) models physically; one would need to adjoin time and velocity to obtain the Galilean group counterpart of that (Ex IV.15) [354].

Another feature of note is that Minkowski spacetime's null cones are preserved by conformal transformations; conformal transformations become causal transformations in the context of indefinite spacetime (Ex III.11).

B.3 Complex Transformations and Geometries

These are for use in various applications, starting with quantum symmetries. $\mathbb{R}^2$ can also be modelled as $\mathbb{C}$ and $\mathbb{R}^{2p}$ as $\mathbb{C}^p$. $GL(p, \mathbb{C})$ and $SL(p, \mathbb{C})$ are defined as expected. We also need *unitary transformations*: such that $U^\dagger U = \mathbb{I}$, which form the *unitary groups* $SU(p)$ *special unitary transformations*, for which $|\det U| = 1$ as well, which form the *special unitary groups* $U(p)$, and *antiunitary transformations*, for which $U^\dagger U = -\mathbb{I}$ instead.

Probably the best-known example of projective group is the *M öbius group* [815] $PGL(2, \mathbb{C})$ (after mathematician August Mobius). This acts upon $\mathbb{C} \cup \infty$ as the *fractional linear transformations*

$$z \longrightarrow \frac{az + b}{cz + d},$$

for $a, b, c, d \in \mathbb{C}$ such that $ad - bc \neq 0$. It is a 6-d group, due to there being one complex restriction on it: for $\lambda \in \mathbb{C}$,

$$\frac{\lambda az + \lambda b}{\lambda cz + \lambda d} = \frac{\lambda}{\lambda} \frac{az + b}{cz + d} = \frac{az + b}{cz + d}. \tag{B.10}$$

Appendix C
Basic Analysis

C.1 Real Analysis

The real numbers $\mathbb{R}$ possess the further *completeness property* . One way of expressing this is through the *Fundamental Axiom of Analysis*: if $x_n \in \mathbb{R}, n \geq 1$ is an increasing sequence and $x_n < B$ for some constant $B \in \mathbb{R}$ for each n, then there is an $x \in \mathbb{R}$ such that $x_n \to x$ as $n \to \infty$. I.e. every increasing sequence that is bounded above has a limit which also belongs to $\mathbb{R}$. Analysis on $\mathbb{R}$ can be built upon this property. On the other hand, e.g. the rational numbers $\mathbb{Q}$ do not have such a benevolent property [822]. Without this, it is much harder to find any kind of substitutes for results that require proving by Analysis (see below for a few more details and e.g. [184] for a systematic exposition).

Analysis often proceeds via an 'ϵ and δ' formulation of concepts, according to which for each ϵ, however small, one can find a corresponding δ of an adequate size for the property under consideration to hold. However, this in itself transcends to $\mathbb{Q}$ rather than being specific to complete spaces.

Completeness for $\mathbb{R}$ renders a number of further concepts useful, in which that do not apply in the absence of completeness such as for $\mathbb{Q}$. One furthermore usually uses a convenient implementation of the completeness property, such as the *Bolzano–Weierstrass Theorem* [184] by which every bounded sequence in $\mathbb{R}$ has a convergent subsequence.

Analysis, as a rigorous theory of limits, for instance brings about the demise of Zeno's paradoxes (outlined in the Introduction). Achilles and the tortoise cross each other at a limit point arrived at within finite time.

A sequence $\{x_n, n \in \mathbb{N}\}$ is a *Cauchy sequence* if $\forall\, \epsilon > 0\, \exists\, N$ such that $|x_p - x_q| < \epsilon\ \forall\, p, q > N$. Completeness can also be very usefully formulated in terms of every Cauchy sequence converging [184].

Many results in Analysis concern functions on some space, for now $f : \mathbb{R} \to \mathbb{R}$. A function f is *continuous* at some point $x \in \mathbb{R}$ if given $\epsilon > 0$, $\exists\, \delta(x, \epsilon)$ such that $|f(x) - f(y)| < \epsilon\ \forall\, |x - y| < \delta(\epsilon, x)$. Intuitively, this means that a function f is continuous if we can get $f(y)$ to be as close to $f(x)$ as we please just by letting y be sufficiently close to x; see Fig. C.1.a). $f : \mathbb{R} \to \mathbb{R}$ is *differentiable* at $x \in \mathbb{R}$ if

E. Anderson, *The Problem of Time*, Fundamental Theories of Physics 190,
DOI 10.1007/978-3-319-58848-3

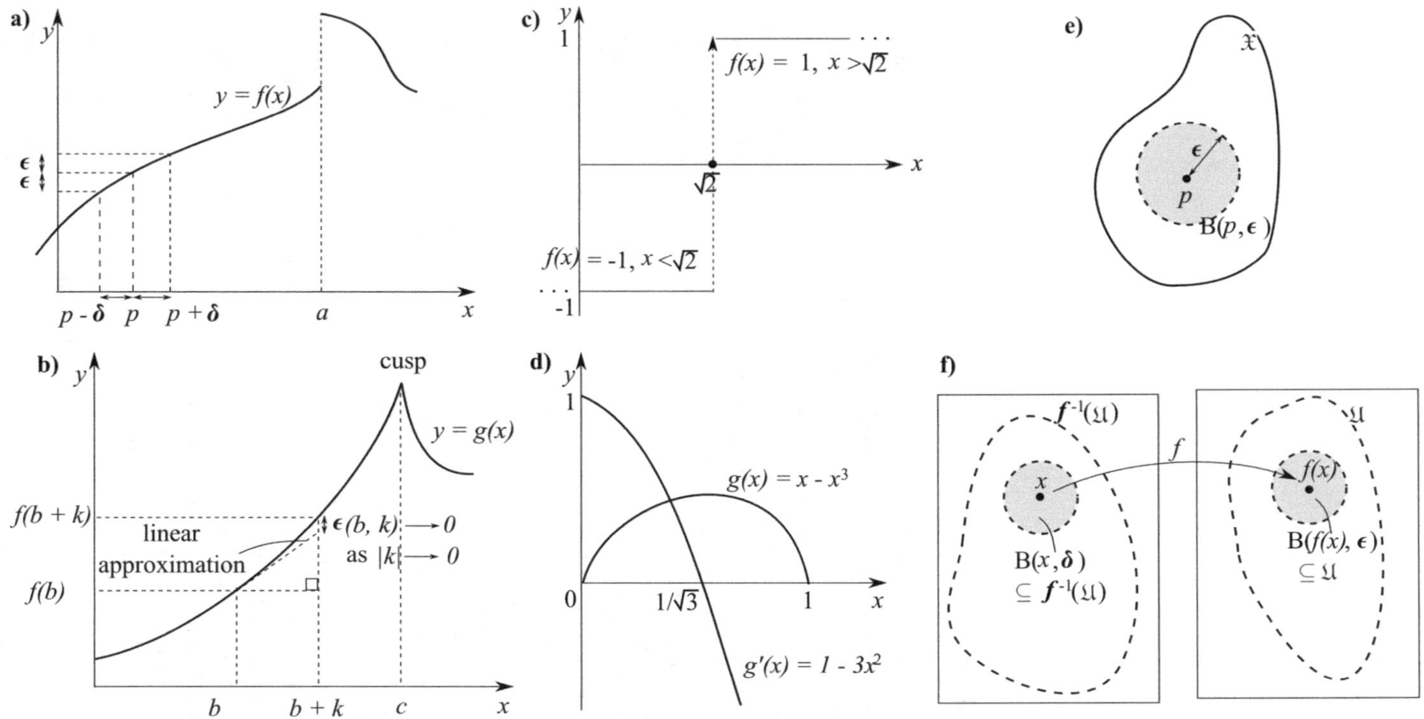

Fig. C.1 **a)** $y = f(x)$ is discontinuous at a as a function on $\mathbb{R}$, but not on $\mathbb{Q}$ if $a = \sqrt{2}$. **b)** A function is differentiable wherever it is locally linear, e.g. at $x = b$; this does not however apply e.g. at a cusp ($x = c$). **c)** A function that is continuous in $\mathbb{Q}$ which does not obey the Intermediate Value Theorem. For e.g. $f(0) = -1 < 0 = v < 1 = f(2)$ but there is no $c \in \mathbb{Q}$ such that $f(c) = 0$. **d)** A function that is differentiable in $\mathbb{Q}$ which does not obey the Mean Value Theorem [339]. For there is no $m \in \mathbb{Q}$ such that $g'(m) = \{g(1) - g(0)\}/\{1 - 0\} = 0$. **e)** An open ball (about p of radius ϵ in $\mathfrak{X} \subset \mathbb{R}^2$). **f)** Continuity in terms of the inverse image of an open set being open

there is a linear map $Df(x)$—the *derivative*—such that when $|h| <$ some δ,

$$f(x+k) = f(x) + Df(x)k + \epsilon(x,k)|k|,$$

where the error term decreases faster than linearly: $|\epsilon(x,k)| \to 0$ as $|k| \to 0$, as per Fig. C.1.b).

Subsequent Analysis theorems of note are [184] firstly the *Intermediate Value Theorem* concerning any continuous function $f : [a,b] \longrightarrow \mathbb{R}$; then for any v such that $f(a) < v < f(b)$, $\exists\, c \in (a,b)$ such that $f(c) = v$. Secondly, the *Mean Value Theorem* concerning any function $g : [a,b] \longrightarrow \mathbb{R}$ continuous on $[a,b]$ and differentiable on (a,b); then there is a point $m \in (a,b)$ such that the derivative

$$g'(m) = \frac{g(b) - g(a)}{b - a}.$$

N.B. that these theorems do not hold for functions $h : \mathbb{Q} \to \mathbb{Q}$; see e.g. Fig. C.1.c)–d). They do however readily extend to functions $k : \mathbb{R}^n \to \mathbb{R}^m$. The real modulus $|\ |$ is here extended to the Euclidean norm $||\ ||$, which furnishes a ball concept: Fig. C.1.e); open balls are convenient in many proofs.

A second definition of continuity of a function—which can indeed readily be proven equivalent to the above one: Ex III.5.iii)—is that f is continuous if the inverse image $f^{-1}(\mathfrak{U})$ of any open set $\mathfrak{U}$ is open. E.g. in the $f : \mathbb{R} \to \mathbb{R}$ case presented in Fig. C.1.f), the region in the domain $\mathbb{R}$ is open whenever the region $\mathfrak{U}$ in the codomain $\mathbb{R}$ is. The latter definition extends to considerably more general cases (see below).

This Sec's results furthermore readily generalize to $\mathbb{R}^n$; some further generalizations are in Sects. C.4 and C.6. Useful notions in this regard, for now presented at the level of $\mathbb{R}^n$, are as follows. A set $\mathfrak{U} \in \mathbb{R}^n$ is i) *open* if whenever $u \in \mathfrak{U}$ there is an $\epsilon(u) > 0$ such that $||u - v|| < \epsilon \Rightarrow v \in \mathfrak{U}$. ii) *Closed* if whenever the sequence of points $\{u_n, n \in \mathbb{N}\}$ lies in $\mathfrak{U}$ and $u_n \to u$, then this *limit point* u also lies in $\mathfrak{U}$.

C.2 Basic Functional Analysis

A *function space* is a space whose constituent points are themselves functions. Some simple examples are 1) the continuous functions $\mathfrak{C}^0$. 2) The once-differentiable functions $\mathfrak{C}^1$ and their further specialization to the k-fold differentiable functions $\mathfrak{C}^k$. 3) The smooth alias infinitely differentiable functions $\mathfrak{C}^\infty$. 4) Finally, the analytic functions $\mathfrak{C}^\omega$ are those that have convergent power series about each point.

Study of function spaces is *Functional Analysis*. Applications of this include infinite-dimensional spaces, PDEs, FDEs, and operators (including quantum ones) acting on function spaces. An example of function spaces that are well-known in Theoretical Physics are the infinite-d *Hilbert spaces*: infinite-d versions of vector spaces that are both equipped with an inner product and are complete. In Quantum Theory, this inner product is usually denoted by Dirac's bra-ket combination $\langle\psi_1|\psi_2\rangle$. See Appendices H.2 and P.5 for a bit more on Hilbert and other function spaces, and [207, 270, 729] if you have need of extensively learning this subject.

C.3 Complex Analysis*

Complex differentiability of a complex function f is equivalent to the *Cauchy–Riemann equations* for real functions $u(x, y)$, $v(x, y)$ such that $f = u + iv$:

$$\partial_x u - \partial_y v = 0, \qquad \partial_x v + \partial_y u = 0. \tag{C.1}$$

These are solved by complex-analytic functions $f(z$ alone) (as opposed to of the complex conjugate $\overline{z}$). Finally, in this case and in sharp contrast to Real Analysis, all functions which are differentiable once are furthermore infinitely differentiable. See e.g. [184, 726] for more about Complex Analysis.

C.4 Metric Spaces

A *metric space*—(see especially [184, 822]) is a set $\mathfrak{X}$ equipped with a *metric function* Dist: $\mathfrak{X} \times \mathfrak{X} \rightarrow \mathbb{R}$ satisfying the following properties.

Metric 1) $\text{Dist}(x, y) \geq 0 \; \forall \, x, y \in \mathfrak{X}$ (non-negativity).
Metric 2) If $\text{Dist}(x, y) = 0$, then $x = y$ (separation).
Metric 3) $\text{Dist}(x, y) = \text{Dist}(y, x)$ (symmetry).
Metric 4) $\text{Dist}(x, y) \leq \text{Dist}(x, z) + \text{Dist}(z, y)$ (triangle inequality).

These properties encapsulate features of the Euclidean notion of distance, which are now applied to a wider range of settings.

Dist—standing for 'distance between'—generalizes the Euclidean norm $\| \; \|$ of $\mathbb{R}^n$, and continues to support the concept of balls. Dist is *translation invariant* if the metric space possesses a '+' operation and $\text{Dist}(x + w, y + w) = \text{Dist}(x, y)$.

An *isometry*—in the metric space sense—is a Dist-preserving transformation, $\iota : \mathfrak{X} \rightarrow \mathfrak{X}$ such that $\text{Dist}(x, y) = \text{Dist}(\iota x, \iota y)$.

For metric spaces, one can still conceive of completeness (in terms of Cauchy sequences), and of continuity and differentiability (in terms of balls).

C.5 Inverse and Implicit Function Theorems*

A *Fixed Point Theorem* holds for metric spaces. The *Implicit Function Theorem*—concerning the existence of implicitly-defined functions—follows from this on metric spaces. The $\mathbb{R}^n \rightarrow \mathbb{R}^n$ *Inverse Function Theorem*—about the existence of an inverse f^{-1} for interior points within sets on which a function f is continuously differentiable, alongside provision of a formula for f^{-1}'s own derivative—then follows as a special case. This is because its inverse function condition is a subcase of implicit function. See e.g. [184] for proofs and further discussion.

C.6 Topological Spaces

These further generalize metric spaces while retaining many key notions of Analysis, such as convergence and continuity; see especially [822] for an introduction. Topological spaces are based on extending the notions of open and closed sets away from Appendix C.1's $\mathbb{R}^n$ context. $\mathbb{R}^n$'s open set notion obviously carries over to metric spaces, but further extension requires reconceptualization. Preliminarily, one associates a collection of subsets $\tau(\mathfrak{X}) = \langle \mathfrak{X}, \tau \rangle$ to a given fixed set $\mathfrak{X}$. The *open subsets* $\mathfrak{U}_{\text{O}}$ are a particular type of collection that is convenient for performing Analysis. *Topological spaces* [68, 613, 822] are collections of open subsets with the following properties.

Topological Space 1) $\mathfrak{X}, \emptyset \in \tau$.
Topological Space 2) The union of any collection of the $\mathfrak{U}_{\text{O}}$ is also in τ.
Topological Space 3) The intersection of any finite number of the $\mathfrak{U}_{\text{O}}$ is also in τ.

Closed sets in $\mathfrak{X}$ are then defined as the complements of sets which are open in $\mathfrak{X}$. The reader should now work through the start of Ex III.5 to establish that the definitions given for $\mathbb{R}^n$ indeed comply with these more general definitions.

As a slight detour [822] useful further on in these Appendices, a point y is a *point of closure* of $\mathfrak{Y} \subseteq \langle \mathfrak{X}, \tau \rangle$ if $\mathfrak{U} \cap \mathfrak{Y} \neq \emptyset$ for any open $\mathfrak{U} \subseteq \mathfrak{X}$ such that $y \in \mathfrak{U}$. The set of points of closure of $\mathfrak{Y}$ in $\mathfrak{X}$ is termed the *closure* of $\mathfrak{Y}$ in $\mathfrak{X}$, and is denoted by Clos($\mathfrak{Y}$). On the other hand, y is an *interior point* of $\mathfrak{Y} \subseteq \langle \mathfrak{X}, \tau \rangle$ if $\exists$ some $\mathfrak{U}$ open in $\mathfrak{X}$ such that $y \in \mathfrak{U} \subseteq \mathfrak{Y}$. The set of interior points, Int($\mathfrak{Y}$), of $\mathfrak{Y}$ in $\mathfrak{X}$ is known as the *interior* of $\mathfrak{Y}$. The *frontier* alias *boundary* of $\mathfrak{Y} \subseteq \langle \mathfrak{X}, \tau \rangle$ is Clos($\mathfrak{Y}$)\Int($\mathfrak{Y}$). $\mathfrak{Y}$ is *dense* in $\mathfrak{X}$ if Clos($\mathfrak{Y}$) $= \mathfrak{X}$ [207].

The open sets version of the definition of continuity remains meaningful in topological spaces' more general setting. Moreover, a *homeomorphism* is a continuous bijection with continuous inverse, i.e. the 'open image of an open set' version of the definition of continuity carries over to topological spaces. *Topological properties* are those properties which are preserved by homeomorphisms; the rest of this Chapter lays out a number of topological properties, supported by Exs III.8 and 9.

Notions of *separation* are topological properties which indeed involve separating two objects (points, certain kinds of subsets) by encasing each in a disjoint subset. A particular such is the *Hausdorffness* property [613, 822] (after mathematician Felix Hausdorff):

$$\text{for } x, y \in \mathfrak{X}, x \neq y, \exists \text{ open sets } \mathfrak{U}_x, \mathfrak{U}_y \in \tau$$
$$\text{such that } x \in \mathfrak{U}_x,\ y \in \mathfrak{U}_y \text{ and } \mathfrak{U}_x \cap \mathfrak{U}_y = \emptyset. \tag{C.2}$$

So this case involves separating points by open sets. Hausdorffness allows for each point to have a neighbourhood without stopping any other point from having one. This is a property of the real numbers, and is additionally permissive of much Analysis. In particular, Hausdorffness secures uniqueness for limits of sequences. Moreover, it extends to how compact sets can be separated by open neighbourhoods, so in Hausdorff spaces 'compact sets behave like points'. Examples in this book that are *not* Hausdorff are the trousers topology and the related branching in Fig. 1.2.e), and the general quotients of Appendix M.

If $\mathfrak{U}_1, \mathfrak{U}_2$ are open sets such that

i) $\mathfrak{U}_1 \cap \mathfrak{U}_2 = \emptyset$,
ii) $\mathfrak{U}_1 \cup \mathfrak{U}_2 = \mathfrak{X}$, and
iii) neither $\mathfrak{U}_1$ nor $\mathfrak{U}_2$ are $\emptyset$,

then $\mathfrak{U}_1, \mathfrak{U}_2$ *disconnect* $\mathfrak{X}$. If $\mathfrak{X}$ is not disconnected by any two sets, $\mathfrak{X}$ is *connected* [68, 613, 822]. This notion is motivated by considering how far the Intermediate Value Theorem can be generalized. On the other hand, $\mathfrak{X}$ is *path-connected* if for $x, y \in \mathfrak{X}$, $\exists$ a path γ from x to y. Finally, $\mathfrak{X}$ is *simply-connected* if it is path-connected and all continuous paths between two given points can be continuously transformed into each other (without leaving $\mathfrak{X}$).

Some notions of countability are concurrently topological properties, due to involving counting of topologically defined entities [613]. *First countability* holds if for each $x \in \mathfrak{X}$, there is a countable collection of open sets such that every open neighbourhood $\mathfrak{N}_x$ of x contains at least one member of this collection. *Second countability* is the stronger condition that there is a countable collection of open sets such that every open set can be expressed as union of sets in this collection. Second-countability is also useful via being a property standardly attributed to manifolds.

A collection of open sets $\{\mathfrak{U}_\mathsf{C}\}$ is an *open cover* for $\mathfrak{X}$ if $\mathfrak{X} = \bigcup_\mathsf{C} \mathfrak{U}_\mathsf{C}$. A subcollection of an open cover that is still an open cover is termed a *subcover*, $\{\mathfrak{V}_\mathsf{D}\}$ for D a subset of the indexing set C. On the other hand, an open cover $\{\mathfrak{V}_\mathsf{D}\}$ is a *refinement* of $\{\mathfrak{U}_\mathsf{C}\}$ if to each $\mathfrak{V}_\mathsf{D}$ there corresponds a $\mathfrak{U}_\mathsf{C}$ such that $\mathfrak{V}_\mathsf{D} \subset \mathfrak{U}_\mathsf{C}$. $\{\mathfrak{V}_\mathsf{D}\}$ is furthermore *locally finite* if each $x \in \mathfrak{X}$ has an open neighbourhood $\mathfrak{N}_x$ such that only finitely many $\mathfrak{V}_\mathsf{D}$ obey $\mathfrak{N}_x \cup \mathfrak{V}_\mathsf{D} \neq \emptyset$.

A topological space $\tau(\mathfrak{X})$ is *compact* [68, 613, 822] if every open cover of $\mathfrak{X}$ has a finite subcover. Compactness generalizes continuous functions being bounded on a closed interval of $\mathbb{R}$.

A topological space $\tau(\mathfrak{X})$ is *paracompact* [613] if every open cover of $\mathfrak{X}$ has a locally finite refinement.

A topological space $\tau(\mathfrak{X})$ has *topological dimension* [672] p if every open cover of $\mathfrak{X}$ has an open refinement such that no point lies in more than $p+1$ subsets. The topological dimension of $\mathbb{R}^p$ is indeed p, so $\mathbb{R}^p$ is not homeomorphic to $\mathbb{R}^q$ for $p \neq q$.

A *topological group* [207, 613] is a set equipped with both a topology and a group operation such that the composition and inverse operations are continuous. Lie groups (Appendix E) are a major example of this. A *topological vector space* [207] is a set equipped with both a topology and the vector space operations such that the addition and scalar multiplication are continuous. Such include some very useful function spaces, as per Appendix H.2.

There is fascinating interplay between topological properties: many combinations of these imply other a priori unrelated properties [672, 822]. Except where explicitly stated, we henceforth assume Hausdorffness and second-countability.

Appendix D
Manifold Geometry

D.1 Topological Manifolds

Passing from topological spaces to topological manifolds is a specialization (rather than an equipping), and is widely useful in Physics and in Mathematics. A topological space $\langle \mathfrak{X}, \tau \rangle$ is *locally Euclidean* if every point $x \in \mathfrak{X}$ has a neighbourhood $\mathfrak{N}_x$ that is homeomorphic to $\mathbb{R}^p$: Euclidean space.[1] I give an arbitrary dimension p treatment of this with coordinates x^{A}, so as to cover a wide range of applications: spacetime (Chap. 7), space (Chaps. 1 and 8) and auxiliary spaces from the Principles of Dynamics (Appendices G to L). $\langle \mathfrak{X}, \tau \rangle$ is a *real topological manifold* if it obeys

Topological Manifold 1) local Euclideanness,
Topological Manifold 2) Hausdorffness, and
Topological Manifold 3) second-countability.

See in particular [613]. I denote the general manifold by $\mathfrak{M}$, and term the above trio of topological space properties 'manifoldness'; moreover, this trio implies paracompactness as well [613]. Second countability ensures sequences suffice to probe most topological properties, whereas Hausdorffness ensures that neighbourhoods retain many of the intuitive properties of their metric space counterparts. In these ways, much of Analysis can be carried over to manifolds.

Riemann's trick for study of manifolds involves introducing the notion of *chart* alias *local coordinate system* for $\mathfrak{M}$: an injective map $\phi : \mathfrak{U} \rightarrow \phi(\mathfrak{U}) \subset \mathbb{R}^n$ for $\mathfrak{U}$ an open subset of $\mathfrak{m}$. Each chart does not in general cover the whole manifold; one gets around this by considering a suitable collection of charts. These serve as homeomorphisms which guarantee the locally Euclidean property. One can loosely think of these as 'deformations of a rubber sheet', with continuous stretching but no guarantee of smoothness. Appendix D.2 then concerns adding in a further level of structure to model the smoothness. One is to compare those charts which overlap,

[1] A topological manifold's topological dimension indeed turns out to be p as well, though this is not straightforward to prove; see [672] for a start on this.

E. Anderson, *The Problem of Time*, Fundamental Theories of Physics 190,
DOI 10.1007/978-3-319-58848-3

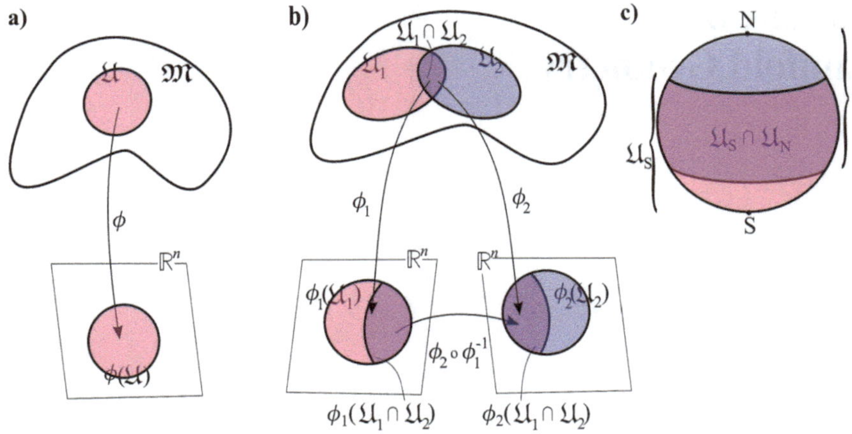

Fig. D.1 **a)** A chart. **b)** Overlapping charts and transition functions. **c)** A 2-chart cover for $\mathbb{S}^2$: from N to the *lower curve* and from S to the *upper curve*

leading to the 2-chart Fig. D.1.b), with $\phi_1 : \mathfrak{U}_1 \to \mathbb{R}^n$, $\phi_2 : \mathfrak{U}_2 \to \mathbb{R}^n$ which do indeed overlap: $\mathfrak{U}_1 \cup \mathfrak{U}_2 \neq 0$. Next, consider a composite map

$$t_{12} := \phi_2 \circ \phi_1^{-1} \tag{D.1}$$

which sends $\mathfrak{U}_1 \cup \mathfrak{U}_2$ to itself. This is a locally defined map of $\mathbb{R}^n \to \mathbb{R}^n$; it is a local coordinate transformation, and is called a *transition functions*. An *atlas* for a topological manifold is a collection of charts that, between them, cover the whole manifold.

Example 1) For some basic intuitions, consider the well-known case of the 2-sphere $\mathbb{S}^2$. No matter which type of flat map of the world one peruses, it is highly distorted in some places. This is a manifestation of multiple charts being required to cover the whole of $\mathbb{S}^2$ (Fig. D.1.c). Moreover, there is interest in comparing charts which partly overlap; atlases (in the colloquial sense) contain multiple part-overlapping charts.

Example 2) Consider the 2-manifolds more generally; these form a simple model arena for which a well-known classification of topological manifolds exists. This is in terms of 1) *genus*: the number of handles. 2) *Non-orientable genus*: number of non-orientability twists. At the topological manifold level, orientability concerns a consistent allocation of handedness of rotation for each simplex in a simplicial complex. For now, we give examples: $\mathbb{S}^2$ and the 2-torus $\mathbb{T}^2$ are orientable with genus 0 and 1 respectively. On the other hand, the 2-*d* real projective space $\mathbb{RP}^2 =$ {set of lines through a point in $\mathbb{R}^3$} has non-orientable genus 1.

Moreover, by the rubber sheet property being all at this stage, it makes no difference whether the handles are 'large' or 'small' or 'near' or 'far'. For those are metric concepts that the current level of structure does not possess.

Example 3) Topological manifolds of dimension ≥ 3 are harder to study in general, with some sensitivities to particular dimensions. Each of dimension 3 and ≥ 4 have rather different mathematical properties at the topological level [848, 876]. Knots (Appendix N.13) also specifically require dimension 3.

Example 4) Some spaces come in series within each of which dimension makes little difference to the properties and presentation of the space. E.g. p-spheres $\mathbb{S}^p$ and p-tori $\mathbb{T}^p$ form such series. As well as the above 2-d models, these include $\mathbb{S}^3$ and $\mathbb{T}^3$, which are each in some senses the simplest 3-d topological manifolds that are compact without boundary, for use as notions of space in GR.

A *manifold with boundary* [606, 614] is locally homeomorphic to some open set in the half-space $\{(x_1, \dots, x_p) \in \mathbb{R}^p \,|\, x_p \geq 0\}$. Charts ending on the half-space's boundaries are describing part of the manifold that is adjacent to its boundary (Fig. 37.5.a). See Appendix M for a substantial generalization of this. Chapter 8's compact without boundary notion makes sense here; moreover, this book takes this to imply connectedness as well as a default.

D.2 Differentiable Manifolds

Charts can furthermore allow for one to tap into the standard $\mathbb{R}^p \longrightarrow \mathbb{R}^q$ Calculus. This allows for manifolds to be equipped with *differentiable structure*. These manifolds possess not only a local differentiable structure in each coordinate patch $\mathfrak{U}_i$ but a notion of global differentiable structure as well. This is due to the '*meshing condition*' on the coordinate patch overlaps (Fig. D.1.b). The transition functions can in this case be interpreted in terms of Jacobian matrices of derivatives for one local coordinate system with respect to another,

$$\mathrm{L}^{\mathsf{A}}{}_{\mathsf{B}} = \frac{\partial(x^{\mathsf{A}})}{\partial(\bar{x}^{\mathsf{B}})}. \tag{D.2}$$

The great mathematician Hassler Whitney [903] showed that the topological manifold notion of atlas can additionally be equipped with differentiable structure. Moreover, our main interest here is really in equivalence classes of atlases. Differentiable structure is then approached using a convenient small atlas [such as in Fig. D.1.d)'s 2-chart approach to the 2-sphere]. In contrast to the previous Section's atlas being $\mathfrak{C}^0$ (the continuous functions) the current Sec's is usually taken to be $\mathfrak{C}^\infty$: the smooth functions. In fact, weakening $\mathfrak{C}^\infty$ to $\mathfrak{C}^k$ $k \geq 1$ makes little difference, since each such differentiable structure is uniquely smoothable [903]. Having Calculus available throughout the manifold, moreover, allows on to study *differential equations* which conventionally represent physical law.

We next introduce *vectors* on the manifold as the tangents to *curves* (mappings $\mathfrak{I} \rightarrow \mathfrak{M}$ for $\mathfrak{I}$ an interval of $\mathbb{R}$: Fig. D.2.b). One can furthermore compose curve and chart maps to make use of standard $\mathbb{R}^p \rightarrow \mathbb{R}^q$ Calculus; moreover one can show straightforwardly that all notions involved are chart-independent.

At this level of structure *orientability* can be taken to involve a continuous assignation of normal vectors. It can be treated in terms of the Jacobians in each sequence

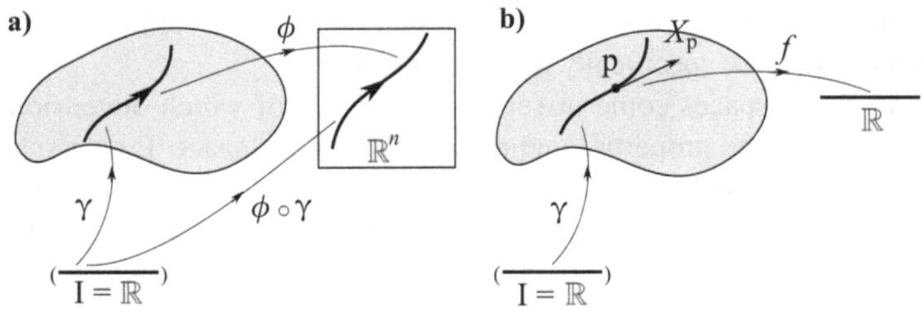

Fig. D.2 **a)** The curve construct. **b)** Basing a notion of vector on the curve construct

of charts that runs across the space (Fig. D.2.a). An orientable differentiable manifold possesses a coordinate atlas all of whose transition functions have positive Jacobians; this consistently maintains orientability throughout the manifold.

We can furthermore apply [814] the basic machinery of (multi)linear algebra to produce notions of 1) covectors as the duals of vectors. At a point p on the manifold, this is a linear map $\mathfrak{T}_p(\mathfrak{M}) \longrightarrow \mathbb{R}$ where $\mathfrak{T}_p(\mathfrak{M})$ is the tangent space at p, a *vector* at p is a linear map $\mathfrak{T}^*_p(\mathfrak{M}) \longrightarrow \mathbb{R}$ where $\mathfrak{T}^*_p(\mathfrak{M})$ is the cotangent space at p (the dual of the tangent space). 2) All the higher-rank tensors on the manifold: *rank (k, l) tensors* at p are the multilinear maps from the product of k copies of $\mathfrak{T}^*_p(\mathfrak{M})$ and l copies of $\mathfrak{T}_p(\mathfrak{M})$ to $\mathbb{R}$. A collection of vectors, one at each $\mathrm{p} \in \mathfrak{M}$, constitutes a *vector field* over $\mathfrak{M}$; *tensor fields* are similarly defined. The more old-fashioned formulation of the definition of a (k, l) tensor (cf. Chaps. 2 and 4) is not in the above 'coordinate-free' language but rather in terms of components. These transform according to

$$\mathrm{T}^{\bar{\mathsf{A}}_1 \ldots \bar{\mathsf{A}}_k}{}_{\bar{\mathsf{B}}_1 \ldots \bar{\mathsf{B}}_l} = \mathrm{L}^{\bar{\mathsf{A}}_1}{}_{\mathsf{A}_1} \ldots \mathrm{L}^{\bar{\mathsf{A}}_k}{}_{\mathsf{A}_k} \mathrm{L}^{\mathsf{B}_1}{}_{\bar{\mathsf{B}}_1} \ldots \mathrm{L}^{\mathsf{B}_l}{}_{\bar{\mathsf{B}}_l} \mathrm{T}^{\mathsf{A}_1 \ldots \mathsf{A}_k}{}_{\mathsf{B}_1 \ldots \mathsf{B}_l} \tag{D.3}$$

in passing between barred and plain coordinate systems; (2.20) is a simple subcase of this.

As a further example, *form fields* [316] are the often encountered totally antisymmetric downstairs-index subcase of tensors. p-volume is the *top form* that a p-space can support. Areas $(\underline{x} \times \underline{y})_3$ in 2-d and scalar triple product volumes $[\underline{x} \times \underline{y} \cdot \underline{z}]$ in 3-d both generalize to $\bigwedge_{I=1}^{p} x_I$ in the language of forms. Finally, a *pseudotensor*'s transformation law additionally features the sign of the determinant of $\mathrm{L}^{\bar{\mathsf{A}}_1}{}_{\mathsf{A}_1}$. The most commonly encountered such in Physics are pseudovectors alias axial vectors, which arise from cross products.

Diffeomorphisms A *diffeomorphism* (see in particular [614]) is a map $\phi : \mathfrak{M} \rightarrow \mathfrak{M}$ that is injective, $\mathfrak{C}^\infty$ (or a bit rougher, e.g. $\mathfrak{C}^k$), and has a an inverse map of matching minimal standard of differentiability. In fact, in 1- to 3-d all homeomorphic smooth manifolds are additionally diffeomorphic. However, in $d \geq 4$, there are examples of pairs that are homeomorphic but not diffeomorphic, including the famous exotic spheres (this is a much harder piece of mathematics; see e.g. [570]).

The *passive* notion of diffeomorphism concerns the coordinate transformation on overlap between two charts, corresponding to the Jacobian matrix from x^{A} around the point p to y^A around $\phi(p)$. This involves representations of the same objects in different coordinate systems. On the other hand, the *active* notion of diffeomorphism involves associating each point in a manifold with another through, at the infinitesimal level, the Lie derivative (see below). This relates different entities on $\mathfrak{M}$ within the one coordinate system. Active and passive diffeomorphisms are *mathematically* equivalent [874]. However, the two differ in physical significance, as per Chap. 10.

Integral Curves An *integral curve* (see e.g. [814]) of a vector field V in a manifold $\mathfrak{M}$ is a curve $\gamma(\nu)$ such that the tangent vector is V_p at each p on γ. It is *complete* if $\gamma(\nu)$ is defined $\forall\, \nu \in \mathbb{R}$. A set of complete integral curves corresponding to a non-vanishing vector field is called a *congruence*. This 'fills' a manifold or region therein upon which the vector field is non-vanishing: the curves go through all points therein.

Lie Derivatives Physics makes plentiful use of *derivatives* acting on vector fields. Such are not straightforward to set up; the flat-space derivatives that one is accustomed to entail taking the limit of the difference between vectors at different points. However, in the context of differentiable manifolds, such vectors belong to different tangent spaces. Whereas in $\mathbb{R}$ one can just move the vectors to the same point, there is no direct counterpart of this procedure on a general manifold (cf. Fig. D.2.c). The usual partial derivation is undesirable since it does not preserve *tensoriality*: the mapping of tensors to tensors. On the other hand, it suffices to construct such a notion of derivatives acting on vectors and acting trivially on scalars. This is because the derivative's action on all the other tensors can then be found by application of the Leibniz rule.

The *Lie derivative* is tensorial, and directional in the sense of involving an additional vector field ξ^i along which the tensors are dragged. It is denoted by $\pounds_{\underline{\xi}}$. Its underlying dragging first principles are worth outlining next; see e.g. [207, 490, 814] for further details in this regard.

There is a map construction between manifolds: $\phi : \mathfrak{M} \rightarrow \mathfrak{N}$, though in this book we concentrate on the $\mathfrak{M} \rightarrow \mathfrak{M}$ case, as per Fig. D.3.a). This induces a *push-forward* $\phi_* : \mathfrak{T}_p(\mathfrak{M}) \rightarrow \mathfrak{T}_{\phi(p)}(\mathfrak{M})$ which maps the tangent vector to a curve γ at p to that at the image of the curve $\phi(\gamma)$ at $\phi(p)$. There is a corresponding *pull-back* ϕ^* : $\mathfrak{T}^*_{\phi(p)}(\mathfrak{M}) \rightarrow \mathfrak{T}^*_p(\mathfrak{M})$ which maps 1-forms in the opposite direction. Finally, $\phi_*\mathbf{T} = \mathbf{T}$ defines a *symmetry* for the general tensor $\mathbf{T}$.

First-principles considerations give the actions of Lie derivation on scalars and vectors as the first equalities below; if required, consult e.g. [814] as regards passage to the second 'computational' forms. For γ the integral curve of $\underline{\xi}$ through p inducing a 1-parameter group of transformations ϕ_ν, the Lie derivative with respect to $\underline{\xi}$ at p of a scalar S is

$$\{\pounds_{\underline{\xi}} S\}_p = \lim_{d\nu \rightarrow 0} \left(\frac{S_{\phi_{d\nu}(p)} - S_p}{d\nu} \right) = \{\xi^A \partial_A S\}_p. \tag{D.4}$$

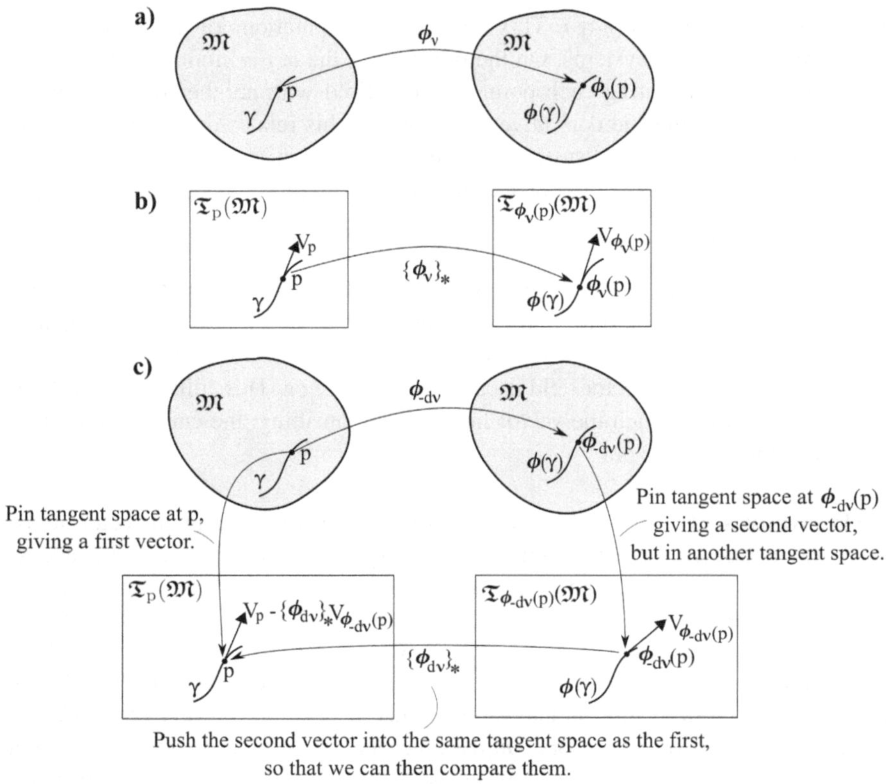

Fig. D.3 **a)** Effect on a curve of mapping a manifold to itself. **b)** The associated induced push–forward on the tangent vector to the curve. **c)** Decomposition of the first-principles construction of the Lie derivative of a vector

For a vector V, it is

$$\{\pounds_{\underline{\xi}} V^{A}\}_{p} = \lim_{d\nu \to 0} \left(\frac{V^{A}_{p} - \{\phi_{d\nu}\}_{*} V^{A}_{\phi_{-d\nu}(p))}}{d\nu} \right) = \{\xi^{B} \partial_{B} V^{A} - \partial_{B} \xi^{A} V^{B}\}_{p}. \tag{D.5}$$

One can then readily obtain the Lie derivatives for tensors of all the other ranks by use of Leibniz's rule (Ex III.16).

D.3 Affine Differential Geometry

To have a non-directional tensorial derivative, one can correct the partial derivative's non-tensoriality by introducing an extra structure: the *affine connection*. This is another non-tensorial object with components denoted by $\Gamma^{A}{}_{BC}$, transforming as

$$\Gamma^{\bar{A}}{}_{\bar{B}\bar{C}} = L^{\bar{A}}_{A} L^{B}{}_{\bar{B}} L^{C}{}_{\bar{C}}\, \Gamma^{A}{}_{BC} + L^{\bar{A}}{}_{A_1} L^{B}{}_{\bar{B}}\, \partial_{B} L^{A}{}_{\bar{C}}. \tag{D.6}$$

By this, the non-tensorial part of its transformation law compensates for that of the partial derivative. The *covariant derivative* is the tensorial derivative obtained in this manner; moreover, this now maps tensors to tensors with downstairs rank increased by one. Let us denote the covariant derivative by $\mathscr{D}_\mathsf{A}$ in general, by $\mathcal{D}_i$ in the spatial version and by ∇_μ in the spacetime version. It is just the partial derivative when acting on scalars, but takes the form

$$\mathscr{D}_\mathsf{A} v^\mathsf{B} = \partial_\mathsf{A} v^\mathsf{B} + \Gamma^\mathsf{A}{}_{\mathsf{BC}} v^\mathsf{C} \tag{D.7}$$

when acting on vectors. For later use, let us note that in cases in which a manifold happens to possess affine structure, its Lie derivatives can be recast in terms of covariant derivatives. E.g. in the case of vectors,

$$\pounds_{\underline{\xi}} V^\mathsf{A} = \xi^\mathsf{B} \mathscr{D}_\mathsf{A} V^\mathsf{A} - \mathscr{D}_\mathsf{A} \xi^\mathsf{A} V^\mathsf{B}. \tag{D.8}$$

The affine connection may be interpreted as giving a notion of straightest possible transport of vectors along curves: *parallel transport*. This holds true to the spirit of Affine Flat Geometry being developed with parallelism at the forefront.

In Affine Geometry, the ‘locally straightest paths’ or *affine geodesics* may be parametrized in terms of some ν so as to have the form

$$\mathscr{D}\dot{x}^\mathsf{C}/\mathscr{D}\nu := \dot{x}^\mathsf{A}\mathscr{D}_\mathsf{A}\dot{x}^\mathsf{C} = \ddot{x}^\mathsf{C} + \Gamma^\mathsf{C}{}_{\mathsf{AB}}\dot{x}^\mathsf{A}\dot{x}^\mathsf{B} = 0; \tag{D.9}$$

here $\dot{}$ denotes $\partial/\partial\nu$. If one passes to a parameter $\mu(\nu)$, the above *geodesic equation* transforms to

$$\dot{\mu}^2 x'^\mathsf{A}\mathscr{D}_\mathsf{A} x'^\mathsf{C} = \dot{\mu}^2\{x''^\mathsf{C} + \Gamma^\mathsf{C}{}_{\mathsf{AB}} x'^\mathsf{A} x'^\mathsf{B}\} = -\ddot{\mu}\, x'^\mathsf{C}, \tag{D.10}$$

where $'$ denotes $\partial/\partial\mu$. This is the *non-affine form* of the geodesic equation; the standard form is preserved iff $\ddot{\mu} = 0$, i.e. under the linear transformations $\mu = A\nu + B$. This is how Chap. 1’s freedom to choose tick-duration and calendar year zero for time is retained in geometrical theories.

The following consequence of the non-tensorial transformation law of the so-called affine connection is of importance for the foundations of GR. For each point, a set of ‘normal coordinates’ can be found in which the affine connection is zero at that particular point [814, 874]; the reader is encouraged to come up with a derivation of this (Ex III.12) prior to consulting the given references.

Moreover, parallel transport along two paths generally depends on the order in which the two paths are traversed (Fig. D.4.a). A combination of derivatives and squares of the affine connection—the *Riemann curvature tensor*— can be associated with this property of the transport of a vector W^A. In terms of components with respect to the usually employed (‘coordinate-induced’ [814]) basis, this takes the form

$$\mathscr{R}^\mathsf{A}{}_{\mathsf{BCD}} := \partial_\mathsf{C}\,\Gamma^\mathsf{A}{}_{\mathsf{BD}} - \partial_\mathsf{D}\,\Gamma^\mathsf{A}{}_{\mathsf{BD}} + \Gamma^\mathsf{E}{}_{\mathsf{BD}}\Gamma^\mathsf{A}{}_{\mathsf{EC}} - \Gamma^\mathsf{E}{}_{\mathsf{BC}}\Gamma^\mathsf{A}{}_{\mathsf{ED}}. \tag{D.11}$$

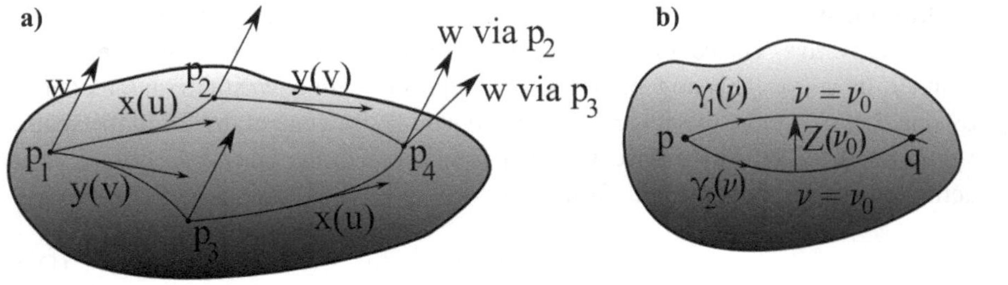

Fig. D.4 **a)** Two different ways of transporting a vector w^{A}. u, v parametrize 2 arbitrary curves with tangents x^{A} and y^{A} respectively. **b)** For two nearby geodesics γ_1, γ_2 in a congruence, each parametrized by λ, the *connecting*—alias *Jacobi—vector field* Z^{A} is the tangent to the curve connecting equal-λ points. In the case depicted, additionally p and q are conjugate points

This object can be arrived at by considering

$$\mathscr{R}^{\mathsf{A}}{}_{\mathsf{BCD}}w^{\mathsf{B}}x^{\mathsf{C}}y^{\mathsf{D}} = \lim_{\Delta\mu,\Delta\nu\longrightarrow 0}\left(\frac{\Delta \mathrm{W}^{\mathsf{A}}}{\Delta\mu\Delta\nu}\right). \tag{D.12}$$

It also follows from the '*Ricci Lemma*' (after mathematician Gregorio Ricci)

$$2\mathscr{D}_{[\mathsf{A}}\mathscr{D}_{\mathsf{B}]}\mathrm{W}^{\mathsf{C}} = \mathscr{R}^{\mathsf{C}}{}_{\mathsf{DAB}}\mathrm{W}^{\mathsf{D}}. \tag{D.13}$$

Finally, if one contemplates two neighbouring geodesics with initially-parallel tangent vectors $\dot{x}^{\mathsf{A}}$, one can arrive at the Riemann curvature tensor by considering the relative acceleration

$$\mathcal{D}^2\mathrm{Z}^{\mathsf{A}}/\mathcal{D}\nu^2$$

of two neighbouring geodesics with connecting vector Z^{A} (Fig. D.4.b). This gives the so-called *geodesic deviation equation*

$$\mathscr{D}^2\mathrm{Z}^{\mathsf{A}}/\mathscr{D}\nu^2 = -\mathscr{R}^{\mathsf{A}}{}_{\mathsf{BCD}}\dot{x}^{\mathsf{B}}\dot{x}^{\mathsf{C}}\mathrm{Z}^{\mathsf{D}}. \tag{D.14}$$

Geodesic deviation is moreover a nonlocal effect (a non-negligibly sized neighbourhood is required for this effect to manifest itself). *Conjugate points* are pairs of points p, q which are linked by a nonzero Jacobi field existing that vanishes at both p and q.

A more general and coordinate-independent definition of the Riemann curvature tensor is (see e.g. [814])

$$\mathscr{R}(\mathrm{X},\mathrm{Y})\mathrm{W} := \mathscr{D}_{\mathrm{Y}}\mathscr{D}_{\mathrm{X}}\mathrm{W} - \mathscr{D}_{\mathrm{X}}\mathscr{D}_{\mathrm{Y}}\mathrm{W} + \mathscr{D}_{[\mathrm{X},\mathrm{Y}]}\mathrm{W} \tag{D.15}$$

In fact (Ex III.14), it is necessary to exclude *torsion*

$$\mathscr{T}(\mathrm{Y},\mathrm{Z}) = \mathscr{D}_{\mathrm{Z}}\mathrm{Y} - \mathscr{D}_{\mathrm{Y}}\mathrm{Z} - [\mathrm{Y},\mathrm{Z}] \tag{D.16}$$

with coordinate adapted basis components $\mathscr{T}^{\mathsf{A}}{}_{\mathsf{BC}} := \Gamma^{\mathsf{A}}{}_{[\mathsf{BC}]} = 0$ in order for it to necessarily be curvature that underlies the symptoms of formulae (D.12)–(D.14).

The Riemann tensor obeys $\mathscr{R}^{\mathsf{A}}{}_{\mathsf{BCD}} = -\mathscr{R}^{\mathsf{A}}{}_{\mathsf{BDC}}$ and the *(first) Bianchi identity* (after mathematician Luigi Bianchi) $\mathscr{R}^{\mathsf{A}}{}_{[\mathsf{BCD}]} = 0$, whereas its derivatives are related by the *(second) Bianchi identity*:

$$\mathscr{D}_{[\mathsf{E}|}\mathscr{R}^{\mathsf{A}}{}_{\mathsf{B}|\mathsf{CD}]} = 0. \tag{D.17}$$

Finally, the *Ricci tensor* is $\mathscr{R}_{\mathsf{BD}} := \mathscr{R}^{\mathsf{A}}{}_{\mathsf{BAD}}$; it is symmetric: $\mathscr{R}_{\mathsf{AB}} = \mathscr{R}_{\mathsf{BA}}$.

D.4 (Semi)Riemannian Manifolds

Riemannian Geometry [207, 814, 874] involves a further encoding of notions of distance and angle, which Affine Geometry does not possess. The *metric tensor* m_{AB} is the structure brought in for this purpose. This is usually taken to be symmetric, non-degenerate and a function of the coordinates alone. In this case, $\langle \mathfrak{M}, \mathbf{m} \rangle$ is a *Riemannian manifold.* Genuine *Riemannian* metrics are those which are positive-definite; those which are indefinite are termed *semi-Riemannian.* [These are also known as *Euclidean-* and *Lorentzian-signature* metrics respectively.] Moreover, semi-Riemannian Geometry loses the separation axiom of distance due to its metric being indefinite. In the spacetime application this is desirable to model how observers perform measurements of both length and time. This indefinite case also splits the orientability concept into separate time and space orientability concepts.

The geometrical interpretation of the Riemannian metric function is that, through this, the metric defines the *length* along a path γ with tangent $\dot{x}^{\mathsf{A}}$ by

$$\text{length} := \int_{\nu_1}^{\nu_2} \sqrt{\mathrm{m}_{\mathsf{AB}}\dot{x}^{\mathsf{A}}\dot{x}^{\mathsf{B}}}\mathrm{d}\nu = \int_{\nu_1}^{\nu_2} \|\dot{\boldsymbol{x}}\|_{\mathbf{m}}\mathrm{d}\nu = \int_{\gamma} \|\mathrm{d}\boldsymbol{x}\|_{\mathbf{m}}.$$

This is a local rendition of the Euclidean notion of distance (B.1). It amounts also to the recovery of a metric in the metric space sense, in the form of a *path metric*, which is the inf over all the geodesics joining the two points in question. N.B. also that 'length' and 'metric space notion of metric' above strictly refer to the positive-definite case. One can now use the $(\cdot)$ to $(\cdot)_{\mathbf{m}}$ version of definitions (B.2)–(B.5). Some applications involve an infinitesimal version rather than one which is integrated over a finite range, as in e.g. 'local angle' in the (semi-)Riemannian geometrical sense.

Examples of positive-definite metrics include 1) the flat metric $\boldsymbol{\delta}$ on $\mathbb{R}^n$, 2) the (hyper)spherical metric (G.7) on $\mathbb{S}^n$, and 3) the Fubini–Study metric (G.8) on $\mathbb{CP}^n$. On the other hand, examples of indefinite metrics include 4) the Minkowski metric $\boldsymbol{\eta}$ of SR the 5) FLRW, 6) Schwarzschild and 7) Kerr–Newman metrics of GR (Chap. 7), and 8) the Minisuperspace metrics of Appendix I.1. Appendix H furthermore considers infinite-dimensional examples.

In studying metric geometries, it can help to cast the metric in block-minimal form. Whereas diagonal is the simplest example of block-minimal, not all metrics are, however, diagonalizable. E.g. the Kerr–Newman metric, the Fubini–Study metric, the non-diagonal Bianchi IX metric [812] and the modewise slightly inhomogeneous metric (Fig. I.2) are not.

The metric and its inverse may be used to lower and raise indices on other tensors. We denote the determinant of the metric by m and its inverse by n^{AB}.

(Semi-)Riemannian Geometry has a *metric connection* (Christoffel symbol, after mathematician Elwin Christoffel)

$$\left\{ {}^{\mathsf{A}}_{\mathsf{BC}} \right\} := \frac{1}{2}\mathrm{m}^{\mathsf{AD}}\{\partial_{\mathsf{C}}\mathrm{m}_{\mathsf{BD}} + \partial_{\mathsf{B}}\mathrm{m}_{\mathsf{CD}} - \partial_{\mathsf{D}}\mathrm{m}_{\mathsf{BC}}\}. \tag{D.18}$$

The corresponding metric geodesics—paths locally of extremal length—are parametrizable as

$$\ddot{x}^{\mathsf{A}} + \left\{ {}^{\mathsf{A}}_{\mathsf{BC}} \right\} \dot{x}^{\mathsf{B}}\dot{x}^{\mathsf{C}} = 0. \tag{D.19}$$

In (semi-)Riemannian Geometry all intrinsic properties follow from the metric: one assumes that the affine connection is the metric connection,

$$\left\{ {}^{\mathsf{A}}_{\mathsf{BC}} \right\} = \Gamma^{\mathsf{A}}{}_{\mathsf{BC}}. \tag{D.20}$$

While Affine Geometry is built upon Euclid's parallel postulate and Absolute Geometry discards this, none the less the two notions can readily coexist in Differential Geometry through Riemannian Geometry carrying a metric connection that can be cast in the affine connection's role.

By the index-lowering use of the metric, one can now consider a version of the Riemann tensor with all its indices downstairs: $\mathscr{R}_{\mathsf{ABCD}} := \mathrm{m}_{\mathsf{AE}}\mathscr{R}^{\mathsf{E}}{}_{\mathsf{BCD}}$ which has the additional symmetry property $\mathscr{R}_{\mathsf{ABCD}} = \mathscr{R}_{\mathsf{CDAB}}$. One can also now obtain the *Ricci scalar* $\mathscr{R} := \mathrm{m}^{\mathsf{AB}}\mathscr{R}_{\mathsf{AB}}$. In dimension p, the curvature tensors contain the following amounts of independent pieces of information:

$$\#\mathscr{R}_{\mathsf{ABCD}} = \frac{p^2\{p^2-1\}}{12}, \qquad \#\mathscr{R}_{\mathsf{AB}} = \frac{p\{p+1\}}{2}, \qquad \#\mathscr{R} = 1. \tag{D.21}$$

This establishes which tensor suffices to describe intrinsic curvature in each dimension. I.e. $\mathscr{R}$ suffices in 2-d, $\mathscr{R}_{\mathsf{AB}}$ in 3-d and all of $\mathscr{R}^{\mathsf{A}}{}_{\mathsf{BCD}}$ is required in all higher-d. It also establishes whether the numbers of equations and unknowns make sense in different possible physical theories based on such geometrical objects. The irreducible part of the information in $\mathscr{R}^{\mathsf{A}}{}_{\mathsf{BCD}}$ that is not contained in $\mathscr{R}_{\mathsf{AB}}$ is the *Weyl tensor* (after noted mathematician Hermann Weyl)

$$\mathscr{C}^{\mathsf{A}}{}_{\mathsf{BCD}} := \mathscr{R}^{\mathsf{A}}{}_{\mathsf{BCD}} - \frac{2}{p-2}\left\{\delta^{\mathsf{A}}{}_{[\mathsf{C}}\mathscr{R}_{\mathsf{D}]\mathsf{B}} - \mathrm{m}_{\mathsf{B}[\mathsf{C}}\mathscr{R}_{\mathsf{D}]}{}^{\mathsf{A}}\right\} - \frac{2}{\{p-1\}\{p-2\}}\delta^{\mathsf{A}}{}_{[\mathsf{D}}\mathrm{m}_{\mathsf{C}]\mathsf{B}}\mathscr{R}, \tag{D.22}$$

which inherits all the symmetry properties of the Riemann tensor.

The *Einstein tensor*

$$\mathscr{G}_{\mathsf{AB}} = \mathscr{R}_{\mathsf{AB}} - \frac{1}{2}\mathscr{R}\,\mathrm{m}_{\mathsf{AB}} \tag{D.23}$$

then has the following significant properties. Firstly, it is divergenceless by the contracted Bianchi identity

$$\mathscr{D}^{\mathsf{A}}\mathscr{G}_{\mathsf{AB}} = 0, \tag{D.24}$$

as used in the construction of the Einstein field equations (7.5). Secondly, it is a symmetric (0, 2) tensor like the metric; this results in (7.5) being well-determined as equations for the metric.

D.5 Some More General Metric Geometries*

If however one does not assume equality (D.20), then the difference of the two connections would constitute an additional tensor. This can in general be split into two pieces: nonmetricity (covariant derivative of the metric) and contorsion (a linear combination of torsion terms in the absence of the previous) [232]. Apart from allowing these, other ways of having more complicated geometry are for the metric to be non-symmetric or degenerate.

A further extension becomes apparent from rephrasing Riemannian Geometry as following from a *metric function*

$$\mathrm{F} := \sqrt{\mathrm{m}_{\mathsf{AB}}\dot{Q}^{\mathsf{A}}\dot{Q}^{\mathsf{B}}}$$

by forming a metric according to

$$\frac{\partial^2\mathrm{F}^2}{\partial\dot{Q}^{\mathsf{A}}\partial\dot{Q}^{\mathsf{B}}}.$$

Now in this Riemannian case, this procedure just returns M_{AB}. However, if $\mathrm{F} = \mathrm{F}(Q^{\mathsf{A}}, \dot{Q}^{\mathsf{A}})$ is a more general homogeneous linear function of the velocities, then further geometries ensue; this gives *Finslerian Geometry* [201] if non-degeneracy continues to be respected. Examples of this include *Riemann's quartic geometry* for

$$\mathrm{F} = \{\mathrm{m}_{\mathsf{ABCD}}\dot{Q}^{\mathsf{A}}\dot{Q}^{\mathsf{B}}\dot{Q}^{\mathsf{C}}\dot{Q}^{\mathsf{D}}\}^{1/4} \tag{D.25}$$

and *Randers geometry* [201, 728] for

$$\mathrm{F} = \sqrt{\mathrm{m}_{\mathsf{AB}}\dot{Q}^{\mathsf{A}}\dot{Q}^{\mathsf{B}}} + \mathrm{l}_{\mathsf{C}}\dot{Q}^{\mathsf{C}}. \tag{D.26}$$

In this book, these are not be considered as options for the geometrization of space or spacetime, but they do occur at the level of configuration space geometry in Sect. 17.2.

D.6 Integration on Manifolds

At the level of Differential Geometry, the Riemann integral is first meaningful in terms of forms [207, 606]. Integration on manifolds relies on being able to form partitions of unity, which in turn is guaranteed by paracompactness. This includes a generalization of both of the familiar Divergence and Stokes' Theorems. One can also define *(n,m)-tensor densities* of weight w at this level, as a further class of geometrical object which, e.g. in components transform as

$$\mathrm{Y}^{\bar{\mathsf{A}}_1 \dots \bar{\mathsf{A}}_n}{}_{\bar{\mathsf{B}}_1 \dots \bar{\mathsf{B}}_m} = (\sqrt{\det \mathrm{L}})^w \mathrm{L}^{\bar{\mathsf{A}}_1}{}_{\mathsf{A}_1} \dots \mathrm{L}^{\bar{\mathsf{A}}_n}{}_{\mathsf{A}_n} \mathrm{L}^{\mathsf{B}_1}{}_{\bar{\mathsf{B}}_1} \dots \mathrm{L}^{\mathsf{B}_1}{}_{\bar{\mathsf{B}}_m} \mathrm{Y}^{\mathsf{A}_1 \dots \mathsf{A}_n}{}_{\mathsf{B}_1 \dots \mathsf{B}_m}. \tag{D.27}$$

At the level of Affine Geometry, one can furthermore re-express the forms version of the above generalization in terms of the covariant derivative,

$$\int_{\partial V} \overline{\mathrm{A}}^{\mathsf{A}} \mathrm{d}\boldsymbol{\Sigma}_{\mathsf{A}} = \int_V \mathscr{D}_{\mathsf{A}} \overline{\mathrm{A}}^{\mathsf{A}} \mathrm{d}\Omega .$$

Here is $\overline{\mathrm{A}}^{\mathsf{A}}$ a (1, 0) vector density of weight 1; $\mathrm{d}\boldsymbol{\Sigma}_{\mathsf{A}}$ and $\mathrm{d}\Omega$ are the obvious (hyper)surface and (hyper)volume elements respectively.

Finally, at the level of Metric Geometry, $\sqrt{|\mathrm{m}|}$ is a particularly significant scalar density of weight 1. Now $\mathrm{Y}^{\mathsf{A}_1 \dots \mathsf{A}_n}{}_{\mathsf{B}_1 \dots \mathsf{B}_m}$ can be expressed as $\sqrt{|\mathrm{m}|}^w \mathrm{T}^{\mathsf{A}_1 \dots \mathsf{A}_n}{}_{\mathsf{B}_1 \dots \mathsf{B}_m}$ for $\mathrm{T}^{\mathsf{A}_1 \dots \mathsf{A}_n}{}_{\mathsf{B}_1 \dots \mathsf{B}_m}$ an unweighted (n,m) tensor, i.e. density can be absorbed into powers of $\sqrt{|\mathrm{m}|}$. In terms of this, Metric Geometry's *(hyper)volume* is given by

$$\mathrm{V} = \int \sqrt{|\mathrm{m}|}\, \mathrm{d}^p x. \tag{D.28}$$

D.7 Conformal Transformations, Metrics and Geometry

In the manifold setting [207, 874], conformal transformations continue to be local-angle-preserving transformations. Consider next the base objects of one's theory to possess *conformal weights*: powers of some suitably smooth positive function which is termed the *conformal factor* ψ. In particular, metrics scale as[2]

$$\mathrm{m}_{\mathsf{AB}} \longrightarrow \overline{\mathrm{m}}_{\mathsf{AB}} := \psi^2 \mathrm{m}_{\mathsf{AB}}. \tag{D.29}$$

Conformal weights can be conceived of as a type of tensor transformation law, the above metric scaling is then the vector case. Consequently, the inverse metric m^{AB} scales as ψ^{-2}: a conformal covector, and $\sqrt{\mathrm{m}}$ as $\psi^{\dim(\mathbf{m})}$: a rank $\dim(\mathbf{m})/2$ conformal tensor. This points to conformal tensors' rank taking values in $\mathbb{R}$, rather than in $\mathbb{N}$ for Cartesian tensors, or in $\mathbb{N} \times \mathbb{N}$ for Differential Geometry's general curved tensors.

---

[2]Moreover, for some applications—such as the GR initial value problem—it is helpful to use powers other than 2 in this definition. Whereas in 1-d there are no angles to preserve, the notion of conformal factor still makes sense, though it is now just a type of reparametrization.

A special status is accorded to *conformally invariant* tensors and operators. Moreover, some more composite objects conformally transform under more complicated relations than power laws—involving derivatives of the conformal factor as well. Since conformal invariants are rare, the objects which do transform under power laws—*conformal covariants*—also merit further attention.

Two useful cases of more complicated transformation laws are [874] 1) that the metric connection conformally transforms as

$$\Gamma^{\mathsf{C}}{}_{\mathsf{AB}} \longrightarrow \overline{\Gamma}^{\mathsf{C}}{}_{\mathsf{AB}} = \Gamma^{\mathsf{C}}{}_{\mathsf{AB}} + 2\delta^{\mathsf{C}}{}_{(\mathsf{A}}\omega_{\mathsf{B})} - \mathrm{m}_{\mathsf{AB}}\mathrm{m}^{\mathsf{CD}}\omega_{\mathsf{D}}, \tag{D.30}$$

for $\omega_{\mathsf{D}} := \partial_{\mathsf{D}}\ln\psi$. Applying this to the affinely-parametrized geodesic equation,

$$\overline{\mathrm{u}^{\mathsf{A}}\mathscr{D}_{\mathsf{A}}\mathrm{u}^{\mathsf{B}}} = \mathrm{u}^{\mathsf{A}}\mathscr{D}_{\mathsf{A}}\mathrm{u}^{\mathsf{B}} + 2\mathrm{u}^{\mathsf{B}}\mathrm{u}^{\mathsf{C}}\omega_{\mathsf{C}} - \|\mathrm{u}\|_{\mathbf{m}}{}^2\omega^{\mathsf{B}}, \tag{D.31}$$

so conformal transformations do not in general preserve geodesics. However, these do preserve null geodesics, since nullness kills the first factor of the third term and the second term merely encodes non-affineness of parametrization. 2) The Ricci scalar $\mathscr{R}$ conformally transforms as

$$\mathscr{R} \longrightarrow \overline{\mathscr{R}} = \psi^{-2}\{\mathscr{R} - \{p-1\}\{2\triangle\ln\psi + \{p-2\}|\mathscr{D}\ln\psi|^2\}\}. \tag{D.32}$$

On the other hand, the vacuum Maxwell equations take a conformally invariant form [874]. Finally, the Weyl curvature tensor is conformally invariant: $\overline{\mathscr{C}}^{\mathsf{A}}{}_{\mathsf{BCD}} = \mathscr{C}^{\mathsf{A}}{}_{\mathsf{BCD}}$. It being zero is useful as a diagnostic for conformal flatness in dimension ≥ 3.

Note that tensor rank in conformal Tensor Calculus is at most relative: there is freedom in how one allots primary transformation laws. This includes being able to swap tensor and co-tensor notions around, since these are just the signs of weights in the conformal Tensor Calculus, and this sign is also part of the aforementioned choice of convention. This means that each conformal Tensor Calculus representation involves making a choice of unit weight and of sign. We choose to approach this by assigning the theory's most basic nontrivially conformally covariant object to be a covector. E.g. (D.29) can be taken to correspond to setting up the metric to be a conformal covector for choice of unit weight +2. But the above freedom can be used to allot the alternative scaling

$$\mathrm{m}_{\mathsf{AB}} \longrightarrow \overline{\mathrm{m}}_{\mathsf{AB}} = \psi^{4/\{p-2\}}\mathrm{m}_{\mathsf{AB}}, \tag{D.33}$$

which gives a useful simplification to the Ricci scalar's conformal scaling

$$\mathscr{R} \longrightarrow \overline{\mathscr{R}} = \psi^{4/\{2-p\}}\left\{\mathscr{R} - 4\frac{p-1}{p-2}\psi^{-1}\triangle\psi\right\}. \tag{D.34}$$

One application of this is in simplifying the conformally-transformed Hamiltonian constraint to the quasilinear elliptic PDE Lichnerowicz–York equation (21.7). This criterion picks out the power 4 for the $p = 3$-d space of standard Geometrodynamics. In close relation to (D.34),

$$\triangle\mathrm{u} - \frac{1}{4}\frac{p-2}{p-1}\mathscr{R}\mathrm{u} \tag{D.35}$$

is a conformally covariant version of the Laplacian operator acting on a scalar field u, provided that u itself scales with weight $\{2 - p\}/2$.

This Sec's approach is ascribing validity to conformal *equivalence classes* of geometries; see Appendix H.6 for more. Moreover, in the indefinite-signature case the conformal structure can also be interpreted as the causal structure., by which the conformal–scale split is also an isolation of the causal structure. This can already be done in the case of SR, as per Chap. 4.

Finally, if Ω is constant, k, the transformation is known as a *homothety* alias *dilation*. The metric is a homothety covector, under the convention that 2 powers of k corresponds to covectors:

$$\mathrm{m}_{\mathsf{AB}} \longrightarrow \overline{\mathrm{m}}_{\mathsf{AB}} := k^2 \mathrm{m}_{\mathsf{AB}}. \tag{D.36}$$

If *just* these conformal factors are allowed, one has the differentiable manifold analogue of Similarity Geometry which preserves ratios of now Riemannian inner products covering both angles and ratios of magnitudes. Homotheties give rise to their own simpler notion of *homothetic tensors* carrying *homothetic weights* $\in \mathbb{R}$. Each conformal invariant gives rise to a corresponding *homothety invariant*, and each conformal covariant to a *homothety covariant*. Moreover, there are many further homothety invariants and covariants due to $\partial_{\mathsf{A}}\phi$ reducing to $\partial_{\mathsf{A}}k = 0$. E.g. the metric connection and the geodesic equation are homothetic scalars, whereas the Ricci scalar is a homothetic vector:

$$\mathscr{R} \longrightarrow \overline{\mathscr{R}} = k^{-2}\mathscr{R}. \tag{D.37}$$

D.8 Exercises III. Basic Mathematics and Geometry

Exercise 0) Establish which of the natural numbers $\mathbb{N}$, the integers $\mathbb{Z}$, the rationals $\mathbb{Q}$, the reals $\mathbb{R}$ and the complex numbers $\mathbb{C}$ are i) groups under each of the usual $+$ and $\times$ operations, ii) fields under both operations, iii) rings, and iv) countable.

Exercise 1) i) Check that projectors $P_{ij} = n_i n_j$ and $\widehat{P} = |\,\psi\rangle\langle\,\psi\,|$ are *idempotent*: $P^2 = P$, and that the latter obeys $P^\dagger = P$ as well, and understand the meanings and implications of these relations. Explain geometrically how projectors are related to the notion of components of a vector and to the general reflection (B.8). ii) Complete $(1,0,0)$ and $\frac{1}{\sqrt{2}}(0,1,1)$ to an orthonormal basis and find the corresponding dual basis. iii) For $\mathfrak{v}$ finite, prove that dual space $\mathfrak{v}^*$ is itself a vector space, that dual bases are indeed bases, that $\mathfrak{v}^{**}$ and $\mathfrak{v}$ are isometric as vector spaces and that $\dim(\mathfrak{v}^*) = \dim(\mathfrak{v})$.

Exercise 2) a) Find Im and Ker for $M : \mathbb{R}^3 \to \mathbb{R}^3$ represented by the matrices with all entries M_{ij} zero except i) $M_{33} = 1$, ii) $M_{13} = 1$, iii) $M_{22} = M_{33} = 1$, and iv) $M_{11} = M_{22} = M_{33} = 1$. b) Consider the exterior derivative d: a map: $\mathbf{\Lambda}^{p-1} \longrightarrow \mathbf{\Lambda}^p$ for $\mathbf{\Lambda}^p$ the space of p-forms. Interpret Ker(d) and Im(d) in terms of closed and exact forms, show that they are groups and that the $(p-1)$th map's Im is a normal

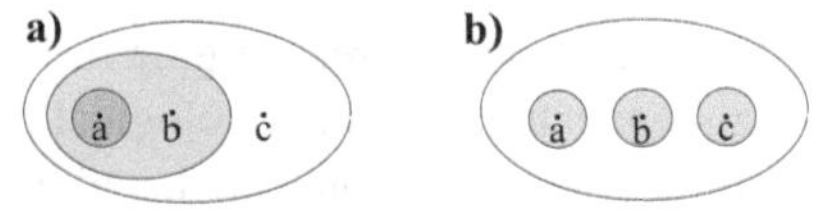

Fig. D.5 Some collections of subsets

subgroup of the pth map's Ker. Evaluate the corresponding quotient groups in the case of forms on $\mathbb{R}$, $\mathbb{R}^2$, $\mathbb{R}/\{0\}$ and $\mathbb{S}^1$.

Exercise 3) i) Show that Euclidean transformations preserve angles and distances, and that similarity transformations preserve angles and ratios of distances. ii) Show that special conformal transformations preserve angles but not ratios more generally. iii) Show that affine transformations in the plane preserve ratios of planar cross-products of differences between vectors. iv) Show that Möbius transformations do not preserve lines, but find a sense in which they do preserve circles. Finally show that Möbius transformations preserve cross-ratios (defined in Fig. G.6).

Exercise 4) i) Prove that if $\|x - y\| < \epsilon$, then $\|y\| - \epsilon < \|x\| < \|y\| + \epsilon$. ii) Prove from the ϵ–δ definition of continuity that if f, g are continuous functions: $\mathbb{R} \to \mathbb{R}$, then so are $f + g$ and fg. iii) Prove that any convergent sequence in $\mathbb{R}$ is a Cauchy sequence.

Exercise 5) i) Prove that a subset $\mathfrak{U}$ of $\mathbb{R}$ is open iff its complement $\mathfrak{U}^c = \mathbb{R}\backslash\mathfrak{U}$ is closed (using the definitions provided at the level of $\mathbb{R}$). ii) Also show that the open sets defined in this context obey the topology axioms. iii) Prove that the ϵ–δ and inverse image of an open set definitions of continuity are equivalent for $f : \mathbb{R}^n \to \mathbb{R}^m$ and f mapping metric spaces to metric spaces, in which cases both definitions are meaningful. [If stuck, use Fig. C.1.e) as a hint.] iv) Study Analysis until you understand at least one proof for each of the Intermediate Value Theorem and the Mean Value Theorem. v)† If proceeding to work through this book's second track, then also cover the Inverse Function Theorem, that once complex-differentiable implies infinitely differentiable, and the converse of Ex III.4.iii) alongside understanding its significance in the metric space and Hilbert space contexts.

Exercise 6) i) How many subsets does a set of n elements contain? How many distinct collections of subsets can be made from it? For $n = 1, 2, 3$, how many inequivalent (i.e. label-independent) collections are there? ii) Show that Fig. D.5.a) is a topology and b) is not. Find all the indistinguishable topologies for iii) $n = 2$ and iv) $n = 3$. v) Show that each of iii) and iv) form a lattice (defined in Appendix S.4).

Exercise 7) [Topological properties entering definition of manifolds] i) Show that the full collection of subsets (the so-called discrete topology) formed from a finite set $\mathfrak{X}$ is Hausdorff and is the only Hausdorff topology thereupon. ii) Prove that all metric spaces are Hausdorff and second-countable. iii) Show that in Appendix M's 'line with 2 origins' is a locally Euclidean, second-countable counter-example to Hausdorffness as claimed. iv) Show that $\mathbb{R}_d \times \mathbb{R}$, for $\mathbb{R}_d$ the real line with discrete topology, is Hausdorff, locally Euclidean but not second-countable. v) Finally give an example of a space which is Hausdorff, second-countable but not locally Euclidean.

Exercise 8) a) Use the open sets definition of continuity to demonstrate that the continuous image of a compact set is compact. Deduce that compactness is a topological property. Also prove that closed subsets of a compact set are compact.

[This only involves rearranging the obvious definitions.] b) Similarly demonstrate that connectedness is a topological property. Also use the definitions to show that path-connected $\Rightarrow$ connected, and find a counter-example to the converse. c) Prove that $\mathbb{R}$ and $\mathbb{R}^2$ are not homeomorphic by considering what happens to each upon removing a point. d) Use $\mathbb{R}$ and a suitable subset thereof to show that completeness is not a topological property.

Exercise 9) What do each of the possible topological identifications of pairs of edges of a square in $\mathbb{R}^2$ give? Which of these spaces are orientable?

Exercise 10) [Spherical Geometry] a) Find the general mathematical form of the *great circles*: the geodesics for $\mathbb{S}^2$), and pick out the particular formulae for a choice of principal axis' equator and meridians. b) Demonstrate Appendix D.3's various interpretations of intrinsic curvature in the case of a sphere. E.g. prove that it contains a spherical triangle with three right angles, and then determine the result of carrying a 'gyroscope' vector around this triangle. Show additionally how that two initially-parallel great circles thereupon converge, quantifying this in terms of the geodesic deviation equation. Which points are conjugate to each other?

Exercise 11) [Conformal Differential Geometry] a) Show that the geodesic equation conformally transforms according to (D.31). Deduce that null geodesics and consequently causal structure, are conformally invariant (then check against Appendix D.7). b)† Show that the equations of motion of RPM are analogously invariant under Appendix L's PPSCT. c) Show that FLRW spacetimes are conformal to (a piece of) Minkowski spacetime $\mathbb{M}^4$. Apply this to construct the Penrose diagrams in Figs. 7.1.c)–d). d) Arrive at the Weyl tensor from the first principles demand of a conformally-invariant curvature tensor in dimension ≥ 4.

Exercise 12) Derive Appendix D.3's normal coordinates.

Exercise 13) In arbitrary dimension p, calculate the number of components of the metric, Riemann tensor, Ricci tensor and Einstein tensor. Also show that the Weyl tensor has $p\{p+1\}\{p+2\}\{p-3\}/12$ components. What is the number of degrees of freedom in the counterpart of GR in each of these dimensions? Deduce a major difference in status of vacuum solutions between $2+1$ and $3+1$ GR.

Exercise 14) Each definition of intrinsic curvature in Appendix D.3 was under the assumption of vanishing torsion. Work out what happens to each of the notions involved in the presence of nonzero torsion.

Exercise 15) Use the Bianchi identity to establish a *short* proof of the closure of classical GR's constraints.

Exercise 16) i) Find the Lie derivative for the arbitrary (p,q) rank tensor ii) Show that $\pounds_{\underline{\xi}}\mathrm{h}_{ab} = 2\,\mathcal{D}_{(a}\xi_{b)}$ and that $\pounds_{\underline{\xi}}\sqrt{\mathrm{h}} = \sqrt{\mathrm{h}}\,\mathcal{D}_a\xi^a$.

Exercise 17) i) Show that the surface of a cube is homeomorphic to the sphere inscribed inside it. ii) Show that the solid sphere is diffeomorphic to $\mathbb{R}^3$.

Exercise 18)† [Simple examples of fibre bundles and Algebraic Topology] i) Provide explicit local charts, projections, transition functions, local sections and structure group for the Möbius strip tangent bundle of Fig. F.2.f). Which of the other notions mentioned in Appendix F are nontrivial for this bundle but trivial for the cylinder bundle? Which are nontrivial for $\mathbb{S}^2$?

Exercise 19)† [Introduction to Symplectic Geometry] i) Derive the symplectic reformulation of the notion of Poisson bracket of Appendix J.9, and the local Darboux Theorem and Poisson tensor properties of Appendix J.12. ii) Work through Chap. 2 of [446] as regards the Symplectic Geometry of constraint surfaces.

Exercise 20)† Work out the geometry of $\mathbb{R}^{3d}/Eucl(d)$ for $d = 1, 2, 3$, interpreted as 3-body problem reduced configuration spaces.

Exercise 21) Show that the GR momentum constraint $\mathcal{M}_i$ can be regarded as conformally covariant, and derive the Lichnerowicz–York equation (21.7).

Appendix E
Lie Groups and Lie Algebras

Lie groups $\mathfrak{g}$ [354] are concurrently groups and differentiable manifolds; additionally their composition and inverse operations are differentiable. Working with the corresponding infinitesimal 'tangent space' around $\mathfrak{g}$'s identity element—the Lie algebra $\mathfrak{g}$—is more straightforward due to vector spaces' tractability, while very little information is lost in doing so. For instance, the representations of $\mathfrak{g}$ determine those of $\mathfrak{g}$. More formally, a *Lie algebra* is a vector space equipped with a product (bilinear map) $|[\ ,\ ]| : \mathfrak{g} \times \mathfrak{g} \longrightarrow \mathfrak{g}$ that is antisymmetric and obeys the Leibniz (product) rule and the *Jacobi identity*

$$|[g_1, |[g_2, g_3]|\,]| + \text{cycles} = 0 \tag{E.1}$$

$\forall\, g_1, g_2, g_3 \in \mathfrak{g}$. This an example of *algebraic structure*: equipping a set with a second or further product operations. Particular subcases of Lie brackets include the familiar Poisson brackets and quantum commutators.

Moreover, a Lie algebra's generators (infinitesimal elements) τ_{p} obey (Fig. E.1)

$$|[\tau_{\mathsf{p}}, \tau_{\mathsf{q}}]| = C^{\mathsf{r}}{}_{\mathsf{pq}}\tau_{\mathsf{r}}, \tag{E.2}$$

where $C^{\mathsf{r}}{}_{\mathsf{pq}}$ are the corresponding *structure constants*.[1] It readily follows that the structure constants with all indices lowered are totally antisymmetric, and also obey

$$C^{\mathsf{o}}{}_{[\mathsf{pq}}C^{\mathsf{r}}{}_{\mathsf{s}]\mathsf{o}} = 0. \tag{E.3}$$

Next suppose that a hypothesis is made about some subset of the generators k_{k} being significant. Denote the rest of the generators by h_{h}. On now needs to check the extent to which the algebraic structure actually complies with this assignation of significance. Such checks place limitations on the generality of intuitions and concepts which hold for simple examples of algebraic structures. A general split

[1] If this role is played by functions instead, then one has strayed into mathematics more complicated than that of Lie algebras; see Appendix V.6 if interested.

E. Anderson, *The Problem of Time*, Fundamental Theories of Physics 190,
DOI 10.1007/978-3-319-58848-3

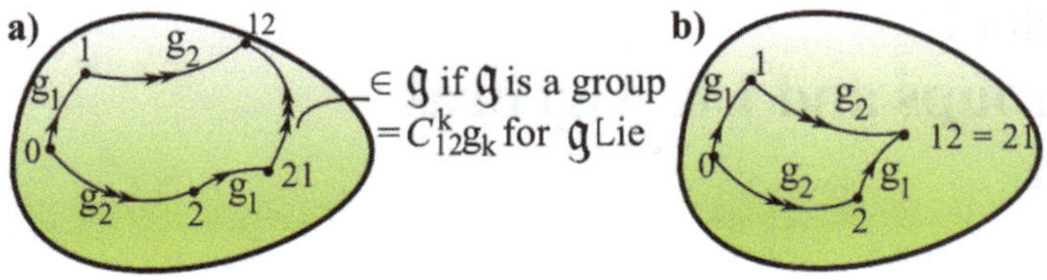

Fig. E.1 **a)** A Lie algebra's commutator. This is a comparison of two triples of objects resulting from applying two transformations g_1, g_2 in either order to a common initial object 0. **b)** The even more straightforward commuting subcase, for which the final objects 12 and 21 coincide as well A lot of instances of **a)** and **b)** occur during investigation of the Problem of Time; these are rendered easy to pick out among the book's figures by all being drawn on *lime-green* egg-shaped spaces

algebraic structure is of the form

$$|[k_{\mathsf{k}}, k_{\mathsf{k}'}]| = C^{\mathsf{k}''}{}_{\mathsf{k}\mathsf{k}'} k_{\mathsf{k}''} + C^{\mathsf{h}}{}_{\mathsf{k}\mathsf{k}'} h_{\mathsf{h}}, \tag{E.4}$$

$$|[k_{\mathsf{k}}, h_{\mathsf{h}}]| = C^{\mathsf{k}'}{}_{\mathsf{k}\mathsf{h}} k_{\mathsf{k}'} + C^{\mathsf{h}'}{}_{\mathsf{k}\mathsf{h}} h_{\mathsf{h}'}, \tag{E.5}$$

$$|[h_{\mathsf{h}}, h_{\mathsf{h}'}]| = C^{\mathsf{k}}{}_{\mathsf{h}\mathsf{h}'} k_{\mathsf{k}} + C^{\mathsf{h}''}{}_{\mathsf{h}\mathsf{h}'} h_{\mathsf{h}''}. \tag{E.6}$$

Denote the second to fifth right hand side terms by 2) to 5). 2) and 5) being zero are clearly subalgebra closure conditions. 3) and 4) are 'interactions between' $\mathfrak{h}$ and $\mathfrak{K}$. The following cases of this are realized in this book.

I) *Direct product*. If 2) to 5) are zero, then $\mathfrak{g} = \mathfrak{K} \times \mathfrak{h}$.
II) *Semidirect product*. If 3) alone is nonzero, then $\mathfrak{g} = \mathfrak{K} \rtimes \mathfrak{h}$.
III) *'Thomas integrability'*. If 2) is nonzero, then $\mathfrak{K}$ is not a subalgebra: attempting to close it leads to some k_{k} are discovered to be integrabilities. Let us denote this by $\mathfrak{K} \ominus \mathfrak{h}$. A simple example of this occurs in splitting the Lorentz group's generators up into rotations and boosts: the group-theoretic underpinning [354] of Thomas precession as per the next Section and Ex IV.9.
IV) *'Two-way integrability'*. If 2) and 5) are nonzero, neither $\mathfrak{K}$ nor $\mathfrak{h}$ are subalgebras, due to their imposing integrabilities on each other. Let us denote this by $\mathfrak{K} \ominus \mathfrak{h}$. In this case, any wishes for $\mathfrak{K}$ to play a significant role by itself are almost certainly dashed by the actual mathematics of the algebraic structure in question.

Note that III) and IV) cover much more diversity of mathematical structure than I) and II) do.

E.1 Examples of Lie Groups and Lie Algebras

Example 1) For Abelian Lie groups, the structure constants are all zero. Subcases of this include $Tr(d)$ and $Dil(d)$—which are both noncompact—and $Rot(2)$ which is compact. The corresponding Lie algebras' generators are

$$P_{\mathsf{A}} := -\frac{\partial}{\partial x^{\mathsf{A}}}, \qquad D := -x^{\mathsf{A}}\frac{\partial}{\partial x^{\mathsf{A}}} \quad \text{and} \quad L := y\frac{\partial}{\partial x} - x\frac{\partial}{\partial y}. \tag{E.7}$$

Example 2) $GL(\mathfrak{v})$ and $SL(\mathfrak{v})$ are also Lie groups (non-Abelian for $\dim(\mathfrak{v}) > 1$). The corresponding Lie algebras are the *general linear algebras* $gl(\mathfrak{v})$ consist of $p \times p$ matrices over $\mathbb{F}$, and the *special linear algebras* $sl(\mathfrak{v})$, which are the zero-trace case of the preceding. The real cases of these are of dimension p^2 and $p^2 - 1$ respectively. The generators for $gl(p, \mathbb{R})$ are, very straightforwardly,

$$G_{\mathsf{AB}} := x^{\mathsf{A}}\frac{\partial}{\partial x^{\mathsf{B}}}, \tag{E.8}$$

and those for $sl(p, \mathbb{R})$ are the tracefree part of the preceding:

$$S_{\mathsf{AB}} := x^{\mathsf{A}}\frac{\partial}{\partial x^{\mathsf{B}}} - \frac{1}{n}\delta^{\mathsf{A}}{}_{\mathsf{B}}x^{\mathsf{C}}\frac{\partial}{\partial x^{\mathsf{C}}}. \tag{E.9}$$

Example 3) $SU(p)$ and $U(p)$ are Lie groups.
Example 4) $SO(p)$ and $O(p)$ are Lie groups.

For each of the pairs in Examples 2) to 4), the version with the S prefix is a Lie subgroup of the other version. 3) share the Lie algebra $su(p) := \{A \in gl(p, \mathbb{C}) \,|\, A + A^{\dagger} = 0\}$ of dimension $p^2 - 1$ (*special unitary algebras*), and 4) share the Lie algebra $so(p) := \{A \in gl(p, \mathbb{R}) \,|\, A + A^{\mathrm{T}} = 0\}$ of dimension $p\{p-1\}/2$(*special orthogonal algebras*).[2] The latter has generators

$$M_{\mathsf{AB}} := x^{\mathsf{A}}\frac{\partial}{\partial x^{\mathsf{B}}} - x^{\mathsf{B}}\frac{\partial}{\partial x^{\mathsf{A}}} \tag{E.10}$$

subject to the nontrivial commutation relation schematically of the form

$$|[M, M]| \sim M. \tag{E.11}$$

Moreover, some (especially smaller) Lie groups conceived of in different manners coincide: so-called 'accidental relations'. As a first example, $Rot(2)$ is mathematically $SO(2) = U(1) = SU(1)$. Thus $so(2) = su(1)$ is Abelian; on the other hand, all the other $so(n)$ and $su(n)$ are non-Abelian. $so(3)$ has the alternating symbol $\epsilon^{i}{}_{jk}$ for its structure constants. The 3-d case of (E.11) also simplifies via duality between M_{AB} and the usual form of for the 3-d angular momenta L_{C}.

[2] $O(p)$ is furthermore a double cover of $SO(p)$: it has two elements per element of $SO(p)$, related by a discrete reflection. For a Lie group $\mathfrak{g}$, the corresponding *(universal) covering group* $\widetilde{\mathfrak{g}}$ is simply-connected and such that $\exists$ a smooth homeomorphism: $\rho : \widetilde{\mathfrak{g}} \rightarrow \mathfrak{g}$ such that $\mathfrak{g} \cong \widetilde{\mathfrak{g}}\,/\ker\rho$. The corresponding Lie algebras then only see the connected component that contains the identity, and so coincide for $O(p)$ and $SO(p)$.

Example 5) the *symplectic groups* $Sp(2p) := \{M \in GL(2p, \mathbb{C}) \,|\, M^T \mathbb{J}_p M = \mathbb{I}\}$ for $\mathbb{J}_p$ given by (A.1), whose corresponding double covers are the *metaplectic groups* $Mp(2p)$. The *symplectic algebras* shared by $Sp(2p)$ and $Mp(2p)$ are $sp(2p) := \{A \in gl(2p) \,|\, A + \mathbb{J}_p A^{\mathrm{T}} \mathbb{J}_p = 0\}$ of dimension $p\{2p+1\}$.
The three preceding examples of families of Lie algebras, alongside the *exceptional Lie algebras* E_6, E_7, E_8, F_4 and G_2 [326] comprise the only semisimple Lie algebras over $\mathbb{C}$. [3] Further accidental relations include $su(2) = so(3)$ and $so(4) = so(3) \times so(3)$. One can also view $SO(p)$ as $O(p)/\mathbb{Z}_2$ and $SU(p)$ as $U(p)/U(1)$. A final case used in this book is that, $SO(3) = SU(2)/\mathbb{Z}_2$, by which $SU(2)$ provides a double cover of $SO(3)$.

Example 6) *Eucl*(p) and *Sim*(p) are some composite Lie groups of particular relevance. These take semidirect product form; e.g. for *Eucl*(p),

$$|[M, P|] \sim P, \tag{E.12}$$

which signifies that P is a *Rot*(p)-vector. On the other hand, *Rot*(p)–*Dil* independence is based upon rotation and dilation generators commuting: a direct product split

$$|[M, D]| = 0 \quad \big(|[\underline{L}, D]| = 0 \text{ in 3-}d \text{ and } |[L, D]| = 0 \text{ in 2-}d\big). \tag{E.13}$$

Let us finally note that the $Tr(p) \rtimes Dil$ combination is based upon

$$[P, D] \sim P. \tag{E.14}$$

Next, returning to Example 2), the $sl(p, \mathbb{R})$ generators can be split into an antisymmetric part, corresponding to the $SO(p)$ subgroup, and a tracefree symmetric part

$$E_{\mathsf{AB}} := x^{\mathsf{A}} \frac{\partial}{\partial x^{\mathsf{B}}} + x^{\mathsf{B}} \frac{\partial}{\partial x^{\mathsf{A}}} - \frac{2}{n} \delta_{\mathsf{AB}} x^{\mathsf{C}} \frac{\partial}{\partial x^{\mathsf{C}}}. \tag{E.15}$$

These are shears and Procrustean stretches (Fig. B.1); each of these are only nontrivial for $p \geq 2$. E.g. corresponding infinitesimal matrices for $sl(2, \mathbb{R})$ are $\begin{pmatrix} 1 & 0 \\ 0 & -1 \end{pmatrix}$, and $\begin{pmatrix} 0 & 1 \\ 1 & 0 \end{pmatrix}$, and the infinitesimal rotation matrix $\begin{pmatrix} 0 & -1 \\ 1 & 0 \end{pmatrix}$. The corresponding generators are

$$x \frac{\partial}{\partial x} - y \frac{\partial}{\partial y} \quad \text{for Procrustean stretches} \quad \text{and} \quad x \frac{\partial}{\partial y} + y \frac{\partial}{\partial x} \quad \text{for shears.} \tag{E.16}$$

The additional trace part of $sl(p, \mathbb{R})$ is just D.

---

[3] The E does not just stand for 'exceptional' but also comes from mathematicians using $A_p := su(p+1)$, $B_p = so(2p+1)$, $C_p = sp(2p)$ and $D_p = so(2p)$. See e.g. [343] for an account of how the even and odd $SO(n)$ indeed behave sufficiently differently to merit treatment as two separate series. Also, *simple* Lie algebras are those which have nontrivial ideals. *Semisimple* Lie algebras are those which are direct sums of simple Lie algebras.

Example 7) The *Heisenberg group Heis*(p)—which occurs in particular in Quantum Theory—adjoins instead an extra $\mathbb{R}^p$ to the Euclidean group. This has the form $Mom(p) \rtimes \{Tr(p) \rtimes Rot(p)\} = \mathbb{R}^p \rtimes \{\mathbb{R}^p \rtimes SO(p)\}$; see Chap. 39 for more advanced mathematical commentary about this group.

Example 8) In Particle Physics, internal gauge groups are restricted to take the form of a direct product of compact simple and $U(1)$ subalgebras by the *Gell-Mann–Glashow Theorem* [886]. Note in particular that the Standard Model's $SU(3) \times SU(2) \times U(1)$ lies within this scope.

Example 9) Further Lie groups and Lie algebras arise upon indefinite signatures becoming significant through SR's means of encoding of the distinction between space and time. The full Lorentz group $O(p, 1)$ has four connected components since both spatial and temporal reflections are involved. Its connected component is the *proper orthochronous Lorentz group Lor*$(p+1) = SO^+(p, 1)$. The xt components of M_{AB} are boosts with generators

$$K_{\mathsf{A}} = t\frac{\partial}{\partial x^{\mathsf{A}}} + x^{\mathsf{A}}\frac{\partial}{\partial t}. \tag{E.17}$$

Since boosts are present, the Lorentz group is noncompact. Schematically, $SO(3, 1)$ decomposes (E.11) into

$$|[J, J]| \sim J, \qquad |[J, K]| \sim K, \qquad |[K, K]| \sim K + J, \tag{E.18}$$

the key Lie bracket being the last one by which the boosts are not a subalgebra. *Thomas precession* (after physicist Llewellyn Thomas) refers to the rotation arising in this manner from a combination of boosts.

Example 10) The Poincaré group *Poin*$(p+1)$ likewise has four connected components to *Poin*$^+(p+1)$'s one. Generators for this are rotations, boosts, spatial translations and temporal translations, $\frac{\partial}{\partial t}$. Equations (6.25)–(6.26) give its nonzero Lie brackets; also clearly $SO(p, 1)$'s Lie algebra can be read off as the subalgebra arising by striking out P_μ.

E.2 Killing Vectors and Isometries

One of the situations in which Lie groups arise is as transformation groups acting on manifolds $\mathfrak{M}$. See e.g. [207] for a basic account, or mathematician Shoshichi Kobayashi's treatise [560] for a detailed exposition including a wider range of transformations in Differential Geometry.

In particular, *isometries* (in the geometrical, not metric space context) are $\mathfrak{M}$-diffeomorphisms that additionally preserve the metric structure $\mathbf{m}$. This is additionally the $\mathbf{T} = \mathbf{m}$ subcase of the aforementioned more general definition $\phi^*\mathbf{T} = \mathbf{T}$ of symmetries for tensors $\mathbf{T}$. Isometries take the infinitesimal form

$$\epsilon_{\mathsf{AB}} \rightarrow \epsilon_{\mathsf{AB}} - 2\,\mathscr{D}_{(\mathsf{A}}\xi_{\mathsf{B})}. \tag{E.19}$$

This exhibits some parallels with the transformations of Gauge Theory $\mathsf{A}_\mathsf{A} \to \mathsf{A}_\mathsf{A} - \partial_\mathsf{A}\Lambda$ (Chap. 6), and is underpinned by the *Killing equation* (after mathematician Wilhelm Killing): the first equality in

$$0 = \pounds_\xi \mathrm{m}_{\mathsf{AB}} = 2\mathscr{D}_{(\mathsf{A}}\xi_{\mathsf{B})} =: (\mathcal{K}\xi)_{\mathsf{AB}}. \tag{E.20}$$

The second equality here is a simple computation, whereas the final definition is for *Killing form* $(\mathcal{K}\xi)_{\mathsf{AB}}$ or *Killing operator* $\mathcal{K}$. (E.20)'s solutions are the *Killing vectors* of $\langle \mathfrak{M}, \mathbf{m} \rangle$.

Killing's Lemma [874] *is that*

$$\mathscr{D}_\mathsf{A}\mathscr{D}_\mathsf{B}\xi_\mathsf{C} = -\mathscr{R}_{\mathsf{BCA}}{}^{\mathsf{D}}\xi_\mathsf{D}. \tag{E.21}$$

One straightforward application of this is in determining that the solution of Killing's equation for flat space ($\mathbb{R}^p$ or $\mathbb{M}^{p+1}$) is the formula at the top of Fig. B.1. **B** is now a 2-tensor that is annihilated by the metric. This means an antisymmetric 2-tensor in the case of $\mathbb{R}^p$, corresponding to the $p\{p-1\}/2$ rotations; it has an additional symmetric piece in the case of $\mathbb{M}^{p+1}$, corresponding to the p boosts. On the other hand, the a^A are the p (spatial) translations in the case of $\mathbb{R}^p$, alongside the single time translation in the case of $\mathbb{M}^{p+1}$. This gives a total of $o\{o+1\}/2$ isometries; $o := p$ for $\mathbb{R}^o$ and $:= p+1$ for $\mathbb{M}^o$; thus in particular 6 for $o = 3$ and 10 for $o = 4$. In this way, one recovers that *Eucl*(o) arises as *Isom*($\mathbb{R}^o$) and *Poin*(o) as *Isom*($\mathbb{M}^o$). Here *Isom*($\langle \mathfrak{M}, \mathbf{m} \rangle$) $:= \{\phi \in$ *Diff*$(\mathfrak{M}) \,|\, \phi^*\mathbf{m} = \mathbf{m}\}$ denotes *isometry group*; these are demonstrated to all be Lie groups in e.g. [560]. Moreover, $o\{o+1\}/2$ is the maximal number [560, 653] of isometries that an o-dimensional $\mathfrak{M}$ can possess, and occurs more widely for the constant curvature spaces. So e.g. $\mathbb{S}^o$ also possesses the maximal number; they are here $\{o+1\}$-*d* rotations: *Isom*($\mathbb{S}^o$) $= SO(o+1)$.

E.3 Conformal and Homothetic Counterparts

The above further generalizes to the notion of *conformal isometry* [207, 874]: a diffeomorphism which additionally preserves the conformal metric structure $\phi^*\mathbf{m} = \Omega^2\mathbf{m}$ for Ω^2 a conformal factor function. It is also the $\mathbf{T} = \mathbf{m}$ subcase of the more general definition $\phi^*\mathbf{T} = \Omega^{2w}\mathbf{T}$ of a *conformal symmetry with weight w* of a tensor $\mathbf{T}$.

In this case, (E.20) generalizes to the *conformal Killing equation*

$$2\,\Omega\,\mathrm{m}_{\mathsf{AB}} = \pounds_\xi \mathrm{m}_{\mathsf{AB}}. \tag{E.22}$$

Moreover, contraction determines that $\Omega = \frac{1}{o} D_\mathsf{C}\xi^\mathsf{C}$, so (E.22) can be rewritten as

$$0 = 2\,\mathscr{D}_{(\mathsf{A}}\xi_{\mathsf{B})} - \frac{2}{o}\mathscr{D}^\mathsf{C} =: \mathcal{L}(\xi)_{\mathsf{AB}}. \tag{E.23}$$

This gives the infinitesimal taken by conformal isometries,

$$\epsilon_{\mathsf{AB}} \longrightarrow \epsilon_{\mathsf{AB}} - 2\mathscr{D}_{(\mathsf{A}}\xi_{\mathsf{B})} + \frac{2}{p}\mathrm{m}_{\mathsf{AB}}\mathscr{D}^{\mathsf{C}}\psi_{\mathsf{C}}, \tag{E.24}$$

and defines the *conformal Killing form* $\mathcal{L}\xi$ or *conformal Killing operator* $\mathcal{L}$. (E.22)'s solutions are the *conformal Killing vectors* of $\langle \mathfrak{M}, \mathbf{m} \rangle$. The conformal Killing vector fields on manifolds also generate a Lie algebra ([560] is an advanced reference covering this).

A *homothety* can now be understood to be a diffeomorphism that additionally preserves the homothetic metric structure $\phi^*\mathbf{m} = c^2\mathbf{m}$. This is also the $\mathbf{T} = \mathbf{m}$ subcase of the more general definition $\phi^*\mathbf{T} = c^{2w}\mathbf{T}$ of a *homothetic symmetry with weight* w of a tensor $\mathbf{T}$. The homothetic Killing vectors solve the *homothety equation*

$$2\,c\,\mathrm{m}_{\mathsf{AB}} = \pounds_{\xi}\mathrm{m}_{\mathsf{AB}}. \tag{E.25}$$

The infinitesimal form taken by homothety is

$$\epsilon_{\mathsf{AB}} \longrightarrow \epsilon_{\mathsf{AB}} + c^2\mathrm{m}_{\mathsf{AB}}. \tag{E.26}$$

Finally, the homothetic Killing vector fields on manifolds also straightforwardly generate a Lie algebra.

The example of flat space—whether $\mathbb{R}^p$ or $\mathbb{M}^{p+1}$, with f_{AB} as the portmanteau metric for δ_{AB} and η_{AB} respectively—now splits into three cases. For $o \geq 3$, $\partial_{\mathsf{A}}\partial_{\mathsf{B}}\phi = 0$ follows as a straightforward integrability from mixed partial equality. By this, ϕ terminates at linear order, giving, upon performing the final integration,

$$\xi^{\mathsf{A}} = a^{\mathsf{A}} + B^{\mathsf{A}}{}_{\mathsf{B}}x^{\mathsf{B}} + cx^{\mathsf{A}} + \{2k^{\mathsf{B}}x^{\mathsf{A}} - k^{\mathsf{A}}x^{\mathsf{B}}\}f_{\mathsf{BC}}x^{\mathsf{C}}. \tag{E.27}$$

c here parametrizes dilations.

$$\xi^{\mathsf{A}} = a^{\mathsf{A}} + B^{\mathsf{A}}{}_{\mathsf{B}}x^{\mathsf{B}} + cx^{\mathsf{A}} \tag{E.28}$$

features as a subcase within, giving a similarity group $Sim(p)$ for $\mathbb{R}^p$, and $Sim(p, 1)$ for $\mathbb{M}^{p+1}$. The dimensionality is $o\{o+1\}/2+1$. One can view (E.28) as being of the form

'complementary function + particular integral'

= {Killing operator kernel: formula atop Fig. B.1.}

+ {outcome of introducing the simple homothetic inhomogeneous term}.

Furthermore, for ϕ not constant k^{A} arises as well. These correspond to *special conformal transformations*

$$x^{\mathsf{A}} \longrightarrow \frac{x^{\mathsf{A}} - k^{\mathsf{A}}x^2}{1 - 2\underline{k}\cdot\underline{x} + k^2x^2} \tag{E.29}$$

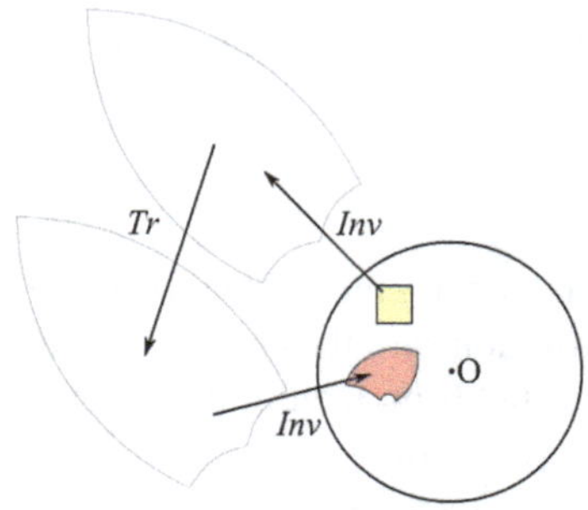

Fig. E.2 Decomposition of special conformal transformation into an inversion, translation and another inversion

formed from an inversion, a translation and then a second inversion (Fig. E.2).

$$C_{\mathsf{A}} := x^2 \frac{\partial}{\partial x^{\mathsf{A}}} - 2\, x_{\mathsf{A}} x^{\mathsf{B}} \frac{\partial}{\partial x^{\mathsf{B}}}. \tag{E.30}$$

The conformal groups $Conf(p)$ and $Conf(p, 1)$ arise for $\mathbb{R}^o$ and $\mathbb{M}^o$ respectively; these are of dimension $\{o + 1\}\{o + 2\}/2$ (in particular 10 for $o = 3$ and 15 for $o = 4$); they are in fact isometric to already-known orthogonal-type groups, e.g. $Conf(p) = SO(p+1, 1)$. (E.27) can be viewed in terms of a larger particular integral arising from the more complicated conformal inhomogeneous term.

An integrability of the form

$$|[C, P]| \sim M + D, \tag{E.31}$$

causes the conformal algebra to be of the Thomas form $(P, C) \ominus (M, D)$. I.e. a translation and an inverted translation compose to give both a rotation ('conformal precession') *and* an overall expansion. Elsewise, C_{A} behaves much like P_{A} does.

The $o = 2$ case is the most interesting. Here $\frac{2}{o} = 1$ in (E.23) paves the way to terms cancelling therein, collapsing it to the Cauchy–Riemann equations (C.1) for $(u, v) = (\xi_x, \xi_y)$. These have an infinity of solutions: any complex-analytic function $f(z)$ will do.

Thus $Conf(2)$ and $Conf(1, 1)$ are infinite-*d* Lie groups. Moreover, one can still consider a finite subgroup of these: the Möbius group of 'global' conformal transformations, or some subgroup of that.

Finally, the $o = 1$ case collapses as well. Preliminarily, Killing's equation collapses to $\mathrm{d}\xi/\mathrm{d}x = 0$, so $\xi = a$, constant, so this case is subsumed within the formula atop Fig. B.1. On the other hand, conformal Killing equation collapses to $\mathrm{d}\xi/\mathrm{d}x = \phi(x)$, amounting to reparametrization by a 1-*d* coordinate transformation $v = \Phi(x) + a$ for $\Phi := \int \phi(x)\mathrm{d}x$. This case is *not* subsumed within (E.27): $Conf(1)$ is also infinite-dimensional, albeit rather less interesting than its 2-*d* counterpart. [1-*d* has no angles to preserve, though conformal factors can be defined for it none the less; note also that **m** *drops out* of the 1-*d* conformal Killing equation.]

E.4 Some Further Groups Acting upon $\mathbb{R}^p$

The above three Sections can be viewed as introducing P_{A}, M_{AB}, D, S_{AB} and C_{A} generators.

Example 11) Combining the first four of these (the second, third and fourth can be jointly packaged as G_{AB}), one arrives at the *affine group* $Aff(p) := Tr(p) \rtimes GL(p, \mathbb{R})$, corresponding to Affine Geometry [222, 644]. $\dim(Aff(p)) = p\{p+1\}$. The nontrivial Lie brackets for this are $|[G, G]| \sim G$ and

$$|[G, P]| \sim P, \tag{E.32}$$

signifying closure of the $GL(p, \mathbb{R})$ subgroup and that P_{A} is a $GL(p, \mathbb{R})$ vector.

Example 12) Dropping D from the preceding, the *equi-top-form group* $Equi(p) := Tr(p) \rtimes SL(p, \mathbb{R})$, which corresponds to the eponymous geometry (equiareal in 2-*d* [222]). $\dim(Equi(p)) = p\{p+1\} - 1$. The nontrivial Lie brackets for this are the S's closing among themselves and

$$|[S, P]| \sim P. \tag{E.33}$$

Note also that

$$|[\text{Shear}, \text{Shear}']| \sim \text{Rotation} \tag{E.34}$$

by which the non-rotational parts of $SL(d, \mathbb{R})$ cannot be included in the absence of the rotations.

Moreover, the K_{A} and $G^{\mathrm{T}}_{(\mathsf{AB})}$ generators are not compatible with each other, as is clear from

$$\begin{gathered}\text{conformal transformations only preserving angles}\\ \text{whereas shears do not preserve angles.}\end{gathered} \tag{E.35}$$

Thus there are two distinct 'apex groups': $Conf(p)$ from including K_{A} and $Aff(p)$ from including S_{AB}. 'Apex' is used here in the sense that the other possibilities are contained within as Lie subgroups. These include a number of subgroups not yet considered (Fig. E.3).

E.5 Yet Further Examples of Lie Groups in Physics

Example 13) The Galileo and Carroll complements of *Poin*(4) in its aspect as a kinematical group are considered in Ex IV.15.

Example 14) *Isom*(AdS) is considered in Ex IV.10, resulting in another interesting accidental relation: Ex IV.17.

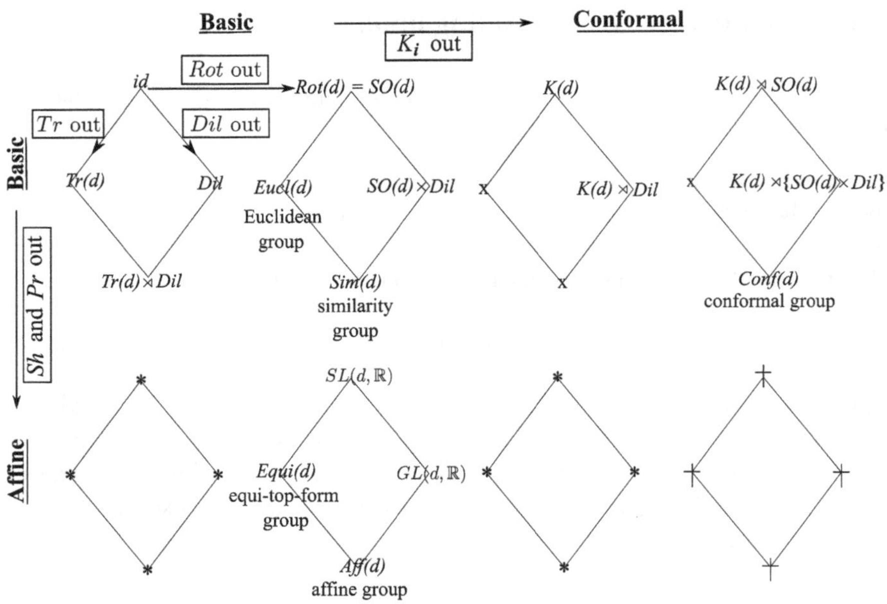

Fig. E.3 Summary sketch, of groups including further groups acting upon $\mathbb{R}^d$. These are arrived at by adding generators as per the labelled arrows. Moreover, the group relations involved do not permit all combinations of generators to be included. In particular, absences marked X are due to integrability (E.31). Absences marked $*$ are due to integrability (E.34). Finally, absences marked $\dagger$ are due to obstruction (E.35). Figures E.4 and G.5 then use a matching layout

E.6 Lie Representations

Representations of Lie groups have much practical relevance to Quantum Theory. Representations are here Lie algebra homomorphisms $\rho : \mathfrak{g} \rightarrow GL(\mathfrak{v})$ for the Lie algebra being on a vector space $\mathfrak{v}$. These are required to preserve the Lie bracket: $\rho(|[g_1, g_2]|) = |[\rho(g_1), \rho(g_2)]|$.

The Representation Theory of compact Lie groups has many parallels with finite Representation Theory (e.g. through Example 2) of Appendix P.2), but the noncompact case does not.

For T_{p} the generators of a brackets algebraic structure $\mathfrak{g}$ with bracket operation $|[\ ,\ ]|$,

$$|[T_{\mathsf{p}}, V_{\mathsf{V}}]| = 0 \tag{E.36}$$

produces an '*associated algebra*' with respect to the same bracket operation.

Example 1) [of associated structure]. When the V_{V} are formed from the generators themselves, this amounts to forming the centre $Z(\mathfrak{uea})$, for $\mathfrak{uea}$ the *universal enveloping algebra* of $\mathfrak{g}$. This has this name due to its encapsulating features which are universal to all representations of the $\mathfrak{g}$ in question. In this case, the resulting V_{V} are known as *Casimirs* [152, 205] (after physicist Hendrik Casimir). These play

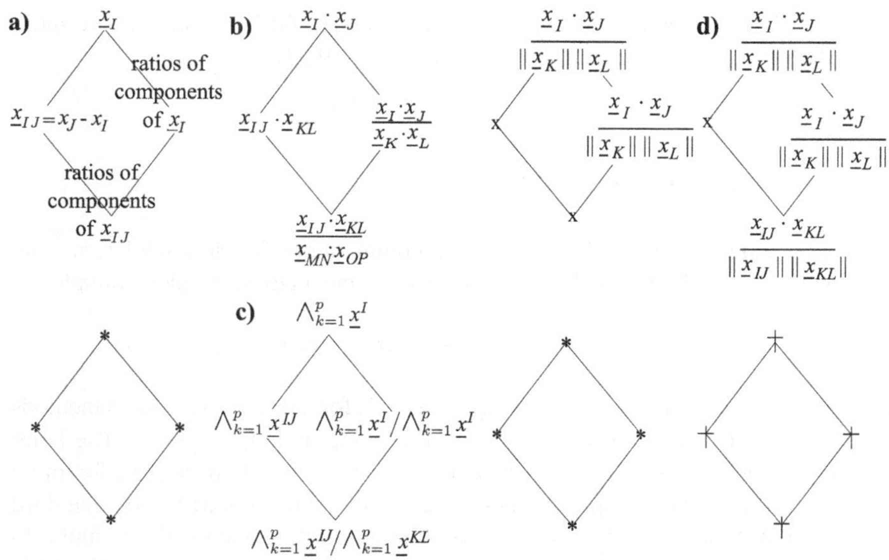

Fig. E.4 Following the preceding figure's layout, we indicate the corresponding invariants. As regards other cases in the figures, the parent potential for *Conf*(p) will continue to hold as a subcase for the subgroups marked with an S, though these have a wider range of good quantities and thus potentials and quantum wavefunctions

a prominent role in Representation Theory, with $SU(2)$'s total angular momentum operator J^2 being the best-known such.

Example 2) [of Lie representations]. $SO(3)$ has representations labelled by $|\mathrm{l}, \mathrm{m}\rangle$. These can be viewed as a ladder. These representation labels can also be viewed as tied to the eigenvalues of the Casimir J^2 (Ex IV.6): $\mathrm{j}\{\mathrm{j}+1\}$, and the eigenvalue of J_3.

Example 3) [of Lie representations]. Wigner [908, 909] showed that this perspective usefully extends to the case of the representations of *Poin*(4). This case has two quadratic Casimirs (Ex IV.7): P^2 and W^2 for

$$W_\mu := \frac{1}{2}\epsilon_{\mu\nu\rho\sigma} M^{\nu\rho} P^\sigma \tag{E.37}$$

the *Pauli–Lubański pseudovector*. Making use of the isotropy subgroups involved [885] and Ex IV.7, the massive particles are labelled by $|\, m, \mathrm{l}, P_\mu, \mathrm{m}\,\rangle$: mass, total spin, momentum and spin's 3-component. On the other hand, the massless particles are labelled by $|\, P_\mu, \lambda\,\rangle$ for λ the *helicity*: component of angular momentum in the direction of motion. From this perspective, spin has passed from being an added-on label in Nonrelativistic QM to being an intrinsic part of Special-Relativistic QM's representations. Moreover these representations can here be taken to be *the types of particle* featuring in the theory (massive or massless, of spins 0, 1/2, 1, ...).

Example 4) [of associated structure]. The observables or beables are a more general class of examples, each associated to a constraint subalgebraic structure. In simpler cases, the latter is still a Lie algebra, though GR's constraints are more complicated: they form the Dirac algebroid (9.31)–(9.33).

E.7 Anticommutator Algebras

Some algebraic structures also involve anticommutators. These model fermionic species; Sect. 6.2's Dirac algebra is of this nature, and an even simpler example is

$$|[\sigma_i, \sigma_j]|_+ := \sigma_i, \sigma_j + \sigma_j, \sigma_i = \delta_{ij}$$

as obeyed by the Pauli matrices. See e.g. [316, 712] for an extensive 'maths methods for physicists' treatment of these and of the ensuing notion of spinors. The latter includes a brief account of these for curved as well as flat spacetime; for more extensive consideration of spinors in GR, see e.g. [75, 232, 706, 814, 868]; the third and fourth of these cover the Ashtekar variables application and the fourth and fifth the Supergravity one. Let us end by noting that this Appendix's coarse split of Lie groups of types I) to IV) readily extends to Lie superalgebras as well [36].

E.8 Exercises IV. Groups and Lie Groups

Exercise 0) i) Give the action of S_4 on the 4 vertices of the tetrahaedron and on the three lines which join opposite pairs of its edges.

ii) Prove that the set of group orbits form a partition, that $\text{Stab}(x) \leq \mathfrak{g}$, that free group action can be re-expressed as $\text{Stab}(\mathfrak{X})$ triviality, and that points on the same group orbit have conjugate stabilizers. Also prove the following.

Orbit-Stabilizer Theorem *If $\mathfrak{g}$ acts on a set $\mathfrak{X}$, there is a bijection between $Orb(x)$ and $\mathfrak{g}/Stab(x)$ given by $gx \leftrightarrow g\ Stab(x)$. Consequently* $|Orb(x)| = |\mathfrak{g}/Stab(x)|$.

iii) Show that all subgroups of an Abelian group are normal and that $Z(\mathfrak{g}) \lhd \mathfrak{g}$. Find the normal subgroups of D_4, A_4 and S_4. Finally show that $SO(d) \lhd O(d)$ and $Tr(d) \lhd Eucl(d)$.

Exercise 1) [Finite Representation Theory]

i) Show that all elements of $\mathfrak{g}$ acting as the identity mapping of $\mathfrak{v}$ produces a rep, 1. [This is known as the *trivial rep*, and counts as one irrep of $\mathfrak{g}$.

ii) The *character* χ of a rep ρ is a map $\chi : \mathfrak{g} \to \mathbb{F}$, $g \mapsto \text{Tr}\,\rho(g)$. Take as facts that 1) the *character table*—with conjugacy classes for columns and characters for rows—is square. 2) Essentially all information about finite group reps is captured by the character Table 3) If $\|\chi\|^2 = 1$, the corresponding rep is an irrep, whereas if $\|\chi\|^2 = n > 1$, this counts how many irreps the corresponding rep ρ is a direct sum of. Additionally for irreps, $\langle \chi, \chi' \rangle = 1$ if $\rho = \rho'$ and 0 otherwise, so characters are orthogonal. [Here,

$$\langle \chi, \chi' \rangle := \frac{1}{|\mathfrak{g}|} \sum_{g \in \mathfrak{g}} \chi(g)\overline{\chi}'(g)$$

—an example of $\mathfrak{g}$-act $\mathfrak{g}$-all construct—and $\|\chi\|^2$ is the corresponding norm.] Show that S_3 has 3 conjugacy classes, and a sign rep σ for which $\chi(\sigma)$ is the sign of the permutation. Finally use orthogonality to finish off the character table.

iii) Obtain the last rep above—ϵ_2—by considering the obvious action of S_3 on the equilateral triangle, noting that this rep's norm is 2, and deducing from orthogonality that this rep is the direct sum of an already-known rep and a new rep.

iv) Show that S_4 has 5 conjugacy classes. Show that iii)'s means of finding ϵ_2 produces an analogous ϵ_3.

v) For reps ρ_1 and ρ_2, $\rho_1 \otimes \rho_2$ is also a rep, though $\otimes$ in does not in general preserve irreducibility, being capable of producing a direct sum of irreps. [This is the *tensor product method*.] Show that $\epsilon_3 \otimes \sigma$ is a new irrep, and so finish off the character table of S_4 by orthogonality. (This question's methods permit many other finite group irreps to be found, though other often more efficient methods exist [326].)

vi)† Justify the above uses of $\otimes$ and orthogonality at the level of modules.

Exercise 2) State the Jacobi identity in a wide enough range of contexts to prove the following. i) The first-class, ii) observables, iii) conserved quantity, and iv) Killing vector properties are preserved by the corresponding brackets. v) That given a constraint algebraic structure, the corresponding notion of observables also closes as an algebraic structure under the same type of brackets. vi) That $[\pounds_U, \pounds_V]W = \pounds_{[U,V]}W$. vii) Eq. (E.3), viii) Eq. (J.29), and ix) that if the Poisson tensor (J.28) is invertible, then its inverse is a closed 2-form (defined in Appendix F.3).

Exercise 3) Find the centres of D_n, A_n, $O(n)$ and $SU(n)$.

Exercise 4) i) Give an action of a permutation group which has only one group orbit, and another which has multiple such. ii) Show that if $\mathbb{Z}$ acts on $\mathbb{R}$ by translations by integers, then the group orbit space $\mathbb{R}/\mathbb{Z} = \mathbb{S}^1$. iii) What is the group orbit space $\mathbb{S}^n/O(n+1)$ for the obvious group action?

Exercise 5) Show that i) $D_n = C_n \rtimes C_2$, ii) $Eucl(d) = Tr(d) \rtimes Rot(d) = \mathbb{R}^d \rtimes SO(d)$ and iii) $Poin(d) = \mathbb{M}^d \rtimes SO(d-1,1)$.

Exercise 6) [Lie Representation Theory] i) Show that $SU(3)$ involves three different sets of ladder operators, each of which individually takes the form familiar from $SU(2)$ in the context of angular momentum addition. ii) Establish that $SU(3)$ has a fundamental 3-irrep, and that $\overline{3}$ is now a distinct rep [unlike for $SU(2)$, where $2 = \overline{2}$]. iii) Show that $3 \otimes \overline{3} = 8 \oplus 1$, for 8 the adjoint irrep (as features in Gell–Mann's eightfold way) and 1 the trivial 'singlet' irrep. [The adjoint rep of $\mathfrak{g}$ is a map $\mathfrak{g} \longrightarrow Aut(\mathfrak{g})$.] Find another tensoring which gives rise to the irrep 10 which also appears approximately in the flavour physics of hadrons. iv) What representations do each of the Riemann and Weyl tensors correspond to?

Exercise 7) [Thomas precession] i) Split the M_{ij} generators into rotations J_i and boosts K_i. ii) Consequently split up the Poincaré group *Poin*(4)'s commutation relations, and deduce that composing two boosts produces a rotation. iii) Give an order of magnitude estimate for the size of the precession in the case of atomic spin–orbit coupling, as well as the size of this correction relative to the 'naive' spin–orbit coupling. iv) Show that upon setting $A_i := \{J_i + i K_i\}/2$ and $B_i := \{J_i - i K_i\}/2$, then each of these quantities close separately. Show similarly that $SO(4) \cong SO(3) \times SO(3)$.

Exercise 8) i) Prove that P^2 and W^2 [cf. Eq./ (E.37)] are quadratic Casimirs for *Poin*(4). ii) Show that $[W_\mu, W_\nu] = i\,\epsilon_{\mu\nu\rho\sigma} W^\rho P^\sigma$. While this is not immediately an algebra, interpret it as a such for P^μ taking a definite value, noting the role played here by stabilizers alias 'little groups'. iii) Show that the stabilizer is $SO(3)$ in the case of massive particles and *Eucl*(2) for massless ones. iv) This provides one motivation for working out the Representation Theory of *Eucl*(2). v) Deduce the massive and massless reps of *Poin*(4), explaining the physical significance of each of their labels.

Exercise 9) a) Prove Killing's Lemma (E.21). b) Find $Isom(\mathbb{R}^p)$, $Isom(\mathbb{M}^p)$, $Isom(\mathbb{S}^p)$, *Isom*(Schwarzschild), *Isom*(Kerr) and $Isom(\text{AdS}_p)$. c) Show that the Killing operator and the conformal Killing operator are self-adjoint on $\mathbf{\Sigma}$ compact without boundary. What are their kernels?

Exercise 10) i) For $p \geq 3$, determine the commutation relations for $Conf(p)$ corresponding to $\mathbb{R}^p$ and $Conf(p-1, 1)$ corresponding to $\mathbb{M}^p$. ii) Reconcile (E.29) and the given form for the special conformal transformation generators.

Exercise 11) Show that $Isom(\mathbb{S}^2) < Isom(\mathbb{R}^3)$ and *Isom*(Schwarzschild), and that $Isom(\mathbb{M}^4) < Conf(p)$. Use these to further exemplify the splits in Appendix E.

Exercise 12) i) Obtain the commutation relations for $SL(p, \mathbb{R})$, $GL(p, \mathbb{R})$, $Equi(p)$ and $Aff(p)$. ii) Show that everything in Fig. E.3 are subalgebras of $Conf(p)$ or $Aff(p)$ and that shears do not close with special conformal transformations.

Exercise 13)† Extend the Killing vector notion to a suitable notion of Killing tensor. Check that your notion is indeed nontrivially realized by the Kerr geometry.

Exercise 14) i) Show that $Conf(3, 1)$ is a symmetry group for Electromagnetism. ii)† Why is relativity not based on this rather than on *Poin*(4)?

Exercise 15) i) Derive the commutations relations of the Galileo and Carroll groups as suitable limits of the Poincaré group *Poin*(4). ii) Show that the Galileo group has semidirect product form. iii)† Work through [354]'s account of the Lie group *contraction* operation in passing from the Poincaré group to the Galileo Group. iv)† Readers with both enthusiasm and some experience in Group Theory and Representation Theory are also invited to find the Casimirs and the reps for the Galileo group, and to compare these with the Poincaré group's.

Exercise 16) i) Show that the Möbius group $\cong SO^+(3, 1)$, with corresponding universal covering group $SL(2, \mathbb{C})$. ii) Deduce that if one SR observer detects circular patterns in the sky, then these will appear circular to all other SR observers within that model universe as well.

Exercise 17)† Show that *Isom*(AdS_5) $\cong$ *Conf*(3, 1) and interpret this result.
Exercise 18)† i) Find commutation relations for i) *Conf*(2). ii) The Virasoro algebra (see Appendix V) corresponding to *Diff*($\mathbb{S}^1$).

Appendix F
More Advanced Topology and Geometry*

F.1 Complex Manifolds

The notion of manifold carries over to the case in which the charts map to open subsets of $\mathbb{C}^m$. The maps involved here are usually *complex-analytic*, i.e. obeying the obvious $\mathbb{C}^m$ extension of the Cauchy–Riemann equations (C.1) for $\mathbb{C}$. E.g. $\mathbb{S}^2$ can be thought of as a complex manifold: the *Riemann sphere*, $\mathbb{C} \cup \infty$. This can be arrived at by the stereographic projection of $\mathbb{S}^2$ onto $\mathbb{C}$ plus the point at infinity that one pole is sent to. $\mathbb{S}^2$ can furthermore be viewed as the simplest *complex projective space* $\mathbb{CP}^1$, whereas $\mathbb{CP}^m$ constitute a 1-parameter family of further examples of complex manifolds. These are the obvious complex analogue of $\mathbb{RP}^m$, i.e. now the spaces of lines through the origin in $\mathbb{C}^{m+1}$. The $\mathbb{C}^{m+1}$ coordinates for these are *homogeneous coordinates*, whereas *inhomogeneous coordinates* are formed by dividing m of these by the remaining one. Since that is only defined for nonzero denominator, these correspond to merely local coordinate charts. See e.g. [673, 891] for a more detailed introduction to complex manifolds, [75, 386, 706, 707] for Theoretical Physics applications and [560, 561] for more advanced results.

F.2 The Hodge-*

The *Hodge-** (after mathematician William Hodge) is a type of form duality map. It is between forms of two generally distinct ranks which add up to the dimension of the underlying manifold, $* : \mathbf{\Lambda}^p \longrightarrow \mathbf{\Lambda}^{n-p}$. This involves metric as well as orientation structure, and its specific form is also affected by the signature of the metric.

Example 1) In 3-d, the Hodge-* interrelates axial vectors and 2-forms; think for instance of two common presentations of the magnetic field.

E. Anderson, *The Problem of Time*, Fundamental Theories of Physics 190,
DOI 10.1007/978-3-319-58848-3

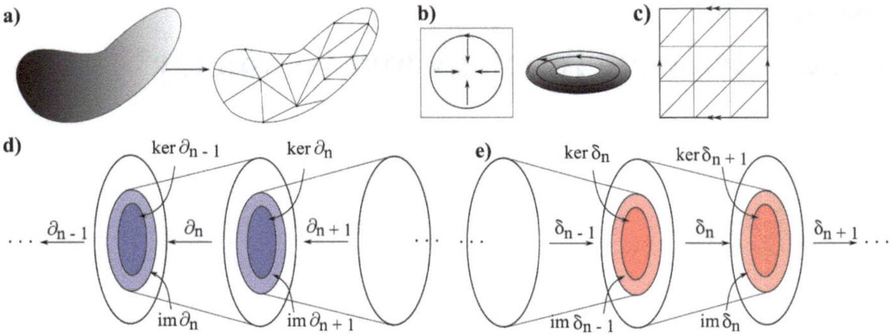

Fig. F.1 **a)** Triangulation as an approximation. **b)** Homotopic triviality of the plane and homotopic nontriviality of the torus $\mathbb{T}^2$. The latter has two distinct types of noncontractible loops, marked in the *bolder ink*, resulting in fundamental group $\pi_1(\mathbb{T}^2) = \mathbb{Z}_2 \times \mathbb{Z}_2$. Spaces, maps, images and kernels in homology and cohomology are depicted in **c)** and **d)** respectively. **c)** Triangulation by a simplicial complex in the case of $\mathbb{T}^2$; see e.g. [68] for what rules this complies with. **d)** and **e)** are useful in envisaging the nature of homology and cohomology

Example 2) In 4-d, the Hodge-* provides an involution[1] of 2-tensors. 2-tensors $\mathrm{T}_{\mu\nu}$ can accordingly be split into self-dual and anti-self-dual parts [707]:

$$\mathrm{T}^{\pm}_{\mu\nu} := \frac{1}{2}\left\{\mathrm{T}_{\mu\nu} \pm \frac{i}{2}\epsilon^{\rho\sigma}{}_{\mu\nu}\mathrm{T}_{\rho\sigma}\right\}. \tag{F.1}$$

Consult e.g. [316, 606, 673] for further information about the Hodge-*; the second of these references includes various further physical applications.

F.3 Algebraic Topology

2-manifolds can be *triangulated*; more generally a p-dimensional counterpart can be represented as a *simplicial complex* (see e.g. [68, 613] for a definition). Figure F.1.c) gives an example of this, but consult the given references and [437] if you require technical details of such. From a physical point of view, this could be viewed either as a discretization of a manifold, or as an inherently discrete version for which the manifold itself is an approximation.

The following constructs each associate series of groups and maps to a given topological manifold (by which they constitute some basic Algebraic Topology). These further structures are in turn useful in classifying topological manifolds.

1) For f, g functions on topological spaces $\mathfrak{X}$ and $\mathfrak{Y}$ respectively, a *homotopy* [68, 613, 673] is a continuous deformation $\Theta : \mathfrak{X} \times [0, 1] \rightarrow \mathfrak{Y}$ such that for $x \in \mathfrak{X}$, $\Theta(x, 0) = f(x)$ and $\Theta(x, 1) = g(x)$. $\mathfrak{X}$ is *contractible* if all loop paths

[1] An *involution* is a self-inverse function.

therein are homotopic to a point. Considering the variety offered by continuously deformable loops on a manifold more generally produces the homotopy groups; these are a type of topological invariant. The *fundamental group*, denoted $\pi_1(\mathfrak{X})$, is the first such. This is trivial for $\mathbb{S}^2$ (Exercise!), whereas Fig. F.1.b) illustrates the case of $\mathbb{T}^2$.

2) *Homology*. One can associate a *chain complex* C_i of Abelian groups to a given topological space $\mathfrak{X}$. Successive members of this complex are related by *boundary operator* homomorphisms $\partial_n : C_n \to C_{n-1}$. Next, *boundaries* are elements of $B_n(\mathfrak{X}) := \text{Im}(\partial_{n+1})$ whereas *cycles* are elements of $Z_n(\mathfrak{X}) := \text{Ker}(\partial_n)$. $\partial_n \partial_{n+1} = 0$ holds: 'the boundary of a boundary is trivial'. It also immediately follows from the C_n being Abelian that all their subgroups are normal. Thereby, it makes sense to define the nth *homology group* as the quotient group $H_n(\mathfrak{X}) := Z_n(\mathfrak{X})/B_n(\mathfrak{X})$; this is a measure of the extent to which each image is a subset of the subsequent kernel (Fig. F.1.c). All in all, homology is a means of constructing topological invariants from cellular arrays [Fig. F.1.e) illustrates a such] which approximate a given $\mathfrak{X}$. If interested in homology, see e.g. [68, 437, 613] for worked examples of how some simple spaces' homology groups can be constructed.

3) *Cohomology* [490, 614, 673, 791] ensues instead in applications in which the maps are taken to go in the opposite direction (Fig. F.1.c): so-called cochains $\delta_n : C_n \to C_{n+1}$. Next, *coboundaries* are elements of $\text{Im}(\delta_{n-1}) := B^n(\mathfrak{X})$, *cocycles* are elements of $\text{Ker}(\delta_n) := Z^n(\mathfrak{X})$. Finally, the quotient $H^n(\mathfrak{X}) := Z^n(\mathfrak{X})/B^n(\mathfrak{X})$ is the nth *cohomology group*.

Example 1) The most commonly encountered type of cohomology evoked in Theoretical Physics is *de Rham cohomology* [490, 673] (after Georges De Rham), which is for a smooth differentiable manifold in the role of $\mathfrak{X}$. Here d is the exterior derivative, so this example concerns closed and exact differential forms. These are respectively those forms f for which $\text{d}f = 0$ and those which can be written as $f = \text{d}g$. $\text{d}^2 = 0$ here means that all exact forms are closed. *Poincaré's Lemma* [207, 673] provides a partial converse to this, for the case in which the domain is contractible.

Example 2) This book also involves the more general *Čech cohomology* [280, 451] (after early 20th century mathematician Eduard Čech). Here a topological space τ is modelled by use of open covers. This turns out to work well for

$$\text{'good covers'} := \text{covers for which every open set and finite intersection thereof are contractible.} \tag{F.2}$$

One next introduces the *nerve* $\mathfrak{N}(\{\mathfrak{U}_\text{C}\})$ of each cover $\{\mathfrak{U}_\text{C}\}$, defined as the simplicial complex which is built as follows.

i) Allotting 1 vertex per element $\mathfrak{U}_\text{C}$ of the cover.
ii) Allotting 1 edge per pair of open covers with non-empty intersection.

iii) Continue through with this pattern to allotting 1 k-simplex per $\{k+1\}$-fold of open covers with non-empty total intersection. In this way, one arrives at

$$\begin{aligned}&(\text{Čech cohomology of } \tau)\\ &\quad = (\text{simplicial cohomology of } \mathfrak{N}(\{\mathfrak{U}_\mathrm{c}\}) \text{ for } \{\mathfrak{U}_\mathrm{c}\} \text{ of form (F.2)}).\end{aligned} \tag{F.3}$$

This is an example of a model space being successful through manifesting properties of the underlying space. In this case, the cohomological operation passing between open covers is refinement of open covers. Finally note that the Čech cohomology coincides with [451] the de Rham one when the topological spaces are additionally smooth differentiable manifolds.

See Appendices F.5, S.2, W.1, and W.3 for mention of further types of cohomology.

Exact sequences are often a useful construct in topology [437]. Here each space in the sequence is linked to the next by a morphism, and these morphisms are such that the image of one map is equal to the kernel of the next map. Finally, *short exact sequences* are of the form $0 \rightarrow A \overset{u}{\rightarrow} B \overset{v}{\rightarrow} C \rightarrow 0$ for u injective and v surjective.

F.4 General and Fibre Bundles

Consider first topological spaces which project down continuously onto lower-d topological spaces, $\pi : \mathfrak{E} \longrightarrow \mathfrak{B}$. Such can be viewed in reverse[2] as higher-d bundle *total spaces* $\mathfrak{E}$, each built over a lower-d *base space* $\mathfrak{B}$; π is a *projection map*. This is the *general bundle* notion; see Fig. F.2.a)–c) and [490] for an outline and [464] for an advanced account.

Suppose that one further introduces a local product structure, in which the total space is made up of identical copies of a *fibre space* (alias just *fibres*) $\mathfrak{F}$, itself for now regarded as a topological space. Then one has a topological-level *fibre bundle*; see Fig. F.2.b), d) and [207, 490, 673, 675] for introductions and [464, 891] for advanced accounts. Moreover, from a global perspective, fibre bundles are in typically 'twisted versions' of product spaces, whereas, conversely, global product spaces are the trivial cases of fibre bundles. Figures F.2.e)–f) are simple examples of these respectively. The inverse image $\pi^{-1}(\mathrm{p})$ is the fibre $\mathfrak{F}_\mathrm{p}$ at p (Fig. F.2.b). That all fibres are the same is mathematically encoded by $\mathfrak{F}_\mathrm{p}$ being homeomorphic to $\mathfrak{F}$, with extra isomorphic equivalence if and when required.

In fact, fibre bundles are also taken to have a *structure group* $\mathfrak{g}$ acting upon the fibres $\mathfrak{F}$, by which they are denoted $\langle \mathfrak{E}, \pi, \mathfrak{B}, \mathfrak{F}, \mathfrak{g} \rangle$.

[2] Bundles were originally considered from a perspective of total space primality in the early 1930s by mathematician Herbert Seifert. Whitney [902], however, switched attention to base space primality, meaning that $\mathfrak{B}$ is an a priori known manifold $\mathfrak{M}$.

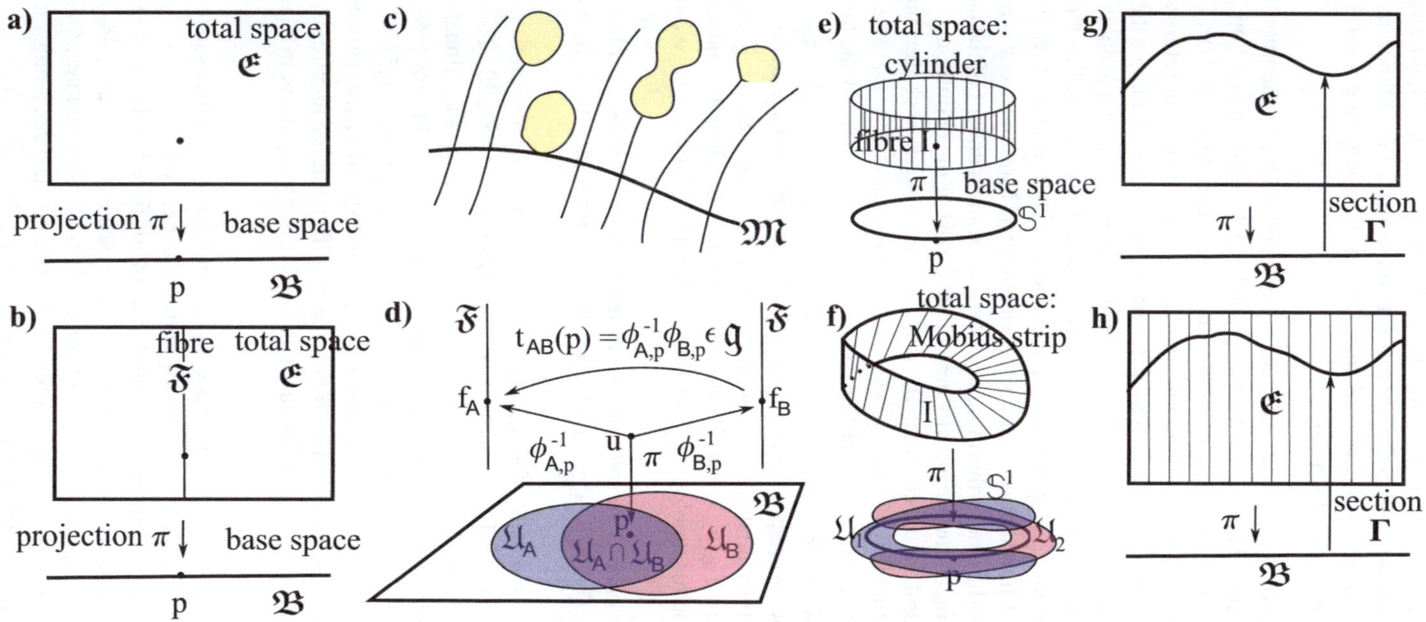

Fig. F.2 **a)** General bundle at the topological level. **b)** For a fibre bundle, the total space $\mathfrak{E}$ is made up of identical disjoint copies of a fibre manifold $\mathfrak{F}$ at each point in the base space $\mathfrak{B}$. **c)** The general bundle structure is a broader concept due to permitting attachment of distinct manifolds to different parts of the base space. **d)** The fibre bundle additionally involves transition functions (cf. manifolds) and a structure group $\mathfrak{g}$. **e)** and **f)** are simple examples of trivial and nontrivial bundle respectively. **g)** and **h)** illustrate the further key notion of (cross-)sections for general and fibre bundles

Example 1) For the significant case of a *principal fibre bundle* $\mathfrak{p}(\mathfrak{M}, \mathfrak{g})$ alias $\mathfrak{g}$-*bundle*, $\mathfrak{g}$ and $\mathfrak{F}$ coincide, so that $\mathfrak{g}$ now just acts on itself. On the other hand, for an *associated fibre bundle*, $\mathfrak{g}$ acts on a distinct type of fibre $\mathfrak{F}$, giving a somewhat more general and complicated structure.

Taking an open cover $\{\mathfrak{U}_A\}$ of $\mathfrak{B}$, each $\mathfrak{U}_A$ is equipped with a homeomorphism $\phi_A : \mathfrak{U}_A \times \mathfrak{F} \to \pi^{-1}(\mathfrak{U}_A)$. This is such that $\pi\phi_A$ sends (p, f)—for f a point on $\mathfrak{F}_p$—down to p. ϕ_A is termed a *local trivialization* since its inverse maps $\pi^{-1}(\mathfrak{U}_A)$ onto $\mathfrak{U}_A \times \mathfrak{F}$ which is a trivial product structure. *Local* triviality here is in reference to *globally* nontrivial fibre bundles encoding information in excess of that in the also globally trivial product space. Figures F.2.e)–f) are, in more detail of a Möbius strip viewed as a nontrivial fibre bundle as compared to the cylinder viewed as a trivial bundle. Both of these have circles for base spaces and line intervals for fibres. In this case, the extra global information is the non-orientability. Moreover, our ongoing definition of fibre bundle can furthermore be shown to be independent of the choice of covering, so we do not enumerate this paragraph as part of the definition.

As a final structural input, consider $\mathfrak{U}_A$ and $\mathfrak{U}_B$—an arbitrarily chosen pair of open sets except that nontrivial overlap between them is guaranteed: $\mathfrak{U}_A \cup \mathfrak{U}_B \neq \emptyset$. Somewhat simplify the notation according to $\phi_A(p, f) = \phi_{A,p}(f)$, $\phi_{A,p}$ is the homeomorphism sending $\mathfrak{F}_p$ to $\mathfrak{F}$. The *transition functions* $t_{AB}(p) := \phi_{A,p}^{-1}\phi_{B,p} : \mathfrak{F} \to \mathfrak{F}$ corresponding to the overlap region as per Fig. F.2.b) are then elements of $\mathfrak{g}$. ϕ_A and ϕ_B are moreover related by a continuous map $t_{AB} : \mathfrak{U}_A \cup \mathfrak{U}_B \to \mathfrak{g}$ according to $\phi_A(p, f) = \phi_A(p, t_{AB}(p^f))$: Fig. F.2.d). Note that this is a bundle analogue of the meshing condition for topological manifolds of Fig. D.1.b), exhibiting a number of parallels with it.

Topological *fibre bundle morphisms* are continuous maps between fibre bundles $\langle \mathfrak{E}_1, \pi_1, \mathfrak{B}_1, \mathfrak{f}_1, \mathfrak{g}_1 \rangle$ and $\langle \mathfrak{E}_2, \pi_2, \mathfrak{B}_2, \mathfrak{F}_2, \mathfrak{g}_2 \rangle$ that map each fibre $\mathfrak{F}_1$ onto a fibre $\mathfrak{F}_2$.

A *section* alias *cross-section* of a topological fibre bundle is a continuous map in the opposite direction to π, $\Gamma : \mathfrak{B} \to \mathfrak{E}$ such that $\pi(\Gamma(x)) = x \ \forall x \in \mathfrak{B}$. This is to cut each fibre precisely once. N.B. that not all fibre bundles possess a global such; whether they do is often insightfully expressible in cohomological terms [464] and gives rise to the theory of characteristic classes (outlined in Appendix F.5).

The above definitions of fibre bundle—and of the corresponding morphisms and sections—can furthermore be elevated to the case of differentiable manifolds, now with smooth maps in place of continuous maps and diffeomorphisms in place of homeomorphisms.

Example 2) Tangent space, cotangent space and the general space of tensors can be thought of as *tangent*, *cotangent* and *tensor bundles* respectively.

Example 3) Gauge Theory can be formulated in terms of fibre bundles (using both principal and more general associated fibre bundles); see e.g. [147, 207, 487, 673, 675] for details. This requires considering *connections* on fibre bundles. One can now indeed interpret Gauge Theory's potential A_μ as a connection, alongside corresponding notions of parallel transport and of covariant derivative D_μ.

The space $\mathfrak{l}_p(\mathfrak{M})$ of *loops* at a point $p \in \mathfrak{p}(\mathfrak{M}, \mathfrak{g})$: curves $\gamma : [0, 1] \rightarrow \mathfrak{M}$ starting and ending at p; these define transformations $\phi_\gamma : \pi^{-1}(p) \rightarrow \pi^{-1}(p)$ on $\mathfrak{f} = \mathfrak{g}$. For $u \in \mathfrak{p}(\mathfrak{M}, \mathfrak{g})$ such that u projects down to p [$= \pi(u)$] the *holonomy group* at u is $\text{Hol}_u := \{g \in \mathfrak{g} \,|\, \phi_\gamma(u) = u\,g, \gamma \in \mathfrak{l}_p(\mathfrak{M})\}$; it is a subgroup of $\mathfrak{g}$.

Finally, the field strength $F_{\mu\nu}$ corresponding to A_μ indeed plays the corresponding role of curvature.

The above references and [446] show how monopoles (Sect. 37.3), Gribov effects (Sect. 37.4) anomalies (Sect. 49.3), and BRST Quantization (Sect. 43.1) can lucidly be studied in fibre bundle terms. Spinors can be as well, in either flat [885] or curved [814, 868, 874] spaces.

Finally returning to the unqualified notion of bundles, these can be viewed as a generalization in which there need no longer be a notion of identical fibre at each point of the base space. This is useful since assuming such identical fibres throughout turns out to be a significantly restrictive assumption in some kinds of modelling required by Theoretical Physics (see e.g. Appendix M).

F.5 Characteristic Classes, Indices and Morse Theory

Characteristic classes (see e.g. [464, 673]) describe obstructions to the presence of global sections in fibre bundles.[3] The examples below indicate that these arise from Cohomology Theory. See Appendix W.1 for a further algebraic interpretation of characteristic classes.

Example 1) For complex vector bundles over a real manifold $\mathfrak{M}$ with $\mathfrak{f} = \mathbb{C}^p$ and $\mathfrak{g} = GL(p, \mathbb{C})$, the *Chern classes* are the $2q$-forms that arise from the expansion of $\det\left(\mathbb{I} + \frac{i}{2\pi}\mathbf{F}\right)$ for $\mathbf{F}$ the curvature 2-form corresponding to the bundle's gauge connection $\mathbf{A}$. The pth Chern class is moreover an element of the $\{p + 1\}$th cohomology group $H^{p+1}(\mathfrak{M}, \mathbb{Z})$. Applications of this include Kinematical Quantization and the theory of anomalies. Finally, the integrand in the Chern–Simons action (38.4) of use in various approaches to spatially 2-d Gravitation is the character associated with the second Chern class [673].

Example 2) For a tangent bundle on a manifold $\mathfrak{M}$, the $H^r(\mathfrak{M}, \mathbb{Z}_2)$-valued characteristic classes are *Stiefel–Whitney classes*. If the first of these is nontrivial, it is a global obstruction to $\mathfrak{M}$ being orientable; if the second is, there is an obstruction to $\mathfrak{M}$ supporting spinorial structure.

Indices are a priori analytic entities, such as $\dim(\ker(\mathcal{D})) - \dim(\ker(\mathcal{D}^\dagger))$ for $\mathcal{D}$ a differential operator on some manifold $\mathfrak{M}$ with adjoint operator $\mathcal{D}^\dagger$. Moreover, by *Index Theorems* these are additionally topological invariants. See e.g. [673, 674] for an outline of the Atiyah–Singer Index Theorem that corresponds to the above

[3]This Section's works are named in part after mathematicians Shiing-Shen Chern, James Simons, Eduard Stiefel and Marston Morse.

example of index; this is useful e.g. for a global perspective on anomalies (Epilogue III.B).

Finally, *Morse Theory* considers relations between functions on a space and (topological notions of) 'the shape' of that space. This is based on consideration of critical points and makes use of further types of cohomology; see e.g. [648, 674] for introductions to this subject. The *Morse functions* are the particular Morse Theory notion required for this book. These are characterized by their Hessians at each critical point: a function f is Morse if its critical points are isolated and non-degenerate.

Appendix G
Configuration Space Geometry: Mechanics*

As well as applying the previous Appendix's Topology and Geometry to modelling each of space and spacetime in Part I, we next consider the further application to configuration space $\mathfrak{q}$ [598]: the space of generalized configurations Q for a physical system. This not only plays a significant role in Classical Dynamics but also underpins Facet 2 of the Problem of Time—Configurational Relationalism—(Part II) and Geometrical Quantization (Part III). See Appendix J for an outline of the also useful notion of phase space, $\mathfrak{P}$hase.

This first Appendix on configurations and configuration spaces considers finite flat-space based Mechanics examples to gain intuition. Some geometrical complications encountered in studying configuration space are then considered in Appendix M, whereas the Field Theory and GR cases of configuration spaces are in Appendices H and N. Notions of distance on configuration spaces are considered in Appendices G.4 (positive-definite $\mathfrak{q}$) and N.8 (indefinite $\mathfrak{q}$ as per GR). This book's other main model arenas' configuration spaces—the Minisuperspace subcase of GR and inhomogeneous perturbations thereabout—are covered in Appendices I (unreduced) and N.10 (reduced).

The morphisms corresponding to $\mathfrak{q}$—i.e. the coordinate transformations of $\mathfrak{q}$—are termed the *point transformations* $Point(\mathfrak{q})$. These straightforwardly induce the transformation theory for the Lagrangian variables $\mathsf{Q}, \dot{\mathsf{Q}}$ or the Machian variables $\mathsf{Q}, \mathrm{d}\mathsf{Q}$. *Point* also admits a time-dependent extension $Point_{\mathrm{t}}$ (termed *rheonomic* point transformations to *Point*'s *scleronomic* ones [598]); this however lies outside the main theme of this book.

G.1 (Relational) Mechanics Configuration Spaces

These models' incipient notion of space is *absolute space* $\mathfrak{a}(d)$ of dimension d. This is usually taken to be $\mathbb{R}^d$ equipped with standard Euclidean inner product alias metric. We consider constellations of N labelled (possibly superposed) material points in $\mathbb{R}^d$ with coordinates q^{iI}. (For simplicity, this book considers just the case of equal

E. Anderson, *The Problem of Time*, Fundamental Theories of Physics 190,
DOI 10.1007/978-3-319-58848-3

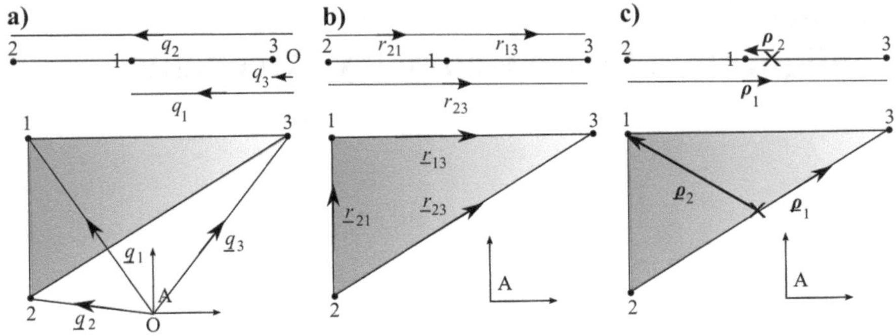

Fig. G.1 Coordinate systems for 3 particles in each of 1- and 2-d. **a)–b)** Absolute particle position coordinates ($\underline{q}_1$, $\underline{q}_2$, $\underline{q}_3$). These are defined with respect to, where they exist, fixed axes A and a fixed absolute origin O. **c)–d)** Relative inter-particle (Lagrange) coordinates $\boldsymbol{r} := \{\underline{r}^{IJ}, I > J\}$. Their relation to the $\underline{q}^I$ is obvious: $\underline{r}^{IJ} := \underline{q}^J - \underline{q}^I$. In the case of 3 particles, any 2 of these form a basis; we use upper-case Latin indices A, B, C for a basis of relative separation labels 1 to n. No absolute origin enters their definition, but reference is still made to fixed coordinate axes A. **e)–f)** Relative particle inter-cluster mass-weighted Jacobi coordinates $\boldsymbol{\rho}$, which are more convenient but still involve A. $\times$ denotes the centre of mass of particles 2 and 3

masses; see [37] for discussion of other cases.) The corresponding (relationally redundant) configuration space $\mathfrak{q}(N, d)$ is just $\mathbb{R}^{Nd}$.

Appendices E.1–E.4 provide a number of $\mathfrak{g}$ that act naturally on $\mathbb{R}^d$, which can subsequently be interpreted as physically redundant transformations acting on $\mathfrak{q}(N, d) = \mathbb{R}^{dN}$. See Fig. G.5 for the corresponding quotient spaces $\mathfrak{q}(N, d)/\mathfrak{g}$, and Appendix M for quotient spaces more generally. Let us denote configuration space dimension by k.[1] Mirror image identifications are in some cases optional, and in other cases obligatory due to being realized by rotations. See Figs. G.9 and G.11 for this, and also concerning the effect of considering indistinguishable particles; both of these matters involve discrete operations being added or removed.

Relative space $\mathfrak{r}(N, d) = \mathfrak{q}(N, d)/Tr(d) = \mathbb{R}^{nd}$ for $n := N - 1$ (Sect. 14.1). *Relative Lagrange coordinates*—some basis of relative inter-particle separation vectors—are conceptually simple natural coordinates for this. Fig. G.1.c)–d) illustrate these for 3 particles in 1- and 2-d, though this notion indeed trivially extends to arbitrary N and d. However, Fig. G.1.e)–f)'s *relative Jacobi coordinates* are more mathematically convenient to work with. These are sets of n inter-particle (cluster) separations chosen such that the kinetic term (or the corresponding arc element) is diagonal. They are widely used in Celestial Mechanics [636] and Molecular Physics [624]. The diagonal form for the kinetic matrix in relative Jacobi coordinates is $\mu_{ijAB} := \mu_A \delta_{ij} \delta_{AB}$, for μ_A the corresponding Jacobi inter-particle cluster reduced

[1] In general this refers to the naïve or largest dimension, since the outcome of quotienting in general produces strata with a variety of dimensions.

masses μ_A. E.g. for the 3-body case,

$$\mu_1 = \frac{m_2 m_3}{m_2 + m_3} \quad \text{and} \quad \mu_2 = \frac{m_1\{m_2 + m_3\}}{m_1 + m_2 + m_3}. \tag{G.1}$$

We furthermore pass to *mass-weighted relative Jacobi coordinates* $\rho^{iA} := \sqrt{\mu_A} R^{iA}$. The kinetic metric is now just an identity array with components $\delta_{ij}\delta_{AB}$. The *unit mass-weighted relative Jacobi coordinates*

$$n^{iA} := \rho^{iA}/\rho =: \rho^{iA}/\sqrt{I} \tag{G.2}$$

are also useful on some occasions. Here, I is the moment of inertia; ρ itself is the *configuration space radius* (alias *hyperradius* [667] in the Molecular Physics literature).

If absolute axes are also to have no meaning, the remaining configuration space is

$$\textit{relational space} \quad \mathfrak{R}(N,d) := \mathfrak{R}(N,d)/Rot(d) \quad [= \mathfrak{q}(N,d)/Eucl(d)]. \tag{G.3}$$

This is of dimension $k = nd—d\{d - 1\}/2$ = d$\{2n + 1—d\}$ /2; in particular, this is $N - 1$ in 2-d, $2N - 3$ in 2-d and $3N - 6$ in 3-d. If, instead, absolute scale is also to have no meaning, the configuration space is Kendall's *preshape space* [539] $\mathfrak{p}(N,d) := \mathfrak{r}(N,d)/Dil = \mathbb{S}^{nd-1}$, with $k = nd - 1$. If both absolute axes and absolute scale are to have no meaning, then the configuration space is Kendall's [539]

$$\textit{shape space} \quad \mathfrak{S}(N,d) := \mathfrak{q}(N,d)/\text{Sim}(d). \tag{G.4}$$

Since the dimension of this plays a recurring role in this book, we give a notation for it,

$$k(N,d) := \dim(\mathfrak{S}(N,d)) = dN - \{d\{d+1\}/2 + 1\} = d\{2n + 1 - d\}/2 - 1. \tag{G.5}$$

In particular, this is $N - 2$ in 1-d, $2N - 4$ in 2-d and $3N - 7$ in 3-d. Finally note that $\mathfrak{p}(N,1) = \mathfrak{S}(N,1)$, since there are no rotations in 1-d.

The above quotient spaces are taken to be not just sets but also normed spaces, metric spaces,, topological spaces, and, where possible, Riemannian geometries. Their analogy with GR's configuration spaces is explained in Fig. G.2.

Useful Lemma [Jacobi Pairs] *Within the subgroups of the affine group, the number of relational configuration spaces requiring independent study is halved, since each version including translations is the same as the version excluding these but with one particle more.*

Proof These subgroups all admit the same trivial notion of taking out the centre of mass. Moreover, the diagonal form in Jacobi's relative $\underline{\rho}^A$ is mathematically identical to that of the mass-weighted point particles $\sqrt{m_I}\underline{q}^I$ bar there being one $\underline{\rho}^A$ less. □

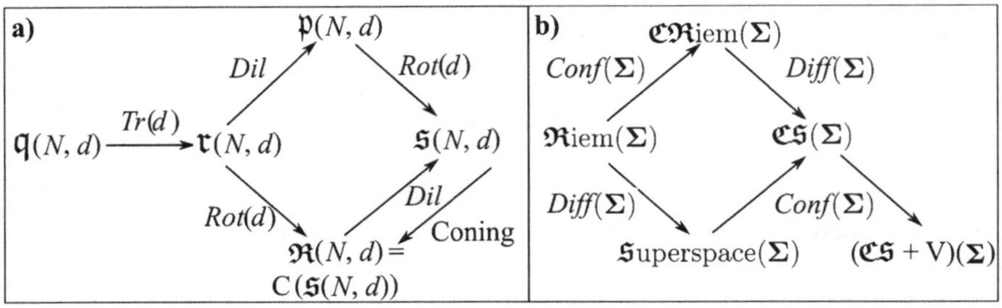

Fig. G.2 This Sec's configuration spaces **a)** are useful model arenas for their GR counterparts **b)** as defined in Appendices H and N

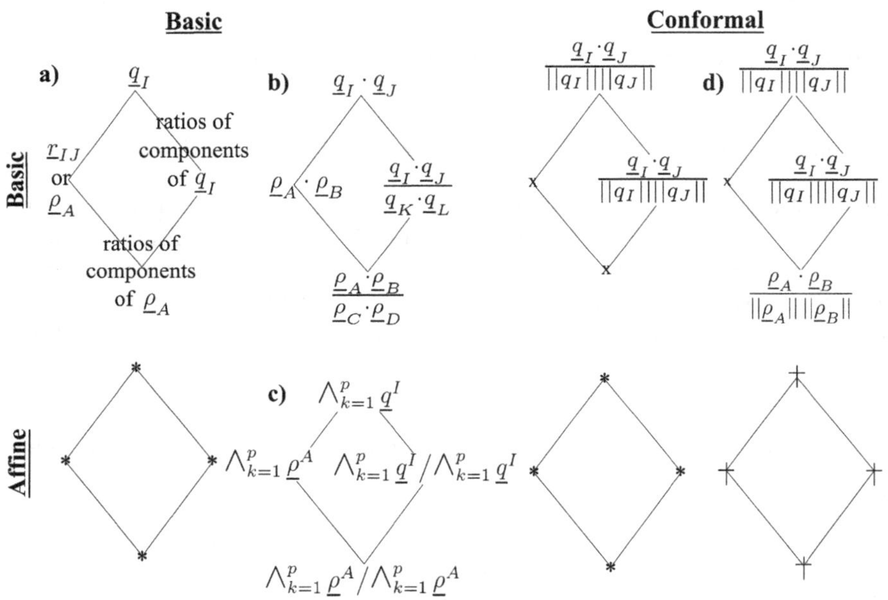

Fig. G.3 Following Fig. E.3, each $\mathfrak{g}$ RPM's invariants are indicated

We next take the form of the invariants of a geometry [Appendix B] and apply these to $\boldsymbol{q}$ and $\boldsymbol{\rho}$ as is appropriate to one's material point particle theory; these are displayed in column 1 of Fig. G.3 and are further exposited in Fig. G.5]. These serve as each corresponding Relational Particle Mechanics' (RPM) potential functional dependence, and end up being its quantum wavefunction dependencies as well. Finally, the corresponding RPM configuration spaces are named in Fig. G.5.[2]

[2]See also Fig. E.3 for an outline of further subgroups of $Conf(d)$, alongside indication of other combinations of generators which fail to close as groups for the reasons stated. Also note that there are no conformal RPM models in 1- or 2-d since a finite space cannot be quotiented by an infinite group.

invariants	group $\mathfrak{g}$	dim($\mathfrak{g}$)	Minimal relationally nontrivial object 1-d	2-d	3-d
none	*id*	0	q^1, q^2	$\underline{q}$ has 2 components	$\underline{q}$ has 3 components
ratios /	*Dil*	1	q^1/q^3, q^2/q^3	$\underline{q}^1, \underline{q}^2$ components support 3 ratios	$\underline{q}$ components support 2 ratios
differences −	*Tr*	d	ρ^1, ρ^2	ρ_x, ρ_y	$\underline{\rho}$ has 3 components
dot products •	*Rot*	$\frac{d\{d-1\}}{2}$	row 1 again	$\|\|\underline{q}^1\|\|, \|\|\underline{q}^2\|\|, (\underline{q}^1 \cdot \underline{q}^2)$	
−/−	*Tr*⋊*Dil*	$d+1$	ρ^1/ρ^3, ρ^2/ρ^3	$\underline{\rho}^1, \underline{\rho}^2$ components support 3 ratios	$\underline{\rho}^1$ components support 2 ratios
•/•	*Rot*×*Dil*	$\frac{d\{d-1\}}{2}+1$	row 2 again	row 4 without scale supports 2 ratios	
−•−	*Tr*⋊*Rot* = *Eucl*	$\frac{d\{d+1\}}{2}$	row 3 again	$\|\|\underline{\rho}^1\|\|, \|\|\underline{\rho}^2\|\|, (\underline{\rho}^1 \cdot \underline{\rho}^2)$	
−•−/−•−	*Tr*⋊{*Rot*×*Dil*} = *Sim*	$\frac{d\{d+1\}}{2}+1$	row 5 again	row 7 without scale supports 2 ratios	
top forms ∧	$SL(d, \mathbb{R})$	d^2-1	row 1 again	3 subtended areas $(q_I \times q_J)_3$	4 subtended volumes $q_I \times q_J \cdot q_K$
∧/∧	$GL(d, \mathbb{R})$	d^2	row 2 again	row 9 supports 2 ratios of subtended areas	row 9 supports 3 ratios of subtended volumes
−∧−	*Tr* ⋊ $SL(d, \mathbb{R})$ = *Equi*	$d\{d+1\}-1$	row 3 again	3 areas $(\underline{\rho}_A \times \underline{\rho}_B)_3$	4 volumes $\underline{\rho}_A \times \underline{\rho}_B \cdot \underline{\rho}_C$
−∧−/−∧−	*Tr* ⋊ $GL(d, \mathbb{R})$ = *Aff*	$d\{d+1\}$	row 5 again	row 11 supports 2 area ratios	row 11 supports 3 volume ratios
∠	*Conf*	$\frac{\{d+1\}\{d+2\}}{2}$ $d>2$; ∞ $d \leq 2$	No freedom for finite point configurations		supports 2 independent local angle freedoms $\frac{\underline{\rho}_A \cdot \underline{\rho}_B}{\|\|\underline{\rho}_A\|\| \, \|\|\underline{\rho}_B\|\|}$
cross ratios ;	$PGL(d, \mathbb{R})$ = *Möbius*	$2\{d^2-1\}$	row 1 again	1 complex cross ratio: $[z_1, z_2; z_3, z_4] = \frac{z_{13} z_{24}}{z_{14} z_{23}}$ degrees of freedom	supports 2 independent projective degrees of freedom

Fig. G.4 Minimal relationally nontrivial units in spatial dimensions 1, 2 and 3 for each group are indicated. The axis and ruler logos denote absolute orientation and absolute scale respectively. $z_{IJ} := z_J - z_I$ are relative particle positions in the complex plane, corresponding to absolute particle positions z_I.
Research Project 115)† Elucidate the topology and geometry of as many of these spaces as possible

The Jacobi pairs simplification moreover does not apply to the configuration spaces of the RPMs corresponding to those further groups which include the special conformal transformations C_i. This is due to the commutation relation (E.31), which causes translations to cease to be trivially removable. Also contrast the conformal case's pure angle information with the similarity case's mixture of angle and ratio information.

Some notions of 2-*d* relational configuration additionally admit a $\mathbb{C}$ formulation. As well as the two components of *Tr*(2) being an obligatory pairing in this setting (keep both or none), *Rot*(2) and *Dil* are also an obligatory pairing: as the modulus and phase parts of a single complex number. The simplest notions of relational configuration that admit a $\mathbb{C}$ formulation have configuration spaces forming the diamond array $\mathbb{C}^N$, $\mathbb{C}^n$, $\mathbb{CP}^n$, $\mathbb{CP}^{n-1}$ corresponding to quotienting out by none, one

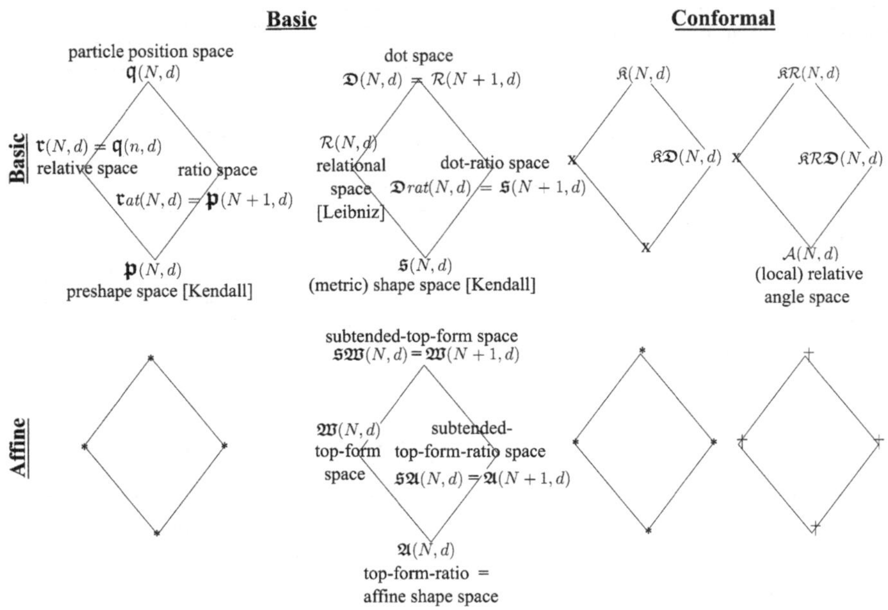

Fig. G.5 Corresponding relational configuration spaces [36]. The equalities given follow from the Jacobi pairs Lemma

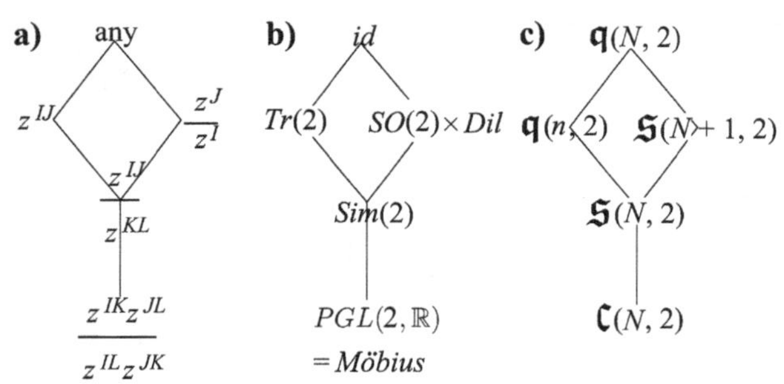

Fig. G.6 Complex suite of **a)** invariants and **b)** the corresponding groups. The expression at the bottom is the *cross-ratio*, for $z^{IJ} := z^I - z^J$. The Möbius group has further subgroups amenable to complex formulation that are not considered here. **c)** The one new kind of relational configuration space here is *cross-ratio space* $\mathfrak{C}(N, 2)$

or both of $Tr(2)$ and $Rot(2) \times Dil$. For sure, the case of Möbius Configurational Relationalism can be modelled in the $\mathbb{C}$ formulation.

G.2 Picking out the Triangleland Example

From here on, we restrict attention to Euclidean and similarity configurations and configuration spaces. Let us start with the latter, due to these being geometrically

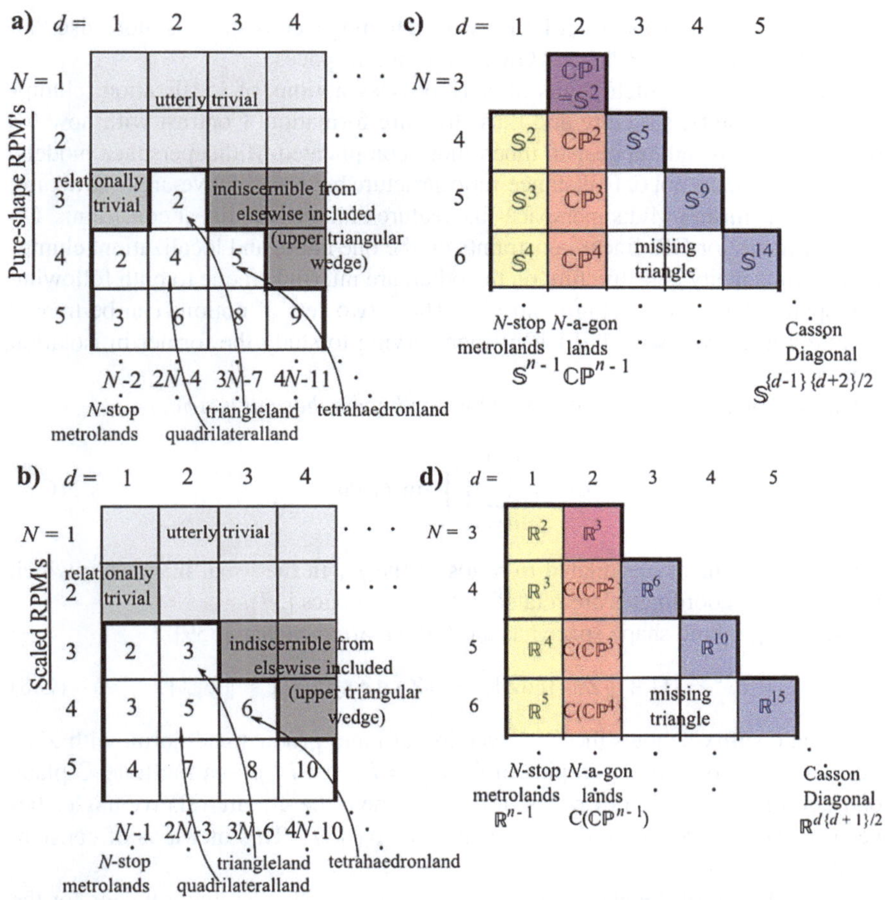

Fig. G.7 **a)** and **b)** are pure-shape and Metric Shape and Scale RPM configuration space dimensions k respectively. **c)** and **d)** are the corresponding topological manifolds ([37] summarizes further topological results about RPM configuration spaces). While this gives 3 tractable series—see [539], including for more about the 'Casson diagonal'—only the two shaded columns admit tractable metrics as well. Let us term 1-d RPM universe models *N-stop metrolands* since their configurations look like underground train lines. We term 2-d RPM universe models *N-a-gon-lands* since their configurations are planar N-sided polygons. The mathematically highly special $N = 3$ case of this is *triangleland*, and the first mathematically-generic $N = 4$ case is *quadrilateralland* [28]. See [37] for the simpler RPM configuration spaces' Algebraic Topology

simpler; furthermore they recur as subproblems within the former. Figure G.7.a)–b) tabulates configuration space dimension k, so as to display inconsistency, triviality, and relational triviality by shading. We follow this up identifying tractable topological manifolds and metric geometries in Fig. G.7.c)–d) [18].

Also note that cases in which

$$\mathfrak{q} = \mathfrak{H}/\mathfrak{g} \text{ is a homogeneous space have more well-understood mathematics} \tag{G.6}$$

[475, 602, 633]; see Appendix M.1 for what a homogeneous space is; note also that this result covers many RPM and GR configuration spaces.

Moreover, N-stop metrolands already possess notions of localization, clumping, inhomogeneity, structure and thus structure formation. Contrast with how for GR these notions only appear in much more complicated Midisuperspace models. N-a-gonlands have not only distance-ratio structure but also relative-angle structure, as well as the further Midisuperspace-like feature of nontrivial linear constraints. On the other hand, for GR, linear constraints on the one hand, and localization, clumping, inhomogeneity and structure on the other, are interlinked due to both following from spatial derivatives being nontrivial. These two sets of notions can be treated separately in RPMs, with 1-d RPMs then serving to study the former in isolation from the latter.

The N-stop metroland shape space has the *hyperspherical metric*

$$\mathrm{d}s^2 = \sum_{p=2}^{n-1} \prod_{m=1}^{p-1} \sin^2\theta_m \mathrm{d}\theta_p^2. \tag{G.7}$$

These $\theta_{\bar{\mathrm{p}}}$ coordinates are related to ratios of the ρ_A in the usual manner in which hyperspherical coordinates are related to Cartesian ones [37].

The N-a-gonland shape space has the *Fubini–Study metric* [539]

$$\mathrm{d}s^2 = \{\{1 + \|\mathbf{Z}\|_{\mathrm{C}}^2\}\|\mathrm{d}\mathbf{Z}\|_{\mathrm{C}}^2 - |(\mathbf{Z}\cdot \mathrm{d}\mathbf{Z})_{\mathrm{C}}|^2\}/\{1 + \|\mathbf{Z}\|_{\mathrm{C}}^2\}^2 \tag{G.8}$$

where the c suffix denotes the $\mathbb{C}^{n-1}$ version of inner product and norm, with $Z_{\bar{p}}$'s indices running over $n-1$ copies of $\mathbb{C}$. $Z_{\bar{p}} = R_{\bar{p}}\exp(i\Phi_{\bar{p}})$—a multiple $\mathbb{C}$ plane polar coordinates version of ratios of the $\underline{\rho}_i$, where the $\Phi_{\bar{p}}$ are relative angles between $\underline{\rho}_A$ and the $R_{\bar{p}}$ are ratios of magnitudes ρ_A [37]. This metric is of constant curvature.

N.B. that each of the above metrics is presented in standard coordinates for the corresponding geometries (hyperspherical angles and inhomogeneous coordinates respectively). Moreover, in the current RPM setting the physical meanings of these coordinates can be traced back to the spatial coordinates describing the particles themselves: see [37]. N.B. also that, as mechanical theories, RPMs have positive-definite kinetic arc elements, which significantly differ from GR's indefinite one. [This book's other principal model arenas—Minisuperspace and inhomogeneous perturbations thereabout—however, inherit GR's indefiniteness.]

Let us next introduce a generalized notion of *cone* over some topological manifold $\mathfrak{M}$. This is denoted by $\mathrm{C}(\mathfrak{M})$ and takes the form

$$\mathrm{C}(\mathfrak{M}) = \mathfrak{M} \times [0, \infty)/\tilde{\ }. \tag{G.9}$$

$\tilde{\ }$ here means that all points of the form $\{\mathrm{p} \in \mathfrak{M}, 0 \in [0, \infty)\}$ are 'squashed' or identified to a single point termed the *cone point*, 0. At the metric level, given a manifold $\mathfrak{M}$ with a metric with line element $\mathrm{d}s$, the corresponding cone has a natural metric of the form

$$\mathrm{d}s_{\mathrm{cone}}^2 := \mathrm{d}\rho^2 + \rho^2\mathrm{d}s^2, \quad \text{(for } \rho \in [0, \infty) \text{ a 'radial' coordinate)}. \tag{G.10}$$

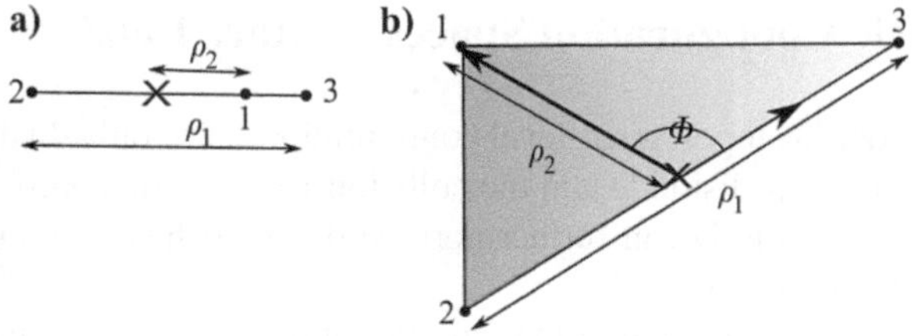

Fig. G.8 **a)** For 3 particles in 1-d, just use the magnitudes of the two 'base' and 'median' Jacobi coordinates. **b)** For 3 particles in 2-d, use the magnitudes of the two Jacobi coordinates and define Φ as the 'Swiss army knife' angle $\arccos(\boldsymbol{\rho}_1 \cdot \boldsymbol{\rho}_3/\rho_1\rho_3)$. This is a relative angle, so, unlike the $\boldsymbol{\rho}$, these three coordinates *do not* make reference to absolute axes A. Next, pure-shape coordinates are the relative angle Φ and some function of the ratio ρ_2/ρ_1. In particular, $\Theta := 2\arctan(\rho_2/\rho_1)$ is the azimuth to Φ's polar angle. Finally, $Z = R\ \exp(i\Phi)$ so as to make contact with the N-a-gon in complex presentation

Relational space is just the cone over shape space [37]; this cone structure renders clear the geometrical meaning of the *scale–shape split* for Metric Shape and Scale RPM. Finally, $C(\mathfrak{S}(N, 1))$ is just $\mathbb{R}^n$.

For triangleland, the additional coincidence $\mathbb{CP}^1 = \mathbb{S}^2$ 'doubles' the amount of geometric and linear methods available (and the spherical ones are both simpler and better-known than complex-projective ones). Here, the Fubini–Study metric simplifies to

$$ds^2 = dZ^2/\{1 + |Z|^2\}^2 = d\Theta^2 + \sin^2\Theta\, d\Phi^2; \tag{G.11}$$

see Fig. G.8 for the meanings of these coordinates.

The scaled case is just the cone over the pure-shape case's configuration space, allowing for that case to be covered as well.

$$ds^2 = d\rho^2 + \rho^2\{d\Theta^2 + \sin^2\Theta_2 d\Phi^2\}/4 = \{dI^2 + I^2\{d\Theta^2 + \sin^2\Theta d\Phi_2^2\}\}/4I; \tag{G.12}$$

$C(\mathfrak{S}(3, 2))$ is also $\mathbb{R}^3$, albeit not with the flat metric. It is, however, conformally flat [37]: just apply the conformal factor $4I$ to the second form of (G.12). The coning construct is moreover independent of which shapes it is being adjoined to, thus constituting a striking examples of Sect. 14.5's 'heterogeneous addendum'.

The corresponding isometry groups include $Isom(\mathfrak{S}(N, 1)) = Isom(\mathbb{S}^{n-1}) = SO(n)$, $Isom(\mathfrak{S}(N, 2)) = Isom(\mathbb{CP}^{n-1}) = SU(n)/\mathbb{Z}_n$ [among which triangleland is further distinguished by $Isom(\mathfrak{S}(3, 2)) = Isom(\mathbb{CP}^1) = SU(2)/\mathbb{Z}_2 = SO(3) = Isom(\mathbb{S}^2)$] and $Isom(\mathfrak{R}(N, 1)) = Isom(\mathbb{R}^n) = Eucl(n)$.

Atomic and Molecular Physics provide a number of useful parallels for the spherical configuration spaces [37]. On the other hand, N-a-gonlands can draw from [28] Geometrical Methods, Shape Statistics, and the standard Representation Theory of $SU(N)$. Finally, The space of N particles on $\mathbb{S}^1$ is straightforwardly just $\mathbb{T}^n$, whereas the space of relational triangles on $\mathbb{S}^2$ itself is hyperbolic 3-space $\mathbb{H}^3$ [539].

G.3 3-Particle Configuration Spaces in More Detail

Let us first consider the topological-level configurations. The only distinct 3-particle shapes are the double collision D and the collision-less generic configuration. If the particles are labelled, the D can furthermore be distinguished by which particle is not involved in the collision.

If the dimension is low enough that configurations cannot be rotated into their mirror images, a first modelling choice is whether the mirror images are to be identified. A second modelling choice is whether the particles are to be labelled or indistinguishable. These considerations combine to give four topological 3-stop metrolands (Fig. G.9) and four topological trianglelands (Fig. G.11).

Consider next the metric-level configurations. 3 particles on a line now have continua of distinguishable non-D configurations. These include a further distinguished notion of *merger* M: a configuration in which the third particle coincides with the centre of mass of the other two: Fig. G.9.a). In configuration space, the M's sit on the mid-points of the arcs between adjacent D's, so e.g. the most extensive 3-stop metroland forms a 'clock-face'.

Triangular configurations (Fig. G.10) include collinear configurations C isosceles configurations I and regular configurations R as distinctive subcases. The last of these are triangles for which $I_{\text{base}} = I_{\text{median}}$: equality of base and median partial moments of inertia, or of the base and median lengths themselves in mass-weighted coordinates.

At the level of the triangleland configuration space shape sphere, C is realized as the equator great circle. This divides the shape sphere into hemispheres of clockwise and anticlockwise ordered triangles: Fig. G.10.b). There are 3 types for each of isosceles I and regular R, corresponding to the 3 ways of labelling 'base' and 'apex'. These are all meridian great circles, alternating between being labelled I and R, and evenly spaced out to form the pattern of the 'zodiac' or 'orange of 12 segments' (Fig. G.11.a). Each I divides the shape sphere into hemispheres of left and right leaning triangles (Fig. G.10.c), and each R into hemispheres of sharp and flat triangles (Fig. G.10.d). All 6 of these great circles intersect at the poles, which correspond to the equilateral triangle configurations E (the orientation-reversed equilateral triangle pole is denoted by $\overline{\text{E}}$). These make for a very natural and significant choice of poles for triangleland, as displayed in Fig. G.11.a). Also, C ∩ I is D at one end and M at the other. On the other hand, C ∩ R has no further special properties, so we denote these points by S for 'spurious'. Finally note that C and E are labelling (or clustering) independent notions, unlike I, R, D, M or S.

The overall pattern [37] is that of the 12-segmented orange cut in half perpendicular to its segments, or of the zodiac additionally split into northern and southern skies. This pattern is, due to its regularity, an example of a *tessellation*: a partition of a space into a number of equal shaped regions: the 'tiles'. Faces, edges and vertices therein being physically significant in the current context, one is really dealing with a *labelled* tessellation (cf. the notion of labelled graph in Appendix A.6). The 4-stop metroland sphere carries an even more elaborate tessellation based on the cube–octahaedron group [37, 59]. Quadrilateralland—much more typical of an

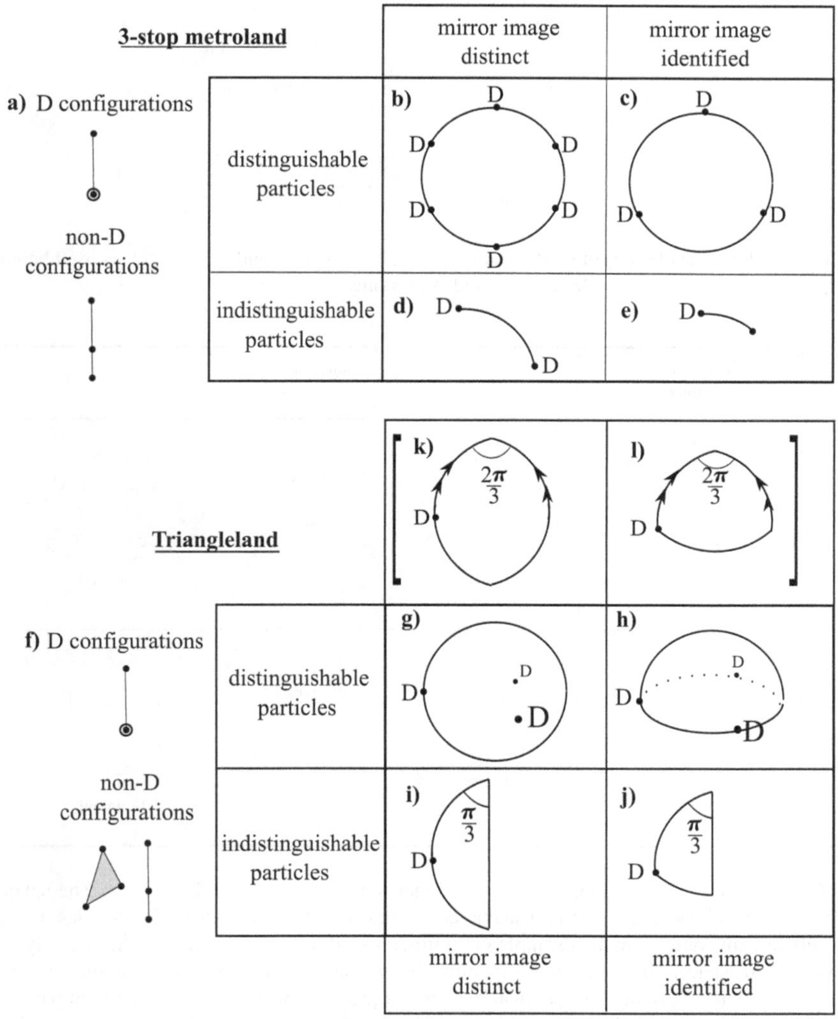

Fig. G.9 Topological-level configurations and configuration spaces for 3 particles (see [56, 57] for more about these). The *double arrows* indicate topological identification. The less used **k**) and **l**), corresponding to quotienting out A_3 instead of S_3, are simple examples of shape space orbifold ($\mathbb{S}^2/\mathbb{Z}_3$) and stratified orbifold. Orbifolds play a significant role in Appendices M. The kinematic geometry itself does not provide any reasons to excise the D points. However if the potential function is singular at the D's—as is the case for the Newtonian gravitational potential—then mathematical study would often begin by excising the D's. Finally, if the D's in **g**) are excised, one obtains the 'pair of pants' [664, 665]

N-a-gonland—is given a comparable configuration space analysis in [28]. Such tessellations provide a useful 'interpretational back-cloth' for the study of dynamical trajectories, probability distributions and quantum wavefunctions. This method was originally applied in the Shape Statistics setting by Kendall [537]: his spherical blackboard, cf. Fig. G.11.f).

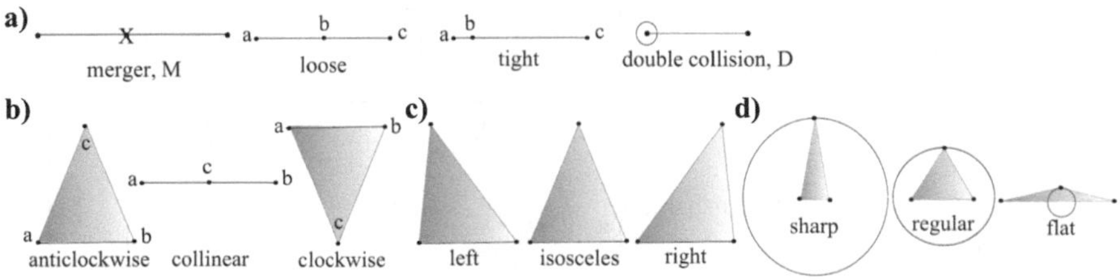

Fig. G.10 Metric level types of configuration for 3 particles in 1- and 2-d. 'Tight' is used here as in 'tight binary' from Celestial Mechanics and Astronomy

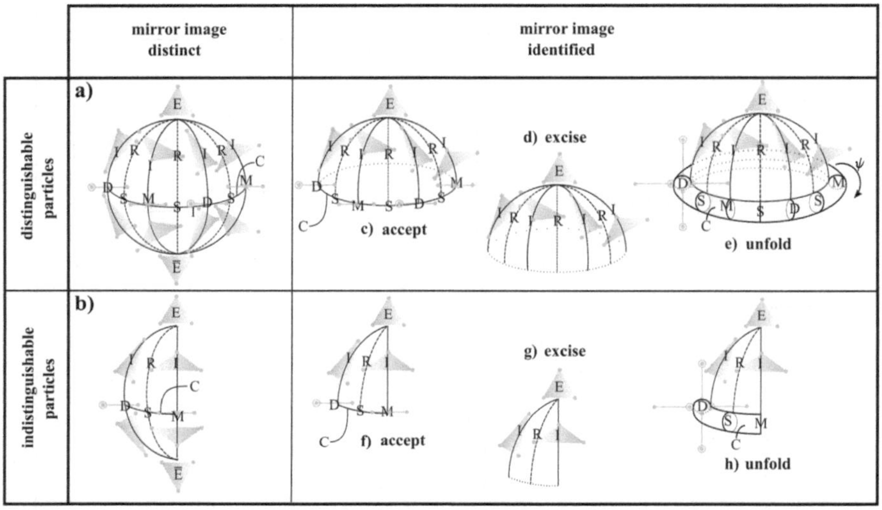

Fig. G.11 3-particle configuration spaces in dimension ≥ 2 at the metric level. **a**) The sphere. **b**) The lune of mirror image distinguishable configurations of indistinguishable particles. **c**) The hemisphere with edge: a simple example of stratified manifold, which occurs in the spatially 3-d version's shape space. This corresponds to mirror image indistinguishable configurations of distinguishable particles. **f**) Kendall's spherical blackboard [537], corresponding to both both particles and mirror images being indistinguishable. Note that stratified manifolds play a significant role in Appendices M and N, alongside Sect. 37.5's deliberation of whether to excise, unfold or accept strata, which is sketched out here as the variants in **b**), **d**), **e**), **f**), **g**) and **h**). Also note that e.g. asymptotically flat GR has analogous additional discrete quotient distinctions corresponding to whether to retain large diffeomorphisms [359]

In 2-d, mirror image identification is optional: a) and b) are both viable options. In 3-d, however, rotation out of the plane sends one mirror image to the other, so a) ceases to be a valid option. As regards stratification, 1-d has no capacity for isotropy groups of different dimension. On the other hand, metric shape spaces for 2-d shapes avoid stratification issues. This is due to only involving $SO(2) = U(1)$, which acts in the same manner same on C and non-C configurations. However, in 3-d the C have only an $SO(2)$ subgroup of the $SO(3)$ acting upon them, so stratification ensues. In 3-d also, the inertia tensor has zero eigenvalues for the C, causing mathematical complications (these prevent inversion of kinetic metric and lead to curvature singularities). This gives one mathematical reason for excision (Fig. G.11.d), along with a physical reason against this: the C are quite clearly physically acceptable configurations.

A second option is to accept the stratification (Fig. G.11.c).

A third possibility is to unfold the equator, by introducing an extra angular coordinate that parametrizes the hitherto unused rotation about the collinearity axis. At the level of configuration space, this has the effect of blowing up the equator into a torus: the 'hemisphere with thick edge' of Fig. G.11.e). However, within the point-particle model setting, the value taken by this extra angular coordinate is not physically meaningful, providing physical and philosophical reasons to not consider this option.

As regards the corresponding relationalspaces, 3-stop metroland's is trivially $\mathbb{R}^2$; indeed N-stop metroland's is $\mathbb{R}^n$. In each case it is entirely clear how to represent an n-sphere within $\mathbb{R}^n$. The $\underline{n}_A$ play the role of Cartesian directions. However, what plays this role for $\mathbb{S}^2$ within $\mathbb{R}^4$? (Ex VI.10). In this case, there are four components of $\underline{n}_A$; how does one relate these to an $\mathbb{R}^3$? It turns out that $\mathbb{R}^4 \rightarrow \mathbb{S}^3 \rightarrow \mathbb{S}^2 \rightarrow \mathbb{R}^3$ handles this, where the second step is the 'Hopf map'. In this manner, the 'Hopf–Dragt' quantities [266, 513, 624] arise:[3]

$$dra_x := \sin\Theta\cos\Phi = 2n_1n_2\,, \quad \cos\Phi = 2\{\underline{n}_1 \times \underline{n}_2\}_3, \tag{G.13}$$

$$dra_y := \sin\Theta\sin\Phi = 2n_1n_2\sin\Phi = 2\underline{n}_1 \cdot \underline{n}_2, \tag{G.14}$$

$$dra_z := \cos\Theta = n_2^2 - n_1^2. \tag{G.15}$$

These appear as 'ubiquitous quantities' [37] in studying the kinematics and the relational dynamics of triangle configurations, and are indeed configurational Kuchař beables for the triangleland RPM [32]. Using $\rho_i = \rho n^i$, these quantities are also available in scaled form:; denote these by Dra_i. The Hopf–Dragt quantities can be interpreted as follows.

dra_x is a quantifier of 'anisoscelesness' *aniso*: departure from the underlying clustering's notion of isoscelesness, cf. anisotropy in Sect. I.1. Specifically, $aniso$ per unit base length in mass-weighted space is the $l_1 - l_2$ indicated in Fig. G.12.a): the amount by which the perpendicular to the base fails to bisect it (which it would do were the triangle isosceles).

dra_y is a quantifier of noncollinearity; this is actually clustering-independent (alias 'democracy invariant' in Molecular Physics [624]). This is furthermore equal to $4 \times$ *area* (the area of the triangle per unit I in mass-weighted space), which is lucid enough to use as notation for this quantity. In comparison, in the equal-mass

[3] 'Hopf' is the name used in Mathematics and 'Dragt' in Molecular Physics, after mathematician Heinz Hopf and physicist Alex Dragt respectively. $n_A := \rho_A/\rho$: the normalized mass-weighted relative Jacobi coordinates, which, by the normalization, are indeed pure metric shape quantities. Finally, the 3-component in the first of these indicates the component in the fictitious third dimension of this cross product.

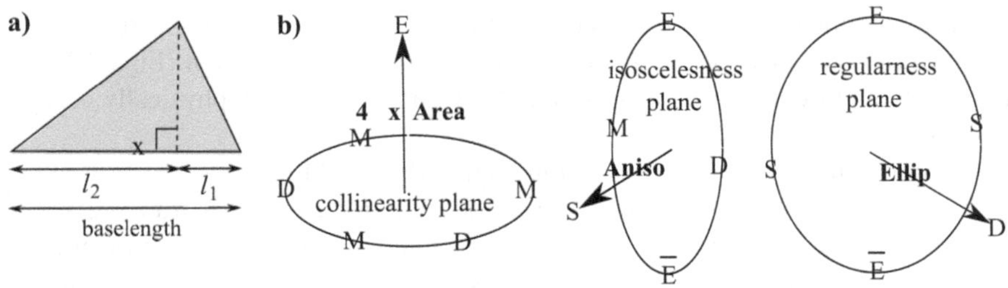

Fig. G.12 **a)** sets up the definition of anisoscelesness quantifier. **b)–d)** interpret the three 'Hopf—Dragt' axes in terms of the physical significance of their perpendicular planes

case

$$(\text{physical area}) = \frac{I\sqrt{3}}{m} area. \tag{G.16}$$

Finally, dra_z is an ellipticity, *ellip*: the difference of the two 'normalized' partial moments of inertia involved in the clustering in question, i.e. that of the base and that of the median. This is clearly a function of pure ratio of relative separations, in contrast to *aniso* being a function of pure relative angle.

Θ itself is also a ratio variable. Moreover, that *ellip* is $\cos\Theta$ subsequently enters the mathematical study of triangleland (as the Legendre variable [220]). On the other hand, Φ is the relative angle 'rightness variable' right corresponding to each clustering. So, in contrast with the pure-ratio variable $dra_z = ellip$, $dra_x = aniso$ and $dra_y = 4 \times area$ provide mixed ratio and relative angle information. The ratio information in both of these of these is a $2n_1n_2 = \sqrt{1 - ellip^2}$ factor. On the other hand, the relative angle information is in the $\cos\Phi$ and $\sin\Phi$ factors.

Figure G.12.b) depicts *Aniso*, *Ellip* and *Area* as vectors in relationalspace, identifying which plane therein each is perpendicular to and which shape space hemispheres this separates.

Maximal collisions are singular for 2- and 3-*d* RPMs. E.g. for scaled triangleland, the Ricci scalar is $R = 6/I$.

Pure-shape triangleland has the maximal three Killing vectors, the 'axial' $\partial/\partial\Phi$ now corresponding to invariance under change of relative angle. Scaled triangleland's topologically and geometrically distinguished origin precludes $\partial/\partial\, dra^i$ from being Killing vectors but still admits the three $dra^j\partial/\partial\, dra^i - dra^i\partial/\partial\, dra^j$. Plain N-stop metroland and N-a-gonland avoid stratification issues due to $SO(2) = U(1)$'s particular straightforwardness, whereas 3-*d* RPMs are not so fortunate. Moreover, mirror image identified triangleland exhibits strata in both its shape space (hemisphere with edge) and relationalspace (half-space with edge), whereas indistinguishable particle versions of RPMs are examples of orbifolds.

G.4 Notions of Distance on Configuration Spaces

Various such [37, 105, 240, 536, 539] can be built from the inner product and norm corresponding to $\mathfrak{q}$'s kinetic metric [4] $\boldsymbol{M}$

$$\text{(Kendall Dist)} = (\boldsymbol{Q},\, \boldsymbol{Q})_{\boldsymbol{M}}, \tag{G.17}$$

$$\text{(Barbour Dist)} = ||\mathrm{d}\boldsymbol{Q}||^2_{\boldsymbol{M}}, \tag{G.18}$$

$$\text{(DeWitt Dist)} = (\mathrm{d}\boldsymbol{Q}, \mathrm{d}\boldsymbol{Q}')_{\boldsymbol{M}}. \tag{G.19}$$

If there is additionally a physically irrelevant $\mathfrak{g}$ acting upon $\mathfrak{q}$,

$$\text{(Kendall } \mathfrak{g}\text{-Dist)} = (\boldsymbol{Q}\cdot \overrightarrow{\mathfrak{g}_g}\, \boldsymbol{Q}')_{\boldsymbol{M}}, \tag{G.20}$$

$$\text{(Barbour } \mathfrak{g}\text{-Dist)} \ = ||\mathrm{d}_g \boldsymbol{Q}||^2_{\boldsymbol{M}} \quad \text{and} \tag{G.21}$$

$$\text{(DeWitt } \mathfrak{g}\text{-Dist)} = (\overrightarrow{\mathfrak{g}}_{\mathrm{d}g}\, \boldsymbol{Q},\, \overrightarrow{\mathfrak{g}}_{\mathrm{d}g}\, \boldsymbol{Q}')_{\boldsymbol{M}}. \tag{G.22}$$

$\mathfrak{g}$-all moves—such as integral, sum, average, inf, sup or extremum—can be applied, after insertion of *Maps* if necessary (cf. Chap. 14). For instance (G.21) subjected to the $\times\sqrt{2\,W}$ and integration maps before a $\mathfrak{g}$-all extremum move gives Best Matching. This can furthermore now be recognized as a subcase of weighted path metric construct (Appendix D.4). On the other hand, (G.22) differs from this in the 'root–sum or integral' ordering manner of Sect. 17.2. (G.20) itself differs from the other two cases in using a *finite* group action to the other two cases' infinitesimal ones. In another sense, it is (G.20) and (G.22)which are akin: these compare two distinct inputs versus (G.21) working around a single input. There is a further issue with 'comparers': if $\boldsymbol{M} = \boldsymbol{M}(\boldsymbol{Q})$, does one use Q_1 or Q_2 in evaluating $\boldsymbol{M}$ itself? This situation does not arise in for Kendall's shapes in $\mathbb{R}^n$, but it does in DeWitt's GR context. DeWitt resolved this (Appendix N.8) in the symmetric manner: using Q_1 and Q_2 to equal extents.

See furthermore Appendix R as regards assessment of how good the best fit is.

In some cases, one might instead be able to work directly with, or reduce down to, $\mathfrak{q}/\mathfrak{g}$ objects, in which case there is no need for the above indirect construct. One would then make use of the relational or reduced $\mathfrak{q}$ geometry $\widetilde{\boldsymbol{M}}$ itself.

A further alternative to the above comparisons of two configurations themselves is to performing intrinsic computations on each

$$\iota : \mathfrak{q} \longrightarrow \mathbb{R}^p, \tag{G.23}$$

[4] This exists independently of whether it is contracted into velocities or changes; e.g. moment of inertia is this metric contracted into mechanical configurations themselves. It only provides a norm if it is positive-definite.

and only then compare the outputs of these computations.[5] In this case, one can consider using norms in the space of computed objects that it is mapped into (the p-dimensional Euclidean metric in the above example). Note however that the outcome of doing this may well depend on the precise quantity under computation. Also ι will in general has a nontrivial kernel; so the candidate ι-Dist would miss out on the separation property of bona fide distances. If this separation fails, one can usually (see e.g. [393]) quotient so as to pass to a notion of distance. [Moreover occasionally this leaves one with a single object so that the candidate notion of distance has collapsed to a trivial one.] Also it is occasionally limited or inappropriate to use such a distance if it is the originally intended space rather than the quotient that has deeper significance attached to it.

ι can again be directly or indirectly $\mathfrak{g}$-invariant; indeed directness is one selection criterion amongst the vast number of possibilities for ι. Other selection criteria include extendibility to unions of configuration spaces, 'physical naturality', and recurrence of the structure used in other physical computations. E.g. a notion of distance that is—or at least shares structural features with—an action, SM partition function, entropy, or notion of information, or a quantum path integral.

[5]This is motivated e.g. by the preceding comparers failing to give distances when $\boldsymbol{M}$ is indefinite—losing the non-negativity and separation properties of bona fide distance—which we know will occur for GR. A range of candidate ι's for the GR case are provided in Appendix N.8.

Appendix H
Field Theory and GR: Unreduced Configuration Space Geometry*

H.1 Field Theory: Unreduced Configuration Space Geometry

Scalar Field Theory's configuration space $\mathfrak{S}$ca the a space of scalar field values $\phi(\underline{x})$ Electromagnetism's configuration space is a space $\mathbf{\Lambda}^1$ of 1-forms $\mathrm{A}_i(\underline{x})$.

Yang–Mills Theory's configuration space is a larger space $\mathbf{\Lambda}$ of 1-forms $\mathrm{A}_i^P(\underline{x})$. [$\mathfrak{S}$ca and $\mathbf{\Lambda}^1$ have implicit dependence ($\mathbb{R}^3$) in many of their more standard uses.]

In modelling the above in more detail, the square-integrable functions $\mathfrak{L}^2$ provide one starting point. One can furthermore pass to e.g. the Fréchet spaces of the next Section, which are subsequently useful in curved-space and GR-coupled versions. Here one has $\mathbf{\Sigma}$ dependence in place of $\mathbb{R}^3$.

H.2 From Hilbert to Banach and Fréchet Spaces

Consider the following ladder of increasingly general topological vector spaces which are infinite-*d* function spaces [207, 522]. A *Hilbert space* $\mathfrak{H}$ilb is a complete inner product space, a *Banach space* $\mathfrak{B}$an is a complete normed space, and a *Fréchet space* $\mathfrak{f}$re is a complete metrizable locally convex topological vector space [426].[1]

Hilbert Spaces are the most familiar in Theoretical Physics due to their use in linear-PDE Fourier Analysis and in Quantum Theory. Functional Analysis has moreover also been extensively developed for Banach spaces [207] (this e.g. underlies Appendices' J and L's treatment of the Calculus of Variations). Major results here are as follows; see [270] for details and proofs.

1) The *Hahn–Banach Theorem.*
2) The *Uniform Boundedness Principle.*
3) The *Open Mapping Theorem.*

[1]These are named after, respectively, mathematicians David Hilbert, Stefan Banach, and Maurice Fréchet.

E. Anderson, *The Problem of Time*, Fundamental Theories of Physics 190,
DOI 10.1007/978-3-319-58848-3

1) is a case of Globalization by Extension of bounded linear functionals on subspaces. Both 2) and 3) apply to continuous linear operators. 2) is self-descriptive, whereas 3) involves surjective such operators being open maps (i.e. maps that preserve the openness property of sets). 2) and 3) follow from *Baire's Category Theorem* concerning the topological space notion of density for complete metric spaces. One consequence of the Open Mapping Theorem is that the Inverse Function Theorem extends to Banach spaces. See e.g. [207] for the form taken by Calculus on Banach spaces.

Treatment of GR configuration spaces moreover involves the even more general Fréchet spaces. Let us first explain their definition. A topological vector space is *metrizable* if its topology can be induced by a metric space metric which is furthermore translation-invariant. This qualification is required since for topological vector spaces, one uses a collection of neighbourhoods of the origin (vector space 0). From this, translation (by the vector space + operation) establishes the collection of neighbourhoods at each other point. A *Hamel basis* itself is a maximal linearly-independent subset of $\mathfrak{v}$ (this is one of various notions of basis in the case of infinite-dimensional spaces). A *base* in a topological vector space $\mathfrak{v}$ is a linearly-independent subset $\mathfrak{b}$ such that $\mathfrak{v}$ is the closure of the linear subspace with Hamel basis $\mathfrak{b}$.

A subset $\mathfrak{Y}$ of a vector space $\mathfrak{v}$ is *convex* if $px + \{1 - p\}y \in \mathfrak{Y}\ \forall x, y \in \mathfrak{Y}, p \in [0, 1]$. A topological vector space $\mathfrak{v}$ is *locally convex* [207] if it admits a base that consists of convex sets. Fréchet spaces are moreover very naturally associated with $\mathfrak{C}^\infty$ smoothness [426].

Let us finally note that many substantial results in Functional Analysis—in particular 1) to 3)—furthermore carry over from Banach spaces to Fréchet spaces [426].

On the other hand, we caution that there is no longer in general an Inverse Function Theorem here, though the *Nash–Moser Theorem* (after mathematicians John Nash and Jurgen Moser) [426] is a replacement for this for a subclass of Fréchet spaces. See footnote 2 for another application, and [426] as regards Calculus on Fréchet spaces more generally.

H.3 Hilbert, Banach and Fréchet Manifolds

Topological manifolds' local Euclideanness and ensuing $\mathbb{R}^p$-portion charts extend well to infinite-*d* cases, for which the charts involve portions of Hilbert, Banach and Fréchet spaces. See e.g. [207, 426, 606] for accounts of Hilbert, Banach and Fréchet manifolds respectively. Banach manifolds are the limiting case as regards retaining a very wide range of analogies with finite manifolds. Fréchet manifolds remain reasonably tractable [207], despite the loss in general of the Inverse Function Theorem, as do Fréchet Lie groups [426].

Finite manifolds' incorporation of differentiable structure also has an analogue in each of the above cases. So e.g. one can consider differentiable functions and tangent vectors for each, and then apply multilinearity to set up versions for tensors of any other rank (p, q) and symmetry type S. In particular, applying this construction to a Fréchet manifold with tangent space $\mathfrak{fre}(\mathfrak{C}^{\infty})$ produces another Fréchet space $\mathfrak{fre}_{S(p,q)}(\mathfrak{C}^{\infty})$.

H.4 Topology of $\mathfrak{Riem}(\Sigma)$

The space of Riemannian geometries $\mathfrak{Riem}(\Sigma)$ can be modelled as an open positive convexcone[2] in the Fréchet space $\mathfrak{fre}_{\mathfrak{S}\mathrm{ym}(0,2)}(\mathfrak{C}^{\infty})$ for $\mathfrak{S}\mathrm{ym}(0, 2)$ the symmetric rank-2 tensors.

$\mathfrak{Riem}(\Sigma)$ can furthermore be equipped [301] with a metric space notion of metric, Dist; this can additionally be chosen to be preserved under *Diff*(Σ). Thus $\mathfrak{Riem}(\Sigma)$ is a metrizable topological space. Consequently $\mathfrak{Riem}(\Sigma)$ obeys all the separation axioms—including in particular Hausdorffness—and it is also paracompact. $\mathfrak{Riem}(\Sigma)$ is additionally second-countable [363], and has an infinite-dimensional analogue of the locally Euclidean property as well; consequently a single type of chart suffices in this case. In this manner, $\mathfrak{Riem}(\Sigma)$ is a manifold that is infinite-dimensional in the sense of Fréchet$(\mathfrak{C}^{\infty})$.

H.5 $\mathfrak{Riem}(\Sigma)$ at Level of Geometrical Metric Structure

Infinite-*d* manifolds can be equipped with connections and metrics [207]. In the dynamical study of GR, $\mathfrak{Riem}(\Sigma)$ is usually taken to carry the infinite-dimensional indefinite Riemannian metric provided by GR's kinetic term, i.e. the inverse DeWitt supermetric M^{abcd} of (8.18). More generally, one might consider other members of the family of *ultralocal supermetrics* [358, 552]

$$\mathrm{M}_{\beta}^{abcd} := \sqrt{\mathrm{h}}\{\mathrm{h}^{ac}\mathrm{h}^{bd} - w\,\mathrm{h}^{ab}\mathrm{h}^{cd}\}. \tag{H.1}$$

These split into 3 cases: the positive-definite $w < 1/3$, the degenerate $w = 1/3$, and the indefinite (heuristically $\{- - + + + + +\}^{\infty}$) $w > 1/3$. Ultralocality readily permits these to be studied pointwise; the more problematic degenerate case is usually dropped from such studies. Pointwise, these supermetrics arise from positive-definite symmetric 3×3 matrices (h_{ab} at that point. The 6-*d* space of these is mathematically [358] $\mathfrak{S}\mathrm{ym}^{+}(3, \mathbb{R})$, which is diffeomorphic to the homogeneous space

$$GL^{+}(3, \mathbb{R})/SO(3) \cong \mathbb{R}_{+} \times \mathbb{R}^{5}. \tag{H.2}$$

[2]This is a Linear Algebra characterization of a space $\mathfrak{S}$ [301, 729], that is not itself linear but obeys $\mathfrak{S} + \mathfrak{S} \subset \mathfrak{S}$ and $m\,\mathfrak{S} \subset \mathfrak{S}$ for $m \in \mathbb{R}_{+}$. See [301] for more on this and for consideration of why Fréchet spaces are appropriate. Do not confuse this use of 'cone' with Appendix G's topological and geometrical uses.

This is the full Minisuperspace; it occurs in Strong Gravity as well [716, 717].

The spatial 3-metric can furthermore be decomposed into scale and unimodular (unit-determinant) parts according to

$$\mathrm{h}_{ab} = \mathrm{h}^{1/3}\mathrm{u}_{ab} \quad \text{for } \mathrm{u}_{ab} := \mathrm{h}^{-1/3}\mathrm{h}_{ab}. \tag{H.3}$$

The pointwise unimodular (unit-determinant) metrics form the 5-d space[3]

$$SL^{+}(3,\mathbb{R})/SO(3) \cong \mathbb{R}^5. \tag{H.4}$$

Ultralocality also implies that these pointwise structures uplift to $\mathfrak{Riem}(\boldsymbol{\Sigma})$ and $\mathfrak{CRiem}(\boldsymbol{\Sigma})$, as per Sect. 43.3. The scale-free part gives rise to 8 Killing vectors and the scale part to a homothety [358]. The corresponding local Riemannian Geometry for this was studied by DeWitt [237], including the form taken by the geodesics. This exhibits various global difficulties: curvature singularities and geodesic incompleteness [874].

H.6 Conformal Variants

The *conformal transformations* $Conf(\boldsymbol{\Sigma})$ are smooth positive functions on $\boldsymbol{\Sigma}$. These form an infinite-d Lie group; moreover this is Abelian under pointwise multiplication. They can be decomposed according to $Conf(\boldsymbol{\Sigma}) = Dil \times VPConf(\boldsymbol{\Sigma})$ for constant scaling Dil and $VPConf(\boldsymbol{\Sigma})$ the global-volume preserving conformal transformations.

The subsequent quotient is *conformal Riem*[4] $\mathfrak{CRiem}(\boldsymbol{\Sigma}) := \mathfrak{Riem}(\boldsymbol{\Sigma})/Conf(\boldsymbol{\Sigma})$. This was first considered by DeWitt [237], who showed it to be simpler and better-behaved than $\mathfrak{Riem}(\boldsymbol{\Sigma})$ at the level of Metric Geometry. This is based firstly on his observation [237] that it is the scale part of the GR configuration space metric which causes the indefiniteness. $\mathfrak{CRiem}(\boldsymbol{\Sigma})$ itself is positive-definite; the natural supermetric thereupon is

$$\mathrm{U}^{abcd} := \mathrm{u}^{ac}\mathrm{u}^{bd}, \tag{H.5}$$

and this is the basis of bona fide notions of distance. Secondly, geodesics are better-behaved upon $\mathfrak{CRiem}(\boldsymbol{\Sigma})$ as compared to $\mathfrak{Riem}(\boldsymbol{\Sigma})$.

On the other hand, $\{\mathfrak{CRiem} + \mathrm{V}\}(\boldsymbol{\Sigma})$'s metric is '$-\{+++++\}^{3\infty}$', which is de facto, rather than just pointwise, hyperbolic. The—direction here corresponds to a global scale variable, such as indeed the global spatial volume, or the cosmological scalefactor a when applicable.

Moreover, many of the configuration spaces have physically-significant singular points both for GR and for RPMs. In particular, $a = 0$ corresponds to the Big Bang

[3]This is the Minisuperspace counterpart of the $\mathfrak{CRiem}(\boldsymbol{\Sigma})$ detailed in the next Section.

[4]This has also been termed 'pointwise conformal superspace' [305]; Chap. 21.4 explains this change of nomenclature.

and $I = 0$ to the maximal collision of Mechanics; these are furthermore analogous through both involving scale variables.

Finally, quotienting out conversely *Dil* alone from $\mathfrak{Riem}(\boldsymbol{\Sigma})$ gives a $\mathfrak{vp}\mathfrak{Riem}(\boldsymbol{\Sigma})$ configuration space: 'volume-preserving $\mathfrak{Riem}(\boldsymbol{\Sigma})$'.

H.7 GR Alongside Minimally-Coupled Matter

This case is useful through including fundamental-field second-order minimally-coupled bosonic matter. The redundant configuration space metric now splits according to the direct sum [835]

$$\mathbf{M} = \mathbf{M}^{\text{grav}} \oplus \mathbf{M}^{\text{mcm}}. \tag{H.6}$$

We use ψ^{Z} to denote fundamental-field second-order minimally-coupled bosonic matter, consisting of blockwise disjoint species ψ^{z}, with $\mathsf{z} \in \mathsf{Z}$. It is usually additionally assumed that $\mathbf{M}$ is independent of the matter fields. This example covers e.g. minimally-coupled scalars, Electromagnetism, Yang–Mills Theory and scalar Gauge Theories, in each case coupled to GR.

In the case of a minimally-coupled scalar field, we denote this configuration space by $\mathfrak{RIEM}(\boldsymbol{\Sigma})$.[5] The (undensitized) metric on this takes the blockwise form $\mathcal{M}(\text{h}) := \begin{pmatrix} 1 & 0 \\ 0 & \mathbf{M}(\text{h}) \end{pmatrix}$ and $\mathbf{M}(\text{h})$ the GR configuration space metric itself. [This immediately extends to the case of N minimally-coupled scalar fields.]

Similar considerations apply throughout to extending $\mathfrak{CRiem}(\boldsymbol{\Sigma})$, $\{\mathfrak{CRiem} + \text{V}\}(\boldsymbol{\Sigma})$ and $\mathfrak{vp}\mathfrak{Riem}(\boldsymbol{\Sigma})$.

H.8 Spaces of Affine Connections

The affine level of structure presents a first 'single-floor' versus 'tower' ambiguity in the sense of Epilogue II.C's nomenclature.

In the single-floor case, if only affine connections are taken to contain physically meaningful information, so $\mathfrak{q} = \mathfrak{a}\text{ff}(\boldsymbol{\Sigma})$, what dynamics do these support? This question can additionally be asked if only 'Weyl connections' —connections which carry out parallel transport up to conformal transformation—are involved, for which $\mathfrak{q} = \mathfrak{ca}\text{ff}(\boldsymbol{\Sigma})$.

In the tower case, one can consider what happens if the spatial affine connection is dynamical independently from the metric. This corresponds to $\mathfrak{q} = \mathfrak{Riem}(\boldsymbol{\Sigma}) \times \mathfrak{a}\text{ff}(\boldsymbol{\Sigma})$, and is to be addressed by varying connections separately, rather than presupposing that the affine connection involved is the metric connection.

[5]We generally use the capped version of a GR configuration space to denote further inclusion of a minimally-coupled scalar field.

Appendix I
GR Model Configuration Space Geometries*

I.1 Minisuperspace: Homogeneous GR

These were introduced in Chap. 9; we now consider a wider range of examples of homogeneous models than in Part I due to the possibility of nontrivial anisotropy. These models are classified by the isometry groups $Isom(\langle \mathbf{\Sigma}, \mathbf{h}\rangle)$ of their spatially homogeneous surfaces.[1] This leads to two cases according to whether $Isom(\langle \mathbf{\Sigma}, \mathbf{h}\rangle)$ acts 'simply transitively' (meaning freely and transitively). If this does not apply, it turns out that [812] there is a single case: $SO(3) \times \mathbb{R}$ acting upon the cylindrical 3-space $\mathbb{S}^2 \times \mathbb{R}$, giving the *Kantowski–Sachs model*. The other case gives the family of *Bianchi models*. The I being Lie groups, they are in turn characterized by the form of their structure constants. They are subdivided according to

$$C^k{}_{ik} = 0 \quad \text{for class A} \quad \text{and} \quad \neq 0 \text{ for class B}. \tag{I.1}$$

A finer classification of $C^k{}_{ij}$ yields a subdivision into Bianchi models labelled by I to IX [812]. The general case's spatial metric can be represented as

$$\mathrm{d}s^2 = h_{ij}\,\mathrm{d}\omega^i\,\mathrm{d}\omega^j \tag{I.2}$$

for 1-forms $\mathrm{d}\omega^k$ obeying $\mathrm{d}^2\omega^k = C^k{}_{ij}\,\mathrm{d}\omega^i \wedge \mathrm{d}\omega^j$.

Example 1) The spatially closed $\mathbb{S}^3$ isotropic case has spatial metric $\mathrm{d}s^2 = a(t)^2\mathrm{d}s^2_{\mathbb{S}^3}$ ($\subset$ Bianchi IX).
Example 2) The Kasner universes ($\subset$ Bianchi I) have

$$\mathrm{d}s^2 = t^{2p_1}\mathrm{d}x_1^2 + t^{2p_1}\mathrm{d}x_1^2 + t^{2p_1}\mathrm{d}x_1^2 \tag{I.3}$$

such that the exponents obey $p_1 + p_2 + p_3 = 1$ and $p_1^2 + p_2^2 + p_3^2 = 1$.

[1]See Appendix N.3 as regards isotropy and isometry turning out to have equivalent groups.

E. Anderson, *The Problem of Time*, Fundamental Theories of Physics 190,
DOI 10.1007/978-3-319-58848-3

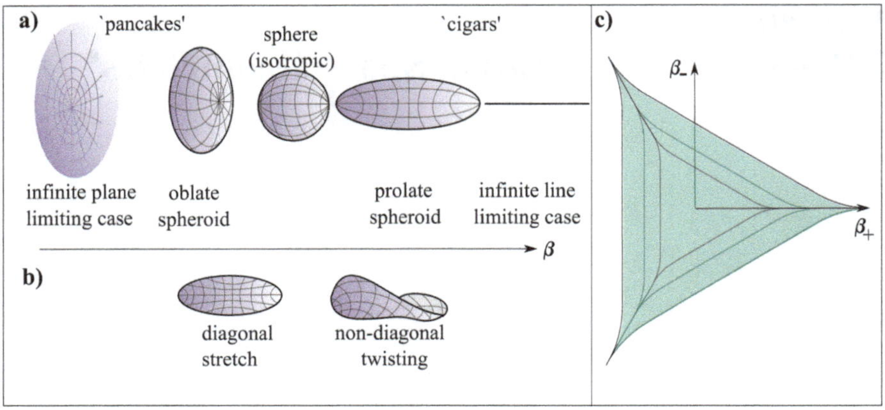

Fig. I.1 **a)** Visualization of diagonal anisotropy in 2-d. **b)** Non-diagonal anisotropy allows for the hypersurface to be 'twisted' as well. In 3-d, there are 2 independent anisotropic stretches and 3 twistings. **c)** The diagonal Bianchi IX model's potential well

Example 3) Diagonal Bianchi IX models have spatial metrics

$$\mathrm{d}s^2 = \exp(2\{-\Omega + \beta_+ + 3\sqrt{3}\,\beta_-\})\mathrm{d}\Omega^2 + \exp(2\{-\Omega + \beta_+ - 3\sqrt{3}\,\beta_-\})\mathrm{d}\beta_+^2 + \exp(2\{-\Omega - 2\,\beta_+\})\mathrm{d}\beta_-^2 \tag{I.4}$$

on $\mathbb{S}^3$. These models are potentially of great importance through being conjectured to be the generic GR behaviour near cosmological singularities [125]. These also have a nontrivial potential term of a specific form inherited from the densitized GR Ricci scalar potential term,

$$\overline{V} = \exp(\Omega)\{V(\boldsymbol{\beta}) - 1\}, \quad \text{for} \tag{I.5}$$

$$V(\boldsymbol{\beta}) = \frac{\exp(-8\,\beta_+)}{3} - \frac{4\exp(-2\,\beta_+)}{3}\cosh(2\sqrt{3}\,\beta_-) + 1 + \frac{2\exp(4\,\beta_+)}{3}\{\cosh(4\sqrt{3}\,\beta_-) - 1\}, \tag{I.6}$$

which is an open-ended well of equilateral triangular cross-section (Fig. I.1.c). For small anisotropies, this takes the approximate form

$$V(\boldsymbol{\beta}) \simeq 8\{\beta_+^2 + \beta_-^2\} =: 8\|\boldsymbol{\beta}\|^2 \tag{I.7}$$

(the lower orders of its Taylor series cancel).
Example 4) The *Taub model* is the $\beta_- = 0$ subcase of the preceding.

To explain the above notation, and to characterize the types of anisotropic configurations in GR more generally, Misner [656] parametrized anisotropy—a type of GR shape variable—by writing the pointwise spatial tracefree metric $u_{ab} = h_{ab}/h^{1/3}$ as $\exp(2\,\beta)_{ab}$ for β_{ab} a tracefree symmetric matrix. In the case of u_{ab} diagonal,

$$\beta_{ab} = \text{diag}(\beta_1, \{\sqrt{3}\,\beta_2 - \beta_1\}/2, -\{\sqrt{3}\,\beta_2 + \beta_1\}/2). \tag{I.8}$$

These are related to $\beta_\pm$ by $\beta_1 = \beta_+ + \sqrt{3}\,\beta_-$, $\beta_2 = \beta_+ - \sqrt{3}\,\beta_-$. Figure I.1 provides a simple conceptual outline of the meaning of anisotropy for a 2-*d* hypersurface.

In the homogeneous case the configuration spaces $\mathfrak{Riem}$, $\mathfrak{CRiem}$ + V, $\mathfrak{S}$uperspace and $\mathfrak{CS}$ + V coincide as *Minisuperspace*, $\mathfrak{Mini}$. On the other hand, $\mathfrak{CRiem}$, $\mathfrak{vpRiem}$, Appendix N.7's $\mathfrak{vpS}$uperspace, and $\mathfrak{CS}$ coincide as *Anisotropy-space*, $\mathfrak{Ani}$, which is yet another example of pure shape space.

As regards configuration space geometries, for diagonal Bianchi class A, $\mathfrak{Mini} = \mathbb{M}^3$ with configuration space metric

$$\mathrm{d}s^2 = -\mathrm{d}\Omega^2 + \mathrm{d}\beta_+^2 + \mathrm{d}\beta_-^2. \tag{I.9}$$

The corresponding $\mathfrak{Ani} = \mathbb{R}^2$, with shape metric

$$\mathrm{d}s^2 = \mathrm{d}\beta_+^2 + \mathrm{d}\beta_-^2. \tag{I.10}$$

Moreover, the Taub solution has $\mathbb{M}^2$ and $\mathbb{R}$ restrictions of these. For the isotropic case, one has $\mathbb{R}$ and a point instead.

Example 5) Upon inclusion of a single minimally-coupled scalar field, we use the corresponding capitalized notation $\mathfrak{MINI}$ and $\mathfrak{ANI}$. In the diagonal Bianchi IX case, $\mathfrak{MINI} = \mathbb{M}^4$ and $\mathfrak{ANI} = \mathbb{R}^3$, going up to $3 + p$ and $2 + p$ in place of 4 and 3 for p minimally-coupled scalar fields.

I.2 Perturbations about Minisuperspace: Unreduced Formulation

As regards incipient (redundant) configuration variables, the 3-metric and scalar field are expanded as [419]

$$\begin{aligned} \mathrm{h}_{ij}(\underline{x}, t) &= \exp(2\Omega(t))\{S_{ij}(t) + \varepsilon_{ij}(\underline{x}, t)\}, \\ \phi(\underline{x}, t) &= N^{-1}\left\{\phi(t) + \sum_{\mathrm{n,l,m}} f_{\mathrm{nlm}}\, \mathrm{Q}^{\mathrm{n}}_{\mathrm{lm}}(\underline{x})\right\}. \end{aligned} \tag{I.11}$$

S_{ij} is here the standard hyperspherical $\mathbb{S}^3$ metric, whereas ε_{ij} are inhomogeneous perturbations of the form

$$\begin{aligned} \varepsilon_{ij} = \sum_{\mathrm{n,l,m}} \Bigg\{ & \sqrt{\frac{2}{3}} a_{\mathrm{nlm}} S_{ij} \mathrm{Q}^{\mathrm{n}}_{\mathrm{lm}} + \sqrt{6}\, b_{\mathrm{nlm}} \{\mathrm{P}_{ij}\}^{\mathrm{n}}_{\mathrm{lm}} \\ & + \sqrt{2}\{c^{\mathrm{o}}_{\mathrm{nlm}}\{\mathrm{S}^{\mathrm{o}}_{ij}\}^{\mathrm{n}}_{\mathrm{lm}} + c^{\mathrm{e}}_{\mathrm{nlm}}\{\mathrm{S}^{\mathrm{e}}_{ij}\}^{\mathrm{n}}_{\mathrm{lm}}\} + 2\{d^{\mathrm{o}}_{\mathrm{nlm}}\{\mathrm{G}^{\mathrm{o}}_{ij}\}^{\mathrm{n}}_{\mathrm{lm}} + d^{\mathrm{e}}_{\mathrm{nlm}}\{\mathrm{G}^{\mathrm{e}}_{ij}\}^{\mathrm{n}}_{\mathrm{lm}}\} \Bigg\}. \end{aligned} \tag{I.12}$$

The superscripts 'o' and 'e' for stand for 'odd' and 'even' respectively. We subsequently use n indices as a shorthand for nlm. Let $x_{\rm n}$ be a collective label for the 6 gravitational modes per n $a_{\rm n}$, $b_{\rm n}$, $c_{\rm n}$ and $d_{\rm n}$, and $y_{\rm n}$ for these alongside the $f_{\rm n}$. The $y_{\rm n}$ are all functions of just the coordinate time (which is also label time for GR) t. The $\mathrm{Q_n}(\underline{x})$ are the $\mathbb{S}^3$ scalar (S) harmonics, $\mathrm{S}^{\rm o}_{i\,\rm n}(\underline{x})$ and $\mathrm{S}^{\rm e}_{i\,\rm n}(\underline{x})$ are the transverse $\mathbb{S}^3$ vector (V) harmonics, and the $\mathrm{G}^{\rm o}_{ij\,\rm n}(\underline{x})$ and $\mathrm{G}^{\rm e}_{ij\,\rm n}(\underline{x})$ are the transverse traceless $\mathbb{S}^3$ symmetric 2-tensor (T) harmonics. The $\mathrm{S}_{ij\,\rm n}(\underline{x})$ are given by $\mathcal{D}_j\mathrm{S}_{i\,\rm n} + \mathcal{D}_i\mathrm{S}_{j\,\rm n}$ (for each of the suppressed o and e superscripts). The $\mathrm{P}_{ij\,\rm n}(\underline{x})$ are traceless objects given by $\mathrm{P}_{ij\,\rm n} := \mathcal{D}_j\mathcal{D}_i\mathrm{Q_n}/\{\mathrm{n}^2-1\} + \mathrm{S}_{ij}\mathrm{Q_n}/3$. $N := \sqrt{2/3\pi}/m_{\rm Pl}$ is a normalization factor.

Additionally, the relational formulation's differential of the frame auxiliary is expanded as

$$\partial \mathrm{F}_i = \exp(\Omega)\sum_{\rm n,l,m}\left\{\mathrm{d}k_{\rm nlm}\{\mathrm{P}_i\}^{\rm n}_{\rm lm}/\sqrt{6} + \sqrt{2}\{\mathrm{d}j^{\rm o}_{\rm nlm}\{\mathrm{S}^{\rm o}_i\}^{\rm n}_{\rm lm} + \mathrm{d}j^{\rm e}_{\rm nlm}\{\mathrm{S}^{\rm e}_i\}^{\rm n}_{\rm lm}\}\right\} \tag{I.13}$$

for $\mathrm{P}_{i\,\rm n} := \mathcal{D}_i\mathrm{Q_n}/\{\mathrm{n}^2-1\}$. The *frame expansion coefficients* $\mathrm{d}j_{\rm n}$, $\mathrm{d}k_{\rm n}$—collectively denoted by $\mathrm{d}u_{\rm n}$—are also functions of t alone. The relational formulation differs from [419] not only in using this distinct formulation of the auxiliary but also in not having a primary lapse to expand. This is because the lapse is not held to have meaningful primary existence, so it is not to be an independent source of perturbations. Consequently the relational formulation has one family of coefficients less than Halliwell–Hawking's (their $g_{\rm nlm}$).

Let us next compare these modewise and SVT splits with the variables encountered in the triangleland model arena. The scaled triangle has a scale–shape split (G.12); this bears some resemblance to the homogeneity–inhomogeneity split of GR, with a closer still resemblance to that in the vacuum case. The pure-shape triangle has a further geometrically meaningful split into θ (ratio) and ϕ (relative angle) variables (G.11). However, due to the presence of the 'spherical polar' $\sin^2\theta\,\pi_\phi^2$ combination, this does not factorize the kinetic arc element (or consequently the Hamiltonian) into polar and azimuthal terms. Consequently, this does not exhibit an analogue of the SVT split itself. On the other hand, the N-a-gon has $n-1$ ratios–relative angle pairs (encoded as $\mathbb{CP}^{n-1}$'s well-known copies of $\mathbb{C}$). Looking at these pairs individually bears some parallel to modewise considerations (probing a large N-a-gon by looking at constituent triangles). RPM perturbations about a Minisuperspace model [46] bear further resemblance but still fall short of having an SVT split.

In the vacuum case, the SIC redundant configuration space is $\mathfrak{Riem}_{0,1,2}(\mathbb{S}^3)$; the 0, 1 and 2 subscripts refer to the orders in perturbation theory that feature in it. This is the $1 + 6 \times \{\text{countable } \infty\}$ space of the scale variable alongside the $x_{\rm n}$. In the minimally-coupled scalar field case, the redundant configuration space is $\mathfrak{RIEM}_{0,1,2}(\mathbb{S}^3)$. This is the $2 + 7 \times \{\text{countable } \infty\}$ space of the scale variable, homogeneous scalar field mode and the $y_{\rm n}$. The first form in Fig. I.2 displays the latter for one mode to second order overall in $y_{\rm n}$, $\mathrm{d}y_{\rm n}$. By (H.6), $\mathfrak{RIEM}_{0,1,2} = \mathfrak{Riem}_{0,1,2}(\mathbb{S}^3) \oplus \mathfrak{Sca}_{0,1,2}(\mathbb{S}^3)$, where $\mathfrak{Sca}$ denotes the scalar field

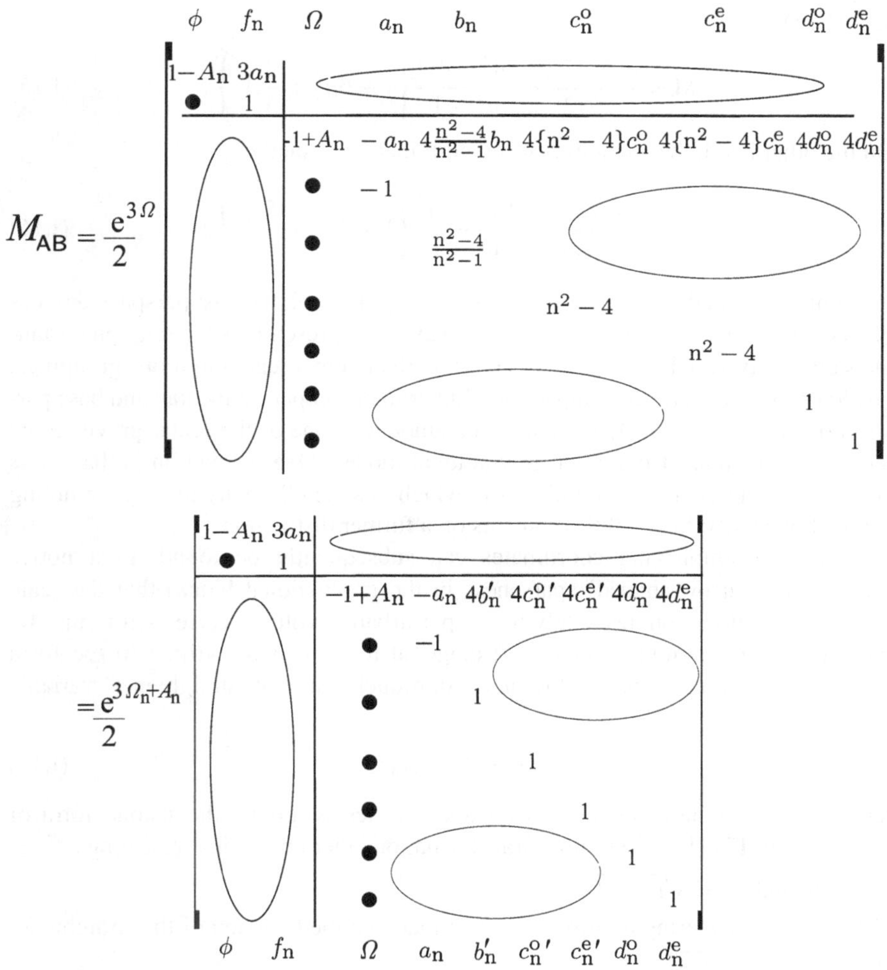

Fig. I.2 **a)** Slightly Inhomogeneous Cosmology (SIC)'s configuration space metric [35]. The heavy dot denotes 'same as the transposed element' since metrics are symmetric. N.B. this is the blockwise corrections' configuration space metric rather than the full one. Moreover, metric variables enter the scalar field sector but scalar field variables do not enter the gravitational sector. This is well-known and held to secure freedom to 'add in' scalar fields in cosmological modelling

configuration space. Thus the former configuration space can readily be read off the figure as a sub-block.

Next introduce the transiently convenient variable

$$A_{\mathrm{n}} := -\frac{3}{2}\left\{a_{\mathrm{n}}^2 - 4\left\{\frac{\mathrm{n}^2-4}{\mathrm{n}^2-1}b_{\mathrm{n}}^2 + \{\mathrm{n}^2-4\}c_{\mathrm{n}}^2 + d_{\mathrm{n}}^2\right\}\right\}. \tag{I.14}$$

This is the gravitational sector configuration space volume correction term: the first perturbative correction to the expansion of the configuration space metric determinant (I.15).

$$M = \frac{\exp(27\,\Omega)}{512}\frac{\{n^2-4\}^3}{n^2-1}\left\{1 - 6a_n^2 + \frac{2}{3}A_n\right\}. \tag{I.15}$$

On the other hand, the vacuum case's determinant is, exactly,

$$M = \frac{\exp(21\,\Omega)}{128}\frac{\{n^2-4\}^3}{n^2-1}\left\{1 + 3a_n^2 + \frac{5}{3}A_n\right\}. \tag{I.16}$$

This enters the study though being the sole coupling to the Minisuperspace degrees of freedom Ω and ϕ. Moreover, in some ways (I.14) resembles the relational triangle's ellipticity variable ${n_2}^2 - {n_1}^2$ (G.15): both are emergent ubiquitous groupings, quadratic and a difference comparison. Rather than comparing median and base partial moments of inertia, A_n compares the amount of one of the scalar gravitational modes against that of the other gravitational modes. One distinction is that A_n is not based on a traceless quadratic form, which ties the ellipticity to the commuting element of $SU(2)$; Sect. 30.5 comments on a further distinction.

Blockwise-simplifying coordinates can subsequently be found. First notice (through the 'arrow-shaped' sparseness in the gravitational block) that the scale degree of freedom couples solely to the perturbative volume correction terms A_n. Indeed, the gravitational sector's off-diagonal terms can be written in the form $dA_n d\Omega_n$ up to a constant. Furthermore, introducing a new 'straightener' variable [34].

$$\Omega_n = \Omega - A_n/3 \tag{I.17}$$

removes the gravitational sector's off-diagonal terms, giving the second form of the metric in Fig. I.2. [The other transformations therein are just rescalings $b'_n := \sqrt{\frac{n^2-4}{n^2-1}}b_n$ and $c'_n := \sqrt{n^2-4}\,c_n$.]

If one is considering multiple n's, the mode-summed version of the straightener variable is [789, 872]

$$\widetilde{\Omega} = \Omega - \sum_n A_n/3. \tag{I.18}$$

The above $\mathfrak{q}$ geometry for SIC is neither flat nor conformally flat. For x_n small, the corresponding Ricci scalar R has no singularities away from the Big Bang. In the minimally-coupled scalar field case, $\partial/\partial\phi$ and $\partial/\partial f_n$ are Killing vectors for SIC's $\mathfrak{q}$ geometry. This corresponds to the 'adding on' status of scalar fields at this level. Additionally, $\partial/\partial\Omega$ is a conformal Killing vector. All of these results are unaffected by passing to multiple scalar fields or to the vacuum case. It remains unknown if this SIC configuration space has further (conformal) Killing vectors.

Let us next introduce scale-free spaces of inhomogeneities for these model arenas. Use w_n as a collective label for the 5 positive-definite gravitational modes b_n, c_n, d_n, and z_n for the 6 positive-definite modes w_n and f_n. $\mathfrak{CRiem}_2(\mathbb{S}^3)$ is the space of the w_n, which is straightforwardly just $\mathbb{R}^5$. Thus it has the obvious 15 Killing vectors built from the Cartesian rescalings of the w_n coordinates 5 $\partial/\partial w_n^{\mathrm{W}}$ and 10

$w_n^W \partial/\partial w_n^{W'} - w_n^{W'} \partial/\partial w_n^W$. On the other hand, $\mathfrak{CRIEM}_{0,1,2}(\mathbb{S}^3)$—the space of homogeneous scalar field modes ϕ alongside the z_n—is neither flat nor conformally flat.

Let us introduce $\mathfrak{M}$odespace as the space of the x_n. This is $\mathbb{M}^6$ and thus it has 21 Killing vectors: 6 $\partial/\partial x_n^X$, 10 $w_n^W \partial/\partial w_n^{W'} - w_n^{W'} \partial/\partial w_n^W$ and 5 $w_n^W \partial/\partial a_n + a_n \partial/\partial w_n^W$. *Positive modespace* $\mathfrak{M}$odespace$^+$ in this case just coincides with $\mathfrak{CR}$iem$_2(\mathbb{S}^3)$. Let us also introduce $\mathfrak{M}$ODESPACE$(\mathbb{S}^3)$ as the space of the x_n. This is $\mathbb{M}^7$ and thus it has 28 Killing vectors: 7 $\partial/\partial y_n^Y$, 15 $z_n^Z \partial/\partial z_n^{Z'} - z_n^{Z'} \partial/\partial z_n^Z$ and 6 $z_n^Z \partial/\partial a_n + a_n \partial/\partial z_n^Z$. Finally, $\mathfrak{M}$ODESPACE$^+$ is the space of the z_n. This is $\mathbb{R}^6$ and thus has 21 Killing vectors: 6 '$\partial/\partial z_n^Z$' and 15 $z_n^Z \partial/\partial z_n^{Z'} - z_n^{Z'} \partial/\partial z_n^Z$.

N.B. that this Sec's analysis readily extends to the case of multiple minimally-coupled scalar fields.

Appendix J
The Standard Principles of Dynamics (PoD). i. Finite Theory

This and the next two Appendices support Facets 1 to 4 of the Problem of Time. Consult [371, 598] as preliminary reading if unfamiliar with this material.

J.1 Lagrangians and Euler–Lagrange Equations

Consider for now a finite second-order classical physical system [371, 598] expressed in Lagrangian variables $\boldsymbol{Q}$, $\dot{\boldsymbol{Q}}$ (named after the great mathematician Joseph-Louis Lagrange). All dynamical information is contained within the *Lagrangian* function $L(\boldsymbol{Q}, \dot{\boldsymbol{Q}}, t)$. The most common form this takes is

$$L = T - V(\boldsymbol{Q}, t) \quad \text{for } T := \| \dot{\boldsymbol{Q}} \|_{\boldsymbol{M}}^2 / 2, \tag{J.1}$$

where $\dot{} = \partial / \partial t$. Next apply the standard prescription of the Calculus of Variations to obtain the equations of motion such that the *action*

$$S = \int \mathrm{d}t \, L \tag{J.2}$$

is stationary with respect to the $\boldsymbol{Q}$. This approach considers the true motion between two particular fixed endpoints e_1 and e_2 alongside the set of varied paths about this motion (subject to the same fixed endpoints). It gives rise to the *Euler–Lagrange equations*,

$$\frac{\mathrm{d}}{\mathrm{d}t}\left\{\frac{\partial L}{\partial \dot{Q}^{\mathsf{A}}}\right\} = \frac{\partial L}{\partial Q^{\mathsf{A}}} \tag{J.3}$$

(named in part after another great mathematician, Leonhard Euler).

These equations simplify in the three special cases below, two of which involve particular types of coordinates. Indeed, one major theme in the Principles of Dynamics is judiciously choosing a coordinate system with as many simplifying coordinates as possible.

E. Anderson, *The Problem of Time*, Fundamental Theories of Physics 190,
DOI 10.1007/978-3-319-58848-3

1) *Lagrange multiplier coordinates* m^{M} are such that L is independent of $\dot{m}^{\mathsf{M}}$,

$$\frac{\partial L}{\partial \dot{m}^{\mathsf{M}}} = 0.$$

The m^{M}-Euler–Lagrange equations then simplify to

$$\frac{\partial L}{\partial m^{\mathsf{M}}} = 0. \tag{J.4}$$

2) *Cyclic coordinates* are such that L is independent of c^{Y} itself,

$$\frac{\partial L}{\partial c^{\mathsf{Y}}} = 0,$$

but features $\dot{c}^{\mathsf{Y}}$: the corresponding *cyclic velocities*. The c^{Y} Euler–Lagrange equations then simplify to

$$\frac{\partial L}{\partial \dot{c}^{\mathsf{Y}}} = const_{\mathsf{Y}}. \tag{J.5}$$

3) *The energy integral type simplification*. If L is free from the independent variable t,

$$\frac{\partial L}{\partial t} = 0,$$

then one Euler–Lagrange equation may be supplanted by the first integral

$$L - \dot{Q}^{\mathsf{A}} \frac{\partial L}{\partial \dot{Q}^{\mathsf{A}}} = \text{constant}. \tag{J.6}$$

Suppose that we can

$$\text{solve} \quad 0 = \frac{\partial L}{\partial m^{\mathsf{M}}}(Q^{\mathsf{O}}, \dot{Q}^{\mathsf{O}}, m^{\mathsf{M}}) \quad \text{as equations for the } m^{\mathsf{M}}. \tag{J.7}$$

Note that these equations come from 1), and that Q^{O} denotes the system's other coordinates which are not multipliers. Moreover, solvability is not in general guaranteed. Firstly, (J.7) can on occasion be not even well-determined due to some of the m^{M} being absent from the equations (e.g. Sect. 19.7) or due to some equations not being independent. Secondly, it is also possible for (J.7) to admit no solution (or only a non-real solution which cannot be applied physically, or a solution that is not in closed form [37]). In the absence of these pathologies, we can pass from $L(Q^{\mathsf{O}}, \dot{Q}^{\mathsf{O}}, m^{\mathsf{M}})$ to a reduced $L_{\text{red}}(Q^{\mathsf{O}}, \dot{Q}^{\mathsf{O}})$, which is known as *multiplier elimination*.

J.2 Conjugate Momenta

These are defined by

$$P_{\mathsf{A}} := \frac{\partial L}{\partial \dot{Q}^{\mathsf{A}}}. \tag{J.8}$$

Explicit computation of this for (J.1) gives the *momentum–velocity relation*

$$P_{\mathsf{A}} = \mathrm{M}_{\mathsf{AA'}} \dot{Q}^{\mathsf{A'}}. \tag{J.9}$$

The definition of P_{A} enables further formulation of the preceding Section's simplifications. Now the preliminary condition in deducing the multiplier condition is $\dot{P}_{\mathsf{Y}} = 0$, the cyclic coordinate condition is

$$P_{\mathsf{Y}} = \text{constant}, \tag{J.10}$$

and the energy integral is

$$L - \dot{Q}^{\mathsf{A}} P_{\mathsf{A}} = \text{constant}. \tag{J.11}$$

J.3 Noether's Theorem

The configurational transformation

$$Q^{\mathsf{A}} \rightarrow Q'^{\mathsf{A}} = Q^{\mathsf{A}} + \epsilon\, \Delta Q^{\mathsf{A}}$$

is a *symmetry of the Lagrangian* if it causes the Lagrangian to remain unchanged modulo a time derivative,

$$L \rightarrow L + \epsilon\, \dot{\theta}.$$

A quantity C is *conserved* if

$$\frac{\mathrm{d}C}{\mathrm{d}t} = 0.$$

Noether's Theorem (*after mathematician Emmy Noether*) *is then that conserved quantities correspond to symmetries. More specifically, the conserved quantity C corresponding to a given symmetry's ΔQ^{A} is*

$$C := P_{\mathsf{A}} \Delta Q^{\mathsf{A}} - \theta. \tag{J.12}$$

E.g. energy, momentum and angular momentum conservation can be viewed in this manner (Ex II.4).

J.4 Legendre Transformations

Suppose we have a function $F(y^{\mathsf{W}}, v^{\mathsf{V}})$ and wish to use

$$z_{\mathsf{W}} := \frac{\partial F}{\partial y^{\mathsf{W}}}$$

as variables in place of the y^{W}. To avoid losing information in the process, we have to apply a *Legendre transformation* (named after 18th and 19th century mathematician Adrien-Marie Legendre). In this way, we pass to a function

$$G(z_{\mathsf{W}}, v^{\mathsf{V}}) = y^{\mathsf{W}} z_{\mathsf{W}} - F(y^{\mathsf{W}}, v^{\mathsf{V}}). \tag{J.13}$$

Moreover, Legendre transformations are symmetric between y^{W} and z_{W}: if one defines

$$y^{\mathsf{W}} := \frac{\partial G}{\partial z_{\mathsf{W}}},$$

the reverse passage yields

$$F(y^{\mathsf{W}}, v^{\mathsf{V}}) = y^{\mathsf{W}} z_{\mathsf{V}} - G(z_{\mathsf{W}}, v^{\mathsf{V}}).$$

In particular, if our function is a Lagrangian $L(\boldsymbol{Q}, \dot{\boldsymbol{Q}})$, we may wish to use some of the conjugate momenta P_{A} as variables in place of the corresponding $\dot{Q}^{\mathsf{A}}$.

J.5 Passage to the Routhian

As a first example of Legendre transformation, start from a Lagrangian with cyclic coordinates c^{Y}, $L(Q^{\mathsf{X}}, \dot{Q}^{\mathsf{X}}, \dot{c}^{\mathsf{Y}})$, and exchange the $\dot{c}^{\mathsf{Y}}$ for the corresponding momenta using (J.10). This is *passage to the Routhian* (after 19th century mathematician Edward Routh)

$$R(Q^{\mathsf{X}}, \dot{Q}^{\mathsf{X}}, p^{c}_{\mathsf{Y}}, t) := L(Q^{\mathsf{X}}, \dot{Q}^{\mathsf{X}}, \dot{c}^{\mathsf{Y}}, t) - P^{c}_{\mathsf{Y}} \dot{c}^{\mathsf{Y}}. \tag{J.14}$$

It amounts to treating the cyclic coordinates as a separate package from the non-cyclic ones; this can be a useful trick, most usually in the context of simplifying the Euler–Lagrange equations.

We proceed by the cyclic velocities analogue of multiplier elimination, known as *Routhian reduction*. This is based upon being able to

$$\text{solve} \quad const_{\mathsf{Y}} = p_{\mathsf{Y}} = \frac{\partial L}{\partial \dot{c}^{\mathsf{Y}}}(Q^{\mathsf{X}}, \dot{Q}^{\mathsf{X}}, \dot{c}^{\mathsf{Y}}) \quad \text{as equations for the } \dot{c}^{\mathsf{Y}}. \tag{J.15}$$

Unlike in the corresponding multiplier elimination, these are not just to be substituted back into the Lagrangian. One additionally needs to apply the Legendre transformation (J.14), thus indeed passing to a Routhian rather than to a naïve reduced Lagrangian. Moreover, the pathologies with (J.7) have counterparts here as well [e.g. Example 5) of Sect. 17.2]. If the above reduction can be performed, we can furthermore use the status of the cyclic momenta as constants (J.10) to free a dynamical problem from its cyclic coordinates. I.e. this completes the corresponding part of the integration of equations of motion. Finally note that the passage from Euler–Lagrange's action principle with no explicit t dependence to Jacobi's action principle (Sect. 15.2) is a subcase of Routhian reduction.

J.6 Passage to the Hamiltonian

A second example of Legendre transformation is passage to the *Hamiltonian* (after the great mathematician and physicist William Rowan Hamilton). In this case, one replaces *all* the velocities $\dot{Q}^{\mathsf{A}}$ by the corresponding momenta P_{A}:

$$H(\boldsymbol{Q}, \boldsymbol{P}, t) := P_{\mathsf{A}}\dot{Q}^{\mathsf{A}} - L(\boldsymbol{Q}, \dot{\boldsymbol{Q}}, t) \tag{J.16}$$

The variables $\boldsymbol{Q}$, $\boldsymbol{P}$ are subsequently termed *Hamiltonian variables*; considerations of whether such a replacement is entirely possible are postponed to Sect. J.15. For instance, for the Lagrangian (J.1),

$$H = \|\boldsymbol{P}\|_N^2/2 + V(\boldsymbol{Q}, t). \tag{J.17}$$

This book concentrates on the t-independent notion of Hamiltonian. The equations of motion are *Hamilton's equations*,

$$\dot{Q}^{\mathsf{A}} = \frac{\partial H}{\partial P_{\mathsf{A}}}, \qquad \dot{P}_{\mathsf{A}} = -\frac{\partial H}{\partial Q^{\mathsf{A}}}; \tag{J.18}$$

in the t-dependent case, these are supplemented by

$$-\frac{\partial L}{\partial t} = \frac{\partial H}{\partial t}.$$

For the Lagrange multiplier coordinates, half of the corresponding Hamilton's equations collapse to just

$$\frac{\partial H}{\partial m^{\mathsf{M}}} = 0. \tag{J.19}$$

On the other hand (J.6) becomes $H = const$ in the t-independent case.

N.B. that further motivations for the Hamiltonian formulation include its admitting a systematic treatment of constraints due to Dirac ([250, 446] and Sect. J.15), and its greater closeness to Quantum Theory.

J.7 Passage to the Anti-Routhian*

The next Appendix requires consideration also of *passage to the anti-Routhian*

$$A(Q^{\mathsf{X}}, P^{\mathsf{X}}, \dot{c}^{\mathsf{Y}}, t) := L(Q^{\mathsf{X}}, \dot{Q}^{\mathsf{X}}, \dot{c}^{\mathsf{Y}}, t) - P_{\mathsf{X}}\dot{Q}^{\mathsf{X}}. \tag{J.20}$$

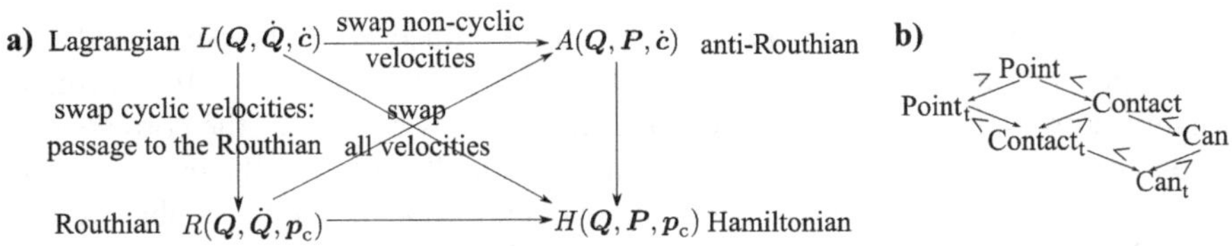

Fig. J.1 **a)** Completion of an elsewise well-known Legendre square by introducing an anti-Routhian that is the diametric opposite of the Routhian in terms of which variables it swaps by Legendre transformation. I.e. it makes the same split into cyclic coordinates and other coordinates, but converts *the other piece's* velocities to momenta. See Fig. Q.1 for the well-known thermodynamical Legendre square counterpart. **b)** Lattice structure of subgroups of Principles of Dynamics morphisms of varying generality

This still involves treating the cyclic coordinates as a separate package, albeit now under the diametrically opposite Legendre transformation. This completes the 'Legendre square' whose other vertices are L, H and R (Fig. J.1.a). Passage to the anti-Routhian turns out to also be a useful trick; a minor use is in Sect. J.15, whereas the major use is in the next Appendix.

J.8 Further Auxiliary Spaces*

The Lagrangian and Hamiltonian variables respectively form the tangent bundle $\mathfrak{T}(\mathfrak{q})$ and cotangent bundle $\mathfrak{T}^*(\mathfrak{q})$ over $\mathfrak{q}$. From a geometrical perspective, the Legendre transformation for passage to the Hamiltonian is thus a map from $\mathfrak{T}(\mathfrak{q})$ to $\mathfrak{T}^*(\mathfrak{q})$.

The Routhian and the anti-Routhian tricks can now be seen to both come at a price. A first part of this price is geometrical: using these requires slightly more complicated *mixed cotangent–tangent bundles* over $\mathfrak{q}$: $\mathfrak{T}(\check{\mathfrak{q}}) \times \mathfrak{T}^*(\mathfrak{C})$ for the Routhian and $\mathfrak{T}^*(\check{\mathfrak{q}}) \times \mathfrak{T}(\mathfrak{C})$ for the anti-Routhian. $\mathfrak{C}$ is here the subconfiguration space of cyclic coordinates and $\check{\mathfrak{q}}$ is the complementary subconfiguration space of the Q^{X}. The second and third parts of the price to pay are in Sects. J.10 and J.13.

J.9 Corresponding Morphisms

The transformation theory for Hamiltonian variables is more subtle than that of the Lagrangian variables' *Point* (Appendix G). This in part reflects the involvement of

$$P_{\mathsf{A}}\dot{Q}^{\mathsf{A}} \tag{J.21}$$

due to its featuring in the conversion from L to H.

Starting from *Point*, one can have the momenta follow suit so as to preserve (J.21) [598]; these transformations indeed preserve H. On the other hand, starting from Point_t induces gyroscopic corrections to H [598]; this illustrates that H itself can change form. More general transformations which mix the $\boldsymbol{Q}$ and the $\boldsymbol{P}$ are also possible. These are however not as unrestrictedly general functions of their $2k$ arguments as *Point*'s transformations are as functions of their k arguments. A first case of these are the transformations which preserve the *Liouville 1-form* (after mathematician Joseph Liouville)

$$P_{\mathsf{A}} \mathrm{d} Q^{\mathsf{A}} \tag{J.22}$$

that is clearly associated with (J.21). These can again be time-independent (termed *scleronomous*) or time-dependent in the sense of parametrization adjunction of t to the $\boldsymbol{Q}$ (termed *rheonomous*). Again, the former preserve H whereas the latter induce correction terms [598]. These are often known as *contact transformations*, so we denote them by *Contact* and $\mathit{Contact}_t$ respectively.

More generally still, preserving the integral of (J.22) turns out to be useful for many purposes [598]. At the differential level, this corresponds to (J.22) itself being preserved up to an additive complete differential dG for G the *generating function*. In this generality, one arrives at the *canonical transformations* alias *symplectomorphisms*, once again in the form a rhenonomous group with a scleronomous subgroup. We denote these by Can_t and *Can* respectively; see Fig. J.1.b) for how this Sec's groups fit together to form a lattice of subgroups.

Note that whereas arbitrary canonical transformations do not permit explicit representation, infinitesimal ones do.

Finally applying Stokes' Theorem to the integral of (J.22) reveals a more basic invariant: the bilinear antisymmetric *symplectic 2-form* [70]

$$\mathrm{d} P_{\mathsf{A}} \wedge \mathrm{d} Q^{\mathsf{A}} . \tag{J.23}$$

We denote this by $\boldsymbol{\omega}$ with components $\omega_{\mathsf{KK}'}$ where the K indices run over 1 to $2k$. This subsequently features in bracket structures (see e.g. two Section further down).

J.10 Morphisms for the (Anti-)Routhian*

Concentrating on the t-independent case that is central to this book, the morphisms for the Routhian formulation are $\mathit{Point}(\check{\mathfrak{q}}) \times \mathit{Can}(\mathfrak{C})$. Moreover, the latter piece is usually ignored due to the c^{Y} being absent and the p^{c}_{Y} being constant. For the t-independent anti-Routhian formulation, the morphisms are $\mathit{Can}(\check{\mathfrak{q}}) \times \mathit{Point}(\mathfrak{C})$. These more complicated morphisms are the second price to pay in considering Routhian or anti-Routhian formulations.

J.11 Poisson Brackets

The *Poisson bracket* (after noted mathematician Siméon Poisson) $\{\,,\,\}$ of quantities $F(\boldsymbol{Q}, \boldsymbol{P})$ and $G(\boldsymbol{Q}, \boldsymbol{P})$ is given by

$$\{F, G\} := \frac{\partial F}{\partial Q^{\mathsf{A}}}\frac{\partial G}{\partial P_{\mathsf{A}}} - \frac{\partial G}{\partial Q^{\mathsf{A}}}\frac{\partial F}{\partial P_{\mathsf{A}}}. \tag{J.24}$$

In terms of these, the equations of motion are

$$\{P_{\mathsf{A}}, H\} = \dot{P}_{\mathsf{A}}, \ \{Q^{\mathsf{A}}, H\} = \dot{Q}^{\mathsf{A}}. \tag{J.25}$$

Furthermore, for any $F(\boldsymbol{Q}, \boldsymbol{P}, t)$, the total derivative

$$\frac{\mathrm{d}F}{\mathrm{d}t} = \{F, H\} + \frac{\partial F}{\partial t},$$

so if F does not depend explicitly on t, the intuitive conserved quantity condition

$$0 = \frac{\mathrm{d}F}{\mathrm{d}t} \text{ becomes } \{F, H\} = 0\,. \tag{J.26}$$

See Sect. 2.13 for some simple and yet significant examples of Poisson brackets. A further example follows from the canonical transformation

$$H \longrightarrow H + \frac{\partial S}{\partial t}$$

for which the new momentum is E. The conjugate to E is a notion of time $Tempus = Tempus(\boldsymbol{Q}, E, t)$, in the sense of there being an energy–time Poisson bracket

$$\{Tempus, E\} = 1; \tag{J.27}$$

this is a classical precursor of Energy–Time Uncertainty Principles. (J.27) is the more general form of energy–time Poisson bracket; moreover, for a conservative system, $Tempus$ simply reduces to the calendar year zero adjusted external background time, $t - t(0)$.

Finally, *Phase space* 𝔓hase [70] is the space of both the $\boldsymbol{Q}$ and the $\boldsymbol{P}$ as equipped with $\{\,,\,\}$ (or the underlying symplectic form $\boldsymbol{\omega}$).

J.12 Poisson Manifolds *

Basic examples of 𝔓hase are often (for simple quadratic 'bosonic' theories) in the form of a cotangent bundle equipped with a Poisson bracket. More generally, 𝔓hase is geometrically a *symplectic manifold*, which is a type of *Poisson manifold* [70, 142, 154, 610]. The difference between these two notions is that the second permits degenerate $\boldsymbol{\omega}$. This generalization plays a significant role in constraint and reduction applications below. These types of manifolds have no local invariants along the lines of Riemannian manifolds' curvature. This can be seen from the trivial form—in which $\boldsymbol{\omega}$ is based on (A.1)'s $\mathbb{J}_p$—always being locally attainable due to *Darboux' Theorem* (after mathematician Gaston Darboux) [70, 614].

Moreover, not all manifolds admit a symplectic structure. For instance, even-dimensionality and orientability are required. In the case of closed manifolds, non-trivial second de Rham cohomology group is also required [614]; this precludes e.g. any sphere other than $\mathbb{S}^2$ from possessing a such.

The *Hamiltonian vector field* is $\mathrm{X_H} := \{H, \ \}$. This can be viewed as a time derivative, a form, or a Lie derivative $\pounds_{\mathrm{X_H}}$. Moreover, being path-connected by segments of the integral curves (Appendix D.2) corresponding to $\mathrm{X_H}$ forms an equivalence. The corresponding equivalence classes constitute *symplectic leaves* (paralleling Chap. 31's treatment of the leaves in foliations). For a Poisson manifold, each leaf carries a natural symplectic structure that is preserved by the $\mathrm{X_H}$. Thus symplectic manifolds are additionally a structure which recurs within the theory of Poisson manifolds [610].

We finally require the *Poisson tensor* $\boldsymbol{C}$ whose components are denoted by $C^{\mathsf{KK'}}$ with K taking values from 1 to $2k$. This is such that the Poisson bracket can be re-expressed as

$$\{F, G\} = C^{\mathsf{KK'}} \partial_{\mathsf{K}} F \, \partial_{\mathsf{K'}} G. \tag{J.28}$$

$\boldsymbol{P}$ is therefore antisymmetric and obeys

$$\epsilon_{\mathsf{KK''K'''}} C^{\mathsf{KK'}} \partial_{\mathsf{K'}} C^{\mathsf{K''K'''}} = 0 \tag{J.29}$$

due to the Jacobi identity. Now clearly $\boldsymbol{\omega} = \boldsymbol{C}^{-1}$ *in cases in which such an inverse exists*, but since it does not always, $\boldsymbol{C}$ is in general a distinct concept. See e.g. [154, 205] for further applications of $\boldsymbol{C}$.

J.13 Peierls Bracket*

A brackets structure can in fact already be built at the Lagrangian tangent bundle level, for all that it is somewhat more complicated than the Poisson bracket. This is the *Peierls bracket* (after physicist Rudolf Peierls) [242, 292, 703]; it is more complicated through its construction involving Green's functions; its explicit form is not required for this book.

The third part of the price to pay if one uses a Routhian or anti-Routhian is that the mixed cotangent–tangent bundle nature of the variables requires in general *mixed Poisson–Peierls brackets*.

J.14 Hamilton–Jacobi Theory

If one carries out the replacement

$$P_{\mathsf{A}} \longrightarrow \frac{\partial S}{\partial Q^{\mathsf{A}}}$$

in H, one obtains the *Hamilton–Jacobi equation*, whose most general form is

$$\frac{\partial S}{\partial t} + H\left(\boldsymbol{Q}, \frac{\partial S}{\partial \boldsymbol{Q}}, t\right) = 0. \tag{J.30}$$

This is to be solved as a PDE for the as-yet undetermined *Hamilton's principal function* $S(\boldsymbol{Q}, t)$. As a particular subcase, if the Hamiltonian is itself time-independent, one can use $S = \chi(\boldsymbol{Q}) - Et$ as a separation ansatz. Here χ is *Hamilton's characteristic function*, which obeys

$$H\left(\boldsymbol{Q}, \frac{\partial \chi}{\partial \boldsymbol{Q}}\right) = E. \tag{J.31}$$

As well as being computationally useful on some occasions [598], the Hamilton–Jacobi formulation is close to the semiclassical approximation to Quantum Theory and the Semiclassical Approach to the Problem of Time and Quantum Cosmology. See [598] for more about Hamilton–Jacobi Theory.

J.15 Hamiltonian Formulation for Constrained Systems

Passage from Lagrangian to Hamiltonian formulation can be nontrivial. The *Legendre (transformation) matrix*

$$\Lambda_{\mathsf{AA'}} := \frac{\partial^2 L}{\partial \dot{Q}^{\mathsf{A}} \partial \dot{Q}^{\mathsf{A'}}} \left(= \frac{\partial P_{\mathsf{A'}}}{\partial \dot{Q}^{\mathsf{A}}} \right) \tag{J.32}$$

—named by its latter form being associated with the Legendre transformation—is in general non-invertible, so the momenta $\boldsymbol{P}$ cannot be independent functions of the velocities $\dot{\boldsymbol{Q}}$. In the case of the action with purely-quadratic kinetic term, the Legendre matrix is just the kinetic matrix $M_{\mathsf{AA'}}$; more generally, it is the metric corresponding to the Lagrangian metric functional. *Constraints* are relations

$$c_{\mathsf{C}}(\boldsymbol{Q}, \boldsymbol{P}) = 0 \tag{J.33}$$

between the momenta $\boldsymbol{P}$ by which these are not independent. Constraints arising at this stage indicates that $\Lambda_{\mathsf{AA'}}$ can be nontrivial: degenerate (thus not even Finsler) or singular. This is quite a general type[1] of constraint considered by Dirac.

[1] As well as the main text's development, these are also restricted in being scleronomous equality constraints, rather than rheonomous constraints. Inequality constraints are also excluded from this

The Euler–Lagrange equations can be rearranged to reveal the explicit presence of the Legendre matrix,

$$\ddot{Q}^{\mathsf{A}'} \frac{\partial^2 L}{\partial \dot{Q}^{\mathsf{A}'} \partial \dot{Q}^{\mathsf{A}}} = \frac{\partial L}{\partial Q^{\mathsf{A}}} - \dot{Q}^{\mathsf{A}'} \frac{\partial^2 L}{\partial Q^{\mathsf{A}} \partial \dot{Q}^{\mathsf{A}'}}. \tag{J.34}$$

The above noninvertibility additionally means that the accelerations are not uniquely determined by $\boldsymbol{Q}$, $\dot{\boldsymbol{Q}}$. Bergmann termed constraints arising from the above non-invertibility of the momentum–velocity relations *primary*. On the other hand, he termed constraints which require input from the variational equations of motion *secondary* [250, 446]. Constraints arising from the propagation of existing constraints using the equations of motion are an intuitively valuable case of this, though see below for limitations on this way of thinking. Let us index primary and secondary constraints by P and S respectively. The constraints $\mathcal{H}$ and $\mathcal{E}$ illustrate that the primary-secondary distinction is artificial insofar as it is malleable by change of formalism: Example 5) of Sect. 24.8.

Dirac introduced the concept of *weak equality* $\approx$, meaning equality up to additive functionals of the constraints. 'Strong equality', on the other hand, means equality in the usual sense.

Dirac also gave a distinct classification of constraints into *first-class* constraints (indexed by F): those whose classical brackets with all the other constraints vanish weakly. *Second-class* constraints are then simply defined by exclusion: as those which are not first-class. The classical brackets involved in the definition are ab initio the Poisson brackets, though we shall see that this can change during the procedure. For the purpose of counting degrees of freedom, it is also useful to note that first-class constraints use up two each whereas second-class constraints use up only one [446].

Dirac began to handle constraints by appending them additively with Lagrange multipliers to a system's incipient or 'bare' Hamiltonian, H. If one additively appends a formalism's primary constraints using a priori any functions F of the $\boldsymbol{Q}$ and $\boldsymbol{P}$, Dirac's generic 'starred' Hamiltonian $H^* := H + F^{\mathsf{P}} c_{\mathsf{P}}$ is formed. On the other hand, Dirac's *total Hamiltonian* is $H_{\text{Total}} := H + u^{\mathsf{P}} c_{\mathsf{P}}$, where the u^{P} are now regarded as unknowns. One begins to consider

$$\dot{c}_{\mathsf{P}} = \{c_{\mathsf{P}}, H\} + u^{\mathsf{P}'}\{c_{\mathsf{P}}, c_{\mathsf{P}'}\} \approx 0. \tag{J.35}$$

The *Dirac Algorithm* [250] then involves checking whether a given set of constraints implies any more constraints or any further types of equation. The equations arising in this manner can be of five types.

---

treatment. *Holonomic* constraints—are ones which are integrable so as to admit a formulation as $c = F(Q, t) = 0$. This is a useful characterization in considering constraints. Note furthermore that, aside from non-integrability, a further way in which constraints can fail to be holonomic is if they are *inequality constraints*. Dirac's treatment, on the one hand, solely considers equality constraints, and, on the other, says nothing about whether these are to be integrable. Moreover, Sects. 39.4 and 43.3 show that Fundamental Physics indeed manifests inequality constraints.

0) *Inconsistencies.*
1) Mere *identities*—equations that reduce to $0 \approx 0$, i.e. $0 = 0$ modulo constraints.
2) Equations independent of the Lagrange multiplier unknowns, which constitute extra secondary constraints.
3) New constraints arise, such that previously known constraints are demonstrated to be in fact second-class, by being second-class with these subsequently encountered constraints.
4) Relations amongst some of the appending Lagrange multipliers functions themselves. These are a further '*specifier equation*' type of equation, i.e. a specification of restrictions on the Lagrange multipliers.[2]

Lest 0) be unexpected, let us supply a basic counter-example to Principles of Dynamics formulations entailing consistent theories. For the Lagrangian $L = \dot{q} + q$, the Euler–Lagrange equations read $0 = 1$. If 0) occurs, the candidate theory is inconsistent. One then either simply gives up on it or one modifies the incipient Lagrangian to pass to a further theory for which this does not happen. Let us call equations of type 1) to 4) the 'consistent equations' arising from the Dirac Algorithm. Moreover, since type 1) are equations with no new content, call types 2) to 4) the 'nontrivial consistent equations. This is a generalization from considering the set of constraints to considering the set of constraints *and specifier equations arising alongside them* in the Dirac Algorithm.

The Dirac Algorithm is to be applied recursively until one of the following three conditions holds.

Termination 0) One has an *inconsistent theory* due to a case of 0) arising.
Termination 1) One has a *trivial theory* due to the iterations of the Dirac Algorithm leaving the system with no degrees of freedom.
Termination 2) *Completion*: the latest iteration of the Dirac Algorithm has produced no new nontrivial consistent equations, indicating that all of these have been found.

If type 2) arises, handle this by defining '$\mathsf{Q} = \mathsf{P} + \mathsf{S}$' as indexing the constraints obtained so far, and restart with a more general form for the problem,

$$\dot{c}_{\mathsf{Q}} = \{c_{\mathsf{Q}}, H\} + u^{\mathsf{P}}\{c_{\mathsf{Q}}, c_{\mathsf{P}}\} \approx 0. \tag{J.36}$$

Indeed, it may be necessary to run the Dirac Algorithm over multiple cycles.

Next suppose that we are handling a set of unknown functions u^{P}. These are of the form [250] $u^{\mathsf{P}} = U^{\mathsf{P}} + V^{\mathsf{P}}$: a split into a complementary function[3] $V^{\mathsf{P}} = v^{\mathsf{Z}} V^{\mathsf{P}}{}_{\mathsf{Z}}$

---

[2]Dirac introduced such entities on p. 14 of [250]; his terminology is 'imposes a condition'. The term 'fixing equations', as in e.g. 'lapse fixing equation', is often used for them in Numerical Relativity. However, this usage is a subcase of gauge-fixing, nor does all gauge-fixing involves specification of Lagrange multipliers. E.g. (6.19) need not be interpreted in this way. On these grounds, the distinct name 'specifier equations' is used in this book.

[3]I.e. the general solution of the corresponding homogeneous system, with Z indexing the number of independent solutions involved.

and a particular solution U^{P}. V^{P} obeys

$$V^{\mathsf{P}}\{c_{\mathsf{C}}, c_{\mathsf{P}}\} \approx 0 \tag{J.37}$$

where v^{A} are the totally arbitrary coefficients of the independent solutions indexed by A. This done, Dirac additionally defined the 'primed Hamiltonian' $H' := H + U^{\mathsf{P}} c_{\mathsf{P}}$, which can be viewed as appending by a determined mixture of free and fixed Lagrange multipliers. Dirac finally defined $H_{\text{Extended}} := H + u^{\mathsf{P}} c_{\mathsf{P}} + u^{\mathsf{S}} c_{\mathsf{S}}$.

We next consider the limitations on thinking in terms of propagating existing constraints. This misses out—or fails to properly identify and handle—second-class constraints and specifier equations. This is especially relevant since many standard quantum procedures are based on just first-class constraints remaining by this stage. This usually entails classical removal of any other nontrivial consistent entities which feature in the original formulation. Four different approaches to this are as follows.

Procedure A) Replace the incipient Poisson brackets with Dirac brackets [250]; this removes second-class constraints.
Procedure B) Extend $\mathfrak{P}$hase with further auxiliary variables so as to 'gauge-unfix' second-class constraints into first-class ones [121, 446].
Procedure C) Classically reduce out the entities in question.
Procedure D) Some approaches make use of gauge-fixing prior to quantizing.

Whereas A) and B) are systematically available, C) is not, though it is solvable for this book's RPM and SIC examples. As regards D), at least in the more standard theories of Physics, first-class secondary constraints can be taken to arise from variation with respect to mathematically disjoint auxiliary variables. Furthermore, the effect of this variation is to additionally use up part of an accompanying mathematically-coherent block of variables that elsewise contains partially physical information. Some constraints are regarded as gauge constraints; however in general exactly which constraints these comprise remains disputed. Moreover, it is agreed upon that second-class constraints are not gauge constraints; all gauge constraints use up two degrees of freedom. Dirac [250] conjectured a fortiori that all first-class constraints are gauge constraints,[4] so that using up two degrees of freedom would conversely imply being a gauge constraint. See however Counter-example 4) of Sect. 24.8). One feature of Gauge Theory is an associated group $\mathfrak{g}$ of transformations that are held to be unphysical. The above-mentioned disjoint auxiliary variables are often in correspondence with such a group. *Gauge-fixing conditions* $\mathcal{F}_{\mathsf{H}}$ may be applied to whatever Gauge Theory (though one requires the final answers to physical questions to be gauge-invariant).

As regards removing second-class constraints prior to quantizing, it is fortunate that *systematic* procedures exist for freeing one's theory of second-class constraints.

[4]This is in Dirac's sense of Gauge Theory [247, 250]: concerning *data at a given time*, so 'gauge' here means *data-gauge*. Contrast this with Bergmann's perspective [133] that Gauge Theory concerns *whole paths* (dynamical trajectories), so 'gauge' there means *path-gauge*.

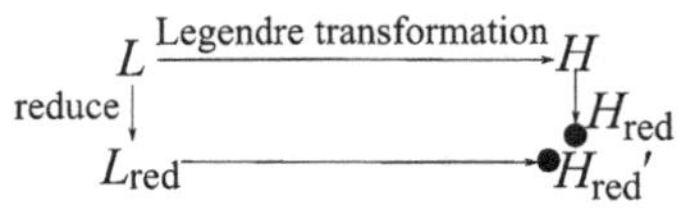

Fig. J.2 The in general noncommuting square of Legendre transformations (*horizontal*) and reductions (*vertical*)

Passage to the Dirac brackets replaces the incipient Poisson brackets with

$$\{F, G\}^* := \{F, G\} - \{F, c_\mathsf{i}\}\{c_\mathsf{i}, c_{\mathsf{i}'}\}^{-1}\{c_{\mathsf{i}'}, G\}. \tag{J.38}$$

Here the -1 denotes the inverse of the given matrix whose i indices run over constraints that are irreducibly second-class [250, 446]. The classical brackets role initially played by the Poisson brackets is then taken over by the Dirac brackets.

Second-class constraints can also always in principle[5] be handled locally by thinking about them instead as 'already-applied' gauge fixing conditions that can be recast as first-class constraints by adding suitable auxiliary variables. By this procedure, a system with first- and second-class constraints *extends* to a more redundant description of a system with just first-class constraints.

Moreover, each of procedures A) to D) render it clear that whether a theory exhibits second-class constraints is in fact a formalism-dependent statement. Also note as regards procedure D) that the square in Fig. J.2 does not in general commute (see e.g. [514]).

Finally, a further useful form for actions is

$$S = \int \mathrm{d}\lambda\{\dot{Q}^\mathsf{A} P_\mathsf{A} - \mathscr{H}_{\mathrm{Total}}\} = \int \mathrm{d}\lambda\{\dot{Q}^\mathsf{A} P_\mathsf{A} - a\,\mathcal{Q}\mathrm{uad} - m^\mathsf{G}\mathcal{G}\mathrm{auge}_\mathsf{G}\}, \tag{J.39}$$

where the second equality is but a common specialization.

J.16 (Anti-)Routhian Analogue of the Legendre Matrix*

With the passage to the Hamiltonian being affected by whether the Legendre matrix is invertible, we should consider whether passage to the (anti-)Routhian is affected as well. The Legendre matrix for the Routhian is

$$\Lambda_{\mathsf{YY}'} := \frac{\partial^2 L}{\partial \dot{c}^\mathsf{Y} \partial \dot{c}^{\mathsf{Y}'}},$$

which is zero by (J.5), so this matrix is an relatively uninteresting albeit entirely obstructive object. The corresponding expressions for acceleration are similarly entirely free of reference to the cyclic variables. On the other hand, the Legendre

[5]This statement follows [446], though we have added the caveat 'locally' since gauge-fixing conditions themselves are not in general global entities.

matrix for the anti-Routhian is

$$\Lambda_{\mathsf{XX'}} := \frac{\partial^2 L}{\partial \dot{Q}^{\mathsf{X}} \partial \dot{Q}^{\mathsf{X'}}},$$

which is in general nontrivial. One can then base a theory of primary constraints on this rather than on the usual larger (J.32). The smaller anti-Routhian trick is the observation that the acceleration of Q^{X} is unaffected by the cyclic variables. I.e. one can take (J.34) again with index X in place of A since the further terms involving the cyclic variables arising from the chain rule are annihilated by (J.5).

J.17 Symplectic Treatment of Constrained Systems*

Given a type of constraint, one can furthermore consider the surface within $\mathfrak{P}$hase that is characterized by that type of constraint holding. See e.g. [142] for the primary constraint surface picked out by the form of the Legendre map; secondary constraint surface is an uncommon subject, but see e.g. [797]. The main thrust of this approach, however, involves first-class and second-class constraint surfaces. See e.g. [152] and for various approaches to these using basic Symplectic Geometry. This includes viewing the Dirac bracket geometrically as a more reduced formulation's notion of Poisson bracket (see e.g. [797] for details), and is both physically insightful and mathematically fruitful.

J.18 Constraint and Beables Algebraic Structures*

The set of constraints in one's possession is entered into ones notion of bracket to form a constraint algebraic structure $\mathfrak{C}$. This may enlarge one's set of constraints, or lead to one adopting a distinct type of bracket. If inconsistency is evaded, the eventual output is an algebraic structure for all of a theory's constraints, symbolically

$$\{\mathcal{C}_{\mathsf{F}}, \mathcal{C}_{\mathsf{F'}}\}_{\text{final}} = C^{\mathsf{F''}}{}_{\mathsf{FF'}} \mathcal{C}_{\mathsf{F''}}. \tag{J.40}$$

In some cases, the $C^{\mathsf{F''}}{}_{\mathsf{FF'}}$ are a Lie algebra's structure constants, whereas in other cases—in particular for GR, they are a Lie algebroid's structure functions. See Appendix V.6 for more about algebroids, and Fig. 24.7 about the lattice $\mathfrak{L}_{\mathfrak{C}}$ of constraint algebraic substructures.

It is now also natural to ask which quantities $\mathcal{B}(\boldsymbol{Q}, \boldsymbol{P})$ form zero brackets with a constraints that form a closed algebraic structure:

$$\{\mathcal{C}_{\mathsf{C}}, \mathcal{B}\}\text{`} =' 0, \tag{J.41}$$

where ‘ $='$ denotes whichever kind of equality (strong, weak, ...). The entities obeying this condition are observables or beables; these are explored further in Chap. 25. More generally, let $|[\ ,\ ]|$ denote whichever type of bracket is used to define observables or beables (e.g. this includes also quantum commutators).

Lemma 1 *Notions of beables can only be meaningfully associated with closed constraint algebraic (sub)structures* [32].

Proof Suppose $\mathcal{B}$ commutes solely with a set of $\mathcal{C}_{\mathsf{C}}$ which is not closed, i.e. it does not include some of the $|[\mathcal{C}_{\mathsf{C}}, \mathcal{C}_{\mathsf{C}'}]|$. However, the Jacobi identity with one $\mathcal{B}$ and two $\mathcal{C}$ as entries and making two uses of (25.1) gives

$$|[\mathcal{B}_{\mathsf{B}}, |[\mathcal{C}_{\mathsf{F}}, \mathcal{C}_{\mathsf{C}'}]|\,]| = -|[\mathcal{C}_{\mathsf{F}}, |[\mathcal{C}_{\mathsf{F}'}, \mathcal{B}_{\mathsf{B}}]|\,]| - |[\mathcal{C}_{\mathsf{F}'}, |[\mathcal{B}_{\mathsf{B}}, \mathcal{C}_{\mathsf{F}}]|\,]| \approx 0, \tag{J.42}$$

which is a contradiction. Thus such a $|[\mathcal{C}_{\mathsf{C}}, \mathcal{C}_{\mathsf{C}'}]|$ in fact has to be included among the quantities $\mathcal{B}$ commutes with. □

Lemma 2 *The* $\mathcal{B}$ *close under whichever* $|[\ ,\ ]|$ *is used to define them*:

$$|[\mathcal{B}_{\mathsf{B}}, \mathcal{B}_{\mathsf{B}'}]|\text{`}=\text{'}\,0. \tag{J.43}$$

Proof Take the Jacobi identity with two $\mathcal{B}$ and one $\mathcal{C}$ as entries

$$|[\mathcal{C}_{\mathsf{F}}, |[\mathcal{B}_{\mathsf{B}}, \mathcal{B}_{\mathsf{B}'}]|\,]| = -|[\mathcal{B}_{\mathsf{B}}, |[\mathcal{B}_{\mathsf{B}'}, \mathcal{C}_{\mathsf{F}}]|\,]| - |[\mathcal{B}_{\mathsf{B}'}, |[\mathcal{C}_{\mathsf{F}}, \mathcal{B}_{\mathsf{B}}]|\,]|\text{`}=\text{'}\,0, \tag{J.44}$$

and make two uses of (25.1). This yields that $|[\mathcal{B}_{\mathsf{B}}, \mathcal{B}_{\mathsf{B}'}]|$ obeys (25.1) as well. □

Note moreover that the beables algebraic structure $\mathfrak{b}$ can be an algebra

$$|[\mathcal{B}_{\mathsf{B}}, \mathcal{B}_{\mathsf{B}'}]| = C^{\mathsf{B}''}{}_{\mathsf{BB}'}\mathcal{B}_{\mathsf{B}''}, \tag{J.45}$$

or an algebroid

$$|[\mathcal{B}_{\mathsf{B}}, \mathcal{B}_{\mathsf{B}'}]| = C^{\mathsf{B}''}{}_{\mathsf{BB}'}(\boldsymbol{Q}, \boldsymbol{P})\mathcal{B}_{\mathsf{B}''}. \tag{J.46}$$

Compare also the rather more familiar case of the Casimirs (Appendix E.6) as a model of associating an algebraic structure with a given algebraic structure via a commutativity condition.

Useful Lemma 3 *If* $\mathcal{B}$ *are beables, then so are the functionals* $\mathcal{F}[\mathcal{B}]$.

Proof The PDE for the beables is linear, so (O.8) applies. □

Corollary [Composition Principle] *In the case of multiple functional dependency restrictions applying, the composition of these restrictions applies. See Sect.* 25.7 *for examples.*

Since classical beables equations are of this form, classical beables form a very sizeable algebraic structure of functions. This also renders 'basis beables' a useful concept. These are a spanning set of linearly independent beables, so that the contents of one's theory can be described entirely in terms of them. A common case is for dim(reduced $\mathfrak{P}$hase) $= 2\{k - g\}$ Kuchař basis beables to be required, where $k = \dim(\mathfrak{q})$ and just g constraints—all first-class linear—involved. See e.g. Sect. 25.7 and [28, 37] for RPM examples of 'basis beables', and Sect. 30.5 for modewise SIC examples.

Finally, see Chaps. 24 and 25 for more features of, and types of, constraint and beables algebraic structures.

J.19 Hamilton–Jacobi Theory in Presence of Constraints

Suppose that one's system has first-class constraints $c_{\mathsf{D}}(\boldsymbol{Q}; \boldsymbol{P})$ other than an 'energy equation'. Then the corresponding Hamilton–Jacobi equation—whether (J.30) or an *incipiently* timeless (J.31) which already takes into account the presence of an 'energy equation' constraint—is supplemented by

$$c_{\mathsf{D}}\left(\boldsymbol{Q}, \frac{\partial \chi}{\partial \boldsymbol{Q}}\right) = 0. \tag{J.47}$$

J.20 Classical Brackets Extended to Include Fermions*

Mixtures of bosonic and fermionic species can be accommodated by introducing physicist Roberto Casalbuoni's brackets [198]

$$\{F, G\}_{\mathrm{C}} := \frac{\partial F}{\partial Q^{\mathsf{A}}} \frac{\partial G}{\partial P_{\mathsf{A}}} - (-)^{\epsilon_F \epsilon_G} \frac{\partial G}{\partial Q^{\mathsf{A}}} \frac{\partial F}{\partial P_{\mathsf{A}}}. \tag{J.48}$$

Here ϵ_S the Grassmann parity of species A (+ for bosons and—for fermions). This bracket also readily generalizes to field-theoretic form. It obeys the Grassmannian generalization of the Jacobi identity,

$$\{\{F, G\}_{\mathrm{C}}, J\}_{\mathrm{C}}(-1)^{\epsilon_F \epsilon_J} + \text{cycles} = 0. \tag{J.49}$$

Appendix K
The Standard Principles of Dynamics. ii. Field Theory

K.1 Classification of Field-Theoretic Versions of the Principles of Dynamics

There are multiple versions of this:

1) Newtonian Field Theory [313, 371].
2) SR Field Theory in flat spacetime form.
3) Field Theory in curved spacetime.
4) GR Field Theory *of* curved spacetime (possibly alongside other fields).

The last three also come in space–time split form, as is useful for Dynamics, conservation, Canonical Approaches and the eventual onset of equal-time commutation relations. On the other hand, the spacetime form has less structures to consider: there are no separated out velocities or momenta, and consequently no notions built upon these such as cyclic coordinates, Routhians or Hamiltonians. Finally, Sect. J.15 and J.18's consideration of constraints and beables carry over well to space–time split field theoretic use.

The book's main applications are scalar Field Theories and Electromagnetism—in both the SR and the GR setting—and vacuum GR itself. All of these are considered in both spacetime and space–time split forms. These examples partly account for the selection of material presented below. Another application is the inherent interest of the change in status of conservation laws and Noether's Theorem in passing from SR to GR This is tied to notions of energy acquiring a more subtle and as yet less fully understood status in GR.

K.2 SR Spacetime Version

Actions are here of the functional form

$$\mathsf{S} = \int \mathrm{d}^4 x \, \mathcal{L}(X^\mu, \eta_{\mu\nu}; \psi^Z(X^\mu)], \tag{K.1}$$

E. Anderson, *The Problem of Time*, Fundamental Theories of Physics 190,
DOI 10.1007/978-3-319-58848-3

which is furthermore taken to be restricted to some Lorentz-invariant combination; see Chap. 6 for examples. Varying with respect to ψ gives the Euler–Lagrange equations, which are now of the form

$$\partial_\mu \left\{ \frac{\delta \mathcal{L}}{\delta\, \partial_\mu \psi^{\mathrm{Z}}} \right\} = \frac{\delta \mathcal{L}}{\delta\, \psi^{\mathrm{Z}}}. \tag{K.2}$$

(K.2) is restricted to the case of second-order theories; (6.8) is a first-order example.

Lagrange multiplier coordinates continue to make sense in Field Theory; they are now in general functions of $\vec{X}$. Their Euler–Lagrange equation collapses to

$$\frac{\delta \mathcal{L}}{\delta \psi^{\mathrm{Z}}} = 0. \tag{K.3}$$

Multiplier elimination continues to make sense as well. Moreover this can now be far more complicated due to the possibility of derivative operators acting on the multipliers. I.e. this is now in general a PDE problem rather than an algebraic one; see Chap. 18 for examples.

Consistency Counter-example 2) to SR or Lorentz invariance guaranteeing consistency. Let $\mathcal{L} = \mathcal{F} + m_\Gamma \mathcal{G}^\Gamma$, for $\mathcal{F}$ and $\mathcal{G}^\Gamma$, $\Gamma = 1$ to p all functionally-independent Lorentz-invariant scalars for a theory with less than p degrees of freedom, and vector-valued Lagrange multiplier m_Γ.

Finally, Noether's Theorem perseveres in the following form. For

$$\psi^{\mathrm{Z}} \to \psi'^{\mathrm{Z}} = \psi^{\mathrm{Z}} + \epsilon\, \Delta\psi^{\mathrm{Z}}$$

preserving $\mathcal{L}$ modulo a divergence,

$$\begin{aligned} \mathcal{L} \to \mathcal{L}' &= \mathcal{L} + \epsilon\, \partial_\mu \theta^\mu, \\ \mathrm{j}^\mu &:= \frac{\delta \mathcal{L}}{\delta\, \partial_\mu \psi^{\mathrm{Z}}} \delta\psi^{\mathrm{Z}} - \theta^\mu \end{aligned} \tag{K.4}$$

is then conserved, i.e. obeys the flat spacetime conservation law

$$\partial_\mu \mathrm{j}^\mu = 0.$$

Ex II.4 contains some examples.

K.3 Space–Time Split SR Version

The split action is of the form

$$\mathsf{S} := \int \mathrm{d}t \int \mathrm{d}^3x\, \mathcal{L} := \int \mathrm{d}t \int \mathrm{d}^3x \{\mathrm{T} - \mathcal{V}\}.$$

The corresponding split Euler–Lagrange equation is

$$-\partial_t \left\{ \frac{\delta \mathcal{L}}{\delta \dot{\psi}^{\mathsf{Z}}} \right\} + \partial_a \left\{ \frac{\delta \mathcal{L}}{\delta\, \partial_a \psi^{\mathsf{Z}}} \right\} = \frac{\delta \mathcal{L}}{\delta \psi^{\mathsf{Z}}}. \tag{K.5}$$

The E–B form of the Maxwell equations (8.4), (3.2) is an example of such.

On occasion, some of the (K.5) will collapse to multiplier equations (K.3), with multiplier elimination in general following the preceding Sec's lead. Furthermore, in the split case there is also a clear-cut notion of cyclic coordinate, for which (K.5) collapses rather to

$$\frac{\delta \mathcal{L}}{\delta \dot{\psi}^{\mathsf{Z}}} = C(x^a) \tag{K.6}$$

The momenta are now

$$\pi^{\psi}_{\mathsf{A}} := \frac{\delta \mathcal{L}}{\delta \dot{\psi}^{\mathsf{A}}}. \tag{K.7}$$

Straightforward analogues of passage to the Routhian and to the Hamiltonian then hold; also phase space and the symplectic approach remain available for Field Theory [142].

It is well-known that a current j_I^{μ} satisfying $\partial_\mu \mathrm{j}_I^{\mu} = 0$ has corresponding conserved charge

$$\mathrm{Q}_I = \int_{\boldsymbol{\Sigma}} \mathrm{d}\boldsymbol{\Sigma}_\mu \mathrm{j}_I^{\mu} = \int_{t=const} \mathrm{d}^3 x\, \mathrm{j}_I^{0}, \tag{K.8}$$

in accord with Noether's Theorem. This includes the case of energy

$$E = \int \mathrm{d}^3 x\, \mathrm{E}^0$$

being straightforwardly conserved modulo introduction of the corresponding momentum flux P^i. What is less widely known is this result's dependence on Minkowski spacetime $\mathbb{M}^n$ possessing a timelike Killing vector, which affects generalizations as per two Secs down.

The Hamiltonian itself is

$$\mathcal{H}(\underline{x}, \pi^{\psi}; \psi] := \pi^{\psi}_{\mathsf{A}} \dot{\psi}^{\mathsf{A}} - \mathcal{L}(\underline{x}, \dot{\psi}; \psi]. \tag{K.9}$$

The field-theoretic Poisson bracket requires smearing,[1] and takes the form

$$\{\mathcal{F}, \mathcal{G}\} := \int_{\mathbb{R}^3} \mathrm{d}^3 x \left\{ \frac{\delta \mathcal{F}}{\delta \mathrm{Q}^{\mathsf{A}}} \frac{\delta \mathcal{G}}{\delta \mathrm{P}_{\mathsf{A}}} - \frac{\delta \mathcal{F}}{\delta \mathrm{P}_{\mathsf{A}}} \frac{\delta \mathcal{G}}{\delta \mathrm{Q}^{\mathsf{A}}} \right\}. \tag{K.10}$$

---

[1] In general, we use $(c_{\mathsf{W}} | \mathrm{A}^{\mathsf{W}}) := \int d^3 x\, c_{\mathsf{W}}(\underline{x}; \psi, \mathbf{h}] \mathrm{A}^{\mathsf{W}}(\underline{x})$ (an 'inner product' notation) for the *smearing* of a W-tensor density-valued constraint c_{W} by an opposite-rank W-tensor smearing with no density weighting: A^{W}. ψ^{A} are the matter fields involved.

The Legendre matrix is now

$$\Lambda_{\mathsf{ZZ'}} = \frac{\delta^2 \mathcal{L}}{\delta \dot{\psi}_{\mathsf{Z}} \delta \dot{\psi}_{\mathsf{Z'}}}.$$

Constraint classification, the Dirac bracket, the extended approach and Dirac's Algorithm carry over to Field Theory. See Chap. 6 for Electromagnetism and Yang–Mills Theory as simple examples of this.

Notions of observables and beables carry over to Field Theory; so do constraint and beables algebraic structures. Useful Lemma 3 on beables continues to hold when beables are determined by an FDE instead of a PDE.

The general field-theoretic Hamilton–Jacobi equation is

$$\frac{\partial \mathcal{S}}{\partial t} + \mathcal{H}\left(t, \underline{x}, \frac{\delta \mathcal{S}}{\delta \psi}; \psi\right] = 0. \tag{K.11}$$

For time-independent $\mathcal{H}$, $\mathcal{S}(\underline{x}, t; \psi] = \chi(\underline{x}; \psi] - Et$ separates this out, leaving an equation

$$\mathcal{H}\left(\underline{x}, \frac{\delta \chi}{\delta \psi}; \psi\right] = E \tag{K.12}$$

to be solved for Hamilton's characteristic function, χ. If first-class constraints c_{D} (other than $\mathcal{H}$) are present, the preceding is to be supplemented by

$$c_{\mathsf{D}}\left(\underline{x}; \psi, \frac{\delta \chi}{\delta \psi}\right] = 0. \tag{K.13}$$

K.4 Curved Spacetime and GR Versions

Actions on curved spacetime are of the form

$$\mathsf{S} = \int \mathrm{d}^4 x \, \mathcal{L}(X^\mu; g_{\mu\nu}, \psi^{\mathsf{Z}}], \tag{K.14}$$

furthermore making use of a generally covariant combination. The 'in curved spacetime' version of this involves just variations with respect to ψ. The GR version—(7.7) plus matter terms—additionally involves variations with respect to **g**, from which the Einstein field equations follow as the corresponding Euler–Lagrange equations. It then follows that GR *adds further field equations*, whereas involving curved space can itself lead to such as Curved Geometry factors in integrals and the absence of Killing vectors.

Consistency Counter-example 3) to GR or spacetime General Covariance guaranteeing consistency. Repeat Counter-example 2)'s multiplier construction but now with spacetime generally covariant objects.

K.5 Space–Time Split GR Version

We next consider the ADM action (8.17); for inclusion of scalar fields and Electromagnetism, see Sect. 18.11; Chap. 32 touches upon canonical treatment of more general matter fields, with references. GR's own momenta are (8.21), whereas the other field momenta are

$$\pi^{\Psi}_{Z} := \frac{\delta \mathcal{L}}{\delta \dot{\psi}^{Z}}. \tag{K.15}$$

Sect. K.3's versions of multiplier coordinates, cyclic coordinates, multiplier elimination, Routhian reduction and passage to the Hamiltonian carry through. ADM lapse α and shift β^i are now examples of multiplier coordinates, with ties to Chap. 8 and 18. The bare GR Hamiltonian is zero, though of course there are constraints $\mathcal{H}$ and $\mathcal{M}_i$, giving Sect. 24.3's total Hamiltonian.

Noether's Theorem carries over to stationary spacetimes [205]; therein time translation symmetry continues to imply energy conservation. Stationary spacetimes are however nongeneric. Once this restriction is dropped, conservation and energy both become rather more involved and unsettled notions within GR's Einsteinian Paradigm of Physics [242, 370]. In more detail, if ξ^{A} is Killing and u^{A} is the tangent to a geodesic γ, then $C = \xi_{A}u^{A}$ is constant along γ. This is because $u^{B}\mathcal{D}_{B}C = u^{B}u^{A}\mathcal{D}_{B}\xi_{A} + \xi_{A}u^{B}\mathcal{D}_{B}u_{A} = 0$ by ξ^{A} Killing and u^{A} geodesic. Moreover, this is only enough to stipulate a conserved quantity if it is compatible with the theory's further structures (e.g. potentials in Mechanics). By this, not all isometries carry over to conserved quantities. On the other hand, conformal Killing vectors correspond to conserved quantities along null (but not in general timelike) geodesics.

In a heuristic sense, the divergence of the Einstein tensor takes the place of the notion of conservation law [370]. The energy–momentum–stress tensor is the variational derivative of the action with respect to the metric (7.8). One format for considering conservation laws in a GR format involves adding gravitational energy–momentum–stress pseudotensors[2] to the matter's energy–momentum–stress tensor. This is so as to form some notion of conserved quantity, at least in some regime. However, general relativists have a number of issues with such approaches [799, 823, 824, 874]. The *Bel–Robinson tensor* (after relativists Lluís Bel and Ivor Robinson) is a gravitational analogue of how $T_{\mu\nu}$ is constructed from $F_{\mu\nu}$ in Electromagnetism, now making use of the Weyl tensor instead:

$$\mathcal{T}_{\mu\nu\rho\sigma} := \mathcal{C}_{\mu\rho\gamma\delta}\mathcal{C}_{\nu}{}^{\gamma}{}_{\sigma}{}^{\delta} + \mathcal{C}^{*}_{\mu\rho\gamma\delta}\mathcal{C}^{*}_{\nu}{}^{\gamma}{}_{\sigma}{}^{\delta}. \tag{K.16}$$

Indeed, out of (especially localized) energy being a contentious issue in GR, numerous energy candidates have been proposed (cf. this book's consideration of numerous time candidates). In fact, the two may be expected to bear some relation to each other, through 'time and energy being conjugate quantities' (see Sect. 35.7).

---

[2]This is meant here in a distinct sense from that of Appendix D.2's.

The Poisson bracket is now

$$\{\mathcal{F}, \mathcal{G}\} := \int_{\Sigma} \mathrm{d}\boldsymbol{\Sigma} \left\{ \frac{\delta \mathcal{F}}{\delta \mathrm{h}_{ij}} \frac{\delta \mathcal{G}}{\delta \mathrm{p}^{ij}} - \frac{\delta \mathcal{F}}{\delta \mathrm{p}^{ij}} \frac{\delta \mathcal{G}}{\delta \mathrm{h}_{ij}} \right\}. \tag{K.17}$$

The GR constraints are first-class (and both secondary in the ADM formulation). The algebraic structure formed by the GR constraints is the Dirac algebroid (9.31)–(9.33). The lapse fixing equations (21.30)–(21.31) are examples of field-theoretic specifier equations.

Finally, the *Peres* alias *Einstein–Hamilton–Jacobi equation* is the FDE

$$\left\| \frac{\delta \mathcal{S}}{\delta \mathbf{h}} \right\|_{\mathbf{M}}^{2} + \mathcal{R} - 2\Lambda = 2\mathcal{E} \ (= 0 \text{ in vacuo}). \tag{K.18}$$

The Hamilton–Jacobi form of the accompanying GR momentum constraint is the FDE

$$-2\,\mathrm{h}^{jk}\mathcal{D}_j \frac{\delta \mathcal{S}}{\delta \mathrm{p}^{ik}} = \mathrm{J}_i \ (= 0 \text{ in vacuo}). \tag{K.19}$$

Appendix L
Temporal Relationalism Implementing Principles of Dynamics (TRiPoD)*

L.1 Finite-Field Theoretic Portmanteau Notation

It is straightforward (Exercise!) to fold the previous Appendix's finite and 'split curved spacetime field' presentations into Chap. 18's portmanteau form. This is based on the suite of analogous calculi (up to the subcase of Fréchet in Appendix H).

The current Appendix summarizes the subsequent Temporal Relationalism implementing (TRi) reformulation of the Principles of Dynamics in portmanteau form. See [44] for an account of the finite case.

L.2 Jacobi–Mach Formulation

We continue to restrict our treatment to second-order physical systems, and now work in the absence of time at the primary level, as per Chap. 15. Consequently, there is no derivative with respect to time and thus no notion of velocity $\dot{\mathsf{Q}}^{\mathsf{A}}$ at the primary level. Instead, we use *change in configuration* $\mathbf{d}\mathsf{Q}^{\mathsf{A}}$ due to being open to resolving primary-level timelessness through Mach's Time Principle: with a secondary notion of time to be abstracted from change. Thus in TRiPoD, *Machian variables* $\mathbf{Q}, \mathbf{dQ}$ supplant the usual Principles of Dynamics's Lagrangian variables $\mathbf{Q}, \dot{\mathbf{Q}}$.

All dynamical information is now contained within the *Jacobi arc element* $\mathbf{d}\mathscr{J}(\mathbf{Q}, \mathbf{dQ})$, which has supplanted the time-independent Lagrangian $\mathscr{L}(\mathbf{Q}, \dot{\mathbf{Q}})$. The action S is itself an *unmodified* concept: it is already in TRi form, albeit now additionally bearing the relation $\mathsf{S} = \int \mathbf{d}\mathscr{J}$ to the TRiPoD formulation's Jacobi arc element $\mathbf{d}\mathscr{J}$. There is clearly also no primary notion of kinetic energy; this has been supplanted by the *kinetic arc element* $\mathbf{ds}$ given by (15.7). Moreover, $\mathbf{d}\mathscr{J} = \sqrt{2\mathscr{W}}\,\mathbf{ds}$ for $\mathscr{W}(\mathbf{Q})$ the usual potential factor, so the kinetic and Jacobi arc elements are related by a conformal transformation. In terms of $\mathbf{d}\mathscr{J}$, Dynamics has been cast in the form of a *geodesic principle* [98], or, in terms of $\mathbf{ds}$ as a *parageodesic principle* [659].

E. Anderson, *The Problem of Time*, Fundamental Theories of Physics 190,
DOI 10.1007/978-3-319-58848-3

We next apply the Calculus of Variations to obtain the equations of motion such that S is stationary with respect to the $\mathbf{Q}$. See Sect. L.3 for comments on the particular form taken by this variation. The resulting equations of motion the '*Jacobi–Mach equations*',

$$\mathbf{d}\left\{\frac{\partial\, \mathbf{d}\mathscr{J}}{\partial\, \mathbf{d}\mathsf{Q}^{\mathsf{A}}}\right\} - \frac{\partial\, \mathbf{d}\mathscr{J}}{\partial \mathsf{Q}^{\mathsf{A}}} = 0, \tag{L.1}$$

in place of the usual Principles of Dynamics's Euler–Lagrange equations.

The Jacobi–Mach equations also admit three simplified cases.

1) *Lagrange multiplier coordinates* $\mathsf{m}^{\mathsf{M}} \subseteq \mathsf{Q}^{\mathsf{A}}$ are such that $\mathbf{d}\mathscr{J}$ is independent of $\mathbf{d}\mathsf{m}^{\mathsf{M}}$,

$$\frac{\partial\, \mathbf{d}\mathscr{J}}{\partial\, \mathbf{d}\mathsf{m}^{\mathsf{M}}} = 0.$$

The corresponding Jacobi–Mach equation is

$$\frac{\partial\, \mathbf{d}\mathscr{J}}{\partial \mathsf{m}^{\mathsf{M}}} = 0. \tag{L.2}$$

2) *Cyclic coordinates* $\mathsf{c}^{\mathsf{Y}} \subseteq \mathsf{Q}^{\mathsf{A}}$ are such that $\mathbf{d}\mathscr{J}$ is independent of c^{Y},

$$\frac{\partial\, \mathbf{d}\mathscr{J}}{\partial \mathsf{c}^{\mathsf{Y}}} = 0,$$

while still featuring $\mathbf{d}\mathsf{c}^{\mathsf{Y}}$: the corresponding *cyclic differential*.[1] The corresponding Jacobi–Mach equation is

$$\frac{\partial\, \mathbf{d}\mathscr{J}}{\partial\, \mathbf{d}\mathsf{c}^{\mathsf{Y}}} = const_{\mathsf{Y}}. \tag{L.3}$$

3) *The energy integral type simplification.* $\mathbf{d}\mathscr{J}$ is independent of what was previously regarded as 'the independent variable t', whereby one Jacobi–Mach equation may be supplanted by the first integral

$$\mathbf{d}\mathscr{J} - \frac{\partial\, \mathbf{d}\mathscr{J}}{\partial\, \mathbf{d}\mathsf{Q}^{\mathsf{A}}}\mathbf{d}\mathsf{Q}^{\mathsf{A}} = \text{constant}. \tag{L.4}$$

Suppose further that the equations corresponding to 1)

$$0 = \frac{\partial\, \mathbf{d}\mathscr{J}}{\partial \mathsf{m}^{\mathsf{M}}}(\mathsf{Q}^{\mathsf{O}}, \mathbf{d}\mathsf{Q}^{\mathsf{O}}, \mathsf{m}^{\mathsf{M}}) \quad \text{can be solved for } \mathsf{m}^{\mathsf{M}}.$$

One can then pass from $\mathbf{d}\mathscr{J}(\mathsf{Q}^{\mathsf{O}}, \mathbf{d}\mathsf{Q}^{\mathsf{O}}, \mathsf{m}^{\mathsf{M}})$ to a reduced $\mathbf{d}\mathscr{J}_{\text{red}}(\mathsf{Q}^{\mathsf{O}}, \mathbf{d}\mathsf{Q}^{\mathsf{O}})$: *multiplier elimination*.

[1]To avoid confusion, note that 'cyclic' in 'cyclic differential' just means the same as 'cyclic' in cyclic velocity. So nothing like 'exact differential' or 'cycle' in Algebraic Topology—which in de Rham's case is tied to differentials—is implied here.

Configuration–change space and configuration–velocity space are conceptually distinct presentations of the same tangent bundle $\mathfrak{T}(\mathfrak{q})$. Formulation in terms of change $\mathrm{d}Q^A$ can furthermore be viewed as introducing a *change covector*. This is in the sense of inducing '*change weights*' to Principles of Dynamics entities, analogously to how introducing a conformal factor attaches conformal weights to tensors. For instance, $\mathrm{d}s$ and $\mathrm{d}\mathcal{J}$ are change covectors as well. On the other hand, S is a change scalar: an entity which remains invariant under passing from the standard Principles of Dynamics to TRiPoD, due to their being already-TRi.

TRiPoD's formulation of momentum is

$$P_A := \frac{\partial\, \mathrm{d}\mathcal{J}}{\partial\, \mathrm{d}Q^A}, \tag{L.5}$$

which is a change scalar as well.

L.3 Free End Notion of Space Variation

Suppose a formulation's multiplier coordinate m is replaced by a cyclic velocity c [20, 64] or a cyclic differential $\mathrm{d}c$ [37, 38]. The zero right hand side of the multiplier equation is replaced by f(notion of space alone) in the corresponding cyclic equation. However, if the quantity being replaced is an entirely physically meaningless auxiliary, in the cyclic formulation, the meaninglessness of its values at the end notion of space becomes nontrivial. I.e. *free end notion of space* variation alias *variation with natural boundary conditions*) [166, 220, 313, 598] is the appropriate procedure. This is a portmanteau of *free end point* variation for finite theories, and *free end spatial hypersurface* variation for Field Theories.[2] Such a variation imposes more conditions than the more usual fixed-end variation does: three conditions per variation,

$$\frac{\partial\, \mathrm{d}\mathcal{J}}{\partial g^G} = \mathrm{d}p_G, \quad \text{alongside } p_G|_{\text{end}} = 0. \tag{L.6}$$

Case 1) If the auxiliaries g^G are multipliers m^G, (L.6) just reduces to

$$p_G = 0, \qquad \frac{\partial \mathcal{J}}{\partial m^G} = 0$$

and redundant equations. So in this case, the end notion of space terms automatically vanish by applying the multiplier equation to the first factor of each. This holds regardless of whether the multiplier is not auxiliary and thus standardly varied, or auxiliary and thus free end notion of space varied. This is because this difference in status merely translates to whether or not the cofactors of the above

[2] To be clear, 'free end' here refers to free value *at* the end notion of space rather than the also quite commonly encountered freedom *of* the end notion of space itself.

zero factors are themselves zero. Consequently the free end notion of space subtlety in no way affects the outcome in the multiplier coordinate case. This probably accounts for the above subtlety long remaining unnoticed.

Case 2) If the auxiliaries g^{G} are considered to be cyclic coordinates c^{G}, (L.6) reduces to

$$\mathsf{p}_{\mathsf{G}}|_{\text{end}-\text{NOS}} = 0 \tag{L.7}$$

alongside

$$\dot{\mathsf{p}}_{\mathsf{G}} = 0 \quad \text{(or equivalently } \mathbf{d}\mathsf{p}_{\mathsf{G}} = 0\text{)}$$
$$\Rightarrow \quad \mathsf{p}_{\mathsf{G}} = C(\text{notion of space}), \text{ invariant along the curve of notion of space.} \tag{L.8}$$

C(notion of space) is now identified as 0 at either of the two end notion of space (L.7). Since this is invariant along the curve of notions of space, it is therefore zero everywhere. So (L.8) and the definition of momentum give

$$\frac{\partial \mathscr{L}}{\partial \dot{\mathsf{c}}^{\mathsf{G}}} := \mathsf{p}_{\mathsf{G}} \quad \text{or equivalently} \quad \frac{\partial \mathbf{d} \mathscr{J}}{\partial \mathbf{d}\mathsf{c}^{\mathsf{G}}} = 0.$$

In conclusion, the above free end point notion of space working ensures that the cyclic and multiplier formulations of auxiliaries in fact give *the same* variational equation. Thus complying with Temporal Relationalism by passing from encoding one's *auxiliaries* as multipliers to encoding them as cyclic velocities or differentials is valid without spoiling the familiar and valid physical equations.

Note that a similar working [20] establishes that passage to the Routhian for an auxiliary formulated in cyclic terms reproduces the outcome of multiplier elimination for that same auxiliary formulated in terms of multipliers.

L.4 TRi Legendre Transformation

One can now apply Legendre transformations that inter-convert changes $\mathbf{d}\mathsf{Q}^{\mathsf{A}}$ and momenta P_{A}.

Example 1) *Passage to the* $\mathbf{d}$*-Routhian*

$$\mathbf{d}\mathscr{R}(\mathsf{Q}^{\mathsf{X}}, \mathbf{d}\mathsf{Q}^{\mathsf{X}}, \mathsf{P}^{\mathsf{Y}}_{\mathsf{c}}) := \mathbf{d}\mathscr{J}(\mathsf{Q}^{\mathsf{X}}, \mathbf{d}\mathsf{Q}^{\mathsf{X}}, \mathbf{d}\mathsf{c}^{\mathsf{Y}}) - \mathsf{P}^{\mathsf{c}}_{\mathsf{Y}}\mathbf{d}\mathsf{c}^{\mathsf{Y}}. \tag{L.9}$$

$\mathbf{d}$-*Routhian reduction* furthermore requires being able to

$$\text{solve } \ const_{\mathsf{Y}} = \frac{\partial \mathbf{d}\mathscr{J}}{\partial \mathbf{d}\mathsf{c}^{\mathsf{Y}}}(\mathsf{Q}^{\mathsf{X}}, \mathbf{d}\mathsf{Q}^{\mathsf{X}}, \mathbf{d}\mathsf{c}^{\mathsf{Y}}) \quad \text{as equations for the } \mathbf{d}\mathsf{c}^{\mathsf{Y}}.$$

This is followed by substitution into (L.9). One application of this is the passage from Euler–Lagrange type actions to the geometrical form of the Jacobi actions, now done without ever introducing a parameter; another is Chap. 16's reduction procedure.

Example 2) *Passage to the* đ*-anti-Routhian,*

$$\text{đ}\mathscr{A}(\mathsf{Q}^{\mathsf{X}}, \mathsf{P}^{\mathsf{X}}, \text{đ}\mathsf{c}^{\mathsf{Y}}) = \text{đ}\mathscr{J}(\mathsf{Q}^{\mathsf{X}}, \text{đ}\mathsf{Q}^{\mathsf{X}}, \text{đ}\mathsf{c}^{\mathsf{Y}}) - \mathsf{P}^{\mathsf{c}}_{\mathsf{X}}\text{đ}\mathsf{Q}^{\mathsf{X}}. \tag{L.10}$$

A subcase of this plays a significant role in the next Section.

Example 3) *Passage to the* đ*-Hamiltonian,*

$$\text{đ}\mathscr{H}(\mathbf{Q}, \mathbf{P}) = \mathsf{P}_{\mathsf{A}}\text{đ}\mathsf{Q}^{\mathsf{A}} - \text{đ}\mathscr{J}(\mathbf{Q}, \text{đ}\mathbf{Q}). \tag{L.11}$$

The corresponding equations of motion are in this case đ*-Hamilton's equations*

$$\frac{\partial\, \text{đ}\mathscr{H}}{\partial \mathsf{P}_{\mathsf{A}}} = \text{đ}\mathsf{Q}^{\mathsf{A}}, \qquad \frac{\partial\, \text{đ}\mathscr{H}}{\partial \mathsf{Q}^{\mathsf{A}}} = -\text{đ}\mathsf{P}_{\mathsf{A}}. \tag{L.12}$$

L.5 TRi-Morphisms and Brackets. i

Suppose we are to keep no cyclic differentials. The usual $\mathfrak{q}$ morphisms apply, except that specifically *Point* rather than *Point$_t$* is involved. Also, the Liouville 1-form (J.22) and the symplectic 2-form (J.23) are already TRi and thus are change 1- and 2-forms respectively. As the inverse of the latter, the Poisson tensor $\boldsymbol{C}$ is recast as a change 2-tensor $\text{đ}^{-2}\boldsymbol{D}$.

Temporal Relationalism also requires use of *Can* rather than *Can$_t$* in the đ-Hamiltonian formulation.

The Poisson bracket portmanteau (24.2) is already-TRi in form, provided that the smearing variables in the field-theoretic case take TRi-smeared form,

$$\{\mathscr{F}, \mathscr{G}\} := \int_{\Sigma} \mathrm{d}\Sigma \left\{ \frac{\delta\mathscr{F}}{\delta \mathsf{Q}^{\mathsf{A}}}\frac{\delta\mathscr{G}}{\delta \mathsf{P}_{\mathsf{A}}} - \frac{\delta\mathscr{F}}{\delta \mathsf{P}_{\mathsf{A}}}\frac{\delta\mathscr{G}}{\delta \mathsf{Q}^{\mathsf{A}}}. \right\} \tag{L.13}$$

Finally, $\mathfrak{P}$hase is already-TRi, since all of $\mathbf{Q}$, $\mathbf{P}$ and the Poisson bracket are.

L.6 đA-Hamiltonians and Phase Spaces, and TRi Dirac-Type Algorithms

The Legendre matrix encoding the non-invertibility of the momentum-velocity relations is now supplanted by the đ^{-1}*-Legendre matrix change vector*

$$\text{đ}^{-1}\Lambda_{\mathsf{A}\mathsf{A}'} := \frac{\partial^2 \text{đ}\mathsf{J}}{\partial\, \text{đ}\mathsf{Q}^{\mathsf{A}} \partial\, \text{đ}\mathsf{Q}^{A'}} \quad \left(= \frac{\partial \mathsf{P}_{\mathsf{A}'}}{\partial\, \text{đ}\mathsf{Q}^{\mathsf{A}}} \right) \tag{L.14}$$

which encodes the non-invertibility of the momentum-change relations. The TRi definition of primary constraint then follows in parallel to how the usual definition of primary constraint follows from the Legendre matrix, with secondary constraint remaining defined by exclusion.

For example, Dirac's argument that Reparametrization Invariance implies at least one primary constraint is now recast in the TRi form of Sect. 15.6's Lemma 5. The specific form of the primary constraint is, of course, $\mathcal{C}$hronos.

The next idea in building a TRi version of Dirac's general treatment of constraints is to append constraints to one's incipient $\mathbf{d}$-Hamiltonian not with Lagrange multipliers—which would break TRi—but rather with cyclic differentials. In this way, a *$\mathbf{d}$A-Hamiltonian* is formed; the 'A' here stands for 'almost', though the $\mathbf{d}$A-Hamiltonian is also a particular case of $\mathbf{d}$-anti-Routhian. Moreover, the $\mathbf{d}$A-Hamiltonian $\mathbf{d}A$ symbol has an extra minus sign relative to the $\mathbf{d}$-anti-Routhian $\mathbf{d}A$ symbol. This originates from the definition of Hamiltonian involving an overall minus sign where the definitions of Routhian and anti-Routhian have none. Furthermore, in the current context, all the cyclic coordinates involved have auxiliary status and occur in best-matched combinations. [In the event of a system possessing physical as well as auxiliary cyclic coordinates, one would use a 'partial' rather than 'complete' anti-Routhian.]

The equations of motion are now *$\mathbf{d}$A-Hamilton's equations*

$$\frac{\partial \mathbf{d}\mathscr{A}}{\partial \mathsf{P}_\mathsf{A}} = \dot{\mathsf{Q}}^\mathsf{A}, \qquad \frac{\partial \mathbf{d}\mathscr{A}}{\partial \mathsf{Q}^\mathsf{A}} = -\dot{\mathsf{P}}_\mathsf{A}, \tag{L.15}$$

augmented by

$$\frac{\partial\, \mathbf{d}\mathscr{A}}{\partial\, \mathbf{d}\mathsf{c}^\mathsf{G}} = 0$$

Let us finally note that Appendix J.15's comment about using the anti-Routhian's own Legendre matrix carries over to the $\mathbf{d}$-anti-Routhian, and thus also to the further identification of a subcase of this as the $\mathbf{d}$A-Hamiltonian.

Examples of the above TRi-Dirac appendings are, firstly, the *starred $\mathbf{d}$A-Hamiltonian* $\mathbf{d}\mathscr{A}^* := \mathbf{d}\mathscr{A} + \mathbf{d}f^\mathsf{P} c_\mathsf{P}$ for arbitrary functions of $\mathbf{Q}, \mathbf{P}$ now represented as cyclic differentials $\mathbf{d}f(\mathbf{Q}, \mathbf{P})$. Secondly, the *total $\mathbf{d}$A-Hamiltonian* $\mathbf{d}\mathscr{A}_{\text{Total}} := \mathbf{d}\mathscr{A} + \mathbf{d}u^\mathsf{P} c_\mathsf{P}$, where $\mathbf{d}u^\mathsf{P}$ are now unknown cyclic differentials.

The next issue to arise is the counterpart of the bracket expression in (J.36). The Best Matched form of the action now ensures the constraints are of the form $c(\mathbf{Q}, \mathbf{P}$ alone), because for these

$$\text{passage from } \mathbf{dQ} \text{ to } \mathbf{P} \text{ absorbs all the } \mathbf{d}\mathsf{g}^\mathsf{G}. \tag{L.16}$$

This is the previously advertised *major* ($\mathbf{d}$-)anti-Routhian trick.

In the case in hand, $c^C = c(\mathsf{Q}^\mathsf{A}, \mathsf{P}_\mathsf{A}$ alone) means that the chain-rule expansion

$$\mathbf{d}c_\mathsf{C} = \frac{\partial\, \mathbf{d}c_\mathsf{C}}{\partial\, \mathsf{Q}^\mathsf{A}} \mathbf{d}\mathsf{Q}^\mathsf{A} + \frac{\partial\, \mathbf{d}c_\mathsf{C}}{\partial \mathsf{P}_\mathsf{A}} \mathbf{d}\mathsf{P}_\mathsf{A}$$

applies. So by (L.15),

$$\mathbf{d}c_\mathsf{C} = \{c_\mathsf{C}, \mathbf{d}\mathscr{A}_{\text{Total}}\} = \{c_\mathsf{C}, \mathbf{d}\mathscr{A}\} + \mathbf{d}u^\mathsf{P}\{c_\mathsf{C}, c_\mathsf{P}\}. \tag{L.17}$$

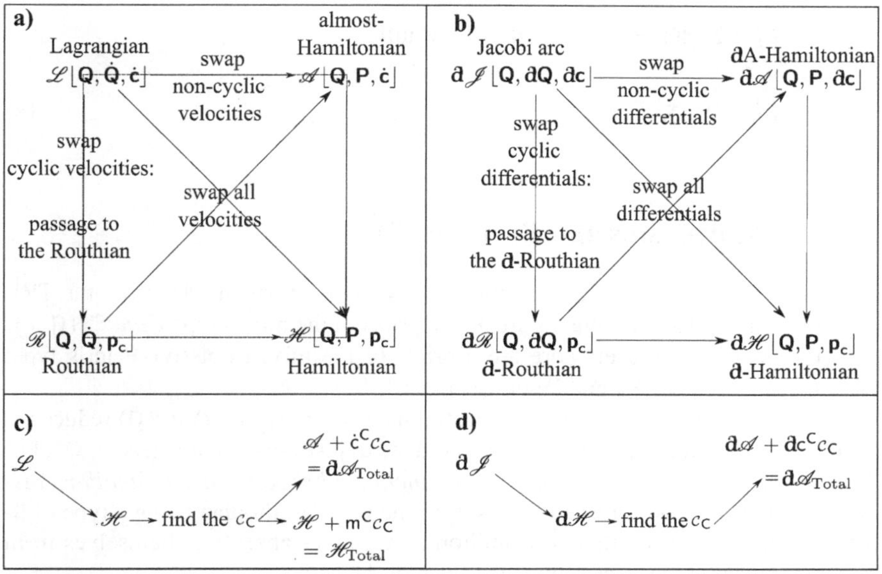

Fig. L.1 **a)** Almost-Hamiltonian subcase of Fig. J.2's Legendre square. **b)** then elevates this square to fully TRi form in terms of ꝺ-Legendre transformations that dually switch momenta and changes. These are between change covectors: $\mathrm{d}\mathcal{J}$, $\mathrm{d}\mathcal{R}$, $\mathrm{d}\mathcal{A}$, $\mathrm{d}\mathcal{H}$, the information-preserving extra terms now being subsystem Liouville forms, which were always change covectors. Routhians go to ꝺ-Routhians; there is no need for 'almost' in this case since Routhians are already allowed to contain velocities, and so already include almost-Routhians as a subset. **c)** and **d)** exhibit the choices by which the total Hamiltonian, A-Hamiltonian and ꝺA-Hamiltonian arise. The starred, primed and extended versions follow suit

Next let this be solved for unknown cyclic differentials under $\mathrm{d}u_{\mathsf{P}} = \mathrm{d}U_{\mathsf{P}} + \mathrm{d}V_{\mathsf{P}}$: the split into the cyclic differential complementary function $\mathrm{d}V^{\mathsf{P}} = v^{\mathsf{Z}}\mathrm{d}V^{\mathsf{P}}{}_{\mathsf{Z}}$ and the cyclic differential particular solution $\mathrm{d}U^{\mathsf{P}}$. Also the *primed* ꝺ*A-Hamiltonian* $\mathrm{d}\mathcal{A}' := \mathrm{d}\mathcal{A} + \mathrm{d}U^{\mathsf{P}}c_{\mathsf{P}}$, where $\mathrm{d}U^{\mathsf{P}}$ are now the cyclic differential particular solution part of $\mathrm{d}u^{\mathsf{P}}$ Finally, the *extended* ꝺ*A-Hamiltonian* $\mathrm{d}\mathcal{A}_{\mathrm{Extended}} := \mathrm{d}\mathcal{A} + \mathrm{d}u^{\mathsf{P}}c_{\mathsf{P}} + \mathrm{d}u^{\mathsf{S}}c_{\mathsf{S}}$. See Fig. L.1 for some context. 𝔓hase is now replaced by A-𝔓hase and ꝺA-𝔓hase; these are all types of bundle twice over: cotangent bundles *and* 𝔤 bundles.

Since only the Poisson bracket part acting on the constraints, the definitions of first- and second-class remain unaffected, as are the Dirac bracket and the extension procedure.

Moreover, Dirac's Algorithm is now supplanted by the TRi-Dirac Algorithm. As regards the five cases this is capable of producing at each step, in any combination, equation types 0), 1) and 3) are as before. On the other hand, equation types 2) and 4) are now phrased in terms of cyclic differentials.

Finally, this book also makes use of the following formulations for actions.

$$\mathsf{S} = \iint_{\Sigma} \mathrm{d}\lambda \mathrm{d}\mathbf{\Sigma}\{\dot{\mathsf{Q}}^{\mathsf{A}}\mathsf{P}_{\mathsf{A}} - \mathcal{A}_{\mathrm{Total}}\} = \iint_{\Sigma} \mathrm{d}\lambda \mathrm{d}\mathbf{\Sigma}\{\dot{\mathsf{Q}}^{\mathsf{A}}\mathsf{P}_{\mathsf{A}} - \mathsf{i}\, \mathcal{Q}uad - \dot{c}^{\mathsf{G}}\mathcal{G}auge_{\mathsf{G}}\}$$

$$= \iint_{\Sigma} \mathrm{d}\boldsymbol{\Sigma}\{\mathbf{d}\mathsf{Q}^{\mathsf{A}}\mathsf{P}_{\mathsf{A}} - \mathbf{d}\mathsf{I}\,\mathcal{Q}\mathrm{uad} - \mathbf{d}\mathsf{c}^{\mathsf{G}}\,\mathcal{S}\mathrm{huffle}_{\mathsf{G}}\}$$

$$= \iint_{\Sigma} \mathrm{d}t^{\mathrm{em}}\mathrm{d}\boldsymbol{\Sigma} \int \{*\mathsf{Q}^{\mathsf{A}}\mathsf{P}_{\mathsf{A}} - \mathcal{Q}\mathrm{uad} - *\mathsf{c}^{\mathsf{G}}\mathcal{G}\mathrm{auge}_{\mathsf{G}}\}. \tag{L.18}$$

L.7 TRi-Morphisms and Brackets. ii)

Suppose there are now cyclic differentials to be kept, or which arise from the TRi-Dirac Algorithm. The morphisms are now a priori of the mixed type $Can(\mathfrak{T}^*(\mathfrak{q})) \times Point(\mathfrak{g})$. Also the brackets here are a priori of the mixed Poisson–Peierls type: Poisson as regards Q^{A}, P_{A} and Peierls as regards $\mathbf{d}\mathcal{g}^{\mathsf{G}}$.

(L.16) implies that, as regards the constraints, $Can(\mathfrak{T}^*(\mathfrak{q})) \times Point(\mathfrak{g})$ reduces to just $Can(\mathfrak{T}^*(\mathfrak{q}))$ and the mixed brackets reduce to just Poisson brackets on Q^{A}, P_{A}. *The physical part of the* **d***A-Hamiltonian's incipient bracket is just a familiar Poisson bracket.* This good fortune follows from the **d**A-Hamiltonian being a type of **d**-anti-Routhian, alongside its non-Hamiltonian variables absenting themselves from the constraints due to the best-matched form of the action.

L.8 TRi Constraint Algebraic Structures and Beables

The algebraic structure of the constraints is unaffected by passing to TRi form: same constraints and same brackets for the purpose of acting on the constraints. This is modulo a minor and physically inconsequential point of difference in formulation of the smearing in the field-theoretic case.

Whenever any of Dirac's Hamiltonians are equivalent, the same applies to A- and **d**A-variants. This includes definitions based on cases of (25.1) as well as the particular δDEs to solve for specific examples, since those are based on the same constraints and the same brackets. As per the end of the previous Section, there remains a minor point of difference in formalism of the smearing in the field-theoretic case.

L.9 TRi Hamilton–Jacobi Theory

This is to be a t-independent version corresponding to a totally constrained **d**A-Hamiltonian. In such a case, only the constraints themselves feature. Moreover, (L.16) implies that this retains the form familiar from the standard Principles of Dynamics' t-independent totally constrained Hamiltonian case. Thus the close parallel to the Semiclassical Approach is also inherited by TRiPoD.

The TRi Hamilton–Jacobi formulation then consists of

$$\mathcal{C}\mathrm{hronos}\left\lfloor \mathbf{Q}, \frac{\delta\chi}{\delta\mathbf{Q}} \right\rfloor = 0, \tag{L.19}$$

with allowed extra dependence on a meaningful constant such as E for Mechanics or Λ for (Minisuperspace) GR. N.B. Hamilton's characteristic function χ is a change scalar. In the case of nontrivial $\mathfrak{g}$ which has been confirmed to act as a gauge group on the configurations, this is supplemented by

$$\mathcal{gauge}\left\lfloor \mathbf{Q}, \frac{\delta\chi}{\delta\mathbf{Q}} \right\rfloor = 0. \tag{L.20}$$

Research Project 116) The Problem of Time strategies require an even wider range of Hamilton–Jacobi Theory formulations than in [598]'s foundational account of Mechanics. Provide a suitable treatise.

L.10 TRiPoD End-Summary

Each entity of Fig. L.2 in the left leg is supplanted by the mirror image entity on the right leg. Moreover, the middle leg is unaffected: its entities are the 'already Temporally Relational' parts of the standard Principles of Dynamics. This mostly consists of change scalars. The new right leg is powered by the free end notion of space variation of Sect. L.3. The text further refines classification of the right and middle leg's entities by homothety weight. N.B. the few asymmetries between the left and right leg: the heavy arrows indicate time assumed versus Machian emergent time, and the right leg does not make use of Lagrange multipliers. Finally, explicitly time-dependent structures such as time-dependent Lagrangians, time-dependent Hamiltonians, $Point_{\mathfrak{t}}$ and $Can_{\mathfrak{t}}$ have no TRi counterparts.

L.11 Parageodesic Principle Split Conformal Transformations

Splitting the relational arc element $\mathbf{d}\mathcal{J}$ into 'kinetic' and 'potential' factors is in fact a nonunique procedure: $\mathbf{d}\widetilde{\mathbf{s}} = \{\mathcal{W}/\Omega\}\{\Omega\mathbf{ds}\}$ will also do. The general such split means that one is dealing with a *parageodesic principle*, so we term the above re-representation a *parageodesic principle splitting conformal transformation* (*PPSCT*). Its first factor involves an ordinary conformal transformation of the kinetic arc element $\mathbf{ds}$; consequently the corresponding kinetic metric scales as a conformal vector. This is now paired with the second factor compensatingly scaling as a conformal covector,

$$\mathcal{W} \longrightarrow \overline{\mathcal{W}} = \mathcal{W}/\Omega^2. \tag{L.21}$$

Also, $*$ is a PPSCT-covector by applying the above to the combination (15.21). This reveals that $*$ is in fact highly nonuniquely defined, at least at this stage in the

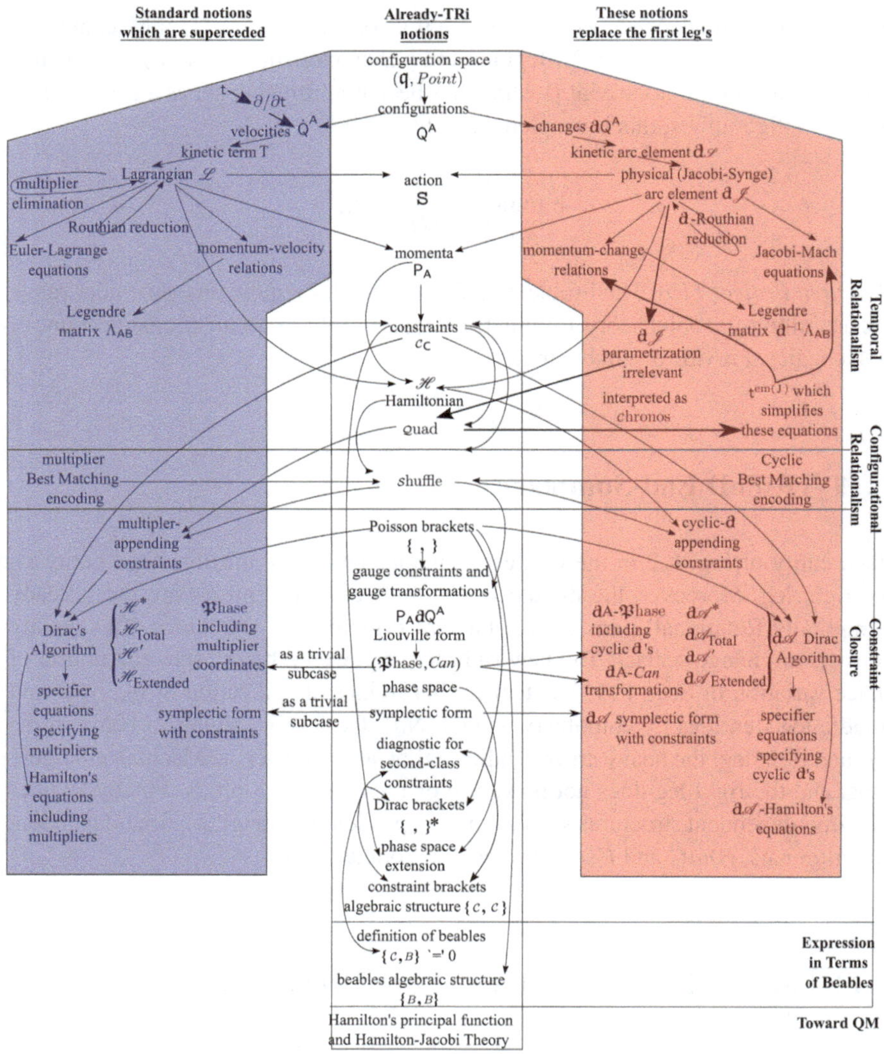

Fig. L.2 The Temporal Relationalism implementation of the Principles of Dynamics (TRiPoD). The powers of $\mathbf{d}$ displayed in the red leg indicate the change-tensor rank of each entity. Parts II and III's subsequent TRi figures are picked out by their matching *blue*, *white* and *red* '*tricolore*' convention, with *white* for already-TRi and the *blue* entities requiring supplanting by the *red* TRi-implementing ones

argument.[3] I.e. integrating,

$$\mathbf{d}t^{em} \longrightarrow \mathbf{d}\bar{t}^{em} = \Omega^2 dt^{em}. \tag{L.22}$$

[3] So as to not confuse '$t^{em(JBB)}$ as features in the previous literature' [37, 98] and the PPSCT-covector explained in the current Appendix, we denote the latter by $\vec{t}$. One can also think of this as N, $\dot{I}$, ∂I scaling as PPSCT-vectors.

Now clearly from the invariance of the action, performing such a transformation should not (and does not) affect the physical content of one's classical equations of motion.[4]

Next, the conjugate momenta are PPSCT-invariant:

$$\overline{\mathsf{P}}_{\mathsf{B}} = \overline{\mathsf{M}}_{\mathsf{AB}} \overline{*}_{\mathsf{g}} \mathsf{Q}^{\mathsf{A}} = \mathsf{M}_{\mathsf{AB}} *_{\mathsf{g}} \mathsf{Q}^{\mathsf{A}} = \mathsf{P}_{\mathsf{B}}. \tag{L.23}$$

Thus PPSCT concerns, a fortiori, $\mathfrak{q}$ rather than $\mathfrak{P}$hase. Moreover, $\mathcal{Q}$uad is a PPSCT-covector; in this way, Misner's conformal covariance of $\mathcal{H}$—underlying his subsequent adoption of conformal operator ordering—is recovered from 'zeroth principles' that are none other than Temporal Relationalism. In cases with nontrivial Configurational Relationalism, the above arc elements pick up Best Matching corrections and **shuffle** constraints, which are conformally invariant. The total $\mathbf{d}$A-Hamiltonian is PPSCT-invariant as well. Finally, the Liouville form $\mathsf{P}_{\mathsf{A}}\dot{\mathsf{Q}}^{\mathsf{A}}$ is PPSCT-invariant, so that the form (L.18) for the action is also indeed PPSCT-invariant.

[4]A more complicated *3-part conformal transformation* $(\mathbf{M}, \mathscr{W}, *) \longrightarrow (\overline{\mathbf{M}}, \overline{\mathscr{W}}, \overline{*}) = (\Omega^2\mathbf{M}, \Omega^{-2}\mathscr{W}, \Omega^{-2}*)$ can also be pinned upon the Euler–Lagrange form of the action, though this would have been harder to find without having seen (L.21) first.

Appendix M
Quotient Spaces and Stratified Manifolds**

We already mentioned in Appendix A.2 that quotienting only works on some occasions in Group Theory, i.e. quotienting out normal subgroups. The current Appendix illustrates that quotienting mathematical structures is, more generally, a rather subtle business, with both limitations on scope of applicability and non-preservation of a number of mathematical properties. This and the next Appendix further support Facet 2 of the Problem of Time: Configurational Relationalism.

M.1 Quotienting out Groups: Further Useful Notions

For g an element and $\mathfrak{H}$ a subgroup of a group $\mathfrak{g}$, then $g\,\mathfrak{H} := \{gh \,|\, h \in \mathfrak{H}\}$ is a *(left) coset*, and the corresponding *(left) coset space* is the set of all of these. The *quotient of the action of a group* $\mathfrak{g}$ on a space $\mathfrak{s}$, denoted by $\mathfrak{s}/\mathfrak{g}$, is the set of all group orbits, which (suitably equipped) is termed the *group orbit space*, $\mathfrak{O}$. Work through Ex IV.4 for some simple examples of this.

A smooth manifold equipped with a transitive smooth action of a Lie group is termed a *homogeneous space*. Readers may wish to convince themselves that $\mathbb{S}^n$ and $\mathbb{CP}^n$ can be thought of in this manner.

M.2 Quotient Topologies

Let us next consider quotienting a topological space by an equivalence relation, $\langle \mathfrak{X}, \tau \rangle / \tilde{}$, so as to produce the corresponding *quotient topology* [613, 672].

N.B. that this does not in general preserve a number of topological properties, including in particular none of the three manifoldness properties. A simple counterexample to preservation of Hausdorffness is as follows. Let $\mathfrak{X} = \{(x, y) \in \mathbb{R}^2 \,|\, y = 0 \text{ or } 1\}$ with the obvious topology, and $(x, y) \sim (z, w)$ iff either $(x, y) = (z, w)$ or $x = z \neq 0$: the line with two origins which cannot be separated. As regards non-preservation of dimension, quotienting is capable of decreasing or increasing topological dimension. Whereas the decreasing case is obvious, space-filling curves [68]

E. Anderson, *The Problem of Time*, Fundamental Theories of Physics 190,
DOI 10.1007/978-3-319-58848-3

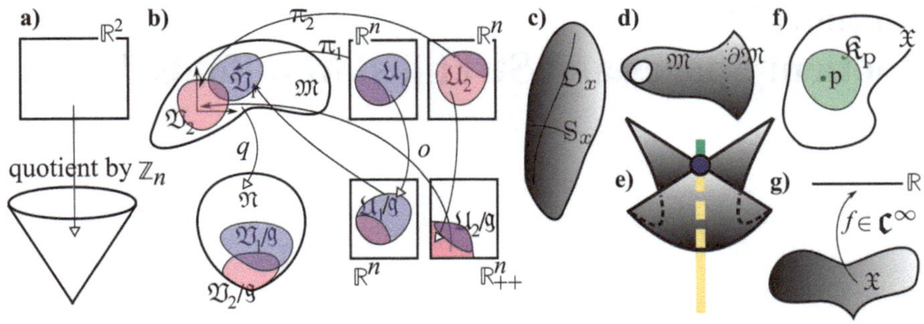

Fig. M.1 **a)** Cones are an example of orbifold; the apex is an orbifold point. **b)** Notion of charts for an orbifold $\mathfrak{N}$ obtained by identifying the two perpendicular arrows at p in the manifold $\mathfrak{M}$, whereby some of the orbifold charts are quarter-spaces $\mathbb{R}^n_{++}$ q is the quotient map corresponding to taking out $\mathfrak{g}$, and o is the group orbit map. **c)** Orbits $\mathfrak{O}_x$ and slices S_x. **d)** A manifold with boundary. **e)** Whitney's umbrella. Its strata are the blue origin, yellow handle, green spoke, and the grey remainder. **f)** Supports the definition of local compactness. **g)** A differential space

provide examples of it increasing. Quotienting can furthermore produce dimension varying from point to point in its quotient; Appendix G already presented simple examples of this. Moreover, in the physical examples below, Hausdorffness and second-countability *are* often retained, so quotienting here leads to entities which are '2/3 of a manifold'.

Quotienting does preserve connectedness, path connectedness and compactness (see [613]), albeit not simple connectedness (e.g. passage to nontrivial universal covering group) or contractibility (e.g. $\mathbb{R}^2/Dil = \mathbb{S}^1$). Moreover, if $\mathfrak{S}/\mathfrak{g}$ arises by a group $\mathfrak{g}$ acting on a space $\mathfrak{S}$ freely and properly, then $\mathfrak{S}/\mathfrak{g}$ is Hausdorff [614]. One application of this result is in guaranteeing the mathematical tractability of 1- and 2-*d* RPM shape spaces.

M.3 Orbifolds

Orbifolds [229, 386] are locally quotients $\mathfrak{M}/\mathfrak{g}$ following from a properly discontinuous action of a finite Lie group $\mathfrak{g}$ on a manifold $\mathfrak{M}$. This construction can moreover be applied to equipped manifolds such as (semi-)Riemannian manifolds. Orbifolds are more general than manifolds, since quotients do not in general preserve manifoldness; consequently some orbifolds carry singularities.

$\mathfrak{M}$ itself admits an open cover $\mathfrak{U}_\mathrm{C}$. Each constituent $\mathfrak{U}_\mathrm{C}$ furthermore possesses an *orbifold chart*: a continuous surjective map $\phi_\mathrm{C} : \mathfrak{V}_\mathrm{C} \rightarrow \mathfrak{U}_\mathrm{C}$ for $\mathfrak{U}_\mathrm{C}$ open $\subseteq \mathbb{R}^p$: for $p = \dim(\mathfrak{M})$ and where $\mathfrak{V}_\mathrm{C}$ and ϕ_C are invariant under the action of $\mathfrak{g}$. In Fig. M.1.b), it is more convenient to use a map π_C in the opposite direction. One can moreover define a notion of meshing between such charts, and finally a notion of orbifold atlas in close parallel to that for manifolds.

The everyday notion of cone can be thought of as a simple example of orbifold (Fig. M.1.a). Another is Fig. G.9.k), in the context of a 3-body problem configuration space. More generally, orbifolds are common in N-body problem configuration spaces, indeed including the generalized sense of cone that applies to relational spaces. The 2-d N-body problem's simplest shape spaces $\mathbb{CP}^{n-1}$ are often best thought of as complex manifolds. There is indeed a notion of complex orbifold as well as of real orbifold, in parallel to how there are real and complex manifolds [673]. Elsewhere in Theoretical Physics, many of the orbifolds which are well-known to occur in String Theory are also complex; in particular, these arise in the study of Calabi–Yau manifolds [229, 386]. Furthermore, some simpler models of this last example are closely related to the preceding one, though both being discrete quotients of $\mathbb{CP}^k$ spaces [37].

M.4 Quotienting by Lie Group Action, and Slices

For the action of a Lie group $\mathfrak{g}$ on a space $\mathfrak{X}$ (e.g. a manifold $\mathfrak{M}$), the generalized *slice* S_x at $x \in \mathfrak{X}$ is a manifold transverse to the group orbit $\mathfrak{O}_x$; see e.g. [466]. This generalizes the fibre bundle notion of local section to the case involving compact transformation groups in place of principal bundles. (The corresponding generalization of the fibre bundle notion of local trivialization—Appendix F.4—in this setting is termed a *tube* [466].)

The slice can be taken to exist in the above compact case. However, in other cases one can occasionally prove *Slice Theorems* to this effect (Appendices N.1 and N.4). Of subsequent relevance below, the Implicit Function Theorem enters these proofs. A slice S_x gives a local chart for $\mathfrak{X}/\mathfrak{g}$; thus the slice notion—when available—is a significant tool for the study of the corresponding group orbit spaces $\mathfrak{O}$.

Slices S_x carry information about the amount of isotropy of points near x [466]. Let us illustrate 'amount of isotropy' using Appendix G's examples. Whereas 2-d mechanical configurations have just the one isotropy group $SO(2)$, 3-d ones have 3 possible isotropy groups: id, $SO(2)$ and $SO(3)$. These have corresponding orbits of the form $SO(3)$, $\mathbb{S}^2$ and 0 respectively. This correspondence follows from the isotropy group also being known as the stabilizer group, and well-known relations between orbits and stabilizers. *Multiple dimensions of isotropy groups point to multiple dimensions of orbits. Thus group orbit spaces $\mathfrak{O}$ are not in general manifolds—entities of unique dimension—but rather collections of manifolds that span various dimensions.* This motivates consideration of further generalizations of manifolds as follows.

M.5 Stratified Manifolds

Manifolds are in general insufficient for the purpose of studying physical reduced or relational configuration spaces $\mathfrak{q}/\mathfrak{g}$. We have already seen that these more generally produce unions of manifolds of in general different dimensions. Moreover, some cases of further physical relevance—such as reduced configuration spaces in

Mechanics and GR, and group orbits spaces in Gauge Theory—'fit together' according to some fairly benevolent rules. The constituent manifolds are here known as strata, and each collection that 'fits together' in this manner is known as a *stratified manifold* .

Historically, the first formulation of stratified manifolds was of differentiable stratified manifolds by Whitney [904] (also reviewed in [905]). Subsequently, noted mathematician René Thom formed a theory of stratified topological manifolds as an arena for dealing with singularities [846].[1] Thom [847] additionally showed that every stratified space in the sense of Whitney is also one of his own stratified spaces and with the same strata.

Let $\mathfrak{X}$ be a topological space that is not presupposed to be a topological manifold. Suppose that this can be split according to $\mathfrak{X} = \mathfrak{X}_\mathrm{p} \cup \mathfrak{X}_\mathrm{q}$ [905]. Here $\mathfrak{X}_\mathrm{p} := \{\mathrm{p} \in \mathfrak{X}, \mathrm{p} \text{ simple}\}$, $\dim_\mathrm{p}(\mathfrak{X}) = \dim(\mathfrak{X})$ where 'simple' means 'regular' and 'ordinary', and $\mathfrak{X}_\mathrm{q} := \mathfrak{X} - \mathfrak{X}_\mathrm{p}$. Proceed to consider a recursion of such splittings, so e.g. $\mathfrak{X}_\mathrm{q}$ can furthermore be split into $\{\mathfrak{X}_\mathrm{q}\}_\mathrm{p}$ and $\{\mathfrak{X}_\mathrm{q}\}_\mathrm{q}$. Then setting $\mathfrak{M}_1 = \mathfrak{X}_\mathrm{p}$, $\mathfrak{M}_2 = \{\mathfrak{X}_\mathrm{q}\}_\mathrm{p}$, $\mathfrak{M}_3 = \{\{\mathfrak{X}_\mathrm{q}\}_\mathrm{q}\}_\mathrm{p}$ etc. gives $\mathfrak{X} = \mathfrak{M}_1 \cup \mathfrak{M}_2 \cup \dots$, $\dim(\mathfrak{X}) = \dim(\mathfrak{M}_1) > \dim(\mathfrak{M}_2) > \cdots$, where each $\mathfrak{M}_\mathrm{I}$ $\mathrm{I} = 1, 2, \dots$ is itself a manifold. The point of this procedure is that it partitions $\mathfrak{X}$ by dimension. Moreover, $\mathfrak{X}$ is only a topological manifold if this is a trivial partition: involving a single piece only. On the other hand, a *strict* partition of a topological space is a (locally finite) partition into strict manifolds. [A manifold $\mathfrak{M}$ within a m-dimensional open set $\mathfrak{U}$ is $\mathfrak{U}$-*strict* if its $\mathfrak{U}$-*closure* $\overline{\mathfrak{M}} := \mathfrak{U} - \mathrm{Clos}\,\mathfrak{M}$ and the $\mathfrak{U}$-*frontier* $\overline{\mathfrak{M}} - \mathfrak{M}$ are topological spaces in $\mathfrak{U}$.]

A set of manifolds in $\mathfrak{U}$ has the *frontier property* if, for any two distinct such, say $\mathfrak{M}$ and $\mathfrak{M}'$,

$$\text{if } \mathfrak{M}' \cap \overline{\mathfrak{M}} \neq \emptyset, \quad \text{then } \mathfrak{M}' \subset \overline{\mathfrak{M}} \text{ and } \dim(\mathfrak{M}') < \dim(\mathfrak{M}). \tag{M.1}$$

A partition into manifolds is itself said to have the frontier property if the corresponding set of manifolds does.

Finally, one definition of a *stratification* of $\mathfrak{X}$ [905] is as a strict partition of $\mathfrak{X}$ which has the frontier property. The corresponding set of manifolds are known as the *strata* of the partition.

The variant that Fischer [301] found in studying $\mathfrak{S}$uperspace($\boldsymbol{\Sigma}$) is based on the *inverse frontier property*. This is (M.1) under interchange of primed and unprimed quantities, which feeds into the corresponding notion of inverted stratification. Another occasionally useful [759] property is the *regular stratification property*,

$$\mathfrak{M}_\mathrm{I} \cup \overline{\mathfrak{M}_\mathrm{J}} \neq \emptyset \quad \Rightarrow \quad \mathfrak{M}_\mathrm{I} \subseteq \overline{\mathfrak{M}_\mathrm{J}} \quad \forall\, \mathrm{I}, \mathrm{J} \in \mathfrak{j}, \tag{M.2}$$

where $\mathfrak{j}$ denotes the set of strata.

Whitney [905] also established that a locally finite partition of $\mathfrak{X}$ with the frontier property is a stratification. Moreover, for each stratum $\mathfrak{M}$, $\overline{\mathfrak{M}} - \mathfrak{M}$ is the union of

[1]The well-known Catastrophe Theory [71] also originates from these authors' works. While this also concerns singular manifolds, it is largely a distinct topic from stratified manifolds.

the other closed strata in $\overline{\mathfrak{M}}$. Indeed, any strict partition of a manifold $\mathfrak{X}$ admits a refinement which is a stratification into connected strata. Take this as a brief indication that refinements of partitions—a type of 'graining'—plays a role in the theory of stratified manifolds.

Simple examples include the following.

Example 0) Manifolds are single-piece stratified manifolds.

Example 1) Appendix D.1's examples of manifolds with boundary can furthermore be interpreted as stratified manifolds. Here the manifold and its boundary are the two constituent strata, the former possessing the full dimension whereas and the latter is of codimension $C = 1$. Figure 37.5 illustrates the types of chart for a particular case of this. Intervals with one or both endpoints regarded as distinct are the simplest subexamples of this.

Example 2) *Manifolds with corners*. These have, in addition to the previous example's strata, the $C = 2$ strata that are the corners themselves. Some but not all [713] of these are stratified manifolds.

Example 3) Cones are stratified manifolds. E.g. for the cone over a compact manifold, the apex and the remainder are the strata.

Example 4) Simplicial complexes are stratified manifolds [713].

Example 5) Whitney's umbrella in Fig. M.1.e).

Since nontrivial stratified manifolds have strata with a range of different dimensions, clearly the locally Euclidean property of manifolds has broken down, and with it the standard notions of charts and how to mesh charts together. These notions still exist for stratified manifolds, albeit in a *more complicated form* (see Fig. 37.5). Also, in general losing Hausdorffness and second-countability leaves stratified manifolds 'further down' than topological manifolds in the diagram of the levels of structure. Moreover, this book considers in any detail only Hausdorff second-countable stratified manifolds, i.e. spaces which are '2/3rds of a manifold'.

Since Whitney, stratified manifolds have additionally been equipped with differentiable structure (see e.g. [798]). Furthermore, individual strata being manifolds, some are metrizable. Additional Riemannian metric structure on stratified spaces is considered by e.g. mathematician Markus Pflaum [713] (Kendall [539] also makes use of this level of structure). Pflaum furthermore sets up a definition of geodesic distance along such lines. His work is also a good source to learn about the morphisms corresponding to stratified manifolds.

Moreover, stratified manifolds and fibre bundles do not fit well together due to stratified manifolds' local structure varying from point to point. Three distinct strategies to deal with this are outlined in Sect. 37.5. Among these, relational considerations point to the strategy of accepting the stratified manifold. In turn, this points to seeking a generalization of Fibre Bundle Theory, for which Sheaf Theory (Appendix W.3) is a strong candidate.

Let us end by noting that stratified orbifolds also make sense, and indeed occur in the study of configuration spaces: the 3-d case of Fig. G.11.f).

M.6 Locally Compact Hausdorff Second-Countable (LCHS) Spaces

Local Euclideanness has been lost, but we consider another local property which also confers much control and understanding on the type of Analysis involved. A topological space $\langle \mathfrak{X}, \tau \rangle$ is *locally compact* [613] if each point $\mathrm{p} \in \mathfrak{X}$ is contained in a neighbourhood $\mathfrak{K}_\mathrm{p} \subseteq \mathfrak{X}$ (Fig. M.1.f).

In particular, these include manifolds, and the outcome of the coning construction. Many of Appendix G's configuration spaces from Mechanics are consequently included.

Furthermore, LCHS spaces are rather well-behaved from an Analysis point of view. LCH spaces have a number of Analysis results in common with complete metric spaces, including an analogue of Baire's Category Theorem (Appendix H.2). See [614] for a first account of further nice Analysis properties of LCHS spaces.

M.7 Differential Spaces and Stratifolds

A *differential space* is a pairing $(\mathfrak{X}, \mathfrak{C})$ of a topological space $\mathfrak{X}$ and a function space $\mathfrak{C}$ equipped with algebraic structure; furthermore, the functions $f \in \mathfrak{C}$ act on $\mathfrak{X}$.

Example 1) The $\mathfrak{C}$ generalizes the standard use of smooth functions in elementary real Differential Topology.

Example 2) *Sikorski spaces* [798] (after mathematician Roman Sikorski) are a prominent, historically early and quite general example of such a pairing. Here $\mathfrak{X}$ is any topological space and $\mathfrak{C}$ is a certain type of subalgebra of the continuous functions $\mathfrak{X} \rightarrow \mathbb{R}$. For instance, mathematical physicist Jedrzej Śniatycki [798] considers the pairing of a type of LCHS manifold with Sikorski spaces. This is the most advanced program as regards providing differential geometric structures thereupon (e.g. use of Marshall forms).

Example 3) *Stratifolds* are differential spaces more recently considered by mathematician Matthias Kreck [570]. These are also rather well-behaved, in part because the $\mathfrak{X}$ half of the pair is LCHS. Moreover, the $\mathfrak{C}$ half of the stratifold's pair receives a sheaf interpretation (outlined in Appendix W.3).

Example 4) Pflaum pairs LCHP stratified manifolds (where the P stands for 'paracompact') with sheaves. These generalize LCHS stratified manifolds because LCHS implies paracompactness. Paracompactness is a desirable property to keep because it protects standard notions of integration (and thus of integral forms of laws and of variational principles). Also (W.1) simplifies some aspects of Hausdorff paracompact spaces.

M.8 Further Stratified Spaces from the Principles of Dynamics

If $\mathfrak{q}$ is stratified, then so are $\mathfrak{T}(\mathfrak{q})$ [713] and the symplectic version of $\mathfrak{T}^*(\mathfrak{q})$ [514]. This is currently available for Pflaum and Śniatycki's treatments, but not for Kreck's.

Appendix N
Reduced Configuration Spaces for Field Theory and GR**

N.1 Gauge Group Orbit Spaces

Gauge Theory does not only involve a configuration space of connections $\mathfrak{Con}$, but also a gauge group $\mathfrak{g}$ acting thereupon. This is a Lie group. In the more usual cases such as Electromagnetism and Yang–Mills Theory, it acts internally. In this way, $\mathbf{\Lambda}^1/\mathfrak{g}$ arise as reduced configuration spaces (more concretely, as group [alias here gauge] orbit spaces $\mathfrak{O}$). Fibre Bundle Theory supports this to some extent, firstly through principal bundles with the above $\mathfrak{g}$ entering as both structure group and fibres. Secondly, a wider range of associated fibre bundles with $\mathfrak{g}$ as structure group and distinct fibres, permit modelling of gauge fields coupled to a number of further (gauged) fields can be modelled. On the other hand, the group orbit space itself is in general heterogeneous, and thus not amenable to fibre bundle description. Moreover, due to the group action in question being smooth and proper,[1] gauge orbit spaces $\mathfrak{O}$ have the separation property—and thus are in particular Hausdorff—as well as second-countable, metrizable and paracompact. See e.g. [564, 759, 775] for more on the topology and geometry of gauge orbit space, and the corresponding symplectic spaces, including in terms of stratified manifolds. Some particular theorems of note here are as follows.

Gauge Theory's Slice Theorem [564] *The action of the gauge group on space* $\mathbf{\Lambda}^1$ *of* A_i *admits a slice at every point*; *this applies in a principal fibre bundle setting*.

Gauge Theory's Stratification Theorem [564] *The decomposition of* $\mathbf{\Lambda}^1/\mathfrak{g}$ *by group orbit type is a regular stratification.*

Let us end by noting that $\mathfrak{L}^2$ mathematics suffices for the above workings, though one can uplift to more general function spaces [759] including so as to attain compatibility with the GR case below.

---

[1] A map is *proper* [613] if the inverse map of each compact set is itself compact.

E. Anderson, *The Problem of Time*, Fundamental Theories of Physics 190,
DOI 10.1007/978-3-319-58848-3

N.2 Loops and Loop Spaces for Gauge Theory

Another approach involves *Wilson loop variables* (after physicist Kenneth Wilson); these contain an equivalent amount of information to the connection variables A_i. Such a formulation is already meaningful for Electromagnetism; this amounts to modelling the space of transverse A_i for which the Gauss constraint has already been taken into account. In this case, the Wilson loop variables are of the form

$$\mathcal{H}_{\mathrm{A}}(\gamma) := \exp\left(i \oint_{\gamma} \mathrm{d}x^i \mathrm{A}_i(x)\right) \tag{N.1}$$

for γ a loop path. The somewhat more involved Yang–Mills Theory version of Wilson loop variables take the form

$$\mathcal{H}_{\mathrm{A}}(\gamma) := \mathrm{Tr}\left(P\ \exp\left(i\mathfrak{g} \oint_{\gamma} \mathrm{d}x^i \mathrm{A}_{iI}(x)\mathrm{g}^I(x)\right)\right), \tag{N.2}$$

for $\mathfrak{g}$ group generators g^I and path-ordering symbol P. $\mathcal{H}_{\mathrm{A}}(\gamma)$ are indeed holonomy variables in the sense of fibre bundles [673]; see [330] for an extensive (if nonrigorous) development of the theory of these.

The curves in question can be taken to be continuous and piecewise smooth, and, for now, to live on $\mathbb{R}^3$. In fact, equivalence classes of curves are required [330].

If modelled in this way, the corresponding loop space is a topological group. It is not however a Lie group, though it is contained within a Lie group: the so-called *extended loop group*. See [330] for a more detailed account.

N.3 Topology of $\mathfrak{g} = \mathit{Diff}(\boldsymbol{\Sigma})$

$\mathit{Diff}(\boldsymbol{\Sigma})$ can be modelled using $\mathfrak{fre}_{(1,0)}(\mathfrak{C}^{\infty})$ [301], which matches the way presented of modelling $\mathfrak{Riem}(\boldsymbol{\Sigma})$. A fortiori, diffeomorphisms are commonly modelled in terms of Fréchet manifolds, a fortiori as Fréchet Lie groups [426].

Next consider the group action $\mathit{Diff}(\boldsymbol{\Sigma}) \times \mathfrak{Riem}(\boldsymbol{\Sigma}) \rightarrow \mathfrak{Riem}(\boldsymbol{\Sigma})$. The group orbits of this are $\mathrm{Orb}(\langle\boldsymbol{\Sigma}, \mathbf{h}\rangle) = \{\phi^*\mathrm{h} \,|\, \phi \in \mathit{Diff}(\boldsymbol{\Sigma})\}$. If two metrics—points in $\mathfrak{Riem}(\boldsymbol{\Sigma})$—lie on the same group orbit, they are isometric. In this manner, the $\mathit{Diff}(\boldsymbol{\Sigma})$ group orbits partition $\mathfrak{Riem}(\boldsymbol{\Sigma})$ into isometric equivalence classes [301]. The corresponding stabilizers $\mathrm{Stab}(\langle\boldsymbol{\Sigma}, \mathbf{h}\rangle) = \{\phi \in \mathit{Diff}(\boldsymbol{\Sigma}) \,|\, \phi^*\mathbf{h} = \mathbf{h}\}$ constitute the isotropy group $\mathit{Isot}(\langle\boldsymbol{\Sigma}, \mathbf{h}\rangle)$. Moreover, $\mathit{Isot}(\langle\boldsymbol{\Sigma}, \mathbf{h}\rangle)$ coincides with [301] $\mathit{Isom}(\langle\boldsymbol{\Sigma}, \mathbf{h}\rangle)$; let us mark this by using $I(\langle\boldsymbol{\Sigma}, \mathbf{h}\rangle)$ to denote this coincident entity. The Lie algebra corresponding to this is isomorphic to that of the Killing vector fields of $\langle\boldsymbol{\Sigma}, \mathbf{h}\rangle$. An interesting result here is that $I(\langle\boldsymbol{\Sigma}, \mathbf{h}\rangle)$ is compact if $\boldsymbol{\Sigma}$ is [653]. Finally, since $I(\langle\boldsymbol{\Sigma}, \mathbf{h}\rangle)$ comes in multiple sizes, there are multiple dimensions of the corresponding group orbits, pointing to the group orbit space not being a manifold.

N.4 Topology of $\mathfrak{S}$uperspace($\boldsymbol{\Sigma}$)

Fischer showed that $\mathfrak{S}$uperspace$(\boldsymbol{\Sigma}) = \mathfrak{Riem}(\boldsymbol{\Sigma})/Diff(\boldsymbol{\Sigma})$ [301] can be taken to possess the corresponding quotient topology. $\mathfrak{S}$uperspace($\boldsymbol{\Sigma}$) additionally admits a metric space metric of the form [301]

$$\text{Dist}([\mathbf{h}_1],[\mathbf{h}_2]) := \inf_{\phi \,\in\, Diff(\boldsymbol{\Sigma})} \left(\text{Dist}(\phi^*\mathbf{h}_1, \phi^*\mathbf{h}_2)\right). \tag{N.3}$$

In this manner, $\mathfrak{S}$uperspace($\boldsymbol{\Sigma}$) is a metrizable topological space and thus obeys all the separation axioms and thus in particular Hausdorffness; it is also second-countable [301]. Thus Superspace is '2/3rds of a manifold' in the sense of Appendix M.

However, unlike $\mathfrak{Riem}(\boldsymbol{\Sigma})$, $\mathfrak{S}$uperspace($\boldsymbol{\Sigma}$) fails to possess the infinite-dimensional analogue of the locally-Euclidean property. Wheeler [899] credited renown American mathematician Stephen Smale with first pointing this out. Fischer [301] subsequently worked out the detailed structure of $\mathfrak{S}$uperspace($\boldsymbol{\Sigma}$) as a stratified manifold. In particular, the appearance of nontrivial strata occurs for $\boldsymbol{\Sigma}$ that admit metrics with non-trivial $I(\langle\boldsymbol{\Sigma},\mathbf{h}\rangle)$. In these cases $Diff(\boldsymbol{\Sigma})$ clearly does not act freely upon these metrics. Rather, the $\mathfrak{S}$uperspace($\boldsymbol{\Sigma}$) quotient space is here a stratified manifold of nested sets of strata ordered by $\dim(I(\langle\boldsymbol{\Sigma},\mathbf{h}\rangle))$.[2] Indeed, Fischer [301] tabulated the allowed isometry groups on various different spatial topologies. In this way, $\mathfrak{S}$uperspace($\boldsymbol{\Sigma}$) is not itself a manifold.

A further useful concept is the *degree of symmetry* of $\boldsymbol{\Sigma}$,

$$\deg(\boldsymbol{\Sigma}) := \sup_{\mathbf{h}\in\mathfrak{Riem}(\boldsymbol{\Sigma})} \left(\dim(I(\langle\boldsymbol{\Sigma},\mathbf{h}\rangle))\right). \tag{N.4}$$

Fischer [301] listed 3-manifolds with $\deg(\boldsymbol{\Sigma}) > 0$, and, in collaboration with mathematical physicist Vincent Moncrief, [304, 305] further characterizes $\deg(\boldsymbol{\Sigma}) = 0$ manifolds. N.B. that for $\deg(\boldsymbol{\Sigma}) = 0$, $\mathfrak{S}$uperspace($\boldsymbol{\Sigma}$) *is* a manifold.

Mathematicians Richard Palais and David Ebin [276] established that $Diff(\boldsymbol{\Sigma})$ is not compact. Moreover, they also showed that $Diff(\boldsymbol{\Sigma})$ acting on $\mathfrak{Riem}(\boldsymbol{\Sigma})$ is one of the cases for which a slice does none the less exist.

Ebin–Palais Slice Theorem [276] *Using* [301]*'s presentation, for each* $\mathbf{h} \in \mathfrak{Riem}(\boldsymbol{\Sigma})$ $\exists$ *a contractible submanifold* $\mathfrak{S}$ *containing* $\mathbf{h}$ *such that*

i) *For* $\phi \in Diff(\boldsymbol{\Sigma})$ *and* $\phi \in I(\langle\boldsymbol{\Sigma},\mathbf{h}\rangle)$, $\phi^*\mathfrak{S} = \mathfrak{S}$.
ii) $\phi \notin I(\langle\boldsymbol{\Sigma},\mathbf{h}\rangle) \Rightarrow \phi^*\mathfrak{S} \cap \mathfrak{S} = \emptyset$.
iii) $\exists$ *in Orb*($\mathbf{h}$) *an open set* $\mathfrak{U}$ *which itself contains* $\mathbf{h}$, *and a local section*

[2] $\mathfrak{S}$uperspacetime($\mathfrak{m}$) also has nested strata and conical singularities corresponding to the geometries with nontrivial Killing vectors.

$\Gamma : \mathfrak{U} \rightarrow Diff(\boldsymbol{\Sigma})$ *such that* $\phi(p, s) = \{\Gamma(p)\}^* s$ *is a diffeomorphism of* $\mathfrak{U} \times \mathfrak{S}$ *onto an open neighbourhood* $\mathfrak{N}_{\mathbf{h}}$ *of* $\mathbf{h}$.[3]

A ready consequence is as follows.

Superspace Decomposition Theorem [301] *The decomposition of* $\mathfrak{S}$*uperspace*$(\boldsymbol{\Sigma})$ *into group orbits is a countable partially-ordered* $\mathfrak{C}^{\infty}$*-Fréchet manifold partition.*

By the preceding and Appendix M.5's definition of inverted stratification, the following holds as well.

Superspace Stratification, Stratum and Strata Theorems [301] *The manifold partition of superspace is an inverted stratification indexed by symmetry type.*

Fischer additionally tabulates classifications of the superspace topologies and of the strata. The Stratum Theorem includes [301] that a stratum of superspace is finite dimensional iff the group action on the manifold is transitive (corresponding to a homogeneous space).

See e.g. [301, 302, 356, 357, 363] for further topological studies of Superspace, and [363] for further difficulties with putting a Riemannian metric on Superspace.

Research Project 117)†† [Hard and long-standing] How satisfactorily can *all* GR singularities be treated from a 'paths in configuration space' perspective? From a space of spacetimes perspective?

N.5 Comparison Between Theories. i. Theorems

Let us next further compare the GR, Gauge Theory and Mechanics cases. Firstly, Slice Theorems are known for each. The previous Section gives the GR case, and Appendix N.1 the Gauge Theory case, whereas e.g. [686, 774] give Mechanics counterparts. Secondly, see the same Sections for the GR and Gauge Theory cases of Stratification Theorems; Mechanics also has a such, at least in the symplectic setting [642]. The above two results provide further directions in which to consider the RPM model arena. Thirdly, the Decomposition Theorem that we have seen arise for GR also has a Gauge Theory version [759]. This is for group orbit spaces $\mathfrak{O}$, and is more mathematically standard: based on a generalization of the *Hodge–de Rham Decomposition Theorem* (see e.g. [207, 316] for a basic outline).

---

[3] Here $s \in \mathfrak{S}$ and $p \in \mathfrak{U}$, which is an open neighbourhood of $I(\langle\boldsymbol{\Sigma}, \mathbf{h}\rangle)$'s identity in the coset space (Appendix M.1): $Diff(\boldsymbol{\Sigma})/I(\langle\boldsymbol{\Sigma}, \mathbf{h}\rangle)$, and 'diffeomorphism' and 'submanifold' are in the sense of Fréchet($\mathfrak{C}^{\infty}$).

N.6 ii. Handling Dynamical Trajectories Exiting a Stratum

For both RPMs and GR, stratification becomes an issue as regards continuations of dynamical trajectories. On the other hand, in Gauge Theory, bounding by strata can be considered within the context of Gribov regions. See e.g. [267] for boundary condition considerations for the Gribov regions of Gauge Theory.

In the GR case, Leutwyler and Wheeler [899] appear to have been the first to ask about initial or boundary conditions on superspace. DeWitt, Fischer and Misner subsequently suggested [240, 301, 658] that when the edge of one of the constituent manifolds—i.e. where the next stratum starts—is reached, the path in Superspace that represents the evolution of the 3-geometry could be *reflected*. Simpler such reflection conditions were also previously considered for Mechanics; Misner's considerations were for Minisuperspace, whereas DeWitt considered a further simple model arena [240].

A subsequent alternative proposal by Fischer [302] concerns extending such motions though working instead with a nonsingular extended space. This no longer encounters the stratified manifold's issues as regards differential equations for motion becoming questionable at the junctions between strata.

Fischer explicitly built such an extended space [302] by use of an unfolding which permits access to Fibre Bundle Methods. The unfolding involved is parametrized by $I(\mathbf{\Sigma})$, as anticipated in the Mechanics case in Appendix G.3 This unfolding improves on previous such constructs by being Generally Covariant. It provides the right amount of information at each geometry—$\mathfrak{S}$uperspace($\mathbf{\Sigma}$)'s notion of point—to make the space of geometries into a manifold. The unfolding attains this by making use of the *bundle of linear frames over* $\mathbf{\Sigma}$, $\mathfrak{F}(\mathbf{\Sigma})$. In this case, no nontrivial isometries fix a frame. Thus the group action on the unfolded space $\mathfrak{Riem}(\mathbf{\Sigma}) \times \mathfrak{F}(\mathbf{\Sigma})$ is free.

Fischer [302] also pointed to $\mathfrak{S}$uperspace($\mathbf{\Sigma}$) possessing a 'natural minimal resolution' of the resultant singularities. This is based upon using the frame bundle quotient space $\mathfrak{Riem}(\mathbf{\Sigma}) \times \mathfrak{F}(\mathbf{\Sigma})/\mathrm{Diff}(\mathbf{\Sigma})$. In this particular case, one can regard $\mathfrak{Riem}(\mathbf{\Sigma})$ as a principal fibre bundle $\mathfrak{P}(\mathfrak{S}\mathrm{uperspace}_{\mathfrak{F}}(\mathbf{\Sigma}), \mathrm{Diff}_{\mathfrak{F}}(\mathbf{\Sigma}))$. I.e. $\mathrm{Diff}_{\mathfrak{F}}(\mathbf{\Sigma}) \xrightarrow{i} \mathfrak{Riem}(\mathbf{\Sigma}) \xrightarrow{\pi} \mathfrak{S}\mathrm{uperspace}_{\mathfrak{F}}(\mathbf{\Sigma})$ for i an inclusion map and π the fibre bundle's projection map (Appendix F.4).

However, the above unfolding runs against Relationalism, due to the $\mathfrak{F}(\mathbf{\Sigma})$ involved being a mathematical construct that does not correspond to more detailed modelling of physical entities.

Within the alternative Accept All Strata strategy toward strata, as a first point, recollect Sect. 37.6's outline of extending geodesics between strata using sheaf methods. As a second point, note that stratifolds (Appendix M.7) happen to further model a number of configuration spaces of interest, as follows. Firstly, use Appendix M.6's statement about Mechanics configuration spaces. Secondly, proceed by Appendix M.7's statement about infinite-dimensional stratifolds moving toward being able to model GR configuration spaces. On the other hand, that the space of spacetimes modulo spacetime diffeomorphisms is not Hausdorff leaves this space outside the scope of stratifolds, as are some loop spaces. Kreck's stratifold, and Śniatycki's and Pflaum's constructs are then mentioned in this book due to their applicability to *a range of* physically interesting examples, rather than as a full resolution for handling all the stratified manifolds that arise in Physics.

Moreover, as Fischer and Moncrief pointed out, the $\deg(\mathbf{\Sigma}) = 0$ case of $\mathfrak{S}$uperspace($\mathbf{\Sigma}$) avoids having strata in the first place, thus not necessitating any boundary conditions or extension procedure. On the one hand, $\deg(\mathbf{\Sigma}) = 0$ carries connotations of genericity, upon which general relativists place much weight. On the other hand, there is considerable interest in studying the simpler superspaces which are based on spaces with Killing vectors—such as $\mathbb{S}^3$ and $\mathbb{T}^3$—for which reduced approaches do encounter stratification.

Another research direction arises from acknowledging that the actual Universe at most involves approximate Killing vectors [218, 806, 807]. This comes at the price of many standard techniques becoming inapplicable.

Example 1) perturbation theory that is *centred about* an exact solution with exact Killing vectors may cease to apply.
Example 2) GR averaging issues enter the modelling.

Example 1) would be covered by modelling the Universe on some specific $\deg(\mathbf{\Sigma}) = 0$ spatial topology: in this case we know there are no Killing vectors for strata to arise from. This would greatly complicate calculations as compared to those we are accustomed to on e.g. $\mathbb{S}^3$. On the other hand, Example 2) would be manifested through us not knowing which $\deg(\mathbf{\Sigma}) = 0$ spatial topology to take; one would now have to average over all plausible such, and quite possibly allow for these to change over evolution. By this stage one would be modelling with 'Big Superspace' (Appendix S.2) and it would be a 'higher level excision' to exclude the superspaces with Killing vectors. One might still hope that sufficiently accurate analysis of the dynamical path would reveal it to avoid $\deg(\mathbf{\Sigma}) \neq 0$ topologies, or at least the solutions with Killing vectors therein. However, issues remain as regards whether each of the nongeneric structures—such that $\deg(\mathbf{\Sigma}) \neq 0$ and h_{ab} possesses one or more Killing vectors—could have a dynamical attractor role, which could force generic *paths* to have endpoints in, or pass arbitrarily close to, non-generic points.

N.7 $\mathfrak{CS}(\mathbf{\Sigma})$ and $\{\mathfrak{CS} + \mathrm{V}\}(\mathbf{\Sigma})$

$Conf(\mathbf{\Sigma})$ and $Diff(\mathbf{\Sigma})$ combine according to $Conf(\mathbf{\Sigma}) \rtimes Diff(\mathbf{\Sigma})$ [303]; similarly $VPConf(\mathbf{\Sigma})$ and $Diff(\mathbf{\Sigma})$ combine according to $VPConf(\mathbf{\Sigma}) \rtimes Diff(\mathbf{\Sigma})$ (Exercise!). Note that $Conf(\mathbf{\Sigma}) \cap Diff(\mathbf{\Sigma}) \neq \emptyset$ due to the conformal isometries outlined in Appendix E.2. However, since quotienting something out twice is clearly the same as quotienting it out once, this does not unduly affect the implementation. Also note that $Conf(\mathbf{\Sigma})$ is contractible, so $Conf(\mathbf{\Sigma}) \rtimes Diff(\mathbf{\Sigma})$ has the same topology as $Diff(\mathbf{\Sigma})$, $Conf(\mathbf{\Sigma}) \rtimes Diff_{\mathfrak{F}}(\mathbf{\Sigma})$ as $Diff_{\mathfrak{F}}(\mathbf{\Sigma})$, $\mathfrak{CS}(\mathbf{\Sigma})$ as $\mathfrak{S}$uperspace($\mathbf{\Sigma}$) and $\mathfrak{CS}_{\mathfrak{F}}(\mathbf{\Sigma})$ as $\mathfrak{S}$uperspace$_{\mathfrak{F}}(\mathbf{\Sigma})$ (see e.g. [363]).

Fischer and mathematical physicist Jerrold Marsden [303] extended Ebin's work by considering the action of the $\mathfrak{C}^\infty$ version of $Conf(\mathbf{\Sigma})$ on $\mathfrak{Riem}(\mathbf{\Sigma})$, as motivated by York's GR initial value problem work [922, 924, 925]. They obtained an analogue of the Ebin–Palais Slice Theorem for $Conf(\boldsymbol{\Sigma}) \rtimes Diff(\boldsymbol{\Sigma})$. They also demonstrated that the cotangent space corresponding to $\mathfrak{CS}(\boldsymbol{\Sigma})$ is an infinite-d weak symplectic manifold near those points corresponding to $(\mathbf{h}, \mathbf{p})$ not possessing any shared conformal Killing vector fields. This signifies topological straightforwardness other than as regards being stratified. [924] includes a linearized version of the stratification. That stratification occurs carries over from $\mathfrak{S}$uperspace($\boldsymbol{\Sigma}$) to $\mathfrak{CS}(\boldsymbol{\Sigma})$, along with many results that follow from contractibility. In fact, Fischer and Marsden [303] already had a $\mathfrak{CS}(\boldsymbol{\Sigma})$ analogue of the Superspace Stratification Theorem. Fischer and Moncrief's Superspace results [305] carry over to $\mathfrak{CS}(\boldsymbol{\Sigma})$ as well. Consequently for the deg($\boldsymbol{\Sigma}$) = 0 case, one gets a second helping of each of orbifolds, manifolds and contractible manifolds.

$\mathfrak{CS}(\boldsymbol{\Sigma})$ must be positive-definite since it is contained within $\mathfrak{CR}$iem($\boldsymbol{\Sigma}$).

$\mathfrak{CR}$iem($\boldsymbol{\Sigma}$) is better-behaved than $\mathfrak{R}$iem($\boldsymbol{\Sigma}$) along lines already established by DeWitt [237]. One might hope that $\mathfrak{CS}(\boldsymbol{\Sigma})$ is better-behaved than $\mathfrak{S}$uperspace($\boldsymbol{\Sigma}$), in parallel to relational space containing a better-behaved shape space.

Conversely, *Dil* alone can be quotiented out of $\mathfrak{S}$uperspace($\boldsymbol{\Sigma}$), giving a $\mathfrak{vp}\mathfrak{S}$uperspace($\boldsymbol{\Sigma}$) configuration space (volume-preserving Superspace). Finally, let us name the further quotient spaces afforded by the unit-determinant diffeomorphisms $UDiff(\boldsymbol{\Sigma})$ by use of extra U-prefixes.

Research Project 118) $\mathfrak{PR}$iem($\mathfrak{m}$) was studied at the geometrical level by DeWitt [241]. Perform a corresponding analysis of $\mathfrak{CpR}$iem($\mathfrak{m}$), [Stern [301] investigated the basic topological properties of $\mathfrak{S}$uperspacetime($\mathfrak{m}$); in particular, this is not Hausdorff.] What are the basic topological properties of conformal superspacetime, $\mathfrak{Css}(\mathfrak{m})$?

N.8 Notions of Distance for Geometrodynamics

Referring back to Sect. G.4, $\|\ \|_{\mathbf{M}}$ is not a notion of distance for $\mathbf{M}$ indefinite, e.g. for GR or its Minisuperspace. The same restriction occurs again for path metrics. Thereby, the Kendall, Barbour and DeWitt comparers do not carry over to GR as notions of distance. [Whereas the DeWitt comparer originates from Geometrodynamics, it did not arise there as a distance, but rather as a *metric functional* from which an indefinite geometry follows by double differentiation.] Four ways to make progress are as follows.

1) Consider $\mathfrak{CR}$iem($\boldsymbol{\Sigma}$) and $\mathfrak{CS}(\boldsymbol{\Sigma})$. These are positive-definite so that the Barbour and DeWitt comparers do carry over here as notions of distance [37].
2) Use an inf implementation instead; cf. (N.3) and the Gromov–Hausdorff notion of distance [393] (see Sect. S.3).
3) Use inhomogeneity quantifiers. To set up a first example, the standard definition of average is (for compact $\boldsymbol{\Sigma}$)

$$\langle \mathrm{A} \rangle := \int_{\boldsymbol{\Sigma}} \mathrm{d}^3 x \sqrt{\mathrm{h}}\, \mathrm{A} \Big/ \int_{\boldsymbol{\Sigma}} \mathrm{d}^3 x \sqrt{\mathrm{h}}. \tag{N.5}$$

Additionally,

$$\overline{\mathrm{A}} := \mathrm{A} - \langle \mathrm{A} \rangle, \tag{N.6}$$

is an 'unnormalized inhomogeneity quantifier' of the 'contrast' type. There is an obvious continuum counterpart of this example. See e.g. [926, 927] for more complicated such quantifiers.

4) Use spectral notions of distance. The basic idea here is to consider the spectrum of some natural differential operator on the manifold. Problems with this include non-uniqueness of such natural operators and the 'isospectral problem' that 'drums of different shapes' can none the less sound exactly the same; in this way, the separation axiom of distance fails.

N.9 Modelling with Infinite-*d* Stratifolds

Work in this direction has started [294, 571, 837], centering around Sheaf Methods and study of cohomology, in Hilbert and Fréchet space settings that do extend to manifolds in these senses.

Research Project 119)† Finish bridging this gap. I.e. can $\mathfrak{S}$uperspace($\boldsymbol{\Sigma}$) and $\mathfrak{CS}(\boldsymbol{\Sigma})$ be modelled as infinite-*d* stratifolds?

N.10 Reduced Treatment of Slightly Inhomogeneous Cosmology

This can be taken to arise from a particular example of the Thin Sandwich, in which the sandwich procedure by itself fails to factor in the *Diff*($\boldsymbol{\Sigma}$) content (Chap. 30). In these models, *Diff*($\mathbb{S}^3$) start to have effect at first order. Let us denote the corresponding space of the $\mathrm{d}u_\mathrm{n}$ by *Diff*$_1$($\mathbb{S}^3$).

Vacuum Case

This case turns out to be more straightforward [34]. Here solving the Thin Sandwich equations gives—for v_n with components $\mathrm{d}^\mathrm{o}_\mathrm{n}$, $\mathrm{d}^\mathrm{e}_\mathrm{n}$, $s_\mathrm{n} := a_\mathrm{n} + b_\mathrm{n}$: the *scalar mode sum* ubiquitous quantity, and A_n given by (I.14): the best matched configuration space metric

$$\frac{2}{\exp(3\,\Omega)}\mathrm{d}s^2_{\mathrm{bm}} = \{-1 + A_\mathrm{n}\}\mathrm{d}\Omega^2 + \frac{2}{3}\mathrm{d}\Omega\mathrm{d}A_\mathrm{n} + \|\mathrm{d}v_\mathrm{n}\|^2. \tag{N.7}$$

This is of dimension 4 + 1: a_n drops out of the line element, so it is short by 1 in removing the *Diff*($\mathbb{S}^3$) degrees of freedom. Moreover, geometrically this is just flat $\mathbb{M}^5$. Indeed,

$$T_\mathrm{n} := \frac{2}{3}\sqrt{A_\mathrm{n} - 1}\cosh\left(\Omega + \frac{1}{3}\ln(A_\mathrm{n} - 1)\right),$$

$$X_n := \frac{2}{3}\sqrt{A_n - 1}\sinh\left(\Omega + \frac{1}{3}\ln(A_n - 1)\right)$$

cast the line element in the familiar form

$$\frac{2}{\exp(3\,\Omega)}\mathrm{d}s^2 = -\mathrm{d}T_n^2 + \mathrm{d}X_n^2 + \|\mathrm{d}v_n\|^2. \quad \text{(N.8)}$$

One can proceed from here by the V part of $\mathcal{H}$ separating out to give an equation [34] for the thus only temporarily convenient mixed-SVT variable $A_n = A_n(\Omega)$. This gives a fully *Diff*($\mathbb{S}^3$)-reduced line element of the form

$$\frac{2}{\exp(3\,\Omega)}\mathrm{d}s^2 = \{-1 + f_n(\Omega)\}\mathrm{d}\Omega^2 + \|\mathrm{d}v_n\|^2, \quad \text{(N.9)}$$

where $f_n(\Omega) := A_n(\Omega) + \frac{2}{3}\mathrm{d}A_n(\Omega)/\mathrm{d}\Omega$. This is conformally flat. Next define a new scale variable $\zeta_n := \int \sqrt{f_n(\Omega) - 1}\,\mathrm{d}\Omega$ to absorb the first term's prefactor. This leaves, up to a conformal factor, the simplified line element

$$\mathrm{d}s^2 = -\mathrm{d}\zeta_n^2 + \|\mathrm{d}v_n\|^2, \quad \text{(N.10)}$$

which is a local-in-time slab within $\mathbb{M}^4$.

$\partial/\partial v_n^V$'s components $\partial/\partial s_n$, $\partial/\partial d_n^o$, $\partial/\partial d_n^e$ are among the 10 conformal Killing vectors; the others are $\partial/\partial\zeta$, $3\, v_n^V\partial/\partial v_n^{V'} - v_n^{V'}\partial/\partial v_n^V$ and $3\, v_n^V\partial/\partial\zeta + \zeta\partial/\partial v_n^V$.

Finally the corresponding shape space is also clearly flat, in this case $\mathbb{R}^3$:

$$\mathrm{d}s^2 = \|\mathrm{d}v_n\|^2. \quad \text{(N.11)}$$

Minimally-Coupled Scalar Field Case

In this case [35], the outcome of the Thin Sandwich elimination is the undisturbed $\mathrm{d}s_0^2$ of the densitized version of (9.12) alongside

$$\mathrm{d}s_{\mathrm{bm}}^{n\,2} = \frac{\exp(3\,\Omega)}{2}\Big\{\|\mathrm{d}v_n\|^2 + \mathrm{d}f_n^2 + \Big\{\Big\{3\mathrm{d}a_n + \sqrt{3\{n^2-4\}}\mathrm{d}s_n\Big\}f_n + 6a_n\mathrm{d}f_n\Big\}\,\mathrm{d}\phi$$
$$+ \frac{2}{3}\mathrm{d}A_n\mathrm{d}\Omega - A_n\{-\mathrm{d}\Omega^2 + \mathrm{d}\phi^2\}\Big\}. \quad \text{(N.12)}$$

However, since this is of dimension $6 + 2$, it is not yet Superspace. In removing the *Diff*($\mathbb{S}^3$) degrees of freedom, the thin-sandwich manoeuvre has now fallen short by 2. In this case, how to progress from here with the reduction is not known.

In this case, the geometry of the currently attained 'half-way house' has been further explored by the Author. Its configuration space block structure can be tidied up by removing as many off-diagonal terms as possible can be done separately in each of the first two blocks. Diagonalize the one by using (I.17) again, now alongside using the *exchanger variable*

$$\phi_{\rm n} = \phi - \frac{3}{2} b_{\rm n} f_{\rm n} \tag{N.13}$$

on the other so as to set the coefficient of $\mathrm{d}a_{\rm n}\mathrm{d}\phi_{\rm n}$ to zero.

[If multiple n's are considered, use instead mode-summed variables $\widetilde{\Omega}$ as per (N.13), alongside

$$\widetilde{\phi} = \phi - \frac{3}{2} \sum_{n} b_{\rm n} f_{\rm n}. \tag{N.14}$$

This example illustrates that the configuration space metric split (H.6) and consequently the Hamiltonian constraint metric–matter split (24.22) are *not in general preserved* by reduction procedures. This gives rise to the complication that even minimally-coupled matter influences the form of the gravitational sector's reduced $\mathfrak{q}$ geometry.

The Ricci scalar for the partly reduced geometry is

$$R = 7\exp(-3\,\Omega)/f_{\rm n}^2. \tag{N.15}$$

Consequently the matter perturbation going to zero—a physically innocuous situation—gives a curvature singularity. This has some parallels [34, 35] with the configuration space singularity corresponding to the collinear configurations in the N-body reduced configuration space.

$\partial/\partial\phi_{\rm n}$, $\partial/\partial d_{\rm n}^{\rm e}$ and $\partial/\partial d_{\rm n}^{\rm o}$ are Killing vectors for this; whereas $\partial/\partial f_{\rm n}$ has lost this status upon reduction, the tensor mode directions have gained this property. As ever, $\partial/\partial\Omega$ is a conformal Killing vector. However, none of the above respect the corresponding potential, so conserved quantities do not ensue.

Also, scaled perturbative Minisuperspace does not have a bona fide configuration space metric based notion of distance. On the other hand, the space of pure inhomogeneities is positive-definite. The vacuum case admits a 3-*d* Euclidean metric with the $\underline{v}_{\rm n}$ as the corresponding coordinates.

All in all, unlike in the vacuum case, it remains unclear how to complete the reduction in the case with minimally-coupled scalar field matter.

N.11 Further Configuration Spaces

N.12 Loops and Loop Spaces

The loops of Appendix N.2 are now taken to involve i) paths embedded in GR's topological notion of space: $\mathbf{\Sigma}$, and ii) the specific gauge group $SU(2)$.

The heuristic outline of the loop spaces (Fig. N.1.c) being loop groups carries over to this case. For more rigorous treatments, see e.g. [77, 179, 307]. The form

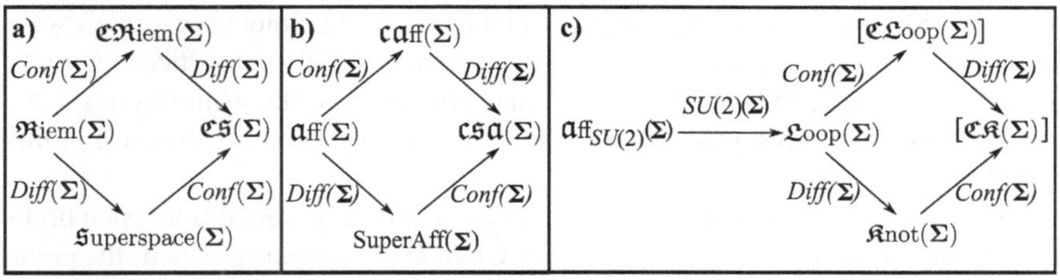

Fig. N.1 Configuration spaces for **a)** GR as Geometrodynamics, **b)** Affine theory and **c)** Nododynamics. The last of these introduces an extra local $SU(2)(\boldsymbol{\Sigma})$ in defining its variables. Taking out these degrees of freedom, one passes to the space of loops, and subsequently quotienting out $Diff(\boldsymbol{\Sigma})$ as well, to the space of knots. The conformal versions of these are presented in square parentheses due to hitherto not usually being studied in Nododynamics

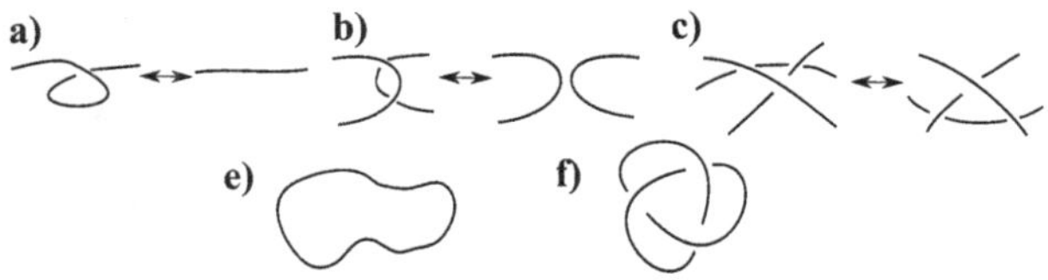

Fig. N.2 Representation of knots as planar graphs with an over-and-under crossing designation. The *Reidemeister moves* that preserve knots are **a)** twist/untwist, **b)** pull back/push under, and **c)** slide string up/down underneath a crossing. **d)** The unknot—alias trivial knot—has no crossings, or can be continuously deformed—by the so-called 'ambient isotopy' notion—into having none. **e)** The trefoil knot is the simplest nontrivial knot

taken by the stratification of the group orbit space in the case of GR is covered in [306].

See e.g. [179] for the LQC equivalent of diagonal anisotropy.

N.13 Knots and Knotspace

A *knot* K is an embedded closed curve in a closed orientable 3-manifold $\boldsymbol{\Sigma}$ (most usually $\mathbb{S}^3$). We restrict attention to smoothly or piecewise linearly embedded curves to avoid 'wild knots' [68]. Two knots K_1, K_2 in a given $\boldsymbol{\Sigma}$ are equivalent if $\exists$ an orientation-preserving automorphism such that $\text{Im}(K_1) = K_2$.

Whereas knot equivalence can be investigated using the Reidemeister moves (Fig. N.2), we do not know of an upper bound on how many such moves are needed to bring knots into obvious equivalence, so these moves are of limited practical use. Rather, we seek characterization in terms of *knot invariants* (a subset of topological invariants). The obvious routes to such are homotopy and homology; these give respectively the *knot group* (fundamental group of the knot complement) [68] and the *Alexander polynomial* [626]. However, neither of these serve to discern between even some of the simplest knots. The advent of the Jones polynomial

[159, 330, 533, 626] revived the subject; a number of further knot polynomials were subsequently discovered at short order. However, these still do not suffice to classify knots. See e.g. [330, 533, 757, 916] for some applications of knots in Physics. Also note the rather obvious topological manifold level Background Dependence in this formulation of knots.

The mathematical form of the corresponding 'Knotspace' remains an open problem. One approach is to view knot space as $\mathfrak{E}mb(\mathbb{S}^1, \mathbb{S}^3)$: embeddings of the circle in the 3-sphere, in the sense of each $\mathfrak{q}$ being a subspace of a more tractable mapping space [181]. This turns the topological problem into one concerning singular maps, which is subsequently *aided* by these forming a *stratified space*. See also [180, 869] for a mathematical account of spaces of knots.

Appendix O
DE Theorems for Geometrodynamics' Problem of Time*

O.1 Types of Global Issues

0) In considering global issues, it is helpful to think complementarily, i.e. to consider not 'globality' but the lack thereof: 'locality'. This could refer to

   i) an infinitesimal neighbourhood (if the type of mathematical space in question possesses such a notion: see Appendix C).
   ii) An extended region that does not cover the whole space, termed *quasilocality* in e.g. [823, 824].
   iii) Topological effects that require the entire space to be taken into consideration—such as the homotopy, homology or cohomology groups outlined in Appendix F.3—remain unaddressed by ii).

Many applications in Classical Physics involve the following set-up, or variants given further below. Analysis can be applied not only to functions on $\mathbb{R}^n$ but to linear and then nonlinear DEs as well. These enter Classical Physics through fundamental physical laws widely taking the form of PDEs. ODEs arise in those simple models that permit separation (as well as directly from Newton's Second Law). Furthermore, a *PDE problem* is a PDE system alongside prescribed data, such as boundary conditions, or initial conditions; prescription of values of fields and their velocities on an initial surface is *Cauchy data* for a *Cauchy problem*. A PDE problem is *well-posed* if there exist unique solutions to it with *continuous dependence* on the initial data.[1] Global issues now enter as follows.

1) Some function spaces are well adapted to this task, including globally. If not using such a function space at the outset, one can pass to it ('Globalization by

[1] Without this last condition, an arbitrarily small change in the data could cause an arbitrarily large immediate, precluding any physical predictability. N.B. this really does mean *immediate*—see e.g. p. 229 of [220]—rather than some issue of chaos or unwanted growing modes. That said, though well-posedness often *also* bounds the growth of such modes [323]. Section O.4's hyperbolic PDE problems require the domain of dependence property as a fourth well-posedness criterion. This enforces a sensible notion of causality, permitting compatibility with Relativity.

E. Anderson, *The Problem of Time*, Fundamental Theories of Physics 190,
DOI 10.1007/978-3-319-58848-3

Replacement'). Which function spaces are adapted to a PDE usually depends on the type of PDE.

Limited Counter-example 1) Use of the analytic functions $\mathfrak{C}^\omega$ is not sensitive to the type of PDE. These admit *analytic continuation* as a Globalization by Extension method . The analytic functions are however also undesirable [732] for modelling Relativistic Physics' spacetime, since analytic continuation precludes independence of causally unlinked regions. Consequently, one needs other function spaces for use in Relativistic Theories; differentiable structure for these is usually taken to be $\mathfrak{C}^\infty$.

Well-posedness criteria can quite often only be established locally. A further 'Globalization by Extension' consideration is whether singular solutions can be included in one's treatment; see Appendix F.5's consideration of Morse Theory for a simple example. Finally, nonlinear PDEs are often harder to handle globally, the Einstein field equations being the main case in question in this book.

2) The above involvement of $\mathbb{R}^n$ is in some ways but a local aspect, to be globalized by passing to some globally-well-defined differentiable manifold $\mathfrak{M}$ with multiple charts $\subset \mathbb{R}^n$. Indeed, meshing charts together is itself a 'Globalization by Replacement' technique. Yet this well-understood technique does not in general carry over to patching together local solutions of PDEs holding on $\mathfrak{M}$, i.e. to patching in a function space context.
3) A further modelling aspect is which role $\mathfrak{M}$ plays. Spacetime versus space is one source of variety here. Another is whether it is a space or a space of spaces, such as configuration space $\mathfrak{q}$ or phase space $\mathfrak{P}$hase.
4) Another local method involves associating with $\mathfrak{M}$ of a Lie algebra $\mathfrak{g}$ acting thereupon. One can then 'Globalize by Replacement' by passing to considering instead a Lie group $\mathfrak{g}$ version (which has its own manifold properties), or, alternatively a Lie algebroid.
5) We have also seen that some physical applications replace $\mathfrak{M}$ by such as a singular manifold or a stratified manifold. In the first case, note the increase in intractability in passing from singularities of solutions on a fixed background manifold versus singularities of manifolds themselves (e.g. GR singularities). In the second case, Appendix M demonstrated that introducing $\mathfrak{g}$ acting upon a bona fide differentiable manifold $\mathfrak{M}$ suffices for stratified manifolds to enter the study.
6) Some physical modelling involves not just $\mathfrak{M}$ but fibre bundles over $\mathfrak{M}$ as well. Indeed, such fibre bundles can serve to encode global properties of $\mathfrak{M}$.

Finally, note that some of the above six features occurring jointly can cause further complications. For instance, stratified manifolds can take one outside of the scope of Fibre Bundle Methods.

O.2 ODEs

Standard existence and uniqueness theorems for equations of the form

$$\dot{x} = f(x, t) \tag{O.1}$$

on metric spaces are based on contraction mappings, in the case of a Lipschitz condition

$$\text{Dist}(f(x), f(x')) \leq k\,\text{Dist}(x, x')$$

holding locally in a neighbourhood $\mathfrak{N}_x$ of x. x' is here an arbitrary point in $\mathfrak{N}_x$, and k is constant.[2] This approach can be extended to establish continuous dependence as well. Finally, we remark that this approach covers not only ODEs on $\mathbb{R}^n$ but e.g. on Riemannian manifolds as well.

Moreover, local ODE solutions cannot always be globally extended, e.g. due to blow-up within finite independent parameter; see e.g. [732, 874]. A global existence result for ODEs for use in Minisuperspace solutions of GR is given in [732]; the same equations also occur in the slightly different context of pointwise in Strong Gravity [716, 717]. This is specific to 3-d and includes all Bianchi types except Bianchi IX.

O.3 First-Order PDEs

There is an integral curves method for first-order quasilinear PDEs [207, 614]

$$\underline{f}(\underline{x}, u) \cdot \underline{\partial} u = g(\underline{x}, u)\ , \tag{O.2}$$

which holds provided that the surface swept out by the integral curves is not itself *characteristic*.[3]

O.4 Second-Order PDEs

For second-order linear equations with constant coefficients, a simple elliptic–hyperbolic distinction can be made in terms of whether or not the underlying quadratic form is positive-definite or indefinite with one negative direction. Wider applicability of such notions, however, requires a fair amount of further abstraction. For a general PDE system

$$F(\underline{x}, \partial^{(1)}\text{u}, \ldots, \partial^{(r)}\text{u}) = 0, \tag{O.3}$$

[2]See e.g. [72] for discussion of the necessity of the Lipschitz condition, and e.g. [426] for further similar ODE theorems on Banach spaces, and their analogues on those Fréchet spaces for which the Nash–Moser Theorem applies.

[3]This is avoided provided that there is no direction in which the directional derivative associated with the flow is tangential to the surface. This involves the same underlying notion of characteristic as in 'method of characteristics'. Consequently e.g. the flat spacetime wave equation has characteristics $x = \pm t$. In contrast, the flat space Laplace equation has no real characteristics [220].

where $\partial^{(i)}$ denotes the ith-order partial derivatives, the *principal symbol* σ_P is obtained by taking the highest-order part and using $\underline{\xi}$ in place of $\underline{\partial} = \partial^{(1)}$. One simple definition of *elliptic system* involves σ_P being positive-definite and invertible. On the other hand, hyperbolicity is more sensitive to lower order terms; this is taken into account by the *Leray notion of hyperbolicity*. See e.g. [204, 732] both for further details of, and variants of, these concepts.

PDE problems are well-known for a range of PDEs linear in the dependent variables. E.g. Electromagnetism is straightforward to treat. Its Maxwell field equations can be split into a constraint equation (2.13) and a system of evolution equations (3.2). On the other hand, GR is nonlinear, by which it requires considerably more work. Its Einstein field equations again split into constraint (8.27)–(8.28) and evolution equation (8.30) systems. In each case the constraints form an elliptic system and the evolution equations can be cast as hyperbolic systems (modulo some caveats in Appendix O.7). The constraint system is to be treated as a *GR initial value problem* and the evolution system as a *GR Cauchy problem*.

O.5 GR Initial Value Problem Theorems. i. Thin Sandwich Approach

The Mechanics analogue of the Thick Sandwich is already not well-posed [308]: a simple periodic example exhibits both nonexistence and nonuniqueness. In fact historically, it was the electromagnetic analogue of the Thick Sandwich [897] not being well-posed (Ex VI.11.i) that led to Thin Sandwich limits being considered instead.

As regards the Thin Sandwich, there are some results [115, 124, 308] concerning well-posedness *in a restricted sense*.

Bartnik–Fodor Theorem *Suppose*

$$\mathrm{h}_{ij} \in \mathfrak{H}^{n+2}(\mathfrak{T}^0_2), \qquad \mathrm{K}_{ij} \in \mathfrak{H}^{n+1}(\mathfrak{T}^0_2), \qquad \mathcal{E} \in \mathfrak{H}^{n+1} \quad \textit{and} \quad \mathrm{J}^i \in \mathfrak{H}^n(\mathfrak{T}^1_0)$$

satisfy the GR constraint equations and the 'potential zeros avoiding condition'

$$2\,\mathcal{E} - \mathcal{R} > 0. \tag{O.4}$$

Here $\mathcal{E}$ is the energy density, and the $\mathfrak{H}^p$ are Sobolev spaces as introduced in Appendix P.5. Additionally, for any choice of shift $\beta^i \in \mathfrak{H}^{n+2}(\mathfrak{T}^1_0)$ and positive lapse $\alpha \in \mathfrak{H}^{n+1}$, $\dot{h}_{ij}$ can be defined by (8.14). *Suppose also that the equation*

$$\mathrm{M}_{(i|j)} = \mu\,\mathrm{K}_{ij} \quad \textit{has only the trivial solution } \mathrm{M}_i \tag{O.5}$$

holds: a locality in configuration space condition involving staying away from Killing vectors. Then there exists a unique continuous map on an open neighbourhood $\mathfrak{N}$ with data

$$D := (\mathrm{h}_{kl}, \dot{\mathrm{h}}_{kl}, \mathcal{E}, \mathrm{J}^i) \in \mathfrak{H}^{n+2}(\mathfrak{T}^0_2) \times \mathfrak{H}^{n+1}(\mathfrak{T}^0_2) \times \mathfrak{H}^{n+1} \times \mathfrak{H}^n(\mathfrak{T}^1_0).$$

This assigns

$$\widetilde{\beta}^i \in \mathfrak{H}^{n+2}(\mathfrak{T}^1_0) \quad \textit{to } \widetilde{D}$$

such that $\widetilde{\beta}^k$ *is a solution of the reduced thin-sandwich equations with data* $\widetilde{D}$.

The proof follows immediately from the Implicit Function Theorem, once this is established to apply to apply because the PDE system is established to be elliptic and with trivial kernel (Exercise!).

Moreover, note that this locality is not just due to a lack of proof to date: a counter-example to global existence was already previously known [124]. Also contrast how in 3-*d* RPMs one has to excise the collinear configurations in order to attain Best Matching with how one has to excise the metrics possessing Killing vectors in order to attain Thin Sandwich proofs.

The local uniqueness, however, can readily be extended to the global *Belasco–Ohanian Theorem* [115] (after physicists Elliot Belasco and Hans Ohanian).

Overall, while the Thin Sandwich historically precedes the next Section's Conformal Approach, and is conceptually interesting and quite strongly tied to a Problem of Time facet, its mathematics remains less worked out, and is quite likely less strong.

O.6 ii. Conformal Approach

$\mathcal{M}_i$ is here taken to be a decoupled equation for the longitudinal potential ζ^i of Sect. 21.6, which is rather better-behaved elliptical PDE than the Thin Sandwich. This is relatively straightforward, out of being linear in the dependent variable. See e.g. [204] for protective theorems.

A *quasilinear* elliptic PDE for unknown u is one with a second-order elliptic part and at most nonlinearity in u itself (no nonlinearity in derivatives of u):

$$\Delta u + f(\underline{x}, u) = 0. \tag{O.6}$$

These afford more theorems than for more generally nonlinear PDEs. The Lichnerowicz–York equation (21.7) of GR is set up to take advantage of this by choosing a conformal scaling (D.33) which renders the equation quasilinear. See e.g. [116, 204, 206, 465] for protective theorems, many of which are global in character; some of these make use of Hölder spaces and some of Sobolev spaces.

O.7 Theorems for the GR Cauchy Problem

The original treatments of Leray and Fourès–Bruhat [311, 312, 617] used $\mathfrak{C}^r$ (or $\mathfrak{C}^\infty$), whereas more modern treatments [206, 440, 460] involve Sobolev spaces. The early works' emphasis on suitable notions of hyperbolicity has proven to be long-standing. Gauge fixing is also involved in the study of the GR evolution equations. The preceding two sentences have to be taken together due to notions of hyperbolicity not being gauge-invariant. E.g. the harmonic gauge (Ex V.13) casts the Einstein field equations into the correct hyperbolic form—due to Leray [617, 732]:

$$\mathcal{M}^{\mathsf{CD}}(\underline{x}, \phi^{\mathsf{A}}, \nabla_{\mathsf{A}}\phi^{\mathsf{B}})\nabla_{\mathsf{C}}\nabla_{\mathsf{D}}\phi^{\mathsf{E}} = \mathcal{F}_{\mathsf{E}}(\underline{x}, \phi^{\mathsf{A}}, \nabla_{\mathsf{A}}\phi^{\mathsf{B}}), \tag{O.7}$$

where $\mathcal{M}^{\mathsf{CD}}$ is a Lorentzian metric functional and both this and $\mathcal{F}_{\mathsf{E}}$ functional are smooth—for *Leray's Theorem* [617, 874] to apply. This guarantees well-posedness.

Hughes–Kato–Marsden Theorem *This further result guarantees existence for the n-d Einstein field equations in harmonic coordinates, if the Sobolev class of the induced metric is no rougher than $\mathfrak{H}^{s+1}$ and that of the extrinsic curvature is no rougher than $\mathfrak{H}^{s}$, for $s > 1.5$.*

Finally, see e.g. [204, 323, 557, 732, 737] for furtherly modernized theorems.

O.8 Basic FDE Theory for GR and QFT

The following results are useful in the theory of observables or beables.

0) By the chain-rule,

If u solves the linear PDE or FDE $\mathcal{L}\mathrm{in}\,\phi$ '=' 0, then so do $\mathrm{F}(u)$ and $\mathcal{F}[u]$. (O.8)

1) **Lemma 1** *The purely configurational gauge-invariant quantities (which coincide with the κ if* $\mathfrak{f}$**lin** = $\mathfrak{g}$**auge***) obey the same equation as that determining which quantities are annihilated by the generators.*

Proof This follows since the constraints involved are homogeneous linear in p_{iI}, so the Poisson bracket removes the p_{iI} factors and replaces them with $\frac{\partial}{\partial q^{iI}}$ factors. [The Poisson bracket term with the other sign is annihilated by the purely configurational restriction.] □

This result is used for instance in the configuration space $\mathfrak{q}$ uplift of the group generators acting upon absolute space $\mathfrak{a}$. In this way, $\mathfrak{a}$-invariants are replaced by ones built out of multiple particle positions q^{iI}.

Also note that the pure-momentum counterpart has no such result since the constraints are not confined to be linear in the $\boldsymbol{Q}$. Those which are linear in both $\boldsymbol{Q}$ and $\boldsymbol{P}$ have pure-$\boldsymbol{P}$ and pure-$\boldsymbol{Q}$ beables in close parallel to each other. On the other hand, those which are not diverge more in the forms of these two contributions to the 'basis beables'.

2) **Kuchař's no-go Theorem** [582] *Nonlocal objects of the form*

$$\int_{\Sigma} \mathrm{d}^3 x \, \mathcal{F}_{ij}(\underline{x}; \mathbf{h}]\mathrm{p}^{ij}(\underline{x}) \tag{O.9}$$

are not Dirac beables, where $\mathcal{F}_{ij}$ is some general spatial tensor-valued mixed function-functional.

This result makes use that metric concomitants are in general built out of covariant derivatives of the Riemann tensor. One then proceeds by proving inductively on the number of covariant derivatives that $\mathcal{F}_{ij}$ cannot contain concomitants with that number of covariant derivatives, by use of algebraic and integrability arguments.

3) **Torre's No-Go Theorem** [854] (see also [63, 194, 853, 858]) *Local functionals*

$$\mathrm{T}(\underline{x}; \mathbf{h}, \mathbf{p}] \tag{O.10}$$

are not Dirac beables either.

This uses that local observables correspond to local 'hidden symmetry'. However, the latter's cohomological classification (in the sense of de Rham for the bundle of metrics over spacetime) leaves no options.

Two further FDE considerations as regards QFT and the Wheeler–DeWitt equation are as follows.

4)

Double functional derivatives acting on functionals are not well-defined. (O.11)

5) At least in the case of such FDEs as have been mastered,

FDEs require regularization. (O.12)

Appendix P
Function Spaces, Measures and Probabilities

P.1 Basic Formulation of Probability

Let us begin by considering a mathematical space $\mathbf{\Omega}$, now cast in the role of *sample space*. Its subsets are now termed *events* $\mathfrak{E}$ (meaning here 'sets of outcomes', not to be confused with the spacetime use of the word 'events'). Each such $\mathfrak{E}$ is associated with a probability $\text{Prob}(\mathfrak{E}) \in [0, 1]$. In the original Probability Theory, these are taken to arise by *sampling*:

$$\text{Prob}(\mathfrak{E}) = (\text{number of times outcomes lying in } \mathfrak{E} \text{ occur in the sample}) \\ /(\text{sample size}) := m/n,$$

for n suitably large. This motivates viewing probability as a *valuation* within the interval $[0, 1]$, since m, n are positive and $m \leq n$.

A simple example of $\mathbf{\Omega}$ is $\{H, T\}$: heads or tails for a coin toss. For an unbiased coin, $\text{Prob}(H) = \frac{1}{2} = \text{Prob}(T)$. More generally, a partition of $\mathbf{\Omega}$ into discrete events $\mathfrak{E}_{\mathsf{E}}$ can be assigned probabilities such that $\sum_{\mathsf{E}} p_{\mathsf{E}} = 1$ (normalization). Then

$$\text{Prob}(\mathfrak{E}_1 \amalg \mathfrak{E}_2) = \frac{m_1 + m_2}{n} = \frac{m_1}{n} + \frac{m_2}{n} = \text{Prob}(\mathfrak{E}_1) + \text{Prob}(\mathfrak{E}_2),$$

while $\sum_{\mathsf{E}} m_{\mathsf{E}} = n$ for a partition. Such an assignation is called a *probability distribution*.

Example 1) For $\{H, T\}$, assigning $\text{Prob}(H) = p$, $\text{Prob}(T) = q := 1 - p$ for an in general biased coin gives a *binomial distribution* $\rho_{n,k,p} = \binom{n}{k} p^k q^{n-k}$ for the number of heads.

A *random variable* is a function $R : \mathbf{\Omega} \to \mathbb{R}$. Each R has a corresponding *probability distribution function* $\rho : \mathbf{\Omega} \to [0, 1]$ such that $\{\text{Prob}(R = r) | r \in \text{Range}(R)\}$. The condition for a finite number of random variables R_{R}, $\mathsf{R} = 1$ to n to be *independent* is that

$$\text{Prob}(R_{\mathsf{R}} = r_{\mathsf{R}}, \mathsf{R} = 1 \text{ to } n) = \Pi_{\mathsf{R}=1}^{n} \text{Prob}(R_{\mathsf{R}} = r_{\mathsf{R}}).$$

E. Anderson, *The Problem of Time*, Fundamental Theories of Physics 190,
DOI 10.1007/978-3-319-58848-3

Expectation

$$\mathrm{E}(R) := \sum_{\omega \in \mathbf{\Omega}} p_\omega R(\omega)$$

for a discrete random variable and

$$\mathrm{E}(R) := \int_{\omega \in \mathbf{\Omega}} \omega \rho(\omega) \mathrm{d}\omega$$

for a continuous one. In either case, *variance* is given by

$$\mathrm{Var}(R) := \mathrm{E}(R - \mathrm{E}(R))^2 = \mathrm{E}(R^2) - \{\mathrm{E}(R)\}^2.$$

A *statistic* is a function $\theta : \mathbf{\Omega} \rightarrow \mathbb{R}$ that is independent of one's sample's data set. θ computes a particular attribute of the data set. We next consider various generalizations of the above.

1) Extend from finite to countable R.

Example 2) The *Poisson distribution* $\rho_{k,\lambda} = \frac{\lambda^k}{k!} \exp(-\lambda)$; its parameter λ is both mean and variance.

2) Further extend R to the uncountable case. Normalization

$$\int_{\mathbf{\Omega}} \rho(\underline{x}) \mathrm{d}\mathbf{\Omega} = 1$$

is also required, where the probability distribution function $\rho : \mathbf{\Omega} \rightarrow \mathbb{R}_0$ assigns e.g.

$$\int_a^b \rho(x) \mathrm{d}x \quad \text{to Prob}(a \le R \le b).$$

Integration is for now meant in the sense of Riemann (so a *p-d* multiple integral for a region of $\mathbf{\Omega}$ generalizes the above 1-*d* probability distribution function's assignment).

Example 3) The *normal* alias *Gaussian distribution* with mean μ and standard deviation $\sigma = \sqrt{\text{variance}}$ is $\rho_{\mu,\sigma}(x) = \frac{1}{\sqrt{2\pi\sigma}} \exp(-\{x - \mu\}^2/2\sigma^2)$.

Example 4) A final example used in this book the is *uniform distribution*, $\rho = 1/n$ for n discrete values or $\rho = 1/\{b - a\}$ on the interval $[a, b]$.

3) One can furthermore view a collection of random variables R_{R} as a *random vector*.

The above notions of independence, expectation, variance and statistic generalize throughout this Appendix and Appendix T. Variance becomes the *co-variance matrix* with components

$$\mathrm{Cov}(R_{\mathsf{R}}, R_{\mathsf{R}'}) := \mathrm{E}(R_{\mathsf{R}} - \mathrm{E}(R_{\mathsf{R}}))\mathrm{E}(R_{\mathsf{R}'} - \mathrm{E}(R_{\mathsf{R}'})).$$

Thus

$$\mathrm{Cov}(R_{\mathsf{R}}, R_{\mathsf{R}}) = \mathrm{Var}(R_{\mathsf{R}}) \quad \text{(no sum)}.$$

In case 3), it is also common to consider random vectors consisting of independent *identically distributed* random variables. *Joint probability distributions* are for two or more random variables considered together, e.g. for Prob($R_1 = r_1$ and $R_2 = r_2$). See e.g. [391] for further reading about this Sec's material.

P.2 Measure Theory*

A *measure* [207] is a map $\mu : \mathfrak{S} \to \mathbb{R}$ such that

μ.1) $\mu(\emptyset) = 0$ (the measure of nothing is zero).
μ.2) $\mu(\bigcup_{\mathsf{A}} \mathfrak{Y}_{\mathsf{A}}) = \sum_{\mathsf{A}} \mu_{\mathsf{A}}(\mathfrak{Y}_{\mathsf{A}})$ for A countable and the subsets $\mathfrak{Y}_{\mathsf{A}}$ pairwise disjoint (countable additivity).

A measure is *normalizable* if $\mu(\mathfrak{S})$ is finite for $\mathfrak{S}$ the whole of the underlying space in question.

Example 1) The *Lebesgue measure* (after mathematician Henri Lebesgue) gives a general theory of integration; in this book, however, we only make detailed use of various more specific subcases. The $\mathfrak{L}^p$ spaces—with norm

$$\mu_f := \|f\|_{\mathfrak{L}^p} = \left\{ \int |f^p(x)| \mathrm{d}x \right\}^{1/p} \tag{P.1}$$

—are measurable spaces; these are also (Exercise!) Banach spaces and the $\mathfrak{L}^2$ case is additionally a Hilbert space.

Example 2) The *Haar measure* [207] is a $\mathfrak{g}$-invariant Borel measures for $\mathfrak{S}$ a compact topological group $\mathfrak{g}$. This permits explicit computational integration over $\mathfrak{g}$. For instance, the compact Lie group counterpart of (A.2) is

$$\int_{g \in \mathfrak{g}} \mathbb{D}g \, O \Big/ \int_{g \in \mathfrak{g}} \mathbb{D}g. \tag{P.2}$$

Example 3) A measure ν is *absolutely continuous* with respect to a second measure μ if $\nu(\mathfrak{X}) = 0$ for each set $\mathfrak{X}$ with $\mu(\mathfrak{X}) = 0$. This requires μ to be a positive measure, but is independent of which type of measure ν is. These are furthermore related by

$$\nu = \int_{\mathfrak{X}} R \, \mathrm{d}\mu:$$

the *Radon–Nikodym Theorem*. Within this, R plays the role of a derivative: the *Radon–Nikodym derivative* [207]; compare the usual Fundamental Theorem of Calculus.

Example 4) Probability measures are considered in Appendix P.3.

Example 5) The theory of *cylindrical measures* [207] is suitable on the locally convex topological vector spaces outlined in Appendix H.2. This is e.g. often applied to more rigorous accounts of loop spaces.

P.3 More Advanced Formulation of Probability*

The great mathematician Andrey Kolmogorov subsequently axiomatized Probability Theory as follows.

Kolmogorov 1) $\text{Prob}(\mathfrak{E}) \in \mathbb{R}_0 \ \forall \ \mathfrak{E} \in \mathbf{\Omega}$ (positivity).

Kolmogorov 2) $\text{Prob}(\mathbf{\Omega}) = 1$ (normalization: the sample space is all).

Kolmogorov 3) $\text{Prob}(\coprod_{\mathsf{E}} \mathfrak{E}_{\mathsf{E}}) = \sum_{\mathsf{E}} \text{Prob}(\mathfrak{E}_{\mathsf{E}})$ for E countable (countable additivity).

These turn out to suffice: $\mathfrak{E}_1 \subseteq \mathfrak{E}_2 \Rightarrow \text{Prob}(\mathfrak{E}_1) \leq \text{Prob}(\mathfrak{E}_2)$, $\text{Prob}(\emptyset) = 0$, and $0 \leq \text{Prob}(\mathfrak{E}) \leq 1$ are then corollaries (Exercise!).

Kolmogorov also formally extended the notion of sample space $\mathbf{\Omega}$ to *probability space* $(\mathbf{\Omega}, \mathfrak{C}, p)$ for $\mathfrak{C}$ a collection of events and $p : \mathfrak{C} \to [0, 1]$ a *probability-assigning map* $\mathfrak{E} \mapsto \text{Prob}(\mathfrak{E})$.

The notion of *σ-algebra*, $\boldsymbol{\sigma}$, arises as a standard answer to the question of what constitutes a suitable collection $\mathfrak{C}$ for use in Probability Theory. This obeys

σ-1) $\mathbf{\Omega} \in \boldsymbol{\sigma}$.

σ-2) If $\mathfrak{E} \in \boldsymbol{\sigma}$, then its complement $\mathfrak{E}^c \in \boldsymbol{\sigma}$ as well (closure under complementation).

σ-3) For $\{\mathfrak{E}_{\mathsf{E}}, \mathsf{E} \text{ countable}\}$ a sequence of elements $\in \boldsymbol{\sigma}$, then $\bigcup_{\mathsf{E}} \mathfrak{E}_{\mathsf{E}} \in \boldsymbol{\sigma}$ (closure under countable unions).

Useful comparison of the Kolmogorov and $\boldsymbol{\sigma}$ axioms is as follows. Firstly, both require $\mathbf{\Omega}$ itself to be the entity to which probabilities can be assigned. Secondly, σ-algebra 2) is implicit in Kolmogorov II) and III). Thirdly, Kolmogorov 3)'s countable additivity is indeed also known as σ-additivity; σ-algebra 3) matches this for countability but not for disjointness.

For $\mathfrak{C}$ any collection of subsets of $\mathbf{\Omega}$, $\cap$ {all σ-algebras containing $\mathfrak{C}$} gives the smallest σ-algebra containing $\mathfrak{C}$, and is known as the *σ-algebra generated by* $\mathfrak{C}$. The *Borel σ -algebra* $\boldsymbol{\sigma}(\mathbf{\Omega})$ is the σ-algebra generated by the set of open subsets of $\mathbf{\Omega}$ [207]. The elements of this algebra are called *Borel subsets*; in outline, these are the sets arrived at by repeatedly applying countable unions and intersections.

A *probability measure* is a measure as befits the necessities of modelling Probability Theory: $\mu : \boldsymbol{\sigma}(\mathbf{\Omega}) \to [0, 1]$ of Probability Theory, and it is a normalized construct. More specifically, a suitable probability map p is provided by the notion of *Borel probability measure* $\mu : \boldsymbol{\sigma}(\mathbf{\Omega}) \to \mathbb{R}$. Such Borel measures are a subcase of Lebesgue measures; given a σ-algebra $\boldsymbol{\sigma}(\mathbf{\Omega})$, $\mathfrak{Y} \subseteq \mathbf{\Omega}$ is *measurable* if $\mathfrak{Y} \in \boldsymbol{\sigma}(\mathbf{\Omega})$.

Finally, suppose that $(\mathbf{\Omega}, \boldsymbol{\sigma}, p)$ is a probability space and $(\mathfrak{S}, \boldsymbol{\sigma}')$ is a measurable space for $\mathfrak{S}$ some equipped set and $\boldsymbol{\sigma}'$ another σ-algebra. Then an $\mathfrak{S}$*-valued stochastic process* is an $\mathfrak{S}$-valued random variable on $\mathbf{\Omega}$ that is indexed by a totally ordered set dubbed 'time'. [In conventionally studied cases, this is a parameter, Newtonian, background time notion and can be either continuous or discrete.] *Markov chains* [391] are a simple case of this, which is memoryless in the sense of $\mathfrak{S}_t$ depending only on the immediately preceding value of t, i.e. on the value taken by that state one time-step previously. *Poisson processes* [391] are a particular continuous-t subcase of this which count occurrences from an underlying exponential distribution; the resulting probability distribution is distributed Poisson. See e.g. [531] for further reading about this Sec's topics.

P.4 Distributions (in the Sense of Functional Analysis)*

A function which is zero outside of some compact set is said to have *compact support*. One can subsequently consider the space of smooth functions with compact support, $\mathfrak{C}_0^\infty$ The space of *distributions* is then the dual of the space of smooth functions with compact support. Distributions make good mathematical sense when put under an integral sign after being multiplied by test functions. Applications of this notion of distribution include placing the Dirac δ-function on a rigorous footing, and use of distributional solutions to PDEs; also measures are a type of distribution. See e.g. [207] for further reading on this topic.

P.5 Further Useful Function Spaces*

For each of the two types of PDE problem discussed in Appendix O.4, there is a useful subcase of Banach spaces.

Hölder spaces (after mathematician Otto Hölder) $\mathfrak{C}^{k,\alpha}$ consist of the following [732]. A function f is *Hölder* if there are constants c and $0 < \alpha < 1$ such that the bound

$$\|f(x) - f(y)\| \leq C\,\|x - y\|^\alpha$$

holds for any points x, y that the function acts upon (so if $\alpha = 1$, f is Lipschitz). The corresponding *Hölder norm* on an open set $\mathfrak{U}$ is then

$$\|f\|_{\mathfrak{C}^{k,\alpha}} := \sup_{x \in \mathfrak{U}} \|f(x)\| + \sup_{x,y \in \mathfrak{U}, x \neq y} \left(\|f(x) - f(y)\|/\|x - y\|^\alpha\right); \quad \text{(P.3)}$$

the k here is as in $\mathfrak{C}^k$. Hölder spaces are well-suited to elliptic PDEs [220].

Sobolev spaces (after mathematician Sergei Sobolev) [204, 206, 207, 440, 460, 557, 732, 874] bear some similarity to $\mathfrak{L}^p$ spaces, but are now built out of the

Sobolev norm which involves up to kth derivatives[1] as well:

$$\| f \|_{\mathfrak{H}^{k,p}} := \sum_{|i| \leq k} \| \partial^{(i)} f \|_{\mathfrak{L}^p} . \tag{P.4}$$

Furthermore the $p = 2$ case $\mathfrak{H}^k := \mathfrak{H}^{k,2}$ are Hilbert spaces (Exercise!). Unlike for $\mathfrak{C}^k$ spaces, Sobolev spaces are set up specifically for PDE Theory (derivative terms now lie within the common function space).

Sobolev spaces are well-suited for hyperbolic PDEs. This can already be envisaged from a simple example such as the flat spacetime Klein–Gordon equation [874]. Consider the values of the scalar field ϕ and its first derivatives on $\mathrm{S}_2 = \mathrm{D}^+(\mathrm{S}_1) \cap \boldsymbol{\Sigma}_2$, for D^+ the future domain of dependence. Then using the construction in Fig. 8.5.b), Gauss' (divergence) Theorem and energy–momentum conservation,

$$\int_{\mathrm{S}_1} \mathrm{T}_{\mu\nu} \mathrm{t}^\mu \mathrm{t}^\nu \mathrm{dS}_1 + \int_{\mathrm{E}} \mathrm{T}_{\mu\nu} \mathrm{n}^\mu \mathrm{t}^\nu \mathrm{d}\mathbf{E} = \int_{\mathrm{S}_2} \mathrm{T}_{\mu\nu} \mathrm{t}^\mu \mathrm{t}^\nu \mathrm{dS}_2 , \tag{P.5}$$

for n^μ the future-directed normal to S_2 and E the 'side edges'. Moreover, the second term ≥ 0 provided that the matter obeys the *dominant energy condition*: [874]

$$-\mathrm{T}^\mu{}_\nu \mathrm{u}^\nu \text{ is a future-directed timelike or null vector } \forall \text{ future-directed timelike } \mathrm{u}^\mu \tag{P.6}$$

and that t^μ is timelike. The definition of the energy-momentum-stress tensor then gives that, for scalar field matter ϕ,

$$\int_{\mathrm{S}_2} \left\{ \dot{\phi}^2 + \{\partial\phi\}^2 + m^2\phi^2 \right\} \mathrm{dS}_2 \leq \int_{\mathrm{S}_1} \left\{ \dot{\phi}^2 + \{\partial\phi\}^2 + m^2\phi^2 \right\} \mathrm{dS}_1 . \tag{P.7}$$

Since each integrand is the sum of squares (which are necessarily positive), this means that control over the data on S_1 gives control of the solution on S_2. The idea of a Sobolev norm is then a generalization of these last two 'energy' integrals to a wider range of *energy estimates* [204].

Finally, as regards e.g. GR applications or quantizing curved configuration spaces, note that both Hölder and Sobolev spaces can indeed also be defined on manifolds [207, 732].

[1] These are meant in a distributional sense to ensure completeness is maintained.

Appendix Q
Statistical Mechanics (SM), Information, Correlation

This and the next Appendix support in particular this book's Timeless Approaches.

Q.1 Thermodynamics

Physical systems with complicated microphysics are none the less observed to admit approximate macroscopic descriptions in terms of a very small number of classical state variables.

Let us distinguish between two types of energy transfer: work w done on a subsystem, which involves directed ordered motion, versus heat q exchanged between a subsystem and its surroundings, which involves random 'thermal' motion. *Temperature* T is a quantifier of thermal motion; an idealized limiting case of no thermal motion corresponding to an absolute zero temperature. Two physical subsystems are in *thermal equilibrium* if heat exchange between them is possible but there is none overall. The *Zeroth Law* (of Thermodynamics) is that

'*in thermal equilibrium with*' is a transitive relation on physical subsystems. (Q.1)

[For comparison with the Zeroth Law of black holes, that corresponds to the somewhat distinct formulation of T being constant throughout a body in thermal equilibrium.]

Temperature T is itself an example of a state variable; in the case of a gas, other examples are its pressure P and the volume V it occupies. P and T are furthermore *intensive* quantities, i.e. ones which do not depend on the amount of matter under consideration (provided that this amount suffices for the approximate macroscopic description to apply in the first place). This is as opposed to *extensive* quantities, which are proportional to the amount of matter, such as V.

Next, the *First Law* is that

$$\mathrm{d}U = \,\unicode{x0111}w + \unicode{x0111}q. \tag{Q.2}$$

E. Anderson, *The Problem of Time*, Fundamental Theories of Physics 190,
DOI 10.1007/978-3-319-58848-3

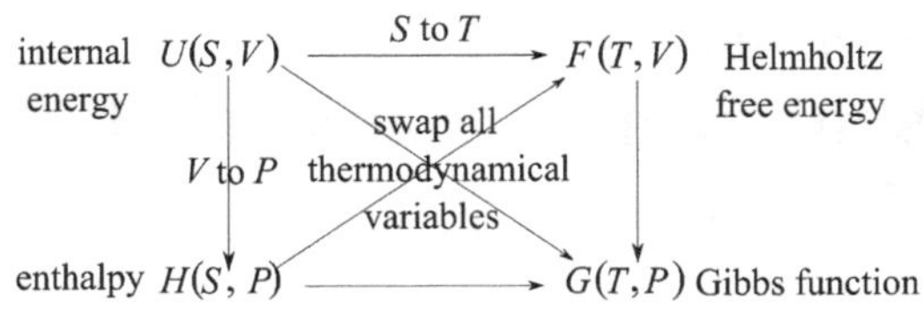

Fig. Q.1 Thermodynamics' Legendre square, all four corners of which are well-known

I.e. the change in internal energy U is the result of the work w done on the subsystem and the heat q exchanged by the system; U is another example of an extensive quantity. Also here đ denotes inexact differential which acts on path-dependent functions, as opposed to d's exact differential which acts on state functions.

Entropy S is another state function [e.g. S can be regarded as $S(V, T)$ for a gas]; this enables assessment of which states are spontaneously accessible from which others. A first historical definition of entropy follows from $\mathrm{d}S = \text{đ}q/T$ for idealized reversible physical processes. The *Second Law* is that entropy is observed to obey

$$S \geq 0 \tag{Q.3}$$

for all physical processes which are sufficiently macroscopic for entropy to be a meaningful notion. At least in simple cases, S is additive upon considering subsystems together (so it is also extensive in these cases). However, in systems involving long-range correlations, entropy can acquire either or both of non-additive and non-extensive character. Gravitational interactions in general, and black hole entropy (7.16) par excellence, are examples of the latter. See e.g. [879] for notions of entropy in general alongside careful detail of what properties these have.

One can furthermore conceive of U as the state function whose dependence is $U(S, V)$. The dependence can be reorganized to involve any two of P, V, T, S. Doing this consistently involves making use of Legendre transformations, leading to the further conceptually and computationally useful state functions (Fig. Q.1): *enthalpy* $H(S, P) := U + PV$, *Gibbs function* (after mathematician J. Willard Gibbs) $G(T, P) := H - TS$ and *Helmholtz free energy* (after mathematician Hermann von Helmholtz) $F(T, V) := U - TS$.

Finally, the *Third Law* is that

absolute zero is not attainable by any finite series of procedures. (Q.4)

Q.2 Thermodynamics for GR More Generally*

The Black Hole Mechanics cases of the Laws of Thermodynamics are outlined in Sect. 7.3). However, there are difficulties with defining gravitational entropy in general—rather than just black hole—situations, with the cosmological setting being of particular interest.

One approach to more general gravitational information candidates bases these on the Weyl tensor $\mathcal{F}[\mathcal{C}_{\mu\nu\rho\sigma}]$ [704]. Monotonicity has caused problems with some such. However, the expression [212] in terms of the Bel–Robinson tensor (K.16)

$$\mathcal{S}^{\text{grav}}_{\text{PL}} := \int \mathcal{T}_{\mu\nu\rho\sigma}\text{u}^\mu\text{u}^\nu\text{u}^\rho\text{u}^\sigma\text{d}\tau \quad \text{(Q.5)}$$

circumvents this problem, and provides an expression that extends to the cosmological context. Moreover, Sect. Q.7 raises two further issues with Weyl tensor based candidates.

On the other hand, in Holographic Approaches, screens in general spacetime have been considered as a possible generalization of event horizons as regards association of entropy [164].

Q.3 Spaces of Substates*

The configuration spaces of Appendices G–H are taken to be for whole-universe models. However in Physics one usually deals with subconfigurations that correspond to subsystems; these are furthermore the primary objects in Records Theory.

For example, subconfigurations of RPMs are not necessarily themselves RPM configurations, e.g. through their carrying a net angular momentum or mutually exchangeable energy. Archetypes of this are the base subsystem of a triangleland or the base alongside the relative angle. Naturally subconfigurations form *subconfiguration spaces* Sub $\mathfrak{q}$; subphase space follows suit.

Q.4 Imperfect Knowledge of a (Sub)system*

Two distinct notions of imperfect knowledge of a (sub)system in use in this book are as follows.

1) *Imperfect knowledge of the system's state*. This can be modelled using localized subsets of a given $\mathfrak{q}$.
2) *Imperfect knowledge of the system's contents*, which could be incorporated by modelling by a union of configuration spaces. Alternatively, one might make use of a space that all the configuration spaces sit in and then use a map $M : \mathfrak{q} \longrightarrow \text{Sub}\,\mathfrak{q}$.

In this setting, the other hitherto ignored coordinates may become relevant in some regions: Fig. Q.2. This is through these other coordinates entering the notion of localization itself. E.g. consider the base of the triangle as a localized subsystem. One notion of localized here is for the apex to have to be outside of the circle of regularity (i.e. with partial moments of inertia obeying $I_{\text{median}} > I_{\text{base}}$). In considering this, one passes from being (in $\mathfrak{q}$) on any open half-line emanating from the triple collision to being on any such which lies within the D-hemisphere. So in this example, one passes from the mathematics of the Dirac string to that of the Iwai string

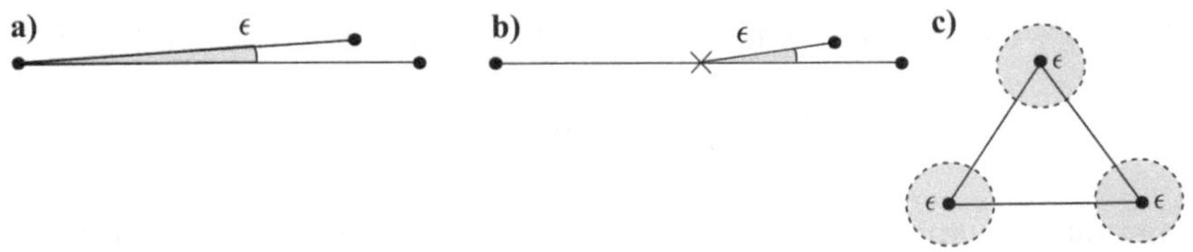

Fig. Q.2 A meaning in space for **a)** Kendall's [539] ϵ-blunt notion of collinearity. In fact, Kendall considered the min for this over all choices of construction, whereas the Author's notion **b)** of ϵ-collinear adapted to the dynamically useful relative Jacobi coordinates [37]. **c)** A notion of ϵ-equilateral [37], for which each vertex is to lie within a shaded disc of radius ϵ

(Sect. 37.3) Finally, note that 'subspaces of configuration space' is not necessarily meant here in any particular mathematical sense. They are physically desirable entities to study but are not necessarily mathematically nice.

Q.5 Classical Statistical Mechanics (SM)

It is well-known that there is a Probability and Statistics basis for Thermal Physics; to some extent, this can be contemplated at the classical level.[1] SM is valuable due to further explaining the nature of macroscopic state variables, and for deriving equations of state and thermodynamical laws.

Firstly, the relation

$$E \sim k_{\mathrm{B}} T \tag{Q.6}$$

gives a rough measure of thermal energy per active mode; the average such is $k_B T/2$; in some physical regimes, some modes are not however active. Here k_{B} is *Boltzmann's constant*, after noted physicist Ludwig Boltzmann. While originally arising in modelling of ideal gases, $E/k_{\mathrm{B}} T$ turns out to be a common grouping in Physics, in the form of Boltzmann factors

$$\exp(\text{`}E\text{'}/k_{\mathrm{B}} T) \tag{Q.7}$$

for various notions of 'E' which have the dimensions of energy.

A second key idea is that of *underlying microstates*. For now, these are in terms of classical phase space regions. The number of microstates W is to be evaluated combinatorially in the discrete case or taken to be proportional to the phase space volume in the continuous case.

In the case of phase space, a pivotal role is played by the *ergodic hypothesis* that the system passes through all of its states that are compatible with thermodynamic equilibrium. Coarse-graining (see the next Sec) is subsequently a means of introducing statistical machinery into Ergodic Theory.

[1] Once QM becomes available—historically in part due to the failings of classical SM prompting its development—one can carefully determine the extent of applicability of classical SM rather than quantum SM.

Approaching the modelling of SM via *ensembles*—not the actual system but rather a group of similar systems which have been suitably randomized—rests on the ergodic hypothesis. The most commonly encountered ensembles are *canonical ensembles*. These involve species number N being fixed and energy E being free.[2] They lead to the simplest version of the thermodynamical laws. On the other hand, *grand ensembles* involve both N and E being free. These are of practical use in Chemistry since chemical reactions in general do not conserve N, as is reflected by the extra pieces entering the ensuing thermodynamical laws including a 'chemical potential' μ. Moreover, in more fundamental physics (for which N is more specifically particle number), grand ensembles are useful as regards handling each of bosons and fermions. *Microcanonical ensembles* are the opposite intermediate stage: now both N and E are fixed. This ensemble is commonly used in Black Hole Physics.

Ensembles provide the constraints involved in extremization of entropy, resulting in the corresponding partition functions. Statistical Mechanics is formulated in terms of *partition functions* for the physical states. These turn out to be built from exponentials; for now in the classical context, these are of the form (Q.7). In this way, this construct generalizes the appearance of 'Boltzmann factors' in a number of physical situations, and also has close ties to the *Maxwell–Boltzmann statistics* of widespread use in Classical Physics. One can then insert the classical Hamiltonian of the system in question into (Q.7) in order to carry out specific calculations.

Classical SM has classical probability distribution function on phase space; each type of ensemble has its own kind of probability distribution function [781].

The usual classical notion of probability density function ρ plays a substantial part in classical SM. Note moreover that the algebraic structure of classical unconstrained observables $\mathcal{U}(\mathbf{Q}, \mathbf{P})$ is additionally the space on which such probability density functions live. Finally, the Wigner functional $\mathcal{W}ig[\mathbf{Q}, \mathbf{P}]$ (cf. Sect. 48.2) is a close analogue of such densities which arises in the semiclassical context.

Q.6 Fine- and Coarse-Graining

In Physics, grainings are principally known in the context of Statistical Mechanics; in the classical case, these are phase space grainings. Moreover, we start with a rather more widely applicable mathematical formulation of grainings. Let $\mathfrak{U} \preceq \mathfrak{V}$ denote '$\mathfrak{U}$ is finer-grained than $\mathfrak{V}$' and $\mathfrak{V} \succeq \mathfrak{U}$ for '$\mathfrak{V}$ is a coarser-grained than $\mathfrak{U}$'; these are ordering relations. The *coarsest* graining is then the whole mathematical space itself. On the other hand, the *finest* graining consists of the individual points that constitute the mathematical space. A more specific partition-based theory of grainings is along the following lines.

---

[2]'Free' here means 'free to fluctuate about equilibrium', which still means that the *average* value of the quantity is fixed.

Graining 1) Refine partitions from $\mathfrak{Y}$ to $\mathfrak{Z}$ so that each $\mathfrak{Y}_\mathsf{Y}$ partition is a subset of a $\mathfrak{Z}_\mathsf{Z}$ one.
Graining 2) A *common refinement* $\mathfrak{W}_\mathsf{W}$ of two partitions $\mathfrak{Y}_\mathsf{Y}$ and $\mathfrak{Z}_\mathsf{Z}$ is concurrently a refinement of both individually.
Graining 3) Moreover, a product of two partitions can be defined in which the new elements are all of the intersections of the two input partitions.

Note that the finest graining's knowledge is attainable in principle, if not in practice, at the classical level: perfect knowledge of state is classically permissible. It is also clear from the mathematical version of the graining concept that there is a configuration space level version of graining as well.

Example 1) Replace a set of particles by just the coordinates of their centre of mass.
Example 2) Replace a set of field values in a region by a mean field value for that region.
Example 3) Replace a set of shape variables by an averaged shape variable. As a first case, replace an N-body configuration by the M-body configuration formed by $M < N$ cluster centres of masses, and then consider shape variables for the M-body configuration rather than the N-body one. As a second case, replace diagonal Minisuperspace GR's two anisotropy parameters by their average value. As a third case, one might define a local inhomogeneity averaging operation that approximates complicated lumps by simple ones, or multiple lumps by single ones.

Examples 1) and 3) involve notions of large scale approximate shape. All of the above are examples of coarse-grainings of subsystem contents, which involves passing from the configuration space of the precise configuration to a distinct configuration space for an approximate configuration. This is known as *collectivization*. Another type of coarse-graining involves replacing a point in the space by a cell in *the same* space, due to only knowing a state's value approximately. Appendices G, H and N's $\mathfrak{q}$ geometry considerations cause it to be straightforward to consider regions within $\mathfrak{q}$ for the spatially 1- or 2-d case of Example 1), for Example 2), and for Example 3) in the modewise perturbative case.

As regards the phase space case which underlies classical SM, coarse-graining here is usually considered in the sense of a cellular approximation.

Example 4) In the case of the continuous phase space of a finite system, the coarse-grained density is given by

$$\rho_n = \int_{\text{cell}} \rho(\boldsymbol{p}, \boldsymbol{q})\mathrm{d}^n\boldsymbol{p}\mathrm{d}^n\boldsymbol{q}\big/W(\text{cell}), \tag{Q.8}$$

where W counts up the for now classical microstates in the cell.
Example 5) The graining concept furthermore extends to Histories Theory as well, as per Chap. 28.

Finally, note that the current Section's examples consider graining at the metric and differentiable manifold levels of structure. However, the given mathematical conceptualization clearly transcends to deeper levels of structure, for which Appendix T provides a number of examples.

Q.7 Classical Microscopic Notions of Entropy

While one can make useful calculations based on Appendix Q.1's notion, it took a further conceptual leap of Boltzmann's to understand the microscopic nature of entropy.

$$S = k_{\mathrm{B}} \ln W \tag{Q.9}$$

for W the number of microstates (i.e. of arrangements of the constituent objects) giving a more primitive combinatorial basis for the entropy concept.

One can subsequently recover entropy from the partition function:

$$S = \left(\frac{\partial k_{\mathrm{B}} T \log Z}{\partial T} \right)_{\mathrm{V}} .$$

Many further expressions for notions of entropy are based on the

$$x \log x \tag{Q.10}$$

function, which arises from maximizing the form (Q.9) of S subject to the corresponding (e.g. canonical) ensemble's constraints. This is through the microstate count giving factorials to which Stirling's approximation can be applied. This $x \log x$ function is, moreover, *the* positive continuous function that is consistent with regraining, monotonicity, and further useful properties for a notion of entropy to have, as exposited in the reviews [221, 879].

Finally, having made these Statistical Mechanics level considerations about notions of entropy, let us caution that entropy candidates for GR in terms of the Weyl tensor bear no known relation to microstate counting arguments or to consistency under regraining.

Q.8 Classical Notions of Information*

One approach to classical information is (see e.g. [129]) that information I is negentropy $I = -S$, so that an incipient classical notion is the Boltzmann-type expression

$$I_{\mathrm{B}} = -\log W. \tag{Q.11}$$

Shannon information (after mathematician Claude Shannon) is likewise related to (Q.10)

$$\begin{aligned} I_{\mathrm{Shannon}}(p_x) &= \sum_x p_x \log p_x \quad \text{(discrete case)}, \\ I_{\mathrm{Shannon}}[\rho] &= \int \mathrm{d}V \rho \log \rho \quad \text{(continuous case)} \end{aligned} \tag{Q.12}$$

for p_x and ρ discrete and continuous probability distributions respectively. This expression arises from (Q.11) by use of (e.g. in the discrete case) $a_i / \sum a_i = p_i$ as partial contents and then applying Stirling's approximation.

Mutual information is defined by (in the portmanteau case)

$$M_{\text{Shannon}}[A, B] = I_{\text{Shannon}}[A] + I_{\text{Shannon}}[B] - I_{\text{Shannon}}[AB] \tag{Q.13}$$

for probability distributions A, B, and AB their joint distribution. Note the parallel with elementary set theory's 'union = sum - intersection' relation. *Relative information* is a conceptually-similar quantity given by

$$\begin{aligned} I_{\text{relative}}[p, q] &= \sum_x p_x \log(p_x/q_x) \quad \text{(discrete case)}, \\ I_{\text{relative}}[\rho_1, \rho_2] &= \int \mathrm{d}V \, \rho_1 \log(\rho_1/\rho_2) \quad \text{(continuous case)}. \end{aligned} \tag{Q.14}$$

This introduced, mutual information can be cast as a relative information, now between a joint probability distribution and the product of the corresponding marginal distributions.

Aside from connotations of the subsystems or perspectival notion of relational, if mutual and relative notions of information are built out of r-formulation objects, they are automatically Configurationally Relational (and timeless). But, as ever, this is a luxury that one cannot afford in the case of GR itself. The more general alternative is to consider $\mathfrak{g}$-act, $\mathfrak{g}$-all versions of these objects.

Q.9 Classical Notions of Correlation*

Again these can be set up for both discrete and continuous cases; these occur in many branches of Mathematics and Physics.

Example 1) The most basic notion from Statistics is

$$\textit{Pearson's correlation coefficient } \rho_{\mathrm{P}} := \mathrm{Cov}(X, Y)/\sigma_X \sigma_Y, \tag{Q.15}$$

in 2-*d* for random variables X, Y. Assessing this relationally, it is scale-invariant and invariant under exchange of dependent and independent variable status of the X and Y; it is not rotationally invariant. Indeed, it is very well-known that the upward and downward pointing lines have the opposite extremes of ρ_{P}.

Example 2) One resolution of this in terms of basic rotational invariants is exemplified by

$$\overline{\rho}_{\mathrm{Rel}} = 2\frac{\sqrt{1 - \rho_{\mathrm{P}}^2}}{\omega + \omega^{-1}} \quad \text{for } \omega := \frac{\sigma_X}{\sigma_Y}. \tag{Q.16}$$

The bar notation here means 'not', since perfect correlation returns the value 0 and not 1, so this quantity is an 'uncorrelation coefficient'. Finally, in terms of the basis of shape quantities in the simplest relationally nontrivial case of a triangle [33],

$$\overline{\rho}_{\mathrm{Rel}} \propto \mathit{area}/\{1 - \mathit{aniso}\}. \tag{Q.17}$$

Examples 3) and 4) ρ_{P} is also more standardly known to be limited through only capturing correlations that approximate a straight line. One can e.g. use [534] Spearman's, or Kendall's τ, rank correlation coefficients to detect non-linear correlations (after statisticians Charles Spearman and *Maurice* Kendall). Rank correlation tests statistics immediately fail to be rotationally relational, since any two data points can be rotated into a tie.

Example 5) Mutual information (Q.13) also serves as a notion of correlation, but one which is in sore need of an n-object extension so as to be able to address most problems. That straightforwardly gives the *total correlation* and *dual total correlation* for any $n > 2$.

Example 6) The n-point function correlators which are well-known to occur in physical Field Theories such as classical Cosmology and QFT. These are Green's functions corresponding to instantaneous operators such as the spatial parts of wave operators. Some such quantities can readily come with [269] translational and rotational invariance directly built in through depending only on $|\underline{x} - \underline{x}'|$. Occasionally they also involve taking the dot product with an arbitrary-direction vector and then integrating over all directions, which is another $\mathfrak{g}$-act, $\mathfrak{g}$-all manoeuvre. For now, we restrict attention to the classical case (see Chap. U.5 for quantum counterparts). For instance, Classical Cosmology has a 2-point function for such as mass density or galaxy number density [215]. This takes the form of a function of inter-particle separation magnitudes.

Example 7) We end by noting that, in the full GR setting, physicist Roustam Zalaletdinov proposed a type of correlation tensor—a comparer—for the study of inhomogeneity [926, 927].

Appendix R
Stochastic Geometry*

Now let the sample space $\mathbf{\Omega}$ (Appendix P.1) be a manifold $\mathfrak{M}$.[1] $\mathfrak{M}$ is furthermore taken to carry a σ-algebra (Appendix P.3) of Borel subsets $\mathfrak{Bo}$, which is generated by the open subsets on $\mathfrak{M}$. A *statistic on a p-d differentiable manifold* $\mathfrak{M}$ is a function $\theta : \mathbf{\Omega} \rightarrow \mathfrak{M}$ such that

$$\theta^{-1}(\mathfrak{U}) = \{\omega \in \mathbf{\Omega} \,|\, \theta(\omega) \in \mathfrak{U}\} \text{ is an event for every open set } \mathfrak{U} \subseteq \mathfrak{M} \,. \tag{R.1}$$

Statistics thereupon are random p-vectors; a given statistic θ induces a probability distribution $\text{Prob}\,\theta^{-1}$ according to $\text{Prob}\,\theta^{-1}(\mathfrak{Bo}) = \text{Prob}\,(\theta^{-1}(\mathfrak{Bo}))$.

R.1 Metric Shape Statistics

Now let the geometry in question a fortiori be a shape space corresponding to a metric Shape Theory. This provides techniques for addressing a wider range of geometrical hypotheses concerning detailed patterns in point data sets upon, most usually, $\mathbb{R}^d$.

Example 1) *Clumping Statistics* investigates hypotheses concerning ratios of relative separations (detailed information which can be attributed both locally and to subsystems). These already exist in 1-*d* and in settings simpler than metric shape spaces, so this topic is well-known. Astrophysical applications include tight binary stars, globular clusters, galaxies and voids: and *absence* of clumping. E.g. Applied Mathematician Stanley Roach [738] provided a discrete statistical study

[1]This is one of the senses in which ‘Geometrical Probability’ is used in the literature [539, 792]. See the first of these for an accessible introduction to probability on manifolds, including for this setting’s notions of independence, expectation, uniform and normal distributions, and binomial and Poisson processes. If you encounter unfamiliar material beyond the scope of Appendix P.1 in the first of these, consult [90]. The second of these is a more advanced reference of particular relevance to the current book.

E. Anderson, *The Problem of Time*, Fundamental Theories of Physics 190,
DOI 10.1007/978-3-319-58848-3

of clumping; this can in turn be interpreted in terms of coarse-grainings of RPM configurations. Geometrical Probability on the shape space $\mathfrak{S}(N, 1) = \mathbb{S}^{n-1}$ is an alternative [878] to this.

Example 2) For spatial dimension ≥ 2, the degrees of freedom include relative angles as well as ratios of relative separations. Now hypotheses concerning relative angle information can be made as well. In particular, angular patterns in 2-*d*—i.e. in N-a-gon constellations—are both nontrivial and mathematically accessible. Spatially 2-*d* case of metric shape geometry are more straightforward due to not being stratified, as per Appendix G. Among these, the minimal relationally nontrivial case concerns alignment in threes; this enters for instance the standing stones and quasar alignment problems outlined in Chap. 26. Such techniques were pioneered in particular by Kendall and collaborators such as Dennis Barden, Keith Carne, Huiling Le and Christopher Small; see [536, 537, 539, 792] for reviews.[2] One is here addressing whether there are a statistically significant number of almost collinear triangles quantified by some (bluntness angle) $< \epsilon \in \mathbb{R}_+$ (Fig. Q.2.a). One then uses probability distributions based on the corresponding shape space geometry (i.e. on Kendall's spherical blackboard in the case of probing in threes. Making use of Borel subsets, Haar measures and Radon–Nikodym derivatives (all defined in Appendix P.1), it can be shown that in 2-*d* N independent identically distributed isotropic Gaussian distributions induce a measure on shape space which coincides with the uniform measure [536]. Most neatly in Appendix G.2's complex presentation of shape coordinates, one obtains a probability measure [539]

$$\mathrm{d}\omega(N, 2) = \frac{\{N-2\}!}{\pi^{N-2}} \frac{\prod_{j=1}^{N-2} \mathrm{d}Z_j}{1 + \sum_{i=1}^{N-2} |Z_i|^2}. \tag{R.2}$$

In the minimal relationally nontrivial case of the triangle, this reduces to

$$\mathrm{d}\omega(3, 2) = \frac{1}{\pi} \frac{\mathrm{d}Z}{1 + |Z|^2}, \tag{R.3}$$

i.e. proportional to the 'square root' of the metric (G.11). See e.g. [536, 539, 792] for further cases.

Furthermore, *how good* a 'best fit'—such as given by Best Matching—is can also be assessed by Shape Statistics. This is by making a set of relational objects out of primed and unprimed vertices (Fig. Q.2.d), to which the corresponding notion of Shape Statistics is to be applied. In the case of metric shapes, this produces triangles that can be tested against the ϵ-bluntness criterion.

Finally, we outline in [36] how the above Metric Shape Statistics is but the first of a larger family, each corresponding to a distinct notion of shape as per Fig. G.4.

[2] Kendall's work is a *solution* to Broadbent's [172] previously posing the standing stones problem as one to be addressed by *some kind of* Geometrical Statistics. Kendall also generalized this to probing in more than threes [536].

R.2 Stratified Manifold Version**

Stochastic Geometry on stratified manifolds is already required for finite theories of Mechanics as per Appendix G. [539] covers an example of this (the Shape Space of tetrahaedrons). Note that assigning the open sets to be Borel subsets $\mathfrak{Bo}$ does not require these to be modelled on $\mathbb{R}^p$ or for the manifold to carry a metric. The same applies to the definition of statistic.

What about 'large' Shape Statistics on GR's $\mathfrak{S}$uperspace($\mathbf{\Sigma}$), $\mathfrak{CS}(\mathbf{\Sigma})$ and $\{\mathfrak{CS} + \mathrm{V}\}(\mathbf{\Sigma})$? In fact, infinite dimensionality is not a concern: stochastical treatment of Banach spaces and Fréchet spaces are known [388, 864, 865, 892], including for Fréchet spaces of the more specific type used to model $\mathfrak{S}$uperspace($\mathbf{\Sigma}$). However, we are not aware of this being extended to Banach or Fréchet *manifolds* much less to a suitable class of Fréchet stratified manifolds. This is one obstacle which keeps us far from having stochastic and statistical treatments of $\mathfrak{S}$uperspace($\mathbf{\Sigma}$) and $\mathfrak{CS}(\mathbf{\Sigma})$ (Research Project 18).

Research Project 120)† Give a stochastic and statistical treatment of gauge group orbit spaces $\mathfrak{O}$, as a model arena with a simpler counterpart of this obstacle.

It is additionally likely that 'large' configuration spaces' statistics will possess a Measure Problem.

Research Project 121)† This gives one reason why investigation of 'medium-sized' examples—Midisuperspaces—are likely to be useful stepping-stones. Moreover, stochastic and statistical theories for the models of anisotropy and modewise small inhomogeneity considered in this book are flat and so are too simple to require Stochastic *Geometry*. One might next turn to analytically tractable Midisuperspaces in this regard.

Appendix S
Deeper Levels. i. Generalized Configuration Spaces**

This and the next Appendix support in particular this book's Epilogues II.C and III.C about Background Independence at deeper levels of mathematical structure and consequent Problem of Time issues.

S.1 Spaces of Differentiable Structures

The amount of differentiability involved turns out to be inconsequential. This is because the configurations in question are spatially 3-d, and differentiable structure on a given topological manifold is unique for dimension $d \leq 3$. [In contrast, see e.g. [614, 674] for the interesting dimensional sensitivity exhibited by the theory of differentiable structures for $d \geq 4$.]

S.2 Spaces of Topological Manifolds

In standard GR, one firstly assumes a fixed spacetime topological manifold (Chap. 7). This is a major restriction on possible spacetime-level solutions. In Geometrodynamics (Chap. 8), one furthermore assumes a fixed $\mathbf{\Sigma} \times \mathbb{R}$. Some commonly used further restrictions are as follows.

i) A fixed spatial dimension.
ii) The orientable $\mathbf{\Sigma}$.
iii) The connected $\mathbf{\Sigma}$ [350, 351].
iv) The compact without boundary $\mathbf{\Sigma}$ [350, 351].
v) Whether the singular spaces are to be included or left out. If these are included, the basic differential geometric notion of chart ceases to suffice to cover all purposes.

The idea of removing such assumptions first appeared in Wheeler's considerations of 'spacetime foam' [896, 897, 899]. This was motivated from the form of the Feynman path integral formulation: i.e. whether Feynman diagrams could be generalized

E. Anderson, *The Problem of Time*, Fundamental Theories of Physics 190,
DOI 10.1007/978-3-319-58848-3

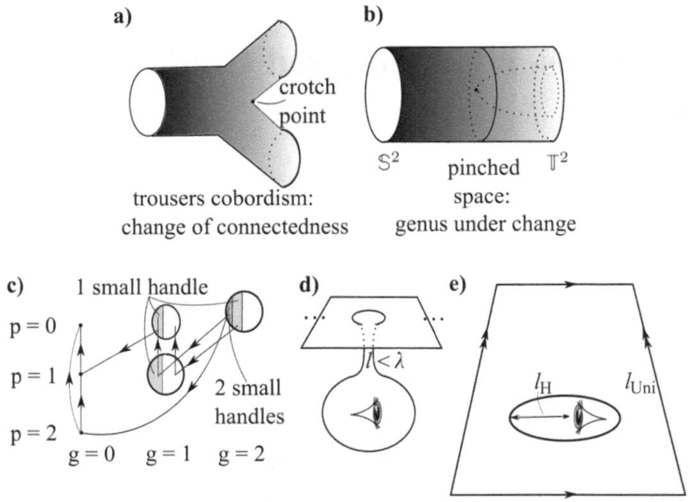

Fig. S.1 **a)** and **b)** are examples of cobordisms. **c)** Depiction of which configurations are 'near' which others in $\mathfrak{BigRiem}$ in the simple case of 2-*d* orientable compact without boundary $\mathbf{\Sigma}$. *g* is genus and *p* is 'pinch number'. **d)** An open universe that looks closed, built by use of one of Hawking's tubes of negligible action, whose aperture l is smaller than the probeable wavelength λ. [This begs the further question of whether such tubes are dynamically stable.] **e)** A closed universe that looks open, due to its characteristic size l_{Uni} greatly exceeding its Hubble radius l_{H}

to topology change diagrams. *Geroch's Theorem* [346] (after mathematical physicist Robert Geroch) then requires choosing between Scylla and Charybdis: singularities or a loss of causality. This result makes use of a combination of mathematics pioneered by Thom and another noted mathematician, John Milnor: singular metric geometries, cobordisms and Morse Theory. At the classical level, one often chooses to keep causality and thus allows for the inclusion of singular metrics.

$\mathbf{\Sigma}_1$ and $\mathbf{\Sigma}_2$ are *cobordant* if there exists a $\mathfrak{M}$ of dimension one greater which interpolates between them (Fig. S.1.a–b). One can view the piece in between as $\mathbf{\Sigma} \times [0, 1]$ with some parameter λ ranging over [0, 1]. This can be viewed as a $\mathfrak{g}_{\text{T}}$-act $\mathfrak{g}_{\text{T}}$-all procedure. To be clear, the $\mathfrak{g}_{\text{T}}$ group is *not* a group of the conceptually simpler kinds that occur in Topology—homotopy, homology or cohomology: *characterizers of* topology. It is, rather, a cobordism group of topology-*altering* 'ripping' operations. For instance, *Surgery Theory* [876] involves such operations.

In the single-floor case—considering the topological level alone—$\mathfrak{Top}$-$\mathfrak{Man}$ is the space of $\mathbf{\Sigma}$'s themselves. This is possibly subject to some restrictions from a menu such as being of a particular dimension, orientable, compact, connected. This supports e.g. the dynamics of a theory with just TFT degrees of freedom.

On the other hand, in the tower case—keeping the upper levels as well—one has 'Geometrodynamics with topology change'. Here the topological manifolds are equipped with differentiable and metric structure. However, this is still dynamically sterile through not providing suitable intermediates, which lie, rather, in $\mathfrak{Riem}(\mathbf{\Sigma})$ where $\mathbf{\Sigma}$ is now allowed to be singular. One gets around this by adjoining pinched manifolds also: a mild case of singularness. We then have (one notion of) $\mathfrak{BigRiem}$ (Fig. S.1), $\mathfrak{BigSuperspace}$ [301] etc, on the set of ordinary and pinched $\boldsymbol{\Sigma}$. An example of this is a geometrodynamics that is additionally a topolodynamics in the sense of admitting changes in topological manifold. These considerations are built upon the following considerations of Fischer [301]. *"Nevertheless, it is hoped that the complete Superspace of all possible topologies can be pieced together from the individual superspaces, so that quantum fluctuations in the topology can be described. This I believe will be possible only after a great deal is known about the individual superspaces."* He then added that *"In such a program, pinched manifolds will play the crucial intermediary role in the topology change."*

Model Arena 1) for 2-*d* orientable closed without boundary manifolds, cobordisms are generated by the genus-changing operation of 'adding handles' (Fig. S.1). In the simplest case, this reduces to $\mathbb{S}$'s indexing set being just over the discrete genus parameter. The next simplest involves indexing over the number of pinchings as well. Consider furthermore just the spheres, tori and the pinched spheres that lie between these. This requires studying the counterparts of $\mathfrak{Riem}(\boldsymbol{\Sigma})$ and $\mathfrak{Superspace}(\boldsymbol{\Sigma})$ for the pinched spheres, and then the $\mathfrak{BigRiem}$ and $\mathfrak{BigSuperspace}$ on the three topologies in question.

Epilogue III.C considers two further model arenas: variable-N RPMs and TFTs.

As regards being more precise *how* singular these pinched spacetimes need to be, *Morse spacetimes* [161] are a nicely tame example. Morse Theory encodes these metrics singularness by a Morse function: $f : \mathfrak{m} \rightarrow [0, 1]$ with $f^{-1}(0) = \boldsymbol{\Sigma}_1$ and $f^{-1}(1) = \boldsymbol{\Sigma}_2$ and r critical points in the interior of $\mathfrak{m}$. These are isolated and non-degenerate for f a Morse function. So in this case one is more specifically interested in e.g. $\mathfrak{BigRiem}$ and $\mathfrak{BigSuperspace}$ restricted to Morse function type singularities. Moreover, Sect. 60.1's TFT example further resembles topology change in GR through its also being linked to Morse Theory.

By Fig. S.1.d)–e), papers nominally concerning 'the topology of the Universe' [219, 597, 618] are really about a large scale approximate notion of topological manifolds that has not necessarily yet been quantified. I.e. some means by which topological manifolds themselves has been equipped with a coarsened length concept. Then handles and tubes which are 'large'—with respect to the probing capacity of the observers—'count', and small ones do not. On some occasions, global effects can serve to discern which universe one is in, though this should not be expected to always be the case. Finally note that spectral notions of distance between metric geometries (Appendix N.8) extend to cases involving comparison between metrics on distinct underlying topological manifolds.

S.3 Spaces of Metric Spaces

In the single-floor case [508, 509], for $\mathfrak{Y}_1$ and $\mathfrak{Y}_2$ subsets of a metric space $\langle \mathfrak{X}, \text{Dist} \rangle$, the *Hausdorff distance* $\text{Dist}_{\text{H}}(\mathfrak{Y}_1, \mathfrak{Y}_2)$ is

$$\max\left(\sup_{y_1 \in \mathfrak{Y}_1} \left(\inf_{y_2 \in \mathfrak{Y}_2} (\text{Dist}(y_1, y_2)) \right), \sup_{y_2 \in \mathfrak{Y}_2} \left(\inf_{y_1 \in \mathfrak{Y}_1} (\text{Dist}(y_1, y_2)) \right) \right). \tag{S.1}$$

The *Gromov–Hausdorff* distance (named in part after mathematician Mikhail Gromov) between any two compact metric spaces $\mathfrak{X}_1$ and $\mathfrak{X}_2$ is then

$$\text{Dist}_{\text{GH}}(\mathfrak{X}_1, \mathfrak{X}_2) := \inf\big(\text{Dist}_{\text{H}}(f_1(\mathfrak{X}_1), f_2(\mathfrak{X}_2)\big). \tag{S.2}$$

This is over all isometric embeddings $f_i : \mathfrak{X}_i \rightarrow \mathfrak{X}$ into all metric spaces $\mathfrak{X}$. [I.e. the embeddings cast $\mathfrak{X}_1$ and $\mathfrak{X}_2$ into the form of subsets of larger metric spaces within which the Hausdorff distance notion applies.] This provides an example of a space of metric spaces which is itself a metric space. Finally note that the above construction is an example of $\mathfrak{g}$-act $\mathfrak{g}$-all method.

On the other hand, in the tower case, as regards how the (positive-definite) Riemannian manifolds sit within the metric spaces, each Riemannian metric induces a metric space metric: the *path metric* (cf. Sect. D.4)

$$\text{Dist}(x, y) = \inf_{\gamma \text{ from } x \text{ to } y} \int_\gamma \sqrt{g_{\mu\nu}(x)\text{d}x^\mu\text{d}x^\nu}. \tag{S.3}$$

Assuming only metric spaces contain physically meaningful information, it is then an interesting question what dynamics they support.

S.4 Lattices

A number of the further spaces of spaces below are lattices, so we introduce these in more general terms. A *lattice* is a poset (Appendix A.1) within which each pair of elements has a least upper bound and a greatest lower bound. In the context of a lattice, these are called *join* $\vee$ and *meet* $\wedge$. $\vee$ and $\wedge$ form an algebra. Each of these operations is furthermore idempotent, commutative and associative, and the pair of them obey the *absorption conditions* $a \vee \{a \wedge b\} = a$ and $a \wedge \{a \vee b\} = a$. An element 1 of $\mathfrak{L}$ is a *unit* if $\forall l \in \mathfrak{L}, l \preceq 1$, and an element 0 of $\mathfrak{L}$ is a *null element* if $\forall l \in \mathfrak{L}, O \preceq l$. A lattice that possesses these is termed a *bounded lattice*. A *lattice morphism* is an order-, join- and meet-preserving map between lattices.

Example 0) Fig. A.2.a), c) and d) can be interpreted as lattices. Moreover d) is one of the minimal-sized nondistributive lattices (Exercise!). Such models are already often used to illustrate the Causal Sets Approach.

Example 1) The *Boolean algebra* (after mathematician George Boole) of classical propositions is a well-known example of a lattice. Here additionally $\vee$ is distributive over $\wedge$ and vice versa, there is a greatest element 1 and a least element 0, identity relations $a \vee 0 = a$ and $a \wedge 1 = a$, a *not operation* $\neg$, and *complement relations* $a \vee \neg a = 1$ and $a \wedge \neg a = 0$.

In fact, the minimal structure required for consideration of propositions is an *orthoalgebra* $\mathfrak{Up}$ [503]. Here, $P \preceq R$ iff $\exists\, S \in \mathfrak{Up}$ such that $R = P \vee S$ in cases in which P and S are disjoint, with $\vee$ not being defined in other cases.

S.5 Spaces of Subgroups, Topological Spaces and Collections

Example 2) The set of subsets of a fixed finite set $\mathfrak{X}$ forms a lattice $\mathfrak{L}_{\mathfrak{X}}$ under the ordering 'is a subset of'. The top and bottom elements here are $\mathfrak{X}$ and $\emptyset$, and the join is the smallest subspace containing a pair of spaces.

Example 3) The subgroups of a group form a lattice $\mathfrak{L}_{\mathfrak{g}}$ under the ordering 'is a subgroup of'. The top and bottom elements here are the whole group $\mathfrak{g}$ and the trivial group *id*, and the join is the subgroup generated by their union. This example generalizes to the subalgebraic structures of an algebraic structure, which the current book uses to model constraints, beables and the selection process underlying Kinematical Quantization. Let us denote the lattices in question by $\mathfrak{L}_{\mathfrak{c}}$ and $\mathfrak{L}_{\mathfrak{b}}$ (exposited in Sect. 24.12).

Example 4) The set of collections $\mathfrak{Collect}(\mathfrak{X})$ of subsets of a fixed finite set $\mathfrak{X}$ forms a lattice $\mathfrak{L}_{\mathfrak{C}}$ under the ordering 'is a subcollection of'. The top and bottom elements here are the power set $\mathfrak{P}(\mathfrak{X})$ and $\emptyset$ (the empty collection, not to be confused with $\{\emptyset\}$: the non-empty collection of subsets consisting of the empty set!)

Example 5) The set of topologies $\mathfrak{Top}$-$\mathfrak{Space}(\mathfrak{X})$ on a fixed finite set $\mathfrak{X}$—a specialization of the previous to collections of subsets which constitute topologies—is itself a lattice $\mathfrak{L}_{\mathfrak{T}}$ [608, 810] under the 'is a finer or coarser topology than' (relative coarseness) operations. The discrete and indiscrete topologies are the top and bottom elements of this: the entirety of $\mathfrak{P}(\mathfrak{X})$ and just $\{\emptyset, \mathfrak{X}\}$ respectively. This particular lattice is complete and complemented; [480, 481, 490, 608, 810] list further properties. Examples 2), 4) and 5) are of further interest as 'spaces of spaces' for some of the sparser levels of mathematical structure.

As regards the tower case, what if all of the topological space and any subsequent emergent differentiable, affine, conformal, and metric structure are dynamical? Now, far from all topological spaces support topological manifolds, so one encounters a problem unless one can consider the structures from here upward as emergent for certain $\tau(\mathfrak{X})$. This does however beg the question of what *other* structures some topological spaces are capable of supporting. This corresponds to a major breakdown here between the sharply characterized tower of Manifold Geometry and the far wider range of topological spaces and of collections more generally (Fig. S.2). A dynamical theory of topological spaces might explain whether and how topological manifolds are prevalent in the set of possible universes as a 'zeroth principles' theory. After that, one would restrict attention to a $\mathfrak{Top}$-$\mathfrak{Man}(\mathfrak{X})$ first-principles theory.

Further discussion involves various useful classes in the nested array of Fig. S.2.

Note also the modelling disjointness between topological spaces on a finite set versus manifolds, which are based on infinite sets. None the less, Čech cohomology based methods can be applied to both sides of this divide.

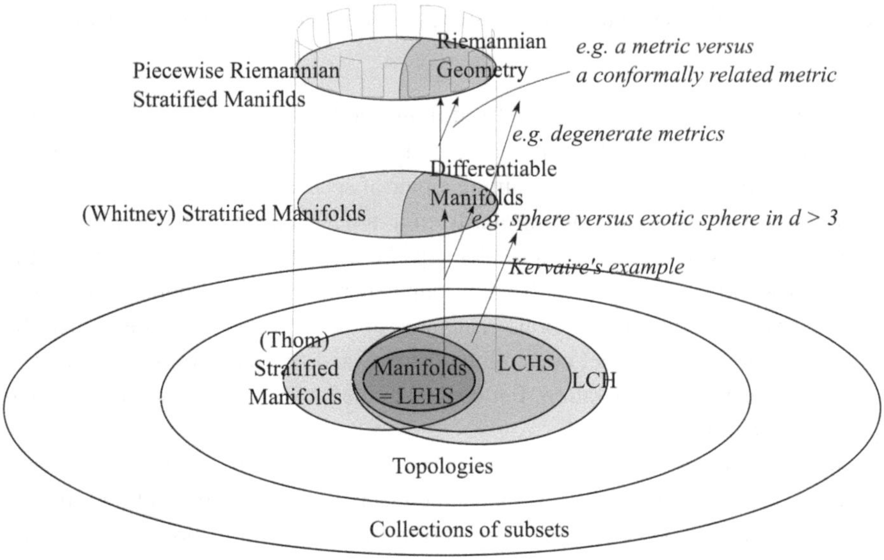

Fig. S.2 The collections, and the principal tower of the standard approach. This figure illustrates by examples that not all cases within a given floor of the tower can be extended up the tower; Kervaire's example refers to [547]. Whereas the Hausdorff second-countable locally Euclidean case just returns the manifolds, their generalization to Hausdorff second-countable topological spaces is also of interest; LCHS spaces are depicted as well. The main purpose is to highlight various well-behaved classes of stratified manifolds: the *HS stratified manifolds* and the *LCHS stratified manifolds*. This last case includes a number of physically relevant stratified manifolds, Kreck's stratifold construct, Śniatycki's differential spaces and Kendall's work on random sets (see Appendix T.4)

S.6 Spaces of Sets?

In the single-floor case, if sets alone contain physically meaningful information, so if $\mathfrak{q}$ is the set of sets, what dynamics do these support?

In the tower case, problems arise since the space of topological spaces becomes unruly if one tries to define it on the set of sets rather than on a fixed set $\mathfrak{X}$. This is rooted in the set of sets suffering from mathematician Bertrand Russell's paradox, limiting the study that far of spaces of space [480, 481]. Yet it is conceivable—if the Universe is built out of sets—that all of the metric, connection, differentiable structure, topological manifold, topological space and set are dynamical (or, a fortiori, undergo quantum fluctuations).

An alternative position, however (Epilogue III.C), is that the approach to set structure (via collections of subsets)—in which simplifications start to occur—is itself a type of background structure.

Appendix T
Deeper Levels. ii. Grainings, Information, Stochastics and Statistics**

Let use now take the sample space $\mathbf{\Omega}$ of Appendix P.1 to be whichever of the deeper levels of structure's $\mathfrak{q}$, and build stochastic and statistical theories thereupon.

T.1 Grainings

Example 1) A more mathematically general starting point than Appendix Q.6's involves replacing metric notions of locality with ones based on subset overlaps. Indeed, Appendix Q.6's partition refinement is a subcase of subset refinement based on Appendix A's ordering by $\subseteq$. Another subcase of note is the condition for whether two neighbourhoods overlap from Čech cohomology as per Appendix F.3, which are based on covers. *Cover refinement* then plays the role of fine-graining. This admits a further sheaf-theoretic generalization as well.

Example 2) Notions of locality can also be defined on lattices, along the chains, including on $\mathfrak{L}_{\mathfrak{X}}$, $\mathfrak{L}_{\mathfrak{C}}$ and $\mathfrak{L}_{\mathfrak{T}}$. Furthermore, graining of subsets, collections of subsets and topologies, and for covers thereover, are lattice concepts.

Example 3) Graining of sets can be considered by use of a variety of sizes of subsets.

T.2 Notions of Information or Entropy

Example 1) In the context of topological spaces, information consists of *which subsets overlap* [2]. This is amenable to using Čech methods, and is further generalizable to Sheaf Methods.

Example 2) Topological entropy or information has been considered on some lattices, though further exploration of this is required in the specific cases of $\mathfrak{L}_{\mathfrak{X}}$, $\mathfrak{L}_{\mathfrak{C}}$ and $\mathfrak{L}_{\mathfrak{T}}$.

Example 3) Entropy or information is a straightforward notion in the case of for unstructured sets, and consequently for $\mathfrak{X}$ and $\mathfrak{Coll}(\mathfrak{X})$.

E. Anderson, *The Problem of Time*, Fundamental Theories of Physics 190,
DOI 10.1007/978-3-319-58848-3

T.3 Stochastic and Statistical Treatment. i. Topological Manifold Level

Some simpler cases have become familiar in the Theoretical Physics literature, namely those probing 'topology of the Universe' i.e. large-scale shape. In more general terms, Smale, Niyogi and Weinberger [681], have worked on stochastic topology, making use of random simplicial complexes, covering set intersections, and (Čech co)homology. This serves to assess the topology of an approximately given configuration.

Sheaves can indeed be applied to Statistics by being a means of presenting and handling local data, (see e.g. [739] for a contemporary outline on this).

T.4 ii. Topological Space Level

Example 1) Kendall's [535] theory of random sets operates within a carrier space $\mathfrak{C}$—taken to be LCHS—which the random set $\mathfrak{X}$ 'sits within'. A *trapping system* $\{\mathfrak{U}_\mathsf{T}\}$ for $\mathfrak{X} \subset \mathfrak{C}$ is then a collection of subsets with the following properties.

Trap 1) $\mathfrak{U}_\mathsf{T} \neq \emptyset$,
Trap 2) $\bigcup_\mathsf{T} \mathfrak{U}_\mathsf{T} = \mathfrak{C}$, so $\{\mathfrak{U}_\mathsf{T}\}$ are a cover of $\mathfrak{C}$.
Trap 3) To each $\mathfrak{U}_\mathsf{T}$ one can associate a countable system $\mathfrak{V}_\mathsf{S}(\mathfrak{U}_\mathsf{T})$ of subtraps (local countability).
Trap 4) If $x \in \mathfrak{U}_\mathsf{T}$, then x belongs to a $\mathfrak{C}$-trap whose $\mathfrak{U}_\mathsf{T}$-closure is covered by $\mathfrak{U}_\mathsf{T}$.

Example 2) One can furthermore contemplate applying this approach to carrier sets that are half of the stratifold construct's pairing. Approaches based on Čech cohomology descend to this level of structure as well.
Example 3) Since the space of topologies on a finite set is a lattice $\mathfrak{L}_\mathfrak{T}$, stochastic treatment of lattices is also to be considered. Well-known examples of this include random points on a square lattice [738] and percolation [781]. This renders plausible a stochastic treatment of $\mathfrak{L}_\mathfrak{T}$ itself (this is one possible approach to Research Project 47).
Example 4) See e.g. [268, 697, 892] for stochastic treatment of metric spaces.

T.5 iii. Set and Collection of Subsets Levels

Once again, the case of sets is more mathematically tractable due to the lack of additional structures.

Research Project 122) Using e.g. that the set of subsets of a finite set $\mathfrak{X}$ also form a lattice, $\mathfrak{L}_\mathfrak{X}$, consider the following sharper version of part of Research Project 49). How probable is it for a random collection of subsets to be a topology? Hausdorff? A cover? A good cover? [In the sense of (F.2).] A basis? A σ-algebra? A trapping system?

Appendix U
Quantum SM, Information and Correlation*

This Appendix supports in particular this book's Timeless Approaches, now at the quantum level.

U.1 Mixed States

Mixed states are more general than pure states. These can be modelled by [487] density matrices

$$\rho = \sum_n o_n \widehat{\mathrm{P}}_n$$

for $o_n = \mathrm{Prob}(\text{state is} \mid \psi_n \rangle)$ and $\widehat{\mathrm{P}}_n$ the corresponding projector, $|\psi_n\rangle\langle\psi_n|$.

$$\mathrm{Prob}(A = a_n \text{ in state } \rho) = \mathrm{Tr}(\widehat{\rho}\,\widehat{P_n}).$$

Note that this gives back the pure state case upon using ρ of the form $|\psi\rangle\langle\psi|$. Density matrices are i) Hermitian, ii) positive semi-definite: $\langle\psi\,|\,\widehat{\rho}\,|\,\psi\rangle \geq 0$, and iii) normalized: $\mathrm{Tr}(\rho) = 1$. In fact, density matrices can be taken to be defined by these properties.

The space of mixed states for a system [130] is much larger than the corresponding spaces of pure states. E.g. the system whose pure states are just an 'up' and 'down' discrete pair (a 'qubit') has a whole $\mathbb{S}^2$'s worth of mixed states. More generally the qu*n*it has a $\mathbb{CP}^{n-1}$ of mixed states, complete with a Fubini–Study metric (G.8), i.e. the extension of the previous sentence upon making the identification $\mathbb{S}^2 = \mathbb{CP}^1$.

E. Anderson, *The Problem of Time*, Fundamental Theories of Physics 190,
DOI 10.1007/978-3-319-58848-3

U.2 Quantum Grainings

Some notions of coarse-graining carry over from the classical level due to remaining defined upon the same classical spaces Sub $\mathfrak{S}$. However, some other notions are new through involving quantum state spaces [130].

In contrast to the classical level's finest graining's knowledge being attainable in principle, a quantum-level distinction is that perfect knowledge of states becomes contentious.

Coarse-graining calculations at the quantum level consist of tracing out modes in ($\mathfrak{S}$ub-)$\mathfrak{H}$ilb,

$$\mathrm{P}_{ij} = \delta_{ij} D_I^{-1} \sum_{A,B=1}^{D_I} \rho_{ii}, \tag{U.1}$$

i.e. a density matrix trace version of (Q.8).

U.3 Quantum Versus Standard Probability Theory

For the purposes of this book, 'quantum probability theory' is approached as follows (other authors may approach this elsewise). At the quantum level, propositions have an inherently probabilistic nature, in the sense of being in terms of 'Prob(A) is' rather than 'A is'. This is embodied by representing propositions by projectors as per Chap. 51.1. Then for state ρ and proposition P implemented by projector $\widehat{\mathrm{P}}$,[1] the assignation of probability is

$$\mathrm{Prob}(P \,|\, \rho) = tr(\widehat{\rho}\widehat{\mathrm{P}}). \tag{U.2}$$

We next consider whether this assignation—a function $Prob : \mathfrak{p}\mathrm{roj}(\mathfrak{H}\mathrm{ilb}) \rightarrow \mathbb{R}$—is unique. (U.2) can be axiomatized as obeying the following [contrast with classical Kolmogorov probability, which uses only ii)].

i) $\mathrm{Prob}(P) \in [0,\ 1] \subset \mathbb{R}_0$ (positivity).
ii) $\mathrm{Prob}(\emptyset) = 0$ and Prob(any outcome) $= 1$.
We need a further criterion along the lines of additivity for disjoint subspaces $\mathfrak{U}$, $\mathfrak{V}$: $\mathrm{Prob}(\mathfrak{U} \cup \mathfrak{V}) = \mathrm{Prob}(\mathfrak{U}) + \mathrm{Prob}(\mathfrak{V})$. Note beforehand that this is less stringent than $\mathrm{Prob}(\mathfrak{U} \cup \mathfrak{V}) = \mathrm{Prob}(\mathfrak{U}) + \mathrm{Prob}(\mathfrak{V}) - \mathrm{Prob}(\mathfrak{U} \cap \mathfrak{V})$, which is not imposed due to the incompatibility in general [487] of quantum observables or beables. Then extend this criterion to the following (compare also now with Kolmogorov's countability axiom).
iii) $\mathrm{Prob}\big(\sum_{n=1}^{\infty} \widehat{P}_i\big) = \sum_{n=1}^{\infty} \mathrm{Prob}(\widehat{P}_i)$ for any finite or countably infinite collection of pairwise mutually orthogonal projectors.

[1] Denote the space of the $\widehat{\mathrm{P}}$ by $\mathfrak{P}\mathrm{roj}(\mathfrak{H}\mathrm{ilb})$.

Note that (U.2) complies with these. Furthermore, *Gleason's Theorem* [487] provides a strong uniqueness criterion for this: for dim($\mathfrak{H}$ilb) > 2, use of a density matrix is the only possibility that fulfils i)–iii).

As regards other violations of classical probability concepts encountered at the quantum level, some proposed wave equations give 'negative probabilities' (Chap. 6), on which grounds they are discarded; some authors term these 'quasiprobability distributions.

Finally note that the difference between classical and quantum probability is great enough for QM density matrices *not* to be 'multivariate probability distributions' in the usual sense familiar from classical probability.

U.4 Quantum SM

A distinct quantum version of SM can be based on Quantum Theory; now imperfect knowledge is required in principle (see e.g. [341]).

The quantum partition function is, for instance, in the case of the canonical formalism for a system with discrete non-degenerate energy spectrum

$$\mathrm{Z}(T) = \mathrm{Tr}(\exp(-\hat{H}/k_{\mathrm{B}}T)).$$

$$\rho(T) = \mathrm{Tr}(\exp(-\hat{H}/k_{\mathrm{B}}T))/\mathrm{Tr}(\exp(-\hat{H}/k_{\mathrm{B}}T))$$

is then the corresponding *thermal density matrix*. [At the quantum level, the sum or integral involved in partition functions takes the form of tracing over states.] See e.g. [781] for further ensembles' partition functions and thermal density matrices.

One major difference comes from quantum level symmetry restrictions on states affecting their count: selections rules giving rise to bosons and fermions respectively. Counting the states in question replaces the classically inspired Maxwell–Boltzmann distribution $\exp(E_i/k_{\mathrm{B}}T)$ with the specifically quantum *Bose–Einstein* and *Fermi–Dirac* distributions, which are respectively,

$$1/\{1 - \exp(E_i/k_{\mathrm{B}}T - \mu)\}, \tag{U.3}$$

$$1/\{1 + \exp(E_i/k_{\mathrm{B}}T - \mu)\}. \tag{U.4}$$

Note that these as given are grand ensemble entities due to the relevance of particle creation and annihilation, which alters particle number N. These distributions can readily be built up into partition functions by inserting degeneracy factors and taking suitable traces.

Finally, from (42.3) it is clear that Configurationally-Relational density matrices can be defined by

$$\rho_{\mathfrak{g}} := \int_{\mathsf{g}\in\mathfrak{g}} \mathbb{D}\mathsf{g}\, \exp\Big(i \sum_{\mathsf{g}\in\mathfrak{g}} \vec{\mathfrak{g}}_{\mathsf{g}^{\mathsf{G}}}\Big)\, \rho\, \exp\Big(-i \sum_{\mathsf{g}\in\mathfrak{g}} \vec{\mathfrak{g}}_{\mathsf{g}^{\mathsf{G}}}\Big). \tag{U.5}$$

Projectors then clearly end up built the same way whether assembled out of states or treated as a subcase of operators.

U.5 Quantum Notions of Entropy and Information

In this setting, quantum states play the role of microstates instead of the classical phase space volumes previously considered.

A suitable quantum analogue of Shannon information (Q.12) is *von Neumann information*,

$$I_{\text{von Neumann}}[\rho] = \text{Tr}(\rho \log \rho). \tag{U.6}$$

See e.g. [633, 879] as regards the classical–quantum correspondence between these. This notion furthermore survives the passage to specially-relativistic QM, and to QFT modulo a short-distance cutoff [874].

The mutual information concept (Q.13) also applies to van Neumann information [130, 725], though this is now of the form

$$M_{\text{von Neumann}}[\rho_A, \rho_B, \rho_{AB}] \\ = I_{\text{von Neumann}}[\rho_A] + I_{\text{von Neumann}}[\rho_B] - I_{\text{von Neumann}}[\rho_{AB}]. \tag{U.7}$$

Relative information at the quantum level is [680]

$$I_{\text{relative}}[\rho_1, \rho_2] = \text{Tr}(\rho_1\{\log\rho_1 - \log\rho_2\}). \tag{U.8}$$

This can be interpreted as a distance on state space. Mutual information can furthermore be seen as a distance between ρ_{AB} and the uncorrelated state $\rho_A \otimes \rho_B$. In Ordinary Quantum Theory, one can view this as a quantifier of entanglement.

Also, a Configurationally Relational version of von Neumann information is

$$I^{\mathfrak{g}}_{\text{von Neumann}} := \int_{\mathsf{g} \in \mathfrak{g}} \mathbb{D}\mathsf{g} \exp\left(i \sum_{\mathsf{g} \in \mathfrak{g}} \vec{\mathfrak{g}_{\mathsf{g}\mathsf{G}}} \right) \rho \exp\left(-i \sum_{\mathsf{g} \in \mathfrak{g}} \vec{\mathfrak{g}_{\mathsf{g}\mathsf{G}}} \right) \\ \times \ln\left(\left(\int_{\mathsf{g} \in \mathfrak{g}} \mathbb{D}\mathsf{g} \exp\left(i \sum_{\mathsf{g} \in \mathfrak{g}} \vec{\mathfrak{g}_{\mathsf{g}\mathsf{G}}} \right) \rho \exp\left(-i \sum_{\mathsf{g} \in \mathfrak{g}} \vec{\mathfrak{g}_{\mathsf{g}\mathsf{G}}} \right) \right) \right). \tag{U.9}$$

Finally, von Neumann entropy has also been used in Black Hole Physics (see e.g. [874]). This gives rise to further difficulties due to the form of the underlying quantum microstates still being a matter of speculation.

U.6 Quantum Correlations

Example 1) Ordinary QFT has n-point functions [712] of the same well-known kind as in Example 1) of Chap. Q.9. $\langle\ \rangle$ here includes inserting the ground-state quantum wavefunction at each end. Giddings–Marolf–Hartle provide a further useful treatment of correlators for Quantum Cosmology in [353]. Some aspects of this [and Example 1)'s relational underpinning] go back to DeWitt [235, 237]. However, we need to proceed with caution because at least some forms of n-point function are not manifestly already-relational. In such a case one could at least formally apply $\mathfrak{g}$-act $\mathfrak{g}$-all with the $\mathfrak{g}$-all move being integration over the $\mathfrak{g}$ in question,

$$\langle \phi(\underline{x}_1) \ldots \phi(\underline{x}_1) \rangle_{\mathrm{g}} := \int_{\mathrm{g} \in \mathfrak{g}} \mathbb{D}\mathrm{g} \int_{\mathfrak{q}} \vec{\mathfrak{g}}_{\mathrm{g}} \left\{ \mathbb{D}\phi \exp(-\mathsf{S}[\phi]) \phi(\underline{x}_1) \ldots \phi(\underline{x}_n) \right\} \Big/ \int_{\mathrm{g} \in \mathfrak{g}} \mathbb{D}\mathrm{g} \int_{\mathfrak{q}} \vec{\mathfrak{g}}_{\mathrm{g}} \left\{ \mathbb{D}\phi \exp(-\mathsf{S}[\phi]) \right\}. \tag{U.10}$$

Example 2) Quantum entanglement means that Quantum Theory has an extra type of correlations that classical theories do not have. This leads to the concept of *discord* = (quantum correlations) − (classical correlations), for which expressions using von Neumann and Shannon informations can be employed. This quantifies how Shannon entropy is insufficient to capture quantum correlations.

Appendix V
Further Algebraic Structures*

This Appendix supports in particular Facets 3 to 6 of the Problem of Time.

Finite groups can be generalized by considering infinite groups which still have a finite number of generators; Lie groups are examples of such. An obvious further generalization is to the case of infinitely many generators. Another generalization involves including fermionic generators in both the finite and infinite generator cases. A further vast generalization occurs upon passing from the structure constants of Lie algebras to the structure functions of Lie algebroids. The rest of this Appendix gives examples of such groups and points to key results in the corresponding Representation Theory.

Applications of this include Kinematical Quantization, classical and quantum constraints and observables or beables. These cover many workings in QFT, Quantum GR, Quantum Supergravity and String Theory.

We next consider representations for various products of groups. Firstly, representations of direct products of groups or of Lie algebras are elementary to handle by tensor products of the individual groups' reps (Exercise!).

V.1 Reps. i. Semidirect Product Groups in General

Quite a lot of the groups involved in Theoretical Physics are semidirect products, such as $Eucl(d) = Tr(d) \rtimes Rot(d) = \mathbb{R}^d \rtimes SO(d)$, $Poin(d) = \mathbb{M}^d \rtimes SO(d-1,1)$, $Conf(\boldsymbol{\Sigma}) \rtimes Diff(\boldsymbol{\Sigma})$, and various groups of the form $\mathfrak{V}^* \rtimes \mathfrak{g}_{\mathrm{can}}$ arising from kinematically quantizing homogeneous spaces $\mathfrak{g}/\mathfrak{H}$ (Appendix M.1). The Representation Theory of semidirect product groups $\mathfrak{H} \rtimes \mathfrak{K}$ is fortunately rendered tractable in terms of that of the constituent groups $\mathfrak{H}$, $\mathfrak{K}$, via *Mackey Theory* [633, 867]. This is a substantially more general version of Wigner's use [908, 909] of the little group, which is very familiar in the Theoretical Physics literature for the Poincaré group $Poin(4)$ in particular and is the subject of Ex III.8. All of this lies within the scope of *induced representation* methods, in which the reps of a subgroup $\mathfrak{H}$ can provide reps for the group $\mathfrak{g}$ itself. [786] is an excellent introduction to induced representa-

E. Anderson, *The Problem of Time*, Fundamental Theories of Physics 190,
DOI 10.1007/978-3-319-58848-3

tions, taking one as far as Mackey's criterion itself. [475, 477, 480, 481] then consider various Theoretical Physics applications of Mackey Theory. Note finally that Sect. 41.5's 'Rieffel induced inner product' is also meant in the sense of induced reps.

On the other hand, the more complicated and diverse Thomas ⊖ and two-way ⊖ integrability structures, have no known systematic means of approaching the corresponding Representation Theory.

V.2 ii. Super-Poincaré Groups

These admit a well-known extension of the approach to Poincaré group Representation Theory. For $\mathrm{N} = 1$ Supersymmetry in 4-d, this is also a semidirect product as per Ex VI.20. Moreover, this differs with increasing N, with multiplets getting larger, up to a maximum $\mathrm{N} = 8$, corresponding to the maximal spacetime dimension being fixed to be 11 (Exercise!).

V.3 iii. Diffeomorphism Groups

We next outline what is known about the reps of diffeomorphism groups, due to their importance in GR.

Example 1) $\mathbb{S}^1$ admits the *Witt algebra*[1]

$$[L_{\mathrm{m}}, L_{\mathrm{n}}] = \{\mathrm{m} - \mathrm{n}\} L_{\mathrm{m+n}}, \tag{V.1}$$

which is an infinite-d Lie algebra. More particularly, this is a central extension of a Lie algebra $\mathfrak{g}$ by an Abelian Lie algebra $\mathfrak{g}$,[2] to the *Virasoro algebra*

$$[L_{\mathrm{m}}, L_{\mathrm{n}}] = \{\mathrm{m} - \mathrm{n}\} L_{\mathrm{m+n}} + c\,\delta_{\mathrm{n+m,0}}\mathrm{n}\{\mathrm{n}^2 - 1\}/12 \tag{V.2}$$

(the c-number c here stands for 'central charge'). These are both examples of the more general class of *Kac–Moody algebras*.

As regards the Representation Theory for these algebras, [368] supply a Weyl–Kac character formula based approach (see e.g. [326] for development of the Weyl character approach for finite (Lie) groups). On the other hand, [674] outlines a distinct Fibre Bundle Method (Borel–Weil construction).

---

[1] The algebras in the current section are named after mathematicians Ernst Witt, Victor Kac and Robert Moody, and physicist Miguel Angel Virasoro. See e.g. [385, 674] for more about the Theoretical Physics involvement of these algebras, in particular in CFT and in String Theory.

[2] For $\mathfrak{H}$ and $\mathfrak{K}$ groups, $\mathfrak{g}$ is an *extension* of $\mathfrak{H}$ by $\mathfrak{K}$ if there is a short exact sequence $1 \to \mathfrak{K} \to \mathfrak{g} \to \mathfrak{h} \to 1$. Then $\mathfrak{K} \lhd \mathfrak{g}$ and $\mathfrak{g}/\mathfrak{K} \cong \mathfrak{H}$. This is furthermore *central* if $\mathfrak{K} \subseteq Z(\mathfrak{g})$. Finally, these additionally admit a cohomological interpretation [214].

Example 2) The more general 3-*d* case is considered e.g. in [475], with [471, 475, 477, 482, 483, 501, 502] giving furthermore some indication of the corresponding Representation Theory.

V.4 iv. Super-diffeomorphism Groups

The extension of the previous Sec's $\mathbb{S}^1$ example is straightforward, giving the *Witt superalgebra* (V.1) alongside

$$[L_{\mathrm{m}}, G_{\mathrm{r}}] = \{\mathrm{m}/2 - \mathrm{r}\} G_{\mathrm{m+r}} \quad \text{and} \tag{V.3}$$

$$\{G_{\mathrm{r}}, G_{\mathrm{s}}\} = 2\, L_{\mathrm{r+s}}. \tag{V.4}$$

This also admits a central extension, to the Virasro superalgebra (V.2), (V.3) and (V.4) with the extra central term $+\frac{c}{3}\left\{\mathrm{r}^2 - 1/4\right\}\delta_{\mathrm{r+s},0}$. The Weyl–Kac character based approach to Representation Theory mentioned above extends nicely to this case as well.

On the other hand, the general case of diffeomorphism supergroup and its corresponding Representation Theory is considerably less well-known than the already limitedly-known general diffeomorphisms; see however e.g. [609].

V.5 v. For Kinematical Quantization of GR

Unitary reps for $\mathfrak{S}\mathrm{ym}(3, \mathbb{R}) \rtimes GL^+(3, \mathbb{R})$, of relevance pointwise to GR, are considered e.g. in [476]. This fortunately still lies within the remit of Mackey Theory as regards construction of representations.

V.6 Algebroids and Their Reps

Lie algebroids are defined as follows [190, 603, 863].

1) Consider a smooth manifold $\mathfrak{M}$, and define a vector bundle $\dot{\mathfrak{j}}$ over this.
2) Then define a Lie algebra structure on the corresponding space of sections of $\dot{\mathfrak{j}}$: $\mathfrak{S}\mathrm{ec}(\dot{\mathfrak{j}})$.
3) The *anchor map* is a bundle map $A : \dot{\mathfrak{j}} \rightarrow \mathfrak{T}(\mathfrak{M})$ such that
    i) $A : \mathfrak{S}\mathrm{ec}(\mathfrak{M}) \rightarrow \mathrm{Vec}(\mathfrak{M})$ (of vectors) is a Lie algebra homomorphism corresponding to the commutator Lie bracket.
    ii) For $f \in \mathfrak{C}^\infty(\mathfrak{M})$, $\Gamma_1, \Gamma_2 \in \mathfrak{S}\mathrm{ec}(\mathfrak{M})$ the derivation rule $|[\Gamma_1, f\Gamma_2]| = f|[\Gamma_1, \Gamma_2]| + (A(\Gamma_1)f)\Gamma_2$ holds.

The algebroid–groupoid interrelation is also more complicated than its algebra–group counterpart.

Example 1) In the case over just a one-point space, one returns to the standard Lie algebra.

Example 2) In the case of tangent bundles, the identity map of $\mathfrak{T}(\mathfrak{M})$ is the anchor map, and the reps are vector bundles over $\mathfrak{M}$ with flat connections. This has applications to the theory of foliations.

Example 3) Lie algebroids arising in the symplectic context are covered in particular in [190].

Example 4) The main algebroid considered in this book is the Dirac alias deformation [454] algebroid (9.31)–(9.33). See in particular [154] for further contemporary coverage of this in a Theoretical Physics setting.

Example 5) The Ashtekar–Dirac algebroid [75].

Example 6) Both Mackey's Kinematical Quantization and its phase space generalization admit interpretation as algebroid mathematics [601, 605].

Representations of algebroids [383] consist of two parts.

I) a vector bundle $\mathfrak{j}$ over $\mathfrak{M}$.

II) A $\mathbb{R}$-bilinear map $\mathfrak{Sec}(\mathfrak{M}) \times \mathfrak{Sec}(\mathfrak{j}) \rightarrow \mathfrak{Sec}(\mathfrak{M}) : A \otimes S \mapsto D_A S$, for a suitable notion of derivative D_A [293].

II) is such that for any $A, B \in \mathfrak{Sec}(\mathfrak{M})$, $S \in \mathfrak{Sec}(\mathfrak{j})$ and $f \in C^\infty(\mathfrak{M})$, $D_{fA}S = f D_A S$, $D_A\{fS\} = f D_A S + \{\rho(A) f\} S$ and $D_A\{D_B S\} - D_B\{D_A S\} = D_{|[A,B]|} S$.

Some Representation Theory methods which extend as far as this case are provided in e.g. [383]. Cohomology suitable for Lie algebroids is covered in [223, 293].

Superalgebroids are treated in outline in e.g. [372]; the above outline readily continues to carry over upon replacing 'Lie algebra' by 'Lie superalgebra'.

Example 7) Supergravity's constraint superalgebroid [232] is probably Theoretical Physics' most salient example of superalgebroid.

V.7 Operator Algebras

Operators on $\mathfrak{H}$ilb form a number of algebras $\mathfrak{B}(\mathfrak{H}\text{ilb}) \supset \mathfrak{C}^*(\mathfrak{H}\text{ilb}) \supset \mathfrak{W}^*(\mathfrak{H}\text{ilb})$. In this book's quantum application of these, they are all taken to be over the field $\mathbb{C}$.

Here the $\mathfrak{B}$ are bounded linear operators T [207] defined on normed spaces. The star superscript $*$ denotes that an involution operation $* : \mathfrak{B} \rightarrow \mathfrak{B}$ enters the algebra; this obeys the following relations.

1) $\{T_1 + T_2\}^* = T_1^* + T_2^*$ and $\{pT\}^* = \bar{p}T^*$ (complex-linearity, for $p \in \mathbb{C}$).
2) $\{T_1 T_2\}^* = T_2^* T_1^*$ (preservation of the algebraic structure itself).
3) $T^{**} = T$ (involution).
4) $\|T^*\| = \|T\|$ (continuity).

T^* is indeed the adjoint operator of T; in this way, Quantum Theory makes active use of this $*$ operation. 1) to 4) characterize $\mathfrak{C}^*$-algebras, which incorporate closure under adjunction. The further $\mathfrak{W}^*$*-algebra* alias *von Neumann algebra* [528] specialization ensures that we are dealing with an algebra of self-adjoint operators, which is also a requisite feature in setting up Quantum Theory. Furthermore, projectors suffice to express all the information in such a model's operators (Exercise!).

See [207] for further outline comments, or [528] for a more detailed account these algebras.

Many physical applications of $\mathfrak{C}^*$ and $\mathfrak{W}^*$ algebras proceed via these being a powerful tool in treating the Representation Theory of quantum-mechanical commutation relations [473, 605, 828]. More specific applications are to Kinematical Quantization algebraic structures [75, 475, 605], quantum constraint algebraic structures, and especially local quantum observables' or beables' algebraic structures [401, 602, 605, 778]. The Rieffel induced inner product is in fact additionally a $\mathfrak{C}^*$ algebra approach construct. A final application of $\mathfrak{W}^*$-algebras outlined in this book is in the sheaf or topos approach to the Kochen–Specker Theorem.

Appendix W
Alternative Foundations for Mathematics**

This Appendix supports in particular the global and deeper levels of structure Epilogues II.B, III.B, II.C and III.C.

Representation Theory unlocks many doors in the study of Quantum Theory. Theoretical physicists may well then ask what 'comes after'—i.e. meaningfully generalizes—Representation Theory; one possibility is Category Theory. This is from the perspective that group representations are a simple, very useful and historically early example of functor category.

And what meaningfully generalizes the fibre bundle based topological methods so useful in global considerations of QFT? One possibility is Sheaf Methods. And what meaningfully generalizes the function spaces that underlie both Mathematical Relativity and Quantum Theory? Functor categories are a possible candidate. And, in thinking about QG, what meaningfully generalizes the General Covariance that is so useful in GR? Perhaps it is Topos Theory!

W.1 Categories

Categories $\underset{\sim}{\mathrm{C}} = (\mathrm{O}, M)$ consist of *objects* O and *morphisms* M: the maps between the objects, $M : \mathrm{O} \longrightarrow \mathrm{O}$, alongside the following.

i) A rule assigning a domain $\mathrm{dom}\, f$ and a codomain $\mathrm{cod}\, f$ to each $f \in M$.
ii) An identity morphism $1_{\mathrm{O}} : \mathrm{O} \longrightarrow \mathrm{O}$.
iii) A composition $f_2 f_1$ for each pair of morphisms f_1, f_2 such that $\mathrm{cod}\, f_1 = \mathrm{dom}\, f_2$.

Functors are maps $\mathcal{F} : \underset{\sim}{\mathrm{C}}_1 \longrightarrow \underset{\sim}{\mathrm{C}}_2$ such that

i) $\mathrm{dom}\, \mathcal{F} f = \mathcal{F}(\mathrm{dom}\, f)$ and $\mathrm{cod}\, \mathcal{F} f = \mathcal{F}(\mathrm{cod}\, f)$.
ii) $\mathcal{F}(1_{\mathrm{O}}) = 1_{\mathcal{F}\mathrm{O}}$.
iii) $\mathcal{F}(f_1 f_2) = \mathcal{F}(f_2)\mathcal{F}(f_1)\ \forall\, f_2 f_1$ defined in $\underset{\sim}{\mathrm{C}}$.

E. Anderson, *The Problem of Time*, Fundamental Theories of Physics 190,
DOI 10.1007/978-3-319-58848-3

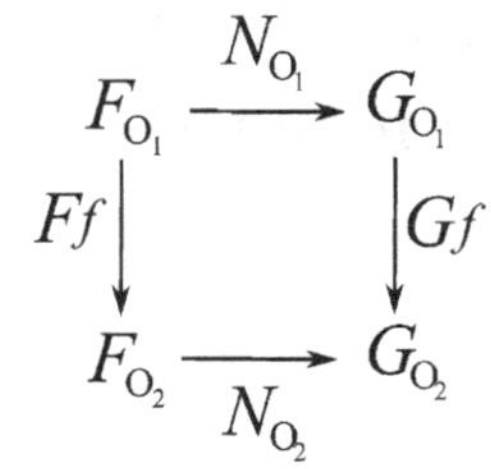

Fig. W.1 Commuting square for natural transformations

Note that 'co' is used in this context to indicate that the map runs in the opposite direction; the two ensuing cases are termed *covariant* and *contravariant* functors.

Categories first appeared in the 1940s work of mathematicians Samuel Eilenberg and Saunders Mac Lane; see in particular [128, 611, 612, 631] for more details on this subject.

Example 1) $\underset{\sim}{\mathrm{Sets}}$ is the category of sets; per se this is foundationally trivial, though indirectly it plays a repeated role in the developments below.
Example 2) $\underset{\sim}{\mathrm{Grp}}$ is the category of groups.
Example 3) $\underset{\sim}{\mathrm{Vec}}$ is the category of vector spaces.
Example 4) The category of topological spaces $\underset{\sim}{\mathrm{Top}}$ played a substantial role in the historical development of this subject in support of the development of Algebraic Topology.
Example 5) Forgetfulness (Appendix A.1) is functorial: involves all spaces within a given category with corresponding morphisms [631].

Functor categories are categories of maps between categories.

A *natural transformation* is a morphism of functor categories themselves: $N : F \rightarrow G$; some authors have argued that these are the core of Category Theory. N maps $\mathrm{O} \longrightarrow M$ such that $N_{\mathrm{O}} : F_{\mathrm{O}} \longrightarrow G_{\mathrm{O}}$ and the diagram in Fig. W.1 commutes for any $F : \mathrm{O}_1 \longrightarrow \mathrm{O}_2$.

Example 6) The representations of Representation Theory are a functor category from $\underset{\sim}{\mathrm{Grp}}$ to $\underset{\sim}{\mathrm{Vec}}$. *Intertwiners*—maps between representations—are then clearly understood as natural transformations. Indeed, understanding intertwiners is often the first benefit for students of Physics beginning to study Category Theory.
Example 7) (Co)homology can be understood as functor categories [631]; cohomology indicates functor contravariance to homology's functor covariance. The characteristic classes of Algebraic Topology provide a second example of natural transformations. This perspective also points clearly to (co)homology having a considerably broader range of applicability than that of the first (de Rham) physical application of these. This is because what (co)homology functors do is to associate Abelian categories to given non-Abelian ones. The latter are then differentiable manifolds in the de Rham case, Poisson manifolds in the Poisson case, covers for topological spaces in the Čech case, singular differentiable manifolds in the Morse

case, maps between differentiable manifolds in the (co)bordism case, and associative algebras over fields and rings in the Hochschild and cyclic cases.[1]

Non-example 8) It would be nice if Quantization itself were in general a clear-cut functor, but it is not in general (Chap. 40).

A final pair of notions that Epilogue III.C makes use of are as follows. A *small category* is one for which both the objects and the morphisms are sets. A *locally small category* is one in which for each pair of objects, the corresponding set of homomorphisms is a set.

W.2 Presheaves

Presheaves are functors $\mathcal{E} : \underset{\sim}{\text{Top}} \rightarrow \underset{\sim}{\text{Sets}}$ (or on occasion some other category such as $\underset{\sim}{\text{Vec}}$), such that the following hold.

Presheaf-1) Each inclusion of open sets $\mathfrak{W} \subseteq \mathfrak{U}$ corresponds to a *restriction morphism* $\text{res}^{\mathfrak{W}}_{\mathfrak{U}} : f(\mathfrak{U}) \rightarrow f(\mathfrak{W})$ in $\underset{\sim}{\text{Sets}}$.[2]

Presheaf-2) $\text{res}^{\mathfrak{U}}_{\mathfrak{U}}$ is the identity morphism.

Presheaf-3) $\text{res}^{\mathfrak{W}}_{\mathfrak{V}} \circ \text{res}^{\mathfrak{V}}_{\mathfrak{U}} = \text{res}^{\mathfrak{W}}_{\mathfrak{U}}$ for open sets $\mathfrak{W} \subseteq \mathfrak{V} \subseteq \mathfrak{U}$. See Fig. W.2 for meanings of these conditions.

For $\mathfrak{U}$ an open subset of $\mathfrak{X}$ (upon which the topological space $\langle \mathfrak{X}, \tau \rangle$ is based), $\mathcal{E}(\mathfrak{U})$ is the *section* of $\mathcal{E}$ over $\mathfrak{U}$. It is a *global section* if it is over the whole of $\mathfrak{X}$ itself. We carry over use of the fibre bundle notation Γ for sections to presheaves. Moreover, we now write $\Gamma(\mathcal{E}, \mathfrak{U})$, which is a useful notation since the case in which $\mathfrak{U}$ rather than $\mathcal{E}$ is fixed is common. This notion of section indeed generalizes that of fibre bundles as regards being the gateway to a more general range of global methods.

W.3 Sheaves

For a presheaf to additionally be a *sheaf* [92, 128, 167, 511, 713]—historically another notion of Leray's—the following further conditions are required.

Sheaf-1) (*local condition*). Let $\{\mathfrak{U}_\mathsf{C}\}$ be an open cover of an open set $\mathfrak{U}$. If $s_1, s_2 \in \mathcal{E}(\mathfrak{U})$ obey $s_1|_{\mathfrak{U}_\mathsf{C}} = s_2|_{\mathfrak{U}_\mathsf{C}}$ for each $\mathfrak{U}_\mathsf{C}$, then $s_1 = s_2$.

Sheaf-2) (*gluing condition*). Let $s_\mathsf{C} \in \mathcal{E}(\mathfrak{U}_\mathsf{C})$ be sections that agree on their pairwise overlaps $s_\mathsf{C}|_{\mathfrak{U}_\mathsf{C} \cap \mathfrak{U}_\mathsf{D}} = s_\mathsf{D}|_{\mathfrak{U}_\mathsf{C} \cap \mathfrak{U}_\mathsf{D}} \ \forall \, \mathsf{C}, \mathsf{D}$. Then there exists a unique $s \in \mathcal{E}(\mathfrak{U})$ such that $\text{res}^{\mathfrak{U}}_{\mathfrak{U}_\mathsf{C}}(s) = s_\mathsf{C}$. See Fig. W.3 for meanings of these conditions.

---

[1]The first of these is named after mathematician Gerhard Hochschild whereas the second is a more recent development by mathematician Alain Connes.

[2]We subsequently use the standard notation for restriction $s|_{\mathfrak{W}}$ to denote $\text{res}_{\mathfrak{W},\mathfrak{U}}(s)$.

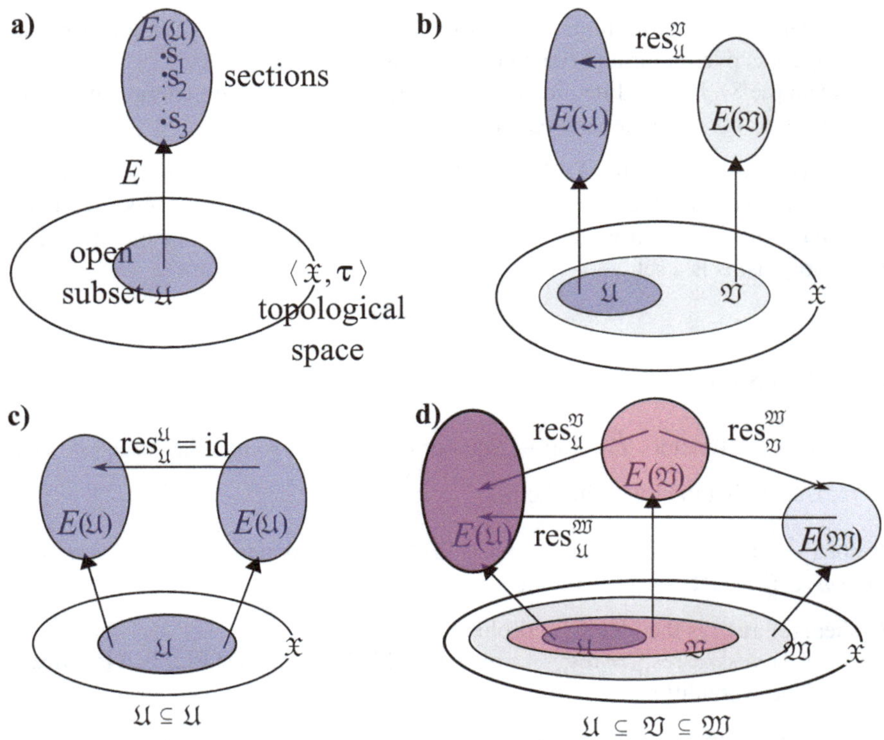

Fig. W.2 **a)** Map from each open subset $\mathfrak{U} \in \langle \mathfrak{X}, \tau \rangle$ to a group of sections over $\mathfrak{U}$. Each section is depicted as a point s_C. **b)** These are equipped with restriction maps $\text{res}^{\mathfrak{V}}_{\mathfrak{U}}$ for each $\mathfrak{U}$ included within each $\mathfrak{V}$. In these last three figures, the (pre)sheaf functor is denoted by a flat backed arrowhead and restriction maps with ordinary arrowheads. **c)** The restriction of an open subset to itself is just the identity. **d)** Restriction is independent of whether one goes via an intermediate subset: the drawn maps form a commuting triangle

Example 1) Each of the sets of: smooth, real-analytic and complex-analytic functions can be viewed as sheaves. This includes in the setting of these being defined over suitable manifolds [891]. The reader might wish to verify this statement and to show that the bounded functions on $\mathbb{C}$ do not form a sheaf.

Example 2) Each of the sets of smooth, real-analytic and complex-analytic sections of a vector bundle form a sheaf [891]. This illustrates how fibre bundles themselves can carry sheaf structure.

The sheaf notion of *section* $s \in E(\mathfrak{U}_C)$ with $s|_{\mathfrak{U}_C} = s_C$ for each C in the cover.

These properties render sheaves adept as a means of formulating more general patching constructs. One can now attach heterogeneous objects to different base space points rather than attaching homogeneous fibres in the formation of a fibre bundle. A simple application of this is to the heterogeneous types of chart involved in the study of a given nontrivial stratified manifold as per Fig. 37.5. See e.g. below and [92, 570, 713] for a wider range of applications to stratified cases of $\mathfrak{q}$ and $\mathfrak{P}$hase.

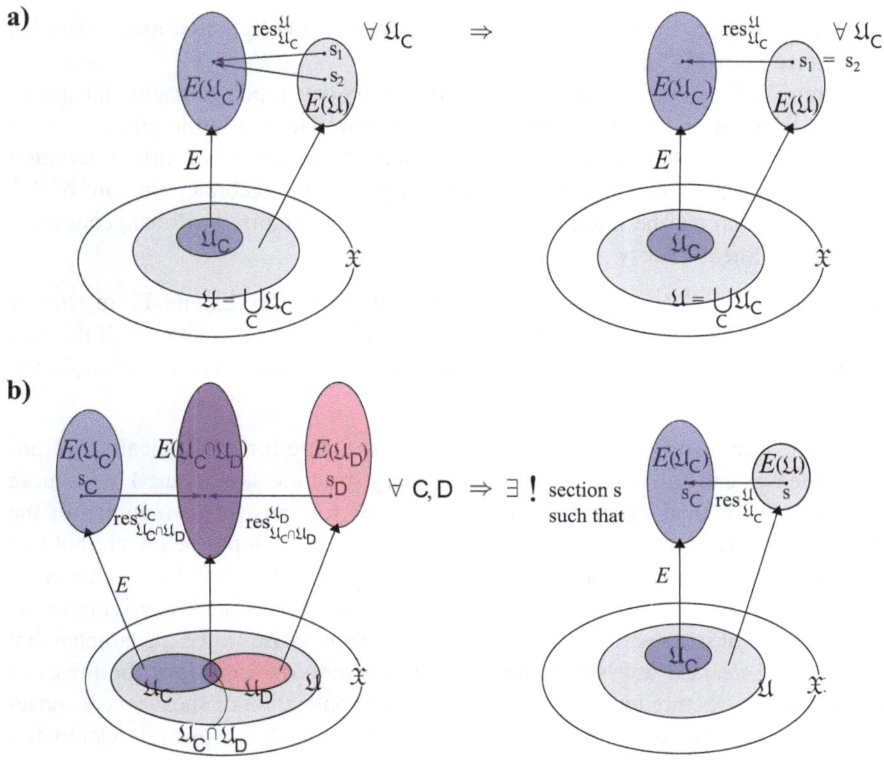

Fig. W.3 Sheaf axioms. **a)** A section is determined by its local restriction in the depicted sense. **b)** A section over all of $\mathfrak{U}$ can be glued together from sections on $\mathfrak{U}_C$ such that $\mathfrak{U} = \bigcup_C \mathfrak{U}_C$ under the depicted circumstances

Sheaves provide a means of formulating obstructions that generalizes the topological treatment using fibre bundles of a number of obstructions that are already familiar in Theoretical Physics. In each case, the notion of section has an associated notion of cohomology concerning obstructions to the presence of global sections. In the case of sheaves, this is given the natural name of *sheaf cohomology*, and indeed turns out to be widely useful from a computational perspective [511]. This ensures the sheaf encodes the topological level of structure of generalized spaces as well as their geometrical structure.

Sheaf cohomology coincides with Čech cohomology on paracompact Hausdorff spaces, (W.1)

which is relevant for the particular simple classes of stratified manifolds considered in Appendix M.7. Furthermore [180],

Sheaf cohomology extends Čech cohomology beyond paracompact Hausdorff spaces. (W.2)

Indeed, this is how sheaf cohomology was first arrived at by noted mathematician Jean-Pierre Serre [785].

Moreover, for all that sheaves were not originally developed with singular spaces in mind, Whitney and Thom's work on the latter proved to be a further place to apply Sheaf Methods. Kreck's subsequent development of stratifolds is a further variation on this theme. The other half of the stratifold pair is an algebraic structure of $\mathfrak{C}^{\infty}$ functions, which can be interpreted as an algebraic structure of global sections in the sense of Sheaf Theory.

Research Project 123)† To what extent does Sheaf Theory extend Fibre Bundle Theory as a *physically useful* theory of global effects and obstructions? If this and Topos Theory do not cover all of the Global Problem of Time's needs, what should the next ports of call be?

Sheaves can also be taken as an underlying structure for topological space notions. Sheaves can be viewed as but a preliminary structure as compared to far more general mathematical structures considered by the foremost mathematician of the 20th century, Alexander Grothendieck [396], of which the topoi below are but one example. In this manner, sheaves and subsequently Grothendieckian mathematics could replace the far simpler collections of subsets, and sets, as lower-lying levels of mathematical structure. Moreover, in this case there ceases to be a guarantee that these are the *deepest* levels of mathematical structure, or, indeed, of the levels of mathematical structure terminating. A very major application of sheaves and subsequent more general structures considered by Grothendieck is Algebraic Geometry. [436].

In Theoretical Physics, sheaves have to date mostly been used in the complex-analytic case in Twistor Theory [459]. They have also been used in modelling algebraic structures of observables [401] and in studying the Complex and Algebraic Geometry arising from String Theory [674, 778]. In this book, we have pointed to a number of further foundational problems which may well become better understood through use of sheaves and the above more general structures.

W.4 Topoi

The original approach leading to these began with equipping categories with the Grothendieck topology, which gives these an open space like structure. The pair formed by a category and the Grothendieck topology thereupon is termed a *site*. Furthermore, a category of sheaves on a site forms a *(Grothendieck) topos*; in this manner, consideration of sheaves offers one route to topoi (Fig. W.4).

Another conceptualization of topoi later considered by mathematicians William Lawvere and Myles Tierney is the *elementary topos*; this furthermore bears some relations to mathematical logic. In this at least a priori much more straightforward approach [611, 612], a topos is envisaged as a category with three extra structures that give it some properties similar to those of sets.

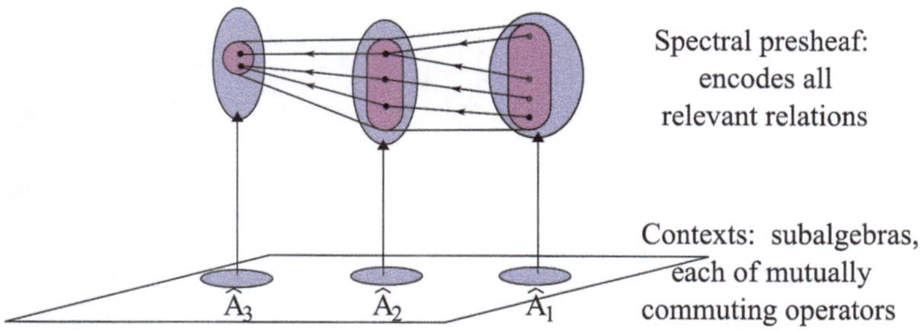

Fig. W.4 Example of a topos presheaf construct, as used in formulating the Kochen–Specker Theorem. This figure illustrates that not all cases within a given floor of the tower can be extended up the tower

1) So-called *finite limits and colimits* in terms of initial and terminal objects and (algebraic-level) pull-backs and push-outs.
2) *Power objects*: for each O_1, O_2 there is an object $O_1^{O_2}$ that acts like the set of functions from O_2 to O_1.
3) *Subobject classifier*: a generalization of the characteristic function from a set to $\{0, 1\}$: 'truth values no and yes', allowing for more general truth values: multi-valued or contextual (terms explained in Epilogue III.B).

See [611, 612] for a basic description of topoi from this point of view, and Isham and Doering's [260] for physical applications based on presheaves on a category $\underset{\sim}{C}$: functors $E : \underset{\sim}{C}^{op} \rightarrow \underset{\sim}{Sets}$, which is often written as $\hat{\underset{\sim}{C}} = \underset{\sim}{Sets}^{\underset{\sim}{C}^{op}}$.

Moreover, being set-like in this manner may render topoi more suitable than categories themselves as regards superceding Set Theory's foundational role.

As regards further breadth of perspectives on what topoi are, see in particular the advanced text [525]. *Indeed, topoi are notorious for being a coincidence of multiple perspectives, in a more mathematically rigorous parallel to how Wheeler argued there to be 'many routes to GR'*. [And yet, two theories being rich in the admission of multiple perspectives is not a sufficient criterion for the two to be related, nor for Nature to realize one just because it realizes the other.]

Let us end by noting that Grothendieck himself viewed topoi as but a step toward further mathematical entities called 'motives', and that he developed numerous further interrelated mathematical structures [66]. On these grounds, future exploration of structures and foundations for use in Physics could benefit from not necessarily being restricted to the sheaf and topos concepts.

Appendix X
Outline of Notation

X.1 For Part I and Unstarred Appendices

We use the italic font for c-numbers and for functions of one variable $F(t)$, and straight font for multivariate functions $\mathrm{F}(\underline{x}, t)$ We also use straight font for functionals based on one variable $\mathrm{F}(t; Q(t)]$, meaning that $\mathrm{d}/\mathrm{d}t$ and $\int \mathrm{d}t$ feature in the dependence. We use calligraphic font for multivariate functionals $\mathcal{F}(\underline{x}, t; \mathcal{Q}(\underline{x}, t)]$, meaning that partial derivatives and integration over more than one variable feature in the dependence. On occasion, operator-valued functionals are allowed, though in such cases they are declared as such. That covers e.g. $\mathcal{F}$ containing derivative operators which act on adjacent expressions to $\mathcal{F}$ rather than solely on other expressions within $\mathcal{F}$ itself.

We correspondingly use d for ordinary derivative, ∂ for partial derivative and δ for functional derivative. I hang 'cov', 'abs' suffixes on these for covariant and absolute derivatives, or I use D in place of ∂ and $\mathcal{D}$ in place of δ in some such contexts. I use $\triangle$ for Laplacians (suitably suffixed to indicate of which type), $\mathbb{D}$ for measures and oversized δ for variational derivative. $* := \mathrm{d}/\mathrm{d}t$ when acting on functions $f(t)$ such as Q^{C}, or $* := \partial/\partial t$ when acting on objects that are functionally dependent on Q^{C}. Underline denotes spatial vector, with double-underline denoting spatial matrix. Bold denotes configuration space vector or tensor. I use the mathfrak font for mathematical spaces, $\mathfrak{ABCDEFGHIJKLMNOPQRSTUVWXYZ}$ $\mathfrak{abcdefghijklmnopqrstuvwxyz}$. I use the bold version of this for spaces of spaces. I also use $\mathcal{ABCDEFGHIJKLMNOPQRSTUVWXYZ}$ (undersized slightly curly calligraphics) for constraints. Finally, I use undersized slanty Latin letters for observables or beables.

E. Anderson, *The Problem of Time*, Fundamental Theories of Physics 190,
DOI 10.1007/978-3-319-58848-3

X.2 Additional Notation for Parts II and III, and Starred Appendices

We now also use the *'finite–field-theoretic portmanteaux'*[1] F of F and F and $\mathscr{F}$ of F and $\mathcal{F}$, denoted by the highly curly font $\mathscr{ABCDEFGHIJKL}$ $\mathscr{MNOPQRSTUVWXYZ}$ $\mathscr{abcdefghijklmnopqrstuvwxyz}$ and the following '*derivative portmanteaux*'. *Ordial derivatives* $\mathbf{d}$: ordinary–partial derivative portmanteau. *Partional derivatives* $\boldsymbol{\partial}$: partial–functional derivative portmanteau.

We use a special font (Large typeface) for such portmanteaux that come integrated over their corresponding notion of space (the action S is a such). The idea is to use these to provide *general cases* of definitions, concepts and results in portmanteau form, to the extent possible. This shorthand embodies the analogies which feature in modelling Field Theories via finite model arenas. These in turn rest on numerous parallels between ordinary and Banach Calculus (Appendix H.2). I use undersized straight Latin letters Q, B, R, T, V, H for Cubert (Chap. 13): how to order Quantization, allotting beables, reduction, allotting a time, use of paths and use of histories. I also use the undersized sans-serif font for conserved quantities. Finally, I use wiggly underlines for categories and oversized italics for functors and natural transformations.

[1]Since GR is central to this book, the Field Theory part is on *curved* spaces. By this later parts of this analogy are more subtle than simply jointly treating particles and flat-spacetime fields, for all that the latter is another plausible use for a distinct portmanteau.

References

1. Abramowitz, M., Stegun, I.A.: Handbook of Mathematical Functions. Dover, New York (1970)
2. Adler, R.L., Konheim, A.G., McAndrew, M.H.: Topological entropy. Trans. Am. Math. Soc. **114**, 309 (1965)
3. Alexander, H.G. (ed.): The Leibnitz–Clark Correspondence. Manchester University Press, Manchester (1956)
4. Alvarez-Gaumé, L., Witten, E.: Gravitational anomalies. Nucl. Phys. B **234**, 269 (1984)
5. Ambjorn, J., Gorlich, A., Jurkiewicz, J., Loll, R., Gizbert-Studnicki, J., Trzesniewski, T.: The semiclassical limit of causal dynamical triangulations. Nucl. Phys. B **849**, 144 (2011). arXiv:1102.3929
6. Ambjorn, J., Jurkiewicz, J., Loll, R.: Causal dynamical triangulations and the quest for quantum gravity? In: Ellis, G., Murugan, J., Weltman, A. (eds.) Foundations of Space and Time. Cambridge University Press, Cambridge (2012). arXiv:1004.0352
7. Anastopoulos, C.: Decoherence and classical predictability of phase space histories. Phys. Rev. E **53**, 4711 (1996). quant-ph/9506031
8. Anastopoulos, C.: Quantum fields in nonstatic background: a histories perspective. J. Math. Phys. **41**, 617 (2000). gr-qc/9903026
9. Anastopoulos, C.: Continuous-time histories: observables, probabilities, phase space. J. Math. Phys. **42**, 3225 (2001). quant-ph/0008052
10. Anastopoulos, C., Hu, B.L.: Intrinsic and fundamental decoherence: issues and problems. Class. Quantum Gravity **25**, 154003 (2008). arXiv:0803.3447
11. Anastopoulos, C., Savvidou, K.N.: Minisuperspace models in histories theory. Class. Quantum Gravity **22**, 1841 (2005). gr-qc/0410131
12. Anderson, J.L.: Relativity principles and the role of coordinates in physics. In: Chiu, H-Y., Hoffmann, W.F. (eds.) Gravitation and Relativity, p. 175. Benjamin, New York (1964)
13. Anderson, J.L.: Principles of Relativity Physics. Academic Press, New York (1967)
14. Anderson, E.: Variations on the seventh route to relativity. Phys. Rev. D **68**, 104001 (2003). gr-qc/0302035
15. Anderson, E.: Strong-coupled relativity without relativity. Gen. Relativ. Gravit. **36**, 255 (2004). gr-qc/0205118
16. Anderson, E.: Leibniz–Mach foundations for GR and fundamental physics. In: Reimer, A. (ed.) Progress in General Relativity and Quantum Cosmology Research. Nova, New York (2005). gr-qc/0405022
17. Anderson, E.: On the recovery of geometrodynamics from two different sets of first principles. Stud. Hist. Philos. Mod. Phys. **38**, 15 (2007). gr-qc/0511070
18. Anderson, E.: Foundations of relational particle dynamics. Class. Quantum Gravity **25**, 025003 (2008). arXiv:0706.3934

E. Anderson, *The Problem of Time*, Fundamental Theories of Physics 190,
DOI 10.1007/978-3-319-58848-3

19. Anderson, E.: Does relationalism alone control geometrodynamics with sources? In: Christiansen, M.N., Rasmussen, T.K. (eds.) Classical and Quantum Gravity Research. Nova, New York (2008). arXiv:0711.0285
20. Anderson, E.: New interpretation of variational principles for gauge theories. I. Cyclic coordinate alternative to ADM split. Class. Quantum Gravity **25**, 175011 (2008). arXiv: 0711.0288
21. Anderson, E.: Records theory. Int. J. Mod. Phys. D **18**, 635 (2009). arXiv:0709.1892
22. Anderson, E.: Relational motivation for conformal operator ordering in quantum cosmology. Class. Quantum Gravity **27**, 045002 (2010). arXiv:0905.3357
23. Anderson, E.: On the semiclassical approach to quantum cosmology. Class. Quantum Gravity **28**, 185008 (2011). arXiv:1101.4916
24. Anderson, E.: The problem of time in quantum gravity. In: Frignanni, V.R. (ed.) Classical and Quantum Gravity: Theory, Analysis and Applications. Nova, New York (2012). arXiv:1009.2157
25. Anderson, E.: Approaching the problem of time with a combined semiclassical-records-histories scheme. Class. Quantum Gravity **29**, 235015 (2012). arXiv:1204.2868
26. Anderson, E.: Problem of time in quantum gravity. Ann. Phys. **524**, 757 (2012). arXiv: 1206.2403
27. Anderson, E.: Quantum cosmology will become a numerical subject. In: Ryutin, R., Petrov, V., Kiselev, V. (eds.) Proceedings of XXIXth International Workshop on High Energy Physics, XXIXth International Workshop on High Energy Physics, Protvino, Moscow Region, Russia, 26–28 June 2013 (2013). arXiv:1306.5812
28. Anderson, E.: Relational quadrilateralland. I. The classical theory. Int. J. Mod. Phys. D **23**, 1450014 (2014). arXiv:1202.4186
29. Anderson, E.: Machian classical and semiclassical emergent time. Class. Quantum Gravity **31**, 025006 (2014). arXiv:1305.4685
30. Anderson, E.: Problem of time: facets and machian strategy. Ann. N.Y. Acad. Sci. **1326**, 42 (2014). arXiv:1306.5816
31. Anderson, E.: Minisuperspace model of machian resolution of problem of time. I. Isotropic case. Gen. Relativ. Gravit. **46**, 1708 (2014). arXiv:1307.1916
32. Anderson, E.: Beables/observables in classical and quantum gravity. SIGMA **10**, 092 (2014). arXiv:1312.6073
33. Anderson, E.: Kendall's shape statistics as a classical realization of Barbour-type timeless records theory approach to quantum gravity. Stud. Hist. Philos. Mod. Phys. **51**, 1 (2015). arXiv:1307.1923
34. Anderson, E.: Origin of structure in the universe: quantum cosmology reconsidered. Gen. Relativ. Gravit. **47**, 101 (2015). arXiv:1501.02443
35. Anderson, E.: Problem of time in slightly inhomogeneous cosmology. Class. Quantum Gravity **33**, 145004 (2016). arXiv:1403.7583
36. Anderson, E.: Six new mechanics corresponding to further shape theories. Int. J. Mod. Phys. D **25**, 1650044 (2016). arXiv:1505.00488
37. Anderson, E.: The problem of time and quantum cosmology in the relational particle mechanics arena. arXiv:1111.1472
38. Anderson, E.: Relationalism. arXiv:1205.1256
39. Anderson, E.: Machian time is to be abstracted from what change? arXiv:1209.1266
40. Anderson, E.: Background independence. arXiv:1310.1524
41. Anderson, E.: Problem of time and background independence: the individual facets. arXiv: 1409.4117
42. Anderson, E.: Where to apply relationalism, write-up of Invited Seminar at the Conferences at Fredericton 2014. arXiv:1411.4316
43. Anderson, E.: Spaces of spaces. arXiv:1412.0239
44. Anderson, E.: TRiPoD (Temporal Relationalism implementing Principles of Dynamics). arXiv:1501.07822
45. Anderson, E.: Configuration spaces in fundamental physics. arXiv:1503.01507

46. Anderson, E.: Spherical relationalism. arXiv:1505.02448
47. Anderson, E.: Explicit partial and functional differential equations for beables or observables. arXiv:1505.03551
48. Anderson, E.: "Shape dynamics": foundations reassessed. arXiv:1505.03553
49. Anderson, E.: Problem of time: temporal relationalism compatibility of other local classical facets completed. arXiv:1506.03517
50. Anderson, E.: Kinematical quantization. arXiv:1604.06248
51. Anderson, E.: On types of observables in constrained theories. arXiv:1604.05415
52. Anderson, E.: Reduced form of slightly inhomogeneous semiclassical quantum cosmology (forthcoming)
53. Anderson, E.: Fibre bundles and algebra analysis of rendering a group physically meaningless (forthcoming)
54. Anderson, E.: Geometry of spacetime reconstruction (forthcoming)
55. Anderson, E.: Limitations on problem of time resolutions for theories with further levels of background independence (2017, forthcoming)
56. Anderson, E.: The smallest shape spaces. I. Shape theory posed, with example of 3 points on the line (forthcoming)
57. Anderson, E.: The smallest shape spaces. III. Triangles in the plane and in 3-*d* (forthcoming)
58. Anderson, E., Barbour, J.B.: Interacting vector fields in relativity without relativity. Class. Quantum Gravity **19**, 3249 (2002). gr-qc/0201092
59. Anderson, E., Franzen, A.: Quantum cosmological metroland model. Class. Quantum Gravity **27**, 045009 (2010). arXiv:0909.2436
60. Anderson, J.L., Gautreau, R.: Operational formulation of the principle of equivalence. Phys. Rev. **185**, 1656 (1969)
61. Anderson, E., Kneller, S.A.R.: Relational quadrilateralland. II. The quantum theory. Int. J. Mod. Phys. D **23**, 1450052 (2014). arXiv:1303.5645
62. Anderson, E., Mercati, F.: Classical machian resolution of the spacetime construction problem. arXiv:1311.6541
63. Anderson, I.M., Torre, C.G.: Symmetries of the Einstein equations. Phys. Rev. Lett. **70**, 3525 (1993). gr-qc/9302033
64. Anderson, E., Barbour, J.B., Foster, B.Z., ó Murchadha, N.: Scale-invariant gravity: geometrodynamics. Class. Quantum Gravity **20**, 157 (2003). gr-qc/0211022
65. Anderson, E., Barbour, J.B., Foster, B.Z., Kelleher, B., ó Murchadha, N.: The physical gravitational degrees of freedom. Class. Quantum Gravity **22**, 1795 (2005). gr-qc/0407104
66. Anel, M., Catren, G. (eds.): New Spaces in Mathematics and Physics, Proceedings of Paris Conference 2015 (forthcoming), for an introduction. https://ncatlab.org/nlab/show/New+Spaces+for+Mathematics+and+Physics
67. Appelquist, T., Chodos, A., Freund, P.G.O.: Modern Kaluza–Klein Theories. Addison–Wesley, Reading (1987)
68. Armstrong, M.A.: Basic Topology. Springer, New York (1983)
69. Armstrong, M.A.: Groups and Symmetry. Springer, New York (1988)
70. Arnol'd, V.I.: Mathematical Methods of Classical Mechanics. Springer, New York (1978)
71. Arnol'd, V.I.: Catastrophe Theory. Springer, Berlin (1984)
72. Arnol'd, V.I.: Ordinary Differential Equations. Springer, Berlin (1992)
73. Arnowitt, R., Deser, S., Misner, C.W.: The dynamics of general relativity. In: Witten, L. (ed.) Gravitation: An Introduction to Current Research. Wiley, New York (1962). arXiv: gr-qc/0405109
74. Ashby, N.: Relativity in the global positioning system. Living Rev. Relativ. **6**, 1 (2003)
75. Ashtekar, A.: Lectures on Nonperturbative Canonical Gravity. World Scientific, Singapore (1991)
76. Ashtekar, A., Horowitz, G.: Phase space of general relativity revisited: a canonical choice of time and simplification of the Hamiltonian. J. Math. Phys. **25**, 1473 (1984)
77. Ashtekar, A., Lewandowski, J.: Representation theory of analytic holonomy $\mathfrak{C}^*$ algebras. In: Baez, J.C. (ed.) Knots and Quantum Gravity, Proceedings of Workshop held at UC River-

side 14–16 May 1993. Oxford Lecture Series in Mathematics and Its Applications, vol. 1. Clarendon/Oxford University Press, Oxford/New York (1994). arXiv:gr-qc/9311010
78. Ashtekar, A., Lewandowski, J.: Background independent quantum gravity: a status report. Class. Quantum Gravity **21**, R53 (2004). gr-qc/0404018
79. Ashtekar, A., Tate, R.S., Uggla, C.: Minisuperspaces: observables and quantization. Int. J. Mod. Phys. D **2**, 15 (1993). gr-qc/9302027
80. Ashtekar, A., Reuter, M., Rovelli, C.: From general relativity to quantum gravity. arXiv: 1408.4336
81. Atkins, P.W., Friedman, R.S.: Molecular Quantum Mechanics. Oxford University Press, New York (1997)
82. Audoin, C., Guinot, B.: The Measurement of Time. Time, Frequency and the Atomic Clock. Cambridge University Press, Cambridge (2001)
83. Bachelard, G.: The Dialectic of Duration (1950), in French; an English translation reprint is: Clinamen Press, Manchester (2000)
84. Bacry, H., Lévy-Leblond, J.-M.: Possible kinematics. J. Math. Phys. **9**, 1605 (1968)
85. Bagger, J., Lambert, N.: Modelling multiple M2's. Phys. Rev. D **75**, 045020 (2007). hep-th/0611108
86. Bagger, J., Lambert, N.: Comments on multiple M2-branes. Phys. Rev. D **77**, 065008 (2008). arXiv:0712.3738
87. Bagger, J., Lambert, N.: Three-algebras and N = 6 Chern-Simons gauge theories. Phys. Rev. D **79**, 025002 (2009). arXiv:0807.0163
88. Bahder, T.B.: Relativity of GPS measurement. Phys. Rev. D **68**, 063005 (2003). gr-qc/0306076
89. Baierlein, R.F., Sharp, D.H., Wheeler, J.A.: Three-dimensional geometry as carrier of information about time. Phys. Rev. **126**, 1864 (1962)
90. Bain, L.J., Engelhardt, M.: Introduction to Probability and Mathematical Statistics. Duxbury Press, Pacific Grove (2000)
91. Balsz, N., Jennings, B.K.: Wigner's function and other distribution functions in mock phase spaces. Phys. Rep. **104**, 347 (1984)
92. Banagl, M.: Topological Invariants of Stratified Spaces. Springer, Berlin (2007)
93. Banks, T.: TCP, quantum gravity, the cosmological constant and all that. Nucl. Phys. B **249**, 322 (1985)
94. Barbero, J.F.: Real Ashtekar variables for Lorentzian space-times. Phys. Rev. D **51**, 5507 (1995). gr-qc/9410014
95. Barbero, J.F., Villaseñor, E.J.S.: Quantization of midisuperspace models. Living Rev. Relativ. **13**, 6 (2010). arXiv:1010.1637
96. Barbour, J.B.: Absolute or Relative Motion? Vol 1: The Discovery of Dynamics. Cambridge University Press, Cambridge (1989)
97. Barbour, J.B.: Time and complex numbers in canonical quantum gravity. Phys. Rev. D **47**, 5422 (1993)
98. Barbour, J.B.: The timelessness of quantum gravity. I. The evidence from the classical theory. Class. Quantum Gravity **11**, 2853 (1994)
99. Barbour, J.B.: The timelessness of quantum gravity. II. The appearance of dynamics in static configurations. Class. Quantum Gravity **11**, 2875 (1994)
100. Barbour, J.B.: GR as a perfectly machian theory. In: Barbour, J.B., Pfister, H. (eds.) Mach's Principle: From Newton's Bucket to Quantum Gravity. Birkhäuser, Boston (1995)
101. Barbour, J.B.: The End of Time. Oxford University Press, Oxford (1999)
102. Barbour, J.B.: Scale-invariant gravity: particle dynamics. Class. Quantum Gravity **20**, 1543 (2003). gr-qc/0211021
103. Barbour, J.B.: Shape dynamics. An introduction. In: Proceedings of the Conference Quantum Field Theory and Gravity, Regensburg (2010). arXiv:1105.0183
104. Barbour, J.B.: The nature of time, fqxi 'nature of time' essay competition: juried first prize. arXiv:0903.3489

105. Barbour, J.B., Bertotti, B.: Mach's principle and the structure of dynamical theories. Proc. R. Soc. Lond. A **382**, 295 (1982)
106. Barbour, J.B., Foster, B.Z.: Constraints and Gauge Transformations: Dirac's Theorem is not Always Valid. arXiv:0808.1223
107. Barbour, J.B., ó Murchadha, N.: Classical and quantum gravity on conformal superspace. gr-qc/9911071
108. Barbour, J.B., ó Murchadha, N.: Conformal superspace: the configuration space of general relativity. arXiv:1009.3559
109. Barbour, J.B., Foster, B.Z., ó Murchadha, N.: Relativity without relativity. Class. Quantum Gravity **19**, 3217 (2002). gr-qc/0012089
110. Bardeen, J.: Gauge-invariant cosmological perturbations. Phys. Rev. D **22**, 1882 (1980)
111. Barrett, J.W., Crane, L.: Relativistic spin networks and quantum gravity. J. Math. Phys. **39**, 3296 (1998)
112. Barrett, J.W., Dowdall, R.J., Fairbairn, W.J., Hellmann, F., Pereira, R.: Lorentzian spin foam amplitudes: graphical calculus and asymptotics. Class. Quantum Gravity **27**, 165009 (2010). arXiv:0907.2440
113. Barrow, J.: Wigner inequalities for a black hole. Phys. Rev. D **54**, 6563 (1996)
114. Bartnik, R.: Remarks on cosmological space-times and constant mean curvature surfaces. Commun. Math. Phys. **117**, 615 (1988)
115. Bartnik, R., Fodor, G.: On the restricted validity of the thin-sandwich conjecture. Phys. Rev. D **48**, 3596 (1993)
116. Bartnik, R., Isenberg, J.: The constraint equations. In: Chruściel, P., Friedrich, H. (eds.) The Einstein Equations and the Large Scale Behavior of Gravitational Fields. Birkhäuser, Basel (2004)
117. Barvinsky, A.O.: Operator ordering in theories subject to constraints of gravitational type. Class. Quantum Gravity **10**, 1985 (1993)
118. Barvinsky, A.O.: Unitary approach to quantum cosmology. Phys. Rep. **230**, 237 (1993)
119. Barvinsky, A.O., Kiefer, C.: Wheeler–DeWitt equation and Feynman diagrams. Nucl. Phys. B **526**, 509 (1998). gr-qc/9711037
120. Barvinsky, A.O., Krykhtin, V.: Dirac and BFV quantization methods in the 1-loop approximation: closure of the quantum constraint algebra and the conserved inner product. Class. Quantum Gravity **10**, 1957 (1993)
121. Batalin, I.A., Tyutin, I.V.: Existence theorem for the effective gauge algebra in the generalized canonical formalism with Abelian conversion of second-class constraints. Int. J. Mod. Phys. A **6**, 3255 (1991)
122. Baumann, D.: TASI lectures on inflation. arXiv:0907.5424
123. Baumgarte, T.W., Shapiro, S.L.: Numerical relativity and compact binaries. Phys. Rep. **376**, 41 (2003). gr-qc/0211028
124. Belasco, E.P., Ohanian, H.C.: Initial conditions in general relativity: lapse and shift formulation. J. Math. Phys. **10**, 1503 (1969)
125. Belinskii, V.A., Khalatnikov, I.M., Lifshitz, E.M.: Oscillatory approach to a singular point in the relativistic cosmology. Adv. Phys. **19**, 525 (1970)
126. Bell, J.S.: Quantum mechanics for cosmologists. In: Isham, C.J., Penrose, R., Sciama, D.W. (eds.) Quantum Gravity 2. A Second Oxford Symposium. Clarendon, Oxford (1981)
127. Bell, J.S.: The theory of local beables. In: Speakable and Unspeakable in Quantum Mechanics, p. 52. Cambridge University Press, Cambridge (1987)
128. Bell, S.e.g.J.L.: Toposes and Local Set Theories. Dover, New York (2008)
129. Ben-Naim, A.: A Farewell to Entropy: Statistical Thermodynamics Based on Information. World Scientific, Singapore (2008)
130. Bengtsson, I., Zyczkowski, K.: Geometry of Quantum States: An Introduction to Quantum Entanglement. Cambridge University Press, Cambridge (2006)
131. Berger, B.K.: Quantum graviton creation in a model universe. Ann. Phys. (N. Y.) **83**, 458 (1974)

132. Berger, B.K.: Quantum effects in the gowdy $\mathbb{T}^3$ cosmology. Ann. Phys. (N. Y.) **156**, 155 (1984)
133. Bergmann, P.G.: “Gauge-invariant” variables in general relativity. Phys. Rev. **124**, 274 (1961)
134. Bergmann, P.G., Komar, A.: The coordinate group symmetries of general relativity. Int. J. Theor. Phys. **5**, 15 (1972)
135. Bergson, H.: The Creative Mind (1907); reprinted by Dover, New York (1998)
136. Berman, D.S., Perry, M.J.: Generalized geometry and M theory. J. High Energy Phys. **1106**, 074 (2011). arXiv:1008.1763
137. Bern, Z.: Perturbative quantum gravity and its relation to gauge theory. Living Rev. Relativ. **5** (2002)
138. Bern, Z., Dixon, L.J., Roiban, R.: Is N = 8 supergravity ultraviolet finite? Phys. Lett. B **644**, 265 (2007). hep-th/0611086
139. Bertlmann, R.A.: In: Anomalies in Quantum Field Theory. Clarendon, Oxford (1996)
140. Bialynicki-Birula, I., Cieplak, M., Kaminski, J.: Theory of Quanta. Oxford University Press, New York (1992)
141. Bičak, J., Kuchař, K.V.: Null dust in canonical gravity. Phys. Rev. D **56**, 4878 (1997). gr-qc/9704053
142. Binz, E., Śniatychi, J., Fischer, H.: Geometry of Classical Fields. Elsevier, Amsterdam (1988)
143. Birrell, N.D., Davies, P.C.W.: Quantum Fields in Curved Space. Cambridge University Press, Cambridge (1982)
144. Bjorken, J.D., Drell, S.D.: Relativistic Quantum Fields. McGraw–Hill, New York (1965)
145. Blackmore, J.T.: Ernst Mach: His Work, Life and Influence. University of California Press, Berkeley (1972)
146. Blaziot, J.-P., Ripka, G.: Quantum Theory of Finite Systems. MIT Press, Cambridge (1986)
147. Bleecker, D.: Gauge Theory and Variational Principles. Dover, New York (2001)
148. Blumenhagen, R., Gmeiner, F., Honecker, G., Lust, D., Weigand, T.: The statistics of supersymmetric D-brane models. Nucl. Phys. B **713**, 83 (2005). hep-th/0411173
149. Blyth, W.F., Isham, C.J.: Quantization of a Friedmann universe filled with a scalar field. Phys. Rev. D **11**, 768 (1975)
150. Bohr, N., Rosenfeld, L.: Zur Frage des Messbarkeit der Electromagnetischen Feldgrössen [On the question of measurability of the electromagnetic field quantities]. K. Dan. Vidensk. **78**, 94 (1933)
151. Bojowald, M.: The semiclassical limit of loop quantum cosmology. Class. Quantum Gravity **18**, L109 (2001). gr-qc/0105113
152. Bojowald, M.: Loop quantum cosmology. Living Rev. Relativ. **8**, 11 (2005). gr-qc/0601085
153. Bojowald, M.: Elements of loop quantum cosmology. In: Ashtekar, A. (ed.) 100 Years of Relativity—Space-Time Structure: Einstein and Beyond. World Scientific, Singapore (2005). arXiv:gr-qc/0505057
154. Bojowald, M.: Canonical Gravity and Applications: Cosmology, Black Holes, and Quantum Gravity. Cambridge University Press, Cambridge (2011)
155. Bojowald, M.: A loop quantum multiverse? AIP Conf. Proc. **15**, 21 (2012). arXiv:1212.5150
156. Bojowald, M., Hossain, G.M., Kagan, M., Shankaranarayanan, S.: Gauge invariant cosmological perturbation equations with corrections from loop quantum gravity. Phys. Rev. D **79**, 43505 (2009). arXiv:0811.1572
157. Bojowald, M., Hoehn, P.A., Tsobanjan, A.: An effective approach to the problem of time. Class. Quantum Gravity **28**, 035006 (2011). arXiv:1009.5953
158. Bojowald, M., Hoehn, P.A., Tsobanjan, A.: Effective approach to the problem of time: general features and examples. Phys. Rev. D **83**, 125023 (2011). arXiv:1011.3040
159. Bollobás, B.: Modern Graph Theory. Springer, New York (1998)
160. Bondi, H.: Relativity. Rep. Prog. Phys. **22**, 97 (1959)
161. Borde, A., Dowker, H.F., Garcia, R.S., Sorkin, R.A., Surya, S.: Causal continuity in degenerate spacetimes. Class. Quantum Gravity **16**, 3457 (1999)

162. Boulware, D.: Quantization of higher derivative theories of gravity. In: Christensen, S. (ed.) Quantum Theory of Gravity. Hilger, Bristol (1984)
163. Boulware, D., Deser, S.: Classical general relativity derived from quantum gravity. Ann. Phys. **89**, 193 (1975)
164. Bousso, R.: The holographic principle. Rev. Mod. Phys. **74**, 825 (2002). hep-th/0203101
165. Brack, M., Bhaduri, R.: Semiclassical Physics. Addison–Wesley, Reading (1997)
166. Brechtken-Manderscheid, U.: Introduction to the Calculus of Variations. T.J. Press, Padstow (1991), English translation of 1983 German text
167. Bredon, G.E.: Sheaf Theory. McGraw–Hill, New York (1997)
168. Bridgman, P.W.: The Logic of Modern Physics. MacMillan, New York (1927)
169. Bridgman, P.W.: A Sophisticate's Primer of Relativity. Routledge, London (1963)
170. Briggs, J.S., Rost, J.M.: On the derivation of the time-dependent equation of Schrödinger. Found. Phys. **31**, 693 (2001). quant-ph/9902035
171. Broad, C.D.: Scientific Thought. Routledge, London (1923)
172. Broadbent, S.: Stimulating the Ley hunter. J. R. Stat. Soc. A **143**, 109 (1980)
173. Bronstein, M.P.: Quantentheories Schwacher Gravitationsfelder [Quantum theories of the weak gravitational field]. Phys. Z. Sowjetunion **9**, 140 (1936)
174. Brout, R., Venturi, G.: Time in semiclassical gravity. Phys. Rev. D **39**, 2436 (1989)
175. Brown, J.D., Kuchař, K.V.: Dust as a standard of space and time in canonical quantum gravity. Phys. Rev. D **51**, 5600 (1995). gr-qc/9409001
176. Brown, J.D., Marolf, D.: On relativistic material reference systems. Phys. Rev. D **53**, 1835 (1996). gr-qc/9509026
177. Brown, J.D., York, J.W.: Jacobi's action and the recovery of time in general relativity. Phys. Rev. D **40**, 3312 (1989)
178. Brunetti, R., Fredenhagen, K.: Time of occurrence observable in quantum mechanics. Phys. Rev. A **66**, 044101 (2002)
179. Brunnemann, J., Fleischhack, C.: On the configuration spaces of homogeneous loop quantum cosmology and loop quantum gravity. arXiv:0709.1621
180. Brylinski, J-L.: Loop Spaces, Characteristic Classes and Geometric Quantization. Springer, New York (2007)
181. Budney, A.R.: Little cubes and long knots. Topology **46**, 1 (2007)
182. Burch, A.: Histories electromagnetism. J. Math. Phys. **45**, 2153 (2004). gr-qc/0311092
183. Burd, A., Tavakol, R.: Invariant Lyapunov exponents and chaos in cosmology. Phys. Rev. D **47**, 5336 (1993)
184. Burkill, J.C., Burkill, H.: A Second Course in Mathematical Analysis. Cambridge University Press, Cambridge (1970)
185. Busch, P.: The time–uncertainty relation, in [669]
186. Butterfield, J., Isham, C.J.: On the emergence of time in quantum gravity. In: Butterfield, J. (ed.) The Arguments of Time. Oxford University Press, Oxford (1999). gr-qc/9901024
187. Butterfield, J., Isham, C.J.: Some possible roles for topos theory in quantum theory and quantum gravity. Found. Phys. **30**, 1707 (2000). gr-qc/9910005
188. Butterfield, J.N., Isham, C.J.: Spacetime and the philosophical challenge of quantum gravity. In: Callender, C., Huggett, N. (eds.) Physics Meets Philosophy at the Planck Scale. Cambridge University Press, Cambridge (2000). gr-qc/9903072
189. Calzetta, E., Hu, B.L.: Wigner distribution function and phase-space formulation of quantum cosmology. Phys. Rev. D **40**, 380 (1989)
190. Cannas da Silva, A., Weinstein, A.: Geometric Models for Noncommutative Algebras. Berkeley Mathematics Lecture Notes Series. Am. Math. Soc., Providence (1999)
191. Carlip, S.: Observables, gauge invariance, and time in $2 + 1$ dimensional quantum gravity. Phys. Rev. D **42**, 2647 (1990)
192. Carlip, S.: Measuring the metric in $(2 + 1)$-dimensional quantum gravity. Class. Quantum Gravity **8**, 5 (1991)
193. Carlip, S.: Quantum Gravity in $2 + 1$ Dimensions. Cambridge University Press, Cambridge (1998)

194. Carlip, S.: Quantum gravity: a progress report. Rep. Prog. Phys. **64**, 885 (2001). gr-qc/0108040
195. Carlip, S.: Challenges for emergent gravity. arXiv:1207.2504
196. Carroll, L.: Through the Looking Glass. MacMillan & Co., London (1871)
197. Carruthers, P., Zachariasen, F.: Quantum collision theory with phase-space distributions. Rev. Mod. Phys. **55**, 245 (1984)
198. Casalbuoni, R.: On the quantization of systems with anticommuting variables. Nuovo Cimento A **33**, 115 (1976)
199. Castagnino, M.: The mathematical structure of superspace as a consequence of time asymmetry. Phys. Rev. D **57**, 750 (1998). gr-qc/9604034
200. Castagnino, M., Laura, R.: Functional approach to quantum decoherence and the classical final limit: the Mott and cosmological problems. Int. J. Theor. Phys. **39**, 1737 (2000). gr-qc/0006012
201. Cheng, X., Shen, Z.: Finsler Geometry. An Approach via Randers Spaces. Springer, Berlin (2012)
202. Choptuik, M.W.: Numerical analysis for numerical relativists. In: Lecture at the VII Mexican School on Gravitation and Mathematical Physics, Playadel Carmen, Mexico, 26 November–1 December 2006 (2006). Available at http://laplace.physics.ubc.ca/People/matt/Teaching/06Mexico/
203. Choquet-Bruhat, Y.: Maximal submanifolds and submanifolds with constant mean extrinsic curvature of a Lorentzian manifold. Ann. Sc. Norm. Super. Pisa **3**, 361 (1976)
204. Choquet-Bruhat, Y.: General Relativity and the Einstein Equations. Oxford University Press, New York (2009)
205. Choquet-Bruhat, Y., DeWitt-Morette, C.: Analysis, Manifolds and Physics, vol. 2. Elsevier, Amsterdam (2000)
206. Choquet-Bruhat, Y., York, J.W.: In: Held, A. (ed.) General Relativity and Gravitation, vol. 1. Plenum Press, New York (1980)
207. Choquet-Bruhat, Y., DeWitt-Morette, C., Dillard-Bleick, M.: Analysis, Manifolds and Physics, vol. 1. Elsevier, Amsterdam (1982)
208. Christodoulou, D.: The chronos principle. Nuovo Cimento B **26**, 67 (1975)
209. Christodoulou, D.: The chronos principle and the interactions of fields of spin 0 and 1. In: Proceedings, Marcel Grossmann Meeting on General Relativity, Trieste, 1975. Oxford University Press, Oxford (1977)
210. Chruściel, P.T.: arXiv:1110.6706
211. Clemence, G.M.: Astronomical time. Rev. Mod. Phys. **29**, 2 (1957)
212. Clifton, T., Ellis, G.F.R., Tavakol, R.: A gravitational entropy proposal. Class. Quantum Gravity **30**, 125009 (2013). arXiv:1303.5612
213. Cohn, P.M.: Classic Algebra. Wiley, Chichester (2000)
214. Cohn, P.M.: Further Algebra and Applications. Springer, London (2003)
215. Coles, P., Lucchin, F.: Cosmology. The Origin and Evolution of Cosmic Structure. Wiley, Chichester (1995)
216. Connes, A.: Noncommutative Geometry. math/0011193
217. Conrady, F., Freidel, L.: On the semiclassical limit of 4d spin foam models. Phys. Rev. D **78**, 104023 (2008). arXiv:0809.2280
218. Cook, G.B., Whiting, B.F.: Approximate Killing vectors on $\mathbb{S}^2$. Phys. Rev. D **76**, 041501 (2007). arXiv:0706.0199
219. Cornish, N., Spergel, D., Starkman, G.: Circles in the sky: finding topology with the microwave background radiation. Class. Quantum Gravity **15**, 2657 (1998). astro-ph/9801212
220. Courant, R., Hilbert, D.: Methods of Mathematical Physics, vols. 1 and 2. Wiley, Chichester (1989)
221. Cover, T.M., Thomas, J.A.: Elements of Information Theory. Wiley, New York (1991)
222. Coxeter, H.S.M.: Introduction to Geometry. Wiley, New York (1989)
223. Crainic, M.: Differentiable and algebroid cohomology, van Est isomorphisms, and characteristic classes. Math. Helv. **78**, 681 (2003)

224. Crane, L.: Clock and category; is quantum gravity algebraic. J. Math. Phys. **36**, 6180 (1995). arXiv:gr-qc/9504038
225. Crane, L.: Categorical physics. hep-th/9301061
226. Cutler, C., Thorne, K.S.: An overview of gravitational-wave sources. In: Bishop, N.T., Maharaj, S.D. (eds.) Proceedings of GR16, Durban, South Africa, 2001 World Scientific, Singapore (2001). gr-qc/0204090
227. Darmois, G.: Les Équations de la Gravitation Einsteinienne [The equations of Einstein's gravity]. Méml. Sci. Math. **25** (1927)
228. Datta, D.P.: Notes on the Born-Oppenheimer approach in a closed dynamical system. Class. Quantum Gravity **14**, 2825 (1997). gr-qc/9706077
229. Davey, J., Hanany, A., Seong, R.-K.: Counting orbifolds. J. High Energy Phys. **06**, 010 (2010). arXiv:1002.3609
230. De Felice, A., Tsujikawa, S.: $f(R)$ theories. Living Rev. Relativ. **13**, 3 (2010)
231. de Sitter, W.: On the secular accelerations and the fluctuations of the longitudes of the Moon, the Sun, Mercury and Venus. Bull. Astron. Inst. Neth. **4**, 21 (1927)
232. D'Eath, P.D.: Supersymmetric Quantum Cosmology. Cambridge University Press, Cambridge (1996)
233. Deser, S., Kay, J.H., Stelle, K.S.: Hamiltonian formulation of supergravity. Phys. Rev. **16**, 2448 (1977)
234. DeWitt, B.S.: Dynamical theory in curved spaces. [A review of the classical and quantum action principles]. Rev. Mod. Phys. **29**, 377 (1957)
235. DeWitt, B.S.: The quantization of geometry. In: Witten, L. (ed.) Gravitation: An Introduction to Current Research. Wiley, New York (1962)
236. DeWitt, B.S.: Gravity: a universal regulator? Phys. Rev. Lett. **13**, 114 (1964)
237. DeWitt, B.S.: Quantum theory of gravity. I. The canonical theory. Phys. Rev. **160**, 1113 (1967)
238. DeWitt, B.S.: Quantum theory of gravity. II. The manifestly covariant theory. Phys. Rev. **160**, 1195 (1967)
239. DeWitt, B.S.: Quantum theory of gravity. III. Applications of the covariant theory. Phys. Rev. **160**, 1239 (1967)
240. DeWitt, B.S.: Spacetime as a sheaf of geodesics in superspace. In: Carmeli, M., Fickler, S.I., Witten, L. (eds.) Relativity, Proceedings of the Relativity Conference in the Midwest, Cincinnati, Ohio, 2–6 June 1969 Plenum, New York (1970)
241. DeWitt, B.S.: Spacetime approach to quantum field theory. In: DeWitt, B., Stora, R. (eds.) Relativity, Groups and Topology II. North-Holland, Amsterdam (1984)
242. DeWitt, B.S.: The Global Approach to Quantum Field Theory, vols. 1 and 2. Oxford University Press, New York (2003)
243. d'Inverno, R.A., Smallwood, J.: Covariant $2+2$ formulation of the initial-value problem in general relativity. Phys. Rev. D **22**, 1233 (1980)
244. Dirac, P.A.M.: The Principles of Quantum Mechanics. Oxford University Press, New York (1930)
245. Dirac, P.A.M.: The theory of magnetic poles. Phys. Rev. **74**, 817 (1948)
246. Dirac, P.A.M.: Quantum theory of localizable dynamical systems. Phys. Rev. **73**, 1092 (1948)
247. Dirac, P.A.M.: Forms of relativistic dynamics. Rev. Mod. Phys. **21**, 392 (1949)
248. Dirac, P.A.M.: The Hamiltonian form of field dynamics. Can. J. Math. **3**, 1 (1951)
249. Dirac, P.A.M.: The theory of gravitation in Hamiltonian form. Proc. R. Soc. Lond. A **246**, 333 (1958)
250. Dirac, P.A.M.: Lectures on Quantum Mechanics. Yeshiva University, New York (1964)
251. Dittrich, B.: Partial and complete observables for canonical general relativity. Class. Quantum Gravity **23**, 6155 (2006). gr-qc/0507106
252. Dittrich, B.: Partial and complete observables for Hamiltonian constrained systems. Gen. Relativ. Gravit. **38**, 1891 (2007). gr-qc/0411013

253. Dittrich, B., Hoehn, P.A.: Constraint analysis for variational discrete systems. J. Math. Phys. **54**, 093505 (2013). arXiv:1303.4294
254. Dittrich, B., Thiemann, T.: Testing the master constraint programme for loop quantum gravity I. General framework. Class. Quantum Gravity **23**, 1025 (2004). gr-qc/0411138
255. Dittrich, B., Hoehn, P.A., Koslowski, T.A., Nelson, M.I.: Chaos, Dirac observables and constraint quantization. arXiv:1508.01947
256. Doering, A., Isham, C.J.: A topos foundation for theories of physics: I. Formal languages for physics. J. Math. Phys. **49**, 053515 (2008). quant-ph/0703060
257. Doering, A., Isham, C.J.: A topos foundation for theories of physics: II. Daseinisation and the liberation of quantum theory. J. Math. Phys. **49**, 053516 (2008). quant-ph/0703062
258. Doering, A., Isham, C.J.: A topos foundation for theories of physics: III. The representation of physical quantities with arrows. J. Math. Phys. **49**, 053517 (2008). quant-ph/0703064
259. Doering, A., Isham, C.J.: A topos foundation for theories of physics: IV. Categories of systems. J. Math. Phys. **49**, 053518 (2008). quant-ph/0703066
260. Doering, A., Isham, C.: 'What is a thing?': topos theory in the foundations of physics. In: Coecke, R. (ed.) New Structures for Physics. Springer Lecture Notes in Physics, vol. 813. Springer, Heidelberg (2011). arXiv:0803.0417
261. Doering, A., Isham, C.J.: Classical and quantum probabilities as truth values. J. Math. Phys. **53**, 032101 (2012). arXiv:1102.2213
262. Douglas, M.R.: The statistics of string/M-theory vacua. J. High Energy Phys. **0305**, 046 (2003). hep-th/0303194
263. Dowker, H.F.: The birth of spacetime atoms as the passage of time. arXiv:1405.3492
264. Dowker, H.F., Kent, A.: Properties of consistent histories. Phys. Rev. Lett. **75**, 3038 (1995). gr-qc/9409037
265. Dowker, H.F., Kent, A.: On the consistent histories approach to quantum mechanics. J. Stat. Phys. **82**, 1575 (1996). gr-qc/9412067
266. Dragt, A.J.: Classification of three-particle states according to SU_3. J. Math. Phys. **6**, 533 (1965)
267. Dudal, D., Sorella, S.P., Vandersickel, N., Verschelde, H.: Gribov no-pole condition, Zwanziger horizon function, Kugo-Ojima confinement criterion, boundary conditions, BRST breaking and all that. Phys. Rev. D **79**, 121701 (2009). arXiv:0904.0641
268. Dudley, R.M. (ed.): Probabilities and Metrics. Convergence of Laws on Metric Spaces, with a View to Statistical Testing. Matematisk Institut, Aarhus Universitat, Aarhus (1976)
269. Duncan, A.: The Conceptual Framework of Quantum Field Theory. Oxford University Press, London (2012)
270. Dunford, N., Schwarz, J.T.: Linear Operators. Part I. General Theory. Wiley, Hoboken (1988)
271. Eardley, D.M., Smarr, L.: Kinematical conditions in the construction of space-time. Phys. Rev. D **19**, 2239 (1979)
272. Earman, J.: Gauge matters. Philos. Sci. **69**, S209 (2002)
273. Earman, J.: Curie's principle and spontaneous symmetry breaking. Int. Stud. Philos. Sci. **18**, 173 (2004)
274. Earman, J.: Reassessing the prospects for a growing block model of the universe. Int. Stud. Philos. Sci. **22**, 135 (2008)
275. Earman, J., Norton, J.D.: What price spacetime substantivalism. Br. J. Philos. Sci. **38**, 515 (1987)
276. Ebin, D.G.: The manifold of Riemannian metrics. Proc. Symp. Pure Math. Am. Math. Soc. **15**, 11 (1970)
277. Echanobe, J., del Campo, A., Muga, J.G.: Disclosing hidden information in the quantum Zeno effect: pulsed measurement of the quantum time of arrival. Phys. Rev. A **77**, 032112 (2008). arXiv:0712.0670
278. Ehlers, J.: Survey of general relativity theory. In: Israel, W. (ed.) Relativity, Astrophysics and Cosmology. Reidel, Dordrecht (1973)

279. Ehlers, J.: Machian ideas and general relativity. In: Barbour, J.B., Pfister, H. (eds.) Mach's Principle: From Newton's Bucket to Quantum Gravity. Birkhäuser, Boston (1995)
280. Eilenberg, S., Steenrod, N.: Foundations of Algebraic Topology. Princeton University Press, Princeton (1952)
281. Einstein, A.: On the electrodynamics of moving bodies. Ann. Phys. (Ger.) **17**, 891 (1905); The English translation is available in e.g. The Principle of Relativity. Dover, New York (1952), formerly published by Methuen, London (1923)
282. Einstein, A.: The field equations of gravitation. Preuss. Akad. Wiss. Berl. Sitzber. **844** (1915); For an English translation, see e.g. Volume 6: The Berlin Years: Writings, 1914–1917 (English translation supplement)
283. Einstein, A.: The foundation of the general theory of relativity. Ann. Phys. (Ger.) **49**, 769 (1916); The English translation is available in The Principle of Relativity. Dover, New York (1952), formerly published by Methuen, London (1923)
284. Einstein, A.: My theory. The Times, 28 November 1919; Republished as Time, space and gravitation. Opt. Br. Opt. J. **58**, 187 (1919)
285. Einstein, A.: Autobiographical notes. In: Schilpp, P.A. (ed.) Albert Einstein: Philosopher–Scientist. Library of Living Scientists, Evanston (1949)
286. Einstein, A.: Lecture before the Prussian Academy of Sciences, 27 January 1921
287. Einstein, A., Grossmann, M.: Outline of a generalized theory of relativity and of a theory of gravitation. I. Physical part by A. Einstein II. Mathematical part by M. Grossmann. Z. Math. Phys. **62**, 225 (1913), 245; For a translation, see e.g. The Collected Papers of Albert Einstein, vol. 4. The Swiss Years by A. Beck and D. Howard
288. Einstein, A.: Dialogue about objections to the theory of relativity. Naturwissenschaften **6**, 697 (1918); An English translation is available in Janssen et al. (2002)
289. Eliot, T.S.: Four Quartets. Harcourt, Orlando (1943)
290. Emmrich, C., Römer, H.: Orbifolds as configuration spaces of systems with gauge symmetries. Commun. Math. Phys. **129**, 69 (1990)
291. Eppley, K., Hannah, E.: The necessity of quantizing the gravitational field. Found. Phys. **7**, 51 (1977)
292. Esposito, G., Marmo, G., Stornaiolo, C.: Peierls brackets in theoretical physics. hep-th/0209147
293. Evens, S., Lu, J.-H., Weinstein, A.: Transverse measures, the modular class and a cohomology pairing for Lie algebroids. Q. J. Math. Oxf. **50**, 417 (1999)
294. Ewald, C.-O.: Hochschild homology and de Rham cohomology of stratifolds. Ph.D. thesis, University of Bonn (2002). http://archiv.ub.uni-heidelberg.de/volltextserver/2918/
295. Fadde'ev, L.D.: The Feynman integral for singular Lagrangians. Theor. Math. Phys. **1**, 1 (1969)
296. Farmer, A.J., Phinney, E.S.: The gravitational wave background from cosmological compact binaries. Mon. Not. R. Astron. Soc. **346**, 1197 (2003). astro-ph/0304393
297. Feynman, R.P.: The quantum theory of gravitation. Acta Phys. Pol. **24**, 697 (1963)
298. Feynman, S.e.g.R.P., Hibbs, A.R.: Quantum Mechanics and Path Integrals. McGraw–Hill, New York (1965)
299. Feynman, R.P. (lecture course given in 1962–1963), published as Feynman, R.P., Morinigo, F.B., Wagner, W.G., Hatfield, B. (eds.): Feynman Lectures on Gravitation. Addison–Wesley, Reading (1995)
300. Fierz, M., Pauli, W.: On relativistic wave equations for particles of arbitrary spin in an electromagnetic field. Proc. R. Soc. Lond. A **173**, 211 (1939)
301. Fischer, A.E.: The theory of superspace. In: Carmeli, M., Fickler, S.I., Witten, L. (eds.) Relativity, Proceedings of the Relativity Conference in the Midwest, Cincinnati, Ohio, 2–6 June 1969 Plenum, New York (1970)
302. Fischer, A.E.: Resolving the singularities in the space of Riemannian geometries. J. Math. Phys. **27**, 718 (1986)
303. Fischer, A.E., Marsden, J.E.: The manifold of conformally equivalent metrics. Can. J. Math. **1**, 193 (1977)

304. Fischer, A.E., Moncrief, V.: The reduced Hamiltonian of general relativity and the σ-constant of conformal geometry. In: Cotsakis, S., Gibbons, G.W. (eds.) Proceedings, Global Structure and Evolution in General Relativity, Karlovassi, 1994. Lecture Notes in Physics, vol. 460. Springer, Berlin (1996)
305. Fischer, A.E., Moncrief, V.: A method of reduction of Einstein's equations of evolution and a natural symplectic structure on the space of gravitational degrees of freedom. Gen. Relativ. Gravit. **28**, 207 (1996)
306. Fleischhack, C.: Stratification of the generalized gauge orbit space. Commun. Math. Phys. **214**, 607 (2000). math-ph/0001006
307. Fleischhack, C.: Loop quantization and symmetry: configuration spaces. arXiv:1010.0449
308. Fodor, G.: The thin sandwich conjecture. Ph.D. thesis, Budapest (1995)
309. For eLISA, ee e.g. https://www.elisascience.org/
310. For pulsar ephemerides, see e.g. http://www.jb.man.ac.uk/pulsar/crab.html
311. Fourès-Bruhat, Y.: Théorè d'Existence pour Certains Systèmes d'Équations aux Dérivées Partielles Non Linéaires. Acta Math. **88**, 141 (1952)
312. Fourès-Bruhat, Y.: Sur l'Intégration des Équations de la Relativité Générale. J. Ration. Mech. Anal. **55**, 951 (1956)
313. Fox, C.: An Introduction to the Calculus of Variations. Oxford University Press, London (1950)
314. Fradkin, E.S., Vasiliev, M.A.: Hamiltonian formalism, quantization and S-matrix for supergravity. Phys. Lett. B **72**, 70 (1977)
315. Franchi, J., Le Jan, Y.: Relativistic diffusions and Schwarzschild geometry. Commun. Pure Appl. Math. **60**, 187 (2007)
316. Frankel, T.: The Geometry of Physics: An Introduction. Cambridge University Press, Cambridge (2011)
317. Fraser, J.T.: The Genesis and Evolution of Time: A Critique of Interpretations in Physics. University of Massachusetts Press, Amherst (1982)
318. Fredehagan, K., Haag, R.: Generally covariant quantum field theory and scaling limits. Commun. Math. Phys. **108**, 91 (1987)
319. Fredenhagen, K., Rejzner, K.: Batalin-Vilkovisky formalism in the functional approach to classical field theory. Commun. Math. Phys. **314**, 93 (2012)
320. Freidel, L., Livine, E.R.: $U(N)$ coherent states for loop quantum gravity. J. Math. Phys. **52**, 052502 (2011). arXiv:1005.2090
321. Freund, P.G.O.: Introduction to Supersymmetry. Cambridge University Press, Cambridge (1986)
322. Friedrich, H., Nagy, G.: The initial boundary value problem for Einstein's vacuum field equations. Commun. Math. Phys. **201**, 619 (1999)
323. Friedrich, H., Rendall, A.: The Cauchy problem for the Einstein equations. Lect. Notes Phys. **540**, 127 (2000). arXiv:gr-qc/0002074
324. Frose Fischer, C.: The Hartree–Fock Method for Atoms. Wiley, New York (1977)
325. Fulling, S.A.: Nonuniqueness of canonical field quantization in Riemannian space-time. Phys. Rev. D **7**, 2850 (1973)
326. Fulton, W., Harris, J.: Representation Theory. A First Course. Springer, New York (2004)
327. Galapon, E.A.: Pauli's theorem and quantum canonical pairs: the consistency of a bounded, self-adjoint time operator canonically conjugate to a Hamiltonian with non-empty point spectrum. Proc. R. Soc. Lond. A **451**, 2671 (2002)
328. Galapon, E.A.: Post-Pauli's theorem emerging perspective on time in quantum mechanics. in [670]
329. Gambini, R., Porto, R.A.: Relational time in generally covariant quantum systems: four models. Phys. Rev. D **63**, 105014 (2001). arXiv:gr-qc/0101057
330. Gambini, R., Pullin, J.: Loops, Knots, Gauge Theories and Quantum Gravity. Cambridge University Press, Cambridge (1996)
331. Gambini, R., Pullin, J.: A First Course in Loop Quantum Gravity. Oxford University Press, New York (2011)

332. Gambini, R., Lewandowski, J., Marolf, D., Pullin, J.: On the consistency of the constraint algebra in spin network quantum gravity. Int. J. Mod. Phys. D **7**, 97 (1998). gr-qc/9710018
333. Gambini, R., Porto, R., Pullin, J.: Consistent discrete gravity solution of the problem of time: a model. In: Kokkotas, K., Stergioulas, N. (eds.) Recent Developments in Gravity. World Scientific, Singapore (2003). gr-qc/0302064
334. Gambini, R., Porto, R., Pullin, J.: A relational solution to the problem of time in quantum mechanics and quantum gravity induces a fundamental mechanism for quantum decoherence. New J. Phys. **6**, 45 (2004). arXiv:gr-qc/0402118
335. Gambini, R., Porto, R., Pullin, J.: Fundamental decoherence from quantum gravity: a pedagogical review. Gen. Relativ. Gravit. **39**, 1143 (2007). gr-qc/0603090
336. Gambini, R., Porto, R., Torterolo, S., Pullin, J.: Conditional probabilities with Dirac observables and the problem of time in quantum gravity. Phys. Rev. D **79**, 041501 (2009). arXiv: 0809.4235
337. Gaul, M., Rovelli, C.: Loop quantum gravity and the meaning of diffeomorphism invariance. Lect. Notes Phys. **541**(277), 277 (2000). gr-qc/9910079
338. Géhéniau, J., Debever, R.: Les Quatorze Invariants de Courbure de l'Espace Riemannien a Quatre Dimensions [The 14 curvature invariants of 4-*d* Riemannian space]. Helv. Phys. Acta Suppl. **4**, 101 (1956)
339. Gelbaum, B.R., Olmsted, J.M.H.: Counterexamples in Analysis. Dover, New York (2003)
340. Gell-Mann, M., Hartle, J.B.: Decoherence as a fundamental phenomenon in quantum dynamics. Phys. Rev. D **47**, 3345 (1993)
341. Gell-Mann, M., Hartle, J.B.: Quasiclassical coarse graining and thermodynamic entropy. Phys. Rev. A **76**, 022104 (2007). quant-ph/0609190
342. Gell-Mann, M., Hartle, J.B.: Decoherent histories quantum mechanics with one 'real' fine-grained history. arXiv:1106.0767
343. Georgi, H.: Lie Algebras in Particle Physics. Addison–Wesley, Reading (1982)
344. Gergely, L.Á., Kovacs, Z.: Gravitational dynamics in $s+1+1$ dimensions. Phys. Rev. D **72**, 064015 (2005). gr-qc/0507020
345. Gerlach, U.H., Sengupta, U.K.: Homogeneous collapsing star: tensor and vector harmonics for matter and field asymmetries. Phys. Rev. D **18**, 1773 (1978)
346. Geroch, R.P.: Topology in general relativity. J. Math. Phys. **8**, 782 (1967)
347. Geroch, R.P.: What is a singularity in general relativity? Ann. Phys. **48**, 526 (1968)
348. Geroch, R.P.: Domain of dependence. J. Math. Phys. **11**, 437 (1970)
349. Geroch, R.P.: General Relativity from A to B. University of Chicago Press, Chicago (1978)
350. Gibbons, G.W.: arXiv:1110.0611 covers further details of this
351. Gibbons, G.W., Hawking, S.W.: Selection rules for topology change. Commun. Math. Phys. **148**, 345 (1992)
352. Gibbons, G.W., Hawking, S.W., Perry, M.: Nucl. Phys. B **138**, 141 (1978)
353. Giddings, S.B., Marolf, D., Hartle, J.B.: Observables in effective gravity. Phys. Rev. D **74**, 064018 (2006). hep-th/0512200
354. Gilmore, R.: Lie Groups, Lie Algebras, and Some of Their Applications. Dover, New York (2006)
355. Ginsburg, V.L., Montgomery, R.: Geometric quantization and no-go theorems. In: Poisson Geometry. Banach Center Publications, vol. 51. Polish Academy of Sciences, Warsaw (2000)
356. Giulini, D.: Properties of 3-manifolds for relativists. Int. J. Theor. Phys. **33**, 913 (1994). gr-qc/9308008
357. Giulini, D.: On the configuration space topology in general relativity. Helv. Phys. Acta **68**, 86 (1995). gr-qc/9301020
358. Giulini, D.: What is the geometry of superspace? Phys. Rev. D **51**, 5630 (1995). gr-qc/ 9311017
359. Giulini, D.: The group of large diffeomorphisms in general relativity. Banach Cent. Publ. **39**, 303 (1997). arXiv:gr-qc/9510022

360. Giulini, D.: The generalized thin-sandwich problem and its local solvability. J. Math. Phys. **40**, 1470 (1999). gr-qc/9805065
361. Giulini, D.: In: Giulini, D., Kiefer, C., Lämmerzahl, C. (eds.) Proceedings of Aspects of Quantum Gravity—from Theory to Quantization, Bad Honnef, 2002. Springer, Berlin (2003). quant-ph/0304202
362. Giulini, D.: Some remarks on the notions of general covariance and background independence. In: Stamatescu, I.O. (ed.) An Assessment of Current Paradigms in the Physics of Fundamental Interactions. Lect. Notes Phys., vol. 721, p. 105 (2007). arXiv:gr-qc/0603087
363. Giulini, D.: The superspace of geometrodynamics. Gen. Relativ. Gravit. **41**(785), 785 (2009). arXiv:0902.3923
364. Giulini, D.: Dynamical and Hamiltonian formulation of general relativity. In: Ashtekar, A., Petkov, V. (eds.) Springer Handbook of Spacetime. Springer, Dordrecht (2014), chapter 17. arXiv:1505.01403
365. Giulini, D.: Equivalence principle, quantum mechanics, and atom-interferometric tests. Talk delivered on October 1st 2010 at the Regensburg conference on Quantum Field Theory and Gravity. arXiv:1105.0749
366. Giulini, D., Joos, E., Kiefer, C., Kupsch, J., Stamatescu, I.-O., Zeh, H.D.: Decoherence and the Appearance of a Classical World in Quantum Theory. Springer, Berlin (1996)
367. Goddard, A.J.: Some remarks on the existence of spacelike hypersurfaces of constant mean curvature. Commun. Math. Phys. **54**, 279 (1977)
368. Goddard, P., Kent, A., Olive, D.: Unitary representations of the Virasoro and super-Virasoro algebras. Commun. Math. Phys. **103**, 105 (1986)
369. Goenner, H.F.M.: On the history of unified field theories. Living Rev. Relativ. **2** (2004)
370. Goldberg, J.: Invariant transformations, conservation laws, and energy-momentum. In: Held, A. (ed.) General Relativity and Gravitation: One Hundred Years After the Birth of Albert Einstein, vol. 1. Plenum Press, New York (1980)
371. Goldstein, H.: Classical Mechanics. Addison–Wesley, Reading (1980)
372. Gorbounov, V., Malikov, F., Schechtman, V.: Gerbes of chiral differential operators. III In: Duval, C., Guieu, L., Ovsienko, V. (eds.) The Orbit Method in Geometry and Physics: In Honor of A.A. Kirillov. Am. Math. Soc. Providence (2001)
373. Goroff, M.H., Sagnotti, A.: The ultraviolet behaviour of Einstein gravity. Nucl. Phys. B **266**, 709 (1986)
374. Gotay, M.J.: Functorial geometric quantization and Van Hove's theorem. Int. J. Theor. Phys. **19**, 139 (1980)
375. Gotay, M.J.: On the Groenewold-Van Hove problem for $\mathbb{R}^{2n}$. J. Math. Phys. **40**, 2107 (1999). math-ph/9809015
376. Gotay, M.J.: Obstructions to quantization. In: Marsden, J., Wiggins, S. (eds.) Mechanics: From Theory to Computation (Essays in Honor of Juan-Carlos Simó), pp. 171–216. Springer, New York (2000). math-ph/9809011
377. Gotay, M.J., Grabowski, J.: On quantizing nilpotent and solvable basic algebras. Can. Math. Bull. **44**, 140 (1999)
378. Gotay, M.J., Grundling, H.B.: On quantizing $\mathfrak{T}^*\mathbb{S}^1$. Rep. Math. Phys. **40**, 107 (1997). quant-ph/9609025
379. Gotay, M.J.: On a full quantization of the torus. In: Antoine, J.-P., et al. (eds.) Quantization, Coherent States and Complex Structures. Plenum, New York (1995)
380. Gotay, M.J., Grundling, H.B., Hurst, C.A.: A Groenewold–Van Hove theorem for $\mathbb{S}^2$. Trans. Am. Math. Soc. **348**, 1579 (1996)
381. Gotay, M.J., Grundling, H.B., Tuynman, G.M.: Obstruction results in quantization theory. J. Nonlinear Sci. **6**, 469 (1996). dg-ga/9605001
382. Gourgoulhon, E.: 3 + 1 Formalism in General Relativity: Bases of Numerical Relativity. Lecture Notes in Physics, vol. 846. Springer, Berlin (2012); An earlier version of this is available as gr-qc/0703035
383. Gracia-Saz, A., Mehta, R.A.: Lie algebroid structures on double vector bundles and representation theory of Lie algebroids. Adv. Math. **223**, 1236 (2010). arXiv:0810.0066

384. Graham, R., Luckock, H.: Cosmological time in quantum supergravity. Phys. Rev. D **55**, 1943 (1997). gr-qc/9603054
385. Green, M., Schwarz, J., Witten, E.: Superstring Theory. Volume 1. Introduction. Cambridge University Press, Cambridge (1987)
386. Green, M., Schwarz, J., Witten, E.: Superstring Theory. Volume 2. Loop Amplitudes, Anomalies and Phenomenology. Cambridge University Press, Cambridge (1987)
387. Greene, R.E., Jakobowitz, H.: Analytic isometric embeddings. Ann. Math. **93**, 189 (1971)
388. Grenander, U.: Probabilities on Algebraic Structures. Wiley, New York (1963)
389. Gribov, V.N.: Quantization of non-Abelian gauge theories. Nucl. Phys. B **139**, 1 (1978)
390. Griffiths, R.B.: Consistent histories and the interpretation of quantum mechanics. J. Stat. Phys. **36**, 219 (1984)
391. Grimmett, G., Welsh, D.: Probability: An Introduction. Oxford University Press, New York (2014)
392. Groenewold, H.: On the principles of elementary quantum mechanics. Physica **12**, 405 (1946)
393. Gromov, M.: Metric Structures for Riemannian and Non-Riemannian Spaces. Birkhäuser, Boston (1999)
394. Gross, D.J., Periwal, V.: String perturbation theory diverges. Phys. Rev. Lett. **60**, 2105 (1988)
395. Gross, S.e.g.D.J., Wess, J.: Scale invariance, conformal invariance, and the high-energy behavior of scattering amplitudes. Phys. Rev. D **2**, 753 (1970)
396. Grothendieck, A.: On some points of homological algebra, I. Tohoku Math. J. **9**, 119 (1957)
397. Grünbaum, A.: In: Philosophical Problems of Space and Time, 2nd edn., p. 405. Reidel, Dordrecht (1973)
398. Gryb, S.B.: Shape dynamics and Mach's principles: gravity from conformal geometrodynamics. Ph.D. thesis, University of Waterloo, Canada (2011). arXiv:1204.0683
399. Guillemin, V., Sternberg, S.: Symplectic Techniques in Physics. Cambridge University Press, Cambridge (1984)
400. Gustavsson, A.: Algebraic structures on parallel M2-branes. Nucl. Phys. B **811**, 66 (2009). arXiv:0709.1260
401. Haag, R.: Local Quantum Physics. Springer, Berlin (1992)
402. Habib, S., Kandrup, H.E.: Wigner functions and density matrices in curved spaces as computational tools. Ann. Phys. **191**, 335 (1989)
403. Hackett, J., Speziale, S.: Grasping rules and semiclassical limit of the geometry in the Ponzano-Regge model. Class. Quantum Gravity **24**, 1525 (2007). arXiv:gr-qc/0611097
404. Hájíček, P.: Group quantization of parametrized systems I. Time levels. J. Math. Phys. **36**, 4612 (1996). gr-qc/9412047
405. Hájíček, P., Isham, C.J.: Perennials and the group-theoretical quantization of a parametrized scalar field on a curved background. J. Math. Phys. **37**, 3522 (1996). gr-qc/9510034
406. Hájíček, P., Kijowski, J.: Covariant gauge fixing and Kuchař decomposition. Phys. Rev. D **61**, 024037 (2000). gr-qc/9908051
407. Hall, B.C.: Quantum Theory for Mathematicians. Springer, New York (2013)
408. Halliwell, J.J.: Correlations in the wave function of the universe. Phys. Rev. D **36**, 3626 (1987)
409. Halliwell, J.J.: Derivation of the Wheeler–DeWitt equation from a path integral for minisuperspace models. Phys. Rev. D **38**, 2468 (1988)
410. Halliwell, J.J.: The interpretation of quantum cosmological models. In: Gleiser, R.J., Kozameh, C.N., Moreschi, O.M. (eds.) Proceedings of the 13th International Conference on General Relativity and Gravitation. IOP Publishing Bristol (1992). gr-qc/9208001
411. Halliwell, J.J.: Somewhere in the universe: where is the information stored when histories decohere? Phys. Rev. D **60**, 105031 (1999). quant-ph/9902008
412. Halliwell, J.J.: Trajectories for the wave function of the universe from a simple detector model. Phys. Rev. D **64**, 044008 (2001). gr-qc/0008046
413. Halliwell, J.J.: The interpretation of quantum cosmology and the problem of time. In: Gibbons, G.W., Shellard, E.P.S., Rankin, S.J. (eds.) The Future of Theoretical Physics and Cos-

mology (Stephen Hawking 60th Birthday Festschrift Volume). Cambridge University Press, Cambridge (2003). gr-qc/0208018
414. Halliwell, J.J.: Probabilities in quantum cosmological models: a decoherent histories analysis using a complex potential. Phys. Rev. D **80**, 124032 (2009). arXiv:0909.2597
415. Halliwell, J.J.: Decoherent histories analysis of minisuperspace quantum cosmology. J. Phys. Conf. Ser. **306**, 012023 (2011). arXiv:1108.5991
416. Halliwell, J.J.: Exact phase space localized projectors from energy eigenstates. Phys. Lett. A **377**, 222 (2013). arXiv:1207.0992
417. Halliwell, J.J.: Path Integral Analysis of Arrival Times with a Complex Potential. arXiv:0801.4308
418. Halliwell, J.J., Dodd, P.J.: Decoherence and records for the case of a scattering environment. Phys. Rev. D **67**, 105018 (2003). quant-ph/0301104
419. Halliwell, J.J., Hawking, S.W.: Origin of structure in the universe. Phys. Rev. D **31**, 1777 (1985)
420. Halliwell, J.J., Louko, J.: Steepest descent contours in the path integral approach to quantum cosmology. 3. A general method with applications to minisuperspace models. Phys. Rev. D **42**, 3997 (1990)
421. Halliwell, J.J., Thorwart, J.: Life in an energy eigenstate: decoherent histories analysis of a model timeless universe. Phys. Rev. D **65**, 104009 (2002). gr-qc/0201070
422. Halliwell, J.J., Wallden, P.: Invariant class operators in the decoherent histories analysis of timeless quantum theories. Phys. Rev. D **73**, 024011 (2006). gr-qc/0509013
423. Halliwell, J.J., Yearsley, J.M.: Pitfalls of path integrals: amplitudes for spacetime regions and the quantum Zeno effect. Phys. Rev. D **86**, 024016 (2012). arXiv:1205.3773
424. Halliwell, J.J., Perez-Mercader, J., Zurek, W.H. (eds.): Physical Origins of Time Asymmetry. Cambridge University Press, Cambridge (1994)
425. Halvorson, H., Mueger, M.: Algebraic quantum field theory. math-ph/0602036
426. Hamilton, R.S.: The inverse function theorem of Nash and Moser. Bull. Am. Math. Soc. **7**, 65 (1982)
427. Hartle, J.B.: The quantum mechanics of closed systems. In: Hu, B.-L., Ryan, M.P., Vishveshwara, C.V. (eds.) Directions in Relativity, vol. 1. Cambridge University Press, Cambridge (1993). gr-qc/9210006
428. Hartle, J.B.: Spacetime quantum mechanics and the quantum mechanics of spacetime. In: Julia, B., Zinn-Justin, J. (eds.) Gravitation and Quantizations: Proceedings of the 1992 Les Houches Summer School. North Holland, Amsterdam (1995). gr-qc/9304006
429. Hartle, J.B.: Spacetime information. Phys. Rev. D **51**, 1800 (1995). gr-qc/9409005
430. Hartle, J.B.: Quantum pasts and the utility of history. Phys. Scr. T **76**, 67 (1998). gr-qc/9712001
431. Hartle, J.B.: Gravity: An Introduction to Einstein's General Relativity. Addison–Wesley, San Francisco (2003)
432. Hartle, J.B.: The physics of 'now'. Am. J. Phys. **73**, 101 (2005). gr-qc/0403001
433. Hartle, J.B., Hawking, S.W.: Wave function of the universe. Phys. Rev. D **28**, 2960 (1983)
434. Hartle, J.B., Marolf, D.: Comparing formulations of generalized quantum mechanics for reparametrization invariant systems. Phys. Rev. D **56**, 6247 (1997)
435. Hartle, J.B., Schleich, K.: The conformal rotation in linearized gravity. In: Batalin, I., Isham, C.J., Vilkovilsky, G.A. (eds.) Quantum Field Theory and Quantum Statistics, vol. 2. Hilger, Bristol (1987)
436. Hartshorne, R.: Algebraic Geometry. Springer, New York (1977)
437. Hatcher, A.: Algebraic Topology. Cambridge University Press, Cambridge (2001)
438. Hawking, S.W.: Zeta function regularization of path integrals in curved spacetime. Commun. Math. Phys. **55**, 133 (1977)
439. Hawking, S.W.: Quantum cosmology. In: DeWitt, B.S., Stora, R. (eds.) Relativity, Groups and Topology II. North-Holland, Amsterdam (1984)
440. Hawking, S.W., Ellis, G.F.R.: The Large-Scale Structure of Space-Time. Cambridge University Press, Cambridge (1973)

441. Hawking, S.W., Page, D.N.: Operator ordering and the flatness of the universe. Nucl. Phys. B **264**, 185 (1986)
442. Hawking, S.W., Page, D.N.: How probable is inflation? Nucl. Phys. B **298**, 789 (1988)
443. Heil, A., et al.: Anomalies from nonfree action of the gauge group. Ann. Phys. **200**, 206 (1990)
444. Heisenberg, W., Pauli, W.: Zer Quantendynamik der Wellenfelder [Quantum dynamics of wave fields]. Z. Phys. **56**, 1 (1929)
445. Henneaux, M.: Geometry of zero signature space-times. Bull. Soc. Math. Belg. **31**, 47 (1979)
446. Henneaux, M., Teitelboim, C.: Quantization of Gauge Systems. Princeton University Press, Princeton (1992)
447. Herrmann, S., Senger, A., Möhle, K., Nagel, M., Kovalchuk, E.V., Peters, A.: Rotating optical cavity experiment testing Lorentz invariance at the 10^{-17} level. Phys. Rev. D **80**, 105011 (2009). arXiv:1002.1284
448. Heunen, C., Landsman, N.P., Spitters, B.: A topos for algebraic quantum theory. Commun. Math. Phys. **291**, 63 (2009). arXiv:0709.4364
449. Heunen, C., Landsman, N.P., Spitters, B.: The principle of general tovariance. http://philsci-archive.pitt.edu/3931/
450. Hillery, M., O'Connell, R.F., Scully, M.O., Wigner, E.P.: Distribution functions in physics: fundamentals. Phys. Rep. **106**, 121 (1984)
451. Hocking, J.G., Young, G.S.: Topology. Dover, New York (1988)
452. Hoehn, P.A.: Effective relational dynamics. J. Phys. Conf. Ser. **360**, 12014–12017 (2012). arXiv:1110.5631
453. Hoehn, P.A., Kubalova, E., Tsobanjan, A.: Effective relational dynamics of a nonintegrable cosmological model. Phys. Rev. D **86**, 065014 (2012). arXiv:1111.5193
454. Hojman, S.A., Kuchař, K.V., Teitelboim, C.: Geometrodynamics regained. Ann. Phys. (N. Y.) **96**, 88 (1976)
455. Holst, S.: Barbero's Hamiltonian derived from a generalized Hilbert–Palatini action. Phys. Rev. D **53**, 5966 (1996)
456. Horowitz, G.T.: Quantum cosmology with a positive-definite action. Phys. Rev. D **31**, 1169 (1985)
457. Horvath, Z., Kovacs, Z., Gergely, L.A.: Geometrodynamics in a spherically symmetric, static crossflow of null dust. Phys. Rev. D **74**, 084034 (2006). gr-qc/0605116
458. Huebschmann, J.: Poisson cohomology and quantization. J. Reine Angew. Math. **408**, 57 (1990). arXiv:1303.3903
459. Huggett, S.A., Tod, K.P.: An Introduction to Twistor Theory. Cambridge University Press, Cambridge (1985)
460. Hughes, T.J.R., Kato, T., Marsden, J.E.: Well-posed quasilinear second order hyperbolic systems with applications to nonlinear elasticity and general relativity. Arch. Ration. Mech. Anal. **63**, 273 (1977)
461. Husain, V., Kuchař, K.V.: General covariance, new variables and dynamics without dynamics. Phys. Rev. D **42**, 4070 (1990)
462. Husain, V., Pawlowski, T.: Dust reference frame in quantum cosmology. Class. Quantum Gravity **28**, 225014 (2011). arXiv:1108.1147
463. Husain, V., Pawlowski, T.: Time and a physical Hamiltonian for quantum gravity. Phys. Rev. Lett. **108**, 141301 (2012). arXiv:1108.1145
464. Husemoller, D.: Fibre Bundles. Springer, New York (1994)
465. Isenberg, J.: Constant mean curvature solutions of the Einstein constraint equations on closed manifolds. Class. Quantum Gravity **12**, 2249 (1995)
466. Isenberg, J., Marsden, J.E.: A slice theorem for the space of solutions of Einstein's equations. Phys. Rep. **89**, 179 (1982)
467. Isenberg, J., Marsden, J.E.: The York map is a canonical transformation. J. Geom. Phys. **1**, 85 (1984)
468. Isenberg, J., Nester, J.: The effect of gravitational interaction on classical fields: a Hamilton Dirac analysis. Ann. Phys. (N. Y.) **107**, 56 (1977)

469. Isenberg, J., Wheeler, J.A.: Inertia here is fixed by mass-energy there in every W model universe. In: Pantaleo, M., de Finis, F. (eds.) Relativity, Quanta and Cosmology in the Development of the Scientific Thought of Albert Einstein. Johnson, New York (1979)
470. Isenberg, J., ó Murchadha, N., York, J.W.: Initial value problem of general relativity. 3. Coupled fields and the scalar-tensor theory. Phys. Rev. D **13**, 1532 (1976)
471. Isham, C.J.: An introduction to quantum gravity. In: Isham, C.J., Penrose, R., Sciama, D. (eds.) Oxford Symposium on Quantum Gravity. Clarendon, Oxford (1975)
472. Isham, C.J.: Some quantum field theory aspects of the superspace quantization of general relativity. Proc. R. Soc. Lond. A **351**, 209 (1976)
473. Isham, C.J.: Quantum field theory in curved spacetimes. A general mathematical framework. In: Proceedings, Differential Geometrical Methods in Mathematical Physics, Bonn, 1977 (1977), Berlin
474. Isham, C.J.: Quantum gravity—an overview. In: Isham, C.J., Penrose, R., Sciama, D.W. (eds.) Quantum Gravity 2: A Second Oxford Symposium. Clarendon, Oxford (1981)
475. Isham, C.J.: Topological and global aspects of quantum theory. In: DeWitt, B., Stora, R. (eds.) Relativity, Groups and Topology II. North-Holland, Amsterdam (1984)
476. Isham, C.J.: Quantum geometry. In: Christensen, S. (ed.) Quantum Theory of Gravity. Hilger, Bristol (1984)
477. Isham, C.J.: Aspects of Quantum Gravity. Lectures Given at Conference: C85–07-28.1 (Scottish Summer School 1985:0001), available on KEK archive
478. Isham, C.J.: Quantum gravity. In: MacCallum, M.A.H. (ed.) Proceedings, General Relativity and Gravitation, Stockholm, 1986. Cambridge University Press, Cambridge (1987)
479. Isham, C.J.: Canonical groups and the quantization of general relativity. Nucl. Phys. B, Proc. Suppl. **6**, 349 (1989)
480. Isham, C.J.: Quantum topology and quantization on the lattice of topologies. Class. Quantum Gravity **6**, 1509 (1989)
481. Isham, C.J.: Quantization on the lattice of topologies. In: Lusanna, L. (ed.) Proceedings, Knots, Topology and Quantum Field Theories, Florence, 1989. World Scientific, Singapore (1989)
482. Isham, C.J.: Canonical groups and the quantization of geometry and topology. In: Ashtekar, A., Stachel, J. (eds.) Conceptual Problems of Quantum Gravity. Birkhäuser, Boston (1991)
483. Isham, C.J.: Canonical quantum gravity and the problem of time. In: Ibort, L.A., Rodríguez, M.A. (eds.) Integrable Systems, Quantum Groups and Quantum Field Theories. Kluwer Academic, Dordrecht (1993). gr-qc/9210011
484. Isham, C.J.: Perennials and the group-theoretical quantization of a parametrized scalar field on a curved background. J. Math. Phys. **35**, 2157 (1994). gr-qc/9308006
485. Isham, C.J.: Prima facie questions in quantum gravity. In: Lect. Notes Phys., vol. 434. (1994). gr-qc/9310031
486. Isham, C.J.: Quantum logic and decohering histories. In: Tchrakian, D.H. (ed.) Topics in Quantum Field Theory. World Scientific, Singapore (1995). quant-ph/9506028
487. Isham, C.J.: Lectures on Quantum Theory. Imperial College Press, London (1995)
488. Isham, C.J.: Structural issues in quantum gravity. In: Francaviglia, M., Longhi, G., Lusanna, L., Sorace, E. (eds.) General Relativity and Gravitation 14: Proceedings. World Scientific, Singapore (1997). gr-qc/9510063
489. Isham, C.J.: Topos theory and consistent histories: the internal logic of the set of all consistent sets. Int. J. Theor. Phys. **36**, 785 (1997). gr-qc/9607069
490. Isham, C.J.: Modern Differential Geometry for Physicists. World Scientific, Singapore (1999)
491. Isham, C.J.: Some reflections on the status of conventional quantum theory when applied to quantum gravity. In: Gibbons, G.W., Shellard, E.P.S., Rankin, S.J. (eds.) The Future of Theoretical Physics and Cosmology (Stephen Hawking 60th Birthday Festschrift Volume). Cambridge University Press, Cambridge (2003). quant-ph/0206090
492. Isham, C.J.: A new approach to quantising space-time: I. Quantising on a general category. Adv. Theor. Math. Phys. **7**, 331 (2003). gr-qc/0303060

493. Isham, C.J.: A new approach to quantising space-time: II. Quantising on a category of sets. Adv. Theor. Math. Phys. **7**, 807 (2003). gr-qc/0304077
494. Isham, C.J.: A new approach to quantising space-time: III. State vectors as functions on arrows. Adv. Theor. Math. Phys. **8**, 797 (2004). gr-qc/0306064
495. Isham, C.J.: A topos perspective on state-vector reduction. Int. J. Theor. Phys. **45**, 1524 (2005). quant-ph/0508225
496. Isham, C.J.: Topos methods in the foundations of physics. In: Halvorson, H. (ed.) Deep Beauty. Cambridge University Press, Cambridge (2010). arXiv:1004.3564
497. Isham, C.J.: An Introduction to General Topology And Quantum Topology, unpublished, Lectures given at Banff in 1989 (and available on the KEK archive)
498. Isham, C.J.: Quantising on a category. quant-ph/0401175
499. Isham, C.J., Kakas, A.: A group theoretic approach to the canonical quantization of gravity. 1. construction of the canonical group. Class. Quantum Gravity **1**, 621 (1984)
500. Isham, C.J., Kakas, A.: A group theoretical approach to the canonical quantization of gravity. 2. Unitary representations of the canonical group. Class. Quantum Gravity **1**, 633 (1984)
501. Isham, C.J., Kuchař, K.V.: Representations of space-time diffeomorphisms. 1. Canonical parametrized field theories. Ann. Phys. (N. Y.) **164**, 288 (1985)
502. Isham, C.J., Kuchař, K.V.: Representations of space-time diffeomorphisms. 2. Canonical geometrodynamics. Ann. Phys. (N. Y.) **164**, 316 (1985)
503. Isham, C.J., Linden, N.: Quantum temporal logic and decoherence functionals in the histories approach to generalized quantum theory. J. Math. Phys. **35**, 5452 (1994). gr-qc/9405029
504. Isham, C.J., Linden, N.: Continuous histories and the history group in generalized quantum theory. J. Math. Phys. **36**, 5392 (1995). gr-qc/9503063
505. Isham, C.J., Linden, N.: Information-entropy and the space of decoherence functions in generalised quantum theory. quant-ph/9612035
506. Isham, C.J., Savvidou, K.N.: Time and modern physics. In: Ridderbos, K. (ed.) Time. Cambridge University Press, Cambridge (2002)
507. Isham, C.J., Savvidou, K.N.: Quantising the foliation in history quantum field theory. quant-ph/0110161
508. Isham, C.J., Kubyshin, Y.A., Renteln, P.: Quantum norm theory and the quantization of metric topology. Class. Quantum Gravity **7**, 1053 (1990)
509. Isham, C.J., Kubyshin, Y.A., Renteln, P.: Quantum metric topology. In: Markov, M.A., Berezin, V.A., Frolov, V.P. (eds.) Proceedings, Quantum Gravity, Moscow, 1990. World Scientific, Singapore (1991)
510. Isham, C.J., Linden, N., Savvidou, K.N., Schreckenberg, S.: Continuous time and consistent histories. J. Math. Phys. **39**, 1818 (1998). quant-ph/9711031
511. Iversen, B.: Cohomology of Sheaves. Springer, Berlin (1986)
512. Iwai, T.: A gauge theory for the quantum planar three-body problem. J. Math. Phys. **28**, 964 (1987)
513. Iwai, T.: A geometric setting for internal motions of the quantum three-body system. J. Math. Phys. **28**, 1315 (1987)
514. Iwai, T., Yamaoka, H.: Stratified reduction of classical many-body systems with symmetry. J. Phys. A, Math. Gen. **38**, 2415 (2005)
515. Jacobson, T.: New variables for canonical supergravity. Class. Quantum Gravity **5**, 923 (1988)
516. Jacobson, T.: Einstein-Aether gravity: a status report. P.o.S.Q.G.-Ph. **020** (2007). arXiv: 0801.1547
517. Jammer, M.: The Philosophy of Quantum Mechanics. Wiley, New York (1974)
518. Jammer, M.: Concepts of Space. The History of Theories of Space in Physics, 3rd, enlarged edn. Dover, New York (1993). Foreword by A. Einstein
519. Jammer, M.: Concepts of Space: The History of Theories of Space in Physics, 3rd edn. Dover, New York (1993)
520. Jammer, M.: Concepts of Mass in Contemporary Physics and Philosophy. Princeton University Press, Princeton (2000)

521. Jammer, M.: Concepts of Simultaneity. From Antiquity to Einstein and Beyond. Johns Hopkins University Press, Baltimore (2006)
522. Jänich, K.: Topology. Springer, New York (1984)
523. Jantzen, R.: The dynamical degrees of freedom in spatially homogeneous cosmology. Commun. Math. Phys. **64**, 211 (1979)
524. Jizba, P., Pons, J.M.: Revisiting the gauge principle: enforcing constants of motion as constraints. J. Phys. A **43**, 205202 (2010). arXiv:0905.3807
525. Johnstone, P.T.: Sketches of an Elephant: A Topos Theory Compendium, vols. 1 and 2. Clarendon, Oxford (2002)
526. Jones, H.F.: Groups, Representations and Physics. Taylor and Francis, New York (1998)
527. Joos, E., Zeh, H.D.: The emergence of classical properties through interaction with the environment. Z. Phys. B **59**, 223 (1985)
528. Kadison, R.V., Ringrose, J.R.: Fundamentals of the Theory of Operator Algebras. Academic Press, Orlando (1983)
529. Kaku, M., Townsend, P.K., van Nieuwenhuizen, P.: Gauge theory of the conformal and superconformal group. Phys. Rev. Lett. B **69**, 304 (1977)
530. Kaku, M., Townsend, P.K., van Nieuwenhuizen, P.: Properties of conformal supergravity. Phys. Rev. D **17**, 3179 (1978)
531. Kallenberg, O.: Foundations of Modern Probability. Springer, New York (1997)
532. Kandrup, H.E.: Generalized Wigner functions in curved spaces: a new approach. Phys. Rev. D **37**, 2165 (1988)
533. Kauffman, L.H.: Knots and Physics. World Scientific, Singapore (1994)
534. Kendall, M.G.: Rank Correlation Methods. Griffin, London (1970)
535. Kendall, D.G.: Foundations of a theory of random sets. In: Harding, E.F., Kendall, D.G. (eds.) Stochastic Geometry (1974)
536. Kendall, D.G.: Shape manifolds, procrustean metrics and complex projective spaces. Bull. Lond. Math. Soc. **16**, 81 (1984)
537. Kendall, D.G.: A survey of the statistical theory of shape. Stat. Sci. **4**, 87 (1989)
538. Kendall, D.G., Kendall, W.S.: Alignments in 2-dimensional random sets of points. Adv. Appl. Probab. **12**, 380 (1980)
539. Kendall, D.G., Barden, D., Carne, T.K., Le, H.: Shape and Shape Theory. Wiley, Chichester (1999)
540. Kent, A.: Consistent sets yield contrary inferences in quantum theory. Phys. Rev. Lett. **78**, 2874 (1997). gr-qc/9604012
541. Kent, A.: Comment on "spacetime information". Phys. Rev. D **56**, 2469 (1997). gr-qc/9610075
542. Kent, A.: Quantum histories. Phys. Scr. T **76**, 78–84 (1998). gr-qc/9809026
543. Kent, A.: Consistent sets and contrary inferences: reply to Griffiths and Hartle. Phys. Rev. Lett. **81**, 1982 (1998). gr-qc/9808016
544. Kent, A.: Quantum histories and their implications. Lect. Notes Phys. **559**, 93 (2000). gr-qc/9607073
545. Kent, A.: One world versus many: the inadequacy of Everettian accounts of evolution, probability, and scientific confirmation. In: Saunders, S., Barrett, J., Kent, A., Wallace, D. (eds.) Many Worlds? Everett, Quantum Theory and Reality. Oxford University Press, Oxford (2010). arXiv:0905.0624
546. Kerman, A.K., Koonin, S.E.: Hamiltonian formulation of time-dependent variational principles for the many-body system. Ann. Phys. **100**, 332 (1976)
547. Kervaire, M.: A manifold which does not admit any differentiable structure. Comment. Math. Helv. **34**, 257 (1960)
548. Kessari, S.: Affine histories in quantum gravity: introduction and the representation for a cosmological model. Class. Quantum Gravity **24**, 1303 (2007). gr-qc/0605105
549. Kiefer, C.: Continuous measurement of minisuperspace variables by higher multipoles. Class. Quantum Gravity **4**, 1369 (1987)

550. Kiefer, C.: Quantum cosmology and the emergence of a classical world. In: Rudolph, E., Stamatescu, I-O. (eds.) FESt-Proceedings on the Concepts of Space and Time. Springer, Berlin (1993). gr-qc/9308025
551. Kiefer, C.: Conceptual issues in quantum cosmology. Lect. Notes Phys. **541**, 158 (2000). gr-qc/9906100
552. Kiefer, C.: Quantum Gravity. Clarendon, Oxford (2004)
553. Kiefer, C.: Concept of time in canonical quantum gravity and string theory. J. Phys. Conf. Ser. **174**, 012021 (2009)
554. Kiefer, C., Singh, T.P.: Quantum gravitational corrections to the functional Schrödinger equation. Phys. Rev. D **44**, 1067 (1991)
555. Kiefer, C., Lueck, T., Vargas Moniz, P.: The semiclassical approximation to supersymmetric quantum gravity. Phys. Rev. D **72**, 045006 (2005). gr-qc/0505158
556. Kittel, C.: Quantum Theory of Solids. Wiley, New York (1963)
557. Klainerman, S., Rodnianski, I.: Causal geometry of Einstein-vacuum spacetimes with finite curvature flux. Invent. Math. **159**, 437 (2005)
558. Klauder, J.R.: Ultralocal scalar field models. Commun. Math. Phys. **18**, 307 (1970)
559. Klauder, J.R.: Overview of affine quantum gravity. Int. J. Geom. Methods Mod. Phys. **3**, 81 (2006). gr-qc/0507113
560. Kobayashi, S.: Transformation Groups in Differential Geometry. Springer, Berlin (1972)
561. Kobayashi, S., Nomizu, K.: Foundations of Differential Geometry, vols. 1 and 2. Wiley, New York (1996)
562. Kochen, S.: The Interpretation of QM. Princeton University Press, Princeton (1979)
563. Komar, A.: New general-relativistic thin-sandwich theorem. Phys. Rev. D **4**, 927 (1971)
564. Kondracki, W., Rogulski, J.: On the stratification of the orbit space for the action of automorphisms on connections. Diss. Math. **250** (1986)
565. Kontsevich, M.: Deformation quantization of Poisson manifolds, I. Lett. Math. Phys. **66**, 157 (2003). q-alg/9709040
566. Kouletsis, I.: Covariance and time regained in canonical general relativity. Phys. Rev. D **78**, 064014 (2008). arXiv:0803.0125
567. Kouletsis, I.: A classical history theory: geometrodynamics and general field dynamics regained. Imperial College preprint TP/97-98/17. gr-qc/9801019
568. Kouletsis, I., Kuchař, K.V.: Diffeomorphisms as symplectomorphisms in history phase space: bosonic string model. Phys. Rev. D **65**, 125026 (2002). gr-qc/0108022
569. Kowalski, M., Rubin, D., et al.: Improved cosmological constraints from new, old and combined supernova datasets. Astrophys. J. **686**, 749 (2008). arXiv:0804.4142
570. Kreck, M.: Differential Algebraic Topology: From Stratifolds to Exotic Spheres. American Mathematical Society, Providence (2010)
571. Kreck, M., Tene, H.: Hilbert stratifolds and a Quillen type geometric description of cohomology for Hilbert manifolds. arXiv:1506.07075
572. Kuchař, K.V.: Canonical quantization of cylindrical gravitational waves. Phys. Rev. D **4**, 955 (1971)
573. Kuchař, K.V.: A bubble-time canonical formalism for geometrodynamics. J. Math. Phys. **13**, 768 (1972)
574. Kuchař, K.V.: Canonical quantization of gravity. In: Israel, W. (ed.) Relativity, Astrophysics and Cosmology. Reidel, Dordrecht (1973)
575. Kuchař, K.V.: Geometrodynamics regained: a Lagrangian approach. J. Math. Phys. **15**, 708 (1974)
576. Kuchař, K.V.: Geometry of hyperspace. I. J. Math. Phys. **17**, 777 (1976)
577. Kuchař, K.V.: Kinematics of tensor fields in hyperspace. II. J. Math. Phys. **17**, 792 (1976)
578. Kuchař, K.V.: Dynamics of tensor fields in hyperspace. III. J. Math. Phys. **17**, 801 (1976)
579. Kuchař, K.V.: Geometrodynamics with tensor sources IV. J. Math. Phys. **18**, 1589 (1977)
580. Kuchař, K.V.: On equivalence of parabolic and hyperbolic superHamiltonians. J. Math. Phys. **19**, 390 (1978)

581. Kuchař, K.V.: Canonical methods of quantization. In: Isham, C.J., Penrose, R., Sciama, D.W. (eds.) Quantum Gravity 2: A Second Oxford Symposium. Clarendon, Oxford (1981)
582. Kuchař, K.V.: General relativity: dynamics without symmetry. J. Math. Phys. **22**, 2640 (1981)
583. Kuchař, K.V.: Dirac constraint quantization of a parametrized field theory by anomaly-free operator representations of space-time diffeomorphisms. Phys. Rev. D **39**, 2263 (1989)
584. Kuchař, K.V.: The problem of time in canonical quantization. In: Ashtekar, A., Stachel, J. (eds.) Conceptual Problems of Quantum Gravity. Birkhäuser, Boston (1991)
585. Kuchař, K.V.: Does an unspecified cosmological constant solve the problem of time in quantum gravity? Phys. Rev. D **43**, 3332 (1991)
586. Kuchař, K.V.: Time and interpretations of quantum gravity. In: Kunstatter, G., Vincent, D., Williams, J. (eds.) Proceedings of the 4th Canadian Conference on General Relativity and Relativistic Astrophysics. World Scientific, Singapore (1992); Reprinted as Int. J. Mod. Phys. Proc. Suppl. D **20**, 3 (2011)
587. Kuchař, K.V.: Canonical quantum gravity. In: Gleiser, R.J., Kozamah, C.N., Moreschi, O.M. (eds.) General Relativity and Gravitation 1992. IOP Publishing, Bristol (1993). gr-qc/9304012
588. Kuchař, K.V.: Geometrodynamics of Schwarzschild black holes. Phys. Rev. D **50**, 3961 (1994). arXiv:gr-qc/9403003
589. Kuchař, K.V.: The problem of time in quantum geometrodynamics. In: Butterfield, J. (ed.) The Arguments of Time. Oxford University Press, Oxford (1999)
590. Kuchař, K.V., Romano, J.D.: Gravitational constraints which generate a Lie algebra. Phys. Rev. D **51**, 5579 (1995). gr-qc/9501005
591. Kuchař, K.V., Ryan, M.P.: Is minisuperspace quantization valid?: Taub in mixmaster. Phys. Rev. D **40**, 3982 (1989)
592. Kuchař, K.V., Torre, C.G.: Gaussian reference fluid and interpretation of quantum geometrodynamics. Phys. Rev. D **43**, 419 (1991)
593. Kuchař, K.V., Torre, C.G.: The harmonic gauge in canonical gravity. Phys. Rev. D **44**, 3116 (1991)
594. Kuchař, K.V., Torre, C.G.: Strings as poor relatives of general relativity. In: Ashtekar, A., Stachel, J. (eds.) Conceptual Problems of Quantum Gravity, p. 326. Birkhäuser, Boston (1991)
595. Lachièze-Rey, M.: Space and observers in cosmology. Astron. Astrophys. **376**, 17 (2001). gr-qc/0107010
596. Lachièze-Rey, M.: In search of relativistic time. arXiv:1312.2866
597. Lachièze-Rey, M., Luminet, J.P.: Cosmic topology. Phys. Rep. **254**, 135 (1995). gr-qc/9605010
598. Lanczos, C.: The Variational Principles of Mechanics. University of Toronto Press, Toronto (1949)
599. Landau, L.D., Lifshitz, E.M.: Quantum Mechanics. Pergamon, New York (1965)
600. Landau, L.D., Lifshitz, E.M.: Fluid Mechanics. Butterworth–Heinemann, Oxford (1987)
601. Landsman, N.P.: Strict deformation quantization of a particle in external gravitational and Yang-Mills fields. J. Geom. Phys. **12**, 93 (1993)
602. Landsman, N.P.: Deformations of algebras of observables and the classical limit of quantum mechanics. Rev. Math. Phys. **05**, 775 (1993)
603. Landsman, N.P.: Mathematical Topics Between Classical and Quantum Mechanics. Springer, New York (1998)
604. Landsman, N.P.: Quantization as a Functor. Quantization, Poisson Brackets and Beyond. Contemp. Math., vol. 315. Amer. Math. Soc., Providence (2002)
605. Landsman, N.P.: Between classical and quantum. In: Handbook of the Philosophy of Physics. Elsevier, Amsterdam (2005). quant-ph/0506082
606. Lang, S.: Differential and Riemannian Manifolds. Springer, New York (1995)
607. Lapchinski, V.G., Rubakov, V.A.: Canonical quantization of gravity and quantum field theory in curved space-time. Acta Phys. Pol. B **10** (1979)

608. Larson, R.E., Andima, S.J.: The lattice of topologies: a survey. Rocky Mt. J. Math. **5**, 177 (1975)
609. Larsson, T.A.: Fock representations of non-centrally extended super-diffeomorphism algebras. physics/9710022
610. Laurent-Gengoux, C., Pichereau, A., Vanhaecke, P.: Poisson Structures. Springer, Berlin (2013)
611. Lawvere, F.W., Rosebrugh, R.: Sets for Mathematics. Cambridge University Press, Cambridge (2003)
612. Lawvere, F.W., Schanuel, S.H.: Conceptual Mathematics. A First Introduction to Categories. Cambridge University Press, Cambridge (2009)
613. Lee, J.M.: Introduction to Topological Manifolds. Springer, New York (2011)
614. Lee, J.M.: Introduction to Smooth Manifolds, 2nd edn. Springer, New York (2013)
615. Lee, J., Wald, R.M.: Local symmetries and constraints. J. Math. Phys. **31**, 725 (1990)
616. Leibniz, G.W.: The Metaphysical Foundations of Mathematics. University of Chicago Press, Chicago (1956)
617. Leray, J.: Hyperbolic Differential Equations. Lecture Notes, Princeton (1953)
618. Levin, J., Scannapieco, E., de Gasperis, G., Silk, J., Barrow, J.D.: How the universe got its spots. Phys. Rev. D **66**, 104010 (2002). astro-ph/9807206
619. Lévy-Leblond, J.-M.: Une Nouvelle Limite Non-Relativiste du Groupe de Poincaré [A new non-relativistic limit of Poincaré's group]. Ann. Inst. Henri Poincaré A **3**, 1 (1965)
620. Lewandowski, J., Marolf, D.: Loop constraints: a habitat and their algebra. Int. J. Mod. Phys. D **7**, 299 (1998). gr-qc/9710016
621. Li, R., Gibble, K., Szymaniec, K.: Improved accuracy of the NPL-CsF2 primary frequency standard: evaluation of distributed cavity phase and microwave lensing frequency shifts. Metrologia **48**, 283 (2011). arXiv:1107.2412
622. Lichnerowicz, A.: L'Intégration des Équations de la Gravitation Relativiste et le Problème des N Corps [The integration of the equations of relativistic gravitation and the N-body problem]. J. Math. Pures Appl. **23**, 37 (1944)
623. Lifshitz, E.M., Khalatnikov, I.M.: Investigations in relativistic cosmology. Adv. Phys. **12**, 185 (1963)
624. Littlejohn, R.G., Reinsch, M.: Gauge fields in the separation of rotations and internal motions in the N-body problem. Rev. Mod. Phys. **69**, 213 (1997)
625. Liu, Z-J., Qian, M.: Gauge invariant quantization on Riemannian manifolds. Trans. Am. Math. Soc. **331**, 321 (1992)
626. Livingston, C.: Knot Theory. Math. Assoc. of America, Washington (1993)
627. Loday, J.-L.: Cyclic Cohomology. Springer, Berlin (1998)
628. Loll, R.: Discrete approaches to quantum gravity in four dimensions. Living Rev. Relativ. **1**, 13 (1998). gr-qc/9805049
629. Lovelock, D.: The Einstein tensor and its generalizations. J. Math. Phys. **12**, 498 (1971)
630. Lusanna, L., Pauri, M.: The physical role of gravitational and gauge degrees of freedom in general relativity—II: Dirac versus Bergmann observables and the objectivity of space-time. Gen. Relativ. Gravit. **38**, 229 (2006). arXiv:gr-qc/0407007
631. Mac Lane, S.: Categories for the Working Mathematician. Springer, New York (1998)
632. Mach, E.: Die Mechanik in ihrer Entwickelung, Historisch-kritisch dargestellt. Barth, Leipzig (1883); An English translation is The Science of Mechanics: A Critical and Historical Account of Its Development. Open Court, La Salle (1960)
633. Mackey, G.: Mathematical Foundations of Quantum Mechanics. Benjamin, New York (1963)
634. Major, S., Rideout, D., Surya, S.: Stable homology as an indicator of manifoldlikeness in causal set theory. Class. Quantum Gravity **26**, 17508 (2009). arXiv:0902.0434
635. Maldacena, J.M.: The large N limit of superconformal field theories and supergravity. Int. J. Theor. Phys. **38**, 1133 (1999). hep-th/9711200
636. Marchal, C.: Celestial Mechanics. Elsevier, Tokyo (1990)

637. Markopoulou, F.: The internal description of a causal set: what the universe looks like from the inside. Commun. Math. Phys. **211**, 559 (2000). gr-qc/9811053
638. Markowitz, W., Hall, R.G., Essen, L., Parry, J.V.L.: Frequency of cesium in terms of ephemeris time. Phys. Rev. Lett. **1**, 105 (1958)
639. Marolf, D.: Poisson bracket on the space of histories. Ann. Phys. **236**, 374 (1994). hep-th/9308141
640. Marolf, D.: Observables and a Hilbert space for Bianchi IX. Class. Quantum Gravity **12**, 1441 (1995). arXiv:gr-qc/9409049
641. Marolf, D.: Solving the problem of time in minisuperspace: measurement of Dirac observables. Phys. Rev. D **79**, 084016 (2009). arXiv:0902.1551
642. Marsden, J.E., Ratiu, T.S.: Introduction to Mechanics and Symmetry. Springer, New York (1999)
643. Marsden, J.E., Tipler, F.J.: Maximal hypersurfaces and foliations of constant mean curvature in general relativity. Phys. Rep. **66**, 109 (1980)
644. Martin, G.E.: Transformation Geometry. Springer, New York (1982)
645. Marzke, R.F., Wheeler, J.A.: Gravitation as geometry—I: the geometry of the space-time and the geometrodynamical standard meter. In: Chiu, H.Y., Hoffman, W.F. (eds.) Gravitation and Relativity. Benjamin, New York (1964)
646. Massar, S., Parentani, R.: Unitary and non-unitary evolution in quantum cosmology. Phys. Rev. D **59**, 123519 (1999). gr-qc/9812045
647. Mathur, S.D.: What exactly is the information paradox? Lect. Notes Phys. **769**, 3 (2009)
648. Matsumoto, Y.: An Introduction to Morse Theory. Am. Math. Soc., Providence (1997)
649. McTaggart, J.M.E.: The unreality of time. Mind **17**, 457 (1908)
650. Mercati, F.: A shape dynamics tutorial. arXiv:1409.0105
651. Merzbacher, E.: Quantum Mechanics. Wiley, New York (1998)
652. Messiah, A.: Quantum Mechanics, vols. 1 and 2. North-Holland, Amsterdam (1965)
653. Meyers, S.B., Steenrod, N.E.: The group of isometries of a Riemannian manifold. Ann. Math. **40**, 400 (1939)
654. Minkowski, H.: Space and Time. Address delivered at 80th Assembly of German Natural Scientists and Physicians at Cologne 21 September 1908; The English translation is available in The Principle of Relativity. Dover, New York (1952), formerly published by Methuen, London (1923)
655. Misner, C.W.: Feynman quantization of general relativity. Rev. Mod. Phys. **29**, 497 (1957)
656. Misner, C.W.: The isotropy of the universe. Astrophys. J. **151**, 431 (1968)
657. Misner, C.W.: Quantum cosmology. I. Phys. Rev. **186**, 1319 (1969)
658. Misner, C.W.: Classical and quantum dynamics of a closed universe. In: Carmeli, M., Fickler, S.I., Witten, L. (eds.) Relativity, Proceedings of the Relativity Conference in the Midwest, Cincinnati, Ohio, 2–6 June 1969 Plenum, New York (1970)
659. Misner, C.W.: Minisuperspace. In: Klauder, J. (ed.) Magic Without Magic: John Archibald Wheeler. Freeman, San Francisco (1972)
660. Misner, C.W., Thorne, K., Wheeler, J.A.: Gravitation. Freedman, San Francisco (1973)
661. Mohr, P.J., Newell, D.B., Taylor, B.N.: CODATA Recommended Values of the Fundamental Physical Constants: 2014 (2015). http://physics.nist.gov/cuu/index.html
662. Moncrief, V., Teitelboim, C.: Momentum constraints as integrability conditions for the Hamiltonian constraint in general relativity. Phys. Rev. D **6**, 966 (1972)
663. Mondragon, M., Montesinos, M.: Covariant canonical formalism for four-dimensional BF theory. J. Math. Phys. **47**, 022301 (2006). arXiv:gr-qc/0402041
664. Montgomery, R.: Infinitely many syzygies. Arch. Ration. Mech. Anal. **164**, 311 (2002)
665. Montgomery, R.: Fitting hyperbolic pants to a 3-body problem. Ergod. Theory Dyn. Syst. **25**, 921 (2005). math/0405014
666. Morris, M.S., Thorne, K.S.: Wormholes in spacetime and their use for interstellar travel: a tool for teaching general relativity. Am. J. Phys. **56**, 395 (1988)
667. Morse, P.M., Feshbach, H.: Methods of Theoretical Physics. Part II. McGraw–Hill, New York (1953)

668. Mott, N.: The wave mechanics of α-ray tracks. Proc. R. Soc. Lond. A **126**, 79 (1929)
669. Muga, G., Sala Mayato, R., Egusquiza, I. (eds.): Time in Quantum Mechanics, vol. 1. Springer, Berlin (2008)
670. Muga, G., Sala Mayato, R., Egusquiza, I. (eds.): Time in Quantum Mechanics, vol. 2. Springer, Berlin (2010)
671. Mukhanov, V.F., Feldman, H.A., Brandenberger, R.H.: Theory of cosmological perturbations. Phys. Rep. **215**, 203 (1992)
672. Munkres, J.R.: Topology. Prentice–Hall, Upper Saddle River (2000)
673. Nakahara, M.: Geometry, Topology and Physics. Institute of Physics Publishing, London (1990)
674. Nash, C.: Differential Topology and Quantum Field Theory. Academic Press, London (1991)
675. Nash, C., Sen, S.: Topology and Geometry for Physicists. Dover, New York (2011)
676. Newton, I.: Philosophiae Naturalis Principia Mathematica [Mathematical Principles of Natural Philosophy]. (1686); For an English translation, see e.g. Cohen, I.B., Whitman, A.: University of California Press, Berkeley, (1999). In particular, see the Scholium on Time, Place, Space and Motion therein
677. Newton-Smith, W.: The Structure of Time. Routledge, London (1980)
678. Nicolai, H.: Supersymmetry and spin systems. J. Phys. A, Math. Gen. **9**, 1497 (1976)
679. Nicolai, H., Peeters, K., Zamaklar, M.: Loop quantum gravity: an outside view. Class. Quantum Gravity **22**, R193 (2005). hep-th/0501114
680. Nielsen, M.A., Chuang, I.L.: Quantum Computation and Quantum Information. Cambridge University Press, Cambridge (2000)
681. Niyogi, P., Smale, S., Weinberger, S.: Finding the homology of submanifolds with high confidence from random samples. Discrete Comput. Geom. **39**, 419 (2008)
682. ó Murchadha, N.: Constrained Hamiltonians and local-square-root actions. Int. J. Mod. Phys. A **20**, 2717 (2002)
683. ó Murchadha, N.: General relativity from the three dimensional linear group. gr-qc/0305038
684. Okawa, H.: Initial conditions for numerical relativity—introduction to numerical methods for solving elliptic PDEs. Int. J. Mod. Phys. A: Special issue **28** (2013), Cardoso, V., Gualtieri, L., Herdeiro, C., Sperhake, U. arXiv:1308.3502
685. Oriti, D.: Disappearance and emergence of space and time in quantum gravity. arXiv: 1302.2849
686. Ortega, P., Ratiu, T.S.: Optimal momentum map. In: Newton, P., Holmes, P., Weinstein, A. (eds.) Geometry, Mechanics and Dynamics: Volume in Honour of the 60th Birthday of Jerrold Marsden. Springer, New York (2002)
687. Osterwalder, K., Schrader, R.: Axioms for Euclidean Green's functions II. Commun. Math. Phys. **42**, 281 (1975)
688. Ovsienko, V., Tabachnikov, S.: Projective Differential Geometry Old and New: From the Schwarzian Derivative to the Cohomology of Diffeomorphism Groups. Cambridge University Press, Cambridge (2005)
689. Padmanabhan, T.: Semiclassical approximations for gravity and the issue of back-reaction. Class. Quantum Gravity **6**, 533 (1989)
690. Padmanabhan, T., Singh, T.P.: On the semiclassical limit of the Wheeler–DeWitt equation. Class. Quantum Gravity **7**, 411 (1990)
691. Page, D.N.: Clock time and entropy. In: Halliwell, J.J., Perez-Mercader, J., Zurek, W.H. (eds.) Physical Origins of Time Asymmetry. Cambridge University Press, Cambridge (1994)
692. Page, D.N.: Sensible quantum mechanics: are probabilities only in the mind? Int. J. Mod. Phys. D **5**, 583 (1996). gr-qc/9507024
693. Page, D.N.: Consciousness and the quantum. arXiv:1102.5339
694. Page, D.N., Wootters, W.K.: Evolution without evolution: dynamics described by stationary observables. Phys. Rev. D **27**, 2885 (1983)
695. Palmer, M.C., Takahashi, M., Westman, H.F.: Localized qubits in curved spacetimes. Ann. Phys. **327**, 1078 (2012). arXiv:1108.3896

696. Parentani, R.: The background field approximation in (quantum) cosmology. Class. Quantum Gravity **17**, 1527 (2000). gr-qc/9803045
697. Parthasarathy, K.R.: Probability Measures on Metric Spaces. Academic Press, New York (1967)
698. Patrangenaru, V., Ellingson, L.: Nonparametric Statistics on Manifolds and Their Applications to Object Data Analysis. Taylor & Francis, Boca Raton (2016)
699. Pauli, W.: Principles of wave mechanics. In: Handbuch der Physik, vol. 24 Springer, Berlin (1933)
700. Pauli, W.: Schlusswort durch den Präsidenten der Konferenz [Opening talk]. Helv. Phys. Acta Suppl. **4**, 261 (1955)
701. Pauli, W.: General Principles of Quantum Mechanics. Springer, Berlin (1980); English translation of work by the same publisher in 1958
702. Peebles, P.J.E.: Principles of Physical Cosmology. Princeton University Press, Princeton (1993)
703. Peierls, R.: The commutation laws of relativistic field theory. Proc. R. Soc. Lond. A **214**, 143 (1952)
704. Penrose, R.: In: Hawking, S.W., Israel, W. (eds.) General Relativity. An Einstein Centenary Survey. Cambridge University Press, Cambridge (1979)
705. Penrose, R.: The Road to Reality. Vintage, London (2005)
706. Penrose, R., Rindler, W.: Spinors and Space-Time: Volume 1, Two-Spinor Calculus and Relativistic Fields. Cambridge University Press, Cambridge (1984)
707. Penrose, R., Rindler, W.: Spinors and Space-Time: Volume 2, Spinor and Twistor Methods in Space-Time Geometry. Cambridge University Press, Cambridge (1988)
708. Peres, A.: Quantum Theory: Concepts and Methods. Kluwer Academic, Boston (1993)
709. Peres, A., Rosen, N.: Queantum limitations on the measurement of gravitational fields. Phys. Rev. **118**, 335 (1960)
710. Perez, A.: Spin foam models for quantum gravity. Class. Quantum Gravity **20**, R43 (2003). gr-qc/0301113
711. Perez, A.: The spin foam approach to quantum gravity. arXiv:1205.2019
712. Peskin, M.E., Schroeder, D.V.: An Introduction to Quantum Field Theory. Perseus Books, Reading (1995)
713. Pflaum, M.J.: Analytic and Geometric Study of Stratified Spaces. Lecture Notes in Mathematics, vol. 1768. Springer, Berlin (2001)
714. Pflaum, M.J.: Smooth structures on stratified spaces. Prog. Math. **198**, 231 (2001)
715. Pilati, M.: The canonical formulation of supergravity. Nucl. Phys. B **132**, 138 (1978)
716. Pilati, M.: Strong coupling quantum gravity 1. Solution in a particular gauge. Phys. Rev. D **26**, 2645 (1982)
717. Pilati, M.: Strong coupling quantum gravity 2. Solution without gauge fixing. Phys. Rev. D **28**, 729 (1983)
718. Poincaré, H.: La Mesure du Temps [The measure of time]. Rev. Métaphys. Morale **6**, 1 (1898)
719. Polchinski, J.: String Theory, vols. I and II. Cambridge University Press, Cambridge (1998)
720. Pons, J.M., Salisbury, D.C.: The issue of time in generally covariant theories and the Komar–Bergmann approach to observables in general relativity. Phys. Rev. D **71**, 124012 (2005). gr-qc/0503013
721. Pons, J.M., Salisbury, D.C., Shepley, L.C.: Gauge transformations in the Lagrangian and Hamiltonian formalisms of generally covariant theories. Phys. Rev. D **55**, 658 (1997). gr-qc/9612037
722. Pons, J.M., Salisbury, D.C., Sundermeyer, K.A.: Revisiting observables in generally covariant theories in the light of gauge fixing methods. Phys. Rev. D **80**, 084015 (2009). arXiv:0905.4564
723. Pons, J.M., Salisbury, D.C., Sundermeyer, K.A.: Gravitational observables, intrinsic coordinates, and canonical maps. Mod. Phys. Lett. A **24**, 725 (2009). arXiv:0902.0401

724. Pons, J.M., Salisbury, D.C., Sundermeyer, K.A.: Observables in classical canonical gravity: folklore demystified In: Proceedings of 1st Mediterranean Conference on Classical and Quantum Gravity. arXiv:1001.2726
725. Preskill, J.: Lecture Notes “Quantum Computation”. http://www.theory.caltech.edu/people/preskill/ph229/index.html
726. Priestley, H.A.: Introduction to Complex Analysis. Oxford University Press, New York (2003)
727. Pullin, J.: In: O’Loughlin, M., Stanič, S., Veberič, D. (eds.) Proceedings of the Second Conference on Time and Matter. University of Nova Gorica Press, Nova Gorica (2008)
728. Randers, G.: On an asymmetrical metric in the four-space of general relativity. Phys. Rev. **59**, 195 (1941)
729. Reed, M., Simon, B.: Methods of Modern Mathematical Physics. II. Fourier Analysis, Self-Adjointness. Academic Press, New York (1975)
730. Reichenbach, H.: The Philosophy of Space and Time. Dover, New York (1950)
731. Reichenbach, H.: The Direction of Time. University of California Press, Berkeley (1956)
732. Rendall, A.D.: Partial Differential Equations in General Relativity. Oxford University Press, New York (2008)
733. Rendall, A.D.: Existence and non-existence results for global constant mean curvature foliations. In: Proceedings of Second World Congress of Nonlinear Analysts. gr-qc/9608045
734. Rideout, D., Wallden, P.: Emergence of spatial structure from causal sets. J. Phys. Conf. Ser. **174**, 012017 (2009). arXiv:0905.0017
735. Riemann, B.: On the hypotheses which lie at the bases of geometry. Nature **8**, 140 (1873); 36 (1873) for a translation
736. Rindler, W.: Relativity. Special, General and Cosmological. Oxford University Press, Oxford (2001)
737. Ringström, H.: The Cauchy Problem in General Relativity. EMS Publishing, Zurich (2009)
738. Roach, S.A.: In: The Theory of Random Clumping. Methuen, London (1968)
739. Robinson, M.: A sheaf-theoretic perspective on sampling. In: Pfander, G.E. (ed.) Sampling Theory, a Renaissance. Birkhäuser, Boston (2015). arXiv:1405.0324
740. Romano, J.D., Tate, R.S.: Dirac versus reduced space quantisation of simple constrained systems. Class. Quantum Gravity **6**, 1487 (1989)
741. Rosenfeld, L.: Über die Gravitationswirkung des Lichtes [Concerning the gravitational effects of light]. Z. Phys. **65**, 589 (1930)
742. Rosenfeld, L.: Zur Quantelung der Wellenfelder [On the quantization of wave fields]. Ann. Phys. **5**, 113 (1930)
743. Rovelli, C.: Is there incompatibility between the ways time is treated in general relativity and in standard quantum mechanics? In: Ashtekar, A., Stachel, J. (eds.) Conceptual Problems of Quantum Gravity, p. 126. Birkhäuser, Boston (1991)
744. Rovelli, C.: Time in quantum gravity: an hypothesis. Phys. Rev. D **43**, 442 (1991)
745. Rovelli, C.: Quantum evolving constants. Phys. Rev. D **44**, 1339 (1991)
746. Rovelli, C.: Group quantization of the Bertotti–Barbour model. In: Ashtekar, A., Stachel, J. (eds.) Conceptual Problems of Quantum Gravity, p. 292. Birkhäuser, Boston (1991)
747. Rovelli, C.: Relational quantum mechanics. Int. J. Theor. Phys. **35**, 1637 (1996). quant-ph/9609002v2
748. Rovelli, C.: Loop quantum gravity. Living Rev. Relativ. **1**, 1 (1998). gr-qc/9710008
749. Rovelli, S.e.g.C.: Quantum spacetime: what do we know? In: Callendar, C., Huggett, N. (eds.) Physics Meets Philosophy at the Planck Scale. Cambridge University Press, Cambridge (2001)
750. Rovelli, C.: GPS observables in general relativity. Phys. Rev. D **65**, 044017 (2002). gr-qc/0110003
751. Rovelli, C.: Partial observables. Phys. Rev. **65**, 124013 (2002). gr-qc/0110035
752. Rovelli, C.: Quantum Gravity. Cambridge University Press, Cambridge (2004)
753. Rovelli, C.: A note on the foundation of relativistic mechanics. I: relativistic observables and relativistic states. gr-qc/0111037

754. Rovelli, C.: A note on the foundation of relativistic mechanics. II: covariant Hamiltonian general relativity. gr-qc/0202079
755. Rovelli, C.: Forget time, fqXi 'Nature of Time' Essay Competition: Community First Prize. arXiv:0903.3832
756. Rovelli, C.: Zakopane lectures on loop gravity. arXiv:1102.3660
757. Rovelli, C., Smolin, L.: Knot theory and quantum gravity. Phys. Rev. Lett. **61**, 1155 (1988)
758. Rovelli, C., Smolin, L.: Loop space representation for quantum general relativity. Nucl. Phys. B **331**, 80 (1990)
759. Rudolph, G., Schmidt, M., Volobuev, I.P.: On the gauge orbit space stratification: a review. J. Phys. A, Math. Gen. **35**, R1 (2002)
760. Ryan, M.P.: Hamiltonian Cosmology. Lec. Notes Phys., vol. 13. Springer, Berlin (1972)
761. Salecker, H., Wigner, E.P.: Quantum limitations of the measurement of space-time distances. Phys. Rev. **109**, 571 (1958)
762. Samuel, J.: Is Barbero's Hamiltonian formulation a gauge theory of Lorentzian gravity? Class. Quantum Gravity **17**, L141 (2000). gr-qc/0005095
763. Sathyaprakash, B.S., Schutz, B.F.: Physics, astrophysics and cosmology with gravitational waves. Living Rev. Relativ. **12**, 2 (2009). arXiv:0903.0338
764. Savitt, S.F. (ed.): Time's Arrow Today. Cambridge University Press, Cambridge (1995)
765. Savvidou, K.N.: Continuous time in consistent histories. Ph.D. thesis, Imperial College London (1999). gr-qc/9912076
766. Savvidou, K.N.: General relativity histories theory: spacetime diffeomorphisms and the Dirac algebra of constraints. Class. Quantum Gravity **18**, 3611 (2001). gr-qc/0104081
767. Savvidou, K.N.: Poincaré invariance for continuous-time histories. J. Math. Phys. **43**, 3053 (2002). gr-qc/0104053
768. Savvidou, K.N.: General relativity histories theory I: the spacetime character of the canonical description. Class. Quantum Gravity **21**, 615 (2004). arXiv:gr-qc/0306034
769. Savvidou, K.N.: General relativity histories theory II: invariance groups. Class. Quantum Gravity **21**, 631 (2004). gr-qc/0306036
770. Savvidou, K.N.: General relativity histories theory. Braz. J. Phys. **35**, 307 (2005). gr-qc/0412059
771. Savvidou, K.N.: Spacetime symmetries in histories canonical gravity. In: Oriti, D. (ed.) Approaches to Quantum Gravity. Cambridge University Press, Cambridge (2009)
772. Schlegel, R.: Time and the Physical World. Dover, New York (1968)
773. Schlosshauer, M.: Decoherence, the measurement problem, and interpretations of quantum mechanics. Rev. Mod. Phys. **76**, 1267 (2005). quant-ph/0312059
774. Schmah, T.: A cotangent bundle slice theorem. Differ. Geom. Appl. **25**, 101 (2007). math/0409148
775. Schmidt, M.: How to study the physical relevance of gauge orbit space singularities? Rep. Math. Phys. **51**, 325 (2003)
776. Schneider, P., Ehlers, J., Falco, E.: Gravitational Lenses. Springer, Berlin (1992)
777. Schouten, J.A.: Ricci Calculus. Springer, Berlin (1954)
778. Schreiber, U.: Differential cohomology in a cohesive infinity-topos. arXiv:1310.7930
779. Schrödinger, E.: Space-Time Structure. Cambridge University Press, Cambridge (1950)
780. Schroeren, D.: Decoherent histories of spin networks. Found. Phys. **43**, 310 (2013). arXiv:1206.4553
781. Schwabl, F.: Statistical Mechanics. Springer, Berlin (2006)
782. Seiberg, N.: Emergent spacetime. In: Gross, D., Henneaux, M., Sevrin, A. (eds.): The Quantum Structure of Space and Time, Proceedings of the 23rd Solvay Conference in Physics. hep-th/0601234
783. Seidelmann, P.K. (ed.): Explanatory Supplement to the Astronomical Almanac. University Science Books, Mill Hill (1992)
784. Senovilla, J.M.M., Garfinkle, D.: The 1965 Penrose singularity theorem. arXiv:1410.5226
785. Serre, J-P.: Faisceaux Algebraiques Coherents [Coherent algebraic sheaves]. Ann. Math. **61**, 197 (1955)

786. Serre, J-P.: Linear Representations of Finite Groups. Springer, New York (1977)
787. Shakespeare, W.: Richard II (1597)
788. Shapiro, L.S.: Affine Analysis of Image Sequences. Cambridge University Press, Cambridge (1995)
789. Shirai, I., Wada, S.: Cosmological perturbations and quantum fields in curved space. Nucl. Phys. B **303**, 728 (1988)
790. Singer, I.M.: Some remarks on the Gribov ambiguity. Commun. Math. Phys. **60**, 7 (1978)
791. Singer, I.M., Thorpe, J.M.: Lecture Notes on Elementary Topology and Geometry. Springer, New York (1967)
792. Small, C.G.S.: The Statistical Theory of Shape. Springer, New York (1996)
793. Smolin, L.: Space and time in the quantum universe. In: Ashtekar, A., Stachel, J. (eds.) Conceptual Problems of Quantum Gravity. Birkhäuser, Boston (1991)
794. Smolin, L.: The classical limit and the form of the Hamiltonian constraint in nonperturbative quantum gravity. gr-qc/9609034
795. Smolin, L.: How far are we from the quantum theory of gravity? hep-th/0303185
796. Smolin, L.: An invitation to loop quantum gravity. hep-th/0408048
797. Śniatycki, J.: Dirac brackets in geometric dynamics. Ann. Inst. Henri Poincaré **20**, 365 (1974)
798. Śniatycki, J.: Differential Geometry of Singular Spaces and Reduction of Symmetry. Cambridge University Press, Cambridge (2013)
799. So, L.L., Nester, J.M., Chen, H.: Energy-momentum density in small regions: the classical pseudotensors. Class. Quantum Gravity **26**, 085004 (2009). arXiv:0901.3884
800. Sorkin, R.D.: Posets as lattice topologies. In: Bertotti, B., de Felice, F., Pascolini, A. (eds.) General Relativity and Gravitation: Proceedings of the GR10 Conference. Volume I, p. 635. Consiglio Nazionale Delle Ricerche, Rome (1983)
801. Sorkin, R.D.: Spacetime and causal sets. In: D'Olivo, J.C., Nahmad-Achar, E., Rosenbaum, M., Ryan, M.P., Urrutia, L.F., Zertuche, F. (eds.) Relativity and Gravitation: Classical and Quantum, p. 150. World Scientific, Singapore (1991)
802. Sorkin, R.D.: Forks in the road, on the way to quantum gravity. Int. J. Theor. Phys. **36**, 2759 (1997). gr-qc/9706002
803. Sorkin, R.D.: Relativity theory does not imply that the future already exists: a counterexample. In: Pektov, V. (ed.) Relativity and the Dimensionality of the World. Springer, Dordrecht (2007)
804. Sorkin, R.D.: Causal sets: discrete gravity. Notes for the Valdivia Summer School. gr-qc/0309009
805. Souriau, J.M.: Structure of Dynamical Systems. A Symplectic View of Physics. Birkhäuser, Basel (1994)
806. Spero, A., Baierlein, R.: Approximate symmetry groups of inhomogeneous metrics. J. Math. Phys. **18**, 1330 (1977)
807. Spero, A., Baierlein, R.: Approximate symmetry groups of inhomogeneous metrics: examples. J. Math. Phys. **19**, 1324 (1978)
808. Stachel, J.: Einstein's search for general covariance, 1912–1915. In: Howard, D., Stachel, J. (eds.) Einstein and the History of General Relativity Einstein Studies, vol. 1. Birkhäuser, Basel (1989)
809. Steen, L.A., Seebach, J.A.: Counterexamples in Topology. Dover, New York (1995)
810. Steiner, A.K.: The lattice of topologies: structure and complementation. Trans. Am. Math. Soc. **122**, 379 (1966)
811. Stelle, K.S.: Renormalization of higher-derivative quantum gravity. Phys. Rev. D **16**, 953 (1977)
812. Stephani, H., Kramer, D., MacCallum, M.A.H., Hoenselaers, C.A., Herlt, E.: Exact Solutions of Einstein's Field Equations, 2nd edn. Cambridge University Press, Cambridge (2003)
813. Stern, A.: Anyons and the quantum hall effect—a pedagogical review. Ann. Phys. **323**, 204 (2007). arXiv:0711.4697
814. Stewart, J.M.: Advanced General Relativity. Cambridge University Press, Cambridge (1991)

815. Stillwell, J.: The Four Pillars of Geometry. Springer, New York (2005)
816. Streater, R.F.: Current commutation relations, continuous tensor products and infinitely divisible group representations. Rend. Sci. Int. Fisica E. Fermi **11**, 247 (1969)
817. Streater, R.F., Wightman, A.S.: PCT, Spin and Statistics, and All That. Benjamin, New York (1964)
818. Strominger, A.: Quantum Gravity and String Theory. http://physics.princeton.edu/strings2014/slides/Strominger.pdf
819. Strominger, A., Vafa, C.: Microscopic origin of the Bekenstein-Hawking entropy. Phys. Lett. B **379**, 99 (1996). hep-th/9601029
820. Sundermeyer, K.: Constrained Dynamics. Springer, Berlin (1982)
821. Surya, S.: Directions in Causal Set Quantum Gravity. arXiv:1103.6272
822. Sutherland, W.A.: Introduction to Metric and Topological Spaces. Clarendon, Oxford (1981)
823. Szabados, L.B.: Quasi-local energy-momentum and angular momentum in GR: a review article. Living Rev. Relativ. **7**, 4 (2004)
824. Szabados, L.B.: Quasi-local energy-momentum and angular momentum in GR: a review article. Living Rev. Relativ. **12**, 4 (2009)
825. Szekeres, P.: The gravitational compass. J. Math. Phys. **6**, 1387 (1965)
826. Szydlowski, M., Szczesny, J.: Invariant chaos in mixmaster cosmology. Phys. Rev. D **50**, 819 (1994)
827. 't Hooft, G., Veltman, M.: One-loop divergences in the theory of gravitation. Ann. Inst. Henri Poincaré **20**, 69 (1974)
828. Tarasov, V.: Quantum Mechanics of Non-Hamiltonian and Dissipative Systems. Elsevier, Amsterdam (2008)
829. Tatarskii, V.I.: The Wigner representation of quantum mechanics. Sov. Phys. Usp. **26**, 311 (1983)
830. Tavakol, R.K., Anderson, E.: We have time because we shall never know. arXiv:1506.07928
831. Taylor, E.F., Wheeler, J.A.: Spacetime Physics. Freeman, San Francisco (1966)
832. Teitelboim, C.: How commutators of constraints reflect spacetime structure. Ann. Phys. (N. Y.) **79**, 542 (1973)
833. Teitelboim, C.: The Hamiltonian structure of spacetime. Ph.D. thesis, Princeton (1973)
834. Teitelboim, C.: Supergravity and square roots of constraints. Phys. Rev. Lett. **38**, 1106 (1977)
835. Teitelboim, C.: The Hamiltonian structure of spacetime. In: Held, A. (ed.) General Relativity and Gravitation: One Hundred Years After the Birth of Albert Einstein, vol. 1. Plenum, New York (1980)
836. Teitelboim, C.: The Hamiltonian structure of two-dimensional space-time and its relation with the conformal anomaly. In: Christensen, S.M. (ed.) Quantum Theory of Gravity. Hilger, Bristol (1984)
837. Tene, H.: Some geometric equivariant cohomology theories. Ph.D. thesis, University of Bonn (2010). arXiv:1210.7923
838. The ATLAS Collaboration: Observation of a new particle in the search for the standard model Higgs boson with the ATLAS detector at the LHC. Phys. Lett. B **716**, 1 (2012)
839. The LIGO Collaboration the Virgo Collaboration: Observation of gravitational waves from a binary black hole merger. Phys. Rev. Lett. **116**, 061102 (2016). arXiv:1602.03837
840. The Planck Collaboration: Planck 2013 results. XVI. Cosmological parameters. Astron. Astrophys. **571**, A16 (2014). arXiv:1303.5076
841. The Planck Collaboration: Planck 2013 results. XXII. Constraints on inflation. Astron. Astrophys. **571**, A22 (2014). arXiv:1303.5082
842. Thiemann, T.: Quantum spin dynamics (QSD). Class. Quantum Gravity **15**, 839 (1998). gr-qc/9606089
843. Thiemann, T.: Lectures on loop quantum gravity. Lect. Notes Phys. **631**, 41 (2003). gr-qc/0210094
844. Thiemann, T.: The Phoenix project: master constraint programme for loop quantum gravity. Class. Quantum Gravity **23**, 2211 (2006). gr-qc/0305080

845. Thiemann, T.: Modern Canonical Quantum General Relativity. Cambridge University Press, Cambridge (2007)
846. Thom, R.: Les Singularités des Applications Différentiables [Singularities in differentiable maps]. Ann. Inst. Fourier (Grenoble) **6**, 43 (1955)
847. Thom, R.: Ensembles et Morphismes Stratifiés [Stratified spaces and morphisms]. Bull. Am. Math. Soc. (N.S.) **75**, 240 (1969)
848. Thurston, W.P.: Three-Dimensional Geometry and Topology. Princeton University Press, Princeton (1997)
849. Tolkien, J.R.R.: The Fellowship of the Ring (1954)
850. Tolstoy, L.: War and Peace (1869)
851. Tomonaga, S.: On a relativistically invariant formulation of the quantum theory of wave fields. Prog. Theor. Phys. **1**, 27 (1946)
852. Torre, C.G.: Is general relativity an 'already parametrized' field theory? Phys. Rev. D **46**, 3231 (1992). hep-th/9204014
853. Torre, C.G.: Symmetries of the Einstein equations. Phys. Rev. Lett. **70**, 3525 (1993)
854. Torre, C.G.: Gravitational observables and local symmetries. Phys. Rev. D **48**, 2373 (1993)
855. Torre, C.G.: In: Dhurandhar, S., Padmanabhan, T. (eds.) Gravitation and Cosmology. Kluwer Academic, Dordrecht (1997)
856. Torre, C.G.: Observables for the polarized Gowdy model. Class. Quantum Gravity **23**, 1543 (2006). gr-qc/0508008
857. Torre, C.G.: Functional time evolution, anomaly potentials, and the geometric phase. Phys. Rev. D **76**, 125012 (2007)
858. Torre, C.G., Anderson, I.M.: Classification of generalized symmetries for the vacuum Einstein equations. Commun. Math. Phys. **176**, 479 (1996). gr-qc/9404030
859. Torretti, R.: Relativity and Geometry. Pergamon, New York (1983)
860. Underhill, J.: Quantization on a manifold with connection. J. Math. Phys. **19**, 1932 (1978)
861. Unruh, W.G.: Notes on black hole evaporation. Phys. Rev. D **14**, 870 (1976)
862. Unruh, W.G., Wald, R.M.: Time and the interpretation of canonical quantum gravity. Phys. Rev. D **40**, 2598 (1989)
863. Vaisman, I.: Lectures on the Geometry of Poisson Manifolds. Birkhäuser, Basel (1994)
864. Vakhania, N.N.: Probability Distributions on Linear Spaces. North-Holland, New York (1981)
865. Vakhania, N.N.: Probability Distributions on Banach Spaces. Reidel, Dordrecht (1987)
866. Van Hove, L.: Sur le Probleme des Relations Entre les Transformations Unitaires de la Mécanique Quantique et les Transformations Canoniques de la Mécanique Classique. Acad. R. Belg. **37**, 610 (1951)
867. Varadarajan, V.S.: George Mackey and his work on representation theory and foundations of physics. Contemp. Math. **449**, 417 (2008). http://www.math.ucla.edu/~vsv/mackey.pdf
868. Vargas Moniz, P.: Quantum Cosmology—The Supersymmetric Perspective, vols. 1 and 2. Springer, Berlin (2010)
869. Vassiliev, V.A.: Cohomology of Knot Spaces. In: Theory of Singularities. Advances in Soviet Math., vol. 1. Am. Math. Soc., Providence (1990)
870. Wada, S.: Consistency of canonical quantization of gravity in the lowest nontrivial order. Preprint UT Komaba 85-8 (1985)
871. Wada, S.: Quantum cosmological perturbations in pure gravity. Nucl. Phys. B **276**, 729 (1986); Erratum-ibid. **284**, 747 (1987)
872. Wada, S.: Consistency of canonical quantization of gravity and boundary conditions for the wave function of the universe. Phys. Rev. D **34**, 2272 (1986)
873. Wainwright, J., Ellis, G.F.R.: Dynamical Systems in Cosmology. Cambridge University Press, Cambridge (2005)
874. Wald, R.M.: General Relativity. University of Chicago Press, Chicago (1984)
875. Wald, R.M.: The formulation of quantum field theory in curved spacetime. In: Rowe, D. (ed.) Proceedings of the 'Beyond Einstein Conference'. Birkhäuser, Boston (2009). arXiv:0907.0416

876. Wall, C.T.C.: In: Ranicki, A. (ed.) Surgery on Compact Manifolds. Mathematical Surveys and Monographs, vol. 69. Am. Math. Soc., Providence (1999)
877. Wallace, D.: The Emergent Multiverse: Quantum Theory According to the Everett Interpretation. Oxford University Press, New York (2012)
878. Watson, G.S.: Statistics on Spheres. Wiley, Chichester (1983)
879. Wehrl, A.: General properties of entropy. Rev. Mod. Phys. **50**, 221 (1978)
880. Weinberg, S.: Derivation of gauge invariance and the equivalence principle from Lorentz invariance of the S-matrix. Phys. Lett. **9**, 357 (1964)
881. Weinberg, S.: Photons and gravitons in S-matrix theory: derivation of charge conservation and equality of gravitational and inertial mass. Phys. Rev. **135**, 1049 (1964)
882. Weinberg, S.: Photons and gravitons in perturbation theory: derivation of Maxwell's and Einstein's equations. Phys. Rev. **138**, 988 (1965)
883. Weinberg, S.: Gravitation and Cosmology: Principles and Applications of the General Theory of Relativity. Wiley, New York (1972)
884. Weinberg, S.: Ultraviolet divergences in quantum theories of gravitation. In: Hawking, S.W., Israel, W. (eds.) General Relativity: An Einstein Centenary Survey, p. 790. Cambridge University Press, Cambridge (1979)
885. Weinberg, S.: The Quantum Theory of Fields. Vol. I. Foundations. Cambridge University Press, Cambridge (1995)
886. Weinberg, S.: The Quantum Theory of Fields. Vol. II. Modern Applications. Cambridge University Press, Cambridge (1995)
887. Weinberg, S.: The Quantum Theory of Fields. Vol III. Supersymmetry. Cambridge University Press, Cambridge (2000)
888. Weinberg, S.: Cosmology. Oxford University Press, New York (2008)
889. Weinberg, S.: Lectures on Quantum Mechanics. Cambridge University Press, Cambridge (2013)
890. Weiss, P.: On the Hamilton–Jacobi theory and quantization of a dynamical continuum. Proc. R. Soc. A **169**, 102 (1939)
891. Wells, R.O.: Differential Analysis on Complex Manifolds. Springer, New York (2008)
892. Weron, A. (ed.): Probability Theory on Vector Spaces (1977: Trzebiesowice, Poland). Springer, Berlin (1978)
893. Weyl, H.: Gravitation und Elektricitat [Gravitation and electricity]. Preuss. Akad. Wiss. Berl. (1918); The English translation is available in e.g. The Principle of Relativity. Dover, New York (1952), formerly published by Methuen, London (1923)
894. Weyl, H.: Quantenmechanik und Gruppentheorie [The Theory of Groups and Quantum Mechanics]. Hirzel, Leipzig (1928); currently available in English translation by Dover, New York (1931)
895. Wheeler, J.A.: On the mathematical description of light nuclei by the method of resonating group structure. Phys. Rev. **52**, 1107 (1937)
896. Wheeler, J.A.: On the nature of quantum geometrodynamics. Ann. Phys. **2**, 604 (1957)
897. Wheeler, J.A.: Geometrodynamics and the issue of the final state. In: DeWitt, B.S., DeWitt, C.M. (eds.) Groups, Relativity and Topology. Gordon & Breach, New York (1964)
898. Wheeler, J.A.: Gravitation as geometry—II. In: Chiu, H.Y., Hoffman, W.F. (eds.) Gravitation and Relativity. Benjamin, New York (1964)
899. Wheeler, J.A.: Superspace and the nature of quantum geometrodynamics. In: DeWitt, C., Wheeler, J.A. (eds.) Battelle Rencontres: 1967 Lectures in Mathematics and Physics. Benjamin, New York (1968)
900. Wheeler, J.A.: Frontiers of time. In: Toraldo di Francia, N., van Fraassen, B. (eds.) Problems in the Foundations of Physics. North-Holland, Amsterdam (1979)
901. Wheeler, J.A.: Geometrodynamic steering principle reveals the determiners of inertia. Int. J. Mod. Phys. A **3**, 2207 (1988)
902. Whitney, H.: Sphere spaces. Proc. Natl. Acad. Sci. USA **21**, 464 (1935)
903. Whitney, H.: Differentiable manifolds. Ann. Math. **37**, 645 (1936)
904. Whitney, H.: Complexes of manifolds. Proc. Natl. Acad. Sci. USA **33**, 10 (1946)

905. Whitney, H.: Tangents to an analytic variety. Ann. Math. **81**, 496 (1965)
906. Whitrow, G.J.: The Natural Philosophy of Time. Nelson, London (1961)
907. Wigner, E.P.: On the quantum correction for thermodynamic equilibrium. Phys. Rev. **40**, 749 (1932)
908. Wigner, E.P.: On unitary representations of the inhomogeneous Lorentz group. Ann. Math. **40**, 149 (1939)
909. Wigner, E.P.: Symmetries and Reflections. Indiana University Press, Bloomington (1967)
910. Will, S.C.M.: The confrontation between general relativity and experiment. Living Rev. Relativ. **17**, 4 (2014). arXiv:1403.7377
911. Williams, R.M., Tuckey, P.A.: Regge calculus: a bibliography and brief review. Class. Quantum Gravity **9**, 1409 (1992)
912. Wiltshire, D.L.: In: Robson, B., Visvanathan, N., Woolcock, W.S. (eds.) Cosmology: The Physics of the Universe. World Scientific, Singapore (1996). gr-qc/0101003
913. Winter, J.: Wigner transformation in curved space-time and the curvature correction of the Vlasov equation for semiclassical gravitating systems. Phys. Rev. D **32**, 1871 (1985)
914. Witten, E.: Search for a realistic Kaluza–Klein theory. Nucl. Phys. B **186**, 412 (1981)
915. Witten, E.: Topological quantum field theory. Commun. Math. Phys. **117**, 353 (1988)
916. Witten, E.: Quantum field theory and the Jones polynomial. Commun. Math. Phys. **121**, 351 (1989)
917. Witten, E.: Anti de Sitter space and holography. Adv. Theor. Math. Phys. **2**, 253 (1998). hep-th/9802150
918. Witten, E.: Perturbative gauge theory as a string theory in twistor space. Commun. Math. Phys. **252**, 189 (2004). hep-th/0312171
919. Woodhouse, N.M.J.: Geometric Quantization. Springer, Berlin (1980); Clarendon Press, Oxford (1991)
920. Wüthrich, C.: Approaching the Planck scale from a generally relativistic point of view: a philosophical appraisal of loop quantum gravity. Ph.D. thesis, Pittsburgh (2006)
921. York Jr., J.W.: Gravitational degrees of freedom and the initial-value problem. Phys. Rev. Lett. **26**, 1656 (1971)
922. York Jr., J.W.: Role of conformal three-geometry in the dynamics of gravitation. Phys. Rev. Lett. **28**, 1082 (1972)
923. York Jr., J.W.: Mapping onto solutions of the gravitational initial value problem. J. Math. Phys. **13**, 125 (1972)
924. York Jr., J.W.: Conformally invariant orthogonal decomposition of symmetric tensors on Riemannian manifolds and the initial-value problem of general relativity. J. Math. Phys. **14**, 456 (1973)
925. York Jr., J.W.: Covariant decompositions of symmetric tensors in the theory of gravitation. Ann. Inst. Henri Poincaré **21**, 319 (1974)
926. Zalaletdinov, R.: Averaging out the Einstein equations and macroscopic spacetime geometry. Gen. Relativ. Gravit. **24**, 1015 (1992)
927. Zalaletdinov, R.: Towards a macroscopic theory of gravity. Gen. Relativ. Gravit. **25**, 673 (1993)
928. Zeh, H.D.: On the interpretation of measurement in quantum theory. Found. Phys. **1**, 69 (1970)
929. Zeh, H.D.: Emergence of classical time from a universal wavefunction. Phys. Lett. A **116**, 9 (1986)
930. Zeh, H.D.: Time in quantum gravity. Phys. Lett. A **126**, 311 (1988)
931. Zeh, H.D.: The Physical Basis of the Direction of Time. Springer, Berlin (1989)
932. Zurek, W.H., Habib, S., Paz, J.P.: Coherent states via decoherence. Phys. Rev. Lett. **70**, 1187 (1992)
933. Zwiebach, B.: A First Course in String Theory. Cambridge University Press, Cambridge (2004)

Index

E. Anderson, *The Problem of Time*, Fundamental Theories of Physics 190,
DOI 10.1007/978-3-319-58848-3

E